International Association of Fire Chiefs

MW00910860

Fundamentals of
Fire Fighter Skills

Canadian Third Edition

JONES & BARTLETT
LEARNING

Jones & Bartlett Learning
World Headquarters
5 Wall Street
Burlington, MA 01803
978-443-5000
info@jblearning.com
www.jblearning.com

National Fire Protection Association
1 Batterymarch Park
Quincy, MA 02169-7471
www.NFPA.org

International Association of Fire Chiefs
4025 Fair Ridge Drive
Fairfax, VA 22033
www.IAFC.org

Jones & Bartlett Learning books and products are available through most bookstores and online booksellers. To contact Jones & Bartlett Learning directly, call 800-832-0034, fax 978-443-8000, or visit our website, www.jblearning.com.

Substantial discounts on bulk quantities of Jones & Bartlett Learning publications are available to corporations, professional associations, and other qualified organizations. For details and specific discount information, contact the special sales department at Jones & Bartlett Learning via the above contact information or send an email to specialsales@jblearning.com.

Production Credits

Chief Executive Officer: Ty Field
President: James Homer
Chief Product Officer: Eduardo Moura
Executive Publisher: Kimberly Brophy
Director of Canadian Sales: Robert Rosenitsch
Senior Acquisitions Editor: Janet Maker
Senior Editor: Jennifer Deforge-Kling
Associate Editor: Olivia MacDonald
Associate Director, Production: Jenny L. Corriveau
Production Editor: Jessica deMartin

Senior Marketing Manager: Brian Rooney
VP, Manufacturing and Inventory Control: Therese Connell
Composition: diacriTech
Cover Design: Kristin E. Parker
Associate Permissions and Photo Researcher: Ashley Dos Santos
Cover Image: Courtesy of Tim Olk
Printing and Binding: Courier Companies
Cover Printing: Courier Companies

To order this product, use ISBN: 978-1-284-03563-6

Library of Congress Cataloging-in-Publication Data
Fundamentals of fire fighter skills / International Association of Fire Chiefs, National Fire Protection Agency.—Canadian Third edition.
    pages cm
Includes index.
ISBN 978-1-284-02690-0
1. Fire prevention. 2. Fire extinction. I. International Association of Fire Chiefs. II. National Fire Protection Association.
TH9146.F855 2014
628.9'25—dc23
                              2013020769
6048
Printed in the United States of America
17 16 15 14 13   10 9 8 7 6 5 4 3 2 1

Brief Contents

Contents

© Greg Henry/ShutterStock, Inc.

© Greg Henry/ShutterStock, Inc.

© Photos.com

© Greg Henry/ShutterStock, Inc.

© Greg Henry/ShutterStock, Inc.

Skill Drills

© Greg Henry/ShutterStock, Inc.

Instructor Resources

■ Instructor's ToolKit

Preparing for class is easy with the resources on this DVD. The DVD includes the following:

- Detailed lesson plans keyed to the PowerPoint presentations with sample lectures, lesson quizzes, and teaching strategies. Teaching tips and ideas to enhance presentations are included.
- Adaptable PowerPoint™ presentations that provide instructors with a powerful way to create presentations that are educational and engaging to their students. These slides can be modified and edited to meet instructors' specific needs.
- Test Bank that contains an abundance of multiple-choice questions and allows instructors to create tailor-made classroom tests and quizzes quickly and easily by selecting, editing, organizing, and printing a test. An answer key with page references to the text is also available.
- Image and Table Bank, which offers a selection of the most important images and tables found in the text. Instructors can use these graphics to incorporate more images into the PowerPoint presentations, make handouts, or enlarge a specific image for further discussion.
- Skill Sheets provide instructors with a resource to track students' skills and conduct skill proficiency exams.

■ Skills and Drills DVD Series

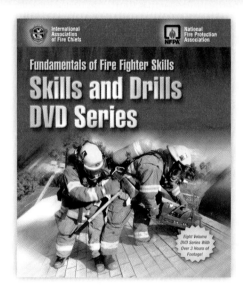

This eight-DVD set walks students through the skills that they must successfully complete to achieve Fire Fighter I and Fire Fighter II certification. Capturing real-life scenes, these DVDs teach students how to successfully perform each skill and offer helpful information, tips, and pointers designed to facilitate progression through practical examinations. This exciting series gives students the chance to witness fire fighters in action and in "real" time.

Digital Curriculum Solution Packages

■ Digital Curriculum Solution Packages

Digital Curriculum Solution Packages allow educators to offer their students cutting-edge digital resources based on high-quality technical content. The innovative, multifaceted online tools facilitate absolute understanding—how fire behaves, the fire-suppression process, and how to be a confident, safe fire fighter at both levels. This total comprehension and command of skills and knowledge is key to success on the fireground.

Gold standard content joins sound instructional design in a user-friendly online interface to give students a truly interactive, engaging learning experience with:

- **Navigate Course Manager**. This easy-to-use and fully hosted online learning platform seamlessly combines authoritative textbook content with interactive tools, assessments, and robust reporting tools that enable instructors to track real-time student progress and engagement.
- **Web Tools**. Using the access code printed inside the front cover, students can unlock additional interactive and mobile education resources such as chapter pre-tests, interactive skill drills, and skill evaluation sheets.
- **Navigate TestPrep**. This dynamic tool is designed to prepare students to sit for examinations by providing practice examinations and simulated certification examinations using multiple-choice questions.
- **Interactive Lectures**. Take the lectures out of the classroom! These interactive lectures provide anytime, anywhere access for students to an innovative educational environment.

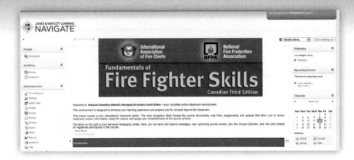

Available packages include:

Advantage Package
- Printed Textbook
- Navigate Course Manager
- Web Tools

Preferred Package
- Printed Textbook
- Navigate Course Manager
- Web Tools
- Navigate TestPrep

Premier Package
- Printed Textbook
- Navigate Course Manager
- Web Tools
- Navigate TestPrep
- Interactive Lectures

Acknowledgements

Jones & Bartlett Learning, the National Fire Protection Association, and the International Association of Fire Chiefs would like to thank the current editors, contributors, and reviewers of *Fundamentals of Fire Fighter Skills, Canadian Third Edition.*

Editorial Board for US Edition

Shawn Kelley
International Association of Fire Chiefs
Fairfax, Virginia

Thomas McGowan
National Fire Protection Association
Quincy, Massachusetts

Glen D. Rudner
Virginia Department of Emergency Management, retired
North Chesterfield, Virginia

Steven Sawyer
National Fire Protection Association
Quincy, Massachusetts

Dave Schottke
Fairfax City Volunteer Fire Department
Fairfax, Virginia

Editorial Board for Canadian Edition

Shawn Kelley
International Association of Fire Chiefs
Fairfax, Virginia

Thomas McGowan
National Fire Protection Association
Quincy, Massachusetts

Walter (Walt) Osellame
Midway Volunteer Fire Department
Midway, British Columbia
College of the Rockies
Cranbrook, British Columbia

Steven Sawyer
National Fire Protection Association
Quincy, Massachusetts

Contributors and Reviewers

Larry E. Badry
Kaslo Volunteer Fire Department
Kaslo, British Columbia

Claude Beauchamp
Ecole Nationale des Pompiers du Quebec
Laval, Quebec

Norman C. Booth
Brantford Fire Department
Brantford, Ontario

Bill Brown
Toronto Fire Services
Toronto, Ontario

Tim Calhoun
Clarington Emergency and Fire Services
Clarington (Bowmanville), Ontario

Robin Chadwick
Georgina Fire Department
Keswick, Ontario

Darren Charlton
Justice Institute of British Columbia Fire & Safety Division
Mission, British Columbia

Dean Colthrop
Justice Institute of British Columbia, Fire and Safety Division
Maple Ridge, British Columbia

Larry Crang
Office of the Fire Marshal
Gravenhurst, Ontario

Craig Dart
Toronto Fire Services
Toronto, Ontario

Dave Dixon
Central York Fire Service
Newmarket, Ontario

Kane Doran
Manitoba Emergency Services College
Brandon, Manitoba

Doug Goodings
Ontario Fire College
North York, Ontario

Bryan Goustos
Brampton Fire and Emergency Services
Brampton, Ontario

Richard Hepditch
Waterloo Fire Rescue
Waterloo, Ontario

Ken Kroeker
Manitoba Emergency Services College
Brandon, Manitoba

Kevin C. Lambert
Ottawa Fire Services
Ottawa, Ontario

Timothy Langille
Thunder Bay Fire Rescue
Thunder Bay, Ontario

Grant Love
North Bay Fire & Emergency Services
North Bay, Ontario

Evan Miller
City of Surrey
Surrey, British Columbia

Douglas R. Moeckl
Lakeland College—Emergency Training
  Centre
Vermillion, Alberta

Bo Moore
Calgary Fire Department
Calgary, Alberta

William (Bill) Reimer
Municipal and Community Affairs
Fort Smith, Northern Territories

David Roe
BFSCO High River Fire Department
High River, Alberta

Ron Sparling
Lambton College
Sarnia, Ontario

Brent Thomas
Hamilton Fire Department
Hamilton, Ontario

Michael Vilneff
Cobourg Fire Department
Cobourg, Ontario

Cody Zebedee
Foothills Fire Department
High River, Alberta

The Orientation and History of the Fire Service

Fire Fighter I

Knowledge Objectives

After studying this chapter, you will be able to:

- List five guidelines for successful fire fighter training. (p 4)
- Describe the general requirements for becoming a fire fighter. (p 5)
- Outline the roles and responsibilities of a Fire Fighter I. (NFPA 5.1 , p 5–6)
- Describe the common positions of fire fighters within the fire department. (NFPA 5.1.1 , p 6–7)
- Describe the specialized response roles within the fire department. (p 7)
- Explain the concept of governance, and describe how the fire department's regulations, policies, and suggested operating guidelines (SOGs), or standard operating procedures (SOPs), affect it. (NFPA 5.1.1 , p 8)
- Locate information in departmental documents and SOGs/SOPs. (NFPA 5.1.2 , p 9)
- List the different types of fire department companies, and describe their functions. (NFPA 5.1.1 p 10)
- Describe how to organize a fire department in terms of staffing, function, and geography. (p 10–11)
- Explain the basic structure of the chain of command within the fire department. (p 11)
- Define the four basic management principles used to maintain organization within the fire department. (p 12–13)
- Explain the evolution of the methods and tools of firefighting from colonial days to the present. (p 13)
- Explain how building codes prevent the loss of life and property. (p 14)
- Describe the evolution of funding for fire department services. (p 17)

Skills Objectives

There are no skills objectives for Fire Fighter I candidates.

© Photos.com

Fire Fighter II   FFII

Knowledge Objectives

- Outline the responsibilities of a Fire Fighter II. (NFPA 6.1 , p 6)
- Describe the roles of a Fire Fighter II within the fire department. (NFPA 6.1.1 , p 6–7)

Skills Objectives

There are no skills objectives for Fire Fighter II candidates. NFPA 1001 contains no Fire Fighter II Job Performance Requirements for this chapter.

Additional NFPA Standards

- NFPA 1500, *Standard on Fire Department Occupational Safety and Health Program*
- NFPA 1582, *Standard on Comprehensive Occupational Medical Program for Fire Departments*

You Are the Fire Fighter

I t is your first day in the station after having completed four months at the recruit academy. Your new crew welcomes you. As you are being shown around the station, the doorbell rings, and you are told your first official duty is to answer it. At the door, you find a young woman and her young son. She sheepishly says that her son has been begging her to meet the fire fighters, because he wants to be one when he grows up.

As you begin the tour, you show the pair the engine room and the boot room. Like most youngsters, the visiting boy is full of questions and asks, "Where are the rest of the fire fighters for all those coats? Do you drive the fire truck? Why does that truck have a big ladder on top?" The mother tells him to give you a chance to explain before asking any more questions. You smile as you ponder your answers.

1. How do you explain the difference between the number of turnout ensembles and the number of personnel at the station?
2. Explain how the duties of the captain, engineer, and fire fighters differ.
3. Explain the types of apparatus used in the fire service and their functions.

Introduction

Training to become a fire fighter is challenging. The art and science of extinguishing fires is much more complex than most people imagine. You will be challenged—both physically and mentally—during this course: You must keep your body in excellent condition, so you can complete your assignments. You must also remain mentally alert to cope with the various conditions you will encounter. Fire fighter training will expand your understanding of fire suppression—that is, the various activities involved in controlling and extinguishing fires.

By the time you complete this course, you will be well equipped to continue a centuries-old tradition of preserving lives and property threatened by fires **FIGURE 1-1**. You will have an understanding of the many duties and responsibilities of

FIGURE 1-1 When you join the fire service, you join a profession with a long and noble history of protecting and serving the community.

© Vince Talotta/Toronto Star/Getty Images

the fire service, and you should have a feeling of personal satisfaction about your accomplishment.

First, this chapter offers five guidelines for you to remember as you progress through this course and take your place as a fire fighter in your community. Next, it presents the progressive responsibilities of a Fire Fighter I, who operates under direct supervision, and a Fire Fighter II, who provides general supervision. The chapter then explains the organization of the fire service and describes the chain of command that connects all members of a department. Finally, it reviews the history and development of the modern-day fire service.

Fire Fighter Guidelines

During this course of study, you will need to practise and work hard. Do your best. The five guidelines listed here will help both to keep you on target to become a proud and accomplished fire fighter and throughout your career as a fire fighter.

1. **Be safe.** Safety should always be top of mind. Keep yourself safe. Keep your team members safe. Keep the public you serve safe.
2. **Follow orders.** Your supervisors have more training and experience than you do. If you can be counted on to follow orders, you will become a dependable member of the team.
3. **Work as a team.** Fighting fires requires the coordinated efforts of all department members. Teamwork is essential to success.
4. **Think!** Lives will depend on the choices you make. Put your brain in gear. Think about what you are studying.
5. **Follow the golden rule.** Treat each team member, victim, or citizen as an important person or as you would treat a member of your own family. Everyone is an important person to someone, and everyone deserves your best efforts.

Fire Fighter Qualifications

Not everyone can become a fire fighter. Those who do succeed in achieving this status understand the vital mission of the fire department: to save lives and protect property through prevention, education, and suppression. A fire fighter must be healthy and in good physical condition, assertive enough to enter a dangerous situation, yet mature enough to work as a member of a team **FIGURE 1-2**. The job requires a person who has the desire to learn, the will to practise, and the ability to apply the skills of the trade. A fire fighter is constantly learning as the body of knowledge about fires expands and the technology used in fighting fires evolves.

The training and performance qualifications for fire fighters are specified in NFPA 1001, *Standard for Fire Fighter Professional Qualifications*. Age, education requirements, medical requirements, and other criteria are established locally.

■ Age Requirements

Most career fire departments require that candidates be at least 18 years of age, although some specify a minimum age of 21. Candidates should possess a valid driver's licence, have a clean driving record, have no criminal record, and be drug-free.

Volunteer fire departments usually have different age requirements, which often depend on insurance considerations. Some volunteer fire departments allow junior members to join at age 16 but restrict their activities until they reach age 18.

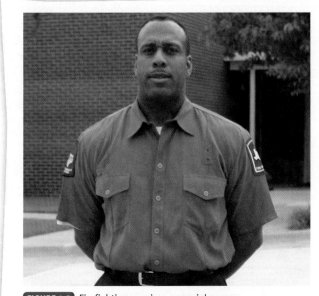

FIGURE 1-2 Firefighting requires a special person.

■ Education Requirements

Most career fire departments require that applicants have at least a high school diploma or the equivalent. Some require candidates to have completed more advanced college-level courses in a field related to fire or Emergency Medical Services (EMS). Most departments require fire fighters who wish to be considered for promotion or extra responsibility to take additional courses.

■ Medical Requirements

Because firefighting is both stressful and physically demanding, fire fighters must undergo a medical evaluation before their training begins. This medical evaluation will identify most medical conditions or physical limitations that could increase the risk of injury or illness either to the candidate or to other fire fighters. Medical requirements for fire fighters are specified in NFPA 1582, *Standard on Comprehensive Occupational Medical Program for Fire Departments*.

■ Physical-Fitness Requirements

Physical-fitness requirements are established to ensure that fire fighters have the strength and stamina needed to perform the tasks associated with firefighting and emergency operations. Several performance-testing scenarios have been validated by qualified organizations. NFPA 1001 allows individual fire departments to choose the physical-fitness testing method that will be used for fire fighter candidates.

■ Emergency Medical Care Requirements

Delivering emergency medical care is an important function of most fire departments. NFPA 1001 allows individual fire departments to specify the level of emergency medical care training required for entry-level personnel. At a minimum, you will need to understand infection-control procedures and be able to perform cardiopulmonary resuscitation (CPR), control bleeding, and manage shock. Many departments require fire fighters to become certified as an Emergency Medical Responder (EMR), an Emergency Medical Technician (EMT)/ Basic Primary Care Paramedic (PCP), or an Advanced Care Paramedic (ACP).

Roles and Responsibilities of a Fire Fighter I and a Fire Fighter II

The first step in understanding the organization of the fire service is to learn your roles and responsibilities as a <u>Fire Fighter I</u> or <u>Fire Fighter II</u>. As you progress through this text, you will learn what to do and how to do it so that you can take your place confidently among the ranks.

■ Roles and Responsibilities of a Fire Fighter I

- <u>Don</u> (put on) and <u>doff</u> (take off) personal protective equipment (PPE) properly.
- Hoist hand tools using appropriate ropes and knots.
- Understand and correctly apply appropriate communications protocols.

- Use self-contained breathing apparatus (SCBA).
- Respond on apparatus to an emergency scene.
- Establish and operate safely in emergency work areas.
- Force entry into a structure.
- Exit a hazardous area safely as a team.
- Set up ground ladders safely and correctly.
- Attack a passenger vehicle fire, an exterior Class A fire, and an interior structure fire.
- Conduct search and rescue in a structure.
- Perform ventilation of an involved structure.
- Overhaul a fire scene.
- Conserve property with salvage tools and equipment.
- Connect a fire department engine to a water supply.
- Extinguish incipient Class A, Class B, Class C, and Class D fires.
- Illuminate an emergency scene.
- Turn off utilities.
- Combat a ground cover fire.
- Perform fire safety surveys.
- Clean and maintain equipment.

FIGURE 1-3 Any contact with the public should be used as an educational opportunity.

Courtesy of Captain David Jackson, Saginaw Township Fire Department

■ Roles and Responsibilities of a Fire Fighter II

- Prepare reports.
- Communicate the need for assistance.
- Coordinate an interior attack line team.
- Extinguish a flammable-liquid fire.
- Control a flammable-gas cylinder fire.
- Protect evidence of fire cause and origin.
- Assess and disentangle victims from motor vehicle accidents.
- Assist special rescue team operations.
- Perform a fire safety survey.
- Present fire safety information.
- Maintain fire equipment.
- Perform annual service tests on fire hoses.

Every member of the fire service will interact with the public. In addition to interacting with citizens at incident scenes, fire fighters will encounter people who visit the fire station requesting a tour or asking questions about specific fire safety issues . Fire fighters should be prepared to assist these visitors and use these opportunities to provide them with additional fire safety information. Use every interaction with the public to deliver positive public relations and an educational message.

Roles Within the Fire Department

■ General Roles

A fire fighter may be assigned any task from placing hose lines to extinguishing fires. Generally, the fire fighter is not responsible for any command functions and does not supervise other personnel—except on a temporary basis when promoted to acting officer. Through additional training and promotion, a fire fighter may fulfill many roles during his or her career,

particularly in smaller departments. Some of the more common positions include the following:

- **Fire apparatus driver/operator**—Often called an engineer or technician, the driver is responsible for getting the fire apparatus to the scene safely, as well as setting up and running the pump or operating the aerial ladder once it arrives on the scene. This function is a full-time role in some departments; in other departments, it is rotated among the fire fighters.
- **Company officer**—The company officer is usually a lieutenant or captain who is in charge of an apparatus. This person leads the company both on scene and at the station. The company officer is responsible for initial firefighting strategy, personnel safety, and the overall activities of the fire fighters on his or her apparatus. Once command is established, the company officer focuses on tactics.
- **Safety officer**—The safety officer watches the overall operation for unsafe practices. He or she has the authority to halt any firefighting activity, allowing it to resume only when it can be done safely and correctly. The senior ranking officer may act as safety officer until the appointed safety officer arrives or until another officer is delegated those duties.
- **Training officer**—The training officer is responsible for updating the training of current fire fighters and for training new fire fighters. He or she must be aware of the most current techniques for firefighting and EMS. The training officer must coordinate various aspects of the fire department's training program and maintain documentation of all training activities.
- **Incident commander (IC)**—The IC is responsible for the management of all incident operations. This position focuses on the overall strategy for a particular

incident and is often assumed by the battalion/district chief.

- **Fire marshal/fire inspector/fire investigator**—The fire marshal inspects businesses and enforces public safety laws and fire codes. He or she may also respond to fire scenes to help investigate the cause of a fire. The investigator may have full police powers to investigate and arrest suspected arsonists and people causing false alarms.
- **Fire and life-safety education specialist**—The fire and life-safety education specialist educates the public about fire safety and injury prevention and presents juvenile fire-safety programs.
- **911 dispatcher/telecommunicator**—Working from the communications centre, the dispatcher takes calls from the public, sends appropriate units to the scene, assists callers with emergency medical information (treatment that can be performed until the medical unit arrives), and assists the IC with obtaining needed resources.
- **Fire apparatus maintenance personnel**—Apparatus mechanics repair and service fire and EMS vehicles, keeping them ready to respond to emergencies. These individuals are usually trained by equipment manufacturers to repair vehicle engines, lights, and all parts of the fire pump and aerial ladders.
- **Fire police**—Fire police are usually fire fighters who control traffic and secure the scene from public access. Many fire police are also sworn peace officers.
- **Information management**—"Info-techs" are fire fighters or civilians who take care of a department's computer and networking systems. Maintaining computerized fire reports, department e-mail, and a functional computer network requires the services of properly trained computer personnel.
- **Public information officer**—The public information officer serves as a liaison between the IC and the news media.
- **Fire protection engineer**—The fire protection engineer reviews plans and works with building owners to ensure that their fire-suppression and detection systems will meet the applicable codes and function as needed. Some fire protection engineers actually design these systems, and most hold an engineering degree.

■ Specialized Response Roles

Many assignments require specialized training. Indeed, most large departments have teams of fire fighters who respond to specific types of calls. Members of these teams are usually required to be fire fighters before they begin additional training. Specialist positions include the following:

- **Aircraft rescue fire fighter (ARFF)**—Aircraft rescue fire fighters are based out of military and civilian airports and receive specialized training in aircraft fires, extrication of victims from aircraft, and extinguishing agents. They wear special PPE and respond in specialized fire apparatus that protect them from high-temperature fires caused by substances such as jet fuel.
- **Hazardous materials technician**—"Haz-mat" technicians have training and certification in chemical identification, leak control, decontamination, and clean-up procedures.
- **Technical rescue technician**—"Tech rescue" technicians are trained in special rescue techniques for incidents involving structural collapse, trench rescue, swiftwater rescue, confined-space rescue, high-angle rescue, and other unusual situations. The units they work in are sometimes called urban search-and-rescue teams.
- **SCUBA dive rescue technician**—Many fire departments, especially those located near waterways, lakes, or an ocean, use SCUBA (self-contained underwater breathing apparatus) technicians who are trained in rescue, recovery, and search procedures in both water and under-ice situations.
- **Emergency Medical Services (EMS) personnel**—EMS personnel administer prehospital care to people who are sick or injured. Prehospital calls account for the majority of responses in many departments, so fire fighters are often cross-trained as EMS personnel. EMS training levels are normally divided into three categories: EMR, PCP, and ACP.
  1. **Emergency Medical Responder (EMR)**—Most EMS providers are EMRs. They have training in basic emergency care, including oxygen therapy, bleeding control, CPR, automated external defibrillation, basic airway devices, and assisting patients with certain medications.
  2. **Primary Care Paramedic (PCP)**—PCPs can perform more procedures than EMRs, but they are not yet ACPs. They have training in specific aspects of advanced life support (ALS), such as intravenous (IV) therapy, interpretation of cardiac rhythms, and defibrillation.
  3. **Advanced Care Paramedic (ACP)**—An ACP has completed the highest level of training in EMS. These personnel have extensive training in ALS, including IV therapy, administering drugs, cardiac monitoring, inserting advanced airways (endotracheal tubes), manual defibrillation, and other advanced assessment and treatment skills.

Working with Other Organizations

Fire departments are a vital part of their communities. To fulfill its mission, each fire department must interact with other organizations in the community. Consider the response to a motor vehicle crash **FIGURE 1-4**. In an area with a centralized 911 call centre, the fire department, EMS, law-enforcement officials, and tow-truck operators will all be notified of this event. If the crash affects utility service, the utility company will be notified as well. If EMS is not part of the fire department, the EMS

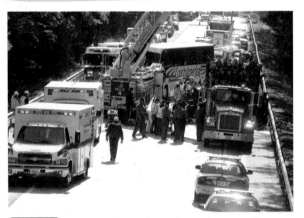

FIGURE 1-4 Many agencies work together at the scene of a motor vehicle crash.

© Patrick Kane/AP Photos

provider will be contacted. In addition, fire fighters frequently interact with hospital personnel who assume care for ill or injured victims who are transported to emergency rooms.

When multiple agencies work together at an incident, a unified command must be established as part of the **Incident Management System (IMS)**. A unified command system eliminates multiple command posts, establishes a single set of incident goals and objectives, and ensures mutual communication and cooperation. Although there can be only one IC at each scene, each agency must have input in handling an emergency. The IMS is discussed more thoroughly in the chapter "Incident Management System."

Other organizations may be involved in different types of incidents. For example, utility companies, The Salvation Army, and the Canadian Red Cross may be part of the team responding to a structure fire. A wildland fire may involve representatives from various government agencies and jurisdictions. Large-scale incidents may call upon several agencies, including the following:

- Public works
- School administrators
- Medical examiners or coroners
- Funeral directors
- Government officials (e.g., mayor, city manager)
- Royal Canadian Mounted Police (RCMP)
- Military
- Public Safety Canada (PS)
- Search-and-rescue teams
- Fire investigators
- Various provincial/territorial agencies

Communications centres keep a list of contact names and phone numbers, so these parties may be notified without delay when their presence at an incident is required.

Fire Department Governance

<u>Governance</u> is the process by which an organization exercises authority and performs the functions assigned to it. The governance of a fire department depends on regulations, policies, and

suggested operating guidelines (SOGs), or standard operating procedures (SOPs). Each of these concepts is discussed in this section.

<u>Regulations</u> are developed by various government or government-authorized organizations to implement laws that have been passed by a government body; for example, federal occupational health and safety laws are adopted as regulations by some jurisdictions. These regulations may apply to activities within the fire department.

<u>Policies</u> are developed to provide definitive guidelines for present and future actions; fire department policies, for example, outline what is expected in stated conditions. These policies often require personnel to make judgments and to determine the best course of action within the stated policy. Policies governing parts of a fire department's operations may be developed by other government agencies, such as personnel policies that cover all employees of a city, county or municipality.

<u>Standard operating procedures (SOPs)</u> provide specific information on the actions that should be taken to accomplish a certain task **FIGURE 1-5**. SOPs are developed within the fire department, are approved by the chief of the department, and ensure that all members of the department perform a given task in the same manner. They provide a uniform way of dealing with emergency situations, enabling fire fighters from different stations or companies to work together smoothly, even if they have never encountered one another before. These procedures are vital, because they enable everyone in the department to function properly and know what is expected for each task. Fire fighters must learn and frequently review departmental SOPs. An example would be your policy and procedure manuals.

In some fire departments, <u>suggested operating guidelines (SOGs)</u>, which are not as strict as SOPs, are utilized because conditions may dictate that the fire fighter or officer use his or her personal judgment in completing the procedure. This flexibility allows the responder to deviate from a set procedure yet still be held accountable for that action.

A practical way to organize a department's SOG/SOP manual is with removable pages collected in a three-ring binder, which ensures that regular updates can be added easily. Each department member should have an SOG/SOP manual and must update it as needed. This manual should be organized in sections—for example, administration, safety, scene operations, apparatus and equipment, station duties, uniforms, and miscellaneous—and use a simple numbering system based on section and policy numbers. Many fire departments maintain their SOGs/SOPs on a computer network, which simplifies the process of providing all employees with up-to-date SOGs/SOPs.

The Organization of the Fire Service

■ Company Types

A fire department includes many different types of companies, each of which has its own job at the scene of an emergency. Companies may be composed of various combinations of people and equipment; in smaller departments, a company may

Anytown Fire Department

Standard Operating Procedure

Date: 1/1/08
Section: 1 – Administration   SOP 01-01   Page 1 of 1
Maintaining Station Logbooks

Purpose

This guideline is provided to ensure that information documented in station logbooks is maintained in a consistent manner, station to station, shift to shift, throughout the department. The station commander shall have discretion in formatting the information.

Scope

This guideline shall be followed by all authorized department personnel entering information into the station logbooks. Any deviations from this guideline will be the responsibility of the individual making the deviation.

Policy

The station log is an important component of the fire department's record-keeping system. All entries must be legible, written in black ink, and the writer identified by name and computer identification number. All members should treat the logbook as a legal document and record all of the station's activities as well as other information deemed important by the station commander.

Procedure

The following guidelines should be used when making logbook notations:

1. The day, date, and shift noted on the top line of the page.
2. A list of each person assigned to that shift to include their computer identification numbers, and the apparatus to which they are assigned for the shift. Personnel on leave should be identified with the type of leave (ie, annual, sick, leave without pay, funeral, military, training) noted. A notation should be made by the name of any member temporarily assigned to another station during that shift.
3. Daily safety talks should be recorded prior to the section dedicated to emergency responses. For more information on safety talks, refer to SOP 31.3.
4. The section of the log dedicated to recording chronological activity should have two (2) columns on the left side of the book: the far left should note time and the second column the incident number. A column to the far right should be used to note the identification number of the person making the entry.
5. Incident numbers for medic unit responses should be noted in blue ink.
6. Incident numbers for fire suppression responses should be noted in red ink.
7. Units responding, the address to which the response is being made, and the type of situation found should be noted behind the message number (ie, E-5 responded to 304 Albemarle Drive/Gas Leak).
8. The term fill-in should be used in the logbook to denote one station standing by in another station to cover a company that has responded to an incident.
9. Entries for medic units should include the hospital to which the patient was transported.
10. Station maintenance/repairs and apparatus maintenance should be documented in the section of the logbook dedicated to that purpose. This documentation may also be maintained in a separate logbook dedicated to those types of entries.

Although the information requirements provided here must be maintained in the station's logbook, the format used for ensuring that documentation is left to the station commander's discretion.

Approved: _____   Date: _____
                          Fire Chief

FIGURE 1-5 A sample SOP.

fill many roles. The most common types of companies and their roles are described here:

- **Engine company**—An engine company is responsible for securing a water source, deploying handlines, conducting search and rescue operations, and putting water on the fire **FIGURE 1-6**. Each fire engine has a pump, carries hoses, and maintains a booster tank of water. Engines also carry a limited quantity of ladders and hand tools.

- **Truck company**—A truck company (also called a ladder company) specializes in forcible entry, ventilation, roof operations, search and rescue, and deployment of ground ladders **FIGURE 1-7**. Trucks carry several ground ladders, ranging from 2.4 to 15.2 metres (8 to 50 feet) in length, as well as an extensive quantity of tools. Trucks are also equipped with an aerial device, such as an aerial ladder, tower ladder, or ladder/platform. These aerial devices can be raised and positioned above a roof to provide fire fighters with a stable, safe work zone.

- **Rescue company**—A rescue company is usually responsible for rescuing victims from fires, confined spaces, trenches, and high-angle situations. Rescue

companies carry an extensive array of regular and specialized tools.

- **Wildland/brush company**—A wildland/brush company is dispatched to wildland and brush fires that larger engines cannot reach. Because they often work in rough terrain, wildland/brush companies use four-wheel drive vehicles. They carry a tank of water and a pump that enables them to pump water while the truck is moving. These companies also carry special firefighting equipment such as portable pumps, rakes, and shovels.

- **Hazardous materials company**—A hazardous materials company responds to and controls scenes involving spilled or leaking hazardous chemicals. These companies have special equipment, PPE, and training to handle most emergencies involving chemicals.

- **Emergency Medical Services (EMS) company**—An EMS company may include medical units such as ambulances or first-response vehicles. These companies respond to and assist in the transport of medical and trauma victims to medical facilities for further treatment. They often carry medications, defibrillators, and other equipment that can stabilize a critical patient during transport. Engine or truck companies may be staffed with EMS providers who act as first responders until a transport ambulance arrives **FIGURE 1-8**.

■ Other Views of Fire Service Organization

There are several other ways to look at the organization of a fire department, such as in terms of staffing, function, and geography.

Staffing

A fire department must have sufficient trained personnel available to respond to an emergency at any hour of the day, every day of the year. Staffing issues affect all fire departments—career departments, combination departments, and volunteer departments. In volunteer departments, it is particularly important to ensure that enough responders are available at all times, particularly during the day. In the past, when many people worked in or near the communities where they lived, volunteer response time was not an issue; today, however,

FIGURE 1-6 An engine company secures a water source and extinguishes the fire.

© Brad Sauter/ShutterStock, Inc.

FIGURE 1-7 A truck company provides aerial support at a fire.

FIGURE 1-8 An EMS company delivers medical services.

many people have longer commutes and work longer hours, so the number of people available to respond during the day may be limited. Some volunteer departments have been forced to hire full-time fire fighters during daytime hours to ensure sufficient personnel are available to respond to an incident.

Function

Fire departments can also be organized along functional lines. For example, the training division is responsible for leading and coordinating department-wide training activities; truck companies have certain functional responsibilities at a fire; hazardous materials squads have different functional responsibilities that support the overall mission of the fire department.

Geography

Each fire department is responsible for a specific geographic area, and each station in the department is assigned the primary responsibility for a geographic area within the community. Fire stations are distributed throughout a community in a manner intended to ensure a rapid response time to every location in that community. This design enables the fire department to distribute and use specialized equipment efficiently throughout the community.

■ Chain of Command

The organizational structure of a fire department consists of a **chain of command**. Although the precise ranks may vary in different departments, the basic concept remains the same across the fire service. The chain of command creates a structure for managing the department and the fire-ground operations.

Fire fighters usually report to a **lieutenant**, who is responsible for a single fire company (e.g., an engine company) on a single shift. Lieutenants can provide a number of practical skills and tips to new recruits.

The next level in the chain of command is the **captain**: A captain is responsible for not only managing a fire company on his or her shift, but also coordinating the company's activities with other shifts. A captain may be in charge of the activities of a fire station.

Captains report directly to chiefs. Several levels of chiefs are designated. A **battalion chief** (also known as a district chief) is responsible for coordinating the activities of several fire companies in a defined geographic area, such as a station or district. A battalion chief is usually the officer in charge of a single-alarm working fire.

Above battalion chiefs are **assistant** or **division chiefs**. An assistant or division chief is usually in charge of a functional area, such as training, within the department. These officers report directly to the chief of the department.

The **chief of the department** has overall responsibility for the administration and operations of the fire department. The chief can delegate responsibilities to other members of the department but is still responsible for ensuring that these activities are carried out properly.

The fire service's chain of command is used to implement department policies. This organizational structure enables a fire department to determine the most efficient and effective way to fulfill its mission and to communicate this information

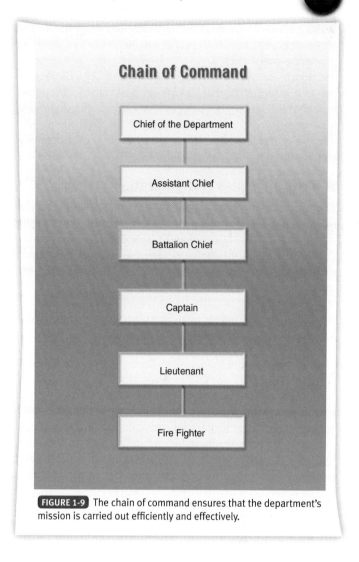

FIGURE 1-9 The chain of command ensures that the department's mission is carried out efficiently and effectively.

to all members of the department **FIGURE 1-9**. Adhering to the chain of command ensures that a given task is carried out in a uniform manner. A variety of documents, including the previously discussed regulations, policies, procedures, and SOGs/SOPs, are used to achieve this goal.

■ Source of Authority

Governments—whether municipal, provincial/territorial, or federal—are charged with protecting the welfare of the public against common threats. Fire is one such peril, because an uncontrolled fire threatens everyone in the community. Citizens accept certain restrictions on their behaviour and pay taxes to protect themselves and the common good. As a consequence, people charged with protecting the public may be given certain privileges to enable them to perform effectively; for example, fire departments can legally enter a locked home without permission to extinguish a fire and protect the public.

The fire service draws its authority from the governing entity responsible for protecting the public from fire—be it a town, city, county, municipality, or special fire district. Federal and provincial/territorial governments also grant authority to fire departments. In some provinces/territories, private corporations have contracts to provide fire protection to municipalities or government agencies. The head of the fire department (i.e., fire chief) is accountable to the leaders of the governing body, such as the city council, the municipality, the mayor, or the city manager. Because of the relationship between a fire department and local government, fire fighters should consider themselves to be civil servants, working for the tax-paying citizens who fund the fire department. The ultimate customers for a fire department are the citizens of a community.

■ Basic Principles of Organization

The fire department uses a paramilitary style of leadership; in other words, fire fighters operate under a rank system, which establishes a chain of command (as discussed earlier in this chapter) in which the fire chief serves as head of the department. Most fire departments use four basic management principles:

1. Discipline
2. Division of labour
3. Unity of command
4. Span of control

Discipline

<u>Discipline</u> is guiding and directing fire fighters to do what their fire department expects of them. Positive discipline consists of providing guidelines for the right way of doing things. Examples of positive discipline are policies, SOPs, training, and education. Corrective discipline consists of actions taken to discourage inappropriate behaviour or poor performance. Examples of corrective discipline are counselling sessions, formal reprimands, or suspension from duty.

Division of Labour

<u>Division of labour</u> is a way of organizing an incident by breaking down the overall strategy into a series of smaller tasks. Some fire departments are divided into units based on function; for example, engine companies establish water supplies and pump water, while truck companies perform forcible entry, rescue, and ventilation functions. Each of these functions can be divided into multiple assignments, which can then be assigned to individual fire fighters. With division of labour, the specific assignment of a task to an individual makes that person responsible for completing the task and prevents duplication of job assignments.

Unity of Command

<u>Unity of command</u> is the concept that each fire fighter answers to only one lieutenant/captain, each lieutenant/captain answers to only one battalion chief/IC, and so on **FIGURE 1-10**. In this way, the chain of command ensures that everyone is answerable to the fire chief and establishes a direct route of responsibility from fire chief to fire fighter.

At a fire ground, all functions are assigned according to incident priorities. A fire fighter with more than one supervising officer during an emergency may be overwhelmed with various assignments, and the incident priorities may not be accomplished in a timely and efficient manner. The concept of unity of command is designed to avoid such conflicts and lead to more effective firefighting.

Span of Control

<u>Span of control</u> is the number of people whom one person can supervise effectively. Most experts believe that span

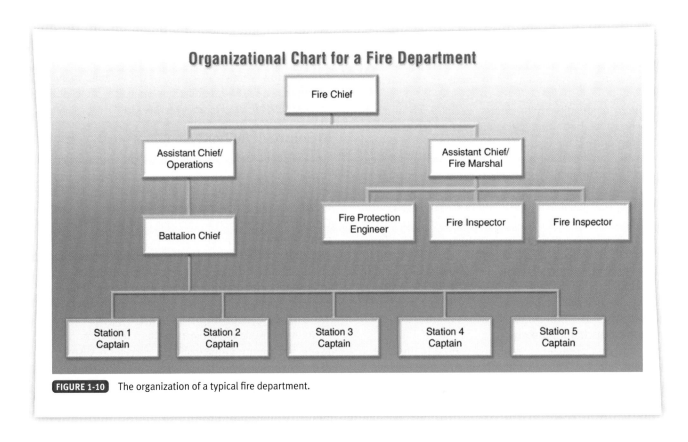

Organizational Chart for a Fire Department

- Fire Chief
  - Assistant Chief/Operations
    - Battalion Chief
      - Station 1 Captain
      - Station 2 Captain
      - Station 3 Captain
      - Station 4 Captain
      - Station 5 Captain
  - Assistant Chief/Fire Marshal
    - Fire Protection Engineer
    - Fire Inspector
    - Fire Inspector

FIGURE 1-10   The organization of a typical fire department.

of control should extend to no more than five people in a complex or rapidly changing environment. This number can change, however, depending on the assignment or task to be completed.

The History of the Fire Service

Since prehistoric times, controlled fire has been a source of comfort and warmth, while uncontrolled fire has brought death and destruction. Historical accounts from the ancient Roman Empire describe community efforts to suppress uncontrolled fire. In 24 BC, the Roman emperor Augustus Caesar created what was probably the first fire department. Called the Familia Publica, it was composed of approximately 600 slaves who were stationed around the city and charged with watching for and fighting fires. Of course, because the Familia Publica consisted of slaves, these conscripts had little interest in preserving the homes of their masters and little desire to take risks, so fires continued to be a problem.

In about 60 AD, under the emperor Nero, the Corps of Vigiles was established as the Roman Empire's fire protectors. This group of 7000 free men were responsible for firefighting, fire prevention, and building inspections. The Corps of Vigiles adopted the formal rank structure of the Roman military, which continues to be used by today's fire departments.

■ The Canadian Fire Service

When early French and British settlers came to Canada, they established communities in forested areas that were vulnerable to fire and did not consider fire protection for their homes. At that time, most structures were built entirely of combustible materials such as straw and wood, as these materials were readily available. Local ordinances soon required the use of less flammable building materials and mandated that fires be **banked** (covered over) throughout the night. In 1630, the city of Boston established the first fire regulations in North America when it banned wood chimneys and thatched roofs. In a short time, these regulations were being applied in Canada as well. There were many fires caused by buildup in chimneys, and because of this, the new fire regulations also required that chimneys be swept out regularly. **Fire wardens** imposed fines on homeowners who did not obey these regulations; the money collected was used to pay for firefighting equipment.

It was not until the late 1700s that organized firefighting methods began to develop. After major fire losses in Montreal and in the Maritimes, the communities decided that they wanted to control the threat of fire. As communities started to grow and prosper, they would organize local councils to address the fire safety issues. This was a lengthy process, and the merchants that suffered the losses determined that they could do a better job.

The earliest firefighting measures required households to have several buckets filled with water close to their front door, and homeowners would respond to fires with these buckets when a fire was identified in the community. They also used axes, ladders, ropes, and hook devices to help with the prevention of fire spread.

FIRE FIGHTER Tips

Early insurance companies marked the homes of their policy holders with a plaque or fire mark that showed the name or logo of the insurance company. The insurance company paid fire companies to respond to those buildings that displayed their fire mark; sadly, others were left to burn.

The Great Fire of London in 1666 prompted the organization and development of fire insurance companies, which would provide their own fire protection with on-call firefighting teams **FIGURE 1-11**. In the late 1700s, insurance policies became available in Canada, and at this time, the British built fire engines to replace some of the primitive tools and equipment from earlier periods. The Royal Navy and the British Army showed the cities in Canada how to use pumping engines, which were already in use on warships. With this technology available, the older communities started to purchase their own fire engines. Some of the engines were donated by British fire insurance companies to help with the fire prevention effort.

Early water supply in communities started with basic wells and then turned to individuals who had containers on wheels to deliver supplies to the households. Soon, the municipal officials began paying these individuals to show up to fires in the community. The first municipal piped-water system opened in Montreal in 1832. This system provided low pressure for the engines. Other cities opened municipal piped-water systems shortly after.

In many countries, the military or police were the individuals who participated in early firefighting efforts. In Canada, the local citizens that had the most property to lose would band together and protect each other's property. They organized into what would ultimately be known as fire companies, and they would each have their own hand engine and a hose reel or a ladder on a cart, which carried salvage materials. This organized group became well known in the various colonies of New England and was looked upon as a positive promotion of volunteer fire companies. Many people who fled the American Revolution came to Canada to settle in its established communities, and they brought this enthusiasm for volunteer fire companies with them. When the gold rush arrived on the west coast, it brought people to Victoria, British Columbia, including many ex-volunteer fire fighters from the United States, who helped to create very strong fire companies.

During the 1800s, the volunteer fire company model would be developed in all Canadian cities and towns. These volunteer fire fighters took their jobs very seriously and were often involved in disputes with municipal officials. These individuals would respond to calls quickly and would take extreme pride in the work they were providing. During this turbulent political time in Canada, these volunteer fire companies were involved in the riots over the Rebellion Losses Bill in Montreal—the then capital city. These fire companies participated in the riots and would ultimately be held responsible for burning the Parliament buildings in Montreal.

FIGURE 1-11 The Great Fire of London in 1666 prompted the organization and development of fire insurance companies, which would provide their own fire protection with on-call firefighting teams.

Today's building codes not only govern construction materials but also frequently require built-in fire prevention and safety measures. Required fire detection equipment notifies both building occupants and the fire department. Built-in fire suppression and sprinkler systems help contain fires to a small area and prevent small fires from growing into major fires. Fire escapes, stairways, and doors that unlock when the alarms sound and that open outward enable occupants to escape a burning building safely. Without modern building code requirements, high-rise buildings and large shopping centres could not be built safely.

FIRE FIGHTER Tips

Modern building codes help to ensure that fire departments receive prompt notification of fires and limit or extinguish fires that might otherwise overwhelm local firefighting resources.

At first, each community established its own building codes. As larger government jurisdictions were established, however, a more uniform code was adopted across Canada, reflecting a minimum standard. Local communities were permitted to make this code stricter, as needed.

Today, Canadian codes and standards are written by national organizations such as the National Research Council (NRC), Canadian Standards Association (CSA), and the National Fire Protection Association (NFPA). Volunteer committees of citizens and representatives of businesses, insurance companies, and government agencies research and develop proposals, which are in turn debated and reviewed by various groups. The final document, known as a **consensus document**, is then presented to the public. Many Canadian jurisdictions adopt selected NFPA codes and standards as law.

FIRE FIGHTER Tips

Alfred Perry was the captain of the volunteer corps of firefighters and was active in the promotion of fire protection in the 18th century for areas such as Montreal. Perry could be considered the Canadian version of Benjamin Franklin, who was the first leader of the volunteer fire company in the United States. Perry might also be considered the father of the Canadian fire service.

■ Building Codes

Throughout history, the threat of fire has served as an impetus for communities to establish building codes. Although the first building codes—which were developed in ancient Egypt—focused on preventing building collapse, building codes were quickly recognized as an effective means of preventing, limiting, and containing fires.

Colonial communities had few building codes. The first settlers had a difficult time erecting even primitive shelters, which often were constructed of wood with thatched straw roofs. The fireplaces used for cooking and heating may have had chimneys constructed of smaller logs. The all-wood construction and use of open fires meant that fires were a constant threat to early settlers. As communities developed, they enacted codes restricting the hours during which open fires were permitted and the materials that could be used for roofs and chimneys. After British forces burned Washington in 1814 (during the War of 1812), codes prohibited the building of wooden houses in the area. Some building codes required the construction of a fire-resistive wall, or fire wall, of brick or mortar between two buildings.

■ Training and Education

Fire fighter training and education have also come a long way over the years. The first fire fighters simply needed sufficient muscular strength and endurance to pass buckets or operate a hand pumper. As equipment became more complex, however, the importance of formalized training and good judgment increased.

Today's fire fighters operate high-tech, costly equipment, including million-dollar apparatus, radios, **thermal imaging cameras**, and **self-contained breathing apparatus (SCBA)** **FIGURE 1-12** . These tools, as well as better fire detection devices, have greatly increased the safety and effectiveness of modern day fire fighters. The most important "machines" on the fire scene, however, remain the intelligent, knowledgeable, well-trained, physically capable fire fighters, who have the ability and determination to attack the fire. A thermal imaging camera may be able to find someone trapped in a burning building, but it takes a smart, able-bodied fire fighter to remove the victim safely.

The increasing complexity of both the world and the science of firefighting requires that fire fighters continually sharpen their skills and increase their knowledge of potential hazards. That need for ongoing education is why training courses such as this one are just as important as good physical fitness to today's fire fighters.

FIRE FIGHTER Tips

You will find many resources on the history of firefighting at your public library, in local bookstores, and on the Internet. These resources make for interesting reading and learning about how things have—or have not—changed over the years.

■ Fire Equipment

Today's equipment and apparatus evolved over a number of years, as new inventions were adapted to the needs of the fire service. Colonial fire fighters, for example, had only buckets, ladders, and **fire hooks** (tools used to pull down burning structures) at their disposal. Homeowners were required to keep buckets filled with sand or water and bring them to the scene of the fire. Some towns also required that ladders be available so that fire fighters could access roofs to extinguish small fires. If all else failed, the fire hook was used to pull down a burning building and prevent the fire from spreading to nearby structures. The hook-and-ladder truck evolved from this early equipment.

Buckets gave way to hand-powered pumpers in 1720, when Richard Newsham developed the first such pumper in London. As many as 16 strong men were used to power the pump, making it possible to propel a steady stream of water from a safe distance. In 1829, more powerful steam-powered pumpers replaced the hand-powered models. Unfortunately, many volunteer fire fighters felt threatened by these steam engines and fought against their use. Steam engines were heavy machines that were pulled to the fire by a team of trained horses. They required constant attention, which limited their use to larger cities that could bear the costs of maintaining the horses and the steamers.

The advent of the internal combustion engine in the early 1900s greatly changed the fire service and enabled even small towns to have machine-powered pumpers. Today, both staffed and unstaffed firehouses keep fire engines ready to respond at any hour of the day or night. Although they require regular maintenance, current equipment does not require the constant attention that horses or steam engines did. Modern fire apparatus carry water, a pumping mechanism, hoses, equipment, and personnel. In this sense, a single apparatus has replaced several single-function vehicles from the past.

The progress in fire protection equipment extends beyond trucks. Without an adequate water supply, modern-day apparatus would be useless. The advent of municipal water systems provides large quantities of water to extinguish major fires.

Romans developed the first municipal water systems, just as they had developed the first fire companies. It was not until the 1800s, however, that water distribution systems were applied to fire-suppression efforts. George Smith, a fire fighter in New York City, developed the first fire hydrants in 1817. He realized that using a valve to control access to the water in the pipes would enable fire fighters to tap into the system whenever a fire occurred. These valves, or **fire plugs**, were used with both above-ground and below-ground piping systems. In Canada, the first municipal piped-water system was opened in Montreal in 1832. In 1861, Toronto and Montreal installed more powerful water main systems that were equipped with hydrants.

Because small fires are more easily controlled, the sooner a fire department is notified, the more likely it will be able to extinguish the fire and minimize losses. The introduction of public call boxes in cities across Canada represented a major advance in this direction. The call boxes, which were placed around the city, enabled citizens to send a coded telegraph signal to the fire department, which received the message as a series of bells. The fire department could determine the location of the call box being used by the number of bells in the signal. When fire fighters arrived at the call box, the caller could then direct them to the exact location of the fire. Similar units are still used in many areas but are being replaced by more immediate and effective communications systems.

FIRE FIGHTER Tips

Buckets made of leather could not hold water for any length of time, so in some parts of the country, fire buckets were kept filled with sand. After the sand was thrown on the fire, the bucket would be refilled with water and used in a bucket brigade.

FIGURE 1-12 A thermal imaging camera is one of the many tools available to modern-day fire fighters.

■ Communications

Good communication is vital for effective firefighting. When a fire is discovered, fire fighters must be summoned and citizens alerted to the danger. During the attack, officers must be able to communicate with fire fighters or summon additional resources. Not surprisingly, then, improvements in communication systems are tied to improvements in the fire service.

During the colonial period, fire wardens or night watchmen patrolled neighbourhoods and sounded the alarm if a fire was discovered. Some towns built a series of fire towers where wardens would watch for fires. In many towns, ringing the community fire bell or church bells alerted citizens to a fire.

In the late 1800s, telegraph fire alarm systems were installed in large cities. These systems enabled more rapid reporting of fires and made it possible for officers to request additional resources or let dispatchers know that the fire was extinguished. In small towns, telegraph fire alarm systems were gradually replaced by community sirens mounted on poles or tall rooftops to signal a fire. Today, most fire departments rely on pagers or two-way radios to summon part-time or volunteer fire fighters to emergencies.

Many of the early communications systems have also been replaced by hard-wired and cellular telephones, which enable citizens to report an emergency from almost anywhere **FIGURE 1-13**. The use of telephones has greatly reduced the time involved in and difficulty of reporting a fire. The introduction of computer-aided dispatch facilities has likewise improved response times, because the closest available fire units can quickly be sent to the site of the emergency.

Communications during an attack also have improved over the years—from simply shouting loud enough to be heard over the chaos to using two-way radios. Before electronic

FIRE FIGHTER Tips

The historic symbol of the chief officer's trumpet is still used on a chief's badge. This series of crossed trumpets is one of the cherished traditions of the fire service.

amplification became available, the chief officer shouted commands through his trumpet. The **chief's trumpet**, or bugle, eventually became a symbol of authority **FIGURE 1-14**. Although chief officers no longer use trumpets for communicating, the use of multiple trumpets to symbolize the rank of chief signifies this person's need to communicate as well as to lead. Today's two-way radios enable fire units and individual fire fighters to remain in contact with one another at all times.

FIGURE 1-14 The chief's trumpet was once used to amplify this commander's voice. Today, it serves as a symbol of authority in the fire service.

FIGURE 1-13 Modern technology allows citizens to report an emergency from almost any location.

Paying for Fire Service

As previously noted, volunteer fire departments were common in colonial Canada and are still used today in many areas. The first fire wardens were employed by communities and paid from community funds. The introduction of hand-powered pumpers hastened the development of a permanent fire service. The question of who would pay for the equipment and the fire fighters, however, was not settled for many years.

Fire insurance companies were established in England soon after the Great Fire of London in 1666 to help victims cope with the financial loss from fires. Such companies collected fees (premiums) from homeowners and businesses and, in return, pledged to repay them for any losses resulting from fire. Fire insurance policies first became available in Canada in the late 1700s.

Because the insurance companies would save money if a fire was put out before much damage was done, they agreed to pay fire companies to try to extinguish fires. Houses that had insurance were designated by a **fire mark** FIGURE 1-15 . Most fire companies were loosely governed and organized, however, and more than one company might show up to fight a fire; if two fire companies arrived at a fire, a dispute would likely arise over which company would collect the money. Consequently, municipalities began assuming the role of providing fire protection.

Today almost all the fire protection in Canada is funded directly or indirectly through tax dollars. Some jurisdictions fund the fire department as a public service; others contract with an independent fire department. Volunteer fire departments frequently conduct fundraising activities. Ultimately, however, all funds for operating fire departments come from the citizens of the community.

Fire Service in Canada Today

The fire service in Canada today is a product of evolution over the past 400 years. As a novice fire fighter, it is helpful for you to learn from the past as well as to study the modern-day Canadian fire service.

The makeup of the Canadian fire services has evolved and grown over the years, but the tradition and overall organization remains the same. Statistics gathered by the NFPA show that most career departments protect communities of 50,000 or more, and volunteer fire fighters protect communities of 50,000 or less. This varies a great deal across Canada but is a good representation of the Canadian fire service today FIGURE 1-16 .

- All career—All of the members of the department are paid, full-time fire fighters.
- Combination departments:
  - Combination—These departments include both paid, full-time fire fighters and either on-call fire fighters or volunteers.
  - Mostly volunteer—More than half of the members in these departments are volunteers, although they do include paid, full-time fire fighters as well.
- All volunteer—All of the members of the fire department are volunteer fire fighters.

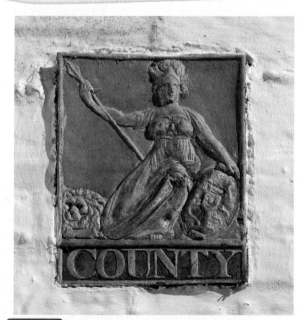

FIGURE 1-15 A fire mark indicated the homeowner had insurance that would pay the fire company for extinguishing the blaze.
© PURPLE MARBLES YORK 1/Alamy Images

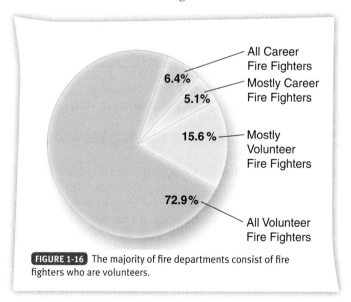

6.4% — All Career Fire Fighters

5.1% — Mostly Career Fire Fighters

15.6% — Mostly Volunteer Fire Fighters

72.9% — All Volunteer Fire Fighters

FIGURE 1-16 The majority of fire departments consist of fire fighters who are volunteers.

Wrap-Up

Chief Concepts

- Throughout your training and career, keep in mind the five fire fighter guidelines:
  1. **Be safe.** Safety for yourself, your teammates, and the public should be your top priority.
  2. **Follow orders.** Remember that your supervisors have more training and experience than you do.
  3. **Work as a team.** Teamwork is essential to safety and successful fire suppression.
  4. **Think!** Always keep your mind sharp and focused.
  5. **Follow the golden rule.** Treat each team member, victim, and citizen as you would like to be treated. Every life is important.
- The training and performance qualifications for fire fighters are specified in NFPA 1001. Age restrictions, educational requirements, medical requirements, and other criteria are established locally.
- A Fire Fighter I works in a team under direct supervision to suppress fires.
- A Fire Fighter II works in a team under general supervision. He or she may assume command, transfer command, and coordinate command within the IMS.
- Throughout your career, you may assume several roles in the fire department. Each role requires additional training:
  - The fire apparatus driver/operator is responsible for getting the fire apparatus to the scene safely as well as setting up and running the pump or operating the aerial ladder.
  - The company officer leads his or her company both on the scene and at the station.
  - The safety officer watches the overall operation for unsafe practices.
  - The training officer is responsible for updating the training of current fire fighters and for training new fire fighters.
  - The IC is responsible for the management of all incident operations.
  - The fire marshal/fire inspector/fire investigator inspects businesses and enforces public safety laws and fire codes. He or she may also respond to fire scenes to help investigate the cause of a fire.
  - The fire and life-safety education specialist educates the public about fire safety and injury prevention.
  - The 911 dispatcher/telecommunicator takes calls from the public and dispatches appropriate units to an emergency.
  - Fire apparatus maintenance personnel repair and service fire and EMS vehicles, keeping them ready to respond to emergencies.
  - Fire police control traffic and secure the scene from the public.
  - Information-management professionals are fire fighters or civilians who take care of a fire department's computer network system.
  - The public information officer serves as a liaison between the IC and the news media.
  - The fire protection engineer reviews plans and works with building owners to ensure that their fire suppression and detection systems will meet the applicable codes and function as needed.
- Many emergencies require specialized skills. Most large fire departments have teams of specialized fire fighters who can respond to specific emergencies:
  - An ARFF works at military and civilian airports and has specialized training in aircraft fires, extricating victims from aircraft, and extinguishing agents.
  - A hazardous materials technician is trained to identify chemicals, control leaks, decontaminate a scene, and clean up a scene.
  - A technical rescue technician is trained in special rescue techniques for incidents involving structural collapse, trench rescue, swiftwater rescue, confined-space rescue, high-angle rescue, and other specialized situations.
  - A SCUBA dive rescue technician is trained in water rescue, recovery, and search procedures.
  - EMS personnel are trained to administer prehospital medical care to victims.
- When multiple agencies, such as police, fire, and EMS, work together at an incident, a unified command must be established as part of the IMS. A unified command establishes a single set of incident goals under a single leader and ensures mutual communication and cooperation.
- Governance is the process by which an organization exercises authority and performs the functions assigned to it. The governance of a fire department depends on regulations, policies, and SOGs/SOPs.
  - Regulations are developed by various government or government-authorized organizations to implement a law that has been passed by a government body.
  - Policies are developed to provide definitive guidelines for present and future actions.
  - SOGs/SOPs provide specific information on the actions that should be taken to accomplish a certain task.

- A fire department includes many different types of companies to perform specific tasks at the scene of an emergency:
  - The engine company is responsible for securing a water source, deploying handlines, conducting search and rescue operations, and putting water on the fire.
  - The truck company is responsible for forcible entry, ventilation, roof operations, search and rescue, and ground ladder deployment.
  - The rescue company is responsible for rescuing victims from fires, confined spaces, trenches, and high-angle situations.
  - A wildland/brush company is dispatched to wildland and brush fires.
  - A hazardous materials company responds to and controls scenes involving spilled or leaking hazardous chemicals.
  - An EMS company includes ambulances and responds to medical emergencies.
- The chain of command may differ from fire department to fire department, but the basic concept remains the same across the fire service. The chain of command, from lowest rank to highest, is as follows:
  - Fire fighter
  - Lieutenant
  - Captain
  - Battalion chief
  - Assistant or division chief
  - Chief of the department
- Four basic management principles apply to the fire service:
  1. Discipline comprises the set of guidelines that a fire department establishes for fire fighters. Regulations, policies, and procedures are all forms of discipline.
  2. Division of labour is a way of organizing an incident by breaking down an overall strategy into a series of smaller tasks.
  3. Unity of command is the concept that each fire fighter answers to only one supervisor.
  4. Span of control is the number of people whom one person can supervise effectively.
- Highly destructive fires spurred communities to enact strict building and fire codes in an effort to prevent large losses of life and property. Today's building codes not only govern construction materials but also frequently require built-in fire prevention and safety measures, such as fire detectors.
- Fire equipment has evolved from leather buckets and wooden ladders in colonial times to today's thermal imaging cameras.
- Communications have evolved from simply shouting to using two-way radios.
- The fire service in Canada today is the product of an evolution over hundreds of years. As a novice fire fighter, it is helpful for you to learn from the past and to study the fire service in Canada today.

Hot Terms

911 dispatcher/telecommunicator Working from the communications centre, the dispatcher takes the calls from the public, sends appropriate units to the scene, assists callers with treatment instructions until the EMS unit arrives, and assists the IC with needed resources.

Advanced Care Paramedic (ACP) EMS personnel with the highest level of training, including cardiac monitoring, administering drugs, inserting advanced airways, manual defibrillation, and other advanced assessment and treatment skills.

Aircraft rescue fire fighter (ARFF) An individual with specialized training in aircraft fires, extrication of victims from aircraft, and extinguishing agents. These fire fighters wear special types of turnout gear and respond in fire apparatus that protect them from high-temperature fires.

Assistant or **division chief** A midlevel chief who often has a functional area of responsibility, such as training, and who answers directly to the fire chief.

Bank To cover a fire to ensure it burns slowly.

Battalion chief Typically the first level of fire chief, these chiefs are often in charge of running calls and supervising multiple stations or districts within a city; also called a district chief. A battalion chief is usually the officer in charge of a single-alarm working fire.

Captain The second rank of promotion in the fire service (between the lieutenant and the battalion chief). A captain is responsible for managing a fire company and for coordinating the activities of that company with the other shifts.

Chain of command A rank structure—spanning from the fire fighter through to the fire chief—for managing a fire department and fire-ground operations.

Chief of the department The top position in the fire department. The fire chief has ultimate responsibility for the fire department and usually answers directly to the mayor or other designated public official.

Chief's trumpet An obsolete amplification device that enabled a chief officer to give orders to fire fighters during an emergency. It was a precursor to the bullhorn and portable radios.

Company officer The officer, or any other position of comparable responsibility in the department, in charge of a fire department company or station. (NFPA 1143)

Consensus document A code document jointly developed by people representing various organizations and interests. NFPA codes and standards are consensus documents.

Discipline The guidelines that a department sets for fire fighters to work within.

Division of labour The breaking down of an incident or task into a series of smaller, more manageable tasks and assignment of personnel to complete those tasks.

Doff To take off an item of clothing or equipment.

Don To put on an item of clothing or equipment.

Emergency Medical Responder (EMR) The first trained individual to arrive at the scene of an emergency to provide initial medical care.

Emergency Medical Services (EMS) company A company that may be made up of medical units and first-response vehicles. Members of this company respond to and assist in the transport of medical and trauma victims to medical facilities. They often have medications, defibrillators, and paramedics who can stabilize a critical patient.

Emergency Medical Services (EMS) personnel Personnel who are responsible for administering prehospital care to people who are sick and/or injured. Prehospital calls make up the majority of responses in most fire departments, so fire fighters are often cross-trained as EMS personnel.

Engine company A group of fire fighters who work as a unit and are equipped with one or more pumping engines with a rated capacity of 2839 litres per minute (750 gallons per minute [gpm]) or more. (NFPA 1410)

Fire and life-safety education specialist A member of the fire department who deals with the public on education, fire safety, and juvenile fire-safety programs.

Fire apparatus driver/operator A member of the fire department who is authorized by the authority having jurisdiction to drive and/or operate fire department vehicles. (NFPA 1451)

Fire apparatus maintenance personnel People who repair and service fire and EMS vehicles, ensuring that they are always ready to respond to emergencies.

Fire Fighter I A person at the first level of progression who has demonstrated the knowledge and skills to function as an integral member of a firefighting team under direct supervision in hazardous conditions. (NFPA 1001)

Fire Fighter II A person at the second level of progression who has demonstrated the skills and depth of knowledge to function under general supervision. (NFPA 1001)

Fire hook A tool used to pull down a burning structure.

Fire mark Historically, an identifying symbol on a building informing fire fighters that the building was insured by a company that would pay them for extinguishing a fire.

Fire marshal/fire inspector/fire investigator A member of the fire department who inspects businesses and enforces laws that deal with public safety and fire codes. A fire investigator may also respond to a fire scene to help the IC investigate the cause of a fire. Investigators may have full police powers and deal directly with investigations and arrests.

Fire plug A valve installed to control water accessed from wooden pipes.

Fire police Members of the fire department who protect fire fighters by controlling traffic and securing the scene from public access. Many fire police are sworn peace officers as well as fire fighters.

Fire protection engineer A member of the fire department who is responsible for reviewing plans and working with building owners to ensure that the design of and systems for fire detection and suppression will meet applicable codes and function as needed.

Fire warden An individual charged with enforcing fire regulations in colonial Canada.

Governance The process by which an organization exercises authority and performs the functions assigned to it.

Hazardous materials company A fire company that responds to and controls scenes where hazardous materials have spilled or leaked. Responders wear special suits and are trained to deal with most chemicals.

Hazardous materials technician A person who responds to hazardous materials or weapons of mass destruction incidents using a risk-based response process by which they analyze the problem at hand, select applicable decontamination procedures, and control a release while using specialized protective clothing and control equipment. (NFPA 472)

Incident commander (IC) The person who is responsible for all decisions relating to the management of an incident and is in charge of the incident site. (NFPA 1500)

Incident Management System (IMS) The combination of facilities, equipment, personnel, procedures, and communications operating within a common organizational structure that has responsibility for the management of assigned resources to effectively accomplish stated objectives pertaining to an incident or training exercise. (NFPA 1670)

Information management Fire fighters or civilians who take care of the computer and networking systems that a fire department needs to operate.

Lieutenant A company officer who is usually responsible for a single fire company on a single shift; the first in line among company officers.

Policy A formal statement that provides guidelines for present and future actions. Policies often require personnel to make judgments.

Primary Care Paramedic (PCP) A member of EMS who can perform limited procedures that usually fall between those provided by an EMR and those provided by an ACP, including IV therapy, interpretation of cardiac rhythms, and defibrillation.

Public information officer An individual who has demonstrated the ability to conduct media interviews and prepare news releases and media advisories. (NFPA 1035)

Regulation A rule, usually issued by a government or other legally authorized agency, that dictates how something must be done. Regulations are often developed to implement a law.

Rescue company A group of fire fighters who work as a unit and are equipped with one or more rescue vehicles. (NFPA 1410)

Safety officer An individual appointed by the authority having jurisdiction as qualified to maintain a safe working environment at all live fire training evolutions. (NFPA 1403)

SCUBA dive rescue technician A responder who is trained to handle water rescues and emergencies, including recovery and search procedures in both water and under-ice situations. (SCUBA stands for *self-contained underwater breathing apparatus*.)

Self-contained breathing apparatus (SCBA) An atmosphere-supplying respirator that supplies a respirable air atmosphere to the user from a breathing air source that is independent of the ambient environment and designed to be carried by the user. (NFPA 1981)

Span of control The maximum number of personnel or activities that can be effectively controlled by one individual (usually three to seven). (NFPA 1006)

Standard operating procedure (SOP) A written organizational directive that establishes or prescribes specific operational or administrative methods to be followed routinely for the performance of designated operations or actions. (NFPA 1521)

Suggested operating guideline (SOG) Another term for SOP that implies that the fire fighter or officer must use his or her personal judgment in completing the procedure.

Technical rescue technician A fire fighter who is trained in special rescue techniques for incidents involving structural collapse, trench rescue, swiftwater rescue, confined-space rescue, and other unusual rescue situations.

Thermal imaging camera An electronic device that detects differences in temperature based on infrared energy and then generates images based on those data. These devices are commonly used in obscured environments to locate victims.

Training officer The person designated by the fire chief as having authority for overall management and control of the organization's training program. (NFPA 1401)

Truck company A group of fire fighters who work as a unit and are equipped with one or more pieces of aerial fire apparatus. (NFPA 1410)

Unity of command The concept that each person within an organization reports to one, and only one, designated person. (NFPA 1026)

Wildland/brush company A fire company that is dispatched to wildland and brush fires where larger engines cannot gain access. Wildland/brush companies have four-wheel drive vehicles and special firefighting equipment.

FIRE FIGHTER
in action

You are with a group of fire fighters from several departments at the province-wide fire school. As usual, you are discussing every topic under the sun; right now, however, the discussion focuses on their various fire departments and how they operate. Fire Fighter Tom Smith explains that his volunteer department has a chief as well as an assistant chief, a captain, and 30 fire fighters, who all report to the chief. You respond that your volunteer department has a chief, 3 assistant chiefs (one to oversee each of training, operations, and fire prevention), 3 captains (one to oversee each station), and 45 fire fighters. Fire fighter Brad Jones says that his department has a chief, 5 district chiefs, and 35 fire fighters.

As the fire fighters continue to discuss the structures of their departments, you find that there are many differences in departmental organization, even though they are all small volunteer departments that respond to comparable types and quantities of incidents.

1. Which of the following is not typically a type of incident to which fire departments respond?
   A. Motor vehicle collisions
   B. Hazardous materials spills
   C. Vicious dogs on the loose
   D. Emergency medical requests

2. What are the three most common staffing methods of fire departments?
   A. Public safety officer
   B. Volunteer
   C. Career
   D. Combination

3. Which role would most likely fall under the Fire Prevention Division?
   A. Fire police
   B. Fire marshal
   C. IC
   D. Company officer

4. If there are an equal number of fire fighters assigned to each station at your department, what is the span of control for each captain?
   A. 1 to 1
   B. 3 to 1
   C. 5 to 1
   D. 15 to 1

5. At Fire Fighter Jones's department, fire fighters report to a district chief as well as the fire chief. Which organizational principle does this violate?
   A. Span of control
   B. Unity of command
   C. Division of labour
   D. Chain of command

6. Which position would Fire Fighter Smith's department most likely add if it wanted a supervisory position between the rank of fire fighter and district chief?
   A. Division chief
   B. Battalion chief
   C. Fire fighter
   D. Assistant chief

You are thrilled to learn that you will be assigned to Ladder Company 34, which is in one of the busiest districts in the city. It covers multiple low-income high-rise apartments in the downtown area and is known to routinely have a working fire each day. It is served by an engine company, a truck company, a rescue company, and a battalion chief.

Generally, it takes years of experience and proving one's skills before getting this coveted assignment. However, your excitement turns to concern as you wonder how well you will perform. Are you up to it? Can you step up to a leadership role within the organization while still a fire fighter? These and other questions race through your mind as you pack your gear to head for the new assignment.

1. What is the benefit of division of labour through the use of engine companies, truck companies, rescue companies, and chiefs?
2. Why are regulations, policies, and SOGs/SOPs important for you to understand and follow?
3. What responsibility do you have for fire prevention in your new district?
4. How can you develop a positive image with your new supervisors and crews?

© Greg Henry/ShutterStock, Inc.

Fire Fighter Safety

© Patricia Marks/ShutterStock, Inc.

© Greg Henry/ShutterStock, Inc.

Fire Fighter I

Knowledge Objectives

After studying this chapter, you will be able to:

- List the major causes of death and injury in fire fighters. (p 26–27)
- List the three groups whom fire fighters must always consider when ensuring safety at the incident scene. (p 27)
- List and describe the four components of a fire fighter safety program. (p 27)
- Describe the 16 fire fighter life-safety initiatives. (p 28)
- Describe the connection between physical fitness and fire fighter safety. (NFPA 5.1.1 , p 29)
- Describe the components of a well-rounded physical-fitness program. (NFPA 5.1.1 , p 29)
- Explain the practices fire fighters should take to promote optimal physical and mental health. (NFPA 5.1.1 , p 29)
- Describe the purpose of an employee assistance program (EAP). (NFPA 5.1.1 , p 30)
- Explain how fire fighter candidates, instructors, and veteran fire fighters work together to ensure safety during training. (p 30)
- Describe the steps to ensure safety when responding to an emergency. (p 30–31)
- Describe the steps to ensure safety when driving to an emergency incident. (p 31)
- List the four general principles that govern emergency-vehicle operation. (p 31)
- List the guidelines for safe emergency-vehicle response. (p 31–34)
- Explain how the teamwork concept is applied during every stage of an emergency incident to ensure the safety of all fire fighters. (p 34)
- Describe how the personnel accountability system is implemented during an emergency incident. (p 35)
- List the common hazards at an emergency incident. (p 35)
- Describe the measures fire fighters follow to ensure electrical safety at an emergency incident. (p 36)
- Describe how to lift and move objects safely. (p 36)
- Explain how rehabilitation is used to protect the safety of fire fighters during an emergency incident. (p 36–37)
- Explain the role of a critical incident stress debriefing (CISD) in preserving the mental well-being of fire fighters. (p 37)
- Describe how to ensure safety at the fire station. (p 37)
- Describe how to ensure safety outside of the workplace. (p 37)

Skills Objectives

There are no skills objectives for Fire Fighter I candidates.

Fire Fighter II   FFII

Knowledge Objectives

There are no knowledge objectives for Fire Fighter II candidates. NFPA 1001 contains no Fire Fighter II Job Performance Requirements for this chapter.

Skills Objectives

There are no skills objectives for Fire Fighter II candidates. NFPA 1001 contains no Fire Fighter II Job Performance Requirements for this chapter.

Additional NFPA Standards

- NFPA 1250, *Recommended Practice in Emergency Service Organization Risk Management*
- NFPA 1451, *Standard for a Fire Service Vehicle Operations Training Program*
- NFPA 1500, *Standard on Fire Department Occupational Safety and Health Program*
- NFPA 1582, *Standard on Comprehensive Operational Medical Program for Fire Departments*

You Are the Fire Fighter

You and your crew are at a large apartment fire that started on the third floor, extended into the attic, and is now venting through the roof. Your captain gave the incident commander (IC) the accountability tags, and you were assigned to Division D to confine the fire on the third floor. You quickly ran through an air cylinder as you pulled ceilings after having stretched the 64-millimetre (2.5-inch) hose to the third floor. Your crew quickly changes cylinders and gets ready to continue the arduous task at hand. Then, you get the word from the IC, "Engine 213, report to rehab."

1. What is the leading cause of fire fighter injury and death?
2. What are the safety measures taken during this incident?
3. What are the potential hazards during this incident?

Introduction

This chapter covers the topics of injury prevention, means of reducing fire fighter injuries and deaths, and safety and health measures needed during all activities performed by fire fighters from training to the fire ground to fire station duties. Firefighting, by its very nature, is dangerous. Each individual fire fighter must learn safe methods of confronting the risks presented during training exercises, on the fire ground, and at other emergency scenes.

Every fire department must do everything it can to reduce the hazards and dangers of the job and to help prevent fire fighter injuries and deaths. Each fire department must have a strong commitment to fire fighter safety, with fire fighters taking the lead for their health and safety. Safety must be fully integrated into every activity, procedure, and job description.

Appropriate safety measures must be applied routinely and consistently. During incidents, safety officers are responsible for evaluating the hazards of various situations and recommending appropriate safety measures to the Incident Commander (IC). Each accident, injury, or near miss must be thoroughly investigated to learn why it happened and how it can be avoided in the future.

Advances in technology and equipment require fire departments to review and revise their safety policies and procedures regularly. Information reviews and research by designated health and safety officers can identify new hazards as well as appropriate risk-management measures. In addition, reports of accidents, fatalities, and near misses from other fire departments can help identify common problems and lead to the development of effective preventive actions.

Causes of Fire Fighter Deaths and Injuries

While Canada does not currently maintain a national database of statistics on fire fighter injuries and line-of-duty deaths, individual provincial/territorial workers' compensation statistics are maintained. Thankfully, the statistical trend shows that the number of annual fire fighter deaths in Canada is quite low. Comparatively, the National Fire Protection Association (NFPA) reports that in the United States each year, 80 to 100 fire fighters are killed, on average, in the line of duty. These deaths occur at emergency incidents, in fire stations, during training, and while responding to or returning from emergency incidents. Approximately the same number of fire fighter deaths occur on the fire ground or emergency scene as during training or while performing other nonemergency duties. Approximately one-fourth of all deaths occur while fire fighters are responding to or returning from alarms. Heart attack and stroke are the highest cause of fire fighter death due to injury.

Vehicle collisions are a major cause of fire fighter fatalities. For every 1000 emergency responses, it is estimated that one vehicle collision involving an emergency vehicle occurs **FIGURE 2-1**. One study found that 39 percent of the fire fighters who died in those incidents were not using seat belts. Fire fighters should never overlook basic safety procedures, such as always fastening seat belts.

FIGURE 2-1 Motor vehicle collisions are a leading cause of death for fire fighters.

© Joseph Kaczmarek/AP Photos

The NFPA estimates that in the United States, 70,090 fire fighters were injured in the line of duty in 2011. Fewer than half of these injuries occurred while fighting fires. The most common injuries were strains, sprains, and soft-tissue injuries. Burns accounted for only 10 percent of the total injuries. Smoke and inhalation injuries accounted for 5 percent of all fire-ground injuries. In Canada, workers' compensation statistics from 2007–2011 injury data indicate that 70 percent of fire fighter injuries were caused by strain-related injuries. Accident statistics indicate that the highest percentage (31 percent) of fire fighter injuries related to accidents is caused by overexertion. In the interest of fire fighter safety, it is important that all fire departments obtain regular information from their provincial/territorial authority regarding injuries and death among fire fighters.

Injury Prevention

Injury prevention is a responsibility shared by each member of the firefighting team. Fire fighters must always consider three things when ensuring safety at an emergency scene:

- Personal safety
- Safety of other team members
- Safety of everyone present at the scene

To reduce the risk of accidents, injuries, occupational illnesses, and fatalities, a successful safety program must include four major components:

- Standards and procedures
- Personnel
- Training
- Equipment

■ Standards and Procedures

Due to the fact that safety is such a high priority, several organizations set the standards intended to ensure a safe working environment for the fire service. NFPA 1500, *Standard on Fire Department Occupational Safety and Health Program*, provides a template for implementing a comprehensive health and safety program. Additional NFPA standards focus on specific subjects directly related to health and safety—for example, NFPA 1582, *Standard on Comprehensive Operational Medical Program for Fire Departments*.

In the United States, the federal **Occupational Safety and Health Administration (OSHA)**, along with a variety of state health and safety agencies, develops and enforces government regulations on workplace safety. In Canada, health and safety agencies within each province/territory develop and enforce workplace safety regulations. It is critical that fire departments ensure that they are cognizant of and compliant with the specific regulations regarding safety in each respective province/territory in Canada. NFPA standards often are incorporated by reference in government regulations.

Each fire department needs to establish and maintain a set of standard operating procedures (SOPs) or standard operating guidelines (SOGs), which outline determined and accepted protocols for the safe performance of all functions and operations specific to that department. These procedures must incorporate safe practices and policies in accordance with the regulatory agencies in each province, territory, or local authority having jurisdiction. Each fire fighter is responsible for understanding and following these procedures.

The fire department chain of command also enforces safety goals and procedures. In particular, the command structure keeps everyone working toward common goals in a safe manner. An incident management system (IMS) is a nationally recognized plan to establish command and control of emergency incidents. The IMS is flexible enough to meet the needs of any emergency situation and should be implemented at every emergency scene—from a routine auto accident to a major disaster involving responders from numerous agencies. Departments should ensure that they are trained in their provincial/territorial standard of IMS. More information on the IMS is presented in the chapter "Incident Management System."

Many fire departments have a health and safety committee that is responsible for establishing policies and monitoring fire fighter safety. Members of this committee should include representatives from every area, component, and level within the department—from fire fighters to chief officers. The health and safety officer and the fire department physician also should be members of the committee.

■ Personnel

A safety program is only as effective as the individuals who implement it. A majority of safety issues occur during emergency operations; this also accounts for the largest portion of most fire department operations and budgets.

Teamwork is an essential element of safe emergency operations. On the fire ground and during any hazardous activity, fire fighters must work together to get the job done. The lives of citizens and fire fighters alike depend on compliance with basic safety concepts and principles of operation.

An overall plan also is essential to coordinate the activities of every team, crew, or unit involved in the operation. The IMS coordinates and tracks the location and function of every individual or work group involved in an operation.

Freelancing is acting independently of a superior's orders or the fire department's SOGs/SOPs. Freelancing has no place on the fire ground; it poses extreme danger to not only the fire fighter who acts independently but also every other fire fighter on the emergency scene. A fire fighter who freelances can easily create a dangerous situation where one did not previously exist by being in the wrong place at the wrong time or by doing the wrong thing; for example, a fire fighter who enters a burning structure without informing a superior could become trapped by rapidly changing conditions. By the time the fire fighter is missed, it may be too late to perform a rescue. Searching for a missing fire fighter exposes others to unnecessary risk.

Safety Officer

Incident safety officers are members of the fire department whose primary responsibility is safety. At the emergency scene, the designated incident safety officer reports directly to the IC and has the authority to correct or stop any action that is judged to be unsafe. Incident safety officers observe operations and conditions, evaluate risks, and work with the IC to identify hazards and ensure the safety of all personnel. They also determine when fire fighters can work without self-contained breathing apparatus (SCBA) after a fire is extinguished.

Incident safety officers can enhance safety in the workplace, at emergency incidents, and at training exercises. Even so, it is important to remember that each and every member of the fire department shares the responsibility for promoting safety, both as an individual and as a member of the team.

Training

Adequate training is essential for fire fighter safety. The initial fire fighter training covers the potential hazards of each skill and outlines the steps necessary to avoid injury. Fire fighters must avoid negligent practices or shortcuts that might potentially contribute to injuries and learn how to identify hazards and unsafe conditions.

The knowledge and skills developed during training classes are essential to maintain safety at actual emergency scenes. The initial training course is just the beginning: fire fighters must continually seek out additional courses to keep their skills current.

Equipment

A fire fighter's equipment ranges from power and hand tools to personal protective equipment (PPE) and electronic instruments. Fire fighters must know how to use equipment in the correct manner and operate it safely at all times. Equipment must also be properly maintained. Poorly maintained equipment may create additional hazards to the user or fail to operate when needed.

Manufacturers usually supply operating instructions and safety procedures for their equipment. These instructions cover proper use of the equipment, its limitations, and warnings about potential hazards. Fire fighters must read and heed these warnings and instructions. In addition, new equipment must meet applicable standards to ensure that it can perform under the difficult and dangerous conditions often encountered on the fire ground.

Reducing Fire Fighter Injuries and Deaths

Reducing fire fighter injuries and deaths requires every fire fighter, every fire department, and the entire fire community to work together. In 2003, the Canadian Fallen Firefighters Foundation (CFFF) was incorporated to construct a national memorial to Canada's fallen fire fighters, conduct an annual national memorial service, and support the families of the fallen.

Most fire fighter injuries and deaths are the result of preventable situations. Recognizing this fact, its counterpart in the United States, the National Fallen Firefighters Foundation (NFFF), has developed programs with the goal of reducing line-of-duty deaths. For example, the National Fire Fighter Near-Miss Reporting System provides a method for reporting situations that could have resulted in injuries or deaths. This system, which is accessible over the Internet, provides a means for all fire fighters to learn from situations that occur both rarely and frequently. Representative near-miss reports are included in most chapters of this text so that you can benefit from the experiences of others without getting in harm's way.

In an effort to do more to prevent line-of-duty deaths and injuries, the NFFF has developed a fire fighter safety initiative called Everyone Goes Home. The goal of this program is to raise awareness of life-safety issues, improve safety practices, and allow everyone to return home at the end of his or her shift. In particular, the Everyone Goes Home program has developed 16 fire fighter life-safety initiatives, which describe steps that need to be taken to change the current culture of the fire service to help make it a safer place for all **TABLE 2-1**. The NFFF has even launched a website, LifeSafetyInitiatives.com, dedicated to information and research on these initiatives.

TABLE 2-1	16 Fire Fighter Life-Safety Initiatives

1. Define and advocate the need for a cultural change within the fire service in regards to safety and incorporating leadership, management, supervision, accountability, and personal responsibility.
2. Enhance personal and organizational accountability for health and safety throughout the fire service.
3. Focus greater attention on the integration of risk management with incident management at all levels, including strategic, tactical, and planning responsibilities.
4. Empower all fire fighters to stop unsafe practices.
5. Develop and implement national standards for training, qualifications, and certification, including regular recertification, that are equally applicable to all fire fighters, based on the duties they are expected to perform.
6. Develop and implement national medical and physical-fitness standards that are equally applicable to all fire fighters, based on the duties they are expected to perform.
7. Create a national research agenda and a data-collection system that are related to the initiatives.
8. Take advantage of available technology whenever it can produce higher levels of health and safety.
9. Thoroughly investigate all fire fighter fatalities, injuries, and near misses.
10. Implement grant programs to support the implementation of safe practices and/or mandate safe practices as an eligibility requirement.
11. Develop and champion national standards for emergency-response policies and procedures.
12. Develop and champion national protocols for response to violent incidents.
13. Ensure that fire fighters and their families have access to counselling and psychological support.
14. Provide more resources for public education, and champion this kind of education as a critical fire- and life-safety program.
15. Advocate for the enforcement of codes and installation of home sprinklers.
16. Make safety a primary consideration in the design of apparatus and equipment.

Safety and Health

Safety and well-being are directly related to personal health and physical fitness. Although fire departments regularly monitor and evaluate the health of fire fighters, each department member is responsible for his or her personal health, conditioning, and nutrition. To be an effective fire fighter, you should eat a healthy diet, maintain a healthy weight, and exercise regularly.

Good health requires that you get an adequate amount of uninterrupted sleep to maintain alertness, prevent stress, and avoid illnesses and injuries. Because it is not always possible to get adequate sleep during long duty shifts, it is important that you get adequate amounts of uninterrupted sleep during your off-duty time. Establish a consistent sleep schedule and routine, such as turning off all electronic devices a half-hour before bedtime to allow your mind to wind down and prepare for sleep.

All fire fighters—whether career or volunteer—should spend at least an hour each day in physical-fitness training. Fire fighters should be examined by either a personal or departmental physician before beginning any new workout routine. An exercise routine that includes weight training, cardiovascular workouts, and stretching, with a concentration on job-related exercises, is ideal; for example, many fire fighters use a stairclimber to focus on the muscle groups used for climbing. This type of exercise builds cardiovascular endurance for the fire ground, but other muscle groups should not be neglected. Physical fitness must be a career-long activity; it is not something to be left at the fire academy upon graduation. Firefighting is a stressful activity and demands you maintain a good fitness level throughout your career.

Fire Fighter Safety Tips

© Photos.com

Maintaining proper hydration is essential to performing at your peak physical level.

FIRE FIGHTER II Tips   FFII

Maintain your physical fitness, because your life and the lives of your crew depend on it!

Hydration is an important part of every workout. A good guideline is to consume 240–300 millilitres (8–10 ounces) of water for every 5–10 minutes of physical exertion. Do not wait until you feel thirsty to start rehydrating. In fact, fire fighters should drink up to 4 litres (1 gallon) of water each day to keep properly hydrated. Proper hydration enables muscles to work longer and reduces the risk of injuries at the emergency scene. More information about hydration is presented in the chapter "Rehabilitation."

Diet is another important aspect of physical fitness. A healthy menu includes fruits, vegetables, low-fat foods, whole grains, and lean protein. Pay attention to portion sizes; unfortunately, most people eat larger portions than their bodies need. Substitute healthy choices (e.g., fruit) for high-calorie desserts.

Heart disease is the leading cause of death both in North America as a whole and among fire fighters in particular. A healthy lifestyle that includes a balanced diet, weight training, and cardiovascular exercises can help reduce many risk factors for heart disease and enable fire fighters to meet the physical demands of the job FIGURE 2-2.

Many fire departments have adopted policies that prohibit the use of tobacco products by fire fighters, both on duty and off duty. Smoking is a major risk factor in cardiovascular disease, reduces the efficiency of the body's respiratory system, and increases the risk of lung and other types of cancer. Fire fighters should avoid tobacco products entirely for both health and insurance reasons.

Alcohol is another substance that fire fighters should avoid. Alcohol is a mood-altering substance, which can be abused. Excessive alcohol use can damage the body and affect performance. Fire fighters who have consumed alcohol within the previous 8 hours must not be permitted to engage in training or emergency operations.

Drug use has absolutely no place in the fire service. Many fire departments have drug-testing programs to ensure that fire fighters do not use or abuse drugs. The illegal use of drugs endangers your life, the lives of your team members, and the public you serve.

FIGURE 2-2 Regular exercise will help you to stay healthy and perform your job effectively.

Of course, everyone is subject to occasional illness or injury. You should not try to work when ill or injured. Operating safely as a member of a team requires both fitness and concentration. Do not compromise the safety of the team or your personal health by trying to work while you are ill or injured.

FIRE FIGHTER II Tips	FFII

Find time for yourself and your family as a mental buffer from the stressors of the job.

■ Employee Assistance Programs

Employee assistance programs (EAPs) provide confidential help with a wide range of problems that might affect performance. Many fire departments have established EAPs so that fire fighters can get counselling, support, or other assistance in dealing with physical, financial, emotional, or substance-abuse problems. A fire officer may refer a fire fighter to an EAP if a problem begins or may begin to affect the individual's job performance. Fire fighters who take advantage of an EAP can do so with complete confidentiality and without fear of retribution.

Safety During Training

During training, fire fighters learn and practise the actual skills that they will use later under emergency conditions. Typically, the patterns that develop during training will continue during actual emergency incidents; thus, developing the proper working habits during training courses helps ensure safety later.

Many of the skills covered during training can be dangerous if they are not performed correctly. According to the NFPA, each year, approximately 10 percent of fire fighter deaths in the United States occur during training exercises. Use of proper protective gear and good teamwork are as important during training as they are on the fire ground.

Instructors and veteran fire fighters are more than willing to share their experiences and advice. They can explain and demonstrate every skill and point out the safety hazards involved, because they have performed these skills hundreds of times and know what to do. But here, too, safety is a shared responsibility. Do not attempt anything you feel is beyond your ability or knowledge. If you see something that you believe is an unsafe practice, bring it to the attention of your instructors or a designated safety officer.

As mentioned earlier, freelancing is a dangerous practice, both on the fire ground and during training exercises. Wait for specific instructions or orders before beginning any task. Do not assume that something is safe and act independently. Follow instructions and learn to work according to the proper procedures.

Teamwork is also important during training exercises. Assignments are given to firefighting teams during most live fire exercises. Teams must stay together. If any member of the team becomes fatigued, is in pain or discomfort, or needs to leave the training area for any reason, notify the instructor or

safety officer. Emergency Medical Services (EMS) personnel should be available to perform an examination and to transport ill or injured personnel to further treatment, if necessary. A fire fighter who is injured during training should not return until medically cleared for duty.

Safety During the Emergency Response

When a company is dispatched to an emergency, fire fighters need to get to the apparatus and don the appropriate PPE quickly before mounting the vehicle and proceeding to the incident. Walk quickly to the apparatus; do not run. Take care not to slip and become injured before you reach the apparatus. Be careful getting into and out of fire apparatus.

Personal protective gear should be properly positioned so that you can don it quickly before getting into the apparatus FIGURE 2-3 . In addition, make sure that seat belts are properly fastened before the fire apparatus begins to move. All personnel responding on fire apparatus must be seated with seat belts fastened, and seat belts should remain fastened until the fire apparatus comes to a complete stop. Fire fighters can don SCBA while seated in some vehicles. Learn how to do this without compromising safety.

Drivers of fire apparatus have a great responsibility: They must get the fire apparatus and crew members to the

FIGURE 2-3 Protective clothing should be properly positioned so that you can quickly don it.

emergency scene without having or causing a traffic accident en route. They must know the streets in their first-due area and any target hazards. They must be able to operate the vehicle skillfully and keep it under control at all times. In addition, they must anticipate all responses from other drivers, who might not see or hear an approaching emergency vehicle or know what to do when confronted by one. Prompt response is a goal, but safe response is a much higher priority. Before you are qualified to drive any piece of fire department apparatus, you must undergo special training that covers the principles of emergency driving, applicable motor vehicle regulations, and characteristics of the equipment you will be driving. You also need to practice driving with a qualified instructor. You need to demonstrate competency in both required knowledge and skills. These driving skills are not required for the Fire Fighter I and Fire Fighter II courses and are beyond the scope of this textbook.

FIRE FIGHTER II Tips   FFII

Developing situational awareness on the emergency scene is essential to your safety and survival. This concept can be summed up by saying that you must have a big picture view of the emergency scene and avoid tunnel vision. This ability comes only through experience and with well-developed competency in basic fire fighter skills. The less you have to concentrate on the components of your task, the more you are able to observe your surroundings. This quality emerges through hours and hours of practice and skill repetition. Do not stop practising until your basic skills become ingrained in your muscle memory. If you can do your job with your eyes closed, you will be able to observe much more when they are open!

Safe Driving Practices

Vehicle occupants who are not wearing seat belts are much more likely to suffer serious injuries or death than occupants who have their seat belts properly fastened. Airbags are most effective when seat belts are properly fastened; they are not effective without properly applied seat belts. Seat belts also greatly reduce the possibility that vehicle occupants will be ejected from the vehicle.

In some departments, where fire fighters are not on duty at the fire station, it may be necessary for members to respond to the station or to the emergency scene in their private vehicles. Driving a private vehicle during an emergency response requires a special degree of safety. Such emergency driving requires added considerations and represents a much different type of driving from the routine driving.

Motor vehicle crashes are the second most common cause of fire fighter deaths. Although some of these crashes involve fire department apparatus, all too many involve fire fighters driving their private vehicles to an emergency. The ensuing collisions sometimes result in injury or death to the fire fighter and to civilians. Even when no one is injured, crashes cause fire fighters to be delayed in getting to the emergency.

Many collisions occur when the motor vehicle operator loses control of the vehicle. These crashes may be caused by driving too fast for the prevailing conditions, braking inappropriately, changing directions too abruptly, or tracking around a curve too fast. In addition, many collisions involving emergency vehicles occur on open roads, where excessive speed is often cited as a primary cause. Intersections are common sites of collisions involving emergency vehicles; most of these collisions occur when the emergency-vehicle operator fails to stop at an intersection and ensure that other traffic has stopped before proceeding through the intersection.

A motor vehicle collision actually consists of a series of separate collision events: the first collision occurs when the vehicle collides with a second vehicle or with a stationary object; the second collision occurs when the occupants of the vehicle collide with the interior of the vehicle.

Fire Fighter Safety Tips   © Photos.com

Always fasten your seat belt each and every time you get into a motor vehicle. It is vital that you follow this rule each and every time you climb into an apparatus.

■ Laws and Regulations Governing Emergency-Vehicle Operation

Four general principles govern emergency-vehicle operation:

- Emergency-vehicle operators are subject to all traffic regulations, unless a specific exemption is made. A specific exemption is a statement that appears in a statute; for example, "The driver of an authorized vehicle may exceed the maximum speed limits so long as he or she does not endanger life or property."
- Exceptions are legal only when the vehicle is operating in emergency mode.
- Even with an exemption, the emergency-vehicle operator can be found criminally or civilly liable if involved in a crash.
- An exemption does not relieve the operator of an authorized emergency vehicle from the duty to drive with reasonable care for all persons using the highway.

Laws governing emergency-vehicle operation vary from one province/territory to another. You must follow the laws and regulations of your province/territory regarding emergency-vehicle operations. Some permit private vehicles to operate as emergency vehicles when responding to a fire station or an emergency scene; others outline a limited use of emergency equipment and dictate certain restrictions. You must understand and follow the specific laws and regulations of your province/territory as well as your department's SOGs/SOPs.

■ Standard Operating Procedures for Personal Vehicles

The use of personal vehicles to respond to fire and EMS calls constitutes a fire department function. For you to use your personal vehicle in a situation requiring an emergency response, this use must be permitted by provincial laws and regulations, and this operation must also be permitted by your local government and fire department. If your local fire department does not permit the use of personal vehicles for emergency responses, then you are not permitted to engage in this behaviour, regardless of what is permitted by your provincial laws. Your fire department should have SOGs/SOPs that spell out whether private vehicles can be used for emergency response. These procedures typically address which kind of training you must complete before you are permitted to use your personal vehicle for emergency response. This training will usually include a course in defensive driving and functioning as an emergency-vehicle operator.

Safe Driving Begins with You

It has often been stated that the biggest cause of vehicle crashes is "a loose nut behind the wheel." In other words, a major factor in crashes is the attitude and ability of the vehicle operator. A competent emergency-vehicle operator needs to have a confident, but not cocky, attitude. The emergency-vehicle operator should possess good judgment, mental and physical fitness, maturity, and good driving habits. It is a requirement to maintain a clean driving record, because a person who has been cited for multiple moving violations in his or her personal vehicle will usually be a poor risk as the driver of an emergency vehicle. It is important for a good driver to know the provincial and local laws relating to motor vehicle operation. In addition, an emergency-vehicle operator needs to understand the reaction time, braking distance, and stopping distance of the vehicle.

Emergency driving requires good reactions and alertness. It is important for the emergency-vehicle operator to be in good physical condition and to have good eyesight. Driving while impaired is illegal and places everyone at risk. It is important to recognize that impairment can result from many different sources—for example, using some prescribed medications, using some over-the-counter drugs, and being overly fatigued. Anyone who has been drinking alcoholic beverages should never think of driving. Driving while eating, texting, or talking on a communications device is a form of distracted driving. Distracted driving during routine driving or when responding to an emergency is to be avoided; it is the cause of many collisions and deaths.

In many cases, one of the best predictors of future performance is past behaviour. Your driving record is an important consideration in your career in the fire service. A person who has past moving violations for speeding, reckless operation, aggressive driving, chargeable crashes, or driving under the influence of alcohol or drugs may be excluded from driving any emergency vehicle. Your driving record, both on duty and off, is important to your career as a fire fighter.

Vehicle Collision Prevention

Safe driving practices will prevent most vehicle collisions. If you are using your personal vehicle to respond to an emergency, it is important to take the characteristics of your vehicle into account while driving. For example, four-wheel drive pickup trucks and large sport-utility vehicles have a higher

centre of gravity than sleek sports cars, and sudden changes in direction are more likely to result in rollover crashes. For this reason, it is critical to learn the characteristics and limitations of your vehicle.

Anticipate the road and road conditions. If you regularly travel the same roads to report to your fire station, learn the characteristics of those roadways, and be able to anticipate when you need to slow down or stop. Always drive at a safe speed for the road; travelling on a limited-access highway is much different from travelling through a school zone, for example. Observe the traffic conditions, and slow down when more traffic congestion is present. Expect that motorists around you may do anything at any time. Upon the approach of an emergency vehicle, for example, some motorists will slow down, some will stop, some will speed up, some will pull to the right, and others will turn left in front of you. Expect the unexpected.

Make allowances for weather conditions. On a clear day, it might seem that you can see forever, but you really cannot. In conditions of rain, snow, fog, dust, or darkness, the distance over which you can see is greatly reduced; reduce your speed to compensate for the limited visibility. At night, your vision is limited by the distance your headlights reach, so reduce your speed accordingly. Recognize that you cannot see objects outside the projection of the headlight beams. Understand that the goal of an emergency response is not to drive as fast as you can, but rather, to arrive on the scene as quickly as you can while maintaining safety.

Adjust your speed of response to accommodate any storm conditions. Rainstorms reduce visibility and produce slippery road surfaces. When water collects on a roadway, your vehicle can hydroplane on the thin film of water that separates your tires from the road's surface; you have no control over your vehicle when hydroplaning occurs. During stormy weather, slow down and be prepared for slippery road surfaces. Be aware of high winds from tornadoes, hurricanes, or other windstorms, which can topple trees onto the roadway. During heavy rainstorms, it is critical that you watch for water flowing over the road: you cannot see how deep the water is, and you cannot tell whether the flood water has washed out part of the roadway. Slow down when driving following an earthquake or tornado, because you will not know where the roadway, bridges, and overpasses have been blocked, damaged, or destroyed.

When operating an emergency vehicle, you are not exempt from the laws of physics, which tell us that when the speed of a vehicle doubles, the force exerted by that vehicle increases by a factor of four; thus, when the speed of a vehicle increases from 30 to 60 kilometres per hour (20 to 40 miles per hour), the force exerted by that vehicle is increased by a factor of four. Higher speeds require more braking power and a longer distance to bring the vehicle to a stop.

To drive safely, you need to understand where the greatest risks are located. As mentioned earlier, studies have shown that many crashes involving emergency vehicles occur at intersections **FIGURE 2-4**. Most fire departments require all emergency vehicles to come to a full stop when the emergency-vehicle operator encounters a stop sign or a red traffic light. If you

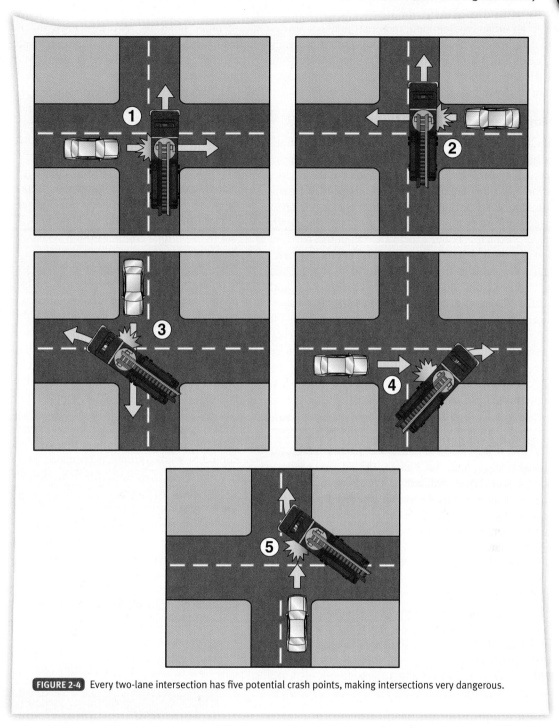

FIGURE 2-4 Every two-lane intersection has five potential crash points, making intersections very dangerous.

enter an intersection without stopping, you may not be able to see approaching traffic, and many motorists do not drive defensively. They assume that if there is no stop sign or red light, they can proceed through the intersection without looking for oncoming vehicles.

In addition, many emergency-vehicle crashes occur on open stretches of straight road during daylight hours. The primary cause of these crashes is excessive speed. Most of us have limited experience with high-speed driving, so it is important to keep within the limits of your driving skills and within the limits of the vehicle you are operating. Many departments limit the speed during emergency responses to 10 kph (6 mph)

over the posted speed limit. Multiple studies have shown that little time is saved by driving at high rates of speed.

During an emergency response, be alert for the presence of other emergency vehicles. These vehicles may carry other fire fighters responding from their homes or fire, EMS, or law-enforcement personnel responding to the same incident. It is difficult to hear the sounds of other apparatus when you are on an emergency response.

Drive with a cushion of safety around you—that is, in such a manner that even if a motorist does the unexpected, you still have enough time and distance to avoid a collision. Remember that the sound of a siren carries only a limited distance. In an urban environment, motorists may not always be able to determine the location of an approaching emergency vehicle, because of surrounding buildings, which distort the sound of the siren. Similarly, emergency lights have limitations and cannot be depended on to clear a path of travel for you.

The Importance of Vehicle Maintenance

Do not underestimate the importance of proper vehicle maintenance for both fire department vehicles and your private vehicle. It is important to perform regular maintenance of the engine, transmission, and other equipment to ensure dependable starting and running. Keep the brakes and tires in proper condition to ensure dependable stopping. In addition, note that tread depth for emergency vehicles should exceed the minimum required for nonemergency vehicles. The suspension system and steering system need to be properly maintained to ensure the safe handling of the vehicle. Likewise, the windshield wipers and washers must be kept in good condition to ensure clear vision. Regularly check all headlights, tail lights, and turn signals for proper operation. Private vehicles that are used for emergency response should be maintained at a higher level than vehicles that are not subject to the stresses of emergency operation.

TABLE 2-2 summarizes the safety guidelines to follow when driving to respond to an emergency call.

Safety at Emergency Incidents

At an emergency incident, fire fighters should never charge blindly into action. The IC will size up the situation, carefully evaluating the conditions to determine whether the burning building is safe to enter or what needs to be done to make the scene safe for action to take place. Wait for instructions and follow directions for the specific tasks to be performed. Do not freelance or act independently of command.

Teamwork

On the fire ground or emergency scene, a firefighting team should always consist of at least two fire fighters who work together and are in constant communication with each other. Some departments call this scheme "the buddy system" **FIGURE 2-5**. In some cases, fire fighters work directly with the company officer, and all crew members function as a team.

In other situations, two individual fire fighters may form a team who are assigned to carry out a specific task. In either case, the company officer must always know where teams are and what they are doing.

Partners or assigned team members should enter together, work together, and leave together. If one member of a team must leave the fire building for any reason, the entire team must leave together, regardless of whether it is a two-person team or an entire crew working as a team.

Before entering a burning building to perform interior search-and-rescue or fire-suppression operations, fire fighters must be properly equipped with approved and appropriate PPE. Partners should check each other's PPE to ensure that it has been donned and is working correctly before they enter a hazardous area.

When working in a hazardous area, team members should maintain visual, vocal, or physical contact with one another at all times. At least one member of each team should have a portable radio to maintain contact with the IC or a designated individual in the chain of command who remains outside the hazardous area. This radio can be used to relay pertinent information and to summon help if the team becomes disoriented, trapped, or injured. In such a situation, the IC can contact the team with new instructions or an evacuation order.

Any fire fighters operating in a hazardous area require backup personnel. The backup team must be able to communicate with the entry team—either by sight or by radio—and should be ready to provide assistance. During the initial stages of an incident, at least two fire fighters must remain outside the hazardous area, properly equipped to respond immediately if the entry team needs to be rescued.

Once crews are assigned to enter the hazardous area, a designated **rapid intervention crew (RIC)**, also known as a rapid intervention team (RIT), should be established and positioned outside the hazardous area. This team's sole responsibility is to be prepared to provide emergency assistance to crews working inside the hazardous area.

TABLE 2-2	Guidelines for Safe Emergency-Vehicle Response

1. Drive defensively.
2. Follow department policy in regard to the posted speed limit.
3. Always maintain a safe distance. Use the 4-Second Rule: stay at least 4 seconds behind another vehicle in the same lane.
4. Maintain an open space or cushion in the lane next to you as an escape route in case the vehicle in front of you stops suddenly.
5. Always assume that other drivers will not hear your siren or see your emergency lights.
6. Select the shortest and least-congested route to the scene at the time of dispatch.
7. Visually clear all directions of an intersection before proceeding.
8. Go with the flow of traffic.
9. Watch carefully for bystanders and pedestrians. They may not move out of your way or could move the wrong way.

FIGURE 2-5 A firefighting team should consist of at least two members who work together.

■ Accountability

Every fire department must have a **personnel accountability system** to track personnel and assignments on the emergency scene. This system should record the individuals assigned to each company, crew, or entry team; the assignments for each team; and the team's current activities. Several kinds of accountability systems are acceptable—from paper assignments or display boards to laptop computers and electronic tracking devices.

Some departments use a "passport" system. In this scheme, each company officer carries a small magnetic board called a passport. Each crew member on duty with that company has a magnetic name tag on the board **FIGURE 2-6**. At the incident scene, the company's passport is given to a designated individual, who uses it to track the assignment and location of every company at the incident.

Other fire departments employ an accountability-tag system. With this approach, each fire fighter carries a name tag and turns it in or places it in a designated location on the apparatus when he or she is on the scene. The tags are then collected and used to track crews who are working together as a unit. This system works well when crews are organized based on the available personnel.

Both passport and accountability-tag systems provide an up-to-date accounting of everyone who is working at the incident and how they are organized. At set intervals, an accountability check is performed to account for everyone. Usually, a company officer reports on the status of each crew. The company officer should always know exactly where his/her crew is and what it is doing. If a crew splits into two or more teams, the company officer should be in contact with at least one member of each team.

An accountability check is also performed when the operational strategy changes or when a situation occurs that could endanger fire fighters. If an accountability check is needed, the list of personnel and assignments should be readily available at the command post.

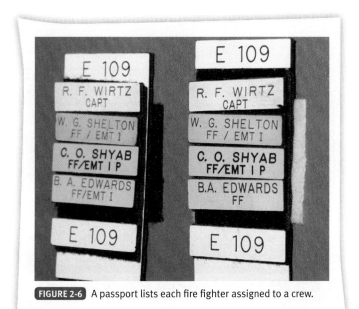

FIGURE 2-6 A passport lists each fire fighter assigned to a crew.

Fire fighters must learn which kind of accountability system their department uses, how to work within it, and how this system works within the ICS. They are responsible for complying with this system and staying in contact with their company officer or assigned supervisor at all times. Most importantly, fire fighters must always remember that a team must stay together.

The accountability system is valuable for identifying fire fighters who are lost or trapped in a fire or other emergency situation. The chapter "Fire Fighter Survival" covers many of these self-survival skills and the steps for calling for help (i.e., mayday). **TABLE 2-3** presents these steps in summary form.

■ Incident Scene Hazards

Fire fighters must be aware of their surroundings when performing their assigned tasks at an emergency scene. At an incident, make a safe exit from the fire apparatus and look at the building or situation for safety hazards, such as traffic, downed utility wires, and adverse environmental conditions. An incident on a street or highway must first be secured with proper traffic- and scene-control devices. Flares, pylons, or barrier tape are all measures that can help keep the public at a safe distance from the scene. Emergency vehicles can be placed to block traffic and protect the incident scene. Always operate within established boundaries and protected work areas.

Changing fire conditions will also affect safety. During the overhaul phase and while picking up equipment, watch out for falling debris, smoldering areas of fire, and sharp objects. If a safety officer is not on the scene, another qualified person should be assigned to monitor the atmosphere to ensure it is safe to remove SCBA or enter the area without SCBA. Because the chance for injury increases when you are tired, do not let your guard down even though the main part of the firefighting operation is over.

■ Using Tools and Equipment Safely

Fire fighters must learn how to use tools and equipment properly and safely before attempting to use them at an actual emergency incident. Follow the proper procedures and safety precautions both in training and at an incident scene. In addition, don protective gear, such as PPE, safety glasses, and hearing protection, as required.

TABLE 2-3	Self-Survival for Lost or Trapped Fire Fighters

Don't get lost or trapped! If you are lost or trapped, take these steps:
- Call for help immediately.
- Activate the radio emergency button.
- Declare a mayday on your portable radio.
- Call on a second radio channel if necessary.
- Activate your personal alert safety system (PASS) device.
- Stay with your crew.
- Follow the hose line or life line out of the building.
- Search for an exit.
- Retreat to a safe refuge.
- Stay calm and conserve air.
- Assume a horizontal position.
- Shine your flashlight at the ceiling and tap with a metal tool to attract attention.

Proper tool maintenance includes sharpening, lubricating, and cleaning each tool. Equipment should always be in excellent condition and ready for use. All fire fighters should be able to do basic repairs, such as changing a saw blade or a hand light battery. Practice these tasks at the fire station until you can perform them quickly and safely at the emergency scene.

Electrical Safety

Electricity is an emergency scene hazard that must always be respected. Electricity is in every building and involved with many vehicle collisions. In some cases, energized power lines may be present on the fire ground. These wires may be energized yet nearly invisible when lying on the ground. This is especially true in conditions of limited visibility, such as darkness, fog, and smoke. Always check for overhead power lines when raising ladders. During any major fire, the electric power supply to the building should be turned off as part of the fire-ground task called *controlling the utilities*.

Fire departments are often called to electrical emergencies, such as those involving downed power lines, fires, arcing wires, and transformer fires. Always have the power disconnected or turn off the power if you are trained and permitted by department policy.

Fire apparatus should be parked outside the area and away from power lines when responding to a call for an electrical emergency. A downed power line should be considered energized until the power company confirms it is dead. Secure the area around the power line and keep the public at a safe distance. Never drive the fire apparatus over a downed line or attempt to move it using tools. If a sparking power line causes a brush fire, if it is safe to do so, attempt to contain the fire from a safe distance, but do not use water near the line.

Lifting and Moving

Lifting and moving objects is part of a fire fighter's daily duties. Do not try to move something that is too heavy alone; ask for help. Never bend at the waist to lift an object; instead, bend at the knees and use your legs to lift the weight. If you must move objects over a long distance, use equipment, such as handcarts, hand trucks, and wheelbarrows.

In addition to moving inanimate objects, fire fighters must often move sick or injured victims. Don't move an injured person, unless his or her life is in danger, until an EMT or other appropriate medical personnel is on scene to stabilize the victim. Discuss and evaluate the options before moving a victim, and then proceed very carefully. If necessary, request help. Never be afraid to call for additional resources, such as another engine or truck company, to assist in lifting and moving a heavy victim.

Working in Adverse Weather Conditions

In adverse weather conditions, fire fighters must dress appropriately. A turnout coat, pants, boots, and helmet can keep you warm and dry in rain, snow, or ice. Firefighting gloves and knit caps will also help retain body heat and keep you warm. If conditions are icy, make smaller movements, watch your step, and keep your balance. During the hot summer months, it is difficult to stay cool in PPE. Many departments have SOGs/SOPs covering PPE use in summer months. Check your department's SOGs/SOPs for guidance on acceptable PPE use in adverse weather. More information on working in adverse weather conditions is presented in the chapter "Extreme Cold Weather Cautions and Operations."

Rehabilitation

<u>Rehabilitation</u> is a systematic process that provides periods of rest and recovery for emergency workers during an incident. It is usually conducted in a designated area away from the hazards of the emergency scene. The rehabilitation area, or "rehab," is usually staffed by fire and EMS personnel.

A fire fighter who is sent to rehabilitation should be accompanied by the other members of the crew. The company officer should inform the IC of their change in location. While in rehabilitation, fire fighters should take advantage of the opportunity to rest, rehydrate, have their vital signs checked, and receive treatment for minor injuries **FIGURE 2-7**. Rehabilitation gives fire fighters the chance to cool off in hot weather and to warm up in cold weather.

Rehabilitation time can be used to replace SCBA cylinders, obtain new batteries for portable radios, and make repairs or adjustments to tools or equipment. Firefighting teams can discuss recently completed assignments and plan their next work cycle. When a crew is released from rehabilitation, its members should be rested, refreshed, and ready for another work cycle. If the crew are too exhausted or unable to return to work or have not been medically cleared to return to work, they should be replaced and released from the incident.

Never be afraid or embarrassed to admit you need a break when on the emergency scene. Heat exhaustion and overexertion are common conditions at firefighting incidents and are characterized by profuse sweating, dizziness, confusion, headache, nausea, cramping, and pain in localized body areas. If a fire fighter shows signs or symptoms of heat exhaustion or overexertion, the company officer should be notified immediately. The company officer will then request approval for rehabilitation so the problem can be treated.

Heat exhaustion is usually not life-threatening and, if identified early, can be remedied by rehydration, cooling, and rest.

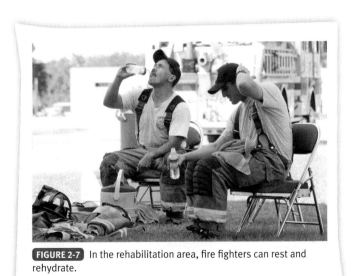

FIGURE 2-7 In the rehabilitation area, fire fighters can rest and rehydrate.

Left untreated, however, heat exhaustion can progress quickly to heatstroke, which is a life-threatening emergency. A lack of sweating, low blood pressure, shallow breathing, and seizures are some of the signs of heatstroke. If you or a member of your team experience these symptoms during training or on the fire ground, place a mayday or fire fighter–in–trouble call. You must report to your superior and seek immediate medical attention.

A fire fighter who experiences chest pain or discomfort should stop and seek medical attention immediately. As noted earlier, heart attack is the leading cause of death among fire fighters. More information on fire fighter rehabilitation is presented in the chapter "Rehabilitation."

■ Violence at the Scene

Fire fighters are sometimes dispatched to calls involving guns, domestic disputes, injuries from an assault, or other violent scenes. In such incidents, the staging area and the fire apparatus should be located at some distance from the scene until the police arrive, investigate, and declare the area safe. Only then should fire fighters proceed to enter the emergency scene. A fire fighter's personal safety should always be paramount. If there is any threat to personal safety, slowly back away from the emergency scene to a safe distance, and request the police to secure the scene. Do not become a victim.

If you are confronted with a potentially violent situation, do not respond violently. Remain calm, speak quietly, and attempt to gain the person's trust and call for police assistance. You might consider taking additional classes to increase your understanding and develop appropriate skills for these situations.

■ Mental Well-Being

Some emergency calls are particularly difficult and emotionally traumatic. After experiencing such an incident, fire fighters who were involved may be encouraged to meet with a counsellor or attend a **critical incident stress debriefing (CISD)**. Usually, a CISD is held as soon as possible after a traumatic call. It provides a forum for firefighting and EMS personnel to discuss the anxieties, stress, and emotions triggered by a difficult call. Follow-up sessions can be arranged for individuals who continue to experience stressful or emotional responses after a challenging incident.

Some fire departments have qualified counsellors available 24 hours a day. The initial counselling may consist of a group session for all fire fighters and rescuers; alternatively, it can also be done on a one-on-one basis or in smaller groups. Everyone handles stress differently, and new fire fighters need time to develop the personal resources necessary to deal with difficult situations. If any call proves emotionally disturbing to you, ask your superior officer for a referral to a qualified counsellor. If you are having trouble sleeping or find yourself having difficulty dealing with your thoughts or feelings, tell a fire officer or contact an EAP.

Safety at the Fire Station

The fire station is just as much a workplace as the fire ground. Indeed, fire fighters will spend much of their time during a 12- to 24-hour shift at the fire station. While there, be careful when working with power tools, ladders, electric appliances, pressurized cylinders, and hot surfaces. Injuries that occur at the firehouse can be just as devastating as those that occur at an emergency incident.

Safety Outside Your Workplace

Continue to follow safe practices when you are off duty. An accident or injury, regardless of where it happens, can end your career as a fire fighter. For example, if you are using a ladder while off duty, follow the same safety practices that you would use on duty. Keep your seat belt fastened in your personal vehicle just as you are required to do when you are on duty FIGURE 2-8 .

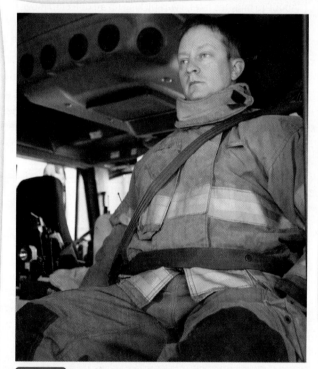

FIGURE 2-8 Always use your seat belt.

Fire Fighter Safety Tips

© Photos.com

Remember these guidelines to stay safe—both on and off the job:

- You are personally responsible for safety. Keep yourself safe. Keep your teammates safe. Keep citizens—your customers—safe.
- Work as a team. The safety of the entire firefighting unit depends on the efforts of each member. Become a dependable member of the team.
- Follow orders. Freelancing can endanger other fire fighters as well as yourself.
- Think! Before you act, think about what you are doing. Many people are depending on you.

Near-Miss REPORT

Report Number: 05-0000535

Synopsis: Inadequate protective clothing contributes to fire fighter injury.

Event Description: At a structure fire in a metal frame and metal siding workshop, I was on a 6-centimetre(2.5-inch) line with a fire fighter from another department. At that time, we were conducting an exterior attack on the fire from the front of the building, in front of a metal roll-up door. The door had fallen down and was lying in the way of us advancing the hose line into the shop to extinguish hot spots. The door was also positioned in such a way that it was difficult for me to get the stream where I wanted it inside the shop. I shut down the nozzle and moved the metal door, which was all the way on the ground now and not connected to the building. We had no problem directing our fire stream after that, and we had a clear access point into the shop.

About 10 minutes later, I was picking up an object to move it out of the way (something smaller than the door) with my right hand, which had just a regular leather work glove on it the whole time, when I noticed blood streaming down my wrist and in my glove. I reported to the ambulance and got it flushed, cleaned, and wrapped; I then went back to work.

If the cut had been any deeper, it would have nicked one of my veins and I could have been in bad shape. I never felt the metal door cut into me. I was wearing full PPE the whole time, with the exception of my leather work gloves. I'll have a pretty good scar to remind me. If I had been wearing my fire gloves, which have extended cuffs on them, and had asked for a hand moving the door, I probably would not have been injured. However, at the time, there were not many fire fighters on scene to help me, and coordination between the four or five departments that were there seemed somewhat nonexistent to me. Four other fire fighters were taken out for rehab, two had IVs started on scene, and two or three were transported to the hospital. The contents of the shop were pretty much a total loss.

Lessons Learned: Wear your PPE so that no skin is exposed. Ask for help when moving heavy objects—don't try to fight the fire by yourself.

Fewer trucks and more personnel on the scene can sometimes lead to less chaos. Likewise, better coordination of tasks and teams leads to more effective operations.

Better training and equipment for volunteer fire fighters would also be helpful. Some fire fighters who were on scene just stood around not knowing what was going on and couldn't do much because they didn't have the knowledge or the equipment to assist safely. Some groups of fire fighters with no PPE were walking around the shop with small portable foam units applying it to the fire through holes in the wall, even as others were washing the foam away with the hose lines from the front of the building and negating what the others were doing.

Chief Concepts

- Every fire fighter and fire department must have a strong commitment to safety and health. Safety must be fully integrated into every activity, procedure, and job description.
- Each year, fire fighters are killed at emergency incidents, in fire stations, during training, and while responding to or returning from an emergency. Even one death is too many.
- The majority of all fire fighter deaths are caused by heart attacks or strokes.
- Motor vehicle collisions are a major cause of fire fighter fatalities. Always wear your seat belt each and every time you are in a motor vehicle.
- Fire fighters must always consider three things when ensuring safety at the scene:
  1. Personal safety
  2. Safety of other team members
  3. Safety of everyone present at the scene
- A successful safety program must have four major components:
  1. Standards and procedures—Regulations, ranging from NFPA standards to provincial/territorial regulations to local laws, govern the fire department's SOPs and SOGs on ensuring a safe work environment.
  2. Personnel—From teams working together to safety officers, personnel ensure that a fire department's safety program is implemented correctly.
  3. Training—Training does not end with this course. To ensure that you attain and maintain a high level of skill and knowledge, you must seek out additional courses and training opportunities throughout your career.
  4. Equipment—To ensure safety, equipment must be properly maintained, and fire fighters must be trained how to use it correctly.
- The 16 fire fighter life-safety initiatives describe the steps that need to be taken to change the current culture of the fire service to help make it a safer place for all fire fighters.
- Safety and well-being are directly related to personal health and fitness. You should eat a healthy diet, maintain a healthy weight, exercise regularly, and sleep for 7 to 8 hours a day, whenever possible.
- A well-rounded fitness routine includes weight training, cardiovascular workouts, and stretching.
- EAPs are available to provide fire fighters with confidential counselling, support, or assistance in dealing with a physical, financial, emotional, or substance abuse problem.
- Each year, up to 10 percent of fire fighter fatalities occur during training exercises. Do not attempt anything you feel is beyond your ability or knowledge. If you see something you feel is unsafe, say something.
- Four general principles govern emergency-vehicle operation:
  1. Emergency-vehicle operators are subject to all traffic regulations, unless a specific exemption is made.
  2. Exceptions are legal only when operating in emergency mode.
  3. Even with an exemption, the emergency-vehicle operator can be found criminally or civilly liable if involved in a crash.
  4. An exemption does not relieve the operator of an authorized emergency vehicle from the duty to drive with reasonable care for all persons using the highway.
- Emergency driving requires good reactions and alertness. Driving while impaired or distracted is unacceptable and illegal. Your driving record, both on duty and off duty, will impact your career as a fire fighter.
- Safe driving practices will prevent most vehicle collisions. Anticipate the road and road conditions. Make allowances for weather conditions. Learn where the greatest risks are located on your routes. Be alert for the presence of other emergency vehicles.
- The personnel accountability system tracks personnel and assignments at the emergency scene. Several types of systems are acceptable—from paper assignments to display boards to laptop computers and electronic tracking devices.
- The electric power supply should be turned off to safely secure the scene for fire fighters.
- Do not try to move a heavy object alone; ask for help.
- A fire fighter who experiences chest pain or discomfort should stop and seek medical attention immediately.
- A CISD is a forum in which firefighting and EMS personnel can discuss anxieties, stress, and emotions triggered by a difficult incident.
- The fire station is just as much a workplace as the fire ground. Be careful when working with power tools, ladders, electric appliances, pressurized cylinders, wet floors, and hot surfaces.
- An accident or injury, regardless of when or where it happens, can end your firefighting career.

Hot Terms

__Critical incident stress debriefing (CISD)__ A postincident meeting designed to assist rescue personnel in dealing with psychological trauma as the result of an emergency. (NFPA 1006)

Employee Assistance Programs (EAPs) Fire service programs that provide confidential help to fire fighters with personal issues.

Freelancing The dangerous practice of acting independently of command instructions.

Occupational Safety and Health Administration (OSHA) The United States federal agency that regulates worker safety and, in some cases, responder safety. OSHA is part of the United States Department of Labor.

Personnel accountability system A system that readily identifies both the locations and the functions of all members operating at an incident scene. (NFPA 1500)

Rapid intervention crew (RIC) A minimum of two fully equipped personnel who are on site, in a ready state, for immediate rescue of injured or trapped fire fighters. Also known as rapid intervention team (RIT).

Rehabilitation An intervention designed to mitigate against the physical, physiological, and emotional stress of firefighting so as to sustain a member's energy, improve performance, and decrease the likelihood of on-site injury or death. (NFPA 1521)

FIRE FIGHTER
in action

You live in a small town, so it is common for the fire fighters to personally know their elected officials. One day, you are eating at the only diner in town, when the mayor comes over and asks if you mind if he sits down with you. You readily say yes to the opportunity to eat with the mayor.

The conversation quickly turns to the fire department. By his line of questioning, you quickly recognize that the mayor does not understand the risks of firefighting and why safety has become so important. He says he is curious why the fire fighters need new safety gear and equipment. You explain that he should contact your chief to discuss this matter further. As the conversation shifts gears, you think about how your chief will answer this question.

1. What is the cause of most fire fighter deaths?
   A. Burns
   B. Building collapse
   C. Heart attack
   D. Smoke inhalation

2. Out of all fire fighter injuries in the line of duty, what percentage occurred while fighting fires?
   A. 25 percent
   B. 45 percent
   C. 65 percent
   D. 85 percent

3. What is the most important action you can take to ensure your safety while responding to an emergency incident?
   A. Remain seated in the fire apparatus.
   B. Fasten your seat belt.
   C. Fasten your SCBA straps.
   D. Wear your helmet.

4. How much water should be consumed to maintain proper hydration for each 5–10 minutes of physical exertion?
   A. 120–180 mL (4–6 oz.)
   B. 240–300 mL (8–10 oz.)
   C. 350–470 mL (12–16 oz.)
   D. 590–710 mL (20–24 oz.)

© Greg Henry/ShutterStock, Inc.

To pass the physical-ability test and get hired as a fire fighter, you had to be in fairly good shape. Even so, during your time at the firefighting academy, you would spend the first hour of each day doing some form of a physical-fitness activity. Each of the recruits made significant improvements and found that the firefighting tasks became even easier with improved physical condition.

You were caught somewhat off guard when you were assigned to the station and found many of the crew members lacking basic physical-fitness levels. You had just completed your one-year probationary period, when the fire chief announced that he was implementing a mandatory fitness program. This decision quickly became the hot topic around the firehouse table. With all of the information you learned about fire fighter safety, you knew the benefits of being physically fit, so you were shocked to find that the crew was opposed to it.

1. What would you say to the other crew members as they discuss this topic around the table?
2. What would you do to create a change in the attitude of the crew?
3. How would you handle it if you were asked by another fire fighter to stop working out while on duty because it was making the rest of them look bad?
4. When should you go with the flow and when should you stand up for your principles?

Personal Protective Equipment and Self-Contained Breathing Apparatus

© Greg Henry/ShutterStock, Inc.

Fire Fighter I

Knowledge Objectives

After studying this chapter, you will be able to:

- List the components of personal protective equipment (PPE) or the structural firefighting ensemble. (NFPA 5.1.2 , p 44–49)
- Describe the type of protection provided by the structural firefighting ensemble. (p 44–45)
- Explain how each design element of a fire helmet works to protect the head, face, and eyes. (p 45–46)
- Explain why protective hoods are a part of the structural firefighting ensemble. (p 46–47)
- Explain how each design element of a turnout coat works to protect the upper body. (p 46–47)
- Describe how each design element of boots works to protect the feet. (p 48)
- Describe how each design element of gloves works to protect the hands and wrists. (p 48)
- Explain how a personal alert safety system (PASS) helps to ensure fire fighter safety. (NFPA 5.3 , p 49)
- List the limitations of PPE. (p 50)
- Explain the role of the fire fighter's work uniform as part of the PPE ensemble. (NFPA 5.1.2 , p 50–51)
- Describe how to inspect the condition of PPE. (NFPA 5.3 , p 51, 54)
- Describe how to properly maintain PPE. (NFPA 5.5.1 , p 51, 54)
- Describe the specialized protective clothing required for vehicle extrication and wildland fires. (p 54)
- List the respiratory hazards posed by smoke and fire. (NFPA 5.3.1A , p 54–56)
- List the conditions that require respiratory protection or self-contained breathing apparatus (SCBA). (NFPA 5.3.1A , p 56)
- Describe the differences between open-circuit breathing apparatus and closed-circuit breathing apparatus. (p 56)
- Describe when a supplied-air respirator is used. (p 56)
- Describe the limitations of SCBA. (NFPA 5.3.1A , p 58–59)
- Describe the physical and psychological limitations of an SCBA user. (NFPA 5.3.1A , p 58–59)
- List and describe the major components of SCBA. (NFPA 5.3.1A , p 59–64)
- Describe the devices on an SCBA that can assist the user in air management. (NFPA 5.3.1 , p 62)
- Describe the pathway that air travels through an SCBA. (p 62)
- Explain how to inspect SCBA to ensure that it is operation ready. (NFPA 5.5.1 , p 62)
- List the complete sequence of donning PPE. (NFPA 5.1.2, 5.3.1A, 5.3.1 , p 73)
- Describe the importance of SCBA inspections and SCBA operational testing. (NFPA 5.5.1 , p 73–78)
- Explain the procedures for refilling SCBA cylinders. (p 78–81)

Skills Objectives

After studying this chapter, you will be able to perform the following skills:

- Don approved personal protective clothing. (NFPA 5.1.2 , p 51–52)
- Doff approved personal protective clothing. (NFPA 5.1.2 , p 51, 53)
- Don an SCBA from an apparatus seat mount. (NFPA 5.3.1B , p 64–65)
- Don an SCBA from a compartment mount. (NFPA 5.3.1B , p 64)
- Don an SCBA from a storage case using the over-the-head method. (NFPA 5.3.1B , p 66–67)
- Don an SCBA from a storage case using the coat method. (NFPA 5.3.1B , p 66, 68)
- Don an SCBA from a seat-mounted position with a safety latch. (NFPA 5.3.1B , p 66, 69)
- Don a face piece. (NFPA 5.3.1B , p 66–72)
- Doff an SCBA. (p 73–74)
- Perform a visible inspection of an SCBA. (NFPA 5.5.1 , p 73–74)
- Perform an operational inspection of an SCBA. (NFPA 5.5.1 , p 75–78)
- Replace an SCBA cylinder. (NFPA 5.3.1B , p 75–76, 79)
- Replace an SCBA cylinder on another fire fighter. (NFPA 5.3.1B , p 78, 80)
- Refill an SCBA cylinder from a cascade system. (p 78–79, 81)
- Clean an SCBA. (NFPA 5.5.1 , p 79–82)

Fire Fighter II   FFII

Knowledge Objectives

There are no knowledge objectives for Fire Fighter II candidates. NFPA 1001 contains no Fire Fighter II Job Performance Requirements for this chapter.

Skills Objectives

There are no skills objectives for Fire Fighter II candidates. NFPA 1001 contains no Fire Fighter II Job Performance Requirements for this chapter.

Additional NFPA Standards

- NFPA 1404, *Standard for Fire Service Respiratory Protection Training*
- NFPA 1500, *Standard on Fire Department Occupational Safety and Health Program*
- NFPA 1582, *Standard on Comprehensive Operational Medical Program for Fire Departments*
- NFPA 1851, *Standard on Selection, Care, and Maintenance of Protective Ensembles for Structural Fire Fighting and Proximity Fire Fighting*
- NFPA 1852, *Standard on Selection, Care, and Maintenance of Open-Circuit Self-Contained Breathing Apparatus (SCBA)*
- NFPA 1971, *Standard on Protective Ensembles for Structural Fire Fighting and Proximity Fire Fighting*
- NFPA 1975, *Standard on Station/Work Uniforms for Emergency Services*
- NFPA 1977, *Standard on Protective Clothing and Equipment for Wildland Fire Fighting*
- NFPA 1981, *Standard on Open-Circuit Self-Contained Breathing Apparatus (SCBA) for Emergency Services*
- NFPA 1982, *Standard on Personal Alert Safety Systems (PASS)*

You Are the Fire Fighter

One cool fall evening, you are dispatched for a smoke odour in the area. It seems like a fairly routine call, because residents are beginning to use their fireplaces for the first time of the year and the weather conditions tend to cause the smoke to hang low. As the engine drives slowly through the neighbourhood of bungalow-style houses, you notice the distinct smell of a working house fire rather than that of firewood.

As you turn the corner, there it is—smoke drifting from the basement windows of a house. Your captain does a quick size-up and requests a full assignment from dispatch. She then tells you and the other fire fighter to pull a cross-lay and attack the fire.

1. Which protective equipment would you need for this situation?
2. What is the most efficient way to put on an SCBA?
3. Once the fire is over, what is the proper way to care for your protective equipment?

Introduction

Two safety components used by the fire fighter require special consideration: personal protective equipment (PPE) and self-contained breathing apparatus (SCBA). For you to remain safe and to do your job effectively, you need to understand the purposes for which this equipment is made, what it can do, and what it cannot do. PPE is designed to protect the body against a *limited* amount of heat. If you exceed the protection offered by this equipment, then severe injuries and death can occur. SCBA enables fire fighters to safely enter some smoky and toxic environments. It provides respiratory protection for a limited period of time. When fire fighters exceed the limits of SCBA, however, the consequences can be deadly.

Personal Protective Equipment

Personal protective equipment (PPE) is an essential component of a fire fighter's safety system. It enables a person to survive under conditions that might otherwise result in death or serious injury. Different PPE ensembles are designed for specific hazardous conditions, such as structural firefighting, wildland firefighting, airport rescue and firefighting, hazardous materials operations, and emergency medical operations.

The various PPE ensembles provide specific protections, so an understanding of their designs, applications, and limitations is critical. For example, a structural firefighting ensemble will protect the wearer from the heat, smoke, and toxic gases present in building fires, but it cannot provide long-term protection from extreme weather conditions and it limits range of motion **FIGURE 3-1**. The more you know about the protection your PPE can provide, the better you will be able to judge conditions that exceed its limitations.

A fire fighter's PPE must provide full-body coverage and protection from a variety of hazards. To be effective, the entire ensemble must be worn whenever potential exposure to those hazards exists. In addition, PPE must be cleaned, maintained, and inspected regularly to ensure that it will continue to

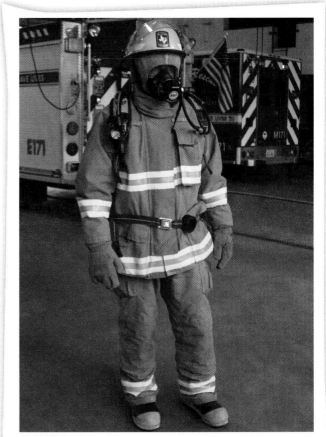

FIGURE 3-1 A protective ensemble for structural firefighting provides protection from multiple hazards.

provide the intended degree of protection when it is needed. Worn or damaged articles must be repaired or replaced.

■ Structural Firefighting Ensemble

Structural firefighting PPE enables fire fighters to enter burning buildings and work in areas with elevated temperatures and concentrations of toxic gases. Without PPE, fire fighters would

TABLE 3-1	Protection Furnished by Personal Protective Equipment
• Provides thermal protection • Repels water • Provides impact protection • Protects against cuts and abrasions • Furnishes padding against injury • Provides respiratory protection	

be unable to conduct search-and-rescue operations or perform fire-suppression activities. A structural firefighting ensemble is designed to cover every square centimetre of the body. It provides protection from the fire, keeps water away from the body, and helps reduce trauma from cuts or falls. This type of PPE is designed to be worn with **self-contained breathing apparatus (SCBA)**, which provides respiratory protection.

The structural firefighting ensemble consists of a protective coat, trousers or coveralls, a helmet, a hood, boots, and gloves. The helmet must have a face shield, goggles, or both. The clothing is worn with an SCBA and a personal alert safety system (PASS) device. All of these elements must be worn together to provide the necessary level of protection **FIGURE 3-2**.

Protection Provided

A structural firefighting ensemble is designed to provide full-body coverage and offers several different types of protection **TABLE 3-1**. The coat and pants have outer shells that can withstand elevated temperatures, repel water, and provide protection from abrasions and sharp objects. The knees may be reinforced with pads for greater protection when crawling. Reflective trim adds visibility in dark or smoky environments. Insulating layers of fire-resistant materials protect the skin from high temperatures. A moisture barrier between different layers of PPE keeps liquids and vapours, such as hot water or steam, from reaching the skin.

The helmet protects the head from falling debris; it includes ear coverings to protect the ears as well. The face shield helps protect the eyes. A fire-retardant hood covers any exposed skin between the coat collar and the helmet. Gloves protect the hands from heat, cuts, and abrasions. Boots protect the feet and ankles from the fire, keep them dry, prevent puncture injuries, and protect the toes from crushing injuries.

An SCBA provides respiratory protection by giving the fire fighter an independent, limited air supply. In this way, the fire fighter's respiratory system remains safe from toxic products and hot gases present in the atmosphere.

Helmet

Fire helmets are manufactured in several designs and shapes using a variety of materials. Nevertheless, each design must meet the requirements specified in NFPA 1971, *Standard on Protective Ensembles for Structural Fire Fighting and Proximity Fire Fighting*. The hard outer shell of the helmet is lined with energy-absorbing material and has a suspension system to provide impact protection against falling objects **FIGURE 3-3**. The helmet shell also repels water, protects against steam, and creates a thermal barrier against heat and cold. Its shape helps to deflect water away from the head and neck.

Face and eye protection can be provided by a face shield, goggles, or both. These components must be attached to the helmet and are used when SCBA is not needed or when the

FIGURE 3-2 The complete structural firefighting ensemble consists of a helmet, coat, trousers, protective hood, gloves, boots, SCBA, and PASS device.

FIGURE 3-3 A helmet is constructed with multiple layers and components.

SCBA face piece is not in place. A chin strap is also required and must be worn to maintain the helmet in the proper position. In addition, this strap helps to keep the helmet on the fire fighter's head during an impact.

Fire helmets have an inner liner for added thermal protection. This liner also provides protection for the wearer's ears and neck. When entering a burning building, the fire fighter should pull down the ear tabs to ensure maximum protection.

Helmets are manufactured with adjustable inner suspension systems that hold the head away from the shell and cushion it against impacts. This suspension system must be adjusted to fit the individual, who should try on the helmet with the SCBA face piece and hood in place. Because helmet shells come in a variety of sizes, fire fighters must be sure to wear the correct-size shell.

In many departments, helmets are colour-coded according to the fire fighter's rank and function. They often carry company numbers, rank insignia, or other markings. Some fire departments use a shield mounted on the front of the helmet to identify the fire fighter's rank and company. Bright, reflective materials are applied to make the fire fighter more visible in all types of lighting conditions.

NFPA 1971 requires that helmets approved for structural firefighting have a label permanently attached to the inside of the shell. This label lists the manufacturer, model, date of manufacture, weight, size, and recommended cleaning procedures.

Protective Hood

Although the helmet's ear tabs cover the ears and neck, this area is still at risk for burns when the head is turned or the neck is flexed. **Protective hoods** provide additional thermal protection for these areas. Such a hood, which is constructed of a flame-resistant material such as **Nomex**®, **PBI**®, or carbon fibre, covers the whole head and neck, except for that part of the face protected by the SCBA face piece **FIGURE 3-4**. The lower

part of the hood, which is called the bib, drapes down inside the turnout coat.

Protective hoods are worn over the face piece but under the helmet. After securing the face piece straps, the fire fighter

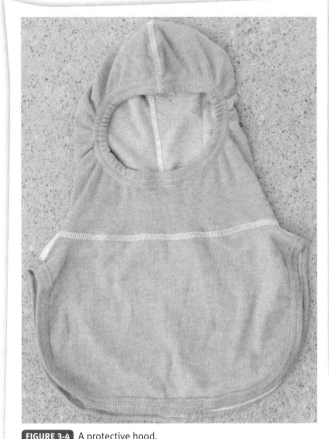

FIGURE 3-4 A protective hood.

Standards for Personal Protective Equipment

The PPE for fire fighters must be manufactured according to exacting standards. Each item must have a permanent label verifying that the particular item meets the requirements of the standard **FIGURE 3-5**. Usage limitations as well as cleaning and maintenance instructions should also be provided.

The requirements for firefighting PPE are outlined in these standards:

- NFPA 1971, *Standard on Protective Ensembles for Structural Fire Fighting and Proximity Fire Fighting*
- NFPA 1977, *Standard on Protective Clothing and Equipment for Wildland Fire Fighting*

The requirements for SCBA and PASS devices are outlined in two standards:

- NFPA 1981, *Standard on Open-Circuit Self-Contained Breathing Apparatus (SCBA) for Emergency Services*
- NFPA 1982, *Standard on Personal Alert Safety Systems (PASS)*

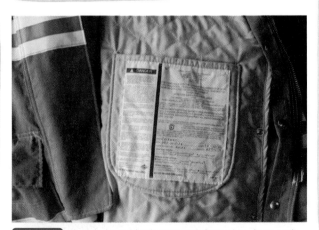

FIGURE 3-5 The label provides important information about each item of PPE.

should carefully fit the hood around the face piece so that no areas of bare skin are left exposed. The hood must fit snugly around the clear area of the face piece so that vision is not compromised and hot gases cannot leak between the face piece and the hood.

Turnout Coat

The coats used for structural firefighting are generally called **bunker coats** or **turnout coats** FIGURE 3-6. Only coats that meet NFPA 1971 requirements should be used for structural firefighting.

Turnout coats consist of three layers. The outer layer or shell is constructed of a sturdy, flame-resistant, water-repellent material such as Nomex®, **Kevlar**®, or PBI®. Fluorescent reflective material applied to the outer shell makes the fire fighter more visible in smoky conditions and at night. The fabric also makes it easier to identify contaminants such as hydrocarbons, blood, and bodily fluids on the coat.

The second layer of the coat is the moisture barrier, which usually consists of a flexible membrane attached to a thermal barrier material (the third layer). The moisture barrier helps prevent the transfer of water, steam, and other fluids to the skin. It is critical because water applied to a fire generates large amounts of superheated steam, which can engulf fire fighters and burn unprotected skin.

The third layer of the turnout coat comprises a thermal barrier, which is made of a multilayered or quilted material that insulates the body from external temperatures. It enables fire fighters to operate in elevated temperatures generated by a fire and keeps the body warm during cold weather.

The front of the turnout coat has an overlapping flap to provide a secure seal. The inner closure is secured first, and then the outer flap is secured, creating a double seal. Several different combinations of D-rings, snaps, zippers, and Velcro can be used to secure the inner and outer closures.

The collar of the turnout coat works with the hood to protect the neck. The collar has snaps or a Velcro closure system in front to keep it in a raised position. The coat's sleeves feature wristlets that prevent liquids or hot embers from getting between the sleeves and the skin. They also prevent the sleeves from riding up on the wrists, so no skin is exposed between the gloves and the sleeves, which could result in wrist burns.

Turnout coats come in two lengths—long and short. Both will protect the body if the matching style of pants or coveralls is worn simultaneously. The coat must be long enough that you can raise your arms over your head without exposing your midsection. The sleeve length should not hinder arm movement, and the coat should be large enough that it does not interfere with breathing or other movements.

Pockets in the coat can be used for carrying small tools or extra gloves. Additional pockets or loops can be installed to hold radios, microphones, flashlights, or other accessories.

Fire Fighter Safety Tips
© Photos.com

The inner thermal liner of most turnout coats can be removed while the outer shell is being cleaned, but the turnout coat should never be used without the thermal liner. Severe injury could occur if the coat is used without the liner.

Bunker Pants

The protective trousers included as part of the structural firefighting ensemble are also called **bunker pants** or **turnout pants** FIGURE 3-7. They can be constructed in a waist-length design or a bib-overall configuration. Like turnout

FIGURE 3-7 Bunker pants.

FIGURE 3-6 A bunker or turnout coat.

coats, bunker pants must satisfy the conditions specified in NFPA 1971.

Also like turnout coats, bunker pants are constructed with three layers: The outer shell resists abrasions and repels water. The second layer is a moisture barrier that protects the skin from liquids and steam burns. The third, inner layer is a quilted, thermal barrier that protects the body from elevated temperatures. Bunker pants are reinforced around the ankles and knees with leather or extra padding.

Bunker pants are manufactured with a double-fastener system at the waist, similar to the front flap of a turnout coat. Fluorescent or reflective stripes around the ankles provide added visibility. Suspenders hold the pants up. Pants should be large enough that you can don them quickly. They should be big enough to allow you to crawl and bend your knees easily, but they should not be bigger than necessary.

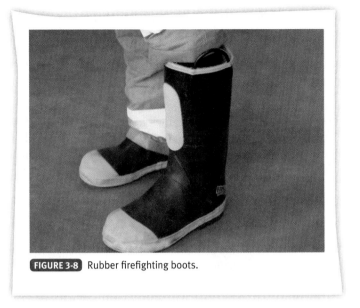

FIGURE 3-8 Rubber firefighting boots.

Boots

Structural firefighting boots can be constructed of rubber or leather; they are available in a variety of lengths. Rubber firefighting boots come in a step-in style without laces **FIGURE 3-8**. Leather firefighting boots are available in a knee-length, pull-on style or in a shorter version with laces **FIGURE 3-9**. Many fire fighters install a zipper on the laced boots to aid in quick donning and doffing of this component of PPE.

Both boot styles must meet the requirements specified in NFPA 1971. The outer layer of these boots repels water and must be both flame and cut resistant. Boots must also have a heavy sole with a slip-resistant design, a puncture-resistant sole, and a reinforced toe to prevent injury from falling objects. An inner liner constructed of a material such as Nomex® or Kevlar® adds thermal protection.

Boots must be the correct size for the fire fighter's feet. The foot should be secure within the boot to prevent ankle injuries and enable secure footing on ladders or uneven surfaces. Improperly sized boots will cause blisters and other problems.

FIGURE 3-9 Leather firefighting boots.

Fire Fighter Safety Tips
© Photos.com

Fire boots must be worn with approved bunker pants. Boots worn without bunker pants leave the legs unprotected and exposed to injury.

Gloves

Gloves are an important part of the firefighting ensemble, because most fire-suppression tasks require the use of the hands **FIGURE 3-10**. Gloves must provide adequate protection, yet permit the manual dexterity needed to accomplish tasks. NFPA 1971 specifies that gloves must be resistant to heat, liquid absorption, vapours, cuts, and penetration. These items must have a wristlet to prevent skin at the wrist from being exposed during normal firefighting activities.

Firefighting gloves are usually constructed of heat-resistant leather. The wristlets are usually made of knitted Nomex® or Kevlar®. The liner adds thermal protection and serves as a moisture barrier.

Many fire fighters carry a second set of gloves in their bunker gear or on the fire apparatus so they can change gloves if one pair becomes wet or is damaged. Do not wring or twist wet gloves, as this motion can tear or damage the inner liners.

Although gloves furnish needed protection, they inevitably reduce manual dexterity. For this reason, fire fighters need to practice manual skills while wearing gloves to become accustomed to them and to learn to adjust their movements accordingly.

Fire Fighter Safety Tips
© Photos.com

Plain leather work gloves or plastic-coated gloves should never be used for structural firefighting. Gloves used for firefighting must be labelled as such and meet the specifications given by NFPA 1971.

Respiratory Protection

The SCBA is an essential component of the PPE used for structural firefighting. Without adequate respiratory protection, fire fighters would be unable to mount an attack. The design, parts, operation, use, and maintenance of this complex piece of equipment are covered later in this chapter.

FIGURE 3-10 Firefighting gloves.

Personal Alert Safety System

A **personal alert safety system (PASS)** is an electronic device that sounds a loud audible signal when a fire fighter becomes trapped or injured. This device will sound automatically if a fire fighter is motionless for a set time period. It can also be manually activated to notify other fire fighters that the user needs assistance.

A PASS device can be either separate from or integrated into the SCBA unit **FIGURE 3-11**. Most PASS devices turn on automatically when the SCBA is activated. The separate PASS devices are often worn on the SCBA harness and must be turned on manually. All fire fighters must check that their

PASS devices are on and working properly before they enter a burning building or hazardous area.

The PASS device combines an electronic motion sensor with an alarm system. If the user remains motionless for 30 seconds, it will produce a low warning tone before sounding a full alarm. The user can reset the device by moving during this warning period. Newer PASS devices may also include a radio transmitter that sends a signal to the command post when the alarm sounds.

Additional Personal Protective Equipment

Some eye protection is provided by the face shield mounted on fire helmets. When additional eye protection is needed, such as when they are using power saws or hydraulic rescue tools, fire fighters can use approved goggles or safety glasses. Each province or territory has specific guidelines for additional eye protection. Follow the requirements of your department or governing body. Goggles can be carried easily in the pockets of the turnout coat.

Fire fighters are often exposed to loud noises such as sirens and engines. Because hearing loss is cumulative, it is important to limit exposure to loud sounds. An intercom system on the apparatus can provide hearing protection **FIGURE 3-12**. A small microphone is incorporated into large earmuffs located at each riding and operating position. Fire fighters don the earmuffs, which reduce engine and siren noise; at the same time, the microphone enables crew members to talk to each other in a normal tone of voice and to hear the apparatus radio. Flexible ear plugs are useful in other situations involving loud sounds. Fire fighters should use the hearing protection supplied by their departments to prevent hearing loss when using power equipment or other equipment that produces noise.

A fire fighter should always carry a **hand light**, given that most interior firefighting takes place in near-dark, zero-visibility conditions. A good working hand light can illuminate your surroundings, mark your location, and help you find your way under difficult conditions.

Two-way radios link the members of a firefighting team. At least one member of each team working inside a burning building or in any hazardous area should always have a radio.

FIGURE 3-11 An integrated PASS device can save a fire fighter's life.

FIGURE 3-12 Some fire apparatus are equipped with an intercom system.

Some fire departments provide a radio for every on-duty fire fighter. Follow your department's suggested operating guideline (SOG) or standard operating procedure (SOP) on radio use. A radio should be considered part of PPE and carried with you whenever it is appropriate.

Any time you are operating in a roadway or are close to traffic, you are required to wear a brightly coloured reflective safety vest that meets the latest iteration of ANSI Standard 207. This standard requires the vest to contain 0.3 square metres (450 square inches) of high-visibility fabric and 0.1 sq. m (201 sq. in.) of retro-reflective material. Each province or territory has regulations that meet or exceed this standard. To be compliant, follow your local requirements.

The final piece of safety equipment is the drag rescue device. The drag rescue device is a fabric handle integrated within the turnout coat that is designed to aid in the rescue of an incapacitated fire fighter. A rescuer can grab onto the drag rescue device and drag the incapacitated fire fighter to safety.

FIRE FIGHTER Tips

Each item of PPE clothing is designed to have overlapping material from one item to the next. For example, boots overlap the pants, coats overlap pants, and coats overlap gloves. This overlapping ensures that the body is protected at all times.

Limitations of the Structural Firefighting Ensemble

The structural firefighting ensemble protects a fire fighter from the hostile environment of a fire. To provide complete protection, each component must be properly donned and worn continuously. Unfortunately, even today's advanced PPE has drawbacks and limitations. Understanding those limitations will help you avoid situations that could result in serious injury or death.

First, this ensemble is not easy to don. It includes several components, all of which must be put on in the proper order and correctly secured. You must be able to don your equipment quickly and correctly, either at the fire station before you respond to an emergency or after you arrive at the scene. Properly donning and doffing PPE takes practice **FIGURE 3-13**.

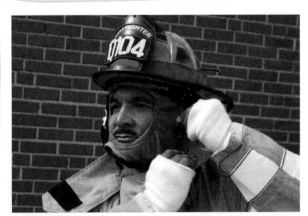

FIGURE 3-13 It takes practice to don the full set of PPE.

PPE is also heavy, accounting for nearly 23 kilograms (50 pounds) of extra weight. This increased weight means that everything you do—even walking—requires more energy and strength. Tasks like advancing an attack line up a stairway or using an axe to ventilate a roof can be difficult, even for a fire fighter in excellent physical condition.

Because PPE retains body heat and perspiration, wearing this equipment makes it difficult for the body to cool itself. Perspiration is retained inside the protective clothing rather than being released through evaporation to cool you. As a consequence, fire fighters in full protective gear can rapidly develop elevated body temperatures, even when the ambient temperature is cool. The problem of overheating is more acute when surrounding temperatures are high, which is one reason why fire fighters must undergo regular rehabilitation and fluid replacement.

Another drawback of PPE is that it limits mobility. Full turnout gear not only limits the range of motion, but also makes movements awkward and difficult. This constraint increases energy expenditure and adds stress.

Wearing PPE also decreases normal sensory abilities; the sense of touch is reduced by wearing heavy gloves, for example. Turnout coats and pants protect skin but reduce its ability to determine the temperature of hot air. Sight is restricted when you wear SCBA. The plastic face piece reduces peripheral vision, and the helmet, hood, and coat make turning the head difficult. Both the earflaps and the protective hood over the ears limit hearing. Speaking becomes muffled and distorted by the SCBA face piece, even if it is equipped with a special voice-amplification system.

Fire Fighter Safety Tips

© Photos.com

Do not don protective clothing inside the apparatus while en route to an emergency incident. Instead, stay in your assigned seat, properly secured by a seat belt or safety harness, while the vehicle is in motion. Either don protective clothing in the firehouse before mounting the apparatus, or put it on at the scene after you arrive.

FIRE FIGHTER Tips

Practise donning your protective clothing ensemble until you can do so quickly and smoothly. Remember that each piece of the ensemble must be in place for maximum protection.

For these reasons, fire fighters must become accustomed to wearing and using PPE. Practising skills while wearing PPE will help you become comfortable with its operation and limitations.

Work Uniforms

A fire fighter's work uniform is also part of the personal protective package. Certain synthetic fabrics can melt at relatively low temperatures and cause severe burns, even when worn under a complete personal protective ensemble. NFPA 1975, *Standard on Station/Work Uniforms for Emergency Services*, defines criteria for selecting appropriate fabrics for work uniforms.

Clothing containing nylon or polyester—even if these materials are blended with natural fibres—should not be worn in a firefighting environment. Clothing made of natural fibres, such as cotton or wool, is generally safer to wear. Nylon and polyester will melt, but cotton and wool may char or even burn without melting. Special synthetic fibres such as Nomex® and PBI®, which are used in turnout clothing, have excellent resistance to high temperatures.

■ Donning Personal Protective Clothing

Donning (i.e., putting on) protective clothing must be done in a specific order to obtain maximum protection; it must also be done quickly. Your goal should be to don PPE properly.

Following a set pattern of donning PPE helps reduce the time it takes to dress. This exercise does not include donning SCBA, however. First become proficient in donning personal protective clothing, and then add the SCBA. Follow the steps in **SKILL DRILL 3-1** to don personal protective clothing:

1. Prepare your equipment for donning. Place equipment in a logical order for donning. (**STEP 1**)
2. Place the legs of your bunker pants over your boots, and fold the pants down around them.
3. Place your protective hood over the top of your boots to remind you to put on your protective hood.
4. Lay out your coat, helmet, and gloves.
5. Place your protective hood over your head and down around your neck. (**STEP 2**)
6. Step into your boots and pull up your bunker pants. Place the suspenders over your shoulders, and secure the front of the pants using the closure system. (**STEP 3**)
7. Put on your turnout coat, and secure the inner and outer closures. (**STEP 4**)
8. Place your helmet on your head with the ear tabs extended, and adjust the chin strap securely. Turn up your coat collar and secure it in front. (**STEP 5**)
9. Put on your gloves. (**STEP 6**)
10. Check all clothing to be sure it is properly secured and in the correct position.
11. Have your partner check your clothing. (**STEP 7**)

■ Doffing Personal Protective Clothing

To **doff** (i.e., remove) your personal protective clothing, reverse the procedure used in getting dressed. Follow the steps in **SKILL DRILL 3-2** to doff personal protective clothing:

1. Remove your gloves. (**STEP 1**)
2. Open the collar of your turnout coat. (**STEP 2**)
3. Release the helmet chin strap and remove your helmet. (**STEP 3**)
4. Remove your turnout coat. (**STEP 4**)
5. Remove your bunker pants and boots. (**STEP 5**)
6. Remove your protective hood. (**STEP 6**)

PPE should be cleaned anytime it becomes dirty—by being exposed to smoke, blood or bodily fluids, or chemicals, for example—to ensure your safety and health. Smoke contains **carcinogens**, substances capable of causing cancer. Bodily fluids can transmit diseases, and chemicals can cause burns.

Once PPE has been properly cleaned, it should be placed in a convenient location until the next response. Such equipment may be kept close to the fire apparatus, on the fire apparatus, or in an equipment locker. Wherever it is stored, personal protective clothing must be properly maintained, organized, and ready for the next emergency response.

■ Care of Personal Protective Clothing

Approved personal protective clothing is built to exacting standards but requires proper care if it is to continue to afford maximum protection. Avoid unnecessary wear on turnout clothing. A complete set of approved turnout clothing (excluding SCBA) costs more than $2000. Keep this expensive equipment in good shape for its intended use—structural firefighting.

Check the condition of your PPE on a regular basis. Clean it when necessary; repair worn or damaged PPE at once. PPE that is worn or damaged beyond repair must be replaced immediately, because it will not be able to protect you.

Avoid unnecessary cuts or abrasions on the outer material. This material already meets NFPA standards—do not look for new opportunities to test its effectiveness. If the fabric is damaged, it must be properly repaired to retain its protective qualities. Follow the manufacturer's instructions for repairing or replacing PPE.

You must also keep personal protective clothing clean so that it will maintain its protective properties. Dirt will build up in the clothing fibres from routine use and exposure to fire environments. Smoke particles will also become embedded in the outer shell material. The interior layers will frequently be soaked with perspiration. Regular cleaning should remove most of these contaminants.

Other contaminants are formed from the by-products of burnt plastics and synthetic products. These residues, which are flammable, can become trapped between the fibres or build up on the outside of PPE, damaging the materials and reducing their protective qualities. A fire fighter who is wearing contaminated PPE is, in essence, bringing additional fuel into the fire on the clothing.

PPE that has been badly soiled by exposure to smoke, other products of combustion, melted tar, petroleum products, or other contaminants needs to be cleaned as soon as possible. Items that have been exposed to chemicals or hazardous materials may have to cleaned by a professional cleaning company or be impounded for decontamination or disposal.

Cleaning instructions are listed on the tag attached to the clothing. Always follow the manufacturer's cleaning instructions. Failure to do so may reduce the effectiveness of the garment and create an unsafe situation for the wearer.

Some fire departments have special washing machines that are approved for cleaning turnout clothing. Other departments contract with an outside firm to clean and repair protective clothing. In either case, the manufacturer's instructions for cleaning and maintaining the garment must be followed. If you

SKILL DRILL 3-1

Donning Personal Protective Clothing
(Fire Fighter I, NFPA 5.1.2)

1 Place your equipment in a logical order for donning.

2 Place your protective hood over your head and down around your neck.

4 Put on your turnout coat and close the front of the coat.

5 Place your helmet on your head and adjust the chin strap securely. Turn up your coat collar and secure it in front.

3 Put on your boots and pull up your bunker pants. Place the suspenders over your shoulders and secure the front of the pants.

6 Put on your gloves.

7 Have your partner check your clothing.

© LiquidLibrary

SKILL DRILL 3-2 Doffing Personal Protective Clothing
(Fire Fighter I, NFPA 5.1.2)

1 Remove your gloves.

2 Open the collar of your turnout coat.

3 Release the helmet chin strap and remove your helmet.

4 Remove your turnout coat.

5 Remove your bunker pants and boots.

6 Remove your protective hood.

FIRE FIGHTER Tips

Follow your fire department's policies in regard to cleaning contaminated PPE (i.e., PPE that comes in contact with bodily fluids, such as blood). Your fire department may require that the PPE be bagged in a biohazard bag and professionally cleaned. Never clean your PPE or work uniform in home washers to prevent cross-contamination with your home laundry.

Fire Fighter Safety Tips   © Photos.com

In hot weather, remove as much of your PPE as possible when you arrive at rehab. This step will help you to cool down quickly. Some fire departments even place fans and misting equipment in the rehabilitation area to provide additional cooling.

Fire Fighter Safety Tips   © Photos.com

Make sure your PPE is dry before using it on the fire ground. If wet protective clothing is exposed to the high temperatures of a structural fire, the water trapped in the liner materials will turn into steam and be trapped inside the moisture barrier. The result may be painful steam burns.

wash your turnout gear in-house, make sure that you follow the manufacturer's instructions for drying these garments properly.

Other PPE may also require regular cleaning and maintenance. For example, the outer shell of your helmet should be cleaned with a mild soap as recommended by the manufacturer. The inner parts of helmets should be removed and cleaned according to the manufacturer's instructions. The chin strap and suspension system must be properly adjusted and all parts of the helmet kept in good repair.

Protective hoods and gloves get dirty quickly and should be cleaned according to the manufacturer's instructions. Most hoods can be washed with appropriate soaps or detergents. Repair or discard gloves or hoods that have holes in them; do not use them. Even a small cut or opening in these items can result in a burn injury.

Boots should be maintained according to the manufacturer's instructions. Rubber boots need to be kept in a place that does not result in damage to the boot. Leather boots must be properly maintained to keep them supple and in good repair. Any type of boot needs to be repaired or replaced if the outer shell is damaged. Do not store these items or any other PPE in direct sunlight.

Correctly storing your PPE is another important step in the proper care and maintenance of your PPE. PPE should only be stored in the proper PPE storage areas within the fire station. PPE should not be stored in living areas or in cars.

■ Specialized Protective Clothing

Vehicle Extrication

Due to the risk of fire at the scene of a vehicle-extrication incident, members of the emergency team will wear full turnout gear. The officer in charge may also designate one or more members to don SCBA and stand by with a charged hose line. A structural firefighting ensemble also protects against many of the hazards present at a vehicle-extrication incident, such as broken glass and sharp metal objects.

Some protective clothing, such as special gloves and coveralls or jumpsuits, has been specifically designed for vehicle extrication. These items are generally lighter in weight and more flexible than structural firefighting PPE, although they may be constructed of the same basic materials.

Fire fighters performing vehicle extrication must always be aware of the possibility of contact with blood or other bodily fluids. Latex gloves should be worn when providing patient treatment. Eye protection should be worn, given the possibilities of breaking glass, contact with bodily fluids, metal debris, and accidents with tools. Proper cleaning of extrication gear is especially important because many of the fluids encountered during vehicle extrication are oil based—for example, gasoline, diesel fuel, transmission fluid, and brake fluid. Because of the flammability of these fluids, they pose a hazard for fire fighters who enter a burning building while wearing oil-contaminated turnout gear.

Wildland Fires

Gear that is designed specifically for fighting wildland or brush fires must meet the requirements outlined in NFPA 1977, *Standard on Protective Clothing and Equipment for Wildland Fire Fighting*. In this type of firefighting ensemble, the jacket and pants are made of fire-resistant materials, such as Nomex® or specially treated cotton, which are designed for comfort and maneuverability while working in the wilderness. Wildland fire fighters wear a helmet of a thermoresistant plastic, eye protection, and pigskin or leather gloves. The boots are designed to provide comfort and sure footing while hiking to a remote fire scene and to provide protection against crush and puncture injuries. Structural firefighting gear is not designed for extended wildland firefighting. It may produce high body temperatures resulting in heat exhaustion or heat stroke.

Respiratory Protection

Respiratory protection equipment is an essential component of the structural firefighting personal protection ensemble. An SCBA is both expensive and complicated; using one confidently requires practice. You must be proficient in using SCBA before you engage in interior fire-suppression activities.

The interior atmosphere of a burning building is considered to be **immediately dangerous to life and health (IDLH)**. Attempting to work in such an atmosphere without proper respiratory protection can cause serious injury or death. Never enter or operate in a fire atmosphere without first ensuring that you have appropriate respiratory protection.

■ Respiratory Hazards of Fires

Fire fighters need respiratory protection because a fire involves a complex series of chemical reactions that can rapidly affect the atmosphere in unpredictable ways. The most readily evident by-product of a fire is smoke. The visible smoke produced by a fire contains many different substances, some of which are dangerous if inhaled. In addition, smoke contains invisible, highly toxic products of combustion. The process of combustion consumes oxygen and can lower the oxygen concentration in the atmosphere below the level necessary to support life. In addition, the atmosphere of a fire may become so hot that one unprotected breath can result in fatal respiratory burns.

These respiratory hazards require fire fighters to use respiratory protection in all fire environments, regardless of whether the environment is known to be contaminated, suspected of being contaminated, or could possibly become contaminated without warning. The use of SCBA allows fire fighters to enter and work in a fire atmosphere with a safe, independent air supply.

Smoke

Most fires do not have a supply of oxygen sufficient to allow them to consume all of the available fuel. As a consequence, most fires result in **incomplete combustion** and produce a variety of by-products, which are then released into the atmosphere. Many of these by-products are extremely toxic. Collectively, the airborne products of combustion are called smoke. Smoke includes three major components: particles, vapours, and gases.

Smoke Particles

Smoke particles consist of unburned, partially burned, and completely burned substances. These particles are lifted in the thermal column produced by the fire and are usually readily visible. The completely burned particles are primarily ash; the unburned and partially burned smoke particles can include a variety of substances. The concentration of unburned or partially burned particles depends on the amount of oxygen that was available to fuel the fire.

Many smoke particles are so small that they can pass through the natural protective mechanisms of the respiratory system and enter the lungs. Some are toxic to the body and can result in severe injuries or death if they are inhaled. These particles also can prove extremely irritating to the eyes and digestive system.

Smoke Vapours

Smoke also contains small droplets of liquids. These smoke vapours are similar to fog, which consists of small water droplets suspended in the air. When oil-based compounds burn, they produce small hydrocarbon droplets that become part of the smoke. If inhaled or ingested, these compounds can affect the respiratory and circulatory systems. Some of the toxic droplets in smoke can cause poisoning if they are absorbed through the skin.

When water is applied to a fire, it creates steam and water droplets that in turn become part of the smoke. These water droplets can absorb some of the toxic substances contained in the smoke.

Toxic Gases

A fire also produces several types of gases. The amount of oxygen available to the fire and the type of fuel being burned determine precisely which gases are produced. A fire fuelled by wood, for example, produces a different mixture of gases than one fuelled by petroleum-based products. (Many common products—such as plastics—are made from petroleum-based compounds.) The concentrations of these gases can change rapidly as the oxygen supply is consumed or as fresh oxygen is introduced to the combustion process.

Many of the gases produced by residential or commercial fires are highly toxic. **Carbon monoxide (CO)**, hydrogen cyanide, and phosgene are three of many gases often present in smoke.

CO is deadly in small quantities. When inhaled, this gas quickly replaces the oxygen in the bloodstream because it binds with hemoglobin molecules (which normally carry oxygen to the body's cells) in the blood 200 times more readily than oxygen does. Even a small concentration of CO can quickly disable and kill a fire fighter. CO is odourless, colourless, and tasteless, which also makes it dangerous. Because this gas cannot be detected without special instruments, fire fighters must always assume that it is present in the atmosphere around a fire.

Hydrogen cyanide is formed when plastic products—such as the polyvinyl chloride pipe used in residential construction—burn. This poisonous gas is quickly absorbed by the blood and interferes with cellular respiration. Just a small amount of hydrogen cyanide can easily render a person unconscious. Recent studies have indicated that cyanide poisoning in fire fighters may be more common than previously recognized.

Phosgene gas is formed from incomplete combustion of many common household products, including vinyl materials. This gas can affect the body in several ways. At low levels, it causes itchy eyes, a sore throat, and a burning cough. At higher levels, phosgene gas can cause pulmonary edema (fluid retention in the lungs) and death. Phosgene gas was used as a weapon in World War I to disable and kill soldiers.

Fires produce other gases that may affect the human body in harmful ways. Among these toxic gases are hydrogen chloride and several compounds containing different combinations of nitrogen and oxygen. Carbon dioxide (CO_2), which is produced when sufficient oxygen is available to allow for complete combustion, is not toxic, but it can displace oxygen from the atmosphere and cause asphyxiation. Inhalation of these substances should be avoided.

Oxygen Deficiency

Oxygen is required to sustain human life. Normal outside or room air contains approximately 21 percent oxygen. A decrease in the amount of oxygen in the air, however, may drastically affect an individual's ability to function **TABLE 3-2**. An atmosphere with an oxygen concentration of 19.5 percent or less is considered oxygen deficient. If the oxygen level drops below 17 percent, people can experience disorientation, an inability to control their muscles, and irrational thinking, which can make escaping a fire much more difficult.

TABLE 3-2	Physiological Effects of Reduced Oxygen Concentration
Oxygen Concentration	**Effect**
21 percent	Normal breathing air
17 percent	Judgment and coordination impaired; lack of muscle control
12 percent	Headache, dizziness, nausea, fatigue
9 percent	Unconsciousness
6 percent	Respiratory arrest, cardiac arrest, death

During compartment fires (i.e., fires burning within enclosed areas), <u>oxygen deficiency</u> occurs in two ways. First, the fire consumes large quantities of the available oxygen, thereby decreasing the concentration of oxygen in the atmosphere. Second, the fire produces large quantities of other gases, which decrease the oxygen concentration by displacing the oxygen that would otherwise be present inside the compartment.

Increased Temperature

Heat is also a respiratory hazard. The temperature of the gases generated during a fire varies, depending on the fire conditions and the distance travelled by the hot gases. Inhaling superheated gases produced by a fire can immediately cause severe burns of the respiratory tract. If the gases are hot enough, a single inhalation can cause fatal respiratory burns. More information about fire behaviour and products of combustion appears in the chapter "Fire Behaviour."

■ Other Toxic Environments

Not all hazardous atmospheric conditions are caused by fires. Indeed, fire fighters may encounter toxic gases or oxygen-deficient atmospheres in numerous types of emergency situations. Respiratory protection is just as important in these situations as in a fire-suppression operation.

For example, toxic gases may be released at hazardous materials incidents from leaking storage containers or industrial equipment, from chemical reactions, or from the normal decay of organic materials. CO can be produced by internal combustion engines or improperly operating heating appliances. Exposure to CO or another toxic gas should be suspected whenever fire fighters find one or more unconscious patients with no evident cause.

Toxic gases can quickly fill confined spaces or below-grade structures. Any confined space or below-grade area must be treated as a hazardous atmosphere until it has been tested to ensure that the concentration of oxygen is adequate and that no hazardous or dangerous gases are present.

■ Conditions That Require Respiratory Protection

More fire deaths are caused by smoke inhalation rather than burns. Clearly, then, adequate respiratory protection is essential to your safety as a fire fighter. In fact, the products of combustion from house fires and commercial fires may be so toxic that just a few breaths can result in death—which is why you must always take respiratory protection precautions before entering the fire. When you first arrive at the scene of a fire, you do not have any way to measure the immediate danger to your life and your health posed by that fire. You must use full PPE and approved breathing apparatus if you will enter and operate within this atmosphere. CSA Standard Z 94.4 regulates and specifies all conditions that require respiratory protection in Canada.

Whenever you are in an area where smoke is present, you must use SCBA—including when you respond to outside fires, such as vehicle and dumpster fires. Use your SCBA at the fire scene and continue to wear it during overhaul until the air has been tested and proven to be safe by the safety officer. Do not remove your SCBA just because a fire has been knocked down.

SCBA must also be used in any situation where there is a possibility of toxic gases being present or oxygen deficiency, such as in a confined space. Always assume that the atmosphere is hazardous until it has been tested and proven to be safe.

■ Types of Breathing Apparatus

The basic respiratory protection hardware used by the fire service is the SCBA. The term *self-contained* in the acronym for this equipment refers to the requirement that the apparatus be the sole source of the fire fighter's air supply. In other words, SCBA is an independent air supply that will last for a predictable duration.

The two main types of SCBA are open-circuit and closed-circuit devices. An <u>open-circuit breathing apparatus</u> is typically used for structural firefighting. A tank of compressed air provides the breathing air supply for the user, and exhaled air is released into the atmosphere through a one-way valve **FIGURE 3-14**. Approved open-circuit SCBA comes in several different models, designs, and options.

A <u>closed-circuit breathing apparatus</u> recycles the user's exhaled air. The air passes through a mechanism that removes CO_2 and adds oxygen within a closed system, where the oxygen is generated from a chemical canister **FIGURE 3-15**.

Many closed-circuit SCBA units include a small oxygen tank as well as chemically generated oxygen. Closed-circuit breathing apparatus are often used for extended operations, such as mine rescue work and operations in long tunnels where breathing apparatus must be worn for a long time.

A <u>supplied-air respirator (SAR)</u> uses an external source for the breathing air **FIGURE 3-16**. In this type of device, a hose line is connected to a breathing-air compressor or to compressed air cylinders located outside the hazardous area. The user breathes air through the line and exhales through a one-way valve, just as with an open-circuit SCBA.

Although SARs are commonly used in industrial settings, they are not used by fire fighters for structural firefighting. Hazardous materials teams and confined-space rescue teams sometimes use SARs for specialized operations. Some fire service SCBA units can be adapted for use as SARs.

FIGURE 3-14 Open-circuit SCBA.

FIGURE 3-15 Closed-circuit SCBA.

FIGURE 3-16 A supplied-air respirator may be needed for special rescue operations.

Fire Fighter Safety Tips

© Photos.com

Always use SCBA before entering a confined space or a below-ground structure. Continue to use respiratory protection until the air is tested and deemed safe by the safety officer. Do not attempt confined-space operations without special training.

FIRE FIGHTER Tips

Do not confuse SCBA and SCUBA. SCBA is used by fire fighters during fire suppression activities. **Self-Contained Underwater Breathing Apparatus (SCUBA)** is used by divers while swimming underwater.

equipment used in the emergency service must meet the design, testing, and certification requirements established by NIOSH.

In the United States, **Occupational Safety and Health Administration (OSHA)** and state agencies are responsible for establishing and enforcing regulations for respiratory protection programs. In some states and in Canadian territories provinces, regulatory occupational safety and health agencies establish and enforce these regulations.

The NFPA has developed three standards directly related to SCBA:

- NFPA 1500, *Standard on Fire Department Occupational Safety and Health Program*, includes the basic requirements for SCBA use and program management.
- NFPA 1404, *Standard for Fire Service Respiratory Protection Training*, sets the requirements for an SCBA training program within a fire department.
- NFPA 1981, *Standard on Open-Circuit Self-Contained Breathing Apparatus (SCBA) for Emergency Services*, includes requirements for the design, performance, testing, and certification of open-circuit SCBA for the fire service.

These standards specify approved SCBA equipment, initial training for each fire fighter, and annual retraining. Each fire department must follow applicable standards and regulations to ensure safe working conditions for all personnel.

■ SCBA Standards and Regulations

In the United States, the **National Institute for Occupational Safety and Health (NIOSH)** sets the design, testing, and certification requirements for SCBA. NIOSH is a federal agency that researches, develops, and implements occupational safety and health programs. It also investigates fire fighter fatalities and serious injuries, and makes recommendations on how to prevent accidents from recurring. NFPA 1981 requires that all SCBA

■ Uses and Limitations of Self-Contained Breathing Apparatus

SCBA is designed to provide pure breathing air to fire fighters who are working in the hostile environment of a fire. It must meet rigid manufacturing specifications so that it can function in the increased temperatures and smoke-filled environments that fire fighters encounter. When properly maintained, such equipment will provide sufficient quantities of air to enable fire fighters to perform rigorous tasks.

Using SCBA requires that fire fighters develop unique skills, including different breathing techniques. This equipment limits normal sensory awareness; the senses of smell, hearing, and sight are all affected by the apparatus. Proficiency in the use of SCBA and other PPE requires ongoing training and practice.

Like any type of equipment, SCBA also has its limitations. Some of these limitations apply to the equipment; others apply to the user's physical and psychological abilities.

Limitations of the Equipment

Because an SCBA carries its own air supply in a pressurized cylinder, its use is limited by the amount of air in the cylinder. SCBA for structural firefighting must carry enough air for a minimum of 30 minutes; cylinders rated for 45 minutes and 60 minutes are also available. These duration ratings, however, are based on ideal laboratory conditions. The realistic useful life of an SCBA cylinder for firefighting operations is usually much less than the rated duration, and actual use time will depend on the size of the user, his or her physical fitness and conditioning, the amount of physical exertion, and the user's degree of calm. An SCBA cylinder will generally have a realistic useful life of no more than 50 percent of the rated time. For example, an SCBA cylinder rated for 30 minutes can be expected to last for a maximum of 15 minutes during strenuous firefighting.

Fire fighters must manage their working time while using SCBA so that they have enough time to exit from the hazardous area before exhausting the air supply. To properly manage the air supply, a fire fighter must consider the following factors:

- The time and the effort it will take to reach the task destination; climbing stairs will take more energy and air than walking across a flat floor, for example.
- The amount of air that will be available upon reaching the task destination.
- The amount of time necessary to complete the task and the air that will be used during that period. Some tasks will take more energy and air.
- The amount of time it will take to reach a safe area. At the end of this time, the fire fighter must have a reserve of air for unexpected emergencies.

SCBA provides a very limited window of time for firefighting and a safe exit from the hazardous conditions of the fire. It is essential that you have a margin of safety built into your air supply for the unexpected. In some cases, you may need to begin exiting from the fire scene before half of your air supply is exhausted.

The weight of an SCBA varies, based on the manufacturer and the type and size of the cylinder. Generally, an SCBA weighs at least 11 kg (25 lb). The size of the unit also makes it more difficult for the user to fit into small places **FIGURE 3-17**. The added weight and bulk decrease the user's flexibility and mobility, and shift the user's centre of gravity.

FIGURE 3-17 SCBA expands a user's profile, making it more difficult to pass through tight spaces.

The design of the SCBA face piece limits the fire fighter's vision—particularly his or her peripheral vision. The face piece may fog up under some conditions, further limiting visibility. SCBA also may affect the user's ability to communicate, depending on the type of face piece and any additional hardware provided, such as voice amplification and radio microphones. The equipment is noisy during inhalation and exhalation, which may limit the user's hearing as well.

Physical Limitations of the User

Conditioning is important for SCBA users. An out-of-shape fire fighter will consume the air supply from this equipment more quickly and will have to exit the fire building long before a well-conditioned fire fighter does so. Overweight or poorly conditioned fire fighters are also at greater risk for heart attack due to physical stress.

Altogether, the protective clothing and SCBA that must be worn when fighting fires weigh more than 23 kg (50 lb). Moving with this extra weight requires additional energy, which in turn increases air consumption and body temperature. Taken collectively, this activity places additional stress on a fire fighter's body. A person with ideal body weight and in good physical condition will be able to perform more work per cylinder of air than a person who is overweight or out of shape.

The weight and bulk of the complete PPE ensemble limits a fire fighter's ability to walk, climb ladders, lift objects, and crawl through restricted spaces. Fire fighters must become accustomed to these limitations and learn to alter their movements accordingly. Practice and conditioning are key to becoming proficient in wearing and using PPE while fighting fires.

Psychological Limitations of the User

In addition to the physical limitations, the user must make mental adjustments when wearing an SCBA. Breathing through an SCBA is different from normal breathing and can be very stressful. Covering your face with a face piece, hearing the air rushing in, hearing valves open and close, and exhaling against positive pressure are all foreign sensations. The surrounding environment, which is often dark and filled with smoke, is foreign as well.

Fire fighters must adjust so that they can operate effectively under these stressful conditions. Practice in donning PPE, breathing through SCBA, and performing firefighting tasks in darkness helps to build confidence, not only in the equipment, but also in the fire fighter's personal skills.

Training generally introduces one skill at a time. Practice each skill as it is introduced and try to become proficient in that skill. As your skills improve, you will be able to tackle tasks characterized by increasing levels of difficulty.

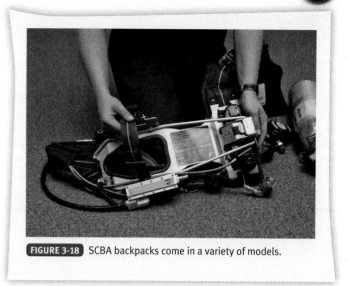

FIGURE 3-18 SCBA backpacks come in a variety of models.

FIRE FIGHTER Tips

The goal of training is to bring you to a level of comfort and proficiency in using your equipment. Start with a friendly environment until you are used to your equipment. Then add stressors such as darkness, smoke, and heat, one at a time.

■ Components of Self-Contained Breathing Apparatus

SCBA consists of four main parts: the backpack and harness, the air-cylinder assembly, the regulator assembly, and the face piece assembly. Although the basic features and operations of all models are similar, you need to become familiar with the specific SCBA used by your department.

Backpack and Harness

The **backpack** provides the frame for mounting the other working parts of the SCBA **FIGURE 3-18**. It is usually constructed of a lightweight metal or composite material.

The **SCBA harness** consists of the straps and fasteners used to attach the SCBA to the fire fighter. Most harnesses have two adjustable shoulder straps and a waist belt. Depending on the specific model of SCBA, the waist belt and shoulder straps will carry different proportions of the weight. The procedures for tightening and adjusting the straps also vary based on the model. The harness must be secure enough to keep the SCBA firmly fastened to the user, but not so tight that it interferes with breathing or movements. The waist belt must be tight enough to keep the SCBA from moving from side to side or getting caught on obstructions. Some SCBA models are equipped with a reinforced harness that can be used to help drag a fallen fire fighter out of danger.

Air-Cylinder Assembly

A compressed **air cylinder** holds the breathing air for an SCBA. This removable cylinder is attached to the backpack harness and can be changed quickly in the field. An experienced fire fighter should be able to remove and replace the cylinder in complete darkness.

Fire fighters should be familiar with the type of cylinders used in their departments. Cylinders are marked with the materials used in their construction, the working pressure, and the rated duration.

The air pressure in filled SCBA cylinders ranges from 15,169 to 31,028 **kilopascals** (2216 to 4500 **pounds per square inch (psi)**). The greater the air pressure, the more air that can be stored in the cylinder.

Low-pressure cylinders, which are pressurized at 15,16 9 kPa (2200 psi) can be constructed of steel or aluminum and are usually rated for 30 minutes of use. Composite cylinders are generally constructed of an aluminum shell wrapped with carbon, Kevlar®, or glass fibres. They are significantly lighter in weight, can be pressurized up to 31,028 kPa (4500 psi), and are rated for 30, 45, or 60 minutes of use.

As previously noted, the rated duration times are established under laboratory conditions. A working fire fighter can quickly use up the air because of exertion, so the ratings should be viewed with caution. Generally, the working time available for a particular cylinder is half the rated duration.

The neck of an air cylinder is equipped with a hand-operated shut-off valve. The **pressure gauge** is located near the shut-off valve; it shows the amount of pressure currently in the cylinder. Be careful not to damage the threads or let any dirt get into the outlet of the cylinder.

FIRE FIGHTER Tips

The pressure gauge on the cylinder is important for checking the amount of air in the cylinder. This gauge cannot be viewed by the user while wearing the SCBA.

Regulator Assembly

<u>SCBA regulators</u> may be mounted on the waist belt or shoulder strap of the harness or attached directly to the face piece **FIGURE 3-19**. The regulator controls the flow of air to the user. Inhaling decreases the air pressure in the face piece. This change in pressure opens the regulator, which in turn releases air from the cylinder into the face piece. When inhalation stops, the regulator shuts off the air supply. Exhaling opens a second valve (the exhalation valve), thereby expelling the exhaled air into the atmosphere. SCBA regulators are capable of delivering large volumes of air to support the strenuous activities required in firefighting. Some units are equipped with a <u>dual-path pressure reducer</u>, a feature that automatically provides a backup method for air to be supplied to the regulator if the primary passage malfunctions.

SCBA regulators maintain a slight positive air pressure to the face piece in relation to the ambient air pressure outside the face piece. This feature helps to prevent the hazardous atmosphere outside the face piece from leaking into the face piece during inhalation. If any leakage occurs in the area where the face piece and the face make the seal, the positive-pressure breathing air inside the face piece will keep the hazardous atmosphere from entering the device. Regardless of the positive pressure, a proper face piece-to-face seal must always be maintained. Breathing with this slight positive pressure may require some practice. New fire fighters often report that it takes more energy to breathe when first using positive-pressure SCBA, but note that this sensation gradually decreases.

To activate the SCBA, it is necessary to open the air-cylinder valve, don the SCBA, attach the regulator to the face piece, and breathe. Some SCBA models require that the user turn a yellow-coloured valve to activate the unit. If the air supply is partially or completely cut off during use, you can fully open the purge valve to secure an emergency supply of air. This action will create a constant flow of air that will rapidly deplete the remaining air supply in the air cylinder. If it is necessary to open the purge valve, you should immediately exit from the IDLH area. During normal use, the purge valve can be momentarily opened to remove condensed air from the face piece.

Like the cylinder, the regulator has a gauge that indicates the pressure of the breathing air remaining in the cylinder. This gauge enables the user to monitor the amount of air remaining in the cylinder. The regulator pressure gauge can be mounted directly on the regulator, or it may be located on a separate hose so it can be attached to a shoulder strap for easier viewing. If the pressure gauges are working correctly, the readings on the regulator and cylinder pressure gauges should be within 690 kPa (100 psi) of each other. The cylinder pressure gauge must be visible to the user during SCBA use.

SCBA is also required to have a second heads-up air-pressure display, which must be visible in the face piece. Some SCBAs contain a total of four <u>light-emitting diode (LED)</u> displays that indicate whether the air cylinder is full, three-fourths full, one-half full, or one-fourth full. Such <u>heads-up displays (HUD)</u> enable the user to constantly monitor the amount of air in the air cylinder. Other LEDs on the display may provide additional information—for example, low batteries or other problems **FIGURE 3-20**.

NFPA standards require that SCBA include an <u>end-of-service-time-indicator (EOSTI)</u>, or low-air alarm. This warning device tells the user that the end of the breathing air supply is approaching. NFPA 1500 requires that an exit strategy be practiced when the SCBA cylinder reaches a level of 600 litres or more. This alarm may take the form of a bell or whistle, a vibration, or a flashing LED. SCBAs are required to have two types of low-air alarms that operate independently of each other and activate different senses; for example, one alarm might ring a bell, whereas the second alarm might vibrate or flash an LED.

Fire fighters should never ignore the EOSTI warning. Most fire departments require fire fighters to exit from the IDLH area before the EOSTI alarm sounds because the low-air alarm does not sound until three-fourths of the air supply has been exhausted. A reserve of one-fourth of the air supply will not be adequate for a safe escape.

Many SCBA models include integrated PASS devices. A PASS is designed to help colleagues locate a downed fire fighter by sending out a loud audible signal. This device can be manually activated by a fire fighter in distress. It is also

FIGURE 3-19 SCBA regulator.

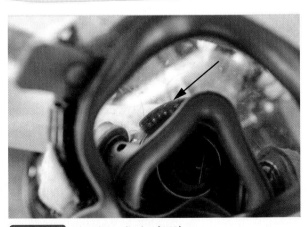
FIGURE 3-20 A heads-up display (HUD).

automatically activated if a fire fighter remains motionless for 30 seconds. Turning on the air supply automatically activates this PASS device, which ensures that a fire fighter does not forget to turn the PASS device on when entering a hazardous area. If your SCBA is equipped with a PASS device, learn how to activate it and how to turn it off **FIGURE 3-21**.

Communications between fire fighters wearing SCBAs is difficult. To facilitate communication, SCBAs are required to be equipped with a voice communication system. This functionality may be as simple as mechanical voice diaphragm, or it may be as sophisticated as an electronic system. Some SCBAs are also equipped with voice communications systems that utilize **two-way radios**.

SCBAs are required to be equipped with a **rapid intervention crew/company universal air connection (RIC UAC)**. This device is connected to the SCBA and used to refill a cylinder if a trapped fire fighter is running out of air. A universal air connection is attached to a hose from a full air cylinder and brought to the downed fire fighter by a rapid intervention team (RIT), who refill the downed fire fighter's cylinder. This connection should not be used for the normal refilling of a cylinder.

Fire department SCBAs are required to be certified to provide protection against certain chemical, biological, radiological, and nuclear agents—that is, agents that could be released as a result of a terrorist attack. SCBAs provide protection against these agents by preventing chemical fumes, disease-causing biological organisms, and radioactive particles from being inhaled and entering the fire fighter's respiratory system. However, an SCBA does not protect from contamination by other means of transmission.

Many accessories are available for SCBAs, including data logging, unit IDs, tracking devices, special rescue-harness connections, corrective eye lenses for face pieces, and electronic communications devices. Data logging and unit IDs aid in keeping track of the use of each SCBA and checking the proper functioning of each unit. Exit locator devices attempt to give the fire fighter some idea of how to exit a hazardous environment. Some of these tracking devices work by sending out sounds of varying pitch. Special rescue harnesses are reinforced straps that enable the SCBA to be used as a rescue harness. Corrective lenses for face pieces enable SCBA users to use their face pieces as corrective eyewear. Electronic communications systems integrate radio communications between SCBA users. If your SCBA is equipped with any of these accessories, you must become competent in the operation of these extra devices. To ensure proper functioning of your SCBA, do not add any devices or accessories that are not approved by the manufacturer of the equipment.

Although many models of SCBA regulators exist, fire fighters must learn how to operate only the particular model that is used in their department. They should be able to operate the regulator in the dark and with gloves on.

Face Piece Assembly

The **face piece** delivers breathing air to the fire fighter and protects the face from high temperatures and smoke **FIGURE 3-22**. It consists of a face mask with a clear lens, an exhalation valve, and—on models with a harness-mounted regulator—a flexible low-pressure supply hose. In some models, the regulator is attached directly to the face piece.

The face piece covers the entire face. The part that comes in contact with the skin is made of special rubber or silicon, because these materials provide for a tight seal. The clear lens allows for better vision. Exhaled air is expelled from the face piece through the one-way exhalation valve, which has a spring mechanism to maintain positive pressure inside the face piece. Because it is difficult to communicate through a face piece, a voice-amplification device or mechanical diaphragm is used to facilitate communications. A mechanical diaphragm uses a vibrating airtight membrane to transmit the fire fighter's voice without the use of electricity.

Face pieces are equipped with **nose cups** to help prevent fogging of the clear lens. In addition, nose cups prevent the buildup of CO_2 by directing exhaled air toward the exhalation valve. Fogging is a greater problem in colder climates. It occurs because the compressed air you breathe is dry, but the air you exhale is moist. The purge valve can be opened slightly for a second or two to clear condensation from the eyepieces. The flow of dry air from the regulator helps to prevent fogging of the lens.

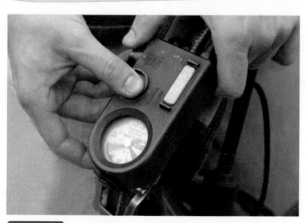

FIGURE 3-21 A PASS device can be integrated into an SCBA.

FIGURE 3-22 SCBA face pieces come in several sizes.

The face piece is held in place with a weblike series of straps, or a net and straps. Face pieces should be stored with the straps in the longest position to make them easier to don. Pull the end of the straps toward the back of the head (not out to the sides) to tighten them and ensure a snug fit.

A leak in the face piece seal may result from an improperly sized face piece, an improper donning procedure, or facial hair around the edge of the face piece. In particular, the following factors can affect the seal on an SCBA face piece:

- Facial hair, sideburns, or beard
- A low hairline that interferes with the sealing surface
- Ponytails or buns that interfere with the smooth and close fit on the head harness
- A skullcap that projects under the face piece or temple pieces
- The absence of teeth
- Improper size face piece

An improperly fitted face piece may lead to exposure to a hazardous environment. Such leaks are dangerous for two reasons: First, a large leak could overcome the positive pressure in the face piece and allow contaminated air to enter the face piece. Second, a leak of any size will deplete the breathing air and reduce the amount of time available for firefighting.

Face pieces are manufactured in several sizes. NFPA 1500 requires that all fire fighters must have their face pieces fit-tested annually to ensure that they are wearing the proper size. Some departments issue individual face pieces to each fire fighter; others provide a selection of sizes on each apparatus. NFPA 1500 also requires that the sealing surface of the face piece be in direct contact with the user's skin; that is, hair or a beard cannot be in the seal area.

■ Pathway of Air Through an SCBA

In an SCBA, the breathing air is stored under pressure in the cylinder. This air passes through the cylinder shut-off valve into the high-pressure **air line** (hose), which then takes it to the regulator. The regulator reduces the high pressure to low pressure.

The regulator opens when the user inhales, reducing the pressure on the downstream side. In an SCBA unit with a face piece-mounted regulator, the air next goes directly into the face piece. In units with a harness-mounted regulator, the air travels from the regulator through a low-pressure hose into the face piece. From the face piece, the air is inhaled through the user's air passages and into the lungs.

When the user exhales, used air is returned to the face piece. The exhaled air is exhausted from the face piece through the exhalation valve. This cycle repeats with every inhalation. As the pressure in the face piece drops, the exhalation valve closes and the regulator opens.

Fire Fighter Safety Tips

© Photos.com

Make sure that your face piece is properly fitted and is the correct size for your face.

FIRE FIGHTER Tips

Controlled-Breathing Technique
Controlled breathing helps extend the SCBA air supply. It consists of a conscious effort to inhale naturally through the nose and to force exhalation from the mouth. Practicing controlled breathing during training will help you to maximize the efficient use of air while you are working.

■ Mounting Breathing Apparatus

SCBA should be located so that fire fighters can don it quickly when they arrive at the scene of a fire. Seat-mounted brackets enable fire fighters to don SCBA en route to an emergency scene, without unfastening their seat belts or otherwise endangering themselves. This approach enables fire fighters to begin work as soon as they arrive.

Several types of apparatus seat-mounting brackets are available. Some hold the SCBA with the friction of a clip. Others are equipped with a mechanical hold-down device that must be released to remove the SCBA. Regardless of which mounting system is employed, it must hold the SCBA securely in the bracket; a collision or sudden stop should not dislodge the SCBA from the brackets, because a loose SCBA can be a dangerous projectile. The fire fighter who dons SCBA from a seat-mounted bracket should not tighten the shoulder straps while seated, so as not to dislodge the SCBA in a sudden-stop situation. Fire fighters must remain securely restrained by a seat belt or combination seat belt/shoulder harness anytime the apparatus is moving.

Compartment-mounted SCBA units also can be donned quickly. These units are used by fire fighters who arrive in vehicles without seat-mounted SCBA or whose seats were not equipped with them. Driver/operators and fire fighters who arrive in their private vehicles often use these types of SCBA. The mounting brackets should be positioned high enough to allow for easy donning of the equipment. Some mounting brackets allow the fire fighter to lower the SCBA without removing it from the mounting bracket. Older apparatus may have the brackets mounted on the exterior of the vehicle. Any exterior-mounted SCBA should be protected from weather and dirt by a secure cover.

SCBA also may be kept in a storage case. This method is most appropriate for transporting extra SCBA units. It should not be used to transport SCBA that will be used during the initial phase of operations at a fire scene.

The SCBA should be stored on the apparatus in ready-for-use condition, with the main cylinder valve closed.

■ Donning Self-Contained Breathing Apparatus

Donning SCBA is an important skill. Fire fighters should be able to don and activate SCBA in 1 minute; both the fire fighter's personal safety and the effectiveness of the firefighting operation depend on mastery of this skill. Fire fighters must already be wearing full PPE before they don SCBA. Before beginning the actual donning process, fire fighters must carefully check the SCBA to ensure it is ready for operation:

- Check whether the air cylinder is at least 90 percent of its rated pressure.
- If the SCBA has an air-saving/donning switch, confirm that it is activated.
- Open the cylinder valve two or three turns, listen for the low-air alarm to sound, and then open the valve fully.
- Check the pressure gauges on both the regulator and the cylinder. The readings for these gauges should be within 10 percent of each other.
- Check all harness straps to ensure that they are fully extended.
- Check all valves to ensure that they are in the correct position. (An open purge valve will waste air.)

Donning SCBA from an Apparatus Seat Mount

Donning SCBA while en route to an emergency can save valuable time. However, this maneuver requires that you don all of your protective clothing before mounting the apparatus. Place your arms through the shoulder straps as you sit down, and then fasten your seat belt. Alternatively, you can fasten your seat belt first, and then slide one arm at a time through the shoulder straps of the SCBA harness. You can partially tighten the shoulder straps while you are seated.

When you arrive at the emergency scene, release your seat belt, activate the bracket release, and exit the apparatus. Be sure to take a face piece with you. Face pieces should be either kept in a storage bag close to each seat-mounted SCBA or attached to the harness. After exiting from the apparatus, attach the waist strap of the SCBA unit, and then tighten and adjust the shoulder and waist straps.

Follow the steps in **SKILL DRILL 3-3** to don SCBA from a seat-mounted bracket. Before beginning this skill drill, inspect the SCBA to ensure it is ready for service.

1. Don your hood, bunker pants, boots, and turnout coat. Safely mount the apparatus and sit in the seat, placing your arms through the SCBA shoulder straps. (**STEP 1**)
2. Fasten your seat belt. Partially tighten the shoulder straps; do not fully tighten them at this time. When the apparatus comes to a complete stop at the emergency scene, release your seat belt and release the SCBA from the mounting bracket.
3. Carefully exit the apparatus. (**STEP 2**)
4. Attach the waist belt of the harness and cinch it down. (**STEP 3**)
5. Adjust shoulder straps until they are tight. (**STEP 4**)
6. Open the main cylinder valve. Activate the air-saver valve if needed. (**STEP 5**)
7. Remove or loosen your helmet and pull the hood back. Don the face piece and check for leaks. Pull the protective hood up over your head, put the helmet on, and secure the chin strap. (**STEP 6**)
8. If necessary, connect the regulator to the face piece. (**STEP 7**)
9. Activate the air flow and PASS alarm. (**STEP 8**)

Donning SCBA from a Compartment Mount

To don a compartment-mounted SCBA, slide one arm through the shoulder harness strap, and then slide the other arm through the other shoulder strap. Release the SCBA from the mounting bracket. Adjust the shoulder straps so that you can carry the SCBA fairly high on your back. Attach the ends of the waist strap and tighten the waist strap.

Follow the steps in **SKILL DRILL 3-4** (Fire Fighter I, NFPA 5.3.1) to don SCBA from a side-mounted compartment or bracket. Before starting, confirm that the SCBA has been inspected and is ready for service. Examine the equipment to see how it is mounted in the compartment. Check the release mechanism and determine how it operates. If the SCBA is mounted on an exterior bracket, remove the protective cover before beginning the donning sequence.

1. Don PPE and hood. Stand in front of the SCBA bracket and fully open the main cylinder valve.
2. Turn your back toward the SCBA, slide your arms through the shoulder straps, and partially tighten the straps.
3. Release the SCBA from the bracket and step away from the apparatus.
4. Attach the waist belt and tighten it.
5. Adjust the shoulder straps.
6. Don the face piece and check for an adequate seal.
7. Pull the protective hood into position, don your helmet, and secure the chin strap.
8. If necessary, connect the regulator to the face piece.
9. Activate the air flow and PASS alarm.

Donning SCBA from the Ground, the Floor, or a Storage Case

Fire fighters must sometimes don an SCBA that is stored in a case or on the ground. Two methods can be used in these situations: the over-the-head method and the coat method.

Over-the-Head Method

To don an SCBA using the over-the-head method, place the SCBA on the ground or on the floor with the cylinder valve facing away from you. Lay the shoulder straps out to each side of the backpack. Grasp the backplate with both hands, and lift the SCBA over your head. Let the backpack slide down your back; the straps will slide down your arms. Balance the unit on your back. First attach and tighten the waist strap, and then tighten the shoulder straps.

Follow the steps in **SKILL DRILL 3-5** to don SCBA using the over-the-head method. Before starting, confirm that the SCBA has been inspected and is ready for service. Don your bunker pants, boots, and turnout coat. Place your protective hood over your neck. Place your helmet and gloves close to the SCBA case.

1. Open the protective case and lay out the SCBA so that the cylinder valve is away from you and the shoulder straps are to the sides. (**STEP 1**)

SKILL DRILL 3-3

Donning SCBA from an Apparatus Seat Mount
(Fire Fighter I, NFPA 5.3.1)

1   Don your head, bunker pants, boots, and turnout coat. Safely mount the apparatus and sit in the seat, placing your arms through the SCBA shoulder straps.

2   Fasten your seat belt. Partially tighten the shoulder straps. When the apparatus stops, release the seat belt and release the SCBA from its brackets. Exit the apparatus.

3   Attach the waist belt of the harness and cinch it down.

4   Adjust shoulder straps until they are tight.

5   Open the main cylinder valve. Activate the air-saver valve if needed.

6   Remove or loosen your helmet and pull the hood back. Don the face piece and check for leaks. Replace the protective hood and helmet, and secure the chin strap.

7   If necessary, connect the regulator to the face piece.

8   Activate the air flow and PASS alarm.

2 Fully open the main cylinder valve. Activate the air-saver valve if needed. (**STEP 2**)

3 Bend down and grasp the SCBA backplate with both hands. Using your legs, lift the SCBA over your head. Once the SCBA clears your head, rotate it 180 degrees so that the waist straps are pointed toward the ground. (**STEP 3**)

4 Slowly slide the pack down your back. Make sure that your arms slide into the shoulder straps. Once the SCBA is in place, tighten the shoulder straps and secure the waist strap. (**STEP 4**)

5 Make sure your hood is pulled down around your neck. Don the face piece and check for an adequate seal. Pull your protective hood into position, don your helmet, and secure the chin strap. (**STEP 5**)

6 If necessary, connect the regulator to the face piece. Activate the air flow and PASS alarm. (**STEP 6**)

Coat Method

To don an SCBA using the coat method, place the SCBA on the ground or on the floor with the cylinder valve facing toward you. Spread out and extend the shoulder straps. Use your left hand to grasp the left shoulder strap close to the backplate; use your right hand to grasp the right shoulder strap farther away from the backplate. Swing the SCBA over your left shoulder. Release your right arm and slide it through the right shoulder harness strap. Tighten both shoulder straps. Attach and tighten the waist belt.

Follow the steps in **SKILL DRILL 3-6** to don SCBA using the coat method. Before starting, confirm that the SCBA has been inspected and is ready for service. Don your bunker pants, boots, and turnout coat. Place your protective hood over your neck. Place your helmet and gloves close to the SCBA case.

1 Open the protective case and lay out the SCBA so that the cylinder valve is facing you and the straps are laid out to the sides. Fully open the main cylinder valve. Open the air-saver valve if needed. Place your dominant hand on the opposite shoulder strap. For safety reasons, be sure to grasp the strap as close to the backplate as possible. (**STEP 1**)

2 Lift the SCBA and swing it over your dominant shoulder, being mindful of people or objects around you. (**STEP 2**)

3 Slide your other hand between the SCBA cylinder and the corresponding shoulder strap. (**STEP 3**)

4 Tighten the shoulder straps. (**STEP 4**)

5 Attach the waist belt and adjust its tightness. (**STEP 5**)

6 Make sure your hood is pulled down around your neck. Don the face piece and check for an adequate seal. (**STEP 6**)

7 Pull the protective hood into position, don your helmet, and secure the chin strap. If necessary, connect the regulator to the face piece. Activate the air flow and PASS alarm. (**STEP 7**)

These instructions will have to be modified for different SCBA models. In particular, the sequence for adjusting the shoulder straps and the waist belt varies with different models.

Modifications must also be made for SCBA models with waist-mounted regulators. Refer to the specific manufacturer's instructions supplied with each unit. Follow the SOGs/SOPs for your department.

To don SCBA from a seat-mounted position with a safety latch, follow the steps in **SKILL DRILL 3-7**:

1 With the SCBA straps fully extended, place your dominant arm between the shoulder strap and the SCBA bottle. Repeat the same process with the other arm. (**STEP 1**)

2 Tighten the shoulder straps. (**STEP 2**)

3 Attach the waist belt and adjust its tightness. (**STEP 3**)

4 If the apparatus has an SCBA locking device, detach the device from the SCBA. *Note:* Fire fighters should never take off their seat belts to perform this action. Stay seated and buckled up while the apparatus is in motion. (**STEP 4**)

■ Donning the Face Piece

Your face piece keeps contaminated air outside and pure, breathable air inside. To perform properly, it must be the correct size and must be adjusted to fit your face. Make sure you have been tested to determine the proper size for you. The requirements for face piece fit testing are described in NFPA 1500. Fit testing needs to be performed before fire fighters are permitted to use SCBA. To ensure a proper fit, such testing should be repeated every year.

No facial hair can be present in the seal area. Eyeglasses that pass through the seal area cannot be worn with a face piece, because they can cause leakage between the face piece and your skin. Your face piece must match your SCBA—you cannot interchange a face piece from a different SCBA model.

Face pieces for various brands and models of SCBAs may differ slightly. Some have the regulator mounted on the face piece; others have it mounted on the harness straps. Fire fighters must learn about the specific face pieces used by their departments.

Follow the steps in **SKILL DRILL 3-8** to don a face piece. Before beginning this procedure, make sure you have donned your PPE and protective hood. Remove your helmet.

1 Fully extend the straps on the face piece. (**STEP 1**)

2 Rest your chin in the chin pocket at the bottom of the mask. (**STEP 2**)

3 Fit the face piece to your face, bringing the straps or webbing over your head. (**STEP 3**)

4 Tighten the lowest two straps. To tighten them, pull the straps straight back, rather than out and away from your head. Check the harness net to make sure it is lying flat against the back of the head. (**STEP 4**)

5 Tighten the pair of straps at your temple, if these straps are present.

6 If your model has additional straps, tighten the top strap(s) last. (**STEP 5**)

7 Check for a proper seal. This process depends on the model and type of face piece you use.

8 Confirm that your nose fits in the chin cup. (**STEP 6**)

SKILL DRILL 3-5

Donning SCBA Using the Over-the-Head Method
(Fire Fighter I, NFPA 5.3.1)

1 Open the protective case and lay out the SCBA so that the cylinder valve is away from you and the shoulder straps are to the sides.

2 Fully open the main cylinder valve. Activate the air-saver valve if needed.

3 Bend down and grasp the SCBA backplate with both hands. Using your legs, lift the SCBA over your head. Once the SCBA clears your head, rotate it 180 degrees so that the waist straps are pointed toward the ground.

4 Slowly slide the pack down your back. Make sure that your arms slide into the shoulder straps. Once the SCBA is in place, tighten the shoulder straps and secure the waist strap.

5 Make sure your hood is pulled down around your neck. Don the face piece and check for an adequate seal. Pull your protective hood into position, don your helmet, and secure the chin strap.

6 If necessary, connect the regulator to the face piece. Activate the air flow and PASS alarm.

© LiquidLibrary

9 Pull the protective hood up so that it covers all bare skin. Make sure it does not get under your face piece or obscure your vision. (**STEP 7**)

10 Replace your helmet and secure the chin strap. (**STEP 8**)

11 Attach the regulator to your face piece or attach the low-pressure air supply hose to the regulator. (**STEP 9**)

SKILL DRILL 3-6

Donning SCBA Using the Coat Method
(Fire Fighter I, NFPA 5.3.1)

1 Open the protective case and lay out the SCBA so that the cylinder valve is facing you and the straps are laid out to the sides. Fully open the main cylinder valve. Open the air-saver valve if needed. Place your dominant hand on the opposite shoulder strap. For safety reasons, be sure to grasp the strap as close to the backplate as possible.

2 Lift the SCBA and swing it over your dominant shoulder, being mindful of people or objects around you.

3 Slide your other hand between the SCBA cylinder and the corresponding shoulder strap.

4 Tighten the shoulder straps.

5 Attach the waist belt and adjust its tightness.

6 Make sure your hood is pulled down around your neck. Don the face piece and check for an adequate seal.

7 Pull the protective hood into position, don your helmet, and secure the chin strap. If necessary, connect the regulator to the face piece. Activate the air flow and PASS alarm.

SKILL DRILL 3-7

Donning SCBA from a Seat-Mounted Position with a Safety Latch
(Fire Fighter I, NFPA 5.3.1)

1 With the SCBA straps fully extended, place your dominant arm between the shoulder strap and the SCBA bottle. Repeat the same process with the other arm.

2 Tighten the shoulder straps.

3 Attach the waist belt and adjust its tightness.

4 If the apparatus has an SCBA locking device, detach the device from the SCBA. *Note*: Fire fighters should never take off their seat belts to perform this action. Stay seated and buckled up while the apparatus is in motion.

© LiquidLibrary

■ Safety Precautions for Self-Contained Breathing Apparatus

As you practice using your SCBA, remember that this equipment is your protection against serious injury or death in hazardous conditions. Practice safe procedures from the beginning.

Before you enter a hazardous environment, make sure that your PASS device is activated and that you are properly logged into your personnel accountability system. Always work in teams of two in hostile environments. In addition, always have at least two fire fighters outside at the ready whenever two fire fighters are working in a hostile environment.

SKILL DRILL 3-8   Donning a Face Piece
(Fire Fighter I, NFPA 5.3.1)

1 Fully extend the straps on the face piece.

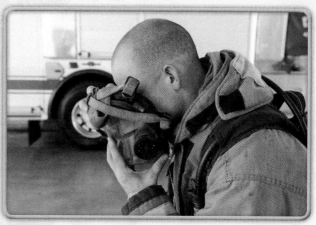

2 Rest your chin in the chin pocket at the bottom of the mask.

3 Fit the face piece to your face, bringing the straps or webbing over your head.

4 Tighten the lowest two straps. To tighten them, pull the straps straight back, rather than out and away from your head. Check the harness net to make sure it is lying flat against the back of the head.

5 Tighten the pair of straps at your temple, if these straps are present. If your model has additional straps, tighten the top strap(s) last.

6 Check for a proper seal. This process depends on the model and type of face piece you use. Confirm that your nose fits in the chin cup.

© LiquidLibrary

(Continues)

SKILL DRILL 3-8

Donning a Face Piece (*Continued*)
(Fire Fighter I, NFPA 5.3.1)

7 Pull the protective hood up so that it covers all bare skin. Make sure it does not get under your face piece or obscure your vision.

8 Replace your helmet and secure the chin strap.

9 Attach the regulator to your face piece or attach the low-pressure air supply hose to the regulator.

© LiquidLibrary

Fire Fighter Safety Tips
© Photos.com

The following method is one way to check the seal of a face mask:

1. Don the SCBA and begin breathing cylinder air.
2. Completely close the air cylinder valve.
3. Breathe on the respirator until all air stops flowing from the breathing regulator.
4. Inhale slowly and hold your breath. Your face piece should be slightly drawn to your face.
5. Listen and feel for air leakage around the face piece seal. Check that the negative pressure in the face piece does not change.
6. If no change occurs, no leaks are present.

■ SCBA Use During Emergency Operations

As a fire fighter, your job is to practice using SCBA until you are confident that you can carry out a variety of tasks in hazardous conditions while depending on your equipment to supply you with safe air to breathe. Because hostile environments are often unpredictable, fire fighters must be prepared to react if an emergency situation occurs while they are using SCBA. In emergencies, follow simple guidelines. Most importantly, keep calm, stop, and think. Panic increases air consumption. Try to control your breathing by maintaining a steady rate of respirations. A calm person has a greater chance of surviving an emergency.

If you experience a problem with your SCBA, try to exit the IDLH area to a safe environment. If your cylinder contains air but no air comes out of your regulator, many SCBAs have a purge value that can be opened slightly to release a constant supply of air. This measure will rapidly empty your cylinder, however. In this circumstance, you must immediately exit from the hazardous environment. If you are in danger, follow the steps for self-survival and calling a mayday. This topic is discussed in the chapter "Fire Fighter Survival."

FIRE FIGHTER Tips

Restricted Spaces
The size and shape of SCBA may make it difficult for you to fit through tight openings when wearing this equipment. Several techniques may help you navigate these spaces:

- Change your body position: Rotate your body by 45 degrees and try again.
- Loosen one shoulder strap and change the location of the SCBA on your back.
- If you have no other choice, you may have to remove your SCBA. In this case, do not let go of the backpack and harness for any reason. Keep the unit in front of you as you navigate through the tight space. Reattach the harness as soon as you are through the restricted space. *Note:* This is an absolutely "last resort" procedure!

You will be using your SCBA in a variety of conditions that most people would consider to be emergencies. Because you are a fire fighter, these activities are predictable for you. Therefore, you need to master a wide variety of firefighting skills while wearing SCBA. Practice these tasks first in conditions of good visibility, and then progress to doing the same tasks in conditions of limited visibility. You need to practice most of the skills you will perform as a fire fighter—for example, advancing hose lines, climbing ladders, crawling through windows, performing rescues, providing medical assistance, and crawling through confined spaces—while wearing SCBA.

FIRE FIGHTER Tips

Many fire departments need to take precautions to ensure proper operation of SCBAs in cold temperatures. SCBAs should be stored in a location that is kept at a temperature greater than 0°C (32°F). After use, this equipment should be dried thoroughly to avoid retained moisture that might freeze moving parts. When SCBA units are doffed in below-freezing temperatures, the face piece and the regulator should be placed inside the turnout coat to keep it warm and prevent any moisture from freezing.

FIRE FIGHTER Tips

Refilling a cylinder using the RIC UAC in below-freezing conditions presents two concerns: The first concern relates to the protective cap on the RIC UAC coupling. Water on the cover of the coupling may freeze, preventing a connection between the filling hose and the universal air connection. The second concern is that if the universal air connection is used to fill an SCBA cylinder in below-freezing temperatures, and the SCBA is then taken into a warmer temperature, the difference in temperature could cause the pressure in the cylinder to expand to unsafe levels. If there is excess pressure, open the air cylinder valve and the purge valve to release the excess pressure.

Exercises for Mastering SCBA

The following are three exercises that may be practiced in order to assist in the mastery of using SCBA properly.

The "Conducting a Primary Search" skill drill in the "Search and Rescue" chapter describes how to perform search and rescue while wearing SCBA. Another skill you need to practice is passing through a restricted space while wearing an SCBA. This activity is illustrated in the "Opening a Wall to Escape" skill drill in the "Fire Fighter Survival" chapter.

A third activity to familiarize you with your SCBA simulates the activities you would perform when ventilating a roof. Once you have mastered donning and doffing your PPE ensemble, you may perform this activity to help you master the use of your SCBA. In doing so, you will work with at least one other team member.

Bring an axe and other appropriate ventilation tools with you to the roof. Don your face piece and activate the air supply. With the axe, assess the construction of the building for safety and sound the rafters for safety.

Next, have a second team member support you. Chop on the roof or simulated roof using short, powerful, controlled strokes. Then switch places with your team member. Be sure to monitor your air supply during this activity. Exit the hazard area in time to be out of hazards before your low-air alarm activates.

■ Doffing Self-Contained Breathing Apparatus

The procedure for doffing SCBA depends on which model you use and whether it has a face piece–mounted regulator or a harness-mounted regulator. Follow the procedures recommended by the manufacturer and your department's SOGs/ SOPs.

In general, when doffing your SCBA, you should reverse the steps you followed to don your SCBA. Follow the steps in **SKILL DRILL 3-9** to doff your SCBA:

1. Remove your gloves. Remove the regulator from your face piece or disconnect the low-pressure air-supply hose from the regulator. (**STEP ①**)
2. Shut off the air-supply valve or fully depress the air-saver/donning switch to stop the flow of air.
3. Remove your helmet and pull your protective hood down around your neck. (**STEP ②**)
4. Loosen the straps on your face piece. (**STEP ③**)
5. Remove your face piece. (**STEP ④**)
6. Release your waist belt. (**STEP ⑤**)
7. Loosen the shoulder straps and remove the SCBA. (**STEP ⑥**)
8. Shut off the air-cylinder valve. (**STEP ⑦**)
9. Bleed the air pressure from the regulator by opening the purge valve. (**STEP ⑧**)
10. If you have an integrated PASS device, turn it off.
11. Place the SCBA in a safe location where it will not get dirty or damaged. (**STEP ⑨**)

Putting It All Together: Donning the Entire PPE Ensemble

The complete PPE ensemble consists of both personal protective clothing and respiratory protection (SCBA). Although donning personal protective clothing and donning and operating SCBA can be learned and practiced separately, you must be able to integrate these skills to have a complete PPE ensemble. Each part of the complete ensemble must be in the proper place to provide whole-body protection.

The steps for donning a complete PPE are summarized here:
1. Place the protective hood over your head and bring it down around your neck.
2. Put on your bunker pants and boots. Adjust the suspenders and secure the front flap of the pants.
3. Put on your turnout coat and secure the front.
4. Open the air-cylinder valve on your SCBA and check the air pressure. Press the air-saver valve to prevent air flow if needed.
5. Put on your SCBA.
6. Tighten both shoulder straps of the SCBA harness.
7. Attach the waist belt of the harness and tighten it. Tighten the chest straps, if present.
8. Fit the face piece to your face.
9. Tighten the face piece straps, beginning with the lowest straps.
10. Check the face piece for a proper seal. (Follow the manufacturer's instructions.)

11. Pull the protective hood up so that it covers all bare skin but does not obscure your vision.
12. Place your helmet on your head with the ear tabs extended, and secure the chin strap.
13. Turn up your coat collar and secure it in front.
14. Put on your gloves.
15. Check your clothing to be sure it is properly secured.
16. Make sure your PASS device is turned on.
17. Attach your regulator or turn it on to start the flow of breathing air.
18. Work safely!

■ SCBA Inspection and Maintenance

SCBA must be properly cleaned, inspected, and prepared for the next use each time it is used, whether in an actual emergency incident or as part of a training exercise. The air cylinder must be changed or refilled, the face piece and regulator must be sanitized according to the manufacturer's instructions, and the unit must be cleaned, inspected, and checked for proper operation. It is the user's responsibility to ensure that the SCBA is in good working order and a ready-to-use condition before it is returned to the fire apparatus.

Each SCBA must be checked on a regular basis to ensure that it meets these requirements. Different procedures can be followed for inspection and for operational testing. Inspection and operational testing should be conducted after a unit has been used, and on a regular schedule. In career departments, inspection and testing are done at the beginning of each shift. In volunteer departments, this step is commonly performed on a weekly schedule.

If an SCBA inspection reveals any problems that cannot be remedied by routine maintenance, the SCBA must be removed from service for repair. Only properly trained and certified personnel are authorized to repair SCBA.

FIRE FIGHTER Tips

Follow the manufacturer's recommendations and avoid the use of aerosol cleaners or any alcohol-containing cleaner, as these materials degrade the rubber material in the face piece. Also, face pieces should be air dried or wiped with a soft, nonabrasive cloth to avoid scratching the lens.

Inspection of SCBA

The purpose of a SCBA inspection is to identify any parts of the SCBA that are visibly damaged and need to be repaired or replaced to ensure continued safe operation. This visible inspection can be done in conjunction with the operational testing sequence described here. Follow the steps in **SKILL DRILL 3-10** for the SCBA inspection:

1. Visually inspect the air cylinder and valve assembly for dents and gouges. Look for black or discoloured areas that indicate exposure to flame. (**STEP ①**)

SKILL DRILL 3-9 Doffing SCBA

1 Remove your gloves. Remove the regulator from your face piece or disconnect the low-pressure air-supply hose from the regulator.

2 Shut off the air-supply valve or fully depress the air-saver/donning switch to stop the flow of air. Remove your helmet and pull your protective hood down around your neck.

3 Loosen the straps on your face piece.

4 Remove your face piece.

5 Release your waist belt.

6 Loosen the shoulder straps and remove the SCBA.

7 Shut off the air-cylinder valve.

8 Bleed the air pressure from the regulator by opening the purge valve.

9 If you have an integrated PASS device, turn it off. Place the SCBA in a safe location where it will not get dirty or damaged.

© LiquidLibrary

2 Check the cylinder for the current hydrostatic test date and date of manufacture. Note that fibre-wrapped cylinders must be replaced every 15 years.

3 Check the cylinder pressure gauge to be sure it is full. (**STEP 2**)

4 Inspect hoses and rubber parts for damage or deterioration. (**STEP 3**)

5 Inspect the harness, webbing, buckles, fasteners, and cylinder-retention system for damage. (**STEP 4**)

6 Verify that the SCBA has been cleaned according to the manufacturer's and department's recommendations.

7 Inspect the regulator for intact gaskets and other visible damage. (**STEP 5**)

8 Inspect the face piece for damage and worn components. Look for damage to the lenses and check for the presence of a nose cup. (**STEP 6**)

9 Inspect the head harness to confirm that all parts are present and working properly. (**STEP 7**)

10 Check the quick disconnects and the universal air connection to make sure they are not damaged and that they are operating properly. (**STEP 8**)

■ Operational Testing

The operational testing sequence is designed to check the function of the many parts of the SCBA to ensure safe use of the device. It concentrates on the working parts of the SCBA. Operational testing should be done after each use as well as at the beginning of each shift in career departments or on a set schedule in volunteer departments. Follow the steps in **SKILL DRILL 3-11** for operational testing of SCBAs.

1 Check the regulator purge valve to be sure it is closed. (**STEP 1**)

2 Depress the air-saver switch, if present. (**STEP 2**)

3 Slowly open the air-cylinder valve. Check for proper operation of the HUD and of the low-battery indicator. Confirm that the low-air warning and PASS devices are working. (**STEP 3**)

4 Check the remote pressure gauge for proper operation. Its reading should be within 10 percent of the air-cylinder pressure gauge reading. (**STEP 4**)

5 Don the face piece. Adjust it to obtain a good seal. Inhale sharply to start the flow of air. Breathe normally to check for proper operation. (**STEP 5**)

6 Remove the regulator or face piece; air should flow freely. (**STEP 6**)

7 Depress the air-saver/donning switch to stop the flow of air. (**STEP 7**)

8 Open the purge valve to check for air flow. (**STEP 8**)

9 Close the purge valve to stop the flow of air. (**STEP 9**)

10 Rotate the air-cylinder valve to close it. (**STEP 10**)

11 Open the purge valve slightly to vent residual air pressure from the system. Watch the HUD to verify its proper operation as the air pressure is exhausted. (**STEP 11**)

12 Once the air flow stops, close the purge valve. Complete any reporting that is required. (**STEP 12**)

Annual Inspection

A complete annual inspection and maintenance procedure must be performed on each SCBA unit. The annual inspection must be performed by a certified manufacturer's representative or a person who has been trained and certified to perform this work. SCBA requires regular inspection and maintenance to ensure that it will perform as intended.

■ Servicing SCBA Cylinders

A pressurized SCBA cylinder contains a tremendous amount of potential energy. Not only does the air within the cylinder exert considerable pressure on its walls, but the cylinder itself is used under extreme conditions on the fire ground. If the cylinder ruptures and suddenly releases this energy, it can cause serious injury or death. For this reason, cylinders must be regularly inspected and tested to ensure they are safe. These devices must be visually inspected during regularly scheduled inspections. A more detailed inspection is required if a cylinder has been exposed to excessive heat, has come into contact with flame, has been exposed to chemicals, or has been dropped.

Transport Canada, under the dangerous goods regulation and CSA Standards Z 180, Z94.4, B339, and B340 regulates cascade systems and requires **hydrostatic testing** for SCBA cylinders on a periodic basis and limits the number of years that a cylinder can be used. For example, composite-fibre overwrapped cylinders must be replaced after 15 years. Hydrostatic testing seeks to identify any defects or damage that might render the cylinder unsafe. Any cylinder that fails a hydrostatic test should be immediately taken out of service and cannot be used.

Cylinders constructed of different materials have different testing requirements. Aluminum, steel, and carbon-fibre cylinders must be hydrostatically tested every 5 years. Cylinders constructed of composite materials such as Kevlar® or fibreglass fibres must be tested every 3 years. Fire fighters must know which types of cylinders are used by their departments and must check each cylinder for a current hydrostatic test date before filling it.

■ Replacing SCBA Cylinders

A used air cylinder can be quickly replaced with a full cylinder in the field to enable you to continue firefighting activities. A fire fighter who is working alone must doff his or her SCBA to replace the air cylinder; two fire fighters who are working together can change each other's cylinders without removing their SCBA. The steps listed in this section outline how a single person makes a cylinder change. This procedure may vary slightly depending on the model of SCBA being used. Follow the procedure recommended by the SCBA manufacturer and by your department's SOGs/SOPs.

Practice changing air cylinders until you become proficient at this task. A fire fighter should be able to change cylinders in the dark and while wearing gloves if necessary. Follow the steps in **SKILL DRILL 3-12** to replace an SCBA cylinder:

1 Place the SCBA on the floor or a bench. (**STEP 1**)

2 Turn off the cylinder valve. (**STEP 2**)

3 Bleed off the pressure by opening the purge valve. (**STEP 3**)

4 Disconnect the high-pressure supply hose. Keep the ends clean. (**STEP 4**)

5 Release the cylinder from the backpack and remove cylinder. (**STEP 5**)

6 Slide a full cylinder into the backpack. Align the outlet to connect the supply hose. Lock the cylinder in place. (**STEP 6**)

7 Check that the O ring is present and in good shape. (**STEP 7**)

SKILL DRILL 3-10

Visible SCBA Inspection
(Fire Fighter I, NFPA 5.5.1)

1 Visually inspect the air cylinder and valve assembly for dents and gouges. Look for black or discoloured areas that indicate exposure to flame.

2 Check the cylinder for the current hydrostatic test date and date of manufacture. Check the cylinder pressure gauge to be sure it is full.

3 Inspect hoses and rubber parts for damage or deterioration.

4 Inspect the harness, webbing, buckles, fasteners, and cylinder-retention system for damage.

5 Verify that the SCBA has been cleaned according to the manufacturer's and department's recommendations. Inspect the regulator for intact gaskets and other visible damage.

6 Inspect the face piece for damage and worn components. Look for damage to the lenses and check for the presence of a nose cup.

7 Inspect the head harness to confirm that all parts are present and working properly.

8 Check the quick disconnects and the universal air connection to make sure they are not damaged and that they are operating properly.

© LiquidLibrary

8 Connect the high-pressure hose to the cylinder. Hand-tighten only. (**STEP 8**)

9 Open the cylinder valve. Check the regulator gauge or remote gauge. Its reading should be within 10 percent of the reading on the cylinder gauge. (**STEP 9**)

■ Replacing an SCBA Cylinder on Another Fire Fighter

If you need to quickly reenter the fire scene, a second person can replace your air cylinder while you continue to wear your SCBA harness. Be sure that you are physically able to fight a second round with the fire. It is better to allow yourself a few minutes in rehab than to get back into the fire immediately and require rescue from other crew members. Do not overtax

---

SKILL DRILL 3-11   SCBA Operational Inspection
(Fire Fighter I, NFPA 5.5.1)

1   Check the regulator purge valve to be sure it is closed.

2   Depress the air-saver switch, if present.

3   Slowly open the air-cylinder valve. Check for proper operation of the HUD and of the low-battery indicator. Confirm that the low-air warning and PASS devices are working.

4   Check the remote pressure gauge for proper operation. Its reading should be within 10 percent of the air-cylinder pressure-gauge reading.

5   Don the face piece. Adjust it to obtain a good seal. Inhale sharply to start the flow of air. Breathe normally to check for proper operation.

6   Remove the regulator or face piece; air should flow freely.

7   Depress the air-saver/donning switch to stop the flow of air.

8   Open the purge valve to check for air flow.

9   Close the purge valve to stop the flow of air.

© LiquidLibrary

SKILL DRILL 3-11   SCBA Operational Inspection (*Continued*)
(Fire Fighter I, NFPA 5.5.1)

10 Rotate the air cylinder valve to close it.

11 Open the purge valve slightly to vent residual air pressure from the system. Watch the HUD to verify its proper operation as the air pressure is exhausted.

12 Once the air flow stops, close the purge valve. Complete any reporting that is required.

© LiquidLibrary

yourself by replacing the cylinder and going back to work without adequate rest when you need it.

Follow the steps in **SKILL DRILL 3-13** to replace an SCBA cylinder on another fire fighter:

1 Remove the regulator from the face piece or remove the face piece so that you can breathe ambient air. (**STEP 1**)

2 Turn off the cylinder valve on the used cylinder. (**STEP 2**)

3 Open the purge valve to bleed off pressure from the high-pressure supply line. (**STEP 3**)

4 Disconnect the high-pressure supply hose. (**STEP 4**)

5 Release the SCBA cylinder from the backpack assembly and set it aside. (**STEP 5**)

6 Slide the full SCBA cylinder into the backpack. (**STEP 6**)

7 Check for the presence and satisfactory condition of the O ring. (**STEP 7**)

8 Lock the SCBA cylinder into place. (**STEP 8**)

9 Connect the high-pressure hose to the SCBA cylinder. (**STEP 9**)

10 Open the cylinder valve and state the pressure reading of the SCBA cylinder to the fire fighter.

11 Notify the fire fighter that the SCBA cylinder change is complete. (**STEP 10**)

■ Refilling SCBA Cylinders

Compressors and **cascade systems** are used to refill SCBA cylinders. A compressor or a cascade system can be permanently located at a maintenance facility or at a firehouse, or it can be mounted on a truck or a trailer for mobile use. Mobile filling units are often brought to the scene of a large fire.

Compressor systems filter atmospheric air, compress it to a high pressure, and transfer it to the SCBA cylinders **FIGURE 3-23**. Cascade systems have several large storage cylinders

of compressed breathing air connected by a high-pressure manifold system. The empty SCBA cylinder is connected to the cascade system, and compressed air is transferred from the storage tanks to the cylinder. The storage cylinder valves must be opened and closed, one at a time, to fill the SCBA cylinder to the recommended pressure. This is done by opening the lowest-pressure SCBA cylinder first and then filling it from cylinders containing more pressure. The specific steps for filling SCBA cylinders vary with different systems. It is important to follow the manufacturer's directions when refilling SCBA cylinders.

Follow the steps in **SKILL DRILL 3-14** to safely fill SCBA cylinders from a cascade system:

1 Complete the fill record form, including the date, hydrostatic test date, and cylinder serial number. (**STEP 1**)

2 Ensure that the cylinder is safe to fill by checking its date of manufacture and the hydrostatic test date. (**STEP 2**)

3 Check the cylinder for visible damage. (**STEP 3**)

FIGURE 3-23 Air compressors are installed at many fire stations to refill SCBA cylinders.

SKILL DRILL 3-12   Replacing an SCBA Cylinder
(Fire Fighter I, NFPA 5.3.1)

1 Place the SCBA on the floor or a bench.

2 Turn off the cylinder valve.

3 Open the purge valve to bleed off the pressure.

4 Disconnect the high-pressure supply hose.

5 Release the cylinder from the backpack and remove cylinder.

6 Slide a full cylinder into the backpack. Align the outlet to the supply hose. Lock the cylinder in place.

7 Check that the O ring is present and in good shape.

8 Connect the high-pressure hose to the air cylinder.

9 Open the cylinder valve. Check the gauge reading.

© LiquidLibrary

4 Follow the cascade system filling procedures.

5 Secure the cascade system after use according to system procedures. (**STEP 4**)

Proper training is required to fill SCBA cylinders. Whether your department has an air compressor or a cascade system, only those fire fighters who have been trained on the safe use of this equipment should refill air cylinders.

■ Cleaning and Sanitizing SCBA

Most SCBA manufacturers provide specific instructions for the care and cleaning of their models. The first step in cleaning the SCBA is to rinse the entire unit using a hose with clean water. The harness assembly and cylinder can be cleaned with a mild soap-water solution. If additional cleaning is needed, the unit can be scrubbed with a stiff brush. After scrubbing, the SCBA harness and cylinder should be rinsed with clean water.

After a fire, face pieces and regulators can be cleaned with a mild soap and warm water or with a disinfectant cleaning solution. The face piece should be fully submerged in the cleaning solution. If additional cleaning is needed, a soft brush can be used to scrub the face piece. During the cleaning process, avoid scratching the lens or damaging the exhalation

SKILL DRILL 3-13

Replacing an SCBA Cylinder on Another Fire Fighter
(Fire Fighter I, NFPA 5.3.1)

1 Remove the regulator from the face piece or remove the face piece so that you can breathe ambient air.

2 Turn off the cylinder valve on the used cylinder.

3 Open the purge valve to bleed off pressure from the high-pressure supply line.

4 Disconnect the high-pressure supply hose.

5 Release the SCBA cylinder from the backpack assembly and set it aside.

6 Slide the full SCBA cylinder into the backpack.

7 Check for the presence and satisfactory condition of the O ring.

8 Lock the SCBA cylinder into place.

9 Connect the high-pressure hose to the SCBA cylinder.

10 Open the cylinder valve and notify the fire fighter that the SCBA cylinder change is complete.

valve. The regulator can be cleaned with the same solution, but should not be submerged. The face piece and regulator should then be rinsed with clean water.

Fire Fighter Safety Tips
© Photos.com

The RIC UAC is designed to be used only for emergency refilling of an air cylinder. It should never be used for routine filling of an air cylinder.

Allow the SCBA time to dry completely before returning it to service. Also check for any damage before returning this equipment to service. Follow the steps in **SKILL DRILL 3-15** to clean and sanitize an SCBA:

1. Inspect the SCBA for any damage that might have occurred before cleaning. (**STEP 1**)

2. Remove the face piece from the regulator. On some models, the regulator also can be removed from the harness. (**STEP 2**)

3. Detach the SCBA cylinder from the harness. (**STEP 3**)

4. Rinse all parts of the SCBA with clean water. Water from a garden hose can be used for this step. (**STEP 4**)

5. Using a stiff brush, along with a soap-water solution, scrub the SCBA cylinder and harness. Rinse and set these pieces aside to dry. (**STEP 5**)

6. In a 19-litre (5-gallon) bucket, make a mixture of mild soap and water; alternatively, use the manufacturer's recommended cleaning and disinfecting solution and water. (**STEP 6**)

7. Submerge the SCBA face piece into the soapy water or cleaning solution. For heavier cleaning, allow the face piece to soak. (**STEP 7**)

SKILL DRILL 3-14   Refilling SCBA Cylinders from a Cascade System

1 Complete the fill record form, including the date, hydrostatic test date, and cylinder serial number.

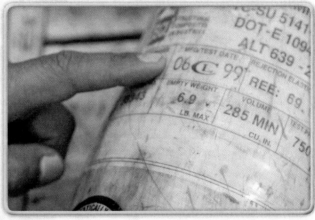

2 Ensure the cylinder is safe to fill by checking its date of manufacture and the hydrostatic test date.

3 Check the cylinder for visible damage.

4 Follow the cascade system filling procedures. Secure the cascade system after use according to system procedures.

© LiquidLibrary

8 Clean the regulator with the soapy water or cleaning solution, following the manufacturer's instructions.

9 Use a soft brush, if necessary, to scrub contaminants from the face piece and regulator. (**STEP 8**)

10 Completely rinse the face piece and the regulator with clean water. Set them aside and allow them to dry.

11 Reassemble and inspect the entire SCBA before placing it back in service. (**STEP 9**)

SKILL DRILL 3-15 | Cleaning SCBA
(Fire Fighter I, NFPA 5.5.1)

1 Inspect the SCBA for any damage that might have occurred before cleaning.

2 Remove the face piece from the regulator. On some models, the regulator also can be removed from the harness.

3 Detach the SCBA cylinder from the harness.

4 Rinse all parts of the SCBA with clean water. Water from a garden hose can be used for this step.

5 Using a stiff brush, along with a soap-and-water solution, scrub the SCBA cylinder and harness. Rinse and set these pieces aside to dry.

6 In a 19-L (5-gal.) bucket, make a mixture of mild soap and water; alternatively, use the manufacturer's recommended cleaning and disinfecting solution and water.

7 Submerge the SCBA face piece into the soapy water or cleaning solution. For heavier cleaning, allow the face piece to soak.

8 Clean the regulator with the soapy water or cleaning solution, following the manufacturer's instructions.

9 Completely rinse the face piece and the regulator with clean water. Set them aside and allow them to dry. Reassemble and inspect the entire SCBA before placing it back in service.

Near-Miss REPORT

Report Number: 09-0000331

Synopsis: SCBA fails during maze exercise.

Event Description: During a Fire Fighter I, Breathing Apparatus 1 class, a student suffered an SCBA failure. The student was in line for the maze exercise. While in line, the fire fighter heard an air leak and loss of air in the mask. The student continued to attempt to breathe in the mask. These attempts led to hyperventilation and a shift in the fire fighter's level of consciousness. The fire fighter's fellow students supported her and removed her mask. Instructors provided initial care (patient assessment and oxygen) and called Emergency Medical Services (EMS). The fire fighter was observed and released by the local Emergency Department Division Chiefs of Training. The Health and Safety Division were notified and an investigation initiated within the hour.

The SCBA failed at the intermediate pressure line between the first stage regulator and the mask-mounted second stage regulator. A product change in the SCBA placed a swivel-nut connection in the middle of the intermediate pressure line. In previous models this had been a one-piece line. The swivel nut became loose and came undone, causing the air leak at 620 kPa (90 psi) and loss of air pressure to the mask.

The fire fighter's SCBA and mask were impounded by Health and Safety. All 40 sets of SCBA at the training academy were inspected. One additional SCBA was found to have the swivel-nut assembly loose. The Air Maintenance Shop was notified. A Safety Alert was sent to all stations. The manufacturer was contacted.

The swivel-nut assembly is on the shoulder strap of the SCBA harness. It is often hidden from view by a cloth channel the line runs through. In attempts to recreate the failure it was found that under pressure, no air leak is heard until the swivel nut is on the last thread. When separation occurs, the O ring is expelled or damaged. Even if the fire fighter could reassemble the line, leakage and loss of integrity would occur. The manufacturer has since developed torque standards for this component.

Lessons Learned:

All product "improvements" need to be critically looked at to determine risk, need for procedural changes, etc. We have added a step in the daily SCBA checklist that addressed the need to visually inspect the intermediate pressure line swivel-nut assembly and attempt to disassemble using finger pressure. Should a swivel-nut assembly be loose, the SCBA is to be removed from service and the Air Maintenance Shop notified. SCBA assigned to the Training Division, because of their high utilization, will receive twice-yearly inspections and testing by the Air Maintenance Shop.

Fire Fighter Safety Tips

© Photos.com

Refilling SCBA cylinders requires special precautions because of the high pressures involved in this procedure. The SCBA cylinder must be placed in a shielded container while it is being refilled **FIGURE 3-24**. Such a container is designed to prevent injury if the cylinder ruptures. In addition, the hydrostatic test date must be checked before the cylinder is refilled to ensure that its certification has not expired. Special procedures must be followed to ensure that the air used to fill the SCBA cylinder is not contaminated.

FIGURE 3-24 SCBA cylinders are refilled in a protective enclosure.

Chief Concepts

- PPE is an essential component of a fire fighter's safety system. It enables a person to survive under conditions that might otherwise result in death or serious injury.
- Structural firefighting PPE enables fire fighters to enter burning buildings and work in areas with elevated temperatures and concentrations of toxic gases. It provides protection from the fire, keeps water away from the body, and helps reduce trauma from cuts or falls. This type of PPE is designed to be worn with SCBA.
- The components of structural PPE include the following items:
  - Bunker coat and pants—Have tough outer shells to withstand temperature extremes and repel water, and insulating layers to protect the skin.
  - Helmet—Protects the entire head and face from trauma; includes ear coverings and a face shield.
  - Protective hood—Constructed of a flame-resistant material to provide additional protection to the ears and neck.
  - Gloves—Constructed of heat-resistant leather to protect the hands from heat, water, vapours, cuts, and puncture wounds.
  - Boots—May be constructed of rubber or leather; designed to protect the feet from heat and trauma.
  - Work uniform—Must be made of natural fibres or special synthetic fibres to prevent burns.
  - SCBA—Provides essential respiratory protection that enables fire fighters to enter the fire atmosphere with a safe, independent air supply.
  - PASS—A device that will automatically sound if a fire fighter is motionless for a set time period.
  - Additional equipment—Goggles, earmuffs, ear plugs, hand light, and two-way radios.
- Structural PPE has several limitations:
  - Weight—PPE adds 23 kg (50 lb.) of extra weight to the fire fighter.
  - Overheating—PPE retains body heat and perspiration, which raises the risk of overheating.
  - Mobility—Full turnout gear limits the range of motion and makes movements awkward and difficult.
- Fire fighters should be able to don PPE in 1 minute or less. Following a set pattern of donning PPE helps reduce the time it takes to dress.
- Check the condition of PPE on a regular basis. Clean it when necessary, and get it repaired at once if it is worn or damaged. Damaged PPE will not protect you.

- Keep PPE clean to maintain its protective properties. The flammable by-products of burnt plastics and synthetic products can become trapped between the fibres, damaging the PPE.
- Cleaning instructions are listed on the tag attached to the clothing. Always follow the manufacturer's cleaning instructions.
- The specialized protective equipment used during vehicle extrication includes special gloves and coveralls or jumpsuits. Latex gloves may be used if the fire fighter provides emergency medical care.
- The PPE for wildland fires is designed for comfort and maneuverability. It includes a jacket and pants made of fire-resistant materials, a helmet, eye protection, and pigskin or leather gloves. The boots are designed to provide comfort and sure footing.
- Respiratory hazards associated with fires include smoke, smoke particles, smoke vapours, toxic gases, oxygen deficiency, and increased temperatures.
- The two main types of SCBA are open-circuit and closed-circuit devices. An open-circuit breathing apparatus is typically used for structural firefighting. A tank of compressed air provides the breathing air supply for the user, and exhaled air is released into the atmosphere through a one-way valve. A closed-circuit breathing apparatus recycles the user's exhaled air and is often used for extended operations, such as mine rescue work.
- One of the biggest limitations of SCBA is the limited amount of air in the cylinder. Fire fighters must manage their working time while using SCBA so that they have enough time to exit from the hazardous area before exhausting the air supply.
- Physical conditioning is important for all SCBA users. A fire fighter with an ideal body weight and in good physical condition will be able to perform more work per cylinder of air than a person who is overweight or out of shape.
- Breathing through an SCBA is different from normal breathing and can be very stressful. Covering your face with a face piece, hearing the air rushing in, hearing valves open and close, and exhaling against a positive pressure are all foreign sensations. Practicing breathing through SCBA in darkness helps to build confidence in the equipment and your skills.
- SCBA consists of four main parts: the backpack and harness, the air-cylinder assembly, the regulator assembly, and the face piece assembly. A compressed air cylinder is removable from the backpack harness and can be changed quickly in the field. SCBA regulators may be mounted on the waist belt or shoulder strap of the

harness or attached directly to the face piece; they control the flow of air to the user. The face piece delivers breathing air to the fire fighter and protects the face from high temperatures and smoke.

■ The pathway of air through an SCBA begins in the cylinder. The air passes through the cylinder shut-off valve into the high-pressure air line that takes it to the regulator. The regulator opens when the user inhales, reducing the pressure on the downstream side. In an SCBA unit with a face piece–mounted regulator, the air next goes directly into the face piece. In units with a harness-mounted regulator, the air travels from the regulator through a low-pressure hose into the face piece. From the face piece, the air is inhaled through the user's air passages and into the lungs. When the user exhales, used air is returned to the face piece. The exhaled air is exhausted from the face piece through the exhalation valve.

■ Each SCBA must be checked on a regular basis to ensure that it is ready for use. Inspection and operational testing should be conducted after a unit has been used, and on a regular schedule. In career departments, inspection and testing is done at the beginning of each shift. In volunteer departments, it is commonly done on a weekly schedule.

■ Compressors and cascade systems are used to refill SCBA cylinders. A compressor or a cascade system can be permanently located at a maintenance facility or at a firehouse, or it can be mounted on a truck or a trailer for mobile use. Mobile filling units are often brought to the scene of a large fire.

■ The steps for donning a complete PPE are as follows:

1. Place the protective hood over your head and bring it down around your neck.
2. Put on your bunker pants and boots. Adjust the suspenders and secure the front flap of the pants.
3. Put on your turnout coat and secure the front.
4. Open the air-cylinder valve on your SCBA and check the air pressure. Press the air-saver valve to prevent air flow if needed.
5. Put on your SCBA.
6. Tighten both shoulder straps of the SCBA harness.
7. Attach the waist belt of the harness and tighten it. Tighten the chest straps, if present.
8. Fit the face piece to your face.
9. Tighten the face piece straps, beginning with the lowest straps.
10. Check the face piece for a proper seal. (Follow the manufacturer's instructions.)
11. Pull the protective hood up so that it covers all bare skin but does not obscure your vision.
12. Place your helmet on your head with the ear tabs extended, and secure the chin strap.
13. Turn up your coat collar and secure it in front.
14. Put on your gloves.
15. Check your clothing to be sure it is properly secured.
16. Make sure your PASS device is turned on.
17. Attach your regulator or turn it on to start the flow of breathing air.
18. Work safely!

Hot Terms

Air cylinder The component of the SCBA that stores the compressed air supply.

Air line The hose through which air flows, either within an SCBA or from an outside source to a supplied air respirator.

Backpack The harness of the SCBA, which supports the components worn by a fire fighter.

Bunker coat The protective coat worn by a fire fighter for interior structural firefighting; also called a turnout coat.

Bunker pants The protective trousers worn by a fire fighter for interior structural firefighting; also called turnout pants.

Carbon monoxide (CO) A toxic gas produced through incomplete combustion.

Carcinogen A cancer-causing substance that is identified in one of several published lists, including, but not limited to, *NIOSH Pocket Guide to Chemical Hazards, Hazardous Chemicals Desk Reference*, and the *ACGIH 2007 TLVs and BEIs*. (NFPA 1851)

Cascade system A method of piping air tanks together to allow air to be supplied to the SCBA fill station using a progressive selection of tanks, each with a higher pressure level. (NFPA 1901)

Closed-circuit breathing apparatus SCBA designed to recycle the user's exhaled air. This system removes carbon dioxide and generates fresh oxygen.

Compressor A device used for increasing the pressure and density of a gas. (NFPA 853)

Doff To take off an item of clothing or equipment.

Don To put on an item of clothing or equipment.

Dual-path pressure reducer A feature that automatically provides a backup method for air to be supplied to the regulator of an SCBA if the primary passage malfunctions.

End-of-service-time-indicator (EOSTI) A warning device on an SCBA that alerts the user that the end of the breathing air is approaching.

Face piece A component of SCBA that fits over the face.

Fire helmet Protective head covering worn by fire fighters to protect the head from falling objects, blunt trauma, and heat.

Hand light A small, portable light carried by fire fighters to improve visibility at emergency scenes; it is often powered by rechargeable batteries.

Heads-up display (HUD) A visual display of information and system conditions status that is visible to the wearer of the SCBA.

Hydrogen cyanide A toxic gas produced by the combustion of materials containing cyanide.

Hydrostatic testing Pressure testing of the extinguisher to verify its strength against unwanted rupture. (NFPA 10)

Immediately dangerous to life and health (IDLH) Any condition that would pose an immediate or delayed threat to life, cause irreversible adverse health effects, or interfere with an individual's ability to escape unaided from a hazardous environment. (NFPA 1670, 2004)

Incomplete combustion A burning process in which the fuel is not completely consumed, usually due to a limited supply of oxygen.

Kevlar® A strong synthetic material used in the construction of protective clothing and equipment.

Kilopascal (kPa) The standard unit of measuring pressure.

Light-emitting diode (LED) An electronic semiconductor that emits a single-colour light when activated.

National Institute for Occupational Safety and Health (NIOSH) The United States federal agency responsible for research and development on occupational safety and health issues.

Nomex® A fire-resistant synthetic material used in the construction of PPE for firefighting.

Nose cups An insert inside the face piece of an SCBA that fits over the user's mouth and nose.

Occupational Safety and Health Administration (OSHA) The United States federal agency that regulates worker safety and, in some cases, responder safety. It is part of the United States Department of Labor.

Open-circuit breathing apparatus SCBA in which the exhaled air is released into the atmosphere and is not reused.

Oxygen deficiency Any atmosphere where the oxygen level is less than 19.5 percent. Low oxygen levels can have serious effects on people, including adverse reactions such as poor judgment and lack of muscle control.

PBI® A fire-retardant synthetic material used in the construction of PPE.

Personal alert safety system (PASS) A device that continually monitors for lack of movement of the wearer and, if no movement is detected, automatically activates an alarm signal indicating the wearer is in need of assistance. The device can also be manually activated to trigger the alarm signal. (NFPA 1982)

Personal protective equipment (PPE) The basic PPE for wildland fire suppression includes a helmet, protective footwear, gloves, and flame-resistant clothing as defined in NFPA 1977, *Standard on Protective Clothing and Equipment for Wildland Fire Fighting*. (NFPA 1051)

Phosgene A chemical agent that causes severe pulmonary damage; it is a by-product of incomplete combustion.

Pounds per square inch (psi) The standard imperial unit for measuring pressure.

Pressure gauge A device that measures and displays pressure readings. In an SCBA, the pressure gauges indicate the quantity of breathing air that is available at any time.

Protective hood A part of a fire fighter's PPE that is designed to be worn over the head and under the helmet; it provides thermal protection for the neck and ears.

Rapid intervention crew/company universal air connection (RIC UAC) A system that allows emergency replenishment of breathing air to the SCBA of disabled or entrapped fire or emergency services personnel. (NFPA 1981)

SCBA harness The part of SCBA that allows fire fighters to wear it as a backpack.

SCBA regulator The part of the SCBA that reduces the high pressure in the cylinder to a usable lower pressure and controls the flow of air to the user.

Self-contained breathing apparatus (SCBA) A respirator with an independent air supply that is used by fire fighters to enter toxic or otherwise dangerous atmospheres.

Self-contained underwater breathing apparatus (SCUBA) A respirator with an independent air supply that is used by underwater divers.

Smoke particles Airborne solid materials consisting of ash and unburned or partially burned fuel released by a fire.

Supplied-air respirator (SAR) An atmosphere-supplying respirator for which the source of breathing air is not designed to be carried by the user. (NFPA 1404)

Turnout coat Protective coat that is part of a protective clothing ensemble for structural firefighting; also called a bunker coat.

Turnout pants Protective trousers that are part of a protective clothing ensemble for structural firefighting; also called bunker pants.

Two-way radio A portable communication device used by fire fighters. Every firefighting team should carry at least one radio to communicate distress, progress, changes in fire conditions, and other pertinent information.

Your crew was first-in to the fully involved house, and you were the lead on the nozzle as you worked to put down the flames. After completing rehab, you were tasked with overhaul, but you did not mind; you were getting to fight fire. You think to yourself, "This is the greatest job in the world."

You are still pumped from the excitement as you pull back into the station. As you step off the engine, you notice the soot covering your gear and your SCBA. You think to yourself that it will be a while before you get to bed.

1. Which method should routinely be used to clean the SCBA?

    A. Soap and water

    B. Clean water only

    C. Nonsolvent degreaser

    D. Solvent-based degreaser

2. When refilling SCBA cylinders from a cascade system, how should it be done?

    A. One cascade cylinder at a time, starting with the one with the lowest pressure

    B. One cascade cylinder at a time, starting with the one with the highest pressure

    C. One cascade cylinder at a time, but the order does not matter

    D. Open all cascade cylinders at the same time

3. In career departments, how frequently should an SCBA unit be operationally tested?

    A. Annually

    B. Quarterly

    C. Monthly

    D. Daily

4. Pressure readings on the cylinder and the regulator should be _____ each other.

    A. the same as

    B. within 69 kPa (10 psi) of

    C. within 345 kPa (50 psi) of

    D. within 690 kPa (100 psi) of

5. Full PPE will weigh nearly _____.

    A. 11 kg (25 lb.)

    B. 16 kg (35 lb.)

    C. 23 kg (50 lb.)

    D. 32 kg (70 lb.)

6. A(n) _____ is an SCBA designed to recycle the user's exhaled air.

    A. supplied-air respirator

    B. open-circuit breathing apparatus

    C. closed-circuit breathing apparatus

    D. cascade system

There have been many changes in your fire department over the past 20 years. You have heard all the stories about how fire fighters use to ride tailboards, wore three-quarter boots rather than bunker pants, used 1.9-centimetre (0.75-inch) booster lines for house fires, and did not wear SCBA. Many of the more senior guys talk fondly about those days and how much better it was.

1. What are the benefits of today's fire service over the fire service of yesteryear when it comes to safety?
2. What kind of conversation would you have with the senior fire fighters?
3. Would you discuss their comments with other recruits?
4. How will you promote today's safety standards?

Fire Service Communications

Fire Fighter I

Knowledge Objectives

After studying this chapter, you will be able to:

- Describe the role of the communications centre in the fire service. (NFPA 5.2 , p 90–91)
- Describe the role and responsibilities of a dispatcher. (NFPA 5.2.1 , p 91)
- List the requirements of a communications centre. (p 91)
- Describe the equipment used in a communications centre. (NFPA 5.2.1 , p 91–92)
- Describe how computer-aided dispatch assists in dispatching the correct resources to an emergency incident. (NFPA 5.2.1 , p 92)
- Describe the basic services provided by the communications centre. (NFPA 5.2.1 , p 93–94)
- List the five major steps in processing an emergency incident. (NFPA 5.2.1A , p 94)
- Describe how dispatchers conduct a telephone interrogation. (NFPA 5.2.2 , p 94–95)
- Describe how municipal fire alarm systems, private and automatic fire alarm systems, and citizens can activate the emergency response system. (NFPA 5.2.1 , p 95–96)
- Describe how location validation systems operate. (NFPA 5.2.1 , p 96–97)
- Describe the three types of fire service radios. (NFPA 5.2.3 , p 100–101)
- Describe how two-way radio systems operate. (NFPA 5.2.3 , p 100–101)
- Explain how a repeater system works to enhance fire service communications. (NFPA 5.2.3 , p 102–103)
- Explain how a trunking system works to enhance fire service communications. (NFPA 5.2.3 , p 103)
- Describe the basic principles of effective radio communication. (NFPA 5.2.3 , p 103–105)
- Describe when to use plain language and how ten-codes are implemented in fire service communications. (NFPA 5.2.3A , p 104–105)
- Outline the information provided in arrival and progress reports. (p 105–106)
- Describe fire department procedures for answering nonemergency business and personal telephone calls. (NFPA 5.2.2, 5.2.2A , p 106)

Skills Objectives

After studying this chapter, you will be able to perform the following skills:

- Initiate a response to a simulated emergency. (NFPA 5.2.1B , p 99)
- Observe the operation of a communications centre. (NFPA 5.2.2 , p 100)
- Display how to use a portable radio. (NFPA 5.2.3B , p 103–104)
- Operate and answer the fire station telephone. (NFPA 5.2.1B, 5.2.2B , p 106–107)

Fire Fighter II — FFII

Knowledge Objectives

After studying this chapter, you will be able to:

- Define emergency traffic. (NFPA 6.2.2, 6.2.2A , p 105)
- Explain how to initiate a mayday call.
- Describe common evacuation signals.
- Explain the importance of an incident report to the entire fire service. (NFPA 6.2, 6.2.1 , p 105–106)
- Describe how to collect the necessary information for a thorough incident report. (NFPA 6.2.1, 6.2.1A , p 106–107)
- Describe the resources that list the codes utilized in incident reports. (NFPA 6.2.1A , p 106–107)
- Explain the consequences of an incomplete or inaccurate incident report.

Skills Objectives

After studying this chapter, you will be able to perform the following skill:

- Display how to use a portable radio. (NFPA 6.2.2B , p 103)

Additional NFPA Standards

- NFPA 901, *Standard Classifications for Incident Reporting and Fire Protection Data*
- NFPA 902, *Fire Reporting Field Incident Guide*
- NFPA 1061, *Standard for Professional Qualifications for Public Safety Telecommunicator*
- NFPA 1221, *Standard for the Installation, Maintenance, and Use of Emergency Services Communications Systems*

You Are the Fire Fighter

A t 3:04 a.m., you are dispatched to a report of a fire at 3256 West Madison. Your captain asks if any additional details are available, and the dispatcher says that the caller said there was smoke coming from the eaves of the building and then hung up. You are getting excited about the prospect of a working fire. Then the dispatcher comes back and says that this may be a false report, because the caller used a cell phone from the east side of town. Your aggravation rises at the prospect of being woken up for another false call—at least until the dispatcher comes back with a frantic sound in her voice and says that a home-security company just reported a fire alarm at 3256 *East* Madison. Thoughts plow through your mind, wondering what is going on.

1. What is the process for receiving 911 calls in an emergency communications centre?
2. How would the dispatcher know where the cell phone call originated?
3. How are alarm systems monitored and then reported to the communications centre?

Introduction

Rapidly developing technology and advanced communications systems are having a tremendous effect on fire department communications. The rapid pace of technology advancement is changing the way many fire services operate. From GPS-embedded self-contained breathing apparatus (SCBA), to heads-up displays in face pieces, technology is redefining the manner in which we provide community service. Understanding these changes and learning to adapt to change is something that all fire personnel need to embrace. As a fire fighter, you must be familiar with the communications systems, equipment, and procedures used in your particular fire department. This chapter provides a basic guide to help you understand how fire department communications systems work and how commonly used systems are configured.

Every fire department depends on a functional communications system. When a citizen requests assistance or an alarm sounds, the communications centre dispatches the appropriate units to the incident, and it continues to maintain communication with those units throughout the entire duration of the incident. The communications system is the link between fire fighters on the scene, the rest of the organization, and others. The communications centre monitors everything that happens at the incident scene and processes all requests for assistance or special resources.

At the scene, fire fighters need to communicate with one another so that the incident commander (IC) can manage the operation efficiently based on progress reports or requests for assistance from fire fighters. The incident management system (IMS) depends on the presence of a functional on-site communications system. During incidents, fire fighters must be able to communicate not only with one another, but also with other emergency response agencies.

The communications centre does more than simply dispatch units and communicate with them during emergency operations. It also tracks the location and status of every other fire department unit. The communications centre must always know which units can be dispatched to an incident, and it must be able to contact those units promptly. The communications centre is also responsible for redeploying units to maintain adequate coverage for all areas.

In addition to these special communications requirements, a fire department must have a communications infrastructure that allows it to function as an effective organization. Basic administration and day-to-day management require an efficient communications network, including telephone and data links with every fire station and work site.

The Communications Centre

The **communications centre** is the hub of the fire department emergency response system. It serves as the central processing point for all information relating to an emergency incident and all of the information relating to the location, status, and activities of fire department units. It connects and controls all of the department's communications systems. The communications centre functions much like the human brain; that is, information comes in via the nerves, is processed, and is then sent back out to be acted upon by the various parts of the body.

The communications centre is a physical location, whose size and complexity will vary depending on the needs of the department. The communications centre for one department may be a small room in the fire station; another department may have a specially designed, highly sophisticated facility with advanced technological equipment. The fire department in a small community may need a simple system, whereas the public safety agencies in a metropolitan area may require a large facility. Some distinct differences between rural and urban dispatch centres are apparent—specifically, the number of dispatchers needed at any given time, the type of training that is required, and, to some extent, the type of equipment used. No matter what their size, however, all communications centres perform the same basic functions **FIGURE 4-1**.

FIGURE 4-1 No matter what their size, all communications centres perform the same basic functions.

Many fire services operate stand-alone communications centres, which serve only a single agency. Others operate or are served by a communications centre that works with several fire departments, and sometimes with all of the fire departments in a province or region. In some areas, the fire department communications centre is co-located with other public safety agencies, including law enforcement and separate Emergency Medical Services (EMS) providers. In Canada, the 911 operations centres are located in the municipal police dispatch centre. 911 dispatchers will answer calls, ascertain the nature of the emergency, and send the call to the appropriate agency. In remote regions within large geographic areas that have small populations, the Royal Canadian Mounted Police and/ or the provincial policing agency, as in Ontario and Quebec, are typically responsible for dispatching emergency vehicles, including fire services.

■ Dispatchers

The employees who staff a communications centre are known as dispatchers. Dispatchers have been professionally trained to work in a public safety communications environment. Their completion of advanced training and professional certification programs ensures that these skilled, competent individuals can fulfill their critical roles in the public safety system.

Just as special qualities can make an individual a good fire fighter, so a dispatcher must possess certain qualities to be successful. The job of a dispatcher can be complicated, demanding, and extremely stressful. Even in the face of these challenges, the successful dispatcher is able to understand and follow complicated procedures, perform multiple tasks effectively, memorize information, and make decisions quickly.

One of the dispatcher's most important skills is the ability to communicate effectively with citizens to obtain critical information, even when the caller is highly stressed or in extreme personal danger. Even if an emotional caller criticizes or insults the dispatcher, the dispatcher must respond professionally and focus on obtaining the essential information. Voice control and

the ability to maintain composure under pressure are important qualities; the dispatcher must always be clear, calm, and in control.

A dispatcher must be skilled in operating all of the systems and equipment in the communications centre. He or she must understand and follow the fire department's operational procedures, particularly those relating to dispatch policies and protocols, radio communications, and incident management. The dispatcher must keep track of the status and location of each unit at all times, and monitor the overall deployment and availability of resources throughout the system. In Canada, there are no established nationwide requirements for the certification of dispatchers. Most municipalities train their personnel in house to perform the duties required for their specific fire service. In municipalities east of of the Ottawa River, dispatchers must be bilingual in French and English. In most areas within Ontario and the western provinces and territories, bilingualism is not mandatory, but is considered an asset.

■ Communications Facility Requirements

The fire service communications centre must be designed and operated to ensure that its critical mission can be performed with a very high degree of reliability. The performance requirements in NFPA 1221, *Standard for the Installation, Maintenance, and Use of Emergency Services Communications Systems*, govern the design and construction of a fire department or public safety communications centre. These requirements apply whether the communications centre serves a small community with only one or two fire stations or a metropolitan area with dozens of stations.

The communications centre should be well protected against natural threats, such as floods, and it should be able to withstand predictable damaging forces such as earthquakes, snowstorms, tornadoes, hurricanes, and other severe storms. It should be located so that it can continue to function in times of civil unrest. In addition, this centre should be able to operate at maximum capacity, without interruption, even when other community services are severely affected. The building should be equipped with emergency generators and other systems so that it can continue to operate for several days in even the most challenging conditions.

Most Canadian municipalities have established contingency plans to ensure communications continuity in the event that their main communications centre goes offline or is compromised.

■ Communications Centre Equipment

Most communications centres are usually equipped with several types of communications systems and equipment. Although the specific requirements will depend on the needs of the service, most centres will have the following equipment:

- Dedicated 911 telephones
- Public telephones
- Direct-line phones to other agencies
- Equipment to receive alarms from public or private fire alarm systems

- Computers and/or hard-copy files and maps to locate addresses and select units to dispatch
- Equipment for alerting and dispatching units to emergency calls
- Two-way radio system(s)
- Recording devices to record phone calls and radio traffic
- Backup electrical generators
- Records and record-management system

■ Computer-Aided Dispatch

Computers are used in almost all communications centres. Large Canadian communications centres use **computer-aided dispatch (CAD)** systems; many smaller centres have smaller-scale versions of CAD systems.

As the name suggests, a CAD system is designed to provide the dispatcher with the ability to multitask, such as reassuring a caller on the phone while simultaneously dispatching and providing updates in real time to responding personnel. Many variations in CAD systems exist—from very simple versions to highly sophisticated systems that perform many different functions. All of the functions performed by a CAD system can be performed manually by trained and experienced dispatchers; the benefit of such a system derives from the fact that the computer enables the dispatcher to perform these functions more quickly and accurately **FIGURE 4-2**.

A well-designed CAD system can shorten the time it takes for a dispatcher to receive and dispatch calls. In this way, a CAD system helps meet the most important objective in processing an emergency call: sending the appropriate units to the correct location as quickly as possible. The increased efficiency provided by a CAD system is particularly important in large communications centres that process a high volume of calls.

Because a computer can process large amounts of information efficiently and accurately, a CAD system can be used to track the status of a large number of units. It will know which units are available to respond to a call, which units are assigned to incidents, and which units are temporarily assigned to cover different areas. The most advanced type of CAD system utilizes

global positioning system (GPS) devices in the department's vehicles, allowing it to track the exact location of each unit. A CAD system can also look up addresses and determine the closest fire stations in order of response for any location. It can then select those units that can respond quickly to an alarm, even if some of the units that would normally respond to the call are currently unavailable.

Some CAD systems transmit dispatch information directly to **mobile data terminals (MDT)** on the apparatus or to computers that are located in the fire station **FIGURE 4-3**. In addition, CAD systems can provide immediate access to information such as preincident plans, hazardous materials lists, lockbox locations, and whether persons with limited mobility reside at the address of the emergency call. By linking the CAD system to other data files, fire fighters gain access to even more useful information such as travel route instructions, maps, and reference materials.

Efficient CAD systems will usually be integrated in the fire services records management system, which is also computerized and becomes the repository for all of the fire services information. In more advanced CAD systems, responding units can book on and off the air with the push of a button on an MDT or vehicle radio. This advancement reduces radio traffic and enables for a more efficient response.

■ Voice Recorders and Activity Logs

Almost everything that happens in a communications centre is recorded, either by a **voice recording system** or by an **activity logging system**. Most communications centres can automatically record everything that is said over the telephone or radio, 24 hours a day. In addition, most have an instant playback unit that allows the dispatcher to replay conversations for the previous 10–15 minutes at the touch of a button. This feature is particularly valuable if the caller talks very quickly, has an unusual accent, hangs up, or is disconnected. The dispatcher can replay the message several times, if necessary, to determine exactly what was said.

The logging system keeps a detailed record of every incident and activity that occurs. These records include every

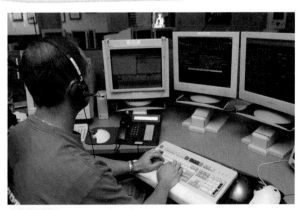

FIGURE 4-2 A CAD system enables a dispatcher to work more quickly and efficiently.

FIGURE 4-3 Some CAD systems transmit dispatch information directly to terminals in fire stations and mobile data terminals in the apparatus.

call that is entered, every unit that is dispatched, and every significant event that occurs in relation to an emergency incident. Times are recorded when a call is received, when the units are dispatched, when they report that they are en route, when they arrive at the scene, when the incident is under control, and when the last unit leaves the scene. One major advantage of a CAD system is that it automatically captures and stores every event as it occurs. Before such systems were developed, someone had to write down and time-stamp every transaction, including every radio transmission; all of these hard-copy logs then had to be retained for future reference. Digital technology allows for a significant amount of data to be stored long-term in CAD and RMS systems. This data can be retrieved for historical, training, or legal purposes.

Voice recorders and activity logs are maintained for several reasons. First, they serve as legal records of the official delivery of a government service by a public agency (i.e., the fire department). These records may be required for legal proceedings, sometimes years after the incident occurred. They may be needed to defend the fire department's actions when questions are raised about an unfortunate outcome. The records provided by voice recorders and activity logs accurately document the events and can often demonstrate that the organization and its employees performed ethically, responsibly, and professionally. They also make it difficult to hide an error, if a mistake was made.

Second, records are valuable in reviewing and analyzing information about department operations. Good record keeping enables a fire department to examine what happened on a particular call as well as to measure workloads, system performance, activity trends, and other factors as part of its planning and budget preparation; for example, analysts and planners can use data from CAD systems to study deployment strategies and to make the most efficient use of fire department resources.

■ Call Response and Dispatch

Critical functions performed by most CAD systems include verifying an address and determining which units should respond to an alarm. This is usually a simple task in a small fire department with only two or three fire stations, but it becomes much more complicated in a large urban system with dozens of fire stations and individual units. When the dispatcher enters an address and an incident description code into a CAD system, the system can make a recommendation on the appropriate units to dispatch in less than a second.

Data links that provide the location of the caller and automatically enter the address can also save time. If duplicate addresses might exist or if the address entered is not a valid location, the CAD system will prompt the dispatcher to ask the caller for more information. Hard-copy files of the most essential information should always be maintained for backup, so that calls can be dispatched even if the system goes down (i.e., is out of service).

The dispatcher's first responsibility is to obtain the information that is required to dispatch the appropriate units to the correct location **FIGURE 4-4**. At that point, the incident must

FIGURE 4-4 Dispatchers must obtain information and relay it accurately to the appropriate responders.

be processed according to standard protocols. The dispatcher must decide which agencies or units should respond and transmit the necessary information to them. The generally accepted performance objective, from the time a call reaches the communications centre until the units are dispatched, is 1 minute or less.

Most requests for fire department response are made by telephone. Although a few communities use a regular seven-digit telephone number or "0" to report an emergency, most areas in Canada and the United States have implemented the 911 system. According to the United States' National Emergency Number Association (NENA), 93 percent of the population in the United States and Canada has access to some type of 911 system to report an emergency.

All calls to 911 are automatically directed to a designated **public safety answering point (PSAP)** for that community or jurisdiction. If all public safety communications are located in the same facility, the call can be answered, processed, and dispatched immediately. If fire and EMS communications operate in separate facilities, those calls must be transferred to a fire-/EMS-trained dispatcher. The call-taking process is described in more detail later in this chapter.

Communications Centre Operations

Several different functions are performed in a fire department or public safety communications centre. All of these activities must be performed accurately and efficiently, even in the most challenging circumstances; for example, the activity level in a communications centre can ratchet up from calm to extreme in less than a minute during an emergency, but the chaos outside must not affect the operations inside. If the communications centre fails to perform its mission, the fire department will not be able to deliver much-needed emergency services.

A communications centre performs the following basic functions:

- Receiving calls for emergency incidents and dispatching fire department units
- Supporting the operations of those fire department units delivering emergency services

- Coordinating fire department operations with other agencies
- Keeping track of the status of each apparatus
- Monitoring the level of coverage and managing the deployment of available apparatus
- Notifying designated individuals and agencies of particular events and situations
- Maintaining records for all emergency and nonemergency responses
- Maintaining information required for dispatch purposes

■ Receiving and Dispatching Emergency Calls

Most fire department responses begin when a call comes in to the communications centre, which then dispatches the appropriate resources. Even if a citizen reports an emergency directly to a fire station or a crew discovers a situation, the communications centre is immediately notified to initiate the emergency response process and generate an incident number. Most emergency response services in Canada generate an incident number for every call for service; this usually includes emergency and nonemergency incident response. The assignment of an incident call number enables effective referencing from the records management system.

There are five major steps in processing an emergency incident:

1. Call receipt
2. Location validation
3. Classification and prioritization
4. Unit selection
5. Dispatch

Call receipt refers to the process of receiving an initial call and obtaining the necessary information to initiate a response. It includes telephone calls from the general public as well as other notification methods, such as automatic fire alarm systems, public fire alarm boxes, requests from other agencies, calls reported directly to fire stations, and calls initiated via radio from fire service units or other public safety agencies.

Location validation ensures that the information received is adequate to dispatch units to the correct location. Potential duplicate addresses—such as two streets with the same or very similar names—must be eliminated from consideration; for example, it is important to differentiate 123 Main Street from 123 Main Road. The information must point to a valid location on a map or in a street index system, and that location must be within the geographic jurisdiction of the potential dispatch units. Without a valid address, a communications centre will be unable to send units to the proper location.

Classification and prioritization is the process of assigning a response category, based on the nature of the reported problem. Most fire services respond to many kinds of situations—from structure fires to high-angle rescues. The nature of the call dictates which units or combination of units should be dispatched. Dispatch policies are usually spelled out in the department's suggested operating guidelines (SOGs), or standard operating procedures (SOPs), and dispatch protocols,

which state the numbers and types of units to dispatch for each type of situation.

Unit selection is the process of determining exactly which unit or units to dispatch, based on the location and classification of the incident. The usual policy is to dispatch the closest available unit that can provide the necessary assistance. When all units are available and in their fire stations, this determination can usually be made quickly based on preprogrammed information. Unit selection becomes more complicated when some units are out of position or unavailable. This step often requires quick decision-making skills, even when the CAD system is programmed to follow set policies for various situations. Some dispatch centres are equipped with automatic vehicle-locator systems that track apparatus by using GPS devices. These systems enable dispatchers to dispatch the closest units that are in service, thereby increasing the efficiency of the system.

Dispatch is the important step of actually alerting the selected units to respond and transmitting the information to them. Fire services use a variety of dispatch systems—from telephone lines to radio systems. The communications centre must have at least two separate ways of notifying each fire station.

FIRE FIGHTER Tips

Dispatchers must be careful about which information is given to the public. Information such as availability of individual fire departments, apparatus placement, and available equipment should be monitored. An SOG/SOP should clarify which information is allowed to be broadcast and which information is not.

■ Call Receipt

Telephones

The public generally uses telephones to report emergency incidents. In most communities, calls to 911 connect the caller with a public service answering point. The PSAP can take the information immediately or transfer the call to the appropriate agency, based on the nature of the emergency. In some systems, calling 911 connects the caller directly with a dispatcher who obtains the required information.

FIRE FIGHTER Tips

Some communities have implemented a 311 system to handle nonemergency calls. This system is often used to link calls with health and human services or other community resources that can assist with a problem when it might not necessarily be life threatening in nature. The underlying premise of the 311 system is to reduce the number of nonemergency calls to 911.

The communications centre has a seven-digit nonemergency telephone number. Some people may use that number to report an emergency, because they are not sure that their particular situation is serious enough to be considered a true emergency.

Any number that is published as a fire department telephone number should be answered at all times, including nights and weekends. The emergency number should always be pronounced as "nine-one-one," not "nine-eleven."

The dispatcher who takes the call must conduct a **telephone interrogation**, asking the caller questions to obtain the required information. Initially, the dispatcher will need to know the location of the emergency and the nature of the situation. Many 911 systems can automatically provide the location of the telephone where a call originates; this information may be unavailable or inaccurate, however, if the call was made on a wireless phone or from someplace other than the location of the emergency incident. The exact location must be obtained so that units can respond directly to the incident scene.

The dispatcher must also interpret the nature of the problem from the caller's description. The caller may be distressed, excited, and unable to organize his or her thoughts. There may be language barriers: the caller might speak a different language or might not know the right words to explain the situation. Dispatchers must follow SOGs/SOPs and use active listening to interpret the information. Many communications centres use a structured set of questions to obtain and classify information on the nature of the situation. The dispatcher must remember that the caller thinks the situation is an emergency, and he or she must treat every call as such until it is determined that no emergency exists.

Dispatchers cannot allow gaps of silence to occur while questioning the caller. If the caller suddenly becomes silent, something may have happened to him or her; the caller might be in personal danger or extremely upset. Conversely, if the dispatcher is silent, the caller might think that the dispatcher is no longer on the line or no longer listening.

Disconnects are another problem, because callers to 911 might hang up accidentally or be disconnected after initially reaching the communications centre. A dispatcher who is unable to return the call and reach the original caller will usually dispatch a police officer to the location to determine if more help is needed.

A dispatcher should never argue with a caller. Although callers may raise their voices, scream, or shout, the dispatcher must continue to speak calmly and remain professional at all times. He or she should remember that the caller is simply reacting to the emergency situation.

With just two critical pieces of information—the location and nature of the problem—the dispatcher can initiate a response; local SOGs/SOPs, however, may require the dispatcher to obtain additional information. Getting the caller's name and contact phone number is useful in case it is necessary to call back for additional information. If the caller is in danger or distress, the dispatcher will try to keep the line open and remain in contact with the caller until help arrives. In many communities, dispatchers are trained to provide self-help instructions for callers, advising them on what to do until the fire service arrives. For example, if a building fire is being reported, the dispatcher will advise the occupants to evacuate

and wait outside; if the caller is reporting a medical incident, the dispatcher may provide first-aid instructions, based on the patient's symptoms.

TDD/TTY/Text Phones

Speech- and/or hearing-impaired citizens can communicate by telephone, using special devices that display text rather than transmitting audio **FIGURE 4-5**. Three such systems are **TDD** (telecommunications device for the deaf), **TTY** (teletype), and **text phones**. Many communications centres have a special telephone line with a seven-digit number set aside for such calls, but the 911 lines must also accommodate text messaging. Any incoming calls without voice contact at the other end should be checked for TDD/TTY/text phone systems.

FIGURE 4-5 TDD/TTY/text phone systems enable speech- and/or hearing-impaired persons to communicate over phone lines.

FIRE FIGHTER Tips

Next generation dispatch systems will permit dispatch centres to receive texts, video clips, and other communications from the general public.

Direct-Line Telephones

A **direct-line** (or ring-down) telephone connects two predetermined points. Picking up the phone at one end causes an immediate ring at the other end. Direct-line connections often link police and fire communications centres or two fire communications centres that serve adjacent areas. Direct lines also may connect hospitals, private alarm companies, utility companies, airports, and similar facilities with the fire department communications centre. These lines are often used in both directions, so that the communications centre can both receive calls for assistance and send notifications and requests for response.

The communications centre may be connected by direct lines to each fire station in its jurisdiction. A direct line also can be linked to the station's public address speakers to announce dispatch messages.

Municipal Fire Alarm Systems

Some communities have installed fire alarm boxes or emergency telephones on street corners or in public places. A fire alarm box transmits a coded signal to the communications centre. Although this signal indicates an alarm and identifies the location of the box, it does not indicate what kind of emergency is occurring. The major drawbacks to this type of reporting system are the high rate of false alarms and the inability to determine the nature of the call. Public fire alarm boxes may also be connected to automatic fire alarm systems in buildings. When the building's alarm system is activated, the public fire alarm box transmits an alarm directly to the communications centre.

Public fire alarm boxes are usually connected by networks of dedicated cables that transmit the alarm signals to a fire department facility and to individual fire stations. In many cases, these hard-wired boxes have been replaced by radio-operated fire alarm boxes. In recent years, many communities have eliminated their public fire alarm systems because of increasing numbers of false alarms, high maintenance costs, and development of alternative notification systems, including private and public hard-wired and cellular telephones. Other communities have converted their public fire alarm systems into call box systems.

A **call box** connects a person directly to a dispatcher. The caller can request a full range of emergency assistance—from police, fire, or emergency medical assistance to a tow truck. Call boxes can be directly connected to a wireless or hard-wired telephone network or can operate on a radio system. Emergency call boxes are infrequently located along major highways and in bridges, tunnels, subways, large complexes such as universities, and in other locations where nearby phones are not available and where mobile phones do not work. In locations without electric or telephone service, call boxes may be solar powered **FIGURE 4-6**.

Private and Automatic Fire Alarm Systems

Private and automatic fire alarm systems use several types of arrangements to transmit alarms to the local fire department communications centre. Many commercial, industrial, and residential buildings have automatic fire alarm systems, which use heat detectors, smoke detectors, or other devices to initiate an alarm, such as water flow alarms on automatic sprinkler systems. The chapter "Fire Detection, Protection, and Suppression Systems" discusses these systems in detail.

The connection used to transmit an alarm from a private system to a public fire communications centre depends on many factors. These systems may be monitored by a privately operated central station alarm service or other alarm monitoring service, which relays the alarm to the fire communications centre. A direct line may connect a central alarm service and the fire department communications centre. Some communications centres provide monitoring services. Finally, private alarm systems may be connected to a public fire alarm box. No matter how the alarm reaches the communications centre, the result is the creation of an incident report and the dispatch of fire department apparatus.

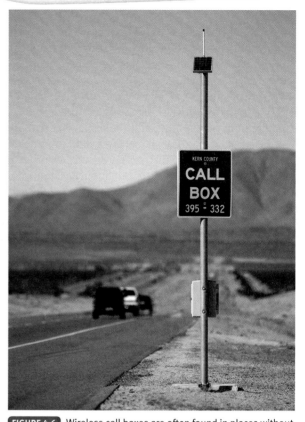

FIGURE 4-6 Wireless call boxes are often found in places without other kinds of telephone service.

© Chase Swift/Corbis

Walk-ins

Although most emergencies are reported by telephone, some people actually come to a fire station seeking assistance **FIGURE 4-7**. When a walk-in, sometimes referred to as a "still alarm," occurs, the station should contact and advise the communications centre of the situation immediately. The communications centre will create an incident record and dispatch any needed additional assistance. Even if the units at the station can handle the situation, they should notify the communications centre that they are occupied with an incident.

■ Location Validation

Enhanced 911 Systems

Most enhanced 911 systems have features that can assist the dispatchers in obtaining and identifying information; for example, **automatic number identification (ANI)** shows the telephone number where the call originated. **Automatic location identification (ALI)** queries a database to show the location of the telephone, the subscriber's name, and other details **FIGURE 4-8**. These features, when combined with an accurate database, will generate a computer display listing all of the information when the call is received. Another feature in an enhanced 911 system ensures that each call will be directed to the appropriate PSAP for that location.

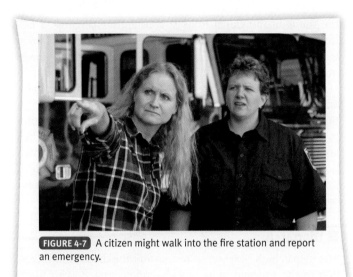

FIGURE 4-7 A citizen might walk into the fire station and report an emergency.

FIGURE 4-9 Calls from wireless telephones are now as common as those from hard-wired telephones.

Using enhanced 911, the dispatcher can often pinpoint the location of an incident, even if the caller is unable to provide that information. Even so, the dispatcher should always confirm that this information is correct and refers to the actual location of the emergency. If the caller has recently moved, the database may contain the old address. Also, sometimes a billing address for the telephone service is provided rather than the physical location of the telephone.

Enhanced 911 features can save both time and lives, but only if they are used in conjunction with an accurate, up-to-date database.

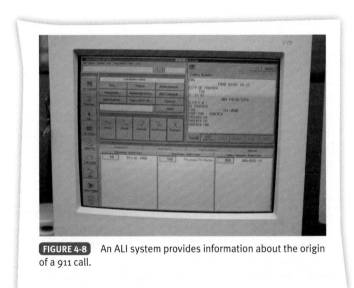

FIGURE 4-8 An ALI system provides information about the origin of a 911 call.

With the widespread use of cell phones, calls from wireless telephones are now as common as those from hard-wired telephones **FIGURE 4-9**.

Although wireless phones have certainly enhanced public access to emergency services, they have also created new challenges. The ANI and ALI enhancements to 911 systems were developed for hard-wired telephone systems. The federal government has mandated that wireless telephone companies work to update their systems to be able to pinpoint the geographic coordinates of a 911 call from a cellular phone and transmit that information to dispatch centres upon request.

Determining the location of the incident may also prove problematic, because people who call 911 on their cellular telephones may not know their exact location. In addition, the dispatch centre may receive numerous calls for the same incident from callers calling from different locations, thus making it even more difficult to determine the correct location.

Initiating a Response to a Simulated Emergency

Fire fighters should understand the steps followed by the personnel who staff the communications centre as they receive emergency calls, solicit needed information from a caller, and dispatch appropriate units to the emergency. Follow the steps in **SKILL DRILL 4-1** to initiate a response to a simulated emergency:

1. Identify your agency.
2. Ask whether there is an emergency.
3. Organize your questions to get the following information:
    A. Incident location
    B. Type of incident
    C. When incident occurred
4. Obtain the following information:
    A. Caller's name
    B. Location of the caller, if different from the incident
    C. Callback number (**STEP 1**)

⑤ Record the information needed.

⑥ Initiate an alarm following the protocols of your communications centre. This may be done by dialing 911 if you are at the fire station, using a mobile phone, or using your mobile or portable radio to transmit the information to the dispatch centre. The protocols in your department may vary from the steps listed here. Follow the protocols of the agency having jurisdiction for your department's communications. (**STEP ❷**)

■ Call Classification and Prioritization

The next step in processing an emergency incident is classifying and prioritizing each call as it is received. The nature of the problem—what is happening at the scene—determines the urgency of the call and its priority. Although most fire service calls are dispatched immediately, it is sometimes necessary to delay dispatch to a lower-priority call if a more urgent situation arises or if several calls come in at the same time. Some calls qualify for emergency response (red lights and siren); others are considered nonemergency. Consult your local SOGs/SOPs to learn the difference in your jurisdiction.

The call classification determines the number and types of units that are dispatched. The standard response to different incidents is established by the fire department's SOGs/SOPs and can range from a single unit to an initial assignment of 10 or more units and dozens of fire fighters.

■ Unit Selection

After verifying the location and classifying the situation, a dispatcher must select the specific units to dispatch to an incident. Generally, the standard assignment to each type of incident in each geographic area is stored on a system of **run cards**. Run cards are prepared in advance and either entered into the CAD system or stored in hard-copy form. They list units in the proper order of response, based on response distance or estimated response time, and often specify the units that would be dispatched through several levels of multiple alarms.

The run card assignments can be used as long as all of the units that would normally respond on a call are available. If any units are unavailable, the dispatcher must adjust the assignments by selecting substitute units in their order of response. In a large, busy system with many simultaneous incidents,

---

FIRE FIGHTER Tips

On occasion, the communications centre may receive calls about issues that cannot be handled by the fire department or other agencies participating in the communications system. In these cases, the dispatcher should make every effort to accommodate the caller. If possible, transfer the caller to the proper agency. In addition, provide the caller with the proper contact information in case the transfer fails. The caller will appreciate the assistance and retain a positive image of the department's service.

---

the status of individual units is constantly changing—a factor that complicates the assignment process unless the system is equipped with an automatic vehicle-locator system that is integrated into the computerized dispatch protocols.

Most CAD systems are programmed to select the units for an incident automatically, based on the location, call classification, and actual status of all units. Such a system will recommend a dispatch assignment, which the dispatcher can then accept or adjust, based on circumstances or special information. The same process is used to dispatch any additional units to an incident, whether it is a multiple alarm or a request for particular units or capabilities.

■ Dispatch

The dispatcher's next step is to transmit the dispatch information to the assigned units quickly and accurately. This information can be transmitted in many different ways, but every department must have at least two separate methods for sending a dispatch message from the communications centre to each fire station. The primary connection can be a hard-wired circuit, a telephone line, a computer-based data link, a microwave system, or a radio system.

Most fire departments dispatch verbal messages to the appropriate fire stations. The dispatch message is broadcast over speakers in the fire station, so that everyone immediately knows the location and nature of the incident and the units that should respond. Radio transmissions are used to contact a unit that is out of the station. The fire station or each vehicle must confirm that the message was received and the units are responding. If the communications centre does not receive the confirmation within a set time period, it must dispatch substitute units.

A CAD system can be programmed to alert the appropriate fire stations automatically. Such a system can send the dispatch information to computer terminals or printers, sound distinctive tones, turn on lights and public address speakers, turn off the stove, and perform additional functions. The dispatch message also can go directly to each individual vehicle that has a computer terminal as well as to the station. The CAD notification often is accompanied by a verbal announcement over the radio and the fire station speakers.

Fire departments with volunteer responders must be able to reach them with a dispatch message. Some departments issue pagers to their individual volunteers. Some pagers can receive a vocal dispatch message; others display an alphanumeric message. Some CAD systems permit text messages to be sent to wireless phones with incident information.Volunteer fire departments in some communities may also rely on outdoor sirens, horns, or whistles to notify their members of an emergency. These audible devices usually can be activated by remote control from the communications centre. Volunteers then call the communications centre by radio or telephone to receive specific instructions.

■ Operational Support and Coordination

After the communications centre dispatches the units, it begins to provide incident support and coordination. Someone in the communications centre must remain in contact with the

SKILL DRILL 4-1	Initiating a Response to a Simulated Emergency
	(Fire Fighter I, NFPA 5.2.1)

1 Identify your agency. Ask whether there is an emergency. Organize your questions to get the following information: incident location, type of incident, and when the incident occurred. Obtain the following information: caller's name, location of the caller, if different from the incident, and callback number.

2 Record the information needed. Initiate an alarm following the protocols of your communications centre.

© LiquidLibrary

responding units throughout the entirety of the incident. The dispatcher must confirm that the dispatched units actually received the alarm, record their en route times, provide any additional or updated information, and record their on-scene arrival times.

Operational support and coordination encompass all communications between the units and the communications centre during an entire incident. Two-way radios, laptop computers, mobile communications devices, wireless phones, and mobile computer terminals may all be used to exchange information.

Generally, the IC will communicate with a dispatcher operating a radio in the communications centre. Progress and incident status reports, requests for additional units or release of extra units, notifications, and requests for information or outside resources are examples of incident communications. The communications centre closely monitors the radio and provides any needed support for the incident.

■ Status Tracking and Deployment Management

The communications centre must know the location and status of every fire department unit at all times. Units should never get lost in the system—whether they are available for dispatch, assigned to an incident, or in the repair shop. As previously stated, the communications centre must always know which units are available and which units are not available for

dispatch. The changing conditions at an incident may require frequent reassignment of units, including ambulances. In addition, units may be needed from outside the normal response area or from other districts.

Tracking the status of the various units is difficult if the dispatcher must rely on radio reports and coloured magnets or tags on a map or status board. CAD systems make this job much easier, because status changes can be entered through digital status units or computer terminals.

Communications centres must continually monitor the availability of units in each geographic area and redeploy units when an area has insufficient coverage. Many fire departments list both unit relocations and multiple response units on the run cards; this information is only valid, however, if major incidents occur one at a time and when all units are available. If a department has a large volume of routine incidents, it may need to redeploy units to balance coverage, even when no major incidents are in progress.

Usually, a supervisor in the communications centre is responsible for determining when and where to redeploy units as well as for requesting coverage from surrounding jurisdictions; for example, units from many different jurisdictions may be redeployed to respond to large-scale incidents under regional or province-/territory-wide plans. These plans must include a system for tracking every unit and a designated communications centre for maintaining contact with all units.

■ Touring the Communications Centre

It is helpful to new fire fighters to tour the communications centre. Observing the actual operation of this centre will give you a much better understanding of the role of the dispatcher. **SKILL DRILL 4-2** lists the steps for touring your local communications centre (Fire Fighter I, NFPA 5.2.2):

1 Arrange for a tour.

2 Conduct yourself in a professional manner.

3 Observe the use of equipment **FIGURE 4-10**.

4 Observe the receipt of a reported emergency.

5 Differentiate the needs of fire, police, and EMS personnel.

6 Understand the dispatcher's job.

FIRE FIGHTER Tips

Vehicle-locator systems and mobile tracking devices are valuable tools to accurately and effectively keep command staff and communication dispatchers up to date on the status of all units. These systems are especially useful in large departments and at large emergency incidents.

FIGURE 4-10 Observing the actual operation of a call centre will give you a much better understanding of the role of dispatchers.

Radio Systems

Fire department communications systems depend on two-way radio systems. Radios link the communications centre and individual units; they also link units at an incident scene. A radio system is an integral component of the IMS, because it links all of the units on an incident—both up and down the chain of command and across the organizational chart. Usually, every fire service vehicle has a mobile radio, and at least one—if not every—member of a team carries a portable radio during an emergency incident. A radio may be the fire fighter's only link to the incident organization and the only means to call for help in a dangerous situation. Radios also are used to transmit dispatch information to fire stations, to page volunteer fire fighters, and to link mobile computer terminals.

Fire departments use many different types of radios and radio systems. Technological advances are rapidly adding new features and system configurations, making it impossible to describe all of the possible features, principles of operation, and systems here. Instead, this section describes common systems and operating features. As a fire fighter, you must know how to operate your assigned radio and learn your own department's radio procedures.

■ Radio Equipment

Three types of fire service radios are distinguished: the portable radio, the mobile radio, and the base station.

A <u>portable radio</u> is a hand-held two-way radio that is small enough for a fire fighter to carry at all times **FIGURE 4-11**. The radio body contains an integrated speaker and microphone,

FIRE FIGHTER Tips

It can be frustrating to the fire fighter on the street when he or she has to deal with an inadequate dispatch centre. Always remember that someone on the other end of the radio transmission is attempting to do his or her job with whatever is available to the dispatcher. Stay professional, and remember that many citizens with scanners are listening to everything you say.

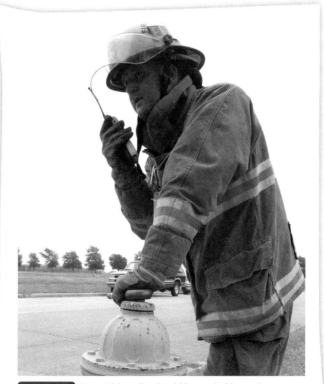

FIGURE 4-11 A portable radio should be carried by each individual fire fighter.

an on/off switch or knob, a volume control, channel select switch, and a push-to-talk (PTT) button.

A portable radio must have an antenna to receive and transmit signals. A popular optional attachment is an extension microphone/speaker unit that can be clipped to a collar or shoulder strap, while the radio remains in a pocket or pouch.

A portable radio is usually powered by a rechargeable battery, which should be checked at the beginning of each shift or prior to each use. Because the battery has a limited capacity, it must be recharged or replaced after extended operations. Battery-operated portable radios also have limited transmitting power. The signal can be heard only within a certain range and is easily blocked or overpowered by a stronger signal.

<u>Mobile radios</u> are more powerful two-way radios that are permanently mounted in vehicles and powered by the vehicle's electrical system **FIGURE 4-12**. Both mobile and portable radios share similar features, but mobile radios usually have a fixed speaker and an attached, hand-held microphone on a coiled cord. The PTT button is on the microphone, and the antenna is usually mounted on the exterior of the vehicle. Fire apparatus often include headsets with a combined intercom/radio system that enables crew members to talk to each other and hear the radio.

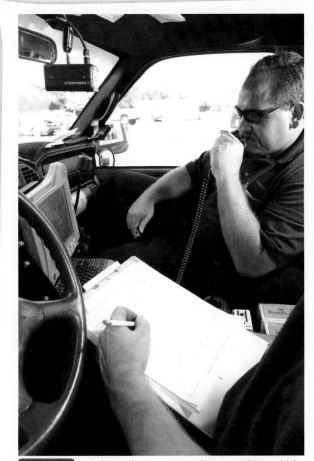
FIGURE 4-12 A mobile radio is permanently mounted in a vehicle.

<u>Base station</u> radios are permanently mounted in a building, such as a fire station, communications centre, or remote transmitter site **FIGURE 4-13**. These kinds of radios are more powerful than either portable or mobile radios. The antenna of the base station is often mounted on a radio tower so the transmissions have a wide coverage area **FIGURE 4-14**. Public safety radio systems often use multiple base stations at different locations to cover large geographic areas. The communications centre operates these stations by remote control.

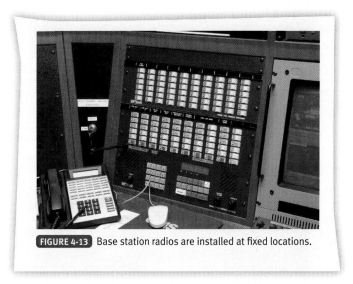
FIGURE 4-13 Base station radios are installed at fixed locations.

Mobile data terminals are computer devices that transmit data by radio. Some departments use them to track unit status, transmit dispatch messages, and exchange different types of information. Some terminals simply include buttons to send status messages such as "responding," "on the scene," and "back in service"; others comprise advanced laptop computers **FIGURE 4-15**.

■ Radio Operation

Radios work by broadcasting electronic signals on certain frequencies, which are designated in units of megahertz (MHz). A frequency is an assigned space on the radio spectrum; only those radios tuned to that specific frequency can hear the message. Every system must be licensed and operated within the guidelines as established by the Canadian Radio-Television and Telecommunications Commission (CRTC).

FIGURE 4-14 A base station radio antenna is often mounted on a radio tower to provide maximum coverage.

FIGURE 4-15 Mobile computers exchange data over a radio channel.

A radio channel uses either one frequency (**simplex channel**) or two frequencies (**duplex channel**). With a simplex channel, each radio transmits signals and receives signals on the same frequency, so a message goes directly from one radio to every other radio that is set to receive that frequency. With a duplex channel, each radio transmits signals on one frequency and receives on the other frequency. Duplex channels are used with repeater systems, which are described later in this chapter.

Canadian fire service frequencies are located in several different ranges. Additional groups of frequencies are allocated in specific geographic areas with a high demand for public safety agency radio channels. Each band has certain advantages and disadvantages relating to geographic coverage, topography (hills and valleys), and penetration into structures. Some bands have a large number of different users, which can cause interference problems, particularly in densely populated metropolitan areas.

Generally, one radio can be programmed to operate on several frequencies in a particular band. This limitation may become a problem if neighbouring fire departments or police, fire, and EMS systems within the same jurisdiction operate on different bands. When different agencies working at the same incident communicate via different bands, they must make complicated arrangements with cross-band repeaters and multiple radios in command vehicles so that they can communicate with one another.

Repeater Systems

Radio messages can be broadcast over only a limited distance, for two reasons: (1) the signal weakens as it travels farther from the source; and (2) buildings, tunnels, and topography create interference. These problems are most significant for hand-held portable radios, which have limited transmitting power. Mobile radios that operate in systems covering large geographic areas face similar problems. To compensate for these shortcomings, fire departments may use **radio repeater systems** **FIGURE 4-16**.

In a repeater system, each radio channel uses two separate frequencies—one to transmit and the other to receive. When a radio transmits over the first frequency, the signal is received by a repeater unit that automatically rebroadcasts it on the second frequency over a more powerful radio. All radios set

Dispatch centre   Field units

A.

Dispatch centre   Remote repeater site   Field units

B.

FIGURE 4-16 Direct and repeater channels send and receive transmissions in different ways. **A.** Direct channel. **B.** Repeater channel.

on the designated channel receive the boosted signal on the second frequency. This approach enables the transmission to reach a wider coverage area.

Many public safety radio systems have multiple receivers (voter receivers), which are geographically distributed over the service area to capture weak signals. The individual receivers forward their signals to a device that selects the strongest signal and rebroadcasts it over the system's base station radio(s) and transmission tower(s). With this configuration, messages that originate from a mobile or hand-held portable radio have as strong a signal as a base station radio at the communications centre. As long as the original signal is strong enough to reach one of the voter receivers, the system is effective. If the signal does not reach a repeater, however, no one will hear it.

Some fire departments switch from a duplex channel to a simplex channel for on-scene communications. This configuration is sometimes called a **talk-around channel**, because it bypasses the repeater system. In this case, the radios transmit and receive on the same frequency, and the signal is not repeated. A talk-around channel often works well for short-distance communications, such as from the command post to crews inside a house fire or from one unit to another inside a building. Conversations on a talk-around channel cannot be monitored by the communications centre; instead, the IC must use a more powerful radio on a repeater channel to maintain contact with the communications centre.

Mobile repeater systems also can boost signals at the incident scene by creating a localized, on-site repeater system. The mobile repeater can be permanently mounted in a vehicle or set up at the command post. It captures the weak signal from a portable radio for rebroadcast. Some fire departments use cross-band repeaters, which boost the power of a weak signal and transmit it over a different radio band. In this way, a fire fighter using a UHF portable radio inside a building can communicate with the IC who is outside on a VHF radio. Similar systems may be used in large buildings or underground structures, so crews working inside can communicate with units on the outside or with the communications centre.

Trunking Systems

Radio systems based on newer technology and digital communications are being introduced to take advantage of the additional frequencies being allocated by Industry Canada for public safety use. Unfortunately, these new systems are generally incompatible with older radios and require expensive infrastructure changes.

The new radio technologies use a **trunking system**, a group of shared frequencies controlled by a computer. The computer allocates the frequencies for each transmission. The radio operator sets the radio on a talk group and communicates with the computer on a control frequency. When a user presses the transmit button, the computer assigns a frequency for that message and directs all of the radios in the talk group to receive the message on that frequency. In this way, a trunking system makes more efficient use of available frequencies and can provide coverage over extensive networks. Using digital

technology is also more efficient than using analog technology, and enables more users to communicate at the same time in a limited portion of the radio spectrum. Trunking systems also allow different radios to be tied together for a given incident. With a trunking system, it is possible to link different agencies on the same system; for instance, public works, the highway department, fire department, and the police department could all be linked for a specific incident if they needed to share radio communications.

■ Using a Radio

As a fire fighter, you must learn how to operate any radio assigned to you and how to work with the particular radio systems used by your fire department. You must know when to use "Channel 3" or "Tac 5" or "Charlie 2," which buttons to press, and which knobs to turn. Study the training materials and SOGs/SOPs provided by your department to learn this information.

When a radio is assigned to you, make sure you know how to operate and maintain it. Check the battery to be sure it is fully charged. Learn the protocol for changing and recharging portable radio batteries. On the fire ground, your radio is your link to the outside world. It must work properly—because your life depends on it.

To use a radio, follow the steps in **SKILL DRILL 4-3** (Fire Fighter I, NFPA 5.2.3, Fire Fighter II, NFPA 6.2.3):

1. Listen to determine that the channel is clear of any other traffic. Depress the PTT button, and wait at least 2 seconds before speaking. This delay enables the system to capture the channel without cutting off the first part of the message. Some systems sound a distinctive tone when the channel is ready.

2. Speak across the microphone at a 45-degree angle and hold the microphone 2.5–5 centimetres (1–2 inches) from your mouth. Never speak on the radio if you have anything in your mouth **FIGURE 4-17**.

3. Know what you are going to say before you start talking. Speak clearly and keep the message brief and to the point.

4. Release the PTT button only after you have finished speaking.

If you hold a portable radio perpendicular to the ground with the antenna pointing toward the sky, you will get better transmission and reception. Range and transmission quality also will improve if you remove the radio from the radio pocket or belt clip before you use it.

You should be familiar with your specific fire service SOGs/SOPs governing the use of radios. All radio transmissions should be pertinent to the situation, with unnecessary radio traffic being avoided. Remember that radio communications are automatically recorded. They also may be heard by everyone with a radio scanner, including the news media and the general public. Given this fact, you should avoid saying anything of a sensitive nature or anything you might later regret. Radio tapes provide a complete record of everything that is said and can be admissible as evidence in a legal case.

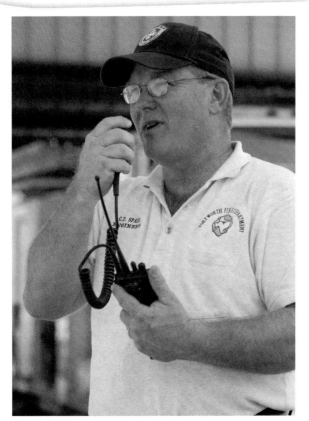

FIGURE 4-17 Speak across the microphone at a 45-degree angle and hold the microphone 2.5–5 cm (1–2 in.) from your mouth.

Most fire departments use plain English for radio communications, but some use codes for standard messages. A few services have developed intricate systems of radio codes to fit any situation. **Ten-codes**, refers to a system of coded messages that begin with the number 10, were once widely used. They can be problematic, however, when units from different jurisdictions have to communicate with one another. To be effective, codes must be understood by all parties and mean the same thing to all users. The United States' National Incident Management System (NIMS) and National Fire Protection Association (NFPA) standards recommend using plain English rather than codes. If your department uses radio codes, you must know all of the codes and understand how they are used. Even plain English may include some generally accepted terminology within a system, such as "code blue" for a cardiac arrest. The objective is to be clearly and easily understood, without having to explain every message in detail. If your department has a few codes or special words for special situations, they should be specified in the SOGs/SOPs.

FIRE FIGHTER Tips

Ten-codes are not approved by NIMS.

Arrival and Progress Reports

Fire fighters and dispatchers must know how to transmit clear, accurate size-up and progress reports. The first-arriving unit at an incident—whether a fire fighter, a company officer, or a chief—should always give a brief initial radio report and establish command as described in the chapter "Incident Management System." This initial report should convey a preliminary assessment of the situation and give other units a sense of what is happening, so that they can anticipate what their assignments may be when they arrive at the scene. The communications centre should repeat the initial size-up information.

FIRE FIGHTER Tips

Many departments have adopted a regional numbering system for apparatus to avoid confusion at major incidents. This numbering often involves assigning a three-digit number for each piece of apparatus. The first number identifies the fire department to which that apparatus belongs. The second and third numbers identify the number of that unit in the department. For example, all apparatus belonging to Everytown Fire Department might start with the number 4. Everytown Engine 3, then, is identified as Engine 403; Everytown Engine 33 is identified by the number 433. When units are dispatched by a regional dispatch centre, this numbering system avoids the need to identify the name of each unit. When fire apparatus from several different departments are dispatched to the same fire, it also avoids the confusion that occurs when two "Engine 3" apparatus are on the same scene. With a three-digit numbering system, Everytown Engine 3 is designated as Engine 433, whereas Anothertown Engine 3 is designated as Engine 633.

For example, on arrival at the scene, a company might make the following report:

Communications, this is Pump 12. Engine 410 is on location at the intersection of High Street and Chambersburg Road. We have a two-storey, wood-frame dwelling with fire showing from two windows on the second floor, side A. Pump 12 is assuming High Street command. This will be an offensive operation.

The communications centre would respond by announcing:

Pump 12, I have you on the scene assuming High Street command. Fire showing from the second floor, side A, of a two-storey dwelling. Offensive operation.

This initial report should be followed by a more detailed report after the IC has had time to gather more information and make initial assignments. An example of such a report follows:

IC: High Street command to communications.

Communications centre: Go ahead, High Street command.

IC: At 176 High Street, we are using all companies on the first alarm for a fire on the second floor, side A, of a two-storey, wood-frame dwelling, approximately 9 by 12 metres (30 by 40 feet). Exposure side 1 is the street, Exposure side 2 is a similar dwelling, Exposure side 3 is an alley, and Exposure side 4 is a vacant lot. Companies are engaged in offensive operations, and the fire appears to be confined. We have an all clear at this time. Car 10 is in command.

Communications centre: Copied High Street command. You have an all-clear for a fire in a dwelling at 176 High Street. The fire appears to be confined to the second floor. Car 10 is IC.

FIRE FIGHTER Tips

Verifying the information received over the radio is vital. Always restate an important message or instruction to confirm that it has been received and understood.

IC: Command to Ladder 207. We need you to open the roof for ventilation, directly over the fire area.

Ladder 207: Ladder 207 copies. We are going to the roof to ventilate directly over the fire.

For the duration of an incident, the IC should provide regular updates to the communications centre. Information about certain events, such as command transfer, "all clear," and "fire under control," should be provided when they occur. Reports also should be made when the situation changes significantly and at least every 10–20 minutes during a working incident.

In some fire departments, the communications centre prompts the IC to report at set intervals, known as **time marks**. Time marking allows the IC to assess the progress of the incident and decide whether the strategy or tactics should be changed. All progress reports are noted in the activity log to document the incident.

If additional resources, such as apparatus, equipment, or personnel, are needed, the IC transmits these requests to the communications centre. The communications centre will advise the IC of additional resources that are dispatched.

Emergency Messages

Sometimes the dispatcher must interrupt normal radio transmissions for **emergency traffic**. Emergency traffic is an urgent message that takes priority over all other communications. When a unit needs to transmit emergency traffic, the dispatcher generates a distinctive alert tone to notify everyone on the frequency to stand by, so the channel is available for the emergency communication. Once the emergency message is complete, the dispatcher notifies all units to resume normal radio traffic.

The most important emergency traffic is a fire fighter's call for help. Most departments use **mayday** to indicate that a fire fighter is lost, is missing, or requires immediate assistance.

If a mayday call is heard on the radio, all other radio traffic should stop immediately. The fire fighter making the mayday call should describe the situation, location, and help needed. Fire fighters should study and practise the procedure for responding to a mayday call. An example of a mayday call follows:

Fire fighter: MAYDAY ... MAYDAY ... MAYDAY.

[All radio traffic stops.]

IC: Unit calling MAYDAY, go ahead.

Fire fighter: This is Pump 403. We are on the second floor and running out of air. Fire has cut off our escape route. We request a ladder to the window on side 3 of the building so we can evacuate.

IC: Command copied. Pump 403, your escape route cut off by fire. I am sending the rapid intervention crew [RIC] to side 3 with a ladder.

The procedure for responding to a mayday call—which is used when fire fighters are in imminent danger—should be studied and practised frequently. The steps for calling a mayday are explained more fully in the chapter "Fire Fighter Survival."

Many fire department portable radios have an emergency button that can be pressed to transmit an emergency signal to the dispatcher. This transmission identifies which radio is sending the emergency signal. If your radios are equipped with this feature, learn the procedure for using it.

Another emergency traffic message is "evacuate the building," which warns all units inside a structure to abandon the building immediately. Most fire departments have specified a standard **evacuation signal** to warn all personnel to pull back to a safe location. The evacuation signal that is commonly used is a sequence of three blasts on an apparatus air horn, repeated several times, or sirens sounded on high–low for 15 seconds. An evacuation warning should be announced at least three times to ensure that everyone hears it; the warning should also be announced on the radio by the IC. Because no universal evacuation signal has been established, fire fighters must learn their department's SOGs/SOP for emergency evacuation.

After the evacuation, the radio airwaves should remain clear so the IC can do a roll call of all units. This step ensures that all personnel have safely exited the building. After the roll call, the IC will allow the resumption of normal radio traffic.

Records and Reporting

The final communications step in an emergency situation is the incident report. Every time a fire department unit responds to an incident, someone must complete the proper reports. The incident report includes where and when the incident occurred, who was involved, and what happened. Incident reports for fires should include details about the origin of the fire, the extent of damage, and any injuries or fatalities. Such reports can be completed and submitted on paper, although many fire departments now enter them on computers and store the information in a computerized database. Several different incident-reporting software packages are available, and many large fire departments use custom-designed software.

In Canada, it is the responsibility of each individual province or territory to compile statistics on fire responses and emergencies. The majority of this information is provided through the municipalities' record management system. However, the insurance industry also compiles useful data for information and financial purposes. In Canada, insurance rates are based on a complex formula that considers fire statistical data, water supplies, and the capacity of the fire service to provide coverage.

■ Obtaining the Necessary Information

After each incident, the officer in charge will need to complete an incident report. The property owner and/or occupant is a primary source of information for the report. Any bystanders or eyewitnesses should also be questioned about what they observed. The officer preparing the report should share information with any authorities—such as a fire marshal, law-enforcement personnel, or insurance investigator—who also are looking into the cause of the fire. The model number and serial number of any equipment damaged by or involved in the fire should be recorded as well.

■ Required Coding Procedures

Codes are predominantly used in incident reports to indicate incident type, actions taken, and property use, as well as other information deemed important by the Authority Having Jurisdiction (AHJ). For the most part, in Canada, Fire Services reports incidents formally in accordance with the guidelines and requirements of the provincial/territorial Office of the Fire Commissioner/Marshal. Training with respect to reporting procedures is usually provided by the AHJ either in person or online. If your department has its own incident reports and coding procedures, you should follow department SOGs/SOPs.

■ Consequences of Incomplete and Inaccurate Reports

From a legal standpoint, records and reports are vital parts of the emergency. Information must be complete, clear, and concise because these records can become admissible evidence in a court case. Improper or inadequate documentation can have long-term negative consequences. Fire reports are considered to be public records under the Freedom of Information Act, so they may be viewed by an attorney, an insurance company, the news media, or the public. If a fatality or loss occurs, incomplete or inaccurate reports may be used to prove that the fire department was negligent. The department, the fire chief, and others may be held accountable.

Taking Calls: Emergency, Nonemergency, and Personal Calls

One of the first things you should learn when assigned to a fire station is how to use the telephone and intercom systems. You must be able to use them to answer a call or to announce a response. Keep your personal calls to a minimum, so incoming phone lines are open to receive emergency calls.

A fire fighter who answers the telephone in a fire station, fire service facility, or communications centre is a representative of the fire service. Use your department's standard greeting when you answer the phone: "Good afternoon, Pleasant Town Fire Department. Captain Smith speaking. How may I help you?" Be prompt, polite, professional, and concise.

Detailed SOGs/SOPs for obtaining information and processing calls should be provided to any fire fighter assigned to answer incoming emergency telephone lines. An emergency call can come in on any fire department telephone line. As mentioned earlier, someone may call your fire station directly instead of dialing 911 or another published emergency number. If this happens, you are responsible for ensuring that the caller receives the appropriate emergency assistance. Your department's SOGs/SOPs should outline exactly which steps you should take in this situation, such as whether you should take the information yourself or connect the caller directly to the communications centre, if possible.

SKILL DRILL 4-4 lists the steps in obtaining essential information from a caller for an emergency response (Fire Fighter I, NFPA 5.2.1):

1 When you answer a call on any incoming line, follow your department's SOGs/SOPs. Answer promptly and professionally. Identify yourself and the department or location; for example, you might say, "Good afternoon. Fire Station 403, Fire Fighter Jones speaking."

2 Determine immediately whether the caller has an emergency **FIGURE 4-18**.

3 If the call involves an emergency, follow your department's SOGs/SOPs. You may have to transfer the call to the communications centre or take the information yourself and relay it to the communications centre.

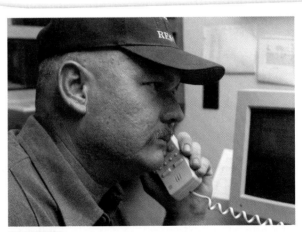

FIGURE 4-18 Determine whether the caller has an emergency.

4 Take the information from the caller, and be sure to get as much information as possible, including:

    a. Type of incident/situation
    b. Location of the incident
    c. Cross streets or identifying landmarks
    d. Indication of scene safety
    e. Caller's name
    f. Caller's location, if different from the incident location
    g. Caller's callback number

5 If your station or your unit will be responding to the call, always advise the communications centre immediately.

6 Be prepared to take accurate information or messages on all calls. Include the date, time, name of the caller, caller's phone number, message, and call taker's name. Have questions already formed and organized to maintain control of the conversation. Never leave someone on hold for any longer than necessary. Always terminate the call in a courteous manner. Let the caller hang up first.

FIRE FIGHTER Tips

Today most people carry cell phones, most of which have the ability to take pictures. Taking a picture of a person in an emergency situation while you are a member of a fire department can violate the privacy rights of that person and subject you and your department to possible legal action. Become familiar with the policy of your department regarding the use of private cell phones for taking pictures of people in emergency situations. Do not ever consider posting any such pictures on the Internet.

Wrap-Up

Chief Concepts

- Every fire service depends on a functional communications system.
- The communications centre is the central processing point for all information relating to an emergency incident and all information relating to the location, status, and activities of fire department units. Depending on the size of the fire department, the communications centre may be a small room or a large building. The communications centre may serve one fire department or several public service agencies.
- Dispatchers receive information from citizens and process that information to correctly dispatch resources to an emergency incident.
- The following equipment is found in communications centres:
  - Dedicated 911 telephones
  - Public telephones
  - Direct-line phones to other agencies
  - Equipment to receive alarms from public or private fire alarm systems
  - Computers and/or hard-copy files and maps to locate addresses and select units to dispatch
  - Equipment for alerting and dispatching units to emergency calls
  - Two-way radio system(s)
  - Recording devices to record phone calls and radio traffic
  - Backup electrical generators
- CAD enables dispatchers to work more effectively. It tracks the status of units and assists the dispatcher in quickly dispatching units to an emergency incident. Some CAD systems transmit dispatch information directly to mobile data terminals.
- Most communications centres automatically record everything that is said over the telephone or radio. This feature can prove valuable if the caller talks very quickly, has a strong accent, hangs up, or is disconnected. The recordings also serve as legal records for the fire department. They can assist in reviewing and analyzing information about department operations as well.
- A communications centre performs the following basic functions:
  - Receiving calls for emergency incidents and dispatching fire department units
  - Supporting the operations of fire department units delivering emergency services
  - Coordinating fire department operations with other agencies
  - Keeping track of the status of each fire department unit at all times
  - Monitoring the level of coverage and managing the deployment of available units
  - Notifying designated individuals and agencies of particular events and situations
  - Maintaining records of all emergency-related activities
  - Maintaining information required for dispatch purposes
- There are five major steps in processing an emergency incident:
  1. Call receipt—The process of receiving an initial call and gathering information.
  2. Location validation—Ensuring that the address is valid.
  3. Classification and prioritization—Assigning a response category based on the nature of the problem.
  4. Unit selection—The process of determining exactly which unit or units to dispatch based on the location and classification of the incident.
  5. Dispatch—The alerting of the selected units to respond and transmit information to them.
- Calls may be received via telephone, municipal fire alarm systems, private and automatic fire alarm systems, and walk-ins.
- Enhanced 911 systems may be able to display information about where a call originated and even the name and address of the caller.
- Fire service communications systems depend on two-way radio systems. A radio system is an integral component of the IMS because it links all of the units on an incident—both up and down the chain of command and across the organizational chart.
- Three types of fire service radios may be used:
  1. Portable radio—A hand-held two-way radio that the fire fighter carries at all times. The battery of such a radio should be checked at the beginning of each shift.
  2. Mobile radio—Two-way radios permanently mounted in vehicles and powered by the vehicle's electrical system.
  3. Base station—Radios permanently mounted in a building, such as a fire station, communications centre, or remote transmitter site. Public safety radio systems often use multiple base stations at different locations to cover large areas.
- Radios work by broadcasting electronic signals on certain frequencies. These frequencies are often programmed into the radio and can be adjusted only by a qualified technician.
- A radio channel uses either one frequency (simplex channel) or two frequencies (duplex channel). With a simplex channel, each radio transmits signals and receives signals on the same frequency, so a message goes directly from one radio to every other radio set to that frequency. With a duplex channel, each radio transmits signals on one frequency and receives messages on another frequency.
- In a repeater system, each radio channel uses two separate frequencies—one to transmit and the other to receive. When a low-power radio transmits over the first frequency, the signal is received by a repeater unit that

automatically rebroadcasts it on the second frequency over a more powerful radio. All radios set on the designated channel receive the boosted signal on the second frequency. This approach enables the transmission to reach a wider coverage area.

- With a trunking system, a group of shared frequencies are controlled by a computer. The computer allocates the frequency for each transmission. The radio operator sets the radio on a talk group and communicates with the computer on a control frequency. When a user presses the transmit button, the computer assigns a frequency for that message and directs all of the radios in the talk group to receive the message on that frequency.
- The first-arriving unit at an incident should always give a brief initial radio report. This report should convey a preliminary assessment of the situation and give other units a sense of what is happening so that they can anticipate what their assignments may be when they arrive at the scene.
- Emergency traffic is an urgent message that takes priority over all other communications.
- When a unit needs to transmit emergency traffic, the dispatcher generates a distinctive alert tone to notify everyone on the frequency to stand by, so the channel is available for the emergency communication. Once the emergency message is complete, the dispatcher notifies all units to resume normal radio traffic.
- The most important emergency traffic is a fire fighter's call for help. Most departments use "mayday" to indicate that a fire fighter is lost, is missing, or requires immediate assistance. If a mayday call is heard on the radio, all other radio traffic should stop immediately. The fire fighter making the mayday call should describe the situation, location, and help needed.
- An incident report describes where and when the incident occurred, who was involved, and what happened. Incident reports for fires should include details about the origin of the fire, the extent of damage, and any injuries or fatalities. Such reports can be completed and submitted on paper, although many fire departments enter them on computers and store the information in a computerized database.
- From a legal standpoint, records and reports are vital parts of the emergency. Information must be complete, clear, and concise, because these records can become admissible evidence in a court case.
- A fire fighter who answers the telephone in a fire station, fire department facility, or communications centre is a representative of the fire department. Use your department's standard greeting when you answer the phone. Be prompt, polite, professional, and concise.

Hot Terms

Activity logging system A device that keeps a detailed record of every incident and activity that occurs.

Automatic location identification (ALI) An enhanced 911 service feature that displays where the call originated or where the phone service is billed.

Automatic number identification (ANI) An enhanced 911 service feature that shows the calling party's phone number on the display screen at the dispatcher's terminal.

Base station A stationary radio transceiver with an integral AC power supply. (NFPA 1221)

Call box A system of telephones connected by phone lines, radio equipment, or cellular technology to a communications centre or fire department.

Communications centre A building or portion of a building that is specifically configured for the primary purpose of providing emergency communications services or PSAP services to one or more public safety agencies under the authority or authorities having jurisdiction. (NFPA 1221)

Computer-aided dispatch (CAD) A combination of hardware and software that provides data entry, makes resource recommendations, notifies and tracks those resources before, during, and after fire service alarms, preserving records of those alarms and status changes for later analysis. (NFPA 1221)

Direct line A telephone that connects two predetermined points.

Dispatch To send out emergency response resources promptly to an address or incident location for a specific purpose. (NFPA 450)

Dispatcher A trained individual who is responsible for answering requests for emergency and nonemergency assistance from citizens. This individual assesses the need for a response and alerts responders to the incident.

Duplex channel A radio system that uses two frequencies per channel: one frequency transmits and the other receives messages. Such a system uses a repeater site to transmit messages over a greater distance than is possible with a simplex system.

Emergency traffic An urgent message, such as a call for help or evacuation, transmitted over a radio that takes precedence over all normal radio traffic.

Evacuation signal A distinctive signal intended to be recognized by the occupants as requiring evacuation of the building. (NFPA 72)

Mayday A code indicating that a fire fighter is lost, missing, or trapped, and requires immediate assistance.

Mobile data terminal (MDT) Technology that allows fire fighters to receive information while in the fire apparatus or at the station.

Mobile radio A two-way radio that is permanently mounted in a fire apparatus.

Portable radio A battery-operated, hand-held transceiver. (NFPA 1221)

Public safety answering point (PSAP) A facility in which 911 calls are answered. (NFPA 1221)

Radio repeater system A radio system that automatically retransmits a radio signal on a different frequency.

Run cards Cards used to determine a predetermined response to an emergency.

Simplex channel A radio system that uses one frequency to transmit and receive all messages.

Talk-around channel A simplex channel used for on-site communications.

TDD/TTY/text phone system User devices that allow speech- and/or hearing-impaired citizens to communicate over a telephone system. TDD stands for telecommunications device for the deaf; TTY stands for teletype; text phones visually display text. The displayed text is the equivalent of a verbal conversation between two hearing persons.

Telephone interrogation The phase in a 911 call during which the dispatcher asks questions to obtain vital information such as the location of the emergency.

Ten-codes A system of predetermined coded messages, such as "What is your 10-20?", used by responders over the radio.

Time marks Status updates provided to the communications centre every 10–20 minutes. Such an update should include the type of operation, the progress of the incident, the anticipated actions, and the need for additional resources.

Trunking system A radio system that uses a shared bank of frequencies to make the most efficient use of radio resources.

Voice recording system Recording devices or computer equipment connected to telephone lines and radio equipment in a communications centre to record telephone calls and radio traffic.

FIRE FIGHTER *in action*

Today is the day you study dispatching in your recruit class. You see that 4 hours is allocated for this topic. You wonder how it can take 4 hours to show how a dispatcher answers a phone, asks some questions, and then pushes a button to dispatch the fire units.

The director of the emergency communications centre arrives and tells the class that you will be going over to the communications centre. You are asked to sit next to a man with a headset on and five computer screens in front of him. He hands you a headset so you can listen in. Within minutes, you are amazed at how hard this job really is.

1. _____ is an enhanced 911 service feature that displays where the call originated or where the phone service is billed.
   A. ALI
   B. ANI
   C. Call box
   D. TDD/TTY/text phone system

2. A _____ is a radio system that uses a shared bank of frequencies to make the most efficient use of radio resources.
   A. trunked system
   B. simplex system
   C. radio repeater system
   D. mobile data system

3. A _____ is a radio system that uses one frequency to transmit and receive all messages.
   A. trunked system
   B. simplex system
   C. radio repeater system
   D. mobile data system

4. A(n) _____ is a code indicating that a fire fighter is lost, missing, or trapped, and requires immediate assistance.
   A. emergency evacuation signal
   B. ten-code
   C. box call
   D. mayday

5. Two-way radios that are permanently mounted in vehicles and powered by the vehicle's electrical system are called _____.

A. portable radios

B. mobile radios

C. base radios

D. simplex radios

6. Which step comes first in the processing of an emergency incident in a communications centre?

A. Classification and prioritization

B. Unit selection

C. Location validation

D. Dispatch

FIRE FIGHTER II
in action

Your engine is dispatched to the report of a fire alarm in a building. It is the third call for the same building during this shift, and the second one since you went to bed. It's difficult to get too excited when you know the building is having trouble with its system. When you reach the site, you get off the engine with your gear, the captain checks the alarm panel, and you catch the elevator to the sixth floor.

As you step off the elevator, you smell smoke and see a haze hanging down approximately 0.6 m (2 ft.) from the ceiling. Clearly, this call is not an alarm malfunction. Your captain tells you to go to the stairwell and connect to the standpipe, but to not charge it, while she checks for the source. She says she thinks the source is food on a stove, so the captain says don't have dispatch wake anyone else up. You get connected to the standpipe, but your captain isn't back. You open the door and find the hallway filled with smoke. You try to call your captain on the radio but do not receive a response. You get a queasy feeling in the pit of your stomach as you realize this isn't good.

1. Would this be a situation where you would initiate emergency traffic?

2. How would you activate a mayday?

3. Which actions would this message precipitate at the communications centre?

4. Which information should you relay to the communications centre?

Incident Management System

INCIDENT
COMMAND

Fire Fighter I

Knowledge Objectives

There are no knowledge objectives for Fire Fighter I candidates. NFPA 1001 contains no Fire Fighter I Job Performance Requirements for this chapter.

Skills Objectives

There are no skills objectives for Fire Fighter I candidates. NFPA 1001 contains no Fire Fighter I Job Performance Requirements for this chapter.

Fire Fighter II · FFII

Knowledge Objectives

After studying this chapter, you will be able to:

- Describe the characteristics of the incident management system (IMS). (NFPA 6.1.1 , p 116–119)
- Explain the organization of the incident management system. (NFPA 6.1.1 , p 119–123)
- Function within an assigned role within the incident management system. (NFPA 6.1.1 , p 119–125)
- Organize and coordinate an incident management system until command is transferred. (NFPA 6.1.1 , p 125–128)
- Transfer command within an incident management system. (NFPA 6.1.1 , p 129–130)

Skills Objectives

After studying this chapter, you will be able to perform the following skills:

- Operate within the incident management system. (NFPA 6.1.2 , p 127)
- Assume command. (NFPA 6.1.2 , p 128)
- Transfer command. (NFPA 6.1.2 , p 129–130)

Note: NFPA 1561 Chapter 3 Definitions recognize IMS and ICS as being the same for the purposes of this Chapter the term Incident Management System (IMS) is understood to have the same meaning as Incident Command System (ICS).

© Photos.com

Additional NFPA Standards

- NFPA 1026, *Standard for Incident Management Personnel Professional Qualifications*
- NFPA 1500, *Standard on Fire Department Occupational Safety and Health Program*
- NFPA 1521, *Standard for Fire Department Safety Officer*
- NFPA 1561, *Standard on Emergency Services Incident Management System*

You Are the Fire Fighter

Just before lunch, you hear the unmistakable sound of the alert tones in the station. Your pump company is dispatched to Highway 401 close to Highway 427 for a motor vehicle collision. The dispatcher informs you that one vehicle is on its side, blocking the eastbound lanes of the 401, while a second vehicle has veered off the westbound lanes and hit a utility pole. In her second transmission, the dispatcher informs you that the Ontario Provincial Police (OPP) are responding as are Emergency Medical Services (EMS). Your backup for this call is another pump, a heavy rescue (squad), and a district chief.

Your pulse quickens as you realize you are the first to arrive on the scene. You know you must assume command and transmit a size-up of the situation. Additionally, you must rapidly develop an incident action plan (IAP). You remain calm as you systematically apply what you have learned with regards to the Incident Management System (IMS). You assume, confirm, and announce command and start gathering information in order to begin the process of creating an IAP.

1. Why is there a need for an incident management system?
2. Who is in charge of a situation like this one, where several different agencies are operating at the same scene?
3. What will your role be?

Introduction

All fire department emergency operations and training exercises should be conducted within the framework of an **Incident Management System (IMS)**. Using an IMS ensures that operations are coordinated within a recognized management system and that operations are conducted with safety as an overriding principle. This type of system provides a standard approach, structure, and operational procedure to organize and manage any operation—from an emergency to a training session. The same principles apply whether the situation involves a single pump company or several responders from more than one agency. It is important to realize that these principles can be applied to emergency and nonemergency events, as the system is flexible and designed to expand and contract according to changing circumstances.

Application of risk-management principles, planning, supervision, and communication are key components of any IMS model. Throughout various North American jurisdictions, there are several different IMS models in use; however, most are based on NFPA 1561, *Standard on Emergency Services Incident Management System*, and the **National Incident Management System (NIMS)**. The incident management structure discussed in this chapter is based on NIMS and National Fire Protection Association (NFPA) standards in the context of Canadian fire services practices.

Within a single fire department or group of fire departments and other public safety agencies (e.g., police or Emergency Medical Service [EMS]) who routinely respond together, the use of a recognized standard system for managing an incident is essential. Enhanced coordination results when every organization that could potentially be involved in a major incident is familiar with and has practised with the same IMS.

As some Canadian fire services may use procedures that vary from the IMS model presented here, it is essential for you to become intimately familiar with the system that is used within your jurisdiction.

The NIMS system was developed as an outgrowth of an American presidential directive issued on February 28, 2003. It provides a consistent national template to enable United States federal, state, and local governments, as well as private sector and nongovernmental organizations, to work together effectively and efficiently to prepare for, prevent, respond to, and recover from domestic incidents, regardless of those incidents' cause, size, or complexity. Such incidents include acts of terrorism and natural disasters. Along with the basic concepts of flexibility and standardization, NIMS principles are now taught in many incident management courses.

In the Canadian context, American IMS principles and concepts have had a profound impact on the development of the IMSs within many Canadian fire services. However, fire fighters should not view any IMS as a static standard; rather, it should be viewed as a living process. Fire service personnel and municipal, provincial/territorial, and federal leaders should continually review their processes to ensure that they remain up to date with the latest developments.

As new methodologies for emergency response are developed, emergency management principles are altered to adapt. This evolution is mainly driven by international information exchanges, which allow governments and emergency responders to learn from one another and collaborate in an effort to improve safety.

History of the Incident Management System

Prior to the 1970s, most fire services in Canada did not have a formalized process for safely managing emergency situations. Often, the organizational structure established to direct on-scene operations at an emergency incident depended on the individual authoritative style of the chief on duty. It became apparent that large, dynamic emergency situations commonly outstripped the functional abilities of a single roving commander, who would—with portable radio in hand—attempt to micromanage tasks, tactics, and strategy. The absence of a single formal recognized process for managing incidents became a glaring weakness within many fire services.

Over the past 30 years, formal IMSs have been developed and refined. Today, IMS structures clearly define roles, responsibilities, and suggested operating guidelines (SOGs), or standard operating procedures (SOPs), that are used to efficiently direct emergency operations. Consequently, the same basic approaches, organizational structures, and terminology are used by thousands of fire departments and emergency response agencies around the world.

The development of a standard system began approximately 40 years ago in the United States, after several large-scale wildland fires in southern California proved disastrous for both the fire service and residents of the region. A number of fire-related agencies at the local, state, and federal levels decided that better organization was necessary to effectively combat these costly fires.

Collectively, these agencies established an organization known as __FIRESCOPE (FIre RESources of California Organized for Potential Emergencies)__ to develop solutions for a variety of problems associated with large, complex emergency incidents during wildland fires. These problems related to the following issues:

- Command and control procedures
- Resource management
- Terminology
- Communications

FIRESCOPE developed the first standard IMS. Originally, the IMS was intended only for large, multijurisdictional or multiagency incidents involving more than 25 resources or operating units. It proved so successful, however, that it was applied to structural firefighting and eventually became an accepted system for managing all emergency incidents **FIGURE 5-1**.

At the same time, the __fire-ground command (FGC)__ system was developed and adopted by many fire departments. The concepts underlying the FGC system were similar to those that formed the foundation of the IMS, albeit with some differences in terminology and organizational structure. The FGC system was initially designed for day-to-day fire department incidents involving fewer than 25 fire-suppression companies, but it could be expanded to meet the needs of larger-scale incidents.

During the 1980s, the IMS developed by FIRESCOPE was adopted by all American federal and most state wildland firefighting agencies. The National Fire Academy (NFA) also used the FIRESCOPE IMS as the model fire service IMS for all of its courses. Additional federal agencies, including the Federal Emergency Management Agency (FEMA) and the Federal Bureau of Investigation (FBI), adopted the same model for use during major disasters or terrorist events. All federal agencies in the United States could now learn and use the same basic system.

In Canada, many fire services adopted American practices and have adjusted them to suit their respective needs. It is important to realize that differences in the federal structure of governance between Canada and the United States have impacted both the development and application of the incident management process.

In Canada, provinces/territories maintain jurisdiction over their respective regions and municipalities for emergency response and management. Constitutionally speaking, municipalities within Canada are created by and fall under the legislative jurisdiction of the province/territory in which they reside. Therefore, Canada is comprised of various emergency response models with no single, overriding IMS doctrine in place; however, the federal government offers a form of IMS training through the Canadian Emergency Management College (CEMC) in Ottawa. Following the terrorist attacks against the United States on September 11, 2001, various levels of government began focusing more on emergency response capabilities and management.

Recently, the Canadian federal government and the country's 10 provincial and 3 territorial governments collaborated to create *An Emergency Management Framework for Canada* (2nd ed., 2011). The intent of this initiative was to align federal emergency response doctrine with that of the provinces and territories. These initiatives notwithstanding, Canadian municipalities' emergency response IMS landscape remains a patchwork quilt of various similar systems and doctrines.

Jurisdictional Authority

Provincial/territorial governments are responsible for emergency management within their respective jurisdictions. The federal government exercises leadership at the national level relating to emergency management responsibilities with regard to areas, lands, and departments that fall under its jurisdictional control.

In an emergency, the first response is almost always by local or provincial/territorial authorities. Should a provincial/territorial government require resources beyond its capacity to cope with an emergency or disaster, the federal government will respond with additional resources and funding; however, it is common for the federal government to wait to be asked by the province/territory for assistance before activating emergency response teams. The need to be asked for assistance is

FIGURE 5-1 IMS was first developed to coordinate efforts during large-scale wildland fires.

© Peter Weber/ShutterStock, Inc.

a common convention, based on the Canadian Constitution, which outlines federal versus provincial/territorial legal jurisdictions.

Jurisdictional authority is usually not a problem at an incident with a single focus or threat, such as a structure fire, where the highest-ranking or designated officer from the local fire service assumes incident command; matters can become more complicated, however, when several jurisdictions are involved or when multiple agencies within a single jurisdiction have authority for various aspects of the incident. Whenever the federal government is directly involved in a disaster response, it is likely that Public Safety Canada (PS), the Royal Canadian Mounted Police (RCMP), and/or the Department of National Defence (DND) will have a presence. Major disaster response is usually directed under a unified command structure, which will be explained in greater detail later in this chapter.

Characteristics of the Incident Management System

Generally speaking, an IMS is designed to provide a defined standard, systematic approach to managing emergency incidents, training, and/or events. In other words, it is designed to break a large, complex situation down into smaller, more manageable parts. The use of an IMS enables any agency to operate more safely and more effectively at the scene of an emergency or other situation. The essence of an IMS is to provide a recognizable, standardized, and flexible process that facilitates cooperation and organization between all agencies, groups, or individuals involved in any given situation.

Effective incident management requires an organizational structure that is highly adaptable to the needs of the incident. IMS provides a hierarchy for decision making and scene management. Under an IMS, specific responsibilities and lines of authority are clearly communicated, and all relationships are clearly defined.

Important characteristics of an IMS include, but are not limited to, the following:

- Hierarchical management structure
- Applicable to all risk and hazard situations
- Unity of command
- Span of control

- Guided by clearly defined risk-management principles
- Flexible, modular organizational structure
- Common terminology
- Integrated communications
- Consolidated incident action plans (IAPs)
- Designated incident facilities
- Resource management

■ All-Risk–All-Hazard System

The IMS has evolved into an all-risk–all-hazard system that can be applied to manage resources at fires, floods, tornadoes, plane crashes, earthquakes, hazardous materials incidents, or any other type of emergency situation **FIGURE 5-2**. This kind of system has also been used to manage many nonemergency events, such as large-scale public events, which have similar requirements for command, control, and communication. The flexibility of the IMS enables the management structure to expand as needed, using whatever components are necessary. The operations of multiple agencies and organizations can be integrated smoothly in order to manage any incident.

■ Applicability

An IMS can and should be used for everyday operations as well as major incidents. Command should be established at every incident—whether it is a trash can fire, an automobile collision, or a building fire **FIGURE 5-3**. Regular use of the system ensures familiarity with standard procedures and terminology. It also increases the users' confidence in the system. Frequent use of an IMS for routine situations makes it easier to apply to larger incidents.

■ Unity of Command

Unity of command is a management concept in which each person has only one direct supervisor. All orders and assignments come directly from that supervisor, and all reports are made to the same supervisor. This approach eliminates the confusion that can occur if a person receives orders from more than one

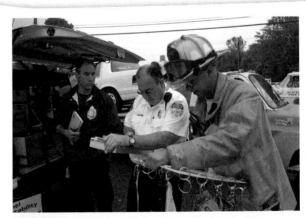

FIGURE 5-2 IMS can be used to manage different types of emergency incidents involving several agencies.

© Mark C. Ide

FIGURE 5-3 Because of effective fire prevention measures, fire fighters respond to more car accidents than fires.

© Rick McClure/AP Photos

You respond to a two-storey detached residential Type V structure. Upon arrival, you find that a section of the main floor is fully involved, and on the second floor, you observe a person gesturing for help from a window. The initial response of two pumps, a ladder company, and a district chief arrive simultaneously.

As IC you size up the situation and believe that you can safely rescue the occupant and extinguish the fire using an aggressive interior fire attack. Your strategy would be offensive, and your tactical objectives would be to rescue the trapped person and extinguish the fire.

The overall strategy for this scenario was to offensively attack the fire and perform a simultaneous rescue. In order to achieve your tactical objectives, you would assign the ladder company to perform a rescue using a ground ladder and order a pump company to advance an attack line to fight the fire.

In this scenario, the rescue itself would not be considered the main incident's strategy, as it is but one component of a much bigger picture, which includes a fire that needs to be contained and extinguished.

supervisor at the same time. Additionally, unity of command reduces confusion and increases accountability.

An IMS is not necessarily a rank-oriented system. The best-qualified person should be assigned at the appropriate level for each situation—even if that means a lower-ranking individual is temporarily assigned to a higher-level position. This can occur when a situation requires expert knowledge in an area such as high-angle rescue, hazardous materials incidents or large-rig rescue, as the rescue team leader may have more experience and knowledge than the ranking officer. This concept is critical for the effective application of the system and must be embraced by all participants.

Span of Control

Span of control refers to the maximum number of subordinates who can report to one supervisor at any level within the organization. Span of control relates to all levels of IMS—from the strategic level to the tactical/operational level to the task level (individual companies or crews).

In most situations, one person can effectively supervise only three to seven people or resources at any given time. Because of the dynamic nature of emergency incidents, an individual who has command or supervisory responsibilities ideally should not directly supervise more than five people. The actual span of control should depend on the complexity of the incident and the nature of the work being performed. For example, at a complex incident involving hazardous materials, the span of control might be only three; during less intense operations, the span of control could be as high as seven.

Risk-Management Principles

A fundamental characteristic of all IMSs is an explanation of the risk-management principles. These principles provide guidance for decision making with regard to risk-determination analysis.

The incident commander (IC) must consider the risks associated with any course of action prior to determining a strategy and assigning tactical objectives to personnel.

The following risk-management principles are taken from the NIMS and reference NFPA 1500, *Standard on Fire Department Occupational Safety and Health*:

- Activities that present a significant risk to safety of members shall be limited to situations where there is a potential to save endangered lives.
- Activities that are routinely employed to protect property shall be recognized as inherent risk to the safety of members, and actions shall be taken to reduce or avoid these risks.
- No risk to the safety of members shall be acceptable when there is no possibility to save lives.

NFPA 1561 states, "Risk-management principles shall be employed routinely by supervisory personnel at all levels of the incident management system," (5.8.1). ICs should constantly and consistently filter all of their decisions through the sieve that is the risk-management doctrine.

Modular Organization

The IMS is designed to be flexible and modular. The IMS organizational structure—command, operations, planning, logistics, and finance/administration—is predefined, ready to be staffed and made operational as needed. Indeed, an IMS has often been characterized as an organizational tool box, where only the tools needed for the specific incident at hand are used. In an IMS, these tools consist of position titles, job descriptions, and an organizational structure that defines the relationships between the positions. Some positions and functions are used frequently, whereas others are needed only for complex or unusually large situations involving a multiagency response.

Common Terminology

An IMS promotes the use of common terminology both within an organization and among all agencies involved in an emergency incident. Common terminology means that each word has a single definition, and no two words used in managing an emergency incident have the same definition. Everyone uses the same terms to communicate the same thoughts, so everyone understands what is meant. Each job comes with one set of responsibilities, and everyone knows who is responsible for each duty.

Common terminology is particularly important for radio communications. For example, in many fire services, the term *tanker* refers to a mobile water supply apparatus; to wildland firefighting agencies, however, a tanker is a fixed-wing aircraft used to drop an extinguishing agent on a fire. Without common terminology, a request for a tanker could result in the operations section chief getting an airplane instead of the truck he or she actually requested from command.

FIRE FIGHTER II Tips   **FFII**

NG 0001, *National Resource Typing Criteria*, is the NIMS document that lists the proper terminology for the common resources used at large-scale emergencies.

■ Integrated Communications

<u>Integrated communications</u> ensure that everyone at an emergency can communicate with both supervisors and subordinates. The IMS must support communication up and down the chain of command at every level. A message must be able to move efficiently through the system from command down to the lowest level and from the lowest level up to the command **FIGURE 5-4**. Within a fire service, the primary means of communicating at the incident scene is by radio. In Canada, the ability to communicate with other agencies at the same incident is often referred to as interoperability. The lack of communications interoperability between New York City's fire and police departments has been recognized as a serious problem that plagued the emergency responders at the 9/11 terrorist attacks.

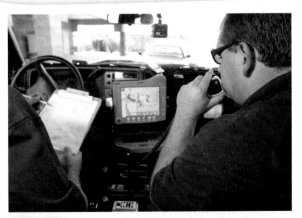

FIGURE 5-4 Integrated communications are essential for successful operations during an emergency.

■ Consolidated Incident Action Plans

<u>Incident action plans (IAPs)</u> are necessary for every incident. For basic and routine incidents, the IAP is usually formulated mentally; however as an incident grows, it is important for the IC to use a formalized means to capture information and assignments. Many Canadian fire service commanders use tactical worksheets as a guide for formulating their action plans. Fire services are increasingly adopting technology such as mobile data terminals and laptop computers as command aids. Regardless of the medium used, the use of a system to capture and organize information is an essential consideration in order for the IMS to be effective.

When an incident grows into one that requires a multiagency response, it is important that the IAP expands in relation to the number of agencies and resources involved. Different components of the organization may perform different functions, but all of their efforts contribute to meeting the same goals and objectives. All actions taken toward the safe mitigation of an incident should be interrelated.

At smaller incidents, command develops an action plan and communicates the incident priorities, objectives, strategies, and tactics to all of the operating units. Representatives from all participating agencies meet regularly to develop and update this plan. In both large and small incidents, all personnel involved in the incident understand what their specific roles are and how they fit into the overall plan.

FIRE FIGHTER Tips

Strategy involves the overall view of the incident; in other words, it considers the high-level aspects of creating an IAP. Tactical objectives are stated goals that are consistent with the strategy (e.g., ventilating the roof area to allow for a safe interior fire attack). Tactics are the steps that are taken to achieve the tactical objectives, such as coordinating the fire attack. Tasks are the specific actions that get the job done—for example, Ladder 1 placing a ladder to access the roof.

■ Designated Incident Facilities

<u>Designated incident facilities</u> are assigned locations where specific functions are always performed; for example, command will always be based at the incident command post (ICP). The staging area, rehabilitation area, casualty collection point, treatment area, base of operations, and heliport are all designated areas where particular functions take place. The facilities required for the specific incident are established according to the specific IAP or a predefined IMS plan.

At large-scale incidents, units are often dispatched to a <u>staging area</u> rather than going directly to the incident location. A staging area is a location close to the incident scene where a number of units can be held in reserve, ready to be assigned if needed

■ Resource Management

<u>Resource management</u> entails the use of a standard system of assigning and tracking the resources involved in an incident.

In structural firefighting, the basic units are companies. Engine companies, ladder companies, and other units are dispatched to the incident and assigned specific functions. In some cases, groups from several different companies are assigned to task forces and strike teams. The resource management system of the IMS keeps track of the various company assignments.

Personnel are the most vital resource the fire service has. Under the IMS, the IC uses a personnel accountability system to track each and every member at the scene. The chapter Fire Fighter Survival discusses the various personnel accountability systems in detail.

The IMS Organization

The IMS structure identifies a full range of duties, responsibilities, and functions that are performed at emergency incidents; it also defines the relationships among those components. Some components are used at almost every incident, whereas others apply to only the largest and most complex situations. The five major components of IMS organization are command, operations, planning, logistics, and finance/administration.

An IMS organization chart may be quite simple or very complex. Each block on an IMS organization chart refers to a function area or a job description.

Positions are staffed as they are needed. The only position that must be filled at every incident is IC. The IC decides which additional components are needed for the given situation and activates those positions by assigning someone to perform those tasks.

Fire fighters must understand the overall structure of the IMS as well as the basic roles and responsibilities of each position within the IMS organization. Knowing how IMS works enables fire personnel to better understand the different roles and responsibilities within the system and how they interrelate to mitigate an incident. The IMS must grow at a pace that is ahead of the incident itself.

ICs must be anticipatory decision makers and request resources before they may be needed or required. Large-scale, rapidly evolving incidents have a way of outstripping resources in a short period of time. The IC should ensure that he or she is time aware and uses time to his or her advantage in the decision-making process by staying ahead of the situation. This is where experience and training come in; a well-trained, experienced IC should be able to anticipate a fire's spread or the need for additional resources at every stage of a fire.

■ Incident Command Levels

There are commonly three levels of command within an IMS:

1. Strategic level
2. Tactical level
3. Task level

The strategic level involves the overall management of the incident. The IC operates at this level and is responsible for establishing strategy, prioritizing tactical objectives, and assigning personnel in accordance with his or her IAP. He or she is also responsible for establishing branches, divisions, groups, and sectors and is ultimately responsible for the safety of all responders throughout the life of the incident.

The tactical level involves officers who will directly supervise groups, divisions, or sectors. In basic or routine incidents, this level will likely involve the IC. At larger and more complex incidents this level may be delegated by command to branch or sector officers.

The task level is where the physical actions take place in order to achieve tactical and strategic outcomes. This level usually involves the work performed by individual crews/companies and/or strike teams.

Incidents are categorized as follows:

■ Basic
■ Routine
■ Complex

A basic response may be one where only one unit (e.g., a pump or ladder) responds and the IC is the officer in charge of the first-responding vehicle and crew—for example, at a gas leak with no fire or injuries reported. The IC will embody strategic, tactical, and task levels of supervision.

A routine incident may be one that includes an initial response to a high-rise alarm in the form of two pumps, a ladder, and a district chief. The IC will operate at the strategic level of the incident from a fixed command location. A complex incident is one that grows beyond the initial response to include other agencies and responders—for example, a major motor vehicle collision involving many casualties and a fire.

■ Command

On an IMS organization chart, the first component is **command** **FIGURE 5-5**. Command is the only position in the IMS that is continually staffed for every incident. As previously discussed, command is established when the first unit arrives on the scene and is maintained until the last unit leaves the scene.

In the IMS structure, command (be it single or unified) is ultimately responsible for managing an incident and has the necessary authority to direct all activities at the incident scene. Command is directly responsible for the following:

■ Creating and establishing an IAP
■ Determining strategy

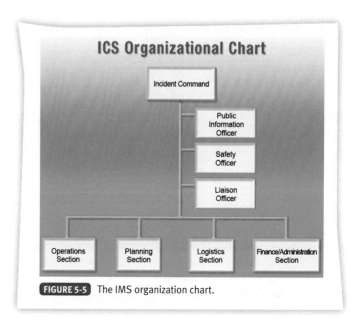

FIGURE 5-5 The IMS organization chart.

- Selecting incident tactics
- Developing the IMS organization
- Managing resources and ensuring that they are accounted for
- Coordinating resource activities
- Providing for personnel and scene safety
- Coordinating with outside agencies
- Ensuring that victim support and assistance is provided
- Releasing information about the incident

There are three primary priorities for the safe and effective mitigation of any incident. The IC is responsible for ensuring that these priorities are completed:

1. Life safety—Ensure that the safety of the public and responders remains a paramount command consideration throughout the life of the incident. This may require rescue by removing people from a place of danger to one of safety.
2. Incident—Ensure that the situation is safely and effectively stabilized.
3. Property—Protect/conserve property and the environment.

These priorities are easily remembered by the acronym LIP. With regard to life safety, the IC should always evaluate the risks associated with any activity and adjust his or her IAP accordingly. This is especially true in the case of a rapidly escalating incident.

Unified Command

Due to the size or nature of the emergency, there may be a need to establish a **unified command**. When multiple agencies with overlapping jurisdictions or legal responsibilities respond to the same incident, a unified command structure provides several advantages. In a unified command situation, representatives from each agency co-operate to share command authority. They work together and are all directly involved in the decision-making process. Unified command helps ensure co-operation, avoids confusion, and guarantees agreement on goals and objectives **FIGURE 5-6**.

FIGURE 5-6 A unified command directly involves multiple agencies in the decision-making process for a large incident.

Most cities in Canada have established trigger mechanisms that signal the need for a multiagency response. These protocols have built-in escalation notifications, which ultimately lead to the activation of an established command centre. Once activated, the command centre will operate under a unified command system, where decisions are made collectively for the safe and effective mitigation of the emergency. If the situation warrants, additional federal and provincial/territorial human and physical resources can be requested, and the unified command structure will expand as required. In order to gain a complete understanding of the command system in your jurisdiction, it is important to acquaint yourself with your city's methodology and incident management practices.

The shopping mall collapse in Elliot Lake, Ontario, in June 2012 is an excellent example of a complex incident that rapidly outstripped the local emergency-response agencies' ability to cope. The Ontario provincial government provided assistance with management coordination through Emergency Management Ontario and activated the City of Toronto Heavy Urban Search and Rescue Team (HUSAR) to coordinate the search for victims. These actions were coordinated with the Elliot Lake fire chief and other local officials. While there were concerns over delays in response, it provides an example over how different levels of government can work together under a recognized IMS. Regardless of the various command mechanisms, it is vital that transfer of command is performed in a seamless and time-sensitive manner.

Incident Command Post

The **incident command post (ICP)** is the headquarters for the incident. Command functions are centred in the ICP; thus, command and all direct support staff should always be located at this site. The location of the ICP should be broadcast to all units as soon as it is established.

Relative to the incident scene, the ICP should be in a nearby, protected location. Often, the ICP for a major incident is located in a special vehicle or building. This choice of location enables the command staff to function without needless distractions or interruptions. For large incidents that are geographically spread out, the command post may be some distance from part of the emergency incident.

■ Command Staff

Individuals on the **Command Staff** perform functions that must report directly to Command and cannot be delegated to work in other major sections of the organization **FIGURE 5-7**. The safety officer, liaison officer, and public information officer are always part of the Command Staff. In addition, aides, assistants, and advisers may be assigned to work directly for members of the Command Staff. An aide is a fire fighter (sometimes an officer) who serves as a direct assistant to a member of the Command or general staff.

Safety Officer

The **safety officer** is responsible for ensuring that safety issues are managed effectively at the incident scene. He or she serves as command's eyes and ears in regard to safety—identifying and evaluating hazardous conditions, watching out for unsafe practices, and ensuring that safety procedures are followed. The safety officer is appointed early during an incident. As the incident becomes more complex and the number of resources present at the scene increases, additional qualified personnel can be assigned as assistant safety officers.

Although the safety officer is an adviser to command, he or she has the authority to stop or suspend operations when unsafe situations occur. This authority is clearly stated in NFPA 1521, *Standard for Fire Department Safety Officer*; as well as NFPA 1026, NFPA 1500 and NFPA 1561. Several provincial/territorial and federal regulations require the assignment of a safety officer at incidents involving hazardous materials and at certain technical rescue incidents.

At a fire scene, the safety officer should be an individual who is knowledgeable in fire behaviour, building construction and collapse, firefighting strategy and tactics, hazardous materials, rescue practices, and departmental safety rules and regulations. He or she should also have considerable experience in incident response and specialized training in occupational safety and health. Many larger fire departments have full-time safety officers who perform administrative functions relating to health and safety when they are not responding to emergency incidents. The Justice Institute of British Columbia's Incident Safety Officer course is an excellent resource for people who are interested in serving in this capacity.

Liaison Officer

The **liaison officer** is command's representative to outside agencies. This member of the command staff is responsible for exchanging information with representatives from those agencies. During an active incident, command may not have time to meet directly with everyone who comes to the ICP. The liaison officer functions as the representative of command under these circumstances, obtaining and providing information or directing people to the proper location or authority. The liaison area should be adjacent to, but not inside, the ICP.

Public Information Officer

The **public information officer (PIO)** is responsible for gathering and releasing incident information to the news media and other appropriate agencies **FIGURE 5-8**. At a major incident, the public will want to know what is being done. Because command must make managing the incident its top priority, the PIO serves as the contact person for media requests, which allows command to concentrate on the

incident. A media headquarters should be established near, but not inside, the ICP. The information presented to the media by the PIO needs to be approved by the IC. Many large Canadian fire services have designated information officers who have the specialized training, knowledge, and preauthorization to liaise with the media.

■ General Staff Functions

Command has the overall responsibility for the entire incident command organization, although some elements of command's responsibilities can be handled by the command staff. When the incident is too large or too complex for just one person to manage effectively, command may appoint someone to oversee parts of the operation. Everything that occurs outside of command at an emergency incident can be divided among the other four major functional components within the IMS **FIGURE 5-9**:

- Operations
- Planning

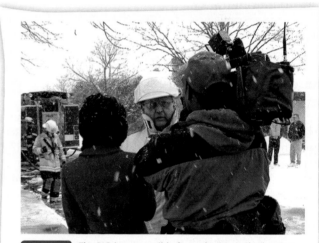

FIGURE 5-8 The PIO is responsible for gathering and releasing incident information to the media and other appropriate agencies.

Courtesy of Captain David Jackson, Saginaw Township Fire Department

FIGURE 5-7 The command staff report directly to command.

FIGURE 5-9 Major functional components of an IMS.

- Logistics
- Finance/administration

The staff of these four sections are known as the **IMS general staff**. Command decides which (if any) of these four positions need to be activated, when to activate them, and who should be placed in each position. (Recall that the blocks on the IMS organizational chart refer to functional areas or job descriptions, not to positions that must always be staffed.)

The four section chiefs on the IMS general staff, when they are assigned, may run their operations from the main ICP, although this structure is not required. At a large incident, the four functional organizations may operate from different locations but will always be in direct contact with command.

Operations

The **Operations Section** is responsible for taking direct action to control the incident **FIGURE 5-10**. The operations section fights the fire, rescues any trapped individuals, treats any injured victims, and does whatever else is necessary to remediate the emergency.

In the case of most structure fires, command directly supervises the functions of the operations section. At complex incidents, a separate **Operations Section Chief** takes on this responsibility, so that command can focus on overall strategy while the operations section chief focuses on the tactics that are required to get the job done.

Operations are conducted in accordance with an IAP, which outlines what the strategic objectives are and how emergency operations will be conducted. At most incidents, the IAP is relatively simple and can be expressed in a few words or phrases; for a large-scale incident, however, it can be a lengthy document that is regularly updated and used for daily briefings of the command staff. Many large Canadian fire services are administratively organized to readily adapt and fill these command positions with assistant deputy and deputy chief positions.

Planning

The **Planning Section** is responsible for the collection, evaluation, dissemination, and use of information relevant to the incident **FIGURE 5-11**. The planning section works with preincident plans, building construction drawings, maps, aerial photographs, diagrams, reference materials, and status boards. It is also responsible for developing and updating the IAP. The planning section outlines what needs to be done by whom and identifies which resources are needed.

Command activates the planning section when information needs to be obtained, managed, and analyzed. The **Planning Section Chief** reports directly to command. Individuals assigned to planning examine the current situation, review available information, predict the probable course of events, and prepare recommendations for strategies and tactics. The planning section also keeps track of resources at large-scale incidents and provides command with regular situation and resource status reports.

Logistics

The **Logistics Section** is responsible for providing supplies, services, facilities, and materials during the incident **FIGURE 5-12**. The **Logistics Section Chief** reports directly to command and serves as the supply officer for the incident. Among the responsibilities of this section are keeping apparatus fuelled,

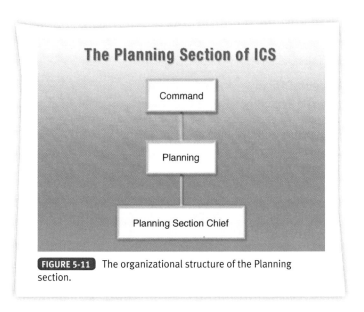

FIGURE 5-11 The organizational structure of the Planning section.

FIGURE 5-12 The organizational structure of the Logistics Section.

FIGURE 5-10 The organizational structure of the operations section.

providing food and refreshments for fire fighters, obtaining the foam concentrate needed to fight a flammable liquid fire, and arranging for specialized equipment to remove a large pile of debris.

In many fire services, these logistical functions are routinely performed by support services personnel. These groups work in the background to ensure that the members of the operations section have whatever they need to get the job done. Resource-intensive or long-duration situations may require the assignment of a logistics section chief, however, because service and support requirements may be so complex or so extensive that they need their own management component.

Finance/Administration

The **Finance/Administration Section** is the fourth major IMS component managed directly by command **FIGURE 5-13**. This section is responsible for the accounting and financial aspects of an incident, as well as any legal issues that may arise in its aftermath. This function is not staffed at most incidents, because cost and accounting issues are typically addressed after the fact. Nevertheless, a finance/administration section may be needed at large-scale and long-term incidents that require immediate fiscal management, particularly when outside resources must be procured quickly.

A Finance/Administration Section may also be established during a natural disaster or during a hazardous materials incident, where reimbursement may come from the shipper, carrier, chemical manufacturer, or insurance company.

Standard IMS Concepts and Terminology

Modern emergency response organizations utilize IMSs in an effort to be efficient and disciplined and to improve safety. Prior to the widespread use of IMSs, emergency scenes were far more disorganized and less effective in ensuring safe outcomes. Above all other emergency-response agencies, fire services tend to be experts in the implementation of IMS, as it is a function of every response. One aspect of utilizing an IMS is its use of standard terminology.

■ Single Resources and Crews

A **single resource** is an individual vehicle and its assigned personnel **FIGURE 5-14**; for example, a pump and its crew would be a single resource, and a ladder company would be a second single resource. A company officer is the individual in charge of a company. A company operates as a work unit with all crew members working under the supervision of the company officer. Companies are assigned to perform tasks such as search and rescue, attacking the fire with a hose line, forcible entry, and ventilation.

A **crew**, or company, is a group of fire fighters who are working without apparatus; for example, members of a pump or ladder company who are assigned to operate inside a building would be considered a crew. Additional personnel at the scene of an incident who are assembled to perform a specific task may also be called a crew. A crew must have an assigned leader or company officer.

■ Divisions, Groups, and Sectors

Organizational units, such as divisions and groups, are established to aggregate single resources and/or crews under one supervisor. The primary reason for establishing divisions and groups is to maintain an effective span of control. The terms *sector* and *division* are interchangeable. Sectors are supervised by sector officers, who function like group or division leaders.

- A **division** usually refers to companies and/or crews working in the same geographic area. In many Canadian fire services, this is commonly referred to as a **sector**; sectors can be geographic and or task oriented.
- A **group** usually refers to companies and/or crews who are working on the same task or function but not necessarily in the same location.

The flexibility of the IMS enables organizational units to be created as required according to the needs of the incident and depending on the size and scope of the incident. In the early stages of an incident, individual companies are often assigned to work in different areas or perform different tasks. As the incident grows and more companies are assigned to it, command can establish divisions and groups. An individual is assigned to supervise each division or group as it is created.

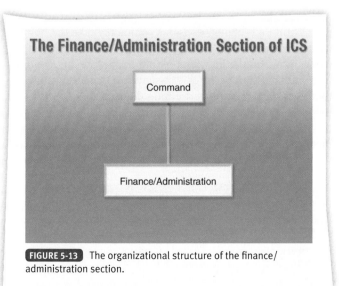

FIGURE 5-13 The organizational structure of the finance/administration section.

FIGURE 5-14 A single resource consists of an individual company and the personnel that arrive on that unit.

These organizational units are particularly useful when several resources are working near one another, such as on the same floor inside a building, or are doing similar tasks, such as ventilation. The assigned supervisor can directly observe and coordinate the actions of several crews.

A division includes those resources responsible for operations within a defined geographic area. This area could be a floor inside a building, the rear of the fire building, or a section of a wildland or brush fire. Divisions are most often employed during routine fire department emergency operations. By assigning all of the units in one area to one supervisor, command is better able to coordinate their activities and ensure that all resources are working to further a similar strategy.

The division structure provides effective coordination of the tactics being employed by different companies working in the same area; for example, the division supervisor would coordinate the actions of a crew that is advancing a hose line into the area, a crew that is conducting search and rescue operations in the same area, and a crew that is performing horizontal ventilation (**FIGURE 5-15**).

An alternative way of organizing resources is by function (rather than by location). A group includes those resources assigned to a specific function, such as ventilation, search and rescue, or water supply. Groups are responsible for performing an assignment, wherever it may be required, and often work across division lines; for example, the officer assigned to supervise the ventilation group would use the radio designation "ventilation group."

Division and group supervisors have the same rank within an IMS. Divisions do not report to groups, and groups do not report to divisions. Instead, the supervisors are required to work together—that is, to coordinate their actions and activities with one another. For example, a **group supervisor** must coordinate with the **division supervisor** when the group enters the division's geographic area, particularly if the group's assignment will affect the division's personnel, operations, or safety; the division supervisor, in turn, must be aware of everything that is happening within that area. Effective communication between divisions and groups is critical during emergency operations.

■ Branches

A **branch** is a higher level of combined resources than a division or group. At a major incident, several different activities may occur in separate geographic areas or involve distinct functions. The span of control might still be a problem, even after the establishment of divisions and groups. In these situations, command can establish branches to place a higher-level supervisor, or **branch director**, in charge of a number of divisions or groups.

■ Location Designators

An IMS uses a standard system to identify the different parts of a building or a fire scene. Every fire fighter must be familiar with this terminology.

The exterior sides of a building are generally identified by letter—A, B, C, and D—in the NIMS; however, many Canadian fire services use a numerical system in which the front of the building is designated side 1 and the sequence continues in a clockwise direction through sides 2 to 4.

In the NIMS, the front of the building is the A side, with B, C, and D following in a clockwise direction around the building. The companies working in front of the building are assigned to division A, and the radio designation for their supervisor is also "division A." Similar terminology is used for companies working at the sides and rear of the building.

The areas adjacent to a burning building are called exposures. Exposures take the same letter as the adjacent side of the building. A fire fighter facing the A side can see both the building to the left (exposure B) and the building to the right (exposure D). If the burning building is part of a row of buildings, the buildings to the left are called exposures B, B1, B2, and so on; the buildings to the right are exposures D, D1, D2, and so on. Other IMSs may number exposures 1–4 but identify large interior spaces by quadrant, starting with Alpha on the corner of sides 1 and 2 then moving clockwise to Bravo, Charlie, and Delta (**TABLE 5-1**).

Within a building, divisions commonly take the number of the floor on which they are working; for example, fire fighters working on the fifth floor would be in division 5, and the radio designation for the chief assigned to that area would also be "division 5." Crews performing different tasks on the fifth floor would all be part of this division.

Some Canadian high-rise fire management systems may not refer to divisions based on numeric floor identifiers,

FIGURE 5-15 A division refers to companies or crews working in the same geographic area.

TABLE 5-1	Location designators in an IMS
Quadrant 1-Alpha	Quadrant 2-Bravo
Quadrant 4-Delta	Quadrant 3-Charlie

Courtesy of Kevin Lambert.

preferring instead to identify groups based on function. This allows for groups like ventilation and stairwell to coordinate their activities as they migrate throughout the building without having to continually change their radio designations. In the event of a mayday, the current location of the person transmitting the mayday is essential; however, if they have been assigned to division 5 and they are on the sixth floor, confusion may occur. Again, it is important for you to be intimately familiar with the adopted process of your particular fire service.

Task Forces and Strike Teams

Task forces and strike teams are groups of single resources that have been assigned to work together for a specific purpose or for a certain period of time. Grouping resources reduces the span of control by placing several units under a single supervisor.

A **task force** includes two to five single resources, which are often different types of units assembled to accomplish a specific task; for example, a task force may be composed of two engines and one truck company, two engines and two brush units, or one rescue company and four ambulances. A task force operates under the supervision of a **task force leader**. All communications for the separate units in the task force are directed to the task force leader.

Task forces may be part of a fire service's standard dispatch philosophy. In the Los Angeles Fire Department, for example, a task force consists of two engines and one ladder truck, staffed by a total of 10 fire fighters. Some departments create task forces that consist of one engine and one brush unit for responses during wildland fire season; the brush unit responds with the engine company wherever it goes.

Canadian fire services may not have task forces as a designation but prefer to utilize an alarm designation for requesting apparatus; for example, a first alarm assignment may consist of a district chief, two pumps, and a ladder, and the second alarm assignment might bring similar additional resources.

In the United Kingdom, ICs request additional resources by identifying the exact number and type of apparatus required—for example, "Make pumps four and ladders two." This approach ensures that the IC continually repeats the number of resources he or she has requested. If the IC wants another pump, he or she simply adds a pump onto the assignment: "Make pumps five."

Notwithstanding, many emergency services adopt the term *task force* for nonfire-related activities; it is a widely used terminology with which you should become familiar.

A **strike team** is comprised of five units of the same type with an assigned leader. A strike team could be five pumps (pump strike team) or five ladders (ladder strike team) **FIGURE 5-16** . A strike team operates under the supervision of a **strike team leader**.

Strike teams are commonly used to combat wildland fires—incidents to which dozens or hundreds of companies may respond. During the wildland fire season, many departments establish strike teams of five pump companies that will be dispatched and work together on major wildland fires **FIGURE 5-17** .

FIGURE 5-16 A strike team is comprised of five units of the same type under one leader.

FIGURE 5-17 The organization of a strike team.

© Keith D. Cullom

The assigned companies rendezvous at a designated location and then respond to the scene together. Each pump has an officer and fire fighters, but only one officer is designated as the strike team leader. All communications for the strike team are directed to the strike team leader.

EMS strike teams, which consist of five ambulances and a supervisor, are often organized to respond to multicasualty incidents or disasters. Rather than requesting 15 ambulances and establishing an organizational structure to supervise 15 single resources, command can request three EMS strike teams and coordinate with three strike team leaders.

In Canada, different jurisdictions will have unique arrangements and agreements in place with their various EMSs. The manner in which EMSs and police are incorporated into an integrated IMS is usually defined through interagency cooperation and by the respective municipal council.

Implementing IMS

The IMS organizes every incident scene in a standard, consistent manner. Fire services usually develop SOGs/SOPs and then train and practise using their IMS to ensure consistency at emergency incidents. This approach increases safety and efficiency. As an incident escalates in size or complexity, the IMS organization expands to fit the situation. The same system can be used at a house fire or a major wildland fire **FIGURE 5-18**.

At a more complex incident, the IC will have to assign duties to other officers to ensure he or she does not exceed his or her span of control. If the span of control is exceeded, the effective allocation of resources and the safety of those on scene will be compromised.

The modular design of an IMS allows the organization to expand, based on the needs of the incident, by activating predetermined components. Command can delegate specific responsibilities and authority to other personnel, thereby creating an effective incident organization. An individual who receives an assignment knows the basic responsibilities of the job, because they are defined in advance.

The most frequently used IMS components in structural firefighting are divisions, sectors, and groups **FIGURE 5-19**. These components place several single resources under one supervisor, effectively reducing command's span of control. Command can also assign individuals to special jobs, such as safety officer and liaison officer, to establish a more effective organization for the incident.

In the largest and most complex incidents, other IMS components can be activated. Command can then delegate responsibility for managing major components of the operation, such as operations or logistics, to general staff positions. Each of these positions has a wide range of responsibilities related to a major component of the organization.

If necessary, branches can be established, creating an additional level within the organization **FIGURE 5-20**. A structural collapse during a major fire, for example, may result in several trapped victims, many of whom need medical treatment and transportation. In such a situation, the operations section chief would have multiple responsibilities that could exceed his or

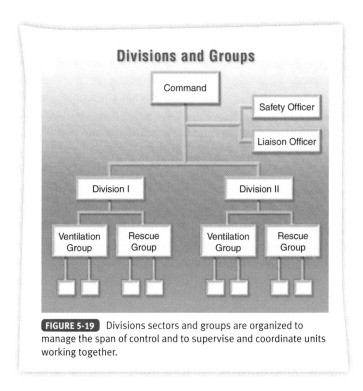

FIGURE 5-19 Divisions sectors and groups are organized to manage the span of control and to supervise and coordinate units working together.

FIGURE 5-18 At a small-scale incident, the typical ICS command structure may consist of only Command and reporting resources.

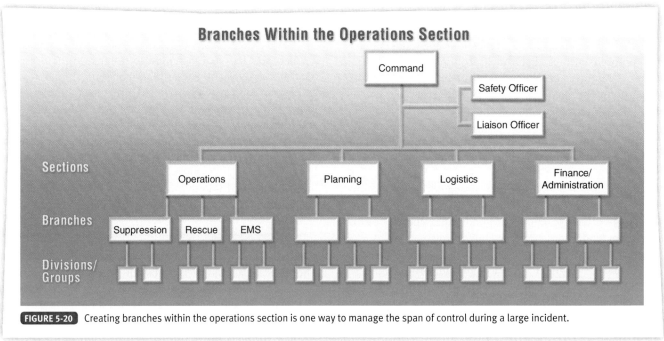

FIGURE 5-20 Creating branches within the operations section is one way to manage the span of control during a large incident.

her span of control. Activating a fire-suppression branch, a rescue branch, and a casualty branch would address this problem. One officer would be responsible for each branch and report to the operations section chief. Within each branch, several divisions, groups, or sectors would report to the branch director.

Standard Position Titles

To help clarify roles within the IMS organization, standard position titles are used for personnel within the organization. Each level of the organization has a different designator for the individual in charge. The position title typically includes the functional or geographic area of responsibility followed by a specific designator TABLE 5-2 .

Division and group supervisors are usually chief officers, but company officers may be assigned to these positions as well. Typically, the company officer from the first unit assigned to a division or group will serve as the supervisor until a chief officer is available. A chief officer with a strong background in operations is usually assigned to the position of operations section chief. The other section chief positions (planning, logistics, and finance/administration) are staffed as needed by command. These individuals should understand their roles and be able to handle the responsibilities associated with these positions.

Most Canadian fire services operate under a paramilitary style of organization. In many instances, their respective IMSs, particularly as they pertain to fire and rescue operations, retain the title *officer* for any and all of those in a supervisory role (e.g., roof division officer or vent group officer). At extremely large and complex incidents, terms like *supervisor* and/or *leader* may come into use as nonfire personnel become more involved in the incident.

Working within the Incident Management System

To an outsider, the IMS might appear to be a large, complicated organizational model that involves a complex set of SOGs/SOPs. By contrast, to the individual fire fighter working within this system, the IMS is really quite simple and uncomplicated. Fire fighters should understand what the IMS is and how it works. More importantly, each individual fire fighter must understand his or her place and role in the system.

This section discusses how every emergency incident is conducted. The following basic principles of the IMS are used at every incident and at every training exercise. In addition, fire fighters generally have specific responsibilities and procedures to follow in most situations.

Three basic components always apply:

- Command is established at every incident and is maintained from the time that the first unit arrives until the time that the last unit leaves. The identity of command may change, but there is always one single function (person) who is in charge of the incident and responsible for everything that happens. SOGs/SOPs may dictate who will be command at any time.
- Each fire fighter always reports to one supervisor (unity of command). A fire fighter's supervisor will usually be a company officer. The company officer directly supervises a small group of fire fighters, such as an engine company or ladder company, who work together. At an incident scene, the company officer provides instructions and must always know where each fire fighter is and what he or she is doing. If the company officer assigns two fire fighters to work together away from the rest of the company, both fire fighters remain under the supervision of the company officer.
- If only one company is present at the scene, the company officer is command until command is transferred. At a basic incident, the company officer may report directly to command; at complex incidents, several layers of supervision may separate a company officer from command.

In order for an IMS to be effective, every fire fighter must understand how to function within the system. To fulfill your role as a member of the incident management team, follow the steps for operating within the IMS SKILL DRILL 5-1 (Fire Fighter II, NFPA 6.1.2):

1. Verify that the IMS is in use.
2. Verify your radio designation and the resources assigned to complete your assignment.
3. Repeat your assignment over the radio to verify it.
4. Complete a size-up to ensure safety.
5. Account for yourself and for other members as is required.
6. Ensure that you conduct operations in accordance with your IMS risk-management principles.
7. Update your supervising officer regularly.
8. Provide personnel accountability reports as necessary.
9. Report the completion of each assignment to your supervising officer.

TABLE 5-2	Levels of an IMS Organization	
IMS Level	**IMS Function/Location**	**Position Designator**
Command	Command and control	Incident commander
Command staff	Safety, liaison, information	Officer
General staff	Operations, planning, logistics, finance/administration	Section chief/officer
Branch	Varies (e.g., operations)	Director/officer
Division/group/sector	Varies (e.g., division A or roof sector)	Supervisor/officer
Unit/crew/strike	Varies (e.g., rehab)	Leader/officer
Team/task force	Varies (e.g., overhaul)	Company officer

■ Responsibilities of the First-Arriving Fire Fighters

The first fire fighters to arrive at an emergency incident are the foundation upon which the IMS is established.

The officer in charge of the first-arriving unit is responsible for taking initial action and assumes command, with all of the authority that accompanies that position. Note: It is important to remember that a ranking officer who is not on scene should not assume command.

The first officer on the scene assumes command and is responsible for managing the operation until he or she is relieved by a senior officer. If there is no officer on the first-arriving unit, the fire fighter with the greatest seniority should assume command until an officer arrives. The position of command must have an unbroken line of succession from the moment the first unit arrives on the scene until the incident is considered finished.

Establishing Command

The IMS establishes that the command function is initiated when the first unit arrives and is transferred as required for the duration of the incident. The individual who is in charge of the first-arriving unit automatically establishes command of the incident until a superior takes over that role.

The individual who initially establishes command must formally announce that fact over the radio. This announcement eliminates any possible confusion over who is in charge. The initial radio reports should include the following information **SKILL DRILL 5-2** (Fire Fighter II, NFPA 6.1.2):

① Confirmed command designation
② Building description and occupancy based on a brief size-up
③ Brief description of the incident
④ Request or release of resources
⑤ Declaration of strategy and initial tactical objectives
⑥ Any immediate safety concerns

For example, the first-arriving unit at a single-family Type V lightweight dwelling fire might deliver the following radio report:

> Pump 4 is on the scene at 309 Central Avenue, with heavy smoke showing from the A side of a Type V single-family dwelling. This is a vent-controlled fire. Pump 4 is laying a 44-millimetre (1.75-inch) attack line and will make an offensive attack through the back door, side C. Pump 4 is Central Avenue command. This may be a lightweight constructed building; use caution.

The officer who initially assumes command must decide whether to take action and directly supervise the initial attack crew or whether to concentrate on managing the incident as command. This decision depends on many factors, including the nature of the situation, the resources available on scene, and the ability of the crew to work safely without direct supervision.

If the incident is large and complicated, the best option for command is to establish an ICP and focus on sizing up the situation, directing incoming units, and requesting additional resources. Command's own crew can be assigned to work with an acting officer or to join forces with another company officer and crew. If the situation is less critical, command might be able to function as both a company officer and command simultaneously, at least temporarily.

If the first-arriving unit is a chief officer, he or she automatically assumes command and begins executing command responsibilities. If a company officer had previously assumed command, the company officer would transfer command to the first-arriving chief officer. The officer relinquishing command should provide an assessment of the situation to the incoming personnel, as described in the "Transfer of Command" section later in this chapter.

Most fire services have written procedures that specify who will assume command in certain situations. If multiple units respond from the same station and arrive together, there should be an established protocol for which officer assumes command. If a company officer arrives only seconds ahead of a chief officer, the chief officer should assume command from the outset.

■ Confirming the Command

The strength of the command system is that it fixes command on an individual who, from the onset of the incident, establishes a strong, direct, and visible presence for the purposes of controlling the incident.

In most Canadian fire services, the establishment of an IC has long been a guiding principle and is used at every incident. The first member to arrive on the scene is to establish command. In order for any member to assume command, they must have a means to communicate—either by mobile radio, portable radio (walkie-talkie)—and assume a radio designation that recognizes that they are in command. The announcement of Command also reinforces Command's personal commitment to the position through a conscious personal act and a standard organizational act. In most Canadian jurisdictions, the command designation is based on a geographic location, such as a street or other well-known location; for example, if an individual assumes command at an incident at the West Edmonton Mall, he or she may establish a radio designation like *West Edmonton Mall Command*.

Strong communication is the key to initial establishment of command. The first-arriving member should identify him- or herself, assume command, and provide the following information in his or her initial radio reports:

■ A description of the situation, building type, and occupancy
■ A description of his or her intended action—to lay an attack line, for example, or to raise a ladder for immediate rescue—only if the commander has the functional capability of a pump, ladder, or rescue company

- A statement of any safety concerns (e.g., "Wall on side 2 is about to collapse.")
- Request or release of resources, as determined by the needs of the incident

Identifying Incident Command

Individual fire service procedures may vary in terms of the specific protocol they use for naming an incident. The first officer to assume command should establish an identity that clearly identifies the location of the incident (e.g., "Pump 10 will be Seventh Avenue command."). This announcement reduces confusion on the radio and establishes a continuous identity for command, regardless of who holds that position during the incident.

There can be only one "Seventh Avenue command" at a time, so there is no confusion about who is in charge of the incident. When anyone needs to talk to command, a call to "Seventh Avenue command" should be answered by the individual who is in command.

Transfer of Command

Transfer of command occurs whenever one person relinquishes command of an incident and another individual becomes Command. For example, the first-arriving company officer would transfer command to the first-arriving chief officer, who would later transfer command to a higher-ranking officer during a major incident. Some services require transfer of command when a higher-ranking officer arrives at an incident; others give the higher-ranking officer the option of assuming command or leaving the existing personnel in the Command assignment.

When a higher-ranking officer arrives at the scene of an emergency incident and takes charge, that officer assumes the moral and legal responsibility for managing the overall operation. Established procedures must be followed whenever command is transferred. One of the most important requirements of command transfer is the accurate and complete exchange of incident information **FIGURE 5-21**. The officer who is relinquishing command needs to give incoming Command

a current situation status report that includes the following information:

- Tactical priorities
- Action plans
- Hazardous or potentially hazardous conditions
- Accomplishments
- Assessment of effectiveness of operations
- Current status of resources
  - Assigned or working
  - Available for assignment
  - En route
  - Additional resource requirements

If transfer of command occurs very early during an incident, the transfer of information may be brief. For example, the first-arriving company officer might have been in command for only a few minutes and have little information to report when command transfers to the first-arriving chief officer. The chief may have heard all of the exchanges over the radio and know what the current situation is. Conversely, if the company officer has been Command for several minutes, there may be a significant amount of information to report, such as the current assignments of all first-alarm companies. Whether the information is minimal or substantial, the transfer must be accurate and complete.

Each fire service establishes specific procedures for transferring command. In some cases, the transfer of command may be done via radio. The most effective method, however, is face-to-face communication. Standard command worksheets and a status board are valuable tools for tracking and transferring information at an ICP.

Command is always maintained for the entire duration of an incident. In the later stages of an incident, after the situation is under control, it might be transferred to a lower-level commander. A downward transfer of command requires the same type of briefing and exchange of information as an upward command transfer. The officer in charge of the last company remaining on the scene would be Command. When that company leaves the scene, the Command function is terminated.

Command Transfer Rationale

Command may be transferred at difference points during an emergency incident for several important reasons. A first-arriving company officer can usually direct the initial operations of two or three additional companies, but as situations become more complex, the problems of maintaining control increase rapidly. A company officer's primary responsibility is to supervise one crew and ensure that the crew members operate safely. When three or more companies are operating at an incident, it is better to have a chief officer assume command.

As more companies are assigned to an incident, the Command structure must also expand. That is, the organization must grow to maintain an effective span of control. Additional chief officers may be assigned to the incident, and a higher-ranking officer may assume the Command role. A command transfer may be required if the situation is beyond the training and experience of the current Command. A more experienced officer may have to assume command to ensure proper management of the incident.

FIGURE 5-21 The incoming Command needs to be briefed before assuming this role.

All fire fighters should know the steps involved in transferring command during an emergency incident **SKILL DRILL 5-3** (Fire Fighter II, NFPA 6.1.2):

1 Assume command.

2 Follow established procedures for transferring command.

3 Transfer command in a face-to-face meeting if possible. If not possible, transfer command over the radio.

4 Communicate to the incoming Command personnel the tactical priorities, action plans, hazardous conditions, effectiveness of operations, status of resources, and need for additional resources.

5 Formally announce the transfer of command over the radio.

Near-Miss REPORT

Report Number: 08-0000111

Synopsis: Poor communications create opposing handlines.

Event Description: My department responded to a report of a garage fire in a large subdivision in our fire district. The structure was a three-stall garage attached to a roughly 370-square-metre (4000-square-foot) house. We responded in an engine with 4740-litre-per-second (1250-gallon-per-minute), 3785-litre (1000-gallon) tank. There were five people on board, whose ranks were as follows: assistant chief, who was the driver (career); chief, who was riding in the passenger seat (career); one fire fighter (volunteer); a captain (career); and me, also a fire fighter (career).

On arrival, we found the garage well involved, as well as two vehicles inside the garage. We stretched a 44-mm (1.75-in.) line to the D side to go into the garage. An attempt to knock down the fire was made, with little success. The next entry was made by me and the captain through the front door on the A side. We encountered heavy smoke and heat. We eventually made our way to the door that went into the garage, which was left open by the occupants upon discovering the fire. At this time, positive-pressure ventilation (PPV) was in place and conditions improved. The captain and I went into the garage and began attacking the fire. At the same time, someone sprayed from an outside line and, in return, pushed heat and smoke onto us inside the garage. The captain immediately got on the radio and told command to shut down that line, and conditions deteriorated rapidly. As a result, we had to pull back into the house and regroup.

Lessons Learned: It is vital to have a good and competent IC with a strong command presence. In this case, the IC did not have any accountability system set in place, and several fire fighters were freelancing. Incident command must keep in constant communication with all fire fighters who are inside a burning structure, and in return, fire fighters must keep incident command updated on their status.

Wrap-Up

Chief Concepts

- An IMS provides a standard, professional, and organized approach to managing emergency incidents. This standardized approach provides common objectives and helps to coordinate resources from multiple agencies.
- Several characteristics are critical to an IMS:
  - Organized approach—An IMS imposes order and discipline on the emergency scene and enables a safer and more efficient operation than would be possible if personnel and units worked independently of one another. In short, the IMS breaks down a large, complex situation into more manageable units.
  - Terminology—An IMS uses a standard terminology for effective communications.
  - All-risk—An IMS can be used at any type of emergency incident.
  - Jurisdictional authority—An IMS enables different jurisdictions, agencies, and organizations to work cooperatively on a single incident.
  - Span of control—An IMS maintains a desired span of control through flexible levels of organization. Span of control ranges from three to seven, with five being the preferred maximum number of resources managed by a single individual.
  - Everyday applicability—The IMS should be used at every single incident.
  - Modular—An IMS is based on standard modules that are activated as needed to manage an incident.
  - Interoperability—Everyone responding to an incident can communicate up and down the chain of command as needed.
  - IAP—Every incident has a plan that outlines the strategic objectives. Large incidents will have formal written plans, whereas smaller incident plans may be mentally formulated.
  - Designated incident facilities—Standardized facilities (such as an ICP, staging area, and rehabilitation area) are established as needed.
  - Resource management—A standard system is used for assigning and keeping track of incident resources.
- Five major functions are part of IMS:
  - Command—Responsible for the entire incident. This is the only function that is always maintained until command is terminated.
  - Operations—Responsible for most fire-ground functions, including suppression, search and rescue, ventilation, salvage, and overhaul.
  - Planning—Responsible for developing the IAP.
  - Logistics—Responsible for obtaining the resources needed to support the incident.
  - Finance/Administration—Responsible for tracking expenditures and managing administrative functions at the incident.
- The command staff assist command at the incident as follows:
  - The safety officer is responsible for the overall safety of the incident. He or she has the authority to stop any action or operation if it creates a safety hazard on the scene.
  - The liaison officer is responsible for coordinating operations between the fire department and other agencies that may be involved in the incident.
  - The PIO is responsible for coordinating media activities and providing the necessary information to the various media organizations.
- An example of a single resource would be a pump company or a ladder company.
- Single resources can be combined into task forces or strike teams and/or sectors.
- Other organizational units that can be established under an IMS include divisions, groups, sectors, branches, task forces, and strike teams.
- The IMS can be expanded infinitely to accommodate any size of incident. Branches can be established to aggregate similar functions, such as suppression, EMS, or hazardous materials.
- Five basic components always apply when working within the IMS:
  - At every incident, command must always be established.
  - Each fire fighter always reports to one supervisor.
  - Risk-management principles should be applied at all emergency incidents.
  - Spans of control should be applied in an escalating situation.
  - The company officer reports to command.
- As the incident grows or continues, it may be necessary to transfer command to another officer. Transfer of command must be done in a seamless manner to ensure continuity of command.

Hot Terms

Branch The organizational level having functional, geographical, or jurisdictional responsibility for major aspects of incident operations. (NFPA 1026)

www.FireFighter.jbpub.com/Canadian3e

Branch director A person in a supervisory position in either the operations or logistics function to provide a span of control. (NFPA 1561)

Command The first component of the IMS. It is the only position in the IMS that must always be staffed.

Command staff The PIO, safety officer, and liaison officer, all of whom report directly to the IC and are responsible for functions in the IMS that are not a part of the function of the line organization. (NFPA 1561)

Crew A team of two or more fire fighters; also known as a company. (NFPA 1500)

Designated incident facilities Assigned locations where specific functions are always performed.

Division A supervisory level established to divide an incident into geographic areas of operations. (NFPA 1561)

Division supervisor A person in a supervisory position who is responsible for a specific geographic area of operations at an incident. (NFPA 1561)

Finance/administration section The section responsible for all costs and financial actions of the incident or planned event, including the time unit, procurement unit, compensation/claims unit, and the cost unit. (NFPA 1026)

Fire-ground command (FGC) An IMS developed in the 1970s for day-to-day fire department incidents generally handled with fewer than 25 units or companies.

FIRESCOPE (FIre RESources of California Organized for Potential Emergencies) An organization of agencies established in the early 1970s to develop a standardized system for managing fire resources at large-scale incidents such as wildland fires.

Group A supervisory level established to divide an incident into functional areas of operation. (NFPA 1561)

Group supervisor A person in a supervisory position who is responsible for a functional area of operation. (NFPA 1561)

IMS general staff The chiefs of each of the four major sections of IMS (outside of command): operations, planning, logistics, and finance/administration.

Incident action plan (IAP) The objectives reflecting the overall incident strategy, tactics, risk management, and member safety that are developed by the IC. IAPs are updated throughout the incident. (NFPA 1500)

Incident command post (ICP) The field location at which the primary tactical-level, on-scene incident command functions are performed. (NFPA 1026)

Incident Management System (IMS) The combination of facilities, equipment, personnel, procedures, and communications operating within a common organizational structure that has responsibility for the management of assigned resources to effectively accomplish stated objectives pertaining to an incident or training exercise. (NFPA 1670)

Integrated communications The ability of all appropriate personnel at the emergency scene to communicate with their respective supervisor and subordinates.

Liaison officer A member of the command staff who serves as the point of contact for assisting or coordinating agencies. (NFPA 1026)

Logistics section The section responsible for providing facilities, services, and materials for the incident or planned event, including the communications unit, medical unit, and food unit within the service branch and the supply unit, facilities unit, and ground support unit within the support branch. (NFPA 1026)

Logistics section chief The general staff position responsible for directing logistical functions. It is generally assigned on complex, resource-intensive, or long-duration incidents.

National Incident Management System (NIMS) A system mandated by the United States Department of Homeland Security Presidential Directive 5 (HSPD-5), *Management of Domestic Incidents*, that provides a systematic, proactive approach guiding government agencies at all levels, the private sector, and nongovernmental organizations to work seamlessly to prepare for, prevent, respond to, recover from, and mitigate the effects of incidents, regardless of cause, size, location, or complexity, so as to reduce the loss of life or property and harm to the environment. (NFPA 1026)

Operations section The section responsible for all tactical operations at an incident or planned event, including up to 5 branches, 25 divisions/groups, and 125 single resources, task forces, or strike teams. (NFPA 1026)

Operations section chief The general staff position responsible for managing all operations activities. It is usually assigned when complex incidents involve more than 20 single resources or when command cannot be involved in the details of tactical operations.

Planning section The section responsible for the collection, evaluation, dissemination, and use of information related to the incident situation, resource status, and incident forecast. (NFPA 1026)

Planning section chief The general staff position responsible for planning functions. It is assigned when command needs assistance in managing information.

Public information officer (PIO) A member of the command staff who is responsible for interacting with the public and media or with other agencies with incident-related information requests. (NFPA 1026)

Resource management A standard system of assigning and tracking the resources involved in an incident. Under NIMS, this includes mutual-aid agreements; the use of special federal, state, local, and tribal teams; and resource-mobilization protocols. (NFPA 1026)

Safety officer A member of the command staff who is responsible for monitoring and assessing safety hazards and unsafe situations and for developing measures for ensuring personnel safety. (NFPA 1561)

Sector A designated geographic or functional area within the IMS hierarchy similar to a group or division as defined by NIMS.

Single resource An individual, a piece of equipment and its personnel, or a crew or team of individuals with an identified supervisor that can be used on an incident or planned event. (NFPA 1026)

Staging area A prearranged, strategically placed area where support response personnel, vehicles, and other equipment can be held in an organized state of readiness for use during an emergency. (NFPA 424)

Strike team Specified combinations of the same kind and type of resources with common communications and a leader. (NFPA 1026)

Strike team leader The person in charge of a strike team. This individual is responsible to the next higher level in the incident organization and serves as the point of contact for said strike team within the organization.

Task force Any combination of single resources assembled for a particular tactical need with common communications and a leader. (NFPA 1051)

Task force leader The person in charge of a task force. This individual is responsible to the next higher level in the incident organization and serves as the point of contact for the task force within the organization.

Transfer of command The formal procedure for transferring the duties of an IC from one person to another at an incident scene. (NFPA 1026)

Unified command An application of the IMS that allows all agencies with jurisdictional responsibility (geographical or functional) for an incident or planned event to manage said incident or event by establishing a common set of objectives and strategies. (NFPA 1561)

FIRE FIGHTER *in action*

At 2:43 P.M., your pump company is dispatched for a report of smoke in the 3700 block of Dunning Road. As you approach the scene, you note a shed approximately 18 metres (60 feet) from a large two-storey farmhouse that is fully involved. As you approach, the pump stops on the road, and you hear your captain's initial report: "Pump 53 is on the scene at 3794 Dunning Road. This is for a fully involved 12- by 18-m [40- by 60-ft.] one-storey garage with a large two-storey exposure. We are laying a line from the road to protect exposures. Pump 53 is Dunning command. Put in a second alarm. Request hydro for live wires down and police for traffic control."

1. What is the best structure for handling this incident?

  A. An IMS using all major functional components

  B. A unified command

  C. A single command

  D. No incident command is needed for a single-building fire.

2. When is it necessary to establish command at a fire?

  A. Only when a chief officer is dispatched

  B. At all incidents

  C. Whenever more than one agency is involved

  D. When the incident requires more resources than the initial response

3. Which phase of the IMS was the captain's report?

  A. Transfer of command

  B. Termination of command

  C. Establishing command

  D. Confirming command

4. Which of the following IMS components must be implemented for this emergency?

  1. Operations

  2. Planning

  3. Logistics

  4. Finance/Administration

  5. Command

     A. 1, 2, 3, 4, and 5

     B. 1, 2, and 3

     C. 2, 3, and 5

     D. 5

5. A second alarm includes two additional pumps, two truck companies, a rescue company, three tankers, two chief officers, and a rehab unit. Which of the following would be an appropriate organizational structure?

  A. IC, safety, division 3, division B, water supply group, and rehab unit

  B. IC, safety, division 3, division 1, water supply group, and rehab unit

  C. IC, safety, exposure division, fire attack group, water supply group, and rehab unit

  D. IC, safety, exposure group, fire attack group, water supply group, and rehab unit

It is 6:00 A.M., and with the exception of your captain, you and the rest of the crew of Pump 413 are sitting on the tailboard talking about the incident from which you just returned. Everyone felt good about having just made a really good stop given the serious conditions upon your arrival. The incident involved a three-storey Type III apartment complex with fire showing from the A side in the third-floor window. Dispatch had reported people trapped in the adjoining units.

Upon arrival, your captain established command and assigned you to confine and attack the fire. Ladder 16 set up ventilation and then checked for extension into the attic, while Rescue 3 did a primary search of the third-floor units. A second alarm was requested, bringing two additional pump companies and two additional ladder companies. The two engines were assigned to exposure protection, one ladder company was assigned as RIC, and the other truck was assigned to salvage.

The discussion eventually turned to complaints by the crew that they could have used the captain's help rather than him standing in the front yard barking orders.

Another chimed in about the waste of crew members being assigned as RIC rather than doing "some real work."

1. You know the importance of having a strong command structure, particularly in serious fires such as this. How do you explain this point to the other crew members?

2. Leadership often requires someone to stand up for what is right, even when it is not popular. How do you get the crew's buy-in for the use of an RIC when others say, "We are paid to risk our lives"?

3. The IMS delegates authority and responsibility down the organizational structure. How do you motivate subordinates to ensure that the overall goal is achieved?

4. Why is a maximum span of control of five to one generally recommended when on the fire ground?

Fire Behaviour

Fire Fighter I

Knowledge Objectives

After studying this chapter, you will be able to:

- Describe the chemistry of fire. (p 138)
- List the three states of matter. (NFPA 5.3.10.A , p 139)
- List the five forms of energy. (p 139–140)
- Explain the concept of the fire triangle. (NFPA 5.3.11.A , p 140)
- Explain the concept of the fire tetrahedron. (NFPA 5.3.11.A , p 140)
- Describe the chemistry of combustion. (NFPA 5.3.11.A , p 140–141)
- Describe the by-products of combustion. (NFPA 5.3.11.A , p 141)
- Explain how fires are spread by direct contact, conduction, convection, and radiation. (NFPA 5.3.12.A , p 141–143)
- Describe the four methods of extinguishing fires. (p 144)
- Define Class A, B, C, D, and K fires. (p 144–145)
- Describe the characteristics of solid-fuel fires. (NFPA 5.3.11.A , p 146)
- Describe the four phases of a solid-fuel fire: ignition phase, growth phase, fully developed phase, and decay phase. (NFPA 5.3.11.A , p 146–147)
- Describe the characteristics of a room-and-contents fire during each of the four phases of a solid-fuel fire. (NFPA 5.3.10A, 5.3.11.A , p 147–149)
- Describe the conditions that cause thermal layering. (NFPA 5.3.12.A , p 150)
- Describe the conditions that lead to flameover. (p 150)
- Describe the conditions that lead to flashover. (p 150)
- Describe the conditions that lead to a backdraft. (NFPA 5.3.11.A , p 150–151)
- Explain how fire behaves in an enclosed compartment. (NFPA 5.3.11.A , p 151)
- Describe how fire behaves in modern structures. (p 151)
- Describe how the wind effect impacts fire behaviour. (p 152)
- Describe the characteristics of liquid-fuel fires. (p 152)
- Define the characteristics of gas-fuel fires. (p 153)
- Explain the concept of vapour density. (p 153)
- Explain the concept of flammability limits. (p 153)
- Describe the causes and effects of a boiling liquid/expanding vapour explosion (BLEVE). (p 153–154)
- Describe the process of reading smoke. (p 154–156)

Skills Objectives

There are no skills objectives for Fire Fighter I candidates. NFPA 1001 contains no Fire Fighter I Job Performance Requirements for this chapter.

© Photos.com

Fire Fighter II — FFII

Knowledge Objectives

There are no knowledge objectives for Fire Fighter II candidates. NFPA 1001 contains no Fire Fighter II Job Performance Requirements for this chapter.

Skills Objectives

There are no skills objectives for Fire Fighter II candidates. NFPA 1001 contains no Fire Fighter II Job Performance Requirements for this chapter.

I t is 3:46 a.m., and you arrive at a house with a fire reported by a neighbour. You pull a 44-millimetre (1.75-inch) hose line and advance it to the front door. Even though there is not any visible smoke, you catch a whiff of that familiar smell of a fire. You notice the windows are stained black with soot. As you prepare to force entry, you notice light-brown smoke pushing through between the threshold and the door. There is a crack across the living room window. The hair on the back of your neck rises as you get an unsettled feeling.

1. Why would this situation make you uneasy?

2. What conditions would you expect in the structure?

3. What type of event is most likely getting ready to occur?

Introduction

This chapter describes the behaviour of fire. Since the days when someone discovered how to use fire for cooking food and keeping warm, fire has fuelled our lives. Although we have progressed from having open fires for cooking and warmth, we continue to depend on fires to fuel the combustion of our society. Most of our electric power is created by burning fossil fuels. In addition, our modern gasoline- and diesel-powered vehicles depend on small explosions (internal combustion) to propel them.

Unfortunately, the destruction of lives and property by uncontrolled fires has also been occurring since ancient times. Despite our advanced technology for detecting and combatting fires, we continue to wage an ongoing battle with fire.

FIRE FIGHTER Tips

When studying fire behaviour, referenced units of measure must be used. In Canada, the metric system is the official system of reference. In the United States, most people use the imperial system of units. Every other country uses the metric system, or international system of units (SI). Whichever system is used, it is important to use it consistently.

This book provides both the metric and imperial systems of reference.

In the metric system:

- Distance is measured in metres (m), centimetres (cm), and millimetres (mm)
- Liquid volume is measured in litres (L) and millilitres (mL)
- Temperature is measured in degrees Celsius (°C)
- Pressure is measured in pascals (Pa) or kilopascals (kPa)

In the imperial system:

- Distance is measured in feet (ft.) and inches (in.)
- Liquid volume is measured in gallons (gal.)
- Temperature is measured in degrees Fahrenheit (°F)
- Pressure is measured in pounds per square inch (psi)

This chapter explores the relationship between various types of fuels and the combustion process. In addition, it details the phases of a fire and fire conditions, such as thermal layering, flameover, flashover, and backdraft. This chapter also describes how to assess the volume, velocity, density, and colour of smoke. By relating the key attributes of smoke to the size of the building, weather conditions, and the rate of change of the smoke, it becomes possible to predict the fire's location and its stage of development.

This book uses the metric system and provides the imperial equivalent as necessary. In the metric system, distance is measured in metres, liquid volume is measured in litres, temperature is measured in degrees Celsius, and pressure is measured in pascals or kilopascals. In the imperial system, distance is measured in feet and inches, liquid volume is measured in gallons, temperature is measured in degrees Fahrenheit, and pressure is measured in pounds per square inch.

The Chemistry of Fire

To be a safe and effective fire fighter, you need to understand the conditions needed for a fire to ignite and grow. Understanding these conditions will increase your effectiveness as a fire fighter. A well-trained and experienced fire fighter can put out more fire with less water because of his or her understanding of the chemistry of fire.

■ What Is Fire?

Fire is a rapid chemical process that produces heat and usually light. Everyone can identify fire when they see it, but few can explain the process that is taking place to produce it. Fire is characterized by the production of a flame, which can be seen in many different colors, depending on what is burning and how much heat is being produced. Fire is neither solid nor liquid but rather is the by-product of combustible vapours being consumed during a chemical process. Solid fuels, such as wood, liquid fuels, such as gasoline, and gaseous fuels, such as propane, can all burn and create flaming combustion, better known as fire.

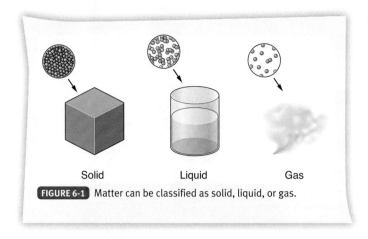

FIGURE 6-1 Matter can be classified as solid, liquid, or gas.

FIGURE 6-2 Energy being converted to work.

■ States of Matter

<u>Matter</u> is made up of atoms and molecules. Matter exists in three states: solid, liquid, and gas **FIGURE 6-1**. An understanding of these states of matter is important in understanding fire behaviour.

A **solid** has a definite shape and size. Most uncontrolled fires are predominantly fed by solid fuels. In structure fires, the building and most of its contents are solids. A solid does not flow under stress. Instead, it has a definite capacity for resisting forces and, under ordinary conditions, retains a definite size and shape. Some solids, such as wax or plastic, may change to a liquid or a gaseous state when heated. Cold makes most solids more brittle, whereas heat makes them more flexible. Because a solid is rigid, only a limited number of molecules are present on its surface; the majority of molecules are cushioned or insulated by the outer surface of the solid.

A <u>liquid</u> will assume the shape of the container in which it is placed. Most liquids expand when heated and will turn into gases when sufficiently heated. Liquids, for all practical purposes, cannot be compressed. This characteristic allows fire fighters to pump water for long distances through pipelines or hoses. A liquid has no independent shape, but it does have a definite volume. Water is the liquid that fire fighters work with the most.

A <u>gas</u> is a type of fluid that has neither independent shape nor independent volume, but rather, tends to expand indefinitely. The gas we most commonly encounter is air—the mix of invisible, odourless, tasteless gases that surrounds the earth. The mixture of gases in air maintains a constant composition—21 percent oxygen, 78 percent nitrogen, and 1 percent other gases, including carbon dioxide. Oxygen is required for us to live, but it is also required for fires to burn. We will explore the reaction of fuels—some of which exist in gaseous form—with oxygen as we look at the chemistry of burning.

■ Fuels

A <u>fuel</u> is a form of energy. Think of the vast amount of heat that is released during a large fire. The energy released in the form of heat and light has been stored in the fuel before it is burned. The release of the energy in 3.8 L (1 gal.) of gasoline, for example, can move a car kilometres down the road **FIGURE 6-2**.

■ Types of Energy

Energy exists in the following forms: chemical, mechanical, electrical, light, and nuclear. Regardless of the form in which the energy is stored, it can be changed from one form to another; for example, electrical energy can be converted to heat or to light. Similarly, mechanical energy can be converted to electrical energy through a generator.

Chemical Energy

<u>Chemical energy</u> is the energy created by a chemical reaction. Some chemical reactions produce heat (<u>**exothermic**</u>); others absorb heat (<u>**endothermic**</u>). The combustion process is an exothermic reaction, because it releases heat energy. Most chemical reactions occur because bonds are established between two substances or bonds are broken as two substances are chemically separated. Heat is also produced whenever oxygen combines with a combustible material. If the reaction occurs slowly in a well-ventilated area, the heat is released harmlessly into the air. If the reaction occurs very rapidly or within an enclosed space, the mixture can be heated to its <u>**ignition temperature**</u> and can begin to burn. A bundle of rags soaked with linseed oil will begin to burn spontaneously, for example, because of the heat produced by oxidation that occurs within the mass of rags. The energy produced during combustion is an example of chemical energy.

Mechanical Energy

<u>Mechanical energy</u> is converted to heat when two materials rub against each other and create friction. For example, a fan belt rubbing against a seized pulley produces heat. Heat is also produced when mechanical energy is used to compress air in a compressor. Water falling over a dam is one example of mechanical energy.

Electrical Energy

<u>Electrical energy</u> is converted to heat energy in several different ways; for example, electricity produces heat when it flows through a wire or any other conductive material. The greater the flow of electricity and the greater the resistance of the material, the greater the amount of heat produced. Examples of electrical energy that can produce enough heat to start a fire include heating elements, overloaded wires, electrical arcs, and lightning. Electrical energy is carried through the electrical

wires inside homes and can be stored in batteries, which convert chemical energy to electrical energy.

Light Energy

Light energy is caused by electromagnetic waves packaged in discrete bundles called photons. This energy travels as thermal radiation, a form of heat. When light energy is hot enough, it can sometimes be seen in the form of visible light. If it is of a frequency that we cannot see, the energy may be felt as heat but not seen as visible light. Candles, fires, light bulbs, and lasers are all forms of light energy. Another example of light energy is the radiant energy we receive from the sun.

Nuclear Energy

Nuclear energy is created by splitting the nucleus of an atom into two smaller nuclei (fission) or by combining two small nuclei into one large nucleus (fusion). Nuclear reactions release large amounts of energy in the form of heat. These reactions can be controlled, as in a nuclear power plant, or uncontrolled, as in an atomic bomb explosion. In a nuclear power plant, the nuclear reaction releases carefully controlled amounts of heat, which then convert water to steam so as to power a steam turbine generator. Both uncontrolled explosions and controlled reactions release radioactive material, which can cause injury or death. Nuclear energy is stored in radioactive materials and converted to electricity by nuclear power–generating stations.

■ Conservation of Energy

The law of conservation of energy states that energy cannot be created or destroyed by ordinary means. Energy can, however, be converted from one form to another. Think of an automobile: chemical energy in the gasoline is converted to mechanical energy when the car moves down the road; if you apply the brakes to stop the car, the mechanical energy is converted to heat energy by the friction of the brakes. Similarly, in a house fire, the stored chemical energy in the wood of the house is converted into heat and light energy during the fire.

■ Conditions Needed for Fire

To understand the behaviour of fire, you need to consider the three basic elements needed for a combustion to occur: fuel, oxygen, and heat. First, a combustible fuel must be present. Second, oxygen must be available in sufficient quantities. Third, a source of ignition (i.e., heat) must be present. If we graphically place these three components together, the result is the **fire triangle** FIGURE 6-3.

A fourth factor, a chemical chain reaction, must result from the first three elements to produce and maintain a self-sustaining fire. That is, the fuel, oxygen, and heat must interact in such a way as to create a self-sustaining chemical chain reaction to keep the combustion process going.

One way of visualizing this process is to show the chemical chain reaction joining the elements of the fuel, oxygen, and heat. This depiction reflects the central role that the chemical reaction plays in maintaining the process of flaming combustion. This relationship is sometimes characterized as the **fire tetrahedron** FIGURE 6-4. A tetrahedron is a four-sided, three-dimensional figure. Each side of the fire tetrahedron represents one of the four elements needed for a fire to occur.

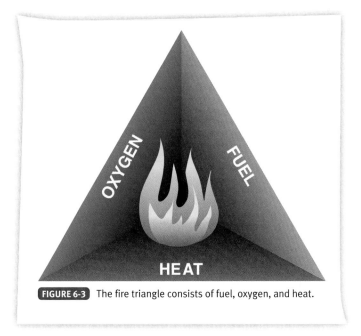

FIGURE 6-3 The fire triangle consists of fuel, oxygen, and heat.

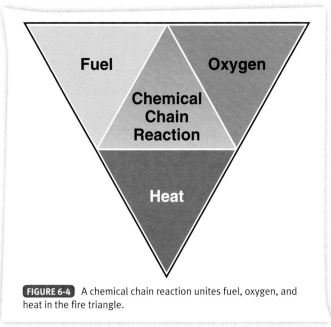

FIGURE 6-4 A chemical chain reaction unites fuel, oxygen, and heat in the fire triangle.

The key point to remember is that fuel, oxygen, and heat must be present for a fire to start and to continue burning. If you remove any of these elements, the fire will go out.

■ Chemistry of Combustion

The smallest unit of matter is an **atom**. Some atoms exist in nature as pairs; for example, two atoms of oxygen are paired up in air. The formula for the oxygen pair is written O_2.

When atoms of one element combine chemically with atoms of another element, they produce a compound that is made up of molecules; for example, when an atom of oxygen (O) combines chemically with two atoms of hydrogen (H_2), the resulting compound is water (H_2O).

Almost all fuels consist of hydrogen (H) and carbon (C) atoms; hence they are called hydrocarbons. Fuels may contain a wide variety of other elements, but for our purpose, we can describe the chemical reaction of combustion simply by considering the fuel to contain carbon and hydrogen. When this fuel combines with oxygen, it produces two by-products: water and carbon dioxide.

Hydrogen–Carbon Fuel + Oxygen = Water and Carbon Dioxide

Graphically, this reaction is depicted as follows:

$$CH \text{ (hydrocarbon fuels)} + O_2 = H_2O + CO_2$$

Of course, the reactions that occur in most actual fires are not as simple as this basic reaction suggests. First, hydrocarbon fuels are very complex molecules, which contain a wide variety of atoms besides hydrogen and carbon. As a consequence, the combustion process produces numerous toxic by-products. Second, the fires encountered by fire fighters burn in the presence of a limited amount of oxygen. The resulting incomplete combustion produces significant quantities of deadly gases and compounds. Also, the production of large quantities of heat and light is an integral part of combustion; thus the reaction for combustion should be rewritten as follows:

$$CH \text{ (hydrocarbon fuels)} + O_2 = H_2O + CO_2 + Light + Heat$$

To understand the reactions involved in fire, it is helpful to differentiate between oxidation, combustion, and pyrolysis. **Oxidation** is the process in which oxygen combines chemically with another substance to create a new compound; for example, steel that is exposed to oxygen results in rust. The process of oxidation can be very slow; indeed, it can take years for oxidation to become evident. Slow oxidation does not produce easily measurable heat. **Combustion**, by contrast, is a rapid chemical process in which the combination of a substance with oxygen produces heat and light. For fire fighters' purposes, the terms *combustion* and *fire* can be used interchangeably. **Pyrolysis** is the decomposition of a material brought about by heat in the absence of oxygen.

■ Products of Combustion

The classic reaction for burning that results in water and carbon dioxide does not occur in the real fire situations fire fighters encounter. Actual fires burn without an adequate supply of oxygen, which results in incomplete combustion and produces a variety of toxic by-products. These by-products are collectively called **smoke**. Smoke includes three major components: particles, which are solids; vapours, which are finely suspended liquids—that is, aerosols; and gases **FIGURE 6-5**.

Smoke particles include unburned, partially burned, and completely burned substances. The unburned particles are lifted in the **thermal column** produced by the fire. Some partially burned particles become part of the smoke because inadequate oxygen is available to allow for their complete combustion. Completely burned particles are primarily ash. Some particles in smoke are small enough that they can get past the protective mechanisms of the respiratory system and enter the lungs, and most of them are very toxic to the body.

Smoke may also contain small droplets of liquids. The fog that is formed on a cool night consists of small water droplets suspended in the air. When water is applied to a fire, small droplets may also be suspended in the smoke or haze that

FIGURE 6-5 Smoke is composed of solids (ashes), vapours (aerosols or small droplets), and gases.

* – ashes
33 – gases
::: – aerosols

forms. Similarly, when oil-based compounds burn, they produce small oil-based droplets that become part of the smoke. Oil-based or lipid compounds can cause great harm to a person when they are inhaled. In addition, some toxic droplets cause poisoning if absorbed through the skin.

Smoke contains a wide variety of gases. The composition of gases in smoke will vary greatly, depending on the amount of oxygen available to the fire at that instant. The composition of the substance being burned also influences the composition of the smoke. In other words, a fire fuelled by wood will produce a different composition of gases than a fire fueled by petroleum-based fuels, including plastics.

Almost all of the gases produced by a fire are toxic to the body, including carbon monoxide, hydrogen cyanide, and phosgene. Carbon monoxide is deadly in small quantities and was used to kill people in gas chambers. Hydrogen cyanide was also used to kill convicted criminals in gas chambers. Phosgene gas was used in World War I as a poisonous gas to disable soldiers. Even carbon dioxide, which is an inert gas, can displace oxygen and cause **hypoxia**. It is important to wear self-contained breathing apparatus (SCBA) at the scene for 24 hours post fire to ensure that all off-gassing has ceased. Tests have shown that dangerous off-gassing can still occur up to 24 hours after a fire has been extinguished.

A discussion of the by-products of combustion would be incomplete without considering heat. Because smoke is the result of fire, it is hot. The temperature of smoke will vary depending on the conditions of the fire and the distance the smoke travels from the fire. Injuries from smoke may occur because of the inhalation of the particles, droplets, and gases that make up smoke. The inhalation of superheated gases in smoke may also cause injuries in the form of severe burns of the skin and the respiratory tract.

■ Fire Spread

Fire fighters work hard to prevent fires and to extinguish them. Fire fighters' enemy is fire, and one way to outsmart this enemy is by learning to think like the enemy. By learning how fires act, fire fighters can learn how to use their suppression forces most effectively. Fires grow and spread by four primary mechanisms: direct contact, conduction, convection, and radiation.

Smoke is a product of incomplete combustion and is primarily composed of hydrocarbons. As a consequence, when smoke is combined with adequate oxygen and heat, it will burn. Smoke contains large amounts of potential energy, which can be released violently under the right conditions.

Smoky environments are deadly! It is essential to use SCBA whenever you are operating in a smoky environment. This rule applies whether you are dealing with a structure fire, a dumpster fire, a car fire, or any other type of fire that generates smoke. Toxic gases can even be present during the overhaul stage of a fire. The following toxic gases are commonly found in smoky environments:

- Carbon monoxide
- Hydrogen chloride
- Hydrogen cyanide
- Carbon dioxide
- Phosgene
- Ammonia
- Chlorine

FIGURE 6-6 Direct contact.

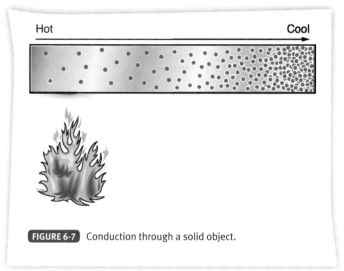

FIGURE 6-7 Conduction through a solid object.

Direct Contact

Direct contact can quickly spread a fire. Direct contact can be as simple as a lit cigarette dropping onto a piece of paper in a wastebasket, embers from a wildland fire dropping onto dry pine needles, or flying embers from a structure fire landing on a wooden roof **FIGURE 6-6** . Many ground and wildland fires are at least partly spread by direct contact.

Conduction

Conduction is the process of transferring heat through matter by movement of the kinetic energy from one particle to another **FIGURE 6-7** . Conduction transfers energy directly from one molecule to another, much as a billiard ball transfers its energy to the next. Objects vary in their ability to conduct energy. Metals generally have a greater ability to conduct heat than wood does, whereas a substance such as fibreglass (used for insulation) has almost no ability to conduct heat or fire.

Objects that are good conductors tend to absorb heat and conduct it throughout; for example, heat applied to a steel beam will be readily conducted along the beam. Because the heat spreads out over the beam, the area of the steel beam to which the heat is being applied will not get as hot as it would if the beam were a poor conductor.

Applying heat to a poor conductor, such as wood, will result in the heat energy remaining in the area of the wood to which the heat is applied. Because the wood has a fairly low ignition temperature, it will catch on fire. Heat applied to the wood acts on a small part of the wood; it is not conducted to other parts of the wood. By comparison, the same amount of heat energy applied to the steel beam will result in less heat at

the point of application, because much of the heat is dissipated to other parts of the steel beam.

This behaviour has serious consequences for fire fighters. If two substances have the same ignition temperature, it will be easier to ignite the substance that is a poor conductor than it is to ignite the substance that is a better conductor. The most important fact to remember about conductivity and fire spread is that poor conductors may ignite more easily but, once ignited, do not spread fire through conduction. Materials that are good conductors are rarely the primary means of spreading a fire.

Fire is rarely spread primarily through conduction.

Convection

Convection is the circulatory movement that occurs in a gas or fluid, such as air or water, with areas of differing temperatures being present in the same medium, owing to the variation of the substance's density and the action of gravity **FIGURE 6-8** . To understand the movement of gases in a fire, consider a container of water being heated on a stove. If you put a drop

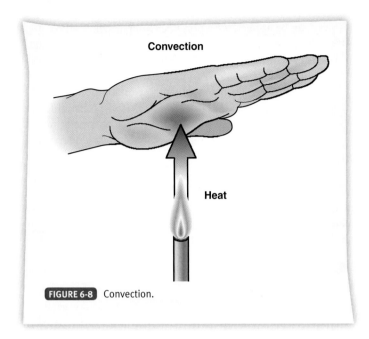

Convection

Heat

FIGURE 6-8 Convection.

FIGURE 6-9 Convection currents in a room-and-contents fire.

of food colouring in one part of the water and start to heat the water, you can readily see the circulation of the water within the container. As the water heats up, it expands and becomes lighter than the surrounding water. This causes a column of water to rise to the top; the cooler water then falls to the bottom. Thus a cycle of warmer water rises to the top, pushing the cooler water to the bottom. The convection currents have the same effect as a small pump pushing the water around in a set pathway.

The convection currents in a fire primarily involve gases generated by the fire. The heat of the fire warms the gases and particles in the smoke. A large fire burning in the open can generate a **plume** of heated gases and smoke that rises high into the air. This convection stream can carry smoke and large bands of burning fuel for several blocks, before the gases cool and fall back to the earth. If winds are present during a large fire, they will influence the direction in which the convection currents travel.

When a fire occurs in a building, the convection currents generated by the fire rise in the room and travel along the ceiling. These currents carry superheated gases, which may ultimately heat flammable materials enough to ignite them **FIGURE 6-9**. If the fire room has openings, convection may then carry the fire outside the room of origin and to other parts of the building.

In structure fires, convection currents are influenced by the layout of the building; for example, air will flow quickly through a structure with an open floor plan versus a structure made up of small, closed rooms. Convection currents also depend on the flow of air currents within the structure. This airflow is greatly influenced by whether doors and windows are closed or open. A final factor that affects convection currents in a structure is the speed and direction of the wind entering the structure. If the wind is strong and is blowing through an opening in the structure, the convection currents will be stronger and move more rapidly. The movement of heated gases and fire can be greatly influenced by wind and ventilation openings.

Radiation

<u>Radiation</u> is the transfer of heat through the emission of energy in the form of invisible waves **FIGURE 6-10**. The sun radiates energy to the earth over the vast expanse of outer space; this electromagnetic radiation readily travels through the vacuum of space. When this energy is absorbed, it becomes converted to heat—that is, the heat you feel as the sun touches your body on a warm day. Of course, the direction in which the radiation travels can be changed or redirected, as when a sheet of shiny aluminum foil reflects the sun's rays and bounces the energy in another direction.

FIGURE 6-10 Radiation.

<u>Thermal radiation</u> from a fire travels in all directions. The effect of thermal radiation, however, is not seen or felt until the radiation strikes an object and heats its surface. Thermal radiation is a significant factor in the growth of a campfire from a small flicker of flame to a fire hot enough to ignite large logs. The growth of a small fire in a wastebasket to a full-blown room-and-contents fire is due in part to the effect of thermal radiation. A building that is fully involved in fire radiates a tremendous amount of energy in all directions. Indeed, the radiant heat from a large building fire can travel one hundred metres to ignite an unattached building.

Methods of Extinguishment

The job of fire fighters is to extinguish fires. Although many variations on the methods used to extinguish fires exist, they boil down to four main methods: cooling the burning material, excluding oxygen from the fire, removing fuel from the fire, and interrupting the chemical reaction with a flame inhibitor **FIGURE 6-11**. There are many variations on the way that these methods can be implemented, and sometimes a combination of these methods is used to achieve fire suppression.

FIGURE 6-11 The four basic methods of fire extinguishment. **A.** Cool the burning material. **B.** Exclude oxygen from the fire. **C.** Remove fuel from the fire. **D.** Interrupt the chemical reaction with a flame inhibitor.

The method most commonly used to extinguish fires is to cool the burning material. The actions fire fighters take to set up a water supply, lay hose lines to the fire, and, in a coordinated manner, "put the wet stuff on the red stuff" are all steps in implementing this method of fire extinguishment.

A second method of extinguishing a fire is to exclude oxygen from the fire. One way to do so is to place the lid on an unvented charcoal grill. Likewise, applying foam to a petroleum fire covers flammable vapours.

Removing fuel from a fire will also extinguish the fire; for example, shutting off the supply of natural gas to a fire being fuelled by the gas will extinguish the fire. In wildland fires, a fire line or control line (i.e., firebreak) cut around a fire puts further fuel out of reach of the fire.

The fourth method of extinguishing a fire is to interrupt the chemical reaction with a flame inhibitor. Halon fire extinguishers work in this way. In the United States, these agents can be applied with portable extinguishers or through a fixed system designed to flood an enclosed space. Halogenated fire-extinguishing systems are often used to protect electronics and computer systems. In Canada, Halon systems are not permitted for use. Their use is contrary to components of the Canadian Environmental Protection Act, 1999. A list of Halon alternatives for use in fire suppression can be obtained through your provincial/territorial authority or through Underwriters Laboratories of Canada (ULC).

Classes of Fire

Fires are generally categorized into one of five classes: Class A, Class B, Class C, Class D, and Class K.

Class A Fires

Class A fires involve ordinary solid combustible materials, such as wood, paper, and cloth **FIGURE 6-12**. Natural vegetation, such as the grass that burns in brush fires, is also considered to be part of this group of materials. The method most commonly used to extinguish Class A fires is to cool the fuel with water to a temperature that is below its ignition temperature.

FIGURE 6-12 A Class A fire involves wood, paper, or other ordinary combustibles.

Class B Fires

Class B fires involve flammable or combustible liquids, such as gasoline, kerosene, diesel fuel, and motor oil **FIGURE 6-13**. Fires involving gases such as propane or natural gas are also

FIGURE 6-13 A Class B fire involves flammable liquids, such as gasoline.

classified as Class B fires. These fires can be extinguished by shutting off the supply of fuel or by using foam to exclude oxygen from the fuel.

■ Class C Fires

Class C fires involve energized electrical equipment FIGURE 6-14. They are listed in a separate class because of the electrical hazard they present. Attacking a Class C fire with an extinguishing agent that conducts electricity can result in injury or death to the fire fighter. Once the power is cut to a Class C fire, the fire is treated as a Class A or Class B fire, depending on the type of material that is burning.

■ Class D Fires

Class D fires involve combustible metals, such as sodium, magnesium, and titanium FIGURE 6-15. These fires are assigned

FIGURE 6-14 A Class C fire involves energized electrical equipment.

FIGURE 6-15 A Class D fire involves metals such as sodium or titanium.

Titanium shavings

to a special class, because the application of water to fires involving these metals will result in violent explosions. Instead, these fires must be attacked with special agents to prevent explosions and to smother the fire and remove the supply of oxygen to the fire.

■ Class K Fires

Class K fires involve combustible cooking oils and fats in kitchens FIGURE 6-16. Heating vegetable or animal fats or oils in appliances such as deep fryers can result in serious fires, which are difficult to fight with ordinary fire extinguishers. Special Class K extinguishers are available to handle this type of fire.

FIGURE 6-16 A Class K fire involves a deep fryer in a kitchen.

Some fires may fit into more than one class; for example, a fire involving a wood building could also involve petroleum products. Likewise, a fire involving energized electrical circuits (a Class C fire) might also involve Class A or Class B materials. If a fire involves live electrical sources, it should be treated as a Class C fire until the source of electricity has been disconnected.

Fire Fighter Safety Tips

© Photos.com

Why do metal fires explode when water is applied to them? When water is applied to certain heated metals, the water (H_2O) reacts with the heated metal and produces hydrogen (H_2), which burns with an explosive force.

Characteristics of Solid-Fuel Fires

Most of the fires encountered by fire fighters involve solid fuels. This section uses wood as an example of a solid fuel, because it is the most commonly encountered solid fuel, and because it shares many characteristics with other solid fuels.

Solid fuels have a definite form and a defined shape. Wood conducts little heat, so the heat acts only on its surface. As a result, the fuel reaches its ignition temperature quickly and ignites. A thin piece of wood burns quickly, because a large surface area is exposed to the heat, but little mass is available to distribute the heat. The greater the ratio of surface area to mass, the faster the wood will ignite and burn—which explains why it is easier to ignite a matchstick than a large log. The speed with which a solid fuel burns is related to the amount of surface that is exposed: the larger the exposed surface, the faster the fuel will burn. Moisture content also influences the speed of combustion: fuels that contain more moisture burn more slowly than fuels that contain less moisture.

Solid fuels do not actually burn in their solid state. Instead, a solid fuel must first be heated or pyrolyzed to decompose it into a vapour before it will burn. Many solid fuels change directly from a solid to a gas; others change from a solid to a liquid before they are vaporized.

Because solid fuels do not generally conduct heat, a heat source applied to the surface of the fuel will cause decomposition of only the surface of the material. The ignition of any solid fuel is related to the temperature of the surface of the material as well as the vapours released through pyrolysis. The temperature of the interior of the material, however, does not affect the speed with which the material can be ignited. If the fuel conducted heat to other parts of the fuel, it would take more heat to raise the temperature of the material to the ignition temperature.

Wood does not have a fixed ignition temperature. Rather, the temperature at which wood will ignite depends on the temperature at which enough flammable vapours to produce the burning reaction are produced. The ignition temperature varies according to the manner and rate at which the heat is applied.

Consider paper as a fuel: Paper is made from wood. The basic ingredient of paper is cellulose, the same ingredient that is found in wood. The ignition temperature of paper is 218°C (425°F). Paper ignites because its surface temperature increases rapidly, which allows for rapid burning and a high heat-release rate.

■ Solid-Fuel Fire Development

Solid-fuel fires progress through four phases: the ignition phase, the growth phase, the fully developed phase, and the decay phase. To illustrate the behaviour of a fire through these phases, we will use a wooden dollhouse as an example. We will make three assumptions:
1. The dollhouse is placed on a sandy surface that will not burn.
2. There are no exposures to which the fire can spread.
3. There is no wind to influence the behaviour of the fire.

Ignition Phase

The **ignition phase** begins as a lighted match is placed next to a crumpled piece of paper. The heat from the match ignites the paper, which sends a small plume of fire upward **FIGURE 6-17**. The heat generated from the paper sets up a small convection current, and the flame produces a small amount of radiated energy. The combination of convection and radiation serves to heat the fuel around the paper. Because the convection acts more in an upward direction, the wood in the dollhouse above the paper will be heated to the ignition temperature first. It is small, so its surface can be easily heated enough to release flammable vapours from the wood.

Growth Phase

The **growth phase** occurs as the dollhouse starts to burn, increasing the convection of hot gases upward **FIGURE 6-18**. The hot gases and the flame both act to raise the temperature of the wood located above the starting point for the fire. Energy generated by the growing fire starts to radiate in all directions. The convection of hot gases and the direct contact with the flame cause major growth to occur in an upward direction. The energy radiated by the flame will also cause some growth of the fire in a lateral direction. The major growth will be in an upward direction, however, because the radiation and

FIGURE 6-17 The ignition phase of a fire.

FIGURE 6-18 The growth phase of a fire.

convection both act in an upward direction, whereas growth beside the flame is primarily a function of radiation alone. The net result of the growth phase is that the fire grows from a tiny flame to a fire that involves the heavier wood in the dollhouse.

Fully Developed Phase

The <u>fully developed phase</u> produces the maximum release of energy **FIGURE 6-19**. All available fuel in the dollhouse has ignited, and heat is being produced at the maximum rate. During the fully developed phase, thermal radiation extends in all directions around the fire. The fire can spread downward by radiation and by flaming material that falls on unburned fuel. The fire can spread upward through the thermal column of hot gases and flame above the fire. In fact, a large free-burning fire can carry a thermal column one hundred metres into the air.

A fully developed fire burning outside is limited only by the amount of fuel available. By comparison, a fully developed fire in an enclosed building is usually limited by the amount of oxygen that is available to it. Inside a structure, the fully developed phase lasts as long as a large supply of fuel and oxygen are available.

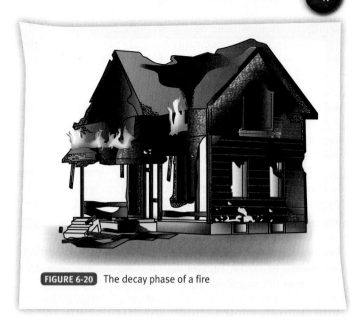

FIGURE 6-20 The decay phase of a fire

FIGURE 6-19 The fully developed phase of a fire.

Fire Fighter Safety Tips

Fire fighters must learn to recognize the warning signs of flashover and use the reach of their hose stream to cool the atmosphere, thereby preventing this condition. A fire fighter—even one who is wearing full turnout gear—has little chance of avoiding severe injury or death if he or she is caught in a flashover.

Decay Phase

The final phase of a fire is the <u>decay phase</u>, the period when the fire is running out of fuel **FIGURE 6-20**. During the decay phase, the rate of burning slows down because less fuel is available. Likewise, the rate of thermal radiation decreases. The amount of hot gases rising above the fire also decreases, because less heat is available to push the gases aloft, and smaller amounts

of gases are being produced by the decreasing volume of fire. Eventually, the flames become smouldering embers, and the fire goes out because of a lack of fuel.

Key Principles of Solid-Fuel Fire Development

Several important principles of fire behaviour are illustrated by the dollhouse fire:

- Hot gases and flame are lighter and tend to rise.
- Convection is the primary factor in spreading the fire upward.
- Downward fire spread occurs primarily from radiation and falling chunks of flaming material.
- If there is not more fuel above or beside the initial flame that can be ignited by convection or radiated heat, the fire will burn out before reaching the growth phase.
- Variations in the direction of fire spread will occur if and when air currents deflect the flame.
- The total material burned reflects the intensity of the heat and the duration of the exposure to the heat.
- An adequate supply of oxygen must be available to fuel a free-burning fire, although some parts of the flame may have a limited supply of oxygen.

The dollhouse example focused on a relatively simple form of fire. Fire behaviour becomes more complex, of course, when fires occur within an actual building.

■ Characteristics of a Room-and-Contents Fire

A fire in a building is not just a fire brought inside; that is, the construction of the building and the contents of its rooms have a major impact on the behaviour of the fire.

Room Contents

Fifty years ago, rooms contained many products made from natural fibres and wood. Today, synthetic products are all around us—not only in buildings but also in all parts of the transportation industry. This change in room contents has changed the behaviour of fires. The synthetic products that are so prevalent today—especially plastics—are usually made from petroleum products. When heated, most plastics will pyrolyze to volatile products. These by-products of the combustion process are usually not only flammable but also toxic. Some will melt and drip; if these droplets are on fire, they can contribute to the spread of the fire. In addition, burning plastics generate dense smoke that is rich in flammable vapours, which makes fire suppression both more difficult and more dangerous. Plastics burn at a higher temperature than wood and plant-based products and generate more heat energy.

Walls and ceilings are often painted. Newer paints are generally emulsions of latex, acrylic, or polyvinyl acetate. When these paints are applied, they form a plastic-like coating. This coating has the characteristics of the plastic from which it was formed and burns readily. Similarly, varnishes and lacquers are very combustible. Most paints and coatings will accelerate the fire present in a building, and their presence may aid in the spread of the fire from one room to another. If walls are not painted, they are usually covered with flammable wallpaper or other plastic-based wall coatings, such as fake-wood panelling.

Carpets made today have low melting points and ignite readily. The backing of carpets is often polypropylene (polyurethane foam is also commonly used for this purpose). Carpet is readily ignitable by radiant heat, even when it is located some distance away from a fire. The huge quantities of black smoke released from burning carpets are similar to the smoke released from burning tires. This smoke consists of unburned fuel, which can burn explosively when exposed to sufficient oxygen.

Furniture manufactured today is made from polyurethane and has improved resistance to heat from glowing sources, such as cigarettes, but has almost no resistance to ignition from flaming sources. Such furniture can be completely involved in fire in 3 to 5 minutes, and it may be reduced to a burning frame in just 10 minutes.

In summary, the increased use of plastics in furniture, bedding, paints, wallcoverings, and carpets sets the stage for hot, rapidly burning, complex fires. To see how such a fire might evolve, we will consider a fire in an ordinary bedroom that has painted walls and wall-to-wall carpeting. The room is furnished with a bed, two nightstands, a lamp, a vanity, and a wastebasket.

Ignition Phase

The ignition phase begins when a fallen candle ignites the contents of the plastic wastebasket next to the vanity **FIGURE 6-21**. The fire starts small, with a localized flame. This phase typically produces an open flame. The concentration of oxygen present at this point is 21 percent—the same as the percentage in the room air.

As more of the contents of the wastebasket are ignited, a plume of hot gases rises from the wastebasket. Combustible materials in the path of the flame begin to ignite, which increases the extent and intensity of the flame. The convection of hot gases is the primary means of fire growth. Some energy is radiated to the area close to the flame, so that the plastic wastebasket starts to melt and may ignite. Oxygen is drawn

FIGURE 6-21 The ignition phase of a typical room-and-contents fire.

in at the bottom of the flame above the wastebasket. At this point, the fire could probably be extinguished with a portable fire extinguisher.

Growth Phase

During the growth phase, additional fuel is drawn into the fire **FIGURE 6-22**. As more fuel is ignited, the size of the fire increases. The plume of hot gases, smoke, and flames creates a convection current, which carries hot gases to the ceiling of the room. Next, flammable materials in the path of this plume ignite. Flames begin to spread upward and outward. Hot gases and smoke rise because they are lighter; they hit the ceiling and spread out to form a layer at the ceiling. These products of combustion are trapped in the room and continue to affect the growth of the fire, unlike in a free-burning fire that occurs outdoors. Radiation starts to play a greater role in the growth of the room-and-contents fire.

The temperature of the room continues to increase as the fire grows. The room temperature is highest at the ceiling and lowest at the floor level. In addition, as the fire intensifies, the visibility will be greatest at the floor level and poorest at the ceiling.

The growth of the fire may be limited by either the amount of fuel available or the amount of oxygen available. If the room

FIGURE 6-22 The growth phase of a typical room-and-contents fire.

in which the fire is burning is noncombustible, the only fuel available will be the contents of the room. Likewise, if the doors and windows of the room are closed or if the house has no doors or windows open to the outside, a limited amount of oxygen will be present. Either of these conditions may limit the growth of the fire. If flames reach the ceiling, however, they are likely to trigger involvement of the whole room.

Fully Developed Phase

As the fire develops, temperatures increase to the point where the flammable materials in the room are undergoing pyrolysis. In turn, large amounts of volatile gases are being released. If the temperature becomes high enough to ignite the materials in the room and sufficient oxygen is present, a condition called **flashover** will occur. In flashover, the temperature in the room reaches a point where the combustible contents of the room ignite all at once. This temperature varies depending on the ignition temperature of the room's contents. Flashover is the final stage in the process of fire growth.

Flashover is not a specific moment, but rather, the transition from a fire that is growing by igniting one type of fuel to a fire where all of the exposed fuel in the room is on fire. If a fire does not have enough ventilation to supply sufficient oxygen, it will not flash over until sufficient oxygen is introduced. If the ventilation openings are too large, the fire may not reach a temperature high enough for the whole room to reach the flashover point. The critical temperature for a flashover to occur is approximately 540°C (1000°F). Once this temperature is reached, all fuels in the room are involved in the fire, including the floor coverings. Indeed, the temperature at the floor level may be as high as the temperature at the ceiling was before flashover. Fire fighters, even when wearing full personal protective equipment (PPE), cannot survive for more than a few seconds in a flashover.

Once the room flashes over, the fire is fully developed. All of the combustible materials are involved in the fire, and the burning fuels are releasing the maximum amount of heat. This condition is sometimes referred to as steady state burning. The burning gases at the ceiling level radiate heat to combustible materials in the room. Carpets and other objects can ignite from this radiant heat; melting plastics can ignite and drop flaming particles onto materials below them. An average-sized furnished room with an open door will be completely involved in a flashover within 5 to 10 minutes of the ignition of the fire.

The amount of fire generated depends on the amount of oxygen available. For the fire to generate the maximum amount of heat, ventilation must be adequate to supply the fire with sufficient oxygen. If openings in the room let oxygen in, those same openings will serve as an escape route for hot gases. In this way, the hot gases can spread the fire outside the original fire room. The total fire damage to the room is the result of the intensity of the heat applied and the amount of time the room is exposed to the heat **FIGURE 6-23**.

Fire-suppression efforts can greatly influence the spread of the fire. In particular, positive-pressure ventilation can force

FIGURE 6-23 The fully developed phase of a typical room-and-contents fire.

the fire into an area that may not have already been involved. The fire will follow the path of the convection currents of smoke and superheated gases.

Decay Phase

The last phase of a fire is the decay phase **FIGURE 6-24**. During the decay phase, open-flame burning decreases to the point where only smouldering fuel is present. A large amount of heat buildup means that the room will remain very hot, even though the amount of heat being produced decreases. Although a large amount of heat is available, the amount of fuel available to be pyrolyzed decreases, so a smaller amount of combustible vapours is available. The amount of fuel available dwindles as the fuel supply is exhausted.

A fire in the decay phase will become a smouldering fire and eventually go out when its fuel supply is exhausted. During this phase, the compartment temperature will start to decrease, though it may remain high for a long period of time. A smouldering fire may continue to produce a large volume of toxic gases, however, and the compartment may remain dangerous, even though the fire appears to be under control.

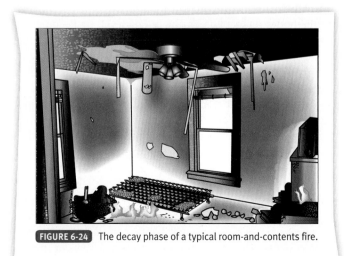

FIGURE 6-24 The decay phase of a typical room-and-contents fire.

■ Special Considerations

Being a fire fighter is analogous to being a sailor: A sailor must understand the wind and waves and work with them to achieve the objective of moving the boat in the desired direction. Likewise, a fire fighter must understand the behaviour of fires and work to control the power of the fire rather than trying to attack it in a reckless manner. Fire fighters need to understand a variety of special conditions to work safely to extinguish fires: thermal layering of gases, flameover, flashover, and backdraft.

Thermal Layering

Thermal layering is the tendency of gases in an enclosed space to form in layers according to their temperature. The gases rise as they are heated and form different layers within a room. The hotter the gases are, the lighter they become, owing to the increased speed of their molecules. As a consequence, the hottest gases travel by convection currents to the top level of the room.

During a fire in a closed room, gases are said to be in thermal balance when they are allowed to seek their own level. The temperature at the floor level is relatively low compared to the hotter temperature at the ceiling level. By keeping low when there is a normal thermal balance (e.g., crawling on the floor), you stay in the area where the temperature is the lowest; put simply, you increase your chance of survival with proper personal protective gear by staying low in the room.

If this normal thermal balance is upset, severe burn injuries can occur. For example, when water is sprayed into the upper part of a room, some of the water will be converted to steam. Steam takes up many times more space than the same amount of liquid water. The steam generated by the application of the water will flood the room and can result in severe burns to fire fighters, even when they are in full personal protective gear. This outcome occurs because the ensuing thermal imbalance replaces the normal layering of gases with superheated steam. The energy in the superheated gases at the top of the room converts the water to steam, and the superheated steam expands and can fill the whole room.

It is important to understand the concept of thermal layering so that you can avoid creating a thermal imbalance. Two actions are important: First, fire fighters must work together to make sure the superheated gases are being vented from the fire room as they are attacking the fire. Second, they must avoid directing water at the ceiling of the fire room except for short bursts that serve to cool the superheated gases. Superheated steam can cause severe burns even through approved personal protective gear. Use proper fire-suppression techniques to avoid creating a thermal imbalance.

Flameover

Flameover, or **rollover**, is the spontaneous ignition of hot gases in the upper levels of a room or compartment. During the growth phase of the fire, the hottest gases rise to the top of the room. When these gases cool, they fall and are pulled back into the fire by convection currents. When one of the gases in this mixture reaches its ignition temperature, it will ignite either from the radiation from the flaming fire or from direct contact with open flames; in other words, the upper layer of flammable vapour catches fire. These flames can flicker across the ceiling and then go out. Alternatively, if they get a few degrees hotter, they can extend throughout the room at ceiling level. Rollover is a sign that the temperature is rising—and if it continues to rise, the temperature throughout the room will soon reach the point where the room and contents will spontaneously and rapidly ignite. Rollover is a sign that flashover is imminent if actions are not taken to change the fire conditions.

Flashover

If the temperatures continue to rise in a room where the flameover is occurring, it will get hot enough to cause a flashover. A flashover is the near-simultaneous ignition of most of the exposed combustible material in an enclosed area or room; in other words, it is as if the entire room erupts into flames within seconds. These flames are not limited to the contents of the room. Because the fire has heated the contents of the room to a point where much of the combustible contents are vaporized, vaporized fuel is present throughout the room. A flashover occurs because the room is filled with a superheated highly flammable gas that suddenly ignites. In this circumstance, there is sufficient heat to cause autoignition of the vaporized and superheated fuel in the room.

In recent controlled experiments, ULC documented that temperatures can rise from 250 to 600°C (480 to 1110°F) in less than 3 minutes in a flashover. Because these conditions can change so rapidly, it may be impossible to exit a structure fast enough to escape injury or death. The temperatures in a flashover are not survivable for fire fighters even when they are wearing full PPE.

Backdraft

A **backdraft** is caused by the introduction of oxygen into an enclosure where the superheated gases and contents are already hot enough for ignition but do not have sufficient oxygen to combust. The development of a backdraft requires a unique set of conditions. When a fire generates quantities of combustible gases, these gases can become heated above their ignition temperatures. Backdrafts require a "closed <u>box</u>"—that is, a room or building that has a limited supply of oxygen. When the fire chamber has a limited amount of oxygen, the oxygen concentration will be reduced, leading to decreased combustion. If a supply of new oxygen is then introduced into the room, explosive combustion can occur, owing to the presence of the superheated gases. This kind of explosive combustion may exert enough force to cause great injury or death to fire fighters.

Signs and symptoms of an impending backdraft include the following:

- Any confined fire with a large heat buildup
- Little visible flame from the exterior of the building
- A "living-fire", where the building appears to be breathing because of smoke puffing out of, and then being sucked back into, the building
- Smoke that seems to be pressurized
- Smoke-stained windows (an indication of a significant fire)
- No smoke showing
- Turbulent smoke
- Thick, yellowish smoke (containing sulphur compounds)

It is important for fire fighters to look out for the conditions that signal a possible backdraft.

What triggers a backdraft when the conditions are right is the introduction of oxygen into a room that was lacking a sufficient oxygen supply to support a fire. A backdraft can occur when a window breaks because of excess heat, when a fire fighter opens a closed door, or when a fire fighter tries to ventilate a fire by opening a door or breaking a window. It is important to reduce the possibility of a backdraft developing. Fire fighters must coordinate the application of water on the fire with efforts to ventilate the fire. More about the techniques of ventilation are presented in the chapter "Ventilation."

FIRE FIGHTER Tips

Environmental conditions may dramatically affect the behaviour of a fire. Solid fuels that are hot and dry represent an increased fire hazard, for example. Increased temperature dehydrates the fuel, so that these kinds of dry materials catch fire more readily.

The heat from a fire can also dry the fuel as part of the burning process. This development has the largest effect during the initial phases of the fire.

Decreased humidity has only a small effect on interior fires but may drive the development of outdoor fires.

Fire Behaviour in Modern Structures

Modern construction techniques have altered the way fire behaves in structures, owing to two major changes. The first has to do with building construction. Houses built in the last 20 or 30 years are constructed to be energy and cost efficient. They contain insulation to prevent heat loss in the winter and the loss of cool air in the summer. In addition, they are sealed to prevent air exchange through the walls and around the

doors and windows. As a consequence, when a fire occurs in the house, only a limited quantity of air can enter the structure while the doors and windows are closed. In many parts of the country, a home's doors and windows are kept closed most of the year, because the building's temperature is being regulated by heating and cooling systems.

Many newer houses are constructed with lightweight manufactured building components. These manufactured building components contain many highly flammable products, such as glues and coatings, most of which are petroleum based. These products are not only flammable but also burn hotter and faster than wood, the primary material used in older houses. The lightweight frames of these houses also fail much faster than the heavier wood frames of older homes. The lightweight construction also provides greater surface area of structural members and provides greater void spaces between floors.

The second major change relates to the widespread use of plastics and petroleum-based products for furniture and accessories. These petroleum-based products contain much more energy than wood and natural-fibre furnishings. They catch fire easily, reach high temperatures quickly, and release huge quantities of thick, dark smoke while burning. This rapid combustion requires huge amounts of oxygen to maintain the growth phase (i.e., to remain burning).

These changes in construction materials and techniques as well as changes in buildings' contents have, in turn, changed the dynamics of fire behaviour inside a structure. Recent tests by ULC documented a pattern of fire behaviour that had been previously observed by veteran fire fighters. When a fire is ignited in a modern structure, it progresses to the fully developed phase quickly until it starts to encounter decreased levels of oxygen. As the concentration of oxygen in the structure declines, the growth of the fire decreases, as reflected by a sharp drop in temperature. It appears as though the fire has entered the decay phase. This apparent decay is not caused by a decrease in fuel, as is the case in older structures, but rather, to an insufficient supply of oxygen to support an active fire.

It is at this point that the fire department often arrives on the scene. One of the first actions taken by the fire department is to open the front door to gain access to the house. It is critical to note that opening the door has the unintentional effect of introducing a fresh supply of oxygen to the fire. When this fresh supply of oxygen comes in contact with superheated flammable fuel, the result is a rapid and explosive growth of the fire. With this violent development, the fire transitions from being ventilation controlled to being fuel controlled.

To prevent this scenario, fire fighters must achieve proper timing in implementing building entry, ventilation, and the application of water to the fire. More information on these topics is presented in the chapters "Forcible Entry," "Ventilation," "Fire Attack Lines and Foam," and "Fire Suppression."

FIRE FIGHTER Tips

A variety of factors—the structure's layout, construction, and contents, as well as the amount of oxygen available in the structure—influence the speed and direction of fire spread.

Wind Effect

One factor that greatly influences the behaviour of a fire is the wind. When wind is present, the behaviour of the fire changes drastically; for example, when the air is relatively still, the best approach to a fire may be through the A side of a building. Conversely, if the wind is blowing at 32–40 kilometres per hour (20–25 miles per hour) from the C side of the same building, entering on the A side may be a deadly miscalculation, because the wind is pushing the fire to the A side of the building **FIGURE 6-25**. The impact of the wind is similar to placing several large ventilation fans on the C side of the building and pushing huge quantities of air and fire toward the A side of the building.

The potential effect of the wind may not be evident during the initial size-up when the structure is closed tightly. However, because the wind can have a huge effect on fire behaviour, the initial size-up of a structure fire must always include an evaluation of the impact of the wind. This evaluation will sometimes change the plan of attack; for example, if the seat of the fire is located on the C side of the building and a strong wind is blowing in that direction, the incident commander might choose not to begin ventilation operations on the C-side windows in case the rescue crew needs to enter through the A side of the building. Opening up a window on the C side could cause the fire to spread rapidly toward the A side.

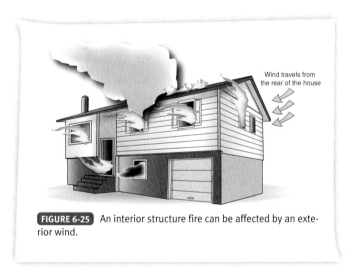

FIGURE 6-25 An interior structure fire can be affected by an exterior wind.

Characteristics of Liquid-Fuel Fires

Fires involving liquid or gaseous fuels have some different characteristics from fires that involve solid fuels. To understand the behaviour of these fires, you need to understand the characteristics of liquid and gaseous fuels.

Recall that solid fuels do not burn in the solid state but instead must be converted to a vapour before they will burn. Liquids share the same characteristic: they must be converted to a vapour and mixed with oxygen in the proper concentration before they will burn. Three conditions must be present for a vapour–air mixture to ignite:

1. The fuel and air must be present at a concentration within a flammable range.
2. There must be an ignition source with enough energy to start ignition.
3. The ignition source and the fuel mixture must make contact for long enough to transfer the energy to the air–fuel mixture.

As liquids are heated, the molecules in the material become more active and the speed of vaporization of the molecules increases. Most liquids will eventually reach their boiling points during a fire. As the boiling point is reached, the amount of flammable vapour generated increases significantly **FIGURE 6-26**. Because most liquid fuels are a mixture of compounds (e.g., gasoline contains approximately 100 different compounds), they do not have a single boiling point. Instead, the flammability of mixture liquid fuel is determined by the compound with the lowest ignition temperature—that is, the temperature at which that compound will spontaneously ignite.

The amount of liquid that will be vaporized is also related to the **volatility** of the liquid: the higher the temperature, the more evaporation. Liquids with a lower molecular weight will tend to vaporize more readily than liquids with a higher molecular weight. As more of the liquid vaporizes, the mixture may reach a point where enough vapour is present in the air to create a flammable vapour–air mixture.

Two additional terms are used to describe the flammability of liquids: flash point and flame point. The **flash point** is the lowest temperature at which a liquid produces a flammable vapour. It is measured by determining the lowest temperature at which a liquid will produce enough vapour to support a small flame for a short period of time (the flame may go out quickly) **TABLE 6-1**. The **flame point** (also known as the **fire point**) is the lowest temperature at which a liquid produces enough vapour to sustain a continuous fire.

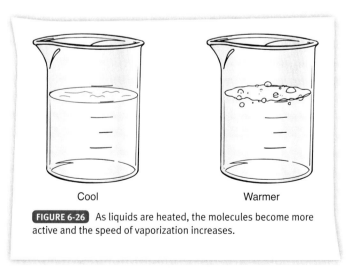

FIGURE 6-26 As liquids are heated, the molecules become more active and the speed of vaporization increases.

TABLE 6-1	Flash Point and Flame Point
Fuel	**Flash Point (°C/°F)**
Gasoline	−43°C (−45°F)
Ethanol	12°C (54°F)
Diesel fuel	40–55°C (104–131°F)
SAE N 10 motor oil	171°C (340°F)

Source: Data from Raymond Friedman, *Principles of Fire Protection Chemistry and Physics*, 3rd ed. (Quincy, MA: National Fire Protection Association, Inc., 1998), 115.

FIRE FIGHTER Tips

Another practical consideration is that liquids flow when not contained. Thus, if the container in which a flammable liquid is kept is damaged, the liquid will spread as the result of spilled or flowing fuel.

Characteristics of Gas-Fuel Fires

By learning about the characteristics of flammable gas fuels, you can help to prevent injuries or deaths in emergency situations and work to mitigate the conditions causing the problem. Two terms are used to describe the characteristics of flammable vapours: vapour density and flammability limits.

■ Vapour Density

<u>Vapour density</u> refers to the weight of a gas fuel and measures the weight of the gas compared to air TABLE 6-2 . The weight of air is assigned the value of 1. A gas with a vapour density less than 1 will rise to the top of a confined space or rise in the atmosphere; for example, hydrogen gas, which has a vapour density of 0.07, is a very light gas. Conversely, a gas with a vapour density greater than 1 is heavier than air and will settle close to the ground; for example, propane gas has a vapour density of 1.52 and will settle to the ground. By comparison, carbon monoxide has a vapour density of 0.97—almost the same as that of air—so it mixes readily with all layers of the air. In situations where a flammable gas is present, fire fighters need to recognize the vapour density of the escaping fuel so that they can take actions to prevent the ignition of the fuel and allow the gaseous fuel to safely escape into the atmosphere.

TABLE 6-2	Vapour Density of Common Gases
Gaseous Substance	**Vapour Density**
Gasoline	>3
Ethanol	1.6
Methane	0.55
Propane	1.52

■ Flammability Limits

Mixtures of flammable gases and air will burn only when they are combined in certain proportions. If too much fuel is present in the mixture, there will not be enough oxygen to support the combustion process; if too little fuel is present in the mixture, there will not be enough fuel to support the combustion process. The range of gas–air mixtures that will burn varies from one fuel to another. Carbon monoxide will burn when mixed with air in concentrations between 12.5 and 74 percent. By contrast, natural gas will burn only when it is mixed with air in concentrations between 4.5 and 15 percent.

The terms <u>**flammability limits**</u> and <u>**explosive limits**</u> are used interchangeably, because, under most conditions, if the flammable gas–air mixture will not explode, it will not ignite. The <u>**lower flammable limit (LFL)**</u> refers to the minimum amount of gaseous fuel that must be present in a gas–air mixture for the mixture to be flammable. In the case of carbon monoxide, the lower limit is 12.5 percent. The <u>**upper flammable limit (UFL)**</u> of carbon monoxide is 74 percent. Test instruments are available to measure the percentage of fuels in gas–air mixtures and to determine when an emergency scene is safe.

■ Boiling Liquid/Expanding Vapour Explosions

One potentially deadly set of circumstances involving liquid and gaseous fuels is a <u>**boiling liquid/expanding vapour explosion (BLEVE)**</u>. A BLEVE occurs when a liquid fuel is stored in a vessel under pressure. If the vessel is filled with propane, for example, the bottom part of the vessel would contain liquid propane and the upper part of the vessel would contain vaporized propane FIGURE 6-27 .

If this sealed container is subjected to heat from a fire, the pressure that builds up from the expansion of the liquid will prevent the liquid from evaporating. Normally, evaporation cools the liquid and allows it to maintain its temperature. If heating continues, the temperature inside the vessel will reach a level that exceeds the boiling point of the liquid. At some point, the vessel will fail, releasing all of the heated fuel in a massive explosion. The released fuel instantly becomes vaporized and ignited as a huge fireball.

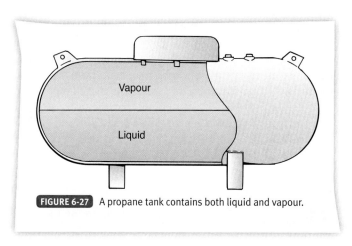

FIGURE 6-27 A propane tank contains both liquid and vapour.

The key to preventing a BLEVE is to cool the top of the tank, which contains the vapour. This action will prevent the fuel from building up enough pressure to cause a catastrophic rupture of the container. Prevention of BLEVEs is covered in more detail in the chapter "Fire Suppression."

Smoke Reading

Learning the principles underlying fire behaviour helps fire fighters to understand the rules that govern the way fires burn. One practical application of these principles is learning how to read the smoke at a fire. Being able to read smoke enables you to learn where the fire is, how big it is, and where it is moving. At a fire, most untrained people look at flames. Flames indicate where the fire is now, but they do not tell you how big the fire is and where it is moving. Fires are dynamic events: what you see this minute will probably change quickly. Before developing a plan of attack, fire fighters need to anticipate where the fire will be in a few minutes. At a structure fire, there are often no visible flames, because the fire is occurring inside the building. The ability to read smoke gives fire fighters the information that they need to mount a more effective attack on a fire and may help to save either their lives or the lives of the building's occupants.

It is helpful to think about smoke as being a fuel. Fuels behave in predictable ways when they are combined with the right mixture of oxygen and heat; for example, flashovers occur when fuels in a room or confined space are heated until they ignite simultaneously. Most of the fuel in a flashover is in the form of smoke. A backdraft is the sudden, explosive ignition of fire gases that occurs when oxygen is introduced into a super-heated space with limited oxygen. As in a flashover, these fire gases are really smoke. Both flashovers and backdrafts are fed by superheated fuel in the form of smoke; thus, as you study the art of smoke reading, think of this exercise as studying the fuel that is all around you at a fire.

As mentioned earlier, smoke is composed of three major components: particles, vapours (aerosols), and gases. The solids consist of carbon, soot, dust, and fibres; the vapours contain suspended hydrocarbons and water; and the gases consist of carbon monoxide and a wide variety of other gases. Taken together, these three components of smoke largely represent fuels that will burn when heated sufficiently in the presence of oxygen; therefore, hot smoke is extremely flammable and will ultimately dictate the fire's behaviour. Smoke today is not the smoke of yesterday. Most of today's building materials and building contents are plastics, which are made from petroleum products. As a consequence, these materials give off more toxic gases and burn at higher temperatures than the materials used for construction and contents in earlier times.

The best place to observe patterns of smoke is from the outside of the fire building. Compare the smoke coming out of openings in different parts of the building. It will tell you more than looking at the flames.

■ Step 1: Determine the Key Attributes of Smoke

The first step in reading smoke is to consider its four key attributes: **smoke volume**, **smoke velocity**, **smoke density**, and **smoke colour**. You also need to factor in the size of the

building, its layout, the wind, and any other characteristics of the building that might change the appearance of these key attributes.

The volume of smoke coming from the fire gives some idea of how much fuel is being heated to the point that it gives off gases (off-gassing). As you assess the smoke volume, also consider the size of the burning building, or box. It takes relatively little smoke to fill up a small building, but a lot of smoke is required to fill up a large building; thus you must consider the size of the box when considering the amount of smoke coming from it. Assessing the volume of smoke alone will not provide a complete picture of the fire, but it sets the stage for a better understanding of the fire.

The velocity (speed) at which smoke is leaving the building suggests how much pressure is accumulating in the building. Smoke is pushed both by heat and by volume. When smoke is pushed by heat, it will rise and then slow down gradually; when smoke is pushed by volume, it will slow down immediately. Assess the smoke velocity to determine whether it is being pushed by heat or by volume.

Next, consider whether the smoke has a laminar flow or a turbulent flow. **Laminar smoke flow** is a smooth or streamlined flow. It indicates that the box (the building and its contents) is absorbing heat and the pressure in the box is not too high **FIGURE 6-28**. By contrast, **turbulent smoke flow** is agitated, boiling, or angry. This type of flow is caused by rapid molecular expansion of the gases within the smoke and the restrictions created by the building **FIGURE 6-29**. This expansion occurs when the box cannot absorb any more heat. In such circumstances, the radiant heat that the box has absorbed is radiated back into the smoke. Turbulent smoke contains an immense amount of energy. When this energy reaches a point where the smoke is heated to its ignition temperature, flashover will occur. Whenever you see turbulent smoke, be aware that flashover is likely to occur very soon.

Next, compare the velocity or speed of the smoke leaving different openings of the building. By comparing the speed and type of flow from similar-sized openings, fire fighters can get a good idea of where the fire is located. A similar-sized opening with a higher velocity of smoke is closer to the location of the fire. Look for the fastest smoke coming from the most restrictive opening. Studying the velocity of smoke provides valuable information as fire fighters prepare to attack the fire.

Evaluate the thickness of the smoke. Smoke density suggests how much fuel is contained in the smoke: the denser the smoke is, the more fuel it contains. Dense smoke can produce a powerful flashover when conditions are right. In fact, thick smoke can flash over even without turbulent flow, because the supply of fuel reaches from the heart of the fire to the location where the smoke is issuing from the building.

When you think of smoke as fuel, you quickly realize that fire fighters who are surrounded by dense smoke are working in a pool of flammable fuel. While no one would send a fire fighter into a pool of petroleum, we are doing just that when we send a fire fighter into dense smoke.

Moreover, dense smoke contains many poisonous substances. When fire fighters are in an environment of dense

The colour of smoke gives you some indication of which stage the fire is in and which substances are burning. When a single substance is burning, the colour of the smoke provides valuable clues about what that substance is. Of course, most of the fires involve multiple substances burning, so using smoke colour alone to determine what is burning may not always yield a clear answer.

When the first fire fighters on the scene evaluate smoke colour, they can determine the stage of heating and obtain information about the location of the fire in the building. Most solid materials will emit a white smoke when they are first heated. This white colour is primarily a result of moisture being released from the material. The same effect is apparent when fire fighters apply water to a fire: the smoke goes from black to white. As the materials dry out, however, the colour of the smoke changes. Smoke from burning wood, for example, changes to tan or brown. Plastics and painted or stained surfaces emit a grey smoke, which is a combination of black from the hydrocarbons and white from the escaping moisture. As materials become hotter, the smoke becomes blacker: the blacker the colour, the hotter the smoke.

The colour of smoke can also help fire fighters determine the location of the fire. As smoke travels, its heat causes the moisture from some of the materials it passes through to evaporate. This added moisture tends to change black smoke to lighter smoke the further it travels from the fire. In addition, the carbon suspended in the black smoke settles out and is filtered by the materials it passes through; therefore, carbon-rich black smoke often becomes lighter in colour as it travels farther from the initial fire site, because of the loss of carbon from the smoke.

How do you differentiate between white smoke caused by early heating and white smoke from a hot fire that has travelled a distance? White smoke that is lazy or slow indicates early heating; white smoke that has its own pressure indicates smoke from a hot fire that has travelled for some distance.

Black Fire

<u>Black fire</u> is a high-volume, high-velocity, turbulent, ultra-dense, black smoke. You can think of it as a form of fire, because in many ways it is. Black fire is so hot (up to nearly 540°C [1000°F]) that it produces as much destruction as flames would. Black fire can produce charring as well as heat damage to steel and concrete. Its presence indicates impending autoignition and flashover; thus black fire is often just seconds away from becoming a deadly environment even for fire fighters wearing full PPE. Because black fire is not tenable for fire fighters in full protective equipment, it certainly is not tenable for unprotected occupants of a building. When black fire occurs, there are usually no lives to be saved in that location.

■ Step 2: Determine What Is Influencing the Key Attributes

Consider the size of the structure that is on fire. How big is the box in which the fire is contained? A small amount of smoke rapidly fills up a small building, so it does not take a very big fire to fill a small building with smoke. Conversely, in a large building such as a big box store or a home-improvement

FIGURE 6-28 Laminar smoke flow.

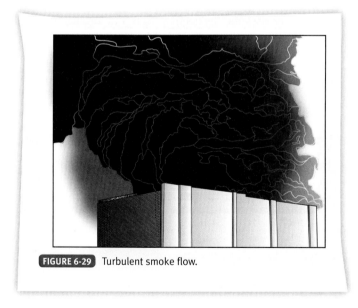

FIGURE 6-29 Turbulent smoke flow.

smoke that has banked down to the floor, they are in an environment that will not support life for a building occupant. Just a few breaths of this toxic brew will result in death. Also, if a flashover occurs, this environment will immediately change to a deadly environment for fire fighters—even those in full turn-out gear and SCBA. There is no sense in a fire fighter risking his or her life to try to save someone who is already dead.

centre, it takes a large fire to produce enough smoke to fill the whole building and pressurize the smoke coming out of the building. A fire fighter who pulls up to a large building that has signs of smoke from the outside should realize that only a significant fire would produce enough smoke to vent to the outside. Always consider the size of the box when examining smoke for clues about the fire.

Also consider how wind, thermal balance, ventilation openings, and an operating sprinkler system might affect the normal characteristics of smoke. Wind can change the direction in which the smoke travels. The normal thermal balance in a room or building can be affected by the heating and air-conditioning systems, especially in a high-rise building. Ventilation openings are made with the intention of changing the normal flow of smoke, so the characteristics of the smoke would naturally be expected to change as the building is ventilated. An operating sprinkler system will cause smoke to be cooled; this cool smoke will then hang close to the bottom of the room, making it difficult to see.

■ Step 3: Determine the Rate of Change

To complete the smoke reading, determine the rate of change for the event. Remember that flames indicate what is happening now, whereas smoke gives a more complete picture of the characteristics of the fire and where it is going. Are the volume of smoke, the velocity of smoke, the density of smoke, and the colour of smoke changing? How are they changing? How rapidly are these changes occurring? What do these changes suggest about the progression of the fire?

■ Step 4: Predict the Event

To assess the size and location of a fire, it is necessary to first consider the four key attributes of smoke—volume, velocity, density, and colour. Then, fire fighters need to consider what might be influencing these key attributes. Next, fire fighters should try to determine the rate of change: Is it getting better or is it getting worse? Finally, it is time to put these pieces of the puzzle together. The previously mentioned information should help fire fighters to better determine the location of the fire, the size of the fire, and the potential for a hostile fire event, such as a backdraft or a flashover. Communicate the key parts of these observations to the company officer. With practice, this assessment process will help you systematically evaluate the information the smoke signals are giving you about the fire.

One way to become more proficient in smoke reading is to review videos of fires. Assess the smoke at the beginning of the tape, and try to identify the location of the fire and the stage of burning. You will be surprised at how much information you can obtain from a short video clip. This information on smoke reading might not have been taught to fire fighters who completed their initial training in the past, but today's smoke is deadly. If you want to stay alive, you need to understand it.

Fire Fighter Safety Tips © Photos.com

When you are in a smoky fire and visibility is poor, stop and shine a light into the smoke. Look at what the smoke is doing. What is the flow of the smoke? If there is no smoke movement, you are probably below the fire.

■ Reading Smoke Through a Door

When you see indications of a hot fire, such as darkened windows, but little visible smoke is coming from around closed doors and windows, you may be dealing with a fire that is in the decay phase because it does not have sufficient oxygen. This is a sign of great danger: as soon as this fire receives a fresh supply of oxygen, it will likely produce a violent backdraft. Fires can be dangerous even when little smoke is showing.

When you open a door, watch what the smoke does. If smoke exits through the top half of the door and clean air enters through the bottom half of the opening, then the fire is probably on the same level. This phenomenon is sometimes called "smoke that has found balance."

If smoke rises and the opening clears out, fresh air is being pulled into the building. This indicates the fire is probably above the level of the opening.

If smoke thins when the door is opened but still fills the door, the fire is probably below the level of the opening.

Fire Fighter Safety Tips © Photos.com

Beware of a fire with little smoke showing initially. It may be an under-ventilated fire that will erupt violently as soon as a supply of fresh oxygen is introduced into the room.

Near-Miss REPORT

Report Number: 06-0000049

Synopsis: Backdraft catches crew by surprise.

Event Description: A call came over our pagers for mutual aid for a working structure fire around 2030 hours or so. Ten to fifteen departments, with crews and all equipment needed, had already been en route to that location. My department responded with our pumpers and ladder truck with full crew in each engine. We arrived on scene and the building was just engulfed in flames. Fire fighters from early responders were attacking it from above in the bucket of their ladder truck. Departments were surrounding the building from north to south and east to west. My department took the rear of the structure with one of our engines and the ladder truck. We were also in company with two other pumpers from different districts. I was on the base of the ladder communicating with the fire fighters up top who were combating the flames.

Meanwhile, the other department accompanying my department was busy sawing and pounding the side of the wall in for ventilation. After a manhole was made, fire fighters started to enter with hoseline in hand. There had to have been at least three (fire fighters) on the main line.

As the crew proceeded towards the fire, the heat got treacherous. Abruptly the blaze turned into a backdraft. As that backdraft progressed towards the crew, they had no idea what was approaching them.

The first fire fighter on hoseline ducked and planted his face in his chest. The second fire fighter, who I believe was not wearing full PPE, was smothered by the fire. After the short experience, the blaze was sucked back into the structure and blew through the top of the building. The fire fighter seemed to be okay but was too shocked from the experience to explain anything that had happened right away.

Lessons Learned: It is very important to wear full PPE while you're fighting any kind of fire whether it be small or large and all types as well. More fire fighters should have also been around for their safety.

Wrap-Up

Chief Concepts

- Fire is a rapid chemical process that produces heat and light.
- Matter is made up of atoms and molecules.
- Matter exists in three states: solid, liquid, and gas.
  - A solid has definite capacity for resisting forces and, under ordinary conditions, retains a definite size and shape.
  - A liquid assumes the shape of the container in which it is placed.
  - A gas is a type of liquid that has neither independent shape nor independent volume, but rather, tends to expand indefinitely.
- Fuels are a form of energy. Energy exists in many forms, including chemical, mechanical, electrical, light, or nuclear.
  - Chemical energy is the energy created by a chemical reaction.
  - Mechanical energy is converted to heat when two materials rub against each other and create friction.
  - Electrical energy is converted to heat energy when it flows through a conductive material. Light energy is caused by electromagnetic waves packaged in photons and travels as thermal radiation.
  - Nuclear energy is stored in radioactive materials and converted to electricity by nuclear power–generating stations.
- The three basic conditions needed for a fire to occur are fuel, oxygen, and heat. A fourth factor, a chemical chain reaction, is required for a self-sustaining fire.
- The by-product of fire is smoke. Smoke includes three major components: particles, vapours, and gases. Smoke consists mainly of unburned forms of hydrocarbon fuels.
- Fire may be spread by direct contact, conduction, convection, and radiation.
- Direct contact is a flame touching a fuel.
- Conduction is the transfer of heat through matter, such as heat travelling up a metal spoon.
- Convection is the circulatory movement that occurs in a gas or fluid. Convection currents in a fire involve gases that are generated by the fire and travel along the ceiling of a room. Such convection currents carry superheated gases, which may ultimately heat flammable materials enough to ignite them.
- Radiation is the transfer of heat through the emission of energy in the form of invisible waves.
- The four principal methods of fire extinguishment are cooling the fuel, excluding oxygen, removing the fuel, and interrupting the chemical reaction.
- Fires are categorized as Class A, Class B, Class C, Class D, and Class K. These classes reflect the type of fuel that is burning and the type of hazard that the fire represents.

- Class A fires involve ordinary solid combustible materials, such as wood, paper, and cloth.
- Class B fires involve flammable or combustible liquids, such as gasoline, kerosene, diesel fuel, and motor oil.
- Class C fires involve energized electrical equipment.
- Class D fires involve combustible metals, such as sodium, magnesium, and titanium.
- Class K fires involve combustible cooking oils and fats in kitchens.
- Most fires encountered by fire fighters involve solid fuels. Solid fuels do not actually burn in a solid state. Instead, they must be heated or pyrolyzed to decompose into a vapour before they will burn.
- Solid-fuel fires develop through four phases: the ignition phase, the growth phase, the fully developed phase, and the decay phase.
- The growth phase occurs as the fuel starts to burn, increasing the convection of hot gases upward.
- The fully developed phase occurs when all available fuel has ignited and heat is being produced at the maximum rate.
- During the decay phase, the rate of burning slows down because less fuel is available.
- The growth of room-and-contents fires depends on the characteristics of the room and the contents of the room. Synthetic products are widely used in today's homes. The by-products of heated plastics are not only flammable but also toxic.
- Special considerations related to room-and-contents fires include thermal layering, flameovers, flashovers, and backdrafts.
- Thermal layering is the property of gases in an enclosed space by which they form layers according to their temperature. The hottest gases travel by convection currents to the top level of the room.
- Flameover is the spontaneous ignition of hot gases in the upper levels of a room.
- Flashover is the near-simultaneous ignition of most of the exposed combustible materials in an enclosed area.
- Backdraft is caused by the introduction of oxygen into an enclosure where superheated gases and contents are hot enough for ignition but the fire does not have sufficient oxygen to cause their combustion.
- Modern structures tend to be more tightly sealed, be constructed of lighter-weight materials, and contain more plastics. These characteristics can lead to a greater risk of backdrafts when a fire occurs.
- Liquid-fuel fires require the proper mixture of fuel and air, an ignition source, and contact between the fuel mixture and the ignition.
- The characteristics of flammable vapours can be described in terms of vapour density and flammability

limits. Vapour density reflects the weight of a gas compared to air. Flammability limits vary widely for different fuels.

- A BLEVE is a catastrophic explosion in a vessel containing both a boiling liquid and a vapour.
- Assessment of smoke volume, velocity, density, and colour enables fire fighters to predict the location of a fire and its stage of development.
- Smoke reading requires fire fighters to evaluate the effect of the building, the weather, and ventilation on the smoke.

Hot Terms

__Atom__ The smallest particle of an element, which can exist alone or in combination.

__Backdraft__ A phenomenon that occurs when a fire takes place in a confined area, such as a sealed aircraft fuselage, and burns undetected until most of the oxygen contained therein is consumed. The heat continues to produce flammable gases, mostly in the form of carbon monoxide. These gases are heated above their ignition temperature, and when a supply of oxygen is introduced (e.g., when normal entry points are opened), the gases ignite with explosive force. (NFPA 402)

__Black fire__ A hot, high-volume, high-velocity, turbulent, ultra-dense black smoke that indicates an impending flashover or autoignition.

__Boiling liquid/expanding vapour explosion (BLEVE)__ An explosion that occurs when a tank containing a volatile liquid at the bottom and a flammable gas at the top is heated to the point of rupture.

__Box__ A burning structure.

__Chemical energy__ Energy that is created or released by the combination or decomposition of chemical compounds.

__Class A fire__ A fire fuelled by ordinary combustible materials, such as wood, cloth, paper, rubber, or many plastics. (NFPA 10)

__Class B fire__ A fire fuelled by flammable or combustible liquids, petroleum greases, tars, oils, oil-based paints, solvents, lacquers, alcohols, or flammable gases. (NFPA 10)

__Class C fire__ A fire that involves energized electrical equipment. (NFPA 10)

__Class D fire__ A fire fuelled by combustible metals, such as magnesium, titanium, zirconium, sodium, lithium, or potassium. (NFPA 10)

__Class K fire__ A fire in a cooking appliance that involves combustible cooking media (i.e., vegetable or animal oils and fats). (NFPA 10)

__Combustion__ A chemical process of oxidation that occurs at a rate fast enough to produce heat and usually light in the form of either a glow or a flame. (NFPA 101)

__Compartment__ A space completely enclosed by walls and a ceiling. The compartment enclosure is permitted to have wall openings to an adjoining space if the openings have a minimum lintel depth of 20 cm (8 in.) from the ceiling and the openings do not exceed 2.4 m (8 ft.) in width. A single opening of 91 cm (36 in.) or less in width without a lintel is permitted when there are no other openings to adjoining spaces. (NFPA 13)

__Conduction__ Heat transfer to another body, or within a body, by direct contact. (NFPA 921)

__Convection__ Heat transfer by circulation within a medium such as a gas or a liquid. (NFPA 921)

__Decay phase__ The phase of fire development in which the fire has consumed either the available fuel or oxygen and is starting to die down.

__Electrical energy__ Heat that is produced by electricity.

__Endothermic__ A reaction that absorbs heat or requires heat to be added.

__Exothermic__ A reaction that results in the release of energy in the form of heat.

__Explosive limits__ The upper and lower concentration limits, at a specified temperature and pressure, of a flammable gas or vapour in the air, expressed as a percentage of the fuel by volume; also known as flammability limits.

__Fire__ A rapid, persistent chemical reaction that releases both heat and light.

__Fire point__ The lowest temperature at which a substance releases enough vapours to ignite and sustain combustion; also known as flame point.

__Fire tetrahedron__ A geometric shape used to depict the four components required for a fire to occur: fuel, oxygen, heat, and chemical chain reactions.

__Fire triangle__ A geometric shape used to depict the three components of which a fire is composed: fuel, oxygen, and heat.

__Flameover__ The condition where unburned fuel (pyrolysate) from the originating fire has accumulated in the ceiling layer to a sufficient concentration (i.e., at or above the lower flammable limit) that it ignites and burns; it can occur without ignition of, or prior to the ignition of, other fuels separate from that which caused the event; also known as rollover. (NFPA 921)

__Flame point__ The lowest temperature at which a substance releases enough vapours to ignite and sustain combustion; also known as fire point.

__Flammability limits__ The upper and lower concentration limits, at a specified temperature and pressure, of a flammable gas or vapour in the air, expressed as a percentage of the fuel by volume; also known as explosive limits.

__Flashover__ A transition phase in the development of a compartment fire in which surfaces exposed to thermal radiation reach their ignition temperature more or less simultaneously and fire spreads rapidly throughout the space, resulting in full-room involvement or total involvement of the compartment or enclosed space. (NFPA 921)

Flash point The minimum temperature of a liquid at which sufficient vapour is given off to form an ignitable mixture with the air, either near the surface of the liquid or within the vessel used. (NFPA 30)

Fuel A material that will maintain combustion under specified environmental conditions. (NFPA 53)

Fully developed phase The phase of fire development in which the fire is free-burning and consuming much of the fuel.

Gas A material that has a vapour pressure greater than 300 kPa (43.5 psi) absolute at 50°C (122°F) or is completely gaseous at 20°C (68°F) at a standard pressure of 101.3 kPa (14.7 psi) absolute. (NFPA 30A)

Growth phase The phase of fire development in which the fire spreads beyond the point of origin and begins to involve other fuels in the immediate area.

Hypoxia A state of inadequate oxygenation of the blood and tissue sufficient enough to cause impairment of function. (NFPA 99B)

Ignition phase The phase of fire development in which the fire is limited to the immediate point of origin.

Ignition temperature Minimum temperature a substance must attain to ignite under specific test conditions. (NFPA 921)

Laminar smoke flow The smooth or streamlined movement of smoke, which indicates that the pressure in the building is not excessively high.

Liquid Any material that (1) has a fluidity greater than that of 300 penetration asphalt when tested in accordance with American Society of Testing and Materials (ASTM) D5, *Standard Test Method for Penetration of Bituminous Materials,* or (2) is a viscous substance for which a specific melting point cannot be determined but that is determined to be a liquid in accordance with ASTM D4359, *Standard Test Method for Determining Whether a Material Is a Liquid or a Solid.* (NFPA 30)

Lower flammable limit (LFL) The concentration of a combustible material in air below which ignition will not occur. Mixtures below this limit are said to be "too lean"; also known as lower explosive limit (LEL). (NFPA 329)

Matter A substance made up of atoms and molecules.

Mechanical energy A form of potential energy that can generate heat through friction.

Oxidation A reaction with oxygen in either its elemental form or the form of one of its compounds. (NFPA 53)

Plume The column of hot gases, flames, and smoke rising above a fire; also called a convection column, thermal updraft, or thermal column. (NFPA 921)

Pyrolysis The destructive distillation of organic compounds in an oxygen-free environment by which the organic matter is converted into gases, liquids, and char. (NFPA 820)

Radiation The emission and propagation of energy through matter or space by means of electromagnetic disturbances that display both wave- and particle-like behaviour. (NFPA 801)

Rollover The condition where unburned fuel (pyrolysate) from the originating fire has accumulated in the ceiling layer to a sufficient concentration (i.e., at or above the lower flammable limit) that it ignites and burns; it can occur without ignition of, or prior to the ignition of, other fuels separate from that which caused the event; also known as flameover. (NFPA 921)

Smoke The airborne solid and liquid particulates and gases evolved when a material undergoes pyrolysis or combustion, together with the quantity of air that is entrained or otherwise mixed into the mass. (NFPA 318)

Smoke colour The attribute of smoke that reflects the stage of burning of a fire and the material that is burning in the fire.

Smoke density The thickness of smoke. (Because it has a high mass per unit volume, smoke is difficult to see through.)

Smoke velocity The speed of smoke leaving a burning building.

Smoke volume The quantity of smoke, which indicates how much fuel is being heated.

Solid A material that has three dimensions and is firm in substance; one of the three phases of matter.

Thermal column A cylindrical area above a fire in which heated air and gases rise and travel upward.

Thermal layering The stratification (layering) that occurs in a room as a result of a fire.

Thermal radiation The means by which heat is transferred to other objects.

Turbulent smoke flow Agitated, boiling, aggressive smoke, which indicates great heat in the burning building. It is a precursor to flashover.

Upper flammable limit (UFL) The highest concentration of a combustible substance in a gaseous oxidizer that will propagate a flame. (NFPA 68)

Vapour density The weight of an airborne concentration or vapour or gas as compared to an equal volume of dry air.

Volatility The ability of a substance to produce combustible vapours.

You are spending your evening searching the Web for firefighting clips. Hours pass by as you watch video after video. Then you run across a clip that makes you drop your mouse: It shows two fire fighters entering a two-storey house with fire coming out of a side window. Smoke is pouring out of the upper half of the front door, and you can see that the upper floor is full of smoke.

As the clip continues, the smoke on the upper floor gets darker and darker and begins boiling out of the window. To your surprise, a fire fighter emerges amidst the turbulent smoke of the window. He is clearly in trouble.

1. What phase is this fire in?
   - **A.** Ignition
   - **B.** Growth
   - **C.** Fully developed
   - **D.** Decay
2. How is the heat from this fire travelling from one room to another on the second floor?
   - **A.** Convection
   - **B.** Direct contact
   - **C.** Conduction
   - **D.** Radiation
3. Which is not a part of the fire triangle?
   - **A.** Fuel
   - **B.** Chemical chain reaction
   - **C.** Oxygen
   - **D.** Ignition source

4. _____ is the speed of the smoke leaving a burning building.
   - **A.** Smoke velocity
   - **B.** Smoke density
   - **C.** Smoke volume
   - **D.** Smoke colour
5. _____ is a hot, high-volume, high-velocity, turbulent, ultra-dense black smoke that indicates an impending flashover or autoignition.
   - **A.** Black fire
   - **B.** Laminar flow
   - **C.** BLEVE
   - **D.** Pyrolysis

You and another fire fighter are working your way through dirty brown smoke in a house fire, trying to find the seat of the fire. The smoke is thick, so you are maintaining physical contact with each other. You can feel the heat on your back, and you hear a faint crackle as you work your way to the back of the structure. Before you can reach the sound, the smoke turns jet black, and the heat gets so intense that you are forced onto your stomach. You know this is a bad situation, but your partner keeps on pushing toward the fire.

1. What type of event is most likely about to occur?
2. What type of protective actions should you take?
3. What do you say, if anything, to your partner?
4. If you feel that you need to retreat, but your partner does not, what should you do?

Building Construction

Fire Fighter I

Knowledge Objectives

After studying this chapter, you will be able to:

- Explain how occupancy classifications affect fire-suppression operations. (p 164–165)
- Explain how the contents of a structure affect fire-suppression operations. (p 165)
- Describe the characteristics of masonry building materials. (p 165–166)
- Describe the characteristics of concrete building materials. (p 166)
- Describe the characteristics of steel building materials. (p 166–167)
- Describe the characteristics of glass building materials. (p 167)
- Describe the characteristics of gypsum building materials. (p 167–168)
- Describe the characteristics of wood building materials. (p 168)
- Describe the characteristics of engineered wood building materials. (p 168)
- Describe the characteristics of plastic building materials. (p 169–170)
- List the five types of building construction. (p 170–171)
- Describe the characteristics of Type I construction. (NFPA 5.3.12.A , p 171–172)
- Describe the effects of fire on Type I construction. (NFPA 5.3.12.A , p 171–172)
- Describe the characteristics of Type II construction. (NFPA 5.3.12.A , p 172)
- Describe the effects of fire on Type II construction. (NFPA 5.3.12.A , p 172)
- Describe the characteristics of Type III construction. (NFPA 5.3.12.A , p 172–173)
- Describe the effects of fire on Type III construction. (NFPA 5.3.12.A , p 172–173)
- Describe the characteristics of Type IV construction. (NFPA 5.3.12.A , p 173–174)
- Describe the effects of fire on Type IV construction. (NFPA 5.3.12.A , p 173–174)
- Describe the characteristics of Type V construction. (NFPA 5.3.12.A , p 174)

- Describe the effects of fire on Type V construction. (NFPA 5.3.12.A , p 174)
- Describe the characteristics of balloon-frame construction. (NFPA 5.3.12.A , p 175)
- Describe the effects of fire on balloon-frame construction. (NFPA 5.3.12.A , p 175)
- Describe the characteristics of platform-frame construction. (NFPA 5.3.12.A , p 175)
- Describe the effects of fire on platform-frame construction. (NFPA 5.3.12.A , p 175)
- Describe the purpose of a foundation in a structure. (p 175–176)
- List the warning signs of foundation collapse. (p 176)
- Explain how floor construction affects fire-suppression operations. (p 176–177)
- Describe the characteristics of fire-resistive floors. (p 176)
- Describe the characteristics of wood-supported floors. (p 176–177)
- Describe the characteristics of ceiling assemblies. (p 177)
- List the three primary types of roofs. (p 177–180)
- Describe the characteristics of trusses. (p 180–181)
- List the types of trusses. (p 180–181)
- Describe the effects of fires on trusses. (p 181–182)
- Describe the characteristics of walls. (p 182–183)
- List the common types of walls in structures. (p 183)
- Describe the characteristics of door assemblies. (p 184)
- Describe the characteristics of window assemblies. (p 184)
- Describe the characteristics of fire doors. (p 184–185)
- Describe the characteristics of fire windows. (p 184–185)
- Explain the effects that interior finishes have on fire-suppression operations. (p 185–186)
- Describe the hazards that buildings under construction or demolition pose to fire fighters. (p 186)
- Describe how building construction factors into preincident planning and incident size-up. (p 186–187)

Skills Objectives

There are no skills objectives for Fire Fighter I candidates. NFPA 1001 contains no Fire Fighter I Job Performance Requirements for this chapter.

Fire Fighter II — FFII

Knowledge Objectives

There are no knowledge objectives for Fire Fighter II candidates. NFPA 1001 contains no Fire Fighter II Job Performance Requirements for this chapter.

Skills Objectives

There are no skills objectives for Fire Fighter II candidates. NFPA 1001 contains no Fire Fighter II Job Performance Requirements for this chapter.

Additional NFPA Standards

- NFPA 80, *Standard for Fire Doors and Other Opening Protectives*
- NFPA 220, *Standard on Types of Building Construction*

You Are the Fire Fighter

Your crew is driving around your district looking at the various buildings. The engineer drives through a historic residential district and points to a two-storey house with a wrap-around porch, tall, narrow windows, and a stone foundation. He then cuts over to the old Main Street and points to a commercial building. It is three storeys tall and made of brick. Every seventh layer of brick is laid crosswise. It has a flat roof and is about 9.1 metres (30 feet) wide and 27.4 m (90 ft.) deep. It has large plate glass windows on the first floor, and there are normal-sized windows on the second and third floors. He then drives to a residential part of town built in the 1940s. Each house is one storey, is about 40 square metres (440 square feet), and has a roof with asphalt shingles. Each home has wooden clapboard siding and a small front porch. He then drives to a newer commercial area where there is a stand-alone convenience store. It was built in 1998 and has a flat roof and metal siding. He turns to you and says, "We just looked at four different buildings. Tell me about each one."

1. Why do fire fighters need to understand building construction?
2. What effect does building construction have on fires?
3. Which building would most likely collapse first if exposed to fire?

Introduction

Knowing the basic types of building construction is vital for fire fighters, because building construction affects how fires grow and spread. Fire fighters must be able to recognize and understand different types of building construction. They must understand how each type reacts when exposed to the effects of heat and fire. This knowledge will help them anticipate a fire's behaviour and respond accordingly. An understanding of building construction will also help fire fighters determine when it is safe to enter a burning building and when it is necessary to evacuate a building. Your safety, the safety of your team members, and the safety of the building's occupants all depend on your knowledge of building construction.

Construction, however, is merely one aspect of a complex situation. Fire fighters also must consider the occupancy of the building and the building's contents. Fire risk factors vary depending on how a building is used. To establish a course of action at a fire, it is necessary to consider the interactions between three factors: the building's construction, the occupancy of the building, and the contents of the building. The following sections relate to occupancy and contents, and the rest of this chapter focuses on those aspects of building construction that are important to a fire fighter.

Occupancy

The term **occupancy** refers to how a building is used. Based on a building's occupancy classification, a fire fighter can predict who is likely to be inside the building, how many people are likely to be there, and what they are likely to be doing. Occupancy also suggests the types of hazards and situations that might be encountered in the building.

Occupancy classifications are used in building, fire, and safety codes to establish construction requirements, including materials and systems that must be used. A change in occupancy may trigger the need to upgrade a structure to meet different code standards—for example, a warehouse space (industrial) that was converted to a bar (assembly). The conversion and accompanying occupancy change would trigger major upgrades to the structure and systems. Occupancy classifications are based on similar types of use, danger to the occupants and risk of fire. For inspection purposes, the National Building Code of Canada (NBC) classifies buildings into six groups: Assembly (Group A), Care and Detention (Group B), Residential (Group C), Business and Personal Services (Group D), Mercantile (Group E), and Industrial (Group F). Some of these groups have multiple divisions.

Occupancy classifications can be used to predict the number of occupants who are likely to be at risk in a fire. For example, hospitals and nursing homes are occupied 24 hours a day by persons who will probably need assistance to evacuate. An office building probably houses a large population during the day, but most of the workers should be able to evacuate the site without assistance. A daycare centre might be filled with young children during the day but is likely to be empty at night. A nightclub is probably empty during the day and crowded at night.

Fire hazards in different types of occupancies also vary greatly. A computer-repair shop has different characteristics than an auto-transmission shop, for example. A factory could be used to manufacture cast iron pipefittings or airplanes. The materials used could be flammable, toxic, or fire resistant. Fire fighters must consider each of these factors when responding to a particular building.

Building codes require that certain types of building construction be used for specific types of occupancies; in most communities, however, buildings that were built for one reason are often being used for some other purpose. Buildings constructed before modern building codes were adopted present significant hazards for fire fighters. The NBC requires a full code upgrade to a structure when it undergoes a major renovation; however, it is possible to perform multiple small renovations without the requisite full code upgrade, which may leave deficient construction in place and present unexpected hazards to both occupants and fire fighters. Often, a building's occupancy use is changed without informing the fire and building authorities. A building may have a temporary use that does not fall under its usual occupancy classification—for example, a warehouse that is periodically used for company celebrations.

Contents

Contents also must be considered when responding to a building. Building contents vary widely but are usually closely related to the occupancy of a building. Some buildings have noncombustible contents that would not feed a fire, whereas others could be so dangerous that fire fighters could not safely attack any fire at the site; for example, an auto-repair shop generally contains many toxic and flammable substances. Whenever possible, fire fighters should prepare a prefire or preincident plan that accounts for such possibilities if buildings present special hazards.

Even similar occupancies can pose different levels of risk, however. For example, three warehouses, although identical from the exterior, could have very different contents that would either increase or reduce the risks to fire fighters. One might contain ceramic floor tiles, the second might be filled with wooden furniture, and the third might serve as a storehouse for swimming pool chemicals. The tiles will not burn, the furniture will burn easily, and the chemicals will create toxic products of combustion and contaminated runoff.

When evaluating the risks at any alarm, consider the occupancy, the contents, and the building construction; they represent an interwoven package of risks to fire fighters.

Types of Construction Materials

An understanding of building construction begins with the materials used. Building components are usually made of different materials. The properties of these materials and the details of their construction determine the basic fire characteristics of the building itself.

Function, appearance, price, and compliance with building and fire codes are all considerations when selecting building materials and construction methods. Architects often place a priority on functionality and aesthetics when selecting materials, whereas builders are often more concerned about price and ease of construction. For their part, building owners are usually interested in durability and maintenance expenses, as well as the materials' initial cost and aesthetics.

Fire fighters, however, use a completely different set of criteria to evaluate construction methods and materials. Their chief concern is the behaviour of the building under fire conditions. In particular, a building material or construction method that is attractive to architects, builders, and building owners might have the potential to create serious problems or deadly hazards for fire fighters.

The most commonly used building materials are wood, engineered wood products, masonry, concrete, steel, aluminum, glass, gypsum board, and plastics. Within these basic categories, hundreds of variations are possible. The key factors that affect the behaviour of each of these materials under fire conditions are outlined here:

- **Combustibility**: Whether or not a material will burn determines its combustibility. Materials such as wood will burn when they are ignited, releasing heat, light, and smoke, until they are completely consumed by the fire. Concrete, brick, and steel are noncombustible materials, which cannot be ignited and are not consumed by a fire.
- **Thermal conductivity**: This characteristic describes how readily a material will conduct heat. Heat flows very readily through metals, such as steel and aluminum. By contrast, brick, concrete, and gypsum board are poor conductors of heat.
- Decrease in strength at elevated temperatures: Many materials lose strength at elevated temperatures; for example, steel loses strength and will bend or buckle when exposed to the high temperatures associated with fires, and aluminum melts in a fire. Some engineered building products are constructed of wood products held together by glue or plastics; the glue in these products begins to fail long before they are hot enough to burn. By contrast, bricks and concrete can generally withstand high temperatures for extended periods of time.
- Thermal expansion when heated: Some materials—steel, in particular—expand significantly when they are heated. A steel beam exposed to fire will stretch (elongate); if it is restrained so that it cannot elongate, it will sag, warp, or twist. As a general rule, a steel beam will elongate at a rate of 2.5 centimetres (1 inch) per 3 metres (10 feet) of length at a temperature of 537.8 degrees °C (1000 degrees °F). This expansion can push out the walls of buildings and cause collapse.

■ Masonry

Masonry includes stone, concrete blocks, and brick. The individual components are usually bonded together into a solid mass by mortar, which is produced by mixing sand, lime, water, and Portland cement. A concrete masonry unit (CMU) is a preassembled wall section of brick or concrete blocks reinforced by steel rods. A complete CMU can be delivered to a construction site, ready to be erected without further preparation.

Masonry materials are inherently fire resistive **FIGURE 7-1**; that is, they do not burn or deteriorate at the temperatures normally encountered in building fires. Masonry is also a poor conductor of heat, so it will limit the passage of heat from one side through to the other side. For these reasons, masonry is often used to construct fire walls to protect vulnerable materials. A fire wall helps prevent the spread of a fire from one side of the wall to the other.

All masonry walls are not necessarily fire walls, however. A single layer of masonry may be placed over a wood-framed building to make it appear more substantial, but if a masonry wall

A. © MitarArt/ShutterStock, Inc.

B. © Ken Durden/ShutterStock, Inc.

FIGURE 7-1 Masonry materials are inherently fire resistive. **A.** Solid masonry wall. **B.** Brick-veneer wall.

contains unprotected openings, a fire can often spread through it. If the mortar has deteriorated or the wall is exposed to fire for a prolonged time, a masonry wall can collapse during a fire.

A masonry structure can also collapse under fire conditions if the roof or floor assembly collapses. In such a case, the masonry falls because of the mechanical action of the collapse. Likewise, aged, weakened mortar can also contribute to a collapse. Regardless of the cause, a collapsing masonry wall can be a deadly hazard to fire fighters.

■ Concrete

Like masonry, concrete is a naturally fire-resistive material. Because it does not burn or conduct heat well, it is often used to insulate other building materials from fire. Concrete does not have a high degree of thermal expansion—that is, it does not expand greatly when exposed to heat and fire—nor will it lose strength when exposed to high temperatures.

Concrete is made from a mixture of Portland cement and aggregates such as sand and gravel. Different formulations of concrete can be produced for specific building purposes. Concrete is inexpensive and easy to shape and form. It is often used for foundations, columns, floors, walls, roofs, and exterior pavement.

Under compression, concrete is strong and can support a great deal of weight; however, it is weak under tension by contrast. When this material is used in building construction, steel reinforcing rods are often embedded in the concrete to strengthen it under tension. In turn, the concrete insulates the steel reinforcing rods from heat.

Although it is inherently fire resistive, concrete can be damaged by exposure to a fire; for example, a fire may convert moisture trapped within the concrete into steam. As the steam expands, it creates internal pressure that can cause surface sections of the concrete to break off in a process called **spalling**. Severe spalling in reinforced concrete can expose the steel rods to the heat of the fire. If the fire is hot enough, the steel might weaken, resulting in a structural collapse, though that is very rare.

Fire Fighter Safety Tips © Photos.com

Heat transfer can occur through exposed steel reinforcing rods. This mode of heat transfer is known as conduction.

■ Steel

Steel is the strongest building material in widespread use. It is considered very strong in terms of both tension and compression. Steel can also be produced in a wide variety of shapes and sizes—from heavy beams and columns to thin sheets. This material is often used in the structural framework of a building to support floor and roof assemblies. It is resistant to age and does not rot; however, most types of steel will rust if they are not protected from exposure to air and moisture.

Steel is an alloy of iron and carbon. Other metals may be added to the mix to produce steels with special properties, such as stainless steel or galvanized steel.

On its own, steel is not fire resistive; it will melt at extremely high temperatures, although such extraordinary temperatures are rarely encountered at structure fires **FIGURE 7-2**. Steel conducts heat well, so it tends to expand and lose strength as the temperature increases. For this reason, other materials—masonry, concrete, or layers of gypsum board—are often used to protect steel from the heat of a fire. Spray-on coatings of mineral or cement-like materials are also used to insulate steel members. The amount of heat absorbed by steel depends on the mass of the object and the amount of protection surrounding it. Smaller, lighter pieces of steel absorb heat more easily than larger and heavier pieces.

Steel both expands and loses strength as it is heated. As a consequence, an unprotected steel roof beam directly exposed to a fire may elongate sufficiently as to cause a supporting wall to collapse. Heated steel beams will often sag and twist, whereas steel columns tend to buckle as they lose strength. The bending and distortion are caused by the uneven heating that occurs in actual fire situations. Many residential occupancies are now constructed using metal studs instead of wood.

Failure of a steel structure depends on three factors: the mass of the steel components, the loads placed upon them, and the methods used to connect the steel pieces. Unfortunately, there are no accurate indicators that would enable fire fighters to

FIGURE 7-2 Steel will lose strength at temperatures of more than 537.8°C (1000°F).

Courtesy of Captain David Jackson, Saginaw Township Fire Department

predict when a steel beam is about to fail. Any sign of bending, sagging, or stretching of steel structural members should be considered a warning of an immediate risk of failure. At the first sign of any of these indicators, all fire fighters should immediately clear the area because of the potential for imminent collapse.

■ Other Metals

A variety of other metals, including aluminum, copper, and zinc, are also used in building construction. Aluminum is often used for siding, window frames, door frames, and roof panels. Copper is used primarily for electrical wiring and piping; it is sometimes used for decorative roofs, gutters, and downspouts as well. Zinc is used primarily as a coating to protect metal components from rust and corrosion.

Aluminum is occasionally used as a structural material in building construction. It is more expensive and not as strong as steel, so its use is generally limited to light-duty applications such as awnings and sunshades. Aluminum expands more significantly than steel when heated and loses strength quickly in the presence of fire. This material also has a lower melting point than steel, so it will often melt and drip in a fire.

■ Glass

Almost all buildings contain glass—in windows, doors, sky-lights, and sometimes walls. Ordinary glass breaks very easily, but glass can be manufactured to resist breakage and to withstand impact or high temperatures.

Although ordinary glass will usually break when exposed to fire, specially formulated fire-rated glass can be used as a fire barrier in certain situations. The thermal conductivity of glass is rarely a significant factor in the spread of fire.

Many different types of glass are available for use in building construction:

- Ordinary window glass will usually break with a loud pop when heat exposure to one side causes it to expand and creates internal stresses, fracturing the glass. The glass breaks in large shards, which usually have sharp edges.

- **Tempered glass** is much stronger than ordinary glass and more difficult to break. Some tempered glass can be broken with a spring-loaded centre punch; it will shatter into small pieces, which do not have the sharp edges of ordinary glass. Tempered or laminated glass is usually required in installations such as storm doors to prevent injuries from shards of broken glass.
- **Laminated glass** is manufactured with a thin sheet of plastic placed between two sheets of glass. The resulting product is much stronger than ordinary glass, is difficult to break with ordinary hand tools, and will usually deform instead of breaking. When exposed to a fire, laminated glass windows are likely to crack, but remain in place. Laminated glass is sometimes used in buildings for soundproofing purposes.
- **Glass blocks** are thick pieces of glass similar to bricks or tiles. They are designed to be built into a wall with mortar, so that light can be transmitted through the wall. Glass blocks have limited strength and are not intended to be used as part of a load-bearing wall, but they can usually withstand a fire. Some glass blocks are approved for use with fire-rated masonry walls.
- **Wired glass** is made by moulding tempered glass with a reinforcing wire mesh **FIGURE 7-3**. When wired glass is subjected to heat, the wire holds the glass together and prevents it from breaking. This material is often used in fire doors and windows designed to prevent fire spread.

■ Gypsum board

Gypsum is a naturally occurring mineral composed of calcium sulphate and water molecules, which is used to make plaster of Paris. Gypsum is a good insulator and noncombustible; it will not burn even in atmospheres of pure oxygen.

Gypsum board (also known as drywall, Sheetrock®, or plasterboard) is commonly used to cover the interior walls and ceilings of residential living areas and commercial spaces. It is also required in some applications, such as residential garages,

FIGURE 7-3 Wired glass.

to create a rated fire-resistant barrier. Gypsum board is manufactured in large sheets consisting of a layer of compacted gypsum sandwiched between two layers of specially produced paper. In a building, these sheets are nailed or screwed in place onto a framework of wood or metal studs, and the edges are secured with a special tape. The nail or screw heads and tape are then covered with a thin layer of plaster.

When a building has a fire-rated assembly, a rated drywall will be used. Rated drywall is often called Type X drywall and is 15 millimetres (0.6 inches) in thickness. Rated drywall has additional reinforcement built into the gypsum matrix to increase the time it takes for the drywall to fail and is thus more predictable in its resistance to fire. Multiple layers of rated drywall are used to increase a building assembly's resistance to fire.

Gypsum board has limited combustibility, because the paper covering will burn slowly when exposed to a fire. It does not conduct or release heat to an extent that would contribute to fire spread, so the material is often used to create a firestop or to protect building components from fire. Notably, gypsum blocks are sometimes used as protective insulation around steel members or to create fire-resistive enclosures or interior fire walls.

When gypsum board is heated, some of the water present in the calcium sulphate will evaporate, causing the board to deteriorate. If it is exposed to fire for a long time, it will fail. Water will also weaken and permanently damage gypsum board.

Even if the gypsum board retains its integrity after a fire, the sections that were directly exposed to the fire should be replaced following the incident. The sections that were heated will probably be unable to serve as effective fire barriers in the future.

Although gypsum board is a good finishing material and an effective fire barrier, it is not a strong structural material. It must be properly mounted on and supported by wood or steel studs. Gypsum board mounted on wood framing will protect the wood from fire for a limited time.

■ Wood

Wood is probably the most commonly used building material in today's environment. It is inexpensive to produce, is easy to use, and can be shaped into many different forms—from heavy structural supports to thin strips of exterior siding. Both soft woods (e.g., pine) and hard woods (e.g., oak) are used for building construction.

FIRE FIGHTER Tips

Older buildings were built with lumber that was true to its stated dimensions. A two-by-four stud actually measured 2 in. by 4 in. In buildings built today, a two-by-four stud measures 1.5 in. (3.8 cm) by 3.4 in. (8.6 cm) or less. In a fire, the true two-by-four stud will resist fire longer than the modern stud.

Solid lumber is squared and cut into uniform lengths. Examples of solid lumber range from the heavy timbers used in mills and barns, to smaller so-called two-by-four studs, to wood siding, interior wood trim, wood floors, and wood shingles on roofs. Solid wood can be used in a wide variety of building applications; this type of lumber may be found in multiple sizes in both older or **legacy construction** and newer or **contemporary construction**.

Engineered Wood Products

Engineered wood products are also called manufactured board, manmade wood, or composite wood. They include plywood, fibreboard, oriented strand board (OSB), and particle board—materials that are used in roofs, walls, floors, countertops, doors, I-beams, trusses, and cabinets. These products are manufactured from small pieces of wood that are held together by glue or adhesive. Such adhesives include urea-formaldehyde resins, phenol-formaldehyde resins, melamine-formaldehyde resins, and polyurethane resins.

Engineered wood products are preferred over natural wood for a variety of reasons: In particular, they are usually less expensive than large, solid planks of natural wood. Such products can also be fabricated in longer lengths and for specific applications that are not possible with natural wood. They can maximize the natural characteristics of wood, thereby producing greater strength and stiffness. These products are easy to work with using ordinary tools, and they can readily be shaped into curved surfaces. With all of these advantages, it is not surprising that the use of engineered wood products has greatly increased in recent years.

Unfortunately, some drawbacks also arise with engineered wood products. One disadvantage is that some of these products may be prone to warping in conditions of high humidity. A second disadvantage is that these products may contain toxic products, such as formaldehyde, which are introduced during manufacture. The biggest disadvantage—at least from the fire fighter's perspective—is that these products burn and fail quickly and without warning under fire conditions and produce large void spaces.

Fire Fighter Safety Tips

© Photos.com

The glue that is used to create **laminated wood** is flammable.

Implications for Fire Fighters

For fire fighters, the most important characteristic of wood and manufactured wood products is their high combustibility. Wood acts as fuel in a structural fire and can provide a path for the fire to spread. It ignites at fairly low temperatures and is gradually consumed by the fire, weakening and eventually causing the collapse of the structure. Great quantities of heat and hot gases are created by the wood-burning process, until eventually, all that remains once the fuel has been consumed is a small quantity of residual ash.

Wood components weaken as they are consumed by fire. A burning wood structure becomes weaker every minute, making it imperative that fire fighters always operate on the side of safety in such environments.

The rate at which wood ignites, burns, and decomposes depends on several factors:

- Ignition—A small ignition source contains less energy and will take a longer time to ignite a fire. Using an accelerant will greatly speed this process.
- Moisture—Damp or moist wood takes longer to ignite and burn. New lumber usually contains more moisture than lumber that has been in a building for many years. A higher relative humidity in the atmosphere also makes wood more difficult to ignite.

- Density—Heavy, dense wood is harder to ignite than lighter or less dense wood.
- Preheating—The more the wood is preheated, the faster it ignites.
- Size and form—The rate of combustion is directly related to the surface area of the wood. During the combustion process, high temperatures cause the release of flammable gases from the surface of the wood. A large solid beam is difficult to ignite and burns relatively slowly; by contrast, a lightweight truss ignites more easily and is consumed rapidly.

Exposure to the high temperatures generated by a fire can also decrease the strength of wood through the process of **pyrolysis**. This chemical change occurs when wood or other materials are heated to a temperature high enough to release some of the volatile compounds in the wood, albeit without igniting these gases. The wood then begins to decompose without combustion.

Fire Retardant–Treated Wood

Wood cannot be treated to make it completely noncombustible. Nevertheless, impregnating wood with mineral salts makes it more difficult to ignite and slows the rate of burning. Fire retardant treatment can significantly reduce the fire hazards of wood construction. On the downside, the treatment process can also reduce the strength of the wood.

In some cases, fire retardant–treated wood can pose a danger for fire fighters. Some fire-retardant chemicals used to treat plywood roofing panels, for example, will cause the wood to rapidly deteriorate and weaken. The plywood may then fail, necessitating the replacement of these panels. Fire fighters should know whether fire retardant–treated plywood is used in the community. If it is, they should avoid standing on roofs made with these panels during a fire, because the extra weight could cause the plywood to collapse.

Pressure-treated preserved wood is a product approved for, and often used for, the foundations of buildings, decks, and areas where wood is exposed to conditions that would cause rot. It is a common foundation material in areas where concrete is not available or is cost-prohibitive. Pressure-treated wood is not resistant to fire but, instead, poses a health risk because the preservative injected into the wood contains toxic materials that can be released during a fire. The preservative used contains arsenic, a heavy metal easily absorbed by the body. Arsenic concentrates in the ash when wood is burned. North American manufacturers of pressure-treated wood reached a voluntary agreement with Health Canada's Pest Management Regulatory Agency and the Environmental Protection Agency (EPA) in the United States to end the manufacture of pressure-treated wood for most consumer applications as of December 31, 2003. However, pressure-treated wood is still allowed for commercial uses, such as wharves and bridges.

Plastics

Plastics are synthetic materials that are used in a wide variety of products today **FIGURE 7-4**. Plastics may be transparent or opaque, stiff or flexible, tough or brittle. They are often combined with other materials to produce building products.

Although plastics are rarely used for structural support, they may be found throughout a building. Building exteriors may include vinyl siding, plastic window frames, and plastic panel skylights. Foam plastic materials may be used as exterior or interior insulation. Plastic pipe and fittings, plastic tub and shower enclosures, and plastic light fixtures are commonly used. Even carpeting and floor coverings often contain plastics.

The combustibility of plastics varies greatly. Some plastics ignite easily and burn quickly, whereas others will burn only when an external source of heat is present. Some plastics can withstand high temperatures and fire exposure without igniting.

Many plastics produce quantities of heavy, dense, dark smoke and release high concentrations of toxic gases as they burn. This smoke resembles smoke from a petroleum fire, which is not surprising, given that most plastics are made from petroleum products.

Thermoplastic materials melt and drip when exposed to high temperatures, even those as low as 260°C (500°F). Although heat can be used to shape these materials into a variety of desired forms, the dripping, burning plastic can rapidly spread a fire. By contrast, **thermoset materials** are fused by heat and will not melt as the plastic burns, although their strength will decrease dramatically.

Insulating materials can be made from ridged plastics, glass fibres and bonding materials, cellulose, mineral wool, asbestos, and numerous other materials. Insulating materials result in several hazards to fire fighters: many are highly combustible with rapid flame spread, and many absorb and retain water, while some are toxic (asbestos). Due to the cold Canadian climate,

FIGURE 7-4 Plastic building materials can be used for siding, window frames, and interior finishes.

© AbleStock

Near-Miss REPORT

Report Number: 07-0000726

Synopsis: Truss roof collapse causes near miss at church fire.

Event Description: We were involved in fire-suppression activities on a working attic fire at a church. Company crew members were attempting to perform a trench cut to ventilate and confine the fire to an area adjacent to the fire wall. Upon laddering the roof, and while crew members were still operating on and from the aerial apparatus, a sudden collapse of the truss roof occurred, endangering the crew because of fire and radiant heat. The collapse involved approximately 12 m (40 ft.) of the roof line, and a fireball extended upward approximately 9 m (30 ft.) into the air. The ventilation crew had to make an immediate descent down the aerial to escape harm. The collapse occurred about 10 minutes after the arrival of first-in companies. All firefighters involved were in full personal protective equipment (PPE) and self-contained breathing apparatus (SCBA). Awareness of wood-truss construction must always be considered during fire-suppression activities.

Lessons Learned: Fire-ground accountability and awareness of wood-truss construction, the fire's impact because of direct impingement, and the duration of the fire must always be considered during fire-suppression activities. In this case, had the crew actually been actively ventilating or on the roof, serious injuries and, most likely, fire fighter fatalities would have occurred.

Awareness training for wood-truss construction should be included in any ventilation and structural fire-suppression training. Command and control, fire-ground accountability, and fire-ground safety-officer training should also be included in such education.

We must always keep in mind that the fire ground is a dynamic, ever-changing scene, and we must adapt to those conditions and stay aware of them. Remember: no structure is worth our lives; everyone goes home.

---

insulating materials are widely used in large volumes. Buildings are often reinsulated, or have had additional insulation added, since their original construction. Under fire conditions, insulating materials present significant risks to fire fighters—from being trapped by large quantities of waterlogged insulation falling from failed ceilings to insulation hiding the presence of slow, smouldering fires. Insulation can also be a toxic hazard if it contains asbestos or other hazardous chemicals.

Types of Construction

Canadian building construction is regulated by the federal government in all federally regulated industries, provinces/territories, and local municipal governments. The NBC and National Fire Code of Canada (NFC) are developed by the National Research Council Canada (NRC) with input from provincial and municipal building officials. The NBC and NFC are model codes that, everywhere except Prince Edward Island, are either adopted by provincial/territorial governments or used as the basis for provincial/territorial codes and standards. Provinces/territories that adopt the national codes have regulatory mechanisms to vary the requirements of the code to meet the needs of their jurisdictions. National and provincial/territorial codes are minimum construction standards. Variations on the code by municipal governments generally only increase the minimum code requirements; they are not allowed to reduce requirements.

Since 2005, the Canadian code system has been an objective-based system, which allows for multiple building solutions to deliver on its four objectives. The objectives under which the codes have been developed are safety, health, accessibility for persons with disabilities, and fire and structural protection of buildings. Prior to that time, the codes were prescriptive in nature. Building codes specify the type of construction to be used based on the height, area, occupancy classification, and location of the building. Other factors, such as the presence of automatic sprinklers, will also affect the required construction classification. National Fire Protection Association (NFPA) 220, *Standard on Types of Building Construction*, provides detailed requirements for each type of building construction.

Building construction is dictated by the occupancy classification of the building. The NBC has nine sections: Two sections dictate the fire safety requirements of buildings. Buildings that are required to be designed by professional architects and engineers, such as assembly occupancies and large buildings, generally fall under sections 1–8. Section 9 of the code relates specifically to housing, small buildings of three storeys or less that do not exceed 600 m² and major occupancies (Groups C, D, E, and F [Divisions 2 and 3)]. The two main sections of the code that deal with fire-safe construction are sections 3 and 9.

Fire fighters need to understand and recognize five types of building construction, designated as Types I through V. These classifications are directly related to fire protection and fire behaviour in a building.

Buildings are classified based on the combustibility of the structure and the fire resistance of its components. Buildings using Type I or II construction are assembled with primarily noncombustible materials and limited amounts of wood and other materials that will burn. This does not mean that Type I and Type II buildings cannot be damaged by a fire; in fact, if the contents of the building burn, the fire can seriously damage or destroy its structure.

In buildings using Type III, IV, or V construction, both the structural components and the buildings' contents will burn. Wood and wood products are used to varying degrees in these buildings. If the wood ignites, the fire will weaken and consume both the structure and its contents. Structural elements of these buildings can be damaged or destroyed in a very short time.

Fire resistance refers to the length of time that a building or building component can withstand a fire before passing flame and the transmission of heat. Fire resistance ratings are stated in hours and minutes, based on the time that a test assembly withstands or contains a standard test fire; for example, walls are rated based on whether they stop the progress of the test fire for periods ranging from 20 minutes to 4 hours. Because ratings are based on a standard test fire but an actual fire could be more or less severe than the test fire, ratings are considered merely guidelines; that is, an assembly with a 2-hour rating will not necessarily withstand a fire for 2 hours.

■ Type I Construction: Fire Resistive

Type I construction (also called fire-resistive construction) is the most fire-resistive category of building construction. It is used for buildings designed for large numbers of people, buildings with a high life-safety hazard, tall buildings, large-area buildings, and buildings containing special hazards. Type I construction is commonly found in schools, hospitals, and high-rise buildings **FIGURE 7-5** .

Buildings with Type I construction can withstand and contain a fire for a specified period of time. The fire resistance and combustibility of all construction materials are carefully evaluated in such cases, and each building component must be engineered to contribute to the fire resistance of the entire building to warrant this classification.

All of the structural members and components used in Type I construction must be made of noncombustible materials, such as steel, concrete, or gypsum board. In addition, the structure must be constructed or protected so that it has at least 2 hours of fire resistance. Building codes specify the fire-resistance requirements for different components; for example, columns and load-bearing walls in multistorey buildings could have a fire-resistance requirement of 3 or 4 hours, whereas a floor might be required to have a fire-resistance rating of only 2 hours. Some codes allow Type I buildings to use a limited amount of combustible materials as part of the interior finishes.

If a Type I building exceeds specific height and area limitations, codes generally require the use of fire-resistive walls and/or floors to subdivide it into compartments. A compartment might consist of a single floor in a high-rise building or a part of a floor in a large-area building. In any event, a fire in one compartment should not spread to any other parts of the building. To ensure that a fire is contained, stairways, elevator shafts, and utility shafts should be enclosed in construction that prevents fire from spreading from floor to floor or from compartment to compartment.

Type I buildings typically use reinforced concrete and protected steel-frame construction. As mentioned earlier, concrete is noncombustible and provides thermal protection around the steel reinforcing rods; reinforced concrete can fail, however, if it is subjected to a fire for a long period of time or if the building's contents create an extreme fire load. Given its tendency to lose strength in the face of high temperatures, structural steel framing must be protected from the heat of a fire. In Type I construction, the structural steel members are generally encased in concrete, shielded by a fire-resistive ceiling, covered with multiple layers of gypsum board, or protected by a spray-on insulating material **FIGURE 7-6** . An unprotected steel beam exposed to a fire can fail and cause the entire building to collapse.

Type I building materials should not provide enough fuel in and of themselves to create a serious fire; thus, it is

FIGURE 7-5 Type I construction is commonly found in schools, hospitals, and high-rise buildings.
© John Foxx/Alamy Images

FIGURE 7-6 Spray-on fireproofing materials are often used to protect structural steel.
© Michael Doolittle/Alamy Images

the contents of the building that determine the severity of a Type I building fire. In theory, a fire could consume all of the combustible contents of a Type I building, yet leave the structure basically intact.

Although Type I construction might provide the highest degree of safety, it does not eliminate all of the risks to occupants and fire fighters. Serious fires can still occur in fire-resistive buildings, because the burning contents may produce copious quantities of heat and smoke. Even though the structure is designed to give fire fighters time to conduct an interior attack, it can be very difficult to extinguish a fire in a Type I building. A fire that is fully contained within fire-resistive construction can be very hot and difficult to ventilate. For this reason, fire-resistive construction is typically combined with an automatic fire detection system and automatic sprinklers in modern high-rise buildings.

Sometimes, the burning contents of a building provide sufficient fuel and generate enough heat to overwhelm the fire-resistant properties of the construction. In these cases, the fire can escape from its compartment and damage or destroy the structure. Similarly, inadequate or poorly maintained construction can undermine the fire-resistive properties of the building materials. Under extreme fire conditions, a Type I building can collapse. Even so, these buildings have a much lower potential for structural failure than do other buildings.

FIGURE 7-7 Type II construction is also referred to as noncombustible construction.

FIRE FIGHTER Tips

A new component in lightweight building construction is the truss joist I-beam, or TGI beam. This framing material is made up of three wooden components. Channels may be drilled through TGI beams, thus providing another avenue for fire to spread.

Engineered lumber and laminated beams are being used more often in modern construction. These products are made up of smaller pieces of wood bound together through an adhesive process. Exposure to fire may break the adhesive bonds and lead to lumber or beam failure.

■ Type II Construction: Noncombustible

Type II construction is also referred to as noncombustible construction **FIGURE 7-7**. All of the structural components in a Type II building must be made of noncombustible materials. The fire-resistive requirements, however, are less stringent for Type II construction than for Type I construction: In some cases, no fire-resistance requirements apply to the building. In other cases (known as protected noncombustible construction), a fire-resistance rating of 1 or 2 hours may be required for certain elements.

Noncombustible construction is most common in single-storey warehouse or factory buildings, where vertical fire spread is not an issue. Unprotected construction of this type is generally limited to a maximum of two storeys. Some multistorey buildings are constructed using protected noncombustible construction.

Steel is the most common structural material in Type II buildings. Insulating materials can be applied to the steel when fire resistance is required. A typical example of Type II construction is a large-area, single-storey building with a steel frame, metal or concrete-block walls, and a metal roof deck. Fire walls are sometimes used to subdivide these large-area buildings to prevent catastrophic losses. Undivided floor areas must be protected with automatic sprinklers to limit the fire risk.

Fire severity in a Type II building is determined by the contents of the building, because the structural components contribute little or no fuel and interior finishes are limited. If the building's contents provide a high fuel load, a fire can collapse and destroy the structure. In this kind of construction, automatic sprinklers should be used to protect combustible and valuable contents. Fires in buildings equipped with monitored fire detection systems are more likely to be reported earlier than fires in buildings without monitored fire detection systems.

■ Type III Construction: Ordinary

Type III construction is also referred to as ordinary construction because it is used in a wide variety of buildings—from commercial strip malls to small apartment buildings **FIGURE 7-8**. Ordinary construction is usually limited to buildings of no more than four storeys, but it can sometimes be found in buildings as tall as six or seven storeys.

Type III buildings have masonry exterior walls, which support the floors and the roof structure. The interior structural and nonstructural members—including the walls, floor, and roof—are all constructed of wood. In most ordinary construction, gypsum board or plaster is used as an interior finish material, covering the wood framework and providing minimal fire protection.

Fire-resistance requirements for interior construction of Type III buildings are limited. The gypsum or plaster coverings over the interior wood components provide some fire

FIGURE 7-8 Type III construction is referred to as ordinary construction because it is used in a wide variety of buildings.

© Brandon Bourdages/ShutterStock, Inc.

resistance. Key interior structural components may be required to have fire-resistance ratings of 1 or 2 hours, or there may be no requirements at all. Some Type III buildings use masonry interior load-bearing walls to meet the requirements for fire-resistant structural support.

A building of Type III construction has two separate fire loads: the contents and the combustible building materials used. Even a vacant building will contain a sufficient quantity of wood and other combustible components to produce a large fire. Given these materials' combustibility, a fire involving both the contents and the structural components can quickly destroy the building.

Fortunately, most fires originate in the building's contents and are extinguished before the flames ignite the Type III structure itself. Once a fire extends into the structure itself and begins to consume the fuel within the walls and above the ceilings, it becomes much more difficult to control and prone to roof or floor collapse.

Type III construction presents several problems for fire fighters; for example, an electrical fire can begin inside the void spaces within the walls, floors, and roof assemblies and extend to the contents. The void spaces also allow a fire to extend vertically and horizontally, spreading from room to room and from floor to floor. In such a case, fire fighters will have to open the void spaces to fight the fire. An uncontrolled fire within the void spaces is likely to destroy the building. Fire fighters need to be particularly aware that fires in void spaces under floors or under the roof can quickly destroy the strength of these components. Do not be surprised by these hidden fires; instead, anticipate them.

The fire resistance of interior structural components often depends on the age of the building and on local building codes. Older Type III buildings were built from solid lumber, which can contain or withstand a fire for a limited time. Newer or renovated buildings might contain lightweight assemblies, which can be damaged much more quickly and are prone to early failure. Floors and roofs in these buildings

collapse suddenly and without warning. Because fire fighters cannot always determine the type of structure during a fire, you should assume that lightweight components may be present in any Type III building.

Exterior walls also can collapse if a fire causes significant damage to the interior structure. Because the exterior walls, floors, and roof are all connected in a stable building, the collapse of the interior structure could make the free-standing masonry walls unstable and prone to collapse.

FIRE FIGHTER Tips

Fire-resistive structures with sprinklers are a great asset in assisting fire fighters with controlling the burning contents. Once they activate, however, the sprinklers produce steam when the water hits the fire. As the sprinklers continue to operate, the steam is cooled and condenses; that condensed water vapour may then fill an enclosed area from floor to ceiling. Without proper ventilation, fire fighters may find themselves with virtually no visibility. Recall that 1 litre of water weighs 1 kilogram (1 gallon weighs 8.3 pounds)—an important reason why fire fighters must calculate the total water load on the building.

■ Type IV Construction: Heavy Timber

Type IV construction is also known as heavy timber construction. A heavy timber building has exterior walls that consist of masonry construction, and interior walls, columns, beams, floor assemblies, and a roof structure that are made of wood. The exterior walls are usually brick and are extra thick to support the weight of the building and its contents.

The wood used in Type IV construction is much heavier than that used in Type III construction. It is more difficult to ignite and will withstand a fire for a much longer time before the building collapses. A typical heavy timber building might have 20-cm-square (8-in.-square) wood posts supporting 36-cm-deep (14-in.-deep) wood beams and 20-cm (8-in.) floor joists. The floors could be constructed of solid wood planks, 5 or 8 cm (2 or 3 in.) thick, with a top layer of wood serving as the finished floor surface.

Heavy timber construction, if it meets building codes, has no concealed spaces or voids. This type of structure helps reduce the horizontal and vertical fire spread that often occurs in ordinary construction buildings. Unfortunately, many heavy timber buildings do contain vertical openings for elevators, stairs, or machinery, which can provide a path for a fire to travel from one floor to another.

In the past, heavy timber construction was used to construct buildings as tall as six to eight storeys with open spaces suitable for manufacturing and storage occupancies. Today, modern buildings of this style are usually built with either fire-resistive (Type I) or noncombustible (Type II) construction. New buildings of Type IV construction are rare, except for special structures that feature the construction components as architectural elements.

The heavy, solid wood columns, support beams, floor assemblies, and roof assemblies used in heavy timber construction will withstand a fire much longer than the smaller wood members used in ordinary and lightweight combustible construction. Nevertheless, once involved in a fire, the structure of a heavy timber building can burn for many hours. A fire that ignites the combustible portions of heavy timber construction is likely to burn until it runs out of fuel and the building is reduced to a pile of rubble. As the fire consumes the heavy timber support members, the masonry walls will become unstable and collapse.

Mill construction was common during the 1800s in the United States, especially in northeastern states. In particular, large mill buildings were used as factories and warehouses. In many communities, dozens of these large buildings were clustered together in industrial areas, creating the potential for huge fires involving multiple multistorey buildings. The radiant heat from a major fire in one of these buildings could spread the fire to nearby buildings, resulting in the loss of several surrounding structures.

Mill construction was state of the art for its time. Automatic sprinkler systems were developed to protect buildings of mill construction; as long as the sprinklers are properly maintained, these buildings have a good safety record; without sprinklers, however, a heavy timber mill building becomes a major fire hazard.

Only a few of these original mill buildings are still used for manufacturing. Many more have been converted into small shops, galleries, office buildings, and residential occupancies. These conversions tend to divide the open spaces into smaller compartments and create void spaces within the structure. The revamped buildings may contain either noncombustible contents or highly flammable contents, depending on their use. Occupancies may range from unoccupied storage facilities to occupied loft apartments and office spaces. Appropriate fire protection and life-safety features, such as modern sprinkler systems, must be built into these conversions to ensure the safety of their occupants.

■ Type V Construction: Wood Frame

In **Type V construction**, all of the major components are constructed of wood or other combustible materials FIGURE 7-9 . Type V construction is often called wood-frame construction and is the most common type of construction used today. Wood-frame construction is used not only in one- and two-family dwellings and small commercial buildings but also in larger structures, such as apartment and condominium complexes and office buildings up to four storeys in height.

Many wood-frame buildings do not have any fire-resistive components. Some codes require a 1-hour fire resistance rating for limited parts of wood-frame buildings, particularly those of more than two storeys. Plaster or gypsum board barriers are often used to achieve this rating, but fire fighters should not assume that these barriers are present or will be effective in all Type V buildings.

Because all of their structural components are combustible, wood-frame buildings can rapidly become totally involved in a fire. In addition, Type V construction usually creates voids and channels that allow a fire within the structure to spread quickly. A fire that originates in a Type V building can easily extend to other nearby buildings.

Wood-frame buildings often collapse and suffer major destruction from fires. Smoke detectors are essential to warn

FIGURE 7-9 Type V (wood-frame) construction is the most common type of construction used today.

a building's residents early if a fire occurs. Although compartmentalization can help limit the spread of a fire, fire detection devices and automatic sprinklers are the most effective way of protecting lives and property in Type V buildings.

Wood-frame buildings are constructed in a variety of ways. Older wood-frame construction employed solid lumber, which relied on its mass to provide strength. To reduce costs and create the largest building with the least amount of material, lightweight construction makes extensive use of engineered wood products, such as wooden I-beams and wooden trusses. With this construction technique, structural assemblies are engineered to be just strong enough to carry the required load. As a result, there is little built-in safety margin, and these buildings can collapse early, suddenly, and completely during a fire.

The presence of flammable building contents may greatly increase the severity of a fire and accelerate the speed with which a fire destroys a wood-frame building. Indeed, the structural elements of such a building can be damaged or destroyed in a very short time. Whenever a fire has been burning under a floor or roof, you should assume that the structure is unsafe and may collapse. Fires that start to consume structural parts of a Type V building present a high risk of building collapse. From a fire fighter's point of view, lightweight wood-frame construction is a campfire waiting to happen.

Wood-frame buildings might be covered with wood siding, vinyl siding, aluminum siding, brick veneer, or stucco. The covering on a building does not reflect the type of construction of that building; just because you see a brick covering does not mean that the building is constructed of solid brick walls.

Two systems are used to assemble wood-frame buildings: balloon-frame construction and platform-frame construction. Because these systems developed in different eras, fire fighters can anticipate the type of construction based on the age of a building.

Fire Fighter Safety Tips

Lightweight wood-frame construction is a campfire waiting to happen. A fire in this type of construction will spread rapidly throughout the building and to adjacent exposures.

Fire Fighter Safety Tips

Many wood-frame buildings have a brick veneer on their exterior. The thin layer of brick or stone might have be applied to enhance the appearance of the building or to reduce the risk of fire spread. Unfortunately, this practice can give the impression that a building is Type III (ordinary) construction when it is really Type V (wood-frame) construction.

During a structural fire, the brick veneer is likely to collapse or peel away as the wall behind it burns or collapses. A single layer of bricks is rarely strong enough to stand independently. Fire fighters should be aware of buildings constructed in this manner and maintain a safe distance from potentially falling bricks.

Balloon-Frame Construction

Balloon-frame construction was popular between the late 1800s and the mid 1900s. In a balloon-frame building, the exterior walls are assembled with wood studs that run continuously from the basement to the roof **FIGURE 7-10**. In a two-storey building, the floor joists that support the first and second floors are nailed to these continuous studs. As a result, an open channel between each pair of studs extends from the foundation to the attic. Each of these channels provides a path that enables a fire to spread from the basement to the attic without being visible on the first- or second-floor levels. Fire fighters must anticipate that a fire originating on a lower level will quickly extend through these voids, and they should open the void spaces to check for hidden fires and to prevent rapid vertical extension.

Platform-Frame Construction

Platform-frame construction is used for almost all modern wood-frame construction. In a building with platform construction, the exterior wall studs are not continuous. Instead, the first floor is constructed as a platform, and the studs for the exterior walls are erected on top of it. The first set of studs extends only to the underside of the second-floor platform, which blocks any vertical void spaces. The studs for the second-floor exterior walls are erected on top of the second-floor platform. At each level, then, the floor platform blocks the path of any fire rising within the void spaces. Platform framing prevents fire from spreading from one floor to another through continuous spaces between the studs. Although a fire can eventually burn through the wood platforms, they will slow the fire spread.

Building Components

The construction classification system gives fire fighters a general idea of the materials used in building and an indication of how the materials will react to fire. In addition, each building

FIGURE 7-10 In a balloon-frame building, the exterior walls house channels that enable a fire to spread easily from the basement to the attic.

also has several different components that should be recognized. A fire fighter who understands how these components function will have a better understanding of the risks involved with a burning building. Some building construction features are safer for fire fighters, as this section explains.

The following major building components are discussed in this section:

- Foundations
- Floors and ceilings
- Roofs
- Trusses
- Walls
- Doors and windows
- Interior finishes and floor coverings

■ Foundations

The primary purpose of a building's foundation is to transfer the weight of the building and its contents to the ground **FIGURE 7-11**. The weight of the building itself is called the dead load; the weight of the building's contents is called the live load. The foundation ensures that the base of the building is planted firmly in a fixed location, which helps keep all other components connected.

Modern foundations are usually constructed of concrete or masonry, although wood pilings and pressure-treated wood are used in some areas. Foundations can be either shallow or deep, depending on the type of building and the soil composition. Some buildings are built on concrete footings or piers; others are supported by steel or concrete piles or wooden posts driven into the ground.

FIGURE 7-11 The foundation supports the entire weight of the building.

© Dorn1530/ShutterStock, Inc.

FIGURE 7-12 The space between the ceiling and the underside of the floor above it often holds electrical and communications wiring and other building systems.

© Lourens Smak/Alamy Images

As long as the foundation remains stable and intact, it is usually not a major concern for fire fighters; if the foundation shifts or is in poor repair, however, it can become a critical concern. A burning building with a weak foundation or inadequate lateral bracing could be in imminent danger of collapse.

Most foundation problems are caused by circumstances other than a fire, such as improper construction, shifting soil conditions, or earthquakes. Although fires can damage timber-post and other wood foundations, most foundations remain intact even after a severe fire has damaged the rest of the building.

When examining a building, take a close look at the foundation. Look for cracks that indicate movement of the foundation. If the building has been modified or remodelled, look for areas where the support could have been compromised.

■ Floors and Ceilings

Floor construction is very important to fire fighters for three major reasons: First, fire fighters who are working inside a building must rely on the integrity of the floor to support their weight. A floor that fails could drop the fire fighters into a fire on a lower level. Second, in a multistorey structure, fire fighters may be working below a floor (or roof), which would fall on them if it collapsed. Third, the floor system influences whether a fire spreads vertically from floor to floor within a building or whether it remains contained on a single level. Some floor systems are designed to resist fires, whereas others have no fire-resistant capabilities.

Fire-Resistive Floors

In multistorey buildings, floors and ceilings are generally considered a combined structural system; that is, the structure that supports a floor also supports the ceiling of the storey below. In a fire-resistive building, this system is designed to prevent a fire from spreading vertically and to prevent a collapse when a fire occurs in the space below the floor–ceiling assembly. Fire-resistive floor–ceiling systems are rated in terms of hours of fire resistance based on a standard test fire.

Concrete floors are common in fire-resistive construction. The concrete can be cast in place or assembled from panels or beams of precast concrete, which are made at a factory and then transported to the construction site for placement.

The thickness of the concrete floor depends on the load that the floor needs to support.

Concrete floors can be either self-supporting or supported by a system of steel beams or trusses. In the latter case, the steel can be protected by spray-on insulating materials or covered with concrete or gypsum board. If the ceiling is part of the fire-resistive rating, it provides a thermal barrier to protect the steel members from a fire in the area below the ceiling.

The ceiling below a fire-resistant floor can be constructed from plaster or gypsum board, or it can be a system of tiles suspended from the floor structure. In many cases, a void space between the ceiling and the floor above it contains building systems and equipment such as electrical or telephone wiring, heating and air-conditioning ducts, and plumbing and fire sprinkler system pipes **FIGURE 7-12**. If the space above the ceiling is not subdivided by fire-resistant partitions or protected by automatic sprinklers, a fire can quickly extend horizontally across a large area.

Wood-Supported Floors

Wood floor structures are common in non-fire-resistive construction. Such wooden floor systems range from heavy timber construction, which is often found in old mill buildings, to modern lightweight construction.

Although heavy timber construction provides a huge fuel load for a fire, it can also contain and withstand a fire for a considerable length of time without collapsing. Heavy timber construction uses posts and beams that are at least 20 cm (8 in.) on the smallest side and often as large as 36 cm (14 in.) in depth. The floor decking is often assembled from solid wood boards, 5 or 8 cm (2 or 3 in.) thick, which are covered by an additional 2.5 cm (1 in.) of finished wood flooring. The depth of the wood in this system will often contain a fire for an hour or more before the floor fails or burns through.

Conventional wood flooring, which was widely used for many decades, is much lighter than heavy timber but uses solid lumber as beams, floor joists, decking, and finished flooring. It burns readily when exposed to a fire but generally takes approximately 20 minutes to burn through or reach structural failure. This time estimate is only a general, unscientific guideline; the actual burning rate depends on many other factors.

Modern lightweight construction uses structural elements that are much less substantial than conventional lumber; for example, lightweight wooden trusses or engineered wooden I-beams are often used as supporting structures, thin sheets of plywood are used as decking, and the top layer often consists of a thin layer of concrete or wood covered by carpet. This floor construction provides little resistance to fire. The lightweight structural elements can fail or the fire can burn through the decking quickly. In the United States, the Department of Health and Human Services National Institute for Occupational Safety and Health (NIOSH) reported that it took 19 minutes for a traditional unprotected residential floor assembly to fail (NIOSH Publication no. 2009-114). Under the same test conditions, an unprotected lightweight engineered I-joist failed in only 6 minutes.

Unfortunately, it is impossible to tell how a floor is constructed by looking at it from above. The important information about a floor can be observed only from below, if it is visible at all. The building's age and local construction methods can provide significant clues to a floor's stability, but you should be aware that many older buildings have been renovated using modern lightweight systems and materials. Fire fighters should use preincident surveys and other planning activities to gather essential structural information about buildings. In some office and computer-room renovations, the floor has been replaced with removable floor panels, so as to hide the different types of cables under the floor.

Ceiling Assemblies

From a structural standpoint, the ceiling is actually considered to be part of the floor assembly. Ceilings are mainly included in buildings for aesthetic reasons. Their primary function is to hold light fixtures and diffuse light. Ceilings also conceal heating, ventilation, and air conditioning (HVAC) distribution systems. In addition, they conceal wiring and fire sprinkler systems.

Ceilings can be part of the fire-resistive package; for example, they can be covered with plaster or gypsum board, or they can consist of dropped ceilings covered with fire-rated tiles. Some ceilings are fire rated as part of the floor assembly.

Fire fighters should be aware that hollow spaces between the floor and the ceiling can contribute to the horizontal spread of fire. Penetrations in the floor can also contribute to the vertical spread of fire.

■ Roofs

Safe interior firefighting operations depend on a stable roof. If the roof collapses, the fire fighters inside the building may be injured or killed. Interior fires generate hot gases that accumulate under the roof; thus, fire fighters who are performing ventilation activities on the roof to release the heated gases must also depend on the structural integrity of the roof.

The primary purpose of a roof is to protect the inside of a building from the weather. Often, roofs are not designed to be

as strong as floors, especially in warmer climates that do not experience heavy snowfall. If the space under the roof is used for storage, or if extra HVAC equipment is mounted onto the roof, the load can exceed the designer's expectations. Adding the weight of several fire fighters to an overloaded roof can be disastrous. Also, this space may contain different types of insulation to help protect the structure from climate changes; in older homes, for example, this insulation often contains ground-up paper that can lead to smouldering and fire spread.

Several methods and materials are used for roof construction. Roofs are constructed in three primary designs: pitched, curved, and flat. The major components of a roof assembly are the supporting structure, the roof deck, and the roof covering.

Pitched Roofs

A **pitched roof** has sloping or inclined surfaces. Pitched roofs are used on many houses and some commercial buildings. The pitch or angle of the roof can vary depending on local architectural styles and climate. Variations of pitched roofs include gable, hip, mansard, gambrel, and lean-to roofs **FIGURE 7-13**.

The slope of a pitched roof can present either a minor inconvenience or a complete lack of secure footing, depending on its angle. Fire fighters working on a pitched roof are always in danger of falling, but the risk is particularly high when the roof is wet or icy, when the roof covering is not secure, or when smoke or darkness obscures vision. Roof ladders and aerial apparatus are used to provide a secure working platform for fire fighters working on a pitched roof.

Pitched roofs are usually supported by either rafters or trusses. **Rafters** are solid wood joists mounted in an inclined position. Pitched roofs supported by rafters usually have solid wood boards as the roof decking. Fire fighters can sometimes detect the impending failure of a roof with solid decking supported by rafters by a spongy feeling underfoot.

Modern lightweight construction uses manufactured wood trusses to construct most pitched roofs. Many lightweight trusses are manufactured using gusset plates that penetrate the wood no more than 0.9 cm (0.4 in.). The decking usually consists of thin plywood or a sheeting material such as OSB. When these trusses and decking are exposed to heat and fire, they often fail after a short period of time with no prior warning. Fire fighters cannot work safely on these roofs when the supporting structures become involved in a fire.

Steel trusses also are used to support pitched roofs. The fire resistance of steel trusses is directly related to how well the steel is protected from the heat of the fire.

Several roof-covering materials are used on pitched roofs, usually in the form of wooden shakes (no longer permitted in most areas after the interface fires of 2003–2004) or shingles, asphalt shingles or tiles **FIGURE 7-14**. Asphalt shingles are usually made from felt or mineral fibres impregnated with asphalt, although metal and fibreglass shingles are also used. Modified bitument membrane (MBM) roofing, which is held to the roof deck by melting a bitumen layer to adhere the material to the roof decking without mechanical fastening by nails or screws, is also available. Wooden shingles and shakes, often made from cedar, are popular in some areas. In some construction, wooden shingles or shakes are mounted on individual wood slats instead of continuous decking; these types of shingles

A. © AbleStock

B. Courtesy of Pacific Northwest National Laboratory

C. Courtesy of Doyal Dum/US Army Corps of Engineers

D. © AbleStock

FIGURE 7-13 Examples of pitched roofs: **A.** Gable roof. **B.** Hip roof. **C.** Mansard roof. **D.** Gambrel roof.

burn rapidly in dry weather conditions. Metal sheet roofing can also be used but may not be installed on wood slats as wooden shingles and shakes often are.

Shingles are generally durable, economical, and easily repaired. A shingle roof that has aged and deteriorated should be removed completely and replaced. Some older buildings may have newer layers of shingles on top of older layers, which can make it difficult to cut an opening for ventilation.

Roofing tiles are usually made from clay or concrete products. Clay tiles, which have been used for roofing since ancient times, can be either flat or rounded. Rounded clay tiles are sometimes called mission tiles. Clay tiles are both durable and fire resistant.

Slate tiles are produced from thin sheets of rock. Slate is an expensive, long-lasting roofing material, but it is very brittle and becomes slippery when wet. Slate tiles also pose a threat

to fire fighters below when a pitched roof is being ventilated, because falling tiles can become airborne missiles.

Metal panels of galvanized or painted steel, copper, and aluminum may be used on pitched roofs as well. Expensive metals such as copper are often used for their decorative appearance, whereas lower-cost galvanized or painted steel is used on barns and industrial buildings but is becoming common on residential and other buildings. Corrugated, or profiled, galvanized or painted metal panels are strong enough to be mounted on a roof without roof decking. Metal roof coverings will not burn, but they can conduct heat to the roof decking. Also, fire-damaged electrical wires can cause a metal roof deck to become energized and present a shock hazard.

A new roofing system using panels created by filling a cavity between two metal panels (or one metal and one wood panel) with a insulating foam core is becoming more common.

A. © Paul Springett/Up the Resolution/Alamy Images

B. © Ron Chapple/Thinkstock/Alamy Images

C. © TH Photo/Alamy Images

FIGURE 7-14 Several roof-covering materials are used on pitched roofs. **A.** A tile roof. **B.** A wood shingle roof. **C.** A slate roof.

These systems use the foam insulation and the metal or metal–wood skins to provide structural integrity. These systems provide a high air-vapour sealing ability and easy installation, but when exposed to fire, the metal or wood can delaminate from the foam insulation at temperatures as low as 500°C (932°F), greatly reducing its structural support capability.

Curved Roofs

Curved roofs are often used for supermarkets, warehouses, industrial buildings, arenas, auditoriums, bowling alleys, churches, airplane hangars, and other buildings that require large, open interiors. These kinds of roofs are usually supported by steel or wood bowstring trusses or arches.

The decking on buildings with curved roofs can range from solid wooden boards or plywood to corrugated steel sheets. The type of material used must be identified before ventilation openings can be made. Often, the roof covering consists of layers that include felt, mineral fibres, and asphalt, although some curved roofs are covered with foam plastics or plastic panels.

Flat Roofs

Flat roofs are found on houses, apartment buildings, shopping centres, warehouses, factories, schools, and hospitals **FIGURE 7-15**. Most flat roofs have a slight slope so that water can drain off the structure. If the roof does not have the proper slope or if the drains are not maintained, water may pool on the roof, overloading the structure and causing a collapse.

The support systems for most flat roofs are constructed of either wood or steel. A wood support structure uses solid **wooden beams** and joists; laminated wood beams may be included to provide extra strength or to span long distances. Lightweight construction uses wood trusses or I-beams as the supporting members. Such lightweight assemblies are much less fire resistant and collapse more quickly than do solid-wood systems.

Steel may also be used to support flat roofs. Open-web steel trusses (sometimes called bar joists) can span long distances and remain stable during a fire if they are not subjected to excessive heat. Automatic sprinklers can protect the steel from the heat; without such protection, however, the steel over a fire will soon weaken and sag. Heated steel support members can even elongate enough to push out the exterior walls of the building. Cracks in the outside walls at the roof level are a warning indicator of this condition.

Flat roof decks can be constructed of wood planking, plywood, corrugated steel, gypsum, or concrete. The material used depends on the building's age and size, the climate, the cost of materials, and the type of roof covering.

The roof covering is applied on top of the deck. Most older flat roofs are covered with multilayer, built-up roofing systems that help prevent leakage. A typical built-up roof covering contains five layers: a vapour barrier, thermal insulation, a waterproof membrane, a drainage layer, and a wear course. The outside of the flat roof reveals only the top layer, which is often gravel; it serves as the wear course and protects the underlying layers from wear and tear. Cutting through all of the layers to open a ventilation hole can be quite a challenge.

Most flat roof coverings contain highly combustible materials, including asphalt, roofing felt, tarpaper, rubber or plastic membranes, and insulation layers. These materials can be

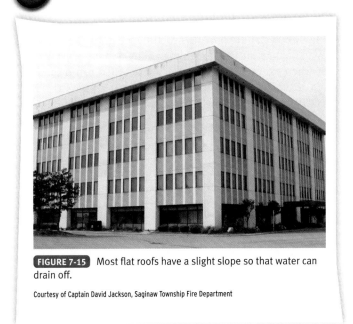

FIGURE 7-15 Most flat roofs have a slight slope so that water can drain off.

Courtesy of Captain David Jackson, Saginaw Township Fire Department

difficult to ignite; once lit, however, they burn readily, releasing great quantities of heat and black smoke in the process.

Gravel-ballasted built-up roof systems are generally being replaced by MBM systems in Canada. This roof system has similar challenges to the older built-up roof system but will not have a gravel-ballast layer. The surface of the MBM roofing will appear similar to an asphalt shingle. The material is adhered to a mechanically fastened deck or insulation by heating and melting it to the surface, so once exposed to heat, it can release from the decking. This roofing system has the similar problems to the built-up system under fire conditions.

Flat roofs can present unique problems during vertical ventilation operations. A roof with a history of leaks could have several patches where additional layers of roofing material make for an extra-thick covering. Sometimes a whole new roof is constructed on top of the old one. After opening a hole in the new roof, fire fighters might discover the old roof and have to cut into that as well. Fire fighters might also discover fire burning in the void space between the two roofs.

A similar problem may be encountered in remodelled buildings or buildings with additions. An old flat roof may be found under a new pitched section, resulting in two separate void spaces. The old roof may also provide a large supply of fuel for the fire. Such a situation is difficult to predict unless the incident commander (IC) has the building plan in advance.

In most of Canada, extreme temperature conditions and high heating/cooling costs have resulted in high insulation levels being installed on most flat roof buildings. Extruded or expanded polystyrene insulation of more than 20 cm (8 in.) is common, and in areas with long winters, insulation thicknesses can exceed 30 cm (12 in.). These insulation levels can disguise significant damage to the supporting structure below a roof without being detectable from the exterior. Fire fighters working on a roof may not be aware the structure is failing by normal means; with high levels of insulation, techniques such as sounding the roof and observing snow melt will not work to provide warning of imminent failure.

Fire Fighter Safety Tips
© Photos.com

When working on a flat roof, avoid walking on skylights and other openings that will not support your weight. Stay away from the edges of the roof.

Fire Fighter Safety Tips
© Photos.com

Flat roofs with parallel chord trusses and corrugated metal decks have been known to roll down into ventilation holes during roof operations. Some fire departments no longer perform vertical ventilation in buildings with this type of roof.

■ Trusses

A **truss** is a structural component that is composed of smaller pieces joined to form a triangular configuration or a system of triangles. Trusses are used extensively in support systems for both floors and roofs. They are common in modern construction for several reasons. The triangular geometry creates a strong, rigid structure that can support a load much greater than its own mass; for example, both a solid beam and a simple truss with the same overall dimensions can support the same load. The truss has several advantages over the solid beam, however. It requires much less material than the beam, is much lighter, and can span a long distance without supports. Trusses are often prefabricated and transported to the building location for installation there.

Trusses are frequently used in residential construction, apartment buildings, small office buildings, commercial buildings, warehouses, fast-food restaurants, airplane hangars, churches, and even firehouses. They may be clearly visible, or they may be concealed within the construction. Trusses are widely used in new construction and often replace heavier solid beams and joists in renovated or modified older buildings.

The strength of a truss depends on both its members and the connections between them. A properly assembled and installed truss is strong; if any member or connector begins to fail, however, the strength of the entire truss will be compromised.

Trusses can be used for many different purposes. In building construction, those made of wood, steel, or combinations of wood and steel are used primarily to support roofs and floors. Three of the most commonly used types of trusses are the parallel chord truss, the pitched chord truss, and the bowstring truss.

A **parallel chord truss** has two parallel horizontal members connected by a system of diagonal and sometimes vertical members **FIGURE 7-16**. The top and bottom members are called the chords, and the connecting pieces are called the web members. Parallel chord trusses are typically used to support flat roofs or floors. In lightweight construction, an engineered **wood truss** is often assembled with wood chords and either wood or light-steel web members. A steel bar joist is another example of a parallel chord truss.

A **pitched chord truss** is typically used to support a sloping roof **FIGURE 7-17**. Most modern residential construction uses

FIGURE 7-16 A parallel chord truss.

FIGURE 7-17 A pitched chord truss.

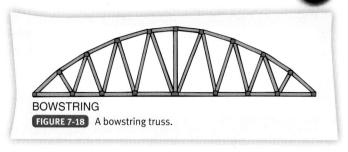

BOWSTRING

FIGURE 7-18 A bowstring truss.

FIGURE 7-19 A building with a bowstring truss.

Courtesy of Captain David Jackson, Saginaw Township Fire Department

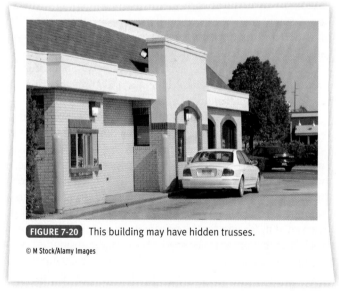

FIGURE 7-20 This building may have hidden trusses.

© M Stock/Alamy Images

a series of prefabricated wood pitched chord trusses to support the roof. The roof deck is supported by the top chords, and the ceilings of the occupied rooms are attached to the bottom chords. In this way, the trusses define the shape of the attic.

A <u>bowstring truss</u> has the same shape as an archery bow, where the top chord represents the curved bow and the bottom chord represents the straight bowstring **FIGURE 7-18**. The top chord resists compressive forces and the bottom chord resists tensile forces. Bowstring trusses are usually quite large and widely spaced in the structure. They are popular in warehouses, supermarkets, and similar buildings with large, open floor areas. The roof of a building with bowstring trusses has a distinctive curved shape.

Trusses are everywhere, so it is important that you learn how to recognize them. A building with a bowstring-truss roof will have a distinctive curved roofline **FIGURE 7-19**. In some buildings, the trusses are exposed and readily visible from the inside of the building; in others, the floor or roof trusses are not visible, because they are enclosed in attic or floor spaces **FIGURE 7-20**.

Fire Fighter Safety Tips

© Photos.com

Checking the floor or roof to see if it is spongy does not necessarily ensure safety with modern lightweight construction.

Trusses Under Fire

Under normal circumstances, a properly engineered and installed truss will do a satisfactory job of supporting the roof or floor assembly for which it was designed; however, like many other building components, trusses display different characteristics when exposed to heat and fire. Because these components are engineered to work with the least amount of wood or steel needed to support a given weight, they provide little margin of safety for fire fighters during a fire.

Trusses consist of a small cross-section of wood with a high surface-to-mass ratio. As a consequence, a fire burning for a very short period of time can destroy enough wood to weaken a truss. All it takes is a failure at one point in a truss to produce a failure in the entire truss. Wooden trusses

constructed with metal gusset plates are especially prone to failure. These gusset plates tie the chords and members of the truss together. They have sharp pointed indentions that are pounded into the wood to hold the chords and web members together. Because the gusset plates are embedded in the wood at a depth of only 0.9 cm (0.4 in.), heating the metal gusset plate will decompose the wood fibres, quickly leading to failure of the truss. Likewise, wooden trusses held together with steel pins can fail in a fire. When heated, the steel pins conduct heat into the wood, which in turn causes decomposition of the wood. When the wood weakens, failure initially occurs at that point, followed by failure of the entire truss. Fire can also weaken the glue in glued trusses, causing the truss to weaken.

Because trusses are constructed with a small margin of safety, failure of one truss usually results in overload and subsequent failure of adjoining trusses. For this reason, failure of a single truss usually results in catastrophic failure of the floor or roof it supports.

In the past, fire fighters were taught that sounding the floor or checking the roof to see if it was spongy would help determine if that part of the structure was safe. This technique has proved unreliable when dealing with structural members built from trusses. In fact, fire fighters have reported cases where roofs and floors have collapsed with no warning signs.

Steel trusses are also prone to failure during fires. When its thin, tubular steel is heated enough to cause a tube in the truss to weaken, the entire truss can fail. If steel trusses are protected from heat with thermal insulation, they may withstand the heat from a fire for a longer period of time; many steel trusses do not have any protection from heat, however, so they often fail during a major fire. Steel has a tendency to sag before it fails completely, which may provide a slight warning of its impending failure.

During a fire, steel trusses and I-beams can fail because of elongation and twisting. Francis L. Brannigan noted that a 30.5-m (100-ft.) beam or truss can elongate by as much as 23 cm (9 in.) when heated to 537.8°C (1000°F). Elongated steel beams can rotate and drop wooden trusses—with deadly consequences. Elongation can also push down masonry walls.

In some cases, trusses may fail because of overload. As buildings are modified or when the use of a building changes, more weight may be placed on floors or roofs than they were intended to carry; for example, air conditioners may be placed on roofs, and attic spaces may be used for storage purposes. This extra weight might exceed the weight for which the trusses were designed. In the event of a fire, the added weight can contribute to the rapid failure of a truss-supported floor or roof.

Truss spaces can also contribute to the spread of fires. Spaces under floors occupied by trusses can serve as conduits for the spread of fires. These fires are often difficult to detect and may spread rapidly, much as fires spread quickly through any open spaces. Because truss-supported floors can be covered with many types of material (including cement), it may be difficult to determine that a fire burning under the floor has weakened it. Be alert for the possibility of a fire burning under you. Fires that occur in the space above a truss-supported ceiling are also challenging to detect. Whenever a truss-supported ceiling has fire above it, be prepared for the possibility of a rapid collapse of the trusses supporting the roof.

Trusses exist in many types of buildings—an important fact for fire fighters to recognize. Identifying the presence of trusses will enable you to better predict the behaviour of a particular building under fire conditions. Understanding that buildings with trusses give fire fighters few signs of impending failure enables you to evaluate the risk and benefit of different approaches to fire suppression in that building.

> ### Fire Fighter Safety Tips
> © Photos.com
>
> Trusses can and will fail without warning early in a fire. Fire fighters should stay off of roofs when fire is known or suspected to be in the truss loft.

> ### FIRE FIGHTER Tips
>
> A new component in lightweight building construction is the truss joist I-beam (or TGI beam). This framing material is made up of three wooden components. Channels may be drilled through TGI beams, thus providing another avenue for fire to spread.
>
> Engineered lumber and laminated beams are being used more often in modern construction. These products are made up of smaller pieces of wood bound together with an adhesive process. Exposure to fire may break the bonds of the adhesive process and lead to lumber or beam failure.

■ Walls

Walls are the most visible parts of a building because they shape the exterior and define the interior. Walls may be constructed of masonry, wood, steel, aluminum, glass, and other materials.

Walls are either load bearing or non-load-bearing. **Load-bearing walls** provide structural support **FIGURE 7-21**. A load-bearing wall supports a portion of both the building's weight (dead load) and its contents (live load), transmitting that load down to the building's foundation. Damaging or removing a load-bearing wall can result in a partial or total collapse of the building. Load-bearing walls can be either exterior or interior walls.

FIGURE 7-21 A load-bearing wall provides structural support.

Non-bearing walls, by contrast, support only their own weight **FIGURE 7-22**. For this reason, most non-bearing walls can be breached or removed without compromising the structural integrity of the building. Many non-bearing walls are interior partitions that divide the building into rooms and spaces. The exterior walls of a building can also be non-bearing, particularly when a system of columns supports the building.

In addition to load-bearing and non-bearing walls, several specialized walls may also be used in construction:

- A fire separation is a construction assembly that acts as a barrier against the spread of fire. Fire separations can be both vertical (walls) or horizontal (floors/ceilings).
- **Party walls** are constructed on the line between two properties or dwelling units and are shared by a building on each side of the line. They are almost always load-bearing walls. A party wall, depending on the occupancy, will be constructed as a fire wall or a fire separation between the two properties or dwellings. The wall separating the dwelling units of a duplex is only required to be a fire separation.
- A **fire wall** is a fire separation of noncombustible construction that subdivides a building or separates adjoining buildings with the required fire rating and has structural stability to remain intact under fire conditions for the required rated time. If a fire wall is rated for 4 hours, it should remain intact and standing for that period of time. Concrete block is often used in the construction of a fire wall.
- **Interior fire separations** are interior walls that extend from the floor to the underside of the rated assembly above. They often enclose fire-rated interior corridors or service rooms or divide a floor area into separate fire compartments. Interior fire separations also enclose vertical shafts, stairwells, and elevator shafts.

- **Curtain walls** are non-bearing exterior walls attached to the outside of a building. These walls often serve as the exterior skin on a steel-frame high-rise building.

Solid, load-bearing masonry walls (at least 15–20 cm [6–8 in.] thick) can be used for buildings up to six storeys high. In contrast, non-load-bearing masonry walls can be almost any height. Masonry walls provide a durable, fire-resistant outer covering for a building, so they are often used as fire walls. A well-designed masonry fire wall is often completely independent of the structure on either side. Even if the building on one side burns completely and collapses, the fire wall should prevent the fire from spreading to the building on the other side.

Older buildings often have masonry load-bearing walls that are more than a metre thick at the bottom but decrease in thickness as the building's height increases. Modern masonry walls are typically reinforced with steel rods or concrete to provide a more efficient and more durable structural system.

When properly constructed and maintained, masonry walls are strong and can withstand a vigorous assault by fire. If the interior structure begins to collapse and exerts unanticipated force on the exterior walls, however, solid masonry walls may fail during a fire.

Even though a building has an outer layer of brick or stone, fire fighters should never assume that the exterior walls are masonry. Many buildings that look like masonry are actually constructed using wood-frame techniques and materials. A single layer of brick or stone, called a veneer, is applied to the exterior walls to give the appearance of a durable and architecturally desirable outer covering. If the wood structure is damaged during a fire, this veneer is likely to collapse. For this reason, it is imperative that a collapse zone be set up around a burning building.

Fire Fighter Safety Tips   © Photos.com

To determine if a brick wall is a solid masonry wall or a brick veneer, look at the pattern of the bricks. If every seventh course (row) of bricks is turned sideways so that the end of the brick is visible (king row), it is usually a solid masonry wall; if all of the courses of brick are the same, it is probably a brick veneer.

Wood framing is used to construct the walls in most houses and many small commercial buildings. The wood framing used for exterior walls can be covered with a variety of materials, including wood siding, vinyl siding, aluminum siding, stucco, and masonry veneer. Moisture barriers, wind barriers, and other types of insulation are usually applied to the outside of the vertical studs before the outer covering is applied.

Vertical wooden studs support the walls and partitions inside the building. If fire resistance is critical, steel studs are used to frame walls. The wood framing is usually covered with gypsum board along with any of a variety of interior finishing materials. The space between a wall's two surfaces may be empty, it may contain thermal and sound insulating materials, or it may contain electrical wiring, telephone wires, and plumbing. These spaces often provide pathways through which fire can spread.

Non bearing wall

FIGURE 7-22 A nonbearing wall supports only its own weight.

■ Doors and Windows

Doors and windows are important components of any building. Although they generally have different functions—doors provide entry and exit, whereas windows provide light and ventilation—in an emergency, doors and windows are almost interchangeable. A window can serve as an entry or an exit, for example, whereas a door can provide light and ventilation.

Hundreds of door and window designs exist with many different applications. Of particular concern to fire fighters are fire doors and fire windows. Even wood doors provide some barrier to fire spread. Closing a door during a fire, then, may give you a few additional minutes to search a room.

Door Assemblies

Most doors are constructed of either wood or metal. Hollow-core wood doors are often used inside buildings. A typical hollow-core door has an internal framework whose outer surfaces are covered by thin sheets of wood. Because hollow-core doors can be easily opened with simple hand tools, they should not be used where security is a concern. A fire can usually burn through a hollow-core door in a short time.

Solid-core wood doors are used where a more substantial door is required. These types of doors are manufactured from solid panels or blocks of wood and are more difficult to force open. A solid-core door provides some fire resistance and can often keep a fire contained within a room for 20 minutes or more, giving fire fighters a chance to arrive on the scene and mount an effective attack.

Metal doors are more durable and fire resistant than wooden doors. Some metal doors have a solid wood core that is covered on both sides by a thin sheet of metal; others are constructed entirely of metal and reinforced for added security. The interior of a metal door can be either hollow or filled with wood, a sound-deadening material, or thermal insulation. Although most approved fire doors are metal, not all metal doors should be considered fire doors.

Fire Fighter Tips

Manufactured (mobile) homes use lightweight building components throughout to reduce their weight. As a result, most parts of these structures are combustible. Such homes typically have few doors and small windows, making entry for fire suppression or rescue difficult. Once a fire starts, especially in an older manufactured (mobile) home, it can destroy the entire structure within a few minutes.

Window Assemblies

Fire fighters frequently use windows as entry points to attack a fire and as emergency exits. Windows also provide ventilation during fires, allowing smoke and heat to escape the building and cooler, fresh air to enter it.

Windows come in many shapes, sizes, and designs for different buildings and occupancies **FIGURE 7-23**. Fire fighters must become familiar with the specific types of windows found in local occupancies, learn to recognize them, and

understand how they operate. It is especially important to know if a particular type of window is very difficult to open or cannot be opened. Due to the high cost of energy in Canada, most windows are double- or triple-paned windows in new buildings, and in most existing housing, the windows have been upgraded. With security a concern in many areas, security bars and coatings or panels are installed in windows that are designed to prevent entry. Energy efficiency and security features make forcible entry more difficult during firefighting operations. The chapter "Forcible Entry" contains more information on window construction.

In new and renovated residential construction, the requirements for bedroom windows as a secondary means of escape maintain that they shall have a unobstructed opening of not less than 0.35 sq. m (3.8 sq. ft.) and no dimension less than 38 cm (15 in.). Another design feature requirement is that the windows will open without the need for keys, tools, or special knowledge.

Fire Doors and Fire Windows

Doors and windows in fire separations are constructed to prevent the passage of flames and heat during a fire. They must be tested and meet the standards specified in NFPA 80, *Standard for Fire Doors and Other Opening Protectives*. Fire doors and fire windows come in many different shapes and sizes, however, and they provide different levels of fire resistance; for example, fire doors can swing on hinges, slide down or across an opening, or roll down to cover an opening. In all nonresidential occupancies, doors are required to open in the direction of travel to the exit.

The fire rating on a door or window covers the actual door or window, the frame, the hinges or closing mechanism, the latching hardware, and any other equipment that is required to operate the door or window. All of these items must be tested and approved as a combined system. All fire doors must have a

FIGURE 7-23 Windows can open in many different ways. This is just one example.

mechanism that keeps the door closed or automatically closes the door when a fire occurs. Exit doors that are required to be closed in the event of a fire but for operational reasons are normally open can be closed by activation of the fire alarm system. The device that closes a door in the fire separation must meet requirements for holding force, opening force, and speed in which it closes the door.

Fire doors and fire windows are rated for a particular duration of fire resistance to a standard test fire. This is similar to the fire-resistance rating system used for building construction assemblies. A 1-hour rating, however, does not guarantee that the door will resist any fire for 60 minutes; rather, it simply establishes that the door will resist the standard test fire for 60 minutes. In any given fire situation, a door rated at 1 hour will probably last twice as long as a door rated 30 minutes.

<u>Fire windows</u> are used when a window is needed in a required fire-resistant wall. These windows are often made of wired glass and tempered, which allows the glass to withstand exposures to high temperatures without breaking. Fire-resistant glass without wires is available for some applications; in such cases, special steel window frames are required to keep the glass firmly in place.

Wired and tempered glass is also used to provide vision panels in or next to fire doors **FIGURE 7-24**. Vision panels allow a person to view conditions on the opposite side of a door without opening the door or to make sure no one is standing in front of the door before opening it. In these configurations, the components making up the entire assembly—including the door, the window, and the frame—must be tested and approved together.

When a window is required only for the passage of light, glass blocks can sometimes be used instead of wired glass. Glass blocks will resist high temperatures and remain in place during a fire, assuming they are properly installed. The size of the opening is limited and depends on the fire-resistance rating of the wall.

■ Interior Finishes and Floor Coverings

The term <u>interior finish</u> is typically used to refer to the exposed interior surfaces of a building. These materials affect how a particular building or occupancy reacts when a fire occurs. Considerations related to interior finishes include whether a material will ignite easily or resist ignition, how quickly a flame will spread across the material's surface, how much energy the material will release when it burns, how much smoke it will produce, and which components the smoke will contain (e.g., toxic particles or gases).

A room with a bare concrete floor, concrete-block walls, and a concrete ceiling has no interior finishes that will increase the fire load. But when the same room has an acrylic carpet with rubber padding on the floor, wooden baseboards, varnished wood panelling on the walls, and foam plastic acoustic insulation panels on the ceiling, the situation will be vastly different: these interior finishes ignite quickly, spread flames quickly across other surfaces, and release significant quantities of heat, flame, and toxic smoke.

FIGURE 7-24 Wired glass is used to provide vision panels in or next to fire doors.

Courtesy of Securalldoors.com

Different interior finish materials contribute in various ways to a building fire. Each individual material has certain characteristics, and fire fighters must evaluate the particular combination of materials in a room or space. Typical wall coverings include painted plaster or gypsum board, wallpaper or vinyl wallcoverings, wood panelling, and many other surface finishes. Floor coverings might include different types of carpet, vinyl floor tiles, or finished wood flooring. All of these products will burn to some extent, and each one involves a different set of fire risk factors. Fire fighters must understand the hazards posed by different interior finish materials if they are to operate safely at the scene of a fire. Keep in mind that most of the furnishings and interior finishes in a house consist of some type of plastic derived from petroleum products.

■ Manufactured Housing

Manufactured housing includes both mobile homes and modular homes. These homes are constructed with lightweight building components throughout the structure to reduce weight. Most parts of older manufactured houses are combustible. Many manufactured houses lack large windows that can be used for rescue purposes and fire suppression, and they often contain few doors. Once a fire has started in an older mobile home, it can destroy the entire structure within just a few minutes. As a consequence, the death rate in older– mobile home fires is three times as high as the death rate in fires involving other types of single-family homes. People who live in mobile homes without a working smoke alarm are at an especially high risk of dying from a fire.

Modern mobile or modular homes are required to be built to the codes and standards that have been adopted in the province/territory in which they are located and thus have similar fire-resistance characteristics to site-built homes.

Relocatable camp buildings that are designed to be set up and moved from work site to work site are given special considerations in some provinces/territories. Older relocatable camp structures were built similarly to older mobile homes and have all the same issues: the buildings deteriorate faster due to frequent moves, poor environmental conditions, and the transient nature of their occupants. These camps are moved every few weeks in some industries.

■ Buildings under Construction or Demolition

Buildings that are under construction or renovation or in the process of demolition present a variety of problems and extra hazards for fire fighters. In many cases, the fire protection features found in finished buildings are missing from these structures. For example, automatic sprinklers might not have been installed yet or might have been disconnected. Smoke detectors might not have been installed or might be inoperable. There may be no coverings on the walls, leaving the entire wood framework fully exposed. Missing doors and windows can provide an unlimited supply of fresh air to feed a fire and allow the flames to spread rapidly. Fire-resistive enclosures may be missing, leaving critical structural components unprotected. Sprinkler and standpipe systems may be inoperative. In addition, construction sites and demolition sites often store large quantities of building materials close to buildings. These materials can add a huge additional fire load and contribute to the rapid spread of fire, often from one building to another. All of these factors can enhance fire spread, leading to quick structural failure.

Many fires at construction and demolition sites are inadvertently caused by workers using torches to weld, cut, or take apart pieces of the structure. Tanks of flammable gases and piles of highly combustible construction materials might be left in locations where they could add even more fuel to a fire. Buildings under construction or demolition are often

left unoccupied for many hours, resulting in delayed discovery and reportage of fires. In some cases, it might prove difficult for fire apparatus to approach the structure or for fire fighters to access working hydrants. All of these problems must be anticipated when considering the fire risks associated with a construction or demolition site.

> ### Fire Fighter Safety Tips
> © Photos.com
>
> Buildings that are under construction and buildings that are being renovated or demolished are at high risk for major fires, for several reasons. For example, workers may be using torches for certain tasks. In addition, these buildings are often unoccupied for many hours during the construction period, so alarms may be delayed. Fire detection and suppression systems may be inoperable. Removal of doors and windows allows the entry of sufficient oxygen for a small fire to grow into a major blaze in a short period of time, and helps the fire spread.
>
> Clearly, buildings under construction or demolition are at high risk for a major fire. Such was the case at the Deutsche Bank building in New York City, where two fire fighters lost their lives in 2007. This building was damaged in the terrorist attacks of September 11, 2001, and was set for demolition. The New York City Fire Department responded to a working fire in the building and attempted to use the standpipe system to fight the fire; a part of the standpipe system had been removed, however, and was inoperable.

> ### Fire Fighter Safety Tips
> © Photos.com
>
> The majority of fire fighter deaths and injuries occur from structural collapse at residential fires. The most frequent cause of death and injury is due to the collapse of floors, followed by the collapse of ceilings and walls. Recent tests by Underwriters Laboratories of Canada (ULC) revealed that common items found in modern construction, such as finger-joined parallel chord trusses, can fail in as little as 4 minutes.

Preincident Planning and Incident Size-Up

When a building is on fire, how do you determine which type of construction was used to build the building? Even experienced fire fighters need to assess a building carefully to make this determination.

The best way to gather this information is to conduct preincident planning. Preincident planning allows fire department personnel to identify the type of building construction as well as to survey the specific characteristics of that building. This type of investigation helps you prepare to quickly, safely, and effectively attack a fire in that building. Preincident planning allows you the luxury of being at the building before a fire breaks out.

During preincident planning, you can document the characteristics of a building before a fire starts there **FIGURE 7-25**.

In your examination of the building, consider the following questions: Which type of building construction is present? Which type of occupancy is this building? What does it contain? Which type of support do the floors have? Is it safe to climb onto the roof? Does the building contain roof trusses, floor trusses, or manufactured wood I-joists that might collapse suddenly? Which types of fire detection and fire-suppression systems are installed in the building? The importance of preplanning cannot be overemphasized. An excellent way to identify building construction is to inspect the building multiple times while it is being built. Your preplanning will be greatly advanced by this exercise. More detailed information about preplanning is presented in the chapter "Preincident Planning."

Of course, it is not possible to do preincident planning for every property in your department. In cases where you have not conducted a preincident inspection, it will be necessary to rely on an incident size-up to determine the type of construction present in the fire building **FIGURE 7-26**. By learning

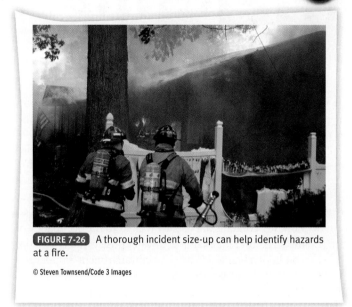

FIGURE 7-26 A thorough incident size-up can help identify hazards at a fire.

© Steven Townsend/Code 3 Images

the general characteristics of the building types in your first-due area, you will get some idea of the type of construction in a given neighbourhood. In a neighbourhood where the houses were built in the 1920s, for example, you are likely to find balloon construction. In a neighbourhood populated by recently built houses, you are likely to encounter lightweight construction with wooden I-beams and trusses.

Use the information you have learned about building construction to make firefighting safer both for you and for building residents. Your knowledge of building construction, preincident inspections, and incident size-up will help you gather the information you need to remain a safe and effective fire fighter.

Over the course of your career as a fire fighter, it is also important to continue to keep up with the changes in building construction. Spend some time at construction sites. Review the latest fire studies from organizations like ULC, which are typically available online. The information you gain may be valuable at your next fire scene.

FIGURE 7-25 Preincident inspections help you determine the type of building construction before a fire occurs there.

FIRE FIGHTER Tips

It is important to know which types of construction are found in your response area. When a building is under construction, contact the builder or site foreperson and arrange for a tour. Ask questions about the structure and its construction. Inquire about any special features and its fire protection system.

Wrap-Up

Chief Concepts

- It is vital for fire fighters to understand the basic types of building construction and to recognize how building construction affects fire growth and spread.
- Based on a structure's occupancy classification, a fire fighter can predict who is likely to be inside the building; for example, a bank is unlikely to have occupants at 3 a.m.
- Building contents are usually related to the occupancy of the structure; for example, a warehouse is likely to store goods or chemicals, whereas an apartment building is likely to be filled with human occupants and furniture.
- The most commonly used building materials are wood, engineered wood products, masonry, concrete, steel, aluminum, glass, gypsum board, and plastics. The key factors that affect the behaviour of each of these materials under fire conditions are combustibility, thermal conductivity, decrease in strength at elevated temperatures, and thermal expansion when heated.
- Masonry includes stone, concrete blocks, and brick. These types of components are usually bonded together with mortar. Masonry is fire resistive and a poor conductor of heat. Even so, a masonry structure can collapse under fire conditions if the roof or floor assembly collapses.
- Concrete is a fire-resistive material that does not conduct heat well. It is often used to insulate other building materials from fire. Concrete does not lose strength when exposed to high temperatures. Steel reinforcing rods are often embedded within this material to strengthen it. The high heat from a fire may convert trapped moisture in the concrete to steam and cause spalling; severe spalling can, in turn, expose steel reinforcing rods to heat.
- Steel is often used in the structural framework of buildings. It is strong and resistant to aging but can rust. Steel is not fire resistive and conducts heat. Steel both expands and loses strength when heated. The risk of failure of a steel structure depends on the mass of the steel components, the loads placed on them, and the methods used to connect the components. Any sign of bending, sagging, or stretching of steel structural members should be viewed as a warning of imminent collapse.
- Glass is noncombustible but not fire resistive. Ordinary glass will usually break when exposed to fire. Tempered glass is stronger and more difficult to break. Laminated glass is likely to crack and remain in place when exposed to fire. Glass blocks have limited strength and are not intended to be used as part of a load-bearing wall. Wired glass incorporates wires that hold the glass together and prevent it from breaking when exposed to heat.

- Gypsum board is commonly used to cover the interior walls and ceilings of residences. It is manufactured in large sheets consisting of a layer of compacted gypsum sandwiched between two layers of specially produced paper. Gypsum board has limited combustibility and does not conduct or release heat to contribute to fire spread. It may fail if exposed to fire for extended periods. Rated drywall is a major component in many rated fire separations.
- Engineered wood products are also called manufactured board, manmade wood, or composite wood. These products are manufactured from small pieces of wood that are held together with glue or adhesives (e.g., urea-formaldehyde resins, phenol-formaldehyde resins, melamine-formaldehyde resins, and polyurethane resins). Engineered wood products may warp under high humidity and release toxic fumes during a fire.
- The most important characteristic of wood and engineered wood products is their high combustibility. The rate at which wood ignites, burns, and decomposes depends on several factors: ignition, moisture, density, preheating, and its size and form.
- Wood cannot be treated to make it completely noncombustible, but it can be made to be more difficult to ignite and burn through the application of a fire-retardant treatment.
- Plastics are rarely used for structural support but may be found throughout a building. The combustibility of plastics varies greatly. Many plastics produce quantities of heavy, dense, dark smoke and release high concentrations of toxic gases.
- There are five types of building construction:
  - Type I construction: fire resistive—Can withstand and contain a fire for a specified period of time. All of the structural members and components used in this construction must be made of noncombustible materials and constructed so that they provides at least 2 hours of fire resistance. Under extreme fire conditions, a Type I building can collapse.
  - Type II construction: noncombustible—Includes structural components made of noncombustible materials. The fire severity is determined by the contents of the building.
  - Type III construction: ordinary—Used in a wide variety of buildings—from strip malls to small apartment buildings. These buildings have masonry exterior walls and interior structural members constructed of wood. The resulting structure may have void spaces that can allow fire spread.

- Type IV construction: heavy timber—Marked by masonry construction and interior walls, columns, beams, floor assemblies, and a roof structure made of heavy wood. It has no concealed spaces or voids. The columns, support beams, floor assemblies, and roof assemblies will withstand a fire much longer than the smaller wood members used in ordinary construction.
- Type V construction: wood frame—All of the major components are constructed of wood or other combustible materials. Type V is the most common type of construction used today. Many of these structures do not have any fire-resistive components. Fire detection devices and automatic sprinklers are the most effective way of protecting lives in these buildings.

- In a balloon-frame building, an open channel that extends from the foundation to the attic can enable hidden fire to spread throughout the structure.
- Platform-frame construction is used in most modern construction. In this building technique, each level is constructed as a separate platform. At each level, the floor platform blocks the path of any fire rising within the void spaces.
- Every building contains the following major components:
  - Foundation—Usually constructed of concrete or masonry and designed to transfer the weight of the building and its contents to the ground.
  - Floors and ceilings—Might be designed to resist fires or might have no fire-resistant capabilities. In a fire-resistive building, the floor is designed to prevent a fire from spreading vertically and to prevent a collapse when a fire occurs in the space below the floor–ceiling assembly. Wood floor structures are common in non-fire-resistive construction. The ceiling is considered to be part of the floor assembly.
  - Roof—Designed to protect the inside of the building from the weather. Roofs are constructed in three primary designs: pitched, curved, and flat. The major components of a roof assembly are the supporting structure, the roof deck, and the roof covering.
  - Trusses—A structural component composed of smaller pieces joined to form a triangular configuration or a system of triangles. Trusses are used as support systems for both floors and roofs. The strength of a truss depends on its members as well as the connections between them. The types of trusses include parallel chord truss, pitched chord truss, and bowstring truss. Depending on their construction, trusses may fail when exposed to heat and fire.
  - Walls—Either load bearing or non-bearing. Load-bearing walls provide structural support. Non-bearing walls support only their own weight.
  - Doors—Typically constructed of either wood or metal. Hollow-core wood doors are often used inside buildings and can be easily opened. Metal doors are more durable and fire resistant.
  - Windows—Used frequently by fire fighters as entry points to attack a fire and as emergency exits.
  - Interior finishes and floor coverings—The exposed interior surfaces of a building. Different interior finish materials contribute in various ways to a structure fire; for example, plastics emit toxic gases when they burn.

- Buildings under construction or demolition often lack fire protection features and present additional hazards to fire fighters.
- The preincident plan should contain information on the type of building construction present in a given structure.

Hot Terms

Balloon-frame construction An older type of wood-frame construction in which the wall studs extend vertically from the basement of a structure to the roof without any fire stops.

Bowstring truss A truss that is curved on the top and straight on the bottom.

Combustibility A property that describes whether a material will burn and how quickly it will burn.

Contemporary construction Buildings constructed since about 1970 that incorporate lightweight-construction techniques and engineered wood components. These buildings exhibit less resistance to fire than older buildings.

Curtain wall A non-load-bearing wall that separates the inside and outside of a building but is not part of the building's support structure.

Curved roof A roof with a curved shape.

Dead load The weight of all materials of construction incorporated into a building, including, but not limited to, walls, floors, roofs, ceilings, stairways, built-in partitions, finishes, cladding, and other similarly incorporated architectural and structural items, as well as fixed service equipment, such as cranes. (NFPA 5000)

Fire wall A fire division assembly with a fire-resistance rating of 3 test hours or longer built to permit the complete burn out and collapse of the structure on one side of the wall without the fire extending through the fire wall or the collapse of the fire wall itself. (NFPA 901)

Fire window A window assembly rated in accordance with NFPA 257, *Standard on Fire Test for Window and Glass Block Assemblies*, and installed in accordance with NFPA 80, *Standard for Fire Doors and Other Opening Protectives*. (NFPA 5000)

Flat roof A horizontal roof; often found on commercial or industrial occupancies.

Glass blocks Thick pieces of glass similar to bricks or tiles.

Gypsum A naturally occurring material consisting of calcium sulphate and water molecules.

Gypsum board The generic name for a family of sheet products consisting of a noncombustible core made primarily of gypsum with paper surfacing. (NFPA 5000)

Interior finish The exposed surfaces of walls, ceilings, and floors within buildings. (NFPA 101)

Interior fire separations Interior walls that extend from the floor to the underside of the rated assembly above. They often enclose fire-rated interior corridors or service rooms or divide a floor area into separate fire compartments. Interior fire separations also enclose vertical shafts, stairwells and elevator shafts.

Laminated glass A type of safety glass comprised of a thin layer of plastic between two layers of glass, which prevents the glass from shattering and falling apart when broken.

Laminated wood Pieces of wood that are glued together.

Legacy construction An older type of construction (prior to 1970) that used sawn lumber.

Live load The load produced by the use and occupancy of the building or other structure, which does not include construction or environmental loads, such as wind load, snow load, rain load, earthquake load, flood load, or dead load. Live loads on a roof are those produced during maintenance by workers, equipment, and materials; and during the life of the structure by movable objects such as planters and by people. (NFPA 5000, ASCE/SEI 7-4.1)

Load-bearing wall A wall designed to provide structural support for a building.

Manufactured (mobile) home A factory-assembled structure, or a structure that can be transported in one or more sections, that is built on a permanent chassis and designed to be used as a dwelling without a permanent foundation when connected to the required utilities, including the plumbing, heating, air-conditioning, and electric systems contained therein.

Masonry A built-up unit of construction or combination of materials such as clay, shale, concrete, glass, gypsum, tile, or stone set in mortar. (NFPA 5000)

Non-bearing wall Any wall that is not a load-bearing wall. (NFPA 5000, ASCE/SEI 7-11.2)

Occupancy The purpose for which a building or other structure, or part thereof, is used or intended to be used. (NFPA 5000)

Parallel chord truss A truss in which the top and bottom chords are parallel.

Party wall A wall constructed on the line between two properties.

Pitched chord truss A type of truss typically used to support a sloping roof.

Pitched roof A roof with sloping or inclined surfaces.

Platform-frame construction A construction technique for building the frame of a structure one floor at a time. Each floor has a top and bottom plate, which act as fire stops.

Pyrolysis The destructive distillation of organic compounds in an oxygen-free environment whereby the organic matter is converted into gases, liquids, and char. (NFPA 820)

Rafters Joists mounted in an inclined position to support a roof.

Spalling The chipping or pitting of concrete or masonry. (NFPA 921)

Tempered glass A type of heat-treated safety glass that, under stress or fire, will break into smaller, less dangerous pieces.

Thermal conductivity A property that describes how quickly a material will conduct heat.

Thermoplastic material A plastic material capable of being repeatedly softened by heating and hardened by cooling and, in its softened state, repeatedly shaped by moulding or forming. (NFPA 5000)

Thermoset material A plastic material that, after having been cured by heat or other means, is substantially infusible and cannot be softened and formed. (NFPA 5000)

Truss A collection of lightweight structural components joined in a triangular configuration that can be used to support either floors or roofs.

Type I construction Buildings with structural members made of noncombustible materials that have a specified fire resistance; also known as fire-resistive construction.

Type II construction Buildings whose structural members are made of noncombustible materials without fire resistance; also known as noncombustible construction.

Type III construction Buildings with exterior walls made of noncombustible or limited combustible materials, but with interior floors and walls made of combustible materials; also known as ordinary construction.

Type IV construction Buildings constructed with noncombustible or limited combustible exterior walls, and interior walls and floors made of large-dimension combustible materials; also known as heavy timber construction.

Type V construction Buildings with exterior walls, interior walls, floors, and roof structures made of wood; also known as wood-frame construction.

Wired glass A glazing material with embedded wire mesh. (NFPA 80)

Wooden beam A load-bearing member assembled from individual wood components.

Wood truss An assembly of small pieces of wood or wood and metal.

A severe wind weather event has ripped through your community in the early hours of the morning, damaging numerous buildings. Your crew has been tasked with conducting a damage assessment of a two-block area and reporting your findings to the incident commander. The first house is slightly damaged with no risk of collapse. The next house has significant damage: it is a wood-frame house with several collapsed load-bearing walls. It is unsafe to enter without shoring. The third wood-frame home is completely destroyed. The fourth building is a noncombustible commercial structure with block walls and steel trusses. It has partially collapsed. As you continue your search, you realize the magnitude of this event.

1. What type of construction is the commercial structure?
  A. Type II
  B. Type III
  C. Type IV
  D. Type V

2. What type of construction were the houses?
  A. Type II
  B. Type III
  C. Type IV
  D. Type V

3. A _____ is a wall that is designed to provide structural support for a building.
  A. non-load-bearing wall
  B. load-bearing wall
  C. curtain wall
  D. fire wall

4. A _____ is usually constructed of concrete or masonry and designed to transfer the weight of the building and its contents to the ground.
  A. foundation
  B. roof
  C. beam
  D. column

5. Type III construction is also known as _____.
  A. heavy timber
  B. noncombustible
  C. ordinary
  D. fire resistive

6. _____ are mounted in an inclined position to support a roof.
  A. Trusses
  B. Rafters
  C. Joists
  D. Beams

You arrive at an old two-storey, wood-frame farmhouse with wood siding and shake shingles. It has very tall, narrow windows. The windows on the second floor are directly above the windows on the first floor. The house is about 12 by 12 m (40 by 40 ft.) and has a basement. There is turbulent, brown smoke coming from the basement windows, and there is light brown, laminar flow smoke coming out the gable end of the roof. You do not see any smoke through the front window. Your captain directs you to pull a 44-mm (1.75-in.) hoseline from the engine and go to the front door to attack the fire in the basement.

1. What type of construction do you suspect this building is?
2. Where is the fire going to travel?
3. What hazards may be associated with this type of construction?
4. What are the signs of a potential building collapse?

Portable Fire Extinguishers

© Glen Henry/ShutterStock, Inc.

Fire Fighter I

Knowledge Objectives

After studying this chapter, you will be able to:

- State the primary purposes of fire extinguishers. (NFPA 5.3.16.A , p 195)
- Define Class A fires. (NFPA 5.3.16.A , p 196)
- Define Class B fires. (NFPA 5.3.16.A , p 196–197)
- Define Class C fires. (NFPA 5.3.16.A , p 197)
- Define Class D fires. (NFPA 5.3.16.A , p 197)
- Define Class K fires. (NFPA 5.3.16.A , p 197–198)
- Explain the classification and rating system for fire extinguishers. (NFPA 5.3.16.A , p 198)
- Describe the types of agents used in fire extinguishers. (NFPA 5.3.16.A , p 202–205)
- Describe the types of operating systems in fire extinguishers. (NFPA 5.3.16.A , p 205–211)
- Describe the basic steps of fire extinguisher operation. (NFPA 5.3.16 , p 211–213)
- Explain the basic steps of inspecting, maintaining, recharging, and hydrostatic testing of fire extinguishers. (NFPA 5.3.16 , p 220–222)
- Select the proper class of fire extinguisher. (NFPA 5.3.16.B , p 211)

Skills Objectives

After studying this chapter, you will be able to perform the following skills:

- Transport the extinguisher to the location of the fire. (NFPA 5.3.16.B , p 211, 213–214)
- Attack a Class A fire with a stored-pressure water-type fire extinguisher. (NFPA 5.3.16.B , p 213, 215)
- Attack a Class A fire with a multipurpose dry chemical fire extinguisher. (NFPA 5.3.16.B , p 214)
- Attack a Class B flammable liquid fire with a dry chemical fire extinguisher. (NFPA 5.3.16.B , p 215–216)
- Attack a Class B flammable liquid fire with a stored-pressure foam fire extinguisher. (NFPA 5.3.16.B , p 216–217)
- Operate a carbon-dioxide extinguisher. (NFPA 5.3.16.B , p 217–218)
- Use a halogenated agent–type extinguisher. (NFPA 5.3.16.B , p 218–219)
- Use dry powder fire-extinguishing agents. (NFPA 5.3.16.B , p 219–220)
- Use a wet chemical (Class K) fire extinguisher. (NFPA 5.3.16.B , p 220–221)

Fire Fighter II — FFII

Knowledge Objectives

There are no knowledge objectives for Fire Fighter II candidates. NFPA 1001 contains no Fire Fighter II Job Performance Requirements for this chapter.

Skills Objectives

There are no skills objectives for Fire Fighter II candidates. NFPA 1001 contains no Fire Fighter II Job Performance Requirements for this chapter.

Additional NFPA Standards

- NFPA 10, *Standard for Portable Fire Extinguishers*
- NFPA 11, *Standard for Low-, Medium-, and High-Expansion Foam*

You Are the Fire Fighter

You are the designated cook for the day, and the kitchen is bare. You go into the grocery store while the rest of the crew waits in the apparatus. As you make your way through the produce section, an employee runs up to you and says that there is a fire in the back room. You instinctively run to the back room and you notify your crew by portable radio. You find a fire in a trash can that is starting to spread. You quickly scan the room looking for a fire extinguisher. Not seeing one, you ask the employee where the closest one is located.

1. What type of fire extinguisher do you need for this fire?
2. How large of a fire will it put out?
3. How do fire inspectors ensure that a fire extinguisher will work when needed?

Introduction

Portable fire extinguishers are required in many types of occupancies as well as in commercial vehicles, boats, aircraft, and various other locations. Fire prevention efforts encourage citizens to keep fire extinguishers in their homes, particularly in their kitchens. Fire extinguishers are used successfully to put out hundreds of fires every day, preventing millions of dollars in property damage as well as saving lives. Most fire extinguishers are easy to operate and can be used effectively by an individual with only basic training.

Fire extinguishers range in size from models that can be operated with one hand to large, wheeled models that contain several hundred kilograms of **extinguishing agent** (material used to stop the combustion process) **FIGURE 8-1**. Extinguishing agents include water, water with different additives, dry chemicals, wet chemicals, dry powders, and gaseous agents. Each agent is suitable for specific types of fires.

Fire extinguishers are designed for different purposes and involve different operational methods. As a fire fighter, you must know which is the most appropriate kind of extinguisher to use for different types of fires, which kinds must not be used for certain fires, and how to use any fire extinguisher you may encounter, especially those used by your department and carried on your apparatus. The selection of the proper fire extinguisher builds on the information presented in the chapter "Fire Behaviour".

Fire fighters use fire extinguishers to control small fires that do not require the use of a hose line. In addition, a portable backpack-type fire extinguisher can be used to control and overhaul a wildland fire located beyond the reach of hoses. Special types of fire extinguishers are appropriate for situations where the application of water would be dangerous, ineffective, or undesirable; for example, applying water to a fire that involves expensive electronic equipment is both dangerous and costly. Using the correct fire extinguisher could control the fire without causing additional damage to the equipment.

This chapter also covers the operation and maintenance of the most common types of portable fire extinguishers. These principles will enable you to use fire extinguishers correctly and effectively, thereby reducing the risk of personal injury and property damage.

FIGURE 8-1 Portable fire extinguishers can be large or small. **A.** A wheeled extinguisher. **B.** A one-hand fire extinguisher.

Courtesy of Amerex Corporation

A.   **B.**

FIRE FIGHTER Tips

Fire fighters are often called out to do fire extinguisher training for the public. It is essential that you understand the characteristics and operations of each type of fire extinguisher.

Fire Fighter Safety Tips

© Photos.com

Do not place yourself in a dangerous situation by trying to fight a large fire with a small fire extinguisher. You cannot fight a fire or protect yourself with an empty extinguisher, an under-sized extinguisher, or the inappropriate type of extinguisher.

© Duncan Noakes/Dreamstime.com

Purposes of Fire Extinguishers

Portable fire extinguishers have two primary uses: to extinguish **incipient-stage** fires (those that have not spread beyond the area of origin) and to control fires where traditional methods of fire suppression are not recommended.

Fire extinguishers are placed in many locations so that they will be available for immediate use on small, incipient-stage fires, such as a fire in a wastebasket. A trained individual with a suitable fire extinguisher could usually control this type of fire **FIGURE 8-2**. As flames spread beyond the wastebasket to other contents of the room, however, such a fire becomes increasingly difficult to control with only a portable fire extinguisher.

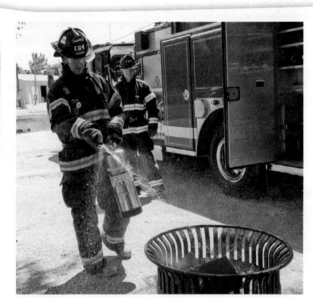

FIGURE 8-2 A trained individual with a suitable fire extinguisher can usually control an incipient-stage fire.

Fire extinguishers are also used to control fires in situations where traditional extinguishing methods are not recommended; for example, using water on fires that involve energized electrical equipment increases the risk of electrocution to fire fighters. Applying water to a fire in a computer or electrical control room could cause extensive damage to the electrical equipment. In these cases, it would be better to use a fire extinguisher filled with the appropriate extinguishing agent. Special extinguishing agents are also required for fires that involve flammable liquids, cooking oils, and combustible metals. Extinguishers are designed to provide an effective discharge pattern for fires involving specific types of fuels. Using an improper type of extinguisher can spread burning material. An appropriate type of fire extinguisher needs to be available in areas containing flammable materials.

■ Use of Portable Fire Extinguishers in Incipient-Stage Fires

Most fire department vehicles carry at least one fire extinguisher, and many vehicles carry two or more extinguishers of different types. Fire fighters often use these portable extinguishers to control incipient-stage fires quickly. At times, a fire fighter may even use an extinguisher from the fire-site premises to control an incipient fire.

One advantage of fire extinguishers is their portability. It may take less time to control a fire with a portable extinguisher than it would to advance and charge a hose line. Fire department vehicles that are not equipped with water or fire hoses usually carry at least one multipurpose fire extinguisher. If you arrive at an incipient-stage fire in one of these vehicles, you might be able to control the fire with the portable extinguisher. When you take this course of action, make sure you are not placing yourself in a dangerous situation by using a fire extinguisher.

The primary disadvantage of fire extinguishers is that they are "one-shot" devices. In other words, once the contents of a fire extinguisher have been discharged, the device is no longer effective in fighting fires until it is recharged. If the extinguisher does not control the fire, some other device or method must be employed. This is a serious limitation when compared to a fire hose with a continuous water supply.

■ Special Extinguishing Agents

As mentioned earlier, some types of fires require special extinguishing agents. As a fire fighter, you must know which fires require special extinguishing agents, which type of extinguisher should be used, and how to operate the different types of special-purpose extinguishers.

Special wet chemical extinguishing agents are used for commercial kitchen fires that involve cooking oils in deep fryers. A dry chemical is used for residential kitchen fires involving vegetable and peanut oils. Specific dry chemicals are used for combustible-metal fires, and certain "clean" agents are used for fires in electronic equipment. Using water or an unsuitable fire-extinguishing chemical to fight these types of fires can cause unnecessary damage and may pose a danger to the fire extinguisher operator.

Portable extinguishers are sometimes used in combination with other techniques; for example, with a pressurized-gas fire, water may be used to cool hot surfaces and prevent reignition, while a dry chemical agent is used to extinguish the flames **FIGURE 8-3**. Certain types of portable extinguishers can be helpful in overhauling a fire. The extinguishing agents contained in these devices break down the surface tension of the water, allowing water to penetrate the materials and reach deep-seated fires.

FIGURE 8-3 Water can cool hot surfaces.

FIGURE 8-4 Most ordinary combustible materials are included in the definition of Class A fires.

Natural vegetation, such as grass and trees, is also Class A material. Water is the most commonly used extinguishing agent for Class A fires, although several other agents can be used effectively.

FIRE FIGHTER Tips

Most residential fires involve Class A materials. Even though the source of the ignition could be Class B or Class C in nature, the resulting fire frequently involves normal combustibles. A fire that begins with an unattended pot of grease on the stove, for example, might ignite the wooden cabinets and other kitchen contents. A short-circuit in an electrical outlet might ignite ordinary combustibles in the immediate vicinity. In both of these cases, the fire would involve primarily Class A materials.

Classes of Fires

It is essential to match the appropriate type of extinguisher to the type of fire. Fires and fire extinguishers are grouped into classes according to their characteristics. Some extinguishing agents work more efficiently than others on certain types of fires. In some cases, selecting the proper extinguishing agent will mean the difference between extinguishing a fire and being unable to control it.

It is dangerous to apply the wrong extinguishing agent to a fire. Using a water extinguisher on an electrical fire, for example, can cause an electrical shock as well as a short-circuit in the equipment. Likewise, a water extinguisher should never be used to fight a grease fire. Burning grease is generally hotter than 100 degrees Celcius (212 degrees Fahrenheit), so when water comes in contact with it, the water converts to steam, which expands very rapidly. If the water penetrates the surface of the grease, the steam is produced within the grease. As the steam expands, the hot grease erupts like a volcano and splatters over everything and everyone nearby, potentially resulting in burns or injuries to people and spreading the fire.

Before selecting a fire extinguisher, ask yourself, "Which class of fire am I fighting?" There are five classes of fires, each of which affects the choice of extinguishing equipment.

■ Class A Fires

<u>Class A fires</u> involve ordinary combustibles such as wood, paper, cloth, rubber, household rubbish, and some plastics FIGURE 8-4.

■ Class B Fires

<u>Class B fires</u> involve flammable or combustible liquids, such as gasoline, oil, grease, tar, lacquer, oil-based paints, and some plastics FIGURE 8-5. Fires involving flammable gases, such as propane or natural gas, are also categorized as Class B fires.

FIGURE 8-5 Class B fires involve flammable liquids and gases.

Examples of Class B fires include a fire in a pot of molten roofing tar, a fire involving splashed fuel on a hot lawnmower engine, and burning natural gas that is escaping from a gas meter struck by a vehicle. Several different types of extinguishing agents are approved for use in Class B fires.

Class C Fires

<u>Class C fires</u> involve energized electrical equipment, which includes any device that uses, produces, or delivers electrical energy **FIGURE 8-6**. A Class C fire could involve building wiring and outlets, fuse boxes, circuit breakers, transformers, generators, or electric motors. Power tools, lighting fixtures, household appliances, and electronic devices such as televisions, radios, and computers could be involved in Class C fires as well. The equipment must be plugged in or connected to an electrical source, but not necessarily operating, for the fire to be classified as Class C.

Electricity does not burn, but electrical energy can generate tremendous heat that could ignite nearby Class A or B materials. As long as the equipment is energized, the incident must be treated as a Class C fire. Agents that will not conduct electricity, such as dry chemicals or carbon dioxide (CO_2), must be used on Class C fires.

Class D Fires

<u>Class D fires</u> involve combustible metals, such as magnesium, titanium, zirconium, sodium, lithium, and potassium. Special techniques and extinguishing agents are required to fight combustible-metal fires **FIGURE 8-7**. Commonly used extinguishing agents can react violently—even explosively—if they come in contact with burning metals. Violent reactions also can occur when water strikes burning combustible metals.

Class D fires are most often encountered in industrial occupancies, such as machine shops, repair shops, and metal recycling plants, as well as in fires involving aircraft and automobiles. Magnesium and titanium—both combustible metals—are used to produce automotive and aircraft parts, because they combine high strength with light weight. Sparks from cutting, welding, or grinding operations could ignite a

FIGURE 8-7 Combustible metals in Class D fires require special extinguishing agents.
© Andrew Lambert Photography/Photo Researchers, Inc.

Class D fire, or the metal items could become involved in a fire that originated elsewhere.

Because of the chemical reactions that could occur during a Class D fire, it is important to select the proper extinguishing agent and application technique for this kind of incident. Choosing the correct fire extinguisher for a Class D fire requires expert knowledge and experience.

Class K Fires

<u>Class K fires</u> involve combustible cooking oils and fats **FIGURE 8-8**. This is a relatively new classification; cooking oil fires were previously classified as Class B combustible liquid fires. The introduction of high-efficiency, modern deep fryers and the trend toward using vegetable oils instead of animal fats to fry foods in recent years have resulted in higher cooking temperatures. These higher temperatures required the

FIGURE 8-6 Class C fires involve energized electrical equipment or appliances.
Courtesy of Tyco Fire and Building Products

FIGURE 8-8 Class K fires involve cooking oils and fats.
© Scott Leman/ShutterStock, Inc.

development of a new class of wet chemical extinguishing agents to combat fires involving cooking oils. Many restaurants continue to use extinguishing agents that were approved for Class B fires, however. Class B extinguishers are not as effective for cooking oil fires as are Class K extinguishers.

Classification of Fire Extinguishers

Portable fire extinguishers are classified and rated based on their characteristics and capabilities. This information is important for selecting the proper extinguisher to fight a particular fire TABLE 8-1. It is also used to determine which types of fire extinguishers should be placed in a given location so that incipient-stage fires can be controlled quickly.

In Canada, the **Underwriters Laboratory of Canada (ULC)** is the organization that developed the standards, classification, and rating system for portable fire extinguishers. This system rates fire extinguishers for both safety and effectiveness. Each fire extinguisher has a specific rating that identifies the classes of fires for which it is both safe and effective.

TABLE 8-1	Types of Fires
Class A	Ordinary combustibles
Class B	Flammable or combustible liquids
Class C	Energized electrical equipment
Class D	Combustible metals
Class K	Kitchen fires involving oils and fats

The classification system for fire extinguishers uses both letters and numbers. The letters indicate the classes of fire for which the extinguisher can be used, and the numbers indicate its effectiveness. Fire extinguishers that are safe and effective for more than one class will be rated with multiple letters. For example, an extinguisher that is safe and effective for Class A fires will be rated with an *A*; one that is safe and effective for Class B fires will be rated with a *B*; and one that is safe and effective for both Class A and Class B fires will be rated with both an *A* and a *B*.

Class A and Class B fire extinguishers also include a number, indicating the relative effectiveness of the fire extinguisher in the hands of a nonexpert user. On Class A extinguishers, this number reflects the amount of water it contains. An extinguisher that is rated 1-A contains the equivalent of 4.7 litres (1.25 gallons) of water. A typical Class A extinguisher contains 9.5 L (2.5 gal.) of water and has a 2-A rating. The higher the number, the greater the extinguishing capability of the extinguisher. An extinguisher that is rated 4-A should be able to extinguish approximately twice as much fire as one that is rated 2-A.

The effectiveness of Class B extinguishers is based on the approximate area of burning fuel that these devices are capable of extinguishing. A 10-B rating indicates that a nonexpert user should be able to extinguish a fire in a pan of flammable liquid that is 0.9 square metres (10 square feet) in surface area. An extinguisher rated 40-B should be able to control a flammable liquid pan fire with a surface area of 3.7 sq. m (40 sq. ft.).

Numbers are used to rate an extinguisher's effectiveness only for Class A and Class B fires. If the fire extinguisher can also be used for Class C fires, it contains an agent proven to be nonconductive to electricity and safe for use on energized electrical equipment. For instance, a fire extinguisher that carries a 2-A:10-B:C rating can be used on Class A, Class B, and Class C fires. It has the extinguishing capabilities of a 2-A extinguisher when applied to Class A fires, has the capabilities of a 10-B extinguisher when applied to Class B fires, and can be used safely on energized electrical equipment.

Standard test fires are used to rate the effectiveness of fire extinguishers. Such testing may involve different agents, amounts, application rates, and application methods. Fire extinguishers are rated for their ability to control a specific type of fire as well as for the extinguishing agent's ability to prevent rekindling. Some agents can successfully suppress a fire but are unable to prevent the material from reigniting. A rating is given only if the extinguisher completely extinguishes the standard test fire and prevents rekindling.

Labelling of Fire Extinguishers

Fire extinguishers that have been tested and approved by an independent laboratory are labelled to clearly designate the classes of fire the unit is capable of extinguishing safely. This traditional lettering system has been used for many years and is still found on many fire extinguishers. More recently, a universal pictograph system, which does not require the user to be familiar with the alphabetic codes for the different classes of fires, has been developed.

■ Traditional Lettering System

The traditional lettering system uses the following labels FIGURE 8-9:

- Extinguishers suitable for use on Class A fires are identified by the letter *A* on a solid green triangle. The triangle has a graphic relationship to the letter *A*.
- Extinguishers suitable for use on Class B fires are identified by the letter *B* on a solid red square. Again, the shape of the letter mirrors the graphic shape of the box.
- Extinguishers suitable for use on Class C fires are identified by the letter *C* on a solid blue circle, which also incorporates a graphic relationship between the letter *C* and the circle.
- Extinguishers suitable for use on Class D fires are identified by the letter *D* on a solid yellow, five-pointed star.
- Extinguishers suitable for use on Class K (combustible cooking oil) fires are identified by a pictograph showing a fire in a frying pan. Because the Class K designation is new, there is no traditional-system alphabet graphic for it.

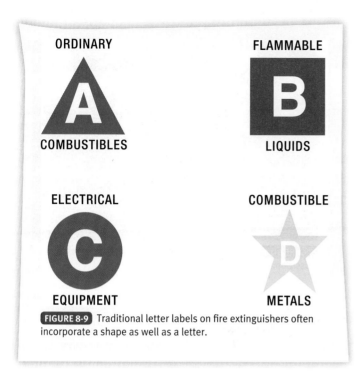

FIGURE 8-9 Traditional letter labels on fire extinguishers often incorporate a shape as well as a letter.

■ Pictograph Labelling System

The pictograph system, such as described for Class K fire extinguishers, uses symbols rather than letters on the labels. This system also clearly indicates whether an extinguisher is inappropriate for use on a particular class of fire. The pictographs are all square icons, each of which is designed to represent a certain class of fire **FIGURE 8-10**. The icon for Class A fires is a burning trash can beside a wood fire. The Class B fire extinguisher icon is a flame and a gasoline can; the Class C icon is a flame and an electrical plug and socket. As noted previously, extinguishers rated for fighting Class K fires are labelled with an icon showing a fire in a frying pan.

Under this pictograph labelling system, the presence of an icon indicates that the extinguisher has been rated for that class of fire. A missing icon indicates that the extinguisher has not been rated for that class of fire. A red slash

FIGURE 8-10 The icons for Class A, B, C, and K fires.

across an icon indicates that the extinguisher must not be used on that type of fire, because doing so would create additional risk.

For example, an extinguisher rated for Class A fires only would show all three icons, but the icons for Class B and Class C would have red diagonal lines through them. This three-icon array signifies that the extinguisher contains a water-based extinguishing agent, making it unsafe to use on flammable liquid or electrical fires.

Certain extinguishers labelled as appropriate for Class B and Class C fires do not include the Class A icon, but may be used to put out small Class A fires. The fact that they have not been rated for Class A fires indicates that they are less effective in extinguishing a common combustible fire than a comparable Class A extinguisher would be.

FIRE FIGHTER Tips

The safest and surest way to extinguish a Class C fire is to turn off the power and treat it like a Class A or B fire. If you are unable to turn off the power, you should be prepared for reignition, because the electricity could reignite the fire after it is extinguished.

Fire Extinguisher Placement

Fire codes and regulations require the installation of fire extinguishers in many areas so that they will be available to fight incipient-stage fires. NFPA 10, *Standard for Portable Fire Extinguishers*, lists the requirements for placing and mounting portable fire extinguishers, as well as the appropriate mounting heights.

The regulations for each type of occupancy specify the maximum floor area that can be protected by each extinguisher, the maximum travel distance from the closest extinguisher to a potential fire, and the types of fire extinguishers that should be provided. Two key factors must be considered when determining which type of extinguisher should be placed in each area: the class of fire that is likely to occur and the potential magnitude of an incipient fire.

Extinguishers should be mounted so they are readily visible and easily accessible **FIGURE 8-11**. Heavy extinguishers should not be mounted high on a wall. If the extinguisher is mounted too high, a smaller person might be unable to lift it off its hook or could be injured in the attempt.

According to NFPA 10, the recommended mounting heights for the placement of fire extinguishers are as follows:

- Fire extinguishers weighing up to 18.1 kilograms (40 pounds) should be mounted so that the top of the extinguisher is not more than 1.5 metres (5 feet) above the floor.
- Fire extinguishers weighing more than 18.1 kg (40 lb.) should be mounted so that the top of the extinguisher is not more than 0.9 m (3 ft.) above the floor.
- The bottom of an extinguisher should be at least 10 centimetres (4 inches) above the floor.

FIGURE 8-11 Extinguishers should be mounted in locations with unobstructed access and visibility.

FIGURE 8-12 Light hazard areas include offices, churches, and classrooms.

■ Classifying Area Hazards

Areas are divided into three risk classifications—light, ordinary, and extra hazard—based on the amount and type of combustibles that are present, including building materials, contents, decorations, and furniture. The quantity of combustible materials present is sometimes called a building's **fire load** and is measured as the average weight of combustible materials per square metre (or per square foot) of floor area. The larger the fire load, the larger the potential fire.

The occupancy use category does not necessarily determine the building's hazard classification. The recommended hazard classifications for different types of occupancies are simply guidelines based on typical situations. The hazard classification for each area should be based on the actual amount and type of combustibles that are present.

Light (Low) Hazard

Light (low) hazard locations are areas where the majority of materials are noncombustible or arranged so that a fire is not likely to spread. Light hazard environments usually contain limited amounts of Class A combustibles, such as wood, paper products, cloth, and similar materials. A light hazard environment might also contain some Class B combustibles (flammable liquids and gases), such as photocopier chemicals or modest quantities of paints and solvents, but all Class B materials must be kept in closed containers and stored safely. Examples of common light hazard environments include most offices, classrooms, churches, assembly halls, and hotel guest rooms **FIGURE 8-12**.

Ordinary (Moderate) Hazard

Ordinary (moderate) hazard locations contain more Class A and Class B materials than do light hazard locations. Typical examples of ordinary hazard locations include retail stores with on-site storage areas, light manufacturing facilities, auto showrooms, parking garages, research facilities, and workshops or service areas that support light hazard locations, such as hotel laundry rooms or restaurant kitchens **FIGURE 8-13**.

FIGURE 8-13 Auto showrooms, hotel laundry rooms, and parking garages are classified as ordinary hazard areas.

Ordinary hazard areas also include warehouses that contain Class I and Class II commodities. Class I commodities include noncombustible products stored on wooden pallets or in corrugated cartons that are shrink-wrapped or wrapped in paper. Class II commodities include noncombustible products stored in wooden crates or multilayered corrugated cartons.

Extra (High) Hazard

Extra (high) hazard locations contain more Class A combustibles and/or Class B flammables than do ordinary hazard locations. Typical examples of extra hazard areas include woodworking shops; service and repair facilities for cars, aircraft, or boats; and many kitchens and other cooking areas that have deep fryers, flammable liquids, or gases under pressure **FIGURE 8-14**. In addition, areas used for manufacturing processes such as painting, dipping, or coating, as well as facilities used for storing or handling flammable liquids, are classified as extra hazard environments. Warehouses containing products that do not meet the definitions of Class I and Class II commodities are also considered extra hazard locations.

■ Determining the Most Appropriate Placement of Fire Extinguishers

Several factors must be considered when determining the number and types of fire extinguishers that should be placed in each area of an occupancy. Among these factors are the types of fuels found in the area and the quantities of those materials.

Some areas may need extinguishers with more than one rating or more than one type of fire extinguisher. Environments that include Class A combustibles require an extinguisher rated for Class A fires; those with Class B combustibles require an extinguisher rated for Class B fires. Areas that contain both Class A and Class B combustibles, however, require either an

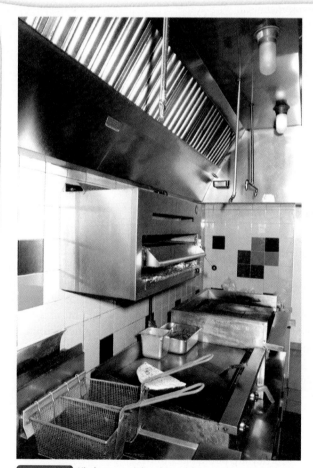

FIGURE 8-14 Kitchens, woodworking shops, and auto repair shops are considered possible extra hazard locations.

extinguisher that is rated for both types of fires or a separate extinguisher for each class of fire.

Most buildings require extinguishers that are suitable for fighting Class A fires because ordinary combustible materials—furniture, partitions, interior finish materials, paper, and packaging products—are so common. Even where other classes of products are used or stored, there is still a need to defend the facility from a fire involving common combustibles.

A single multipurpose extinguisher is generally less expensive than two individual fire extinguishers and eliminates the problem of selecting the proper extinguisher for a particular fire. However, it is sometimes more appropriate to install Class A extinguishers in general-use areas and to place extinguishers that are especially effective in fighting Class B or Class C fires near those specific hazards.

In some facilities, a variety of conditions are present. In these occupancies, each area must be individually evaluated and the extinguisher installation tailored to its particular circumstances. A restaurant is a good example of this situation. The dining areas contain common combustibles, such as furniture, tablecloths, and paper products, that would require an extinguisher rated for Class A fires. In the restaurant's kitchen, where the risk of fire involves cooking oils, a Class K extinguisher would provide the best defence.

Similarly, within a hospital, extinguishers for Class A fires would be appropriate in hallways, offices, lobbies, and patient rooms. Class B extinguishers should be mounted in laboratories and areas where flammable anesthetics are stored or handled. Electrical rooms should have extinguishers that are approved for use on Class C fires, whereas hospital kitchens would need Class K extinguishers.

Methods of Fire Extinguishment

Understanding the nature of fire is key to understanding how extinguishing agents work and how they differ from one another. All fires require three basic ingredients: fuel, heat, and oxygen. In scientific terminology, burning is called **rapid oxidation**. This chemical process occurs when a fuel is combined with oxygen, resulting in the formation of ash or other waste products and the release of energy as heat and light.

The combustion process begins when the fuel is heated to its **ignition point**—the temperature at which it begins to burn. The energy that initiates this process can come from many different sources, including a spark or flame, friction, electrical energy, or a chemical reaction. Once a substance begins to burn, it will generally continue burning as long as adequate supplies of oxygen and fuel to sustain the chemical reaction are present, unless something interrupts the process.

Most extinguishers stop the burning by cooling the fuel below its ignition point, by cutting off the supply of oxygen, or by performing both actions. Some extinguishing agents interrupt the complex system of molecular chain reactions that occur between the heated fuel and the oxygen. Modern portable fire extinguishers contain agents that use one or more of these methods.

Cooling the Fuel

If the temperature of the fuel falls below its temperature, the combustion process will stop. Water extinguishes a fire using this method.

Cutting Off the Supply of Oxygen

Creating a barrier that interrupts the flow of oxygen to the flames will also extinguish a fire. Putting a lid on a pan of burning food is an example of this technique **FIGURE 8-15**; applying a blanket of foam to the surface of a burning liquid is another example. Surrounding the fuel with a layer of CO_2 can also cut off the supply of oxygen necessary to sustain the burning process.

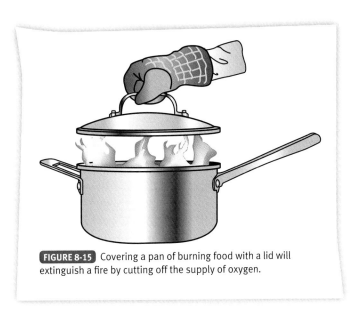

FIGURE 8-15 Covering a pan of burning food with a lid will extinguish a fire by cutting off the supply of oxygen.

Interrupting the Chain of Reactions

Some extinguishing agents work by interrupting the molecular chain reactions required to sustain combustion. In some cases, just a very small quantity of the agent can accomplish this objective.

Types of Extinguishing Agents

An extinguishing agent is the substance contained in a portable fire extinguisher that puts out a fire. A variety of chemicals, including water, are used in portable fire extinguishers. The best extinguishing agent for a particular hazard depends on several factors, including the types of materials involved and the anticipated size of the fire. Portable fire extinguishers use seven basic types of extinguishing agents:

1. Water
2. Dry chemicals
3. CO_2
4. Foam
5. Halogenated agents
6. Dry powder
7. Wet chemicals

Water

Water is an efficient, plentiful, and inexpensive extinguishing agent. When it is applied to a fire, it is quickly converted from a liquid into steam, absorbing great quantities of heat in the process. As the heat is removed from the combustion process, the fuel cools below its ignition temperature, and the fire stops burning.

Water is an excellent extinguishing agent for Class A fires. Many Class A fuels will absorb liquid water, which further lowers the temperature of the fuel. This also prevents rekindling.

Water is a much less effective extinguishing agent for other classes of fire; applying water to hot cooking oil, for example, can cause explosive splattering, which can spread the fire and endanger the operator of the fire extinguisher. Many burning flammable liquids will simply float on top of water. Because water conducts electricity, it is dangerous to apply a stream of water to any fire that involves energized electrical equipment. If water is applied to a burning combustible metal, a violent reaction can occur. Because of these limitations, plain water is used only in Class A fire extinguishers.

One notable disadvantage of water is that it freezes at 0°C (32°F). In areas that are subject to below-freezing temperatures, **loaded-stream fire extinguishers** can be used to counteract this limitation. These extinguishers combine an alkali metal salt with water. The salt lowers the freezing point of water, so the extinguisher can be used in much colder areas.

Wetting agents can also be added to the water in a fire extinguisher. These agents reduce the surface tension of the water, allowing it to penetrate more effectively into many fuels, such as baled cotton and fibrous materials.

Another method of applying water is through the use of water mist. Water mist fire extinguishers contain distilled water, which is discharged from the device as a fine spray. These extinguishers are used where regular extinguishers might cause excessive damage. Typical uses include museums, rare-book collections, and hospital operating rooms.

Dry Chemicals

Dry chemical fire extinguishers deliver a stream of very finely ground particles onto a fire. Different chemical compounds are used to produce extinguishers of varying capabilities and characteristics. The dry chemical extinguishing agents work in two ways. First, the dry chemicals interrupt the chemical chain reactions that occur as part of the combustion process. Second, the tremendous surface area of the finely ground particles allows them to absorb large quantities of heat.

Dry chemical extinguishing agents offer several advantages over water extinguishers:

- They are effective on Class B (flammable liquids and gases) fires.
- They can be used on Class C (energized electrical equipment) fires, because the chemicals are nonconductive.
- They are not subject to freezing.

The first dry chemical extinguishers were introduced during the 1950s and were rated only for Class B and C fires. The industry term for these B:C-rated units is *ordinary dry chemical* extinguishers.

During the 1960s, **multipurpose dry chemical fire extinguishers** were introduced. These extinguishers are rated for Class A, B, and C fires. The chemicals in these extinguishers form a crust over Class A combustible fuels, thereby preventing rekindling FIGURE 8-16 .

FIGURE 8-16 Multipurpose dry chemical extinguishers can be used for Class A, B, and C fires.

Multipurpose dry chemical extinguishing agents take the form of fine particles that are treated with other chemicals to help maintain an even flow when the extinguisher is being used. These additives are intended to prevent the particles from absorbing moisture, which could cause packing or caking and interfere with the extinguisher's discharge.

One disadvantage of dry chemical extinguishers is that the chemicals—particularly the multipurpose dry chemicals—are corrosive and can damage electronic equipment, such as computers, telephones, and photocopiers. The fine particles are carried in air and settle like a fine dust inside the equipment. Over a period of months, this residue can corrode metal parts, causing considerable damage. If electronic equipment is exposed to multipurpose dry chemical extinguishing agents, it should be cleaned professionally within 48 hours after exposure.

Five compounds are used as the primary dry chemical extinguishing agents:

- Sodium bicarbonate (rated for Class B and C fires only)
- Potassium bicarbonate (rated for Class B and C fires only)
- Urea-based potassium bicarbonate (rated for Class B and C fires only)
- Potassium chloride (rated for Class B and C fires only)
- Ammonium phosphate (rated for Class A, B, and C fires)

Sodium bicarbonate is often used in small household extinguishers. Potassium bicarbonate, potassium chloride, and urea-based potassium bicarbonate all have greater fire-extinguishing capabilities (per unit volume) for Class B fires than does sodium bicarbonate. Potassium chloride is more corrosive than the other dry chemical extinguishing agents.

Ammonium phosphate is the only dry chemical extinguishing agent that is rated as suitable for use on Class A fires. Although ordinary dry chemical extinguishers can be used against Class A fires, a water dousing is also needed to extinguish any smouldering embers and prevent rekindling.

The selection of which dry chemical extinguisher to use depends on the compatibility of different agents with one another and with any products that they might contact. Some dry chemical extinguishing agents cannot be used in combination with particular types of foam.

■ Carbon Dioxide

Carbon dioxide is a gas that is one-and-a-half times heavier than air. When CO_2 is discharged on a fire, it forms a dense cloud that displaces the air surrounding the fuel. This effect interrupts the combustion process by reducing the amount of oxygen that can reach the fuel. The placement of a blanket of CO_2 over the surface of a liquid fuel can also disrupt the fuel's ability to vaporize.

In portable **carbon-dioxide (CO_2) fire extinguishers**, CO_2 is stored under pressure as a colourless and odourless liquid. It is discharged through a hose and expelled on the fire through a horn. The CO_2 is very cold when it is released; it forms a visible cloud of dry ice when moisture in the air freezes as it comes into contact with the CO_2.

CO_2 fire extinguishers are rated for Class B and C fires only. This extinguishing agent does not conduct electricity and has two significant advantages over dry chemical agents: it is not corrosive and it does not leave any residue.

CO_2 also has several limitations and disadvantages:

- Weight—CO_2 extinguishers are heavier than similarly rated extinguishers that use other extinguishing agents FIGURE 8-17 .
- Range—CO_2 extinguishers have a short discharge range, which requires the operator to be close to the fire, increasing the risk of personal injury.
- Weather—CO_2 does not perform well at temperatures below –17.8°C (0°F) or in windy or drafty conditions, because it dissipates before it reaches the fire.
- Confined spaces—When used in confined areas, CO_2 dilutes the oxygen in the air. If it is diluted enough, people in the space can begin to suffocate.
- Suitability—CO_2 extinguishers are not suitable for use on fires involving pressurized fuel or on cooking-grease fires.

FIGURE 8-17 CO_2 extinguishers are heavy due to the weight of the container and the large quantity of agent needed to extinguish a fire. They also have a large discharge nozzle, making them easily identifiable.

FIGURE 8-18 An FFFP extinguisher produces an effective foam for use on Class B fires.
© craftvision/iStockphoto

Foam

Foam fire extinguishers discharge a water-based solution to which a measured amount of foam concentrate has been added. The nozzles on foam extinguishers are designed to introduce air into the discharge stream, thereby producing a foam blanket. Foam extinguishing agents are formulated for use on either Class A or Class B fires.

Class A foam extinguishers for ordinary combustible fires extinguish fires in the same way that water extinguishes fires. This type of extinguisher can be produced by adding Class A foam concentrate to the water in a standard 9.5-L (2.5-gal.), stored-pressure extinguisher. The foam concentrate reduces the surface tension of the water, allowing for better penetration of the foam into the burning materials.

Class B foam extinguishers discharge a foam solution that floats across the surface of a burning liquid, creating a blanket that separates the fuel from oxygen. This blanket prevents the fuel from vaporizing. It forms a barrier between the fuel and the oxygen, extinguishing the flames and preventing reignition. These foaming agents are not suitable for Class B fires that involve pressurized fuels or cooking oils, however. The most commonly encountered Class B additive is __film-forming fluoroprotein (FFFP) foam__ **FIGURE 8-18**.

Some Class B foam extinguishing agents are approved for use on **polar solvents**—that is, water-soluble flammable liquids, such as alcohols, acetone, esters, and ketones. Only extinguishers that are specifically labelled for use with polar solvents should be used if these products are present.

Although they are not specifically intended to extinguish Class A fires, most Class B foams can also be used on ordinary combustibles. The reverse is not true, however: Class A foams are *not* effective on Class B fires. Foam extinguishers are not suitable for use on Class C fires and cannot be stored or used at freezing temperatures.

Wet Chemicals

__Wet chemical fire extinguishers__ are the only type of extinguisher to qualify under the new Class K rating requirements. They use __wet chemical extinguishing agents__, which are chemicals applied as water solutions. Before Class K extinguishing agents were developed, most fire-extinguishing systems for kitchens used dry chemicals. The minimum requirement for a commercial kitchen was a 40-B-rated sodium-bicarbonate or potassium-bicarbonate extinguisher. These systems required extensive clean up after their use, which often resulted in serious business interruptions.

All new fixed extinguishing systems in restaurants and commercial kitchens now use wet chemical extinguishing agents. These agents are specifically formulated for use in commercial kitchens and food-product manufacturing facilities, especially where food is cooked in a deep fryer. The fixed systems discharge the agent directly over the cooking surfaces. There is no numeric rating of their efficiency in portable fire extinguishers.

The Class K wet chemical agents include aqueous solutions of potassium acetate, potassium carbonate, and potassium citrate, either singly or in various combinations. These wet agents convert the fatty acids in cooking oils or fats to a soap or foam, a process known as **saponification**.

When wet chemical agents are applied to burning vegetable oils, they create a thick blanket of foam that quickly smothers the fire and prevents it from reigniting while the hot oil cools. Such agents are discharged as a fine spray, which reduces the risk of splattering. They are very effective at extinguishing cooking oil fires, and clean up afterward is much easier, allowing a business to reopen sooner.

Halogenated Agents

Halogenated extinguishing agents are produced from a family of liquefied gases, known as halogens, that includes fluorine, bromine, iodine, and chlorine. Hundreds of different formulations can be produced from these elements; these myriad versions have many different properties and potential uses. Although several of these formulations are very effective for extinguishing fires, only a few of them are commonly used as extinguishing agents.

Halogenated extinguishing agents are commonly called **clean agents** because they leave no residue and are ideally suited for areas that contain computers or sensitive electronic equipment. Per kilogram, they are approximately twice as effective at extinguishing fires as is CO_2.

Two categories of halogenated extinguishing agents are distinguished: Halons and halocarbons. A 1987 international agreement known as the Montreal Protocol limited Halon production, because these agents damage the earth's ozone layer. Halons have since been replaced by a new family of extinguishing agents, halocarbons.

The halogenated agents are stored as liquids and are discharged under relatively high pressure. They release a mist of vapour and liquid droplets that disrupts the molecular chain reactions within the combustion process, thereby extinguishing the fire. These agents dissipate rapidly in windy conditions, as does CO_2, so their effectiveness is limited in outdoor locations. Because halogenated agents also displace oxygen, they should be used with care in confined areas.

Halon 1211 (bromochlorodifluoromethane) should be used judiciously and only in situations where its clean properties are essential, owing to its detrimental environmental impact. Small Halon 1211 extinguishers are rated for Class B and C fires, but are unsuited for use on fires involving pressurized fuels or cooking grease. Larger Halon extinguishers are also rated for Class A fires.

Currently, four types of halocarbon agents are used in portable extinguishers: hydrochlorofluorocarbon, hydrofluorocarbon, perfluorocarbon, and fluoroiodocarbon.

Dry Powder

Dry powder extinguishing agents are chemical compounds used to extinguish fires involving combustible metals (Class D fires). These agents are stored in fine granular or powdered form and are applied to smother the fire. They form a solid crust over the burning metal, which both blocks out oxygen (the fuel for the fire) and absorbs heat.

The most commonly used dry powder extinguishing agent is formulated from finely ground sodium chloride (table salt) plus additives to help it flow freely over a fire. A thermoplastic material mixed with the agent binds the sodium-chloride particles into a solid mass when they come into contact with a burning metal.

Another dry powder agent is produced from a mixture of finely granulated graphite powder and phosphorus-containing compounds. This agent cannot be expelled from fire extinguishers; instead, it is produced in bulk form and applied by hand, using a scoop or a shovel. When applied to a metal fire, the phosphorus compounds release gases that blanket the fire and cut off its supply of oxygen; the graphite absorbs heat from the fire, allowing the metal to cool below its ignition point. Other specialized dry powder extinguishing agents are available for fighting specific types of metal fires; for details, see the National Fire Protection Association's (NFPA) *Fire Protection Handbook*.

Class D agents must be applied very carefully so that the molten metal does not splatter. No water should come in contact with the burning metal, because even a trace quantity of moisture can cause a violent reaction.

Fire Extinguisher Design

All portable fire extinguishers use pressure to expel their contents onto a fire, and many rely on pressurized gas to eject the extinguishing agent. This gas can be stored either with the extinguishing agent in the body of the extinguisher or externally in a separate cartridge or cylinder. With external storage, the extinguishing agent is put under pressure only as it is used.

Some extinguishing agents, such as CO_2, are called **self-expelling agents**. Most of these agents are normally gases, but are stored as liquids under pressure in the fire extinguisher. When the confining pressure is released, the agent rapidly expands, causing it to self-discharge. Hand-operated pumps are used to expel the agent when water or water with additives is the extinguishing agent.

Portable Fire Extinguisher Components

Most hand-held portable fire extinguishers have six basic parts **FIGURE 8-19**:

- A cylinder or container that holds the extinguishing agent
- A carrying handle
- A nozzle or horn
- A trigger and discharge valve assembly
- A locking mechanism to prevent accidental discharge
- A pressure indicator

Locking mechanism

Pressure indicator

Trigger

Handle

Cylinder

Nozzle

FIGURE 8-19 Portable fire extinguishers have six basic parts.
Courtesy of Amerex Corporation

Cylinder or Container

The body of the extinguisher, known as the **cylinder**, or container, holds the extinguishing agent. Nitrogen, compressed air, or CO_2 can be used to pressurize the cylinder to expel the agent. **Stored-pressure fire extinguishers** hold both the extinguishing agent, in wet or dry form, and the expeller gas under pressure in the cylinder. **Cartridge/cylinder-operated fire extinguishers** rely on an external cartridge of pressurized gas, which is released only when the extinguisher is to be used.

Handle

The **handle** is used to carry a portable fire extinguisher and, in many cases, to hold it during use. The actual design of the handle varies from model to model, but all extinguishers that weigh more than 1.4 kg (3 lb.) have handles. In many cases, the handle is located just below the trigger mechanism.

Nozzle or Horn

The extinguishing agent is expelled through a **nozzle** or **horn**. In some extinguishers, the nozzle is attached directly to the valve assembly at the top of the extinguisher. In other models, the nozzle is found at the end of a short hose.

Foam extinguishers have a special aspirating nozzle that introduces air into the extinguishing agent, thereby creating the foam. Carbon-dioxide extinguishers have a tubular or conical horn, which is often mounted at the end of a short hose.

Pump tank fire extinguishers, which are nonpressurized, manually operated water extinguishers, usually have a nozzle at the end of a short hose. The manually operated pump may be mounted directly on the cylinder or it may be part of the nozzle assembly.

Trigger

The **trigger** is the mechanism that is squeezed or depressed to discharge the extinguishing agent. In some models, the trigger is a button positioned just above the handle; on most portable fire extinguishers, however, the trigger is a lever located above the handle. To release the extinguishing agent, the operator lifts the extinguisher by the handle and simultaneously squeezes down on the discharge lever.

Cartridge/cylinder-operated extinguisher models usually have a two-step operating sequence. First, a handle or lever is pushed to pressurize the stored agent; then, a trigger-type mechanism incorporated in the nozzle assembly is used to control the discharge.

Locking Mechanism

The **locking mechanism** is a simple quick-release device that prevents accidental discharge of the extinguishing agent. The simplest form of locking mechanism is a stiff pin, which is inserted through a hole in the trigger to prevent it from being depressed. The pin usually has a ring at the end so that it can be removed quickly.

A special plastic tie, called a **tamper seal**, is used to secure the pin. This seal is designed to break easily when the pin is pulled. Removing the pin and tamper seal is best accomplished with a twisting motion. The tamper seal makes it easy to see whether the extinguisher has been used and not recharged. It also discourages people from playing or tinkering with the extinguisher.

Pressure Indicator

The **pressure indicator** or gauge shows whether a stored-pressure extinguisher has sufficient pressure to operate properly. Over time, the pressure in an extinguisher may dissipate. Checking the gauge first will tell you whether the extinguisher is ready for use.

Pressure indicators vary in terms of both design and sophistication. Most extinguishers use a needle gauge. Pressure may be shown in kilopascals (kPa) or pounds per square inch (psi), or on a three-step scale (i.e., too low, proper range, too high). Pressure gauges are usually colour-coded, with a green area indicating the proper pressure zone. CO_2 extinguishers do not have pressure indicators or gauges but, rather, are weighted to determine the remaining agent. Likewise, extinguishers that are pressurized by a cartridge lack a pressure gauge.

Some disposable fire extinguishers intended for home use have an even simpler pressure indicator—that is, a plastic pin built into the cylinder. Pressing on the pin tests the pressure within the extinguisher. If the pin pops back up, the extinguisher has enough pressure to operate; if it remains depressed, the pressure has dropped below an acceptable level.

Wheeled Fire Extinguishers

Wheeled fire extinguishers are large units that are mounted on wheeled carriages and typically contain between 68 and 158.8 kg (150 and 350 lb.) of extinguishing agent. The wheeled design lets one person transport the extinguisher to the fire. If a wheeled extinguisher is intended for indoor use, doorways and aisles must be wide enough to allow for its passage to every area where it could be needed.

Wheeled fire extinguishers usually have long delivery hoses, so the unit can stay in one spot as the operator moves around to attack the fire from more than one side. Usually, a separate cylinder containing nitrogen or some other compressed gas provides the pressure necessary to operate the extinguisher.

Wheeled fire extinguishers are most often installed in special hazard areas, such as at military bases, airports, and industrial settings. Models with rubber tires or wide-rimmed wheels are available for outdoor installations.

Fire Extinguisher Characteristics

Portable fire extinguishers vary according to their extinguishing agent, capacity, effective range, and the time it takes to completely discharge the extinguishing agent. They also have different mechanical designs. This section describes the basic characteristics of seven types of extinguishers, which are organized by type of extinguishing agent:

- Water extinguishers
- Dry chemical extinguishers
- CO_2 extinguishers
- Class B foam extinguishers
- Halogenated agent extinguishers
- Dry powder extinguishing agents
- Wet chemical extinguishers

Water Extinguishers

Water extinguishers are used to cool the burning fuel below its ignition temperature. They are intended for use primarily on Class A fires. Class B foam extinguishers, which are a specific type of water extinguisher, are intended for fires involving flammable liquids. Water extinguishers include stored-pressure, loaded-stream, and wetting-agent models.

Stored-Pressure Water Extinguishers

The most popular kind of **stored-pressure water-type fire extinguisher** is the 9.5-L (2.5-gal.) model with a 2-A rating **FIGURE 8-20**. Many fire department vehicles carry this type of fire extinguisher for use on incipient-stage Class A fires. Such an extinguisher expels water in a solid stream with a range of 10.7–12.2 m (35–40 ft.) through a nozzle at the end of a short hose. The discharge time is approximately 55 seconds if the extinguisher is used continuously. A full extinguisher weighs about 13.6 kg (30 lb.).

Because the contents of these extinguishers can freeze, these devices should not be installed in areas where the temperature is expected to drop below 0°C (32°F). Antifreeze models

FIGURE 8-20 Most fire departments use stored-pressure water extinguishers.

of stored-pressure water extinguishers, called loaded-stream extinguishers, are available. The loaded-stream agent will not freeze at temperatures as low as –40°C (–40°F).

The recommended procedure for operating a stored-pressure water extinguisher is to set it on the ground, grasp the handle with one hand, and pull out the ring pin or release the locking latch with the other hand. At this point, the extinguisher can be lifted and used to douse the fire. Use one hand to aim the stream at the fire, and squeeze the trigger with the other hand. The stream of water can be turned into a spray by putting a thumb at the end of the nozzle; this technique is often used after the flames have been extinguished as part of the effort to thoroughly soak the fuel.

Stored-pressure water extinguishers can be recharged at any location that provides water and a source of compressed air. Follow the manufacturer's instructions to ensure proper and safe recharging.

Loaded-Stream Water Extinguishers

Loaded-stream water extinguishers discharge a solution of water containing an alkali metal salt that prevents freezing at temperatures as low as –40°C (–40°F). The most common model is the 9.5-L (2.5-gal.) unit, which is identical to a typical stored-pressure water extinguisher. Hand-held models are available with capacities of 3.8–9.5 L (1–2.5 gal.) of water and are rated from 1-A to 3-A.

Larger units, including a 64-L (17-gal.) unit rated 10-A and a 125-L (33-gal.) unit rated 20-A, are also available. Pressure for these extinguishers is supplied by a separate cylinder of CO_2.

Wetting-Agent and Class A Foam Water Extinguishers

Wetting-agent water-type fire extinguishers expel water that contains a solution intended to reduce its surface tension (the physical property that causes water to bead or form a puddle on a flat surface). Reducing the surface tension allows water to spread over the fire and penetrate more efficiently into Class A fuels.

Class A foam extinguishers contain a solution of water and Class A foam concentrate. This agent has foaming properties as well as the ability to reduce surface tension.

Both wetting-agent and Class A foam extinguishers are available in the same configurations as water extinguishers, including hand-held stored-pressure models and wheeled units. These extinguishers should not be exposed to temperatures below 4.4°C (40°F).

Pump Tank Water Extinguishers

Pump tank water fire extinguishers come in sizes ranging from 1-A-rated, 6-L (1.5-gal.) units to 4-A-rated, 19-L (5-gal.) units. The water in these devices is not stored under pressure. Instead, the pressure needed to expel the water is provided by a hand-operated, double-acting, vertical piston pump, which moves water out through a short hose on both the up and the down strokes. This type of extinguisher sits upright on the ground during use. A small bracket at the bottom allows the operator to steady the extinguisher with one foot while pumping.

Pump tank extinguishers can be used with antifreeze agents. The manufacturer should be consulted for details, because some antifreeze agents (such as common salt) can corrode the extinguisher or damage the pump. Extinguishers with steel shells corrode more easily than those with copper or nonmetallic shells.

Backpack Water Extinguishers

Backpack water extinguishers are used primarily outdoors for fighting brush and grass fires **FIGURE 8-21**. Most of these units have a tank capacity of 19 L (5 gal.) and weigh approximately 22.7 kg (50 lb.) when full. Backpack extinguishers are listed by ULC but do not carry numeric ratings.

The water tank can be made of fibreglass, stainless steel, galvanized steel, nylon, canvas, or brass. Backpack extinguishers are designed to be refilled easily in the field—from a lake or a stream, for example—through a wide-mouth opening at the top. A filter keeps dirt, stones, and other contaminants from entering the tank. Antifreeze agents, wetting agents, or other special water-based extinguishing agents can also be used with backpack water extinguishers.

Most backpack extinguishers are operated via hand pumps. The most common design has a trombone-type, double-acting piston pump located at the nozzle, which is attached to the tank by a short rubber hose. To discharge the extinguisher, the operator holds the pump in both hands and moves the piston back and forth.

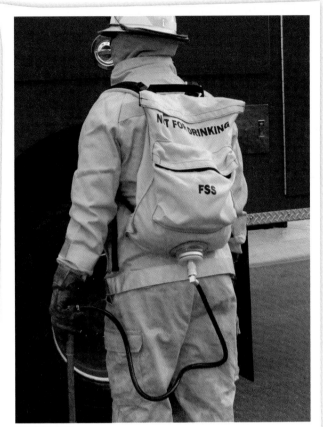

FIGURE 8-21 A backpack extinguisher can be easily refilled from a natural water source such as a lake or river.

Some models have a compression pump built into the side of the tank. On these devices, it takes about 10 strokes of the pump handle to build up the initial pressure, which is maintained through continuous slow strokes. The operator uses the other hand to control the discharge. A lever-operated shut-off nozzle is provided at the end of a short hose.

■ Dry Chemical Extinguishers

Dry chemical extinguishers contain a variety of chemical extinguishing agents in granular form. Hand-held dry chemical extinguishers are available with capacities ranging from 0.4 to 13.6 kg (1 to 30 lb.) of agent. Wheeled fire extinguishers are available with capacities up to 158.8 kg (350 lb.) of agent. Ordinary dry chemical models can be used to extinguish Class B and C fires, whereas multipurpose dry chemical units are rated for use on Class A, B, and C fires. Some large dry chemical extinguishers are mounted on fire apparatus to deal with special risks.

All dry chemical extinguishing agents can be used on Class C fires that involve energized electrical equipment. Unfortunately, the residue left by the dry chemical can be very damaging to computers, electronic devices, and electrical equipment.

Stored-pressure units expel the dry chemical agent in the same manner as stored-pressure water extinguishers. The dry chemical in a cartridge/cylinder-operated extinguisher is not stored under pressure. Instead, these extinguishers have a sealed, pressurized cartridge connected to the storage cylinder. They are activated by pushing down on a lever that punctures the cartridge and pressurizes the cylinder.

Most small, hand-held dry chemical extinguishers are designed to discharge their contents completely in as little as 8–20 seconds. Larger units may discharge for as long as 30 seconds.

Depending on the extinguisher's size, the horizontal range of the discharge stream can be from 1.5 to 9.1 m (5 to 30 ft.). Some models have special nozzles that allow for a longer range. The long-range nozzles are useful when the fire involves burning gas or a flammable liquid under pressure or when the operator is working in a strong wind.

The trigger allows this type of extinguisher to be discharged intermittently, starting and stopping the agent flow. Releasing the trigger stops the flow of agent; however, this does not mean that the extinguisher can be put aside and used again later. These extinguishers do not retain their internal pressure for extended periods, because the granular dry chemical causes the valve to leak.

Dry chemical fire extinguishers can be stored and used in areas where temperatures fall below the freezing point.

Ordinary Dry Chemical Extinguishers

Ordinary dry chemical extinguishers are available in handheld models with ratings up to 160-B:C. Larger, wheeled units carry ratings up to 640-B:C.

Fire Fighter Safety Tips   © Photos.com

Discharging a dry-chemical fire extinguisher in a confined space can create a cloud of very fine dust that can impair vision and cause difficulty breathing. Self-contained breathing apparatus (SCBA) should be used to protect fire fighters from both toxic gases from the fire and the dry chemical dust discharged from the fire extinguisher.

FIRE FIGHTER Tips

A dry chemical extinguisher will usually continue to lose pressure after a partial discharge. Pressure loss can occur even when only a very small amount of agent has been discharged and the pressure gauge still indicates that the extinguisher is properly charged. This loss occurs because the agent leaves residue in the valve assembly and allows the stored pressure to leak out slowly.

Multipurpose Dry Chemical Extinguishers

Ammonium phosphate is commonly called a multipurpose dry chemical agent because it can be used on Class A, B, and C fires. When it is used on Class A fires, the chemical coats the surface of the fuel to prevent continuing combustion.

Multipurpose dry chemical extinguishers are available in hand-held models with ratings ranging from 1-A to 20-A and from 10-B:C to 120-B:C. Larger, wheeled models

have ratings ranging from 20-A to 40-A and from 60-B:C to 320-B:C.

Multipurpose dry-chemical extinguishers should never be used on cooking oil (Class K) fires, such as those involving deep fryers located in commercial kitchens. The ammonium phosphate–based extinguishing agent is acidic and will not react with cooking oils to produce the smothering foam needed to extinguish this type of fire. Even worse, the acid will counteract the foam-forming properties of any alkaline extinguishing agent that is applied to the same fire.

■ Carbon-Dioxide Extinguishers

CO_2 extinguishers are rated to fight Class B and C fires. CO_2 extinguishes a fire by enveloping the fuel in a cloud of inert gas, which reduces the oxygen content of the surrounding atmosphere and smothers the flames. Because the CO_2 discharge is very cold, it also helps to cool the burning materials and surrounding areas.

CO_2 gas is one-and-a-half times heavier than air, colourless, odourless, nonconductive, and inert. It is also noncorrosive and does not leave any residue. These factors are important in areas where costly electronic components or computer equipment must be protected. CO_2 is also used around food preparation areas and in laboratories.

CO_2 is both an expelling agent and an extinguishing agent. It is stored under a pressure of 5675 kPa (823 psi), which keeps the CO_2 in liquid form at room temperature. When the pressure is released, the liquid CO_2 rapidly converts to a gas, and the expansion of the gas forces the agent out of the container.

The CO_2 is discharged through a siphon tube that reaches to the bottom part of the storage cylinder. It is forced through a hose to a horn or cone-shaped applicator that is used to direct the flow of the agent. When discharged, the agent is very cold and contains a mixture of CO_2 gas and solid CO_2, which is quickly converted to a gas.

Compared to other types of extinguishers, CO_2 extinguishers have a relatively short discharge range, 0.9–2.4 m (3–8 ft.). This limitation presents a safety issue because the operator must be close to the fire. Depending on their size, CO_2 extinguishers can discharge completely in 8–30 seconds. CO_2 extinguishers are not toxic when used around food, and their use does not require clean up, because the CO_2 gas dissipates into the air.

CO_2 fire extinguishers are not recommended for outdoor use or for use in locations with strong air currents. In such settings, the gas will be rapidly dissipated and will not be effective in smothering the flames.

CO_2 extinguishers have a trigger mechanism that can be operated intermittently to preserve the remaining agent. The pressurized CO_2 will remain in the extinguisher, but the extinguisher must be recharged after use. The extinguisher can be weighed to determine how much agent is left in the storage cylinder.

The smaller CO_2 extinguishers contain from 0.9 to 2.3 kg (2 to 5 lb.) of agent. These units are designed to

Fire Fighter Safety Tips

© Photos.com

Do not aim the CO_2 extinguisher discharge at anyone or allow it to come in contact with exposed skin, because frostbite could result. CO_2 discharged into a confined space will reduce the oxygen level in that space. For this reason, anyone entering the confined area must wear SCBA.

be operated with one hand. The horn is attached directly to the discharge valve on the top of the extinguisher by a hinged metal tube. In larger models, the horn is attached at the end of a short hose. These models require two-handed operation.

Class B Foam Extinguishers

Class B foam extinguishers are very similar in appearance and operation to water extinguishers; instead of plain water, however, they discharge a solution of water and FFFP foam concentrate. The agent is discharged through an aspirating nozzle, which mixes air into the stream. The result is a foam blanket that will float over the surface of a flammable liquid.

Class B foam agents are very effective in fighting Class A fires but are not suitable for Class C fires or for fires involving flammable liquids or gases under pressure. They are not intended for use on cooking oil fires, and only certain foam extinguishers can be used on fires involving polar solvents. Detailed information on the use of FFFP is available in NFPA 11, *Standard for Low-, Medium-, and High-Expansion Foam.*

FFFP hand-held stored-pressure extinguishers are available in two sizes: 6 L (1.6 gal.), rated 2-A:10-B, and 9.5 L (2.5 gal.), rated 3-A:20-B. A wheeled model with a 125-L (33-gal.) capacity is rated 20-A:160-B.

Foam extinguishing agents are not effective at freezing temperatures. Consult the extinguisher manufacturer for information on using foam agents effectively at low temperatures.

Wet Chemical Extinguishers

Wet chemical extinguishers are used to protect Class K (kitchen) installations, which are equipped with deep fryers, cooking oils, and grills. Many commercial cooking installations use fixed, automatic fire-extinguishing systems as their first line of defence. Portable Class K wet chemical extinguishers are currently available in two sizes, 6 L (1.5 gal.) and 9.5 L (2.5 gal.). There are no numerical ratings for these extinguishers.

FIRE FIGHTER Tips

Dry chemical agents and dry powder agents have very different meanings in relation to fire extinguishers. Dry powder fire extinguishers are designed for use in suppressing Class D fires. Dry chemical fire extinguishers are rated for Class B and C fires or for Class A, B, and C fires. Do not confuse these two terms.

Halogenated Agent Extinguishers

Halogenated agent fire extinguishers include both Halon agents and halocarbon agents. Because Halon agents can destroy the Earth's protective ozone layer, their use is strictly controlled. The halocarbons, however, are not subject to the same environmental restrictions. Both types of agents are available in hand-held extinguishers rated for Class B and C fires. Fire departments must check to ensure which type of halogenated agent fire extinguishers are permitted in their province or territory. Larger-capacity models are also rated for use on Class A fires.

In this type of extinguisher, the agent is discharged as a streaming liquid, which can be directed at the base of a fire. The discharges from halogenated agent extinguishers have a horizontal stream range of 2.7–4.6 m (9–15 ft.).

The halogenated agents are nonconductive and leave no residue that could damage electrical equipment.

Halon 1211 (bromochlorodifluoromethane) is available in hand-held stored-pressure extinguishers with capacities that range from 0.4 kg (1 lb.), rated 1-B:C, to 10 kg (22 lb.), rated 4-A:80-B:C. Wheeled Halon 1211 models are available with capacities up to 68 kg (150 lb.) with a rating of 30-A:160-B:C. The wheeled fire extinguishers use a nitrogen booster charge from an auxiliary cylinder to expel the agent.

Dry Powder Extinguishing Agents

Dry powder extinguishing agents are intended for fighting Class D fires involving combustible metals. The extinguishing agents and techniques required to put out Class D fires vary greatly, depending on the specific fuel, the quantity involved, and the physical form of the fuel (e.g., grindings, shavings, or solid objects). The agent and the application method must be suited to the particular situation.

Each dry powder extinguishing agent is listed for use on specific combustible-metal fires. This information and instructions related to the recommended application method are printed on the container. Consult the manufacturer's recommendations for information about Class D agents and extinguishers, as well as NFPA's *Fire Protection Handbook.*

Dry Powder Fire Extinguishers

Dry powder fire extinguishers using sodium chloride–based agents are available with 13.6-kg (30-lb.) capacity in either stored-pressure or cartridge/cylinder-operated models. Wheeled models are available with 68- and 158.8-kg (150- and 350-lb.) capacities.

Dry powder extinguishers have adjustable nozzles that allow the operator to vary the flow of the extinguishing agent. When the nozzle is fully opened, the hand-held models have a range of 1.8–2.4 m (6–8 ft.). Extension wand applicators are available to direct the discharge from a more distant position.

Bulk Dry Powder Agents

Bulk dry powder agents are available in 18.1- and 22.7-kg (40- and 50-lb.) pails and 158.8-kg (350-lb.) drums. The same sodium chloride–based agent that is used in portable fire extinguishers can be stored in bulk form and applied by hand.

Another dry powder extinguishing agent for Class D fires, graded granular graphite mixed with phosphorus-containing compounds, cannot be expelled from a portable fire extinguisher. Instead, this agent must be applied manually from a pail or other container using a shovel or scoop.

> **Fire Fighter Safety Tips**   © Photos.com
>
> Before deciding to use a fire extinguisher, size up the fire to ensure that the extinguisher has an adequate capacity and holds the proper extinguishing agent.

> **Fire Fighter Safety Tips**   © Photos.com
>
> Always test a fire extinguisher before using it. Activate the trigger long enough to ensure that the agent discharges properly.

Use of Fire Extinguishers

Fire extinguishers should be simple to operate. Indeed, an individual with only basic training should be able to use most fire extinguishers safely and effectively. Every portable extinguisher should be labelled with printed operating instructions.

There are six basic steps in extinguishing a fire with a portable fire extinguisher:

1. Locate the fire extinguisher.
2. Select the proper classification of extinguisher.
3. Transport the extinguisher to the location of the fire.
4. Activate the extinguisher to release the extinguishing agent.
5. Apply the extinguishing agent to the fire for maximum effect.
6. Ensure your personal safety by having an exit route.

Although these steps are not complicated, practice and training are essential for effective fire suppression with a fire extinguisher. In particular, tests have shown that the effective use of Class B portable fire extinguishers depends heavily on user training and expertise. A trained expert can extinguish a fire up to twice as large as a nonexpert can, using the same extinguisher.

As a fire fighter, you should be able to operate any fire extinguisher that you might be required to use, whether it is carried on your fire apparatus, hanging on the wall of your firehouse, or placed in some other location in your community.

■ Locating a Fire Extinguisher

As a fire fighter, you should know which types of fire extinguishers are carried on your department's apparatus and where each type of extinguisher is located. You should also know where fire extinguishers are located in and around the fire station and other workplaces. You should have at least one

fire extinguisher in your home and another in your personal vehicle, and you should know exactly where they are located. Knowing the exact locations of extinguishers can save valuable time in an emergency.

■ Selecting the Proper Fire Extinguisher

Selecting the proper extinguisher requires an understanding of the classification and rating system for fire extinguishers. Knowing which types of agents are available, how they work, which ratings the fire extinguishers carried on your fire apparatus have, and which extinguisher is appropriate for a particular fire situation is also important.

Fire fighters should be able to assess a fire quickly, determine whether the fire can be controlled by an extinguisher, and identify the appropriate extinguisher. Using an extinguisher with an insufficient rating may not completely extinguish the fire, which can place the operator in danger of being burned or otherwise injured. If the fire is too large for the extinguisher, consider other options, such as obtaining additional extinguishers or making sure that a charged hose line is ready to provide backup.

Understanding the fire extinguisher rating system and the different types of agents will enable a fire fighter who must use an unfamiliar extinguisher to determine whether it is suitable for a particular fire situation. A quick look at the label should be all that is needed in such a case.

Fire fighters should also be able to determine the most appropriate type of fire extinguisher to place in a given area, based on the types of fires that could occur and the hazards that are present `TABLE 8-2`. In some cases, one type of extinguisher might be preferred over another. An extinguishing agent such as CO_2 or a halogenated agent is better for a fire involving electronic equipment, for example, because it leaves no residue. A dry chemical extinguisher is generally more appropriate than a CO_2 extinguisher to fight an outdoor fire, because wind will quickly dissipate CO_2. A dry chemical extinguisher would also be the best choice for a fire involving a flammable liquid leaking under pressure from a pipe, whereas foam would be a better choice for a fire involving a liquid spill on the ground.

■ Transporting a Fire Extinguisher

The best method of transporting a hand-held portable fire extinguisher depends on the extinguisher's size, weight, and design. Hand-held portable models can weigh as little as 0.4 kg (1 lb.) or as much as 22.7 kg (50 lb.). The ability to handle the heavier extinguishers depends on an individual operator's personal strength.

Extinguishers with a fixed nozzle should be carried in the favoured or stronger hand. This approach enables the operator to depress the trigger and direct the discharge easily. Extinguishers that have a hose between the trigger and the nozzle should be carried in the weaker or less-favoured hand so that the favoured hand can grip and aim the nozzle.

Heavier extinguishers may have to be carried as close as possible to the fire and placed upright on the ground. The operator can then depress the trigger with one hand, while holding the nozzle and directing the stream with the other hand.

TABLE 8-2 Fire Extinguisher Types

Extinguishing Agent	Method of Operation	Capacity	Horizontal Range of Stream	Approximate Time of Discharge	Protection Required Below 4.4°C (40°F)	UL/ULC Classifications[a]
Water	Stored-pressure Stored-pressure or pump Pump Pump	6 L 9.5 L 15 L 19 L	9.1–12.2 m 9.1–12.2 m 9.1–12.2 m 9.1–12.2 m	40 sec 1 min 2 min 2–3 min	Yes Yes Yes Yes	1-A 2-A 3-A 4-A
Water (wetting agent)	Stored-pressure Stored-pressure Stored-pressure Stored-pressure	6 L 95 L (wheeled) 170 L (wheeled) 227 L (wheeled)	6.1 m 10.7 m 10.7 m 10.7 m	30 sec 1 ½ min 2 min 2 ½ min	Yes Yes Yes Yes	2-A 10-A 30-A 40-A
Loaded stream	Stored-pressure Stored-pressure	9.5 L 125 L (wheeled)	9.1–12.2 m 15.2 m	1 min 3 min	No No	2-A 20-A
Water mist	Stored-pressure Stored-pressure	6 L 9.5 L	NA NA	60–70 sec 70–80 sec	Yes Yes	2-A:C 2-A:C
FFFP	Stored-pressure Stored-pressure Nitrogen cylinder	9.5 L 6 L 125 L	6.1–7.6 m 6.1–7.6 m 9.1 m	50 sec 50 sec 1 min	Yes Yes Yes	3-A:20 to 40-B 2A:10B 20-A:160-B
Carbon dioxide[b]	Self-expelling Self-expelling Self-expelling Self-expelling	1.1–2.3 kg 4.5–6.8 kg 9.1 kg 22.7–45.4 kg (wheeled)	0.9–2.4 m 0.9–2.4 m 0.9–2.4 m 0.9–3 m	8–30 sec 8–30 sec 10–30 sec 10–30 sec	No No No No	1–5-B:C 2–10-B:C 10-B:C 10 to 20-B:C
Regular dry chemical (sodium bicarbonate)	Stored-pressure Cartridge or stored-pressure Cartridge or stored-pressure Stored-pressure Nitrogen cylinder or stored-pressure	0.4–1.1 kg 1.2–2.3 kg 2.7–13.6 kg 22.7 kg (wheeled) 34–158.8 kg (wheeled)	1.5–2.4 m 1.5–6.1 m 1.5–6.1 m 6.1 m 4.6–13.7 m	8–12 sec 8–25 sec 10–25 sec 35 sec 20–105 sec	No No No No No	2–10-B:C 5–20-B:C 10–160-B:C 160:B-C 40–320-B:C
Purple K dry chemical (potassium bicarbonate)	Cartridge or stored-pressure Cartridge or stored-pressure Cartridge or stored-pressure Cartridge or stored-pressure Nitrogen cylinder or stored-pressure	0.9–2.3 kg 2.5–4.5 kg 7.3–13.6 kg 21.8–22.7 kg (wheeled) 56.7–142.9 kg (wheeled)	1.5–3.7 m 1.5–6.1 m 3–6.1 m 6.1 m 4.6–13.7 m	8–10 sec 8–20 sec 8–25 sec 30–35 sec 30–80 sec	No No No No No	5–30-B:C 10–80-B:C 40–120-B:C 120–160-B:C 80–640-B:C
Super K dry chemical (potassium chloride)	Cartridge or stored-pressure Cartridge or stored-pressure Cartridge or stored-pressure Cartridge or stored-pressure Cartridge or stored-pressure	0.9–2.3 kg 2.3–4.1 kg 4.3–9.1 kg 8.8–13.6 kg 56.7–90.7 kg (wheeled)	1.5–2.4 m 2.4–3.7 m 3–4.6 m 1.5–6.1 m 4.6–13.7 m	8–10 sec 10–15 sec 15–20 sec 10–25 sec 30–40 sec	No No No No No	5–10-B:C 20–40-B:C 40–60-B:C 60–80-B:C 160-B:C
Multipurpose/ ABC dry chemical (ammonium phosphate)	Stored-pressure Stored-pressure or cartridge Stored-pressure or cartridge Stored-pressure or cartridge Stored-pressure or cartridge Nitrogen cylinder or stored-pressure	0.4–2.3 kg 1.1–4.1 kg 4.1–7.7 kg 7.7–13.6 kg 20.4–22.7 kg (wheeled) 49.9–142.9 kg (wheeled)	1.5–3.7 m 1.5–3.7 m 1.5–6.1 m 1.5–6.1 m 6.1 m 4.6–13.7 m	8–10 sec 8–15 sec 10–25 sec 10–25 sec 25–35 sec 30–60 sec	No No No No No	1–3-A[c] and 2–10-B:C 1–4-A and 10–40-B:C 2–20-A and 10–80-B:C 3–20-A and 30–120-B:C 20–30-A and 80–160-B:C 20–40-A and 60–320-B:C
Dry chemical (foam-compatible)	Cartridge or stored-pressure Cartridge or stored-pressure Cartridge or stored-pressure Nitrogen cylinder or stored-pressure	2.1–4.1 kg 4.1–12.2 kg 8.2–13.6 kg 68–158.8 kg (wheeled)	1.5–6.1 m 1.5–6.1 m 1.5–6.1 m 4.6–13.7 m	8–10 sec 10–25 sec 10–25 sec 20–150 sec	No No No No	10–20-B:C 20–30-B:C 40–60-B:C 80–240-B:C
Dry chemical (potassium bicarbonate urea based)	Stored-pressure Stored-pressure Nitrogen cylinder or stored-pressure	2.3–5 kg 4.1–10.4 kg 79.4 kg (wheeled)	3.3–6.7 m 4.6–9.1 m 21.3 m	18 sec 17–33 sec 62 sec	No No No	40–80-B:C 60–160-B:C 480-B:C
Wet chemical	Stored-pressure Stored-pressure Stored-pressure	3 L 6 L 9.5 L	2.4–3.7 m 2.4–3.7 m 2.4–3.7 m	30 sec 35–45 sec 75–85 sec	No No No	K K K

TABLE 8-2	Fire Extinguisher Types *(Continued)*						
Extinguishing Agent	Method of Operation	Capacity	Horizontal Range of Stream	Approximate Time of Discharge	Protection Required Below 4.4°C (40°F)	UL/ULC Classifications[a]	
Halon 1211 (bromochloro-difluromethane)	Stored-pressure	0.4–0.9 kg	1.8–3 m	8–10 sec	No	1–2-B:C	
	Stored-pressure	0.9–1.4 kg	1.8–3 m	8–10 sec	No	5-B:C	
	Stored-pressure	2.5–4.1 kg	2.7–4.6 m	8–15 sec	No	1-A:10-B:C	
	Stored-pressure	5.9–10 kg	4.3–4.9 m	10–18 sec	No	2–4-A and 20–80-B:C	
	Stored-pressure	22.7 kg	10.7 m	30 sec	No	10-A:120-B:C	
	Stored-pressure	68 kg (wheeled)	6.1–10.7 m	30–44 sec	No	30-A:160–240-B:C	
Halon 1211/1301 (bromochlorodi-fluromethane and bromotrifluoro-methane) mixtures	Stored-pressure or self-expelling	0.4–2.3 kg	0.9–3.7 m	8–10 sec	No	1–10-B:C	
	Stored-pressure	4.1–9.1 kg	3–5.5 m	10–22 sec	No	1-A:10-B:C–4-A:80-B:C	
Halocarbon type	Stored-pressure	0.6–68 kg	1.8–10.7 m	9–23 sec	No	1B:C to 10A:80-B:C	

Source: Data from Arthur E. Cote, *Fire Protection Handbook*, 20th ed., vol. II (Quincy, MA: NFPA, 2008).
Note: Halon should be used only where its unique properties are deemed necessary.
[a]UL and ULC ratings checked as of July 24, 1987. Readers concerned with subsequent ratings should review the pertinent lists and supplements issued by these laboratories: Underwriters Laboratories, Inc., 333 Pfingsten Road, Northbrook, IL 60062, U.S.A., or Underwriters Laboratories of Canada, 7 Crouse Road, Scarborough, ON M1R 3A9.
[b]Carbon-dioxide extinguishers with metal horns do not carry a C classification.
[c]Some small extinguishers containing ammonium phosphate–based dry chemical do not carry an A classification.
NA = Not available.

To transport a fire extinguisher, follow the steps in **SKILL DRILL 8-1**:

1 Locate the closest appropriate fire extinguisher. (**STEP 1**)

2 Assess that the extinguisher is safe and effective for the type of fire being attacked. Release the mounting bracket straps. (**STEP 2**)

3 Lift the extinguisher using good body mechanics. Lift small extinguishers with one hand, and large extinguishers with two hands. (**STEP 3**)

4 Walk briskly—do not run—toward the fire. If the extinguisher has a hose and nozzle, carry the extinguisher with one hand, and grasp the nozzle with the other hand. (**STEP 4**)

■ Basic Steps of Fire Extinguisher Operation

Activating a fire extinguisher to apply the extinguishing agent is a single operation that involves four steps. The **PASS** acronym is a helpful way to remember these steps:

1. Pull the safety pin.
2. Aim the nozzle at the base of the flames.
3. Squeeze the trigger to discharge the agent.
4. Sweep the nozzle across the base of the flames.

Most fire extinguishers have very simple operation systems. Practise discharging different types of extinguishers in training situations to build confidence in your ability to use them properly and effectively.

■ Ensure Your Personal Safety

When using a fire extinguisher, always approach the fire with an exit behind you. If the fire suddenly expands or the extinguisher fails to control it, you must have a planned escape route. Never let the fire get between you and a safe exit. After suppressing a fire, do not turn your back on it. Always watch it, and be prepared for it to rekindle until the fire has been fully overhauled.

When ordinary civilians use fire extinguishers on incipient fires, they are probably wearing their normal clothing. As a fire fighter, however, you should wear your personal protective clothing and use appropriate personal protective equipment (PPE). Take advantage of the protection they provide.

If you must enter an enclosed area where an extinguisher has been discharged, wear full PPE and use SCBA. The atmosphere within the enclosed area will probably contain a mixture of combustion products and extinguishing agents, and the oxygen content within the space may be dangerously low.

To attack a Class A fire with a stored-pressure water fire extinguisher, follow the steps in **SKILL DRILL 8-2**:

1 Size up the fire to determine whether a stored-pressure water extinguisher is safe and effective for this fire. Ensure that extinguisher is large enough to be safe and effective. (**STEP 1**)

2 Ensure your safety. Make sure you have an exit route from the fire. Do not turn your back on a fire. (**STEP 2**)

3 Remove the hose and nozzle. Quickly check the pressure gauge to verify that the extinguisher is adequately charged. (**STEP 3**)

4 Pull the pin to release the extinguisher control valve. You must be within 10.7–12.2 m (35–40 ft.) of the fire to be effective. (**STEP 4**)

5 Aim the nozzle and sweep the water stream at the base of the flames. (**STEP 5**)

6 Overhaul the fire; take steps to prevent rekindling, break apart tightly packed fuel, and summon additional help if needed. (**STEP 6**)

Note: This fire extinguisher is safe and effective to use on Class A fires. It is not effective on Class B fires, and it is not safe to use on Class C fires.

SKILL DRILL 8-1   Transporting a Fire Extinguisher
(Fire Fighter I, NFPA 5.3.16)

1 Locate the closest fire extinguisher.

2 Assess that the extinguisher is safe and effective for the type of fire being attacked. Release the mounting bracket straps.

3 Lift the extinguisher using good body mechanics. Lift small extinguishers with one hand, and large extinguishers with two hands.

4 Walk briskly—do not run—toward the fire. If the extinguisher has a hose and nozzle, carry the extinguisher with one hand and grasp the nozzle with the other hand.

© LiquidLibrary

To attack a Class A fire with a multipurpose dry chemical fire extinguisher, follow the steps in **SKILL DRILL 8-3** (Fire Fighter I, 5.3.16):

1 Size up the fire to determine whether a multipurpose dry chemical fire extinguisher is safe and effective for this fire. Ensure that extinguisher is large enough to be safe and effective.

2 Ensure your safety. Make sure you have an exit route from the fire. Do not turn your back on a fire.

3 Remove the hose and nozzle. Quickly check the pressure gauge to verify that the extinguisher is adequately charged.

4 Pull the pin to release the extinguisher control valve. You must be within 3–9.1 m (10–30 ft.) of the fire to be effective.

5 Aim the nozzle and sweep the dry chemical discharge at the base of the flames. Coat the burning fuel with dry chemical.

6 Overhaul the fire; take steps to prevent rekindling, break apart tightly packed fuel, and summon additional help if needed.

Note: This fire extinguisher is safe and effective on Class A, B, and C fires.

SKILL DRILL 8-2

Attacking a Class A Fire with a Stored-Pressure Water-Type Fire Extinguisher
(Fire Fighter I, NFPA 5.3.16)

1 Size up the fire to determine whether a stored-pressure water extinguisher is safe and effective for this fire. Ensure that extinguisher is large enough to be safe and effective.

2 Ensure your safety. Make sure you have an exit route from the fire. Do not turn your back on a fire.

3 Remove the hose and nozzle. Quickly check the pressure gauge to verify that the extinguisher is adequately charged.

4 Pull the pin to release the extinguisher control valve. You must be within 10.7–12.2 m (35–40 ft.) of the fire to be effective.

5 Aim the nozzle and sweep the water stream at the base of the flames.

6 Overhaul the fire; take steps to prevent rekindling, break apart tightly packed fuel, and summon additional help if needed.

© LiquidLibrary

To attack a Class B flammable liquid spill or pool fire with a dry chemical extinguisher, follow the steps in **SKILL DRILL 8-4**:

1 Size up the fire to determine whether a dry chemical fire extinguisher is safe and effective for this fire. Ensure that extinguisher is large enough to be safe and effective.

2 Ensure your safety. Make sure you have an exit route from the fire. Do not turn your back on a fire.

3 Remove the hose and nozzle. Quickly check the pressure gauge to verify that the extinguisher is adequately charged. (**STEP 1**)

4 Pull the pin to release the extinguisher control valve. You must be within 3–9.1 m (10–30 ft.) of the fire to be effective. (**STEP 2**)

5 Aim the nozzle and sweep the dry chemical discharge across the surface of the burning liquid. Start at the

SKILL DRILL 8-4   Attacking a Class B Flammable Liquid Fire with a Dry Chemical Fire Extinguisher
(Fire Fighter I, NFPA 5.3.16)

1 Check the pressure gauge.

2 Pull the pin to release the extinguisher value.

3 Aim the nozzle and sweep the dry chemical discharge across the surface of the burning liquid. Start at the near edge of the fire and work toward the back.

4 Overhaul the fire; take steps to prevent rekindling, keep a blanket of dry chemical over the fuel, and summon additional help if needed.

© LiquidLibrary

near edge of the fire and work toward the back. (**STEP 3**)

6 Overhaul the fire; take steps to prevent rekindling, keep a blanket of dry chemical over the fuel, and summon additional help if needed. (**STEP 4**)

Note: This fire extinguisher is safe to use on Class B and C fires. It is effective on Class B fires, but is not rated for Class A fires.

To attack a Class B flammable liquid spill or pool fire with a stored-pressure fire extinguisher containing an FFFP extinguishing agent, follow the steps in **SKILL DRILL 8-5**:

1 Size up the fire to determine if an FFFP fire extinguisher is safe and effective for this fire. Ensure

that extinguisher is large enough to be safe and effective.

2 Ensure your safety. Make sure you have an exit route from the fire. Do not turn your back on a fire.

3 Remove the hose and nozzle. Quickly check the pressure gauge to verify that the extinguisher is adequately charged.

4 Pull the pin to release the extinguisher control valve. You must be within 6.1–7.6 m (20–25 ft.) of the fire to be effective. (**STEP 1**)

5 Aim the nozzle and discharge the stream of foam so the foam drops gently onto the surface of the burning liquid at the front or the back of the

SKILL DRILL 8-5 Attacking a Class B Flammable Liquid Fire with a Stored-Pressure Foam
Fire Extinguisher (FFFP)
(Fire Fighter I, NFPA 5.3.16)

1 Size up the fire and ensure your safety. Remove the hose and nozzle. Quickly check the pressure gauge to verify that the extinguisher is adequately charged. Pull the pin to release the extinguisher control valve. You must be within 6.1–7.6 m (20–25 ft.) of the fire to be effective.

2 Aim the nozzle and discharge the stream of foam so the foam drops gently onto the surface of the burning liquid at the front or the back of the container. Let the foam blanket flow across the surface of the burning liquid. Avoid splashing foam on the burning liquid, because it can cause the burning fuel to splatter.

3 Overhaul the fire; keep a thick blanket of foam intact over the hot liquid, reapply foam over any hotspots, and summon additional help if needed.

© LiquidLibrary

container. Let the foam blanket flow across the surface of the burning liquid. Avoid splashing foam on the burning liquid, because it can cause the burning fuel to splatter. (**STEP 2**)

6 Overhaul the fire; keep a thick blanket of foam intact over the hot liquid, reapply foam over any hotspots, and summon additional help if needed. (**STEP 3**)

Note: This fire extinguisher is safe to use on Class A, B, and C fires. It is effective on Class B fires by forming a foam blanket. It is effective on Class A fires by cooling and penetrating the fuel.

To operate a CO_2 extinguisher, follow the steps in **SKILL DRILL 8-6** :

1 Size up the fire to determine whether CO_2 is a safe and effective agent for this fire. Ensure that extinguisher is large enough to be safe and effective.

2 Ensure your safety. Make sure you have an exit route from the fire. Do not turn your back on a fire.

3 Remove the horn or nozzle. Quickly squeeze to verify that the extinguisher is charged; CO_2 extinguishers do not have pressure gauges. (**STEP 1**)

SKILL DRILL 8-6 Operating a Carbon-Dioxide Extinguisher
(Fire Fighter I, NFPA 5.3.16)

1 Size up the fire to determine whether CO_2 is a safe and effective agent for this fire. Ensure that extinguisher is large enough to be safe and effective. Ensure your safety. Make sure you have an exit route from the fire. Do not turn your back on a fire. Pull the pin to release the extinguisher control valve.

2 You must be within 1.5–2.4 m (5–8 ft.) of the fire to be effective. Remove the horn or nozzle. Quickly squeeze to verify that the extinguisher is charged; CO_2 extinguishers do not have pressure gauges.

3 Aim the horn or nozzle, and sweep at the base of the flames.

4 Overhaul the fire; take steps to prevent rekindling, and summon additional help if needed.

© LiquidLibrary

4 Pull the pin to release the extinguisher control valve. You must be within 1.5–2.4 m (5–8 ft.) of the fire to be effective. (**STEP 2**)

5 Aim the horn or nozzle, and sweep at the base of the flames. (**STEP 3**)

6 Overhaul the fire; take steps to prevent rekindling, and summon additional help if needed. (**STEP 4**)

Note: The fire extinguisher is most effective when used in areas without wind. It is designed for Class B and C fires only.

To apply a halogenated extinguishing agent to a fire in an electrical equipment room, follow the steps in **SKILL DRILL 8-7** (Fire Fighter I, 5.3.16):

1 Size up the fire to determine whether a halogenated stored-pressure fire extinguisher is safe and effective for this fire. Ensure that extinguisher is large enough to be safe and effective.

2 Ensure your safety. Turn off electricity if possible. Make sure you have an exit route from the fire. Do not turn your back on a fire.

③ Remove the hose and nozzle. Quickly check the pressure gauge to verify that the extinguisher is adequately charged.

④ Pull the pin to release the extinguisher control valve. You must be within 0.9–5.5 m (3–18 ft.) of the fire to be effective.

⑤ Aim the nozzle at the base of the flames to sweep the flames off the surface starting at the near edge of the flames.

⑥ Overhaul the fire; take steps to prevent rekindling, continue to apply the extinguishing agent to cool the fuel, and summon additional help if needed.

Note: This fire extinguisher is safe to use on Class A, B, and C fires. It is effective on Class B fires, but requires large quantities to be effective on Class A fires.

To apply a dry powder extinguishing agent to an incipient-stage fire involving combustible-metal filings or shavings (Class D fire), follow the steps in **SKILL DRILL 8-8**:

① Size up the fire to determine whether a dry powder fire extinguisher is safe and effective for the metal involved in this fire. Ensure that extinguisher is large enough to

be safe and effective. Avoid the use of water and other extinguishing agents that may react explosively.

② Ensure your safety. Burning metals produce very high temperatures and can burn explosively. Make sure you have an exit route from the fire. Do not turn your back on a fire.

③ Remove the hose and nozzle. Quickly check the pressure gauge to verify that the extinguisher is adequately charged.

④ Pull the pin to release the extinguisher control valve. You must be within 1.8–2.4 m (6–8 ft.) of the fire to be effective.

⑤ Aim the nozzle and fully open the valve to provide a maximum range, and then reduce the valve to produce a soft, heavy flow down to completely cover the burning metal. The method of application may vary depending on the type of metal burning and the extinguishing agent. (**STEP ①**)

⑥ Overhaul the fire; take steps to prevent rekindling, continue to place a thick layer of the extinguishing

---

SKILL DRILL 8-8   Using a Dry Powder Fire Extinguisher (Class D)
(Fire Fighter I, NFPA 5.3.16)

1 Aim the nozzle and fully open the valve to provide a maximum range, and then reduce the valve to produce a soft, heavy flow down to completely cover the burning metal. The method of application may vary depending on the type of metal burning and the extinguishing agent.

2 Overhaul the fire; take steps to prevent rekindling, continue to place a thick layer of the extinguishing agent over the hot metal to form an airtight blanket, allow the hot metal to cool, and summon additional help if needed.

agent over the hot metal to form an airtight blanket, allow the hot metal to cool, and summon additional help if needed. (**STEP ❷**)

Note: This fire extinguisher is safe to use only on Class D fires. It is effective only on the metals for which it is intended.

To apply a wet chemical (Class K) extinguishing agent on a fire in a deep fryer, follow the steps in **SKILL DRILL 8-9** :

❶ Size up the fire to determine whether a wet chemical (Class K) fire extinguisher is safe and effective for this fire. Ensure that extinguisher is large enough to be safe and effective.

❷ Ensure your safety. Make sure you have an exit route from the fire. Do not turn your back on a fire.

❸ Remove the hose and nozzle. Quickly check the pressure gauge to verify that the extinguisher is adequately charged.

❹ Pull the pin to release the extinguisher control valve. You must be within 2.4–3.7 m (8–12 ft.) of the fire to be effective.

❺ Aim the nozzle and discharge the stream of wet chemical drops gently into the surface of the burning liquid at the front or the back of the deep fryer. (**STEP ❶**)

❻ Let the deep foam blanket flow across the surface of the burning liquid. Avoid splashing foam on the burning liquid, as it can cause the burning fuel to splatter. (**STEP ❷**)

❼ Continue to apply the agent until the foam blanket has extinguished all of the flames. (**STEP ❸**)

❽ Do not disturb the foam blanket even after all of the flames have been suppressed. If reignition occurs, repeat these steps.

❾ Overhaul the fire; keep a thick blanket of foam intact over the hot liquid, reapply wet chemical over any hot spots or thin spots, and summon additional help if needed. (**STEP ❹**)

Note: This fire extinguisher is designed to be used on Class K fires—that is, fires in deep fryers and on grills. It is not rated for other classes of fire.

The Care of Fire Extinguishers

Fire extinguishers must be regularly inspected and properly maintained to ensure that they will be available for use in an emergency. Records must be kept to confirm that the required inspections and maintenance have been performed on schedule. The individuals assigned to perform these functions must be properly trained and must always follow the manufacturer's recommendations for inspecting, maintaining, recharging, and testing the equipment. Persons performing maintenance on or recharging portable fire extinguishers should be qualified according to NFPA 10.

■ Inspection

According to NFPA 10, an inspection is a quick check to verify that a fire extinguisher is available and ready for immediate

use. Fire extinguishers on fire apparatus should be inspected as part of the regular equipment checks mandated by your department. Inspection should take place each month. The fire fighter charged with inspecting the extinguishers should perform the following tasks:

- Ensure that all tamper seals are intact.
- Determine fullness by weighing or hefting the extinguisher.
- Examine all parts for signs of physical damage, corrosion, or leakage.
- Check the pressure gauge to confirm that it is in the operable range.
- Ensure that the extinguisher is properly identified by type and rating.
- Check the hose and nozzle for damage or obstruction by foreign objects.
- Check the hydrostatic test date of the extinguisher.

If an inspection reveals any problems, the extinguisher should be removed from service until the required maintenance procedures are performed. Spare extinguishers should be used until the problem is corrected.

The pressure gauge on a stored-pressure extinguisher indicates whether the pressure is sufficient to expel the entire agent. Some stored-pressure extinguishers use compressed air, whereas others use compressed nitrogen. The weight of the extinguisher and the presence of an intact tamper seal should indicate that the unit is full of extinguishing agent.

Cartridge-type extinguishers contain a predetermined quantity of a pressurizing gas that expels the agent. This gas is released only when the cartridge is punctured. A properly charged extinguisher will be full of extinguishing agent and will have a cartridge that has not been punctured.

Self-expelling agents, such as CO_2 and some halogenated agents, do not require a separate gas cartridge. The only way to determine whether a CO_2 extinguisher is properly charged is to weigh it. The proper weight should be indicated on the outside of the extinguisher.

■ Maintenance

The maintenance requirements and intervals for various types of fire extinguishers are outlined in NFPA 10. Maintenance includes an internal inspection as well as any repairs that may be required. These procedures must be performed periodically, depending on the type of extinguisher. An inspection may also reveal the need to perform maintenance procedures.

Maintenance procedures must always be performed by a qualified person. Some procedures can be performed only at a properly licensed facility. The specific qualifications and training requirements are determined by the manufacturer and the jurisdictional authority. Untrained personnel should never be allowed to perform fire extinguisher maintenance.

Common indications that an extinguisher needs maintenance include the following findings:

- The pressure gauge reading is outside the normal range.
- The inspection tag is out of date.

SKILL DRILL 8-9

Using a Wet Chemical (Class K) Fire Extinguisher
(Fire Fighter I, NFPA 5.3.16)

1 Size up the fire. Make sure you have an exit route from the fire. Remove the hose and nozzle. Check the pressure gauge. Pull the pin to release the extinguisher control valve. Aim the nozzle and discharge the stream of wet chemical drops gently onto the surface of the burning liquid at the front or the back of the deep fryer.

2 Let the deep foam blanket flow across the surface of the burning liquid. Avoid splashing foam on the burning liquid.

3 Continue to apply the agent until the foam blanket has extinguished all of the flames.

Courtesy of Tyco (ANSUL)

4 Do not disturb the foam blanket even after all of the flames have been suppressed. If reignition occurs, repeat these steps.

© LiquidLibrary

- The tamper seal is broken, especially in extinguishers with no pressure gauge.
- The extinguisher does not appear to be full of extinguishing agent.
- The hose or nozzle assembly is obstructed.
- There are signs of physical damage, corrosion, or rust.
- There are signs of leakage around the discharge valve or nozzle assembly.

■ Recharging

A fire extinguisher must be recharged after each and every use, even if it was not completely discharged. The only exceptions are nonrechargeable extinguishers, which should be replaced after any use. All performance standards assume that an extinguisher will be fully charged when it has to be used. Immediately after any use, the extinguisher should be taken out of service until it has been properly recharged. This guideline also applies to any extinguisher that leaks or has a pressure gauge reading above or below the proper operating range. Most rechargeable extinguishers must be recharged by qualified personnel.

When an extinguisher is recharged, the extinguishing agent is refilled, and the system that expels the agent is fully charged. Both the quantity of the agent and the pressurization must be verified. After recharging is complete, a tamper seal is installed; this seal provides assurance that the extinguisher has not been fully or even partially discharged since the last time it was recharged.

Typical 9.5-L (2.5-gal.), stored-pressure water extinguishers can be recharged by fire fighters using water and a source of compressed air. Before the extinguisher is refilled, all remaining stored pressure must be discharged so that the valve assembly can be safely removed. Water is then added up to the water-level indicator, and the valve assembly is replaced. Finally, compressed air is introduced to raise the pressure to the level indicated on the gauge.

■ Hydrostatic Testing

Most fire extinguishers are pressurized vessels, meaning that they are designed to hold a steady internal pressure. The ability of an extinguisher to withstand this internal pressure is measured by periodic **hydrostatic testing**. The hydrostatic testing requirements for fire extinguishers are established by Transport Canada and can be found in NFPA 10 (**TABLE 8-3**). Such tests are conducted in a special test facility and involve filling the extinguisher with water and applying above-normal pressure.

Each extinguisher has an assigned maximum interval between hydrostatic tests, usually 5 or 12 years, depending on the construction material and vessel type (**TABLE 8-4**). The date of the most recent hydrostatic test must be indicated on the outside of the extinguisher. An extinguisher may not be refilled if the date of the most recent hydrostatic test is not within the prescribed limit. Any extinguisher that is out of date should be removed from service and sent to the appropriate maintenance facility for hydrostatic testing.

TABLE 8-3 Hydrostatic Test Pressure Requirements

Carbon-dioxide extinguishers Carbon-dioxide and nitrogen cylinders (used with wheeled extinguishers)	5/3 service pressure stamped on cylinder
Carbon-dioxide extinguishers with cylinder specification ICC3	20,685 kPa (3000 psi)
All stored-pressure and bromochlorofluoromethane (Halon 1211)	Factory test pressure not to exceed three times the service pressure
Carbon-dioxide hose assemblies	8619 kPa (1250 psi)
Dry chemical and dry powder hose assemblies	2068 kPa (300 psi) or at service pressure, whichever is higher

TABLE 8-4 Hydrostatic Test Interval for Extinguishers

Extinguisher Type	Test Interval (Years)
Stored-pressure water-loaded steam and/or anti-freeze agent	5
Wetting agent	5
FFFP (film-forming fluoroprotein foam)	5
Dry chemical with stainless-steel shells	5
Carbon dioxide	5
Wet chemical	5
Dry chemical stored-pressure with mild steel shells, brazed brass shells, or aluminum shells	12
Dry chemical cartridge/cylinder-operated with mild steel shells	12
Halogenated agents	12
Dry powder stored-pressure cartridge/cylinder-operated with mild steel shells	12

Chief Concepts

- Fire extinguishers are used successfully to put out hundreds of fires every day, preventing millions of dollars in property damage and saving uncounted lives.
- Fire fighters use fire extinguishers to control small fires that do not require the use of a hose line.
- Portable fire extinguishers have two primary uses: to extinguish incipient-stage fires and to control fires where traditional methods of fire suppression are not recommended.
- The primary disadvantage of fire extinguishers is that they are one-shot devices.
- The five classes of fires affect the choice of extinguishing equipment:
  - Class A fires involve ordinary combustibles such as wood, paper, cloth, rubber, household rubbish, and some plastics.
  - Class B fires involve flammable or combustible liquids, such as gasoline, oil, grease, tar, lacquer, oil-based paints, and some plastics.
  - Class C fires involve energized electrical equipment, which includes any device that uses, produces, or delivers electrical energy.
  - Class D fires involve combustible metals, such as magnesium, titanium, zirconium, sodium, lithium, and potassium. Special techniques and extinguishing agents are required to fight combustible-metal fires.
  - Class K fires involve combustible cooking oils and fats.
- Portable fire extinguishers are classified and rated based on their characteristics and capabilities.
- The classification system for fire extinguishers uses both letters and numbers. The letters indicate the classes of fire for which the extinguisher can be used, and the numbers indicate its effectiveness. Fire extinguishers that are safe and effective for more than one class will be rated with multiple letters. Numbers are used to rate an extinguisher's effectiveness for only Class A and Class B fires.
- The traditional lettering system uses the following labels:
  - Extinguishers suitable for use on Class A fires are identified by the letter A on a solid green triangle. The triangle has a graphic relationship to the letter A.
  - Extinguishers suitable for use on Class B fires are identified by the letter B on a solid red square. The shape of the letter mirrors the graphic shape of the box.
  - Extinguishers suitable for use on Class C fires are identified by the letter C on a solid blue circle, which also incorporates a graphic relationship between the letter C and the circle.
  - Extinguishers suitable for use on Class D fires are identified by the letter D on a solid yellow, five-pointed star.
  - Extinguishers suitable for use on Class K (combustible cooking oil) fires are identified by a pictograph showing a fire in a frying pan.
- The pictograph system uses symbols rather than letters on the labels. Under this system, the presence of an icon indicates that the extinguisher has been rated for that class of fire.
- NFPA 10 lists the requirements for placing and mounting portable fire extinguishers as well as the appropriate mounting heights:
  - Fire extinguishers weighing up to 18.1 kg (40 lb.) should be mounted so that the top of the extinguisher is not more than 1.5 m (5 ft.) above the floor.
  - Fire extinguishers weighing more than 18.1 kg (40 lb.) should be mounted so that the top of the extinguisher is not more than 0.9 m (3 ft.) above the floor.
  - The bottom of an extinguisher should be at least 10 cm (4 in.) above the floor.
- Areas are divided into three risk classifications based on the fire risks associated with the materials in those areas:
  - In light (low) hazard locations, the majority of materials are noncombustible or arranged to limit fire spread, such as in offices.
  - Ordinary (moderate) hazard locations contain more Class A and Class B materials than do light hazard locations, such as auto showrooms and parking garages.
  - Extra (high) hazard locations contain more Class A combustibles and/or Class B flammables than do ordinary hazard locations, such as woodworking shops.
- Most extinguishers stop fires by cooling the fuel below its ignition point, by cutting off the supply of oxygen, or by combining the two techniques.
- Portable fire extinguishers use seven basic types of extinguishing agents:
  - Water—Used to extinguish Class A fires.
  - Dry chemicals—Used mainly to extinguish Class B and C fires. This type of extinguisher can be used to suppress Class A fires, but water is also needed to fully extinguish any embers.
  - Carbon dioxide—Used to extinguish Class B and C fires only.
  - Foam—Used to extinguish Class A or B fires. Be sure to double-check the type of foam in the extinguisher.

Most Class B foams can be used on Class A fires, but Class A foams are not effective on Class B fires.

- Halogenated agents—Include Halons and halocarbons. Halons are limited-use chemicals due to their propensity to damage the environment. These agents are used to extinguish Class B and C fires.
- Dry powders—Chemical compounds used to extinguish Class D fires.
- Wet chemicals—Used to extinguish Class K fires.

■ Most hand-held portable fire extinguishers have six basic parts:
  1. A cylinder or container that holds the extinguishing agent
  2. A carrying handle
  3. A nozzle or horn
  4. A trigger and discharge valve assembly
  5. A locking mechanism to prevent accidental discharge
  6. A pressure indicator or gauge that shows whether a stored-pressure extinguisher has sufficient pressure to operate properly

■ Portable fire extinguishers vary according to their extinguishing agent, capacity, and effective range, along with the time it takes to completely discharge the extinguishing agent. They also have different mechanical designs:
  - Stored-pressure water-type fire extinguishers—Expel water in a solid stream with a range of 10.7–12.2 m (35–40 ft.) through a nozzle at the end of a short hose.
  - Loaded-stream water extinguishers—Discharge a solution of water containing a chemical to prevent freezing.
  - Wetting-agent water-type fire extinguishers—Expel water that contains a solution to reduce its surface tension, which enables the water to penetrate more efficiently into Class A fuels.
  - Pump tank water-type fire extinguishers—Do not store water under pressure, but instead expel it through use of a hand-operated, double-acting, vertical piston pump.
  - Backpack water extinguishers—Nineteen-litre (five-gallon) water tanks used primarily outdoors for fighting brush and grass fires.
  - Ordinary dry chemical extinguishers—Available in hand-held models and larger, wheeled units to extinguish Class B and C fires.
  - Multipurpose dry chemical extinguishers—Available in hand-held models and larger, wheeled units to extinguish Class A, B, and C fires.
  - Carbon-dioxide extinguishers—Include a siphon tube and horn or cone-shaped applicator that is used to direct the flow of the agent. These extinguishers have a relatively short discharge range of 0.9–2.4 m (3–8 ft.).

- Class B foam extinguishers—Very similar in appearance and operation to water extinguishers, except these extinguishers have an aspirating nozzle.
- Wet chemical extinguishers—Available in two sizes, 6 L (1.5 gal.) and 9.5 L (2.5 gal.).
- Halogenated agent extinguishers—Available in handheld extinguishers for Class B and C fires and larger-capacity models for use on Class A fires.
- Dry powder fire extinguishers—Rely on sodium chloride–based agents. Models with 13.6-kg (30-lb.) capacity are available in either stored-pressure or cartridge/cylinder-operated versions. Wheeled models are available with 68- and 158.8-kg (150- and 350-lb.) capacities.

■ There are six basic steps in extinguishing a fire with a portable fire extinguisher:
  - Locate the fire extinguisher.
  - Select the proper classification of extinguisher.
  - Transport the extinguisher to the location of the fire.
  - Activate the extinguisher to release the extinguishing agent.
  - Apply the extinguishing agent to the fire for maximum effect.
  - Ensure your personal safety by having an exit route.

■ Activating a fire extinguisher to apply the extinguishing agent is a single operation that involves four steps. The PASS acronym is a helpful way to remember these steps:
  - Pull the safety pin.
  - Aim the nozzle at the base of the flames.
  - Squeeze the trigger to discharge the agent.
  - Sweep the nozzle across the base of the flames.

■ When using a fire extinguisher, always approach the fire with an exit behind you. If you are outside, make sure the wind is at your back.

■ Fire extinguishers must be regularly inspected and properly maintained to ensure that they will be available for use in an emergency.

■ The fire fighter charged with inspecting the extinguishers should perform the following tasks:
  - Ensure that all tamper seals are intact.
  - Determine fullness by weighing or hefting the extinguisher.
  - Examine all parts for signs of physical damage, corrosion, or leakage.
  - Check the pressure gauge to confirm that it is in the operable range.
  - Ensure that the extinguisher is properly identified by type and rating.
  - Shake dry chemical extinguishers to mix or redistribute the agent.
  - Check the hose and nozzle for damage or obstruction by foreign objects.
  - Check the hydrostatic test date of extinguisher.

- Common indications that an extinguisher needs maintenance include the following findings:
  - The pressure gauge reading is outside the normal range.
  - The inspection tag is out of date.
  - The tamper seal is broken, especially in extinguishers with no pressure gauge.
  - The extinguisher does not appear to be full of extinguishing agent.
  - The hose or nozzle assembly is obstructed.
  - There are signs of physical damage, corrosion, or rust.
  - There are signs of leakage around the discharge valve or nozzle assembly.

- A fire extinguisher must be recharged after each and every use, even if it was not completely discharged. The only exceptions are nonrechargeable extinguishers, which should be replaced after any use.
- Most fire extinguishers are pressurized vessels, meaning that they are designed to hold a steady internal pressure. The ability of an extinguisher to withstand this internal pressure is measured by periodic hydrostatic testing. These tests, which are conducted in a special test facility, involve filling the extinguisher with water and applying above-normal pressure.

Wrap-Up, continued

Hot Terms

Ammonium phosphate An extinguishing agent used in dry chemical fire extinguishers that can be used on Class A, B, and C fires.

Carbon-dioxide (CO$_2$) fire extinguisher A fire extinguisher that uses CO$_2$ gas as the extinguishing agent. It is rated for use on Class B and C fires.

Cartridge/cylinder-operated fire extinguisher A fire extinguisher that has the expellant gas in a separate container from the extinguishing agent storage container. The storage container is pressurized by a mechanical action that releases the expellant gas.

Class A fire A fire in ordinary combustible materials such as wood, cloth, paper, rubber, and many plastics. (NFPA 10)

Class B fire A fire in flammable liquids, combustible liquids, petroleum greases, tars, oils, oil-based paints, solvents, lacquers, alcohols, and flammable gases. (NFPA 10, 2007)

Class C fire A fire that involves energized electrical equipment. (NFPA 10)

Class D fire A fire in combustible metals, such as magnesium, titanium, zirconium, sodium, lithium, and potassium. (NFPA 10)

Class K fire A fire in a cooking appliance that involves combustible cooking media (vegetable or animal oils and fats). (NFPA 10)

Clean agent Electrically nonconducting, volatile, or gaseous fire extinguishant that does not leave a residue upon evaporation. (NFPA 2001)

Cylinder The body of the fire extinguisher where the extinguishing agent is stored.

Dry chemical fire extinguisher An extinguisher that uses a mixture of finely divided solid particles to extinguish fires. The agent is usually sodium bicarbonate, potassium bicarbonate, or ammonium phosphate based, with additives being included to provide resistance to packing and moisture absorption and to promote proper flow characteristics. These extinguishers are rated for use on Class B and C fires, although some are also rated for Class A fires.

Dry powder extinguishing agent An extinguishing agent used in putting out Class D fires. Common examples include sodium chloride and graphite-based powders.

Dry powder fire extinguisher A fire extinguisher that uses an extinguishing agent in powder or granular form. It is designed to extinguish Class D combustible-metal fires by crusting, smothering, or heat-transferring means.

Extinguishing agent A material used to stop the combustion process. Extinguishing agents may include liquids, gases, dry chemical compounds, and dry powder compounds.

Extra (high) hazard locations Occupancies where the total amounts of Class A combustibles and Class B flammables are greater than expected in occupancies classed as ordinary (moderate) hazards.

Film-forming fluoroprotein (FFFP) foam A protein-based foam concentrate incorporating fluorinated surfactants that forms a foam capable of producing a vapour-suppressing, aqueous film on the surface of hydrocarbon fuels. (NFPA 403)

Fire load The weight of combustibles in a fire area (measured in sq. m or sq. ft) or on a floor in buildings and structures, including either contents or building parts, or both. (NFPA 914)

Halogenated agent fire extinguisher An extinguisher that uses a halogenated extinguishing agent.

Halogenated extinguishing agent A liquefied gas extinguishing agent that puts out fires by chemically interrupting the combustion reaction between the fuel and oxygen.

Halon 1211 A halogenated agent whose chemical name is bromochlorodifluoromethane, CBrClF$_2$; it is a multipurpose, Class ABC-rated agent effective against flammable liquid fires. (NFPA 408)

Handle The grip used for holding and carrying a portable fire extinguisher.

Horn The tapered discharge nozzle of a CO$_2$–type fire extinguisher.

Hydrostatic testing Pressure testing of a fire extinguisher to verify its strength against unwanted rupture. (NFPA 10)

Ignition point The minimum temperature at which a substance will burn.

Incipient stage The initial or beginning stage of a fire, in which it can be controlled or extinguished by portable extinguishers or small amounts of dry extinguishing agents, without the need for protective clothing or breathing apparatus. (NFPA 484)

Light (low) hazard locations Occupancies where the total amount of combustible materials is less than expected in an ordinary hazard location.

Loaded-stream fire extinguisher A water-based fire extinguisher that uses an alkali metal salt as a freezing-point depressant.

Locking mechanism A device that locks an extinguisher's trigger to prevent its accidental discharge.

Multipurpose dry chemical fire extinguisher A fire extinguisher rated to fight Class A, B, and C fires.

Nozzle A constricting appliance attached to the end of a fire hose or monitor to increase the water velocity and form a stream. (NFPA 1965)

Ordinary (moderate) hazard locations Occupancies that contain more Class A and Class B materials than are found in light hazard locations.

PASS Acronym for the steps involved in operating a portable fire extinguisher: pull pin, aim nozzle, squeeze trigger, sweep across burning fuel.

Polar solvent A water-soluble flammable liquid, such as alcohol, acetone, ester, and ketone.

Pressure indicator A gauge on a pressurized portable fire extinguisher that indicates the internal pressure of the expellant.

Pump tank fire extinguisher A nonpressurized, manually operated water extinguisher that is rated for use on Class A fires. Discharge pressure is provided by a hand-operated, double-acting piston pump.

Rapid oxidation A chemical process that occurs when a fuel is combined with oxygen, resulting in the formation of ash or other waste products and the release of energy as heat and light.

Saponification The process of converting the fatty acids in cooking oils or fats to soap or foam; the action caused by a Class K fire extinguisher.

Self-expelling agents The agents in a fire extinguisher that have sufficient vapour pressure at normal operating temperatures to expel themselves. (NFPA 10)

Stored-pressure fire extinguisher A fire extinguisher in which both the extinguishing material and the expellant gas are kept in a single container and that includes a pressure indicator or gauge. (NFPA 10)

Stored-pressure water-type fire extinguisher A fire extinguisher in which water or a water-based extinguishing agent is stored under pressure.

Tamper seal A retaining device that breaks when the locking mechanism is released.

Trigger The button or lever used to discharge the agent from a portable fire extinguisher.

Underwriters Laboratories of Canada (ULC) The Canadian organization that tests and certifies that fire extinguishers (among many other products) meet established standards. The United States equivalent is Underwriters Laboratories, Inc.

Wet chemical extinguishing agent An extinguishing agent for Class K fires. It commonly consists of solutions of water and potassium acetate, potassium carbonate, potassium citrate, or any combination thereof.

Wet chemical fire extinguisher A fire extinguisher for use on Class K fires that contains a wet chemical extinguishing agent.

Wetting-agent water-type fire extinguisher An extinguisher that expels water combined with a chemical or chemicals to reduce its surface tension; it is used for Class A fires.

Wheeled fire extinguisher A portable fire extinguisher equipped with a carriage and wheels, which is intended to be transported to the fire by one person. (NFPA 10)

You and your crew are at a local high school giving a tour of the engine. You get to the compartment that has three fire extinguishers in it. You explain that the first is a wetting-agent water-type extinguisher, the second is a dry chemical extinguisher, and the third is a CO_2 extinguisher. One of the students asks if there are other types of fire extinguishers and you respond that there are. The teacher then tells you that they have been studying the chemistry of fire over the past two weeks and asks you to explain each type of fire extinguisher and how each works to extinguish a fire. The other crew members chuckle to themselves as they watch you get tested over fire extinguishers.

1. Which class of fire involves combustible cooking media such as vegetable oils, animal oils, and fats?
   A. Class A
   B. Class B
   C. Class J
   D. Class K

2. Which extinguisher would you use on a Class C fire?
   A. Dry powder
   B. FFFP foam
   C. Halon 1211
   D. Wet chemical

3. Which type of extinguisher can typically be used to extinguish fires in ordinary combustibles, combustible liquids, and charged electrical equipment?
   A. Wet chemical fire extinguisher
   B. Multipurpose dry chemical
   C. Wetting-agent water extinguisher
   D. CO_2

4. What does the acronym *PASS* stand for?
   A. Press, aim, squeeze, sweep
   B. Pull, aim, squeeze, sweep
   C. Peek, aim, squeeze, sweep
   D. Poke, aim, squeeze, sweep

5. Hydrostatic testing confirms the extinguisher _____.
   A. has sufficient strength to withstand the internal pressures
   B. will properly expel the propellant when activated
   C. can be used on electrical fires
   D. contains a water-based extinguishing agent

6. Which of the following is a polar solvent?
   A. Alcohol
   B. Gasoline
   C. Diesel fuel
   D. Water

7. What is the term for the retaining device that breaks when the locking mechanism is released?
   A. Handle
   B. Trigger
   C. Tamper seal
   D. Horn

You are in your local fabric store with your wife when the store owner recognizes you. He asks if you work for the fire department, and you respond proudly that you do. Your inner pride quickly fades when the owner begins to blame you for the demise of his business. He begins to rant about all of the needless rules and regulations that he must comply with as a business owner. He points to a man replacing the fire extinguishers and exclaims, "In my 35 years of doing business, I have never had a fire, but every year I have to pay a company to maintain those extinguishers. You guys nickel and dime me to death for something that isn't even needed."

1. How do you calm the irate owner?
2. Which points of his comments would you attempt to correct and which ones would you simply let go?
3. How do you explain the value of fire extinguishers?
4. What would you recommend to the fire chief as a result of this encounter?

Fire Fighter Tools and Equipment

Fire Fighter I

Knowledge Objectives

After studying this chapter, you will be able to:

- Describe the general purposes of tools and equipment. (NFPA 5.5.1.B , p 232–233)
- Describe the safety considerations for the use of tools and equipment. (NFPA 5.5.1.B , p 232)
- Describe why it is important to use tools and equipment effectively. (NFPA 5.5.1.B , p 232–233)
- Describe why it is important for you to know where tools are stored. (NFPA 5.5.1.B , p 233)
- List and describe tools and equipment that are used for rotating. (p 233–235)
- List and describe tools and equipment that are used for pushing or pulling. (p 235–236)
- List and describe tools and equipment that are used for prying or spreading. (p 236–238)
- List and describe tools and equipment that are used for striking. (p 238–240)
- List and describe tools and equipment that are used for cutting. (p 240–243)

- Describe the tools used in response and scene size-up activities. (p 243)
- Describe the tools used in a forcible entry. (NFPA 5.3.4 , p 244–246)
- Describe the tools used during an interior attack. (NFPA 5.3.8 , p 245–246)
- Describe ventilation tools. (NFPA 5.3.11 , p 245–246)
- Describe the hand tools needed during an overhaul assignment. (NFPA 5.3.13 , p 246)
- Explain how tools and equipment are staged for rapid access. (NFPA 5.5.1.B , p 247)
- Describe the importance of properly maintaining tools and equipment. (NFPA 5.5.1, 5.5.1.A , p 247–248)
- Describe how to clean and inspect hand tools. (NFPA 5.5.1, 5.5.1.A, 5.5.1.B , p 248)

Skills Objectives

There are no skills objectives for Fire Fighter I candidates. NFPA 1001 contains no Fire Fighter I Job Performance Requirements for this chapter.

© Photos.com

Fire Fighter II — FFII

Knowledge Objectives

After studying this chapter, you will be able to:

- Describe the tools used in search and rescue operations. (NFPA 6.4.2.A, 6.4.2.B , p 245)
- Explain how tools and equipment are staged for rapid access. (NFPA 6.5.4.B , p 247)
- Describe how to maintain power equipment and power tools. (NFPA 6.5.4, 6.5.4.A , p 247–248)

Skills Objectives

There are no skills objectives for Fire Fighter II candidates. NFPA 1001 contains no Fire Fighter II Job Performance Requirements for this chapter.

You Are the Fire Fighter

You are excited about getting to travel to another station to work on an engine company. Ever since you were in recruit class, engine work has appealed to you. You have seen the engine crew dismount the engine with their tools in hand, heading to the structure to force open a door. You are pumped knowing that today that fire fighter could be you.

It isn't long before the alarm sounds, "Person trapped, motor vehicle collision, 4th and Elm," directs the dispatcher. You slip into your turnout gear, take your seat, and put on your seat belt. As the engine pulls from the station, you mull over how this changes your plan for what you would be doing today.

1. What is considered engine work?
2. What kinds of tools will you need for this job?
3. What other kinds of knowledge will you need?

Introduction

Fire fighters use tools and equipment to perform a wide range of activities. A fire fighter must know how to use tools effectively, efficiently, and safely, even when it is dark or visibility is limited. This chapter provides an overview of the general functions of the most commonly used tools and equipment and discusses how they are used during fire suppression and rescue operations. As you will see, the same tools may be used in different ways during each phase of fire suppression or rescue operations. This chapter also explains how to maintain tools and equipment so that they will always be ready for emergency use.

General Considerations

Tools and equipment are used in almost all fire suppression and rescue operations. As you progress through your training, you will learn how to operate the different types of tools and equipment used by your department.

Hand tools are used to extend or multiply the actions of your body and to increase your effectiveness in performing specific functions. Most of these tools operate using simple machine principles: A pike pole extends your reach, allows you to penetrate through a ceiling, and enables you to apply force to pull down ceiling material; an axe multiplies the cutting force you can exert on a given area. By contrast, power tools and equipment use an external source of power, such as an electric motor or an internal combustion engine. In certain cases, they are faster and more efficient than hand tools.

■ Safety

Safety is a prime consideration when using any kind of tool or equipment. You need to operate the equipment so that you, your fellow fire fighters, victims, and bystanders are not accidentally injured. Personal safety also requires the use of the proper personal protective equipment (PPE):

- Approved helmet
- Firefighting hood
- Eye protection
- Face shield
- Firefighting gloves
- Turnout coat
- Bunker pants
- Boots
- Self-contained breathing apparatus (SCBA)
- Personal alert safety system (PASS)

■ Conditions of Use/Operating Conditions

The best way to learn how to use tools and equipment properly is under optimal conditions of visibility and safety. In the beginning, you must be able to see what you are doing and practise without endangering yourself and others. As you become more proficient, you should practise using tools and equipment under more difficult working conditions. Eventually, you should be able to use tools and equipment safely and effectively even when darkness or smoke decreases visibility. You must be able to work safely in hazardous areas, or from a ladder or pitched roof, where you are surrounded by noise and other activities, while wearing all of your protective clothing and using your SCBA. For this reason, some departments require fire fighters to practise certain skills and evolutions in total darkness or with their face masks covered to simulate the darkness of actual fires.

■ Effective Use

Effective, efficient use of tools and equipment means using the least amount of energy to accomplish the task. Being effective means that you achieve the desired goal; being efficient means that you produce the desired effect without wasting time or energy. When you are assigned a task on the fire ground, your objective is to complete that task safely and quickly. If you waste energy by working inefficiently, you will not be able to perform additional tasks. If you know which tools and equipment are needed for each phase of firefighting, you will be able to achieve the desired objective quickly and have the energy needed to complete the remaining tasks.

New fire fighters are often surprised by the strength and energy required to perform many tasks. An aggressive, continuous program of physical fitness will enable you to maintain your body in the optimal state of readiness. It does little good to practise using a tool for hours if you are not in good physical condition.

As your training continues, you will learn which tools and equipment are used during different phases of fire-ground operations; for example, the tools needed for forcible entry differ from the tools needed for overhaul. Knowing which tools are needed for the work that must be done will help you prepare for the different tasks that unfold on the scene of a fire. The specific steps for properly using tools and equipment are presented in this chapter and throughout this text; your fire department instructors will demonstrate them as well.

Most fire departments have suggested operating guidelines (SOGs), or standard operating procedures (SOPs), that specify which tools and equipment are needed in various situations. As a fire fighter, you must know where every tool and piece of equipment is carried on your apparatus. Knowing how to use a piece of equipment does you no good if you cannot find it quickly. Your company officer and your department's SOGs/SOPs will assist you in knowing which tools to bring to different situations.

Some fire fighters carry a selection of small tools and equipment in the pockets of their turnout coats or bunker pants. Check whether your department requires you to carry certain tools and equipment at all times. Ask senior fire fighters for recommendations about which tools and equipment you should carry.

Functions

An engine or truck company carries a number of tools and different types of equipment. Often, the easiest way to learn and remember these tools is to group them by the function that each performs TABLE 9-1 and TABLE 9-2 . Most of the tools carried by fire departments fit into the following functional categories:

- Rotating (assembly or disassembly)
- Prying or spreading
- Pushing or pulling
- Striking
- Cutting
- Multiple use

■ Rotating Tools

Rotating tools apply a rotational force to make something turn. The most commonly used rotating tools are screwdrivers, wrenches, and pliers, which are used to assemble (fit together)

FIGURE 9-1 Rotate, assemble, or disassemble.

FIGURE 9-2 Pry or spread.

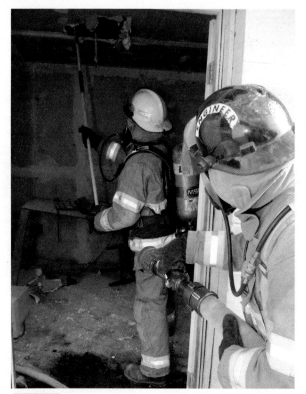
FIGURE 9-3 Push or pull.

TABLE 9-1	General Tool Functions

- Rotate, assemble, or disassemble **FIGURE 9-1**
- Pry or spread **FIGURE 9-2**
- Push or pull **FIGURE 9-3**
- Strike **FIGURE 9-4**
- Cut **FIGURE 9-5**
- A combination of functions **FIGURE 9-6**

FIGURE 9-4 Strike.

FIGURE 9-5 Cut.

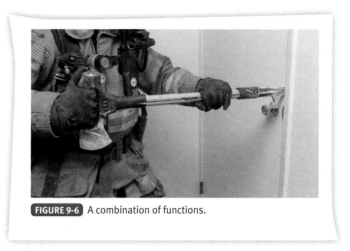

FIGURE 9-6 A combination of functions.

TABLE 9-2	Tool Functions
Function	**Tool**
Rotate	Box-end wrench Gripping pliers Hydrant wrench Open-end wrench Pipe wrench Screwdriver Socket wrench Spanner wrench
Pry or spread	Claw bar Crowbar Flat bar Halligan tool Hux bar Hydraulic spreader Kelly tool Pry bar Rabbet tool
Push or pull	Ceiling hook Clemens hook Drywall hook K tool Multipurpose hook Pike pole Plaster hook Roofman's hook San Francisco hook
Strike	Battering ram Chisel Flat-head axe Hammer Mallet Maul Pick-head axe Sledgehammer Spring-loaded centre punch
Cut	Axe Bolt cutter Chainsaw Cutting torch Hacksaw Handsaw Hydraulic shears Reciprocating saw Rotary saw Seat belt cutter

or disassemble (take apart) parts that are connected with threaded fasteners TABLE 9-3 .

Assembling and disassembling are basic mechanical skills that are routinely employed by fire fighters to solve problems. Most fire apparatus carries a tool kit with a selection of screwdrivers with different heads, open-end wrenches, box wrenches, socket wrenches, pliers, adjustable wrenches, and pipe wrenches.

TABLE 9-3	Tools for Assembly and Disassembly

- Box-end wrenches
- Pipe wrenches
- Gripping pliers
- Screwdrivers
- Hydrant wrenches
- Socket wrenches
- Open-end wrenches
- Spanner wrenches

Various sizes and types of screw heads are available, including slotted head, Phillips head, Roberts head, and others. A screwdriver with interchangeable heads is sometimes more useful than a selection of different screwdrivers. Pliers and wrenches also come in a variety of shapes and sizes for different applications.

A spanner wrench is used to tighten or loosen fire hose **couplings** (one set or pair of connection devices that allow a fire hose to be interconnected with additional lengths of hose). If the couplings are not too tight, they can be loosened by hand, without the use of a spanner wrench. A **hydrant wrench** is used to open or close a hydrant by rotating the valve stem, and to remove the caps from the hydrant outlets.

Examples of rotating tools include the following items **FIGURE 9-7** :

- **Box-end wrench**—A hand tool with a closed end that is used to tighten or loosen bolts. Some styles of box-end wrenches have ratchets for easier use.
- **Gripping pliers**—A hand tool with a pincer like working end that can also be used to bend wire or hold smaller objects.
- Hydrant wrench—A hand tool used to operate the valves on a hydrant. Some versions are plain wrenches, whereas others have a ratchet (a mechanism that will rotate only one way) feature **FIGURE 9-8** .
- **Open-end wrench**—A hand tool with an open end that is used to tighten or loosen bolts.
- Combination wrench—A hand tool with an open-end wrench on one end and a box-end wrench on the other.

FIGURE 9-8 A hydrant wrench in action.

- **Pipe wrench**—A wrench with one fixed grip and one movable grip that can be adjusted to fit securely around pipes and other tubular objects.
- **Screwdriver**—A tool that is used to turn screws.
- **Socket wrench**—A wrench that fits over a nut or bolt and uses the action of an attached handle to tighten or loosen the nut or bolt.
- **Spanner wrench**—A special wrench used to tighten or loosen hose couplings. Spanners come in several sizes.

Fire Fighter Safety Tips © Photos.com

When you are learning to operate tools and equipment, use them in areas without hazards and with adequate light. Gradually increase the amount of PPE worn and decrease the light to the levels typically found on the fire ground. Eventually, you should be so familiar with your tools and equipment that you can use them under adverse conditions while wearing full PPE.

Pushing/Pulling Tools

Tools used for pushing or pulling can extend the reach of the fire fighter as well as increase the power the fire fighter can

FIGURE 9-7 Hydrant wrench, spanner wrench, pipe wrench, open-end wrench, box-end wrench, gripping pliers, screwdriver, and socket wrench.

exert upon an object TABLE 9-4. Pushing and pulling tools have many different uses in fire department operations.

An example of a tool that extends the fire fighter's reach is a <u>pike pole</u> FIGURE 9-9. This wood or fibreglass pole has a metal head attached to one end. It is used primarily to pull down a ceiling in an effort to get to the seat of a fire burning above. The metal head has a sharpened point that can be punched through the ceiling and a hook that can grab and pull the ceiling down. The proper use of a pike pole is discussed in depth in the chapter "Salvage and Overhaul".

Pike poles come in several different sizes and with a variety of heads FIGURE 9-10. The most common length of 1.2–1.8 metres (4–6 feet) enables a fire fighter to stand on a floor and pull down a 3-m-high (10-ft.) ceiling. **Closet hooks**, which are intended for use in tight spaces, are commonly 0.6–1.2 m (2–4 ft.) long. Some pike poles are equipped with handles as long as 3.7–4.3 m (12–14 ft.) for use in rooms with

TABLE 9-4	Tools for Pushing and Pulling

- Ceiling hook
- Multipurpose hook
- Roofman's hook
- Clemens hook
- Pike pole
- San Francisco hook
- Drywall hook
- Plaster hook
- K tool

FIGURE 9-9 A pike pole in action.

FIGURE 9-10 Pike poles come in several different sizes.

very high ceilings; others may have D-type handle for better pulling power. It is important to bring the right-sized pike pole to a fire. It the pike pole is too short, you will not be able to reach the ceiling. If it is too long, you will not be able to use it in a room with a low ceiling.

The different head designs of pike poles are intended for different types of ceilings and come in a variety of configurations. Many fire departments use one type of pike pole for plaster ceilings and another type for drywall ceilings.

Commonly encountered types of pulling poles and hooks include the following:

- **Ceiling hook**—A tool consisting of a long wood or fibreglass pole and a metal point with a spur at right angles that can be used to probe ceilings and pull down plaster–lath material.
- **Clemens hook**—A multipurpose tool that can be used for forcible entry and ventilation applications because of its unique head design.
- **Drywall hook**—A specialized version of a pike pole designed to remove drywall.
- **Multipurpose hook**—A long pole with a wooden or fibreglass handle and a metal hook.
- **Plaster hook**—A long pole with a pointed head and two retractable cutting blades on the side.
- **Roofman's hook**—A long pole with a solid metal hook.
- **San Francisco hook**—A multipurpose tool used for forcible entry and ventilation applications. It includes a built-in gas shut-off and directional slot.

A <u>K tool</u> is another type of pushing or pulling tool FIGURE 9-11. It is used to pull the lock cylinder out of a door, exposing the locking mechanism so it can be unlocked easily.

■ Prying/Spreading Tools

Tools used for prying or spreading may be as simple as a pry bar or as mechanically complex as a hydraulic spreader TABLE 9-5. They come in several versions that are designed for different applications.

FIGURE 9-11 Components of a K tool.
Courtesy of Donald M. Colarusso, AllHandsFire.com

TABLE 9-5	Tools for Prying and Spreading

- Claw bar
- Halligan tool
- Kelly tool
- Crowbar
- Hux bar
- Pry bar
- Flat bar
- Hydraulic spreader
- Rabbet tool

A simple pry bar consists of a hardened steel rod with a tapered end that can be inserted into a small area. The bar acts as a lever to multiply the force that a person can exert to bend or pry objects apart. A properly positioned pry bar can apply an enormous amount of force.

Fire departments use a wide range of prying tools in different sizes and with different features. One of the most popular is a Halligan tool, which was designed by a New York City fire fighter. This tool incorporates a sharp pick, a flat prying surface, and a forked claw. It can be used for forcible entry applications, as can several other prying tools. The specific uses and applications of these tools are covered in the chapters on the various phases of fire suppression.

The following tools are used for prying **FIGURE 9-12**:

- **Claw bar**—A tool with a pointed claw-hook on one end and a forked- or flat-chisel pry on the other end that can be used for forcible entry.
- **Crowbar**—A straight bar made of steel or iron with a forked chisel on the working end.
- **Flat bar**—A specialized prying tool made of flat steel with prying ends suitable for performing forced entry.
- **Halligan tool**—A prying tool designed for use in the fire service that also incorporates a pick, an adze, and a fork; sometimes known as a Hooligan tool. The difference is that a Halligan tool is a forged piece of steel.
- **Hux bar**—A multipurpose tool that can be used for forcible entry and ventilation applications because of its unique design. A Hux bar also can be used as a hydrant wrench.

- **Kelly tool**—A steel bar with two main features: a large pick and a large chisel or fork.
- **Pry bar**—A specialized prying tool made of a hardened steel rod with a tapered end that can be inserted into a small area **FIGURE 9-13**.

Hydraulic spreaders are an example of machine-powered rescue tools that can be used for prying and spreading **FIGURE 9-14**. The use of hydraulic power enables you to

FIGURE 9-13 A pry bar in action.

FIGURE 9-12 Tools used for prying and spreading.
© 2003, Berta A. Daniels

FIGURE 9-14 The components of a hydraulic rescue tool.

apply several tons of force on a very small area. You must have special training to operate these machines safely, however. Fire and rescue departments most commonly use this equipment for extrication of victims from motor vehicles and machinery **FIGURE 9-15**.

Hand-powered hydraulic tools are also used for prying and spreading. One of these, the **rabbet tool**, is designed to quickly open doors **FIGURE 9-16**.

■ Striking Tools

Striking tools are used to apply an impact force to an object **TABLE 9-6**. They are often employed to gain entrance to a

building or a vehicle or to make an opening in a wall or roof. This equipment can also be used to force the end of a prying tool into a small opening. Specific use of these tools is covered in the chapter "Forcible Entry."

Striking tools include the following items **FIGURE 9-17**:

- **Hammer**—A hand tool constructed of solid material with a long handle and a head affixed to the top of the handle, with one side of the head used for striking and the other side used for prying.
- **Mallet**—A short-handled hammer with a round head.
- **Sledgehammer**—A long, heavy hammer that requires the use of both hands.
- **Maul**—A specialized striking tool (weighing 2.7 kilograms [6 pounds] or more) with an axe on one end and a sledgehammer on the other end.
- **Flat-head axe**—A tool with a head that has an axe blade on one end and a flat head that can be used for striking objects on the opposite end **FIGURE 9-18A**.

TABLE 9-6	Striking Tools
• Battering ram • Hammer • Pick-head axe • Chisel • Mallet • Sledgehammer • Flat-head axe • Maul • Spring-loaded centre punch	

FIGURE 9-15 A hydraulic spreader is often used for extricating victims from vehicles.

FIGURE 9-16 A rabbet tool can open doors quickly.

FIGURE 9-17 Striking tools (from top): hammer, mallet, maul, sledgehammer.

A.　　　　　　　　　　　　　　　　　　**B.**

FIGURE 9-18 Two types of axes: **A.** Flat-head axe. **B.** Pick-head axe.

- **Pick-head axe**—A tool with a head that has an axe blade on one end and a pointed pick on the opposite end **FIGURE 9-18B**.
- **Battering ram**—A heavy metal bar used to break down doors **FIGURE 9-19**.
- **Chisel**—A metal tool with one sharpened end that can be used to break apart material when used in conjunction with a hammer, mallet, or sledgehammer **FIGURE 9-20**.
- **Spring-loaded centre punch**: A spring-loaded punch that is used to break tempered automobile glass.

Among the most frequently used tools in the fire service are axes, including both pick-head axes and flat-head axes. Both types of axes have a wide cutting blade that can be used to chop into a wall, roof, or door. A pick-head axe has a point or pick that can be used for puncturing, pulling, and prying **FIGURE 9-21**. A flat-head axe can be used as a striking tool for forcible entry, usually in combination with a prying tool, such as a Halligan tool. Together, the flat-head axe and the Halligan tool are sometimes referred to as *the irons*; this combination is very effective in most forcible entry situations **FIGURE 9-22**. Forcible entry is covered in depth in the chapter " Forcible Entry."

FIGURE 9-19 Using a battering ram.

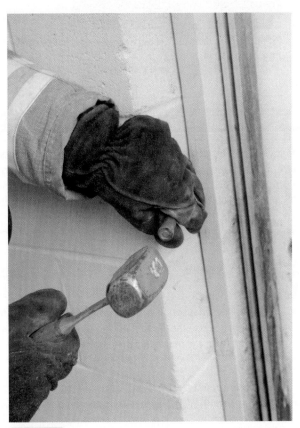

FIGURE 9-20 Using a chisel with a hammer.

FIGURE 9-21 A pick-head axe can be used to pry up boards.

FIGURE 9-22 A flat-head axe can be used with a Halligan tool to force open a door.

FIRE FIGHTER Tips

Many fire departments carry a striking tool and a prying bar strapped together, a combination sometimes referred to as *the irons*. A flat-head axe and a Halligan tool are carried together on many vehicles and are taken into a fire building by one of the crew members.

Fire Fighter Safety Tips

© Photos.com

Some tools require a considerable amount of movement and room to operate. Always confirm that no one is in danger of being injured before you use these tools. Look around and make sure you can operate an axe or sledgehammer safely and effectively.

The spring-loaded centre punch is a striking tool used primarily on cars **FIGURE 9-23**. It can exert a large amount of force on a pinpoint-sized portion of tempered automobile glass. This action disrupts the integrity of tempered glass and causes the window to shatter into small, uniform-sized pieces. A spring-loaded centre punch is often used in vehicular

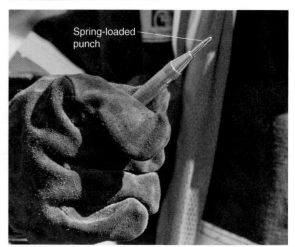

Spring-loaded punch

FIGURE 9-23 A spring-loaded centre punch can be used to break a car window safely.

crashes to gain access to a victim who needs care. Vehicle extrications are covered in depth in the chapter "Vehicle Rescue and Extrication."

■ Cutting Tools

Cutting tools have sharp edges that sever objects. They come in several forms and are used to cut a wide variety of substances **FIGURE 9-24**. Cutting tools used by fire fighters range from knives and wire cutters that can be carried in the pockets of turnout coats to **seat belt cutters**, bolt cutters (a scissor-like tool used to cut through items such as chains or padlocks), saws, cutting torches (a torch that produces a high-temperature flame capable of melting metal), and hydraulic shears **TABLE 9-7**. Each of these tools is designed to work on certain types of materials. Fire fighters can be injured and cutting tools can be ruined if the tools are used incorrectly.

Bolt cutters are most often used to cut through chains or padlocks to open doors or gates. By concentrating the cutting force on a small area, it is possible to break through many chains in just a few seconds.

Fire departments often carry several different types of saws. This equipment can be classified into two main categories, based on the power source. **Handsaws** are manually powered, whereas **mechanical saws** are usually powered by electric motors or gasoline-powered engines. Handsaws include hacksaws, carpenter's handsaws, keyhole saws, and coping saws.

Hacksaws are designed to cut metal **FIGURE 9-25**. Different blades can be used, depending on the type of metal being cut. Hacksaws are useful when metal needs to be cut under closely controlled conditions.

Carpenter's handsaws are designed for cutting wood. Saws with large teeth are effective in cutting large timber or tree branches; they are often useful at motor vehicle crashes where tree limbs may hamper the rescue effort. Saws with finer teeth are designed for cutting finished lumber. A **coping saw** is used to cut curves in wood; it consists of a handsaw with a narrow blade set between the ends of a U-shaped frame.

FIGURE 9-24 Cutting tools. **A.** Combination tool. **B.** Seat belt cutter. **C.** Bolt cutters. **D.** Handsaws.

TABLE 9-7	Cutting Tools

- Axes
- Bolt cutters
- Chainsaws
- Cutting torches
- Hacksaws
- Handsaws
- Hydraulic shears
- Reciprocating saws
- Rotary saws
- Seat belt cutters

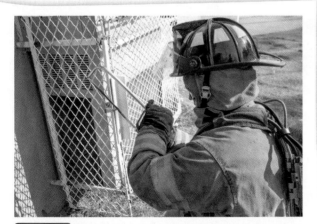

FIGURE 9-25 A fire fighter using a hacksaw to cut through metal.

A **keyhole saw**, a specialty saw, is narrow and slender and can be used to cut keyholes in wood.

Three primary types of mechanical saws are chainsaws, rotary saws, and reciprocating saws. Although handsaws have a valuable role in firefighting operations, power saws have the advantage of accomplishing more work in a shorter period of time. They also enable fire fighters to conserve energy, resulting in less fatigue. Because mechanical saws are powerful, they must be used only by trained operators. Nevertheless, they offer some distinct disadvantages relative to handsaws. Specifically, power saws are heavy to carry and sometimes can be difficult to start. They may also require an electrical connection, although cordless models are becoming more widely available.

Most people are familiar with the gasoline-powered or electric **chainsaws** commonly used to cut wood, particularly trees. Fire fighters often use saws with special chains to cut ventilation openings in roofs constructed of wood, metal, tar, gravel, or insulating materials **FIGURE 9-26** .

Rotary saws may be powered by either electric motors or gasoline engines **FIGURE 9-27** . In some rotary saws, the cutting part of the saw is a round metal blade with teeth, with different blades being used depending on the type of material being cut. Other rotary saws use a flat, abrasive disc for cutting. The discs are made of composite materials and are designed to wear down as they are used. It is important to match the appropriate saw blade or saw disc to the material being cut.

FIGURE 9-26 A fire fighter can use a chainsaw to cut through a roof.

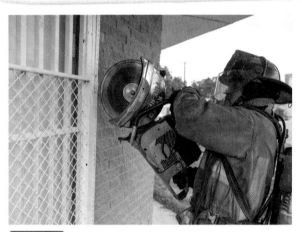

FIGURE 9-27 A fire fighter using a rotary saw.

Reciprocating saws are powered by either an electric motor or a battery motor that rapidly pulls a saw blade back and forth **FIGURE 9-28**. As with rotary saws, different blades are used to cut different materials. Reciprocating saws are most commonly used to cut metal during extrication of a victim from a motor vehicle.

FIRE FIGHTER Tips

Ventilation saws have a depth gauge on the bar and are specifically designed for roof ventilation.

The cutting tools that require the most training are hydraulic shears and cutting torches. **Hydraulic shears** are often used along with hydraulic spreaders and rams in extrication of victims from automobiles; the same hydraulic power source can be used with all three types of tools. Hydraulic shears can cut quickly through metal posts and bars.

Cutting torches produce an extremely high-temperature flame and are capable of heating steel until it melts, thereby cutting through an object **FIGURE 9-29**. Because these torches

FIGURE 9-29 A cutting torch can be used to cut through a metal door.

© 2003, Berta A. Daniels

A.

B.

FIGURE 9-28 **A.** The components of a reciprocating saw. **B.** A reciprocating saw being used during the extrication of a victim from a vehicle.

produce such high temperatures (3148 degrees Celcius [5700 degrees Fahrenheit]), operators must be specially trained before using them. Cutting torches are sometimes used for rescue situations and for cutting through heavy steel objects. One drawback of this equipment is that it cannot be used in situations where flammable fuels are present.

■ Multiple-Function Tools

Certain tools are designed to perform multiple functions, thereby reducing the total number of tools needed to achieve a goal; for example, a flat-head axe can be used as either a cutting tool or a striking tool. Some combination tools can be used to cut, pry, strike, and turn off utilities.

■ Special-Use Tools

Some fire situations require special-use tools that perform other functions; for example, fire departments located in areas where brush and ground fires occur frequently may need to carry rakes, brooms, shovels, and combination tools that can be used for raking, chopping, cutting, and leaf blowing. These tools are described in the chapter "Wildland and Ground Fires."

Rescue squads may also use specialized equipment such as jacks and air bags for lifting heavy objects, **come alongs** (small, hand-operated winches) for lifting or moving heavy objects or bending objects, and tripods **FIGURE 9-30**. You can learn more about the proper use of this special equipment by taking special rescue courses or during in-service training.

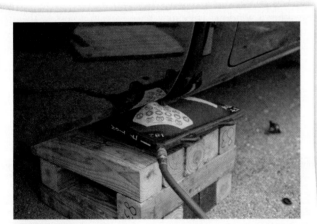

FIGURE 9-30 Heavy-duty air bags can be used to lift vehicles in rescue situations.

Phases of Use

The process of extinguishing a fire usually involves a sequence of steps or stages. Each phase of a fire-ground operation may require the use of certain types of tools and equipment.

The basic steps of fire suppression are summarized here:

- **Response/size-up**—This phase begins when the emergency call is received and continues as the units travel to the incident scene. The last part of this phase involves the initial observation and evaluation of factors used to determine which strategy and tactics will be employed.
- **Forcible entry**—This phase begins when entry to buildings, vehicles, aircraft, or other confined areas is locked or blocked, requiring fire fighters to use special techniques to gain access.
- **Interior attack**—During this phase, a team of fire fighters is assigned to enter the fire structure, locate the fire, and extinguish the fire.
- **Search and rescue**—This phase involves searching for any victims trapped by the fire and extricating them from the building.
- Rapid intervention—During this phase, a **rapid intervention crew (RIC)** provides immediate assistance to injured or trapped fire fighters.
- **Ventilation**—This step involves changing air within a compartment by natural means.
- **Overhaul**—The final phase is to ensure that all hidden fires are extinguished after the main fire has been suppressed.

■ Response/Size-Up

The response and size-up phase enables you to anticipate emergency situations. At this time, you should consider the information from the dispatcher along with preincident plan information about the location. This information can provide you with an idea of the nature and possible gravity of the situation, as well as the types of problems that might arise during firefighting operations; for example, an automobile fire on the highway will present different problems and require different tools than a call for smoke coming from a single-family house. A different thinking process occurs when you are dispatched at midnight to a house fire that may have trapped a family inside than when you respond to a report of a kitchen stove fire at suppertime. Even though information about the incident is limited at this point, the response and size-up phase is the time to start thinking about the types of tools and equipment that you might need.

Most fire departments have SOGs/SOPs that specify the tools and equipment required for different types of fires. Each crew member is expected to bring specific tools and equipment from the apparatus. These requirements take into account the roles that different units have within the fire department. A member of an engine company, for example, will not bring the same tools that a member of a truck company will bring. Likewise, the tools carried for an interior fire attack differ from those carried for an outside or defensive attack.

Upon arrival at the scene, the company officers will size up the situation and develop the action plans for each company, following SOGs/SOPs. As a crew member, you are responsible for following the directions of your company officer.

■ Forcible Entry

Gaining entrance to a locked building or structure can present a challenge to even the most seasoned fire fighter. Buildings are often equipped with security devices designed to keep unwanted people out, but these same devices can make it very difficult for fire fighters to gain access to the building. Forcible entry is the process of entering a building by overcoming these barriers. The skills involved in this phase are discussed in depth in the chapter "Forcible Entry."

Several types of tools can be used for forcible entry, including an axe, a prying tool, or a K tool. A flat-head axe and a Halligan tool, collectively called the irons, are used in combination to pry open a door, although they may permanently damage both the door and the frame. Prying tools used for gaining access include pry bars, crowbars, Halligan tools, Hux bars, and the hydraulic-powered rabbet tool.

A K tool can be used to pull out a cylinder lock mounted in a wood or heavy metal door, so that the lock can be released **FIGURE 9-31**. This comparatively nondestructive process leaves the door and most of the locking mechanism undamaged. The building owner can then have the lock cylinder replaced at a relatively low cost.

A variety of striking tools can be used for forcible entry when brute force is needed to break into a building. These items include flat-head axes, hammers, sledgehammers, and battering rams.

Sometimes the easiest—or only—way to gain access is to use cutting tools. An axe can be used to cut out a door panel. A power saw can be used to cut through a wood wall. Bolt cutters can be used to remove a padlock. Cutting torches or rotary

FIGURE 9-31 Use a K tool to pull out and release a cylinder lock.

power saws can be used to cut through metal security bars or the hinges on metal security doors. Although cutting is a destructive process, it is justified when required to save lives or property.

Many techniques and types of tools can be used to gain entry into secured structures **TABLE 9-8**. The exact tool needed will depend on the method of entry and the type of obstacle. Because having greater experience usually suggests the best way to gain entry in each situation, rely on the orders and advice of your officer and fellow fire fighters. More techniques used for forcible entry are described in the chapter "Forcible Entry."

■ Interior Firefighting Tools and Equipment

The process of fighting a fire inside a building involves several tasks that are usually performed simultaneously or in rapid succession by teams of fire fighters. While one crew is advancing a hose line to attack the fire, another crew may be searching for occupants and a third crew may be performing ventilation tasks. This is referred to as combined operations, or combined ops. An additional company may be standing by as an RIC, in case a fire fighter needs to be rescued.

Some basic tools and equipment should be carried by every crew working inside a burning building. Crews may also carry specialized tools and equipment needed for their particular assignment. The basic tools enable them to solve problems they may encounter while performing interior operations. For example, crew members may encounter obstacles such as locked doors, or they may need to open an emergency escape route. They may need to establish horizontal ventilation by forcing, opening, or breaking a window. They may have to gain access to the space above the ceiling by using a pike pole or making a hole in a wall or floor with an axe. A powerful light is also important, because smoke can quickly reduce interior visibility to just a few inches.

The basic set of tools for interior firefighting includes the following items:

- A prying tool, such as a Halligan tool
- A striking tool, such as a flat-head axe or a sledgehammer

TABLE 9-8	Forcible Entry Tools
Type of Tool	**Use in Forcible Entry**
Prying tools	Prying tools can include a Halligan tool, a flat bar, and a crowbar, among other options. They can be used to break windows or force open doors.
Axe, flat-head axe, pick-head axe	An axe can be used to cut through a door or break a window. A flat-head axe can be used in conjunction with a Halligan tool to force open a door.
Sledgehammer	A sledgehammer can be used to breach walls or break a window. It can also be used in conjunction with other tools, such as a Halligan tool, to force open a door or break off a padlock.
Hammer, mallet	These tools may be used in conjunction with other tools such as chisels or punches to force entry through windows or doors.
Chisel, punch	These tools can be used to make small openings through doors or windows.
K tool	The K tool provides a "through-the-lock" method of opening a door, thereby minimizing damage to the door.
Rotary saw, chainsaw, reciprocating saw	Power saws can cut openings through obstacles including doors, walls, fences, gates, security bars, and other barriers.
Bolt cutter	A bolt cutter can be used to cut off padlocks or cut through obstacles such as fences.
Battering ram	A battering ram can be used to breach walls.
Hydraulic door opener	A hydraulic door opener can be used to force open a door.
Hydraulic rescue tool	A hydraulic rescue tool can be used in a variety of situations to break, breach, or force openings in doors, windows, walls, fences, or gates.

- A cutting tool, such as an axe
- A pushing/pulling tool, such as a pike pole
- A **hand light** or portable light

The specific tools that must be carried by each crew are usually defined in the fire department's training manuals and SOGs/SOPs. These requirements are based on local conditions and preferences and may differ depending on the type of company and the assignment.

The interior attack team is responsible for advancing a hose line, finding the fire, and applying water to extinguish the flames. The members of this team need the basic tools that will enable them to reach the seat of the fire.

Search and Rescue Tools and Equipment

Search and rescue needs to be carried out quickly, shortly after arrival on the fire ground. A search team should carry the same basic hand tools as the interior attack team:

- Pushing tool (short pike pole)
- Prying tool (Halligan tool)
- Striking tool (sledgehammer or flat-head axe)
- Cutting tool (axe)
- Portable hand light

In addition to being equipped for forcible entry and emergency exit, a search and rescue team may use tools to probe under beds for unconscious victims. A short pike pole or closet hook is relatively light and may reduce the time needed to search an area by extending the fire fighter's reach. An axe handle can also be used for this purpose.

Other types of tools used for search and rescue include **thermal imaging cameras**, portable lights, and **life lines** **FIGURE 9-32**, **FIGURE 9-33**, and **FIGURE 9-34**. The techniques of search and rescue are discussed in the chapter "Search and Rescue."

Rapid Intervention Tools and Equipment

The RIC is designated to stand by to provide immediate assistance to any fire fighters who become lost, trapped, or injured during an incident or training exercise. Members of this team should carry the standard set of tools for interior firefighting as well as extra tools and equipment that are particularly important for search and rescue tasks. The extra tools and equipment should help the RIC find and gain access to a fire fighter who is in trouble, extricate a fire fighter who is trapped under debris, provide breathing air for a fire fighter who has experienced an SCBA failure or run out of air, and remove an injured or unconscious fire fighter from the building. All of this equipment should be gathered and staged with the RIC so that it will be immediately available if needed.

An RIC should carry the following special equipment:

- Thermal imaging device
- Additional portable lighting
- Life lines
- Prying tools
- Striking tools
- Cutting tools, including a power saw
- SCBA and spare air cylinders with rapid intervention crew/company universal air connection (RIC UAC)
- Litter or patient packaging device

Ventilation Tools and Equipment

The objective of ventilation is to provide openings so that fresh air can enter a burning structure and hot gases and products

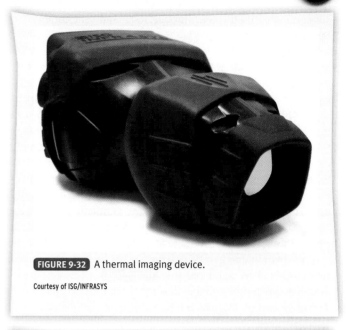

FIGURE 9-32 A thermal imaging device.

Courtesy of ISG/INFRASYS

FIGURE 9-33 A fire fighter using a thermal imaging device.

FIGURE 9-34 An image produced by a thermal imaging device.

of combustion can escape from the building. See the chapter "Ventilation" for more detailed information on ventilation operations. Many of the same tools used for forcible entry are also applied to provide ventilation; for example, power saws and axes are commonly used to cut through roofs and vent combustion by-products.

Fans are used to remove smoke from a building or to introduce fresh air into a structure. With positive-pressure ventilation, fresh air is blown *into* a building through selected openings to force contaminated air out through other openings. This kind of ventilation allows fire fighters to work quickly, safely, and effectively to locate and suppress the fire.

Negative-pressure ventilation uses fans placed at selected openings to draw contaminated air *out of* a building. This technique is used when there are no suitable openings for positive-pressure ventilation or when introducing pressurized air could accelerate the fire. Negative-pressure ventilation may also be the best option when positive-pressure ventilation might potentially spread the products of combustion.

Ventilation fans can be powered either by electric or gasoline motors or by water pressure **FIGURE 9-35**. A gasoline-powered fan may not be suitable in some situations, because it can introduce carbon monoxide into a structure if an exhaust hose is not available. If the building atmosphere contains potentially dangerous levels of flammable gases, a water-powered fan may be the best choice.

Horizontal ventilation usually involves opening outer doors and windows to allow fresh air to enter and to remove contaminated air. When ventilation fans are used, they are placed at these openings. Positive-pressure fans blow in fresh air, whereas negative-pressure fans draw out smoke and contaminants. To provide for adequate ventilation, unlocked or easily released windows and doors should be opened normally. Locked or jammed windows and doors may have to be broken or forced open using basic interior firefighting tools.

It may also be necessary to create interior openings within the building so that contaminated air can reach the exterior openings. Opening interior doors is the easiest, quickest, and often most effective way to ventilate interior spaces. A series of electrically powered fans may be placed throughout a large structure to help move smoke in the desired direction.

In vertical ventilation, openings in the roof or the highest part of a building allow smoke and hot gases to escape.

Whenever possible, existing openings, such as doors, windows, roof hatches, and skylights, should be used for vertical ventilation. It may be necessary to force them open or to break them using forcible entry tools.

In some circumstances, it may be necessary to cut through a roof to make an effective vertical ventilation opening. Cutting tools, such as axes and power saws, are used for this purpose. Pike poles will also be needed to pull down ceilings after the roof covering is opened.

The following special equipment is needed for ventilation:

- Positive-pressure fans
- Negative-pressure (exhaust) fans
- Pulling and pushing tools (long pike poles)
- Cutting tools (power saws and axes)

■ Overhaul Tools and Equipment

The purpose of overhaul is to examine the fire scene carefully and ensure that all hidden fires are extinguished. During this phase, burned debris must be removed and potential hotspots in enclosed spaces behind walls, above ceilings, and under floors exposed. Both tasks can be accomplished using simple hand tools.

As part of overhaul, you need a tool that can shut off activated sprinkler heads so as to avoid excess water damage. In buildings equipped with sprinkler systems, fire fighters should shut off any sprinkler heads that have been activated by the fire to reduce water damage and to restore the fire protection system in the building. Whenever possible, a sprinkler system should be restored to service. A single sprinkler head can be shut off with a wooden wedge—a technique that leaves the sprinkler system with an inoperable sprinkler head when it is turned on. A sprinkler head can also be shut off with a special tool that contains a fusible link. This shut-off tool allows an activated sprinkler head to be placed back in service until a replacement head can be installed.

Pike poles are commonly used for pulling down ceilings and opening holes in walls. Axes, and sometimes power saws, are used to open walls and floors. Prying and striking tools are also used to open closed spaces. Shovels, brooms, rakes, and buckets are used to clear away debris. Water vacuums are used to remove water from buildings.

The widespread introduction of infrared thermal imaging devices has made it possible to see hotspots behind walls without physically cutting into them **FIGURE 9-36**. This technology has reduced the risk of missing dangerous hotspots as well as curtailed the amount of time and effort it takes to overhaul a fire scene. The process of overhauling a fire scene and the tools used to accomplish this task are covered more fully in the chapter "Salvage and Overhaul."

The following tools are used during overhaul operations:

- Pushing tools (pike poles of varying lengths)
- Prying tools (Halligan tool)
- Striking tools (sledgehammer, flat-head axe, hammer, mallet)
- Cutting tools (axes, power saws)
- Debris-removal tools (shovels, brooms, rakes, buckets, carryalls)
- Water-removal equipment (water vacuums)
- Ventilation equipment (electric-, gas-, or water-powered fans)

FIGURE 9-35 Different types of fans used in ventilation.

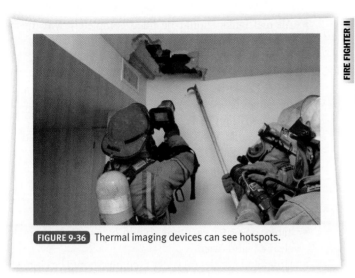

FIGURE 9-36 Thermal imaging devices can see hotspots.

- Portable lighting
- Sprinkler shut-off tools
- Thermal imaging devices

Fire Fighter Safety Tips   © Photos.com

There is a valuable saying in the fire service: "Hand tools— never leave your apparatus without them."

Fire Fighter Safety Tips   © Photos.com

Keep the manufacturers' manuals for all of the department's tools and equipment in a safe and easily accessible location, and refer to them whenever you have any questions about that device.

Tool Staging

Many fire departments have SOGs/SOPs for staging necessary equipment nearby during a fire. This procedure often involves placing a tarp or salvage cover on the ground at a designated location and laying out commonly used tools and equipment where they can be accessed readily. A similar procedure may be used for rescue operations, where tools that are likely to be needed can be laid out, ready for use. This kind of tool staging saves valuable time because fire fighters do not have to return to their own apparatus or search several different vehicles to find a particular tool.

A department's SOGs/SOPs usually specify the types of tools and equipment to be staged. The tool staging area can be located outside the building or, in the case of a high-rise or large building, at a convenient location inside the structure. Additional personnel may be directed to bring particular items to the tool staging location at working fires.

Maintenance

All tools and equipment must be properly maintained so that they will be ready for use when they are needed. That means you must keep equipment clean and free from rust, keep cutting blades sharpened, and keep fuel tanks filled. Every tool and piece of equipment must be ready for use before you respond to an emergency incident. Follow the manufacturer's instructions for cleaning and maintaining each piece of equipment. In addition, use equipment only for its intended purposes. For example, a pike pole is made for pushing and pulling; it is not a lever and will break if used inappropriately. Use the right tool for the job.

■ Cleaning and Inspecting Hand Tools

All hand tools should be completely cleaned and inspected and their conditions documented after each use (TABLE 9-9). Remove all dirt and debris from the tools. If appropriate, use water streams to remove the debris and mild soap to clean the equipment thoroughly. Learn how to safely use the cleaning solutions that your department and the manufacturer specify for cleaning tools and equipment.

To prevent rust, metal tools must be dried completely, either by hand or by air, before being returned to the apparatus. Coating unpainted metal surfaces with a light film of oil will help prevent rusting.

Cutting tools should be sharpened after each use. Before any tool is placed back into service, it should be inspected for damage.

Avoid painting tools, because paint may hide defects or visible damage. Keep the number of markings on a tool to a minimum.

Wooden and fibreglass handles should be inspected for cracks, splinters, gouges, and other defects. Repair or replace any damaged handle.

■ Cleaning and Inspecting Power Equipment and Tools

Power equipment and tools are used for lighting, ventilation, salvage, and overhaul. Use power equipment and tools only after you have received training in their use. Read and heed the instructions supplied with the equipment.

TABLE 9-9	Cleaning and Inspecting Hand Tools
Metal parts	All metal parts should be clean and dry. Remove rust with steel wool. Do not oil the striking surface of metal tools, as this treatment may cause them to slip.
Wood handles	Inspect for damage such as cracks and splinters. Sand the handle if necessary. Do not paint or varnish the handle; instead, apply a coat of boiled linseed oil. Check that the tool head is tightly fixed to the handle.
Fibreglass handles	Clean with soap and water. Inspect for damage. Check that the tool head is tightly fixed to the handle.
Cutting edges	Inspect for nicks or other damage. File and sharpen as needed. Power grinding may weaken some tools, so hand sharpening may be required.

Test power equipment and tools frequently, and have these items serviced regularly by a qualified shop. Keep records of all inspections and maintenance performed on power equipment and tools. In addition, preventive maintenance will help ensure that equipment and tools operate properly when needed. If a power tool or equipment is defective, report this fact to your officer.

Fill each tool with the proper fuel, remembering that some tools operate on gasoline, whereas others use various types of gasoline and oil mixtures. Many small engines operate on gasoline, for example, but others are designed to burn diesel fuel. Fuels have a limited storage life, especially fuels containing ethanol. If the fuel is not used in a certain period of time, it may be necessary to drain and refill equipment with fresh fuel. Check the manufacturer's recommendations before fuelling a particular piece of equipment.

After returning from a fire, clean, inspect, and record maintenance data for all of your overhaul tools to ensure that they are in a ready state.

■ Steps for Cleaning and Inspecting Power Tools

All power equipment should be left in a ready state for immediate use at the next incident:

- All debris should be removed, and the tool should be clean and dry.

- All fuel tanks should be filled completely with fresh fuel.
- Any dull or damaged blades and chains should be replaced.
- Belts should be inspected to ensure they are tight and undamaged.
- All guards should be securely in place.
- All hydraulic hoses should be cleaned and inspected.
- All power cords should be inspected for damage.
- All hose fittings should be cleaned, inspected, and tested to ensure a tight fit.
- Tools should be started to ensure that they operate properly.
- Tanks on water vacuums should be emptied, washed, cleaned, and dried.
- Hoses and nozzles on water vacuums should be cleaned and dried.

It is essential to read the manufacturer-provided manuals and follow all instructions on the care and inspection of power tools and equipment. Keep all manufacturers' manuals in a safe and easily accessible location, and refer to these documents when cleaning and inspecting the tools and equipment.

Learn the proper procedure for reporting a problem with a power tool and taking it out of service. Remember—your safety depends on the quality of your tools and equipment.

Near-Miss REPORT

Report Number: 06-0000499

Synopsis: Hydraulic ram fails during field check.

Event Description: There was a dramatic failure of a long [brand deleted] hydraulic ram at a station today during routine equipment exercising. The weld down the spine of the ram split open, which caused hydraulic fluid to spray far and wide from a longitudinal crack. The excursion of the ram's outer housing caused the heads to shear off the two Allen head machine bolts holding the bracket for the front handgrip of the ram. The fire fighter involved was wearing full PPE and eye protection. He was uninjured, and his person was not contaminated.

Risk management paperwork was completed and submitted by the lieutenant for the damaged gear and ram. The fire fighter turned in that gear and received replacement turnouts on the same day from our warehouse. Digital photographs have been taken of the mechanical equipment by a deputy fire marshal. The deputy chief will receive copies as well as the members of the safety committee made up of management and union members. The logistics utility worker was instructed to take the ram and the new power plant back to the vendor for evaluation.

My observation was that this ram looked very old—almost all of its paint had worn off. I never even contemplated the fact that these housings are not one milled piece. They appear to be two halves welded together; the weld is then ground down flush and the finish applied to give a uniform appearance. None I'd ever seen before looked like this one. We have dozens of pieces of such equipment of different ages and states of wear.

Lessons Learned: All personnel should be reminded to always wear turnout PPE whenever operating hydraulic power equipment under all circumstances. Eye protection is paramount. Routinely, all of our hydraulic power tools should be carefully looked over before operation. Obviously worn equipment should be removed from service and replacements sought through appropriate channels. Command staff as well as rank and file personnel should be sensitive to this issue.

Chief Concepts

- Tools and equipment are used in almost all fire suppression and rescue operations.
- Hand tools are used to extend or multiply the actions of your body and to increase your effectiveness in performing specific functions.
- Power tools and equipment use an external source of power and are faster and more efficient than hand tools.
- Always wear PPE when using tools or equipment.
- Most tools fit into the following functional categories:
  - Rotating tools—Apply a rotational force to make something turn; used to assemble and diassemble items. Rotating tools include screwdrivers, wrenches, and pliers.
  - Prying or spreading tools—Used to pry and spread. May be as simple as a pry bar or as mechanically complex as a hydraulic spreader. Prying and spreading tools include claw bars, crowbars, and Halligan tools.
  - Pushing or pulling tools—Extend your reach and increase the power you can exert upon an object. Pushing and pulling tools include pike poles, closet hooks, K tools, and ceiling hooks.
  - Striking tools—Used to apply an impact force to an object. Striking tools include hammers, mallets, axes, and spring-loaded centre punches.
  - Cutting tools—Tools with sharp edges that sever objects. Cutting tools include bolt cutters, chainsaws, cutting torches, and handsaws.
  - Multiple-use tools—Designed to perform multiple functions, thereby reducing the total number of tools needed to achieve a goal. Some combination tools can be used to cut, pry, strike, and turn off utilities.
- Forcible entry tools are used to gain entrance to a locked building:
  - Prying tools—Used to break windows or force open doors.
  - Axes—Used to cut through a door or break a window.
  - Sledgehammer—Used to breach walls or break a window.
  - Hammer or mallet—May be used with other tools such as chisels or punches to force entry through windows or doors.
  - Chisel or punch—Used to make small openings through doors or windows.
  - K tool—Provides a "through-the-lock" method of opening a door, minimizing damage to the door.
  - Saws—Can cut openings through obstacles including doors, walls, fences, gates, security bars, and other barriers.
  - Bolt cutter—Used to cut off padlocks or cut through obstacles such as fences.
  - Battering ram—Used to breach walls.
  - Hydraulic door opener—Used to force open a door.
  - Hydraulic rescue tool—Used in a variety of situations to break, breach, or force openings in doors, windows, walls, fences, or gates.
- The interior attack team is responsible for advancing a hose line, finding the fire, and applying water to extinguish the flames. The members of this team need the basic tools that will allow them to reach the seat of the fire:
  - A prying tool, such as a Halligan tool
  - A striking tool, such as a flat-head axe or a sledgehammer
  - A cutting tool, such as an axe
  - A pushing/pulling tool, such as a pike pole
  - A strong hand light or portable light
- A search and rescue team should carry the same basic hand tools as the interior attack team, as well as a short pike pole, thermal imaging devices, portable lights, and life lines.
- An RIT should carry the following special equipment:
  - Thermal imaging device
  - Additional portable lighting
  - Life lines
  - Prying tools
  - Striking tools
  - Cutting tools, including a power saw
  - SCBA or spare air cylinders with RIT UAC
- The objective of ventilation is to provide openings so that fresh air can enter a burning structure and hot gases and products of combustion can escape from the building. The following special equipment is needed for ventilation:
  - Positive-pressure fans
  - Negative-pressure (exhaust) fans
  - Pulling and pushing tools (long pike poles)
  - Cutting tools (power saws and axes)
- The purpose of overhaul is to examine the fire scene carefully and ensure that all hidden fires are extinguished.
- The following tools are used during overhaul operations:
  - Pushing tools (pike poles of varying lengths)

- Prying tools (Halligan tool)
- Striking tools (sledgehammer, flat-head axe, hammer, mallet)
- Cutting tools (axes, power saws)
- Debris-removal tools (shovels, brooms, rakes, buckets, carryalls)
- Water-removal equipment (water vacuums)
- Ventilation equipment (electric-, gas-, or water-powered fans)
- Portable lighting
- Sprinkler shut-off tools
- Thermal imaging devices

- Tool staging often involves placing a tarp or salvage cover on the ground at a designated location and laying out commonly used tools and equipment so that they can be readily accessed.
- All tools and equipment must be properly maintained so that they will be ready for use when they are needed. That means you must keep equipment clean and free from rust, keep cutting blades sharpened, and keep fuel tanks filled.
- Test power equipment and tools frequently, and have them serviced regularly.
- Read and follow the manufacturer's manuals, and follow all instructions on the care and inspection of power tools and equipment.

Hot Terms

Battering ram A tool made of hardened steel with handles on the sides used to force doors and to breach walls. Larger versions may be used by as many as four people; smaller versions are made for one or two people.

Bolt cutter A cutting tool used to cut through thick metal objects such as bolts, locks, and wire fences.

Box-end wrench A hand tool used to tighten or loosen bolts. The end is enclosed, as opposed to an open-end wrench. Each wrench is a specific size, and most have ratchets for easier use.

Carpenter's handsaw A saw designed for cutting wood.

Ceiling hook A tool with a long wooden or fibreglass pole that has a metal point with a spur at right angles at one end. It can be used to probe ceilings and pull down plaster–lath material.

Chainsaw A power saw that uses the rotating movement of a chain equipped with sharpened cutting edges. It is typically used to cut through wood.

Chisel A metal tool with one sharpened end that is used to break apart material in conjunction with a hammer, mallet, or sledgehammer.

Claw bar A tool with a pointed claw-hook on one end and a forked- or flat-chisel pry on the other end. It is often used for forcible entry.

Clemens hook A multipurpose tool that can be used for several forcible entry and ventilation applications because of its unique head design.

Closet hook A type of pike pole intended for use in tight spaces, commonly 0.6–1.2 m (2–4 ft.) in length.

Come along A hand-operated tool used for dragging or lifting heavy objects that uses pulleys and cables or chains to multiply a pulling or lifting force.

Coping saw A saw designed to cut curves in wood.

Coupling One set or pair of connection devices attached to a fire hose that allow the hose to be interconnected to additional lengths of hose or adapters and other fire fighting appliances. (NFPA 1963)

Crowbar A straight bar made of steel or iron with a forked chisel on the working end that is suitable for performing forcible entry.

Cutting torch A torch that produces a high-temperature flame capable of heating metal to its melting point, thereby cutting through an object. Because of the high temperatures (3148°C [5700°F]) that these torches produce, the operator must be specially trained before using this tool.

Drywall hook A specialized version of a pike pole that can remove drywall more effectively because of its hook design.

Flat bar A specialized type of prying tool made of flat steel with prying ends suitable for performing forcible entry.

Flat-head axe A tool that has a head with an axe on one side and a flat head on the opposite side.

Forcible entry Techniques used by fire personnel to gain entry into buildings, vehicles, aircraft, or other areas of confinement when normal means of entry are locked or blocked. (NFPA 402)

Gripping pliers A hand tool with a pincer-like working end that can be used to bend wire or hold smaller objects.

Hacksaw A cutting tool designed for use on metal. Different blades can be used for cutting different types of metal.

Halligan tool A prying tool that incorporates a pick and a fork, specifically designed for use in the fire service.

Hammer A striking tool.

Hand light A small, portable light carried by fire fighters to improve visibility at emergency scenes. It is often powered by rechargeable batteries.

Handsaw A manually powered saw designed to cut different types of materials. Examples include hacksaws, carpenter's handsaws, keyhole saws, and coping saws.

Hux bar A multipurpose tool that can be used for several forcible entry and ventilation applications because of its unique design. It may also be used as a hydrant wrench.

Hydrant wrench A hand tool that is used to operate the valves on a hydrant; it may also be used as a spanner wrench. Some models are plain wrenches, whereas others have a ratchet feature.

Hydraulic shears A lightweight, hand-operated tool that can produce up to 4536 kg (10,000 lb.) of cutting force.

Hydraulic spreader A lightweight, hand-operated tool that can produce up to 4536 kg (10,000 lb.) of prying and spreading force.

Interior attack The assignment of a team of fire fighters to enter a structure and attempt fire suppression.

Irons A combination of tools, usually consisting of a Halligan tool and a flat-head axe, that are commonly used for forcible entry.

Kelly tool A steel bar with two main features: a large pick and a large chisel or fork.

Keyhole saw A saw designed to cut circles in wood for keyholes.

K tool A tool that is used to remove lock cylinders from structural doors so the locking mechanism can be unlocked.

Life line A rope secured to a fire fighter that enables the fire fighter to retrace his or her steps out of a structure.

Mallet A short-handled hammer.

Maul A specialized striking tool, weighing 2.7 kg (6 lb.) or more, with an axe on one end and a sledgehammer on the other end.

Mechanical saw A saw that is usually powered by an electric motor or a gasoline engine. The three primary types of mechanical saws are chainsaws, rotary saws, and reciprocating saws.

Multipurpose hook A long pole with a wooden or fibreglass handle and a metal hook on one end used for pulling.

Open-end wrench A hand tool that is used to tighten or loosen bolts. The end is open, as opposed to a box-end wrench. Each wrench is a specific size.

Overhaul The process of final extinguishment after the main body of a fire has been knocked down. All traces of fire must be extinguished at this time. (NFPA 402, 2002)

Pick-head axe A tool that has a head with an axe on one side and a pointed end (pick) on the opposite side.

Pike pole A pole with a sharp point (pike) on one end coupled with a hook. It is used to make openings in ceilings and walls. Pike poles are manufactured in different lengths for use in rooms of different heights.

Pipe wrench A wrench having one fixed grip and one movable grip that can be adjusted to fit securely around pipes and other tubular objects.

Plaster hook A long pole with a pointed head and two retractable cutting blades on the side.

Pry bar A specialized prying tool made of a hardened steel rod with a tapered end that can be inserted into a small area.

Rabbet tool A hydraulic spreading tool designed to pry open doors that swing inward.

Rapid intervention team crew/company (RIC) A minimum of two fully equipped personnel on site, in a ready state, who are assigned to immediate rescue of injured or trapped fire fighters.

Reciprocating saw A saw that is powered by an electric motor or a battery motor, and whose blade moves back and forth.

Response Immediate and ongoing activities, tasks, programs, and systems to manage the effects of an incident that threatens life, property, operations, or the environment. (NFPA 1600)

Roofman's hook A long pole with a solid metal hook used for pulling.

Rotary saw A saw that is powered by an electric motor or a gasoline engine, and that uses a large rotating blade to cut through material. The blades can be changed depending on the material being cut.

San Francisco hook A multipurpose tool that can be used for several forcible entry and ventilation applications because of its unique design, which includes a built-in gas shut-off and directional slot.

Screwdriver A tool used for turning screws.

Search and rescue The process of searching a building for a victim and extricating the victim from the building.

Seat belt cutter A specialized cutting device that cuts through seat belts.

Size-up The observation and evaluation of existing factors that are used to develop objectives, strategy, and tactics for fire suppression. (NFPA 1051)

Sledgehammer A long, heavy hammer that requires the use of both hands.

Socket wrench A wrench that fits over a nut or bolt and uses the ratchet action of an attached handle to tighten or loosen the nut or bolt.

Spanner wrench A type of tool used to couple or uncouple hoses by turning the rocker lugs on the connections.

Spring-loaded centre punch A spring-loaded punch used to break automobile glass.

Thermal imaging camera An electronic camera that detects differences in temperature based on infrared energy and then generates images based on those data. It is commonly used in smoke-filled environments to locate victims as well as to search for hidden fire during size-up and overhaul.

Ventilation The changing of air within a compartment by natural or mechanical means. Ventilation can be achieved by introduction of fresh air to dilute contaminated air or by local exhaust of contaminated air. (NFPA 302, 2004)

You have really settled into your role as a volunteer fire fighter. You know you are helping your community, and you feel the satisfaction when you make a good stop on a fire. It doesn't take long for you to get noticed by others in the department. In fact, at a recent training course, the training officer approaches you on break and asks you to teach a class on basic tools and equipment. You feel great about his confidence in your skills, but you also wonder whether you really know enough about the tools to teach it to others.

1. A _____ is a tool that is used to remove lock cylinders from structural doors so the locking mechanism can be unlocked.
   A. K-tool
   B. kelly tool
   C. rabbet tool
   D. keyhole saw

2. A _____ is a prying tool that incorporates a pick and a fork, designed for use in the fire service.
   A. ceiling hook
   B. pry bar
   C. Halligan tool
   D. crowbar

3. A _____ is a tool made of hardened steel with handles on the sides used to force doors and to breach walls.
   A. battering ram
   B. Halligan tool
   C. maul
   D. mallet

4. The irons usually consists of what two tools?
   A. Pick-head axe and Halligan
   B. Flat-head axe and Halligan
   C. Flat-head axe and Clemens hook
   D. Pick-head axe and Clemens hook

5. Gripping pliers are included in which functional category?
   A. Cutting
   B. Pushing or pulling
   C. Prying or spreading
   D. Rotating

It is your first day after your recruit academy. The captain welcomes you to your new home. He goes over his expectations with you and how you are to operate as part of the crew. He wants you to know that he and the crew want you to be successful and you can count on their support. He introduces you to the senior fire fighter and asks him to show you around the station. When you get to the apparatus, he begins drilling you about every tool on the vehicle: what it's called and what it's for. You are able to answer every question, so you are feeling pretty good about things.

1. How will you react to being quizzed by a fellow crew member?
2. Will you have any questions for your fellow crew member?
3. Does this introduction to the fire station match your expectations of the first day on the job?
4. If you were unable to answer any of the questions, how would you react?

Ropes and Knots

Fire Fighter I

Knowledge Objectives

After studying this chapter, you will be able to:

- Describe the three primary types of fire service rope.
  (NFPA 5.3.20.A, p 256)
- List the two types of life safety rope and their minimum breaking strength. (NFPA 5.3.20.A, p 257)
- Describe the characteristics of escape rope.
  (NFPA 5.3.20.A, p 257)
- Describe the characteristics of utility ropes.
  (NFPA 5.3.20.A, p 257–258)
- List the advantages of synthetic fibre ropes.
  (NFPA 5.3.20.A, p 258–259)
- List the disadvantages of synthetic fibre ropes.
  (NFPA 5.3.20.A, p 259)
- List the types of synthetic fibres that are used in fire service rope. (NFPA 5.3.20.A, p 259)
- Describe how twisted ropes are constructed.
  (NFPA 5.3.20.A, p 259)
- Describe how braided ropes are constructed.
  (NFPA 5.3.20.A, p 259–260)
- Describe how kernmantle ropes are constructed.
  (NFPA 5.3.20.A, p 260)
- Explain the differences between dynamic kernmantle rope and static kernmantle rope. (NFPA 5.3.20.A, p 260)
- List the four components of the rope maintenance formula.
  (NFPA 5.3.20.A, p 263)
- Describe how to preserve rope strength and integrity.
  (NFPA 5.3.20.A, p 263)
- Describe how to clean rope. (NFPA 5.5.1, p 263–264)
- Describe how to inspect rope. (NFPA 5.5.1, p 263–265)
- Describe how to keep an accurate rope record.
  (NFPA 5.5.1, p 265)
- Describe how to store rope properly. (NFPA 5.5.1, p 265–266)
- List the terminology used to describe the parts of a rope when tying knots. (NFPA 5.3.20.A, p 266)
- List the common types of knots that are used in the fire service. (NFPA 5.3.20.A, p 266–267)
- Describe the characteristics of a safety knot.
  (NFPA 5.3.20, p 267–268)
- Describe the characteristics of a hitch. (NFPA 5.3.20, p 267–271)

- Describe the characteristics of a half hitch.
  (NFPA 5.3.20, p 267)
- Describe the characteristics of a clove hitch.
  (NFPA 5.3.20, p 267–268)
- Describe the characteristics of a figure eight knot.
  (NFPA 5.3.20, p 268–269)
- Describe the characteristics of a bowline knot.
  (NFPA 5.3.20, p 272)
- Describe the characteristics of a bend.
  (NFPA 5.3.20, p 272, 279)

Skills Objectives

After studying this chapter, you will be able to perform the following skills:

- Care for life safety ropes. (NFPA 5.5.1.A, 5.5.1.B, p 263)
- Clean fire department ropes. (NFPA 5.5.1.A, p 263–264)
- Inspect fire department ropes. (NFPA 5.5.1, p 265)
- Place a life safety rope in a rope bag. (NFPA 5.5.1, p 265–266)
- Tie a safety knot. (NFPA 5.1.2, 5.3.20.B, p 267–268)
- Tie a half hitch. (NFPA 5.1.2, 5.3.20.B, p 267–269)
- Tie a clove hitch in the open.
  (NFPA 5.1.2, 5.3.20.B, p 268, 271–272)
- Tie a clove hitch around an object.
  (NFPA 5.1.2, 5.3.20.B, p 268–269, 272–273)
- Tie a figure eight knot. (NFPA 5.1.2, 5.3.20.B, p 274–275)
- Tie a figure eight on a bight. (NFPA 5.1.2, 5.3.20.B, p 274, 276)
- Tie a figure eight follow-through.
  (NFPA 5.1.2, 5.3.20.B, p 274, 277)
- Tie a figure eight bend. (NFPA 5.1.2, 5.3.20.B, p 274, 278)
- Tie a bowline. (NFPA 5.1.2, 5.3.20.B, p 274, 279)
- Tie a sheet or Becket bend.
  (NFPA 5.1.2, 5.3.20.B, p 274, 280–281)
- Tie a water knot. (NFPA 5.1.2, 5.3.20.B, p 281)
- Hoist an axe. (NFPA 5.1.2, 5.3.20.B, p 282–283)
- Hoist a pike pole. (NFPA 5.1.2, 5.3.20.B, p 283–284)
- Hoist a ladder. (NFPA 5.1.2, 5.3.20.B, p 284–285)
- Hoist a charged hose line. (NFPA 5.1.2, 5.3.20.B, p 284, 286)
- Hoist an uncharged hose line. (NFPA 5.1.2, 5.3.20.B, p 284, 287)
- Hoist an exhaust fan or power tool.
  (NFPA 5.1.2, 5.3.20.B, p 285–286, 288)

Fire Fighter II — FFII

Knowledge Objectives

After studying this chapter, you will be able to:

- Describe the hardware components used during a rope rescue. (NFPA 6.4.2, p 260–262)
- Describe the characteristics of a carabiner. (NFPA 6.4.2, p 260)
- Describe the characteristics of a harness. (NFPA 6.4.2, p 260–261)
- List the types of incidents that might require a rope rescue. (p 261–262)

Skills Objectives

There are no separate Fire Fighter II skills objectives for this chapter.

Additional NFPA Standard

- NFPA 1983, *Standard on Life Safety Rope and Equipment for Emergency Services*

You Are the Fire Fighter

You and your partner are on the fourth floor of a multistorey apartment building and find yourself trapped by a rapidly progressing fire. You get to a bedroom at the front of the building and know your only escape route is out the window. You don't have time to wait for the aerial ladder to get to you, so you quickly consider your training for self-rescue. You pull out your rope and begin the most important task you have ever done in your life.

1. What type of rope would you want to use for this type of activity?
2. What rope construction method would be used for this rope?
3. What type of knot would you use?

Introduction

In the fire service, ropes are widely used to hoist or lower tools, appliances, or people; to pull a person to safety; or to serve as a life line in an emergency. A rope might be your only means of accessing a trapped person or your only way of escaping from a fire.

Learning about ropes and knots is an important part of your training as a fire fighter. This chapter gives you a basic understanding of the importance of ropes and knots in the fire service. You can then build on this foundation as you develop skills in handling ropes and tying knots. You must be able to tie simple knots accurately without hesitation or delay.

This chapter discusses different types of rope construction and the materials used in making ropes. It covers the care, cleaning, inspection, and storage of ropes. It also shows how to tie eight essential knots and explains how to secure tools and equipment so that they can be raised or lowered using ropes. Finally, it discusses additional uses for ropes in various rescue situations.

Types of Rope

Three primary types of rope are used in the fire service, each of which is dedicated to a distinct function. **Life safety rope** is used solely for supporting people. It must be used whenever a rope is needed to support a person, whether during training or during firefighting, rescue, or other emergency operations **FIGURE 10-1**. **Escape rope** is a single-purpose, emergency self-escape, self-rescue rope. **Utility rope** is used in most other cases, when it is not necessary to support the weight of a person, such as when hoisting or lowering tools or equipment.

■ Life Safety Rope

The life safety rope is a critical tool used *only* for life-saving purposes; it must never be used for utility purposes **FIGURE 10-2**. Life safety rope must be used in every situation where the rope must support the weight of one or more persons. In these situations, rope failure could result in serious injury or death.

Because a fire fighter's equipment must be extremely reliable, the criteria for design, construction, and performance

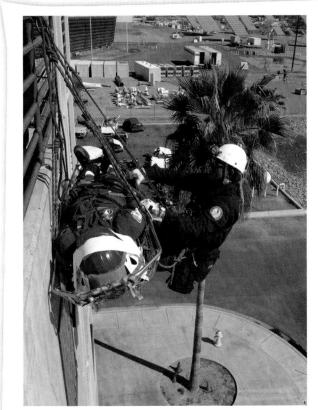

FIGURE 10-1 A life safety rope is a critical tool for fire fighters.

FIGURE 10-2 A life safety rope.

of life safety rope and related equipment are specified in NFPA 1983, *Standard on Life Safety Rope and Equipment for Emergency Services.*

NFPA 1983 lists very specific standards for the construction of life safety rope. This standard also requires the rope manufacturer to include detailed instructions for the proper use, maintenance, and inspection of the life safety rope, including the conditions for removing the rope from service. In addition, the manufacturer must supply a list of criteria that must be reviewed before a life safety rope that has been used in the field can be used again. If the rope does not meet all of the criteria, it must be retired from service.

Types of Life Safety Ropes

NFPA 1983 defines the performance requirements for two types of life safety rope: __technical use life safety rope__ and __general use life safety rope__. Technical use life safety rope by definition has a diameter that is 9.5 mm (0.4 in.) or greater, but is less than 16 mm (0.7 in.). In addition to its smaller diameter, technical use life safety ropes are also weaker than general use life safety ropes. With a minimum breaking strength of 20 kN (4496 pounds force [lbf]), technical use life safety ropes are used by highly trained rescue teams that deploy to very technical environments such as mountainous and/or wilderness terrain TABLE 10-1 .

The most common life safety rope carried by the fire service in the United States is the general use life safety rope. With a minimum breaking strength of 40 kN (8992 lbf), the general use life safety rope allows the technical rescuer a much greater margin of safety. With respect to its diameter, a general use life safety rope can be no larger than 16 mm (0.7 in.) and no smaller than 11 mm (0.5 in.).

■ Escape Ropes

An __escape rope__ is intended to be used by a fire fighter only for self-rescue from an extreme situation. This rope is designed to carry the weight of only one person and to be used only one time FIGURE 10-3 . Its purpose is to provide the fire fighter with a method of escaping from a life-threatening situation. The escape rope should be replaced by a new rope if the escape rope is exposed to an immediate danger to life and health environment.

When you are fighting a fire, you should always have a safe way to get out of a situation and reach a safe location. You might be able to go back through the door that you entered, or you might have another exit route, such as through a different door, through a window, or down a ladder. If conditions suddenly change for the worse, having an escape route can save your life.

Sometimes, however, you can find yourself in a situation where conditions deteriorate so quickly that you cannot use your planned exit route. For example, the stairway you used might collapse behind you, or the room you are in might suddenly flash over (a phase in the development of a contained fire in which exposed surfaces reach ignition temperature more or less simultaneously and fire spreads rapidly throughout the space), blocking your planned route out. In such a situation, you might need to take extreme measures to get out of the building. The escape rope was developed specifically for this type of emergency self-rescue situation. It can support the weight of one person and fits easily into a small packet or pouch FIGURE 10-4 .

■ Utility Rope

Utility rope is used when it is not necessary to support the weight of a person. Fire department utility rope is used for hoisting or lowering tools or equipment, for making __ladder halyards__

FIRE FIGHTER Tips

Escape ropes are not classified as life safety ropes.

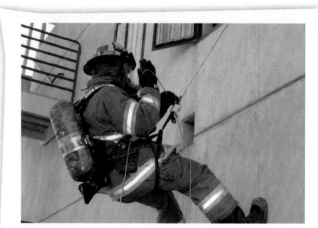

FIGURE 10-3 An escape rope is designed to be used by only one person.

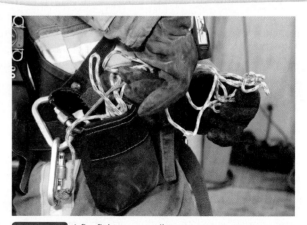

FIGURE 10-4 A fire fighter can easily carry a escape rope.

TABLE 10-1	Life Safety Rope Strength
Rope Classification	**Minimum Breaking Strength**
Escape rope (NFPA 1983)	13.5 kN (3304 lbf)
Technical use life safety rope (NFPA 1983)	20 kN (4496 lbf)
General use life safety rope (NFPA 1983)	40 kN (8992 lbf)

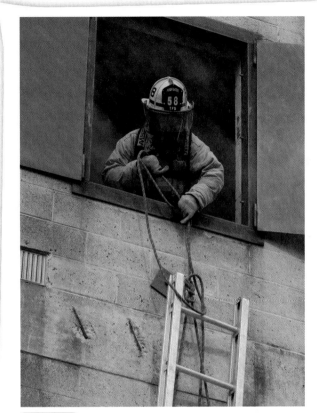

FIGURE 10-5 Utility ropes are used for hoisting and lowering tools.

(rope used on extension ladders to raise a fly section), for marking off areas, and for stabilizing objects **FIGURE 10-5**. Utility ropes also require regular inspection.

Utility ropes must not be used in situations where life safety rope is required. Conversely, life safety rope must not be used for utility applications. A fire fighter must be able to recognize instantly the type of rope from its appearance and markings.

Fire Fighter Safety Tips
© Photos.com

Many fire departments use color coding or other visible markings to identify different types of rope. This allows a fire fighter to determine very quickly if a rope is a life safety rope or a utility rope. The length of each rope should also be clearly marked by a tag or a label on the rope bag.

Rope Materials

Ropes can be made from many different types of materials. The earliest ropes were made from naturally occurring vines or fibres that were woven together. Today, ropes are made of synthetic materials such as nylon or polypropylene. Because ropes have many different uses, certain materials can work better than others in particular situations.

Natural Fibres

In the past, fire departments used ropes made from natural fibres, such as manila, because there were no alternatives. In these ropes, the natural fibres are twisted together to form strands. A strand can contain hundreds of individual fibres of different lengths. Today, ropes made from natural fibres are still used as utility ropes but are no longer acceptable as life safety ropes **TABLE 10-2**. Natural fibre ropes can be weakened by mildew and deteriorate with age, even when properly stored. A wet manila rope can absorb 50 percent of its weight in water, making it very susceptible to deterioration. A wet natural fibre rope is also very difficult to dry.

TABLE 10-2	Drawbacks to Using Natural Fibre Ropes

- Lose their load-carrying ability over time
- Subject to mildew
- Absorb 50 percent of their weight in water
- Degrade quickly

Synthetic Fibres

Since the introduction of nylon in 1938, synthetic fibres have been used to make ropes. In addition to nylon, several newer synthetic materials—such as polyester, polypropylene, and polyethylene—have been used in rope construction **FIGURE 10-6**. Synthetic fibres have several advantages over natural fibres **TABLE 10-3**. For example, synthetic fibres are generally stronger than natural fibres, so it is possible to use a smaller diameter rope without sacrificing strength. Synthetic materials can also produce very long fibres that run the full length of a rope to provide greater strength and added safety.

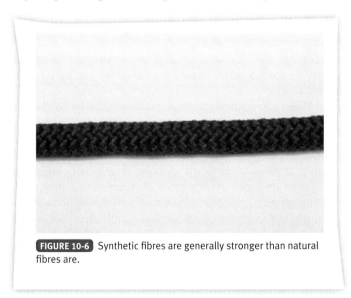

FIGURE 10-6 Synthetic fibres are generally stronger than natural fibres are.

Synthetic ropes are more resistant to rotting and mildew than natural fibre ropes are and do not degrade as rapidly. Depending on the material, they might also provide more resistance to melting and burning than natural fibre ropes do. In addition, synthetic ropes absorb much less water when wet and can be washed and dried. Some types of synthetic rope can float on water, which is a major advantage in water rescue situations.

TABLE 10-3	Advantages to Using Synthetic Fibre Ropes

- Strength-to-diameter ratio
- Difficult to meet block creel construction requirement with natural fibres
- Longevity over natural fibres

However, ropes made from synthetic fibres do have some drawbacks. Prolonged exposure to ultraviolet light as well as exposure to strong acids or alkalis can damage some synthetic ropes and decrease its life expectancy.

Life safety ropes are always made of synthetic fibres. Before any rope can be used for life safety purposes, it must meet the manufacturing requirements outlined in the current NFPA 1983. These standards specify that life safety rope must be woven of **block creel construction** (without knots or splices in the yarns, ply yarns, strands, braids, or rope). Rope of any other material or construction may not be used as a life safety rope.

The most common synthetic fibre used in life safety ropes is nylon. It has a high melting temperature with good abrasion resistance and is strong and lightweight. Nylon ropes are also resistant to most acids and alkalis. Polyester is the second most common synthetic fibre used for life safety ropes. Some life safety ropes are made of a combination of nylon and polyester or other synthetic fibres.

Polypropylene is the lightest of the synthetic fibres. Because it does not absorb water and floats, polypropylene rope is often used for water rescue situations (**FIGURE 10-7**). However, polypropylene is not as suitable as nylon for fire department life-safety uses because it is not as strong, it is hard to knot, and it has a low melting point.

FIGURE 10-7 Polypropylene rope is often used in water rescues.

Rope Construction

Several types of rope construction are possible. The best choice of rope construction depends on the specific application.

■ Twisted and Braided Rope

Twisted ropes, which are also called laid ropes, are made of individual fibres twisted into strands. The strands are then twisted together to make the rope (**FIGURE 10-8A**). This method of rope construction has been used for hundreds of years. Both natural and synthetic fibres can be used to make twisted rope.

This method of construction does have a disadvantage in that it exposes all of the fibres to the outside of the rope, where they are subject to abrasion. Abrasion can damage the rope fibres and reduce rope strength. Twisted ropes tend to stretch and are prone to unravelling when a load is applied.

Braided ropes are constructed by weaving or intertwining strands—typically synthetic fibres—together in the same way that hair is braided (**FIGURE 10-8B**). This method of construction also exposes all of the strands to the outside of the rope where they are subject to abrasion. Braided rope stretches under a load, but it is not prone to twisting. A double-braided rope has an inner braided core covered by a protective braided sleeve so that only the fibres in the outer sleeve are exposed to the

FIRE FIGHTER Tips

Ropes used for water rescue are often kept in special throw bags. The rescuer holds onto one end of the rope and throws the bag for the victim to catch. The rescuer then uses the rope to pull the victim to shore.

A.

B.

FIGURE 10-8 **A.** Twisted rope. **B.** Braided rope.

outside; the inner core remains protected from abrasion in this construction.

Kernmantle Rope

Kernmantle rope consists of two distinct parts: the kern and the mantle. The kern, which is the centre or core of the rope. The mantle or sheath is a braided covering that protects the core from dirt and abrasion. Although both parts of a kernmantle rope are made with synthetic fibres, different fibres can be used for the kern and the mantle.

Each fibre in the kern extends for the entire length of the rope without knots or splices. This block creel construction is required under NFPA 1983 for all life safety ropes. The continuous filaments produce a core that is stronger than one constructed of shorter fibres that are twisted or braided together.

Kernmantle construction produces a very strong and flexible rope that is relatively thin and lightweight **FIGURE 10-9**. This construction is well suited for rescue work and is very popular for life safety rope.

Dynamic and Static Rope

A rope can be either dynamic or static, depending on how it reacts to an applied load. A **dynamic rope** is designed to be elastic and stretch when it is loaded. A **static rope** has a limited range of elasticity. The differences between dynamic and static ropes result from both the fibres used and the construction method.

Dynamic rope is typically used in safety lines for mountain climbing because it stretches and cushions the shock if a climber falls for a long distance. A static rope is more suitable for most fire rescue situations, where falls from great heights are not anticipated. Teams that specialize in rope rescue often carry both static and dynamic ropes for use in different situations.

Dynamic and Static Kernmantle Ropes

Kernmantle ropes can be either dynamic or static. A dynamic kernmantle rope is constructed with overlapping or woven fibres in the core. When the rope is loaded, the core fibres are pulled tighter, which gives the rope its elasticity.

In the core of a static kernmantle rope, all of the fibres are laid parallel to each other. Such a rope has very little elasticity and limited elongation under an applied load. Most fire department life safety ropes use static kernmantle construction. This type of rope is well suited for lowering a person and can be used with a pulley system for lifting individuals. It can also be used to create a bridge between two structures. In this type of rope, the manufacturer includes an imbedded trailer. This trailer includes the name of the manufacturer, model number, make number, serial number, and date of manufacture.

Technical Rescue Hardware

During technical rescue incidents, ropes are often used to access and extricate individuals. In addition to the rope itself, several hardware components might be used. Fire fighters, for example, often use a **carbiner** **FIGURE 10-10**. This device is used to connect one rope to another rope or to other hardware such as an anchor plate, swivel, or pulley. Several types of carabiners are available, and you should know how to operate the type used by your department. Only a few are recognized for use in the fire service and rescue. Refer to NFPA 1983.

Harnesses

A **harness** is a piece of rescue or safety equipment made of webbing and worn by a person. It is used to secure the person to a rope or to a solid object. Two types of harnesses—Class II and Class III—can be used by rescuers, depending on the circumstances encountered:

- Class II harness (seat harness) fastens around the rescuer's waist and legs and has a design load of 272.2 kg (600 lbs). It is used to support a fire fighter, particularly in rescue situations **FIGURE 10-11**.
- Class III harness (full body harness) fastens around the rescuer's waist and thighs as well as secures the rescuer's waist and shoulders. It is the most secure type of harness and is often used to support a fire fighter who is being raised or lowered on a life safety rope **FIGURE 10-12**.

FIGURE 10-9 The parts of a kernmantle rope.

FIGURE 10-10 Carabiners.

FIRE FIGHTER II

FIGURE 10-11 Class II harness (seat harness).
RescueTECH1, Inc.

FIGURE 10-12 Class III harness (full body harness).

Harnesses need to be cleaned and inspected regularly, just as life safety ropes do. Follow the manufacturer's instructions for cleaning and inspecting harnesses.

■ Rope Rescue

Rope rescue involves raising and lowering rescuers to access injured or trapped individuals, as well as raising or lowering victims who are rescued so that they can be given appropriate medical treatment. This chapter outlines the basics of rope rescue and gives you a foundation for learning the more

FIRE FIGHTER II

complex parts of rope rescue. An approved rope rescue course is required to attain true proficiency in rope rescue skills.

Rope rescue courses cover the technical skills needed to raise or lower people using mechanical advantage systems and to remove someone from a rock ledge or a confined space **FIGURE 10-13**. They also cover the equipment and skills needed to accomplish these rescues safely.

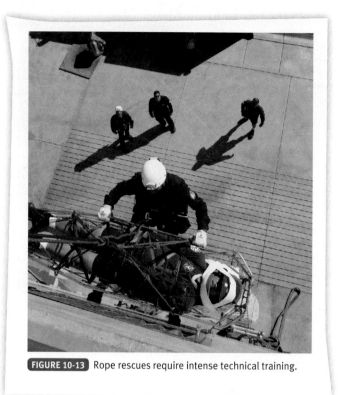

FIGURE 10-13 Rope rescues require intense technical training.

Rope Rescue Incidents

Most rope rescue incidents involve people who are trapped in normally inaccessible locations, such as a mountainside or the outside of a building **FIGURE 10-14**. Rescuers often have to lower themselves using a system of anchors, webbing, ropes, carabiners, and other devices to reach the trapped person. Once rescuers reach the person, they then have to stabilize him or her and determine how to get the person to safety. Sometimes the victim has to be lowered or raised to a safe location. Extreme cases could involve more complicated operations, such as transporting the victim in a basket lowered by a helicopter.

The type and number of ropes used in a rope rescue depend on the situation. There is almost always a primary rope that bears the weight of the rescuer (or rescuers) as he or she attempts to reach the victim. The rescuers often have a second line attached to them, known as a belay line, which serves as a backup if the main line fails. Additional lines might be needed to raise or lower the trapped individual, depending on the circumstances.

Trench Rescue

Rescues in collapsed trenches often are complicated and involve a number of different skills, such as shoring, air quality monitoring, confined-space operations, and rope rescue.

FIGURE 10-14 Ropes are invaluable when a person is trapped in an inaccessible area, such as outside a high building.

FIGURE 10-15 Ropes are often used to remove an injured or unconscious victim from a confined space.

Courtesy of Captain David Jackson, Saginaw Township Fire Department

Ropes are often used to stabilize and remove the trapped person. After the rescuers shore the walls of the trench and remove the dirt covering the victim, they place the person in a Stokes basket or on a backboard and lift him or her to the surface. If the trench is deep, ropes might be used to raise the victim to the surface.

Confined-Space Rescue

A confined-space rescue can take place in locations as varied as tanks, silos, underground electrical vaults, storm drains, and similar structures. It is often very difficult to extricate an unconscious or injured person from these locations because of the poor ventilation and limited entry or exit area. For this reason, ropes are often used to remove an injured or unconscious victim **FIGURE 10-15**.

Water Rescue

Ropes can be used in a variety of ways during water rescue operations. The simplest situation involves a rescuer on the shore throwing a rope to a person in the water and pulling that individual to shore. A more complicated situation could involve a rope stretched across a stream or river **FIGURE 10-16**. A boat might be tethered to the rope, and rescuers on shore might maneuver the boat using a series of ropes and pulleys.

FIGURE 10-16 Ropes ensure rescuer safety during water rescues.

© Mike Meadows/AP Photos

Rope Maintenance

All ropes—but especially life safety ropes—need proper care to perform in an optimal manner. Maintenance is necessary for all kinds of equipment and all types of rope, and it is absolutely essential for life safety ropes. Your life and the lives of others depend on the proper maintenance of your life safety ropes.

There are four parts to the maintenance formula:

- Care
- Clean
- Inspect
- Store

Care for the Rope

You must follow certain principles to preserve the strength and integrity of rope **TABLE 10-4** :

- Protect the rope from sharp and abrasive surfaces. Use edge protectors when the rope must pass over a sharp or unpadded surface.
- Protect the rope from rubbing against another rope or webbing. Friction generates heat, which can damage or destroy the rope.
- Protect the rope from heat, chemicals, and flames.
- Protect the rope from prolonged exposure to sunlight. Ultraviolet radiation can damage rope.
- Do not step on a rope! Your footstep could force shards of glass, splinters, or abrasive particles into the core of the rope, damaging the rope fibres.
- Follow the manufacturer's recommendations for rope care.

TABLE 10-4	Principles to Preserve Strength and Integrity of Rope

- Protect the rope from sharp, abrasive surfaces.
- Protect the rope from heat, chemicals, and flames.
- Protect the rope from rubbing against another rope.
- Protect the rope from prolonged exposure to sunlight.
- Follow the manufacturer's recommendations.

FIRE FIGHTER Tips

A **shock load** can occur when a rope is suddenly placed under unusual tension—for example, when someone attached to a life safety rope falls until the length of the rope or another rescuer stops the drop. A utility rope can be shock-loaded in a similar manner if a piece of equipment that is being raised or lowered drops suddenly.

Any rope that has been shock-loaded should be inspected and might have to be removed from service. Although there might not be any visible damage, shock loading can cause damage that is not immediately apparent. Repeated shock loads can severely weaken a rope so that it can no longer be used safely. Keeping accurate rope records helps identify potentially damaged rope.

To care for life safety ropes properly, follow the steps in **SKILL DRILL 10-1** (Fire Fighter I, NFPA 5.5.1):

1. Protect the rope from sharp and abrasive edges; use edge protectors.
2. Protect the rope from rubbing against other ropes.

Clean the Rope

Many ropes made from synthetic fibres can be washed with a mild soap and water. In addition, a special rope washer can be attached to a garden hose **FIGURE 10-17** . Some manufacturers recommend placing the rope in a mesh bag and washing it in a front-loading washing machine.

When washing a rope, use a mild detergent. Do not use bleach because it can damage rope fibres. Follow the

FIGURE 10-17 Some fire departments use a rope washer to clean their ropes.
Courtesy of Captain David Jackson, Saginaw Township Fire Department

manufacturer's recommendations for specific care of your rope. Do not pack or store wet or damp rope. Air drying is usually recommended. The drying rope should be suspended and should not lie on the floor. The use of mechanical drying devices is not usually recommended. Rope should never be dried or stored in direct sunlight.

Follow the steps in **SKILL DRILL 10-2** to clean fire department ropes:

1. Wash the rope with mild soap and water. (**STEP ①**)
2. Use a rope washer or machine, if recommended by the rope's manufacturer. (**STEP ②**)
3. Air dry the rope out of direct sunlight.
4. Inspect the rope and replace it in the rope bag so that it is ready for use. (**STEP ③**)

Inspect the Rope

Life safety ropes must be inspected after each use, whether the rope was used for an emergency incident or in a training exercise **TABLE 10-5** . Unused rope should also be inspected on a regular schedule. Some departments inspect all rope, including

SKILL DRILL 10-2   Cleaning Fire Department Ropes
(Fire Fighter I, NFPA 5.5.1)

1 Wash the rope with mild soap and water.

2 Use a rope washer or machine if recommended by the rope's manufacturer.

3 Air dry the rope out of direct sunlight. Inspect the rope and replace it in the rope bag so that it is ready for use.

© LiquidLibrary

life safety and utility ropes, every three months. Obtain the inspection criteria from the rope manufacturer.

Inspect the rope visually, looking for cuts, frays, or other damage as you run it through your fingers. Make sure that your grasping hand is gloved to protect yourself from sharp objects imbedded in the rope. Because you cannot see the inner core of a kernmantle rope, feel for any **depressions** (flat spots or lumps on the inside). Examine the sheath for any discolourations, abrasions, or flat spots **FIGURE 10-18**. If you have any doubt about whether the rope has been damaged, consult your company officer **TABLE 10-6**.

A life safety rope that is no longer usable must be pulled from service and either destroyed or marked as a utility rope. A downgraded rope must be clearly marked so that it cannot be confused with a life safety rope.

TABLE 10-5	Questions to Consider When Inspecting Life Safety Ropes

- Has the rope been exposed to heat or flame?
- Has the rope been exposed to abrasion?
- Has the rope been exposed to chemicals?
- Has the rope been exposed to shock loads?
- Are there any depressions, discolourations, or lumps in the rope?

FIGURE 10-18 Rope inspection is a critical step.

TABLE 10-6	Signs of Possible Rope Deterioration

- Discolouration
- Shiny markings from heat or friction
- Damaged sheath
- Core fibres poking through the sheath

Follow the steps in **SKILL DRILL 10-3** (Fire Fighter I, NFPA 5.5.1) to inspect fire department ropes:

1. Inspect the rope after each use and at regular intervals.
2. Examine the core by looking for depressions, flat spots, or lumps.
3. Examine the sheath by looking for discolourations, abrasions, flat spots, and imbedded objects.
4. Remove the rope from service if it is damaged.

Rope Record

Each piece of rope must be marked for identification. In addition, a **rope record** must be kept for each piece of life safety rope. This record should include a history of when the rope was purchased, when it was used, how it was used, and what kinds of loads were applied to it. Each inspection should also be recorded. Fire departments should maintain rope records for both utility ropes and life safety ropes.

■ Store the Rope

Proper care ensures a long life for your rope and reduces the chance of equipment failure and accident. Store ropes away from temperature extremes, out of sunlight, and in areas where there is some air circulation. Avoid placing ropes where fumes from gasoline, oils, or hydraulic fluids might damage them. Apparatus compartments used to store ropes should be separated from compartments used to store any oil-based products or machinery powered by gasoline or diesel fuel. Do not place any heavy objects on top of the rope **FIGURE 10-19**.

Rope bags are used to protect and store ropes. Each bag should hold only one rope. Rope may also be coiled for storage **FIGURE 10-20**. Very long pieces of rope are sometimes stored on spools **FIGURE 10-21**.

Follow the steps in **SKILL DRILL 10-4** to place a life safety rope into a rope bag:

1. Inspect rope to be certain it is fit for service.
2. Carefully load the life safety rope into the rope bag. **(STEP 1)**
3. Do not try to coil the rope in the bag because this action will cause the rope to kink and become tangled when it is pulled out.
4. Return the rope and the rope bag to its proper location for storage or service. **(STEP 2)**

Knots

Knots are prescribed ways of fastening lengths of rope or webbing to objects or to each other. As a fire fighter, you must know how to tie certain knots and when to use each type of knot. Knots can be used for one or more particular purposes. Hitches, such as the clove hitch, are used to attach a rope

FIGURE 10-19 Ensure that ropes are stored safely.

FIGURE 10-20 Ropes may be coiled for storage.
© AbleStock

FIGURE 10-21 Long sections of rope are sometimes stored on spools.

SKILL DRILL 10-4

Placing a Life Safety Rope in a Rope Bag
(Fire Fighter I, NFPA 5.5.1)

1 Inspect rope to be certain it is fit for service. Carefully load the life safety rope into the rope bag.

2 Do not try to coil the rope in the bag because this action will cause the rope to kink and become tangled when it is pulled out. Return the rope and the rope bag to its proper location for storage or service.

© LiquidLibrary

TABLE 10-7	Effect of Knots on Rope Strength	
Group	**Knot**	**Reduction in Strength**
Loop knots	Figure eight on a bight	20%
	Figure eight follow-through	19%
	Bowline	33%

around an object. Knots, such as the figure eight and the bowline, are used to form loops. Bends, such as the sheet bend or Becket bend, are used to join two ropes together. Safety knots, such as the overhand knot, are used to secure the ends of ropes to prevent them from coming untied.

Any knot reduces the load-carrying capacity of the rope in which it is placed by a certain percentage **TABLE 10-7**. You can avoid an unnecessary reduction in rope strength if you know which type of knot to use and how to tie the knot correctly.

■ Terminology

Specific terminology is used to refer to the parts of a rope in describing how to tie knots **FIGURE 10-22**:

- The **working end** is the part of the rope used for forming the knot.
- The **running end** is the part of the rope used for lifting or hoisting.
- The **standing part** is the rope between the working end and the running end.
- A **bight** is formed by reversing the direction of the rope to form a U bend with two parallel ends **FIGURE 10-23**.

- A **loop** is formed by making a circle in the rope **FIGURE 10-24**.
- A **round turn** is formed by making a loop and then bringing the two ends of the rope parallel to each other **FIGURE 10-25**.

A fire fighter should know how to tie and properly use these simple knots:

- Safety knot (overhand knot)
- Half hitch
- Clove hitch
- Figure eight
- Figure eight on a bight
- Figure eight follow-through

FIGURE 10-22 The sections of a rope used in tying knots.

FIGURE 10-23 A bight.

FIGURE 10-24 A loop.

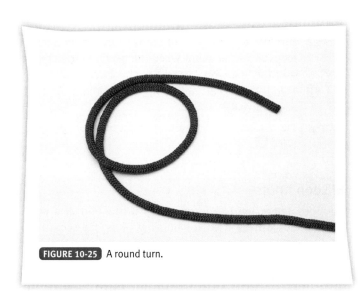
FIGURE 10-25 A round turn.

- Figure eight bend
- Bowline
- Bend (sheet or Becket bend)

■ Safety Knot

A **safety knot** (also referred to as an overhand knot or a keeper knot) is used to secure the leftover working end of the rope. It provides a degree of safety to ensure that the primary knot will not become undone. A safety knot should always be used to finish the other basic knots.

A safety knot is simply an overhand knot in the loose end of the rope that is made around the standing part of the rope. It secures the loose end and prevents it from slipping back through the primary knot.

Follow the steps in **SKILL DRILL 10-5** to tie a safety knot:

1. Take the loose end of the rope, beyond the knot, and form a loop around the standing part of the rope. (**STEP ❶**)
2. Pass the loose end of the rope through the loop. (**STEP ❷**)
3. Tighten the safety knot by pulling on both ends at the same time.
4. Test whether you have tied a safety knot correctly by sliding it on the standing part of the rope. A knot that is tied correctly will slide. (**STEP ❸**)

■ Hitches

Hitches are knots that wrap around an object, such as a pike pole or a fencepost. They are used to secure the working end of a rope to a solid object or to tie a rope to an object before hoisting it.

Half Hitch

The half hitch is not a secure knot by itself, which is why it is used only in conjunction with other knots. For example, when hoisting an axe or pike pole, you use the half hitch to keep the hoisting rope aligned with the handle. On long objects, you might need to use several half hitches.

Follow the steps in **SKILL DRILL 10-6** to tie a half hitch:

1. Grab the rope with your palm facing away from you. (**STEP ❶**)
2. Rotate your hand so that your palm is facing you. This will make a loop in the rope. (**STEP ❷**)
3. Pass the loop over the end of the object. (**STEP ❸**)
4. Finish the half-hitch knot by positioning it and pulling tight. (**STEP ❹**)

Clove Hitch

A clove hitch is used to attach a rope firmly to a round object, such as a tree or a fencepost. It can also be used to tie a hoisting rope around an axe or pike pole. A clove hitch can be tied anywhere in a rope and will hold equally well if tension is applied to either end of the rope or to both ends simultaneously.

There are two different methods of tying this knot. A clove hitch tied in the open is used when the knot can be formed and then slipped over the end of an object, such as an axe or pike pole. It is tied by making two consecutive loops in the rope. Follow the steps in **SKILL DRILL 10-7** to tie a clove hitch in the open:

1. Starting from left to right on the rope, grab the rope with crossed hands with the left positioned higher than the right. (**STEP ❶**)

SKILL DRILL 10-5 Tying a Safety Knot
(Fire Fighter I, NFPA 5.5.2)

1 Take the loose end of the rope, beyond the knot, and form a loop around the standing part of the rope.

2 Pass the loose end of the rope through the loop.

3 Tighten the safety knot by pulling on both ends at the same time. Test whether you have tied a safety knot correctly by sliding it on the standing part of the rope. A knot that is tied correctly will slide. Note: When complete, the safety knot should sit directly next to any accompanying knots.

© LiquidLibrary

2 Holding onto the rope, uncross your hands. This will create a loop in each hand. (**STEP 2**)

3 Slide the right-hand loop behind the left-hand loop. (**STEP 3**)

4 Slide both loops over the object. (**STEP 4**)

5 Pull in opposite directions to tighten the clove hitch.

6 Tie a safety knot in the working end of the rope. (**STEP 5**)

If the object is too large or too long to slip the clove hitch over one end, the same knot can be tied around the object. Follow the steps in **SKILL DRILL 10-8** to tie a clove hitch around an object:

1 Place the working end of the rope over the object. (**STEP 1**)

2 Make a complete loop around the object, working end down. (**STEP 2**)

FIRE FIGHTER Tips

When tying a safety knot in a clove hitch, the end of the rope should pass around the object the clove hitch is being tied to and the overhand safety knot is tied onto the rope.

3 Make a second loop around the object a short distance above the first loop.

4 Pass the working end of the rope under the second loop, above the point where the second loop crosses over the first loop. (**STEP 3**)

5 Tighten the knot and secure it by pulling on both ends. (**STEP 4**)

6 Tie a safety knot in the working end of the rope. (**STEP 5**)

■ Loop Knots

Loop knots are used to form a loop in the end of a rope. These loops can be used for hoisting tools, for securing a person during a rescue, for securing a rope to a fixed object, or for identifying the end of a rope stored in a rope bag. When tied properly, these knots will not slip, yet are easy to untie.

Figure Eight Knot

A figure eight is a basic knot used to produce a family of other knots, including the figure eight on a bight, the figure eight

SKILL DRILL 10-6 Tying a Half Hitch
(Fire Fighter I, NFPA 5.5.2)

1 Grab the rope with your palm facing away from you.

2 Rotate your hand so that your palm is facing you. This will make a loop in the rope.

3 Pass the loop over the end of the object.

4 Finish the half-hitch knot by positioning it and pulling tight.

© LiquidLibrary

with a follow-through, and the figure eight bend. Follow the steps in **SKILL DRILL 10-9** to tie a figure eight knot:

1 Form a bight in the rope. (**STEP 1**)

2 Loop the working end of the rope completely around the standing end of the rope. (**STEP 2**)

3 Thread the working end through the opening of the bight. (**STEP 3**)

4 Tighten the knot by pulling on both ends simultaneously. When you pull the knot tight, it will have the shape of a figure eight. (**STEP 4**)

SKILL DRILL 10-7   Tying a Clove Hitch in the Open
(Fire Fighter I, NFPA 5.5.2)

1 Starting from left to right on the rope, grab the rope with crossed hands with the left positioned higher than the right.

2 Holding onto the rope, uncross your hands. This will create a loop in each hand.

3 Slide the right-hand loop behind the lefthand loop.

5 Pull in opposite directions to tighten the clove hitch. Tie a safety knot in the working end of the rope.

4 Slide both loops over the object.

© LiquidLibrary

Figure Eight on a Bight

The figure eight on a bight knot creates a secure loop at the working end of a rope. This loop can be used to attach the end of the rope to a fixed object or a piece of equipment or to tie a life safety rope around a person. Follow the steps in **SKILL DRILL 10-10** to tie a figure eight on a bight:

1 Double over a section of the rope to form a bight. The closed end of the bight becomes the working end of the rope. (**STEP ❶**)

2 Hold the two sides of the bight as if they were one rope.

SKILL DRILL 10-8   Tying a Clove Hitch Around an Object
(Fire Fighter I, NFPA 5.5.2)

1   Place the working end of the rope over the object.

2   Make a complete loop around the object, working end down.

3   Make a second loop around the object a short distance above the first loop. Pass the working end of the rope under the second loop, above the point where the second loop crosses over the first loop.

4   Tighten the knot and secure it by pulling on both ends.

5   Tie a safety knot in the working end of the rope.

Near-Miss REPORT

Report Number: 10-1038

Synopsis: Knot overlooked during rope training.

Event Description: Our rescue unit was staffed one day with untrained personnel. The apparatus was scheduled for rope training and the untrained member was going to rappel. All safety equipment was checked and appropriate rigging and knots were tied and checked for safe operation. As the untrained member stepped into position to rappel, someone noticed that the knot wasn't properly attached to the member's harness. The member was pulled away from the edge and the equipment was checked again.

Lessons Learned: Untrained personnel should not be allowed to participate in training if they are not educated in all areas of the training.

---

3 Form a loop in the doubled section of the rope. (**STEP ②**)

4 Pass the working end of the bight through the loop. (**STEP ③**)

5 Pull the knot tight. Pulling the knot tight locks the neck of the bight and forms a secure loop. (**STEP ④**)

6 Use a safety knot to tie the loose end of the rope to the standing part. (**STEP ⑤**)

Figure Eight Follow-Through

A figure eight follow-through knot creates a secure loop at the end of the rope when the working end must be wrapped around an object or passed through an opening before the loop can be formed. It is very useful for attaching a rope to a fixed ring or a solid object with an "eye." Follow the steps in **SKILL DRILL 10-11** to tie a figure eight with a follow-through:

1 Tie a simple figure eight in the standing part of the rope, far enough back to make a loop. Leave this knot loose. (**STEP ①**)

2 Thread the working end through the opening or around the object, and bring it back through the original figure eight knot in the opposite direction. (**STEP ②**)

3 Once the working end has been threaded through the knot, pull the knot tight. (**STEP ③**)

4 Secure the loose end with a safety knot. (**STEP ④**)

Figure Eight Bend

The figure eight bend or tracer 8 is used to join two ropes together. Follow the steps in **SKILL DRILL 10-12** to tie a figure eight bend:

1 Tie a figure eight near the end of one rope. (**STEP ①**)

2 Thread the end of the second rope completely through the knot from the opposite end.

3 Pull the knot tight. (**STEP ②**)

4 Tie a safety knot on the loose end of each rope to the standing part of the other. (**STEP ③**)

Bowline

A bowline knot also can be used to form a loop. This type of knot is frequently used to secure the end of a rope to an object or anchor point. It can also be used to hoist equipment. Follow the steps in **SKILL DRILL 10-13** to tie a bowline:

1 Make the desired-sized loop and bring the working end back to the standing part. (**STEP ①**)

2 Form another small loop in the standing part of the rope with the section close to the working end on top.

3 Thread the working end up through this loop from the bottom. (**STEP ②**)

4 Pass the working end over the loop, around and under the standing part, and back down through the same opening. (**STEP ③**)

5 Tighten the knot by holding the working end and pulling the standard part of the rope backward. (**STEP ④**)

6 Tie a safety knot in the working end of the rope. (**STEP ⑤**)

■ Bends

<u>Bends</u> are used to join two ropes together. The sheet bend or Becket bend can be used to join two ropes of unequal size. A sheet bend knot also can be used to join rope to a chain or to pull up a rope of a different diameter. It should not be relied upon to support a human load. Follow the steps in **SKILL DRILL 10-14** to tie a sheet or Becket bend:

1 Form a bight in the working end of one rope. If the ropes are of unequal size, the bight should be made in the larger rope. (**STEP ①**)

2 Pass the end of the second rope up through the opening of the bight, between the two parallel sections of the first rope. (**STEP ②**)

SKILL DRILL 10-9   Tying a Figure Eight Knot
(Fire Fighter I, NFPA 5.5.2)

1 Form a bight in the rope.

2 Loop the working end of the rope completely around the standing part of the rope.

3 Thread the working end back through the bight.

4 Tighten the knot by pulling on both ends simultaneously.

SKILL DRILL 10-10 Tying a Figure Eight on a Bight
(Fire Fighter I, NFPA 5.5.2)

1 Form a bight and identify the closed end of the bight as the working end of the rope.

2 Holding both sides of the bight together, form a loop.

3 Feed the working end of the bight back through the loop.

4 Pull the knot tight.

5 Secure the loose end of the rope with a safety knot.

© LiquidLibrary

SKILL DRILL 10-11 Tying a Figure Eight Follow-Through
(Fire Fighter I, NFPA 5.5.2)

1 Tie a simple figure eight in the standing part of the rope, far enough back to make a loop. Leave this knot loose.

2 Thread the working end through the opening or around the object, and bring it back through the original figure eight knot in the opposite direction.

3 Once the working end has been threaded through the knot, pull the knot tight.

4 Secure the loose end with a safety knot.

SKILL DRILL 10-12 Tying a Figure Eight Bend
(Fire Fighter 1, NFPA 5.1.2)

1 Tie a figure eight near the end of one rope.

2 Thread the end of the second rope completely through the knot from the opposite end. Pull the knot tight.

3 Tie a safety knot on the loose end of each rope to the standing part of the other.

SKILL DRILL 10-13 Tying a Bowline
(Fire Fighter I, NFPA 5.5.2)

1 Make the desired-sized loop and bring the working end back to the standing part.

2 Form another small loop in the standing part of the rope with the section close to the working end on top. Thread the working end up through this loop from the bottom.

3 Pass the working end over the loop, around and under the standing part, and back down through the same opening.

4 Tighten the knot by holding the working end and pulling the standard part of the rope backward.

5 Tie a safety knot in the working end of the rope.

SKILL DRILL 10-14 Tying a Sheet or Becket Bend
(Fire Fighter I, NFPA 5.5.2)

1 Using your left hand, form a bight at the working end of the first (larger) rope.

2 Thread the working end of the second (smaller) rope up through the bight.

3 Loop the second (smaller) rope completely around both sides of the bight. Pass the working end of the second (smaller) rope between the original bight and under the second rope.

4 Tighten the knot.

5 Tie a safety knot in the working end of each rope.

SKILL DRILL 10-15 | Tying a Water Knot
(Fire Fighter I, NFPA 5.5.2)

1 In one end of the webbing, approximately 152 mm (6 in.) from the end, tie an overhand knot.

2 With the other end of the webbing, start retracing from the working end through the knot until approximately 152 mm (6 in.) is left on the other end.

3 Tie an overhand knot on each tail as a safety.

© LiquidLibrary

3 Loop the second rope completely around both sides of the bight.

4 Thread the second rope under itself and on top of the two sides of the bight. (**STEP 3**)

5 To tighten the knot, hold the first rope firmly while pulling back on the second rope. (**STEP 4**)

6 Use a safety knot to secure the loose ends of both ropes. (**STEP 5**)

Water Knot

The water knot or ring bend is used to join webbing of the same or different sizes together. When a single piece of webbing is used and the opposite ends are tied to each other, a loop or sling is created. Follow the steps in **SKILL DRILL 10-15** to tie a water knot:

1 In one end of the webbing, approximately 152 mm (6 in.) from the end, tie overhand knot. (**STEP 1**)

2 With the other end of the webbing, start retracing from the working end through the knot until approximately 152 mm (6 in.) is left on the other end. (**STEP 2**)

3 Tie an overhand knot on each tail as a safety. (**STEP 3**)

There are many ways to tie each of these knots. Find one method that works for you and use it all the time. In addition, your department might require that you learn how to tie other

knots. It is important to become proficient in tying knots. With practice, you should be able to tie these knots in the dark, with heavy gloves on, and behind your back.

A knot should be properly "dressed" by tightening and removing twists, kinks, and slack from the rope. The finished knot is then firmly fixed in position. The configuration of a properly dressed knot should be evident so that it can be easily inspected.

Knot-tying skills can be quickly lost without adequate practice. Practise tying knots while you are on the telephone or watching television **FIGURE 10-26**. You never know when you will need to use these skills in an emergency situation. For added practice, try tying these knots with your gloves on or in darkness. These are conditions you will encounter often as a fire fighter.

Hoisting

Tying knots is not an idle exercise but rather a practical skill that you will frequently use on the job. In emergency situations, you might have to raise or lower a tool to other fire fighters. It is important for you to learn how to raise and lower an axe, a pike pole, a ladder, a charged hose line, an uncharged hose line, and an exhaust fan.

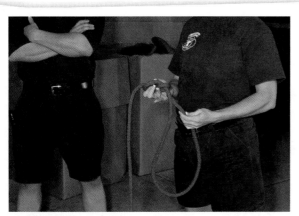

FIGURE 10-26 To maintain your knot-tying skills, practise tying different knots frequently.

It is essential to ensure that the rope is tied securely to the object being hoisted so that the tool does not fall. Additionally, your coworkers must be able to remove the rope and place the tool into service quickly. When you are hoisting or lowering a tool, make sure no one is standing under the object. Keep the scene clear of people to avoid any chance of an accident.

Hoisting an Axe

An axe should be hoisted in a vertical position with the head of the axe down. After receiving the end of the utility rope from the fire fighter above, follow the steps in **SKILL DRILL 10-16** to hoist an axe:

1. Tie the end of the hoisting rope around the handle near the head using either a figure eight on a bight or a clove hitch. Slip the knot down the handle from the end to the head. (**STEP 1**)

2. Loop the standing part of the rope under the head. (**STEP 2**)

3. Place the standing part of the rope parallel to the axe handle. (**STEP 3**)

4. Use one or two half hitches along the axe handle to keep the handle parallel to the rope.

5. Communicate with the fire fighter above that the axe is ready to raise. (**STEP 4**)

To release the axe, hold the middle of the handle, release the half hitches, and slip the knot up and off.

Hoisting a Pike Pole

A pike pole should be hoisted in a vertical position with the head at the top. After receiving the end of the utility rope from the fire fighter above, follow the steps in **SKILL DRILL 10-17** to hoist a pike pole:

1. Place a clove hitch over the bottom of the handle and secure it close to the bottom of the handle.

2. Leave enough length of rope below the clove hitch for a tag line while raising the pike pole. (**STEP 1**)

3. Place a half hitch around the handle above the clove hitch to keep the rope parallel to the handle. (**STEP 2**)

4. Slip a second half hitch over the handle and secure it near the head of the pike pole. (**STEP 3**)

5. Communicate with the fire fighter above that the pike pole is ready to raise. (**STEP 4**)

Hoisting a Ladder

A ladder should be hoisted in a vertical position. A tag line should be attached to the bottom to keep the ladder under control as it is hoisted. If it is a roof ladder, the hooks should be in the retracted position. After receiving the end of the utility rope from the fire fighter above, follow the steps in **SKILL DRILL 10-18** to hoist a ladder:

1. Tie a figure eight on a bight to create a loop, approximately 1–1.3 m (3 or 4 ft.) in diameter, that is large enough to fit around both ladder beams. (**STEP 1**)

2. Pass the rope between two beams of the ladder, three or four rungs from the top.

3. Pull the end of the loop under the rungs and toward the tip at the top of the ladder. (**STEP 2**)

4. Place the loop around the top tip of the ladder and pull on the running end of the rope to remove the slack from the rope. (**STEP 3**)

5. Attach a tag line to the bottom rung of the ladder to stabilize it as it is being hoisted.

6. Communicate with the fire fighter above that the ladder is ready to be raised.

7. Stabilize the bottom of the ladder as it is being hoisted by holding on to the tag line. (**STEP 4**)

Hoisting a Charged Hose Line

It is almost always preferable to hoist a dry hose line because water adds considerable weight to a charged line. Water weighs 3.79 kilograms (8.33 pounds) per gallon, which can make hoisting a charged line much more difficult. However, there can be occasions when it is necessary to hoist a charged hose line. After receiving the end of the utility rope from the fire fighter above, follow the steps in **SKILL DRILL 10-19** to hoist a charged hose line:

1. Make sure that the nozzle is completely closed and secure. A charged hose line should have the nozzle secured in a closed position as it is hoisted because an unsecured handle (known as the bale) could get caught on something as the hose is being hoisted. The nozzle could open suddenly, resulting in an out-of-control hose line suspended in midair, with the potential for causing serious injuries and damage.

2. Use a clove hitch, 0.3–0.6 m (1 or 2 ft.) behind the nozzle, to tie the end of the hoisting rope around a charged hose line.

3. Use a safety knot to secure the loose end of the rope below the clove hitch. (**STEP 1**)

SKILL DRILL 10-16 | Hoisting an Axe
(Fire Fighter I, NFPA 5.5.2)

1 Tie the end of the hoisting rope around the handle near the head using either a figure eight on a bight or a clove hitch. Slip the knot down the handle from the end to the head.

2 Loop the standing part of the rope under the head.

3 Place the standing part of the rope parallel to the axe handle.

4 Use one or two half hitches along the axe handle to keep the handle parallel to the rope. Communicate with the fire fighter above that the axe is ready to raise.

© LiquidLibrary

4 Make a bight in the rope even with the nozzle shut-off handle, which must be in the forward (completely off) position.

5 Insert the bight through the handle opening and slip it over the end of the nozzle. When the bight is pulled tight, it will create a half hitch and secure the handle in the off position while the charged hose line is hoisted.

6 Communicate with the fire fighter above that the hose line is ready to hoist. (**STEP 2**)

The knot can be released after the line is hoisted by removing the tension from the rope and slipping the bight back over the end of the nozzle.

■ Hoisting an Uncharged Hose Line

Before hoisting a dry hose line, you should fold the hose back on itself and place the nozzle on top of the hose. This action ensures that water will not reach the nozzle if the hose is accidentally charged while being hoisted. It also eliminates an unnecessary stress on the couplings by ensuring that the rope pulls on the hose and not directly on the nozzle. After receiving the end of the utility rope from the fire fighter above, follow the steps in **SKILL DRILL 10-20** to hoist an uncharged hose line:

1 Fold about 1 m (3 ft.) of hose back on itself and place the nozzle on top of the hose. (**STEP 1**)

SKILL DRILL 10-17   Hoisting a Pike Pole
(Fire Fighter I, NFPA 5.5.2)

1 Place a clove hitch over the bottom of the handle and secure it close to the bottom of the handle. Leave enough length of rope below the clove hitch for a tag line while raising the pike pole.

2 Place a half hitch around the handle above the clove hitch to keep the rope parallel to the handle.

3 Slip a second half hitch over the handle and secure it near the head of the pike pole.

4 Communicate with the fire fighter above that the pike pole is ready to raise.

© LiquidLibrary

2 Make one or two half hitches in the rope and slip it over the nozzle. Move the half hitch along the hose and secure it about 15 centimetre (about 6 in.) from the fold. (**STEP 2**)

3 Tie a clove hitch near the end of the rope, wrapping the rope securely around both the nozzle and the hose. The clove hitch should hold both the nozzle and the hose. (**STEP 3**)

4 Communicate with the fire fighter above that the hose line is ready to hoist.

5 Hoist the hose with the fold at the top and the nozzle pointing down.

6 Before releasing the rope, the fire fighters at the top must pull up enough hose so that the weight of the hanging hose does not drag down the hose. (**STEP 4**)

SKILL DRILL 10-18 Hoisting a Ladder
(Fire Fighter I, NFPA 5.5.2)

1 Tie a figure eight on a bight to create a loop, approximately 0.9 to 1.2 m (3 or 4 ft.) in diameter, that is large enough to fit around both ladder beams.

2 Pass the rope between two beams of the ladder, three or four rungs from the top. Pull the end of the loop under the rungs and toward the tip at the top of the ladder.

3 Place the loop around the top tip of the ladder. Pull on the running end of the rope to remove the slack from the rope.

4 Attach a tag line to the bottom rung of the ladder to stabilize it as it is being hoisted. Communicate with the fire fighter above that the ladder is ready to be raised. Stabilize the bottom of the ladder as it is being hoisted by holding on to the tag line.

© LiquidLibrary

FIRE FIGHTER II Tips   **FFII**

Ropes and knots are challenging in the best of circumstances. To be proficient and to work in an effective and efficient manner when tying knots and moving equipment, you should consistently work on your rope and knot skills. Keep a short piece of rope close at hand so that you can practise your skills during down times.

■ Hoisting an Exhaust Fan or Power Tool

Several types of tools and equipment—including an exhaust fan, a chainsaw, a circular saw, or any other object that has a strong closed handle—can be hoisted using the same technique. In this situation, the hoisting rope is secured to the object by passing the rope through the opening in the handle. A figure eight with a follow-through knot is then used to close the loop.

SKILL DRILL 10-19   Hoisting a Charged Hose Line
(Fire Fighter I, NFPA 5.5.2)

1 Make sure that the nozzle is completely closed and secure. Use a clove hitch, 0.3 to 0.6 m (1 or 2 ft.) behind the nozzle, to tie the end of the hoisting rope around a charged hose line. Use a safety knot to secure the loose end of the rope below the clove hitch.

2 Make a bight in the rope even with the nozzle shut-off handle. Insert the bight through the handle opening and slip it over the end of the nozzle. When the bight is pulled tight, it will create a half hitch and secure the handle in the off position while the charged hose line is hoisted. Communicate with the fire fighter above that the hose line is ready to hoist.

© LiquidLibrary

Some types of equipment require that you use additional half hitches to balance the object in a particular position while it is being hoisted. For example, power saws are hoisted in a level position to prevent the fuel from leaking out.

You should practise hoisting the actual tools and equipment used in your department. You should be able to perform this task automatically and in adverse conditions.

Remember that you always use utility rope for hoisting tools. You do not want to get oil or grease on designated life safety ropes. If a life safety rope becomes oily or greasy, it should be taken out of service and destroyed so that it will not be mistakenly used again as a life safety rope. The damaged rope can be cut into short lengths and used for utility rope.

After receiving the end of the utility rope from the fire fighter above, follow the steps in **SKILL DRILL 10-21** to hoist an exhaust fan:

1 Tie a figure eight knot in the rope about 1 m (3 ft.) from the working end of the rope. (**STEP ❶**)

2 Loop the working end of the rope around the fan handle and back to the figure eight knot. (**STEP ❷**)

3 Secure the rope by tying a figure eight with a follow-through by threading the working end back through the first figure eight in the opposite direction.

4 Attach a tag line to the fan for better control.

5 Communicate with the fire fighter above that the exhaust fan is ready to hoist. (**STEP ❸**)

SKILL DRILL 10-20	Hoisting an Uncharged Hose Line
	(Fire Fighter I, NFPA 5.5.2)

1 Fold about 1 m (3 ft.) of hose back on itself and place the nozzle on top of the hose.

2 Make one or two half hitches in the rope and slip it over the nozzle. Move the half hitch along the hose and secure it about 15 cm (6 in.) from the fold.

3 Tie a clove hitch near the end of the rope, wrapping the rope securely around both the nozzle and the hose. The clove hitch should hold both the nozzle and the hose.

4 Communicate with the fire fighter above that the hose line is ready to hoist. Hoist the hose with the fold at the top and the nozzle pointing down. Before releasing the rope, the fire fighters at the top must pull up enough hose so that the weight of the hanging hose does not drag down the hose.

© LiquidLibrary

SKILL DRILL 10-21 | Hoisting an Exhaust Fan or Power Tool
(Fire Fighter I, NFPA 5.5.2)

1 Tie a figure eight knot in the rope about 1 m (3 ft.) from the working end of the rope.

2 Loop the working end of the rope around the fan handle and back to the figure eight knot.

3 Secure the rope by tying a figure eight with a follow-through by threading the working end back through the first figure eight in the opposite direction. Attach a tag line to the fan for better control. Communicate with the fire fighter above that the exhaust fan is ready to hoist.

© LiquidLibrary

Chief Concepts

- The three primary types of rope in the fire service are
  - *Life safety rope.* Used to support people during a rescue. This type of rope is used only for life-saving purposes.
  - *Escape rope.* Used as a single-purpose emergency self-escape rope
  - *Utility rope.* Used to perform all other tasks, such as hoisting equipment.
- Life safety ropes are rated as either technical use life safety rope or general use life safety rope.
- An escape rope is designed to be used once by one fire fighter.
- Ropes can be made of natural or synthetic fibres. Natural fibres can be weakened by mildew and age, so they cannot be used to make life safety ropes. Synthetic fibres are generally stronger than natural fibres and are used in life safety ropes.
- Synthetic fibres can be damaged by ultraviolet light.
- Nylon is the most common synthetic fibre used in life safety ropes, followed by polyester and polypropylene. Polypropylene does not absorb water, so it is often used for water rescues.
- Common rope construction includes the following types:
  - *Twisted ropes.* Individual fibres twisted into strands that are twisted together to form the rope.
  - *Braided ropes.* Constructed by weaving or intertwining strands together in the same way that hair is braided.
  - *Kernmantle ropes.* Consists of the kern and the mantle. The kern is the core of the rope and the mantle is the braided covering that protects the core. These ropes can be either dynamic or static.
- During technical rescue incidents, several hardware components are used in addition to rope to rescue victims:
  - *Carabiner.* Connects one rope to another rope or a harness
  - *Harness.* Webbing that secures a person to a rope or solid object
- Most rope rescue incidents involve people who are trapped in normally inaccessible locations such as trenches, confined spaces, and open water.
- All ropes need proper care to perform in an optimal manner. The four parts to the rope maintenance formula are the following:
  - Care
  - Clean
  - Inspect
  - Store

- The principles of caring for a rope include the following:
  - Protect the rope from sharp and abrasive surfaces.
  - Protect the rope from rubbing against another rope or webbing.
  - Protect the rope from heat, chemicals, and flames.
  - Protect the rope from prolonged exposure to sunlight.
- When inspecting life safety rope, consider these questions:
  - Has the rope been exposed to heat or flame?
  - Has the rope been exposed to abrasion?
  - Has the rope been exposed to chemicals?
  - Has the rope been exposed to shock loads?
  - Are there any depressions, discolourations, or lumps in the rope?
- A rope record for a life safety rope includes a history of when the rope was purchased, when it was used, how it was used, each inspection, and what kinds of loads were applied to it.
- Ropes should be stored away from temperature extremes, out of sunlight, and in areas with good air circulation. Rope bags can be used to protect and store ropes.
- Knots can be used for one or more particular purposes. Hitches, such as the clove hitch, are used to attach a rope around an object. Knots, such as the figure eight and the bowline, are used to form loops. Bends, such as the sheet bend or Becket bend, are used to join two ropes together. Safety knots, such as the overhand knot, are used to secure the ends of ropes to prevent them from coming untied.
- Specific terminology is used to refer to the parts of a rope in describing how to tie knots:
  - The working end is the part of the rope used for forming the knot.
  - The running end is the part of the rope used for lifting or hoisting.
  - The standing part is the rope between the working end and the running end.
  - A bight is formed by reversing the direction of the rope to form a U bend with two parallel ends.
  - A loop is formed by making a circle in the rope.
  - A round turn is formed by making a loop, and then bringing the two ends of the rope parallel to each other.
- The knots that a fire fighter should know how to tie are as follows:
  - *Safety knot.* Used to finish other basic knots
  - *Half hitch.* Knots that wrap around an object
  - *Clove hitch.* Used to attach a rope to a round object

- *Figure eight.* Basic knot used to produce a family of other knots
- *Figure eight on a bight.* Used to create a secure loop at the working end of a rope
- *Figure eight follow-through.* Used to create a secure loop at the end of the rope when the working end must be wrapped around an object or passed through an opening before the loop can be formed

- *Bowline.* Used to create a loop
- *Water Knot.* Used to join the ends of webbing together
- The bends that a fire fighter should know how to create in order to join two ropes together are the sheet bend and the Becket bend.

Hot Terms

Bend A knot used to join two ropes together.

Bight A U shape created by bending a rope with the two sides parallel.

Block creel construction Rope constructed without knots or splices in the yarns, ply yarns, strands or braids, or rope. (NFPA 1983)

Braided rope Rope constructed by intertwining strands in the same way that hair is braided.

Carabiner An auxiliary equipment system item; load-bearing connector with a self-closing gate used to join other components of life safety rope. (NFPA 1983)

Depressions Indentations felt on a kernmantle rope that indicate damage to the interior (kern) of the rope.

Dynamic rope A rope generally made from synthetic materials that is designed to be elastic and stretch when loaded. Dynamic rope is often used by mountain climbers.

Escape rope A system component; a single-purpose, one-time use, emergency self-escape (self-rescue) rope; not classified as a life safety rope. (NFPA 1983)

Harness A piece of equipment worn by a rescuer that can be attached to a life safety rope.

Hitch A knot that attaches to or wraps around an object so that when the object is removed, the knot will fall apart. (NFPA 1670, *Standard on Operations and Training for Technical Search and Rescue Incidents*)

General use life safety rope A life safety rope that is no larger than 16 mm (0.7 in.) and no smaller than 11 mm (0.5 in.), with a minimum breaking strength of 40 kN (8992 lbf).

Kernmantle rope Rope made of two parts—the kern (interior component) and the mantle (the outside sheath).

Knot A fastening made by tying together lengths of rope or webbing in a prescribed way. (NFPA 1670)

Ladder halyards Rope used on extension ladders to raise a fly section.

Life safety rope Rope dedicated solely for the purpose of supporting people during rescue, firefighting, other emergency operations, or during training evolutions. (NFPA 1983)

Loop A piece of rope formed into a circle.

Rope bag A bag used to protect and store rope so that the rope can be easily and rapidly deployed without kinking.

Rope record A record for each piece of rope that includes a history of when the rope was placed in service, when it was inspected, when and how it was used, and which types of loads were placed on it.

Round turn A piece of rope looped to form a complete circle with the two ends parallel.

Running end The part of a rope used for lifting or hoisting.

Safety knot A knot used to secure the leftover working end of the rope.

Shock load An instantaneous load that places a rope under extreme tension, such as when a falling load is suddenly stopped as the rope becomes taut.

Standing part The part of a rope between the working end and the running end.

Static rope A rope generally made out of synthetic material that stretches very little under load.

Technical use life safety rope A life safety rope with a diameter that is 9.5 mm (0.4 in.) or greater, but is less than 12.5 mm (0.5 in.), with a minimum breaking strength of 20 kN (4496 lbf). Used by highly trained rescue teams that deploy to very technical environments, such as mountainous and/or wilderness terrain.

Twisted rope Rope constructed of fibres twisted into strands, which are then twisted together.

Utility rope Rope used for securing objects, for hoisting equipment, or for securing a scene to prevent bystanders from being injured. Utility rope must never be used in life safety operations.

Water knot A knot used to join the ends of webbing together.

Working end The part of the rope used for forming a knot.

You are normally assigned to a truck company, but today you are assigned to the rescue company at another station. The captain is scheduled to teach the Intro to Tech Rescue course to the fire fighters from several departments, so you quickly familiarize yourself with the location of the equipment, climb on, and head to the training centre. It quickly becomes apparent that one particular student lacks basic rope skills that the remainder of the class already possesses. The lead instructor asks you to review the basics with the fire fighter while the rest get everything laid out.

1. What type of rope is designed to be used for securing objects, hoisting equipment, and blocking off scenes?
   A. Technical use life safety rope
   B. General use life safety rope
   C. Utility rope
   D. Escape rope

2. The part of a rope used for lifting or hoisting is called the _____.
   A. standing end
   B. running end
   C. bight
   D. round turn

3. The knot used to secure the leftover working end of the rope is called the _____.
   A. safety
   B. figure eight
   C. bowline
   D. half hitch

4. A _____ rope is a rope generally made out of synthetic material that stretches very little under load.
   A. dynamic
   B. twisted
   C. braided
   D. static

5. Once rope has gotten wet, what should be done with it?
   A. Lay it in the sunlight so that it does not mould.
   B. Air dry it out of direct sunlight.
   C. Place it in a clothes dryer on high heat.
   D. The rope can no longer be used.

6. Which knot is used to create a secure loop at the end of the rope when the working end must be wrapped around an object or passed through an opening before the loop can be formed?
   A. Figure eight follow-through
   B. Figure eight on a bight
   C. Half hitch
   D. Clove hitch

7. A _____ is when a piece of rope is looped to form a complete circle with the two ends parallel.
   A. bight
   B. hitch
   C. round turn
   D. loop

You are dispatched to a scene where a young boy has climbed up on a cell tower. On your arrival, you find that he is clinging tightly to an I-beam about 15 m (50 ft.) up the tower and the parents of the boy are on-scene and they are frantic. The terrain does not allow for the use of an aerial ladder, so the decision is made to use a high-angle rescue. The crowd is getting very upset because of the delay. You will remain on the ground while the rope rescue technician makes the ascent.

1. What type of hardware will be needed for this rescue?
2. What type of harness will be needed for this rescue?
3. How will the parents be reassured?
4. How will emotions of the crowd be managed?

Response and Size-Up

© Keith Muratori/ShutterStock, Inc.

Fire Fighter I

Knowledge Objectives

After studying this chapter, you will be able to:

- Explain how fire stations receive and process dispatch information. (p 295)
- Identify the information in a typical dispatch message. (p 295)
- Describe how to safely ride a fire apparatus to an emergency scene. (NFPA 5.3.2 , p 295–296)
- List the NFPA standards that require fire fighters to wear safety belts while riding in a fire apparatus. (NFPA 5.3.2 , p 296)
- List the prohibited practices when riding in a fire apparatus to an emergency scene. (NFPA 5.3.2.A , p 297)
- Describe how to manage traffic safely at an emergency scene. (NFPA 5.3.3, 5.3.3.A , p 297–298)
- Describe the actions fire fighters must take when arriving at an emergency scene. (NFPA 5.3.3 , p 298)
- Explain how to shut off a structure's electrical service. (NFPA 5.3.3, 5.3.18, 5.3.18.A, 5.3.18.B , p 299–300)

- Explain how to shut off a structure's gas service. (NFPA 5.3.3, 5.3.18, 5.3.18.A, 5.3.18.B , p 300)
- Explain how to shut off a structure's water service. (NFPA 5.3.3, 5.3.18, 5.3.18.A, 5.3.18.B , p 300)
- Describe the process of performing an initial size-up. (p 300–304)
- List the two basic categories of information used in the size-up process. (p 301–303)
- Explain how the size-up process determines the resources required at the emergency incident. (p 303–304)
- List the five objectives of an incident action plan (IAP). (p 304–305)
- Describe the two methods of attack used to extinguish a fire. (p 306)

Skills Objectives

After studying this chapter, you will be able to perform the following skills:

- Mount an apparatus safely. (NFPA 5.3.2.A, 5.3.2.B , p 295–296)
- Dismount from an apparatus safely. (NFPA 5.3.2.B, 5.3.3.B , p 297)

© Photos.com

Fire Fighter II · FFII

Knowledge Objectives

There are no knowledge objectives for Fire Fighter II candidates. NFPA 1001 contains no Fire Fighter II Job Performance Requirements for this chapter.

Skills Objectives

There are no skills objectives for Fire Fighter II candidates. NFPA 1001 contains no Fire Fighter II Job Performance Requirements for this chapter.

Additional NFPA Standards

- NFPA 1500, *Standard on Fire Department Occupational Safety and Health Program*

You Are the Fire Fighter

You are dispatched to a report of a house fire. As you pull out of the station, the dispatcher reports multiple calls and that occupants are trapped in the building. Upon your arrival, there is black smoke boiling out of a back window, with brown smoke coming from the eaves and upstairs windows. As the fire officer gives a size-up of the building and conditions, a neighbour runs up and says that he heard someone screaming from the upstairs. Without missing a beat, the fire officer develops a mental incident action plan (IAP) and makes assignments to the crews. You are amazed by how she is able to calmly coordinate such a chaotic scene.

1. What type of attack will be made on this fire?
2. What is the highest priority for this fire?
3. What objectives must be assigned for this incident?

Introduction

Response involves a series of actions that begin when a crew is dispatched to an alarm and end with the crew's arrival at the emergency incident. Response actions include receiving the alarm, donning protective clothing and equipment, mounting the apparatus, and transporting equipment and personnel to the emergency incident quickly and safely.

Because fire fighters must be ready to react immediately to an alarm, preparations for response begin long before the alarm is sounded. These preparations include checking personal equipment, ensuring that the fire apparatus is ready, and making sure that all equipment carried on the apparatus is ready for use. Fire fighters also should be familiar with their response district, know the buildings under their protection, and understand their department's suggested operating guidelines (SOGs), or standard operating procedures (SOPs).

Response actions for the apparatus driver also include considering road and traffic conditions, determining the best route to the incident, identifying nearby hydrant locations or water sources, and selecting the best position for the apparatus at the incident scene.

Size-up is a systematic process of information gathering and situation evaluation that begins when an alarm is received. It continues during the response and includes the initial observations made upon arrival at the incident scene. Size-up information is essential for determining the appropriate strategy and tactics for each situation.

The incident commander (IC) and company officers are ultimately responsible for obtaining the necessary information to manage the emergency incident. Each fire fighter should also be involved in the process of gathering and processing information. The observations made by individual fire fighters will help them anticipate which actions might be necessary at the scene and will provide company officers with critical information.

Response

Trained fire fighters must be ready to respond to an emergency at any time during their tour of duty. This process begins by ensuring that your personal protective equipment (PPE) is complete, ready for use, and in good condition. At the beginning of each tour of duty, place your PPE in its designated location, which depends on your assigned riding position on the apparatus **FIGURE 11-1**.

You should also conduct a daily inspection of the self-contained breathing apparatus (SCBA) for your riding position at the beginning of each tour of duty **FIGURE 11-2**. The air

FIGURE 11-1 Check your PPE and place it in the designated location for your riding position.

cylinder should be full, the face piece clean, and the personal alert safety system (PASS) operable. Also check the availability and operation of your hand light and any hand tools you might require, based on your assigned position on the apparatus **FIGURE 11-3** . Recheck your PPE and tools thoroughly when you return from each emergency response, and clean them whenever they are used. This routine will help ensure that your gear and equipment are fully functional when the next alarm comes.

■ Alarm Receipt

The emergency response process begins when an alarm is received at the fire station. Fire fighters should be familiar with the dispatch method or methods used by their departments. In many fire departments, a local or regional communications centre dispatches individual units. In smaller departments, the dispatcher may be located at the fire station or other location. Departments should have both a primary method and a backup method of transmitting alarms to stations.

FIGURE 11-2 Conduct the daily check of your SCBA at the beginning of each tour of duty.

FIGURE 11-3 Check the tools assigned to your riding position.

Although radio, telephone, and public address systems are often used to transmit information to fire stations, the use of computer terminals and printers to transmit dispatch messages is increasing. Some fire departments still use a system of bells to transmit alarms. Although volunteer or rural departments may use outdoor sirens or horns to summon fire fighters to an emergency, most volunteer fire fighters receive dispatch messages over pagers or cell phones.

Dispatch messages contain varying amounts of useful information, based on what the dispatcher learns from the person initially reporting the incident. At a minimum, the dispatch information will include the location of the incident, the type of emergency (e.g., vehicle fire, structure fire, medical call), and the units that are due to respond. Computer-aided dispatch systems often provide additional information about the building or premises on the dispatch printout.

As more information becomes available, the telecommunicator will include it in dispatch messages to later-responding units or advise the responding units by radio while they are en route. Each additional piece of information can help responding fire fighters plan their strategies and prepare themselves for the incident.

When an alarm is received, the response should be prompt and efficient. Responding fire fighters should walk briskly to the apparatus. There is no need to run; the objective is to respond quickly, without injuring anyone or causing any damage. Follow established procedures to ensure that stoves, faucets, and other appliances at the station are shut off. Wait until the apparatus bay doors are fully open before leaving the station.

FIRE FIGHTER Tips

Volunteer fire fighters who are not assigned to specific tours of duty or riding positions should check their PPE, SCBA, and associated tools and equipment on a regular basis to ensure that these items are ready for use whenever an emergency response becomes necessary. After each use, all PPE should be carefully cleaned and checked before it is put away.

■ Riding the Apparatus

Don your PPE before mounting the apparatus; do not attempt to dress while the apparatus is on the road. Your seat belt should be fastened whenever the apparatus is moving. Wait until you dismount at the incident scene to don any protective clothing that was not donned prior to mounting the apparatus. Don SCBA only after the apparatus stops at the scene, unless the SCBA is seat mounted.

All equipment should be properly mounted, stowed, or secured on the fire apparatus. Unsecured equipment in the crew compartment can prove dangerous if the apparatus must stop or turn quickly, because a flying tool, map book, or PPE can seriously injure a fire fighter.

Be careful when mounting and dismounting apparatus, because the steps on fire apparatus are often high and can be slippery. Follow the steps in **SKILL DRILL 11-1** to mount an apparatus properly:

1 When mounting (climbing aboard) fire apparatus, always maintain three points of contact. (**STEP 1**)

2 Fasten your seat belt, and leave it fastened until the apparatus is stopped at its destination. Don any other required safety equipment for the response, such as hearing protection and intercom. (**STEP 2**)

All fire fighters must be seated in their assigned riding positions with seat belts and/or harnesses fastened before the apparatus begins to move. NFPA 1500, *Standard on Fire Department Occupational Safety and Health Program,* and NFPA 1001, *Standard for Fire Fighter Professional Qualifications,* require all fire fighters to be in their seats, with seat belts secured, whenever the vehicle is in motion. Do not unbuckle your seat belt to don any clothing or equipment while the apparatus is en route to an incident.

The noise produced by sirens and air horns can have long-term, damaging effects on a fire fighter's hearing. For this reason, your department should provide hearing protection for personnel riding on fire apparatus. Use it. Some of these devices include radio and intercom capabilities so that fire fighters can talk to one another and hear information from the dispatcher or IC.

During transport, limit conversation to the exchange of pertinent information. Listen for instructions from the IC, for instructions from your company officer, and for additional information about the incident over the radio. The vehicle operator's attention should be focused on safely driving the apparatus to the scene of the incident.

The ride to the incident is a good time to consider any relevant factors that could affect the situation. These factors could include the time of day or night, the temperature, the presence of precipitation or wind, the type of occupancy, the type of construction, and the location and type of incident. Using this time to think ahead will help you mentally prepare for the various possibilities that you might encounter at the scene.

■ Emergency Response

The fire apparatus driver/operator must always exercise caution when driving to an incident. Fire apparatus can be very large, heavy, and difficult to maneuver. Operating an emergency vehicle without the proper regard for safety can endanger the lives of both the fire fighters on the vehicle and any civilian drivers and pedestrians encountered along the way. Although the impulse to respond as speedily as possible is understandable, safety should never be compromised for a faster response time.

Fire fighters who drive emergency vehicles must have special driver training and know the laws and regulations that apply to emergency responses. Many jurisdictions require a special driver's licence to operate fire apparatus. The rules that apply to emergency vehicles are very specific, and the driver/operator is legally responsible for the safe operation of the vehicle and the safety of the vehicle occupants at all times.

SKILL DRILL 11-1 Mounting Apparatus
(Fire Fighter I, NFPA 5.3.2)

1 When mounting (climbing aboard) fire apparatus, always maintain three points of contact.

2 Fasten your seat belt and then don any other required safety equipment for the response, such as hearing protection.

© LiquidLibrary

An emergency vehicle must always be operated with due regard for the safety of everyone on the road. Although most provinces/territories permit drivers of emergency vehicles to take exception to specific traffic regulations when responding to emergency incidents, driver/operators must always consider the potential actions of other drivers before making such a decision. For example, traffic laws require other drivers to yield the right-of-way to an emergency vehicle; there is no assurance, however, that other drivers will do so when an emergency vehicle approaches. The driver/operator must also anticipate which routes other units responding to the same incident will take.

Fire fighters who respond to emergency incidents in their personal vehicles must follow the specific laws and regulations of their province/territory and follow departmental SOGs/SOPs related to this issue. Some jurisdictions allow fire fighters to equip their privately owned vehicles with warning devices and to operate them as emergency vehicles; others grant no special status to privately owned vehicles driven during an emergency response. In some areas, volunteer fire fighters responding to an emergency incident can use coloured lights to request the right-of-way from other drivers. Additional information about emergency driving is presented in the chapter "Fire Fighter Safety."

Prohibited Practices

For safety's sake, SOGs/SOPs often prohibit specific actions during response. As previously noted, fire fighters must remain seated with their seat belts securely fastened while the emergency vehicle is in motion. Never unfasten your seat belt to retrieve or don equipment. Do not dismount the apparatus until the vehicle comes to a complete stop.

In addition, you should never stand up while riding on apparatus. Do not hold on to the side of a moving vehicle or stand on the rear step. When a vehicle is in motion, everyone aboard must be seated and belted in an approved riding position.

Dismounting a Stopped Apparatus

When the apparatus arrives at the incident scene, the driver/operator will park it in a location that is both safe and functional. Wait until the vehicle comes to a complete stop before dismounting. Always check for traffic before opening the doors or stepping out of the apparatus. During the dismount, watch for other hazards—for example, ice and snow, downed power lines, uneven terrain, or hazardous materials—which could be present.

Be careful when dismounting apparatus: the increased weight of PPE and adverse conditions can contribute to slips, strains, and sprains. Use handrails when mounting or dismounting the apparatus. Follow the steps in **SKILL DRILL 11-2** to dismount an apparatus safely:

1. Become familiar with your riding position and the safest way to dismount. (**STEP** **1**)
2. Keep one hand grasping a handhold and one foot placed firmly on a flat surface when leaving the apparatus, especially on wet or potentially icy roadway surfaces. (**STEP** **2**)

Traffic Safety on the Scene

An emergency incident scene presents several risks to fire fighters in addition to the hazards of fighting fires and performing other duties. One of these dangers is traffic, particularly when the incident scene is on a street or highway. Traffic safety should be a major concern for the first-arriving units, because approaching drivers might not see emergency workers or realize how much room fire fighters need to work safely.

SKILL DRILL 11-2 Dismounting a Stopped Apparatus
(Fire Fighter I, NFPA 5.3.2, 5.3.3)

1 Become familiar with your riding position and the safest way to dismount.

2 Maintain the one-hand and one-foot placement when leaving the apparatus, especially on wet or potentially icy roadway surfaces.

© LiquidLibrary

The first unit or units to arrive at the incident scene has a dual responsibility: not only must the fire fighters focus on the emergency situation facing them, but they must also consider approaching traffic, including other emergency vehicles and other less-obvious hazards. Always check for traffic before opening doors and dismounting the apparatus, and watch out for traffic when working in the street. Follow departmental SOGs/SOPs to close streets quickly and to block access to areas where operations are being conducted.

One of the most dangerous work areas for fire fighters is on the scene of a highway incident, where traffic can be approaching at high speeds. Place pylons, flares, emergency scene signage, the traffic safety officer, and other warning devices far enough away from the incident to slow approaching traffic and direct it away from the work area. Placement of emergency vehicles on the scene is also critical. With proper placement, such vehicles can act as a barrier between oncoming traffic and the scene. All fire fighters working at highway incidents should wear high-visibility safety vests in addition to their normal PPE. These vests should meet CSA Z96, *High-Visibility Safety Apparel*. Many fire departments have specific SOGs/SOPs covering required safety procedures for these incidents.

FIRE FIGHTER Tips

- Do not attempt to mount or dismount a moving vehicle.
- Do not remove your seat belt until the apparatus comes to a complete stop.
- Do not stand directly behind an apparatus that is backing up. Instead, stand off to one side, where the driver can see you in the rear-view mirror. All fire apparatus should have working, audible alarms when in reverse gear.

Arrival at the Incident Scene

From the moment that fire fighters arrive at an emergency incident scene, their department's SOGs/SOPs and the structured incident management system (IMS) must guide all of their actions. Fire fighters should always work in assigned teams (companies or crews) and be guided by an incident action plan (IAP). Teamwork and disciplined action are essential to provide for the safety of all fire fighters and the effective, efficient conduct of operations.

The command structure plans, coordinates, and directs operations. Fire fighters responding to an incident on an apparatus constitute the crew assigned to that vehicle and take direction from their company officer, who ensures that their actions are coordinated with the overall plan. A fire fighter who arrives at the scene independently of an apparatus must report to the IC and receive an assignment to a company or crew under the supervision of a company officer.

Fire fighters who take action on their own, without regard to SOGs/SOPs, the command structure, or the strategic plan, are **freelancing**. This type of activity—whether it is done by individual fire fighters, groups of fire fighters, or full companies—is unacceptable. Freelancing cannot be tolerated at any emergency incident. The safety of every person on the scene can be compromised by fire fighters who do not work within the system.

Personnel must not respond to an emergency incident scene unless they have been dispatched or have an assigned duty to respond. Unassigned units and individual personnel arriving on the scene can overload the IC's ability to manage the incident effectively. Individuals who simply show up and find something to do are likely to compromise their own safety and create more problems for the command staff. All personnel must operate within the established system, reporting to a designated supervisor under the direction of the IC.

■ Personnel Accountability System

A **personnel accountability system** should be used to track every fire fighter at every incident scene. This system maintains an updated list of the fire fighters assigned to each vehicle or crew and tracks each crew's assignment at the fire scene. If any fire fighters are reported missing, lost, or injured, the accountability system can identify who is missing and what his or her last assignment was.

Different fire departments use different types of accountability systems. Consult your agency's SOGs/SOPs for more specific information about the system being used by your department and always follow the required procedures.

Many fire departments use **personnel accountability tags (PATs)** to track individual fire fighters. An accountability tag generally includes the fire fighter's name, ID number, and photograph and may include additional information, such as an individual's medical history, medications, allergies, and other factors. When this system is used, fire fighters who respond to an incident on an apparatus deposit their PATs in a designated location on the vehicle **FIGURE 11-4**. The PATs are then collected from each vehicle and taken to the command post. Fire fighters who respond directly to the scene must report to the command post to receive an assignment and deposit their PATs.

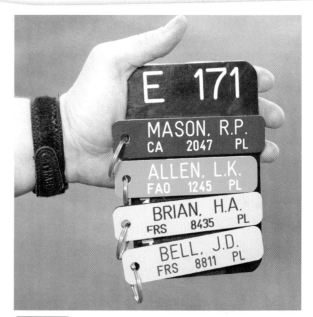

FIGURE 11-4 There should be a designated location on the fire apparatus for the fire fighters to deposit their PATs.

■ Controlling Utilities

Controlling utilities is one of the first tasks that must be accomplished at many working structure fires. Most departments have written SOGs/SOPs that define when the utilities are to be shut off. Although this responsibility is often assigned to a particular company or crew, all fire fighters should know how to shut off a building's electrical, gas, and water services.

Electrical and gas utilities to a burning building may be disconnected for any of several reasons; for example, if faulty electrical equipment or a gas appliance caused the fire, shutting off the supply will help alleviate the problem. Disconnecting electrical service can also prevent problems such as short-circuits and electrical arcing that could result from fire or water damage. In addition, disconnecting electrical service reduces the risk of electrical injury to fire fighters. Shutting down gas service eliminates the potential for explosion due to damaged, leaking, or ruptured gas piping.

Controlling utilities is particularly important if fire fighters need to open walls or ceilings to look for fire in the void spaces, to cut ventilation holes in roofs, or to penetrate through floors. Because these spaces often contain both electrical lines and gas pipes, the danger of electrocution or explosion exists unless these utilities are disconnected.

For example, a fire fighter using an axe to open a void space could be electrocuted if the tool contacts energized electrical wires. Fire fighters operating a hose could be injured if the water conducts electricity back to the nozzle. Fire fighters might also be electrocuted if they come in contact with wires that have been pulled loose or exposed by a structural collapse.

Gas pipes in walls, ceilings, roofs, or floors can also be inadvertently damaged by fire fighters. A power saw slicing through a gas pipe can cause a rapid release of gas and create the potential for an explosion. A structural collapse can rupture gas lines or create leaks in the stressed piping.

Controlling utilities also alleviates the risk of an additional fire or explosion. Shutting off electrical service, for example, eliminates potential ignition sources that might cause an explosion if leaking gas has accumulated inside the building; shutting off the gas supply will prevent any further leakage. The electrical supply must be disconnected at a location outside the area where gas might be present. Interrupting the power at a remote location alleviates the risk of an electrical arc that could cause a gas explosion.

If a serious water leak has occurred inside the building, shutting off the water supply may prevent electrical problems and help minimize additional water damage to the structure and contents. Shutting off the water supply to a fire-damaged building is often necessary to control leaking pipes, which could cause additional damage to the building.

Electrical Service

Electric delivery service arrangements depend on the utility company providing that service and the age of the system. The most common installation is a service drop from above-ground utility wires to the electric meter, which is typically mounted on the outside of the building **FIGURE 11-5**. Other types of electric service installations include underground connections, inside meters, and multiple services or separate electric meters for different tenants or areas inside a building.

Fire departments must work with their local electric utility companies to identify the different types of electric services and arrangements used in the area and to determine the proper procedures for dealing with each type. Often, a main disconnect switch can be operated to interrupt the power. Most utility companies will train fire fighters to identify and operate shut-off devices on typical installations. Other utility companies request that the fire department call a company representative to shut off the power. Shutting off large systems that involve high-voltage equipment should be done by a company technician or a trained individual from the premises.

A utility company representative should be called to interrupt power from a remote location, such as a utility pole. This step may be necessary if the outside wires have been damaged by fire, if fire fighters are working with ladders or aerial apparatus, or if an explosion is possible. In urban areas, the electric utility company can usually dispatch a qualified technician or crew to respond quickly to a fire or emergency incident. In many cases, qualified technicians are dispatched automatically to all working fires. Some utility companies can interrupt the service to an area from a remote location in response to a fire department request.

In general, to turn off an electrical service to a structure, you must first identify the circuit breaker box that supplies the

FIGURE 11-5 The electric service often has an exterior meter connected to above-ground utility wires.

© Carlos Arroyo/ShutterStock, Inc.

building or part of the building in which you need to disconnect the power. You can then turn off the main circuit breaker that supplies the whole breaker box. It is important to leave a fire fighter at the circuit breaker box or to use a lockout system that tags the box to ensure that the power does not get turned on accidentally.

Gas Service

Both natural gas and liquefied petroleum gas (LPG) are used for heating and cooking. Generally, natural gas is delivered through a network of underground pipes. By contrast, LPG is usually delivered by a tanker truck and stored in a container on the premises. In some areas, LPG is distributed from one large storage tank through a local network of underground pipes to several customers.

A single valve usually controls the natural gas supply to a building. This valve is generally located outside the building at the entry point of the gas piping; natural gas service has a distinctive piping arrangement. In older buildings, the shut-off valve for a natural gas system may be located in the basement.

The shut-off valve for a natural gas system is usually a quarter-turn valve with a locking device, so it can be secured in the off position **FIGURE 11-6** . When the handle is in line with the pipe, the valve is open; when the handle is at a right angle to the pipe, the valve is closed. A special key, an adjustable wrench, or a spanner wrench can be used to turn the handle.

The valve for an LPG system is usually located at the storage tank. The more common type of LPG valve, however, has a distinctive handle that indicates the proper rotation direction to open or close the valve. To shut off the flow of gas, rotate the handle to the fully closed position.

A gas valve that has been shut off must not be reopened until the system piping has been inspected by a qualified person. Air must be purged from the system, and any pilot lights must be reignited to prevent gas leaks.

Water Service

Water service to a building can usually be shut off by closing one valve at the entry point. Many communities permit the water service to be turned off at the connection between the utility pipes and the building's system. This underground valve, which is often accessed through a curb box, is located outside the building and can be operated with a special wrench or key. In most cases, another valve is found inside the building, usually in the basement (if there is one), where the water line enters. In warmer climates, water supply valves are sometimes located above ground.

Size-Up

Size-up is the process of initially evaluating an emergency situation to determine which actions need to be taken and which actions can be undertaken safely. It is always the first step in making plans to bring the situation under control. The initial size-up of an incident is conducted when the first unit arrives on the scene and identifies the appropriate actions for that unit or units. Size-up is also used to determine whether additional resources are needed. Size-up is one of the most important steps that needs to be taken at an emergency scene. It is often the key factor in determining whether an emergency operation is a success or a failure.

> ### Fire Fighter Safety Tips
> © Photos.com
>
> Size-up is often a key factor in determining whether an emergency operation is a success or a failure.

The initial size-up is often conducted by the first-arriving company officer, who serves as the IC until a higher-ranking officer arrives on the scene and assumes command. The IC uses the size-up results to develop an initial plan and to set the stage for the actions that follow.

As more complete and detailed information becomes available, the IC will improve both the initial size-up and the initial plan. At a major incident, the size-up process might continue through several stages. The ongoing size-up must consider the effectiveness of the initial plan, the impact that fire fighters are having on the problem, and any changing circumstances at the incident.

Although officers usually perform size-up, all fire fighters must understand how to formulate an operational plan, how to gather and process information, and how to use this information to change plans during the operation. If no officer is present on the first-arriving unit, a fire fighter becomes responsible for assuming command and conducting the preliminary size-up until an officer arrives. Individual fire fighters are often asked to obtain information and to report their observations for ongoing size-up. They should routinely make observations during incidents to maintain their personal awareness of the situation and to develop their personal competence. One factor in an effective size-up relies on the appropriate use of the senses of each fire fighter.

■ Managing Information

The size-up process requires a systematic approach to managing information. Emergency incidents are often complicated, chaotic, and rapidly changing. The IC must look at a

FIGURE 11-6 A typical natural gas service includes a pressure-reducing valve and a gas meter, along with the shut-off valve.

complicated situation, identify the key factors that apply to the specific incident, and develop an action plan based on known facts, observations, realistic expectations, and certain assumptions. As the operation unfolds, the IC must continually revise, update, supplement, and reprocess the size-up information to ensure that the plan is still valid or to identify when the plan needs to be changed.

Size-up relies on two basic categories of information: facts and probabilities. Facts are data elements that are accurate and based on prior knowledge, a reliable source of information, or an immediate, on-site observation. Probabilities are factors that can be reasonably assumed, predicted, or expected to occur but are not necessarily accurate. It is often necessary to rely on probabilities, because fire fighters rarely have all the facts that are necessary to create an action plan based on facts alone.

Often, the initial size-up is based on a limited amount of factual information and a larger amount of probable information. It is refined as more factual information is obtained and as probable information is either confirmed or revised. Effective size-up requires a combination of training, experience, and good judgment.

FIRE FIGHTER Tips

Building maintenance personnel can often provide valuable information about a structure and its mechanical systems.

Facts

Facts are bits of accurate information obtained from various sources. For example, the communications centre will provide some facts about the incident during dispatch. A preincident plan will contain many facts about the structure. Maps, manuals, and other references could provide additional information. Mobile computer data terminals, laptop computers, and tablets have the ability to call up previously entered facts at a touch of a few buttons. Individuals with specific training, such as a building engineer or a utility representative, can add even more specific information. In addition, a seasoned officer will have built up a knowledge bank of valuable information based on experience, training, and direct observations.

The initial dispatch information will contain facts such as the location and nature of the situation. This information could be very general (e.g., "a building reported on fire in the vicinity of Central Avenue and Main Street") or very specific (e.g., "a smoke alarm activation in room 3102 of the First National Bank Building at 173 South Main Street").

The time of day, temperature, and weather conditions are other factors that can easily be determined and incorporated into the initial size-up. Based on these basic facts, an officer might have certain expectations about the incident; for example, whether a building is likely to be occupied or unoccupied, whether the occupants are likely to be awake or sleeping, and whether traffic will delay the arrival of additional units may all be inferred from the time of day. Weather conditions such as snow and ice might delay the arrival of fire apparatus, create operational problems with equipment, and require additional fire fighters to perform basic functions such as stretching hose lines and raising ladders. Strong winds can cause rapid **extension** (spread) of a fire to exposed buildings or push a fire in unexpected ways. High heat and humidity will affect fire fighters' performance and may cause heat casualties.

An experienced fire officer will also have a basic knowledge of the community and the department's available resources. The officer may remember the types of occupancies, construction characteristics of typical buildings, and specific information about particular buildings from previous incidents or preincident planning visits.

A **preincident plan** can be especially helpful during size-up, because it contains significant information about a structure. A preincident plan generally provides details about a building's construction, layout, contents, special hazards, and fire protection systems. The chapter "Preincident Planning" contains specific information about the contents and development of such plans. As you study that chapter, consider how preincident plans feed into the size-up process.

Basic facts about a building's size, layout, construction, and occupancy can often be observed upon arrival, if they are not known in advance. In particular, the officer must consider the size, height, and construction of the building during the size-up. The action plan for a single-storey, wood-frame dwelling on 4047 square metres (1 acre) of land will be quite different than that for a steel-frame, high-rise tower on a crowded urban street.

The age of the building is often an important consideration in size-up, because building and fire codes change over time. Older wooden buildings may have utilized **balloon-frame construction**, which can provide a path for fire to spread rapidly into uninvolved areas. Floor and roof systems in newer buildings typically use trusses for support; these trusses are susceptible to early failure and collapse. Lightweight and composite engineered building systems tend to fail quickly and without warning under fire conditions. In addition, the weight of air conditioning or heating units on the roof may contribute to early structural collapse. Knowledge of building construction is vital when performing an initial size-up. Review the chapter "Building Construction" for construction details that will enable you to perform an accurate size-up.

A plan for rescuing occupants and attacking a fire must also incorporate information about the building's layout, such as the number, locations, and construction of stairways. The plan for a building with an open stairway that connects several floors might identify keeping the fire out of the stairway as a priority. This focus enables fire fighters to use the stairway for rescue and prevents the fire from spreading to other floors. Ladder placement, use of aerial or ground ladders, and emergency exit routes all depend on the building layout.

Any special factors that will assist or hinder operations must be identified during size-up as well; for example, bars on the windows will limit access and complicate the rescue of trapped victims. Fire detection and suppression systems play a vital part in initial size-up. A fire in a building that contains a properly designed and installed fire detection system with external notification will be reported sooner to the fire department than an unoccupied building with no alarm system. Likewise, an operational sprinkler system may confine or extinguish the fire. The presence of a standpipe system makes getting a hose line in operation on an upper floor of a burning building much quicker and easier. The presence of fire walls may also confine a fire to the location of origin. The information presented in the chapter "Fire Detection, Protection, and Suppression Systems" will help you become more skillful in performing a size-up.

The occupancy of a building also represents critical information. An apartment building, a restaurant, an automotive repair garage, an office building, a warehouse filled with concrete blocks (often referred to as a concrete–masonry unit), and a chemical manufacturing plant all present fire fighters with different sets of problems that must be addressed.

It is important for the IC to visualize the building from the side where the IC is located, but it is equally important to try to visualize the other three sides of the building. The IC should initially perform a size-up by walking completely around all four sides of the building, a process often referred to as a "walk-around" or "doing a 360." By completing a loop around a building, the IC can observe fire and smoke conditions on all sides of the building. In addition, the IC can look for signs of occupancy, locations of windows and doors, building construction, fire-suppression devices, exposures, and locations of utility entrances, such as electrical and gas. It will not always be possible for the IC to complete this lap around the fire building. In those cases where the commander cannot complete a walk-around, the companies assigned to the sides and back can give a limited size-up report to the IC, who will usually be located on the front (A side) of the building.

The size of the fire and its location within the building often dictate hose line size, placement, and ventilation sites. The fire's location is also critical in determining which occupants are in immediate need of rescue; for example, someone directly above a burning apartment should be rescued immediately, whereas a person at the opposite end of the building, far from the fire, is probably not in immediate danger.

Direct visual observations will give the best information about the size and location of a fire, particularly when combined with information about the building. Although visible flames can indicate where the fire is located and how intensely it is burning, they might not tell the whole story. Flames issuing from only one window suggest that the fire is confined to that room, but it could also be spreading through void spaces to other parts of the building.

Often, the location of a fire within a building cannot be determined from the exterior, particularly when visibility is obscured by smoke **FIGURE 11-7**. An experienced fire fighter will observe where smoke is visible, how much is apparent, what colour it is, and how it moves. Smoke reading, as described in the chapter "Fire Behaviour," is a vital part on the scene size-up, because this process helps the IC predict where the fire might go. Sometimes even the smell of smoke from afar can be a clue. A burning pot of food, an overheated electrical motor, and burning wood all have distinctive odours, for example.

Inside a building, fire fighters can use their observations and sensations to help them work safely and effectively. If smoke limits visibility, a crackling sound or the sensation of heat coming from one direction may indicate the **seat of the fire**

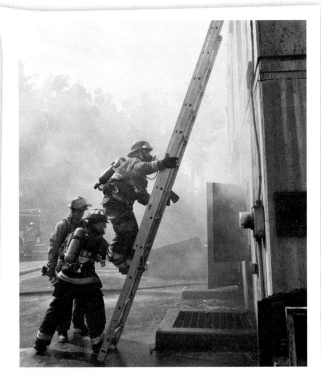

FIGURE 11-7 It can be difficult to determine the location and extent of a fire inside the building.
© 2003, Berta A. Daniels

(i.e., the main area of the fire). Blistering paint and smoke seeping through cracks can lead fire fighters to a hidden fire burning within a wall. Thermal imaging cameras are an effective device for locating fires, as they can sometimes pinpoint the location of a fire that otherwise would be hard to find. In this way, thermal imagers help fire fighters determine the location of a fire quickly and without putting fire fighters at much risk.

The IC needs to gather as much factual information as possible about a fire. Because the IC often is located at a command post outside the building, company officers will be requested to report their observations from different locations. A company that is operating a hose line inside the building can report on interior conditions, whereas the ventilation crew on the roof will have a very different perspective. The IC may request that an officer or a fire fighter prepare a **reconnaissance report**, which entails the inspection and exploration of a specific area to gather information for the IC. The IC assembles, interprets, and bases decisions on this information.

Regular progress reports from companies working in different areas will offer updated information about the situation. Progress reports enable an IC to judge whether an operational plan is effective or needs to be changed.

Probabilities

Probabilities refer to events and outcomes that can be predicted or anticipated, based on facts, observations, common sense, and previous experiences. Fire fighters frequently use probabilities to anticipate or predict what is likely to happen in various situations. The attack plan is also based on probabilities, predicting where the fire is likely to spread and anticipating potential problems.

Near-Miss REPORT

Report Number: 10-0000966

Synopsis: Good entry practice alerts crew to danger.

Event Description: Our department responded to a large four-storey (three up and one down) home involved in fire and smoke. The home was in a very rural part of our district, which significantly delayed our arrival, due to both weather and travel time.

I was a member of the first unit to arrive. Upon arrival, I found a female occupant trapped on the second-floor balcony. I threw up a ground ladder from the first-arriving engine company, and the shift captain accessed the victim and assisted her down the ladder.

With the victim removed, I completed another 360-degree inspection of the home. I could see a glow in the basement but no fire. There was now turbulent smoke coming from the eaves of the home. I instructed the shift captain to enter from the front door on side A and assess fire activity and the percentage of involvement. I told him what I had observed from the exterior and to be aware of fire activity in the basement.

The captain forced the door and began to enter with three other members. He was crawling on the floor, even though there were only mild to moderate smoke conditions on the first floor. Visibility was not an issue. The shift captain was carrying a pick-head axe and sounded the floor as he crawled. Each of the other team members carried a tool. In total, the crew had two pike poles, a drywall hook, and the axe carried by the captain into the home.

Upon sounding the floor a few times, the captain reported that one of the floorboards had given way under the stress of the axe. He immediately ordered his team to retreat and evacuate. He then notified me that when the floor gave way after sounding, he could see fire throughout the basement. Soon after evacuating, the entire first floor became involved in fire.

Lessons Learned: In my opinion, this is a situation that went right. We communicated as a team and used our training and experience to ensure that everyone went home. Due to the clear visibility in the first floor, it could have been a time when complacency led us to walk in and put out the fire; however, we crawled in, and as a result, we crawled out. We could have very easily been carried out in a bag. Our shift captain recognized the integrity of the floor had been compromised and retreated to safety with his team. We later learned that the home had a dumb waiter system that spread the fire rapidly from the basement to the attic area. A good size-up, communication, and training allowed us to clock out at the end of shift.

An IC must be able to quickly identify the probabilities that apply to a given situation; for example, a fire in an apartment building in the middle of the night will probably involve occupants who need to be rescued. In such circumstances, the IC would assign additional crews to search for potential victims, even if no factual information indicates that any occupants need to be rescued. Similarly, a fire burning on the top floor of a structure in a row of attached buildings has a high probability of spreading to adjoining buildings through the cockloft. In this case, the IC's plan would include opening the roof above the fire and sending crews into the exposed occupancies to check for fire extension in the cockloft.

The concepts of convection, conduction, and radiation enable an IC to predict how a fire will extend in a particular situation. By observing a particular combination of smoke and fire conditions in a particular type of building, the IC can identify a range of possibilities for fire extension within the structure or to other exposed buildings. Using these probabilities, the IC can predict what is likely to happen and develop a plan to control the situation effectively. Effective size-up requires a good knowledge of fire behaviour to evaluate the probabilities for the spread of a fire.

The IC must also evaluate the potential for collapse of a burning structure. The building's construction, the location and intensity of the fire, and the length of time the structure has been burning are all factors that must be considered in making this determination; houses built in the last 30 years using lightweight construction techniques, for example, tend to collapse in a much shorter period of time than houses built in earlier eras. When the possibility of collapse exists, risk management dictates that the IC not permit fire fighters to enter the building or immediately order all fire fighters to exit the building.

■ Resources

Resources include all of the means that are available to fight a fire or conduct emergency operations at any other type of emergency incident. Resource requirements depend on the size and type of incident. Resource availability depends on the capacity of a fire department to deliver fire fighters, fire apparatus, equipment, water, and other items that can be used at the scene of an incident.

A fire department's basic resources consist of its personnel and apparatus. Firefighting resources are usually defined as the number of engine companies, ladder companies, special units, and command officers required to control a particular fire. An IC should be able to request the required number of companies and know that each unit will arrive with the appropriate equipment and the necessary fire fighters to perform a standard set of functions at the emergency scene. The IC usually will have a good idea of how many and which types of companies are available to respond, how they are staffed and equipped, and how long they should take to arrive. This information might have to be updated at the incident, particularly in areas served by volunteer fire fighters, because the number of fire fighters available to respond may vary at different times.

Water supply is another critical resource. It is rarely a problem if the fire occurs in a district that has hydrants and a strong, reliable municipal water system with adequately sized water mains. In an area without hydrants, however, water supply may be severely limited. Even when a static water supply is available, it takes time to establish a water supply from such a source. If the water supply is limited to tankers or tenders, the amount of water that can be delivered by a tanker shuttle is limited. In this situation, the IC would need to call for an additional supply of tankers.

Sometimes, resources include more than fire fighters, fire apparatus, and a water source. A fire in a flammable-liquids storage facility will require large quantities of foam and the equipment to apply it effectively. A hazardous materials incident might require special monitoring equipment, chemical-protective clothing, and bulk supplies to neutralize or absorb a spilled product. A building collapse might require heavy equipment to move debris and a structural engineer to determine where fire fighters can work safely. The fire department must have these supplies available or be able to obtain them quickly when they are needed. In addition, resources for a large-scale incident must include food and fluids for the rehydration of fire fighters, fuel for the apparatus, and other supplies. The Red Cross, Salvation Army, and law-enforcement agencies often provide support resources.

The size-up process enables the IC to determine which resources will be needed to control the specific situation and to ensure their availability at an appropriate time. An action plan to control an incident can be effective only if the necessary resources can be assembled on a timely basis. If a delay occurs, the IC must anticipate how much the fire will grow and where it will spread. If the desired resources cannot be obtained, the IC must develop a realistic plan using the available resources to gain eventual control of the situation.

Ideally, a fire department will be able to dispatch enough fire fighters and apparatus to control any situation within its jurisdiction. More practically, most departments have established mutual aid agreements with surrounding jurisdictions that require them to assist one another if a situation requires more resources than the local community can provide. In some

areas, hundreds of fire fighters and apparatus can be assembled to respond to a large-scale incident. In more remote areas, the resources available to fight a fire can be very limited.

If resources are insufficient or delayed, a fire can become too large to be controlled by available personnel; for example, if only 20 fire fighters and four apparatus are available, they will have to bring the fire under control before it gets too big, limit actions to protecting exposures, or wait until the fire burns itself down to a manageable size before tackling it. The IC must always determine which actions can be taken safely with the resources that are currently available.

Resources, in turn, must be organized to support efficient emergency operations. Fire fighters must be properly equipped and organized in companies. Equipment and procedures must be standardized. The IMS must be designed to manage all of the resources that could be used at a large-scale incident, and the communications system must enable the IC to coordinate operations effectively.

Incident Action Plan

Based on information gathered during size-up, the IC develops an incident action plan (IAP) that outlines the steps needed to control the situation. The initial IC develops a basic plan for beginning operations. If the situation expands or becomes more complicated, this plan may be revised and expanded as additional information is obtained, more resources become available, and the IMS structure grows.

The IAP should be based on the five basic fire-ground objectives, which are listed here in order of priority:

1. Save lives. This includes keeping fire fighters safe and rescuing victims.
2. Protect exposures.
3. Confine the fire.
4. Extinguish the fire.
5. Salvage property and overhaul the fire.

This system of priorities clearly establishes that the highest priority in any emergency situation is saving lives. The remaining priorities focus on saving property. Making sure that the fire does not spread to any **exposure** (an area adjacent to the fire that may become involved if not protected) is a higher priority than confining the fire within the burning building. After the fire is confined, the next priority is to extinguish it. The final priority is to protect property from additional damage and make sure the fire is completely out.

These priorities are not separate and exclusive, of course. That is, one objective does not have to be accomplished before fire fighters tackle the next one. Often, more than one objective can be addressed simultaneously, and certain activities help achieve more than one objective. For example, if a direct attack on the fire will bring it under control very quickly, the objectives of protecting exposures, confining the fire, and extinguishing the fire might all be addressed simultaneously. Extinguishing the fire might also be the best way to protect the lives of building occupants. Frequently, salvage crews may be working on lower floors even as fire-suppression crews continue attacking a fire on an upper floor.

RECEO VS

Fire departments and their ICs employ different strategies and tactics to meet incident objectives. Lloyd Layman was first credited with developing a tactical model that identified seven basic tactics that some fire departments follow on the fireground. This model is commonly referred to by the acronym RECEO VS:

1. **R**escue—The removal of victims from a life-threatening situation
2. **E**xposures—The protection of surrounding structures from a fire
3. **C**onfinement—The containment of a fire to one area or the prevention of further areas from becoming involved in the fire
4. **E**xtinguishment—The suppression of the fire
5. **O**verhaul—The process of ensuring that the fire is suppressed completely
6. **V**entilation—The process of removing smoke, heat, and the products of combustion from a structure
7. **S**alvage—The removal or protection of property that could be damaged during firefighting or overhaul operations

Fire Officer: Principles and Practice, 2nd edition (Jones and Bartlett Publishers, LLC, and NFPA, 2010), covers the strategy and tactics employed by fire officers when creating an IAP in greater detail. Note that the steps in this acronym are not listed in the order that they are performed on the fire ground. It is only intended to aid in recalling what needs to be done—not when it needs to be done.

These priorities guide the IC in making difficult decisions, particularly if not enough resources are available to address every priority. If a decision must be made between saving lives and saving property, saving lives comes first. After rescue is completed and if the fire is still spreading, the IC should then place exposure protection ahead of salvage and overhaul.

A similar set of priorities can be established for any emergency situation: saving lives is always more important than protecting property. The IC must always place a higher priority on bringing the problem under control than on cleaning up after the problem.

■ Rescue

Protecting lives is the first consideration at a fire or any other emergency incident. The need for rescue depends on many circumstances. Notably, the number of people in danger is likely to vary based on the type of occupancy and the time of day. A commercial building that is crowded during the workday might have few, if any, occupants at night. At night, rescue is more likely needed in a residential occupancy, such as a house, an apartment building, or a hotel.

The degree of risk to the lives of the building's occupants must also be evaluated. A fire that involves one apartment on the 10th floor of a high-rise building could threaten the lives of both the residents on that floor and the residents who live directly above the fire. In contrast, residents below the fire are probably not in any danger, and residents several floors above the fire might be safer staying in their apartments until the fire is extinguished instead of walking down smoke-filled stairways.

If the size-up suggests that people may be trapped in a burning building, the IC must make a determination of whether it is safe to send fire fighters into the burning building. The IC should use survivability profiling to make this decision. **Survivability profiling** entails examining the situation and making an informed decision based on the presenting circumstances of whether civilians could survive the existing fire and smoke conditions. Through this process, the IC decides whether to commit fire fighters to life-saving and interior suppression operations to save live victims. Fire fighters should engage in interior operations only if there is a reasonable probability that trapped civilians might be alive.

The upper limit in which people can survive has been shown to be 100°C (212°F). Death occurs at temperatures of 176.7°C (350°F) in approximately 3 minutes. Lethal first-degree respiratory burns occur in 230 seconds (less than 4 minutes). An analysis of burn victims has demonstrated that respiratory burns from heat kill victims first, followed closely by toxic smoke.

In today's fires, temperatures can rise higher than 260°C (500°F) in 3 to 4 minutes. Flashovers occur at 593.3°C (1100°F). We must always remember that if an environment is not tenable for a fire fighter in full PPE, it is unlikely that trapped victims will survive either. It is not acceptable to risk the lives of fire fighters to try to save the lives of already dead victims.

Fire Fighter Safety Tips © Photos.com

It is not acceptable to risk fire fighter safety if fire conditions support the probability that victims are already deceased.

Often, the best way to protect lives is to extinguish the fire quickly. For this reason, efforts to control the fire are usually initiated at the same time as rescue operations; for example, hose lines may be used to protect exit paths and keep the fire away from victims during search and rescue operations. The chapter "Search and Rescue" discusses specific tactics and techniques for these operations.

■ Exposure Protection

The secondary objective of protecting property encompasses multiple priorities. The first priority in protecting property is to keep the fire from spreading beyond the area of origin or involvement when the fire department arrives. The IC must start by making sure the fire is not expanding. If the fire is burning in only one room, the objective should be to ensure that it does not spread beyond that room. If more than one room is involved, the objective might be to contain the fire to one apartment or one floor level. Sometimes the objective is to confine the fire to one building, particularly if multiple buildings are attached or closely spaced to the site of the fire.

In some cases, the IC must look ahead of the fire and identify a place to stop its spread. If flames are extending quickly through the cocklofts in a row of attached buildings, for example, the IC might place companies with hose lines ahead of the fire to stop its progress.

The IC must sometimes weigh potential losses when deciding where to attack a fire. If a fire in a vacant building threatens to spread to an adjacent occupied building, the IC will usually assign companies to protect the exposure before attacking the main body of fire. If, however, a fire in an occupied building might spread to a vacant building, the IC's decision might be to attack the fire first and worry about controlling the spread later.

Confinement

After ensuring that the fire is not actively extending into any exposed areas, the IC will focus on confining it to a specific area. To do so, the IC will define a perimeter and plan operations so that the fire does not expand beyond that area. Fire fighters on the perimeter must be alert to any indications that the fire is spreading to those limits.

Thermal imaging cameras, which can detect sources of heat, are valuable tools for finding fires in void spaces and hence for containing the spread of fire. The principles and use of thermal imaging devices are covered more fully in the chapter "Search and Rescue."

Extinguishment

Depending on the size of the fire and the risk involved, the IC will mount either an offensive attack or a defensive strategy to extinguish a fire. An **offensive attack** is used with most small fires. With this approach, fire fighters advance into the fire building with hose lines or other extinguishing agents and overpower the fire. If the fire is not too large and the attacking fire fighters can apply enough extinguishing agent, the fire can usually be terminated quickly and efficiently. Extinguishing the fire in this way often satisfies several priorities at the same time, including exposure protection and confinement.

At regular intervals during an offensive attack, the IC must evaluate the progress being made. If the progress is inadequate, the IC must either adjust the tactics or alter the overall strategy; for example, the fire might produce more heat than the water from hose lines can absorb, or fire fighters might be unable to penetrate far enough to make a direct attack on the fire.

Sometimes the problem can be resolved by adding resources and intensifying the offensive attack. Larger hose lines or more fire fighters with additional hose lines, for instance, might be able to extinguish the fire successfully. Coordinating ventilation with an offensive attack might enable fire fighters to get close enough to extinguish the fire. Special extinguishing agents might also be used to extinguish the fire.

When the fire is too large or too dangerous to extinguish with an offensive attack, the IC will implement a **defensive attack**. In these situations, all fire fighters are ordered out of the building, and heavy streams are operated from outside the fire building. A defensive strategy is required when the IC determines that the risk to fire fighters' lives is excessive, when rescue is impossible, and when structural collapse is possible. The IC who adopts a defensive strategy has determined that there is no property left to save or that the potential for saving property does not justify the risk to fire fighters. Sometimes a defensive strategy is effective in extinguishing the fire; at other times, it simply keeps the fire from spreading to exposed properties **FIGURE 11-8**.

In some situations, all fire fighters may be withdrawn from the area, and the fire allowed to burn itself out. These situations generally involve potentially explosive or hazardous materials that represent an extreme danger to fire fighters.

Each strategy poses some risk to fire fighters, so it is important that the IC consider both the risks and the benefits of each strategy. Sending fire fighters into an unoccupied building may pose a large risk to fire fighters, with the only potential benefit being to save property. Sending fire fighters into an occupied building to save lives has great benefit, but only if there is a chance to save the occupants and the operation does not pose a huge risk to the fire fighters. Each IC must evaluate the risk versus benefit of each action to be taken at an emergency scene.

FIGURE 11-8 A defensive strategy involves an exterior fire attack with heavy streams.
© Gregory Price, Lewiston Son Journal/AP Photos

Ventilation

Ventilation operations include all activities to remove smoke, heat, and the products of combustion from the interior of the structure. Ventilation can be a stand-alone strategy to prevent the spread of heat and smoke in the structure, or it may be used in combination with other tactics to support the safe removal of occupants. Ventilation operations are covered in depth in the chapter "Ventilation."

■ Salvage and Overhaul

Salvage operations are conducted to save property by preventing avoidable property losses. <u>Salvage</u> is the removal or protection of property that could be damaged during firefighting or overhaul operations **FIGURE 11-9**. Salvage operations are often aimed at reducing smoke and water damage to the structure and contents, once the fire is under control.

The <u>overhaul</u> process is conducted after a fire is under control, with the goal being to completely extinguish any remaining pockets of fire **FIGURE 11-10**. The IC is responsible for ensuring that the fire is completely out before terminating operations. Floors, walls, ceilings, and attic spaces should be checked for signs of heat, smoke, or fire. Window casings, wooden door jambs, baseboards, electrical outlets, and heating or air conditioning vents can often hide small,

smouldering fires. To mitigate this risk, debris from the burned contents of the structure should be removed and thoroughly doused to reduce the potential for <u>rekindle</u>, or a reignition of the fire. For large fires, a fire department unit may be assigned to stay at the scene, or to return to the scene every few hours to check for indications of residual fire. These operations are covered in depth in the chapter "Salvage and Overhaul."

FIGURE 11-10 Overhaul is conducted after a fire is under control to completely extinguish any remaining pockets of fire.

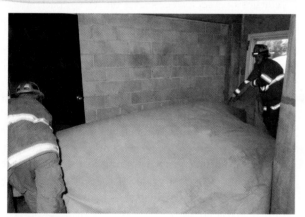

FIGURE 11-9 Salvage operations are conducted to save property by preventing avoidable property losses.

Wrap-Up

Chief Concepts

- Response actions include receiving the alarm, donning protective clothing and equipment, mounting the apparatus, and transporting equipment and personnel to the emergency incident quickly and safely.
- Size-up is the systematic process of gathering information and evaluating the incident. It is essential for determining the appropriate strategy and tactics to handle the incident.
- You must be prepared to respond to an emergency at all times. Inspect your PPE and SCBA daily to ensure that it is ready for operations.
- The response process begins when the alarm is received at the fire station. Dispatch messages will include the location of the incident, the type of emergency, and the units that are due to respond.
- Don your PPE before mounting the apparatus. While riding in the apparatus, your seat belt should be secure, and you should be concentrating on mentally preparing for the emergency. Consider any relevant factors that could affect the emergency, such as the weather and the time of day.
- Upon arrival at the scene, traffic safety should be a major concern. Always check for traffic before exiting the apparatus. Follow departmental SOGs/SOPs to close streets quickly and block access for civilian vehicles to the incident.
- Fire fighters should always work in assigned teams and be guided by a strategic plan for the incident. Teamwork and disciplined action are essential to provide for the safety of all fire fighters and the effective, efficient conduct of operations.
- Upon arriving at the scene, check into the personnel accountability system.
- To protect the safety of fire fighters, controlling utilities is one of the first tasks to be accomplished. The electrical service and gas supply should be shut off and locked.
- Size-up is the first step in making plans to bring the emergency incident under control.
  - The initial size-up is often conducted by the first-arriving company officer, who serves as the IC until a higher-ranking officer arrives at the scene and assumes command.
  - The IC uses the initial size-up information to develop an initial plan. This plan is revised as additional information is gathered.
- Size-up relies on two basic categories of information: facts and probabilities.
  - Facts are data elements that are accurate and based on prior knowledge, a reliable source of information, or an immediate, on-site observation.

- Probabilities are factors that can be reasonably assumed, predicted, or expected to occur, but are not necessarily accurate.
- The preincident plan contains facts that can be essential in creating a plan. It provides details about a building's construction, layout, contents, special hazards, and fire protection systems. This information can be used in determining how to rescue occupants and attack a fire.
- The IC often operates out of a command post outside the structure and relies on company officers to provide reconnaissance reports. Progress reports from companies enable an IC to judge whether an operational plan is effective or needs to be changed.
- Probabilities refer to events and outcomes that can be predicted or anticipated, based on facts, observations, common sense, and previous experiences. Fire fighters frequently use probabilities to anticipate or predict what is likely to happen in various situations. The attack plan is also based on probabilities, predicting where the fire is likely to spread and anticipating potential problems.
- Resources for fire fighters include all of the means that are available to fight a fire or conduct emergency operations at any other type of emergency incident. Resource requirements reflect the size and type of incident. Resource availability depends on the capacity of a fire department to deliver fire fighters, fire apparatus, equipment, water, and other items that can be used at the scene of an incident.
- Based on information gathered during size-up, the IC develops an IAP that outlines the steps needed to control the situation. The IAP should be based on the five basic fire-ground objectives:
  - Save lives—Protecting lives is the first consideration at a fire or any other emergency incident. Often the best way to protect lives is to quickly extinguish the fire.
  - Protect exposures—The first priority in protecting property is to keep the fire from spreading.
  - Confine the fire—Once the exposures are protected, the IC will focus on confining the fire to a specific area.
  - Extinguish the fire—Depending on the size of the fire and the risk involved, the IC will mount either an offensive attack or a defensive strategy to extinguish a fire.
  - Salvage property and overhaul the fire—Salvage is the removal or protection of property that could be damaged during firefighting operations. Overhaul operations completely extinguish any remaining pockets of fire.

Hot Terms

Balloon-frame construction An older type of wood-frame construction in which the wall studs extend vertically from the basement of a structure to the roof without any fire stops.

Defensive attack Exterior fire-suppression operations directed at protecting exposures.

Exposure The heat effect from an external fire that might cause ignition of, or damage to, an exposed building or its contents. (NFPA 80A)

Extension Fire that moves into areas not originally involved in the incident, including walls, ceilings, and attic spaces; also, the movement of fire into uninvolved areas of a structure.

Freelancing The dangerous practice of acting independently of command instructions.

Offensive attack An advance into the fire building by fire fighters with hose lines or other extinguishing agents that are intended to overpower the fire.

Overhaul The process of final extinguishment after the main body of a fire has been knocked down. All traces of fire must be extinguished at this time. (NFPA 402)

Personnel accountability system A system that readily identifies both the locations and the functions of all members operating at an incident scene. (NFPA 1500)

Personnel accountability tag (PAT) An identification card used to track the location of a fire fighter on an emergency incident.

Preincident plan A written document resulting from the gathering of general and detailed information to be used by public emergency response agencies and private industry for determining the response to reasonable anticipated emergency incidents at a specific facility.

Reconnaissance report The inspection and exploration of a specific area for the purpose of gathering information for the incident commander.

Rekindle A return to flaming combustion after apparent but incomplete extinguishment. (NFPA 921)

Response The deployment of an emergency service resource to an incident. (NFPA 901)

Salvage A firefighting procedure for protecting property from further loss following an aircraft accident or a fire. (NFPA 402)

Seat of the fire The main area of the fire.

Size-up The observation and evaluation of existing factors that are used to develop objectives, strategy, and tactics for fire suppression. (NFPA 1051)

Survivability profiling The process of using available data and observations to determine if there is a reasonable chance that victims trapped inside might still be alive. This process is used to decide whether to send fire fighters into a fire or other hazardous environment.

Thermal imaging cameras Electronic cameras that detect differences in temperature based on infrared energy and then generate images based on those data. These devices are commonly used in obscured environments to locate victims.

It is the weekend. You and your buddies are ready for some downtime. You ask them over to your place for a barbecue. Most of them are fire fighters, so it does not take long for the conversation to centre around what you learned during training this week. When you tell them that you are learning about response and size-up, they quickly begin quizzing you to see if they can stump you.

1. A(n) _____ is a written document resulting from the gathering of general and detailed information to be used by public emergency response agencies and private industry for determining the response to reasonable anticipated emergency incidents at a specific facility.
   A. preincident plan
   B. reconnaissance report
   C. incident action plan
   D. size-up

2. _____ is conducted after a fire is under control, with the goal being to completely extinguish any remaining pockets of fire.
   A. Overhaul
   B. Salvage
   C. Exposure protection
   D. Extinguishment

3. What is the systematic process of information gathering and situation evaluation that begins when an alarm is received?
   A. Preincident plan
   B. Reconnaissance report
   C. Incident action plan
   D. Size-up

4. When should you don your PPE?
   A. Prior to mounting the apparatus
   B. While en route to the scene
   C. While en route to the scene, but only if you can remain in your seat belt
   D. Upon arrival at the scene

5. How should fire fighters cut the electricity to a building?
   A. Turn off the main circuit breaker serving the building.
   B. Pull the electrical meter to the building.
   C. Cut the loops at the electrical pole feeding the building.
   D. Only electrical system personnel should ever cut any electricity to a building or area.

6. When does the emergency response process begin?
   A. When the dispatcher answers the phone
   B. When the fire chief receives information from the dispatcher
   C. When an alarm is received at the fire station
   D. When the apparatus leaves the fire station

Your crew is the first to arrive at a one-car rollover on a major highway. The car is in a ditch, so the driver/operator pulls the fire engine onto the shoulder at a slight angle. You see the car's driver slumped over the steering wheel, blood dripping down his face. Despite the presence of the engine with lights flashing, vehicles still zoom past the scene at full speed.

1. Who is responsible for setting out safety cones?
2. What is the first item of PPE every member of your crew should don?
3. What additional resources can assist in maintaining your safety?
4. What is your first priority at this incident?

Forcible Entry

Fire Fighter I

Knowledge Objectives

After studying this chapter, you will be able to:

- Describe the situations and circumstances that require forcible entry into a structure. (NFPA 5.3.4 , p 314–315)
- List the general safety rules to follow when utilizing forcible entry tools. (p 315)
- List the general carrying tips when utilizing forcible entry tools. (NFPA 5.3.4.B , p 315)
- List the general maintenance tips when utilizing forcible entry tools. (p 315–316)
- List the types of tools used in forcible entry. (p 316)
- List the striking tools used in forcible entry. (p 316)
- Describe the tasks that striking tools are used for in forcible entry. (p 316)
- List the prying and spreading hand tools used in forcible entry. (p 316–317)
- Describe the tasks that prying and spreading hand tools are used for in forcible entry. (p 317)
- List the cutting tools used in forcible entry. (p 317–318)
- Describe the tasks that cutting tools are used for in forcible entry. (p 317–318)
- List the lock and specialty tools used in forcible entry. (p 318–319)
- Describe the tasks that lock and specialty tools are used for in forcible entry. (p 318–319)
- Describe the basic components of a door. (NFPA 5.3.4.A , p 319)
- Explain the differences between a solid-core and a hollow-core door. (NFPA 5.3.4.A , p 320)
- Describe the basic classifications of doors. (NFPA 5.3.4.A , p 320–321)
- Explain how the door classification affects forcible entry operations. (NFPA 5.3.4.A , p 321–325)
- Describe the basic configurations of window construction. (NFPA 5.3.4.A , p 325–327)
- Describe the common styles of window frames. (NFPA 5.3.4.A , p 327–331)
- Explain how the style of window frame affects forcible entry operations. (NFPA 5.3.4.A , p 327–331)

- Describe the major components of a door lock. (NFPA 5.3.4.A , p 331–332)
- Describe the major components of a padlock. (NFPA 5.3.4.A , p 332–333)
- Describe the four major types of locks. (NFPA 5.3.4.A , p 332–333)
- Explain how the type of lock affects forcible entry operations. (NFPA 5.3.4.A , p 332–333)
- Describe the tools used to force entry through locks. (p 334–335)
- Describe how to force entry through security gates and windows. (p 335–337)
- Explain the differences between load-bearing and nonbearing walls. (NFPA 5.3.4.A , p 337)
- Describe the materials used in exterior and interior walls. (NFPA 5.3.4.A , p 337)
- List the basic steps and considerations in forcible entry operations. (p 340, 342)
- Describe how forcible entry operations affect salvage operations. (NFPA 5.3.14.A , p 342)

Skills Objectives

After studying this chapter, you will be able to:

- Force entry into an inward-opening door. (NFPA 5.3.4.B , p 321–322)
- Force entry into an outward-opening door. (NFPA 5.3.4.B , p 322–324)
- Open an overhead garage door using the triangle method. (NFPA 5.3.4.B , p 325–326)
- Force entry through a wooden double-hung window. (NFPA 5.3.4.B , p 327–328)
- Force entry through a casement window. (NFPA 5.3.4.B , p 330)
- Force entry through a projected window. (NFPA 5.3.4.B , p 331)
- Force entry using a K tool. (NFPA 5.3.4.B , p 334–335)
- Force entry using an A tool. (NFPA 5.3.4.B , p 335–336)
- Force entry by unscrewing a lock. (NFPA 5.3.4.B , p 335–336)
- Breach a wall frame. (NFPA 5.3.4.B , p 337–338)
- Breach a masonry wall. (NFPA 5.3.4.B , p 338–339)
- Breach a metal wall. (NFPA 5.3.4.B , p 338, 340)
- Breach a floor. (NFPA 5.3.4.B , p 338–340)

© Photos.com

Knowledge Objectives

After studying this chapter, you will be able to:

- Describe the connection between tools and forcible entry tasks. (NFPA 6.3.2.A, 6.3.2.B , p 316–319)

Skills Objectives

There are no skills objectives for Fire Fighter II candidates. NFPA 1001 contains no Fire Fighter II Job Performance Requirements for this chapter.

You Are the Fire Fighter

There is a working fire on the second floor of a local hotel. Your crew is assigned to search for occupants who may still be in the building. The hotel is a three-storey, wood-frame building with a hallway down the centre. It was built in 1981. As you dismount the fire apparatus, your captain tells you to grab the tools for forcible entry.

1. What kind of access issues would you expect with this type of structure?
2. What kind of tools would you want to take with you?
3. How would the forcible entry challenges for this structure differ from other types of structures?

Introduction

One of a fire fighter's most dynamic and challenging tasks, <u>forcible entry</u>, is defined as gaining access to a structure when the normal means of entry are locked, secured, obstructed, blocked, or unable to be used for some other reason. This chapter examines forcible entry into structures.

Forcible entry requires strength, knowledge, proper techniques, and skill. Because it often causes damage to the property, fire fighters must consider the results of using different forcible entry methods and select the one appropriate for the situation. If rapid entry is needed to save a life or prevent a more serious loss of property, it is appropriate to use maximum force. When the situation is less urgent, fire fighters can take more time and use less force.

Another factor to consider when making forcible entry is the need to secure the premises after operations are completed. Fire fighters must never leave the premises in a condition that would allow unauthorized entry. If a door or a lock is destroyed during an urgent forcible entry, the building owner should be called to secure the building. When the situation is not urgent, consider using entry methods that result in less damage and can be repaired more easily.

A fire fighter who is skilled in forcible entry should be able to get the job done quickly, with as little damage as possible. As part of this task, the fire fighter must consider the type of building construction, possible entry points, the types of securing devices that are present, and the best tools and techniques for the specific situation. Selecting the right tool and using the proper technique save valuable time and could save lives.

Doors, windows, locks, and security devices can be combined in countless variations to prevent easy entry. Given this diversity, fire fighters must be familiar with new technology, including new styles of windows, doors, locks, and security devices that are common in the local response area. In addition, they must understand how these devices operate. Learn to recognize different types of doors, windows, and locks. The best time to examine these components is during inspection and preincident planning visits. Touring buildings under construction or renovation is also an excellent way to learn about building construction and to examine different devices. Talking with construction workers and locksmiths can provide valuable information about how to best approach a particular lock, window, or door.

This chapter reviews door, window, and lock construction. In addition, it discusses the tools and techniques used for forcible entry through doors, windows, walls, and floors. The skill drills cover the most common forcible entry techniques.

Forcible Entry Situations

Forcible entry is usually required at emergency incidents where time is a critical factor. A rapid forcible entry, using the right tools and methods, might result in a successful rescue or allow an engine company to make an interior attack and control a fire before it extends farther into the structure or to adjoining structures. Fire fighters must study and practise forcible entry techniques so that their proficiency with these skills increases with every experience.

Company officers usually select both the point of entry and the method to be used. They ensure that the efforts of different companies are properly coordinated for safe, effective operations. For example, forcible entry actions must be coordinated with hose teams, because the entry must be made before engine companies can get inside the structure to attack a fire. Opening a door before the hose lines are in place and ready to advance allows fresh air to enter the structure, possibly resulting in rapid fire spread or a backdraft. Making an opening at the wrong location can undermine a well-planned attack. Anytime you open a door or window, that action may provide ventilation for a fire. Consider forcible entry to be part of ventilation, and make sure it is coordinated with the rest of the fire attack.

Before beginning forcible entry, remember this simple rule: try before you pry. In other words, always check doors or windows to confirm that forcible entry is truly needed. Be absolutely sure that you are at the address of the emergency incident. There is nothing harder to explain than why you broke down the door of a house located three doors down from the structure that is on fire. Check for the presence of a lockbox if your community has a lockbox program. The presence of a lockbox will give you immediate access to a key to open that door. An unlocked door requires no force; a window that can be opened does not need to be broken. Checking first takes only a few seconds, but it could save several minutes of effort and unnecessary property damage. It is equally important to look for alternative entry points; do not spend time working on a locked door when, for example, a nearby window provides easy access to the same room.

Unusually difficult forcible entry situations may require the expertise of specialty companies, such as technical rescue teams. These companies have experience working with specialized tools and equipment to gain access to almost any area.

FIRE FIGHTER Tips

Many communities have installed lockbox systems. With these systems, a master key carried by fire department members opens a small lockbox at the incident structure. This lockbox contains the keys needed to open the structure's door. If your community has a lockbox system, always look for a lockbox before beginning any type of forcible entry. The fastest way to gain access to a building is with a key.

Forcible Entry Tools

Choosing the right forcible entry tool can be a very important decision. Fire departments use many different types of forcible entry tools—from basic cutting, prying, and striking tools to sophisticated mechanical and hydraulic equipment. Fire fighters must be familiar with the following information:

- Which tools the department uses
- What the uses and limitations of each tool are
- How to select the proper tool for the job
- How to safely operate each tool
- How to carry the tools safely
- How to inspect and maintain each tool, thereby ensuring its readiness for service and safe operation

■ General Tool Safety

Any tool that is used incorrectly or is maintained improperly can be dangerous to both the operator and other persons in the vicinity. Anyone who uses a tool should understand the proper operating and maintenance procedures before beginning the task. Always follow the manufacturer's recommendations for operation and maintenance.

General safety tips for using tools include the following points:

1. Always wear the appropriate protective equipment. When conducting forcible entry during fire-suppression operations, this includes a full set of structural firefighting personal protective equipment (PPE).

- Goggles or approved eye protection are required when working with cutting or striking tools to prevent eye injuries from pieces of metal, glass, or other materials.
- Gloves provide protection from sharp cutting blades, sharp edges, and broken glass. Good gloves can mean the difference between a minor injury and a severed finger.
- A helmet provides essential protection from falling debris.
- A turnout coat and pants help to protect skin from sharp metal edges or broken glass.
- Boots protect the feet from sharp objects and from falling objects.

2. Learn to recognize the materials used in building and lock construction, and become familiar with the appropriate tools and techniques used for each kind of material; for example, many locks are made of **case-hardened steel**, which combines carbon and nitrogen to produce a very hard outer coating that cannot be penetrated by most ordinary cutting tools. Using the wrong tool to cut case-hardened steel could break the tool and potentially injure the user and other fire fighters.

3. Keep all tools clean, properly serviced according to the manufacturer's guidelines, and ready for use. Immediately report any tool that is damaged or broken and take it out of service.

4. Do not leave tools lying on the ground or floor; instead, return them to a tool staging area or the apparatus. Tools that are left lying around are tripping hazards that may increase the number of injuries on the fire ground.

■ General Carrying Tips

Tools can cause injuries even when they are not in use. Carrying a tool improperly, for example, can result in muscle strains, abrasions, or lacerations. The following general carrying tips apply to all tools:

- Do not try to carry a tool or piece of equipment that is too heavy or designed to be used by more than one person FIGURE 12-1. Instead, request assistance from another fire fighter to carry this type of tool.
- Always use your legs—not your back—when lifting heavy tools.
- Keep all sharp edges and points away from your body at all times. Cover or shield them with a gloved hand to protect those around you.
- Carry long tools with the head down toward the ground. Be aware of overhead obstructions and wires, especially when using pike poles and ladders.

■ General Maintenance Tips

All tools should be kept in a ready state; that is, the tool should be in proper working order, in its proper storage place, and ready for immediate use. Forcible entry tools are generally stored in a designated compartment on the apparatus. Hand tools should be clean, and cutting blades should be sharp. Power tools should be completely fuelled and treated with a

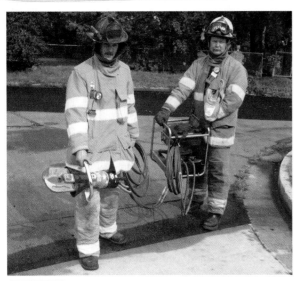

FIGURE 12-1 Request assistance to help carry a heavy load.

fuel-stabilizing product, if necessary, to ensure easy starting. Also remember that metal cutting blades can be damaged by gasoline vapours. Every crew member should be able to locate the right tool immediately, confident that it is ready for use **FIGURE 12-2** .

FIGURE 12-2 All tools should be kept in a neat and ready state.

All tools require regular maintenance and cleaning to ensure that they will be ready for use in an emergency. Thorough, conscientious daily or weekly checks of these items should be performed, particularly with infrequently used tools. Always follow the manufacturer's instructions and guidelines, and store maintenance manuals in an easily accessible location. Keep proper records to track maintenance, repairs, and any warranty work that is performed on tools.

■ Types of Forcible Entry Tools

Forcible entry tools include hand tools and power tools. They can be classified into the following categories:

- Striking tools
- Prying/spreading tools
- Cutting tools
- Lock tools

Many of the basic firefighting tools are described and pictured in the chapter "Fire Fighter Tools and Equipment."

Striking Tools

<u>Striking tools</u> generate an impact force directly on an object or another tool. They are generally hand tools that are powered by human energy. The head of a striking tool is usually made of hardened steel.

Flat-Head Axe

The flat-head axe was one of the first tools developed by humans and is still widely used for a variety of purposes. One side of the axe head is a cutting blade; the other side is a flat striking surface. Fire fighters often use the flat side to strike a Halligan tool and drive a wedge into an opening. Most fire apparatus carry both flat-head axes and pick-head axes. The pick-head axe is a cutting tool, not a striking tool.

Battering Ram

The **battering ram** is used to force doors and breach walls. Originally, it was a large log used to smash through enemy fortifications. Today's battering rams are usually made of hardened steel and have handles; two to four people are needed to use a battering ram.

Sledgehammer

Sledgehammers come in a variety of weights and sizes. The head of the hammer can weigh from 0.9 to 9.1 kilograms (2 to 20 pounds). The handle may be short like a carpenter's hammer or long like an axe handle. A sledgehammer can be used by itself to break down a door or in conjunction with other striking tools, such as the Halligan bar.

Prying/Spreading Hand Tools

Hand tools designed for prying and spreading are often used by fire fighters to force entry into buildings. This section describes the most commonly used forcible entry tools.

Halligan Tool (Bar)

The Halligan tool, or bar, is widely used by the fire service **FIGURE 12-3**. Pairing a Halligan tool with a flat-head axe creates a tool set often referred to collectively as "the **irons**." These two tools are commonly used to perform forcible entry. The Halligan tool incorporates three different tools—the **adz**, the **pick**, and the **claw**. The adz end is used to pry open doors and windows; the pick end is used to make holes or break glass; and the claw is used to pull nails and to pry apart wooden slats. A fire fighter can use a flat-head axe to strike the Halligan bar into place, thereby creating a better **bite** (a small opening that allows better tool access in forcible entry) between a door or window and its frame.

FIGURE 12-3 A Halligan tool has multiple purposes.

Pry Bar/Hux Bar/Crowbar

Pry bars, Hux bars, and crowbars are made from hardened steel cast into a variety of shapes and sizes. These tools are most commonly used to force doors and windows, to remove nails, or to separate building materials. The different shapes allow fire fighters to exert different amounts of leverage in diverse situations.

Pry Axe

The **pry axe** is a multipurpose tool that can be used both to cut and to force open doors and windows **FIGURE 12-4**. It includes an adz, a pick, and a claw. The tool consists of two parts: the body, which has the adz and pick and looks similar to a miniature pick axe, and a handle with a claw at the end, which slides into the body. The handle of the pry axe can be extended to provide extra leverage when prying, or it may be removed and inserted into the head of the adz to provide rotational leverage. Extreme caution should be used when handling this tool. Over time, the mechanism that locks the handle into position may become worn, allowing the handle to slip out and potentially injure fire fighters.

Hydraulic Tools

Hydraulic-powered tools, such as spreaders, cutters, and rams, are often used for forcible entry. These tools require hydraulic pressure, which can be provided by a high-pressure, motor-operated pump or a hand pump.

FIGURE 12-4 The pry axe is a multipurpose tool that can be used to cut and pry.

Courtesy of Paratech Inc.

Hydraulic cutters and spreaders are typically used in vehicle extrication but can also be used in some forcible entry situations **FIGURE 12-5**. Hydraulic rams come in a variety of lengths and sizes and can be used to apply a powerful force in one direction. The operation of these tools is covered in the chapter "Vehicle Rescue and Extrication."

A rabbet tool is a small hydraulic spreader operated by a hand-powered pump. This tool is designed with teeth that will fit into a door jamb or **rabbet** (a type of door frame that has the door stop cut into the frame). As the spreader opens, it applies a powerful force that can open many doors.

Cutting Tools

Cutting tools are primarily used for cutting doors, roofs, walls, and floors. Although they do not work as quickly as power tools, hand-operated cutting tools are proven and reliable in many situations. Because they do not require a power source, these items often can be deployed more quickly than power tools.

FIGURE 12-5 A hydraulic spreader.

Several power cutting tools are available for different applications. These tools can be powered by batteries, electricity, gasoline, or hydraulics, depending on the amount of power required. Each type of tool and power source has both advantages and disadvantages; for example, battery-powered tools are portable and can be placed in operation quickly but have limited power and operating times.

Axe

Many different types of axes are available, including flat-head, pick-head, pry, and multipurpose axes. The cutting edge of an axe can be used to break into plaster and wood walls, roofs, and doors **FIGURE 12-6**.

FIGURE 12-6 The cutting edge of an axe can be used to break into plaster and wood walls, roofs, and doors.

Although a flat-head axe was described earlier in this chapter as a striking tool, it is also classified as a cutting tool. The pick-head axe is similar to a flat-head axe but has a pick instead of a striking surface opposite the blade. This pick can be used to make an entry point or a small hole if needed.

Specialty axes, such as the pry axe and the multipurpose axe, can also be used for purposes other than cutting. The multipurpose axe, for example, can be used for cutting, striking, or prying. It includes a pick, a nail puller, a hydrant wrench, and a gas-main shut-off wrench.

Bolt Cutters

Bolt cutters are used to cut metal components such as bolts, padlocks, chains, and chain-link fences. These tools are available in several different sizes, based on the blade opening and the handle length. The longer the handle, the greater the cutting force that can be applied. Note that bolt cutters may not be able to cut into some heavy-duty padlocks that are constructed from case-hardened metal.

Circular Saw

Most fire departments use gasoline-powered circular saws both for making forcible entry and for cutting ventilation holes. These tools are light, powerful, and easy to use, with blades that can be changed quickly. The different blades enable the saw to cut several materials.

Fire departments generally carry several different circular saw blades so that the proper blade can be used in different situations.

- Carbide-tipped blades are specially designed to cut through hard surfaces or wood. They stay sharp for long periods, so more cuts can be made before the blade needs to be changed.
- Metal-cutting blades consist of a composite material made with aluminum oxide. They are used to cut metal doors, locks, or gates.
- Masonry-cutting blades are abrasive and made of a composite material that includes silicon carbide or steel. They can cut concrete, masonry, and similar materials. Because masonry-cutting blades resemble metal-cutting blades, the operator must check the label on the blade before using it. Blades with missing labels should be discarded. Do not store masonry-cutting blades near gasoline, because the gasoline vapours will cause the composite materials in the blade to decompose. In such a case, when the blade is later used, it could disintegrate.

FIRE FIGHTER Tips

Almost all cutting tools have built-in safety features. Always take advantage of the safety features on every tool. To avoid injury, never remove safety guards from tools—not even for "just one cut."

FIRE FIGHTER Tips

Caution should be used when employing gas-powered equipment in flammable atmospheres, because they are likely to produce sparks. Caution should also be used when using such equipment in enclosed spaces, because these tools produce carbon monoxide in their exhaust fumes.

Lock Tools/Specialty Tools

Lock and specialty tools are used to disassemble the locking mechanism on a door. These devices will cause minimal damage to the door and the door frame. If they are used properly, the door and frame should be undamaged, although the lock generally must be replaced. An experienced user can usually gain entry in less than a minute with these tools.

K Tool, A Tool, and J Tool

The **K tool** is designed to cut into a lock cylinder **FIGURE 12-7A**. To shear the cylinder from a lock, you place the cutting edge

A.

B.

C.

of the K tool on the top of the lock, insert a pry bar into the K tool, and strike the pry bar with a flat-head axe. If the lock has a protective ring, the K tool might not be able to cut through it. Once the lock cylinder is removed, another tool can then be used to open the locking mechanism.

The **A tool** is similar to the K tool except that the pry bar is built into the cutting part of the tool **FIGURE 12-7B**. This tool gets its name from its A-shaped cutting edges. To use an A tool, you put the cutting head on the lock cylinder and use a striking tool to force it down into the cylinder until the lock can be forced out of the door. Once the lock cylinder is removed, you would then use another tool to open the locking mechanism.

A **J tool** will fit between double doors that have panic bars **FIGURE 12-7C**. Slide the J tool between the doors and pull to engage the panic bars.

Shove Knife

A shove knife is an old tool yet is frequently used by modern-day fire fighters. Slipping the knife between the door and the frame at the latch forces the latch back and opens the door. Most new doors have latches that will not respond to this tool.

Duck-Billed Lock Breakers

To open a padlock using a **duck-billed lock breaker**, drive the point into the shackles (the U-shaped part) of the lock. The increasing size of the point forces the shackles apart until they break. While this causes no damage to the lock body, the lock will be inoperable after use of the duck-billed lock breaker.

Locking Pliers and Chain

The locking pliers and chain are used to clamp a padlock securely in place so that the shackles can be cut safely with a circular saw or cutting torch. To use this tool, one fire fighter clamps the pliers to the lock body while a second fire fighter maintains a steady tension on the chain as the lock is being cut.

Bam-Bam Tool

The **bam-bam tool** contains a case-hardened screw, which is inserted and secured in the keyway of a lock. After the screw is in place, you drive the handle of the tool into the keyway and pull the tumbler out of the lock.

Doors

In most cases, the best point to attempt forcible entry to a structure is the door or a window. Doors and windows are constructed as entry points, so they are made generally of weaker materials than are walls or roofs. An understanding of the basic construction of doors will help you select the proper tool and increase your likelihood of successfully gaining entry.

■ Basic Door Construction

Doors can be categorized both by their construction material and by the way they open. Both interior and exterior doors have the same basic components **FIGURE 12-8**:

- **Door** (the entryway itself)
- **Jamb** (the frame)
- **Hardware** (the handles, hinges, and other components)
- **Locking mechanism**

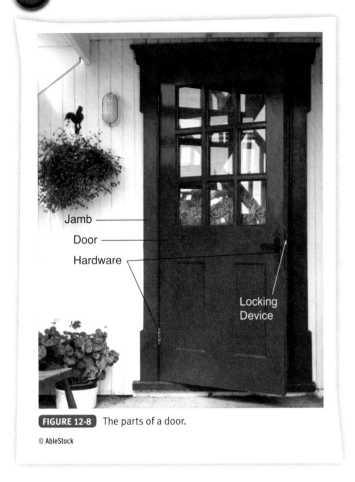

FIGURE 12-8 The parts of a door.
© AbleStock

FIGURE 12-9 A slab door may have either solid-core or hollow-core construction.

■ Construction Materials

Doors are generally constructed of wood, metal, or glass. The design of the door and the construction material used will determine the difficulty of forcing entry through this point.

Wood

Wood doors are commonly used in residences and may be found in some commercial buildings. Three types of wood swinging doors are distinguished: slab, ledge, and panel. Slab doors come in solid-core or hollow-core designs and are attached to wood-frame construction with normal hardware **FIGURE 12-9**.

Solid-core doors are constructed of solid wood core blocks covered by a face panel. Some have a fire rating. Such doors are typically used for entrance doors. Solid-core doors are heavy and may be difficult to force, but their construction enables them to contain fire better than hollow-core doors do.

Hollow-core doors have a lightweight, honeycomb interior, which is covered by a face panel. They are often used as interior doors, such as for bedrooms. Hollow-core doors are easy to force and will burn through quickly.

Ledge doors are simply wood doors with horizontal bracing. These doors, which are often constructed of tongue-and-groove boards, may be found on warehouses, sheds, and barns.

Panel doors are solid wood doors that are made from solid planks to form a rigid frame, with solid wood panels set into the frame. Panel doors are used as both exterior and interior

doors and may be made from a variety of types of wood. These doors resist fire longer than hollow-core slab doors do and are typically easier to breach than solid-core slab doors if entry is attempted at the panels.

Metal

Metal doors may be either decorative (for residential use) or utilitarian (for warehouses and factories). Like wood doors, they may have either a hollow-core or a solid-core construction. Hollow-core metal doors have a metal framework interior so they are as lightweight as possible. By contrast, solid-core metal doors have a foam or wood interior that is intended to reduce the door's weight without affecting its strength. Metal doors may be set in either a wood frame or a metal frame. Residential metal doors may appear to be panel doors and are often used as entry doors.

Glass

Glass doors generally have a steel frame with tempered glass; alternatively, they may be simply tempered glass and not require a frame, but have metal supports to attach hardware. Glass doors are easy to force, but can be dangerous owing to the large number of small broken pieces that are produced when glass is broken.

Fire Fighter Safety Tips
© Photos.com

Opening a door can allow air to enter the structure, potentially resulting in fire spread or a backdraft. Be careful and listen to instructions when performing forcible entry. Everything needs to be coordinated!

A.

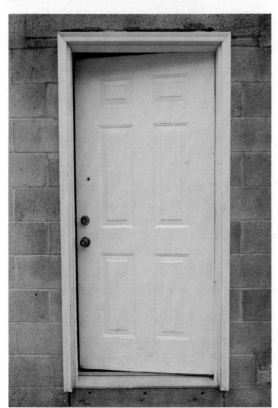

B.

FIGURE 12-10 **A.** An outward-opening door will have hinges showing. **B.** A door with the hinges not showing will open inward.

■ Types of Doors

Doors are also classified based on how they open. The five most common ways that doors open are inward, outward, sliding, revolving, and overhead. Overhead, sliding, and revolving doors are the most readily identified. Inward-opening and outward-opening doors can be differentiated based on whether the hinges are visible. If you can see the hardware, the door will swing toward you (outward) **FIGURE 12-10A** ; if the hinges are not visible, the door will swing away from you (inward) **FIGURE 12-10B** .

Door frames may be constructed of either wood or metal. Wood-framed doors come in two styles: stopped and rabbet. Stopped door frames have a piece of wood attached to the frame that stops the door from swinging past the latch. Rabbeted door frames are constructed with the stop cut built into the frame so it cannot be removed.

Metal-framed doors are more difficult to force than wood-framed doors. Metal frames have little flexibility, so when a metal frame is used with a metal door, forcing entry will be very difficult. Metal frames look like rabbeted door frames.

Inward-Opening Doors
Design
Inward-opening doors of wood, steel, or glass can be found in most structures. They have an exterior frame with a stop or rabbet that keeps the door from opening past the latch. The locking mechanisms may range from standard doorknob locks to deadbolt locks or sliding latches.

Forcing Entry
A simple solution to gaining entry through inward-opening doors may be to break a small window on the door or adjacent to it, reach inside, and operate the locking mechanism. Remember: try before you pry.

If no window is available and the door is locked, stronger measures are required. To force an inward-opening door, first ascertain which type of frame it has. If the door has a stopped frame, use a prying tool near the locking mechanism to pry the stop away from the frame. After removing the stop, reinsert the prying tool near the latch, and pry the door away from the frame. Once the latch clears the strike plate, push the door inward. Use a striking tool to force the prying tool farther into the jamb.

To force entry into an inward-opening door, follow the steps in **SKILL DRILL 12-1** :

1. Look for any safety hazards as you evaluate the door. Inspect the door for the location and number of locks and their mechanisms. **(STEP ①)**
2. Place the adz of the Halligan tool into the door frame between the door jamb and the door stop, near the lock, with the bevelled end against the door. **(STEP ②)**
3. Once the Halligan tool is in position, have your partner, on your command, drive the tool farther into the gap between the rabbeted jamb or stop and the door. Make sure that the tool is not driven into the door jamb itself. **(STEP ③)**
4. Once the tool is past the stop but between the door and the jamb, push the Halligan tool toward the door to force it open.

SKILL DRILL 12-1 Forcing Entry into an Inward-Opening Door
(Fire Fighter I, NFPA 5.3.4)

1 Size up the door, looking for any safety hazards. Inspect the door for the location and number of locks and mechanisms.

2 Place the adz of the Halligan tool into the door frame between the door jamb and the door stop, near the lock, with the bevelled end of the tool against the door.

3 Once the Halligan tool is in position, have your partner, on your command, drive the tool farther into the gap between the rabbeted jamb or stop and the door. Make sure that the tool is not driven into the door jamb itself.

4 Once the tool is past the stop but between the door and the jamb, push the Halligan tool toward the door to force it open. If more leverage is needed, your partner can slide the axe head between the bevel of the Halligan tool and the door. It may be necessary to push in on the door. Secure the door to prevent it from closing behind you.

© LiquidLibrary

5 If more leverage is needed, your partner can slide the axe head between the bevel of the Halligan tool and the door. It may be necessary to push in the door.

6 Once the door is open, secure it to prevent it from closing behind you. (STEP **4**)

Outward-Opening Doors

Design

Outward-opening doors are used in commercial occupancies and for most exits **FIGURE 12-11**. They are designed so that people can leave a building quickly during an emergency.

Outward-opening doors may be constructed of wood, metal, or glass. They usually have exposed hinges, which may present an entry opportunity. More frequently, however, these hinges will be sealed so that the pins cannot be removed. Several types of locks, including handle-style locks and deadbolts, may be used with these doors.

Forcing Entry

Before forcing entry to an outward-opening door, check the hinges to see if they can be disassembled or the pins removed. If that would take too long or cannot be done, place the adz

FIGURE 12-11 An outward-opening door enables rapid exit from a building.

FIRE FIGHTER Tips

Self-closing doors may work via a means as simple as a mechanism above the door or as complex as a magnet or electrically charged release that functions by closing the doors when the fire alarm sounds. Either way, when these doors are forced, the mechanism that closes the doors should be disabled, along with using some type of door chock or wedge to keep the doors from closing behind you once entry is made.

end of a prying tool into the door frame near the locking mechanism. Use a striking tool to drive it farther into the door jamb and get a good bite on the door. Then leverage the tool to force the door outward, away from the jamb.

To force entry into an outward-opening door, follow the steps in **SKILL DRILL 12-2** :

1. Size up the door, looking for any safety hazards. Determine the number and location of locks, and identify the locking mechanism.
2. Place the adz end of the Halligan tool between the door and the frame, near the locking mechanism, or between the mechanism and a secondary lock. **(STEP** 1**)**
3. Once the Halligan tool is in position, give the command to your partner to strike the Halligan tool and drive the adz end farther into the gap. **(STEP** 2**)**

SKILL DRILL 12-2   Forcing Entry into an Outward-Opening Door
(Fire Fighter I, NFPA 5.3.4)

1 Size up the door, looking for any safety hazards. Place the adz end of the Halligan tool between the door and the frame either near the locking mechanism, or between the mechanism and a secondary lock.

2 Once the Halligan tool is in position, have your partner strike the Halligan tool on your command and drive the adz end farther into the gap.

3 Pry in a downward direction with the fork end of the tool, and then force the door outward. Secure the door to prevent it from closing behind you.

© LiquidLibrary

④ Pry down with the fork end of the tool, and then force the door outward.

⑤ Always secure the door to prevent it from closing behind you. **(STEP ③)**

Sliding Doors
Design

Most sliding doors are constructed of tempered glass in a wooden or metal frame. Such doors are commonly found in residences and hotel rooms that open onto balconies or patios **FIGURE 12-12**. Sliding doors generally have two sections and a double track; one side is fixed in place, while the other side slides. A weak latch on the frame of the door secures the movable side. Many people prop a wood or metal rod in the track to provide additional security. If this has been done, look for another means of entry, if possible, to avoid destroying the door and leaving the premise open.

FIGURE 12-12 A reinforced sliding glass door may be difficult to open without breaking the glass.

Forcing Entry

Forcing open sliding glass doors may be either very easy or very difficult. If the doors are not reinforced with a rod in the track, they should be easy to force. The locking mechanisms are not strong, and any type of prying tool can be used. Place the tool into the door frame near the locking mechanism, and force the door away from the mechanism. If a rod in the track prevents the door from moving and there is no other way to enter, the only choice is to break the glass to gain forcible entry. This may create a hazard if fire fighters must drag hose lines or

victims through the sharp glass debris. Clean all broken glass from the opening to avoid such problems.

Revolving Doors
Design

Revolving doors are most commonly found in upscale buildings and buildings in large cities **FIGURE 12-13**. They are usually made of four glass panels with metal frames. The panels are designed to collapse outward with a certain amount of pressure to allow for rapid escape during an emergency. Revolving doors are generally secured by a standard cylinder lock or slide-latch lock.

FIGURE 12-13 Building codes may require outward-opening doors next to a revolving door.

Courtesy of Horton Automatics

Because of NFPA Life Safety Code® fire and building code requirements, standard outward-opening doors are often found adjacent to the revolving doors. It may be easier to attempt to force entry through these doors instead of going through the revolving doors. Repairing any damage to the outward-opening doors will be less expensive for the building owner, too.

Forcing Entry

Forcible entry through revolving doors should be avoided whenever possible. Even if the doors can be forced open, the opening created will not be large enough to allow many people to exit quickly and easily. Forcible entry through a revolving door can be achieved by attacking the locking mechanism directly or by breaking the glass.

Overhead Doors
Design

Overhead doors come in many different designs, ranging from standard residential garage doors to high-security commercial roll-up doors **FIGURE 12-14** . Most residential garage doors have three or four panels, which may or may not include windows. Some residential overhead garage doors come in a single section and tilt rather than roll up. These doors, which may be made of wood or metal, usually have a hollow core that is filled with insulation or foam. By contrast, commercial security overhead doors are made of metal panels or hardened steel rods. They may use solid-core or hollow-core construction, depending on the amount of security needed. Overhead doors can be secured with cylinder-style locks, padlocks, or automatic garage door openers.

A.

B.

FIGURE 12-14 **A.** Commercial overhead doors are made of metal panels or steel rods. **B.** Residential overhead garage doors may tilt or roll up.

© Dave White/iStockphoto.com

© Photodisc/Creatas

Forcing Entry

Before forcing entry through an overhead door, perform a careful size-up of the door. Most residential garage doors are not very sturdy; breaking a window or panel and manually operating the door lock or pulling the emergency release on the automatic opener from the inside may be all that is needed. If the fire is located behind the overhead door, however, it might have weakened the door springs, making it impossible to raise the door. In such a case, either the door must be cut or another means of entry found.

Secure any raised door with a pike pole or other support to ensure that it does not close on either fire fighters or their attack lines. Put the prop under the door near the track. If the door has an emergency release cord, it might be able to serve as a second safety measure.

The quickest way to force entry through a security roll-up door is to cut the door with a torch or saw. Make the cut in the shape of a triangle, with the point at the top. Pad this opening to prevent injury to those entering and exiting through the opening. To open an overhead garage door using the triangle method, follow the steps in **SKILL DRILL 12-3** :

1. Before cutting, check for any safety hazards during the size-up of the garage door.
2. Select the appropriate tool to make the cut. The best choice is a power saw with a metal-cutting blade.
3. Wearing full turnout gear and eye protection, start the saw and ensure it is in proper working order. (**STEP 1**)
4. Be aware of the environment behind the door. If necessary, cut a small inspection hole, large enough to insert a hose nozzle. (**STEP 2**)
5. Starting at a centre high point in the door, make a diagonal cut to the right, down to the bottom of the door. (**STEP 3**)
6. From the same starting point, make a second diagonal cut to the left, down to the bottom of the door.
7. Fold the door down to form a large triangle. (**STEP 4**)
8. If necessary, pad or protect the cut edges of the triangle and the bottom panel to prevent injuries as fire fighters enter or leave the premises. (**STEP 5**)

Fire Fighter Safety Tips
© Photos.com

Always secure doors in the open position before entering a building. Prop open overhead doors with a pike pole. Use the emergency release cord if one is available. If this step is not taken, the overhead doors could come down on fire fighters and on their attack lines, trapping them and cutting off their water supply. This problem could complicate the attack and lead to injuries or deaths.

Windows

Windows provide both air flow and light to the inside of buildings; they also can provide emergency entry or exits. Windows are often easier to force than doors. Understanding how to force entry into a window requires an understanding of both window-frame construction and glass construction. Window frames are made of the same materials used in doors—wood, metal, vinyl, or combinations of these materials—and will often match the door construction of a building.

SKILL DRILL 12-3   Opening an Overhead Garage Door Using the Triangle Method
(Fire Fighter I, NFPA 5.3.4)

1 Before cutting, check for any safety hazards. Select the appropriate tool to make the cut. Wearing full turnout gear and eye protection, start the saw and ensure it is in proper working order.

2 Be aware of the environment behind the door. If necessary, cut a small inspection hole, large enough to insert a hose nozzle.

3 Starting at a centre high point in the door, make a diagonal cut to the right, down to the bottom of the door.

4 From the same starting point, make a second diagonal cut to the left, down to the bottom of the door. Fold the door down to form a large triangle.

5 If necessary, pad or protect the cut edges of the triangle and the bottom panel to prevent injuries as fire fighters enter or leave the premises.

© LiquidLibrary

This section reviews window construction, glass construction, and forcible entry techniques. The goal of forcible entry using windows is to gain entry with minimal or no damage to the window. As with doors, you should try to open the window normally before using any force. Although breaking the glass is the easiest way to force entry, it is also dangerous. Glass may be the least expensive part of the window, but replacement costs are increasing as energy-efficient windows are being used more widely.

■ Safety

Always take proper protective measures, including wearing full PPE, when forcing entry through windows. Ensure that the area around the window, both inside and outside the building and below the window, is clear of other personnel. Broken pieces of wood, shattered glass, and sharp metal can cause serious injury.

Remember that forcing entry through windows during a fire situation might cause the fire to spread. Do not attempt forcible entry through a window unless a proper fire attack is in place. Always stand to the windward side, with your hands higher than the breaking point, when breaking windows. This positioning ensures that the broken glass will fall away from your hands and body. Placing the tip of the tool in the corner of the window will give you more control in breaking the window. After the window is broken, clear glass from the entire frame, so that no glass shards will stick out and cause injury to fire fighters who enter or exit through the window.

■ Glass Construction

The **glazed** (transparent) part of the window is most commonly made of glass. Window glass comes in several configurations: regular glass, double-triple-pane glass, plate glass (for large windows), laminated glass, and tempered glass. Plastics such as Plexiglas® and Lexan may also be used in windows. The window may contain one or more panes of glass. Insulated glass, for example, usually consists of two or more pieces of glass in the window and will be discussed in more detail later.

Regular or Annealed Glass

Single-pane regular or **annealed** glass is often used in construction because it is relatively inexpensive; larger pieces are called plate glass. This type of glass is easily broken with a pike pole. When broken, plate glass creates long, sharp pieces called shards, which can penetrate helmets, boots, and other protective gear, causing severe lacerations and other injuries.

Double-Triple-Pane Glass (Insulated Windows)

Double-triple-pane glass is used in many homes because it improves home insulation by using two panes of glass with an air pocket between them. Some double-triple-pane windows may include an inert gas, such as argon, between the panes for additional insulation value. These windows are sealed units, which makes them more expensive to replace. However, replacing the glass alone is less expensive than replacing the entire window assembly. Forcing entry through insulated windows is basically the same as forcing entry through single-pane windows, except that the two panes may need to be broken separately. These kinds of windows also will produce dangerous glass shards.

Plate Glass

Commercial **plate glass** is a stronger, thicker glass used in large window openings. Although it is being replaced by tempered glass in modern construction for safety reasons, commercial plate glass can still be found in older large buildings, storefronts, and residential sliding doors. It can easily be broken with a sharp object such as a Halligan tool or a pike pole. When broken, commercial plate glass will create many large, sharp pieces.

Laminated Glass

Laminated glass, also known as safety glass, is used to prevent windows from shattering and causing injury. Laminated glass is moulded with a sheet of plastic between two sheets of glass. This type of glass is most commonly used in vehicle windshields, but it may also be found in other applications such as doors or building windows.

Tempered Glass

Tempered glass is specially heat treated, making it four times stronger than regular glass. This type of glass is commonly found in side and rear windows in vehicles, in commercial doors, in newer sliding glass doors, and in other locations where a person might accidentally walk into the glass and break it. Tempered glass breaks into small pellets without sharp edges to help prevent injury during accidents. The best way to break tempered glass is by using a sharp, pointed object in the corner of the frame. During vehicle extrication, a centre-punch is often used to break a vehicle window.

Wired glass is tempered glass that has been reinforced with wire. This kind of glass may be clear or frosted, and it is often used in fire-rated doors that require a window or sight line from one side of the door to the other. Wired glass is difficult to break and force.

■ Frame Designs

Window frames come in many styles. This chapter covers the most common ones, but fire fighters must be familiar with both the types in their response area and the forced entry techniques for each type.

Double-Hung Windows
Design

Double-hung windows contain two movable sashes, usually made of wood or vinyl, that move up and down **FIGURE 12-15**. These kinds of windows are frequently found in residences and have wood, plastic, or metal runners. Newer double-hung window sashes may be removed or swung in for cleaning. They may have either one locking mechanism that is found in the centre of the window or two locks on each side of the lower sash that prevent the sashes from moving up or down.

FIGURE 12-15 Double-hung windows allow the inner and outer sashes to move freely up and down.

Forcing Entry

Forcing entry through double-hung windows involves opening or breaking the locking mechanism. Place a prying tool under the lower sash, and force it up to break the lock or remove it from the track. This technique must be done carefully; otherwise, it may cause the glass to shatter. Because forcing the locks causes extensive damage to the window, breaking the glass and opening the locks may be a less expensive method of gaining entry.

To force entry through a wooden double-hung window, follow the steps in **SKILL DRILL 12-4**:

1 Size up the window to check for any safety hazards and locate the locking mechanism. (**STEP 1**)

2 Place the pry end of the Halligan tool under the bottom sash in line with the locking mechanism. (**STEP 2**)

3 Pry the bottom sash upward to displace the locking mechanism.

4 Secure the window in the open position so that it cannot close. (**STEP 3**)

Fire Fighter Safety Tips
© Photos.com

When forcing entry through older, double-hung windows, try sliding a hacksaw blade or similar blade up between the sashes. The teeth on the blade will grip the locking mechanism and can often move it into the unlocked position. This technique will not work on newer double-hung windows because of their newer latch design.

FIGURE 12-16 A single-hung window has only one movable sash.

Single-Hung Windows
Design
Single-hung windows are similar to double-hung windows, except that the upper sash is fixed and only the lower sash moves **FIGURE 12-16**. The locking mechanism is the same as on double-hung windows as well. It may be difficult to distinguish between single-hung and double-hung windows from the exterior of a building.

Forcing Entry
Forcing entry through a single-hung window involves the same technique as forcing entry through a double-hung window. Place a prying tool under the lower sash, and force it up, which will usually break the lock or remove the sash from the track. Be careful, because the glass will probably shatter. Breaking the glass and opening the window is generally easier with a single-hung window.

Jalousie Windows
Design
A **jalousie window** is made of adjustable sections of tempered glass encased in a metal frame that overlap each other when closed **FIGURE 12-17**. This type of window is often found in mobile homes and is operated by a small hand wheel or crank located in the corner of the window.

Forcing Entry
Forcing entry through jalousie windows can be difficult because they are made of tempered glass and have several panels. Forcing open the panels or breaking a lower one to operate the crank is possible but does not leave a large-enough

---

SKILL DRILL 12-4
Forcing Entry through a Wooden Double-Hung Window
(Fire Fighter I, NFPA 5.3.4)

1 Size up the window to check for any safety hazards and locate the locking mechanism.

2 Place the pry end of the Halligan tool under the bottom sash in line with the locking mechanism.

3 Pry the bottom sash upward to displace the locking mechanism. Secure the window so that it does not close.

© LiquidLibrary

FIGURE 12-17 Jalousie windows are opened and closed with a small hand crank.

opening for a person to enter the building. Removing or breaking the panels one at a time is time-consuming and does not always leave a clean entry point with the framing system still in place. The best strategy is to avoid these windows.

Awning Windows
Design
Awning windows are similar in operation to jalousie windows, except that they usually have one large or two medium-sized glass panels instead of many small panes **FIGURE 12-18**. Awning

FIGURE 12-18 An awning window has larger panels than a jalousie window does.

windows are operated by a hand crank located in the corner or in the centre of the window. These kinds of windows can be found in residential, commercial, and industrial structures. Residential awning windows may be framed in wood, vinyl, or metal, whereas commercial and industrial windows will usually have metal frames.

Commercial and industrial awning windows often use a lock and a notched bar to hold the window open, rather than a crank. These so-called projected windows are discussed later in this chapter.

Forcing Entry
Forcing entry through an awning window requires the same technique as is used for jalousie windows; that is, break or force the lower panel and operate the crank, or break out all the panels. Depending on the size of the panels and the window frame, it may be easier to access an awning window. The larger opening of an awning window makes entry easier as well.

Horizontal-Sliding Windows
Design
Horizontal-sliding windows are similar to sliding doors **FIGURE 12-19**. The latch is similar as well and attaches to the window frame. People often place a rod or pole in the track to prevent break-ins. Newer sliding windows have latches between the windows, similar to those found on double-hung windows.

Sliding sash

Window frame

FIGURE 12-19 Horizontal-sliding windows work like sliding doors.

Forcing Entry
Forcing entry through sliding windows is just like forcing entry through sliding doors: Place a pry bar near the latch to break the latch or the plate. If a rod has been inserted into the track, look for another entry point or break the glass, although the latter step should be the last resort.

Casement Windows
Design
Casement windows have a steel or wood frame and open away from the building with a crank mechanism **FIGURE 12-20**. Although they are similar to jalousie or awning windows,

casement windows have a side hinge, rather than a top hinge. Several types of locking mechanisms can be used with these windows. Like jalousie windows, these windows should be avoided when forcible entry is necessary, because they are difficult to force open.

FIGURE 12-20 Casement windows open to the side with a crank mechanism.

Forcing Entry

The best way to force entry through a casement window is to break out the glass from one or more of the panes, locate the locking mechanisms, and crank them manually. To force entry through a casement window, follow the steps in **SKILL DRILL 12-5**:

1. Size up the window to check for any safety hazards and locate the locking mechanism.
2. Select an appropriate tool to break out a windowpane.
3. Stand to the windward side of the window, and break out the pane closest to the locking mechanism, keeping the tool lower than your hands. (**STEP 1**)
4. Remove all of the broken glass in the pane to prevent injuries. (**STEP 2**)
5. Reach in and manually operate the window crank to open the window. (**STEP 3**)

Projected Windows

Design

Projected windows (also called factory windows) are usually found in older warehouse or commercial buildings **FIGURE 12-21**. They can project inward or outward on an upper hinge. Screens are rarely used with these windows, but forcing entry may not be easy, depending on the integrity of the frame, the type of locking mechanism used, and the window's distance off the ground. These windows may have fixed, metal-framed wire glass panes above them.

Forcing Entry

Avoid forcing entry through a projected window when possible. These windows are often difficult to force open, and they may be difficult for a person to enter. To force entry, access one

---

SKILL DRILL 12-5

Forcing Entry through a Casement Window
(Fire Fighter I, NFPA 5.3.4)

1 Size up the window to check for any safety hazards and locate the locking mechanism. Select an appropriate tool to break out a windowpane. Stand to the windward side of the window, and break out the pane closest to the locking mechanism, keeping the tool lower than your hands.

2 Remove all of the broken glass in the pane to prevent injuries.

3 Reach in and manually operate the window crank to open the window.

FIGURE 12-21 Projected windows may open inward or outward.

pane, unlock the mechanism, and open the window by hand. If the opening created is not large enough, break out the entire window assembly.

To force entry through a projected or factory window, follow the steps in **SKILL DRILL 12-6**:

1. Size up the window to check for any safety hazards and locate the locking mechanism.
2. Select an appropriate tool, such as a pike pole, to break a pane of glass.
3. Stand to the windward side of the window and break the pane closest to the locking mechanism, keeping the tool head lower than your hands. (**STEP 1**)

4. Remove all of the broken glass in the frame to prevent injuries. (**STEP 2**)
5. Reach in and manually open the locking mechanism and the window.
6. If necessary, use a cutting tool, such as a torch, to remove the window frame and enlarge the hole. (**STEP 3**)

Locks

Locks have been used for hundreds of years. They range from the very simple push-button locks found in most homes to the complex computer-operated locks found in banks and other high-security areas.

■ Parts of a Door Lock

All door locks have the same basic parts. Gaining entry will be easier with an understanding of what these parts are and how they operate **FIGURE 12-22**. There are three major parts of a door lock:

- **Latch**—The part of the lock that catches and holds the door frame
- **Operator lever**—The handle, doorknob, or keyway that turns the latch to lock it or unlock it
- **Deadbolt**—A second, separate latch that locks and reinforces the regular latch

■ Parts of a Padlock

All padlocks, like all door locks, have similar parts **FIGURE 12-23**:

- **Shackles**—The U-shaped top of the lock that slides through a hasp and locks in the padlock itself

---

SKILL DRILL 12-6   Forcing Entry through a Projected Window
(Fire Fighter I, NFPA 5.3.4)

1 Size up the window to check for any safety hazards and locate the locking mechanism. Select an appropriate tool, such as a pike pole, to break a pane of glass. Stand to the windward side of the window, and break the pane closest to the locking mechanism, keeping the tool head lower than your hands.

2 Remove all of the broken glass in the frame to prevent injuries.

3 Reach in and manually open the locking mechanism and the window. If necessary, use a cutting tool, such as a torch, to remove the window frame and enlarge the hole.

FIGURE 12-22 All door locks have the same basic parts.

FIGURE 12-23 All padlocks have four basic parts.

Courtesy of IR Safety and Security – Americas

- **Unlocking mechanism**—The keyway, combination wheels, or combination dial used to open the padlock
- **Lock body**—The main part of the padlock that houses the locking mechanisms and the retention part of the lock

■ Safety

Tool safety is an important consideration when executing forcible entry through a door or padlock. The tools used for cutting the cylinders must be kept sharp, and fire fighters must be careful to avoid being cut when using these tools. Care also must be taken when using striking tools when forcing a lock to avoid pinching or crushing fingers or hands. Fire fighters

must wear proper protective equipment, such as gloves, and use tools carefully to minimize the risk of injury.

■ Types of Locks

The four major lock categories are cylindrical locks, padlocks, mortise locks, and rim locks.

Cylindrical Locks
Design
<u>Cylindrical locks</u> are the most common fixed lock in use today **FIGURE 12-24**. They are relatively inexpensive and are standard in most doors. The locks and handles are set into predrilled holes in the doors. One side of the door usually has a key-in-the-knob lock; the other side will have a keyway, a button, or some type of locking/unlocking mechanism.

Forcing Entry
Forcing entry into a cylindrical lock is usually very easy, because most mechanisms are not very strong. Place a pry bar near the locking mechanism, and lever it to force the lock.

Padlocks
Design
<u>Padlocks</u> are the most common locks on the market today and come in many designs and strengths. Both regular-duty and heavy-duty padlocks are available. Padlocks come with a variety of unlocking mechanisms, including keyways, combination wheels, or combination dials. Operating the unlocking

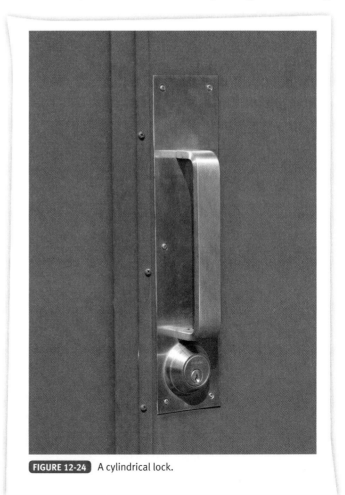

FIGURE 12-24 A cylindrical lock.

mechanism opens one side of the lock to release the shackle and allow entry. Shackles for regular padlocks generally have a diameter of 6 millimetres (0.25 inch) or less and are not made of case-hardened metal. Shackles for heavy-duty padlocks are 6 mm (0.25 in.) or larger in diameter and made of case-hardened steel or some other case-hardened metal. The design of some padlocks hides the latch, making forced entry difficult.

One of these locks is the Pro Series Hidden Shackle Padlock, also referred to as the **hockey puck lock** FIGURE 12-25 . These types of locks have hidden shackles and cannot be forced through conventional methods. A large pipe wrench with a cheater bar can be used to twist the locks from their mounting tabs; however, if the lock has a mounting bracket, this technique will be ineffective. In such a case, you will need a rotary saw or a cutting torch to force entry. When using this method, you should focus your effort approximately two-thirds of the way up from the keyway to cut the pin in half.

Forcing Entry

Several techniques can be used to force entry through padlocks without causing extensive property damage. Before breaking the padlock, consider cutting the shackle or hasp first. This approach saves the lock and makes securing the building easier. Breaking the shackle is the best method for forcing entry through padlocked doors. If the padlock is made of case-hardened steel, however, many conventional methods of breaking the lock will prove ineffective.

The tools most commonly used to force entry through a padlock are bolt cutters, duck-billed lock breakers, a bam-bam tool, locking pliers and chain, and a torch:

- Bolt cutters can quickly and easily break regular-duty padlocks, but cannot be used on heavy-duty, case-hardened steel padlocks. To use bolt cutters on a padlock, open the jaws as wide as possible. Close the jaws around one side of the lock shackles to cut through the shackle. Once one shackle of a regular-duty lock is cut, the other side will spin freely and allow access.
- The duck-billed lock breaker has a large metal wedge attached to a handle. Place the narrow end of the wedge into the centre of the shackle, and force it through with another striking tool. The wedge will spread the shackle until it breaks, freeing the padlock and allowing access to the building.
- The bam-bam tool can pull the lock cylinder out of a regular-duty padlock, but it will not work on higher-end padlocks that include security rings to hold the cylinder in place. This tool contains a case-hardened screw that is placed in the keyway. Once the screw is set, the sliding hammer pulls the tumblers out of the padlock so that the trip-lever mechanism inside the lock can be opened manually.
- The locking pliers and chain are attached to a padlock to secure it in a fixed position so that the lock can be cut. Use a rotary saw with a metal-cutting blade or a torch to cut the padlock after it has been secured. Securing the padlock is necessary because it is too dangerous to hold the lock with a gloved hand or to cut into it while it is loose.

Mortise Locks
Design

Mortise locks, like cylindrical locks, are designed to fit in pre-drilled openings inside a door; they are commonly found in hotel rooms FIGURE 12-26 . Mortise locks have both a latch and a bolt built into the same mechanism, each of which operates independently of the other. The latch will lock the door, but the bolt can also be deployed for added security.

Forcing Entry

The design and construction of mortise locks make them difficult to force. A door with a mortise lock may be forced with conventional means but will probably require the use of a through-the-lock technique.

Rim Locks and Deadbolts
Design

Rim locks and deadbolts are different types of locks that can be surface mounted on the interior of the door frame FIGURE 12-27 . They are commonly found in residences as secondary locks to support the through-the-handle locks. Such locks can be identified from the outside by the keyway that has been bored into the door. These locks have a bolt that extends at least 2.5 centimetres (1 in.) into the door frame, making the door more difficult to force.

Forcing Entry

Rim locks and deadbolts are the most challenging locks to break. They can be very difficult to force with conventional methods, such that through-the-lock methods may be the only option. Skill Drills 12-7 through 12-9 present various through-the-lock techniques.

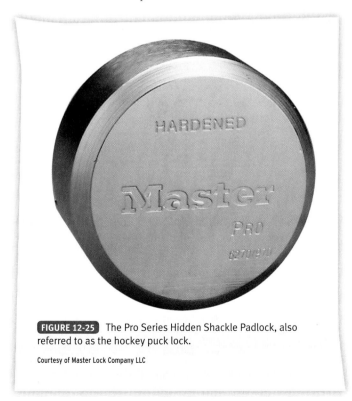

FIGURE 12-25 The Pro Series Hidden Shackle Padlock, also referred to as the hockey puck lock.

Courtesy of Master Lock Company LLC

FIGURE 12-26 Mortise locks have a latch and a bolt, which operate independently of each other.

Courtesy of IR Safety and Security - Americas

FIGURE 12-27 Rim locks and deadbolts are mounted on the inside of a residence door.

Courtesy of IR Safety and Security - Americas

To force entry using a K tool, follow the steps in **SKILL DRILL 12-7**:

1. Size up the lock area, checking for any safety hazards. Determine which type of lock is used, whether it is regular or heavy-duty construction, and whether the cylinder has a case-hardened collar, which may hamper proper cutting.

2. Place the K tool over the face of the cylinder, noting the location of the keyway.

3. Using a Halligan or similar pry tool, tap the K tool onto the cylinder. (**STEP 1**)

4. Place the adz end of the Halligan tool into the slot on the K tool, and strike the end of the Halligan tool with a flat-head axe to drive the K tool farther into the lock cylinder. (**STEP 2**)

5. Pry up on the Halligan tool to pull out the lock and expose the locking mechanism.

6. Using the small tools that come with the K tool, turn the mechanism to open the lock.

7. The lock should release, allowing you to open the door. (**STEP 3**)

FIRE FIGHTER Tips

If the K tool does not fit over a residential lock, try an A tool.

SKILL DRILL 12-7   Forcing Entry Using a K Tool
(Fire Fighter I, NFPA 5.3.4)

1 Place the K tool over the face of the cylinder, noting the location of the keyway. Using a Halligan or similar pry tool, tap the K tool onto the cylinder.

2 Place the adz end of the Halligan tool into the slot on the K tool, and strike the end of the Halligan tool with a flat-head axe to drive the K tool farther into the lock cylinder.

3 Pry up on the Halligan tool to pull out the lock and expose the locking mechanism. Using the small tools that come with the K tool, turn the mechanism to open the lock. The lock should release, allowing you to open the door.

© LiquidLibrary

To force entry using an A tool, follow the steps in **SKILL DRILL 12-8**:

1 Size up the door and lock area, checking for any safety hazards.

2 Place the cutting edges of the A tool at the top of the cylinder, between the cylinder and the door frame. (**STEP 1**)

3 Using a flat-head axe or similar tool, drive the A tool into the cylinder.

4 Pry up on the A tool to remove the cylinder from the door.

5 Insert a key tool into the hole to manipulate the locking mechanism and open the door. (**STEP 2**)

To force entry by unscrewing the lock, follow the steps in **SKILL DRILL 12-9**:

1 Size up the door and lock area, checking for any safety hazards.

2 Using a set of vise grips, lock the pliers onto the outer housing of the lock with a good grip. (**STEP 1**)

3 Unscrew the housing until it can be fully removed.

4 Manipulate the locking mechanism with lock tools to disengage the arm, and then open the door. (**STEP 2**)

■ Forcing Entry through Security Gates and Windows

Many homes are equipped with metal security gates and security bars over the windows, which are intended to provide protection from break-ins for the building residents. Security gates are usually equipped with a lock that must be unlocked using a key from both the inside and the outside. As a consequence, if a key is not available to the building occupants, they can become trapped in their own house. Likewise, security bars should be equipped with a locking mechanism that enables the residents to quickly open the security bars from the inside and use the window for an exit in case of a fire. Unfortunately, many security bars are permanently mounted into the window frame and cannot be removed quickly.

FIRE FIGHTER Tips

Electromagnetic locks use powerful magnets to keep doors shut. Doors with electromagnetic locks may be forced open with prying tools or may require hydraulic tools. Follow the suggested operating guidelines (SOGs), or standard operating procedures (SOPs) of your fire department.

Thus, these security devices present a dilemma: the same devices that were designed to keep criminals out can, in the event of a fire, make it difficult for the building's occupants and fire fighters to escape from the fire. Such devices also complicate fire fighters' efforts to gain access to the building to perform search and rescue operations and fire-suppression activities. Given these facts, it is important for fire fighters to know how to rapidly force entry when these devices are in place.

SKILL DRILL 12-8 — Forcing Entry Using an A Tool
(Fire Fighter I, NFPA 5.3.4)

1 Place the cutting edges of the A tool at the top of the cylinder, between the cylinder and the door frame.

2 Using a flat-head axe or similar tool, drive the A tool into the cylinder. Pry up on the A tool to remove the cylinder from the door. Insert a key tool into the hole to manipulate the locking mechanism and open the door.

© LiquidLibrary

SKILL DRILL 12-9 — Forcing Entry by Unscrewing the Lock
(Fire Fighter I, NFPA 5.3.4)

1 Using a set of vise grips, lock the pliers onto the outer housing of the lock with a good grip.

2 Unscrew the housing until it can be removed. Manipulate the locking mechanism with lock tools to disengage the arm, and then open the door.

© LiquidLibrary

FIRE FIGHTER Tips

Some gates that secure driveways may be siren activated. In such a case, no forcible entry technique other than sounding the siren on the fire apparatus is required to open the gates.

One method of breaching a security gate is to remove the lock cylinder using a K tool, an A tool, or a bam-bam tool. If the lock cylinder can be accessed and removed, it is a simple matter to turn the locking mechanism and open the gate. If the lock cylinder has been protected to minimize the chance of break-in, however, it may be necessary to cut through part of the gate to open it. This can be done by using a circular saw that is equipped with an appropriate blade or, in some cases, by using a hydraulic cutter. An alternative is to use a hydraulic spreader to force the anchor from the masonry or wood to which it is attached. Be sure to wear PPE, including adequate eye protection, when operating these tools. Plan your cuts to minimize the amount of metal that needs to be cut to open the gate.

Breaching security bars on a window can sometimes be accomplished by removing the lock cylinder, if one is present on the outside of the security bars. If no lock cylinder is present, it will be necessary to cut through the bars using either a circular saw or hydraulic cutters. If the bars are not well anchored, you might be able to pry them from the window. A hydraulic spreader can sometimes be used to force the anchor from the masonry or wood to which it is attached. This method will not always work, however, when the bars are securely anchored in masonry or concrete.

Look for the types of security gates and bars that are present in your community. Talking with the installers of these products will give you more knowledge about how to gain access through them.

Breaching Walls and Floors

On occasion, forced entry through a window, lock, or door might not be possible or might take too long. In these cases, consider the option of breaching a wall or floor. Breaching a wall can also be an option for removal of injured people or for emergency escape. An understanding of basic construction concepts is required to execute this technique safely and quickly.

■ Load-Bearing/Nonbearing Walls

Before breaching a wall, first consider whether the wall is load bearing. A load-bearing wall supports the building's ceiling and/or rafters, so removing or damaging such a wall could cause the building or wall to collapse. By contrast, a nonbearing wall can be removed safely and without danger. Nonbearing walls are also called partition walls or simply **partitions**. Refer to the chapter "Building Construction" for a review of wall construction.

■ Exterior Walls

Exterior walls can be constructed of one or more materials. Many residences have both wood and brick, aluminum siding, or masonry-block construction **FIGURE 12-28**. Commercial buildings usually have concrete, masonry, or metal exterior walls. Commercial buildings also may have steel I-beam construction; older commercial structures may have heavy timber construction. Because there are exceptions to every rule, it is important to know the construction methods and materials used for buildings in your specific response area.

Deciding whether to breach an exterior wall is a difficult choice. Masonry, metal, and brick are formidable materials, and breaking through them can be very difficult. The best tools to use in breaching a concrete or masonry exterior wall are a battering ram, a sledgehammer, and a rotary saw with a concrete blade.

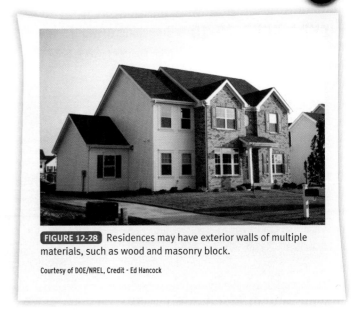

FIGURE 12-28 Residences may have exterior walls of multiple materials, such as wood and masonry block.

Courtesy of DOE/NREL, Credit - Ed Hancock

■ Interior Walls

Interior walls in residences are usually constructed of wood or metal studs covered by plaster, gypsum, or Sheetrock®. Some newer residential construction contains a laminate drywall, which is extremely difficult to penetrate. Commercial buildings may have concrete-block interior walls. Breaching an interior wall can be dangerous to fire fighters for several reasons; for example, many interior walls contain electrical wiring, plumbing, cable wires, and phone wires—all of which present hazards to fire fighters. Interior walls may also be load bearing. Load-bearing walls can be breached without undue hazard as long as studs are not removed; extreme care should be taken, however, if any studs are removed.

After determining whether the wall is load bearing, sound it to locate a stud away from any electrical outlets or switches. Tap on the wall; the area between studs will make a hollow sound compared to the solid sound directly over the stud.

After locating an appropriate site, make a small hole (which can also serve as an inspection hole to check for fire) to check for any obstructions. If the area is clear, expand the opening to reveal the studs. Walls should be breached as close as possible to the studs, because this choice makes a large opening and cutting is easier. If possible, enlarge the opening by removing at least one stud to enable quick escapes.

To breach a wall frame, follow the steps in **SKILL DRILL 12-10**:

1. Size up the wall, checking for safety hazards such as electrical outlets, wall switches, or any signs of plumbing. Inspect the overall scene to ensure that the wall is not load bearing.
2. Using a striking or cutting tool, sound the wall to locate any studs, and then make a hole between the studs. (**STEP 1**)
3. Cut the Sheetrock® as close to the studs as possible. (**STEP 2**)
4. Enlarge the hole by extending it from stud to stud and as high as necessary. (**STEP 3**)

SKILL DRILL 12-10 Breaching a Wall Frame

(Fire Fighter I, NFPA 5.3.4)

1 Size up the wall, checking for safety hazards such as electrical outlets, wall switches, or any signs of plumbing. Inspect the overall scene to ensure that the wall is not load bearing. Using a striking or cutting tool, sound the wall to locate any studs, and then make a hole between the studs.

2 Cut the Sheetrock® as close to the studs as possible.

3 Enlarge the hole by extending it from stud to stud and as high as necessary.

© LiquidLibrary

To breach a masonry wall, you create an upside-down *V* in the wall. Follow the steps in **SKILL DRILL 12-11**:

1 Size up the wall, checking for any safety hazards such as electrical outlets, wall switches, and plumbing. Inspect the overall area to ensure that the wall is not load bearing.

2 Select a row of masonry blocks at 0.6–0.9 metres (2–3 feet) above the floor.

3 Using a sledgehammer, knock two holes each in five masonry blocks along the selected row. Each hole should pierce into the hollow core of the masonry blocks.

4 Repeat the process on four masonry blocks above the first row.

5 Repeat the process on three masonry blocks above the second row, on two masonry blocks above the third row, and on one masonry block above the fourth row. (**STEP 1**)

6 An upside-down V has now been created. (**STEP 2**)

7 Begin knocking out the remaining portion of the masonry blocks by hitting the masonry blocks parallel to the wall rather than perpendicular to it. (**STEP 3**)

8 Clear any reinforcing wire using bolt cutters, and enlarge the hole as needed. (**STEP 4**)

To breach a metal wall, follow the steps in **SKILL DRILL 12-12**:

1 Size up the wall, checking for any safety hazards such as electrical outlets, wall switches, or plumbing. Inspect the overall area to ensure that the wall is not load bearing.

2 Select an appropriate tool such as a rotary saw with a metal-cutting blade or a torch.

3 Open the wall using a diamond-shaped or inverted-V cut. (**STEP 1**)

4 Make the cut so the opening piece falls or is bent into the building so no sharp bottom edges are exposed. (**STEP 2**)

■ Floors

The two most popular floor materials found in residences and commercial buildings are wood and poured concrete. Both are very resilient and may be difficult to breach. Conduct a thorough size-up of the site before considering forcible entry through this route; breaching a floor should be a last-resort measure. A rotary saw with an appropriate blade is the best

SKILL DRILL 12-11 | Breaching a Masonry Wall

(Fire Fighter I, NFPA 5.3.4)

1 Select a row of masonry blocks at 0.6–0.9 metres (2–3 feet) above the floor. Using a sledgehammer, knock two holes each in five masonry blocks along the selected row. Each hole should pierce into the hollow core of the masonry blocks; repeat the process on four masonry blocks above the first row; repeat the process on three masonry blocks above the second row, on two masonry blocks above the third row, and on one masonry block above the fourth row.

2 An upside-down *V* has now been created.

3 Begin knocking out the remaining portion of the masonry blocks by hitting the masonry blocks parallel to the wall rather than perpendicular to it.

4 Clear any reinforcing wire using bolt cutters, and enlarge the hole as needed.

© LiquidLibrary

tool to breach a floor; a chainsaw may be a better choice when used on a wood floor.

To breach a floor, follow the steps in **SKILL DRILL 12-13**:

1 Size up the floor area, checking for hazards in the area to be cut.

2 Sound the floor with an axe or similar tool to locate the floor joists. (**STEP 1**)

3 Use an appropriate cutting tool to cut one side of the hole.

4 Cut the opposite side. (**STEP 2**)

5 Remove any flooring, such as carpet, tile, or floorboards, that has been loosened.

SKILL DRILL 12-12 Breaching a Metal Wall
(Fire Fighter I, NFPA 5.3.4)

1 Select an appropriate tool such as a rotary saw with a metal-cutting blade or a torch. Open the wall using a diamond-shaped or inverted-V cut.

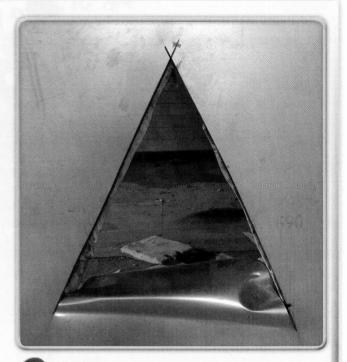

2 Make the cut so the opening piece falls or is bent into the building so no sharp bottom edges are exposed.

© LiquidLibrary

6 Cut a similar opening into the subfloor, until the proper size of hole is achieved. **(STEP 3)**

7 Secure the area around the hole to prevent others working in the area from falling through the hole. **(STEP 4)**

■ Vehicle Entry

Some fire departments receive training from a locksmith in the use of slim jims and other devices to unlock vehicles. If you are not trained and equipped to gain entry into cars, in an emergency, you can quickly gain entry by breaking a side window as far away from the trapped child or pet as possible. This is easily done using a spring-loaded centre punch or other sharp object as described in the chapter "Vehicle Rescue and Extrication."

FIRE FIGHTER Tips

Fire departments are regularly called because people have locked their car keys in their cars. Many fire departments do not respond to these routine calls. Instead, they may refer individuals to locksmiths or emergency lockout companies that can promptly respond to such requests. These routine calls are very different from calls where a child or a pet has been locked in a vehicle, however; this type of situation represents an emergency that requires a fire department response.

Systematic Forcible Entry

The first step in forcible entry is to think. Several issues need to be evaluated before deciding on a course of action. Double-check the address for the incident; it is embarrassing to try to explain to homeowners why you have destroyed their back-door when the burning food on the stove was reported three houses down the street. Finding the right address can be a challenge, especially if rear entrances off an alley are not clearly marked.

Second, look for the presence of a lockbox. Lockboxes are attached to buildings, usually close to a main entrance. These boxes can be unlocked with a master key that is carried by the fire department. The master key is coded to open all lockboxes in that jurisdiction. Once entry is made into the lockbox, the fire department officer can remove keys that will open the specific building. With a lockbox system, rapid entry can be obtained by the fire department into any building equipped with a lockbox, and the building can be secured after the incident by the fire department. To maintain security over the lockbox master keys, some systems are equipped with a device that keeps the master key locked in the fire apparatus unless released by the emergency dispatcher. This approach creates a record of each time the master key is used, allows the dispatcher to record the time and

SKILL DRILL 12-13 Breaching a Floor

(Fire Fighter I, NFPA 5.3.4)

1 Sound the floor with an axe or similar tool to locate the floor joists.

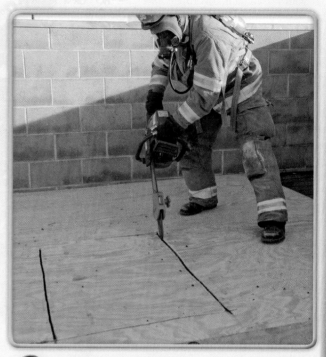

2 Use an appropriate cutting tool to cut one side of the hole. Cut the opposite side.

3 Remove any flooring, such as carpet, tile, or floorboards, that has been loosened. Cut a similar opening into the subfloor, until the proper size of hole is achieved.

4 Secure the area around the hole to prevent others working in the area from falling through the hole.

place the key was used, and identifies the officer requesting the use of the master key. You need to determine if your department has a lockbox program, and if it does, how it works.

Third, evaluate the threat level in the incident. When flames and smoke are visible, the need for forcible entry is much higher than when a fire alarm activation occurs in the middle of a summer thunderstorm with no sign of fire. Life safety is a major consideration. A fire alarm activation in a closed business in the middle of the night is much less potentially life threatening than a call where people are reportedly in distress or trapped.

Fourth, consider how to make entry with the least amount of damage. This objective must be balanced with the severity of the emergency. Generally speaking, you want to make entry with the least damage in the shortest amount of time, but this assessment must be balanced with the type of emergency present.

Do not forget that you need to work in a coordinated fashion with other members of the team. When a forcible entry team opens a door or window, the breach produces as much ventilation to the fire as if the door or window were opened in ventilation operations.

Forcible Entry and Salvage

■ Before Entry

The decision of which type of forcible entry to use will have a direct impact on salvage operations. In a nonemergency situation, through-the-lock methods will minimize damage, thereby saving the property from excessive repairs. In an emergency situation, more urgent methods of forcible entry may cause massive amounts of destruction, requiring expensive repairs to a door, window, or wall.

■ After Entry

When the operation is over and normal salvage and overhaul have begun, the structure must be secured before fire fighters leave the scene. The use of conventional forcible entry techniques and specialized tools requires that doors and windows be secured to prevent break-ins or theft after the call. If through-the-lock methods are used, a hasp and padlock may be all that are needed to secure a site. A resourceful crew might also screw a forced door shut from the inside and leave from a door that has not been forced. Broken windows or doors can be boarded up with sheets of plywood. Police will often increase patrols around a damaged building, when needed, to deter potential thieves. The building owner should be called to secure the building.

FIRE FIGHTER Tips

Many building owners are grateful to fire departments for explaining to them the types of assistance that board-up companies can provide. These companies are available in most communities to secure buildings after fire, water, or storm damage. The average homeowner has few opportunities to use such a company, and many people are not aware of the role these companies play in reducing the damage and in helping to prevent further damage to the building. Some fire departments offer building owners a list of licenced or approved companies that provide emergency board-up services in their community.

The fire department is charged not only with saving lives, but also with protecting property. Do everything possible to reduce the amount of damage caused by both the fire and your own actions. Homeowners will be especially appreciative of support from the fire department after the fire is extinguished.

Chief Concepts

- Forcible entry is required at emergency incidents where time is a critical factor. Company officers usually select both the point of entry and the method to be used. They also ensure that the efforts of different companies are coordinated for safe and effective operations.
- Before beginning forcible entry operations, remember to try before you pry.
- Fire departments use many different types of forcible entry tools—ranging from basic cutting, prying, and striking tools to sophisticated mechanical and hydraulic equipment.
- Four types of forcible entry tools are used:
  - Striking tools—Generate an impact force directly on an object or another tool. They include the flat-head axe, battering ram, and sledgehammer.
  - Prying/spreading tools—Designed for prying and spreading. They include the Halligan tool, pry bar/hux bar/crowbar, pry axe, spreaders, cutters, and rams.
  - Cutting tools—Primarily used for cutting doors, roofs, walls, and floors. They include the axe, bolt cutters, and circular saw.
  - Lock tools and specialty tools—Used to disassemble the locking mechanism on a door. They include the K tool, A tool, J tool, shove knife, duck-billed lock breakers, locking pliers and chain, and bam-bam tool.
- Every door contains four major components:
  - Door—The entryway itself
  - Jamb—The frame
  - Hardware—The handles, hinges, and other components
  - Locking mechanism
- Doors are generally constructed of wood, metal, or glass. Wood and metal doors can be solid-core or hollow-core types.
- Doors are classified by how they open:
  - Inward—Can be made of wood, steel, or glass and can be found in most structures. The locking mechanisms range from standard doorknob locks to deadbolt locks or sliding latches.
  - Outward—Used in commercial occupancies and for most exits. The hinges are often exposed. Several types of locks, including handle-style locks and deadbolts, may be used with these doors.
  - Sliding—Constructed of tempered glass in a wooden or metal frame. A weak latch on the frame of the door secures the movable side.
  - Revolving—Made of four glass panels with metal frames. Generally secured by a standard cylinder lock or slide-latch lock.

- Overhead—Range from standard residential garage doors to high-security commercial roll-up doors. Made be secured with cylinder-style locks, padlocks, or automatic garage door openers.
- Window construction includes glazed glass, regular or annealed glass, double-pane glass, plate glass, laminated glass, and tempered glass.
- Window frame designs include the following types:
  - Double-hung windows—Made of wood or vinyl and two movable sashes
  - Single-hung windows—Similar to double-hung windows except that the upper sash is fixed and only the lower sash moves
  - Jalousie windows—Made of adjustable sections of tempered glass encased in a metal frame that overlap each other when closed
  - Awning windows—Operate like jalousie windows except that awning windows are one large or two medium-sized glass panels instead of many small panes
  - Horizontal-sliding windows—Similar to sliding doors
  - Casement windows—Have a steel or wood frame, and open away from the building with a crank mechanism
  - Projected windows—Usually found in older warehouses, and project either outward or inward on a hinge
- Locks range in sophistication from the very simple push-button locks to complex computer-operated locks.
- There are three major parts of a door lock:
  - Latch—The part of the lock that catches and holds the door frame
  - Operator lever—The handle, doorknob, or keyway that turns the latch to lock it or unlock it
  - Deadbolt—A second, separate latch that locks and reinforces the regular latch
- All padlocks, like all door locks, have similar parts:
  - Shackles—The U-shaped top of the lock that slides through a hasp and locks in the padlock itself
  - Unlocking mechanism—The keyway, combination wheels, or combination dial used to open the padlock
  - Lock body—The main part of the padlock that houses the locking mechanisms and the retention part of the lock
- Locks can be classified into four major categories:
  - Cylindrical locks—Most common fixed lock in use. One side of the door usually has a key-in-the-knob lock; the other side will have a keyway, a button, or some type of locking mechanism.

- Padlocks—Most common locks on the market. Available with a variety of unlocking mechanisms, including keyways, combination wheels, or combination dials.
- Mortise locks—Designed to fit in predrilled openings and have both a latch and a bolt into the same mechanism, each of which operates independently of the other
- Rim locks—Include deadbolts that can be surface mounted on the interior of the door frame. These locks have a bolt that extends at least 2.5 cm (1 in.) into the door frame.

■ On occasion, breaching a wall or floor may be necessary. Before breaching a wall, first consider whether the wall is load bearing. Removing or damaging this type of wall could cause the building or wall to collapse.

■ The steps in systematic forcible entry are as follows:
- Think.
- Look for a lockbox.
- Evaluate the situation.
- Make entry with the least amount of damage possible.
- Recognize that forcible entry operations will require you to take measures to secure the property during salvage operations.

Hot Terms

Adz The prying part of the Halligan tool.

Annealed The process of forming standard glass.

A tool A cutting tool with a pry bar built into the cutting part of the tool.

Awning windows Windows that have one large or two medium-size panels that are operated by a hand crank from the corner of the window.

Bam-bam tool A tool with a case-hardened screw, which is secured in the keyway of a lock and used to remove the keyway from the lock.

Battering ram A tool made of hardened steel with handles on the sides, which is used to force doors and to breach walls. Larger versions may be used by as many as four people; smaller versions are made for one or two people.

Bite A small opening made to enable better tool access in forcible entry.

Bolt cutter A cutting tool used to cut through thick metal objects, such as bolts, locks, and wire fences.

Case-hardened steel Steel created in a process that uses carbon and nitrogen to harden the outer core of a steel component, while the inner core remains soft. Case-hardened steel can be cut only with specialized tools.

Casement windows Windows in a steel or wood frame that open away from the building via a crank mechanism.

Claw The forked end of a tool.

Cutting tools Tools that are designed to cut into metal or wood.

Cylindrical locks The most common fixed locks in use today. The locks and handles are placed into predrilled holes in the doors. One side of the door will usually have a key-in-the-knob lock; the other will have a keyway, a button, or some other type of locking/unlocking mechanism.

Deadbolt Surface- or interior-mounted lock on or in a door with a bolt that provides additional security.

Door An entryway; the primary choice for forcing entry into a vehicle or structure.

Double-hung windows Windows that have two movable sashes that can go up and down.

Double-pane glass A window design that traps air or inert gas between two pieces of glass to help insulate a house.

Duck-billed lock breaker A tool with a point that can be inserted into the shackles of a padlock. As the point is driven farther into the lock, it gets larger and forces the shackles apart until they break.

Exterior wall A wall—often made of wood, brick, metal, or masonry—that makes up the outer perimeter of a building. Exterior walls are often load bearing.

Forcible entry Techniques used by fire personnel to gain entry into buildings, vehicles, aircraft, or other areas of confinement when normal means of entry are locked or blocked. (NFPA 402)

Glazed A glass or transparent or translucent plastic sheet used in windows, doors, skylights, or curtain walls. [ASCE/SEI 7:6.2] (NFPA 5000)

Hardware The parts of a door or window that enable it to be locked or opened.

Hockey puck lock A type of lock with hidden shackles that cannot be forced open through conventional methods.

Hollow-core door A door made of panels that are honeycombed inside, creating an inexpensive and lightweight design.

Horizontal-sliding windows Windows that slide open horizontally.

Interior wall A wall inside a building that divides a large space into smaller areas; also known as a partition.

Irons A combination tool, normally consisting of the Halligan tool plus a flat-head axe.

Jalousie windows Windows made of small slats of tempered glass, which overlap each other when the window is closed. Often found in trailers and mobile homes, jalousie windows are held together by a metal frame and operated by a small hand wheel or crank found in the corner of the window.

Jamb The part of a doorway that secures the door to the studs in a building.

J tool A tool that is designed to fit between double doors equipped with panic bars.

K tool A tool that is used to remove lock cylinders from structural doors so the locking mechanism can be unlocked.

Laminated glass Safety glass. The lamination process places a thin layer of plastic between two layers of glass, so that the glass does not shatter and fall apart when broken.

Latch A spring-loaded latch bolt or a gravity-operated steel bar that, after release by physical action, returns to its operating position and automatically engages the strike plate when it is returned to the closed position. (NFPA 80)

Lock body The part of a padlock that holds the main locking mechanisms and secures the shackles.

Locking mechanism A standard doorknob lock, deadbolt lock, or sliding latch.

Mortise locks Door locks with both a latch and a bolt built into the same mechanism; the two locking mechanisms operate independently of each other. Mortise locks are often found in hotel rooms.

Operator lever The handle of a door that turns the latch to open it.

Padlocks The most common types of locks on the market today, built to provide regular-duty or heavy-duty service. Several types of locking mechanisms are available, including keyways, combination wheels, and combination dials.

Partition A nonstructural interior wall that spans horizontally or vertically from support to support. The supports may be the basic building frame, subsidiary structural members, or other portions of the partition system. [ASCE/SEI 7:11.2] (NFPA 5000)

Pick The pointed end of a pick axe, which can be used to make a hole or bite in a door, floor, or wall.

Plate glass A type of glass that has additional strength so it can be formed in larger sheets but will still shatter upon impact.

Projected windows Windows that project inward or outward on an upper hinge; also called factory windows. They are usually found in older warehouses or commercial buildings.

Pry axe A specially designed hand axe that serves multiple purposes. Similar to a Halligan bar, it can be used to pry, cut, and force doors, windows, and many other types of objects. Also called a multipurpose axe.

Rabbet A type of door frame in which the stop for the door is cut into the frame.

Rim locks Surface- or interior-mounted locks located on or in a door with a bolt that provide additional security.

Shackles The U-shaped part of a padlock that runs through a hasp and then is secured back into the lock body.

Solid-core door A door design that consists of wood filler pieces inside the door. This construction creates a stronger door that may be fire rated.

Striking tools Tools designed to strike other tools or objects such as walls, doors, or floors.

Tempered glass A type of safety glass that is heat treated so that it will break into small pieces that are not as dangerous.

Unlocking mechanism A keyway, combination wheel, or combination dial.

You are on the first-in engine for an alarm in a large industrial building. The company manufactures rubber belts and hoses. It is early Sunday morning, so you are pretty sure this is going to be an accidental alarm trip. Dispatch advises that the key holder is en route with a 25-minute estimated time of arrival. There is no evidence of fire showing from the outside of the structure. Your captain has you set a ground ladder to a first floor window to take a look across the factory floor. From the ladder, you can see a thick haze across the factory floor.

1. The window you peer through is called a factory window and a _____ window.
   A. casement
   B. projected
   C. double-hung
   D. single-hung

2. The front door to the office area is a glass door. What type of glass would most likely be used in this door?
   A. Tempered
   B. Plate
   C. Double-pane
   D. Laminate

3. Which type of door is most commonly used as a front exit?
   A. Inward swinging
   B. Outward swinging
   C. Sliding
   D. Revolving

4. On a(n) _____ door, you can check the hinges to see if they can be disassembled or the pins removed before more destructive methods are used.
   A. inward swinging
   B. outward swinging
   C. sliding
   D. revolving

5. Which tool is designed to fit between double doors equipped with panic bars so it can be forced open?
   A. Halligan tool
   B. Bam-bam tool
   C. K tool
   D. J tool

6. A _____ is a second, separate latch that locks and reinforces the regular latch.
   A. deadbolt
   B. jamb
   C. latch
   D. rabbet

You have just been transferred into a new station. You are assisting another fire fighter as he is cleaning out the compartments on the fire apparatus. He pulls a variety of equipment from the forcible entry compartment. As he pulls out the K tool, he says, "This is covered with grime from sitting in this tray for years. I can't remember when I last used it. It's faster to use the sledge to break the door down."

1. How would you address the issue of the K tool being covered with grime?
2. How do you get others to do what is best, rather than what is fastest?
3. How do you determine the best method to force entry into a structure?
4. What should you say to this fire fighter if he asks, "Don't you agree?"

Ladders

Fire Fighter I

Knowledge Objectives

After studying this chapter, you will be able to:

- List and describe the parts of a ladder. (NFPA 5.3.6.A , p 351–353)
- Categorize the different types of ladders. (NFPA 5.3.6.A , p 353–356)
- Inspect ladders. (NFPA 5.5.1 , p 356–357)
- Maintain ladders. (NFPA 5.5.1 , p 357–358)
- Clean ladders. (NFPA 5.5.1 , p 358–359)
- Describe when, where, and who performs service testing on ladders. (NFPA 5.5.1 , p 359–360)
- Specify the hazards associated with ladders. (NFPA 5.3.6 , p 360–361)
- Itemize the measures fire fighters should take to ensure safety when working with and on ladders. (NFPA 5.3.6 , p 360–362)
- Cite the factors and guidelines used to select the appropriate ladder from the fire apparatus. (NFPA 5.3.6.A , p 363–364)
- Describe how to remove a ladder from the apparatus. (p 364)
- Describe how to lift ladders. (p 364–365)

Skills Objectives

After studying this chapter, you will be able to perform the following skills:

- Inspect, clean, and maintain a ladder. (NFPA 5.5.1 , p 357–359)
- Carry a portable ladder using the one–fire fighter carry. (NFPA 5.3.6.B , p 365–366)
- Carry a portable ladder using the two–fire fighter shoulder carry. (NFPA 5.3.6.B , p 366–367)
- Carry a portable ladder using the three–fire fighter shoulder carry. (NFPA 5.3.6.B , p 366–367)
- Carry a portable ladder using the two–fire fighter suitcase carry. (NFPA 5.3.6.B , p 368)
- Carry a portable ladder using the three–fire fighter suitcase carry. (NFPA 5.3.6.B , p 368–369)
- Carry a portable ladder using the three–fire fighter flat carry. (NFPA 5.3.6.B , p 368, 370)
- Carry a portable ladder using the four–fire fighter flat carry. (NFPA 5.3.6.B , p 369, 371)
- Carry a portable ladder using the three–fire fire fighter flat-shoulder carry. (NFPA 5.3.6.B , p 370, 372)
- Carry a portable ladder using the four–fire fighter flat-shoulder carry. (NFPA 5.3.6.B , p 371–373)
- Raise a portable ladder using the one–fire fighter rung raise for ladders shorter than 4.3 metres (14 feet). (NFPA 5.3.6.B , p 374–375)
- Raise a portable ladder using the one–fire fighter rung raise for ladders taller than 4.3 m (14 ft.). (NFPA 5.3.6.B , p 374, 376)
- Tie the halyard. (NFPA 5.3.6.B , p 374–375, 377)
- Raise a portable ladder using the two–fire fighter beam raise. (NFPA 5.3.6.B , p 374, 378)
- Raise a portable ladder using the two–fire fighter rung raise. (NFPA 5.3.6.B , p 377, 379–380)
- Raise a portable ladder using the three–fire fighter rung raise. (NFPA 5.3.6.B , p 380–381)
- Raise a portable ladder using the four–fire fighter rung raise. (NFPA 5.3.6.B , p 380, 383)
- Climb a ladder. (NFPA 5.3.12B , p 384–385)
- Use a leg lock to work from a ladder. (NFPA 5.3.6.B , p 386–387)
- Deploy a roof ladder. (NFPA 5.3.12B , p 386, 388)
- Inspect a chimney. (NFPA 5.3.6.B , p 389)

© Photos.com

Fire Fighter II | FFII

Knowledge Objectives

There are no knowledge objectives for Fire Fighter II candidates. NFPA 1001 contains no Fire Fighter II Job Performance Requirements for this chapter.

Skills Objectives

There are no skills objectives for Fire Fighter II candidates. NFPA 1001 contains no Fire Fighter II Job Performance Requirements for this chapter.

Additional NFPA Standards

- NFPA 1901, *Standard for Automotive Fire Apparatus*
- NFPA 1931, *Standard for Manufacturer's Design of Fire Department Ground Ladders*
- NFPA 1932, *Standard on Use, Maintenance, and Service Testing of In-Service Fire Department Ground Ladders*
- NFPA 1983, *Standard on Life Safety Rope and Equipment for Emergency Services*

You Are the Fire Fighter

A long with your crew, you find yourself at the scene of a fast-moving fire in an apartment building. When your crew reports to the command post, the incident commander (IC) tells your captain that a truck company was sent inside to conduct a primary search so he needs your crew to set ground ladders to the building. The IC points to the aerial ladder and tells you that he wants you and the other fire fighters to ladder the first, second, and third floors of the B and D sides of the building while he does a 360-degree walk-around. You head to the truck, quickly making estimates of the ladders you will need.

1. Which types of ladders will work best for this situation?
2. How would you estimate the ladder lengths you will need?
3. What concerns do you have when you are laddering the building?

Introduction

Despite the many technological advances that have occurred in the fire service, ladders remain a fundamental piece of equipment that have not changed much. Ladders are one of the fire fighter's basic tools, carried on nearly every piece of structural firefighting apparatus.

Portable ladders are among the most functional, versatile, durable, inexpensive, easy-to-use, and rapidly deployable tools used by fire fighters. No advanced technology can substitute for a ladder as a means of rapid, safe vertical access for fire rescue, ventilation, suppression, and escape. Every fire fighter must be proficient in the basic skills of working with ladders and physically capable of performing ladder evolutions.

Functions of a Ladder

Ladders provide a vertical path, either up or down, from one level of a structure to another. They can provide access to an area or **egress** (a method to exit or escape) from an area **FIGURE 13-1**. At times, a ladder can be used as a work platform so that a fire fighter can perform various functions in locations that could not be reached otherwise.

Ladders are most often used to provide access to and exit from areas above **grade** (the level at which the ground intersects the foundation of a structure). When used outside a building, they can enable fire fighters to reach the roof for ventilation operations, enter a window for an interior search, or rescue a victim. Exterior ladders also provide an emergency exit for crews working inside a structure. When used within a building, ladders can allow fire fighters to access the attic or cockloft or may provide a safe path between floors, enabling fire fighters to avoid a damaged stairway. A response team can use a ladder to access and exit from the top of a rail car during a hazardous materials incident.

Fire fighters also use ladders to access and exit areas below grade. Lowering a ladder into a trench or manhole

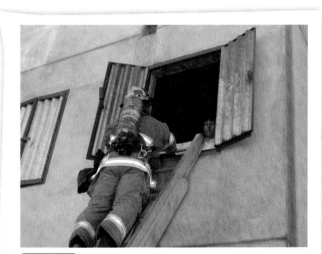

FIGURE 13-1 Ladders can provide access to or exit from a structure.

allows them to mark the location of a victim in case of secondary collapse while the rescue operation is in progress, to reach the level of an incident, or to escape from a below-grade site safely. In addition, a ladder can be used to reach an injured person in a ravine or to access a highway accident from an overpass.

Ladders are useful even at grade level. They can be deployed to create a bridge across a small opening, for example, or to enable fire fighters to climb over a fence or obstruction to reach an emergency scene.

Finally, ladders can be used as work platforms. For example, a fire fighter who has climbed a ladder to a window can stand on the ladder to open the window for ventilation or to direct a hose stream into the building or when working on a roof. Ladders are also used during overhaul operations, where they may serve as a platform for a fire fighter to remove exterior trim in a wood-frame building.

CHAPTER 13 Ladders 351

Secondary Functions of Ladders

Ladders serve several other secondary functions in the fire service. Roof ladders provide stable footing and distribute the weight of fire fighters during pitched-roof operations. A <u>ladder gin</u> is an A-shaped structure formed with two ladder sections. Ladder gins are used to lower a rescuer into a trench or manhole or to raise a victim safely from a below-grade site.

Ladders provide elevated platforms for equipment as well as for fire fighters. They are especially useful when elevation would extend the reach or enhance the operations of the equipment. For example, a hose line attached to a ladder section can be deployed to protect exposures or confine a fire. A smoke ejector may be mounted on a ladder to provide ventilation. Portable lighting also may be secured to ladders to illuminate working areas.

Ladders can be used as ramps for moving equipment, or for hoisting or lowering victims. They can help shore up a damaged wall, support a hose line over an opening, or guide a rope up, down, or over an obstacle. By using a tarp or salvage cover, fire fighters can turn a ladder into a channel or chute for water and debris.

Ladder Construction

In its most basic design, a ladder consists of two beams connected by a series of parallel rungs. Fire service ladders, however, are specialized tools with several different parts. To use and maintain ladders properly, fire fighters must be familiar with the various types of ladders, as well as with the different parts and terms used to describe them.

Basic Ladder Components

The basic components of a straight ladder are found in most other types of ladders as well **FIGURE 13-2**.

Beams

A <u>beam</u> is one of the two main structural components that run the entire length of most ladders or ladder sections. The

beams support the rungs and carry the load of a person from the rungs down to the ground.

Three basic types of ladder beam construction are used:

- **Trussed beam**: A trussed beam ladder has a top rail and a bottom rail, which are joined by a series of smaller pieces called truss blocks. The rungs are attached to the truss blocks. Trussed beams are usually constructed of aluminum or wood **FIGURE 13-3A**. This construction is often used for longer extension ladders.
- **I-beam**: An I-beam ladder has thick sections at the top and the bottom, which are connected by a thinner section. The rungs are attached to the thinner section of the beam. This type of beam is usually made from fibreglass **FIGURE 13-3B**.
- **Solid beam**: A solid beam ladder has a simple rectangular cross-section. Many wooden ladders have solid beams. Rectangular aluminum beams, which are usually hollow or C-shaped, are also classified as solid beams.

Rail

The <u>rail</u> is the top or bottom section of a trussed beam. Each trussed beam has two rails. The term *rail* can also be used to refer to the top and bottom surfaces of an I-beam.

Truss Block

A <u>truss block</u> is a piece or assembly that connects the two rails of a trussed beam. The rungs are attached to the truss blocks. Truss blocks can be made from metal or wood.

Rung

A <u>rung</u> is a crosspiece that spans the two beams of a ladder. The rungs serve as steps and transfer the weight of the user to the beams. Most portable ladders used by fire departments have aluminum rungs, but wooden ladders are still constructed with wood rungs.

Tie Rod

A <u>tie rod</u> is a metal bar that runs from one beam of the ladder to the other and keeps the beams from separating. Tie rods are typically found in wooden ladders.

Tip

The <u>tip</u> is the very top of the ladder.

Butt

The <u>butt</u> is the end of the ladder that is placed against the ground when the ladder is raised. It is sometimes called the heel or base.

Butt Spurs

<u>Butt spurs</u> are metal spikes that are attached to the butt of a ladder. They prevent the butt from slipping out of position.

Butt Plate

A <u>butt plate</u> is an alternative to a simple butt spur. This swivelling plate is attached to the butt of the ladder and incorporates both a spur and a cleat or pad.

Roof Hooks

<u>Roof hooks</u> are spring-loaded, retractable, curved metal pieces that are attached to the tip of a roof ladder. These

FIGURE 13-2 The basic components of a straight ladder.

A.   **B.**

FIGURE 13-3 Two types of beam construction found in ladders: **A.** Trussed beam. **B.** I-beam.

hooks are used to secure the tip of the ladder to the peak of a pitched roof.

Heat Sensor Label

A **heat sensor label** identifies when the ladder has been exposed to a specific amount of heat conditions that could damage its structural integrity; such a label changes colours when it is exposed to a particular temperature.

Protection Plates

Protection plates are reinforcing pieces that are placed on a ladder at chaffing and contact points to prevent damage from friction or contact with other surfaces.

■ Extension Ladder Components

An extension ladder is an assembly of two or more ladder sections that fit together and can be extended or retracted to adjust the ladder's length. Extension ladders have additional parts **FIGURE 13-4**.

Bed (Base) Section

The **bed (base) section** is the widest section of an extension ladder. It serves as the base; all other sections are raised from the bed section. The bottom of the bed section rests on the supporting surface.

Fly Section

A **fly section** is the part of an extension ladder that is raised or extended from the bed section. Extension ladders often have more than one fly section, each of which extends from the previous section.

Guides

Guides are strips of metal or wood that guide a fly section of an extension ladder as it is being extended. Channels or slots in the bed or fly section may also serve as guides.

FIGURE 13-4 Extension ladders have additional parts that are not found in straight ladders.

Halyard

The **halyard** is the rope or cable used to extend or hoist the fly sections of an extension ladder. The halyard runs through the pulley.

Pawls

Pawls are the mechanical locking devices that are used to secure the extended fly sections of an extension ladder. They are sometimes called dogs, ladder locks, or rung locks.

Pulley

The **pulley** is a small grooved wheel that is used to change the direction of the halyard pull. A downward pull on the halyard creates an upward force on the fly sections, extending the ladder.

Stops

Stops are pieces of wood or metal that prevent the fly sections of a ladder from overextending and collapsing the ladder. They are also referred to as stop blocks.

Staypole

Staypoles (also called tormentors) are long metal poles that are attached to the top of the bed section and that help stabilize the ladder as it is being raised and lowered. One pole is attached to each beam of a long (12.2 m [40 ft.] or longer) extension ladder with a swivel joint. Each pole has a spur on the other end. Ladders with staypoles are typically referred to as Bangor ladders. In some parts of the United States, Bangor ladders may be referred to as tormentor ladders.

Types of Ladders

Ladders used in the fire service can be classified into two broad categories: aerial and portable. **Aerial ladders** are permanently mounted and operated from a piece of motorized fire apparatus. **Portable ladders** are carried on fire apparatus but are designed to be removed and used in other locations.

■ Aerial Apparatus

Aerial apparatus vary greatly in terms of their design and function. This discussion is limited to a brief overview of the basic styles of aerial apparatus. Detailed requirements for aerial apparatus are documented in NFPA 1901, *Standard for Automotive Fire Apparatus*. This standard sets minimum performance requirements for apparatus referred to as either aerial ladders or elevating platforms.

Aerial ladders are permanently mounted, power-operated ladders with a working length of between 15.2 m (50 ft.) and 41.7 m (137 ft.). Aerial ladders contain two or more sections **FIGURE 13-5**. Most aerial ladders have a permanently mounted waterway and monitor nozzle. Those built on a straight chassis are often referred to as straight-stick aerials; those built on a tractor-trailer chassis are called tillered aerials or tiller trucks. A tillered ladder truck requires a second operator at the back of

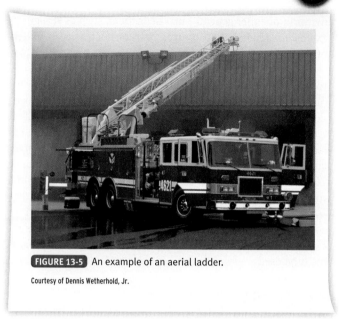

FIGURE 13-5 An example of an aerial ladder.

Courtesy of Dennis Wetherhold, Jr.

the trailer to steer the back of the trailer. Tillered ladder trucks can be maneuvered into tight places that straight chassis trucks cannot reach.

An elevating platform apparatus includes a passenger-carrying platform (bucket) attached to the tip of a ladder or boom **FIGURE 13-6**. The ladder or boom must have at least two sections, which may be telescoping or articulating (jointed). An elevating platform apparatus that is 33.5 m (110 ft.) or less in length must also have a prepiped waterway and a permanently mounted monitor nozzle. An elevated platform supported by a boom may have an aerial ladder for continuous access to the platform.

■ Portable Ladders

Most fire apparatus carry portable ladders (also called ground ladders). Portable ladders are designed to be removed from the apparatus and used in different locations. They may take

FIGURE 13-6 An elevated platform serves as a secure working platform for fire fighters.

© SteveStone/iStockphoto

the form of either general-purpose ladders, such as straight or extension ladders, or specialized ladders, such as attic ladders or roof ladders.

Fire department engines are required to carry at least one roof ladder, one extension ladder, and one folding ladder. Most fire department engines carry 7.3- or 8.5-m (24- or 28-ft.) ladders, which can reach the roof of a typical two-storey building. Ladder trucks are required to carry a total of 35 m (115 ft.) of ladders. They must carry at least one roof ladder, two straight ladders, and two extension ladders. Ladder trucks usually carry 10.7- or 12.2-m (35- or 40-ft.) ladders, which can reach the roofs of most three-storey buildings.

Generally, fire service portable ladders are limited to a maximum length of 15.2 m (50 ft.). If a building's height exceeds the capabilities of portable ladders, aerial apparatus must be used.

Straight Ladder

A straight ladder is a single-section, fixed-length ladder. Straight ladders may also be called wall ladders or single ladders. These lightweight ladders can be raised quickly, and can reach the windows and roofs of one- and two-storey structures. Straight ladders are commonly 3.7 to 4.2 m (12 to 14 ft.) long, but can be as much as 6.1 m (20 ft.) long.

FIRE FIGHTER Tips

Fire service portable ladders should be constructed and certified as compliant with the most recent edition of NFPA 1931, *Standard for Manufacturer's Design of Fire Department Ground Ladders*.

Roof Ladder

A **roof ladder** (sometimes called a hook ladder) is a straight ladder that is equipped with retractable hooks at one end. The hooks secure the tip of the ladder to the peak of a pitched roof, when the ladder lies flat on the roof. A roof ladder provides stable footing and distributes the weight of fire fighters and their equipment, thereby helping reduce the risk of structural failure in the roof assembly. Roof ladders can be used on flat roofs as a work platform to distribute weight. Roof ladders are usually 3.7 to 5.5 m (12 to 18 ft.) long **FIGURE 13-7**.

Fire Fighter Safety Tips

© Photos.com

Roof ladders are not free-hanging ladders. The roof hooks will not support the full weight of the ladder and anyone on it when the ladder is in a vertical position.

Extension Ladder

An **extension ladder** is an adjustable-length ladder with multiple sections. The bed or base section supports one or more fly sections. Pulling on the halyard extends the fly sections along

FIGURE 13-7 Roof ladders are commonly 3.7 to 5.5 m (12 to 18 ft.) long.

Courtesy of Duo-Safety Ladder Corporation

a system of brackets or grooves **FIGURE 13-8**. The fly sections, in turn, lock in place at set intervals so that the rungs are aligned to facilitate climbing.

An extension ladder is usually heavier than a straight ladder of the same length and requires more than one person to set it up. Because the length is adjustable, an extension ladder can replace several straight ladders, and it can be stored in places where a longer straight ladder would not fit.

Bangor Ladder

Bangor ladders (tormentor) are extension ladders with staypoles that offer added stability during raising, lowering, and climbing operations. Staypoles are required on ladders of 12.2 m (40 ft.) or greater length, but can often be found on 10.7-m (35-ft.) ladders. The poles help keep these heavy

FIGURE 13-8 An extension ladder can be used at any length, from fully retracted to fully extended.

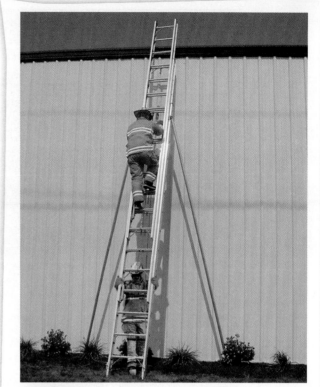

FIGURE 13-9 Staypoles are used to stabilize a Bangor ladder while it is being raised or lowered.

Courtesy of Duo-Safety Ladder Corporation

ladders under control while fire fighters are maneuvering them into place **FIGURE 13-9**. When the ladder is positioned correctly, the staypoles are planted in the ground on either side for additional stability.

Combination Ladder

A **combination ladder** can be converted from a straight ladder to a stepladder configuration (A-frame), or from an extension ladder to a stepladder configuration. Such ladders are convenient for indoor use and for maneuvering in tight spaces. They are generally 1.8 to 3 m (6 to 10 ft.) in length in the A-frame configuration and 3 to 4.6 m (10 to 15 ft.) in length in the extension configuration **FIGURE 13-10**.

Folding Ladder

A **folding ladder** (also called an attic ladder) is a narrow, collapsible ladder that is designed to allow access to attic scuttle holes and confined areas **FIGURE 13-11**. The two beams fold in to enhance the ladder's portability. Folding ladders are commonly available in 2.4- to 4.3-m (8- to 14-ft.) lengths.

Fresno Ladder

A **Fresno ladder** is a narrow, two-section extension ladder that is designed to provide attic access. The Fresno ladder is

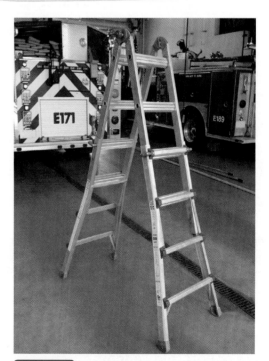

FIGURE 13-10 Combination ladders can be used in several different configurations. This is a combination stepladder and straight ladder.

A.

B.

FIGURE 13-11 Folding ladders are designed to be used in narrow spaces or restrictive passages. **A.** In the portable position. **B.** Fully extended.

generally short—just 3 to 4.3 m (10 to 14 ft.)—so it has no halyard; it is extended manually **FIGURE 13-12**. A Fresno ladder can be used in tight-space applications, such as bridging over a damaged section of an interior stairway.

Inspection, Maintenance, and Service Testing of Portable Ladders

Portable ladders used by the fire service must be able to withstand extreme conditions. NFPA 1931, *Standard for Manufacturer's Design of Fire Department Ground Ladders*, establishes requirements for the construction of new ladders. These requirements are based on probable conditions during emergency operations.

Ladders can be dropped, overloaded, or exposed to temperature extremes during use. NFPA 1932, *Standard on Use, Maintenance, and Service Testing of In-Service Fire Department Ground Ladders*, provides general guidance for ground ladder users. Portable ladders must be regularly inspected, maintained, and service tested, following the NFPA standard. In addition, ladders should always be inspected and maintained in accordance with the manufacturer's recommendations.

FIGURE 13-12 A Fresno ladder is extended manually.

Courtesy of Duo-Safety Ladder Corporation

- Pawls: Check the pawl (dog) assemblies for proper operation.
- Butt spurs: Check the butt spurs for excessive wear or other defects.
- Heat sensors: Check the labels to see whether the sensors indicate that the ladder has been exposed to excessive heat.
- Bolts and rivets: Check all bolts and rivets for tightness; bolts on wood ladders should be snug and tight without crushing the wood.
- Welds: Check all welds on metal ladders for cracks or apparent defects.
- Roof hooks: Check the roof hooks for sharpness and proper operation.
- Metal surfaces: Check metal surfaces for signs of surface corrosion.
- Fibreglass and wood surfaces: Check fibreglass ladders for loss of gloss on the beams. Check for damage to the varnish finish on wooden ground ladders. Check for signs of wood rot—that is, dark, soft spots.

If the inspection reveals any deficiencies, the ladder must be removed from service until repairs are made. Minor repairs that require simple maintenance can often be performed by properly trained fire fighters at the fire station. By contrast, repairs involving the structural or mechanical components of a ladder must be performed only by qualified personnel at a properly equipped repair facility.

Ladders that have been exposed to excessive heat must be removed from service and tested before being used again. A ladder that fails the structural stability test should be permanently removed from service.

Fire Fighter Safety Tips

© Photos.com

Fire fighters should perform routine ladder maintenance. Repairs to portable ladders should be performed only by trained repair personnel.

Inspection

Ground ladders must be visually inspected at least monthly. A ladder should also be inspected after each use. A general visual inspection should, at a minimum, address the following items:

- Beams: Check the beams for cracks, splintering, breaks, gouges, checks, wavy conditions, or deformations.
- Rungs: Check all rungs for snugness, tightness, punctures, wavy conditions, worn serrations, splintering, breaks, gouges, checks, or deformations.
- Halyard: Check the halyard for fraying or kinking; ensure that it moves smoothly through the pulleys.
- Wire–rope halyard extensions: Check the wire–rope halyard extensions on three- and four-section ladders for snugness. This check should be performed when the ladder is in the bedded position to ensure that the upper sections will align properly during operation.
- Ladder slides: Check the ladder slide areas for chafing. Also check for adequate wax, if the manufacturer requires wax.

Maintenance

All fire fighters should be able to perform routine ladder maintenance. Maintenance is simply the regular process of keeping the ladder in proper operating condition. Fundamental maintenance tasks include the following:

- Clean and lubricate the dogs, following the manufacturer's instructions **FIGURE 13-13**.
- Clean and lubricate the slides on extension ladders in accordance with the manufacturer's recommendations.
- Replace worn halyards and wire rope on extension ladders when they fray or kink **FIGURE 13-14**.
- Clean and lubricate hooks. Remove rust and other contaminants, and lubricate the folding roof hook assemblies on roof ladders to keep them operational **FIGURE 13-15**.

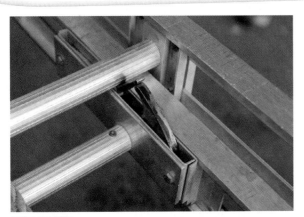

FIGURE 13-13 Extension ladder dogs (pawls) must operate smoothly.

FIGURE 13-14 Replace the halyard if it is worn or damaged.

FIGURE 13-15 Roof hooks must operate smoothly.

- Check the heat sensor labels. Replace the sensors when they reach their expiration date. Remove a ladder that has been exposed to high temperatures from service for testing.
- Maintain the finish on fibreglass and wooden ladders in accordance with the manufacturer's recommendations.

- Ensure that portable ladders are *not* painted except for the top and bottom 46 centimetres (18 inches) of each section, because paint can hide structural defects in the ladder. The tip and butt are painted for purposes of identification and visibility.
- Maintenance records should be kept.

Ladders that are in storage should be placed on racks or in brackets and protected from the weather. Fibreglass ladders can be damaged by prolonged exposure to direct sunlight **FIGURE 13-16**.

FIGURE 13-16 Portable ladders should be stored on racks or in brackets, out of the weather or direct sunlight.

■ Cleaning

Ladders must be cleaned regularly to remove any road grime and dirt that have built up on the apparatus during storage. Ladders should be cleaned before each inspection to ensure that any hidden faults can be observed. They should also be cleaned after each use to remove dirt and debris.

Use a soft-bristle brush and water to clean ladders. A mild, diluted detergent may be used, if allowed by the manufacturer's recommendations. Remove any tar, oil, or grease deposits with a safety solvent as recommended by the manufacturer.

Rinse and dry the cleaned ladder before placing it back on the apparatus.

Fire Fighter Safety Tips
© Photos.com

Do not get any solvent on the halyard of an extension ladder. Contact with solvents can damage halyard ropes.

The steps for inspecting, cleaning, maintaining, and storing ladders are listed in **SKILL DRILL 13-1**:

1. Clean all components following manufacturer and national standards.
2. Visually inspect the ladder for wear and damage. (**STEP 1**)

3 Lubricate the ladder dogs, guides, and pulleys using the recommended material. (**STEP 2**)

4 Perform a functional check of all components. (**STEP 3**)

5 Complete the maintenance record for the ladder. Tag and remove the ladder from service if deficiencies are found.

6 Return the ladder to the apparatus or storage area. (**STEP 4**)

■ Service Testing

Service testing is performed periodically to evaluate the continued usefulness of a ladder during its life. Service tests should be performed on both new ladders and ladders that have been in use for some time. Such tests measure the structural integrity of a portable ladder, ensuring that ladders are safe to use for their intended purposes. Service testing of portable ladders must follow NFPA 1932, *Standard on Use,*

SKILL DRILL 13-1

Inspect, Clean, and Maintain a Ladder
(Fire Fighter I, NFPA 5.5.1)

1 Clean all components following manufacturer and national standards. Visually inspect the ladder for wear and damage.

2 Lubricate the ladder dogs, guides, and pulleys using the recommended material.

3 Perform a functional check of all components.

4 Complete the maintenance record for the ladder. Tag and remove the ladder from service if deficiencies are found. Return the ladder to the apparatus or storage area.

Maintenance, and Service Testing of In-Service Fire Department Ground Ladders.

Service tests should be conducted before a new ladder is used and annually while the ladder remains in service. A ladder that has been exposed to extreme heat, has been overloaded, has been impact or shock loaded, is visibly damaged, or is suspected of being unsafe for any other reason must be removed from service until it has passed a service test. A repaired ladder must also undergo a service test before it can be returned to service, unless a halyard replacement was the only repair.

One part of the ladder service test is the horizontal-bending test, which evaluates the structural strength of a ladder. To perform this test, the ladder is placed in a horizontal position across a set of supports. A weight is then put on the ladder, and the amount of deflection or bending caused by the weight is measured to evaluate the ladder's strength **FIGURE 13-17**. Additional tests are performed on the extension hardware of extension ladders. The hooks on roof ladders are tested as well.

FIGURE 13-17 A horizontal-bending test evaluates the structural strength of a ladder.

Courtesy of Underwriters Laboratories

Service testing of ground ladders and aerial ladders requires special training and equipment and must be conducted only by qualified personnel. Many fire departments use outside contractors to perform these tests. The results of all service tests must be recorded and kept for future reference.

Ladder Safety

Several potential hazards are associated with ladder use, but unfortunately these risks are easily overlooked during emergency operations. As a consequence, many fire fighters have been seriously injured or killed in ladder accidents, both on the fire ground and during training sessions.

Ladders must be used with caution; follow standard procedures and regularly reinforce your skills through training. Many safety precautions should be followed from the time a ladder is removed from the apparatus until it is returned to the apparatus. Basic safety issues include the following concerns:

- General safety
- Lifting and moving ladders
- Placement of ground ladders
- Working on a ladder
- Rescue operations
- Ladder damage

These issues are explained in general here and will be revisited in the discussions of using specific types of ladders.

General Safety Requirements

In any firefighting operation, full turnout clothing and protective equipment are essential for working with ground ladders. Turnout gear provides protection from mechanical injuries as well as from fire and heat injuries. Helmets, coats, pants, gloves, and protective footwear provide protection from falling debris, impact injuries, and pinch injuries. Fire fighters must be able to work with and on ladders while wearing self-contained breathing apparatus (SCBA). You should practise working with ladders while wearing all of your personal protective equipment (PPE), including gloves and eye protection.

Lifting and Moving Ladders

Teamwork is essential when working with ladders. Some ladders are heavy and can be very awkward to maneuver, particularly when they are extended. Crew members must use proper lifting techniques and coordinate all movements.

Never attempt to lift or move a ladder that weighs more than you are capable of lifting safely. Because of their shape, ladders may be awkward to carry, especially over uneven terrain, through gates, or over snow and ice. It is better to ask for help in moving a ladder than to risk injury or delay the placement of the ladder.

Placement of Ground Ladders

Fire fighters working on the fire ground should survey the area where a portable ladder will be used before placing the ladder or beginning to raise it. If possible, inspect the area before retrieving the ladder from the apparatus. If you note any hazards, consider changing the position of the ladder. Work in the safest available location.

Sometimes a ladder must be placed in a potentially hazardous location. If you are aware of the hazard, you can take corrective actions and implement special precautions to avoid accidents.

The most important check assesses the location of overhead utility lines. If a ladder comes in contact with or close to an electrical wire, the fire fighters who are handling it can be electrocuted. If the line falls, it can electrocute other fire fighters as well. Avoid placing a ladder against a surface that has been energized by a damaged or fallen power line.

Do not assume that it is safe to use wood or fibreglass ladders around power lines. Metal ladders will certainly conduct electricity, but a wet or dirty wood or fibreglass ladder can also conduct electricity.

If possible, do not raise ladders anywhere near overhead wires. At a minimum, keep ladders at least 3 m (10 ft.) away from any power lines while using or raising them. If a power line is nearby, make sure that enough fire fighters are present to keep the ladder under control as it is raised. One fire fighter should watch to make sure that the ladder does not come too close to the line. If the ladder falls into a power line, the results can be deadly.

Other types of overhead obstructions may also be hazards during a ladder raise. If the ladder hits something as it is being raised, the weight and momentum of the ladder will shift, which can cause fire fighters to lose control and drop the ladder.

Ladders must be placed on stable and relatively level surfaces **FIGURE 13-18**. The shifting weight of a climber can easily tip a ladder placed on a slope or unstable surface. Always evaluate the stability of the surface before placing the ladder, and check it again during the course of the firefighting operations. A ladder placed on snow or mud, for example, may become unstable as the snow melts or as the base of the ladder sinks into the mud. In addition, water from the firefighting operations can make soft ground even more unstable.

It is also important to ensure that the tip of the ladder is placed against a stable surface. Placing the tip of a ladder against an unstable surface can result in the ladder falling forward into a fire or being pushed backward. The failure of an unstable wall, for example, can occur during firefighting operations or during overhaul. Avoid injury or death—do not place the tip of a ladder against any unstable surface.

Ladders should not be exposed to direct flames or extreme temperatures, because excessive heat can cause permanent damage to or catastrophic failure of such equipment. Check the heat sensor labels immediately after a ladder has been exposed to high temperatures.

■ Working on a Ladder

Before climbing a ladder, make sure it is set at the proper climbing angle (approximately 75 degrees) for maximum load

capacity and safety. To accomplish this, position the base of the ladder out from the building at a distance equal to one-fourth of the ladder's height **FIGURE 13-19**. For example, if a ladder is being raised to a window 6.1 m (20 ft.) above the ground, the base of the ladder should be placed 1.5 m (5 ft.) from the building. Achieving a proper angle is important to your safety and the safety of people being rescued.

When using an extension ladder, ensure that the pawls (dogs) are locked and the halyard tied before anyone climbs the ladder.

To prevent slipping, another fire fighter should secure the base of the ladder by heeling (footing) it **FIGURE 13-20**. This technique uses the fire fighter's weight to keep the base of the ladder from slipping. The base of the ladder can also be mechanically secured to a solid object. An alternative to securing the base is to use a rope or strap to secure the tip of the ladder to the building.

All ladders have weight limits. Most fire department portable ladders are designed to support a weight of 340 kilograms (750 pounds). In a rescue situation, this weight limit is equivalent to one victim and two fire fighters wearing full protective

Fire Fighter Safety Tips
© Photos.com

Always maintain at least 3 m (10 ft.) of clearance between a ladder and utility lines to prevent electrocution.

FIGURE 13-18 Ladders must be placed on stable and relatively level surfaces.

FIGURE 13-19 Place the base of a ladder out from the building at a distance equal to one-fourth of the ladder's height to achieve a climbing angle of 75 degrees.

6 m

1.5 m

75°

FIGURE 13-20 To prevent a ladder from slipping, fire fighters should secure the base by heeling the ladder.

Fire Fighter Safety Tips

© Photos.com

Many ladders used for construction work and home repair are rated for a working load of only 102 to 136 kg (225 to 300 lb.) — significantly less than the 340-kg (750-lb.) rating of approved fire department ladders. Remember the difference between these two types of ladders. If you encounter a situation where a construction ladder is already present, you still need to set up and use a fire department–approved ladder to avoid the possibility of a catastrophic ladder failure.

clothing and equipment. It would also accommodate two fire fighters climbing or working on the ladder along with their equipment. In contrast, ladders designed for everyday use may be rated for as little as 102 kg (225 lb.). Given the potential difference in ratings, it is important to avoid using unapproved ladders for fire department operations.

The weight should be distributed along the length of the ladder. Only one fire fighter should be on each section of an extension ladder at any time.

While climbing the ladder, be prepared for falling debris, misguided hose streams, or people falling from the building. Knowing these risks, it is important for fire fighters to wear full turnout gear when working on ladders—such gear will help to protect you.

If fire conditions change rapidly while you are working on a ladder, you must be prepared to climb down quickly; for example, if the fire suddenly flashes over or if flames break through a window located near the ladder, you must be prepared to move quickly out of danger. Turnout gear will not protect you from direct exposure to flames for more than a few seconds.

A fire fighter who is working from a ladder is in a less stable position than one who is working on the ground. While on a ladder, you must constantly adjust your balance, especially when swinging a hand tool or reaching for a trapped occupant. There is always the danger of falling as well as a risk to people below if something falls or drops. To minimize your risk of falling, use a safety belt or a leg lock to secure yourself to the ladder.

■ Rescue

Ladders can be an essential tool for rescue. For example, portable ladders are often used to reach and remove trapped occupants from the upper storeys of a building. When positioning a ladder for rescue, place the tip of the ladder at or slightly below the window sill. This position provides a direct pathway for the fire fighter entering the window and a direct route for exiting the structure. The ladder should not occlude any part of the window opening. Fire fighters need to address several important safety concerns before going up a ladder to rescue someone.

A person who is in extreme danger may not wait to be rescued. Jumpers not only risk their own lives, but may also endanger the fire fighters trying to rescue them. Fire fighters have been seriously injured by persons who jumped before a rescue could be completed.

A trapped person might try to jump onto the tip of an approaching ladder or to reach out for anything or anyone nearby. As a fire fighter, you might be pulled or pushed off the ladder by the person you are trying to rescue. If several people are trapped, they might all try to climb down the ladder at the same time.

It is important to make verbal contact with any person you are trying to rescue. You must remain in charge of the situation and not let the individual panic. Tell the person to remain calm and wait to be rescued. If enough fire fighters are present, one could maintain contact with the person while others raise the ladder.

After you reach the person, you still have to get him or her down to the ground safely. Try to make the person realize that any erratic movements could cause the ladder to tip over. It is often helpful to have one fire fighter assist the person, while a second fire fighter acts as a guide and backup to the rescuer. Although modern fire ladders are designed to support this load, a ladder that is overloaded during a rescue operation must be removed from service for inspection and service testing. See the chapter "Search and Rescue" for specific ladder rescue techniques.

■ Ladder Damage

Ladders are easily damaged while in use. For example, a ladder might be overloaded or used at a low angle during a rescue. An unexpected shift in fire conditions could bring the ladder in direct contact with flames. Whenever a ladder is used outside of its recommended limits, it should be taken out of service for inspection and testing, even if there is no visible damage.

Using Portable Ladders

Portable ladders are often urgently needed during emergency incidents. Because an accident or error in handling or using a ladder can result in death or serious injury, all fire fighters must know how to work with ladders.

Using a ladder requires that fire fighters complete a series of consecutive tasks. The first step is to select the best ladder for the job from those available. Fire fighters must then remove the ladder from the apparatus and carry it to the location where it will be used. The next step is to raise and secure the ladder. At the end of the operation, the ladder must be lowered and returned to the apparatus. Each of these tasks is important to the safe and successful completion of the overall objective.

■ Selecting the Ladder

The first step in using a portable ladder is to select the appropriate ladder from those available. Fire fighters must be familiar with all of the ladders carried on their apparatus. Engine and ladder company apparatus usually carry several portable ladders of various lengths. Many other types of apparatus, such as tankers (water tenders) and rescue units, often carry additional portable ladders. NFPA 1901, *Standard for Automotive Fire Apparatus*, identifies minimum requirements for portable ladders for each type of apparatus (TABLE 13-1).

Selecting an appropriate ladder requires that fire fighters estimate the heights of windows and rooflines. General guidelines have been developed for estimating the heights for residential and commercial buildings (TABLE 13-2).

Ladder placement will also affect the size and length of the ladder needed. When the ladder is used to access a roof, for example, the tip of the ladder should extend more than a metre (several feet) above the roofline to provide a convenient handhold and secure footing for fire fighters as they mount and dismount. This extra length also makes the tip of the ladder visible to fire fighters who are working on the roof

(FIGURE 13-21). A common rule of thumb is to ensure that at least five ladder rungs show above the roofline.

Ladders used to provide access to windows must be longer than those used in rescue operations. During access operations, the ladder is placed next to the window, with the ladder tip even with the top of the window opening (FIGURE 13-22). During rescue operations, however, the tip of the ladder should be immediately below the windowsill, which prevents the ladder from obstructing the window opening while a trapped occupant is removed (FIGURE 13-23).

The final factor in determining the correct length of ladder is the angle formed by the ladder and the placement surface (ground). A portable ladder should be placed at an angle of approximately 75 degrees for maximum strength and stability. To satisfy this criterion, the ladder must be slightly longer than the vertical distance between the ground and the target point. Generally, a ladder requires an additional 0.3 m (1 ft.) in length for every 4.6 m (15 ft.) of vertical height.

FIGURE 13-21 The tip of the ladder should extend above the roofline during roof operations.

TABLE 13-1	Minimum Ladder Complement for Apparatus as Specified in NFPA 1901, *Standard for Automotive Fire Apparatus*
Pumper	1 folding ladder 1 roof ladder, straight and equipped with hooks 1 extension ladder
Quick attack	1 extension ladder must be 3.7 m (12 ft.) or longer
Aerial/ladder	Portable ladders that have a total length of 35 m (115 ft.) or more and contain a minimum of: • 1 folding ladder • 2 roof ladders, straight and equipped with hooks • 2 extension ladders
Quint	Portable ladders that have a total length of 26 m (85 ft.) or more and contain a minimum of: • 1 folding ladder • 1 roof ladder, straight and equipped with hooks • 1 extension ladder

TABLE 13-2	Approximate Distances for Residential and Commercial Construction
Residential floor-to-floor height	2.4–3 m (8–10 ft.)
Residential floor-to-windowsill height	1 m (3 ft.)
Commercial floor-to-floor height	3–3.7 m (10–12 ft.)
Commercial floor-to-windowsill height	1.2 m (4 ft.)

FIGURE 13-22 To provide access through a window, the tip of the ladder should be on the upwind side of the window and even with the top of the window opening.

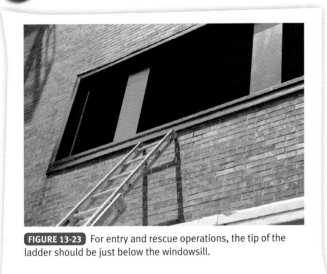
FIGURE 13-23 For entry and rescue operations, the tip of the ladder should be just below the windowsill.

FIGURE 13-24 This hydraulic mechanism lowers the ladders to a convenient height when they are needed.

Courtesy of Dennis Wetherhold, Jr.

Thus, to reach a window that is 9.1 m (30 ft.) above grade level, a ladder would have to be at least 9.7 m (32 ft.) long. Because ladders used in roof operations need to extend at least above the roofline, accessing a roof that is 9.1 m (30 ft.) above grade level requires a ladder at least 10.7 m (35 ft.) long.

■ Removing the Ladder from the Apparatus

Ladders are mounted on apparatus in various ways. Such equipment should be stored in locations where it will not be exposed to excessive heat, engine exhaust, or mechanical damage. Fire fighters must be familiar with how ladders are mounted on their apparatus and practise removing them safely and quickly.

Ladders are often nested one inside another and mounted on brackets on the side of the engine. Fire fighters should note the nesting order and location of the ladders relative to the brackets when removing or replacing them. Ladders that are not needed should not be placed on the ground in front of an exhaust pipe because they could be damaged by the hot exhaust.

Ladders can also be stored on overhead hydraulic lifts **FIGURE 13-24**. The hydraulic mechanism keeps them out of the way until they are needed, when they can be lowered to a convenient height. It is important to review the operation of the hydraulic lift used on your fire department's apparatus so that you can operate it efficiently and safely during an emergency situation.

On some vehicles, portable ladders are stored in compartments under the hose bed or aerial device. The ladders may lie flat or vertically on one beam **FIGURE 13-25**. To remove or replace them, fire fighters slide the ladders out the rear of the vehicle. For this reason, it is important to leave adequate room behind such apparatus so that ladders can be removed at a fire scene.

■ Lifting Ladders

Many ladders are heavy and awkward to handle. To prevent lifting injuries while handling ladders, fire fighters must work together to lift and carry long or heavy ladders.

FIGURE 13-25 Ladders in rear compartments can be stored either flat or vertically. This configuration requires adequate space behind the truck to remove ladders at an emergency scene.

When fire fighters are working as a team to lift or carry a ladder, one fire fighter must act as the leader, providing direction and coordinating the actions of all team members. The lead fire fighter needs to call out the intended movements clearly, using standard terminology and clear two-step commands. For example, when lowering a ladder, the leader should say, "Prepare to lower," followed by the command, "Lower." Team members must communicate in a clear and concise fashion, using the same terminology for commands; there should be no confusion about the specific meaning of each command. Because terminology may differ among jurisdictions, fire fighters who are newly or temporarily assigned to an apparatus should verify which commands are used and what they mean.

A prearranged method should be in place for determining the lead fire fighter for ladder lifts. In some fire departments, the fire fighter on the right side at the butt end of the ladder is always the leader. In other departments, the fire fighter at the

Near-Miss REPORT

Report Number: 07-854

Synopsis: Fire fighters trapped on a roof without a second means of egress.

Event Description: We were called to the scene of a large fire in a two-storey apartment building in mid-December. It was cold out, and while we normally run four fire fighters, this particular night we had only three. The roof was slick from the water and ice. We were assigned to the ventilation sector by the IC. We threw our ground ladder up as well as placed a roof ladder on the roof. We made a ventilation hole that after a few seconds allowed the fire to come through. The roof conditions had deteriorated so badly by this point that we were unable to make it back to our ladders. An aerial operator from a neighbouring department saw that we were having trouble and moved the stick over just below where we were. Our only option at this point was to slide down the roof and hope we hit the ladder. Luckily, all three of us made it to the aerial without injury.

Lessons Learned: From that fire on, we constantly train on putting up two means of egress. It also made us firm believers in having a four–fire fighter truck crew. We believe if we would have had four firefighters, the extra fire fighter could have been throwing other ladders, giving us second and third points of egress.

tip of the ladder on the right side may be the leader. The departmental policy should be consistent for all ladder operations.

Additionally, you must use good lifting techniques when handling ladders. When bending to pick up a ladder, bend at the knees and keep the back straight **FIGURE 13-26**. Lift and lower the load with the legs, rather than the back. Take care to avoid twisting motions during lifting and lowering, as these motions often lead to back strains.

■ Carrying Ladders

Once a ladder has been removed from the apparatus or lifted from the ground, it must be carried to the placement site. Ladders can be carried at shoulder height or at arm's length. They can be carried either on edge or flat. The most common carries are explained in the following sections.

One–Fire Fighter Carry

Most straight ladders and roof ladders less than 5.5 m (18 ft.) long can be safely carried by a single fire fighter. The steps involved in the one–fire fighter carry are outlined in **SKILL DRILL 13-2**:

1. Start with the ladder mounted in a bracket or standing on one beam. Locate the centre of the ladder.
2. Place an arm between two rungs of the ladder just to one side of the middle rung. (**STEP 1**)
3. Lift the top beam of the ladder and rest it on your shoulder. (**STEP 2**)
4. Walk carefully with the butt end first. Carry roof ladders with the hook end in the front. (**STEP 3**)

A straight ladder is carried with the butt end first and pointed slightly downward; a roof ladder is carried tip first. Most ladders are carried with the butt end forward, because

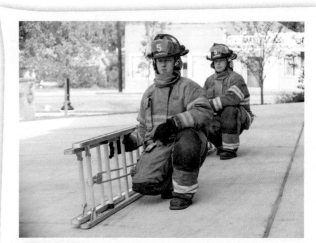

FIGURE 13-26 Bend your knees and keep your back straight when lifting or lowering a ladder.

the placement of the butt determines the positioning of the ladder. Roof ladders are carried with the hooks toward the front. A roof ladder usually must be carried up another ladder to reach the roof. The roof ladder is then usually pushed up the pitched roof slope from the butt end until the hooks engage the peak.

Two–Fire Fighter Shoulder Carry

The two–fire fighter shoulder carry is generally used with extension ladders up to 10.7 m (35 ft.) long. It can also be used to carry straight ladders or roof ladders that are too long for one person to handle. To perform the two–fire fighter shoulder carry, follow the steps in **SKILL DRILL 13-3**:

SKILL DRILL 13-2   One–Fire Fighter Carry
(Fire Fighter I, NFPA 5.3.6)

1 Start with the ladder mounted in a bracket or standing on one beam. Locate the centre of the ladder. Place an arm between two rungs of the ladder just to one side of the middle rung.

2 Lift the top beam of the ladder and rest it on your shoulder.

3 Walk carefully with the butt end first.

© LiquidLibrary

---

1 Start with the ladder mounted in a bracket or standing on one beam. Both fire fighters are positioned on the same side of the ladder.

2 Facing the butt end of the ladder, one fire fighter is positioned near the butt end of the ladder and the second fire fighter is positioned near the tip of the ladder. (STEP **1**)

3 Both fire fighters place one arm between two rungs and, on the leader's command, lift the ladder onto their shoulders.

4 The ladder is carried butt end first. (STEP **2**)

5 The fire fighter closest to the butt covers the sharp butt spur with a gloved hand to prevent injury to other fire fighters in the event of a collision. (STEP **3**)

Three–Fire Fighter Shoulder Carry

Three fire fighters may be needed to carry a heavy ladder. This carry is similar to the two–fire fighter shoulder carry, with the additional fire fighter positioned at the middle of the ladder. All three fire fighters stand on the same side of the ladder. To perform the three–fire fighter shoulder carry, follow the steps in SKILL DRILL 13-4 :

1 Start with the ladder mounted in a bracket or standing on one beam. Three fire fighters stand on the same side of the ladder.

2 Facing the butt end of the ladder, one fire fighter is positioned at the butt end of the ladder, one fire fighter is at the middle of the ladder, and one fire fighter is at the tip of the ladder. (STEP **1**)

SKILL DRILL 13-3 — Two–Fire Fighter Shoulder Carry
(Fire Fighter I, NFPA 5.3.6)

1 Start with the ladder mounted in a bracket or standing on one beam. Both fire fighters are positioned on the same side of the ladder. Facing the butt end of the ladder, one fire fighter is positioned near the butt end of the ladder and a second fire fighter is positioned near the tip of the ladder.

2 Both fire fighters place one arm between two rungs and, on the leader's command, lift the ladder onto their shoulders. The ladder is carried butt end first.

3 The butt spurs are covered with a gloved hand while the ladder is transported.

© LiquidLibrary

SKILL DRILL 13-4 — Three–Fire Fighter Shoulder Carry
(Fire Fighter I, NFPA 5.3.6)

1 All three fire fighters are positioned on the same side of the ladder. Facing the butt end of the ladder, one fire fighter is positioned at the butt end of the ladder, one fire fighter is at the middle of the ladder, and one fire fighter is at the tip of the ladder.

2 Each fire fighter places an arm between two rungs and, on the leader's command, all fire fighters hoist the ladder onto their shoulders.

3 The ladder is carried butt end first. The fire fighter closest to the butt covers the sharp butt spur with a gloved hand.

© LiquidLibrary

3 Each fire fighter places an arm between two rungs and, on the leader's command, all fire fighters hoist the ladder onto their shoulders. (**STEP 2**)

4 The ladder is carried butt end first. The fire fighter closest to the butt covers the sharp butt spur with a gloved hand to prevent injury to other fire fighters in the event of a collision. (**STEP 3**)

Two–Fire Fighter Suitcase Carry

The two–fire fighter suitcase carry is commonly used with straight and extension ladders. With this technique, the ladder is carried at arm's length. To perform the two–fire fighter suitcase carry, follow the steps in **SKILL DRILL 13-5**:

1. Start with the ladder resting on the ground on one beam. Both fire fighters are positioned on the same side of the ladder.

2. Facing the butt end of the ladder, one fire fighter is positioned at the butt end of the ladder and the other fire fighter is positioned at the tip of the ladder. (**STEP 1**)

3. Each fire fighter reaches down and grasps the upper beam of the ladder. (**STEP 2**)

4. On the leader's command, both fire fighters pick the ladder up from the ground and carry it with the butt end forward. The fire fighter closest to the butt covers the sharp butt spur with a gloved hand. (**STEP 3**)

Three–Fire Fighter Suitcase Carry

A three–fire fighter suitcase carry can be used for heavier ladders. This technique is similar to the two–fire fighter suitcase carry, with the addition of a third fire fighter at the centre of the ladder. All three fire fighters remain on the same side of the ladder. To perform the three–fire fighter suitcase carry, follow the steps in **SKILL DRILL 13-6**:

1. Start with the ladder resting on the ground on one beam. All three fire fighters are positioned on the same side of the ladder.

2. Facing the butt end of the ladder, one fire fighter is positioned at the butt end of the ladder, one fire fighter is at the middle of the ladder, and one fire fighter is at the tip of the ladder. (**STEP 1**)

3. All three fire fighters reach down and grasp the upper rung of the ladder. (**STEP 2**)

4. On the leader's command, the fire fighters pick the ladder up from the ground and carry it at arm's length. (**STEP 3**)

5. The fire fighter closest to the butt covers the sharp butt spur with a gloved hand. (**STEP 4**)

Three–Fire Fighter Flat Carry

The three–fire fighter flat carry is typically used with extension ladders up to 10.7 m (35 ft.) long. To perform the three–fire fighter flat carry, follow the steps in **SKILL DRILL 13-7**:

1. The carry begins with the bed section of the ladder flat on the ground. Two fire fighters stand on one side of the ladder, and the third fire fighter stands on the opposite side of the ladder.

2. Facing the butt end of the ladder, one fire fighter is positioned at the butt end of the ladder with the second fire fighter at the tip end of the ladder. The third fire fighter is positioned at the centre of the ladder on the opposite side. (**STEP 1**)

3. All three fire fighters kneel down and grasp the closer beam at arm's length. (**STEP 2**)

4. On the leader's command, all three fire fighters pick up the ladder and carry it butt end forward. The fire fighter closest to the butt covers the sharp butt spur with a gloved hand. (**STEP 3**)

SKILL DRILL 13-5 | Two–Fire Fighter Suitcase Carry
(Fire Fighter I, NFPA 5.3.6)

1 Start with the ladder resting on the ground on one beam. Both fire fighters are positioned on the same side of the ladder. Facing the butt end of the ladder, one fire fighter is positioned at the butt end of the ladder and the other fire fighter is positioned at the tip of the ladder.

2 Each fire fighter reaches down and grasps the upper beam of the ladder.

3 On the leader's command, both fire fighters pick the ladder up from the ground and carry it with the butt end forward. The fire fighter closest to the butt covers the sharp butt spur with a gloved hand.

SKILL DRILL 13-6 | Three–Fire Fighter Suitcase Carry
(Fire Fighter I, NFPA 5.3.6)

1 Start with the ladder resting on the ground on one beam. All three fire fighters are positioned on the same side of the ladder. Facing the butt end of the ladder, one fire fighter is positioned at the butt end of the ladder, one fire fighter is at the middle of the ladder, and one fire fighter is at the tip of the ladder.

2 All three fire fighters reach down and grasp the upper beam of the ladder.

3 On the leader's command, the fire fighters pick the ladder up from the ground and carry it at arm's length.

4 The fire fighter closest to the butt covers the sharp butt spur with a gloved hand.

© LiquidLibrary

Four–Fire Fighter Flat Carry

The four–fire fighter flat carry is similar to the three–fire fighter flat carry described in the previous section, except that two fire fighters are positioned on each side of the ladder. On each side, one fire fighter is at the tip and the other is at the butt end of the ladder. To perform the four–fire fighter flat carry, follow the steps in **SKILL DRILL 13-8**:

1 The carry begins with the bed section of the ladder flat on the ground. Two fire fighters stand on each side of the ladder.

2 All four fire fighters face the butt end of the ladder. One fire fighter is positioned at each corner of the ladder, with two fire fighters at the butt end of the ladder and two fire fighters at the tip end of the ladder. (**STEP 1**)

3 On the leader's command, all four fire fighters kneel down and grasp the closer beam at arm's length. (**STEP 2**)

4 On the leader's command, all four fire fighters pick up the ladder and carry it butt end forward. (**STEP 3**)

5 The fire fighter closest to the butt covers the sharp butt spur with a gloved hand. (**STEP 4**)

SKILL DRILL 13-7   Three–Fire Fighter Flat Carry
(Fire Fighter I, NFPA 5.3.6)

1   The bed section of the ladder is flat on the ground. Two fire fighters stand on one side of the ladder, and the third fire fighter stands on the opposite side of the ladder. Facing the butt end of the ladder, one fire fighter is positioned at the butt end of the ladder with the second fire fighter at the tip end of the ladder. The third fire fighter is positioned at the centre of the ladder on the opposite side.

2   All three fire fighters kneel down and grasp the closer beam at arm's length.

3   On the leader's command, all three fire fighters pick up the ladder and carry it butt end forward. The fire fighter closest to the butt covers the sharp butt spur with a gloved hand.

© LiquidLibrary

Three–Fire Fighter Flat Shoulder Carry

The three–fire fighter flat shoulder carry is similar to the three–fire fighter flat carry, except that the ladder is carried on the shoulders instead of at arm's length. The fire fighters face the tip of the ladder as they lift it, but then pivot into the ladder as they raise it to shoulder height.

This carry is useful when fire fighters must carry the ladder over short obstacles. Raising the ladder to shoulder height increases the potential for back strain, however, so fire fighters must follow proper lifting techniques.

To perform the three–fire fighter flat shoulder carry, follow the steps in **SKILL DRILL 13-9** :

1   The carry begins with the bed section of the ladder flat on the ground. Two fire fighters are positioned on one side of the ladder, and the third fire fighter is on the opposite side of the ladder. Facing the tip end of the ladder, one fire fighter is positioned at the butt end of the ladder, and the second fire fighter is at the tip end of the ladder. The third fire fighter is at the centre of the ladder on the opposite side. **(STEP ❶)**

2   All three fire fighters kneel and grasp the closer beam. **(STEP ❷)**

3   On the leader's command, the fire fighters stand, raising the ladder. **(STEP ❸)**

4   As the ladder approaches chest height, the leader instructs the fire fighters to pivot into the ladder. **(STEP ❹)**

5   The fire fighters place the beam of the ladder on their shoulders. **(STEP ❺)**

6   The ladder is carried in this position with the butt moving forward. The fire fighter closest to the butt covers the sharp butt spur with a gloved hand. **(STEP ❻)**

Fire Fighter Safety Tips
© Photos.com

When fire fighters of different heights are working together and carrying ladders, you can make these operations safer and smoother by considering the placement of fire fighters. When performing three-person carries, place shorter fire fighters at the butt end of the ladder and taller fire fighters at the tip end of the ladder. This strategy prevents a shorter fire fighter from being placed between two taller fire fighters and having to carry the ladder at an awkward and unsafe height. When performing four-person carries, placing shorter fire fighters at the butt end of the ladder prevents the ladder from being tipped to one side while it is being carried.

SKILL DRILL 13-8 Four–Fire Fighter Flat Carry
(Fire Fighter I, NFPA 5.3.6)

1 Two fire fighters stand on each side of the ladder. All four fire fighters face the butt end of the ladder. One fire fighter is positioned at each corner of the ladder, with two fire fighters at the butt end of the ladder and two fire fighters at the tip end of the ladder.

2 On the leader's command, all four fire fighters kneel down and grasp the closer rung at arm's length.

3 On the leader's command, all four fire fighters pick up the ladder and carry it butt end forward.

4 The fire fighter closest to the butt covers the sharp butt spur with a gloved hand.

© LiquidLibrary

Four–Fire Fighter Flat Shoulder Carry

The four–fire fighter flat shoulder carry is similar to the three–fire fighter flat shoulder carry described in the previous section, except that two fire fighters are positioned on each side of the ladder. On each side, one fire fighter is at the tip and the other at the butt end of the ladder. To perform the four–fire fighter flat shoulder carry, follow the steps in **SKILL DRILL 13-10** :

1 The carry begins with the bed section of the ladder flat on the ground. Two fire fighters are positioned on each side of the ladder.

2 All four fire fighters face the tip end of the ladder. One fire fighter is positioned at each corner of the ladder, with two fire fighters at the butt end and two fire fighters at the tip end of the ladder. (**STEP 1**)

3 On the leader's command, all four fire fighters kneel and grasp the closer beam. (**STEP 2**)

4 On the leader's command, the fire fighters stand, raising the ladder. (**STEP 3**)

5 As the ladder approaches chest height, the fire fighters all pivot toward the ladder. (**STEP 4**)

SKILL DRILL 13-9 Three—Fire Fighter Flat Shoulder Carry
(Fire Fighter I, NFPA 5.3.6)

1 Two fire fighters are positioned on one side of the ladder, and the third fire fighter is on the opposite side of the ladder. Facing the tip end of the ladder, one fire fighter is positioned at the butt end of the ladder, and the second fire fighter is at the tip end of the ladder. The third fire fighter is at the centre of the ladder on the opposite side.

2 All three fire fighters kneel and grasp the closer beam.

3 On the leader's command, the fire fighters stand, raising the ladder.

4 As the ladder approaches chest height, the leader instructs the fire fighters to pivot into the ladder.

5 The fire fighters place the beam of the ladder on their shoulders.

6 The ladder is carried in this position with the butt moving forward. The fire fighter closest to the butt covers the sharp butt spur with a gloved hand.

© LiquidLibrary

6 The fire fighters place the beam of the ladder on their shoulders. All four fire fighters face the butt of the ladder.

7 The ladder is carried in this position, with the butt moving forward. The fire fighter closest to the butt covers the sharp butt spur with a gloved hand. **(STEP 5)**

■ Placing a Ladder

The first step in raising a ladder is selecting the proper location for the ladder. Generally, an officer or senior fire fighter will select the general area for ladder placement, and the fire fighter at the butt of the ladder will determine the exact site at which to position it. Both the officer ordering the ladder and the fire fighter placing the ladder need to consider several factors in their decisions.

A raised ladder should be positioned at an angle of approximately 75 degrees to provide the best combination of strength, stability, and vertical reach. This angle also creates a comfortable climbing angle. When a fire fighter is standing on a rung, the rung at shoulder height will be approximately one arm's length away. The ratio of the ladder height (vertical reach) to the distance from the structure should be 4:1. For example, if the vertical reach of a ladder is 6.1 m (20 ft.), the butt of the ladder should be placed 1.5 m (5 ft.) out from the wall. Most new ladders have an inclination guide on their beams to help with ladder placement.

When calculating vertical reach, remember to add extra footage for rooftop operations. If the butt of the ladder is placed too close to the building, the climbing angle will be too steep. In such a case, the tip of the ladder could pull away from the

SKILL DRILL 13-10 Four–Fire Fighter Flat Shoulder Carry
(Fire Fighter I, NFPA 5.3.6)

1 Two fire fighters are positioned on each side of the ladder. All four fire fighters face the tip end of the ladder. One fire fighter is positioned at each corner of the ladder, with two fire fighters at the butt end and two fire fighters at the tip end of the ladder.

2 On the leader's command, all four fire fighters kneel and grasp the closer beam.

3 On the leader's command, the fire fighters stand, raising the ladder.

4 As the ladder approaches chest height, the fire fighters all pivot toward the ladder.

5 The fire fighters place the beam of the ladder on their shoulders. All four fire fighters face the butt of the ladder. The ladder is carried in this position, with the butt moving forward. The fire fighter closest to the butt covers the sharp butt spur with a gloved hand.

© LiquidLibrary

building as the fire fighter climbs, making the ladder unstable. Conversely, if the butt is too far from the structure, the climbing angle will be too shallow, reducing the load capacity of the ladder and increasing the risk that the butt could slip out from under the ladder.

The ladder should be placed on a stable and relatively level surface, as ladders placed on uneven surfaces are prone to tipping **FIGURE 13-27**. As a fire fighter climbs the ladder, the moving weight will cause the load to shift back and forth between the two beams. Unless both butt ends are firmly placed on solid ground, this uneven load can start a rocking motion that will tip the ladder over.

Ladders should not be placed on top of unstable structures such as manholes or trap doors. The combined weight of

the ladder, fire fighters, and equipment could cause the cover or door to fail, injuring the fire fighters.

As mentioned earlier, ladders should be placed in areas with no overhead obstructions. If a ladder comes into contact with overhead utility lines—and particularly electric power lines—both the fire fighters working on the ladder and those stabilizing it could be injured or killed. This is true for all ladders, regardless of their composition (fibreglass, wood, or metal).

Be aware that a ladder can become energized even if it does not actually touch an electric line. For example, a ladder that enters the electromagnetic field surrounding a power line can become energized. For this reason, ladders should be placed at least 3 m (10 ft.) away from energized power lines.

FIGURE 13-27 A ladder placed on an uneven surface is unsafe because it can easily tip over.

Finally, portable ladders should not be placed in high-traffic areas unless no other alternative is available. For example, because the main entrance to a structure is heavily used, a ladder should not be placed where it would obstruct the door.

Fire Fighter Safety Tips

© Photos.com

Ladders can become energized without making contact with overhead electric lines. An electric force field surrounds electric lines, with the size of this field increasing in tandem with both the voltage in the line and the relative humidity in the air. Always keep ladders at least 3 m (10 ft.) away from electric lines.

■ Raising a Ladder

Once the position has been selected, the ladder must be raised for climbing. Two commonly used techniques for raising portable ladders are the beam raise and the rung raise. A beam raise is typically used when the ladder must be raised parallel to the target surface. A rung raise is often used when the ladder can be raised from a position perpendicular to the target surface.

The number of fire fighters required to raise a ladder depends on the length and weight of the ladder, as well as on the available clearance from obstructions. A single fire fighter can safely raise many straight ladders and lighter extension ladders. Two or more fire fighters are required to raise longer and heavier ladders.

One–Fire Fighter Rung Raises

There are two variations of the one–fire fighter rung raise. One is used by a single fire fighter to raise a small, straight ladder, typically 4.3 m (14 ft.) or less in length. Always check for overhead hazards before raising a ladder. To perform this raise, follow the steps in **SKILL DRILL 13-11**:

1 Place the ladder flat on the ground. Position the heel of the ladder where it will be when the ladder is raised.

2 Stand at the tip of the ladder and check for overhead hazards.

3 Raise the ladder to hip level.

4 Walk hand-over-hand down the rungs until the ladder is vertical. **(STEP 1)**

5 Place one foot against the beam of the ladder and lean it into place.

6 Check the ladder for a 75-degree climbing angle. Reposition it if necessary. Check the tip of the ladder and the heel of the ladder to ensure safety before climbing. **(STEP 2)**

The other variation of the one–fire fighter rung raise is generally used with straight ladders longer than 4.3 m (14 ft.) and with extension ladders that can be safely handled by one fire fighter. Each fire fighter will have a different safety limit, depending on his or her strength and the weight of the ladder. To perform this variation of the one–fire fighter rung raise, follow the steps in **SKILL DRILL 13-12**:

1 Place the butt of the ladder on the ground directly against the structure so that both spurs contact the ground and the structure. Lay the ladder on the ground. If the ladder is an extension ladder, place the base section on the ground.

2 Stand at the tip of the ladder and check for overhead hazards. **(STEP 1)**

3 Take hold of a rung near the tip, bring that end of the ladder to chest height, and then step beneath the ladder. **(STEP 2)**

4 Raise the ladder using a hand-over-hand motion as you walk toward the structure until the ladder is vertical and against the structure. **(STEP 3)**

5 If an extension ladder is being used, hold the ladder vertical against the structure and extend the fly section by pulling the halyard smoothly, with a hand-over-hand motion, until the desired height is reached and the dogs are locked. **(STEP 4)**

6 Pull the butt of the ladder out from the structure to create a 75-degree climbing angle. To move the butt away from the structure, grip a lower rung and lift slightly while pulling outward. At the same time, apply pressure to an upper rung to keep the tip of the ladder against the structure. **(STEP 5)**

7 If the ladder is an extension ladder, it will be necessary to rotate the ladder so the fly section is out.

8 The halyard should be tied as described in the "Tying the Halyard" skill drill. **(STEP 6)**

9 Check the tip and the heel of the ladder to ensure safety before climbing. **(STEP 7)**

Tying the Halyard

The halyard of an extension ladder should always be tied after the ladder has been extended and lowered into place. A tied halyard stays out of the way and provides a safety backup to the dogs for securing the fly section. To tie the halyard, follow the steps in **SKILL DRILL 13-13**:

1 Wrap the excess halyard rope around two rungs of the ladder and pull the rope tight across the upper of the two rungs. **(STEP 1)**

SKILL DRILL 13-11 — One–Fire Fighter Rung Raise for Ladders Less Than 4.3 m Long
(Fire Fighter I, NFPA 5.3.6)

1 Place the ladder flat on the ground. Position the heel of the ladder where it will be when the ladder is raised. Stand at the tip of the ladder and check for overhead hazards. Raise the ladder to hip level. Walk hand-over-hand down the rungs until the ladder is vertical.

2 Place one foot against the beam of the ladder and lean it into place. Check the ladder for a 75-degree climbing angle. Reposition it if necessary. Check the tip of the ladder and the heel of the ladder to ensure safety before climbing.

© LiquidLibrary

2 Tie a clove hitch around the upper rung and the vertical section of the halyard. Refer to the chapter "Ropes and Knots" to review how to tie a clove hitch. (**STEP 2**)

3 Pull the clove hitch tight. (**STEP 3**)

4 Place an overhand safety knot as close to the clove hitch as possible to prevent slipping. (**STEP 4**)

Two–Fire Fighter Beam Raise

The two–fire fighter beam raise is used with midsized extension ladders up to 10.7 m (35 ft.) long. To perform the two–fire fighter beam raise, follow the steps in **SKILL DRILL 13-14**:

1 The two–fire fighter beam raise begins with a shoulder or suitcase carry. One fire fighter stands near the butt of the ladder, and the other fire fighter stands near the tip. (**STEP 1**)

2 The fire fighter at the butt of the ladder places that end of the ladder on the ground, while the fire fighter at the tip of the ladder holds that end. (**STEP 2**)

3 The fire fighter at the butt of the ladder places a foot on the butt of the beam that is in contact with the ground and grasps the upper beam. (**STEP 3**)

4 The fire fighter at the tip of the ladder checks for overhead hazards and then begins to walk toward the butt, while raising the lower beam in a hand-over-hand fashion until the ladder is vertical. (**STEP 4**)

5 The two fire fighters pivot the ladder into position as necessary. (**STEP 5**)

6 The fire fighters face each other, one on each side of the ladder, and heel the ladder by each placing the toe of one boot against the opposing beams of the ladder. (**STEP 6**)

7 One fire fighter extends the fly section by pulling the halyard smoothly with a hand-over-hand motion until the fly section is at the height desired and the pawls (dogs) are locked.

8 The other fire fighter stabilizes the ladder by holding the outside of the two beams, so that if the fly comes down suddenly it will not strike the fire fighter's hands. (**STEP 7**)

9 The fire fighter facing the structure places one foot against one beam of the ladder, and then both fire fighters lean the ladder into place.

10 The halyard is tied.

11 The fire fighters check the ladder for a 75-degree climbing angle and for stability at the top and the bottom of the ladder. (**STEP 8**)

SKILL DRILL 13-12 | One–Fire Fighter Rung Raise for Ladders More Than 4.3 m Long
(Fire Fighter I, NFPA 5.3.6)

1 Place the butt of the ladder on the ground directly against the structure so that both spurs contact the ground and the structure. Lay the ladder on the ground. If the ladder is an extension ladder, place the base section on the ground. Stand at the tip of the ladder and check for overhead hazards.

2 Take hold of a rung near the tip, bring that end of the ladder to chest height, and then step beneath the ladder.

3 Raise the ladder using a hand-over-hand motion as you walk toward the structure until the ladder is vertical and against the structure.

4 If an extension ladder is being used, hold the ladder vertical against the structure and extend the fly section by pulling the halyard smoothly, with a hand-over-hand motion, until the desired height is reached and the dogs are locked.

5 Pull the butt of the ladder out from the structure to create a 75-degree climbing angle. To move the butt away from the structure, grip a lower rung and lift slightly while pulling outward. At the same time, apply pressure to an upper rung to keep the tip of the ladder against the structure.

6 If the ladder is an extension ladder, it will be necessary to rotate the ladder so the fly section is out. The halyard should be tied as described in the "Tying the Halyard" skill drill.

7 Check the tip and the heel of the ladder to ensure safety before climbing.

SKILL DRILL 13-13 Tying the Halyard
(Fire Fighter I, NFPA 5.3.6)

1 Wrap the excess halyard rope around two rungs of the ladder and pull the rope tight across the upper of the two rungs.

2 Tie a clove hitch around the upper rung and the vertical section of the halyard.

3 Pull the clove hitch tight.

4 Place an overhand safety knot as close to the clove hitch as possible to prevent slipping.

© LiquidLibrary

Two–Fire Fighter Rung Raise

The two–fire fighter rung raise is also commonly used with midsized extension ladders up to 10.7 m (35 ft.) long. To perform the two–fire fighter rung raise, follow the steps in **SKILL DRILL 13-15**:

1 The two–fire fighter rung raise begins from a shoulder carry or suitcase carry, with one fire fighter near the butt of the ladder and the other near the tip.

2 The fire fighter at the butt of the ladder places the butt of the lower beam on the ground, while the fire fighter at the tip holds the other end.

3 The fire fighter at the tip rotates the ladder so that both beams are in contact with the ground. (**STEP 1**)

4 The fire fighter at the butt of the ladder stands on the bottom rung, grasps a higher rung with both hands, crouches down, and leans backward to heel the ladder.

5 The fire fighter at the tip of the ladder checks for overhead hazards, then swings under the ladder and walks toward the butt, advancing down the ladder and lifting the rungs in a hand-over-hand fashion until the ladder is vertical. (**STEP 2**)

6 The two fire fighters stand on opposite sides of the ladder and pivot it into position as necessary. (**STEP 3**)

7 The fire fighters face each other, one on each side of the ladder, and heel the ladder by each placing the toe

SKILL DRILL 13-14   Two–Fire Fighter Beam Raise
(Fire Fighter I, NFPA 5.3.6)

1 The two–fire fighter beam raise begins with a shoulder or suitcase carry. One fire fighter stands near the butt of the ladder, and the other fire fighter stands near the tip.

2 The fire fighter at the butt of the ladder places that end of the ladder on the ground, while the fire fighter at the tip of the ladder holds that end.

3 The fire fighter at the butt of the ladder places a foot on the butt of the beam that is in contact with the ground and grasps the upper beam.

4 The fire fighter at the tip of the ladder checks for overhead hazards and then begins to walk toward the butt, while raising the lower beam in a hand-over-hand fashion until the ladder is vertical.

5 The two fire fighters pivot the ladder into position as necessary.

6 The fire fighters face each other, one on each side of the ladder, and heel the ladder by each placing the toe of one boot against the opposing beams of the ladder.

7 One fire fighter extends the fly section by pulling the halyard smoothly with a hand-over-hand motion until the fly section is at the height desired and the pawls (dogs) are locked. The other fire fighter stabilizes the ladder by holding the outside of the two beams, so that if the fly comes down suddenly it will not strike the fire fighter's hands.

8 The fire fighter facing the structure places one foot against one beam of the ladder, and then both fire fighters lean the ladder into place. The halyard is tied. The fire fighters check the ladder for a 75-degree climbing angle and check for stability at the top and the bottom of the ladder.

© LiquidLibrary

SKILL DRILL 13-15   Two–Fire Fighter Rung Raise
(Fire Fighter I, NFPA 5.3.6)

1 The two–fire fighter rung raise begins from a shoulder carry or suitcase carry, with one fire fighter near the butt of the ladder and the other near the tip. The fire fighter at the butt of the ladder places the butt of the lower beam on the ground, while the fire fighter at the tip holds the other end. The fire fighter at the tip rotates the ladder so that both beams are in contact with the ground.

2 The fire fighter at the butt of the ladder stands on the bottom rung, grasps a higher rung with both hands, crouches down, and leans backward to heel the ladder. The fire fighter at the tip of the ladder checks for overhead hazards, then swings under the ladder and walks toward the butt, advancing down the ladder and lifting the rungs in a hand-over-hand fashion until the ladder is vertical.

3 The two fire fighters stand on opposite sides of the ladder and pivot it into position as necessary.

4 The fire fighters face each other, one on each side of the ladder, and heel the ladder by each placing the toe of one boot against the opposite beams of the ladder.

5 If using an extension ladder, one fire fighter extends the fly section by pulling the halyard smoothly with a hand-over-hand motion until the tip is at the desired height and the ladder locks (dogs) are locked. The other fire fighter stabilizes the ladder by holding the outside of the base section's beams, so that if the fly comes down suddenly it will not strike the fire fighter's hands.

6 The fire fighter facing the structure places one foot against one beam of the ladder, and then both fire fighters lean the ladder into place. The halyard is tied. The fire fighters check the ladder for a 75-degree climbing angle and for stability at the bottom and at top of the ladder.

© LiquidLibrary

of one boot against the opposite beams of the ladder. (**STEP 4**)

8 If using an extension ladder, one fire fighter extends the fly section by pulling the halyard smoothly with a hand-over-hand motion until the tip is at the desired height and the ladder locks (dogs) are locked. The other fire fighter stabilizes the ladder by holding the outside of the base section's beams, so that if the fly comes down suddenly it will not strike the fire fighter's hands. (**STEP 5**)

9 The fire fighter facing the structure places one foot against one beam of the ladder, and then both fire fighters lean the ladder into place.

10 The halyard is tied.

11 The fire fighters check the ladder for a 75-degree climbing angle and for stability at the bottom and top of the ladder. (**STEP 6**)

Three- and Four–Fire Fighter Rung Raises

The three- and four–fire fighter rung raises are used for very heavy ladders. The basic steps in these raises are similar to the two–fire fighter rung raise. In a three–fire fighter rung raise, the third fire fighter assists with the hand-over-hand raising of the ladder. When four fire fighters are raising the ladder, two anchor the butt of the ladder while the other two raise the ladder. Each of the two fire fighters at the butt places his or her inside foot on the bottom rung of the ladder and bends over, grasping a rung in front of him or her.

To perform the three–fire fighter rung raise, follow the steps in **SKILL DRILL 13-16**:

1 The raise begins from a shoulder carry or suitcase carry, with one fire fighter near the butt of the ladder, one fire fighter in the middle, and one fire fighter near the tip.

2 The fire fighter at the butt of the ladder places the butt of the lower beam on the ground, while the fire fighter at the tip holds that end. The fire fighter in the middle moves to the tip. The fire fighters at the tip rotate the ladder so that both butts are in contact with the ground.

3 The fire fighter at the butt of the ladder stands on the bottom rung, grasps a higher rung with both hands, crouches down, and leans backward to heel the ladder. (**STEP 1**)

4 The fire fighters at the tip of the ladder check for overhead hazards, then begin to walk toward the butt, advancing down the ladder and lifting the rungs in a hand-over-hand fashion until the ladder is vertical. (**STEP 2**)

5 All three fire fighters pivot the ladder into position as necessary. (**STEP 3**)

6 Two fire fighters face each other, one on each side of the ladder, grasp the outsides of the beams, and heel the ladder by each placing the toe of one boot against the opposite beams of the ladder. (**STEP 4**)

7 The third fire fighter extends the fly section by pulling the halyard smoothly with a hand-over-hand motion until the tip is at the desired height and the dogs are locked. (**STEP 5**)

8 One fire fighter heels the ladder while the other two lean the ladder into place.

9 The halyard is tied.

10 The fire fighters check the ladder for a 75-degree climbing angle and for security at the butt end and at the tip of the ladder. (**STEP 6**)

To perform the four–fire fighter rung raise, follow the steps in **SKILL DRILL 13-17**:

1 The raise begins with a four–fire fighter flat carry. Two fire fighters are at the butt of the ladder, and two fire fighters are at the tip. (**STEP 1**)

2 The two fire fighters at the butt of the ladder place the butt end of the ladder on the ground, while the two fire fighters at the tip hold that end of the ladder. (**STEP 2**)

3 The two fire fighters at the butt of the ladder stand side by side, facing the ladder. Each fire fighter places the inside foot on the bottom rung and the other foot on the ground outside the beam. Both crouch down, grab a rung and the beam, and lean backward. (**STEP 3**)

4 The two fire fighters at the tip of the ladder check for overhead hazards and then begin to walk toward the butt of the ladder, advancing down the rungs in a hand-over-hand fashion until the ladder is vertical. (**STEP 4**)

5 The fire fighters pivot the ladder into position, as necessary. (**STEP 5**)

6 Two fire fighters heel the ladder by placing a boot against each beam. Each fire fighter places the toe of one boot against one of the beams. The third fire fighter stabilizes the ladder by holding it on the outside of the rails. (**STEP 6**)

7 The fourth fire fighter extends the fly section by pulling the halyard smoothly with a hand-over-hand motion until the tip reaches the desired height and the dogs are locked. (**STEP 7**)

8 The two fire fighters facing the structure each place one foot against one beam of the ladder while the other two fire fighters lower the ladder into place.

9 The halyard is tied.

10 The fire fighters check the ladder for a 75-degree climbing angle and for security at the butt end and at the tip of the ladder. (**STEP 8**)

Fire Fighter Safety Tips

© Photos.com

When raising the fly section of an extension ladder, do not wrap the halyard around your hand. If the ladder falls or if the fly section retracts unexpectedly, your hand could be caught in the halyard and you could be seriously injured. Also, do not place your foot under the fly sections. If the rope slips, your foot will be crushed.

Fly Section Orientation

When raising extension ladders, fire fighters must know whether the fly section should be placed toward the building (fly in) or away from the building (fly out). The ladder manufacturer and department suggested operating guidelines (SOGs)/standard operating procedures (SOPs) will specify whether the fly should be facing in or out. In general, manufacturers of fibreglass and metal ladders recommend that the fly sections be placed away (fly out) from the structure. In contrast, wooden fire-service ladders are often designed to be used with the fly in.

SKILL DRILL 13-16 | Three–Fire Fighter Rung Raise
(Fire Fighter I, NFPA 5.3.6)

1 The fire fighter at the butt of the ladder places the butt of the lower beam on the ground, while the fire fighter at the tip holds that end. The fire fighter in the middle moves to the tip. The fire fighters at the tip rotate the ladder so that both butts are in contact with the ground. The fire fighter at the butt of the ladder stands on the bottom rung, grasps a higher rung with both hands, crouches down, and leans backward to heel the ladder.

2 The fire fighters at the tip of the ladder check for overhead hazards, then begin to walk toward the butt, advancing down the ladder and lifting the rungs in a hand-over-hand fashion until the ladder is vertical.

3 All three fire fighters pivot the ladder into position as necessary.

4 Two fire fighters face each other, one on each side of the ladder, grasp the outsides of the beams, and heel the ladder by each placing the toe of one boot against the opposite beams of the ladder.

5 The third fire fighter extends the fly section by pulling the halyard smoothly with a hand-over-hand motion until the tip is at the desired height and the dogs are locked.

6 One fire fighter heels the ladder while the other two lean the ladder into place. The halyard is tied. The fire fighters check the ladder for a 75-degree climbing angle and for security at the butt end and at the tip of the ladder.

© LiquidLibrary

■ Securing the Ladder

Several approaches may be used to prevent a ladder from moving once it is in place. One option is to have a fire fighter stand between the ladder and the structure, grasp the beams, and lean back to pull the ladder into the structure—a process called heeling the ladder **FIGURE 13-28**. A fire fighter on the outside of the ladder, facing the structure, can also heel

the ladder by placing a boot against the beam of the ladder **FIGURE 13-29**.

Each of these methods has its own advantages and disadvantages:

- Heeling from under the ladder: The ladder protects the heeling fire fighter from debris that may fall from the structure, but this approach can cause an eye injury

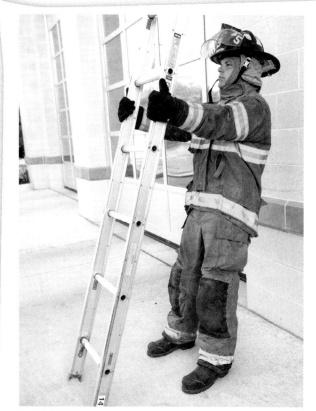

FIGURE 13-28 The fire fighter's weight can pull the ladder against the structure to secure the ladder.

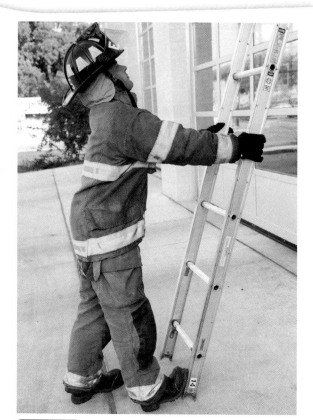

FIGURE 13-29 A fire fighter can push against the ladder and place his or her foot on the butt end of the ladder to heel it.

if the fire fighter looks up during his or her partner's climb (owing to falling debris).

- Heeling from outside the ladder: This technique allows the heeling fire fighter to survey the building and warn the climber of possible hazards and avoid falling debris.

A rope, a rope–hose tool, or webbing also can be used to secure a ladder in place. The lower part of the ladder can be tied to any solid object to keep the base from kicking out; in fact, the base should always be secured if the ladder is used at a low angle. The tip of the ladder can be tied to a secure object near the top to keep it from pulling away from the building. The best method is to secure both the tip and the base.

■ Climbing the Ladder

Climbing a ladder on the fire or emergency scene should be done in a deliberate and controlled manner. Always make sure the ladder is secure (tied or heeled). Before climbing an extension ladder, verify that the dogs are locked and the halyard is secured.

The proper climbing angle should be checked as well. See the section "Placing a Ladder" in this chapter for more information. A raised ladder should be positioned at

an angle of approximately 75 degrees to provide the best combination of strength, stability, and vertical reach. This angle also creates a comfortable climbing angle. When a fire fighter is standing on a rung, the rung at shoulder height will be approximately one arm's length away. The ratio of the ladder height (vertical reach) to the distance from the structure should be 4:1. For example, if the vertical reach of a ladder is 6.1 m (20 ft.), the butt of the ladder should be placed 1.5 m (5 ft.) out from the wall. Some ladder beams include level guide stickers (placed there by the ladder's manufacturer), so that the bottom line of the indicator will be parallel to the ground when the ladder is positioned properly. Standing at the base of the ladder and extending the arms straight out is another way to check the climbing angle **FIGURE 13-30**. The hands should comfortably reach the beams or rungs if the angle is appropriate. Finally, the angle can be measured by ensuring that the butt of the ladder is one-fourth of the working height out from the base of the structure.

Keep any bouncing and shifting to a minimum while climbing the ladder. Your eyes should be focused forward, with only occasional glances upward. This approach will prevent debris such as falling glass from injuring your face or eyes. Lower the protective face shield for additional protection.

SKILL DRILL 13-17 | Four–Fire Fighter Rung Raise
(Fire Fighter I, NFPA 5.3.6)

1 The raise begins with a four–fire fighter flat carry. Two fire fighters are at the butt of the ladder, and two fire fighters are at the tip.

2 The two fire fighters at the butt of the ladder place the butt end of the ladder on the ground, while the two fire fighters at the tip hold that end of the ladder.

3 The two fire fighters at the butt of the ladder stand side-by-side, facing the ladder. Each fire fighter places the inside foot on the bottom rung and the other foot on the ground outside the beam. Both crouch down, grab a rung and the beam, and lean backward.

4 The two fire fighters at the tip of the ladder check for overhead hazards and then begin to walk toward the butt of the ladder, advancing down the rungs in a hand-over-hand fashion until the ladder is vertical.

5 The fire fighters pivot the ladder into position, as necessary.

6 Two fire fighters heel the ladder by placing a boot against each beam. Each fire fighter places the toe of one boot against one of the beams. The third fire fighter stabilizes the ladder by holding it on the outside of the rails.

7 The fourth fire fighter extends the fly section by pulling the halyard smoothly with a hand-over-hand motion until the tip reaches the desired height and the dogs are locked.

8 The two fire fighters facing the structure each place one foot against one beam of the ladder while the other two fire fighters lower the ladder into place. The halyard is tied. The fire fighters check the ladder for a 75-degree climbing angle and for security at the butt end and at the tip of the ladder.

Use a hand-over-hand motion on the rungs of the ladder, or slide both hands along the underside of the beams while climbing. Sliding the hands along the beams is more secure because it maintains three contact points with the ladder at all times (both hands and one foot).

If tools must be moved up or down, it is better to hoist them by rope than to carry them up a ladder. Carrying tools on a ladder reduces the fire fighter's grip and increases the potential for injury if a tool slips or falls. Sometimes you will need to carry a tool up a ladder, however, so you should practise doing it safely. To climb a ladder while holding a tool, hold the tool against one beam with one hand and maintain contact with the opposite beam with the other hand. Tools such as pike poles can be hooked onto the ladder and moved up every few rungs. One method of climbing a ladder while carrying a tool is described in **SKILL DRILL 13-18** :

1 Check the ladder to ensure that it is at a safe angle to climb and is secure.

2 Grasp the tool comfortably and securely in one hand, and hold that hand and the tool against the beam of the ladder. (**STEP 1**)

3 Wrap the other hand around the opposite beam and begin climbing. (**STEP 2**)

4 Climb smoothly and safely by maintaining contact between your free hand and the beam and by sliding the tool along the opposite beam. (**STEP 3**)

5 Be sure not to overload the ladder. No more than two fire fighters should be on a ladder at the same time. As mentioned earlier, a properly placed ladder should be able to support two rescuers (with their protective clothing and equipment) and one victim.

■ Dismounting the Ladder

When a ladder is used to reach a roof or a window entry, the fire fighter will have to dismount the ladder. Fire fighters can minimize their risk of slipping and falling by making sure that the surface is stable—that is, testing its stability with a tool—before dismounting; for example, fire fighters who dismount from a ladder onto a roof typically sound the roof with an axe before stepping onto the surface **FIGURE 13-31** .

FIGURE 13-30 The climbing angle can be checked by standing on the bottom rung and holding the arms straight out.

FIGURE 13-31 Check the stability of the roof before dismounting from the ladder.

SKILL DRILL 13-18 Climbing a Ladder While Carrying a Tool
(Fire Fighter I, NFPA 5.3.12)

1 Check the ladder to ensure that it is at a safe angle to climb and is secure. Grasp the tool comfortably and securely in one hand, and hold that hand and the tool against the beam of the ladder.

2 Wrap the other hand around the opposite beam and begin climbing.

3 Climb smoothly and safely by maintaining contact between your free hand and the beam and by sliding the tool along the opposite beam.

© LiquidLibrary

Try to maintain contact with the ladder at three points when dismounting; for example, a fire fighter who steps onto a roof should keep two hands and one foot on the ladder while checking the footing. This consideration is particularly important if the surface slopes or is covered with rain, snow, or ice. Do not shift your weight onto the roof until you have tested the footing.

It is also important to check the footing when entering a window from a ladder positioned to the side of the window. Before stepping into the window, confirm that the interior surface is structurally sound and offers secure footing. The three-points-of-contact rule also applies in this situation.

Under heavy fire and smoke conditions, fire fighters sometimes dismount from a ladder by climbing over the tip and sliding over the windowsill into the building. Sound the floor inside the window before entering to confirm that it is solid and stable. To remount a ladder under these conditions, back out the window feet first and rest your abdomen on the sill until you can feel the ladder under your feet. Under better conditions, sit on the windowsill with legs out and roll onto the ladder.

■ Working from a Ladder

Fire fighters must often work from a ladder. To avoid falling while doing so, the fire fighter must be secured to the ladder. Fire fighters use two different methods to secure themselves to a ladder.

The first method involves the use of a **ladder belt**—a piece of equipment specifically designed to secure a fire fighter to a ladder or elevated surface **FIGURE 13-32**. Fire fighters should use only ladder belts designed and certified under the requirements specified by NFPA 1983, *Standard on Life Safety Rope and Equipment for Emergency Services*. Utility belts designed to carry tools should never be used as ladder belts.

The second method used to secure a fire fighter to a ladder is to apply a leg lock. The leg lock is simple and secure, and requires no special equipment.

To apply a leg lock, follow the steps in **SKILL DRILL 13-19**:

1 Climb to the desired work height and step up to one more rung. (**STEP 1**)

FIGURE 13-32 A ladder belt has a large hook that is designed to secure a fire fighter to a ladder.

2 Note the side of the ladder where the work will be performed. Extend your leg between the rungs on the side opposite the side you will be working. (**STEP 2**)

3 Once your leg is between the rungs, bend your knee and bring your foot back under the rung and through to the climbing side of the ladder. (**STEP 3**)

4 Secure your foot against the next lower rung or the beam of the ladder. Use your thigh for support and step down one rung with the opposite foot. (**STEP 4**)

5 The use of a leg lock enables you to have two hands free for a variety of tasks, including placing a roof ladder or controlling a hose stream. (**STEP 5**)

Placing a Roof Ladder

Several methods can be used to properly position a roof ladder on a sloping roof. One commonly used method is described in **SKILL DRILL 13-20**:

1 Carry the roof ladder to the base of the climbing ladder that is already in place to provide access to the roofline. (**STEP 1**)

2 Place the butt end of the roof ladder on the ground, and rotate the hooks of the roof ladder to the open position. (**STEP 2**)

3 Use a one–fire fighter beam or rung raise to lean the roof ladder against one beam of the other ladder with the hooks oriented outward. (**STEP 3**)

4 Climb the lower ladder until you reach the midpoint of the roof ladder, and then slip one shoulder between two rungs of the roof ladder and shoulder the ladder. (**STEP 4**)

5 Climb to the roofline of the structure, carrying the roof ladder on one shoulder.

6 Secure yourself to the ladder by using a ladder belt or applying a ladder leg lock. (**STEP 5**)

7 Place the roof ladder on the roof surface with hooks down. Push the ladder up toward the peak of the roof with a hand-over-hand motion. (**STEP 6**)

8 Once the hooks have passed the peak, pull back on the roof ladder to set the hooks and check that they are secure. (**STEP 7**)

9 To remove a roof ladder from the roof, reverse the process just described. After releasing the hooks from the peak, it may be necessary to turn the ladder on one of its beams or turn it so the hooks are pointing up to slide the ladder down the roof without catching the hooks on the roofing material. Carrying a roof ladder in this manner requires strength and practice.

SKILL DRILL 13-19 Using a Leg Lock to Work from a Ladder
(Fire Fighter I, NFPA 5.3.6)

1 Climb to the desired work height and step up to one more rung.

2 Note the side of the ladder where the work will be performed. Extend your leg between the rungs on the side opposite the side you will be working.

3 Once your leg is between the rungs, bend your knee and bring your foot back under the rung and through to the climbing side of the ladder.

4 Secure your foot against the next lower rung or the beam of the ladder. Use your thigh for support and step down one rung with the opposite foot.

5 The use of a leg lock enables you to have two hands free for a variety of tasks.

SKILL DRILL 13-20 Deploying a Roof Ladder
(Fire Fighter I, NFPA 5.3.12)

1 Carry the roof ladder to the base of the climbing ladder that is already in place to provide access to the roofline.

2 Place the butt end of the roof ladder on the ground, and rotate the hooks of the roof ladder to the open position.

3 Use a one–fire fighter beam or rung raise to lean the roof ladder against one beam of the other ladder with the hooks oriented outward.

4 Climb the lower ladder until you reach the midpoint of the roof ladder, and then slip one shoulder between two rungs of the roof ladder and shoulder the ladder.

5 Climb to the roofline of the structure, carrying the roof ladder on one shoulder. Secure yourself to the ladder by using a ladder belt or applying a ladder leg lock.

6 Place the roof ladder on the roof surface with hooks down. Push the ladder up toward the peak of the roof with a hand-over-hand motion.

7 Once the hooks have passed the peak, pull back on the roof ladder to set the hooks and check that they are secure.

When deploying a roof ladder, make sure that the hooks are pointed away from you as you raise and lower the ladder on the roof. Should the roof ladder slip or fall from the roof, the hooks could strike you if they are positioned toward you.

Inspect a Chimney

Each year, fire departments receive many calls for chimney fires and other chimney-related problems. Many times it is necessary to inspect the chimney to determine whether a problem is truly present. To access the chimney, it is necessary to safely ladder the building and proceed to the roof. The steps for inspecting a chimney are described in **SKILL DRILL 13-21**:

1. Make sure you are equipped with a complete set of PPE, including SCBA, and are operating under the direction of your officer.
2. Remove the proper-size extension ladder and the roof ladder from the apparatus.
3. Carry the ladders to the building and raise the extension ladder so that five rungs project above the roofline.
4. Confirm that the roof is safe to climb on. If it is not, use an aerial ladder.
5. When possible, place the roof ladder on the upwind side of the chimney.
6. Working with a partner, climb to the chimney while carrying the appropriate equipment.
7. Inspect the chimney, and report your findings to the IC.
8. Remove the tools and ladder when the task is complete.

As you work on learning the skills needed to perform vital ladder operations, do not lose sight of the dangers inherent in working with ladders. As you practise the skills described in this chapter, follow these safety rules:

- Wear appropriate PPE. This equipment should include SCBA when you are operating on the fire ground.
- Choose the proper ladder for the job.
- Choose the proper number of people for the job. Move smoothly, safely, and efficiently, keeping the ladder under your control at all times.
- Lift with your legs, keeping your back straight.
- Use proper hand and foot placement during carries, raises, climbing, and lowering of ladders.
- Look overhead for wires and obstructions before raising or lowering a ladder. Communicate the results of this check by stating, "Clear overhead," to your partner.
- Allow the heel person to take charge of the processes of raising and lowering the ladder.
- After extending an extension ladder, check the pawls (ladder locks) to confirm that they are set and make sure the halyard is tied off.
- Check the ladder for the proper 75-degree angle.
- Ensure that the wall or roof will support the ladder before climbing it.
- Remain aware of hazards to yourself and to others.
- Maintain hand contact with the ladder. Watch your hands when climbing up, and watch your legs when climbing down.
- Use an appropriate leg lock or ladder belt when working on a ladder.

Wrap-Up

Chief Concepts

- The primary function of ladders is to provide safe access to and egress from otherwise inaccessible areas.
- Ladders can be used for several secondary functions, including channelling debris, serving as a lift point, and holding other firefighting equipment.
- The two main structural components of the ladder are the beam and the rung. The beams support the rungs and carry the load of the fire fighter from the rungs down to the ground.
- There are three basic types of ladder beam construction: trussed beam, I-beam, and solid beam. Trussed beams are usually constructed of aluminum or wood. I-beam ladders are usually made from fibreglass.
- Many wooden ladders have solid beams. Rectangular aluminum beams, which are usually hollow or C-shaped, are also classified as solid beam ladders.
- The rung is a crosspiece that spans the two beams of the ladder. Most portable ladders in fire departments have aluminum rungs, but wooden rungs are also in use.
- Additional parts of a typical portable ladder include the following components:
  - Rail—The top or bottom section of a trussed beam or the top and bottom surfaces of an I-beam
  - Truss block—A piece or assembly that connects the two rails of a trussed beam
  - Tie rod—A metal bar that connects one beam to the other
  - Tip—The very top of the ladder
  - Butt—The end of the ladder
  - Butt spurs—Metal spikes designed to help stabilize the ladder in place
  - Butt plate—An alternative to butt spurs; a swivelling plate that incorporates both a spur and a cleat
  - Protection plates—Reinforcing plates placed at chafing or contact points
- An extension ladder is an assembly of two or more ladder sections that fit together and can be extended or retracted to adjust the length of the ladder. Additional parts of an extension ladder include the following components:
  - Bed section—The base of the extension ladder
  - Fly section—The section of the ladder that is raised from the bed section
  - Guides—Strips of metal or wood or channels that guide the fly section during extension
  - Halyard—The rope or cable used to extend the fly sections
  - Pawls—Mechanical locking devices that secure the fly sections
  - Pulley—A small wheel used to change the direction of the halyard pull

- Stops—Pieces of wood or metal that prevent the fly sections from becoming overextended
- Staypoles—Long metal poles used to stabilize the extension ladder
- There are two general categories of ladders: aerial and portable. Aerial ladders are permanently mounted and operated from a fire apparatus. Portable ladders are carried on fire apparatus and removed from the apparatus to be used in other locations.
- Portable ladders include general-purpose ladders such as straight or extension ladders and specialized ladders such as attic or roof ladders.
- Fire apparatus are required to carry at least one roof ladder, one extension ladder, and one folding ladder.
- A straight ladder is a single-section, fixed-length ladder.
- A roof ladder is a straight ladder with retractable hooks at one end.
- An extension ladder is an adjustable-length ladder with multiple sections.
- Bangor ladders are extension ladders with staypoles for added stability.
- A combination ladder can be converted from a straight ladder to a stepladder configuration, or from an extension ladder to a stepladder configuration.
- A folding ladder is a narrow, collapsible ladder designed to allow access to attic scuttle areas and confined areas.
- A Fresno ladder is a narrow, two-section extension ladder designed to provide attic access.
- Portable ladders must be regularly inspected, maintained, and service tested following NFPA 1932.
- A ladder should be inspected visually after every use. If the inspection reveals any deficiencies, the ladder must be removed from service until repairs are made.
- All fire fighters should be able to perform routine ladder maintenance, including cleaning the ladders to remove any road grime and dirt. Ladders should be cleaned after every use.
- Service testing of portable ladders must follow NFPA 1932.
- Safety precautions should be followed from the time a ladder is removed from the apparatus until it is returned to the apparatus. Full turnout gear and PPE are essential for working with portable and aerial ladders. Additional general safety rules include the following:
  - Lift with your legs, keeping your back straight.
  - Look overhead for wires and obstructions before raising or lowering a ladder.
  - After extending an extension ladder, check the pawls (ladder locks) to confirm that they are set and make sure the halyard is tied off.
  - Check the ladder for the proper 75-degree angle. Ensure that the wall or roof will support the ladder before climbing it.

- When positioning a ladder for rescue, place the tip of the ladder at or slightly below the windowsill. This provides a direct pathway to enter and exit the window.
- Communication is key to coordinating efforts when working with ladders. One leader should be designated, and his or her commands should be followed during carries and raises.

Hot Terms

Aerial ladder A self-supporting, turntable-mounted, power-operated ladder of two or more sections permanently attached to a self-propelled automotive fire apparatus and designed to provide a continuous egress route from an elevated position to the ground. (NFPA 1901)

Bangor ladder A ladder equipped with tormentor poles or staypoles that stabilize the ladder during raising and lowering operations.

Base (bed) section The lowest or widest section of an extension ladder. (NFPA 1931)

Beam The main structural side of a ground ladder. (NFPA 1931)

Butt The end of the beam that is placed on the ground, or other lower support surface, when ground ladders are in the raised position. (NFPA 1931)

Butt plate An alternative to a simple butt spur; a swivelling plate with both a spur and a cleat or pad that is attached to the butt of the ladder.

Butt spurs That component of ground ladder support that remains in contact with the lower support surface to reduce slippage. (NFPA 1931)

Combination ladder A ground ladder that is capable of being used both as a stepladder and as a single or extension ladder. (NFPA 1931)

Egress To go or come out; to exit from an area or a building.

Extension ladder A non-self-supporting ground ladder that consists of two or more sections travelling in guides, brackets, or the equivalent arranged so as to allow length adjustment. (NFPA 1931)

Fly section Any section of an aerial telescoping device beyond the base section. (NFPA 1901)

Folding ladder A single-section ladder with rungs that can be folded or moved to allow the beams to be brought into a position touching or nearly touching each other. (NFPA 1931)

Fresno ladder A narrow, two-section extension ladder that has no halyard. Because of its limited length, it can be extended manually.

Grade A measurement of the angle used in road design and expressed as a percentage of elevation change over distance. (NFPA 1901)

Guides Strips of metal or wood that serve to guide a fly section during extension. Channels or slots in the bed or fly section may also serve as guides.

Halyard Rope used on extension ladders for the purpose of raising a fly section(s). (NFPA 1931)

Heat sensor label A label that changes colour at a preset temperature to indicate a specific heat exposure. (NFPA 1931)

I-beam A ladder beam constructed of one continuous piece of I-shaped metal or fibreglass to which the rungs are attached.

Ladder belt A compliant equipment item that is intended for use as a positioning device for a person on a ladder. (NFPA 1983)

Ladder gin An A-shaped structure formed with two ladder sections. It can be used as a makeshift lift when raising a trapped person. One form of the device is called an A-frame hoist.

Pawls Devices attached to a fly section(s) to engage ladder rungs near the beams of the section below for the purpose of anchoring the fly section(s). (NFPA 1931)

Portable ladder A ladder carried on fire apparatus but designed to be removed from the apparatus and deployed by fire fighters where needed.

Protection plates Reinforcing material placed on a ladder at chafing and contact points to prevent damage from friction and contact with other surfaces.

Pulley A device with a free-turning, grooved metal wheel (sheave) used to reduce rope friction. Side plates are available for a carabiner to be attached. (NFPA 1670)

Rail The top or bottom piece of a trussed beam assembly used in the construction of a trussed ladder. Also, the top and bottom surfaces of an I-beam ladder. Each beam has two rails.

Roof hooks The spring-loaded, retractable, curved metal pieces that allow the tip of a roof ladder to be secured to the peak of a pitched roof. The hooks fold outward from each beam at the top of a roof ladder.

Roof ladder A single ladder equipped with hooks at the top end of the ladder. (NFPA 1931)

Rung The ladder crosspieces on which a person steps while ascending or descending. (NFPA 1931)

Solid beam A ladder beam constructed of a solid rectangular piece of material (typically wood), to which the ladder rungs are attached.

Staypoles Poles attached to each beam of the base section of extension ladders, which assist in raising the ladder and help provide stability of the raised ladder. (NFPA 1931)

Stop A piece of material that prevents the fly sections of a ladder from becoming overextended, leading to collapse of the ladder.

Tie rod A metal rod that runs from one beam of the ladder to the other to keep the beams from separating. Tie rods are typically found in wood ladders.

Tip The very top of the ladder.

Truss block A piece of wood or metal that ties the two rails of a trussed beam ladder together and serves as the attachment point for the rungs.

Trussed beam A ladder beam constructed of top and bottom rails joined by truss blocks that tie the rails together and support the rungs.

FIRE FIGHTER *in action*

You just got back to the station from the largest fire of your career. The site of the fire was a large, old, two-and-a-half-storey wood-frame house that had been converted into apartments. Your engine was the first to arrive and found people trapped on the second floor with heavy smoke conditions throughout. You pulled the 44.5-millimetre (1.75-in.) hose, and your captain fell in behind you. You made the stairs and confined the fire while the rescue company rescued the trapped occupants. One victim was rescued down the 10.7-m (35-ft.) extension ladder from your engine.

You can still feel the aftereffects of the adrenaline as you look at the dirt and grime covering virtually every piece of equipment on the engine. You know it will take some time to get everything cleaned up and put back into a ready state. With the help of your crewmate, you start by pulling off the extension ladder and go to work cleaning.

1. When possible, which is the preferred method of cleaning the ladder?
   A. Stiff-bristle brush, mild detergent, and water
   B. Soft-bristle brush and water
   C. Stiff-bristle brush and water
   D. Soft-bristle brush, mild solvent, and water

2. How should wire–rope halyard extensions be checked?
   A. With the fly section fully extended
   B. With the fly section partially extended
   C. With the fly section fully bedded
   D. The position of the fly section does not matter

3. In addition to after each use, when should a ladder normally be visually inspected?
   A. Daily
   B. Weekly

   C. Monthly
   D. Yearly

4. What are the spring-loaded, retractable, curved metal pieces that allow the tip of a roof ladder to be secured to the peak of a pitched roof?
   A. Halyard
   B. Tie rod
   C. Pawls
   D. Roof hooks

5. What are the two main structural components that run the entire length of most ladders or ladder sections?
   A. Rails
   B. Truss blocks
   C. Beams
   D. Guides

It is the last Saturday morning of the month, and you are performing your monthly visual inspection. You place one ladder at a time on the engine-room floor and get the spray nozzle. You collect your soft-bristle brush and go to work carefully cleaning each component. Once the ladder is clean, you dry off the moisture and then carefully inspect each part of the ladder, ensuring nothing is overlooked. You then lubricate the moving parts as recommended by the manufacturer, before placing the ladder back on the apparatus.

While you continue to work, a family has stopped by the station. Fire Fighter Smith, who is filling in from one of the busier ladder companies, is giving them a tour. The group briefly stops to observe you working diligently to ensure that your ladders operate like a well-oiled machine. You give them a brief overview of what you are doing, and the visitors tell you that they are surprised at the level of care that goes into cleaning. As they move on, you overhear Smith comment to the family that you work at a slow station and that real fire fighters do not have time to spend their days cleaning ladders.

1. Would you address the comment with the guests?
2. How would you address the issue with Smith?
3. Is this an issue you would address with your captain?
4. If you were transferred to a busy station, would you continue to clean, inspect, and maintain the ladders the same as you do at the slow station?

© Glen E. Ellman

Search and Rescue

Fire Fighter I

Knowledge Objectives

After studying this chapter, you will be able to:

- Describe the mission of search operations. (NFPA 5.3.9 , p 396)
- Describe the mission of rescue operations. (NFPA 5.3.9 , p 396)
- Explain how search and rescue operations are coordinated with other fire-suppression operations. (NFPA 5.3.9 , p 396–399)
- Identify the factors to evaluate during a search and rescue size-up. (NFPA 5.3.9 , p 397–399)
- Describe how to perform a risk–benefit analysis. (NFPA 5.3.9 , p 397)
- Describe the factors that determine the level of risk faced by occupants. (NFPA 5.3.9.A , p 397–398)
- Explain how search operations are coordinated. (NFPA 5.3.9 , p 399)
- List the priorities of search operations. (NFPA 5.3.9 , p 399)
- Describe the objectives of a primary search. (NFPA 5.3.9.A , p 399–400)
- Describe the search patterns commonly used in search operations. (NFPA 5.3.9.A , p 401–402)
- Explain how thermal imaging devices are used during search operations. (NFPA 5.3.9.A , p 402)
- Describe how and when search ropes are utilized during search operations. (NFPA 5.3.9.A , p 403)
- Describe the role of a fire officer during search operations. (NFPA 5.3.9.A , p 403)
- Explain how a vent–entry–search is commonly performed. (NFPA 5.3.9.A , p 403)
- Describe the objectives of a secondary search. (NFPA 5.3.9.A , p 404–405)
- Explain how fire fighters maintain safety through risk management. (NFPA 5.3.9 , p 405–406)
- List the tools and equipment used in search and rescue operations. (NFPA 5.3.9.A , p 406)
- Describe the methods fire fighters use to determine whether an area is tenable. (NFPA 5.3.9.A , p 406–407)
- List the major types of rescue. (NFPA 5.3.9.A , p 407)

- Describe the concept of sheltering-in-place. (NFPA 5.3.9.A , p 407–408)
- Describe how to assist a victim to an exit. (NFPA 5.3.9.A , p 408)
- List the common types of simple victim carries performed during rescue operations. (NFPA 5.3.9.A , p 408)
- List the five emergency drags performed during rescue operations. (NFPA 5.3.9.A , p 412–413)
- Describe the conditions that may require a ground ladder rescue. (NFPA 5.3.9.A , p 420)

Skills Objectives

After studying this chapter, you will be able to:

- Conduct a primary search. (NFPA 5.3.9B , p 403–404)
- Conduct a secondary search. (NFPA 5.3.9B , p 405)
- Perform a one-person walking assist. (NFPA 5.3.9B , p 408–409)
- Perform a two-person walking assist. (NFPA 5.3.9B , p 408, 410)
- Perform a two-person extremity carry. (NFPA 5.3.9B , p 408, 409, 412)
- Perform a two-person seat carry. (NFPA 5.3.9B , p 409, 412)
- Perform a two-person chair carry. (NFPA 5.3.9B , p 409–410, 413)
- Perform a cradle-in-arms carry. (NFPA 5.3.9B , p 411, 413)
- Perform a clothes drag. (NFPA 5.3.9B , p 412)
- Perform a blanket drag. (NFPA 5.3.9B , p 412–414)
- Perform a standing drag. (NFPA 5.3.9B , p 414, 416)
- Perform a webbing sling drag. (NFPA 5.3.9B , p 414, 417)
- Perform a fire fighter drag. (NFPA 5.3.9B , p 414, 417)
- Perform a one-person emergency drag from a vehicle. (NFPA 5.3.9B , p 415, 418)
- Perform a long-backboard rescue. (NFPA 5.3.9B , p 415–420)
- Rescue a conscious victim from a window. (NFPA 5.3.9B , p 421–422)
- Rescue an unconscious victim from a window. (NFPA 5.3.9B , p 421–423)
- Rescue an unconscious child or small adult from a window. (NFPA 5.3.9B , p 421, 424)
- Rescue a large adult. (NFPA 5.3.9B , p 424–426)

© Photos.com

Fire Fighter II

Knowledge Objectives

There are no knowledge objectives for Fire Fighter II candidates. NFPA 1001 contains no Fire Fighter II Job Performance Requirements for this chapter.

Skills Objectives

There are no skills objectives for Fire Fighter II candidates. NFPA 1001 contains no Fire Fighter II Job Performance Requirements for this chapter.

Additional NFPA Standards

- NFPA 1500, *Standard on Fire Department Occupational Safety and Health Program*

You arrive on the scene of a fast-moving apartment fire. The structure is a three-storey, centre-hallway, ordinary-construction apartment building with fire on the second floor. The second- and third-floor hallways are smoke charged. Dispatch is reporting occupants trapped. You are assigned to conduct a primary search on the third floor while another crew begins the fire attack. This is a call that many fire fighters will rarely, if ever, experience in their careers.

1. How will you go about making the search?
2. What will you do if you find a victim in the first unit you search?
3. Would you take a hose line with you on the search?

Introduction

The mission of the fire department is to save lives and protect property. Saving lives is always the highest priority at a fire scene, which explains why search and rescue are such important operations. The first fire fighters to arrive must always consider the possibility that lives could be in danger and act accordingly.

Saving lives remains the highest priority until it is determined that every occupant who was in danger has been found and moved to a safe location, or until it is no longer possible to rescue anyone successfully. The first situation occurs when a thorough search is performed and confirms that there are no occupants remaining to be rescued. The second situation occurs when fire conditions make it unlikely or impossible that any occupants could still be alive to be rescued.

Search and rescue are almost always performed in tandem, yet they are actually separate actions: The purpose of **search** operations is to locate living victims. The purpose of **rescue** operations is to physically remove an occupant or victim from a dangerous environment. Rescue occurs when a fire fighter leads an occupant to an exit or carries an unconscious victim out of a burning building and down a ladder. Both are examples of rescues because both the occupant and the victim were physically removed from imminent danger through the actions of a fire fighter.

Any fire department company or unit can be assigned to search and rescue operations. Many fire departments routinely assign this responsibility to ladder (or truck) companies and rescue companies. All fire fighters must be trained and prepared to perform search and rescue functions.

Search and rescue operations must be conducted quickly and efficiently. A systematic approach will ensure that every occupant who can possibly be saved is successfully located and removed from danger. Fire fighters should practise the specific search and rescue procedures used by their departments.

Often, the first-arriving fire fighters will not know whether any occupants are inside a burning building or how many occupants might be inside. Fire fighters should never assume that a building is unoccupied. The only way to confirm that every occupant has safely evacuated a building is to conduct a thorough search. The decision to enter a burning building to search for living victims is one that is made by the incident commander (IC). Fire fighters should not risk their lives when the conditions are such that no victim or fire fighter could survive.

Search and Rescue Operations

■ Coordinating Search and Rescue Operations with Fire Suppression

Although search and rescue is a top priority, it is rarely the only action taken by first-arriving fire fighters. Instead, the IC and fire fighters must plan and coordinate all fire-suppression operations—from ventilation to fire attack—to support the search and rescue priority.

At times, fire fighters must take action to confine or control a fire before search and rescue operations can begin; for example, it might be necessary to position hose lines to keep the fire away from potential victims or to protect the entry and exit paths, so that the victims can be found and safely removed **FIGURE 14-1**. In some cases, the best way to save lives is to control the fire and eliminate the danger quickly.

FIGURE 14-1 To support search and rescue operations, fire fighters may first have to position hose lines to protect the means of egress.

Courtesy of Donald M. Colarusso, AllHandsFire.com

Other fire scene activities also must be coordinated with search and rescue. For example, forcible entry is sometimes needed to provide entrances and exits for search and rescue teams. Well-placed ventilation may reduce interior temperatures and improve visibility, thereby enabling search teams to locate victims more rapidly. Portable lighting can also provide valuable assistance to interior search crews.

Often, the search for potential victims provides valuable information about the location and extent of the fire within a building. In essence, the searchers act as a reconnaissance team to determine which areas are involved and where a fire might spread. They report this information to the IC, who develops the overall fire-suppression plan.

■ Search and Rescue Size-Up

The size-up process at every fire should include a specific evaluation of the critical factors for search and rescue—namely, the number of occupants in the building, their location, the degree of risk to their lives, and their ability to evacuate by themselves. Usually, this information is not immediately available, so fire fighters' actions must be based on a combination of observations and expectations. The type of occupancy; the size, construction, and condition of the building; the apparent smoke and fire conditions; and the time of day and the day of the week are all important observations that could indicate whether and how many people may need to be rescued.

A search and rescue plan can then be developed based on this information. Such a plan identifies the areas to be searched, the priorities for searching different areas, the number of search teams required, and any additional actions needed to support the search and rescue activities. One search team might be sufficient for a small building, whereas multiple teams might be needed to search a large building.

FIGURE 14-2 The IC must weigh the risk to the fire fighters against the possibility of saving any occupant inside before authorizing an interior search.

© Irene Teesalu/ShutterStock, Inc.

FIRE FIGHTER Tips

Although search teams both search for and rescue victims, the preferred term for these teams is simply *search team*. During a technical rescue incident, teams that search and rescue are called search and rescue teams.

Risk–Benefit Analysis

Plans for search and rescue must take into consideration both the risks and the benefits of the operation. In some situations, search and rescue operations must be limited or cannot be performed at all; for example, if a building demonstrates potential backdraft conditions, the risk to fire fighters might be too high and the possibility of saving any occupants inside the structure might be too low to justify sending a search and rescue team into the building **FIGURE 14-2**. A similar decision might be made if the fire occurs in an abandoned building or a lightweight construction building in danger of structural collapse. In other situations, it might not be possible to conduct an interior search until the fire has been extinguished.

Occupancy Factors

When a building is known or believed to be occupied, fire fighters should first rescue the occupants who are in the most immediate danger, followed by those who are in less danger. Search teams should be assigned on the basis of these priorities. Several factors determine the level of risk faced by the occupants of a burning building—namely, the location of the fire within the building, the direction of the fire spread, the volume and intensity of the fire, and smoke conditions in different areas of the structure.

In establishing search priorities, fire fighters should also consider where occupants are most likely to be located. Occupants who are close to the fire, above the fire, or in the path of fire spread are usually at greater risk than occupants who are located farther away from the fire. An occupant in the apartment where the fire started is probably in immediate danger, whereas the occupants of a lower floor may be relatively safe.

Building occupants who are at the windows or on balconies calling for help obviously realize that they are in danger and want to be rescued. There may be other occupants in the building who cannot be seen from the exterior, however—perhaps because they are asleep, unconscious, incapacitated, or trapped. A search must be conducted to locate these individuals before they can be rescued.

Occupants who are asleep are at greater risk than those who are awake. Young children and elderly people, who are more likely to need assistance to evacuate, are at greater risk than adolescents and adults. Individuals who are confined to a bed or wheelchair are at greater risk than occupants who can move freely. Consequently, more fire fighters will be needed for search and rescue operations in a nursing home than in an office building, where the occupants can usually make their way out unassisted.

A fire in a single-family dwelling generally places the members of one family at risk, whereas a fire in a large apartment building may endanger several families. Generally, the life risk in residential occupancies is higher at night and on weekends, when more people are at home. Risk is greatest late at night, when the occupants are probably asleep. A fire that occurs in a residential occupancy on a weekday afternoon would present a significantly different rescue problem from a fire in the same location at 2:00 a.m.

An office building that is fully occupied on a weekday afternoon is typically unoccupied at midnight. A warehouse fire probably presents a direct risk to few occupants. Conversely, a nightclub fire on a Friday or Saturday night could endanger hundreds of lives. The risk posed to victims may also increase depending on the building construction: occupants of an unprotected, wood-frame building are in greater danger than those in a building with fire-resistive construction and automatic sprinklers.

Fire Fighter Safety Tips

© Photos.com

Search and rescue size-up considerations include the following factors:

- Occupancy
- Size of the building
- Construction of the building
- Time of day and day of week
- Number of occupants
- Degree of risk to the occupants presented by the fire
- Ability of occupants to exit on their own

Observations

Fire fighters should never automatically assume that a building is unoccupied. An observant fire fighter notices clues that indicate whether a building is occupied and how many occupants are likely to be present. The initial size-up at an incident can provide valuable information on the probability of finding occupants, the number of occupants, and the most likely location of any occupants.

For example, the first-responding unit at a residential fire might see two cars in the driveway and numerous toys on the front lawn. These signs indicate that the house is probably occupied by a family with children. In such a situation, the IC would likely assign several search teams and emphasize the importance of quickly searching every room where victims might be located. Conversely, an absence of cars, locked doors, and an overflowing mailbox would suggest that the house is vacant. If there is a chance that savable victims might be present, the IC might order a search, in case someone is inside, but the assignment might go to a single search team **FIGURE 14-3** .

Similar observations can be made for other structures. Although an empty parking lot does not guarantee that a building is empty, the presence of a car probably indicates that at least one maintenance worker or security guard is inside, but it might also mean that a car is simply parked in the lot. Boarded-up windows and doors on a building surrounded by a chain-link fence are indications that the building is probably unoccupied.

Occupant Information

Fire fighters must make sure that everyone is accounted for. An important question to ask is, "Are you sure that no one else was home?" The dispatcher may relay information on the status of the occupants. Neighbours are also another source of information about the status of the occupants.

It is often difficult to obtain accurate information from occupants who have just escaped from a burning building, especially if they believe that family members, close friends, or even a beloved pet might still be inside. Occupants may be too emotional to speak or even to think clearly. If someone says that he or she thinks occupants are still inside, fire fighters should obtain as much information as possible. Ask specific questions, such as "Who is still inside?" or "Where is your son's room located?" and "Where was he when you last saw him?" It is always easier to find someone if you know for whom you are looking and where to look. Many fire fighters can tell stories about searching for a missing baby, only to discover that "Baby" is the family dog.

A.

B.

FIGURE 14-3 Exterior observations can often provide a good indication of whether a building is occupied. **A.** Parked cars indicate the residence is occupied. **B.** Boarded-up windows indicate the residence is unoccupied.

Courtesy of Captain David Jackson, Saginaw Township Fire Department

FIRE FIGHTER Tips

Do not consider all of the information gathered from neighbours and bystanders to be accurate. In the stress of the moment, well-intentioned people may make incorrect statements. Do not rush into an empty burning house because of a bad tip from a bystander.

Building Size and Arrangement

The size and arrangement of the building are other important factors to consider in planning and conducting a search. A large building with many rooms must be searched in a systematic fashion. Access to an interior layout or floor plan is often helpful when planning and assigning teams to search a building. To ensure that each area is searched promptly and that no areas are omitted, specific areas must be assigned to different search teams in an organized manner. Assignments are often based on the stairway locations, corridor arrangements, and the apartment or room numbering system. Teams must report when they have completed searching each area.

Because it is difficult to determine this information in the midst of an emergency, fire fighters often conduct preincident surveys of buildings such as apartment complexes and offices. They assemble information about the building during the preincident survey and use it to prepare preincident plans, so they will be prepared when an emergency occurs.

Preincident plans can include a variety of valuable information:

- Corridor layouts
- Exit locations
- Stairway locations
- Apartment layouts
- Number of bedrooms in apartments
- Locations of handicapped residents' apartments
- Special function rooms or areas

Fire fighters should note how the floors of a building are numbered. Buildings constructed on a slope may appear to have a different number of levels when viewed from different sides; for example, a fire may appear to be on the third floor to a fire fighter standing in front of the building, while a fire fighter at the rear of the building might report that it is on the fifth floor. Without knowing how floors are numbered, fire fighters can be sent to work above a fire instead of below it.

■ Search Coordination

The overall plan for the incident must focus on the life-safety priority as long as search and rescue operations are still under way. As soon as all searches are complete, the priority can shift to controlling and extinguishing the fire.

The IC makes search assignments and designates a fire officer to be in charge of the search effort (search officer). The search officer keeps the IC informed of the status and results of the search effort. As the search team completes its search of each area, the search officer must notify the IC of the results. An "all clear" report indicates that an area has been searched and all victims have been removed.

If fire fighters who have been assigned to perform some other task discover a victim or come upon a critical rescue situation, they must notify the IC immediately. Because life safety is always the highest priority, the IC may send a rescue team to meet the search team so that the search team does not lose the continuity of its search.

Another critical aspect of search coordination is keeping track of everyone who was rescued or who escaped without assistance. This information should be tracked at the command post so that reports of missing occupants can be matched to reports of rescued victims. Fire fighters should also conduct an exterior search for any missing occupants. An occupant who escaped from the fire may be lying unconscious on the ground nearby or in the care of neighbours, for example. An occupant who jumped from a window could be injured and unable to move.

■ Search Priorities

A search begins in the areas where victims are at the greatest risk. One or two search teams can usually go through all of the rooms in a single-family dwelling in 15 minutes or less. Multiple search teams and a systematic division of the building are needed to cover larger structures, such as apartment buildings. Area search assignments should be based on a system of priorities:

- The first priority is to search those areas where live victims may be located immediately around the fire, and then the rest of the fire floor.
- The second priority is to search the area directly above the fire and the rest of that floor.
- The next priority is higher-level floors, working from the top floor down, because smoke and heat are likely to accumulate in these areas.
- Generally, areas below the fire floor are a lower priority.

These priorities are guidelines and need to be modified based on the circumstances at that incident.

At a high-rise building fire, the IC might assign two or more search teams to each floor. The search teams must work together closely and coordinate their searches to ensure that all areas are covered.

Primary Search

Two types of searches are performed in buildings: primary and secondary. A **primary search** is a quick attempt to locate any potential victims who are in danger. This search should be as thorough as time permits and should cover any places where victims are likely to be found. Fire conditions might make it impossible to conduct a primary search in some areas, or they may limit the time available for an exhaustive search. A primary search should be completed in 15 minutes or less.

A **secondary search** is conducted after the fire has been suppressed. At this point, any remaining occupants have most likely been exposed to conditions that cannot support life, such as flashover. The main purpose of the secondary search is to find deceased victims. If possible, the secondary search should be conducted by a different search team, so that each area of the building is examined with a fresh set of eyes.

During the primary search, fire fighters rapidly search the accessible areas of a burning structure where conditions may still permit human life to exist. The objective of this search is to find any potential victims as quickly as possible and to remove them from danger. When fire fighters complete the primary search, they have gone as far as they can and have removed anyone who was rescued. The phrase *primary all clear* is used to report that the primary search has been completed.

By necessity, the primary search is conducted quickly and gives priority to those areas where victims are most likely to be located. Time is always a critical concern, because fire fighters must reach potential victims before they are burned, overcome by smoke and toxic gases, or trapped by a structural collapse. Search teams may have only a few minutes to conduct a primary search. In that limited amount of time, fire fighters must try to find anyone who could be in danger and remove those individuals to a safe area. Often, active fire conditions may limit the areas that can be searched quickly as well as the time that can be spent in each area.

Fire fighters should try to check all of the areas where victims might be, such as beds, cribs, chairs, and sofas. Adults who tried to escape on their own are often found near doors or windows. Some people—particularly children—may try to hide in a closet, in a bathtub or shower enclosure, or under a bed or another piece of furniture. Be aware that young children may be frightened by the appearance of a fire fighter wearing full personal protective gear.

The primary search is frequently conducted in conditions that expose both fire fighters and victims to the risks presented by heavy smoke, heat, structural collapse, and entrapment by the fire. Search teams must often work in conditions of zero visibility and may have to crawl along the floor to stay below layers of hot gases. Because the beam from a powerful hand light might be visible for only a few centimetres inside a smoke-filled building, search teams must practise searching for victims in total darkness. They must know how to keep track of their location and how to get back to their entry point. Practising these skills in a controlled environment will enable fire fighters to perform confidently under similar conditions at a real emergency incident.

Fire fighters must rely on their senses when they search a building:

- Sight—Can you see anything?
- Sound—Can you hear someone calling for help, moaning, or groaning?
- Touch—Do you feel a victim's body?

If smoke and fire restrict visibility, fire fighters must use sound and touch to find victims. Every few seconds a member of the search team should yell out, "Is anyone in here? Can anyone hear me?" Then listen. Hold your breath to quiet your breathing regulator and stop moving. People who have suffered burns or are semiconscious may not be able to speak intelligibly, so you will need to listen for guttural sounds, cries, faint voices, groans, and moans. Focus on the direction of any sounds you hear.

As you search, feel around for the hands, legs, arms, and torsos of potential victims. Use side crawls to extend your reach sideways; sweep ahead and to the sides as you crawl. Practise until you can tell the difference between a piece of furniture and a human body. The type of furniture you encounter may also provide clues for finding victims. Spindle-type legs in a bedroom may indicate a crib that needs to be searched. An unusually low bed may be the bottom berth of a children's bunk bed.

Zero-visibility conditions can be very disorienting. In such circumstances, fire fighters must follow walls and note turns and doorways to avoid becoming lost. After locating a victim or completing a search of an area, the search team must be able to retrace its path and return to its entry point. Search teams must also identify a secondary escape route, in case fire conditions change and block their planned exit route; thus, during the search, fire fighters should note the locations of stairways, doors, and windows. Fire fighters should always be aware of the nearest exit and an alternate exit.

If the situation deteriorates rapidly, a window can serve as an emergency exit from a room. Fire fighters should keep track of the exterior walls and window locations, even reaching up to feel for the windows if conditions force them to crawl.

Fire fighters must remain in voice contact with someone outside the room and must be able to state their location at all times, particularly if they need assistance. The search team members must be able to give their location, including the building section and floor, to the IC, if they need assistance. If the search team needs a ladder for the rescue, the IC will need to know where the ladder should be placed (which side of the building) and how long a ladder will be needed (which floor). Fire fighters should use standard incident management system (IMS) terminology to describe their location over the radio.

Fire Fighter Safety Tips © Photos.com

While searching for victims, you also need to focus on your safety. Continuously monitor your air supply and the conditions of the fire and the structure. You should also be aware of the status of other crews, such as the fire attack crew and the ventilation crew.

Search Techniques

Fire fighters should employ standard search techniques to search assigned areas quickly, efficiently, and safely. They should operate in teams and should always stay close together **FIGURE 14-4**. Partners must remain in direct visual, voice, or physical contact with each other.

At least one member of each search team should also have a radio to maintain contact with the command post or someone outside the building. The search team can use this radio to call for help if they become disoriented, trapped by fire, or need assistance. If the search team finds a victim, they can

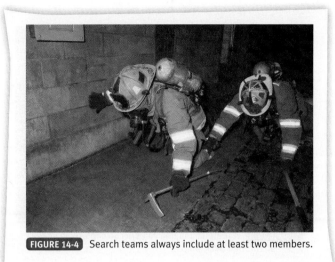

FIGURE 14-4 Search teams always include at least two members.

FIGURE 14-5 To use a clockwise search pattern, turn left when you enter, then turn right at each corner of the room.

notify the IC so that help will be available to help remove the victim from the building and begin medical treatment. The search team must notify the IC when the search of each area has been completed, so the IC can make informed decisions about which steps to take next.

■ Search Patterns

Each room should be searched using a standard pattern. If the rooms are relatively small, fire fighters may follow the walls around the perimeter of each room and reach toward the middle to feel for victims. In larger rooms, one fire fighter can maintain contact with the wall while the other maintains contact with the first fire fighter, and the two can then work their way around the room in tandem. Fire fighters should regularly practise and use the standard search system adopted by their department.

Some fire departments use a clockwise pattern to search a room, whereas others use a counter-clockwise pattern. In a clockwise search (also known as a left-hand search), fire fighters turn left at the entry point, keep the left hand in contact with the wall, and use the right arm to sweep the room. At each corner, the fire fighters make a right turn, eventually returning to the entry door FIGURE 14-5 . A counter-clockwise search moves around the room in the opposite direction FIGURE 14-6 .

While conducting the search, paint a picture in your mind of everything that you are touching. In particular, everything that was on your left side should be on your right side when you are exiting the room.

Fire fighters may use the handle of a hand tool to extend their reach while sweeping across the floor and under furniture. They should also check on top of beds, in chairs, and in other locations where victims may hide, such as closets, bathrooms, bathtubs, or shower enclosures.

The same clockwise/counter-clockwise search pattern applies to searches of areas that are divided up into small spaces, such as an apartment or an apartment building. Fire fighters should work their way around the apartment from room to room in a standard direction, searching each individual room

FIGURE 14-6 In a counter-clockwise search pattern, turn right when you enter, then turn left at each corner around the room.

in the same manner. At the end of the search, they will be back at the entry point.

If a door is closed, fire fighters should check the temperature of the door to determine whether an active fire is present on the other side. They should follow their department's suggested operating guidelines (SOGs), or standard operating procedures (SOPs), for performing such a check. A thermal imaging device, which shows heat images instead of visual images, can be used to determine whether a fire is present behind the door. A hot door should not be opened unless a hose line is readily available to douse the fire. It is usually better to leave a hot door closed and move on to search adjacent rooms.

Fire fighters must always keep track of their position in relation to the entry door, so they can find their way out of the room. To keep the door open during the search of a small room, one team member should remain at the door while the second team member performs the search. A chock can be used to keep the door from closing, or a flashlight can be placed in the doorway to serve as a beacon and help fire fighters maintain their orientation. Fire fighters should always exit through the same door that they entered.

The searched room or apartment should be marked so other personnel will know that it has been searched. Chalk, crayons, felt-tipped markers, or masking tape can be used to mark the door for this purpose. Some fire departments use a two-part marking system to indicate when a search is in progress and when it has been completed: in this system, a slash (/) indicates that a search is in progress, and an X indicates that the search has been completed **FIGURE 14-7**. Other fire departments place an object in the doorway or attach a tag or latch strap to the doorknob to indicate that the room has been searched.

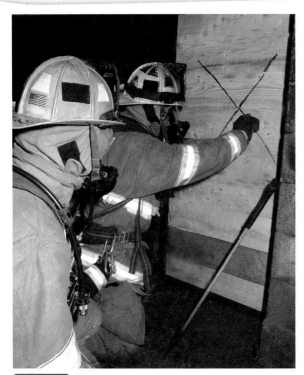

FIGURE 14-7 In large buildings, doors should be marked after the rooms are searched.

Thermal Imaging Devices

A **thermal imaging device** is a valuable tool for conducting a search in a smoke-filled building. This piece of equipment is similar to a television camera, except that it captures heat images instead of visible-light images. The images appear on a display screen and show the relative temperatures of different objects. The thermal imaging device can be set to distinguish small temperature differences, enabling fire fighters to conduct a search quickly and thoroughly.

The major benefit of using a thermal imaging device is that it can "see" a person in conditions of total darkness or through smoke that completely obscures normal vision **FIGURE 14-8**. Fire fighters will be able to identify the shape of a human body with a thermal image scan because the body will be either warmer or cooler than its surroundings.

Temperature differences also mean that the thermal imaging device can show furniture, walls, doorways, and windows.

FIGURE 14-8 A thermal imaging camera can capture an image of a person through thick smoke.

This information may enable a fire fighter to navigate through the interior of a smoke-filled building. The thermal imaging camera can sometimes be used to locate a fire in a smoke-filled building or behind walls or ceilings **FIGURE 14-9**. A thermal image scan of the exterior of a building can sometimes locate the fire source and the direction of fire spread. Scanning a door before opening it may indicate whether the room is safe to enter.

FIGURE 14-9 A thermal imaging camera can be used to check for hidden fire.

Several types of thermal imaging cameras are available. Some are hand-held cameras, while others are helmet-mounted devices that produce an image in front of the face piece of the user's self-contained breathing apparatus (SCBA). Fire fighters need training and practise to become proficient in using these devices, interpreting the images, and maneuvering confidently through a building while wearing a thermal imaging camera.

FIRE FIGHTER Tips

Learn the features of your thermal imaging camera and practise with it regularly.

FIRE FIGHTER Tips

Not all rooms have just four walls; some may have five or more. To determine how many walls a room has, check for walls on each side of the door as you enter the room. If there is a wall parallel to the door, on the right of the door, and to the left of the door, then you will need to search five walls. If there is a parallel wall only to the right or to the left of the door, then the room has four walls. Noting the number of walls in a room about to be searched is a good way to ensure that the entire room has been searched and to maintain orientation.

Search Ropes

Fire fighters must use special techniques to search large open areas, such as warehouses or gymnasiums, when visibility is limited. **Search ropes** should be used when it is impossible to cover the interior by following the walls. Some fire departments may use a hose line instead of a rope. They also should be used in areas with interconnected rooms or spaces, or in areas with multiple aisles created by display counters or storage racks. Without a search rope, the search teams might not be able to find their way out of the area.

Search ropes should be anchored at the entry point to provide a direct path to this location for each fire fighter. One fire fighter should remain at the anchor point to tend the rope and maintain accountability for the fire fighters who enter. Another fire fighter stretches a large-diameter rope down the centre of a room. The other fire fighters then attach their individual search ropes to the main line and branch off to cover the area. The search ropes always provide a reliable return path to the entry point.

Some fire departments may modify how they utilize search ropes. For example, knots may be placed in the search ropes to help fire fighters keep track of distance. Some fire departments may attach ropes perpendicular to the main search rope so that the fire fighters may extend their reach farther. Finally, some fire departments may permit their fire fighters to briefly slide one or two body lengths away from the search rope to investigate an area.

Officer-Led Search

In some fire departments, the search team may consist of a fire officer and one to three fire fighters. With this approach, one or more fire fighters enter and systematically search the room, while the fire officer remains at the door, monitoring the conditions of the room. If the conditions become threatening, the fire officer will give the order to evacuate. The goal is to allow the fire fighters to focus on searching the room, while the fire officer focuses on their safety. This strategy is referred to as an "oriented search" by Retired Assistant Chief Skip Coleman of the Toledo, Ohio, fire department.

One search method that may be used during this type of search is a parallel–crawl–return. With this technique, the fire fighter's body remains parallel to the wall at all times. If the fire fighter feels something during a tool or gloved-hand sweep, the fire fighter may slide away from the wall, investigate further, and then slide back into position, parallel to the wall **FIGURE 14-10**.

FIGURE 14-10 One method of searching a room requires the fire fighter to be parallel to the wall to maintain a proper orientation.

Vent–Entry–Search

In some circumstances, a search team must enter and exit a room through a window instead of a door. Fire departments may practise different variations of the vent–entry–search method, but the basic process is as follows:

- Vent—Open a window in an area that is removed from the fire. Before opening the window, determine that the conditions in that room are tenable for any victims. This action clears some of the smoke from that room, but it may also introduce a supply of oxygen to the fire.
- Entry—When the conditions permit, as soon as possible after venting, the fire fighter should enter through the window and immediately locate and close the door that leads to the hallway. This step isolates the room from the fire and reduces the supply of oxygen to the fire, decreasing the chance of a flashover for a few minutes. Never enter the structure if there are any signs of a potential flashover or backdraft.
- Search—As soon as the hall door is closed, the fire fighter should begin a quick and systematic search of the room, paying special attention to the bed if the fire occurs at night. If a victim is found, the fire fighter transports the victim to the window to be rescued.

Vent–enter–search is a high-risk activity that should be undertaken only in emergency situations where there is a potential to save a life. This procedure is extremely risky, because it places the fire fighter between the location of the fire and the ventilation hole (window) that was created. Fire fighters should use this procedure only as directed by the IC and as part of their department's SOGs/SOPs.

Conducting a Primary Search

Follow the steps in **SKILL DRILL 14-1** to perform a primary search of a structure and rescue victims:

1. Don your personal protective equipment (PPE), including SCBA, and enter the personnel accountability system.
2. Bring hand tools, a hand light, a radio, a thermal imaging device, and search ropes, if indicated.

Fire Fighter Safety Tips

© Photos.com

There are two additions to the classic vent–enter–search method that can make this procedure safer:

1. Isolate the fire. Be sure to close the door between the room and the hallway as soon as you enter the room. According to recent National Institute of Technology and Standards studies, this step will help to keep the room more tenable for a few minutes while you perform a quick search of the room.

2. Use a team of two fire fighters when performing VES. One fire fighter enters the room to isolate it and perform the search. The second fire fighter remains outside the window through which entry was made; this fire fighter monitors the fire conditions and can warn the inside fire fighter of changing fire conditions. The second fire fighter can also assist in rescuing any victims who are found.

3 Notify command that the search is starting, and indicate the area to be searched and the direction of the search.

4 Use hand tools or ground ladders if needed to gain access to the site.

5 Conduct a quick and systematic search by staying on an outside wall and searching from room to room.

6 Maintain contact with an outside wall. (**STEP 1**)

7 Maintain team integrity using visual, voice, or direct contact.

8 Use the most efficient movement based on the hazard encountered: duckwalk, crawl, stand only when you can see your feet and it is not hot, and use tools to extend your reach.

9 Clear each room visually or by touch.

10 Search the area, including stairs, up to the landing on the next floor. (**STEP 2**)

11 Periodically listen for victims and sounds of fire.

12 Communicate the locations of doors, windows, and inside corners to the other team members.

13 Observe fire, smoke, and heat conditions; update command on this information.

14 Locate and remove victims; notify the IC.

15 When the search is complete, conduct a personnel accountability report.

16 Report the results of the search to your officer. (**STEP 3**)

■ Secondary Search

A secondary search is conducted after the fire is under control or fully extinguished. During the slower secondary search, fire fighters search to locate any deceased victims and search any areas not covered in the primary search. Conditions in the building are usually better during the secondary search,

SKILL DRILL 14-1   Conducting a Primary Search
(Fire Fighter I, NFPA 5.3.9)

1 Notify command that the search is starting, and indicate the area to be searched and the direction of the search. Conduct a quick and systematic search by staying on an outside wall and searching from room to room. Maintain contact with an outside wall.

2 Maintain team integrity using visual, voice, or direct contact. Use the most efficient movement based on the hazard encountered: duckwalk, crawl, stand only when you can see your feet and it is not hot, and use tools to extend your reach. Clear each room visually or by touch. Search the area, including stairs, up to the landing on the next floor.

3 Periodically listen for victims and sounds of fire. Communicate the locations of doors, windows, and inside corners to other team members. Observe fire, smoke, and heat conditions; update command on this information. Locate and remove victims; notify the IC. When the search is complete, conduct a personnel accountability report. Report the results of the search to your officer.

© LiquidLibrary

FIRE FIGHTER Tips

Some fire departments have members of their search teams carry hand tools but use their gloved hands to actually sweep the room for victims. Some fire departments believe that a gloved hand can assess an area more accurately than sweeping a hand tool about to feel if it hits an object.

because the fire is under control and ventilation should have cleared away the smoke. Lights may be set up to improve visibility. As a consequence, fire fighters will be able to see better and move around without having to crawl under layers of hot gases FIGURE 14-11 .

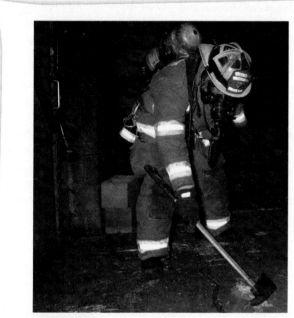

FIGURE 14-11 A secondary search is conducted after the fire is under control.

Even though the fire may be under control, safety remains a primary consideration during a secondary search. SCBA must be used until the air is tested and pronounced safe to breathe, because levels of carbon monoxide and other poisonous gases often remain high for a long time after a fire is extinguished. In burned areas, fire fighters must have a hose line available in case the fire **rekindles** (starts up again). The structural stability of the building also must be evaluated before beginning a secondary search. Portable lighting should be used, and holes in floors and other possible hazards should be marked.

The secondary search should start as soon as the fire is under control and sufficient resources are available. It is conducted slowly and methodically to ensure that no areas are overlooked. Whenever possible, a different team of fire fighters should perform the secondary search. They should check

all places where someone—especially a child—could hide, including closets, bathtubs, shower enclosures, behind doors, under windows, under beds and furniture, and inside toy chests. After a secondary search is completed, fire fighters should notify the IC by reporting, "Secondary all clear."

Follow the steps in SKILL DRILL 14-2 (Fire Fighter, I NFPA 5.3.9) to perform a secondary search:

1. Don PPE, including SCBA, and enter the personnel accountability system.
2. Bring hand tools, a hand light, a radio, a thermal imaging device, and search ropes, if indicated.
3. Notify command that the search is starting, and indicate the area to be searched and the direction of the search.
4. Conduct a systematic, thorough search for any possible victims.
5. Check for hidden fire, salvage needs, and indicators of the fire's cause, its origin, or arson.
6. Protect any evidence.
7. Maintain team integrity using visual, voice, or direct contact.
8. Conduct a personnel accountability report and notify command of the results of the search.

Search Safety

Search and rescue operations present a high risk to fire fighters. During searches, fire fighters are exposed to the same hazards that may endanger the lives of potential victims. Even though fire fighters have the advantages of protective clothing, protective equipment, training, teamwork, and SOGs/SOPs, they can still be seriously or fatally injured during these operations. Safety is an essential consideration in all search and rescue operations.

■ Risk Management

Search and rescue situations require a very special type of risk management. Although every emergency operation involves a degree of unavoidable, inherent risk to fire fighters, during search and rescue operations, fire fighters will probably encounter situations that involve a significantly higher degree of personal risk. Assuming this level of risk is acceptable only if there is a reasonable probability of saving a life.

The IC is responsible for managing the level of risk during emergency operations. He or she must determine which actions will be taken in each situation. Actions that present a high level of risk to the safety of fire fighters are justified only if lives can be saved. In contrast, only a limited risk level is acceptable to save property. When there is no possibility of saving either lives or property, no risk is acceptable.

The acceptance of unusual levels of risk can be justified only when victims are known or believed to be in immediate danger, and there is a reasonable probability that they can be rescued. It is not acceptable to risk the lives of fire fighters when rescuing any victims is not a possibility; for example, if a building is fully involved in flames, a room has flashed over, or the smoke has banked all the way down to the floor, there is little or no chance that any occupants can still be alive, so there is no reason to send fire fighters inside to search for them. The risk involved in conducting search and rescue operations must always be weighed against the probability of finding someone who is savable—that is, who is still alive to be rescued.

A primary search is conducted during the active stages of a fire, when time is a critical factor. During a primary search, it may be necessary to accept a high level of risk to save a life. A secondary search, which is conducted after the fire is extinguished, is designed to locate any deceased victims and should not expose fire fighters to a high level of risk.

The IC must decide whether to conduct a primary search based on a risk–benefit evaluation of the situation. In making this decision, the IC must consider the stage of the fire, the condition of the building, and the presence of any other hazards. These risks must be weighed against the probability of finding any occupants who could be rescued. Such an on-scene evaluation is often guided by established policies; for example, many fire departments have policies that prohibit fire fighters from entering abandoned structures known to be in poor structural condition. This kind of policy reflects the fact that such a building has a high risk of collapse and the low probability that anyone would be inside.

The IC may decide not to conduct a primary search because the risk to fire fighters is too great or because the possibility of making a successful rescue is too remote. Such a decision might be based on advanced fire conditions, potential backdraft conditions, imminent structural collapse, or other circumstances **FIGURE 14-12**. The IC may be able to identify these conditions from the exterior, or he or she may learn of them from a team assigned to conduct a search. A search team that encounters conditions that make entry impossible should report their findings back to the IC. In these situations, a secondary search is conducted when it is safe to enter.

FIGURE 14-12 A primary search cannot be safely conducted in a building that is fully involved in a fire.

Courtesy of District Chief Chris E. Mickal/New Orleans Fire Department, Photo Unit

Search and Rescue Equipment

To perform search and rescue properly, fire fighters must have the appropriate equipment:

- PPE
- Portable radio
- Hand light or flashlight
- Forcible entry/exit tools
- Hose lines
- Thermal imaging devices
- Ladders
- Long rope(s)
- A piece of tubular webbing or short rope (4.9–7.3 metres [16–24 feet])

Full PPE, which is always required for structural firefighting, is essential for interior search and rescue operations. The proper attire includes a helmet, protective hood, bunker coat, turnout pants, boots, and gloves. Each fire fighter must use SCBA and carry a flashlight or hand light. Before entering the building, fire fighters must activate their personal alert safety system (PASS) devices.

At least one member of each search team should be equipped with a portable radio. Ideally, each individual should have a radio. If a fire fighter gets into trouble, the radio is the best means to obtain assistance.

Each fire fighter assigned to search and rescue should carry at least one forcible entry hand tool, such as an axe, Halligan tool, or short ceiling hook. These tools can be used both to open an area for a search and, if necessary, to open an emergency exit path. A hand tool can also be used to extend the fire fighter's reach during a sweep for unconscious victims.

A search team that is working close to the fire should carry a hose line or be accompanied by another team with a hose line. A hose line can protect the fire fighters and enable them to search a structure more efficiently. This measure is essential when a search team is working close to or directly above the fire. The hose line can be used to knock down the fire, to protect a means of egress (stairway or corridor), and to protect the victims as they are escaping.

Methods to Determine Whether an Area Is Tenable

Fire fighters must make rapid, accurate, and ongoing assessments about the safety of the building for fire fighters while working at a structure fire. When they arrive, fire fighters must quickly determine the type of structure involved, the possibility of collapse, and the life-safety risk involved. They also must evaluate the stability of the structure and the potential for backdraft or flashover. Sagging walls, chipped or cracked mortar or cement, warped or failing structural steel, and any partial collapse are significant indicators of an impending collapse. Dark, black, turbulent smoke, blackened windows, the appearance of a "breathing" building, and intense heat are signs of possible flashover or backdraft.

Even after the decision has been made to enter a burning structure, fire fighters must continually reevaluate the safety of the operation. Those working on the search team rely on fire officers, team members, and the IC to notify them of changing fire conditions. Fire officers must be alert for changing conditions inside the building and for any information about changing conditions from the other parts of the fire scene. The IC may know or see things that fire fighters working inside might not know, for example, and may call for an evacuation as the fire situation changes. For their part, fire fighters in the interior of the building should check and recheck the surfaces they are working on. Floors that have burned through, that are sagging, or that are warm should be avoided. A rapid rise in the amount of heat or flame rollover may indicate the potential for flashover.

Fire Fighter Safety Tips

© Photos.com

During search and rescue operations, fire fighters should follow these guidelines:
- Work from a single plan.
- Maintain radio contact with the IC, both through the chain of command and via portable radios.
- Monitor fire conditions during the search.
- Coordinate ventilation operations with search and rescue activities.
- Adhere to the personnel accountability system.
- Stay with a partner.

FIRE FIGHTER Tips

The IC must always balance the risks involved in an emergency operation with the potential benefits:
- Actions that present a high level of risk to the safety of fire fighters are justified only if there is a potential to save lives.
- Only a limited level of risk is acceptable to save valuable property.
- It is not acceptable to risk the safety of fire fighters when there is no chance of saving lives or property.

An exception to this rule is permitted in an imminent life-threatening situation, where immediate action can prevent the loss of life or serious injury. Only under such specific circumstances are fire fighters allowed to take actions at a higher level of risk. The initial IC must evaluate the situation and determine whether this level of risk is justified. The IC also must be prepared to explain his or her decision.

Rescue Techniques

Rescue is the removal of a located person who is unable to escape from a dangerous situation. Fire fighters rescue people not only from fires but also from a wide variety of accidents and mishaps. Although this section refers primarily to rescuing occupants from burning buildings, some of the techniques discussed here can be used for other types of situations. As a fire

FIRE FIGHTER Tips

Specific safety requirements for search and rescue operations are defined in NFPA 1500, *Standard on Fire Department Occupational Safety and Health Program,* and in regulations enforced by Public Safety Canada. The NFPA requirements state that a team of at least two fire fighters must enter together, and at least two other fire fighters must remain outside the danger area, ready to rescue the fire fighters who are inside the building. This policy is sometimes called the two-in–two-out rule.

fighter, you must learn and practise the various types of assists, carries, drags, and other techniques used to rescue people from fires. After mastering these techniques, you will be able to rescue victims from life-threatening situations.

Rescue is the second component of search and rescue. When you locate a victim during a search, you must direct, assist, or carry that person to a safe area. Rescue can be as simple as verbally directing an occupant toward an exit, or it can be as demanding as extricating a trapped, unconscious victim and physically carrying that person out of the building. The term *rescue* is generally applied to situations where the rescuer physically assists or removes the victim from the dangerous area.

Most people who realize that they are in a dangerous situation will attempt to escape on their own. Children and elderly persons, physically or developmentally handicapped persons, impaired people, and ill or injured persons, however, may be unable to escape and will need to be rescued. People who are sleeping or under the influence of alcohol or drugs may not become aware of the danger in time to escape from the structure on their own. Toxic gases may incapacitate even healthy individuals before they can reach an exit. Victims also may become trapped when a rapidly spreading fire, an explosion, or a structural collapse cuts off potential escape routes.

Because fires threaten the lives of both victims and rescuers, the first priority is to remove the victim from the fire building or dangerous area as quickly as possible. It is sometimes better to move the victim to a safe area first and then provide any necessary medical treatment. The assists, lifts, and carries described in this chapter should not be used if you suspect that the victim has a spinal injury, however, unless there is no other way to remove the person from the life-threatening situation.

Always use the safest and most practical means of egress when removing a victim from a dangerous area. A building's normal exit system, such as interior corridors and stairways, should be used if it is open and safe. If the regular exits cannot be used, an outside fire escape, a ladder, or some other method of egress must be found. Ladder rescues, which are covered in this chapter, can be both difficult and dangerous, whether the victim is conscious and physically fit or is unconscious and injured FIGURE 14-13 .

Shelter-in-Place

In some situations, the best option is to shelter the occupants in place instead of trying to remove them from a fire building. This option should be considered when the occupants

FIGURE 14-13 Bringing an unconscious victim or a person who needs assistance down a ladder can be difficult and dangerous.

are conscious and are found in a part of the building that is adequately protected from the fire by fire-resistive construction or fire-suppression systems. If smoke and fire conditions block the exits, victims might be safer staying in the sheltered location than attempting to evacuate through a hazardous environment.

Such a situation occurs in a high-rise apartment building when a fire is confined to one apartment or one floor. In this scenario, the stairways and corridors may be filled with smoke, but the occupants who are remote from the fire would be very safe on their balconies or in their apartments with the windows open to provide fresh air. They would be exposed to greater levels of risk if they attempted to exit than if they remained in their apartments until the fire is extinguished. The decision to follow a shelter-in-place strategy in this case must be made by the IC. Potential locations to shelter-in-place occupants should be identified on the preincident plan.

Fire Fighter Safety Tips
© Photos.com

To exit from a fire, follow the hose line in the direction of the male coupling.

■ Exit Assist

The simplest rescue is the exit assist, in which the victim is responsive and able to walk without assistance or with very little assistance. The fire fighter may simply need to guide the person to safety or to provide a minimal level of physical support. Even if the victim can walk without assistance, the fire fighter should take the person's arm or use the one-person

walking assist (discussed later in this section) to make sure that the victim does not fall or become separated from the rescuer.

Two types of assists can be used to help responsive victims exit a fire situation:
- One-person walking assist
- Two-person walking assist

One-Person Walking Assist

The one-person walking assist can be used if the person is capable of walking. To perform a one-person walking assist, follow the steps in **SKILL DRILL 14-3**:

1 Help the victim stand next to you, facing the same direction. (**STEP 1**)

2 Have the victim place his or her arm behind your back and around your neck.

3 Hold the victim's wrist as it drapes over your shoulder.

4 Put your free arm around the victim's waist and help the victim to walk. (**STEP 2**)

Two-Person Walking Assist

The two-person walking assist is useful if the victim cannot stand and bear weight without assistance. With this assist technique, the two rescuers completely support the victim's weight. It may be difficult to walk through doorways or narrow passages using this type of assist. To perform a two-person walking assist, follow the steps in **SKILL DRILL 14-4**:

1 Two fire fighters stand facing the victim, one on each side of the victim. (**STEP 1**)

2 Both fire fighters assist the victim to a standing position. (**STEP 2**)

3 Once the victim is fully upright, place the victim's right arm around the neck of the fire fighter on the right side; place the victim's left arm around the neck of the fire fighter on the left side. The victim's arms should drape over the fire fighters' shoulders. The fire fighters hold the victim's wrist in one hand. (**STEP 3**)

4 Both fire fighters put their free arms around the victim's waist, grasping each other's wrists for support and locking arms together behind the victim. (**STEP 4**)

5 Both fire fighters slowly assist the victim to walk. Fire fighters must coordinate their movements and move slowly. (**STEP 5**)

■ Simple Victim Carries

Four simple carry techniques can be used to move a victim who is conscious and responsive, but incapable of standing or walking:
- Two-person extremity carry
- Two-person seat carry
- Two-person chair carry
- Cradle-in-arms carry

Two-Person Extremity Carry

The two-person extremity carry requires no equipment and can be performed in tight or narrow spaces, such as the corridors of mobile homes, small hallways, and narrow spaces between buildings. The focus of this carry is on the victim's extremities. To perform a two-person extremity carry, follow the steps in **SKILL DRILL 14-5**:

1. Two fire fighters help the victim to sit up. **(STEP ①)**
2. One fire fighter kneels behind the victim, reaches under the victim's arms, and grasps the victim's wrists. **(STEP ②)**
3. The second fire fighter backs in between the victim's legs, reaches around, and grasps the victim behind the knees. **(STEP ③)**
4. At the command of the first fire fighter, both fire fighters stand up and carry the victim away, walking straight ahead. The fire fighters must coordinate their movements. **(STEP ④)**

Two-Person Seat Carry

The two-person seat carry is used with victims who are disabled or paralyzed. This type of carry requires the assistance of two fire fighters, and moving through doors and down stairs may be difficult. To perform a two-person seat carry, follow the steps in **SKILL DRILL 14-6**:

1. Two fire fighters kneel near the victim's hips, one on each side of the victim. **(STEP ①)**

2. Both fire fighters raise the victim to a sitting position and link arms behind the victim's back. **(STEP ②)**
3. The fire fighters place their free arms under the victim's knees and link them together. **(STEP ③)**
4. If possible, the victim puts his or her arms around the fire fighters' necks for additional support. **(STEP ④)**

Two-Person Chair Carry

The two-person chair carry is particularly suitable when a victim must be carried through doorways, along narrow corridors, or up or down stairs. In this technique, two rescuers use a chair to transport the victim. A folding chair cannot be used for this purpose, because folding chairs can collapse when used in this way to carry a victim. The chair must be strong enough to support the weight of the victim while he or she is being carried. The victim should feel much more secure with this carry than with the two-person seat carry, and he or she should be encouraged to hold on to the chair. To perform a two-person chair carry, follow the steps in **SKILL DRILL 14-7**:

1. Tie the victim's hands together or have the victim grasp his or her hands together.
2. One fire fighter stands behind the seated victim, reaches down, and grasps the back of the chair. **(STEP ①)**

SKILL DRILL 14-3
Performing a One-Person Walking Assist
(Fire Fighter I, NFPA 5.3.9)

① Help the victim stand.

② Have the victim place his or her arm around your neck, and hold on to the victim's wrist, which should be draped over your shoulder. Put your free arm around the victim's waist, and help the victim to walk.

© LiquidLibrary

SKILL DRILL 14-4

Performing a Two-Person Walking Assist
(Fire Fighter I, NFPA 5.3.9)

1 Two fire fighters stand facing the victim, one on each side of the victim.

2 The fire fighters assist the victim to a standing position.

3 Once the victim is fully upright, drape the victim's arms around the necks and over the shoulders of the fire fighters, each of whom holds one of the victim's wrists.

4 Both fire fighters put their free arm around the victim's waist, grasping each other's wrists for support and locking their arms together behind the victim.

5 Assist walking at the victim's speed.

© LiquidLibrary

3 The fire fighter tilts the chair slightly backward on its rear legs, so that the second fire fighter can step back between the legs of the chair and grasp the tips of the chair's front legs. The victim's legs should be between the legs of the chair. (**STEP 2**)

4 When both fire fighters are correctly positioned, the fire fighter behind the chair gives the command to lift and walk away.

5 Because the chair carry may force the victim's head forward, watch the victim for airway problems. (**STEP 3**)

SKILL DRILL 14-5 Performing a Two-Person Extremity Carry
(Fire Fighter I, NFPA 5.3.9)

1 Two fire fighters help the victim to sit up.

2 The first fire fighter kneels behind the victim, reaches under the victim's arms, and grasps the victim's wrists.

3 The second fire fighter backs in between the victim's legs, reaches around, and grasps the victim behind the knees.

4 The first fire fighter gives the command to stand and carry the victim away, walking straight ahead. Both fire fighters must coordinate their movements.

© LiquidLibrary

Cradle-in-Arms Carry

The cradle-in-arms carry can be used by one fire fighter to carry a child or a small adult. With this technique, the fire fighter should be careful of the victim's head when moving through doorways or down stairs. To perform the cradle-in-arms carry, follow the steps in **SKILL DRILL 14-8** :

1. Kneel beside the child, and place one arm around the child's back and the other arm under the thighs. **(STEP 1)**
2. Lift slightly and roll the child into the hollow formed by your arms and chest. **(STEP 2)**
3. Be sure to use your leg muscles to stand. **(STEP 3)**

Fire Fighter Safety Tips   © Photos.com

Keep your back as straight as possible and use the large muscles in your legs to do the lifting!

■ Emergency Drags

The most efficient method to remove an unconscious or unresponsive victim from a dangerous location is a drag. Five types of emergency drags can be used to remove unresponsive victims from a fire situation:

- Clothes drag
- Blanket drag

SKILL DRILL 14-6 Performing a Two-Person Seat Carry
(Fire Fighter I, NFPA 5.3.9)

1 Kneel beside the victim, near the victim's hips.

2 Raise the victim to a sitting position, and link arms behind the victim's back.

3 Place your free arms under the victim's knees, and link arms.

4 If possible, the victim puts his or her arms around your necks for additional support.

© LiquidLibrary

- Webbing sling drag
- Fire fighter drag
- Emergency drag from a vehicle

When using an emergency drag, the rescuer should make every effort to pull the victim in line with the long axis of the body to provide as much spinal protection as possible. The victim should be moved head first to protect the head.

Clothes Drag

The clothes drag is used to move a victim who is on the floor or the ground and is too heavy for one rescuer to lift and carry alone. In this technique, the rescuer drags the person by pulling on the clothing in the neck and shoulder area. The rescuer should grasp the clothes just behind the collar, use the

arms to support the victim's head, and drag the victim away from danger. To perform the clothes drag, follow the steps in **SKILL DRILL 14-9**:

1 Crouch behind the victim's head, grab the shirt or jacket around the collar and shoulder area, and support the head with your arms. (**STEP 1**)

2 Lift with your legs until you are fully upright. Walk backward, dragging the victim to safety. (**STEP 2**)

Blanket Drag

The blanket drag can be used to move a victim who is not dressed or who is dressed in clothing that is too flimsy for the clothes drag (e.g., a nightgown). This procedure requires the use of a large sheet, blanket, curtain, or rug. Place the

SKILL DRILL 14-7 Performing a Two-Person Chair Carry
(Fire Fighter I, NFPA 5.3.9)

1 One fire fighter stands behind the seated victim, reaches down, and grasps the back of the chair.

2 The fire fighter tilts the chair slightly backward on its rear legs, so that the second fire fighter can step back between the legs of the chair and grasp the tips of the chair's front legs. The victim's legs should be between the legs of the chair.

3 When both fire fighters are correctly positioned, the fire fighter behind the chair gives the command to lift and walk away. Because the chair carry may force the victim's head forward, watch the victim for airway problems.

© LiquidLibrary

SKILL DRILL 14-8 Performing a Cradle-in-Arms Carry
(Fire Fighter I, NFPA 5.3.9)

1 Kneel beside the child, and place one arm around the child's back and the other arm under the thighs.

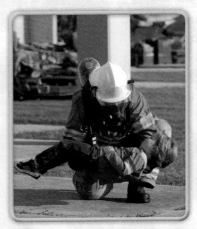

2 Lift slightly and roll the child into the hollow formed by your arms and chest.

3 Be sure to use your leg muscles to stand.

© LiquidLibrary

item on the floor, roll the victim onto it, and then pull the victim to safety by dragging the sheet or blanket.

To perform a blanket drag, follow the steps in **SKILL DRILL 14-10** :

1 Lay the victim supine (face up) on the ground. Stretch out the material for dragging next to the victim. (**STEP 1**)

2 Roll the victim onto the right or left side. Neatly bunch one-third of the material against the victim so the victim will lie approximately in the middle of the material. (**STEP 2**)

3 Lay the victim back down (supine) on the material. Pull the bunched material out from underneath the victim, and wrap it around the victim. (**STEP 3**)

4 Grab the material at the head and drag the victim backward to safety. (**STEP 4**)

The standing drag is performed by following the steps in **SKILL DRILL 14-11** :

1 Kneel at the head of the supine victim. (**STEP 1**)

2 Raise the victim's head and torso by 90 degrees, so that the victim is leaning against you. (**STEP 2**)

3 Reach under the victim's arms, wrap your arms around the victim's chest, and lock your arms. (**STEP 3**)

4 Stand straight up using your legs.

5 Drag the victim out. (**STEP 4**)

Webbing Sling Drag

The webbing sling drag provides a secure grip around the upper part of a victim's body, allowing for a faster removal from the dangerous area. In this drag, a sling is placed around the victim's chest and under the armpits, then used to drag the victim. The webbing sling helps support the victim's head and neck. A webbing sling can be rolled and kept in a turnout coat pocket. A carabiner can be attached to such a sling to secure the straps under the victim's arms and provide additional protection for the victim's head and neck.

To perform the webbing sling drag, follow the steps in **SKILL DRILL 14-12** :

1 Using a prepared webbing sling, place the victim in the centre of the loop, so the webbing is behind the victim's back in the area just below the armpits. (**STEP 1**)

2 Take the large loop over the victim, and place it above the victim's head. Reach through, grab the webbing behind the victim's back, and pull through all the excess webbing. This creates a loop at the top of the victim's head and two loops around the victim's arms. (**STEP 2**)

3 Adjust your hand placement to protect the victim's head while dragging. (**STEP 3**)

Fire Fighter Drag

The fire fighter drag can be used if the victim is heavier than the rescuer, because it does not require lifting or carrying the victim. To perform the fire fighter drag, follow the steps in **SKILL DRILL 14-13** :

1 Tie the victim's wrists together with anything that is handy—a cravat (a folded triangular bandage), gauze, a belt, or a necktie. (**STEP 1**)

2 Get down on your hands and knees, and straddle the victim. (**STEP 2**)

3 Pass the victim's tied hands around your neck, straighten your arms, and drag the victim across the floor by crawling on your hands and knees. (**STEP 3**)

SKILL DRILL 14-9   Performing a Clothes Drag
(Fire Fighter I, NFPA 5.3.9)

1 Crouch behind the victim's head, and grab the shirt or jacket around the collar and shoulder area.

2 Lift with your legs until you are fully upright. Walk backward, dragging the victim to safety.

© LiquidLibrary

SKILL DRILL 14-10 · Performing a Blanket Drag
(Fire Fighter I, NFPA 5.3.9)

1 Stretch out the material you are using next to the victim.

2 Roll the victim onto one side. Neatly bunch one-third of the material against the victim's body.

3 Lay the victim back down (supine). Pull the bunched material out from underneath the victim, and wrap it around the victim.

4 Grab the material at the head, and drag the victim backward to safety.

© LiquidLibrary

Emergency Drag from a Vehicle

An emergency drag from a vehicle is performed when the victim must be quickly removed from a vehicle to save his or her life. The drags described in this section might be used, for example, if the vehicle is on fire or if the victim requires cardiopulmonary resuscitation.

One Rescuer

There is no effective way for one person to remove a victim from a vehicle without some movement of the neck and spine. Preventing excess movement of the victim's neck, however, is important. To perform an emergency drag from a vehicle with only one rescuer, follow the steps in **SKILL DRILL 14-14**:

1 Grasp the victim under the arms, and cradle his or her head between your arms. (**STEP 1**)

2 Gently pull the victim out of the vehicle. (**STEP 2**)

3 Lower the victim down into a horizontal position in a safe place. (**STEP 3**)

Long-Backboard Rescue

If four or more fire fighters are present, one fire fighter can support the victim's head and neck, while the second and third fire fighters can move the victim by lifting under the victim's arms. The victim can then be moved in line with the long axis of the body, with the head and neck stabilized in a neutral position. Whenever possible, a long backboard should be used to remove a victim from the vehicle. Protocol will vary among provinces/territories. Follow your department's SOGs/SOPs for the long-backboard rescue. Follow the steps in **SKILL DRILL 14-15** to perform this type of rescue:

1 The first fire fighter supports the victim's head and cervical spine from behind. Support may be applied

SKILL DRILL 14-11   Performing a Standing Drag
(Fire Fighter I, NFPA 5.3.9)

1 Kneel at the head of the supine victim.

2 Raise the victim's head and torso by 90 degrees, so that the victim is leaning against you.

3 Reach under the victim's arms, wrap your arms around the victim's chest, and lock your arms.

4 Stand straight up using your legs. Drag the victim out.

© LiquidLibrary

from the side, if necessary, by reaching through the driver's-side doorway. (**STEP ①**)

② The second fire fighter serves as team leader and, as such, gives commands until the patient is supine on the backboard. Because the second fire fighter lifts and turns the victim's torso, he or she must be physically capable of moving the victim. The second fire fighter works from the driver's-side doorway. If the first fire fighter is also working from that doorway, the second fire fighter should stand closer to the door hinges toward the front of the vehicle. The second fire fighter applies a cervical collar. (**STEP ②**)

③ The second fire fighter provides continuous support of the victim's torso until the victim is supine on the backboard. Once the second fire fighter takes control of the torso, usually in the form of a body hug, he or she should not let go of the victim for any reason. Some type of cross-chest shoulder hug usually works well, but you will have to decide which method works best for you with any given victim. You cannot simply reach into the car and grab the victim, because this will merely twist the victim's torso. Instead, you must rotate the victim as a unit.

④ The third fire fighter works from the front passenger's seat and is responsible for rotating the victim's legs and feet as the torso is turned, ensuring that they remain free of the pedals and any other obstruction. With care, the third fire fighter should first move

SKILL DRILL 14-12 Performing a Webbing Sling Drag
(Fire Fighter I, NFPA 5.3.9)

1 Place the victim in the centre of the loop, so the webbing is behind the victim's back.

2 Take the large loop over the victim and place it above the victim's head. Reach through, grab the webbing behind the victim's back, and pull through all the excess webbing. This creates a loop at the top of the victim's head and two loops around the victim's arms.

3 Adjust your hand placement to protect the victim's head while dragging.

© LiquidLibrary

SKILL DRILL 14-13 Performing a Fire Fighter Drag
(Fire Fighter I, NFPA 5.3.9)

1 Tie the victim's wrists together with anything that is handy.

2 Get down on your hands and knees and straddle the victim.

3 Pass the victim's tied hands around your neck, straighten your arms, and drag the victim across the floor by crawling on your hands and knees.

© LiquidLibrary

the victim's nearer leg laterally, without rotating the victim's pelvis and lower spine. The pelvis and lower spine rotate only as the third fire fighter moves the second leg during the next step. Moving the nearer leg early makes it much easier to move the second leg in concert with the rest of the body. After the third fire fighter moves the legs together, both legs should be moved as a unit. (**STEP 3**)

5 The victim is rotated 90 degrees, so that his or her back is facing out the driver's door and his or her feet are on the front passenger's seat. This coordinated movement is done in three or four one-eighth turns. The second fire fighter directs each quick turn by saying, "Ready, turn," or "Ready, move." Hand position changes should be made between moves.

SKILL DRILL 14-14 Performing a One-Person Emergency Drag from a Vehicle
(Fire Fighter I, NFPA 5.3.9)

1 Grasp the victim under the arms, and cradle his or her head between your arms.

2 Gently pull the victim out of the vehicle.

3 Lower the victim down into a horizontal position in a safe place.

© LiquidLibrary

6 In most cases, the first fire fighter will be working from the back seat. At some point—either because the door post is in the way or because he or she cannot reach farther from the back seat—the first fire fighter will be unable to follow the torso rotation. At that time, the third fire fighter should assume temporary support of the victim's head and neck until the first fire fighter can regain control of the head from outside the vehicle. If a fourth fire fighter is present, he or she stands next to the second fire fighter. The fourth fire fighter takes control of the victim's head and neck from outside the vehicle, without involving the third fire fighter. As soon as the change has been made, the rotation can continue. (**STEP 4**)

7 Once the victim has been fully rotated, the backboard is placed against the victim's buttocks on the seat. Do not try to wedge the backboard under the victim. If only three fire fighters are present, place the backboard within arm's reach of the driver's door before the move, so that the board can be pulled into place when needed. In such cases, the far end of the board can be left on the ground. When a fourth fire fighter is available, the first fire fighter exits the rear seat of the car, places the backboard against the victim's buttocks, and maintains pressure in toward the vehicle from the far end of the board. (Note: when the door opening allows, some fire fighters prefer to insert the backboard onto the car seat before rotating the victim.)

8 As soon as the victim has been rotated and the backboard is in place, the second and third fire fighters lower the victim onto the board while supporting the head and torso so that neutral alignment is maintained. The first fire fighter holds the backboard until the victim is secured. (**STEP 5**)

9 The third fire fighter moves across the front seat to be in position at the victim's hips. If the third fire fighter stays at the victim's knees or feet, he or she will be ineffective in helping to move the body's weight, because the knees and feet follow the hips.

10 The fourth fire fighter maintains support of the head and now takes over giving the commands. The second fire fighter maintains direction of the extrication. This fire fighter stands with his or her back to the door, facing the rear of the vehicle.

11 The backboard should be immediately in front of the third fire fighter. The second fire fighter grasps the victim's shoulders or armpits. Then, on command, the second and third fire fighters slide the victim 20–30 centimetres (8 to 12 inches) along the backboard, repeating this slide until the victim's hips are firmly on the backboard. (**STEP 6**)

12 The third fire fighter gets out of the vehicle and moves to the opposite side of the backboard, across from the second fire fighter. The third fire fighter now takes control at the victim's shoulders, and the second fire fighter moves back to take control of the hips. On command, these two fire fighters move the victim along the board in 20- to 30-cm (8- to 12-in.) slides until the victim is placed fully on the board. (**STEP 7**)

13 The first (or fourth) fire fighter continues to maintain support of the victim's head. The second and third fire fighters grasp their sides of the board, and then carry it and the victim away from the vehicle and toward the prepared cot nearby. (**STEP 8**)

These steps must be considered a general procedure to be adapted as needed based on the type of vehicle, type of victim, and rescue crew available. Two-door cars differ from four-door models; larger cars differ from smaller compact models; pickup trucks differ from full-size sedans and four-wheel-drive vehicles. Likewise, you will handle a large, heavy adult differently from a small adult or child. Every situation will be different—a different car, a different patient, and a different crew. Your resourcefulness and ability to adapt are necessary elements to successfully perform this rescue technique.

Assisting a Person Down a Ground Ladder

Using a ground ladder to rescue a trapped occupant is one of the most critical, stressful, and demanding tasks performed by fire fighters. Assisting someone down a ladder carries a considerable risk of injury to both fire fighters and victims. Whenever it is possible to use a stairway, fire escape, or aerial tower for rescue, these options should be considered before using ground ladders. Fire fighters must use proper techniques to safely accomplish a ladder rescue. In addition, they must have the physical strength and stamina needed to accomplish the rescue without causing injury to anyone involved. Although circumstances could require an individual fire fighter to work alone, at least two fire fighters should work as a team to rescue a victim whenever possible.

Time is a critical factor in many rescue situations—someone waiting to be rescued is often in immediate danger and may be

SKILL DRILL 14-15 Performing a Long-Backboard Rescue
(Fire Fighter I, NFPA 5.3.9)

1 The first fire fighter provides in-line manual support of the victim's head and cervical spine.

2 The second fire fighter gives commands and applies a cervical collar.

3 The third fire fighter frees the victim's legs from the pedals and moves the legs together without moving the victim's pelvis or spine.

4 The second and third fire fighters rotate the victim as a unit in several short, coordinated moves. The first fire fighter (relieved by the fourth fire fighter, as needed) supports the victim's head and neck during rotation (and later steps).

© LiquidLibrary

(Continues)

SKILL DRILL 14-15   Performing a Long-Backboard Rescue (*Continued*)
(Fire Fighter I, NFPA 5.3.9)

5 The first (or fourth) fire fighter places the backboard on the seat against the victim's buttocks. The second and third fire fighters lower the victim onto the long backboard.

6 The third fire fighter moves to an effective position for sliding the victim. The second and third fire fighters slide the victim along the backboard in coordinated, 20- to 30-cm (8- to 12-in.) moves until the victim's hips rest on the backboard.

7 The third fire fighter exits the vehicle and moves to the backboard opposite the second fire fighter. Working together, they continue to slide the victim until the victim is fully on the backboard.

8 The first (or fourth) fire fighter continues to stabilize the victim's head and neck while the second, third, and fourth fire fighters carry the victim away from the vehicle.

Courtesy of AAOS

© LiquidLibrary

preparing to jump. As a consequence, fire fighters may have only a limited time to work. They must quickly and efficiently raise a ladder and assist the victim to safety.

Ladder rescue begins with proper placement of the ladder. As described in the chapter "Ladders," a ladder used to rescue a person from a window should have its tip placed just below the windowsill. This positioning makes it easier for the victim to mount the ladder. If possible, one or more fire fighters in the interior of the building should help the victim onto the ladder, and one fire fighter should stay on the ladder to assist the individual down.

Any ladder used for rescue should be heeled or tied in. The weight of an occupant and one or two fire fighters, all moving on the ladder at the same time, can easily destabilize a ladder that is not adequately secured.

Rescuing a Conscious Person from a Window

When a rescue involves a conscious person, fire fighters should establish verbal contact as quickly as possible to reassure the victim that help is on the way. Many people have jumped to their deaths just seconds before a ladder would have been raised to a window.

All ladder rescues should be performed with two fire fighters whenever possible. **SKILL DRILL 14-16** presents a technique that could, if necessary, be performed by a single fire fighter:

1 The rescue team places the ladder into the rescue position, with the tip of the ladder just below the windowsill, and secures the ladder in place. (**STEP 1**)

2 The first fire fighter climbs the ladder, makes contact with the victim, and climbs inside the window to assist the victim. Make contact as soon as possible to calm the victim, and encourage the victim to stay at the window until the rescue can be performed. (**STEP 2**)

3 The second fire fighter climbs up to the window, standing one rung below the windowsill. This leaves at least one rung available for the victim. The fire fighter on the ladder should apply a leg lock or use a safety belt. When ready, the fire fighter advises the victim to slowly come out onto the ladder, feet first and facing the ladder. (**STEP 3**)

4 The fire fighter forms a semicircle around the victim, with both hands on the beams of the ladder. (**STEP 4**)

5 The fire fighter and victim proceed slowly down the ladder, one rung at a time, with the fire fighter always staying one rung below the victim.

6 If the victim slips or loses his or her footing, the fire fighter's legs should keep the victim from falling.

7 The fire fighter can take control of the victim at any time by leaning in toward the ladder and squeezing the victim against the ladder. (**STEP 5**)

Rescuing an Unconscious Person from a Window

If the trapped person is unconscious, one or more fire fighters will have to climb inside the building and pass the person out of the window to a fire fighter on the ladder. Caution should be used when lowering an unconscious victim down a ladder, because it is very easy for the victim's arms or legs to get caught in the ladder. To rescue an unconscious person via a ladder, follow the steps in **SKILL DRILL 14-17** :

1 The rescue team places and secures the ladder in the rescue position, with the tip of the ladder just below the windowsill. (**STEP 1**)

2 One fire fighter climbs up the ladder and into the window to assist from the inside. The second fire

---

SKILL DRILL 14-16   Rescuing a Conscious Victim from a Window
(Fire Fighter I, NFPA 5.3.9)

1 The rescue team places the ladder into the rescue position, with the tip of the ladder just below the windowsill, and secures the ladder in place.

2 The first fire fighter climbs the ladder, makes contact with the victim, and climbs inside the window to assist the victim. Make contact as soon as possible to calm the victim, and encourage the victim to stay at the window until the rescue can be performed.

3 The second fire fighter climbs up to the window, standing one rung below the windowsill. This leaves at least one rung available for the victim. When ready, the fire fighter advises the victim to slowly come out onto the ladder, feet first and facing the ladder.

© LiquidLibrary

(Continues)

SKILL DRILL 14-16 | Rescuing a Conscious Victim from a Window (*Continued*)
(Fire Fighter I, NFPA 5.3.9)

4 The fire fighter forms a semicircle around the victim, with both hands on the beams of the ladder.

5 The fire fighter and victim proceed slowly down the ladder, one rung at a time, with the fire fighter always staying one rung below the victim. If the victim slips or loses his or her footing, the fire fighter's legs should keep the victim from falling. The fire fighter can take control of the victim at any time by leaning in toward the ladder and squeezing the victim against the ladder.

© LiquidLibrary

fighter climbs up to the window opening and waits for the victim. (**STEP 2**)

3 The fire fighter on the ladder keeps a firm grip on the ladder with both hands on the rungs. One leg should be straight, and the other should be bent so that the thigh is horizontal to the ground, with the knee at a 90-degree angle. The foot of the straight leg should be one rung below the foot of the bent leg.

4 When both fire fighters are ready, the interior fire fighter passes the victim out through the window and onto the ladder. The victim's back should be toward the ladder, so that the victim is face to face with the fire fighter on the ladder. (**STEP 3**)

5 The victim is lowered so that his or her groin rests on the horizontal leg of the fire fighter. The fire fighter places his or her arms under the arms of the victim and holds on to the rungs. It is important to keep the balls of both feet on the rungs of the ladder, because it is much more difficult to move your feet in this position if the heels are close to the rungs.

6 The fire fighter climbs down the ladder slowly, one rung at a time. The victim is always supported at the groin by one of the fire fighter's legs. The fire fighter's arms remain under the victim's arms to support the upper torso. Optionally, the fire fighter in the interior of the building may tie the victim's hands together and place them over the neck of the fire fighter on the ladder. (**STEP 4**)

Rescuing an Unconscious Child or a Small Adult from a Window

Small adults and children can be cradled across a fire fighter's arms during a rescue. To use this rescue technique, the child must be light enough that the fire fighter can descend safely, using only arm strength to support the victim. To carry an unconscious child down a ladder, follow the steps in **SKILL DRILL 14-18**:

1 The rescue team sets up and secures the ladder in the rescue position, with the tip of the ladder just below the windowsill. (**STEP 1**)

SKILL DRILL 14-17

Rescuing an Unconscious Victim from a Window
(Fire Fighter I, NFPA 5.3.9)

1 Place the tip of the ladder just below the windowsill.

2 One fire fighter enters to rescue the victim. The second fire fighter climbs to the window.

3 The fire fighter waiting on the ladder places both hands on the rungs, with one leg straight and the other horizontal to the ground, with the knee at an angle of 90 degrees. The interior fire fighter passes the victim through the window and onto the ladder, keeping the victim's back toward the ladder.

4 The victim is lowered so that he or she straddles the fire fighter's leg. The fire fighter's arms should be positioned under the victim's arms, holding on to the rungs. Step down one rung at a time, transferring the victim's weight from one leg to the other. The victim's arms can also be secured around the fire fighter's neck.

2 The first fire fighter climbs the ladder and enters the window to rescue from the interior. The second fire fighter climbs the ladder to the window opening and waits for the victim. Both of the second fire fighter's arms should be level with his or her hands on the beams. (**STEP 2**)

3 When ready, the interior fire fighter passes the victim to the fire fighter on the ladder so the victim is cradled across the fire fighter's arms. (**STEP 3**)

4 The fire fighter climbs down the ladder slowly, with the victim being held in his or her arms. The fire fighter's hands should slide down the beams. (**STEP 4**)

Rescuing a Large Adult

Three fire fighters using two ladders may be needed to rescue very tall or heavy adults. To rescue a large adult using a ladder, follow the steps in **SKILL DRILL 14-19**:

1 The rescue team places and secures two ladders, side by side, in the rescue position. The tips of the two ladders should be just below the windowsill. (**STEP 1**)

SKILL DRILL 14-18 Rescuing an Unconscious Child or a Small Adult from a Window
(Fire Fighter I, NFPA 5.3.9)

1 Place the ladder in rescue position, with the tip below the windowsill.

2 One fire fighter enters the window to assist the victim. The second fire fighter stands on the ladder to receive the victim, with both arms level and hands on beams.

3 The victim is placed in the fire fighter's arms.

4 The fire fighter descends, keeping arms level and sliding the hands down the beams.

© LiquidLibrary

SKILL DRILL 14-19 Rescuing a Large Adult
(Fire Fighter I, NFPA 5.3.9)

1 The rescue team places and secures two ladders, side by side, in the rescue position. The tips of the two ladders should be just below the windowsill.

2 Multiple fire fighters may be required to enter the window to assist from the inside.

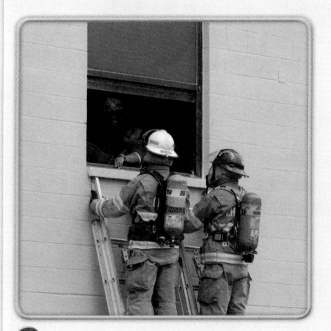

3 Two fire fighters, one on each ladder, climb up to the window opening and wait for the victim.

4 When ready, the victim is lowered down across the arms of the fire fighters, with one fire fighter supporting the victim's legs and the other supporting the victim's arms. Once in place, the fire fighters can slowly descend the ladder, using both hands to hold on to the ladder rungs.

② Multiple fire fighters may be required to enter the window to assist from the inside. (**STEP ②**)

③ Two fire fighters, one on each ladder, climb up to the window opening and wait for the victim. (**STEP ③**)

④ When ready, the victim is lowered down across the arms of the fire fighters, with one fire fighter supporting the victim's legs and the other supporting the victim's arms. Once in place, the fire fighters can slowly descend the ladder, using both hands to hold on to the ladder rungs. (**STEP ④**)

■ Removal of Victims by Ladders

Ladders should be used to remove victims only when it is not possible to use interior stairways, fire escapes, or aerial towers. A ladder rescue is often frightening to conscious victims. Rescuing an unconscious victim by ladder is dangerous and difficult, but it may be the best way to save a life.

Aerial Ladders and Platforms

An aerial ladder or platform can be used for rescue operations. The same basic rescue techniques are used with both aerial and ground ladders, but aerial ladders offer several advantages over ground ladders: Aerial ladders are much stronger and have a longer reach. In addition, they are wider and more stable than ground ladders, with side rails that provide greater security for both rescuers and victims **FIGURE 14-14**.

Aerial platforms are even more suitable than aerial ladders for rescue operations. These devices reduce the risk of slipping and falling because the victim is lowered to the ground mechanically. An aerial platform is usually preferred for rescue work if one is available.

FIGURE 14-14 An aerial ladder is stronger and more stable than a ground ladder.

Ground Ladders

Before a ground ladder can be used in a rescue, it must be properly positioned and secured. Positioning and securing ground ladders are covered in the chapter "Ladders." Additional personnel will be needed to secure the ladder and to assist in bringing the victim down the ladder. It is not often that fire fighters need to perform a rescue using a ground ladder. This type of rescue is physically taxing and requires carefully executed skills. It is important to practise the skills to be prepared for this event, as those skills may be needed at any time.

Chief Concepts

- Search and rescue are almost always performed in tandem, yet they are separate operations. The purpose of search operations is to locate living victims. The purpose of rescue operations is to physically remove an occupant or victim from a dangerous environment.
- The IC and fire fighters must plan and coordinate all fire-suppression operations—from ventilation to fire attack—to support the search and rescue priority.
- Sometimes fire fighters must confine or control a fire before search and rescue operations can begin.
- The size-up process at every fire should include a specific evaluation of the critical factors for search and rescue—the number of occupants in the building, their location, the degree of risk to their lives, and their ability to evacuate by themselves. The type of occupancy; the size, construction, and condition of the building; the apparent smoke and fire conditions; and the time of day and the day of the week are all important observations that could indicate whether and how many people may need to be rescued. A search and rescue plan can then be developed based on this information.
- Plans for search and rescue must consider the risks and benefits of the operations. In some cases, such as when conditions are ripe for a backdraft, search and rescue operations may not be performed.
- Search and rescue size-up considerations include the following factors:
  - Occupancy
  - Size of the building
  - Construction of the building
  - Time of day and day of week
  - Number of occupants
  - Degree of risk to the occupants presented by the fire
  - Ability of occupants to exit on their own
- Never assume that a building is occupied or unoccupied. Look for clues, such as cars parked in the driveway or a light on in the window, that indicate the presence of an occupant. However, the decision to perform search and rescue operations should not be made solely on these observations.
- Dispatch, neighbours, bystanders, and occupants who escaped on their own are sources of information on the status of a structure's occupants.
- A search begins in the areas where victims are at the greatest risk. Area search assignments should be based on a system of priorities:
  - First priority—Search the areas where live victims may be located immediately around the fire, and then the rest of the fire floor.
  - Second priority—Search the area directly above the fire and the rest of that floor.
  - Third priority—Search the higher-level floors, working from the top floor down, because smoke and heat are likely to accumulate in these areas.
  - Lowest priority—Search areas below the fire floor.
- A primary search is a quick attempt to locate victims who are in danger. It should be completed in 15 minutes or less. Fire fighters must rely on their senses, such as touch, hearing, and sight, as they search for victims in smoky conditions.
- When searching a room, fire fighters should use the walls to orient themselves and as guides around the room. The room should be searched in a systematic pattern.
- Thermal imaging devices may be used to search for victims in darkness or smoke-obscured conditions. The thermal imaging device detects the temperature differential of the victim's body from the room.
- Search ropes may be used in large, open areas so fire fighters may maintain their orientation while searching for victims.
- The vent–entry–search method uses a window as the entry and exit point to a room.
- A secondary search is conducted after the fire is under control or fully suppressed. It is a slower search that is designed to locate any deceased victims and search any areas not covered in the primary search.
- Although every emergency operation involves a degree of unavoidable, inherent risk to fire fighters, during search and rescue operations, fire fighters will probably encounter situations that involve a significantly higher degree of personal risk. Assuming this level of risk is acceptable only if there is a reasonable probability of saving a life. In some cases, the IC may decide that the risk to fire fighters is too great and a primary search may not be conducted.
- Sagging walls, chipped or cracked mortar or cement, warped or failing structural steel, and any partial collapse are significant indicators of an impending collapse. Dark, black, turbulent smoke, blackened windows, the appearance of a "breathing" building, and intense heat are signs of possible flashover or backdraft.
- Rescue is the removal of a located person who is unable to escape from a dangerous situation; it is the second component of search and rescue. When you locate a victim during a search, you must direct, assist, or carry that person to a safe area. Rescue can be as simple as verbally directing an occupant toward an exit, or it can be as demanding as extricating a trapped, unconscious victim and physically carrying that person out of the building.
- Children and elderly persons, physically or developmentally handicapped persons, impaired people, and ill or

www.FireFighter.jbpub.com/Canadian3e

injured persons may be unable to escape from a structure and, therefore, may need to be rescued. Victims also may become trapped when a rapidly spreading fire, an explosion, or a structural collapse cuts off potential escape routes. Because fires threaten the lives of both victims and rescuers, the first priority is to remove the victim from the fire building or dangerous area as quickly as possible.

- In some situations, the best option is to shelter the occupants in place instead of trying to remove them from a fire building. This option should be considered when the occupants are conscious and are found in a part of the building that is adequately protected from the fire by fire-resistive construction and/or fire-suppression systems. Such a situation occurs in a high-rise apartment building when a fire is confined to one apartment or one floor.
- The simplest rescue is the exit assist, in which the victim is responsive and able to walk without assistance or with very little assistance. Two types of assists can be used to help responsive victims exit a fire situation:
  - One-person walking assist
  - Two-person walking assist
- Four simple carry techniques can be used to move a victim who is conscious and responsive, but incapable of standing or walking:
  - Two-person extremity carry
  - Two-person seat carry
  - Two-person chair carry
  - Cradle-in-arms carry
- The most efficient method to remove an unconscious or unresponsive victim from a dangerous location is a drag. Five emergency drags can be used to remove unresponsive victims from a fire situation:
  - Clothes drag
  - Blanket drag
  - Webbing sling drag
  - Fire fighter drag
  - Emergency drag from a vehicle

- When using an emergency drag, make every effort to pull the victim in line with the long axis of the body to provide as much spinal protection as possible. The victim should be moved head first to protect the head.
- Assisting someone down a ground ladder carries a considerable risk of injury to both fire fighters and victims. Whenever it is possible to use a stairway, fire escape, or aerial tower for rescue, these options should be considered before using ground ladders.

Hot Terms

<u>Primary search</u> A rapid search to locate living victims.

<u>Rekindle</u> A return to flaming combustion after apparent but incomplete extinguishment. (NFPA 921)

<u>Rescue</u> Those activities directed at locating endangered persons at an emergency incident, removing those persons from danger, treating the injured, and providing for transport to an appropriate health care facility. (NFPA 402)

<u>Search</u> The process of locating victims who are in danger.

<u>Search rope</u> A guide rope used by fire fighters that allows them to maintain contact with a fixed point.

<u>Secondary search</u> A methodical search designed to locate deceased victims after the fire has been suppressed, searching areas not covered during the primary search.

<u>Thermal imaging device</u> An electronic device that detects differences in temperature based on infrared energy and then generates images based on those data. This device is used in smoky environments to locate victims.

<u>Two-in-two-out rule</u> A safety procedure that requires a minimum of two personnel to enter a hazardous area and a minimum of two backup personnel to remain outside the hazardous area during the initial stages of an incident.

You are working the night shift on a truck company at a dual-company station. Shortly after 1:15 a.m., the gong sounds, indicating a fire in a house located just down the street from the station. You slide down the pole and quickly mount the apparatus, when the dispatcher indicates that neighbours are reporting they hear screaming coming from the house. When you arrive, you find a two-storey, wood-frame house, with heavy smoke showing. The engine begins pulling handlines, while two members of your crew are assigned ventilation duties and you and the other fire fighter are assigned primary search responsibility.

1. The main purpose of _____ is to find deceased victims.
- **A.** search
- **B.** rescue
- **C.** primary search
- **D.** secondary search

2. When rescuing a victim from a window, the ladder tip should be placed _____.
- **A.** just below the windowsill
- **B.** just above the windowsill
- **C.** beside the window
- **D.** above the window header

3. Where should a search begin?
- **A.** Search the areas where live victims may be located immediately around the fire.
- **B.** Search the area directly above the fire and the rest of that floor.
- **C.** Search the higher-level floors, working from the top floor down.
- **D.** Search the fire floor, working your way toward the fire.

4. The minimum number of rescuers required to properly remove a victim from a vehicle and onto a longboard is _____.
- **A.** two
- **B.** three
- **C.** four
- **D.** five

5. The most efficient method to remove an unconscious or unresponsive victim from a dangerous location is a _____.
- **A.** two-person walking assist
- **B.** clothes drag
- **C.** two-person extremity carry
- **D.** two-person chair carry

You are at the training centre, waiting for the class to begin. A number of the fire fighters are standing around outside discussing search and rescue. One of the more outspoken individuals makes the comment, "We should search every structure regardless whether it is vacant." He goes on to say, "As far as I am concerned, we should search no matter what because we might save a life."

1. What would you do if every other person there agreed with his comments?

2. What does the fire service generally recommend for firefighting in vacant structures?

3. Is this fire fighter's suggestion in these circumstances correct?

4. How would you address this situation?

© James Quigg/AP Photos

Ventilation

Fire Fighter I

Knowledge Objectives

After studying this chapter, you will be able to:

- Describe the two components of ventilation operations. (p 432)
- Explain the conditions that lead to an underventilated fire. (NFPA 5.3.11A , p 432–433)
- Describe the potential hazards posed by an underventilated fire. (p 432–434)
- Describe the potential hazards posed by the products of combustion. (p 433–434)
- Describe how ventilation operations affect the behaviour of the products of combustion. (p 434)
- Describe how convection aids in fire spread. (p 433–434)
- List the benefits of ventilation. (p 434–435)
- Explain how proper ventilation operations can aid in preventing backdraft. (p 435)
- Explain how proper ventilation operations can aid in preventing flashover. (p 435)
- Explain how the principles of heat transfer affect ventilation operations. (NFPA 5.3.12A , p 436)
- Explain how wind and atmospheric forces affect ventilation operations. (NFPA 5.3.11A , p 436–437)
- Explain how fire-resistive construction affects ventilation operations. (p 437–438)
- Describe how the characteristics of ordinary construction affect ventilation operations. (p 438)
- Describe how the characteristics of wood-frame construction affect ventilation operations. (p 438–439)
- List the two basic types of ventilation. (p 440)
- Explain how horizontal ventilation removes contaminated atmosphere from a structure. (p 440–441)
- List the two methods of horizontal ventilation. (NFPA 5.3.11 , p 441–448)
- Explain how natural ventilation removes contaminated atmosphere from a structure. (NFPA 5.3.11A , p 441)
- Describe the techniques used to provide natural ventilation to a structure. (NFPA 5.3.11 , p 441–445)
- Explain how mechanical ventilation removes contaminated atmosphere from a structure. (NFPA 5.3.11A , p 445–448)
- Describe the techniques used to provide mechanical ventilation to a structure. (NFPA 5.3.11 , p 445–448)

- Describe how negative-pressure ventilation removes contaminated atmosphere from a structure. (p 445–446)
- Describe the techniques used to provide negative-pressure ventilation to a structure. (p 446–448)
- Describe how positive-pressure ventilation removes contaminated atmosphere from a structure. (NFPA 5.3.11A , p 446–448)
- Describe the advantages of positive-pressure ventilation. (p 447–448)
- Describe the disadvantages of positive-pressure ventilation. (p 447–448)
- Describe the techniques used to provide positive-pressure ventilation to a structure. (NFPA 5.3.11 , p 448)
- Describe how hydraulic ventilation removes contaminated atmosphere from a structure. (NFPA 5.3.11A , p 448–449)
- Describe how vertical ventilation removes contaminated atmosphere from a structure. (NFPA 5.3.12 , p 449–453)
- Describe how to ensure fire fighter safety during vertical ventilation operations. (p 450–453)
- Identify the warning signs of roof collapse. (NFPA 5.3.12A , p 453)
- Describe the components and characteristics of roof support structures. (p 453)
- Describe the components and characteristics of roof coverings. (p 453)
- Explain how roof construction affects fire resistance. (p 453–454)
- List the differences in solid-beam construction and truss construction in roofs. (p 454–456)
- List the basic types of roof design. (p 455–456)
- Describe the characteristics of flat roofs. (p 455)
- Describe the characteristics of pitched roofs. (p 455)
- Describe the characteristics of arched roofs. (p 455–456)
- Describe the techniques of vertical ventilation. (NFPA 5.3.12 , p 456–457)
- List the tools utilized in vertical ventilation. (NFPA 5.3.12 , p 457)
- List the types of roof cuts utilized in vertical ventilation operations. (NFPA 5.3.12A , p 459–464)
- Describe the characteristics of a rectangular or square cut. (p 459–460)
- Describe the characteristics of a louvre cut. (p 460–462)
- Describe the characteristics of a triangular cut. (p 461–462)
- Describe the characteristics of a peak cut. (p 462–463)
- Describe the characteristics of a trench cut. (p 462, 464)

- Describe the special considerations in ventilating concrete roofs. (p 464)
- Describe the special considerations in ventilating metal roofs. (p 464–465)
- Describe the special considerations in ventilating basements. (p 465)
- Describe the special considerations in ventilating high-rise buildings. (p 465–466)
- Describe the special considerations in ventilating windowless buildings. (p 467)
- Describe the special considerations in ventilating large buildings. (p 467)
- Explain how to ensure that ventilation equipment is in a state of readiness. (p 467–469)

Skills Objectives

After studying this chapter, you will be able to:

- Break glass with a hand tool. (NFPA 5.3.11B , p 441–442)
- Break a window with a ladder. (p 443–444)
- Break a window with a Halligan tool. (p 443, 445)
- Deliver negative-pressure ventilation. (p 446, 447)
- Deliver positive-pressure ventilation. (p 448–449)
- Perform hydraulic ventilation. (p 449)
- Sound a roof. (NFPA 5.3.12B , p 452)
- Operate a power saw. (p 457–458)
- Perform a rectangular cut. (p 459–460)
- Make a louvre cut. (p 460–462)
- Make a triangular cut. (p 461–462)
- Make a peak cut. (p 462–463)
- Make a trench cut. (p 464)
- Perform a readiness check on a power saw. (p 467–468)
- Maintain a power saw. (p 467–469)

Fire Fighter II   FFII

Knowledge Objectives

After studying this chapter, you will be able to:

- List the three tactical priorities in structural firefighting operations and how the tactical priorities affect ventilation operations. (NFPA 6.3.2, 6.3.2B , p 439)
- Identify the warning signs of roof collapse. (NFPA 6.3.2A , p 453)

Skills Objectives

There are no skills objectives for Fire Fighter II candidates. NFPA 1001 contains no Fire Fighter II Job Performance Requirements for this chapter.

You Are the Fire Fighter

It is a routine day at the grocery store where you work full-time. You are busy stocking shelves, when your pager goes off, saying there is a fire in a house only a few blocks away. Because the fire is between you and the fire station, your departmental policy allows you to go directly to the scene. On your arrival, you find a one-storey, wood-frame house with dirty, brown smoke pushing out through the eaves, siding, and other cracks. The windows are stained black. The neighbours say they do not think anyone is home, because the occupants work during the day.

1. What is the best way to ventilate this structure?
2. What concerns do you have about ventilating this structure?
3. When would you want to ventilate this structure?

Introduction

Fire service **ventilation** is the process of removing smoke, heat, and toxic gases from a burning building and replacing them with cooler, cleaner, oxygen-rich air. The ventilation process has two different components. The first is the removal of smoke, toxic gases, and hot air. This removal is helpful in fire-suppression efforts and in improving the conditions for any trapped victims by increasing the visibility and reducing the air temperature. The second is the addition of cooler, cleaner, and oxygen-rich air. Although the removal of heat and smoke is beneficial, the sudden addition of oxygen-rich air to a smoke-filled and fuel-rich atmosphere can result in a sudden flashover or a backdraft.

Modern construction techniques use lightweight and manufactured building components that contain resins and glues. Modern construction is designed to keep the wind, wet, and cold out by using tightly sealed windows, vapour shields, and insulation. Structures with heating and cooling systems also tend to have closed windows year-round. Finally, structures are filled with petroleum-based furnishings. The cumulative result is that today's fires often quickly reach high temperatures and use up the available oxygen rapidly. The fire has a limited supply of oxygen while the doors and windows remain shut. It also has an abundance of fuel, even though it is an **underventilated** fire.

Fire Fighter Safety Tips
© Photos.com

Effective ventilation removes smoke, gases, and superheated air, but it also brings in cooler, cleaner, and oxygen-rich air. This can have a dramatic effect on a fire if ventilation is not coordinated with fire attack operations.

Fire Fighter Safety Tips
© Photos.com

Today, fire in a room can reach the point of flashover in a few minutes.

If fire fighters arrive when a fire is in an underventilated state, they might see little smoke and assume that they are confronting a small fire, a fire that has exhausted most of its fuel supply, or a fire in a state of decay. An underventilated fire simply needs a supply of oxygen to become a raging fire, a flashover, or a backdraft. This supply can occur if heat breaks a window or if a fire fighter forces open a door. The phenomenon of underventilated fires has led to modifications in ventilation techniques used by the fire service.

Fire Behaviour and Ventilation

When ventilation is coordinated with fire attack and the application of water, it can save lives and reduce property damage. Proper ventilation assists in locating and rescuing victims. It can enable hose teams to advance and locate the source of the fire and can prevent fire spread. A lack of ventilation or use of improper ventilation techniques can lead to violent backdrafts, delay in extinguishing the fire, and unnecessary extension of the fire. Improper ventilation can injure fire fighters and civilians and increase property damage.

During its normal progression and growth, a fire gives off smoke, heat, and toxic gases. As long as fuel and oxygen are available to feed it, the fire will continue to burn and produce these three products of combustion. When there is fire inside a building, the structure acts as a container or box, trapping the products of combustion. As the fire grows and develops, the smoke, heat, and toxic gases spread throughout the structure, where they present a direct risk to the lives of occupants and fire fighters, reduce visibility, and increase property damage. These trapped products of combustion create a potential for explosion. Ventilation removes some of these products from the interior atmosphere and allows them to escape in a controlled manner.

The primary principle that controls the spread of smoke, heat, and toxic gases within a room or a building is convection. Heated gases expand, which means that they become less dense than cooler gases. As a result, the hot gases produced by a fire in a closed room will rise to the ceiling and spread outward, displacing cooler air and pushing it toward the floor **FIGURE 15-1**. As the fire continues to burn, the hot layer of gas banks (curves) down closer to the floor.

If the heated products of combustion escape from the room, the same principles apply as they spread throughout a building. Smoke, heat, and toxic gases spread horizontally, along the ceiling, until they find a path such as a stairway, elevator shaft, pipe **chase** (an open space within a wall where wires and pipes can run), or an outside opening that allows them to reach a higher level. They will then flow upward through the vertical opening, until they reach another horizontal obstruction, such as the ceiling of the highest floor or the underside of the roof **FIGURE 15-2**. At that point, they again spread out horizontally and bank down

as they accumulate. This process of spreading out and banking down is called **mushrooming** **FIGURE 15-3**.

As long as the products of combustion remain trapped within the structure, they present a series of risks and dangers both to the building's occupants and to fire fighters. Most of the gases produced by a fire—most notably, carbon monoxide and cyanide—are toxic and flammable. The contaminated atmosphere they create poses a life-threatening condition to occupants as well as to anyone who enters without self-contained breathing apparatus (SCBA) **FIGURE 15-4**. The gases can be so hot that simply breathing them causes fatal respiratory burns. As the fire continues to consume oxygen from the atmosphere, the oxygen deficiency poses yet another respiratory hazard. The particulate matter in smoke can severely obscure

FIGURE 15-1 Convection currents cause heated products of combustion to rise within a room and spread along the ceiling.

FIGURE 15-2 The heated products of combustion spread horizontally along the ceiling, until they find a path that allows them to flow upward.

FIGURE 15-3 Mushrooming occurs as the heated products of combustion flow outward and downward.

FIGURE 15-4 SCBA is essential when entering an area contaminated by products of combustion.

© Keith D. Cullom

visibility, making it impossible for occupants to find their way to an exit or for rescuers to locate trapped occupants. Smoke also irritates the eyes and the mucous membranes of the respiratory system.

Convection—the flow of heated gases produced by the fire—is one of the primary mechanisms of fire spread. These gases can be hot enough to ignite combustible materials along their path; thus, the fire spreads along the paths taken by the heated gases and in the areas where they accumulate. If the hot gases spread into additional areas through vertical or horizontal openings, there is a high probability that those areas will become involved in the fire.

In addition to igniting combustible materials, the hot gases, smoke, and other products of combustion can themselves explode. In many cases, they include a rich supply of partially burned fuels that are hot enough to ignite but lack sufficient oxygen to support combustion. When these products are mixed with clean air, this flammable atmosphere can ignite or explode in a backdraft (a sudden explosive ignition of fire gases that occurs when oxygen is introduced into a superheated space). Even if they do not ignite, the products of combustion are often hot enough to cause thermal burns to exposed skin.

Benefits of Proper Ventilation

Ventilation is the process of controlling the flow of smoke, heat, and toxic gases, so that they are released safely and effectively from a building. Ventilation must be closely coordinated with all other activities being carried out at a fire scene. We do not send fire fighters into buildings with gas leaks without ventilation, yet every day, we send fire fighters into smoke-filled buildings—sometimes before ventilation has taken place.

The life-safety benefits of ventilation are of primary importance. Properly performed ventilation allows the smoke to lift slightly, so that fire fighters can locate trapped occupants more rapidly. In addition, proper ventilation can keep smoke away from people who are away from the fire area.

The role of ventilation in removing heat is important for fire fighters who are advancing an initial attack hose line. As

the fire fighters advance, ventilation can relieve the pressure and intense heat of the fire and create a much safer environment. A vent in front of the attack hose line can help to pull away the steam created when water cools the flames. It also cools the atmosphere and prevents the steam from banking down on top of the attack line. In addition, the ventilation opening causes the smoke to lift and allows fire fighters to aim the hose stream directly at the **seat of the fire** (the main area of the fire). By reducing the potential for a backdraft or a flashover (the sudden ignition of all combustible objects in a room), effective ventilation makes conditions safer for the attack team.

Proper ventilation also helps to limit fire spread within the structure. Ventilating near the source of the fire can limit the area of involvement and reduce the spread of heat and toxic gases throughout the structure **FIGURE 15-5**. Initial ventilation allows more oxygen to reach the fire. As mentioned previously, this can result in a rekindling of the fire in the fire room, flashover, or a violent backdraft. For this reason, it is essential to have a hose line in place and to be ready to hit the fire prior to ventilation. Careful timing of ventilation enables fire fighters to remain safe while confining and extinguishing the fire quickly.

Releasing trapped heat and smoke from the upper levels of a building can prevent the fire from spreading horizontally into adjoining areas or neighbouring buildings. A well-placed opening releases heat and smoke to the outside, so that fire fighters can operate effectively inside the building. This ventilation should be as close to the fire as possible. Ventilating a fire far from the fire can set up convection currents that carry the fire between its source and the vent opening.

Another concern is the property damage caused by exposure to smoke and heat. Soot and other residues of smoke often cause as much property damage as do flames. Significant heat damage can occur in areas that were exposed only to

A.   **B.**

FIGURE 15-5 Proper ventilation enables fire fighters to control a fire rapidly. **A.** Vented structure. **B.** Unvented structure.

products of combustion, not to the fire itself. Even when a fire is contained and under control, prompt and effective ventilation of the accumulated heat and smoke can be an important factor in limiting property losses.

After the fire is under control, ventilation should be continued until all of the heat, smoke, and toxic products of combustion have been removed from the building. Complete ventilation creates a clear atmosphere for overhauling the fire and limits property losses caused by smoke damage.

Backdraft and Flashover Considerations

Ventilation is a major consideration in two significant fire-ground phenomena: backdraft and flashover. Both can be deadly situations, and fire fighters should exercise great caution when conditions indicate that either is possible.

■ Backdraft

A backdraft can occur when a building is charged with hot gases and most of the available oxygen has been consumed. Few flames might be apparent, but the hot gases can contain rich amounts of unburned or partially burned fuel. If clean air is introduced to the mixture, the fuel can ignite and explode. To reduce the chance of a backdraft, fire fighters try to release as much heat and unburned products of combustion as possible.

Placing a ventilation opening as high as possible within the building or area can help to eliminate potential backdraft conditions. A roof opening can draw the hot gases up and relieve the interior pressure. As the mixture rises into the open atmosphere and is exposed to oxygen, it might ignite. The ventilation crew should ensure that a charged hose line is in place to attack the fire before they ventilate the building.

FIRE FIGHTER Tips

It might also be necessary to place a hose line to protect exposures.

The attack crews should charge their hose lines outside the building. The ventilation crew should not ventilate until the attack crew is ready to advance the attack line. It is very important to remember that opening the front door of a structure to begin a fire attack is an act of ventilation. The door should remain closed until the attack team is ready to advance the hose line into the building.

Once the attack team sees flames inside the structure, they should open their hose streams to extinguish the fire and cool the interior atmosphere as quickly as possible. In cases where flames are beginning to move across the ceiling, a short burst of water on the ceiling helps to cool the upper area of the room without upsetting the thermal balance.

■ Flashover

Both ventilation and cooling are needed to relieve potential flashover conditions. Flashover can occur when the air in a room is superheated and exposed combustibles in the space are near their ignition point TABLE 15-1. In a typical room-and-contents fire, the ceiling temperature can reach 648.8 degrees Celsius (1200 degrees Fahrenheit). In this situation, applying short bursts of water cools the upper atmosphere, while ventilation draws the heated smoke and gases out of the space and oxygen-rich air enters the space. Ventilation openings should be placed close to the fire to vent the heat and flames away from the attack crew. Venting provides an air flow path to the fire and a pathway from the fire. The fire can travel along the pathway of the venting gases if these gases are not cooled. The steam produced by the water stream moves toward the ventilation opening. Flame conditions might intensify briefly, but the heat and steam should be directed away from the attack crew.

TABLE 15-1	Chemical Properties of Fire Gases	
Chemical Name	Ignition Temperature	Flammable Range
Propane	480°C (896°F)	2.1–9.5 percent
Methane	580°C (1076°F)	5–15 percent
Carbon monoxide	609°C (1128°F)	12.5–74 percent
Hydrogen cyanide	538°C (1000°F)	5.6–40 percent

Fire Fighter Safety Tips

© Photos.com

Because the potential harm from a flashover is so great, you need to be able to recognize the signs of a potential flashover.

FIRE FIGHTER Tips

Fire fighters trained in flashover containers are taught to use small hose streams with a 38- or 44-millimetre (1.5- or 1.75-inch) nozzle to cool the upper atmosphere to reduce the flashover. These containers are about 36.2 cubic metres (1280 cubic feet), or approximately the size of a room at a residence; therefore, this technique might not be effective in larger commercial structures.

Factors Affecting Ventilation

To ventilate a structure successfully, fire fighters must consider how fire behaviour dictates the movement of the products of combustion. Armed with this knowledge, fire fighters can develop a plan that specifies where and when to create openings to release these products and limit smoke and fire spread. Fire behaviour is directly related to the principles of heat transfer—namely, convection, conduction, and radiation. The natural movement of smoke and heat within a structure is primarily attributable to convection.

■ Convection

Convection is the transfer of heat through a circulating medium of liquid or gas. Heated air naturally expands and rises, carrying smoke and gases as it flows up and outward from a fire. This heated mixture flows across the underside of ceilings, through doorways, and up stairways, elevator shafts, and vertical concealed spaces, such as pipe chases. Unless it is released to the exterior, the heat accumulates in any accessible space in the upper levels of a building.

As the heated air rises, it displaces cooler air, which is then drawn toward the seat of the fire. The cooler air brings oxygen to support combustion, causing the fire to intensify. As the fire continues to burn, the products of combustion spread throughout the building, eventually banking down to lower levels. Within any room or space, the hottest gases are found at the highest level, with cooler gases being stratified below.

Convection currents can carry smoke and superheated gases to uninvolved areas within the structure or into adjoining spaces, if an opening is available. The flow of heated gases always follows the path of least resistance, so it can be altered by making an opening that offers less resistance. Often structures have openings and natural voids that might have been in the original construction or were added following the original construction. These features contribute to the convection-based fire spread in the building. Fire fighters must be able to read the building from the outside to try to determine where the fire will travel, whether through false ceilings, multiple roofs, or concealed spaces.

Fire fighters take advantage of this basic principle by making ventilation openings that cause the convective flow to draw the heated products out of the building. Doorways, windows, or roof openings can all be used for ventilation. To accomplish this goal, fire fighters must identify the most appropriate type of openings as well as the most feasible locations for openings in the structure.

Thermopane windows increase the energy efficiency of a structure. The windows fit tightly so that little air leakage occurs, and the glass resists heat transfer through it. According to the Underwriter's Laboratory (UL) report *Impact of Ventilation on Fire Behavior in Legacy and Contemporary Residential Construction* by Stephen Kerber (2010), Thermopane windows keep more heat and smoke in the building than conventional windows. This increased heat and smoke may lead to faster flashover times.

Fire fighters should evaluate the structure and fire conditions for signs of a backdraft before creating any type of ventilation opening **FIGURE 15-6**.

■ Mechanical Ventilation

In addition to making or controlling openings to influence convective flow, fire fighters can use mechanical ventilation to

FIRE FIGHTER Tips

Recent National Institute of Standards and Technology (NIST) and UL studies indicate that the action of applying water to the fire does not push the fire or make conditions untenable in other parts of a residence. The movement of the fire is primarily caused by the flow of the smoke and hot gases. This movement is influenced by natural convection currents, mechanical ventilation, and wind.

FIRE FIGHTER Tips

Concerns about saving energy have led to increased use of insulated thermal glass, both in new construction and as a replacement for conventional glass in older windows. Thermopane windows have multiple layers of glass sealed together, with a small air space between the layers. These windows are much more difficult to break than conventional glass windows are. In fact, a fire fighter might have to use multiple strikes and additional force to break a Thermopane window.

direct the flow of combustion gases. Fans can be used to draw or pull smoke through openings (**negative-pressure ventilation**) or to introduce fresh air to displace smoke and other products of combustion (**positive-pressure ventilation**). Hose streams can be used to create a hydraulic ventilation to remove hot gases and smoke from a room. Some buildings have ventilation systems that are designed to remove smoke or prevent smoke from entering certain areas.

Fire Fighter Safety Tips

© Photos.com

If the fire is located in structural spaces, do not use positive-pressure ventilation until access to these spaces is available and attack crews are in place.

■ Wind and Atmospheric Forces

Wind and other atmospheric forces can play a significant role in ventilation. The direction and force of the wind should always be considered when determining where and how to ventilate a building, because even a slight breeze can have a dramatic influence on the effectiveness of a ventilation opening. The force of the wind can turn an underventilated fire in a decay state into a flashover.

For example, wind blowing against one side of a building can prevent smoke, heat, or other products of combustion from escaping through a ventilation opening on that side. If a door or window on that side of the building is opened, or breaks because of heat, a strong current of oxygen-rich air will rush into the structure and lead to the rapid acceleration of the fire. The wind can push the heat and other products of combustion throughout the structure, creating serious life-safety risks for those inside the building.

Conversely, opening a ventilation outlet on the downwind side of a building can produce effective ventilation. As the wind blows around the building, it creates a positive-pressure zone on the upwind side and a negative-pressure zone on the downwind side. After the fire has been extinguished, opening additional doors and windows on the upwind side creates a positive-pressure zone that can rapidly clear residual smoke if there is an exit vent on the other side of the building **FIGURE 15-7**.

Temperature and humidity also can affect ventilation, particularly in tall buildings and large-area structures. On a

FIGURE 15-6 **A.** Building structure and fire conditions should be evaluated for signs of backdraft before any type of ventilation opening is made. **B.** Improperly venting a fire in an oxygen-deficient atmosphere can cause a backdraft.

FIGURE 15-7 On a windy day, proper ventilation helps to remove smoke and heat from a structure.

cold day, there will probably be a strong updraft through a heated high-rise building. A downdraft is more likely on a hot day in an air-conditioned building. Because humidity makes the atmosphere denser, removing cool smoke from a building could prove difficult on a humid day.

Wind, temperature, and humidity are powerful, persistent forces of nature. It is safer, more efficient, and more effective to work with them than to try to work against them. Weather conditions should be a major part of the morning briefing and part of the building size-up before attacking a fire.

Building Construction Considerations

The way a building is constructed affects ventilation operations. Each construction type presents its own set of problems.

■ Fire-Resistive Construction

__Fire-resistive construction__ refers to a building in which all of the structural components are made of noncombustible or limited-combustible materials. This type of construction generally has spaces divided into compartments, which limit potential fire spread. A fire is most likely to involve the combustible contents of an interior space within such a building.

Although fire-resistive construction is intended to confine a fire, it does contain potential avenues for fire spread. Openings for mechanical systems, such as heating and cooling ducts, plumbing and electrical chases, elevator shafts, and stairwells, all provide paths for fire spread. A fire can also spread from one floor to another through exterior windows, a phenomenon called __leap-frogging__ (or auto-exposure) **FIGURE 15-8**.

Smoke can travel through a fire-resistive building using the same routes. Stairways, elevator shafts, and heating, ventilation, and air conditioning (HVAC) systems are common pathways. Because the windows in these buildings are often sealed and are not broken easily, there can be limited opportunities for creating ventilation openings. In some cases, one stairway can be used as an exhaust shaft, while other stairways are cleared for rescue and access. Ventilation fans are often required to direct the flow of smoke. Some windowless buildings contain a number of glass panels with an indicator (sometimes a Maltese cross) in a lower corner, indicating that the panel is made of tempered safety glass and can be easily broken with a sharp, pointed object.

The roof on a fire-resistive building is usually supported by a steel or concrete roof deck. It can be difficult or impossible to make vertical ventilation holes in such a roof. These roofs might, however, have openings for skylights or for HVAC

FIGURE 15-8 Leap-frogging occurs when the fire jumps from floor to floor through exterior windows.

FIGURE 15-9 Vertical ventilation is essential when a fire is burning in an attic or cockloft.

© Steven Townsend/Code 3 Images

ducts. Similar roofing materials can be found in buildings with noncombustible construction.

■ Ordinary Construction

Ordinary construction buildings have exterior walls made of noncombustible or limited-combustible materials that support the roof and floor assemblies. The interior walls and floors are usually wood construction; the roof usually has a wood deck and a wood structural support system. In this type of construction, the roof can be cut with power saws or axes to open vertical ventilation holes. Ordinary construction buildings usually have windows and doors that can be used for horizontal ventilation.

In this type of construction, the walls and floors might include numerous openings for plumbing and electrical chases. If a fire breaks through the interior finish materials (plaster or drywall), it can spread undetected within these void spaces to other portions of the structure. Interior stairwells generally allow a fire to extend vertically within the building. Often, ordinary construction buildings have been remodelled and have multiple roof ceilings, making ventilation more difficult. Although ventilation can be performed either horizontally or vertically in such a case, it can be hard to ventilate in the area close to where the fire is burning.

The vertical openings in ordinary construction frequently provide a path for fire to extend into an attic or **cockloft** (open space between the ceiling of the top floor and the underside of the roof). Heat and smoke can accumulate within these spaces and then spread laterally between sections of a large building or into attached buildings. Vertical ventilation is essential when fire is burning in an attic or cockloft **FIGURE 15-9**.

■ Wood-Frame Construction

Wood-frame construction has many of the same features as ordinary construction. The primary difference is that the exterior walls in a wood-frame building are not required to be constructed of masonry or noncombustible materials.

Wood-frame buildings often contain many void spaces where fire can spread, including attics or cocklofts. Lightweight wood-truss roofs and manufactured I-beam floors, which can fail quickly under fire conditions, are common in newer buildings.

Older buildings of wood-frame construction were often assembled with balloon-frame construction. This type of construction includes direct vertical channels within the exterior walls, so a fire can spread very quickly from a lower level to the attic or cockloft in such a building. These buildings were constructed with ready-made ventilation ducts that can carry superheated gases and smoke to the attic. Heated smoke and gases can accumulate under the roof, requiring rapid vertical roof ventilation.

Modern wood-frame construction uses platform-frame techniques. In these buildings, the structural frame is built one floor at a time. Between each floor, a plate at the floor and the ceiling acts as a fire stop. Its presence limits the fire from spreading up and helps contain the fire on a single floor.

Modern wood-frame construction typically uses lightweight components to save money and provide more longer and open areas, which presents problems when a fire occurs. Lightweight trusses are subject to failure after only a few minutes of exposure to a fire. Wooden I-beams contain much less wood than the solid beams used in older construction, burn more rapidly, and might fail quickly.

Tactical Priorities

Ventilation is directly related to the three major tactical priorities in structural firefighting operations: life safety, fire containment, and property conservation.

■ Venting for Life Safety

Life safety is the primary goal of the fire service, whether it involves rescuing civilians or protecting fire department personnel. Good ventilation practices help to clear smoke, heat, and toxic gases from the structure, which gives occupants a better chance to survive. It provides firefighting crews with increased visibility and makes the structure more tenable, enabling more rapid searches for victims.

Ventilation also limits fire spread and allows fire fighters to advance hose lines more safely and rapidly to attack the fire. The rapid control of the fire reduces the risk to fire fighters as well as to building occupants.

■ Venting for Fire Containment

A fire fighter's second priority is to contain the fire and gain control of the situation. Fire spread can often be controlled or limited through effective ventilation. Releasing smoke and superheated gases to the exterior prevents these products of combustion from spreading throughout the interior of a building or into adjoining spaces.

Not surprisingly, attack teams can be more effective when there is less smoke and heat, because they can advance attack lines more quickly to extinguish the fire. A coordinated fire attack and ventilation effort must consider the timing and the direction of the attack as well as the type and location of possible ventilation openings.

■ Venting for Property Conservation

The third priority of the fire service is property conservation, which involves reducing the losses caused by means other than direct involvement in the fire. Ventilation can play a significant role in limiting property damage. If a structure is ventilated rapidly and correctly, the damages associated with smoke, heat, water, and overhaul operations can all be reduced.

Location and Extent of Smoke and Fire Conditions

Fire fighters must be able to recognize when ventilation is needed and where it should be provided, based on the circumstances of each fire situation. Many factors must be considered, including the size of the fire, the stage of combustion, the location of the fire within a building, and the available ventilation options. An experienced fire fighter learns to recognize these significant factors immediately and to interpret each situation quickly.

Ventilation should occur as close to the fire as possible. The most desirable options are to provide a ventilation opening directly over the seat of the fire or to open a door or window that will release heat and smoke from the fire directly to the exterior. If it is not possible to create a ventilation opening in the immediate area of the fire, fire fighters must be able to predict how ventilation from another location will affect the fire. Fire travels toward a ventilation opening, following the flow of heat and smoke.

The colour, location, movement, and amount of smoke can provide valuable clues about the fire's size, intensity, and fuel **FIGURE 15-10** . Thin, light-coloured smoke moving lazily out of the building usually indicates a small fire involving ordinary combustibles. Thick, dark grey smoke pushing out of a structure suggests a larger, more intense fire. A fire involving petroleum products produces a large quantity of black, rolling smoke that rises in a vertical column. Remember that the initial appearance of a limited amount of smoke can also indicate an underventilated fire.

Smoke movement is a good indicator of the fire's temperature. A very hot fire produces smoke that moves quickly, rolling and forcing its way out through an opening. The hotter the fire, the faster the smoke's movement. Cooler smoke moves

FIGURE 15-10 The colour, location, and amount of smoke can provide valuable clues for fire fighters.

Courtesy of District Chief Chris E. Mickal/New Orleans Fire Department, Photo Unit

In many recent fires where fire fighters have been injured, the initial crews described the smoke as being light and light coloured. This seemingly benign appearance should not fool fire fighters: such smoke can be just as dangerous as thick black smoke. Always remember that these unburned products of combustion are producing toxic gases and vapours, no matter the smoke's colour.

more slowly and gently. On a cool, damp day with very little wind, this type of smoke might hang low to the ground (a phenomenon known as **smoke inversion**).

A different phenomenon occurs in buildings with automatic sprinkler systems. The water discharged from the sprinklers cools the smoke and produces a cold smoke that hardly moves within the building. This type of smoke can fill a large warehouse space from floor to ceiling with an opaque mixture that behaves much like fog on a damp day. Because this cold smoke has a lower temperature than the surrounding atmosphere, it sits at the lower levels of the room, making it hard to see and hard to locate the seat of a fire. Mechanical ventilation is often needed to clear this cold smoke from the building. A similar situation occurs when smoke becomes trapped within a building long enough to cool it to the ambient temperature.

Types of Ventilation

To remove the products of combustion and other airborne contaminants from a structure, fire fighters use two basic types of ventilation: horizontal and vertical.

Horizontal ventilation takes advantage of the doors, windows, and other openings at the same level as the fire. In some cases, fire fighters might make additional openings in a wall to provide horizontal ventilation.

Vertical ventilation involves making openings in roofs or floors, so that heat, smoke, and toxic gases can escape from the structure in a vertical direction. Pathways for vertical ventilation can include ceilings, stairwells, exhaust vents, and roof openings, such as skylights, scuttles, or monitors. Additional openings can be created by cutting holes in the roof or the floor and making sure that the opening extends through every layer of the roof or floor.

Ventilation can be either natural or mechanical. Natural ventilation depends on convection currents and other natural forces, such as the wind, to move heat and smoke out of a building and to allow clean air to enter. Mechanical ventilation uses fans or other powered equipment to introduce clean air or to exhaust heat and smoke.

When referring to ventilation techniques, fire fighters use the term *contaminated atmosphere* to describe the products of combustion that must be removed from a building and the term *clean air* to refer to the outside air that replaces them. The contaminated atmosphere can include any combination of heat, smoke, and gases produced by combustion, including toxic gases. A contaminated atmosphere also can be defined as a dangerous or undesirable atmosphere that was not caused by a fire; for example, natural gas, carbon monoxide, ammonia, and many other products can contaminate the atmosphere as a result of a leak, spill, or similar event. The same basic techniques of ventilation can be applied to many of these situations as well.

■ Horizontal Ventilation

Horizontal ventilation uses horizontal openings in a structure, such as windows and doors, and can be employed in many situations, particularly in small fires **FIGURE 15-11**. Horizontal ventilation is commonly used in residential fires, room-and-contents fires, and fires that can be controlled quickly by the attack team.

Horizontal ventilation can be a rapid, generally easy way to clear a contaminated atmosphere. Often, outlets for horizontal ventilation can be made by simply opening a door or window. In other cases, fire fighters might need to break a window or to use forcible entry techniques. Search and attack teams operating inside a structure as well as fire fighters operating outside the building use horizontal ventilation to reduce heat and smoke conditions.

Horizontal ventilation is most effective when the opening goes directly into the room or space where the fire is located. Opening an exterior door or window, for example, allows heat and smoke to flow directly outside and eliminates the problems that can occur when these products travel through the interior spaces of a building.

FIGURE 15-11 Windows are frequently used in horizontal ventilation.

Courtesy of District Chief Chris E. Mickal/New Orleans Fire Department, Photo Unit

Horizontal ventilation is more difficult if there are no direct openings to the outside or if the openings are inaccessible. In these situations, it might be necessary to direct the air flow through other interior spaces or to use vertical ventilation; for example, fire fighters need to use horizontal ventilation if an inaccessible or damaged roof makes vertical ventilation impossible.

Horizontal ventilation also can be used in less-urgent situations if the building can be ventilated without additional structural damage. Low-urgency situations might include residual smoke caused by a stovetop fire or a small natural gas leak.

Horizontal ventilation tactics include both natural and mechanical methods.

Natural Ventilation

<u>Natural ventilation</u> depends on convection currents, wind, and other natural air movements to allow a contaminated atmosphere to flow out of a structure. The heat of a fire creates convection currents that move smoke and gases up toward the roof or ceiling and out and away from the fire source. Opening or breaking a window or door allows these products of combustion to escape through natural ventilation.

Natural ventilation can be used only when the natural convection air currents or wind are adequate to move the contaminated atmosphere out of the building and replace it with fresh air. Mechanical devices (discussed later in this chapter) can be used if natural forces do not provide adequate ventilation.

Natural ventilation is often used when quick ventilation is needed, such as when attacking a room-and-contents fire or a first-floor residential fire with people trapped on the second floor. For search and attack teams to enter these buildings and act quickly, smoke and heat must be ventilated immediately.

Wind speed and direction play an important role in natural ventilation. If possible, windows on the downwind side of a building should be opened first so that the contaminated atmosphere flows out. Openings on the upwind side can then be used to bring in clean air. Conversely, opening a window on the upwind side first could push the fire into uninvolved areas of the structure, particularly on a windy day.

FIRE FIGHTER Tips

Opening windows on the upwind side of a building before opening downwind-side windows can force the fire into uninvolved areas of the structure.

Breaking Glass

Natural ventilation uses existing or created openings in a building. The methods used to create horizontal openings employ a variety of tools and techniques. For example, some windows can simply be opened by hand. If the window cannot be opened and the need for ventilation is urgent, however, fire fighters should not hesitate to break the glass. Breaking a window is a fast and simple way to create a ventilation opening.

When breaking glass, the fire fighter should always use a hand tool—Halligan tool, axe, or pike pole—and keep his or her hands above or to the side of the falling glass. This tactic prevents pieces of glass from sliding down the tool and potentially injuring the fire fighter. The tool should then be used to clear the entire opening of all remaining pieces of glass, thereby creating the largest opening possible and providing a way for fire fighters to enter or exit through the window in the event of an emergency.

Clearing the broken glass also reduces the risk of injury to anyone who might be in the area and eliminates the danger of sharp pieces falling out later. Before breaking glass—and particularly if the window is above the ground floor—fire fighters must look out to ensure that no one will be struck by the falling glass.

To break glass on a ground floor with a hand tool, follow the steps in **SKILL DRILL 15-1**:

1 Wear full personal protective equipment (PPE), including eye protection and SCBA.

2 Select a hand tool and position yourself to the side of the window. (**STEP 1**)

3 With your back facing the wall, swing backward forcefully with the tip of the tool, striking the top one-third of the glass. (**STEP 2**)

4 Clear the remaining glass from the opening with the hand tool. (**STEP 3**)

Breaking a Window from a Ladder

A similar technique can be used to break a window on an upper floor, with the fire fighter working from a ladder. The ladder should be positioned to the upwind, or windward, side of the window, with the tip even with the top of the window, so that smoke is carried away from the fire fighter. The fire fighter should climb to a position level with the window and lock into the ladder for safety. Using a hand tool, the fire fighter should strike and break the window, and then clear the opening completely. The fire fighter should not be positioned below the window, where glass could slide down the handle of the tool **FIGURE 15-12**.

Breaking a Window with a Ladder

Fire fighters on the ground can use the tip of a ladder to break and clear a window when immediate ventilation is needed on an upper floor. Dropping the tip of the ladder into the upper half of the glass breaks it. Moving the tip of the ladder back and forth inside the opening usually clears the remaining glass.

This technique requires proper ladder selection. Usually, second-floor windows can be reached by 4.9- or 6.1-metre (16- or 20-foot) roof ladders, as well as by bedded 7.3- or 8.5-m (24- or 28-ft.) extension ladders. Extension ladders are needed to break windows on the third or higher floors.

To perform this maneuver, the ladder can be raised directly into the top half of the window, or it can be raised next to the window to determine the proper height for the tip. The ladder is then rolled into the window, drawn back at the tip, and forcibly dropped into the top third of the window. The objective is to push the broken glass into the window opening, although

SKILL DRILL 15-1

Breaking Glass with a Hand Tool
(Fire Fighter I, NFPA 5.3.11)

1 Position yourself to the side of the window.

2 With your back facing the wall, swing backward forcefully with the tip of the tool striking the top one-third of the glass.

3 Clear the remaining glass from the opening with the hand tool.

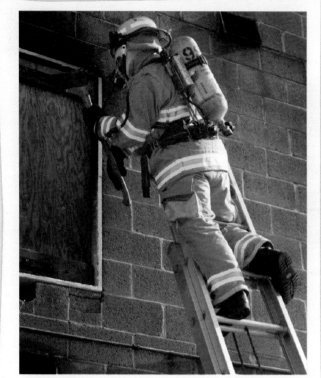

FIGURE 15-12 When breaking an upper-storey window, the fire fighter should place the ladder to the side of the window.

there is always a risk that some of the glass will fall outward. For this reason, fire fighters working below the window must wear full PPE to shield themselves from falling glass. A word of caution: the higher the window, the greater the danger of falling glass.

When a ladder is used to break a window, shards of glass might be left hanging from the edges of the opening. If the opening will be used for access, it needs to be cleared with a tool before anyone enters through it.

To break a window with a ladder, follow the steps in **SKILL DRILL 15-2**:

1. Wear full PPE, including eye protection.
2. Select the proper size of ladder for the job.
3. Check for overhead lines. Use standard procedures for performing a ladder raise using one, two, or three fire fighters, depending on the size of the ladder. (**STEP 1**)
4. Raise the ladder next to the window. Extend the tip so that it is even with the top third of the window. If a roof ladder is used, extend the hooks toward the window. (**STEP 2**)
5. Position the ladder in front of the window. (**STEP 3**)
6. Forcibly drop the ladder into the window.
7. Exercise caution: falling glass can cause serious injury. Glass might slide down the beams of the ladder. (**STEP 4**)
8. Raise the ladder from the window, and move it to the next window to be ventilated. Either carry the ladder vertically, or pivot the ladder on its feet. (**STEP 5**)

Breaking a Window from Above

In some instances, fire fighters might be located above the window that needs to be broken. For example, fire fighters could be on the roof, performing roof ventilation; they could be on a floor above the fire; or they might not have a ladder available that can reach the window.

One way to break a window from above is to attach a Halligan tool to a rope or chain, toss it over the roof edge, and allow it to swing into the window. An alternative method is to use a long-handled tool, such as a pike pole, and to reach down from the edge of the roof, from a balcony, or from a window on the floor above the fire. Smoke and heat will be released when the window is broken, so be sure to stand to the side to keep from being enveloped by these products of combustion.

To break a window from above with a Halligan tool, follow the steps in **SKILL DRILL 15-3**:

1. Look over the roof edge to locate the window to be broken. Always be sure to check if crews are operating below you when you do this maneuver, because glass and debris might fall down from the broken window.
2. Attach the Halligan tool securely to the rope or chain, using a clip or hook.
3. Lower the Halligan tool over the edge of the roof until the tool is located in the middle of the window. Mark this spot on the rope or chain; it is the pivot point for the toss. (**STEP 1**)
4. Raise the tool back to the roof. (**STEP 2**)
5. Toss the Halligan tool over the roofline and out approximately the distance down to the window. This action allows the tool to swing into the window with enough force to break the glass.
6. Be sure to be in a safe position, because this technique puts you above and in the direct path of the products of combustion that are being vented. Always be aware of the location of the roof's edge as well as your safety zone on the roof.
7. If the tool becomes lodged in the window, give it a few shakes and pulls. It should free itself, breaking whatever it is caught on. Always check whether crews are operating below you when you perform this maneuver, because glass and debris might fall down. (**STEP 3**)

Opening Doors

Like windows, door openings can be used for natural ventilation operations. Doorways have certain advantages when used for ventilation. Specifically, they are large openings—often twice the size of an average window. In fact, some doorways can be large enough to drive a truck through. A locked door can be opened using forcible entry techniques.

The problem with using doorways as ventilation openings is that their use for this purpose compromises entry and exit for interior fire attack teams as well as search teams. If heat and smoke are pouring through a doorway, fire fighters cannot enter or exit through that portal. Because protecting exit paths is a priority, doorways are more suitable as entries for clean air. The contaminated atmosphere should, therefore, be discharged through a different opening.

SKILL DRILL 15-2 | Breaking Windows with a Ladder
(Fire Fighter I, NFPA 5.3.11)

1 Check for overhead lines. Use standard procedures for performing a ladder raise.

2 Raise the ladder next to the window. Extend the tip so that it is even with the top third of the window. If a roof ladder is used, extend the hooks toward the window.

3 Position the ladder in front of the window.

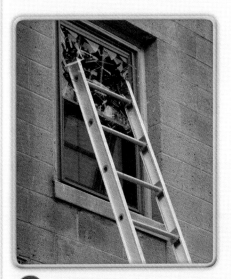

4 Forcibly drop the ladder into the window. Exercise caution: falling glass can cause serious injury.

5 Raise the ladder from the window and move it to the next window to be ventilated.

© LiquidLibrary

FIRE FIGHTER Tips

Watch for glass sliding down the beams of the ladder. Wear full protective clothing and SCBA and gloves when performing this type of ventilation, because this procedure can cause injury to an unwary fire fighter.

Fire fighters often approach a fire through the interior of a building to attack from the unburned side. Opening an exterior door that leads directly to the fire can create an opening for oxygen-rich air to enter the building or a vent that pushes the heat and smoke out of the building. Treat the opening of a door as a major act of ventilation. It should be opened only when the hose line is charged and the attack team is ready to advance.

Before opening an exterior door, examine the area around the door. A small amount of smoke can be the sign of a small fire or a sign of an underventilated fire. Proceed with extreme caution. Once the exterior door is opened, note how the air moves through the exterior door. A rapid movement of air into the structure can indicate an underventilated fire. If the fire is actively burning, you might see clean air entering through the bottom half of the exterior door and smoke escaping through the top half.

SKILL DRILL 15-3   Breaking Windows on Upper Floors: Halligan Toss
(Fire Fighter I, NFPA 5.3.11)

1 Lower the Halligan tool over the edge of the roof to measure the distance.

2 Raise the Halligan tool back to the roof.

3 Toss the Halligan tool over the roofline, so that it swings back to break the glass.

© LiquidLibrary

Doorways are also good places to put mechanical ventilation devices when natural ventilation needs to be augmented. Fans can be used to push clean air in (positive-pressure ventilation) or to pull contaminated air out (negative-pressure ventilation).

Fire Fighter Safety Tips   © Photos.com

Opening an exterior door is an act of ventilation. In an underventilated fire, this act can quickly produce a backdraft.

■ Mechanical Ventilation

Mechanical ventilation uses large high-powered fans to augment natural ventilation; that is, it involves the use of fans or other powered equipment. There are three different methods of mechanical ventilation: Negative-pressure ventilation uses fans called **ejectors** to exhaust smoke and heat from a structure. Positive-pressure ventilation uses fans to introduce clean air into a structure and push the contaminated atmosphere out. **Hydraulic ventilation** moves air by using fog or broken-pattern fire streams to create a pressure differential behind and in front of the nozzle.

In some buildings, HVAC systems can be used to provide mechanical ventilation. To perform this function, such a system needs to be configured so that the zones can be controlled correctly and effectively. The building engineer or maintenance staff might have the specific knowledge required to assist in this effort. Some HVAC systems can be used to exhaust smoke or supply fresh air if a fire occurs, whereas others should be shut down immediately. Information about HVAC systems should be obtained during preincident planning surveys.

Negative-Pressure Ventilation

The basic principles of air flow are used in negative-pressure ventilation. Fire fighters locate the source of the fire and, using a fan or blower, exhaust the products of combustion out through a window or door. The fan creates a negative pressure that draws the heat, smoke, and fire gases out of the building. In turn, fresh, clean air must enter the structure, to replace the exhausted contaminated air.

Smoke ejectors are usually 41–61 centimetres (16–24 inches) in diameter, although larger models are available. They can be powered by electricity, gasoline, or water pressure **FIGURE 15-13**.

Negative-pressure ventilation can be used to clear smoke out of a structure after a fire, particularly if natural ventilation

FIGURE 15-13 A smoke ejector pulls products of combustion out of a structure. Gasoline-powered smoke ejectors might have problems running because of carbon particles in the smoke.

Courtesy of Super Vacuum Mfg. Co., Inc.

would be too slow or if no natural cross-ventilation exists. In large buildings, negative-pressure ventilation can be used to pull heat and smoke out while fire fighters are working, although it is not generally used before or during a fire attack. Although this type of ventilation is usually associated with horizontal ventilation, an ejector can be used in a roof-vent operation to pull heated gases out of an attic.

Negative-pressure ventilation has several limitations related to its positioning, power source, maintenance, and air flow control. Often, fire fighters must enter the heated and smoke-filled environment to set up the ejector, and they might need to use braces and hangers to position it properly. Because most ejectors run on electricity, a power source and a cord are required for them to function. The ejectors also get very dirty while they are running, so they must be taken apart and cleaned after each use.

Air flow control with a negative-pressure system is also difficult. The fan must be completely sealed so that the exhausted air is not immediately drawn back into the building. This phenomenon, which is called churning, reduces the effectiveness of the ejector. **Churning** can be eliminated by completely blocking the opening around the fan.

Smoke ejectors usually have explosion-proof motors, which makes them excellent choices for venting flammable or combustible gases and other hazardous environments.

To perform negative-pressure ventilation, follow the steps in **SKILL DRILL 15-4** :

① Determine the area to be ventilated and the outside wind direction.

② If possible, place the smoke ejector so that it will exhaust on the downwind side of the building.

③ Hang the fan as high in the selected opening as possible.

④ Remove any obstructions from the area used to ventilate smoke, including curtains, screens, and other debris.

⑤ Use a salvage cover to prevent churning, if the structure's windows are not intact. (**STEP ①**)

⑥ Provide an opening on the windward side of the structure to provide cross-ventilation. (**STEP ②**)

FIRE FIGHTER Tips

Negative-pressure ventilation is not frequently used in initial fire attacks; however, you should understand how it works, because it can be a useful tool to exhaust contaminated air caused by smoke or other toxic products.

Positive-Pressure Ventilation

Positive-pressure ventilation uses large, powerful fans to force fresh air into a structure. The fans create a positive pressure inside the structure, which displaces the contaminated atmosphere and pushes heat and other products of combustion out. Positive-pressure ventilation can be used to reduce interior temperatures and smoke conditions in coordination with an initial fire attack or used to clear a contaminated atmosphere after a fire has been extinguished.

Positive-pressure fans are usually set up at exterior doorways, often at the same opening used by the attack team **FIGURE 15-14** . These fans are most effective if placed about 0.9–1.2 m (3–4 ft.) in front of the exterior door. This creates a cone of air that improves the flow of fresh air entering the fire building. This pushes clean air in the same direction as the

FIGURE 15-14 Positive-pressure fans are usually set up at exterior doorways, often at the same opening used by the attack team.

SKILL DRILL 15-4 Delivering Negative-Pressure Ventilation
(Fire Fighter I, NFPA 5.3.11)

1. Hang the fan in the upper part of the opening. Use salvage covers to prevent churning.

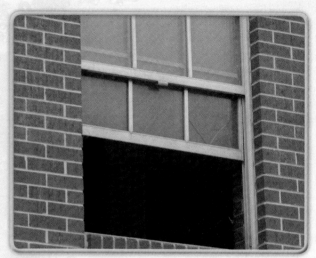

2. Provide openings on the windward side for cross-ventilation.

© LiquidLibrary

advancing hose line, so that fire fighters move toward the fire enveloped in a stream of clean, cool air.

While the fan pushes fresh air into the building, it is equally important to have an opening on the other side of the building to exhaust the superheated air and smoke. This opening should be as close to the fire as possible. Improper placement of the fan or the exhaust opening can help to spread the fire. When performed properly, positive-pressure ventilation can push heat and smoke out of the building through an exhaust vent on the opposite side of the building.

For this kind of system to work effectively, the integrity of the building must be intact, and ventilation openings kept to a minimum. Positive-pressure ventilation does not work properly if there are too many openings. There must be an exhaust opening near the seat of the fire to allow the heat and smoke to escape FIGURE 15-15 . The entry and exit openings should

be approximately the same size to create the desired positive pressure within the structure. If there is no exhaust opening or if it is too small, the heat and smoke can flow into other areas of the building or back toward the attack team. If the exhaust opening is too large, it does not create the buildup of pressure inside the structure that would otherwise create the velocity needed to remove the smoke effectively.

To increase the efficiency of positive-pressure ventilation and to prevent smoke and fire from reaching other parts of the building, fire fighters should close doors to unaffected areas of the structure. This is an effective action to maintain a tenable environment for people trapped in other parts of the building.

When planning positive-pressure ventilation, the force and direction of the wind need to be taken into consideration. Positive-pressure fans should never be placed in a position where they are blowing against the wind. Check for the direction and speed of the wind before placing fans, and monitor the wind for any changes in direction or intensity.

Because smaller areas can be ventilated rapidly, venting one room at a time or one section of the building at a time can quickly clear trapped smoke out of a building after a fire. In a high-rise building, positive-pressure fans can blow fresh air

FIGURE 15-15 There must be an opening near the seat of the fire to allow heat and products of combustion to exhaust.

Safety Tips

© Photos.com

The importance of close coordination among members of the ventilation crew and the attack team cannot be overstressed. Good communication and proper coordination of operations can prevent backdraft.

FIRE FIGHTER Tips

In high-rise fires, a large number of people need to be safely evacuated, often by a single stairway. The stairway must be free of smoke and toxic gases. Positive-pressure fans can be used to keep smoke and products of combustion from entering the stairwell. In emergencies, they can also be used to introduce clean, oxygen-rich air into the potentially toxic atmosphere of the stairwell. When using positive-pressure fans for this purpose, keep in mind that the engines can increase carbon-monoxide levels, so ensure that the fans are properly ventilated.

up through a stair shaft. By opening the doors, fire fighters can clear one floor at a time. Very large structures can be ventilated by placing multiple fans side by side or one behind the other.

Positive-pressure ventilation has several advantages over negative-pressure ventilation. For instance, a single fire fighter can set up a fan very quickly. Because the fan is positioned outside the structure, the fire fighter does not have to enter a hazardous environment, and the fan does not interfere with interior operations. Positive pressure is both quick and efficient because the entire space within the structure is under pressure.

When used with a well-placed exhaust opening, positive-pressure ventilation can help keep a fire confined to a smaller area and increase safety for interior operating crews, as well as any building occupants. In addition, positive-pressure fans do not require as much cleaning and maintenance as negative-pressure fans, because the products of combustion never move through the fan.

Positive-pressure ventilation does have some disadvantages, however. The most important concern is the potential for improper use of positive pressure to spread a fire. If the fire is burning in a structural space such as in a pipe chase or above a ceiling, positive pressure can push it into unaffected areas of the structure. If the fire is located in structural void spaces, positive-pressure ventilation should not be used until access to these spaces is available and attack crews are in place. Ineffective positive-pressure ventilation occurs if positive pressure is used without an adequate exhaust opening for the heat and smoke.

Positive-pressure fans operate at high velocity and can be very noisy. Most are powered by internal combustion engines, so they can increase carbon-monoxide levels if they are run for significant periods of time after the fire is extinguished. Natural ventilation should be used after the structure is cleared to prevent carbon-monoxide buildup. Because positive-pressure fan motors can get hot, they are unsafe to use if combustible vapours are present.

SKILL DRILL 15-5 outlines the steps for performing positive-pressure ventilation:

➊ Determine the location of the fire within the building and the direction of attack.

➋ Place the fan 0.9–1.2 m (3–4 ft.) in front of the opening to be used for the fire attack. (**STEP** ➊)

➌ Provide an exhaust opening at or near the fire. This opening can be made either before the fan is started or when the fan is started. (**STEP** ➋)

➍ Check for interior openings that could allow the products of combustion to be pushed into unwanted areas.

➎ Start the fan and check the cone of air produced; it should completely cover the opening. This can be checked by running a hand around the door frame to feel the direction of air currents.

➏ Allow smoke to clear—usually 30 seconds to 1 minute, depending on the size of the area to be ventilated and the smoke conditions. (**STEP** ➌)

FIRE FIGHTER Tips

Doors used for ventilation must be kept open to ensure that ventilation is sustained. Depending on the time available and the type of door and lock, fire fighters might remove the door, disable the closure device, or place a wedge in the hinge side or under the door. Overhead doors must be blocked in the open position, thereby ensuring that they cannot accidentally roll down and close. Check the manufacturer's recommendations.

On some fans, if the fan was tipped down while getting it out of the apparatus, the oil sensor will cause the fan not to start. Fire fighters need to know the procedure to use for the fan to get it started.

Hydraulic Ventilation

Hydraulic ventilation uses the water stream from the hose line to exhaust smoke and heated gases from a structure. To use this ventilation technique, the fire fighter working the hose line directs a narrow fog stream or a broken stream from a smooth-bore nozzle out of the building through an opening, such as a window or doorway. The contaminated atmosphere is drawn into a low-pressure area behind the nozzle. An induced draft created by the high-pressure stream of water then pulls the smoke and gas out through the opening **FIGURE 15-16**. A well-placed fog stream can move a tremendous volume of air through an opening.

Hydraulic ventilation is most useful in clearing smoke and heat out of a room after the fire is under control. Hydraulic ventilation can move several hundred cubic metres of air per minute and is effective in clearing both heat and smoke from a fire room. Because it does not require any specialized equipment, it can be performed by the attack team with the same hose line that was used to control the fire.

Some disadvantages are also associated with the use of hydraulic ventilation. Fire fighters must enter the heated, toxic environment and remain in the path of the products of combustion as they are being exhausted. If the water supply is limited, using water for ventilation must be balanced with using water for fire attack. This technique can cause excessive water damage if it is used improperly or for long periods of time.

SKILL DRILL 15-5 Delivering Positive-Pressure Ventilation
(Fire Fighter I, NFPA 5.3.12)

1 Place the fan in front of the opening to be used for the fire attack.

2 Provide an exhaust opening at or near the fire.

3 Start the fan and allow the smoke to clear.

© LiquidLibrary

FIGURE 15-16 Hydraulic ventilation can be used to draw smoke out of a building after the fire is controlled.

Also, in cold climates, ice buildup can occur on the ground, creating a safety hazard.

Ideally, the ventilation opening should be created before the hose line advances into the fire area. The hose stream also can be used to break a window before it is directed on the fire. The steam created when the water hits the fire pushes the heat and smoke out through the vent. Hydraulic ventilation can then be used to clear the remaining heat and smoke from the room.

To perform hydraulic ventilation, follow the steps in **SKILL DRILL 15-6**:

1 Enter the room and remain close to the ventilation opening.

2 Place the nozzle through the opening, and open the nozzle to a narrow fog or broken-spray pattern.

3 Keep directing the stream outside and back into the room until the fog pattern almost fills the opening. The nozzle should be 0.6–1.2 m (2–4 ft.) inside the opening.

4 Stay low, out of the heat and smoke, or to one side to keep from partially obstructing the opening.

■ Vertical Ventilation

Vertical ventilation refers to the release of smoke, heat, and other products of combustion into the atmosphere in a vertical direction. Vertical ventilation occurs naturally, owing to convection currents, if an opening is available above a fire. Convection causes the products of combustion to rise and flow through the opening. In some situations, vertical ventilation can be assisted by mechanical means such as fans or hose streams.

Although vertical ventilation refers to any opening that allows the products of combustion to travel up and out, this term is most often applied to describe operations on the roof of a structure. The roof openings can be existing features, such as skylights or bulkheads, or holes created by fire fighters who cut through the roof covering. The choice of roof openings depends primarily on the building's roof construction.

A vertical ventilation opening should be made as close as possible to the seat of the fire **FIGURE 15-17**. Smoke issuing from the roof area, melted asphalt shingles, or steam coming from the roof surface are all signs that fire fighters can use to identify the hottest point.

Safety Considerations in Vertical Ventilation

Vertical ventilation should be performed only when it is necessary and can be done safely. Rooftop operations involve significant inherent risks. If horizontal ventilation can sufficiently control the situation, vertical ventilation might be unnecessary. Before performing vertical ventilation, fire fighters must evaluate all pertinent safety issues and avoid unnecessary risks. In particular, they should assess the roof for roof scuttles, heat vents, plumbing vents, louvre ventilation, and fan shafts.

Near-Miss REPORT

Report Number: 10-0000745

Event Description: Brackets [] denote reviewer de-identification.

Tower [1] responded to a residential structure fire. Tower [1] was the first-arriving unit to the scene, where light smoke was showing from the front door. The homeowner reported the fire in the basement of the home. Crews opted for an interior attack with a hand-line from Tower [1]. Crews located the basement access from the kitchen and proceeded to make their way to the basement of the home. Upon entering the basement, fire was found in the laundry room of the home. Interior attack teams extinguished the visible flames and then called for ventilation to help with low visibility. After requesting ventilation, heat intensified at a rapid rate downstairs, forcing attack teams to pull out of the basement to the main level of the home. After reaching the top of the stairs, interior crews noted extreme heat now on the main level. Heavy, thick smoke conditions blinded attack teams, causing them to take shelter in the kitchen of the home. Fire was observed to be above firefighters. Original egress was no longer an option, because of dark smoke and fire. The interior teams' attack line was burned and rendered not usable. Crews recognized the rapid change and bailed out the kitchen window because of extreme heat and smoke conditions. Two of the three crew members suffered minor injuries. Approximately 20 seconds after crews bailed out, the kitchen floor where they were standing collapsed into the basement. The flashover was believed to be caused by extreme heat building in a couch on the main level of the home. When the ventilation fan was placed to the front door, it is possible that the rush of fresh air helped to fuel the smouldering couch and ignited a fire.

Lessons Learned: Our department teaches us to recognize rapid change. In this instance, our bail-out training helped with confidence not to hesitate and to bail out. Another important point is that the flashover occurred approximately 5 minutes after fire was deemed out. Crews were venting the structure when the flashover occurred. There is a rule of thumb that 15 seconds of positive-pressure ventilation will help to reduce flashover conditions, but in this case, 3 minutes of fresh air helped to fuel the flashover. It's important to never let your guard down when on the fire ground.

FIGURE 15-17 The vertical ventilation opening should be as close as possible to the seat of the fire.

The most obvious risk in a rooftop operation is the possibility that the roof will collapse. Many fire fighters have died during vertical ventilation attempts because the fire had compromised the structural integrity of the roof support system. If

the structural system supporting the roof fails, both the roof and the fire fighters standing on it will fall into the fire. Fire fighters also can fall through an area of the roof deck that has been weakened by the fire.

Determine the type of roof construction before beginning roof ventilation operations. If the roof is constructed with lightweight trusses, ventilate the roof from an aerial ladder or from an aerial basket. Roofs built using lightweight manufactured trusses can fail quickly and without warning.

Falling from a roof—either off of the building or through a ventilation opening, open shaft, or skylight into the building—are risks fire fighters encounter when operating on roofs. Smoke reduces visibility and can cause disorientation, increasing the risk of falling off the edge of a roof. Factors such as darkness, snow, ice, or rain can also create hazardous situations. Not

Thermal imaging cameras can be used to determine if there are changes in the temperature as a result of a fire underneath the roof. This can help you avoid unsafe roof operations and determine the location of heated gases under the roof.

Ventilate roofs constructed with lightweight construction from an aerial ladder or aerial tower whenever possible **FIGURE 15-18**. Do not attempt to stand on these roofs if there are signs of heat or fire underneath.

FIGURE 15-18 Fire fighters operating from the safety of aerial apparatus.

Courtesy of Captain David Jackson, Saginaw Township Fire Department

escape routes should be separate from each other, preferably in opposite directions from the operation site. The team working on the roof should always know where these exit routes are located, and they should plan ventilation operations to work toward these locations.

The ventilation opening should never be located between the fire fighters and their escape route. A ventilation opening releases smoke and hot gases from the fire. If this column of smoke (and possibly fire) separates the fire fighters from their exit from the roof, they could be in serious danger **FIGURE 15-19**.

Once a ventilation opening has been made, fire fighters should withdraw to a safe location. There is no good reason to stand around the vent hole and look at it; this behaviour simply exposes the fire fighters to an unnecessary risk.

Before fire fighters climb or walk onto any roof, they should test the roof to ensure that it will support their weight. Older roofs constructed of heavy beams can become soft or spongy before they fail. Modern roofs constructed with lightweight trusses can fail suddenly and without warning.

FIGURE 15-19 Be prepared for smoke and flames that are released when the roof is opened.

© craig robinson/iStockphoto

surprisingly, a sloping roof presents even greater risks than a flat roof. Safe operations on steep roofs require fire fighters to operate from an aerial ladder or basket or to use a roof ladder.

Effective vertical ventilation in prompt fire control reduces the risks to fire fighters operating inside the building. In some situations, prompt vertical ventilation can save the lives of building occupants by relieving interior conditions, opening exit paths, and allowing fire fighters to enter and perform search and rescue operations.

Vertical ventilation should always be performed as quickly and efficiently as possible. If they provide an adequate opening, skylights, ventilators, bulkheads, and other existing openings should be opened first. If fire fighters must cut through the roof covering to create a hole, the location should be identified and the operation should be performed promptly and efficiently.

Fire fighters who are working on the roof should always have two safe exit routes. A second ground ladder or aerial device should be positioned to provide a quick alternative exit, in case conditions on the roof deteriorate rapidly. The two

Sounding is one way to test the stability of roofs that are not constructed with lightweight trusses, using a tool such as an axe or pike pole. Simply strike the surface of the roof with the blunt end of the tool. If the roof is in good condition, the impact should produce a firm rebound and a reassuring sound; if the impact does not produce this sound or if the tool penetrates the roof, the structure is not safe. During ventilation operations, fire fighters should continually sound the roof, probing ahead and around themselves with the handle of an axe or a pike pole to confirm that the area is solid.

Modern lightweight construction frequently uses thin plywood for roof decking, with a variety of waterproof and insulating coverings then being applied to the plywood. A relatively minor fire under this type of roof can cause the plywood to delaminate, with no visible indication of damage or weakness from above. The weight of a fire fighter on a damaged area can be enough to open a hole and cause the fire fighter to fall through the roof.

Sounding is not a foolproof method to test the condition of a roof. It can give fire fighters a false sense of security on some types of roofs. For example, slate and tile roofs can appear solid even if the supporting structure is ready to fail, owing to the rigid surface of the tiles. Sounding lightweight construction truss roofs is not a reliable test of roof safety. Always look for other visible indicators of possible collapse and take the roof construction into account.

Determine the type of roof construction before beginning roof ventilation operations. If the roof is constructed with lightweight trusses, operate from an aerial ladder or from an aerial basket. To sound a roof for safety, follow the steps in **SKILL DRILL 15-7** :

1 Before stepping onto a roof, use a hand tool, such as an axe, pike pole, or sledge, to strike the roof decking with considerable force. Listen and feel for rebound. Be aware that sounding lightweight-construction truss roofs is not a reliable test of roof safety. (**STEP 1**)

2 Sound the area ahead and to each side of your path. The material should sound solid and remain intact. If the tool penetrates, the roof is not safe.

3 Locate support members by listening to the sound and watching the bounce of the tool. Support members create a solid sound and make the tool bounce from the roof. By contrast, the spaces between supports will sound hollow, and the tool will not bounce.

4 Continue to sound the area around you periodically to monitor conditions. (**STEP 2**)

5 Sound the roof as you leave the area. You are not completely safe until your feet are back on solid ground. (**STEP 3**)

Fire Fighter Safety Tips

 © Photos.com

Sounding a roof to see if it is stable is not a safe test with a truss or lightweight roof.

A fire fighter's path to a proposed vent-hole site should follow the areas of greatest support and strength. These locations include the roof edges, which are supported by bearing walls, and the hips and valleys of the roof, where structural materials are doubled for strength. These areas should be stronger than interior sections of the roof.

Fire fighters should always be aware of their surroundings when working on a roof; knowing where the roof's edge is located, for example, can help avoid accidentally walking or falling off the roof. Maintaining this knowledge is especially important—although difficult—in darkness, adverse weather,

---

SKILL DRILL 15-7   Sounding a Roof
(Fire Fighter I, NFPA 5.3.12)

1 Use a hand tool to check the roof before stepping onto it.

2 Use the tools to sound ahead and on both sides of you as you walk. Locate the support members by sound and rebound. Check conditions around your work area periodically.

3 Sound the roof along your exit path.

© LiquidLibrary

or heavy smoke conditions. It is equally important to watch for openings in the roof, such as skylights and ventilation shafts.

The order of the cuts in making a ventilation opening should be planned carefully. Fire fighters should be upwind, have a clear exit path, and be standing on a firm section of the roof or using a roof ladder. If fire fighters are upwind from the open hole, the wind will push the heat and smoke away from them; thus, the escaping heat and smoke should not block their exit path. To prevent an accidental fall into the building, fire fighters should stand on a portion of the roof that is firmly supported, particularly when making the final cuts.

Basic Indicators of Roof Collapse

Roof collapse poses the greatest risk to fire fighters performing vertical ventilation. Some roofs—particularly truss roofs—give little or no warning that they are about to collapse.

Fire fighters who are assigned to vertical ventilation tasks should always be aware of the condition of the roof. They should immediately retreat from the roof if they notice any of the following signs:

- Any visible indication of sagging roof supports
- Any indication that the roof assembly is separating from the walls, such as the appearance of fire or smoke near the roof's edges
- Any structural failure of any portion of the building, even if it is some distance from the ventilation operation
- Any sudden increase in the intensity of the fire from the roof opening
- High heat indicators on a thermal imaging camera

Fire Fighter Safety Tips
© Photos.com

Signs of imminent roof collapse include fire showing around roof vents, melting snow, and evaporating or steaming water. These signs indicate that the roof is being compromised by fire below it and is not safe.

Roof Construction

Roofs can be constructed of many different types of materials in several configurations. Nevertheless, all roofs have two major components: a support structure and a roof covering.

■ Roof Support Structures

The roof support system provides the structural strength to hold the roof in place. It must also be able to bear the weight of any rain or snow accumulation, any rooftop machinery or equipment, and any other loads placed on the roof, such as fire fighters or other people walking on the roof. The support system can be constructed of solid beams of wood, steel, or concrete, or a system of trusses. Trusses are produced in several configurations and are made of wood, steel, or a combination of wood and steel.

■ Roof Coverings

The roof covering constitutes the weather-resistant surface of the roof and can have several layers. Roof-covering materials include shingles and composite materials, tar and gravel, rubber, foam plastics, and metal panels. The roof decking is a rigid layer made of wooden boards, plywood sheets, or metal panels. In addition to the decking, the roof covering typically includes a waterproof membrane and insulation to retain heat in winter and limit solar heating in the summer.

Vertical ventilation operations often involve cutting a hole through the roof covering. The selection of tools, the technique used, the time required to make an opening, and the personnel requirements all depend on the roofing materials and the layering configurations.

■ Effects of Roof Construction on Fire Resistance

Fire can very quickly burn through some roofs, whereas others retain their integrity for long periods. The difference arises because each type of material used in roof construction is affected differently by fire. These reactions either increase or decrease the time available to perform roof operations. For example, solid beams are generally more fire resistant than truss systems, because the former materials are larger and heavier. Wooden beams are affected by fire more quickly than concrete beams. Steel, unlike wood, does not burn but elongates and loses strength when heated.

Most roofs eventually fail as a result of fire exposure—some very quickly and others more gradually. In some situations, the fire weakens or burns through the supporting structure so that it collapses, while the roof covering remains intact. Structural failure usually results in sudden and total collapse of the roof. A structural failure involving other building components could also cause the roof to collapse.

In other cases, the fire burns through the roof covering, even though the structure is still sound. This type of roof failure usually begins with a "burn through" close to the seat of the fire or at a high point above the fire. As combustible products in the roof covering become involved in the fire, the opening spreads, eventually causing the roof to collapse.

The inherent strength and fire resistance of a roof are often determined by local climate conditions. In areas that receive winter snow loads, roof structures can usually support a substantial weight and might include multiple layers of insulating material. A serious fire could burn under this type of roof with very little visible evidence from above. Heavily constructed roofs supported by solid structural elements can burn for hours without collapsing.

In warmer climates, however, roof construction is often very light. Roofs in these geographic areas might simply function as umbrellas. The supporting structure can be just strong enough to support a thin deck with a waterproof, weather-resistant membrane. This type of roof can burn through very quickly, and the supporting structure can collapse after a short fire exposure.

Solid-Beam versus Truss Construction

The two major structural support systems for roofs use either solid-beam or truss construction. It can be impossible to determine which system is used simply by looking at the roof from the exterior of the building, particularly when the building is on fire. Given this fact, information on the roof support system should be obtained during preincident planning surveys.

The basic difference between solid-beam and truss construction is in the way individual load-bearing components are made. Solid-beam construction uses solid components, such as girders, beams, and rafters. Trusses are assembled from smaller, individual components. In most cases, it makes no difference whether solid-beam or truss construction is used. For fire fighters, however, there is an important difference: a truss system can collapse quickly and suddenly when it is exposed to a fire.

Trusses are constructed by assembling relatively small and lightweight components in a series of triangles. The resulting system can efficiently span long distances while supporting a load. In most cases, a solid beam that spanned an equal distance and carried an equivalent load would be much larger and heavier than a truss **FIGURE 15-20**. The disadvantage of truss construction is that some trusses fail completely when only one of the smaller components is weakened or when only one of the connections between components fails.

A truss is not necessarily bad or inherently weak construction. Indeed, some trusses are very strong and are assembled with substantial components that can resist fire just as well as solid beams. Many trusses, however, are made of lightweight materials and are designed to carry as much load as possible with as little mass as possible. These trusses can be expected to fail more quickly when exposed to fire.

Truss construction can be used with almost any type of roof or floor, including flat roofs, pitched roofs, arched roofs, or overhanging roofs. Fire fighters should assume that any modern construction uses truss construction for the roof support system until proven otherwise.

The individual components of a lightweight truss used for roof support are often wood 20 × 40 (50.8 mm by 101.6 mm) (2-inch × 4-inch) sections or a combination of wood and lightweight steel bars or tubes. Because these components are small, they can be weakened quickly by a fire. Even more critical, however, are the points where the individual components join together. These connection points are often the weakest portions of the truss and the most common points of failure.

For example, the components of a truss constructed with 20 × 40 wood pieces can be connected by heavy-duty staples or by **gusset plates** (connecting plates made of wood or lightweight metal) **FIGURE 15-21**. These connections can fail quickly in a fire. When one connection fails, the truss loses its ability to support a load.

If only one truss fails, other trusses in the system might be able to absorb the additional load, but the overall strength of the system is still compromised. A fire that causes one truss to fail probably weakens additional trusses in the same area. These trusses also can fail, sometimes as soon as the additional load is transferred. In most cases, a series of trusses fails in rapid succession, resulting in a total collapse of the roof. Sometimes just the weight of a fire fighter walking on the roof is enough to trigger a collapse.

Some trusses are made of metal, usually individual steel bars or angle sections that are welded together. Lightweight steel trusses, known as bar joists, often support flat roofs on commercial or industrial buildings. When exposed to the heat of a fire, these metal trusses expand and lose strength. A roof

FIGURE 15-20 A lightweight truss can carry as much weight as a much heavier solid beam but might fail quickly and without warning during a fire.

FIGURE 15-21 Gusset plates can fail in a fire, destroying the truss.

Courtesy of Captain David Jackson, Saginaw Township Fire Department

supported by bar joists will probably sag before it collapses, a warning that fire fighters should heed. The expansion can stretch the trusses, which causes the supporting walls to collapse suddenly. Horizontal cracks in the upper part of a wall indicate that the steel roof supports are pushing outward.

◼ Roof Designs

Fire fighters must be able to identify the various types of roofs and the materials used in their construction. Creating ventilation openings requires the use of many different tools and techniques, depending on the type of roof decking.

Flat Roof

Flat roofs can be constructed with many different support systems, roof decking systems, and materials. Although they are classified as flat, most roofs have at least a slight slope so that water can flow to roof drains or scuppers.

Flat roof construction is generally very similar to floor construction. The roof structure can be supported by solid components, such as beams and rafters (the elements that support the roof), or by trusses. The horizontal beams or trusses often run from one exterior load-bearing wall (a wall that supports the weight of a floor or roof) to another load-bearing wall. In some buildings, the roof is supported by a system of vertical columns and/or interior load-bearing walls.

The roof deck is usually constructed of multiple layers, beginning with wooden boards, plywood sheets, or metal decking. Then come one or more layers of roofing paper, insulation, tar and gravel, rubber, gypsum, lightweight concrete, or foam plastic. Some roof decks have only a single layer, consisting of metal sheets or precast concrete sections.

Flat roofs often have vents, skylights, scuttles (small openings or hatches with movable lids), or other features that penetrate the roof deck. Removing the covers from these openings provides vertical ventilation without the need to make cuts through the roof deck.

Flat roofs also can have **parapet walls**, freestanding walls that extend above the normal roofline. A parapet can be an extension of a firewall, a division wall between two buildings, or a decorative addition to the exterior wall.

Pitched Roofs

Pitched roofs have a visible slope that provides for rain, ice, and snow runoff. The pitch (angle) of the roof usually depends on both local weather conditions and the aesthetics of construction. A pitched roof can be supported either by trusses or by a system of rafters and beams. The rafters usually run from one load-bearing wall up to a centre ridge pole and back to another load-bearing wall.

Most pitched roofs have a layer of solid sheeting, which can be metal, plywood, or wooden boards. This layer is often covered by a weather-resistant membrane and an outer covering such as shingles, slate, or tiles. Some pitched roofs have a system of **laths** (thin, parallel strips of wood) instead of solid sheeting to support the outer covering.

As with a flat roof, the type of construction material will dictate how to ventilate a pitched roof. For example, a slate or tile pitched roof can be opened by breaking the tiles and pushing them through the supporting laths. A tin roof can be cut and peeled back like the lid on a can. Opening a wooden roof usually requires cutting, chopping, or sawing.

Pitched roofs have either steep or gradual slopes. Roof ladders should be used to provide a stable support while working on this kind of roof. A ground or aerial ladder can be used to access the lower part of the roof; in such a case, the roof ladder is placed on the sloping surface and hooked over the centre ridge pole.

Arched Roofs

<u>Arched roofs</u> are generally found in commercial structures, because they create large open spans without requiring the use of columns **FIGURE 15-22**. Arched roofs are common in warehouses, supermarkets, bowling alleys, and similar buildings. These roofs are supported by trusses or by other types of construction, including large wood, steel, or concrete arches.

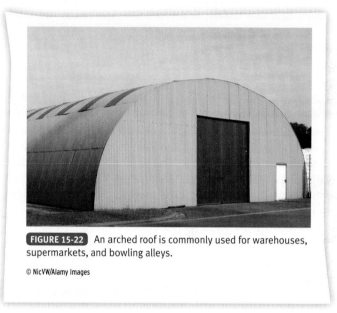

FIGURE 15-22 An arched roof is commonly used for warehouses, supermarkets, and bowling alleys.

© NicVW/Alamy Images

Arched roofs are often supported by bowstring trusses, which give the roof its distinctive curved shape. Bowstring trusses are usually constructed of wood and spaced 1.8–6.1 m (6–20 ft.) apart. They support a roof deck of wooden boards or plywood sheeting and a covering of waterproof membrane. Although these roofs are quite distinctive when seen from above or outside, their structure might not be evident from inside the building, because a flat ceiling is attached to the bottom chords of the trusses. This creates a huge attic space that is often used for storage. A hidden fire within this space can severely and quickly weaken the bowstring trusses.

The collapse of a bowstring-truss roof is usually very sudden. A large area can collapse owing to the long spans and wide spaces between trusses. Fire fighters should be wary of any fire involving the truss space in these buildings. Many fire fighters have been killed when a bowstring-truss roof collapsed.

Buildings with this type of roof construction should be identified and documented during preincident planning surveys. Parapet walls, smoke, and other obstructions make it difficult to recognize the distinctive arched shape of a bowstring-truss roof during an incident.

Vertical Ventilation Techniques

■ Roof Ventilation

The objective of any roof ventilation operation is simple: to provide the largest opening in the appropriate location, using the least amount of time and the safest technique. Among the different roof openings that can provide vertical ventilation are built-in roof openings, inspection openings to locate the optimal place to vent, primary expandable openings located directly over the fire, and defensive secondary openings intended to prevent fire spread (TABLE 15-2).

TABLE 15-2	Types of Vertical Ventilation Openings
• Built-in roof openings • Inspection openings • Primary (expandable) openings • Secondary (defensive) openings	

Before starting any vertical ventilation operation, fire fighters must make an initial assessment. They should note construction features and indications of possible fire damage, establish safety zones and exit paths, and identify built-in roof openings that can be used immediately. The sooner a building is ventilated, the safer it becomes for hose teams working inside the structure.

Ventilation operations must not be conducted in unsafe locations. A less efficient opening in a safe location is better than an optimal opening in a location that jeopardizes the lives of the ventilation team. In some cases, it might not be safe to conduct any rooftop operations, unless they can be performed from the safety of an aerial ladder or an aerial basket. If vertical ventilation cannot be performed, fire fighters have to rely on horizontal ventilation techniques.

Vertical ventilation is most effective when the opening is placed at the highest point, directly over the fire. Information on the fire's location as well as visible clues from the roof can be used to pinpoint the best location to vent. The IC or attack crews might be able to provide some direction, and the ventilation team should look for signs such as smoke issuing through the roof, tar bubbling up, or other indications of heat. If the roof is wet, rising steam or melted snow could indicate the hottest area.

Examination holes allow the team members to evaluate conditions under the roof and to verify the proper location for a ventilation opening. Sometimes a single slit cut (a **kerf cut**) into the roof with a power saw gives fire fighters some indication of the location of the fire. If a single cut is not effective, a triangular examination hole can be created very quickly by making three small cuts in the roof. Examination holes are used to determine how large an area is involved,

whether a fire is spreading, and in which direction it is moving (FIGURE 15-23).

Once the spot directly over the fire is located, the ventilation team should determine the most appropriate type of opening to make. Built-in rooftop openings provide readily available ventilation openings (FIGURE 15-24). Skylights, rooftop stairway exit doors, louvres, and ventilators can quickly be transformed into ventilation openings simply by removing a cover or an obstruction. This approach also results in less property damage to the building. Many residential roofs have a vent installed along the ridgeline of the roof to keep the attic space ventilated. Often, these can be removed quickly because they are attached with small nails.

FIGURE 15-23 A triangular cut can be used as an examination hole.

FIGURE 15-24 Built-in rooftop openings often can provide vertical ventilation very quickly.

Courtesy of Four Seasons Solar Products LLC

If the roof must be cut to provide a ventilation opening, cutting one large hole is better than making several small ones. The original hole should be rectangular and at least 1.2 m long by 1.2 m wide (4 ft. long by 4 ft. wide). If necessary, this hole can be expanded by continuing the cuts to make a larger opening **FIGURE 15-25**. If the crews inside the building report that smoke conditions are lifting and temperature levels are dropping, the vent is effective and probably large enough. If interior crews do not see a difference, the opening might be obstructed or need to be expanded.

Once the roof opening is made, a hole of the same size should be made in the ceiling material below to allow heat and smoke to escape from the interior of the building. If the ceiling is not opened, only the heat and smoke from the space between the ceiling and the roof will escape through the roof opening. The blunt end of a pike pole or hook can be used to push down as much of the ceiling material as possible. A roof opening is much less effective if there is no ceiling hole or if the ceiling hole is too small.

■ Tools Used in Vertical Ventilation

Several tools can be used in roof ventilation. Fire fighters commonly use power saws to cut vent openings but also use many hand tools as well; for example, axes, Halligan tools, pry bars, tin cutters, pike poles, and other hooks can all be used to remove coverings from existing openings, cut through the roof decking, remove sections of the roof, and punch holes in the interior ceiling. Usually the specific tools needed can be determined by looking at the building and the roof construction features.

Ventilation team members should always carry a standard set of tools as well as a utility rope for hoisting additional equipment, if needed. All personnel involved in roof operations should use SCBA and wear full PPE. In this scenario, ground ladders or aerial apparatus can provide access to the roof.

Power saws effectively cut through most roof coverings. A rotary saw with a wood- or metal-cutting blade or a chainsaw can be used for this purpose, for example. Special carbide-tipped saw blades can cut through typical roof construction materials. To operate a power saw, follow the steps in **SKILL DRILL 15-8**:

1. Make sure that the saw is checked during the daily apparatus inspection.
2. Confirm that the cutting device or blade is appropriate for the material anticipated. If it is not, put the correct blade on before going to the roof.
3. Briefly inspect the blade or chain for obvious damage.
4. Ensure that your proper protective gear is in place, including eye protection, hearing protection, SCBA, and full PPE. (**STEP 1**)
5. Before going to the roof, start the saw to ensure that it runs properly. Set the choke to the halfway position, or as recommended by the manufacturer. (**STEP 2**)
6. Stay clear of any moving parts of the saw, use a foot or knee to anchor the saw to the ground, and pull the starter cord as recommended to start the saw. (**STEP 3**)
7. Run the saw briefly at full throttle to verify its proper operation. (**STEP 4**)
8. Shut down the saw, wait for the blade to stop completely, and then carry the saw to the roof. (**STEP 5**)
9. If possible, work off of a roof ladder or aerial platform for added safety. (**STEP 6**)
10. Start the saw in an area slightly away from where you intend to cut. (**STEP 7**)
11. Always run the saw at maximum throttle when cutting. The saw should be running at full speed before the blade touches the roof decking. Keep the throttle fully open while cutting and removing the blade from the cut to reduce the tendency for the blade to bind. (**STEP 8**)

Wind direction

1.4 m²
1.2 m   1.2 m

Total 1.4 m²

1.4 m²
1.4 m²
1.2 m   1.2 m

Total 2.8 m²

1.4 m²   1.2 m
1.4 m²   1.4 m²
1.2 m   1.2 m

Total 4.2 m²

FIGURE 15-25 A ventilation hole can be quickly expanded by extending the cuts.

SKILL DRILL 15-8
Operating a Power Saw
(Fire Fighter I, NFPA 5.3.12)

1 Confirm that the cutting device or blade is appropriate for the material anticipated. Briefly inspect the blade or chain for obvious damage. Ensure that your proper protective gear is in place.

2 Before going to the roof, start the saw to ensure that it runs properly. Set the choke to the halfway position, or as recommended by the manufacturer.

3 Stay clear of any moving parts of the saw, use a foot or knee to anchor the saw to the ground, and pull the starter cord as recommended to start the saw.

4 Run the saw briefly at full throttle to verify its proper operation.

5 Shut down the saw, wait for the blade to stop completely, and then carry the saw to the roof.

6 If possible, work off of a roof ladder or aerial platform for added safety.

7 Start the saw in an area slightly away from where you intend to cut.

8 Always run the saw at maximum throttle when cutting. The saw should be running at full speed before the blade touches the roof decking.

■ Types of Roof Cuts

Roof construction is a major consideration in determining which type of cut to use. Some roofs are thin and easy to cut with an axe or a power saw; others have multiple layers that are difficult to cut. Being familiar with the roof types in the local area and the best methods for opening them increases your efficiency and effectiveness in ventilation operations.

Fire Fighter Safety Tips

© Photos.com

Always carry and handle tools and equipment in a safe manner. When carrying tools up a ladder, hold the beam of the ladder instead of the rungs, and run the hand tool up the beam. This technique works for most hand tools, including axes, pike poles and hooks, and tin-roof cutters. Hoist tools using a rope with an additional tag line to keep them away from the building. Carry power saws and equipment in a sling across your back. Always turn off power equipment before you carry it.

Rectangular or Square Cut

A 1.2- by 1.2-m (4- by 4-ft.) rectangular or square cut is the most common vertical ventilation opening. It requires the fire fighter to make four cuts completely through the roof decking, using an axe or power saw. When using a power saw, the fire fighter must carefully avoid cutting through the structural supports. After the four cuts are completed, a section of the roof deck can be removed.

The fire fighter should stand upwind of the opening, with an unobstructed exit path. The first and last cuts should be made parallel to and just inside the roof supports. The fire fighter making the cuts must always stand on a solid portion of the roof. A triangular cut in one corner of the planned opening can be used as a starting point for prying the deck up with a hand tool.

Depending on the roof construction, it might be possible to lift out the entire section at once. If several layers of roofing material are present, fire fighters might have to peel them off in layers. The decking itself could consist of plywood sheets or individual boards that have to be removed one at a time.

SKILL DRILL 15-9
Making a Rectangular or Square Cut
(Fire Fighter I, NFPA 5.3.12)

1 Locate the roof supports by sounding. Make the first cut parallel to the roof support.

2 Make a triangular cut at the first corner.

3 Make two cuts perpendicular to the roof supports, and then make the final cut parallel to another roof support.

4 Pull out or push in the triangular cut.

5 Punch out the ceiling below. Be wary of a sudden updraft of hot gases or flames.

© LiquidLibrary

To perform the rectangular cut, follow the steps in **SKILL DRILL 15-9**:

1 Check the roof for soundness. Use a tool to locate the roof supports.

2 Make the first cut parallel to a roof support. The cut should be made so that the support is outside the area that will be opened. (**STEP 1**)

3 Make a small triangular cut at the first corner of the opening. (**STEP 2**)

4 Make two cuts perpendicular to the roof supports. Do not cut through roof supports. Rock the saw over them to avoid damaging the integrity of the roof structure. (**STEP 3**)

5 Make the final cut parallel to and slightly inside a roof support. Be sure to stand on the solid portion of the roof when making this cut.

6 Use a hand tool to pull out the corner triangle or push it through to create a small starter hole. (**STEP 4**)

7 If possible, use the hand tool to pull the entire roof section free, and flip it over onto the solid roof. It might be necessary to pull the decking out, one board at a time.

8 After opening the roof deck, use a pike pole to punch out the ceiling below. This hole should be the same size as the opening in the roof decking. Be wary of a sudden updraft of hot gases or flames. (**STEP 5**)

SKILL DRILL 15-10 Making a Louvre Cut
(Fire Fighter I, NFPA 5.3.12)

1 Locate the roof supports by sounding.

2 Make two parallel cuts perpendicular to the roof supports.

3 Cut parallel to the supports and between pairs of supports in a rectangular pattern.

4 Tilt the panel to a vertical position.

Louvre Cut

Another common cut is the <u>louvre cut</u>, which is particularly suitable for flat or sloping roofs with plywood decking. To make a louvre cut, fire fighters use a power saw or axe to make two parallel cuts, approximately 1.2 m (4 ft.) apart, perpendicular to the roof supports. The fire fighters then make cuts parallel to the roof supports, approximately halfway between each pair of supports. Using each roof support as a fulcrum, the cut sections are tilted to create a series of louvred openings.

Louvre cuts can quickly create a large opening. Continuing the same cutting pattern in any direction creates additional louvre sections. To make a louvre cut, follow the steps in **SKILL DRILL 15-10** :

1 Sound the roof with a tool to locate the roof supports. (**STEP** **1**)

2 Make two parallel cuts perpendicular to the roof supports. Do not cut through the roof supports. Rock the saw over them to avoid damaging the integrity of the roof structure. (**STEP** **2**)

3 Make cuts parallel to the supports and between pairs of supports in a rectangular pattern. (**STEP** **3**)

4 Strike the nearest side of each section of the roofing material with an axe or maul, pushing it down on one side; use the support at the centre of each panel as a fulcrum. Tilt the panel over the middle support.

SKILL DRILL 15-11 Making a Triangular Cut
(Fire Fighter I, NFPA 5.3.12)

1 Locate the roof supports.

2 Make the first cut from just inside a support member in a diagonal direction toward the next support member.

3 Begin the second cut at the same location as the first, and make it in the opposite diagonal direction, forming a V shape.

4 Make the final cut along the support member so as to connect the first two cuts. Cutting from this location allows fire fighters the full support of the member directly below them while performing ventilation.

© LiquidLibrary

5 Moving horizontally along the roof, make additional louvre openings.

6 Open the interior ceiling area below the opening by using the butt end of a pike pole. This hole should be the same size as the opening made in the roof decking. (**STEP 4**)

Triangular Cut

The **triangular cut** works well on metal roof decking because it prevents the decking from rolling away as it is cut. Using saws or axes, the fire fighter removes a triangular section of decking. Smaller triangular cuts can be made between supports (so that the decking falls into the opening) or over supports (to create a louvre effect). Because triangular cuts are generally smaller than other types of roof ventilation openings, several might be needed to create an adequately sized vent.

To make a triangular cut, follow the steps in **SKILL DRILL 15-11**:

1 Locate the roof supports. (**STEP 1**)

2 Make the first cut from just inside a support member in a diagonal direction toward the next support member. (**STEP 2**)

3 Begin the second cut at the same location as the first, and make it in the opposite diagonal direction, forming a V shape. (**STEP 3**)

4 Make the final cut along the support member so as to connect the first two cuts. Cutting from this location allows fire fighters the full support of the member directly below them while performing ventilation. (**STEP 4**)

Peak Cut

Peak cuts are limited to peaked roofs sheeted with plywood. In these structures, the 1.2- by 2.4-m (4- by 8-ft.) sheets of plywood are applied starting at the bottom edge of the roof and working toward the top. A partial sheet is usually placed at the top, along the roof peak. A small gap is left between the two sides of the roof to allow the plywood sheeting to expand and contract with temperature changes.

To make a peak cut, the fire fighter first uses a hand tool to clear the outer covering along the peak (roof cap), which reveals the roof supports through the gap in the sheeting. Using a power saw or axe, the fire fighter makes a series of vertical cuts between the supports from the top to the bottom of the plywood sheet. The individual panels are then struck with an axe and louvred or pried up with a hook.

Repeating this technique on both sides of the roof essentially removes the entire peak. Unless multiple layers of roof decking are present, there is no need to make horizontal cuts. The opening can be expanded either vertically or horizontally with a minimum of cutting.

To make a peak cut, follow the steps in **SKILL DRILL 15-12**:

1 Sound the roof with a tool to locate the roof supports. (**STEP 1**)

2 Clear the roofing materials away from the roof cap. (**STEP 2**)

3 Make the first cut vertically, at the farthest point away. Start at the roof peak in the area between the support members, and cut down to the bottom of the first plywood panel. (**STEP 3**)

4 Make parallel downward cuts between supports, moving horizontally along the roofline to make additional ventilation openings. (**STEP 4**)

5 Strike the nearest side of the roofing material with an axe or maul, pushing it in, using the support located at the centre as a fulcrum. This causes one end of the roofing material to go downward into the opening and the other to rise up.

6 If necessary, repeat this process on both sides of the peak, horizontally across the peak, or vertically toward the roof edge. (**STEP 5**)

7 Open the interior ceiling area below the opening by using the butt end of a pike pole. This hole should be the same size as the vent opening made in the roof decking. (**STEP 6**)

Trench Cut

Trench cut (or **strip cut**) ventilation is used to stop fire spread in long narrow buildings, such as strip malls and small storage complexes. A trench cut creates a large opening ahead of the fire, removing a section of fuel and letting heated smoke and gases flow out of the building. Essentially, it is a firebreak in the roof.

Trench cuts are a defensive ventilation tactic intended to stop the progress of a large fire, particularly one that is advancing through an attic or cockloft. The IC who chooses this tactic is writing off part of the building and identifying a point where crews will be able to stop the fire.

A trench cut is made from one exterior wall across to the other. It begins with two parallel cuts, spaced 0.6–1.2 m (2–4 ft.) apart. Approximately every 1.2 m (4 ft.), fire fighters make short perpendicular cuts between the two parallel cuts. They can then lift the roof covering out in sections, completely opening a section of the roof. On a pitched roof, the trench should run from the peak down. On a sloping roof, it should start at the higher end and work toward the lower end.

A trench cut is a **secondary cut**, used to limit the fire spread, rather than a **primary cut**, located over the seat of the fire. A primary vent should still be made before crews start working on the trench cut.

A trench cut must be made far in advance of the fire. The ventilation crew must be able to complete the cut, and

SKILL DRILL 15-12 | Making a Peak Cut
(Fire Fighter I, NFPA 5.3.12)

1 Locate the roof supports.

2 Clear the roofing materials away from the roof cap.

3 Make the first cut vertically, at the farthest point away. Start at the roof peak in the area between the support members, and cut down to the bottom of the first plywood panel.

4 Make parallel downward cuts between supports, moving horizontally along the roofline to make additional ventilation openings.

5 Strike the nearest side of the roofing material with an axe or maul, pushing it in, using the support located at the centre as a fulcrum. This causes one end of the roofing material to go downward into the opening and the other to rise up. If necessary, repeat this process on both sides of the peak, horizontally across the peak, or vertically toward the roof edge.

6 Open the interior ceiling area below the opening by using the butt end of a pike pole. This hole should be the same size as the vent opening made in the roof decking.

© LiquidLibrary

the interior crew get into position before the fire passes the trench. Inspection holes should be made to ensure that the fire has not already passed the chosen site before the trench cut is completed. The crew members' lives will be in danger, and the tactic is useless if the fire advances beyond the trench before it is opened.

Although trench cuts are effective, they require both time and personnel to complete. As with all types of vertical ventilation, careful coordination between the ventilation crew and the interior attack crew is essential. While the trench cut is being made, attack teams should be deployed inside the building to defend the area in front of the cut.

To make a trench cut, follow the steps in **SKILL DRILL 15-13** (Fire Fighter I, NFPA 5.3.12) **FIGURE 15-26**:

1. After a primary cut has been made over the seat of the fire, cut a number of small inspection holes to identify a point sufficiently far ahead of the fire.

2. Make two parallel cuts, 0.6–1.2 m (2–4 ft.) apart, across the entire roof, starting at the ridge pole (for peaked roofs) or a load-bearing wall (for flat roofs).

3. Cut between the two long cuts to make a row of rectangular sections.

4. Remove the rectangular panels to open the trench.

FIGURE 15-26 Making a trench cut.

Special Considerations

Many obstacles can be encountered during ventilation operations. Fire fighters must be creative and remember the basic objectives of ventilation: to create high openings as rapidly as possible so that hose lines can be advanced into the building.

Poor access or obstructions such as trees, fences, electric lines, or tight exposures can prevent fire fighters from getting close enough to place ladders. Many residential and commercial roofs have multiple layers; some even have a new roof built on top of an older one. Window openings in abandoned buildings might be boarded or sealed.

Steel bars, shutters, and other security features can also hinder ventilation efforts. A building that requires forcible entry is also likely to present ventilation challenges. To enhance their security, some commercial buildings have roofs made of steel plating. In these buildings, fire fighters should not attempt vertical ventilation through the roof but rather should use alternative ventilation techniques.

■ Ventilating Concrete Roofs

Some commercial or industrial structures have concrete roofs. Concrete roofs can be constructed with poured-in-place concrete, with precast concrete sections of roof decking placed on a steel or concrete supporting structure, or with T-beams. Such roofs are generally flat and difficult to breach. The roof decking is usually very stable, but fire conditions underneath could weaken the supporting structural components or load-bearing walls, leading to failure and collapse.

There are few options for ventilating concrete roofs. Even special concrete-cutting saws are generally ineffective against these structures. In such cases, fire fighters should use alternative ventilation openings, such as vents, skylights, and other roof penetrations or horizontal ventilation.

■ Ventilating Metal Roofs

Metal roofs and metal roof decks also present many challenges for the ventilation crew. Because metal conducts heat more quickly than other roofing materials, discolouration and warping of this material can indicate the seat of the fire. Tin-cutter hand tools can be used to slice through thin metal coverings, whereas special saw blades might be needed to cut through metal roof decking. In many cases, the metal is on the bottom and supports a built-up or composite roof covering.

Metal roof decking is often supported by lightweight steel-bar joists, which can sag or collapse when exposed to a fire. Because the metal decking is lightweight, the supporting structure can be relatively weak, with widely spaced bar joists. The resulting assembly can fail quickly with only limited fire exposure.

As the fire heats the metal deck, the tar roof covering can melt and leak through the joints into the building, where it releases flammable vapours. When this sequence of events occurs, it can quickly spread the fire over a wide area under the roof decking. Fire fighters should look for indications of dripping or melting tar and begin rapid ventilation to dissipate the flammable vapours before they can ignite. Hose streams should be used to cool the roof decking from below to stop the tar from melting and producing vapours.

When a metal roof deck is cut, the metal can roll down and create a dangerous slide directly into the opening. The triangular cut prevents the decking from rolling away as easily, so it is the preferred option, even though several cuts can be needed to create an adequately sized vent.

■ Ventilating Basements

Basement fires are especially difficult to ventilate. Basements generally have just a few small windows, if they have any at all. Some basement stairways lead only to the ground floor interior. Some basements have no exterior exit.

If a basement fire occurs, fire fighters should open or break windows or exterior doorways into the basement to provide as much ventilation as possible. If the basement has few or no exterior windows or doors, the interior stairways and other vertical openings act as chimneys. Heat, smoke, and gases rise up through these conduits and into the rest of the structure. In buildings with balloon-frame construction, a basement fire can travel through unprotected wall spaces directly up to the underside of the roof.

A combination of vertical and horizontal ventilation can sometimes be used in attacking a basement fire. A direct path is created from the basement to the first floor, where the heat and smoke are pushed out through a door or window. A stairway or some other opening can be used for vertical ventilation, or a hole can be cut in the floor directly over the fire.

The opening from the basement to the ground floor should be placed near a window or door that can be used for horizontal ventilation. Heat and smoke can then be exhausted using a hose line (hydraulic ventilation) or a negative-pressure smoke ejector. If this approach is taken, a hose line must be ready in case the fire spreads into the ground floor.

A basement fire presents problems to fire fighters working both in the basement and above it. To attack the fire, fire fighters must enter the basement, but smoke and hot gases moving up the stairway as the fire fighters descend can make entry difficult or impossible. For their part, fire fighters operating above the fire are in danger if the floor or stairway collapses.

The preferred method of attacking a basement fire is to create as many ventilation openings as possible on one side of the basement, which draws the heat and smoke in that direction. Fire fighters can then enter the basement from the opposite side, along with clean air. The direction of the attack must be determined, and the entry and ventilation points must be clearly identified before this operation begins.

Ventilation openings in basement fires help push heat and smoke away from the attacking fire fighters. As the products of combustion rise, fire fighters can sometimes crawl beneath them to reach the basement. If the entrance stairway is the only ventilation opening when the water hits the fire, the steam that is produced will fill the basement with heat and smoke. A ventilation opening through the floor into the basement provides a second exit for the products of combustion. Sometimes when adequate ventilation is not possible, basement fires need to be fought using a cellar nozzle or hose lines from the outside.

■ Ventilating High-Rise Buildings

Ventilating a high-rise building can be challenging. A high-rise building resembles a stack of individual floor compartments connected by stairways, elevator shafts, and other vertical passages. Most high-rise buildings have sealed windows that are difficult to break. In addition, high-rise buildings have unique patterns of smoke movement, such that smoke might be trapped on individual floors or move up or down within the vertical shafts.

Many newer high-rise buildings have incorporated smoke-management capabilities into their HVAC systems. Use of this type of system enables different areas to be pressurized with fresh air and contaminated air to be exhausted directly to the outside. If the HVAC system does not have this capacity, it can complicate problems by circulating smoke to different areas of the building.

A phenomenon called the **stack effect** can occur in high-rise structures. The stack effect is a response to the differences in temperature inside and outside a building. A cold outer atmosphere and a heated interior cause smoke to rise quickly through stairways, elevator shafts, and other vertical openings, filling the upper levels of the building **FIGURE 15-27**.

The opposite situation can occur on a hot summer day, when the interior temperature is much cooler than the outside atmosphere. In this scenario, the heavy cooler air pushes the smoke down the vertical openings, toward a lower-level exit **FIGURE 15-28**.

The situation can change if the fire produces enough heat to alter the temperature profile within the building. The air currents within a tall building might, for example, be stronger than the convection currents that normally govern the flow of smoke and other products of combustion. A strong wind through an open or broken window can change the direction of smoke. Contaminated air might suddenly move toward the opening or be pushed back as a strong draft of fresh air enters the building.

After smoke mixes with fresh air or is hit by water from sprinklers or hose streams, it cools and can sit in one location. This cold smoke can fill several floors and usually needs to be cleared by mechanical means.

A key objective in ventilating a high-rise building is to manage the air movement in stairways and elevator shafts. Positive-pressure ventilation can help keep stairways

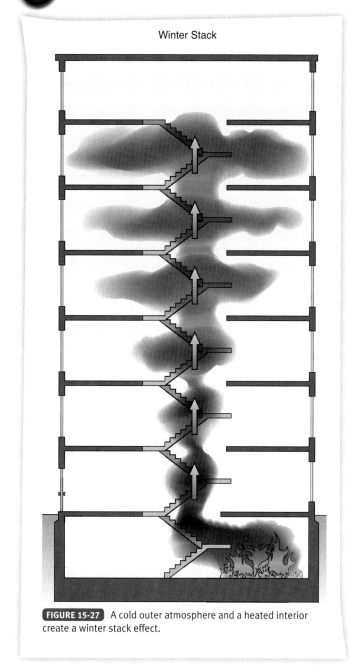

FIGURE 15-27 A cold outer atmosphere and a heated interior create a winter stack effect.

FIGURE 15-28 A summer stack effect occurs when the interior temperature is much cooler than the outside air.

clear of smoke and ventilate individual floors in high-rise structures.

At least one stairwell should be kept well ventilated and designated as an occupant and rescue route. These stairs must be used by rescuers and escaping victims only. Some newer buildings have smoke-proof stair towers or pressurized stair shafts that are designed to keep smoke out of the stairway. Otherwise, positive-pressure fans can be used in a stairway to keep smoke

out. Place the positive-pressure fans at a ground-floor doorway, positioning them so that fresh air blows into the stairway. Make certain that the engines of the positive-pressure fans are well ventilated, because they can increase carbon-monoxide levels.

A pressurized stairway also can be used to clear smoke from a floor. Opening a door from the stairway allows fresh air to enter the floor. The contaminated atmosphere can then be vented out through a window or another stairway.

■ Ventilating Windowless Buildings

Many structures do not have windows **FIGURE 15-29**. Some buildings are designed without windows; others have bricked-up or covered windows. These buildings pose two significant risks to fire fighters: heat and products of combustion are trapped, and fire fighters have no secondary exit route.

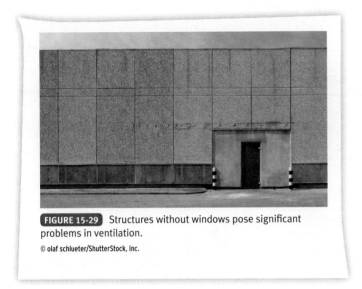

FIGURE 15-29 Structures without windows pose significant problems in ventilation.

© olaf schlueter/ShutterStock, Inc.

Windowless buildings are similar to basements in terms of the ventilation approach to be used. Any ventilation needs to be as high as possible and probably requires mechanical assistance. Using existing rooftop openings, cutting openings in the roof, reopening boarded-up windows or doors, and making new openings in exterior walls are all possible ways to ventilate windowless structures.

■ Ventilating Large Buildings

Providing adequate ventilation is more difficult in large buildings than in smaller ones. In a large building, a ventilation hole placed in the wrong location can draw the fire toward the opening, spreading the fire to an area that had not previously been involved. This underscores the importance of coordinating ventilation operations with the overall fire attack strategy.

Smoke cools as it travels into unaffected portions of a large building. A fire sprinkler system also cools the smoke, causing it to stratify. As the cold smoke fills the area, it becomes more difficult to clear.

If possible, fire fighters should use interior walls and doors to create several smaller compartments in a large building, thereby limiting the spread of heat and smoke. The smaller areas can be cleared one at a time with positive-pressure fans. Several fans can be used in a series or in parallel lines to clear smoke from a large area.

Equipment Maintenance

It is important that all equipment used for ventilation is kept in good repair and ready to operate at peak efficiency. It is important to read and follow the manufacturer's instructions for properly maintaining power equipment at regular and properly documented intervals. Fuel must be rotated regularly if not used. Fuel tanks need to be filled to the recommended levels. All fire fighters who use ventilation tools need to practise using them regularly to ensure safe and effective operation when needed.

To perform a readiness check on a power saw, see **SKILL DRILL 15-14**:

1. Don the proper PPE.
2. Follow the manufacturer's and your department's recommendations.
3. Make certain that the fuel tank is full. Make certain that the bar and chain oil reservoirs are full. (**STEP ❶**)
4. Check the throttle trigger for smooth operation. (**STEP ❷**)
5. Ensure that the saw, blade, air filter, and chain brake are clean and working.
6. Inspect the blade for even wear, and lubricate the sprocket tip, if needed. (**STEP ❸**)
7. Check the chain for wear, missing teeth, or other damage. Check the chain end for proper tension.
8. Check the chain catcher. (**STEP ❹**)
9. Check for loose nuts and screws, and tighten them, if needed.
10. Check the starter and starter cord for wear. (**STEP ❺**)
11. Inspect the spark plugs. (**STEP ❻**)
12. Start the saw.
13. Make certain that both the bar and the chain are being lubricated while the saw is running.
14. Check the chain brake. (**STEP ❼**)
15. Make certain that the stop switch functions.
16. Record the results of the inspection. (**STEP ❽**)

To maintain a power saw, see **SKILL DRILL 15-15**:

1. Don the proper PPE.
2. Follow the manufacturer's and your department's recommendations.
3. Remove the clutch cover, bar, and chain, and inspect them for damage and wear. Replace these components, if necessary. (**STEP ❶**)
4. Clean and inspect the clutch cover, drive, bar, chain, and power head using a cloth, cleaning solvent, and compressed air.
5. Inspect the air filter, and clean or replace it, as needed. (**STEP ❷**)
6. Coat the components with a substance recommended by the manufacturer to prevent rusting.
7. Place the bar and chain back on, flipping the bar over each time to help wear the bar evenly.
8. Replace the clutch cover. (**STEP ❸**)
9. Adjust the chain tension (make sure the bar and chain are cool before adjusting them). (**STEP ❹**)

10 Tighten the clutch cover nuts and other nuts and screws.

11 Fill the power saw with fuel. Fill the bar and chain oil reservoirs.

12 Perform a periodic check.

13 Document the maintenance on the departmental log. (**STEP 5**)

SKILL DRILL 15-14

Performing a Readiness Check on a Power Saw
(Fire Fighter I, NFPA 5.3.12)

1 Make certain that the fuel tank is full. Make certain that the bar and chain oil reservoirs are full.

2 Check the throttle trigger for smooth operation.

3 Ensure that the saw, blade, air filter, and chain brake are clean and working. Inspect the blade for even wear, and lubricate the sprocket tip, if needed.

4 Check the chain for wear, missing teeth, or other damage. Check the chain end for proper tension. Check the chain catcher.

5 Check for loose nuts and screws, and tighten them, if needed. Check the starter and starter cord for wear.

6 Inspect the spark plugs.

7 Start the saw. Make certain that both the bar and the chain are being lubricated while the saw is running. Check the chain brake.

8 Make certain that the stop switch functions. Record the results of the inspection.

SKILL DRILL 15-15 Maintaining a Power Saw
(Fire Fighter I, NFPA 5.3.12)

1 Remove, clean, and inspect the clutch cover, bar, and chain for damage and wear. Replace if necessary.

2 Inspect the air filter and clean/replace it, as needed.

3 Lubricate components as recommended by the manufacturer. Place the bar and chain back on, flipping the bar over each time to help wear bar evenly. Replace the clutch cover.

4 Adjust the chain tension (make sure the bar and chain cool before adjusting).

5 Fill the power saw with fuel. Fill the bar and chain oil reservoirs.

Wrap-Up

Chief Concepts

- Ventilation is the process of removing smoke, heat, and toxic gases from a burning building and replacing them with cooler, cleaner, oxygen-rich air. The ventilation process has two different components. The first is the removal of smoke, toxic gases, and hot air. The second is the addition of cooler, cleaner, and oxygen-rich air.

- The sudden addition of oxygen-rich air to a smoke-filled and fuel-rich atmosphere can result in a sudden flashover or a backdraft.

- Because of modern construction techniques, fires have a large supply of fuel in furnishings and a limited supply of oxygen as a result of tightly shut windows and doors. This can lead to an underventilated fire. An underventilated fire simply needs a supply of oxygen to become a raging fire, a flashover, or a backdraft.

- When ventilation is coordinated with fire attack and the application of water, it can save lives and reduce property damage.

- As a fire grows and develops, the products of combustion—smoke, heat, and toxic gases—spread throughout the structure and present a life-threatening risk to occupants and fire fighters, reduce visibility, and increase property damage. These trapped products of combustion create a potential for explosion. Ventilation removes some of these products from the interior atmosphere and allows them to escape in a controlled manner.

- The principle of convection controls the spread of the products of combustion. Convection is one of the primary mechanisms of fire spread. These gases can be hot enough to ignite combustible materials along their path.

- With convection, when heated gases expand, they become less dense than cooler gases. As a result, the hot gases produced by a fire in a closed room rise to the ceiling and spread outward, displacing cooler air and pushing it toward the floor. As the fire continues to burn, the hot layer of gas banks down closer to the floor. If the heated products of combustion escape from the room, they spread horizontally along the ceiling until they find a path or an outside opening that allows them to reach a higher level. They then flow upward through the vertical opening until they reach another horizontal obstruction, such as a ceiling.

- The benefits of proper ventilation include improved visibility, removal of heat, removal of steam created by the attack lines, reduced risk of backdraft, and limited fire spread.

- To reduce the risk of backdraft, ventilation must release as much heat and unburned products of combustion as possible. Placing a ventilation opening as high as possible within a building can help to eliminate potential backdraft conditions. When it is safe, fire attack crews should extinguish the fire and cool the interior as quickly as possible.

- Both ventilation and cooling are needed to relieve potential flashover conditions. Short bursts of water are applied to the ceiling by the attack crews, and the ventilation openings are placed close to the fire.

- Convection currents can carry smoke and superheated gases to uninvolved areas within the structure. Fire fighters must be able to predict where the fire might travel by convection currents. Fire fighters can also take advantage of convection by making ventilation openings that cause the convective flow to draw the heated products out of the structure.

- The direction and force of the wind should be considered when determining where and how to ventilate a building. Even a slight breeze can have a dramatic influence on the effectiveness of a ventilation opening.

- All of the structural components in fire-resistive construction are made of noncombustible or limited-combustible materials. It generally has spaces divided into compartments, which limit potential fire spread. However, the products of combustion can spread through stairways and HVAC systems. The roof is usually supported by a steel or concrete roof deck that is difficult to impossible to cut vertical ventilation holes in.

- Ordinary-construction buildings have exterior walls made of noncombustible or limited-combustible materials that support the roof and floor assemblies. The walls and floors are usually wood construction. The roof can be cut with power saws or axes to open vertical ventilation holes. These buildings usually have windows and doors that can be used for horizontal ventilation. Vertical openings frequently provide a path for fire to extend into an attic or cockloft.

- Wood-frame buildings often contain many void spaces. Lightweight construction components, found in newer buildings, can fail quickly under fire conditions.

- Ventilation is related to the three major tactical priorities in firefighting operations: life safety, fire containment, and property conservation.
- Fire fighters must be able to recognize when ventilation is needed and where it should be provided, based on the circumstances of each fire situation. They must consider many factors, including the size of the fire, the stage of combustion, the location of the fire within a building, and the available ventilation options.
- Horizontal ventilation takes advantage of the doors, windows, and other openings at the same level as the fire. In some cases, fire fighters make additional openings in a wall to provide horizontal ventilation. Horizontal ventilation is commonly used in residential fires, room-and-contents fires, and fires that can be controlled quickly by the attack team.
- Vertical ventilation involves making openings in roofs or floors so that heat, smoke, and toxic gases can escape from the structure in a vertical direction. Pathways for vertical ventilation can include ceilings, stairwells, exhaust vents, and roof openings, such as skylights, scuttles, or monitors. Additional openings can be created by cutting holes in the roof or the floor and making sure that the opening extends through every layer of the roof or floor.
- Horizontal ventilation tactics include natural and mechanical methods. Natural ventilation includes opening or breaking a window or opening a door and allowing convection currents, wind, and other natural air movements to help flow contaminated atmosphere out of a structure. Mechanical ventilation uses large high-powered fans to augment natural ventilation.
  - Mechanical ventilation includes negative-pressure ventilation, positive-pressure ventilation, and hydraulic ventilation.
  - Negative-pressure ventilation uses smoke ejectors to exhaust smoke and heat from a structure. It can be used to move smoke out of a structure after a fire.
  - Positive-pressure ventilation uses fans to introduce clean air into a structure and push the contaminated atmosphere out. It can be used to reduce interior temperatures and smoke conditions in coordination with a fire attack or to clear a contaminated atmosphere after a fire has been extinguished.
  - Hydraulic ventilation moves air using fog or broken-pattern fire streams to create a pressure differential behind and in front of the nozzle. It is most useful in clearing smoke and heat out of a room after the fire is under control.

- Vertical ventilation refers to any opening that allows the products of combustion to travel up and out. This term is most often applied to describe operations on the roof of a structure. The roof openings can be existing features, such as skylights or bulkheads, or holes created by fire fighters who cut through the roof covering. The choice of roof openings depends primarily on the building's roof construction.
- Before performing vertical ventilation, fire fighters must evaluate all pertinent safety issues and avoid unnecessary risks. The biggest risk is roof collapse. Assess the roof for roof scuttles, heat vents, plumbing vents, louvre ventilation, and fan shafts to prevent tripping or falling from the roof.
- When working on a roof, have two safe exit routes. A second ground ladder or aerial device should be positioned to provide a quick escape route. The ventilation opening should never be located between the exit route and the ventilation crew.
- Fire fighters who are assigned to vertical ventilation tasks should always be aware of the condition of the roof. They should immediately retreat from the roof if they notice any of the following signs:
  - Any visible indication of sagging roof supports
  - Any indication that the roof assembly is separating from the walls, such as the appearance of fire or smoke near the roof edges
  - Any structural failure of any portion of the building, even if it is some distance from the ventilation operation
  - Any sudden increase in the intensity of the fire from the roof opening
  - High heat indicators on a thermal imaging camera
- The two major components of roof construction are roof support structures and roof coverings. The roof support system provides the structural strength to hold the roof in place. The structural system is either solid-beam or truss construction. The roof covering is the weather-resistant surface and may consist of many layers.
- Roof designs include flat roofs, pitched roofs, and arched roofs.
  - Flat roof construction is similar to floor construction. It can be supported by solid components or by trusses. Flat roofs often have vents, skylights, scuttles, or other features that penetrate the roof deck. Removing the covers from these openings provides vertical ventilation without the need to make cuts through the roof deck.

- Pitched roofs have a visible slope. They can be supported by trusses or a system of rafters and beams. Many of these roofs have a layer of solid sheeting covered by a weather-resistant membrane and outer covering. The roof construction material dictates how to ventilate the roof.
- Arched roofs create large open spans without the use of columns. They are often supported with bowstring trusses. The collapse of a bowstring truss is usually very sudden.

■ The types of vertical ventilation openings include the following:
- Built-in roof openings
- Inspection openings
- Primary (expandable) openings
- Secondary (defensive) openings

■ Some commercial or industrial structures have concrete roofs. There are few options for ventilating these structures. Use alternative ventilation openings such as vents or skylights.

■ Metal roofs conduct heat and are often supported by lightweight steel metal joists.

■ Both horizontal and vertical ventilation can be required to vent a basement.

■ HVAC systems may be used to ventilate high-rise buildings.

■ To ensure successful ventilation operations, all equipment and tools must be in a ready state and properly maintained.

Hot Terms

Arched roof A rounded roof usually associated with a bow-string truss.

Chase Open space within walls for wires and pipes.

Churning Recirculation of exhausted air that is drawn back into a negative-pressure fan in a circular motion.

Cockloft The concealed space between the top-floor ceiling and the roof of a building.

Ejectors Also called smoke ejectors, electric fans used in negative-pressure ventilation.

Fire-resistive construction Buildings that have structural components of noncombustible materials with a specified fire resistance. Materials can include concrete, steel beams, and masonry-block walls. Fire-resistive construction is also known as Type I building construction, as defined in NFPA 220, *Standard on Types of Building Construction*.

Gusset plate A connecting plate used in trusses, typically made of wood or lightweight metal.

Horizontal ventilation The process of making openings on the same level as the fire so that smoke, heat, and gases can escape horizontally from a building through openings such as doors and windows.

Hydraulic ventilation Ventilation that relies on the movement of air caused by a fog stream that is placed 0.6–1.2 m (2–4 ft.) in front of the open window.

Kerf cut A cut that is only the width and depth of the saw blade. It is used to inspect cockloft spaces from the roof.

Lath Thin strips of wood used to make the supporting structure for roof tiles.

Leap-frogging A condition in which fire spreads from one floor to the other through exterior windows (auto-exposure).

Louvre cut A cut that is made using power saws and axes to cut along and between roof supports so that the sections created can be tilted into the opening.

Mechanical ventilation A process of removing heat, smoke, and gases from a fire area by using exhaust fans, blowers, air-conditioning systems, or smoke ejectors. (NFPA 402)

Mushrooming The process in which rising smoke, heat, and gases encounter a horizontal barrier such as a ceiling and begin to move out and back down.

Natural ventilation The flow of air or gases created by the difference in the pressures or gas densities between the outside and inside of a vent, room, or space. (NFPA 853)

Negative-pressure ventilation Ventilation that relies on electric fans to pull or draw the air from a structure or area.

Ordinary construction Buildings whose exterior walls are made of noncombustible or limited-combustible materials, but whose interior floors and walls are made of combustible materials. Ordinary construction is also known as Type III building construction, as defined in NFPA 220, *Standard on Types of Building Construction*.

Parapet walls Walls on a flat roof that extend above the roofline.

Peak cut A ventilation opening that runs along the top of a peaked roof.

Positive-pressure ventilation Ventilation that relies on fans to push or force clean air into a structure.

Primary cut The main ventilation opening made in a roof to allow smoke, heat, and gases to escape.

Roof covering The membrane, which can also be the roof assembly, that resists fire and provides weather protection to the building against water infiltration, wind, and impact. (NFPA 5000)

Roof decking The rigid component of a roof covering.

Seat of the fire The main area of the fire.

Secondary cut An additional ventilation opening made for the purpose of creating a larger opening or limiting fire spread.

Smoke inversion The condition in which smoke hangs low to the ground because of the presence of cold air.

Sounding The process of striking a roof with a tool to determine where the roof supports are located.

Stack effect The vertical air flow within buildings caused by the temperature-created density differences between the building interior and exterior or between two interior spaces. (NFPA 92B)

Strip cut Another term for a trench cut.

Trench cut A roof cut that is made from one load-bearing wall to another load-bearing wall and that is intended to prevent horizontal fire spread in a building.

Triangular cut A triangle-shaped ventilation cut in the roof decking that is made using a saw or an axe.

Underventilated fire A fire that has an adequate supply of fuel available but is not receiving an adequate supply of oxygen.

Ventilation The changing of air within a compartment by natural or mechanical means. Ventilation can be achieved by introduction of fresh air to dilute contaminated air or by local exhaust of contaminated air. (NFPA 402)

Vertical ventilation The process of making openings so that smoke, heat, and gases can escape vertically from a structure.

Wood-frame construction Buildings whose exterior walls, interior walls, floors, and roofs are made of combustible wood materials. Wood-frame construction is also known as Type V building construction, as defined in NFPA 220, *Standard on Types of Building Construction*.

FIRE FIGHTER
in action

You arrive at work to find that you are being assigned to the truck company for the day. Your captain hardly gets the words out of his mouth when you are dispatched to a fire in a building. The structure is a newer commercial building constructed of metal-clad, structural steel. The caller indicated that a construction crew was digging a trench when it hit a gas main. The gas ignited, impinged on the building, and caught it on fire. The IC indicated that this would be an offensive attack with an objective to limit fire spread from the current suite that is involved. Your crew is assigned to ventilate the structure to support fire control.

1. You decide to vertically ventilate the structure. What is your greatest concern?
   A. Falling from the roof
   B. The roof collapsing
   C. Cutting the hole in the wrong location
   D. Crews putting a hose line through the ventilation hole

2. What is a trench cut?
   A. A cut that is only the width and depth of the saw blade; it is used to inspect cockloft spaces from the roof
   B. A ventilation opening that runs along the top of a peaked roof
   C. An additional ventilation opening made for the purpose of creating a larger opening or limiting fire spread
   D. A roof cut that is made from one load-bearing wall to another load-bearing wall and that is intended to prevent horizontal fire spread in a building

3. What is hydraulic ventilation?
   A. The process of making openings on the same level as the fire so that smoke, heat, and gases can escape horizontally from a building through openings such as doors and windows
   B. A process of removing heat, smoke, and gases from a fire area by using exhaust fans, blowers, air conditioning systems, or smoke ejectors
   C. Ventilation that relies on fans to push or force clean air into a structure
   D. Ventilation that relies on the movement of air caused by a fog stream that is placed 0.6–1.2 m (2–4 ft.) in front of the open window

4. What is smoke inversion?
   A. The condition in which smoke hangs low to the ground due to the presence of cold air
   B. The vertical airflow within buildings caused by temperature-created density differences between the building's interior and exterior or between two interior spaces
   C. The process in which rising smoke, heat, and gases encounter a horizontal barrier such as a ceiling and begin to move out and back down
   D. Recirculation of exhausted air that is drawn back into a negative-pressure fan in a circular motion

5. A cut that is made using power saws and axes along and between roof supports so that the sections created can be tilted into the opening is a
   _____.
   A. louvred cut
   B. kerf cut
   C. triangular cut
   D. secondary cut

6. When setting up positive-pressure ventilation, how far should the fan be placed in front of the opening?
   A. Inside the building, 0.9–1.2 m (3–4 ft.) from the doorway
   B. In the doorway
   C. 0.9–1.2 m (3–4 ft.) in front of the doorway
   D. 2.4–3.1 m (8–10 ft.) in front of the doorway

You were dispatched to a fire in a 167.2-sq.m (1800-sq.ft.) wood-frame house. Smoke was venting out a broken B-side window. Your crew was the first to arrive and began fire attack, the second-arriving engine was assigned to the B-side exposure protection, and the truck company arrived and set up positive-pressure ventilation in support of fire attack. As the truck company started the fan, visibility inside the structure immediately improved, and the heat decreased significantly; however, unbeknownst to you, the smoke coming from the window burst into flames and was being forcefully driven toward the exposure.

After the fire was extinguished, the exposure protection crew came over and was visibly mad about the use of the fan. The seasoned captain turned toward the truck crew and said, "You guys and those fans are going to get someone killed! You nearly burned down the house next door. You have got to wait to start the fans until the fire is out." You were shocked because you knew how much faster and safer you were able to do fire attack.

1. How do you weigh the pros and cons of using positive-pressure ventilation during attack operations?
2. At which times should you *not* use positive-pressure ventilation?
3. What would you say to the captain?
4. What would you say to the truck crew?

Water Supply

Knowledge Objectives

After studying this chapter, you will be able to:

- Describe the equipment and procedures that are used to access static sources of water. (NFPA 5.3.15.A , p 479–480)
- Describe the characteristics of a mobile water supply apparatus. (p 480)
- Describe the advantages of a portable tank system. (NFPA 5.3.15.A , p 480–481)
- Describe how municipal water systems supply water to communities. (p 482–484)
- List the types of fire hydrants. (p 484–485)
- Describe the characteristics of wet barrel hydrants. (NFPA 5.3.15.A , p 484)
- Describe the characteristics of dry barrel hydrants. (NFPA 5.3.15.A , p 484–485)
- Describe the common guidelines that govern the location of fire hydrants. (p 486)
- Describe how to inspect a fire hydrant. (p 488–489)
- Describe how to test a fire hydrant. (p 489–490)
- Explain the principles of fire hydraulics. (p 490–492)
- Describe how water flow is measured. (p 490)
- Describe how water pressure is measured. (p 490–491)
- Explain how friction loss affects water pressure. (p 491)
- Explain how elevation pressure affects water pressure. (p 491)
- Describe how to prevent water hammer. (NFPA 5.3.1.B , p 491–492)
- List the two types of fire hose. (NFPA 5.3.15.A , p 492)
- Describe the characteristics of small-diameter hose. (NFPA 5.3.15.A , p 492)
- Describe the characteristics of medium-diameter hose. (NFPA 5.3.15.A , p 492)
- Describe the characteristics of large-diameter hose. (NFPA 5.3.15.A , p 492)
- Explain how fire hose is constructed. (NFPA 5.3.15.A , p 492–493)
- Describe the characteristics of double-jacket hose. (NFPA 5.3.15 , p 493)
- Describe the characteristics of rubber-jacket hose. (NFPA 5.3.15 , p 493)
- Describe the characteristics of couplings. (NFPA 5.3.15 , p 493)
- List the common types of couplings. (NFPA 5.3.15 , p 493)
- Describe the two types of supply hose. (NFPA 5.3.15 , p 499)
- List the common types of hose damage. (NFPA 5.5.2.A , p 500–501)
- Describe how to clean and maintain hose. (NFPA 5.5.2 , p 502)
- Describe how to perform a hose inspection. (NFPA 5.5.2 , p 502)
- List the common hose appliances used in conjunction with fire hoses. (NFPA 5.3.15 , p 503)
- Describe the characteristics of wyes. (NFPA 5.3.15 , p 503)
- Describe the characteristics of water thieves. (NFPA 5.3.15 , p 503–504)
- Describe the characteristics of Siamese connections. (NFPA 5.3.15 , p 504)
- Describe the characteristics of reducers. (NFPA 5.3.15 , p 504)
- Describe the characteristics of hose jackets. (NFPA 5.3.15 , p 505)
- Describe the characteristics of hose rollers. (NFPA 5.3.15 , p 505)
- Describe the characteristics of hose clamps. (NFPA 5.3.15 , p 505)
- Describe the types of valves used to control water in pipes or hose lines. (NFPA 5.3.15 , p 506)
- List the common types of hose rolls used to organize supply hose. (p 507–509)
- Describe the procedures used to connect supply lines to a fire hydrant. (NFPA 5.3.15 , p 509–510)
- Describe the common types of supply line evolutions. (NFPA 5.3.15 , p 510–518)
- Describe the common techniques used to load supply hose. (NFPA 5.3.15 , p 518–523)
- Describe the common techniques used to carry and advance supply hose. (NFPA 5.3.15 , p 525–529)
- Describe the two types of standpipe systems. (p 530)
- Describe how to unload a fire hose. (NFPA 5.3.15 , p 531)

Skills Objectives

After studying this chapter, you will be able to perform the following skills:

- Set up a portable tank. (NFPA 5.3.15.B , p 480–481)
- Operate a fire hydrant. (NFPA 5.3.15.A , p 486–487)
- Shut down a fire hydrant. (NFPA 5.3.15.B , p 486–488)
- Replace the swivel gasket on a fire hose. (NFPA 5.3.15.B , p 495)
- Perform the one–fire fighter foot-tilt method of coupling a fire hose. (NFPA 5.3.15 , p 496)
- Perform the two–fire fighter method of coupling a fire hose. (NFPA 5.3.15 , p 496–497)

- Perform the one–fire fighter knee-press method of uncoupling a fire hose. (NFPA 5.3.15 , p 496, 498)
- Perform the two–fire fighter stiff-arm method of uncoupling a fire hose. (NFPA 5.3.10.B , p. 498–499)
- Uncouple a fire hose with a spanner wrench. (NFPA 5.3.10.B , p 498, 500)
- Clean and maintain fire hose. (NFPA 5.5.2.B , p 502)
- Mark a defective hose. (NFPA 5.5.2.A , p 503)
- Perform a straight hose roll. (p 507)
- Perform a single-doughnut roll. (p 507–508)
- Perform a twin-doughnut roll. (p 509–510)
- Perform a self-locking twin-doughnut roll. (p 509, 511)
- Perform a forward hose lay. (NFPA 5.3.15.B , p 512–513)
- Attach a fire hose to a four-way hydrant valve. (NFPA 5.3.15B , p 514–516)
- Perform a reverse hose lay. (NFPA 5.3.15B , p 516–517)
- Perform a split hose lay. (NFPA 5.3.15B , p 518)
- Perform a flat hose load. (NFPA 5.3.15B , p 519–520)
- Perform a horseshoe hose load. (NFPA 5.3.15B , p 519, 521)
- Perform an accordion hose load. (NFPA 5.3.15B , p 522–523)
- Attach a soft suction hose to a fire hydrant. (NFPA 5.3.15B , p 523–525)
- Attach a hard suction hose to a fire hydrant. (NFPA 5.3.15B , p 525–526)
- Perform a working hose drag. (NFPA 5.3.10.B , p 525, 527)
- Perform a shoulder carry. (NFPA 5.3.10.B , p 525–527)
- Advance an accordion load. (NFPA 5.3.10.B , p 526, 528–529)
- Connect a hose line to supply a fire department connection. (p 530–531)
- Replace a damaged hose line. (NFPA 5.5.2.A , p 530–531)
- Drain a fire hose. (p 531–532)

Fire Fighter II   FFII

Knowledge Objectives

After studying this chapter, you will be able to:

- Describe how to perform a service test on a fire hose. (NFPA 6.5.5.A , p 503)
- List the information that should be noted on a hose record. (NFPA 6.5.5.A , p 503)

Skills Objectives

After studying this chapter, you will be able to perform the following skills:

- Perform an annual service test. (NFPA 6.5.5.B , p 503)

Additional NFPA Standards

- NFPA 24, *Standard for the Installation of Private Fire Service Mains and Their Appurtenances*
- NFPA 1142, *Standard on Water Supplies for Suburban and Rural Fire Fighting*
- NFPA 1962, *Standard for the Care, Use, Inspection, Service Testing, and Replacement of Fire Hose, Couplings, and Nozzles and Fire Hose Appliances*

You arrive at a fire in a barn that is fully involved. The attack is clearly going to be a defensive operation, but with a house and other buildings nearby, aggressive actions will be required to protect exposures. The house and other buildings are located approximately 30 metres (98 feet) down a narrow, tree-lined lane. While crews are setting up exposure protection, you are assigned to be the water supply officer (WSO). The incident commander (IC) says he needs a 1893-litre-per-minute (L/min.) (500 gallon-per-minute [gpm]) flow to be established.

1. Which water sources could you consider?

2. Which components need to be included to establish a water supply?

3. How will you deal with the narrow lane?

Water Supply

Water supply is essential to achieving extinguishment of any structure fire. To accomplish this objective, many factors must come together. This chapter covers several topics related to **water supply**. It describes the hydraulics or behaviour of flowing water, rural and municipal water supplies, the operation of fire hydrants, the construction and care of fire hose, the roles of hose appliances, and supply line evolutions; thus this chapter covers the water flow during the first half of its journey—from the source to the pumper. The chapter "Fire Attack Lines and Foam" covers the second half of the journey, from the pumper to the fire.

The importance of a dependable and adequate water supply for fire-suppression operations is self-evident. The hose line is not only the primary weapon for fighting fire but also the fire fighter's primary defence against being burned or driven out of a burning building. The basic plan for fighting most fires depends on having an adequate supply of water to safely confine, control, and extinguish the fire.

If the water supply is interrupted while crews are working inside a building, fire fighters can be trapped, injured, or killed. Fire fighters entering a burning building need to be confident that their water supply is both reliable and adequate to operate hose lines for their protection and to extinguish the fire.

Ensuring a dependable water supply is a critical fire-ground operation, which must be accomplished as soon as possible. A water supply should be established at the same time as other initial fire-ground operations are conducted. At many fire scenes, size-up, forcible entry, raising ladders, search and rescue, ventilation, and establishing a water supply all occur concurrently.

Fire fighters can obtain water from one of two sources: **municipal water systems** furnish water under pressure through fire hydrants **FIGURE 16-1**. In contrast, rural areas may depend on **static water sources**, such as lakes and streams. These bodies of water serve as drafting sites for fire department apparatus to obtain and deliver water to the fire scene.

FIGURE 16-1 The water that comes from a hydrant is provided by a municipal or private water system.

Often, the water that is carried in a tank on one of the first-arriving vehicles is used in the initial attack. Although many fires are successfully controlled using tank water, this tactic does not ensure the adequacy and reliability of the water supply. Establishing an adequate, continuous water supply then becomes the primary objective to support the fire attack **FIGURE 16-2**. The operational plan must ensure that an adequate and reliable water supply is available before the tank becomes empty.

Water Supply in Rural Nonhydrant Areas

Many fire departments protect areas that are not serviced by municipal water systems. In these areas, residents usually depend on individual wells or cisterns to supply water for their domestic uses. Because there are no fire hydrants in these areas, fire fighters must depend on water from other sources for firefighting activities. Fire fighters in rural areas must know how to get water from the sources that are available.

In Canada, many municipalities attempt to maintain sufficient water volume to meet the Fire Underwriters Survey (FUS) requirement for Accredited Superior Tanker Shuttle Service, a recognized equivalency to hydrant protection. A fire service's ability to provide a continuous and reliable water supply is directly related to the insurance ratings in the municipality: the better the water supply, the better the insurance rates, and the safer the community.

■ Static Sources of Water

Several potential static water sources can be used for fighting fires in rural areas. Both natural and man-made bodies of water, such as rivers, streams, lakes, ponds, oceans, canals, reservoirs, swimming pools, and cisterns, can be used to supply water for fire suppression **FIGURE 16-3**. Some areas have many different static sources, whereas others have few or none at all.

Water from a static source can be used to fight a fire directly if it is close enough to the fire scene. Otherwise, it must be transported to the fire using long hose lines, engine relays, or mobile water supply tankers/tenders.

To be useful, static water sources must be accessible to a fire engine or portable pump. If a road or hard surface is located within 6.1 m (20 ft.) of the water source, a fire engine can drive close enough to draft water directly into the pump through a hard suction hose. Rural fire departments should identify these areas and practise establishing drafting operations at all of these locations. Some fire departments construct special access points, so engines can approach the water source.

Dry hydrants, also called drafting hydrants, provide quick and reliable access to static water sources. A dry hydrant is a pipe with a strainer on one end and a connection for a hard suction hose on the other end. The strainer end should be located below the water's surface and away from any silt or potential obstructions. The other end of the pipe should be accessible to fire apparatus, with the connection at a convenient height for an engine hookup **FIGURE 16-4**. When a hard suction hose is connected to a drafting hydrant, the engine can draft water from the static source.

Dry or drafting hydrants are often installed in lakes and rivers and close to clusters of buildings where there is a recognized need for fire protection. They may also be installed in farm cisterns or connected to swimming pools on private property to make water available for the local fire department. In some areas, dry or drafting hydrants are used to enable fire fighters to reach water under the frozen surface of a lake or river. National Fire Protection Association (NFPA) 1142, *Standard on Water Supplies for Suburban and Rural Fire Fighting*, has more information about dry hydrants.

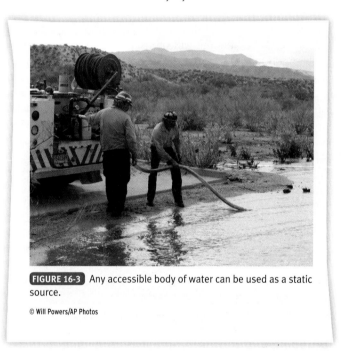

FIGURE 16-3 Any accessible body of water can be used as a static source.

© Will Powers/AP Photos

FIGURE 16-2 Mobile water supply apparatus can deliver limited quantities of water to the scene of a fire.

© Peter Kim/ShutterStock, Inc.

Connection

Water level →

Strainer

FIGURE 16-4 A dry hydrant, or drafting hydrant, can be placed at an accessible location near a static water source.

The portable pump is an alternative means of accessing water in areas that are inaccessible to fire apparatus **FIGURE 16-5**. The portable pump can be hand-carried or transported by an off-road vehicle to the water source. Portable pumps can deliver water as fast as 1893 L/min. (500 gpm).

Mobile Water Supply Apparatus

Mobile water supply apparatus (also called tankers or water tenders) can deliver water to fight fires in rural areas as well. These trucks are designed specifically to carry large volumes of water. Fire department engines usually carry at least 1893 litres (500 gallons) of water in the booster tank, whereas fire department tankers generally carry 3785–13,247 L (1000–3500 gal.) of water **FIGURE 16-6**. Some tankers can transport as much as 18,925 L (5000 gal.) of water.

If a mobile water supply apparatus is the only source of water for fighting a structural fire, the attack must be carefully planned. Enough water must be available on the scene to supply the required hose lines. If the water supply is exhausted before the fire is extinguished, the attack team will be in serious danger, and the building will probably be lost. Conversely, if water use is too conservative, fire-suppression efforts will probably be unsuccessful. To be both safe and successful, the fire department must be able to deliver adequate amounts of water and to maintain a reliable supply of water to the fire scene.

FIGURE 16-5 A portable pump can be used if the water source is inaccessible to a fire department engine.

© Maran Garai/ShutterStock, Inc.

FIGURE 16-6 A fire department tanker or tender.

© IAN MARLOW/Alamy Images

Some fire departments begin the attack using water from the tank of the first-arriving unit. Tenders then pump additional water directly into the attack pumper to keep it full. In rural areas, several tankers may be dispatched for a structural fire.

These tenders can deliver water to the attack engine in sequence, refilling their own supplies at a remote location and returning to the fire scene as needed.

Portable Tanks

Portable tanks carried on fire apparatus can be quickly set up at a fire scene. These tanks typically hold between 2271 and 18,925 L (600 and 5000 gal.) of water and should be placed so that they can be accessed from multiple directions. With this supply method, one engine drafts water from the tank using a hard suction hose, just as it would if the water came from any other static source **FIGURE 16-7**. A tanker is used to fill the portable tank. The pump driver/operator primes the pump and begins drafting water out of the portable tank, while the tanker leaves to get another load of water.

Speed is a primary advantage when using a portable tank system, in that tankers do not have to hook up to the attack pumper and slowly transfer the water. Instead, a dump valve enables the tankers to offload as much as 11,355 L (3000 gal.) of water in 1 minute into a portable tank **FIGURE 16-8**. The faster the tanker can offload its water, the more quickly it can return to the fill site for another load.

Another advantage of the portable tank system is its ability to expand rapidly. Additional portable tanks can be set up and linked together to increase water storage capacity, additional pumpers can be used to draft water from the portable tanks, and additional tankers can be used to deliver more water at a faster rate. A series of several tankers may also be used as shuttles, dumping water either simultaneously or in sequence. When using multiple tanks, a jet siphon assists in keeping the water level at the maximum capacity in the tank closest to the pumper.

To set up a portable tank, follow the steps in **SKILL DRILL 16-1**:

1 Two fire fighters lift the portable tank off the apparatus. This tank may be mounted on a side rack or on a hydraulic rack that lowers it to the ground. (**STEP 1**)

FIGURE 16-7 An engine may be set up to draft water from a portable tank.

Courtesy of Captain David Jackson, Saginaw Township Fire Department

2 Place the portable tank on level ground beside the pump. The pump driver/operator will indicate the best location. (**STEP** **2**)

3 Expand the tank (metal-frame type) or lay it flat (self-expanding type). (**STEP** **3**)

4 Place a strainer on the end of the suction hose, put the suction hose into the tank, and connect it to the engine. (**STEP** **4**)

5 A second fire fighter helps the tanker driver discharge water into the portable tank. If the tank is self-expanding, the fire fighters may need to hold the collar until the water level is high enough for the tank to support itself. (**STEP** **5**)

FIGURE 16-8 A dump valve allows a tanker to quickly discharge water into the portable tank.

SKILL DRILL 16-1 Setting Up a Portable Tank
(Fire Fighter I, NFPA 5.3.15)

1 Two fire fighters lift the portable tank off the apparatus. This tank may be mounted on a side rack or on a hydraulic rack that lowers it to the ground.

2 Place the portable tank on level ground beside the engine. The pump driver/operator will indicate the best location.

3 Expand the tank (metal-frame type) or lay it flat (self-expanding type).

4 One fire fighter helps the pump operator place the strainer on the end of the suction hose, put the suction hose into the tank, and connect it to the engine.

5 The second fire fighter helps the tanker driver discharge water into the portable tank. If the tank is self-expanding, the fire fighters may need to hold the collar until the water level is high enough for the tank to support itself.

© LiquidLibrary

■ Tender Shuttles

When a large volume of water is needed for an extended period at a fire, a **tender shuttle**, or tanker shuttle, can be used to deliver water from a fill site to the fire scene. The number of tenders needed will depend on the distance between the fill site and the fire scene, the time it takes them to dump and refill, and the flow rate required at the fire scene.

All of the components of a tender shuttle must be set up so water moves efficiently from the fill site to the fire scene. At the fill site, the tenders must be refilled without delay.

The routes in both directions must be planned so that tenders make efficient round trips, without having to back up or make U-turns. At the fire scene, the tenders should be able to drive up, dump their water into the portable tanks, and immediately return to the fill site. The portable tanks, in turn, must be large enough to receive the full load of each tanker as it arrives. An effective water shuttle can deliver more than 1000 L (264 gal.) of water per minute without interruption.

A separate component of the incident command structure titled "water supply" is usually established to coordinate a tanker-shuttle operation. Departments must practise these operations to ensure proper coordination and effective water delivery.

Every municipality has its own specific practises for supplying water in nonhydrant and/or rural areas. Ensure that you fully understand the suggested operating guidelines (SOGs), or standard operating procedures (SOPs), in your jurisdiction.

Fire Fighter Safety Tips
© Photos.com

Never get between two apparatus or tanks!

Municipal Water Systems

Municipal water systems make clean water available to people in populated areas and provide water for fire protection. As the name suggests, most municipal water systems are owned and operated by a local government agency, such as a city, county, or special water district. Some municipal water systems are privately owned; however, the basic design and operation of both private and government systems are very similar. Municipal water is supplied to homes, commercial establishments, and industries.

Hydrants make the same water supply available to the fire department. In addition, most automatic sprinkler systems and many standpipe systems are connected directly to a municipal water source. A municipal water system has three major components: the water source, the treatment plant, and the distribution system.

■ Water Sources

Municipal water systems can draw water from wells, rivers, streams, lakes, or man-made storage facilities called **reservoirs**. The source will depend on the geographic and hydrologic features of the area. Many municipal water systems draw water from several sources to ensure a sufficient supply. Underground pipelines or open canals supply some cities with water from sources that are many miles away.

The water source for a municipal water system needs to be large enough to meet the demands of the service area. Most of these systems include large storage facilities, thereby ensuring that they will be able to meet the community's water supply demands if the primary water source becomes unavailable for some reason. The backup supply for some systems can provide water for several months or years. By contrast, in some other systems, the supply may last only a few days.

■ Water Treatment Facilities

Municipal water systems each also include a water treatment facility, where impurities are removed from the water **FIGURE 16-9**. The nature of the treatment system depends on the quality of the untreated source water. Water that is clean and clear from the source requires little treatment. Other systems must use extensive filtration to remove impurities and foreign substances. Some treatment facilities use chemicals to remove impurities and improve the water's taste. All of the water in the system must be suitable for drinking.

Chemicals and ultraviolet (UV) radiation are used to kill bacteria and harmful organisms and to keep the water pure as it moves through the distribution system to individual homes or businesses. After the water has been treated, it enters the distribution system.

■ Water Distribution System

The distribution system delivers water from the treatment facility to the end-users and fire hydrants through a complex network of underground pipes, known as **water mains**. In most cases, the distribution system also includes pumps, storage tanks, control valves, reservoirs, and other necessary components to ensure that the required volume of water can be delivered where and when it is needed, at the required pressure.

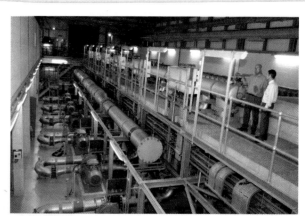

FIGURE 16-9 Impurities are removed at the water treatment facility.

© Paul Glendell/Alamy Images

Water pressure requirements differ, depending on how the water will be used. Generally, water pressure ranges from 138 to 552 kilopascals (kPa; 20 to 80 pounds per square inch [psi]) at the delivery point. The recommended minimum pressure for water coming from a fire hydrant is 138 kPa (20 psi), but it is possible to operate with lower hydrant pressures, under some circumstances.

Most water distribution systems rely on an arrangement of pumps to provide the required pressure, either directly or indirectly. In systems that use pumps to supply direct pressure, if the pumps stop operating, the pressure will be lost, and the system will be unable to deliver adequate water to the end-users or to hydrants. Most municipal systems maintain multiple pumps and backup power supplies to reduce the risk of a service interruption due to a pump failure. These extra pumps can sometimes be used to boost the flow for a major fire or during a high-demand period.

In a pure **gravity-feed system**, the water source, treatment plant, and storage facilities are located on high ground, while the end users live in lower-lying areas, such as a community in a valley **FIGURE 16-10**. This type of system may not require any pumps, because gravity, through the elevation differentials, provides the necessary pressure to deliver the water. In some systems, the elevation pressure is so high that pressure-control/reduction devices are needed to keep parts of the system from being subjected to excessive pressure.

Most municipal water supply systems use both pumps and gravity to deliver water. Pumps can be used to deliver water from the treatment plant to **elevated water-storage towers** or to reservoirs located on hills or high ground. These elevated storage facilities then maintain the desired water pressure in the distribution system, ensuring that water can be delivered under pressure even if the pumps are not operating **FIGURE 16-11**. When the elevated storage facilities need refilling, large supply pumps are used. Additional pumps may be installed to increase the pressure in particular areas, such as a booster pump that provides extra pressure for a hilltop neighbourhood.

A combination pump- and gravity-feed system must maintain enough water in the elevated storage tanks and reservoirs to meet anticipated demands. If more water is being used than the pumps can supply, or if the pumps are out of service, some systems will be able to operate for several days by relying solely

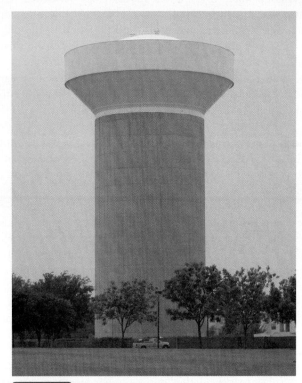

FIGURE 16-11 Water that is stored in an elevated tank can be delivered under pressure to the end-users.

on their elevated storage reserves. Other systems may be able to function for only a few hours in such scenarios.

The underground water mains that deliver water to the end-users come in several different sizes. Large mains, known as **primary feeders**, carry large quantities of water to a section of the town or city. Smaller mains, called **secondary feeders**, distribute water to a smaller area. The smallest pipes, called **distributors**, carry water to the users and hydrants along individual streets.

The size of the water mains required depends on the amount of water needed both for normal consumption and for fire protection in that location. Most jurisdictions specify the minimum-size main that can be installed in a new municipal water system to ensure an adequate flow; some municipal water systems, however, may include undersized water mains in older areas of the community. The volume of water delivered through a water main may decrease if the pipe becomes corroded or partly filled with sediment. Fire fighters must know the arrangement and capacity of the water systems in their response areas.

Water mains in a well-designed system will follow a grid pattern. A grid arrangement provides water flow to a fire hydrant from two or more directions and establishes multiple paths from the source to each area. This kind of organization helps to ensure an adequate flow of water for firefighting. The grid design also helps to minimize downtime for the other portions of the system if a water main breaks or needs maintenance work. With a grid pattern, the water flow can be diverted around the affected section.

FIGURE 16-10 A gravity-feed system can deliver water to a low-lying community without the need for pumps.

Older water distribution systems may have dead-end water mains, which supply water from only one direction. Such dead-end water mains may still be found in the outer reaches of a municipal system. Hydrants on this type of main will have a limited water supply. If two or more hydrants on the same dead-end main are used to fight a fire, the upstream hydrant will have more water and water pressure than the downstream hydrants.

Control valves installed at intervals throughout a water distribution system allow different sections of the water supply to be turned off or isolated. These valves are used when a water main breaks or when work must be performed on a section of the system.

Shut-off valves are located at the connection points where the underground mains meet the distributor pipes. These valves control the flow of water to individual customers or to individual fire hydrants FIGURE 16-12 . If the water system in a building or to a fire hydrant becomes damaged, the shut-off valves can be closed to prevent further water flow.

The fire department should notify the water department when fire operations will require prolonged use of large quantities of water. The water department may, in turn, be able to increase the normal volume and/or pressure by starting additional pumps. In some systems, the water department can open valves to increase the flow to a certain area in response to fire department operations at major fires.

Types of Fire Hydrants

Fire hydrants provide water for firefighting purposes. Public hydrants are part of the municipal water distribution system and draw water directly from the public water mains. Hydrants are also installed on private water systems supplied by the municipal water system or from a separate source. The water source, as well as the adequacy and reliability of the supply to private hydrants, must be noted and identified to ensure that sufficient water supplies will be available when fighting fires.

Most fire hydrants consist of an upright steel casing (barrel) attached to the underground water distribution system. The two major types of municipal hydrants are the dry barrel hydrant and the wet barrel hydrant. Hydrants are equipped with one or more valves to control the flow of water through the hydrant. In addition, one or more outlets are provided to connect fire department hoses to the hydrant. These outlets are sized to fit the 64-millimetre (2.5-inch) or larger fire hoses used by the local fire department. Threads utilized on fire hydrants are usually of the national standard type, although some jurisdictions have their own thread type.

■ Wet Barrel Hydrants

Wet barrel hydrants are used in locations where temperatures do not drop below freezing. These hydrants always have water in the barrel and do not have to be drained after each use.

Wet barrel hydrants usually have separate valves that control the flow to each individual outlet FIGURE 16-13 . A fire fighter can hook up one hose line and begin flowing water, and later attach a second hose line and open the valve for that outlet without shutting down the hydrant.

■ Dry Barrel Hydrants

Dry barrel hydrants are used in climates where temperatures fall below freezing. The valve that controls the flow of water into the barrel of the hydrant is located at the base, several feet below ground (i.e., below the frost line), to keep the hydrant from freezing FIGURE 16-14 .

FIGURE 16-13 A wet barrel hydrant has a separate valve for each outlet.

Courtesy of American AVK Company

FIGURE 16-12 A shut-off valve controls the water supply to an individual user or fire hydrant.

The length of the barrel depends on the climate and the depth of the valve. Water enters the barrel of the hydrant only when it is needed. Turning the nut on the top of the hydrant rotates the operating stem, which opens the valve so that water flows up into the barrel of the hydrant.

Whenever this type of hydrant is not in use, the barrel must remain dry. If the barrel contains standing water, it will freeze in cold weather and render the hydrant inoperable. After each use, the water drains out through an opening at the bottom of the barrel. This drain is fully open when the hydrant valve is fully shut. When the hydrant valve is opened, the drain closes, thereby preventing water from being forced out of the drain when the hydrant is under pressure; if the drain becomes clogged, however, the hydrant may not drain. In this case, it may be necessary to pump water out of the hydrant before it freezes.

A partially opened valve means that the drain is also partially open, so pressurized water can flow out. This leakage can erode (undermine) the soil around the base of the hydrant and may damage the hydrant. For this reason, a hydrant should always be either fully opened or fully closed. A fully opened hydrant also makes the maximum flow available to fight a fire.

Most dry barrel hydrants contain only one large valve that controls the flow of water **FIGURE 16-15**. Each outlet must be connected to a hose or an outlet valve or have a hydrant cap firmly in place before the valve is turned on.

Many fire departments use special hydrant valves that allow additional connections to be made after water is flowing through the first hose. If additional outlets are needed later, separate outlet valves can be connected before the hydrant is opened.

In some areas, vandals may dispose of trash or place foreign objects in the empty barrels of dry barrel hydrants. These materials can obstruct the water flow or damage a fire department pumper if they are drawn into the pump. It is a good practice to check the operation of the hydrant and flush out debris before connecting a hose to a hydrant. To do so, the fire fighter making the connection opens the large outlet cap and then opens the hydrant valve just enough to ensure that water flows into the hydrant and flushes out any foreign matter; this step takes only a few seconds. The fire fighter then closes the hydrant valve, connects the hose, and reopens the valve all the way. Fire departments should ensure that inspections and tests to keep hydrants operating smoothly are performed regularly.

A.

B.

Valve

FIGURE 16-14 A dry barrel hydrant (A) is controlled by an underground valve (B).

Courtesy of American AVK Company

FIGURE 16-15 Most dry barrel hydrants have only one large valve at the bottom of the barrel that controls the flow of water.

FIRE FIGHTER Tips

The terms *dry barrel hydrant* and *dry hydrant* are very easy to confuse, but they are very different. Dry barrel hydrants are connected to a pressurized municipal water system. Dry hydrants, also known as drafting hydrants, are a permanent piping system that are connected to a static water source, such as a stream, a pond, or a lake. To get water from a dry or drafting hydrant, you must draft (pump) the water using a fire or portable pump.

Fire Hydrant Locations

Fire hydrants are located according to local standards and nationally recommended practices. They may be placed a certain distance apart, perhaps every 150 m (492 ft.) in residential areas and every 90 m (295 ft.) in high-value commercial and industrial areas. In many communities, hydrants are located at every street intersection, with mid-block hydrants being installed if the distance between intersections exceeds a specified limit.

In some cases, the requirements for locating hydrants are based on the occupancy, construction, and size of a building. A builder may be required to install additional hydrants when a new building is constructed so that no part of the building will be more than a specified distance from the closest hydrant.

Knowing the plan for installing fire hydrants makes them easier to find in emergency situations. Fire fighters who perform fire inspections or develop preincident plans should identify the locations and flow rates of nearby fire hydrants for each building or group of buildings.

Fire Hydrant Operation

■ Operate a Hydrant

Fire fighters must be proficient in operating fire hydrants. **SKILL DRILL 16-2** outlines the steps for getting water from a dry barrel hydrant efficiently and safely. These same steps, with the modifications noted, apply to wet barrel hydrants as well.

1 Remove the cap from the outlet you will be using. (**STEP 1**)

2 Quickly look inside the hydrant opening for any objects that may have been thrown into the hydrant. (Omit this step for a wet barrel hydrant.) (**STEP 2**)

3 Check that the remaining hydrant caps are snugly attached. (Omit this step for a wet barrel hydrant.) (**STEP 3**)

4 Place the hydrant wrench on the stem nut. Check the top of the hydrant for the arrow indicating which direction to turn the nut to open the hydrant valve. (**STEP 4**)

5 Open the hydrant just enough to determine that there is a good flow of water and to flush out any objects that may have been put into the hydrant. (Omit this step for a wet barrel hydrant.) (**STEP 5**)

6 Shut off the flow of water. (Omit this step for a wet barrel hydrant.) (**STEP 6**)

7 Attach the hose or valve to the hydrant outlet. (**STEP 7**)

8 When instructed to do so by your officer or the pump driver/operator, start the flow of water. Turn the hydrant wrench to fully open the valve. This may take 12–14 turns, depending on the type of hydrant. (**STEP 8**)

9 Open the hydrant slowly to avoid a pressure surge. Once the flow of water has begun, you can open the hydrant valve more quickly. Make sure that you open the hydrant valve completely; if the valve is not fully opened, the drain hole will remain open. (**STEP 9**)

Note: This skill drill applies to a dry barrel hydrant. For a wet barrel hydrant, omit steps 2, 3, 5, and 6. Open the valve for the particular outlet that will be used.

Individual fire services may have their own variations on this procedure; for example, some departments specify that the wrench be left on the hydrant. Other departments require that the wrench be removed and returned to the fire apparatus so that an unauthorized person cannot interfere with the operation. Always follow the SOGs/SOPs for your department.

FIRE FIGHTER Tips

Fire fighters should treat hose cabinets in the same manner as they would a fire hydrant. In many ways, a hose cabinet (standpipe) is simply an extension of the domestic water supply. Standpipe systems should be flushed prior to attaching your hose. This will assist in flushing slag and possible debris. Some may dismiss this as creating unnecessary water damage; however, when compared to the possibility of a clogged nozzle, the damage will be comparatively minimal. Notwithstanding, most standpipe systems are installed in Type I and Type II buildings where the floor plates are comprised of concrete.

■ Shutting Down a Hydrant

Shutting down a hydrant properly is just as important as opening a hydrant properly. If the hydrant is damaged during shutdown, it cannot be used until it has been repaired. To shut down a hydrant efficiently and safely, follow the steps in **SKILL DRILL 16-3**:

1 Turn the hydrant wrench slowly until the valve is closed. (**STEP 1**)

2 Allow the hose to drain by opening a drain valve or disconnecting a hose connection downstream. Slowly disconnect the hose from the hydrant outlet, allowing any remaining pressure to escape. (**STEP 2**)

3 On dry barrel hydrants, leave one outlet open until the water drains from the hydrant. (**STEP 3**)

4 Replace the hydrant cap. (**STEP 4**)

Note: Do not leave or replace the caps on a dry barrel hydrant until you are sure that the water has completely

SKILL DRILL 16-2 | Operating a Fire Hydrant
(Fire Fighter I, NFPA 5.3.15)

1 Remove the cap from the outlet you will be using.

2 Quickly look inside the hydrant opening for foreign objects (dry barrel hydrant only).

3 Check that the remaining caps are snugly attached (dry barrel hydrant only).

4 Attach the hydrant wrench to the stem nut. Check for an arrow indicating the direction to turn to open.

5 Open the hydrant enough to verify flow and flush the hydrant (dry barrel hydrant only).

6 Shut off the flow of water (dry barrel hydrant only).

7 Attach the hose or valve to the hydrant outlet.

8 When instructed, turn the hydrant wrench to fully open the valve.

9 Open the hydrant slowly to avoid a pressure surge.

© LiquidLibrary

drained from the barrel. If you feel suction on your hand when you place it over the opening, the hydrant is still draining. In very cold weather, you may have to use a hydrant pump to remove all of the water and prevent freezing.

In many communities, hydrants are painted in bright reflective colours for increased visibility. The bonnet (the top of the hydrant) and or caps may also be colour-coded to indicate the available flow rate of a hydrant **TABLE 16-1**. Coloured reflectors are sometimes mounted next to hydrants or placed in the pavement in front of them to make them more visible at night. In areas that experience high volumes of snow, hydrants are sometimes marked with a small flag, which is mounted on top of a pole attached to the hydrant **FIGURE 16-16**.

SKILL DRILL 16-3 — Shutting Down a Hydrant
(Fire Fighter I, NFPA 5.3.15)

1 Turn the wrench slowly to close the hydrant valve.

2 Drain the hose line. Slowly disconnect the hose from the hydrant outlet.

3 Leave one hydrant outlet open until the hydrant is fully drained (dry barrel hydrant only).

4 Replace the hydrant cap.

© LiquidLibrary

TABLE 16-1	Fire Hydrant Colours

NFPA 24, *Standard for the Installation of Private Fire Service Mains and Their Appurtenances*, recommends that fire hydrants be colour-coded to indicate the water flow available from each hydrant at 138 kPa (20 psi). It is recommended that the top bonnet and the hydrant caps be painted according to the following system. The colours give you an idea of how much water can be obtained from a hydrant during a fire.

Class	Flow Available at 138 kPa (20 psi)	Colour
Class C	Less than 1893 L/min. (500 gpm)	Red
Class B	1893–3781 L/min. (500–999 gpm)	Orange
Class A	3785–5674 L/min. (1000–1499 gpm)	Green
Class AA	5678 L/min. (1500 gpm) and higher	Light blue

Hydrants should be installed at an appropriate height above the ground. Their outlets should not be so high or so low that fire fighters have difficulty connecting hose lines to them. NFPA 24 requires a minimum of 46 centimetres (18 in.) from the centre of a hose outlet to the finished grade. Hydrants should be positioned so that the connections—especially the large steamer connection—are facing the street.

During a hydrant inspection, check the exterior of the hydrant for signs of damage. Open the **steamer port** of dry barrel hydrants to ensure that the barrel is dry and free of

Before you leave the fire scene, make sure that dry barrel hydrants are completely drained, even if the weather is warm. During winter, any water left in the hydrant can freeze. Fire fighters may lose valuable time connecting a hose to a frozen hydrant, only to discover that it will not operate. If this happens, fire fighters will be without water until they can locate a working hydrant and can reposition and reconnect the hose lines.

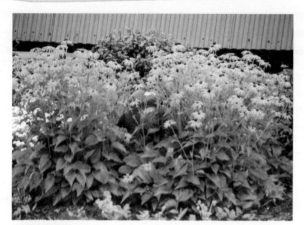

FIGURE 16-16 Hydrants should not be hidden or obstructed.

Courtesy of Captain David Jackson, Saginaw Township Fire Department

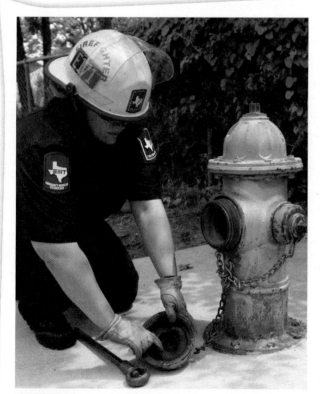

FIGURE 16-17 All hydrants should be checked at least annually.

debris. Make sure that all caps are present and that the outlet hose threads are in good, working order **FIGURE 16-17**.

The second part of the inspection ensures that the hydrant works properly. Open the hydrant valve just enough to confirm that water flows out and flushes any debris out of the barrel. After flushing, shut down the hydrant. Leave the cap off dry barrel hydrants to ensure that they drain properly. A properly draining hydrant will create suction against a hand placed over the outlet opening. When the hydrant is fully drained, replace the cap.

If the threads on the discharge ports need cleaning, use a steel brush and a small triangular file to remove any burrs in the threads. Also, check the gaskets in the caps to make sure they are not cracked, broken, or missing. Replace worn gaskets with new ones, which should be carried on each apparatus. Follow the manufacturer's recommendations for maintaining any parts that require lubrication.

■ Testing Fire Hydrants

The amount of water available to fight a fire at a given location is a crucial factor in planning an attack. Urban water supply usually falls under the authority of public works or the local water authority. Generally speaking, fire fighters are not responsible for maintaining, testing, or repairing fire hydrants in Canada. The only exception to this might be under exceptional circumstances such as a massive snowstorm, when fire personnel may be utilized to clear snow from around fire hydrants.

Flow and Pressure

To understand the procedures for testing fire hydrants, fire fighters must remember the terminology that is used. The flow or quantity of water moving through a pipe, hose, or nozzle is measured in terms of its volume, usually specified in units of litres per minute or gallons per minute. Water pressure refers to an energy level and is measured in units of kilopascals (kPa) or psi. Volume and pressure are two different, but mathematically related, measurements.

Water that is not moving has potential (static) energy. When the water is moving, it has a combination of potential energy and kinetic (in-motion) energy. Both the quantity of water flowing and the pressure under a specific set of conditions must be measured as part of testing any water system, including hydrants.

Static pressure is the pressure in a system when the water is not moving. Static pressure is potential energy, because it would cause the water to move if there were some place the water could go. This kind of pressure causes the water to flow out of an opened fire hydrant. If no static pressure is present, nothing happens when a fire fighter opens a hydrant.

Static pressure is generally created by **elevation pressure** or pump pressure or both. An elevated storage tank creates elevation pressure in the water mains. Gravity also creates elevation pressure (sometimes referred to as head pressure) in a water system as the water flows from a hilltop reservoir to the water mains in the valley below. Pumps create pressure by bringing the energy from an external source into the system.

To measure static pressure in a water distribution system, place a pressure gauge on a hydrant port, and open the hydrant valve. No water can be flowing out of the hydrant when static pressure is measured.

When measured in this way, the static-pressure reading assumes that there is no flow in the system. Of course, because municipal water systems deliver water to thousands of users, there is almost always at least some water flowing within the system; thus, in most cases, a static-pressure reading actually measures the normal operating pressure of the system.

Normal operating pressure refers to the amount of pressure in a water distribution system during a period of normal consumption; in a residential neighbourhood, for example, people are constantly using water to care for lawns, wash clothes, bathe, and perform other normal household activities.

In an industrial or commercial area, normal consumption occurs during a normal business day as water is used for various purposes. The water distribution system uses some of the static pressure to deliver this water to residents and businesses. A pressure gauge connected to a hydrant during a period of normal consumption will indicate the normal operating pressure of this system.

Fire fighters need to know how much pressure is in the system when a fire occurs. Because the regular users of the system will be drawing off a normal amount of water even during firefighting operations, the normal operating pressure is sufficient for measuring available water. *Note:* The normal operating pressure may change according to the time of day in some areas, owing to high demand during certain hours.

Residual pressure is the amount of pressure that remains in the system when water is flowing. When fire fighters open a hydrant and start to draw large quantities of water out of the system, some of the potential energy of still water is converted to the kinetic energy of moving water. However, not all of the potential energy turns into kinetic energy; some of it is used to overcome friction in the pipes. The pressure remaining while the water is flowing constitutes the residual pressure.

Residual pressure is important because it provides the best indication of how much more water is available in the system: the more water flowing, the less residual pressure in the system. In theory, when the maximum amount of water is flowing, the residual pressure is zero, and there is no more potential energy to push more water through the system. In reality, 138 kPa (20 psi) is considered the minimum usable residual pressure necessary to reduce the risk of damage to underground water mains or pumps.

At the scene of a fire, the pump driver/operator uses the difference between static pressure and residual pressure to determine how many more attack lines or appliances can be operated from the available water supply. The pump driver/operator can refer to a set of tables to calculate the maximum amount of available water; these tables are based on the static- and residual-pressure readings taken during hydrant testing. **Flow pressure** measures the quantity of water flowing through an opening during a hydrant test. When a stream of water flows out through an opening (orifice), all of the pressure is converted to kinetic energy. To calculate the volume of water flowing, fire fighters measure the pressure at the centre

of the water stream as it passes through the opening and then factor in the size and flow characteristics of the orifice. A **Pitot gauge** is used to measure pressure (in kPa or psi) and to calculate the rate of flow (in litres per minute or gallons per minute).

Knowing the static pressure, the flow in litres per minute (or gallons per minute), and the residual pressure enables fire fighters to calculate the amount of water that can be obtained from a hydrant or a group of hydrants on the same water main.

During pre-fire planning for buildings equipped with sprinkler and standpipe connections, fire fighters should inspect the threads and inlets to ensure that they have not been compromised.

FIRE FIGHTER Tips

You have learned about the initial steps needed to establish a water supply. Now you can appreciate how important these tasks are for the entire firefighting team and understand how identifying the locations of fire hydrants and static water sources during preincident planning can save time and lives.

FIRE FIGHTER II Tips   FFII

It is important for fire fighters to recognize that caps denote an outlet and intakes, or water-inflow appliances, are equipped with plugs. The only exception is the "Steamer Port" connection, where one would connect a female soft suction to the pump's male main intake.

Fire Hydraulics

To understand how a water supply system works, it is necessary to first understand the principles of water movement. Water is a liquid, which means that it cannot be compressed. The study of **fire hydraulics** deals with the properties of energy, pressure, and water flow as related to fire suppression. When operating supply lines or attack hose lines at a fire, it is important to understand some basic principles of hydraulics—namely, friction loss in different sizes of hose lines, elevation-related changes in pressure, and the development of water hammers. Fire fighters who advance to the position of pump driver/operator will learn more about fire hydraulics.

■ Flow

Flow refers to the volume of water that is being moved through a pipe or hose. In fire hydraulics, flow is measured in gpm. Canadian fire departments use the metric system and measure flow in litres per minute (L/min.). The conversion factor is 1 gpm = 3.785 L/min.

■ Pressure

The amount of energy in a body or stream of water is measured as pressure. In fire hydraulics, pressure is measured in psi.

Near-Miss REPORT

Report Number: 11-0000246

Synopsis: Hydrant cap blows off during checks.

Event Description: I was sent to a local apartment complex to investigate reported damage to a fire hydrant. Pump [number removed] was on the air for area familiarization and was in the apartment complex. The apartment complex consists of 32 12-unit, three-storey, lightweight-construction apartment buildings with limited access. The buildings are all equipped with an NFPA 13R–compliant sprinkler system (*Standard for the Installation of Sprinkler Systems in Low-Rise Residential Occupancies*). The complex is served by a private fire hydrant system, which is maintained by the complex owners.

While in the complex, fire fighters stopped at several hydrants to see which pressures and flows were available. After checking one hydrant, the crew moved to an adjacent hydrant and was charging the hydrant when the 64-mm (2.5-in.) discharge cap came off. The cap was stopped by its chain. The fire fighter opening the hydrant was not injured, as he was standing to the rear of the hydrant, out of the way of the cap.

An examination of the hydrant revealed damage to the threads, which prevented the cap from being replaced properly. The fire fighters stated they noticed the cap appeared to have been backed off the threads prior to charging the hydrant. Both fire fighters stated neither one of them attempted to tighten the cap prior to charging the hydrant. Both fire fighters stated they noticed the damage to the threads after they attempted to replace the cap, when the cap would not screw back onto the connection.

Lessons Learned: This example demonstrates the need to slow down and check equipment prior to using it—especially privately owned and maintained equipment. Some of the private hydrant systems are not properly maintained due to budgetary issues faced by their owners. Municipal budget cuts also affect the enforcement of regulations related to not maintaining fire protection equipment by property owners.

Pressure is required to push water through a hose, to expel water through a nozzle, or to lift water up to a higher level. A pump adds energy to a water stream, causing an increase in pressure. In the metric system, pressure is measured in kPa. The conversion factor is 1 psi = 6.895 kPa.

■ Friction Loss

<u>Friction loss</u> is the decrease in pressure that occurs as water moves through a pipe or hose. This loss of pressure represents the energy required to push the water through the hose. Friction loss is influenced by the diameter of the hose, the volume of water travelling through the hose, and the distance the water travels. In a given size hose, a higher flow rate produces more friction loss. In a 64-mm (2.5-in.) hose, for example, a 1140-L/min. (300-gpm) flow causes much more friction loss than a 760-L/min. (200-gpm) flow. At a given flow rate, the smaller the diameter of the hose, the greater the friction loss. At a flow of 1890 L/min. (500 gpm), the friction loss in a 64-mm (2.5-in.) hose is much greater than the friction loss in a 102-mm (4-in.) hose. With any combination of flow and diameter, the friction loss is directly proportional to the distance; at a flow of 950 L/min. (250 gpm) in a 64-mm (2.5-in.) hose, for example, the friction loss in 61 m (200 ft.) of hose is double the friction loss in 30.5 m (100 ft.) of hose.

■ Elevation Pressure

Elevation affects water pressure. An elevated water tank supplies pressure to a municipal water system because of the difference in height between the water in the water tank and the underground delivery pipes. If a fire hose is laid down a hill, the water at the bottom will have additional pressure owing to the change in elevation. Conversely, if a fire hose is advanced upstairs to the third floor of a building, it will lose pressure as a result of the energy that is required to lift the water. The pump driver/operator must take elevation changes into account when setting the discharge pressure.

■ Water Hammer

<u>Water hammer</u> is a surge in pressure caused by suddenly stopping the flow of a stream of water. A fast-moving stream of water has a large amount of kinetic energy.

If the water suddenly stops moving when a valve is closed, all of the kinetic energy is converted to an instantaneous increase in pressure. Because water cannot be compressed, the additional pressure is transmitted along the hose or pipe as a shock wave.

Water hammer can rupture a hose, cause a coupling to separate, or damage the plumbing on a piece of fire apparatus. Severe water hammer can even damage an underground piping system. In addition, fire fighters have been injured by equipment that was damaged by a water hammer.

A similar situation can occur if a valve is opened too quickly, and a surge of pressurized water suddenly fills a hose. The surge in pressure can damage the hose or cause the fire fighter at the nozzle to lose control of the stream.

To prevent water hammer, always open and close fire hydrant valves slowly. Pump driver/operators should open and close the valves on pumps slowly. When you are operating the nozzle on an attack line, open the nozzle slowly. Most importantly, when you close the shut-off valve on an attack line, do it slowly.

Fire Fighter Safety Tips   © Photos.com

Open and close hydrants and nozzles slowly to prevent water hammer.

Fire Hoses

■ Functions of Fire Hoses

Fire hoses are used for two main purposes: as supply hoses and as attack hoses. **Supply hoses**, or supply lines, are used to deliver water from a static source or from a fire hydrant to an attack engine. The water can come directly from a hydrant, or it can come from another engine that is being used to provide a water supply for the attack engine. Supply line sizes are 64 mm (2.5 in.), 76 mm (3 in.), 102 mm (4 in.), 127 mm (5 in.), and 152 mm (6 in.). Supply hoses are designed to carry larger volumes of water at lower pressures than attack lines.

Attack hoses, or attack lines, are used to discharge water from an attack engine onto the fire. Most attack hoses carry water directly from the attack engine to a nozzle that is used to direct the water onto the fire. In some cases, an attack line is attached to a deck gun, an aerial device, or some other type of master stream appliance. Attack hoses usually operate at higher pressures than do supply lines. These lines can also be attached to the outlet of a standpipe system inside a building. The steps for using attack lines are covered in the chapter "Fire Attack Lines and Foam."

■ Sizes of Hose

Fire hoses range in size from 25 to 152 mm (1 to 6 in.) in diameter **FIGURE 16-18** . The nominal hose size refers to the inside diameter of the hose when it is filled with water. Small-diameter hoses are used as attack lines; large-diameter hoses are almost always used as supply lines. Medium-diameter hoses can be used as either attack lines or supply lines.

Small-diameter hose (SDH) ranges in size from 25 to 51 mm (1 to 2 in.) in diameter. Many fire department vehicles are equipped with a reel of 25-mm (1-in.) hard rubber hose called a **booster hose**, or booster line, which is used for fighting small outdoor fires. A lightweight, collapsible 25-mm (1-in.) hose, known as forestry hose, is often used to fight brush fires.

FIGURE 16-18 Fire hose comes in a wide range of sizes for different uses and situations.

The hoses that are most commonly used to attack interior fires are either 38 or 44 mm (1.5 or 1.75 in.) in diameter. They are usually connected directly to a handline nozzle. Some fire departments also use 51-mm (2-in.) attack lines. Each section of attack hose is usually 15.2 m (50 ft.) long.

Medium-diameter hose (MDH) has a diameter of 64 or 76 mm (2.5 or 3 in.). Hoses in this size range can be used as either supply lines or attack lines. Large handline nozzles are often attached to 64-mm (2.5-in.) hose to attack larger fires. When used as an attack hose, the 76-mm (3-in.) size is more often used to deliver water to a **master stream device** or a fire department connection. These hose sizes also come in 15.2-m (50-ft.) lengths.

Large-diameter hose (LDH) has a diameter of 89 mm (3.5 in.) or more. Standard LDH sizes include 102- and 127-mm (4- and 5-in.) diameters, which are used as supply lines by many fire departments. The largest LDH size is 152 mm (6 in.) in diameter. Standard lengths of either 15.2 m (50 ft.) or 30.5 m (100 ft.) are available for LDH.

Fire hose is designed to be used as either attack hose or supply hose. Attack hose must withstand higher pressures and is designed to be used in a fire environment where it can be subjected to high temperatures, sharp surfaces, abrasion, and other potentially damaging conditions. Large-diameter supply hose is constructed to operate at lower pressures than attack hose and in less severe operating conditions; however, it must still be durable and resistant to external damage. Attack hose can be used as supply hose, but large-diameter supply hose must never be used as attack hose.

Attack hose must be tested annually at a pressure of at least 2068 kPa (300 psi) and is intended to be used at pressures up to 1896 kPa (275 psi). Supply hose must be tested annually at a pressure of at least 1379 kPa (200 psi) and is intended to be used at pressures up to 1276 kPa (185 psi.) Most LDH is constructed as supply line; however, some fire departments use special LDH that can withstand higher pressures.

■ Hose Construction

Most fire hose is constructed with a waterproof inner liner surrounded by either one or two outer layers. The outer layers provide the strength needed to withstand the high pressures exerted by the water inside the hose. This strength is provided by a woven mesh made from high-strength synthetic fibres, such as nylon, which are resistant to high temperatures, mildew, and many chemicals. These fibres can also withstand some mechanical abrasion.

<u>Double-jacket hose</u> is constructed with two layers of woven fibres. The outer layer serves as a protective covering, while the inner layer provides most of the strength. The tightly woven outer jacket can resist abrasion, cutting, hot embers, and other external damage. The woven fibres are treated to resist water and provide added protection from many common hazards that are likely to be encountered at the scene of a fire.

Instead of a double jacket, some fire hoses are constructed with a durable rubberlike compound as the outer covering. This material is bonded to a single layer of strong woven fibres, which provide the strength needed to keep the hose from rupturing under pressure. This type of construction is called <u>rubber-covered hose</u>, or rubber-jacket hose, **FIGURE 16-19** .

Both types of hose are designed to be stored flat and to fold easily when there is no water inside the hose. This allows a much greater length of hose to be stored in the hose compartments on a fire apparatus.

The <u>hose liner</u> (also called the hose inner jacket) is the inner part of the hose **FIGURE 16-20** . This liner prevents water from leaking out of the hose and provides a smooth inside surface for water to move against. Without this smooth surface, excessive friction would arise between the moving water and inside of the hose thus reducing the amount of pressure that could reach the nozzle. The inner liner is usually made of a synthetic rubber compound or a thin, pliable membranous material, which can be flexed and folded without developing leaks. In a double-jacket hose, the inner liner is bonded to the inner woven jacket. In a rubber-covered hose, the inner and

FIGURE 16-20 The liner inside a fire hose can be made from synthetic rubber or a variety of other membranous materials.

outer layers are usually bonded together, and the woven fibres are contained within these layers.

■ Hose Couplings

Couplings are used to connect individual lengths of fire hose together. They are also used to connect a hose line to a hydrant, an intake or discharge valve on an engine, or a variety of nozzles, fittings, and appliances. A coupling is permanently attached to each end of a section of fire hose. The two most common types of fire hose couplings are <u>threaded hose couplings</u> and <u>Storz-type hose couplings</u> (unthreaded).

Threaded Couplings

Threaded couplings are used on most hoses up to 76 mm (3 in.) in diameter and on both soft and hard suction hose. A set of threaded couplings consists of a male coupling, which has the threads on the outside, and a female coupling, which has matching threads on the inside **FIGURE 16-21** . A length of fire hose has a male coupling on one end and a female coupling on the other end. The female coupling has a swivel, so the male and female ends can be attached together without twisting the hose.

Male hose couplings are manufactured in a single piece, whereas female hose couplings are manufactured in two pieces; thus the combination of the male and female hose couplings together is sometimes called a three-piece coupling set. Occasionally, when fire fighters need to reduce the size of a hose coupling to make it fit a smaller coupling or when they need to attach two hoses that possess different threads,

FIGURE 16-19 Rubber-covered hose.

FIGURE 16-21 A set of threaded couplings includes one male coupling and one female coupling. The female coupling (on the left) has threads on the inside of the coupling. The male coupling (on the right) has exposed threads on the outside of the coupling.

a reducer or adaptor will be added to both the male and female couplings; this arrangement is called a five-piece coupling set.

Most fire departments use standardized hose threads, which allow fire hose from different departments to be connected together. The use of standardized hose threads is not universal among all jurisdictions, however, so some fire departments carry special adaptors on their apparatus for use during mutual aid firefighting operations.

When connecting fire hoses with threaded couplings, the threads must be properly aligned so that the male and female couplings will engage fully. When the couplings are properly aligned, the two ends should attach together with minimal resistance. The swivel on the female coupling should be turned until the connection is snug, but only hand-tight, so that the couplings can be easily disconnected.

If any leakage occurs after the hose is filled with water, further tightening may be needed. You can use a **spanner wrench** to gently tighten the couplings until the leakage is stopped. Spanner wrenches are also used to connect and disconnect hose couplings **FIGURE 16-22**. In particular, you may need to use a spanner wrench to uncouple the hose, after it has been pressurized with water. Normally, two spanner wrenches are used together to rotate the two couplings in opposing directions.

Fire hose couplings are constructed with lugs to aid in the coupling and uncoupling of the hose. Three types of lugs are used for this purpose: pin lugs, rocker lugs, and recessed lugs **FIGURE 16-23**. Pin lugs look like small cylinders that extend outward from the hose coupling. They are rarely found on fire hose today, because the pins tend to snag as the hose is being pulled over rough surfaces. Instead, pin lugs have largely been replaced by rocker lugs. **Rocker lugs**, or rocker pins, are rectangular extensions on hose couplings. The edges of rocker lugs are bevelled to prevent them from catching on objects as the hose is dragged. Recessed lugs are circular indentations in hose couplings that are shaped to hold a specially designed spanner wrench. They are usually found on the couplings for 19- or 25-mm (0.75- or 1-in.) booster hose.

Higbee indicators (sometimes called Higbee notches or Higbee cuts) show the position on a pair of couplings where

FIGURE 16-23 Pin lugs, recessed lugs, and rocker lugs.

the ends of the threads are properly aligned with each other. Using the Higbee indicators will help you to couple hose more quickly. When the indicators on the male and female couplings are aligned, the two couplings should connect quickly and easily **FIGURE 16-24**.

An important part of a threaded coupling is the rubber gasket. The gasket is an O-shaped piece of rubber that sits inside the swivel section of the female coupling. When the male coupling is tightened against it, a seal is formed that stops water from leaking. If the gasket is damaged or missing, the coupling will leak. These gaskets can deteriorate with time. Also, using a wrench to tighten couplings on an empty hose or to overtighten couplings on a filled hose can damage the gaskets and cause them to leak. The gaskets must be changed periodically as part of the maintenance on the hose.

Although a leaking coupling is not a critical problem during most firefighting operations, it can result in unnecessary water damage. During cold weather, a leaking coupling can cause ice to form and create a significant safety hazard. The best way to prevent leaks is to make sure the gaskets are in

FIGURE 16-22 Spanner wrenches are used to tighten or loosen hose couplings.

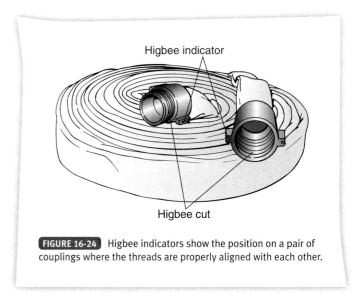

FIGURE 16-24 Higbee indicators show the position on a pair of couplings where the threads are properly aligned with each other.

good condition and to replace any gaskets that are missing or damaged. To replace the swivel gasket, follow the steps in **SKILL DRILL 16-4**:

1. Fold the new gasket, bringing the thumb and forefinger together and creating two loops. (**STEP ①**)
2. Place either of the two loops into the coupling and against the gasket seat. (**STEP ②**)
3. Using the thumb, push the remaining unseated portions into the coupling until the entire gasket is properly positioned against the coupling seat. (**STEP ③**)

FIRE FIGHTER Tips

Pump operators should have a ready supply of gaskets available, so they can replace damaged gaskets on the scene, when required.

FIGURE 16-25 Storz-type couplings are designed so that the couplings on both ends of a length of hose are the same.

Storz-Type Couplings

Storz-type hose couplings are designed so that the couplings on both ends of a length of hose are the same; in other words, there is no male or female end to the hose. With this system, a coupling can be attached to any other coupling of the same diameter **FIGURE 16-25**. Storz-type couplings are made for all hose sizes; however, in North America, they are most often used on LDH.

Storz-type couplings are connected by mating the two couplings face to face and then turning them clockwise one-third of a turn. To disconnect a set of couplings, the two parts are rotated counterclockwise one-third of a turn. A spanner wrench can be used to tighten a leaking coupling or to release a connection that cannot be loosened by hand. Some Storz-type couplings have a small pin or lever that must be released before uncoupling the hose—a design that prevents accidental uncoupling of the hose.

Adaptors are used to connect Storz-type couplings to threaded couplings or to connect couplings of different sizes. Many fire departments use LDH with Storz-type couplings as a supply line between a hydrant and an engine.

SKILL DRILL 16-4 Replacing the Swivel Gasket
(Fire Fighter I, NFPA 5.5.4)

1. Fold the new gasket, bringing the thumb and the forefinger together and creating two loops.

2. Place either of the two loops into the coupling and against the gasket seat.

3. Using the thumb, push the remaining unseated portions into the coupling until the entire gasket is properly positioned against the coupling seat.

© LiquidLibrary

Storz-type couplings can also be found on fire department connections. They can be used to supply a standpipe or sprinkler system. In addition, some hydrant manufacturers place Storz-type connections on hydrants—a practice that allows fire departments to rapidly make hydrant connections.

Coupling and Uncoupling Hose

Several techniques are used for attaching and releasing hose couplings. Depending on the circumstances, one technique may be more effective than another.

A fire fighter should learn how to perform each skill. To perform the one–fire fighter foot-tilt method of coupling fire hose, follow the steps in **SKILL DRILL 16-5**:

1. Place one foot on the hose behind the male coupling.
2. Push down with your foot to tilt the male coupling upward. (**STEP 1**)
3. Place one hand behind the female coupling and grasp the hose. (**STEP 2**)
4. Place the other hand on the coupling swivel.
5. Bring the two couplings together and align the Higbee indicators. Rotate the swivel in a clockwise direction to connect the hoses. (**STEP 3**)

To perform the two–fire fighter method for coupling a fire hose, follow the steps in **SKILL DRILL 16-6**:

1. Pick up the male end of the coupling. Grasp it directly behind the coupling and hold it tightly against the body. (**STEP 1**)
2. The second fire fighter holds the female coupling firmly with both hands. (**STEP 2**)

3. The second fire fighter brings the female coupling to the male coupling. (**STEP 3**)
4. The second fire fighter aligns the female coupling with the male coupling. Use the Higbee indicators for easy alignment. (**STEP 4**)
5. The second fire fighter turns the female coupling counter-clockwise until it clicks, which indicates that the threads are aligned. (**STEP 5**)
6. Turn the female coupling clockwise to couple the hoses. (**STEP 6**)

Charged hose lines should never be disconnected while the water inside the hose is under pressure. The loosened couplings can flail around wildly and cause serious injury to fire personnel or bystanders in the vicinity. Instead, always shut off the water supply, and bleed off the pressure before uncoupling hoses. The water pressure will make it difficult to uncouple a charged hose line. If the coupling resists an attempt to uncouple it, check to make sure the pressure is relieved before using spanner wrenches to loosen the coupling.

To perform the one–fire fighter knee-press method of uncoupling a fire hose, follow the steps in **SKILL DRILL 16-7**:

1. Pick up the connection by the female coupling end. (**STEP 1**)
2. Turn the connection upright, resting the male coupling on a firm surface. (**STEP 2**)
3. Place a knee on the female coupling, and press down on it with your body weight to compress the gasket.
4. Turn the female swivel counterclockwise and loosen the coupling. (**STEP 3**)

SKILL DRILL 16-5 Performing the One–Fire Fighter Foot-Tilt Method of Coupling a Fire Hose
(Fire Fighter I, NFPA 5.3.10)

1 Place one foot on the hose behind the male coupling. Push down with your foot to tilt the male coupling upward.

2 Place one hand behind the female coupling and grasp the hose.

3 Place the other hand on the coupling swivel. Bring the two couplings together and align the Higbee indicators. Rotate the swivel in a clockwise direction to connect the hoses. © LiquidLibrary

SKILL DRILL 16-6 Performing the Two–Fire Fighter Method of Coupling a Fire Hose
(Fire Fighter I, NFPA 5.3.10)

1 Pick up the male end of the coupling. Grasp it directly behind the coupling and hold it tightly against the body.

2 The second fire fighter holds the female coupling firmly with both hands.

3 The second fire fighter brings the female coupling to the male coupling.

4 The second fire fighter aligns the female coupling with the male coupling. Use the Higbee indicators for easy alignment.

5 The second fire fighter turns the female coupling counterclockwise until it clicks, which indicates that the threads are aligned.

6 Turn the female coupling clockwise to couple the hoses.

SKILL DRILL 16-7

Performing the One–Fire Fighter Knee-Press Method of Uncoupling a Fire Hose
(Fire Fighter I, NFPA 5.3.10)

1 Pick up the connection by the female coupling end.

2 Turn the connection upright, resting the male coupling on a firm surface.

3 Place a knee on the female coupling, and press down on it with your body weight. Turn the female swivel counterclockwise and loosen the coupling.

© LiquidLibrary

To perform the two–fire fighter stiff-arm method of uncoupling a hose, follow the steps in **SKILL DRILL 16-8**:

1 Two fire fighters face each other and firmly grasp their respective couplings. (**STEP 1**)

2 With elbows locked straight, they push toward each other. (**STEP 2**)

3 While pushing toward each other, the fire fighters turn the coupling counterclockwise, loosening the coupling. (**STEP 3**)

To uncouple a hose with spanner wrenches, follow the steps in **SKILL DRILL 16-9**:

1 With the connection on the ground, straddle the connection above the female coupling. (**STEP 1**)

2 Place one spanner wrench on the female coupling, with the handle of the wrench to the left. (**STEP 2**)

3 Place the second spanner wrench on the male coupling, with the handle of the wrench to the right. (**STEP 3**)

4 Push both spanner wrench handles down toward the ground, loosening the connection. (**STEP 4**)

To connect two lines with a damaged coupling, follow the steps in **SKILL DRILL 16-10**:

1 Open a hose jacket and place the damaged coupling in one end.

2 Place the second coupling in the other end of the jacket.

3 Close the hose jacket, ensuring that the latch is secure. Slowly bring the hose line up to pressure, allowing the gaskets to seal around the hose ends.

Fire Fighter Safety Tips

© Photos.com

Never attempt to uncouple charged hose lines.

■ Attack Hose

Attack hose is designed to be used for fire suppression. This type of hose carries the water from the attack pumper to the fire. The common sizes of attack hose are 38- or 44-mm (1.5- or 1.75-in.) lines, 25-mm (1-in.) booster lines, and 25- or 38-mm (1- or 1.75-in.) forestry lines. Attack hose is discussed in detail in the chapter "Fire Attack Lines and Foam."

■ Supply Hose

Supply hose is used to deliver water to an attack engine from a pressurized source, such as a fire hydrant or supply pump. Supply lines range from 64 to 152 mm (2.5 to 6 in.) in diameter. The choice of diameter is based on the preferences and operating requirements of each fire department. It also depends on the amount of water needed to supply the attack engine, the distance from the source to the attack pump, and the pressure that is available at the source.

Fire service pumps are normally loaded with at least one bed of hose that can be laid out as a supply line. When threaded couplings are used, this hose can be laid out from the hydrant to the fire (straight lay) or from the fire to the hydrant (reverse lay). Sometimes engines are loaded with two beds of hose so they can easily drop a supply line in either direction. If Storz-type couplings are used or the necessary adaptors are provided, hose from the same bed can be laid in either direction.

A 64-mm (2.5-in.) hose may be used as either a supply line or an attack line. Such hose has a limited flow capacity as supply hose, but it can be effective at low- to moderate-flow rates and over short distances. Sometimes two parallel lines of 64-mm (2.5-in.) hose are used to provide a more effective water supply.

SKILL DRILL 16-8

Performing the Two–Fire Fighter Stiff-Arm Method of Uncoupling a Fire Hose
(Fire Fighter I, NFPA 5.3.10)

1 Two fire fighters face each other and firmly grasp their respective couplings.

2 With elbows locked straight, they push toward each other.

3 While pushing toward each other, the fire fighters turn the coupling counterclockwise, loosening the coupling.

© LiquidLibrary

Large-diameter supply lines are much more efficient than 64-mm (2.5-in.) hose for moving larger volumes of water over longer distances. Given this fact, many fire departments use 102- or 127-mm (4- or 5-in.) hose as their standard supply line. A single 127-mm (5-in.) supply line can deliver flows exceeding 5678 L/min. (1500 gpm) under some conditions; however, LDH is heavy and difficult to move after it has been charged with water. This hose comes in 15.2- and 30.5-m (50- and 100-ft.) lengths. A typical pump may carry anywhere from 228.6 to 381 m (750 to 1250 ft.) of supply hose.

Soft Suction Hose

A **soft suction hose** is a short section of LDH that is used to connect a fire department engine directly to the large steamer outlet on a hydrant FIGURE 16-26 . This type of hose is used to allow as much water as possible to flow from the hydrant to the pump through a single line. A soft suction hose has a female connection on each end, with one end matching the local hydrant threads and the other end matching the threads on a large-diameter inlet to the engine. The couplings have large handles to allow for quick tightening by hand. The hose can range from 102 to 152 mm (4 to 6 in.) in diameter and is usually between 3 and 7.6 m (10 and 25 ft.) in length.

Hard Suction Hose

A **hard suction hose** is a special type of supply hose that is used to draft water from a static source, such as a river, lake, or portable drafting basin FIGURE 16-27 . The water is drawn through the hose into the pump on a fire department engine or into a portable pump. This type of line is called a hard suction hose, because it is designed to remain rigid and will not collapse when a vacuum is created in the hose to draft the water into the pump. It can also be used to carry water from a fire hydrant to the pumper.

SKILL DRILL 16-9 Uncoupling a Hose with Spanner Wrenches
(Fire Fighter I, NFPA 5.3.10)

1 With the connection on the ground, straddle the connection above the female coupling.

2 Place one spanner wrench on the female coupling, with the handle of the wrench to the left.

3 Place the second spanner wrench on the male coupling, with the handle of the wrench to the right.

4 Push both spanner wrench handles down toward the ground, loosening the connection.

© LiquidLibrary

Hard suction hose normally comes in 3- or 6.1-m (10- or 20-ft.) sections. The diameter is based on the capacity of the pump but can be as large as 152 mm (6 in.). Hard suction hose can be made from either rubber or plastic; however, the newer plastic versions are much lighter and more flexible.

Long handles are provided on the female couplings of hard suction hose to assist in tightening the hose. In order to draft water, it is essential to have an airtight connection at each coupling. Sometimes it may be necessary to gently tap the handles on the female couplings with a rubber mallet to tighten or disconnect the hose; tapping these handles with anything metal, however, could cause damage to the handles or the coupling. To create an airtight seal between couplings, it is important to have a flexible gasket to facilitate the seal. These gaskets need to be replaced regularly.

Hose Care, Maintenance, and Inspection

Fire hose should be regularly inspected and tested following the procedures in NFPA 1962, *Standard for the Care, Use, Inspection, Service Testing, and Replacement of Fire Hose, Couplings, Nozzles, and Fire Hose Appliances*. Hoses that are not properly maintained can deteriorate over time and eventually burst. In addition, the gaskets in female couplings need to be checked regularly and replaced when they are worn or damaged.

■ Causes and Prevention of Hose Damage

Fire hose is a lifeline for fire fighters. Every time fire fighters respond to a fire, they rely on fire hose to deliver the water needed to attack the fire and protect themselves from it. Fire

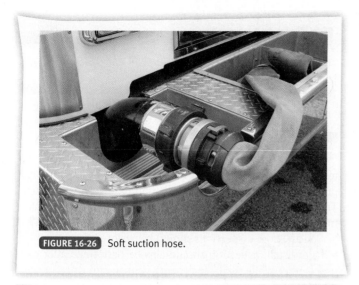

FIGURE 16-26 Soft suction hose.

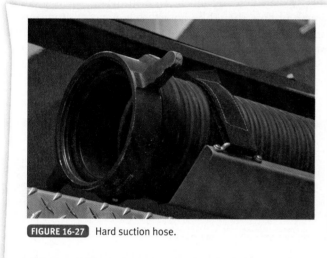

FIGURE 16-27 Hard suction hose.

hose is a highly engineered product designed to perform well under adverse conditions. Fire fighters must be careful to prevent damage to the hose that could result in premature or unexpected failure. The factors that most commonly damage fire hose include mechanical causes, UV radiation, chemicals, heat, cold, and mildew.

Mechanical Damage

Mechanical damage can occur in many ways; hose that is dragged over rough objects or along a roadway can be damaged by abrasion, for example. Broken glass and sharp objects can cut through the hose. Be especially careful if you need to place a hose line through a broken window. Remove any protruding sharp pieces of glass from the edge first. Grit particles caught in the fibres can damage the jacket or puncture holes in the liner. Reloading dirty hose can cause damage to the fibres in the hose jacket.

Fire hose is likely to be damaged if it is run over by a vehicle. For this reason, hose ramps should be used if traffic must drive over a hose that is in the roadway. Hose couplings can also be damaged by dropping them on the ground. In particular, the exposed threads on male couplings are easily damaged if they are dropped. Avoid dragging hose couplings, as this practice can cause damage to the threads and to the swivels.

Heat and Cold

Hoses can be damaged by heat and cold, as well as by prolonged exposure to sunlight. Heat is an obvious concern when fighting a fire. A hose that is directly exposed to a fire can burn through and burst quickly. Burning embers and hot coals can also damage the hose, causing small leaks or weakening the hose, so that it is likely to burst under pressure. Always visually inspect any hose that has come into direct contact with a fire.

Avoid storing a hose where it will come into contact with a hot surface, such as a heating unit or the exhaust pipe on a vehicle. If the apparatus is parked outside, use a hose cover to protect the hose from sunlight.

In cold weather, freezing is a threat to hoses. Freezing can rupture the inner liner and break fibres in the hose jacket. When fire fighters are working in below-freezing temperatures, water should be kept flowing through the hose to prevent freezing. If a line must be shut down temporarily, the nozzle should be left partly open to keep the water moving, and the stream should be directed to a location where it will not cause additional water damage. When a line is no longer needed, the hose should be drained and rolled before it freezes.

Hose that is frozen or encased in ice can often be thawed out with a steam generator. Another option is to carefully chop the hose out using an axe, being careful not to cut the hose itself. The hose can then be transported back to the fire station to thaw. Do not attempt to bend a section of frozen hose. In situations where the hose is frozen solid, it may be necessary to transport the hose back to the fire station on a flatbed truck.

Chemicals

Many chemicals can damage fire hoses. Such chemicals may be encountered at incidents in facilities where chemicals are manufactured, stored, or used, as well as in locations where their presence is not anticipated. Most vehicles contain a wide variety of chemicals that can damage fire hose, including battery acid, gasoline, diesel fuel, antifreeze, motor oil, and transmission fluid. Hose may come into contact with these chemicals at vehicle fires or at the scene of a collision where chemicals are spilled on the roadway. In particular, supply hoses often come in contact with chemical residue when lines are laid in the roadway. Fire fighters should remove chemicals from the hose as soon as possible, wash the hose with an approved detergent, thoroughly rinse it, and let it dry completely.

Mildew

Mildew is a type of fungus that can grow on fabrics and materials in warm, moist conditions. A fire hose that has been packed away while it is still wet and dirty is a natural breeding ground for mildew. This fungus feeds on nutrients found in many natural fibres, which can cause the fibres to rot and deteriorate. In the days when cotton fibres were used in fire hose jackets, mildew was a major problem. Hose had to be washed and completely dried after every use, before it could be placed back on the apparatus.

Modern fire hose is made from synthetic fibres that are resistant to mildew, and most types can be repacked without drying. Nevertheless, mildew may still grow on exposed fibres if they are soiled with contaminants that will provide mildew with the necessary nutrients. The fibres in rubber-covered hose are protected from mildew.

■ Cleaning and Maintaining Hoses

To clean hose that is dirty or contaminated, follow the steps in **SKILL DRILL 16-11**:

1. Lay the hose out flat.
2. Rinse the hose with water. (**STEP ❶**)
3. Gently scrub the hose with mild detergent, paying attention to soiled areas. (**STEP ❷**)
4. Turn over the hose and repeat steps 1 and 2.
5. Give a final rinse to the hose with water.
6. Lay out or hang the hose and allow it to dry before properly storing it. (**STEP ❸**)

FIRE FIGHTER Tips

To prevent UV radiation damage, do not dry hose in sunlight.

■ Hose Inspections

Each length of hose should be tested at least annually, according to the procedures listed in NFPA 1962. The hose testing procedures are complicated and require special equipment. This equipment must be operated according to the manufacturer's instructions.

Visual hose inspections should be performed at least quarterly. A visual inspection should also be performed after each use—either while the hose is being cleaned and dried or once it is reloaded onto the apparatus. If any defects are found, that length of hose should be immediately removed from service and tagged with a description of the problem. Hose that has not been used in 30 days should be unpacked, inspected, cleaned, and reloaded. The appropriate notifications must be made to have the hose repaired.

SKILL DRILL 16-11   Cleaning and Maintaining Hoses
(Fire Fighter I, NFPA 5.5.2)

1 Lay the hose out flat. Rinse the hose with water.

2 Gently scrub the hose with mild detergent, paying attention to soiled areas.

3 Turn over the hose and repeat steps 1 and 2. Give a final rinse to the hose with water. Lay out or hang the hose and allow it to dry before properly storing it.

© LiquidLibrary

FIRE FIGHTER II

To clearly mark a defective hose, follow the steps in **SKILL DRILL 16-12** **FIGURE 16-28** :

1. Inspect the hose for defects.

2. Upon finding a defect, mark the area on the hose and remove the hose from service.

3. Tag the hose as defective and provide a description of the defect, take it out of service, and notify your superiors.

To perform an annual service test on fire hose, follow the steps in **SKILL DRILL 16-13**

1. Don turnout gear.

2. Connect up to 91.4 m (300 ft.) of hose to a hose-testing gate valve on the discharge valve of a fire department pumper or hose tester.

3. Attach a nozzle to the end of the hose.

4. Slowly fill each hose with water at 345 kPa (50 psi), and remove kinks and twists in the hose.

5. Open the nozzle to purge air from the hose, discharging the water away from the test area. Close the nozzle.

6. Measure and record the length of each section of hose.

7. Mark the position of each hose coupling on the hose.

8. Check each coupling for leaks. When leaks are found, remove the hose from service if the leak is behind the coupling. If the leak is in front of the coupling, tighten the leaky couplings. Replace gaskets if necessary after shutting down the hose line.

9. Close the hose-testing gate valve.

10. Ensure that all fire fighters are clear of the test area.

11. Increase the pressure on the hose to the pressure required by NFPA 1962, and maintain that pressure for 5 minutes.

12. Monitor the hose and couplings for leaks as the pressure increases during the test.

13. Shut the gates and open the nozzle to bleed off the pressure.

14. Uncouple the hose and drain it.

15. Inspect the marks placed on the hose jacket near the couplings to determine whether slippage occurred.

FIGURE 16-28 Upon finding a defect, mark the area on the hose and remove the hose from service.

FIRE FIGHTER II

16. Tag hose that failed.

17. Mark hose that passed.

18. Record the results in the departmental logs.

Hose Records

A hose record is a written history of each individual length of fire hose. Each length of hose should be identified with a unique number stenciled or painted on it. A hose record will contain information such as the following:

- Hose size, type, and manufacturer
- Date when the hose was manufactured
- Date when the hose was purchased
- Dates when the hose was tested
- Any repairs that have been made to the hose

Some fire departments keep hose records on cards or paper files. Others keep the records in a database in a computer system.

Hose Appliances

A **hose appliance** is any device used in conjunction with a fire hose for the purpose of delivering water. You should be familiar with wyes, water thieves, Siamese connections, double-male and double-female adaptors, reducers, hose clamps, hose jackets, and hose rollers. It is important to learn how to use those hose appliances and tools required by your fire department. In particular, you should understand the purpose of each device and be able to use each appliance correctly. Some hose appliances are used primarily with supply lines, whereas others are most often used with attack lines; many hose appliances have applications with both supply lines and attack lines. Both types are discussed in this section.

■ Wyes

A **wye** is a device that splits one hose stream into two hose streams. The word *wye* refers to a Y-shaped part or object. When threaded couplings are used, a wye has one female connection and two male connections.

The wye that is most commonly used in the fire service splits one 64-mm (2.5-in.) hose line into two 38-mm (1.5-in.) hose lines. This appliance is used primarily on attack lines. A **gated wye** is equipped with two quarter-turn ball valves, so that the flow of water to each of the split lines can be controlled independently **FIGURE 16-29**. A gated wye enables fire fighters to initially attach and operate one hose line and then to add a second hose later, if necessary. The use of a gated wye avoids the need to shut down the hose line supplying the wye in order to to attach the second line. Some gated wyes are used on LDH. These wyes have an inlet for LDH and several 64-mm (2.5-in.) outlets.

■ Water Thief

A **water thief** is similar to a gated wye but includes an additional 64-mm (2.5-in.) outlet **FIGURE 16-30**. It is used primarily on attack lines and larger ones are used with LDH. The water that comes from a single 64-mm (2.5-in.) inlet can be directed to two 38-mm (1.5-in.) outlets and one 64-mm (2.5-in.) outlet. Under most conditions, it will not be possible to supply all three outlets at the same time, because the capacity of the supply hose is limited.

FIGURE 16-29 A gated wye is used to split one 64-mm (2.5-in.) hose line into two 38-mm (1.5-inch) lines.

Courtesy of Akron Brass Company

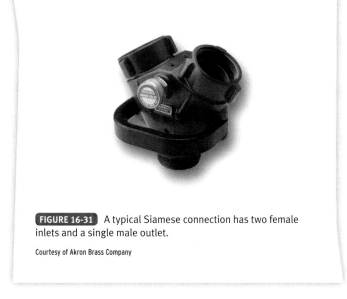

FIGURE 16-31 A typical Siamese connection has two female inlets and a single male outlet.

Courtesy of Akron Brass Company

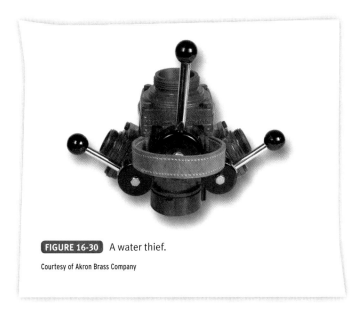

FIGURE 16-30 A water thief.

Courtesy of Akron Brass Company

A water thief can be placed near the entrance to a building to provide the water for interior attack lines. One or two 38-mm (1.5-in.) attack lines can be used in such a case. If necessary, they can then be shut down and a 64-mm (2.5-in.) line substituted for them. Sometimes, the 64-mm (2.5-in.) line is used to knock down a fire, and the two 38-mm (1.5-in.) lines are used for overhaul.

■ Siamese Connection

A **Siamese connection** is a hose appliance that combines two hose lines into one. The most commonly used type of Siamese connection combines two 64-mm (2.5-in.) hose lines into a single line **FIGURE 16-31**. This scheme increases the flow of water on the outlet side of the Siamese connection. A Siamese connection that is used with threaded couplings has two female connections on the inlets and one male connection on the outlet. It can be used with supply lines and with some attack lines.

A Siamese connection is sometimes used on an engine inlet to allow water to be received from two different supply lines. These kinds of connections are also used to supply master stream devices and ladder pipes. Siamese connections are commonly installed on the fire department connections that are used to supply water to standpipe and sprinkler systems in buildings. Fire fighters should go over their department's policies on the correct procedures for supplying fire department Siamese connections.

■ Adaptors

Adaptors are used for connecting hose couplings that have the same diameter but dissimilar threads. Dissimilar threads may be encountered when different fire departments are working together or in industrial settings where the hose threads of the building's equipment do not match the threads of the municipal fire department. Some private fire hydrants may have different threads from the municipal system and require an adaptor. Adaptors are also used to connect threaded couplings to Storz-type couplings. They are useful for both supply lines and attack lines.

Adaptors can also be used when it is necessary to connect two female couplings or two male couplings. A **double-female adaptor** is used to join two male hose couplings. A **double-male adaptor** is used to join two female hose couplings. Double-male and double-female adaptors are often employed when performing a reverse hose lay **FIGURE 16-32**.

■ Reducers

A **reducer** is used to attach an SDH to an LDH **FIGURE 16-33**. Usually the larger end has a female connection, and the smaller end has a male connection. One type of reducer is used to attach a 64-mm (2.5-in.) hose line to a 38-mm (1.5-in.) attack line. Many 64-mm (2.5-in.) nozzles are constructed with a built-in reducer, so that a 38-mm (1.5-in.) line can be attached for overhaul. Reducers are also used to attach a 64-mm (2.5-in.) supply line to a larger suction inlet on a fire pumper.

FIGURE 16-32 Double-female and double-male adaptors are used to join two couplings of the same sex.

FIGURE 16-34 A hose jacket is used to repair a leaky hose line.

© 2003, Berta A. Daniels

FIGURE 16-33 A reducer is used to connect a smaller-diameter hose line to the end of a larger-diameter line.

© 2003, Berta A. Daniels

FIGURE 16-35 A hose roller is used to protect a hose when it is hoisted over the sharp edge of a roof or a windowsill.

© 2003, Berta A. Daniels

■ Hose Jacket

A <u>hose jacket</u> is a device that is placed over a leaky section of hose to stop a leak **FIGURE 16-34**. The best way to handle a leak in a section of hose is to replace the defective section. A hose jacket can provide a temporary fix until the section of hose can be replaced. This device should be used only in cases where it is not possible to quickly replace the leaky section of hose. It can be used for both supply lines and attack lines.

The hose jacket consists of a split metal cylinder that fits tightly over the outside of a hose line. This cylinder is hinged on one side to allow it to be placed over the leak. A fastener is then used to clamp the cylinder tightly around the hose. Gaskets on each end of the hose jacket prevent water from leaking out the ends of the hose jacket.

■ Hose Roller

A <u>hose roller</u> is used to protect a hose line that is being hoisted over the edge of a roof or over a windowsill **FIGURE 16-35**. This appliance keeps the hose from chafing or kinking at the sharp edge. A hose roller is sometimes called a hose hoist, because it makes it easier to raise (i.e., hoist) a hose over the edge of a building. They are typically used with attack lines. Hose rollers are also used to protect ropes while hoisting an object over the edge of a building and during rope rescue operations.

■ Hose Clamp

A <u>hose clamp</u> is used to temporarily stop the flow of water in a hose line. These devices are often applied to supply lines, allowing a hydrant to be opened before the line is hooked up to the intake of the attack engine. The fire fighter at the hydrant does not have to wait for the pump operator to connect the line to the pump intake before opening the hydrant. As soon as the line is connected, the clamp is released. A hose clamp can also be used to stop the flow in a line if a hose ruptures or if an attack line needs to be connected to a different appliance **FIGURE 16-36**.

FIGURE 16-36 A hose clamp is use to temporarily interrupt the flow of water in a hose line.

FIGURE 16-37 **A.** Ball valve.

Valves

Valves are used to control the flow of water in a pipe or hose line. Several different types of valves are used on fire hydrants, fire apparatus, standpipe and sprinkler systems, and attack hose lines. The important thing to remember when opening and closing any valve or nozzle is to do so *slowly* so as to prevent water hammer.

Common valves include the following types:

- **Ball valves**—Ball valves are used on nozzles, gated wyes, and engine discharge gates. They consist of a ball with a hole in the middle. When the hole is lined up with the inlet and the outlet, water flows through it. As the ball is rotated, the flow of water is gradually reduced until it is shut off completely **FIGURE 16-37A**.
- **Gate valve**—Gate valves are found on hydrants and on sprinkler systems. Rotating a spindle causes a gate to move slowly across the opening. The spindle can be rotated with a wrench or a wheel-type handle **FIGURE 16-37B**.
- **Butterfly valve**—Butterfly valves are often found on the large pump intake connections where a hard suction hose or soft suction hose is joined. They are opened or closed by rotating a handle one-quarter turn **FIGURE 16-37C**.
- **Four-way hydrant valve**—This appliance, which is attached to a fire hydrant, enables water to flow directly from the hydrant. It can boost pressure in a supply line without interrupting the flow of water. The use of a four-way hydrant valve is covered in detail later in this chapter.
- **Remote-controlled hydrant valve**—Remote-controlled hydrant valves are attached to fire hydrants. They allow the operator to turn a hydrant on without flowing water into the hose line. These skalves are operated by the pump operator using a radio control. They free up the hydrant person for other tasks and give the pump operator control over the hydrant from a distance.

FIGURE 16-37 **B.** Gate valve.

FIGURE 16-37 **C.** Butterfly valve.

Courtesy of Akron Brass Company

Hose Rolls

An efficient way to transport a single section of fire hose is in the form of a roll. Rolled hose is both compact and easy to manage. A fire hose can be rolled many different ways, depending on how it will be used. Follow the SOGs/SOPs of your department when rolling hose.

■ Straight or Storage Hose Roll

The straight roll is a simple and frequently used hose roll. It is used for general handling and transporting of hose, as well as for rack storage of hose. To perform a straight roll, follow the steps in **SKILL DRILL 16-14**:

1 Lay the length of hose to be rolled flat and straight. (**STEP 1**)

2 Begin by folding the male coupling overtop of the hose. (**STEP 2**)

3 Roll the hose to the female coupling. (**STEP 3**)

4 Set the hose roll on its side, and tap any protruding hose flat with your foot. (**STEP 4**)

■ Single-Doughnut Hose Roll

The single-doughnut roll is used when the hose will be put into use directly from its rolled state. In this arrangement, both couplings are on the outside of the roll. The hose can be connected and extended by one fire fighter. As the hose is extended, it unrolls. To perform a single-doughnut roll, follow the steps in **SKILL DRILL 16-15**:

1 Place the hose flat and in a straight line. (**STEP 1**)

2 Locate the midpoint of the hose. (**STEP 2**)

SKILL DRILL 16-14

Performing a Straight Hose Roll
(Fire Fighter I, NFPA 5.5.2)

1 Lay the length of hose to be rolled flat and straight.

2 Begin by folding the male coupling over on top of the hose.

3 Roll the hose to the female coupling.

4 Set the hose roll on its side, and tap any protruding hose flat with a foot.

© LiquidLibrary

SKILL DRILL 16-15 Performing a Single-Doughnut Roll
(Fire Fighter I, NFPA 5.5.2)

1 Place the hose flat and in a straight line.

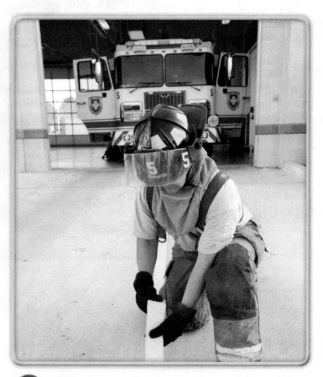

2 Locate the midpoint of the hose.

3 From the midpoint, move 1.5 m (5 ft.) toward the male coupling end. Start rolling the hose toward the female coupling.

4 At the end of the roll, wrap the excess hose of the female end over the male coupling to protect the threads.

3 From the midpoint, move 1.5 m (5 ft.) toward the male coupling.

4 Start rolling the hose toward the female coupling. (**STEP 3**)

5 At the end of the roll, wrap the excess hose on the female end over the male coupling to protect the threads. (**STEP 4**)

■ Twin-Doughnut Hose Roll

The twin-doughnut roll is primarily used to make a small compact roll that can be carried easily. To perform a twin-doughnut hose roll, follow the steps in **SKILL DRILL 16-16**:

1 Lay the hose flat and in a straight line. (**STEP 1**)

2 Bring the male coupling alongside the female coupling. (**STEP 2**)

3 Fold the far end over and roll toward the couplings, creating a double roll. (**STEP 3**)

4 The roll can be carried by hand, rope, or strap. (**STEP 4**)

■ Self-Locking Twin-Doughnut Hose Roll

The self-locking twin-doughnut hose roll is similar to the twin-doughnut roll, except that it forms its own carrying loop. To perform a self-locking twin-doughnut roll, follow the steps in **SKILL DRILL 16-17**:

1 Lay the hose flat and bring the couplings alongside each other. (**STEP 1**)

2 Move one side of the hose over the other, creating a loop. This creates the shoulder-carrying loop. (**STEP 2**)

3 Bring the loop back toward the couplings to the point where the hose crosses. (**STEP 3**)

4 From the point where the hose crosses, begin to roll the hose toward the couplings, with the loop at the centre. This creates a loop on each side of the roll. (**STEP 4**)

5 On completion of rolling, position the couplings on the top of the rolls. (**STEP 5**)

6 Position the loops so that one is larger than the other. Pass the larger loop over the couplings and through the smaller loop, which will secure the rolls together and form the shoulder loop. (**STEP 6**)

Fire Hose Evolutions

Fire hose evolutions are standard methods of working with fire hose to accomplish different objectives in a variety of

---

SKILL DRILL 16-16   Performing a Twin-Doughnut Roll
(Fire Fighter I, NFPA 5.5.2)

1 Lay the hose flat and in a straight line.

2 Bring the male coupling alongside the female coupling.

© LiquidLibrary

(Continues)

SKILL DRILL 16-16 Performing a Twin-Doughnut Roll (*Continued*)
(Fire Fighter I, NFPA 5.5.2)

3 Fold the far end over and roll toward the couplings, creating a double roll.

4 The roll can be carried by hand, by a rope, or by a strap.

© LiquidLibrary

situations. Most fire departments set up their equipment and conduct regular training so that fire fighters will be prepared to perform a set of standard hose evolutions. Hose evolutions involve specific actions that are assigned to specific members of a crew, depending on their riding positions on the apparatus. Every fire fighter should know how to perform all of the standard evolutions quickly and proficiently. When an officer calls for a particular evolution to be performed, each crew member should know exactly what to do.

Fire Fighter Safety Tips

© Photos.com

When performing a forward lay, the fire fighter who is connecting the supply hydrant must not stand between the hose and the hydrant. When the apparatus starts to move, the hose could become tangled and suddenly be pulled taut. Anyone standing between the hose and the hydrant could be seriously injured.

Hose evolutions are divided into supply line operations and attack line operations. Supply line operations involve laying hose lines and making connections between a water supply source and an attack engine. Attack line operations involve advancing hose lines from an attack engine to apply water onto a fire. Attack line evolutions are covered in the chapter "Fire Attack Lines and Foam."

■ Supply Line Evolutions

The objective of laying a supply line is to deliver water from a hydrant or a static water source to an **attack engine**. In most cases, this operation involves laying a continuous hose line out of the bed of the fire apparatus as the vehicle drives forward. It can be done using either a **forward lay** or a **reverse lay**. A forward lay starts at the hydrant and proceeds toward the fire; the hose is laid in the same direction as the water flows—from the hydrant to the fire. A reverse lay involves laying the hose from the fire to the hydrant; the hose is laid in the opposite direction to the water flow. Each fire department will determine its own preferred methods and procedures for supply line operations, based on available apparatus, water supply, and regional considerations.

Forward Hose Lay

The forward hose lay is most often used by the first-arriving engine company at the scene of a fire **FIGURE 16-38**. This method allows the engine company to establish a water supply without assistance from another company. A forward lay also

SKILL DRILL 16-17 Performing a Self-Locking Twin-Doughnut Roll
(Fire Fighter I, NFPA 5.5.2)

1 Lay the hose flat and bring the couplings alongside each other.

2 Move one side of the hose over the other, creating a loop. This creates the carrying shoulder loop.

3 Bring the loop back toward the couplings to the point where the hose crosses.

4 From the point where the hose crosses, begin to roll the hose toward the couplings, with the loop as its centre. This creates a loop on each side of the roll.

5 On completion of the rolling, position the couplings on the top of the rolls.

6 Position the loops so that one is larger than the other. Pass the larger loop over the couplings and through the smaller loop.

© LiquidLibrary

FIGURE 16-38 A forward hose lay is made from the hydrant to the fire.

places the attack engine close to the fire, providing easy access to the additional hoses, tools, and equipment that are carried on the apparatus.

A forward hose lay can be performed using MDH (64 mm [2.5 in.] or 76 mm [3 in.]) or LDH (89 mm [3.5 in.] or larger). The larger the hose's diameter, the more water that can be delivered to the attack engine through a single supply line. When MDH is used and the beds are arranged to lay dual lines, a company can lay two parallel lines from the hydrant to the fire.

If the fire hydrant is located close to the fire, it may supply a sufficient quantity of water from the hydrant alone; for example, a 127-mm (5-in.) hose can supply 2650 L/min. (700 gpm) over a distance of 152 m (500 ft.), losing only some 138 kPa (20 psi) to friction.

To perform a forward hose lay, follow the steps in **SKILL DRILL 16-18**:

1 Stop the fire apparatus 3 m (10 ft.) from the hydrant. (**STEP 1**)

2 Grasp enough hose to reach to the hydrant and to loop around the hydrant. Step off the apparatus, carrying the hydrant wrench and all necessary tools.

3 Loop the end of the hose around the hydrant, or secure the hose as specified in the local SOGs/SOPs. Do not stand between the hose and the hydrant. Never stand on the hose. (**STEP 2**)

4 Signal the pump driver/operator to proceed to the fire once the hose is secured. (**STEP 3**)

SKILL DRILL 16-18 Performing a Forward Hose Lay
(Fire Fighter I, NFPA 5.3.15)

1 Stop the fire apparatus 10 feet (3 m) from the hydrant.

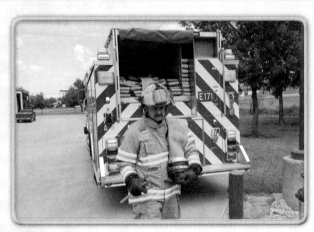

2 Grasp enough hose to reach to the hydrant and to loop around the hydrant. Step off the apparatus, carrying the hydrant wrench and all necessary tools. Loop the end of the hose around the hydrant or secure the hose as specified in the local SOGs/SOPs. Do not stand between the hose and the hydrant. Never stand on the hose.

3 Signal the pump driver/operator to proceed to the fire once the hose is secured.

4 Once the apparatus has moved off and a length of supply line has been removed from the apparatus and is lying on the ground, remove the appropriate-size hydrant cap nearest to the fire. Follow the local SOGs/SOPs for checking the operating condition of the hydrant.

© LiquidLibrary

(Continues)

SKILL DRILL 16-18 Performing a Forward Hose Lay (*Continued*)
(Fire Fighter I, NFPA 5.3.15)

5 Attach the supply hose to the outlet. An adaptor may be needed if a large-diameter hose with Storz-type couplings is used.

6 Attach the hydrant wrench to the hydrant.

7 The pump driver/operator uncouples the hose and attaches the end of the supply line to the pump inlet or clamps the hose close to the pump, depending on the local SOGs/SOPs. When the pump driver/operator signals to charge the hose by prearranged hand signal, radio, or air horn, open the hydrant slowly and completely.

8 Follow the hose back to the engine and remove any kinks from the supply line.

© LiquidLibrary

5 Once the apparatus has moved off and a length of supply line has been removed from the apparatus and is lying on the ground, remove the appropriate-size hydrant cap nearest to the fire. Follow the local SOGs/SOPs for checking the operating condition of the hydrant. (**STEP 4**)

6 Attach the supply hose to the outlet. An adaptor may be needed if an LDH with Storz-type couplings is used. (**STEP 5**)

7 Attach the hydrant wrench to the hydrant. (**STEP 6**)

8 The pump driver/operator uncouples the hose and attaches the end of the supply line to the pump inlet or clamps the hose close to the pump, depending on the local SOGs/SOPs.

9 When the pump driver/operator signals to charge the hose by prearranged hand signal, radio, or air horn, open the hydrant slowly and completely. (**STEP 7**)

10 Follow the hose back to the engine and remove any kinks from the supply line. (**STEP 8**)

Four-Way Hydrant Valve

In cases where long supply lines are needed or when MDH is used, it is often necessary to place a supply engine at the hydrant. The supply engine pumps water through the supply line to increase the pressure and volume of water to the attack engine. Some departments use a four-way hydrant valve to connect the supply line to the hydrant, ensuring that the supply line can be charged with water immediately yet still allowing for a supply engine to connect to the line later.

When the four-way valve is placed on the hydrant, the water flows initially from the hydrant through the valve to the supply line, which delivers the water to the attack engine. The second engine can hook up to the four-way valve and redirect the flow by changing the position of the valve. The water then flows from the hydrant to the supply engine. The supply engine increases the pressure and discharges the water into the supply line, boosting the flow of water to the attack engine. This operation can be accomplished without uncoupling any lines or interrupting the flow.

To use a four-way valve, follow the steps in **SKILL DRILL 16-19**:

1. Stop the attack engine 3 m (10 ft.) past the hydrant to be used.
2. Grasp the four-way valve, the attached hose, and enough hose to reach and loop around the hydrant. Carry the four-way valve from the apparatus, along with the hydrant wrench and any other needed tools.
3. Loop the end of the hose around the hydrant, or secure the hose with a rope, as specified in the local SOGs/SOPs. Do not stand between the hydrant and the hose.
4. Signal the attack engine driver/operator to proceed to the fire. (**STEP 1**)
5. Once enough hose has been removed from the apparatus and is lying in the street, remove the steamer port from the fire hydrant. Follow the local SOGs/SOPs for checking the operating condition of the hydrant.

SKILL DRILL 16-19 Attaching a Hose to a Four-Way Hydrant Valve
(Fire Fighter I, NFPA 5.3.15)

1 Stop the attack engine 3 m (10 ft.) past the hydrant to be used. Grasp the four-way valve, the attached hose, and enough hose to reach to and loop around the hydrant. Carry the four-way valve from the apparatus, along with the hydrant wrench and any other needed tools. Loop the end of the hose around the hydrant or secure the hose with a rope as specified in the local SOGs/SOPs. Signal the attack engine driver/operator to proceed to the fire.

2 Once enough hose has been removed from the apparatus and is lying in the street, remove the steamer port from the fire hydrant. Follow the local SOGs/SOPs for checking the operating condition of the hydrant. Attach the four-way valve to the hydrant outlet. Attach the hydrant wrench to the hydrant. The attack engine driver/operator uncouples the hose and attaches the end of the supply line to the pump inlet. The attack engine driver/operator signals charge the supply line. Open the hydrant slowly and completely.

SKILL DRILL 16-19 Attaching a Hose to a Four-Way Hydrant Valve (*Continued*)
(Fire Fighter I, NFPA 5.3.15)

3 Initially, the attack engine is supplied with water from the hydrant. When the supply engine arrives at the fire scene, the supply engine driver/operator stops at the hydrant with the four-way valve.

4 The supply engine driver/operator attaches a hose from the four-way valve outlet to the intake side of the engine.

5 Attach a second hose to the inlet side of the four-way valve, and connect the other end to the pump discharge.

6 Change the position of the four-way valve to direct the flow of water from the hydrant, through the supply engine, and into the supply line.

© LiquidLibrary

6 Attach the four-way valve to the hydrant outlet (an adaptor may be needed).

7 Attach the hydrant wrench to the hydrant.

8 The attack engine driver/operator uncouples the hose and attaches the end of the supply line to the pump inlet.

9 The attack engine driver/operator signals by prearranged hand signal, radio, or air horn to charge the supply line.

10 Open the hydrant slowly and completely. (**STEP 2**)

11 Initially, the attack engine is supplied with water from the hydrant. When the supply engine arrives at the fire scene, the supply engine driver/operator stops at the hydrant with the four-way valve. (**STEP 3**)

12 The supply engine driver/operator attaches a hose from the four-way valve outlet to the intake side of the engine. (**STEP 4**)

13 Attach a second hose to the inlet side of the four-way valve, and connect the other end to the pump discharge. (**STEP 5**)

14 Change the position of the four-way valve to direct the flow of water from the hydrant, through the supply engine, and into the supply line. (**STEP 6**)

FIRE FIGHTER Tips

When connecting a supply line to a hydrant, wait until you get the signal from the pump driver/operator to charge the line. If the hydrant is opened prematurely, the hose bed could become charged with water, or a loose hose line could discharge water at the fire scene. Either situation will disrupt the water supply operation and could cause serious injuries. Make sure that you know your department's signal to charge a hose line, and do not become so excited or rushed that you make a mistake.

FIRE FIGHTER Tips

When laying out supply hose with threaded couplings, you may find that the wrong end of the hose is on top of the hose bed. Double-male and double-female connectors will allow you to attach a female coupling to a female coupling or a male coupling to a male coupling, respectively. A set of adaptors (one double-male and one double-female) should be easily accessible for these situations. Some departments place a set of adaptors on the end of the supply hose for this purpose.

Reverse Hose Lay

The reverse hose lay is the opposite of the forward lay **FIGURE 16-39**. In the reverse lay, the hose is laid out from the fire to the hydrant, in the opposite direction to the flow of the water. This evolution can be used when the attack engine

FIGURE 16-39 A reverse hose lay is made from the fire to a fire hydrant.

arrives at the fire scene without a supply line. It may be a standard tactic in areas where sufficient hydrants are available and additional companies that can assist in establishing a water supply will arrive quickly. In this scenario, one of these companies will be assigned to lay a supply line from the attack engine to a hydrant.

In this situation, the attack engine focuses on immediately attacking the fire using water from the onboard tank. The supply engine stops close to the attack engine, and hose is pulled from the bed of the supply engine to an intake on the attack engine. The supply engine then drives to the hydrant (or alternative water source) and pumps water back to the attack engine. Usually the supply engine parks in such a way that hose can be pulled from the supply engine, to the inlet, and to the attack engine.

To perform a reverse lay, follow the steps in **SKILL DRILL 16-20** :

1 Pull sufficient hose to reach from the supply engine to the inlet of the attack engine.

2 Anchor the hose to a stationary object, if possible. Wrapping a portion of the hose around the front wheels of the attack engine is one possible strategy; wrapping it around a tree is another. Do not stand between the hose and the stationary object. (**STEP 1**)

3 The supply engine driver/operator drives away, laying out hose from the attack engine to the fire hydrant or static water source. (**STEP 2**)

4 Connect the supply line to the inlet of the attack engine. (**STEP 3**)

5 The supply engine driver/operator uncouples the supply hose from the hose remaining in the hose bed and attaches the supply hose to the discharge side of the supply engine. (**STEP 4**)

6 The supply engine driver/operator connects the supply engine to the hydrant or water source. The four-way hydrant valve can be used here to boost the supply pressure as needed.

7 Upon the signal from the attack engine, the supply engine driver/operator charges the supply line. (**STEP 5**)

SKILL DRILL 16-20 Performing a Reverse Hose Lay
(Fire Fighter I, NFPA 5.3.15)

1 Pull sufficient hose to reach from the supply engine to the inlet of the attack engine. Anchor the hose to a stationary object, if possible. Do not stand between the hose and the stationary object.

2 The supply engine driver/operator drives away, laying out hose from the attack engine to the fire hydrant or other static water source.

3 Connect the supply line to the inlet of the attack engine.

4 The supply engine driver/operator uncouples the supply hose from the hose remaining in the hose bed and attaches the supply hose to the discharge side of the supply engine.

5 The pump driver/operator connects the supply engine to the hydrant or water source. The four-way hydrant valve can be used here to boost the supply pressure as needed. Upon the signal from the attack engine, the supply engine driver/operator charges the supply line.

© LiquidLibrary

Split Hose Lay

A **split hose lay** is performed by two engine companies in situations where hose must be laid in two different directions to establish a water supply **FIGURE 16-40**. This evolution could be used when the attack engine must approach a fire either along a dead-end street with no hydrant or down a long driveway. To perform a split hose lay, the attack engine drops the end of its supply hose at the corner of the street and performs a forward lay toward the fire. The supply engine stops at the same intersection, pulls off enough hose to connect to the end of the supply line that is already there, and then performs a reverse hose lay to the hydrant or static water source. With the two lines connected together, the supply engine can pump water to the attack engine.

A split lay often requires coordination by two-way radio, because the attack engine must advise the supply engine of the plan and indicate where the end of the supply line is being dropped. In many cases, the attack engine is out of sight when the supply engine arrives at the split point.

FIGURE 16-40 Two engines performing a split hose lay. Engine 1 performs a forward hose lay from the corner to the fire; Engine 2 performs a reverse hose lay from the hose at the corner of a hydrant.

A split hose lay does not necessarily require split hose beds. It can be performed with or without split beds if the necessary adaptors are used.

To perform a split hose lay, follow the steps in **SKILL DRILL 16-21** (Fire Fighter I, NFPA 5.3.15):

1. The attack engine driver/operator stops at the intersection or driveway entrance.
2. A fire fighter from the attack engine removes the end of the supply line from the hose bed and anchors it.
3. The attack engine driver/operator lays out a hose line while proceeding slowly toward the fire.
4. The attack engine fire fighter either proceeds by foot to the fire or waits at the intersection for the supply engine, according to local SOGs/SOPs.
5. When the supply engine arrives at the intersection, it stops and connects the end of its supply hose to the hose end laid in the street by the attack engine. If threaded couplings are used, a double-male adaptor may be required.
6. A fire fighter anchors the end of its supply line and lays out a hose line to the hydrant or static water source.
7. The attack engine driver/operator can start pumping from the booster tank and then switch over to the supply line once the supply engine is ready to pump.
8. The supply engine driver/operator positions the apparatus at the water supply according to the local SOGs/SOPs.
9. The supply engine driver/operator pulls off hose from the hose bed until the next coupling. The hose is broken at this connection and connected to the pump discharge or the four-way hydrant valve, following the local SOGs/SOPs.
10. Upon the signal from the attack engine, the supply engine driver/operator charges the supply line.

■ Loading Supply Hose

This section describes the basic procedures for loading supply line hose into the hose beds on a fire apparatus. Hose can be loaded in several different ways, depending on how it will be laid out at the fire scene. The hose must be easily removable from the hose bed, without kinks or twists, and without the possibility of becoming caught or tangled. The ideal hose load would be easy to load, avoid wear and tear on the hose, have few sharp bends, and allow the hose to play out of the hose bed smoothly and easily.

A **split hose bed** is a hose bed that is divided into two or more sections. This division is made for several purposes:

- One compartment in a split hose bed can be loaded for a forward lay with the female coupling out; and the other side can be loaded for a reverse lay with the male coupling out. This allows a line to be laid in either direction without adaptors.
- Two parallel hose lines can be laid at the same time (laying dual lines). Dual lines are beneficial if the situation requires more water than one hose line can supply.
- The split beds can be used to store hoses of different sizes. For example, one side of the hose bed could be loaded with 64-mm (2.5-in.) hose, which can be used as a supply line or an attack line; the other side of the hose bed could be loaded with 127-mm (5-in.) hose for use as a supply line. This setup enables the use of the most appropriate-size hose for any given situation.
- All of the hose from both sides of the hose bed can be laid out as a single hose line. This is done by coupling the end of the hose in one bed to the beginning of the hose to the other bed.

In a variation of the split hose bed known as a combination load, the last coupling in one bed is normally connected to the first coupling in the other bed. When one long line is needed, all of the hose plays out of one bed first, and then the hose continues to play out from the second bed. To lay dual lines, the connection between the two hose beds is uncoupled, and the hose can play out of both beds simultaneously. When the two sides of a split bed are loaded with hose in opposite directions, either a double-female or double-male adaptor is used to make the connection between the two hose beds.

When referring to the hose bed, the end closest to the cab is called the front of the hose bed. The end closest to the tailboard is called the rear of the hose bed.

When a split hose bed is loaded for a combination load, the end of the last length of hose in one bed is coupled to the beginning of the first length in the opposite bed. Begin loading the first bed with the initial coupling hanging out. This coupling will be the last to be deployed from the first bed when the hose is laid out. Leave enough of the end of the hose hanging down to reach the other hose bed. Load the second bed in the normal manner. When both beds have been loaded, connect the hanging coupling from the bottom of the first hose bed to the end coupling on the top of the adjoining bed.

You must learn the specific hose loads used by your fire department. When loading hose, always remember that the time and attention that go into loading the hose properly will prove valuable when it becomes necessary to use the hose at a fire. You should wear appropriate safety gear when loading hose—at a minimum, your helmet, gloves, and boots. Remember that many modern hose beds are a metre (several feet) above the ground, and that a fall from a hose bed could result in significant injuries.

Three basic hose loads are commonly used for supply line hose: the flat load, the horseshoe load, and the accordion load. Any one of these methods can be used to load hose for either a straight or reverse lay.

Flat Hose Load

The <u>flat hose load</u> is the easiest loading technique to implement and can be used for any size of hose, including LDH. Because the hose is placed flat in the hose bed, it should lay out flat without twists or kinks. With this type of load, wear and tear on the edges of the hose from the movement and vibration of the vehicle during travel are minimized. The flat hose load can be used with a single hose bed or a split bed. Many variations of the flat hose load exist, so follow your department's SOGs/SOPs.

To load hose in a flat load, follow the steps in **SKILL DRILL 16-22**:

1 If you are loading hose with threaded couplings, determine whether the hose will be used for a forward lay or a reverse lay.

2 To set up the hose for a forward lay, place the male hose coupling in the hose bed first; to set up the hose for a reverse lay, place the female hose coupling in the hose bed first. (**STEP 1**)

3 Start the hose load with the coupling at the front end of the hose compartment. (**STEP 2**)

4 Fold the hose back on itself at the rear of the hose bed. (**STEP 3**)

5 Run the hose back to the front end on top of the previous length of hose.

6 Fold the hose back on itself so that the top of the hose is on the previous length. (**STEP 4**)

7 When laying the hose back to the rear of the hose bed, angle the hose to the side of the previous fold. (**STEP 5**)

8 Continue to lay the hose in neat folds until the whole hose bed is covered with a layer of hose.

9 To make this hose load neat, make every other layer of hose slightly shorter, or alternate the folds. This will keep the ends from getting too high at the folds.

10 Continue to load the layers of hose until the required amount of hose is loaded. (**STEP 6**)

Horseshoe Hose Load

The <u>horseshoe hose load</u> is accomplished by placing the hose on its edge and positioning it around the perimeter of the hose bed in a U shape. At the completion of the first U shape, the hose is folded inward to form another U shape in the opposite direction. This continues until a complete layer is filled; then another layer is started above the first. When the hose load is completed, the hose in each layer forms the shape of a horseshoe. A major advantage of the horseshoe load is that it contains fewer sharp bends than the other hose loads.

A horseshoe load cannot be used for LDH, because the hose tends to fall over when it stands on edge. This kind of

load also leads to more wear on the hose, because the weight of the hose is supported by the edges. Also, when laying out a horseshoe load, the hose tends to lay out in a wavelike manner from one side of the street to the other.

Many variations of the horseshoe hose load exist, so follow your department's SOGs/SOPs. To perform a horseshoe hose load, follow the steps in **SKILL DRILL 16-23**:

1 If you are loading threaded hose, determine whether you want to load the hose for a forward lay or a reverse lay. For a forward lay, start with the male coupling in the rear corner of the hose bed; for a reverse lay, start with the female coupling in the rear corner of the hose bed. (**STEP 1**)

2 Lay the first length of hose on its edge against the right or left wall of the hose bed. (**STEP 2**)

FIRE FIGHTER Tips

The following tips will help you to do a better job when loading hose:

- Drain all of the water out of the hose before loading it.
- Rolling the hose first will result in a flatter hose load, because there will be no air in the hose.
- Do not load hose too tightly. Leave enough room so that you can slide a hand between the folds of hose. If hose is loaded too tightly, it may not lay out properly.
- Load hose so that couplings do not have to turn around as the hose is pulled out of the hose bed. Make a short fold in the hose close to the coupling to keep the hose properly oriented. This short fold is called a **Dutchman** **FIGURE 16-41**.
- Couple sections of hose so that the flat sides are oriented in the same direction.
- Check gaskets before coupling hose.
- Tighten couplings by hand only. With a good gasket, the hose should not leak.

FIGURE 16-41 A Dutchman fold ensures that a coupling will not have to turn around and possibly become stuck—as the hose is laid out.

Courtesy of Scott Dornan, ConocoPhillips Alaska

SKILL DRILL 16-22 Performing a Flat Hose Load
(Fire Fighter I, NFPA 5.5.2)

1 To set up the hose for a forward lay, place the male hose coupling in the hose bed first; to set up the hose for a reverse lay, place the female hose coupling in the hose bed first.

2 Start the hose load with the coupling at the front end of the hose compartment.

3 Fold the hose back on itself at the rear of the hose bed.

4 Run the hose back to the front end on top of the previous length of hose. Fold the hose back on itself so that the top of the hose is on the previous length.

5 When laying the hose back to the rear of the hose bed, angle the hose to the side of the previous fold.

6 Continue to lay the hose in neat folds until the whole hose bed is covered with a layer of hose. Continue to load the layers of hose until the required amount of hose is loaded.

SKILL DRILL 16-23 Performing a Horseshoe Hose Load
(Fire Fighter I, NFPA 5.5.2)

1 For a forward lay, start with the male coupling in the rear corner of the hose bed; for a reverse lay, start with the female coupling in the rear corner of the hose bed.

2 Lay the first length of hose on its edge against the right or left wall of the hose bed.

3 At the front of the hose bed, lay the hose across the width of the bed, and continue down the opposite side toward the rear.

4 When the hose reaches the rear of the hose bed, fold the hose back on itself, and continue laying it back toward the front of the hose bed. Keep the hose tight to the previous row of hose around the hose bed until it is back to the rear on the starting side. Fold the hose back on itself again, and continue packing the hose tight to the previous row.

5 Continue to pack the hose on the first layer. Each fold of hose will decrease the amount of space available inside the horseshoe. Once the centre of the horseshoe is filled in, begin a second layer by bringing the hose from the rear of the hose bed and laying it around the perimeter of the hose bed. Complete additional layers using the same pattern used for the first layer. Finish the hose load with any adaptors or appliances used by your department.

© LiquidLibrary

3 At the front of the hose bed, lay the hose across the width of the bed, and continue down the opposite side toward the rear. (**STEP 3**)

4 When the hose reaches the rear of the hose bed, fold the hose back on itself, and continue laying it back toward the front of the hose bed. Keep the hose tight to the previous row of hose around the hose bed until it is back to the rear on the starting side. Fold the hose back on itself again, and continue packing the hose tight to the previous row. (**STEP 4**)

5 Continue to pack the hose on the first layer. Each fold of hose will decrease the amount of space available inside the horseshoe. Once the centre of the horseshoe is filled in, begin a second layer by bringing the hose from the rear of the hose bed and laying it around the perimeter of the hose bed. Complete additional layers using the same pattern used for the first layer. Finish the hose load with any adaptors or appliances used by your department. (**STEP 5**)

Accordion Hose Load

The <u>accordion hose load</u> is performed with the hose placed on its edge. In this loading technique, the hose is laid side-to-side in the hose bed. One advantage of the accordion load is that it is easy to implement in the hose bed. One layer is loaded from left to right, and then the next layer is loaded above it from left to right.

The accordion hose load also has some disadvantages. Because the hose is stacked on its side, the hose experiences more wear than with a flat load. The accordion hose load is not recommended for LDH, because the hose tends to collapse when placed on its side.

Many variations of the accordion hose load exist, so follow your department's SOGs/SOPs. To perform the accordion hose load, follow the steps in **SKILL DRILL 16-24**:

1 Determine whether the hose will be used for a forward lay or a reverse lay.

2 To set up hose for a forward lay, place the male hose coupling in the hose bed first; to set up the hose for a reverse lay, place the female hose coupling in the hose bed first.

3 Start the hose lay with the coupling at the front end of the hose compartment.

4 Lay the first length of hose on its edge against the side of the hose bed. (**STEP 1**)

5 Double the hose back on itself at the rear of the hose bed. Leave the female end extended so that the two hose beds can be cross-connected. (**STEP 2**)

6 Lay the hose next to the first length, and bring it to the front of the hose bed.

SKILL DRILL 16-24   Performing an Accordion Hose Load
(Fire Fighter I, NFPA 5.5.2)

1 Lay the first length of hose on its edge against the side of the hose bed.

2 Double the hose back on itself at the rear of the hose bed. Leave the female end extended so that the two hose beds can be cross-connected.

3 Lay the hose next to the first length and bring it to the front of the hose bed. Fold the hose at the front of the hose bed so that the bend is even to the edge of the hose bed. Continue to lay folds of hose across the hose bed.

4 Alternate the length of the hose folds at each end to allow more room for the folded ends. When the bottom layer is completed, angle the hose upward to begin the second tier. Continue the second layer by repeating the steps used to complete the first layer.

7 Fold the hose at the front of the hose bed so that the bend is even to the edge of the hose bed. Continue to lay folds of hose across the hose bed. (**STEP 3**)

8 Alternate the length of the hose fold at each end to allow more room for the folded ends.

9 When the bottom layer is completed, angle the hose upward to begin the second tier.

10 Continue the second layer by repeating the steps used to complete the first layer. (**STEP 4**)

■ Connecting a Fire Department Engine to a Water Supply

When an engine is located at a hydrant, a supply hose must be used to deliver the water from the hydrant to the engine.

This special type of supply line is intended to deliver as much water as possible over a short distance. In most cases, a soft suction hose is used to connect directly to a hydrant. Alternatively, the connection can be made with a hard suction hose or with a short length of large-diameter supply hose.

Attaching a Soft Suction Hose to a Hydrant

Large-diameter soft suction hose is normally used to connect an engine directly to a hydrant. To attach a soft suction hose to a hydrant, follow the steps in **SKILL DRILL 16-25**:

1 The pump driver/operator positions the apparatus so that its inlet is the correct distance from the hydrant. (**STEP 1**)

2 Remove the hose, any needed adaptors, and the hydrant wrench. (**STEP 2**)

SKILL DRILL 16-25 Attaching a Soft Suction Hose to a Fire Hydrant
Fire Fighter I (NFPA 5.3.15)

1 The pump driver/operator positions the apparatus so that its inlet is the correct distance from the hydrant.

2 Remove the hose, any needed adaptors, and the hydrant wrench.

3 Attach the soft suction hose to the inlet of the engine if it is not already attached to the engine intake.

4 Unroll the hose.

(Continues)

SKILL DRILL 16-25 Attaching a Soft Suction Hose to a Fire Hydrant (*Continued*)
Fire Fighter I (NFPA 5.3.15)

5 Remove the large hydrant cap. Check the hydrant for proper operation.

6 Attach the soft suction hose to the hydrant.

7 Ensure that there are no kinks or sharp bend in the hose that might restrict the flow of water.

8 Open the hydrant slowly when so indicated by the driver/operator. Check all connections for leaks. Tighten the couplings if necessary.

9 Where required, place chafing blocks under the hose where it contacts the ground to prevent mechanical abrasion.

3 Attach the soft suction hose to the inlet of the engine. In some departments, this end of the hose is preconnected. (**STEP 3**)

4 Unroll the hose. (**STEP 4**)

5 Remove the large hydrant cap, and check the hydrant for proper operation. (**STEP 5**)

6 Attach the soft suction hose to the hydrant. Sometimes it will be necessary to use an adaptor. (**STEP 6**)

7 Ensure that there are no kinks or sharp bends in the hose that might restrict the flow of water. (**STEP 7**)

8 Open the hydrant slowly when so indicated by the pump driver/operator. Check all connections for leaks. Tighten the couplings if necessary. (**STEP 8**)

9 Where required, place chafing blocks under the hose where it contacts the ground to prevent mechanical abrasion. (**STEP 9**)

Attaching a Hard Suction Hose to a Hydrant

Although it is not commonly done, a hard suction hose can be used to connect an engine to a hydrant. Implementing this connection can be a difficult task, because the hard suction is heavy and has limited flexibility, so additional personnel may be needed to lift and attach the hose. The apparatus must be carefully positioned to make this connection properly.

To attach a hard suction hose to a hydrant, follow the steps in **SKILL DRILL 16-26**:

1 The pump driver/operator positions the apparatus so that the intake on the apparatus is located the correct distance from the hydrant. (**STEP 1**)

2 Remove the pump inlet cap on the apparatus.

3 Remove the hydrant steamer-outlet cap, which is the largest one. Check the hydrant for proper operation.

4 Place an adaptor on the hydrant if needed.

5 With a partner's help, remove a section of hard suction hose from the engine. (**STEP 2**)

6 Connect the hard suction hose to the large intake on the engine. (**STEP 3**)

7 Connect the opposite end to the hydrant, using a double-female adaptor with the hydrant thread on one side and the suction hose thread on the other side.

8 The pump driver/operator may need to move the apparatus slightly to make the final connection.

9 Slowly open the hydrant when instructed to do so by the pump driver/operator. (**STEP 4**)

■ Supply Hose Carries and Advances

Several different techniques are used to carry and advance supply hose. The best technique for a particular situation will depend on the size of the hose, the distance over which it must be moved, and the number of fire fighters available to perform the task. The same techniques can be used for both supply and attack lines.

Whenever possible, a hose line should be laid out and positioned as close as possible to the location where it will be operated before it is charged with water. A charged line is

---

> **Fire Fighter Safety Tips** © Photos.com
>
> When loading hose on an apparatus, always use caution in climbing up and down on the apparatus. If you are loading hose at a fire scene, watch out for wet, slippery surfaces, ice, or other hazards. Also, wet hose can be heavy, and you may need to reach, stretch, or lift the hose to get it into the hose bed. Use caution!

much heavier and more difficult to maneuver than a dry hose line. A suitable amount of extra hose should be available to allow for maneuvering after the line is charged.

Working Hose Drag

A working hose drag is used to deploy hose from a hose bed and advance the line over a relatively short distance to the desired location. Depending on the size and length of the hose, several fire fighters may be required to perform this task.

To perform a working hose drag, follow the steps in **SKILL DRILL 16-27**:

1 Place the end of the hose over your shoulder. (**STEP 1**)

2 Hold onto the coupling with one hand. (**STEP 2**)

3 Walk in the direction you want to advance the hose. (**STEP 3**)

4 As the next hose coupling is ready to come off the hose bed, have a second fire fighter grasp the coupling and place the hose over his or her shoulder. (**STEP 4**)

5 Continue this process until enough hose has been pulled out of the hose bed. (**STEP 5**)

Shoulder Carry

The shoulder carry is used to transport full lengths of hose over a longer distance than it is practical to drag the hose. It is also useful when a hose line must be advanced around obstructions; for example, this technique can be used to stretch an attack line from the front of a building, around to the rear entrance, and up to the second floor. The shoulder carry can also be employed to stretch a supply line to an attack engine in a location where the hose cannot be laid out by another engine.

This technique requires practice and good teamwork to be successful. By working together to complete tasks efficiently, fire fighters can achieve their goal of extinguishing a fire in the shortest period of time. To perform a shoulder carry, follow the steps in **SKILL DRILL 16-28**:

1 Stand at the tailboard of the engine. Grasp the end of the hose, and place it over your shoulder so that the coupling is at chest height.

2 Have a second fire fighter place additional hose on your shoulder so that the ends of the folds reach to about knee level. Continue to place folds on your shoulder—but only as many as you can safely carry. (**STEP 1**)

3 Hold the hose to prevent it from falling off your shoulder. Continue to hold the hose and move forward about 4.5 m (15 ft.).

SKILL DRILL 16-26 Attaching a Hard Suction Hose to a Fire Hydrant
Fire Fighter I (NFPA 5.3.15)

1 The pump driver/operator positions the apparatus so that the intake on the apparatus is located the correct distance from the hydrant.

2 Remove the pump inlet cap. Remove the hydrant steamer outlet cap. Place an adaptor on the hydrant if needed. With a partner's help, remove a section of hard suction hose from the engine.

3 Connect the hard suction hose to the large intake on the engine.

4 Connect the opposite end to the hydrant. Slowly open the hydrant when instructed to do so by the pump driver/operator.

© LiquidLibrary

4 A second fire fighter stands at the tailboard to receive a load of hose and then move forward. (**STEP 2**)

5 When enough fire fighters have received hose loads, the hose can be uncoupled from the hose bed. (**STEP 3**)

6 The pump driver/operator connects the coupling to the engine.

7 All of the fire fighters start walking toward the fire, coordinating their movements. The last fire fighter should not start offloading hose from his or her shoulder until all of the hose has been laid out. The next fire fighter in line then starts laying out hose from his or her shoulder.

8 Each fire fighter lays out his or her supply of hose until the entire length is laid out. (**STEP 4**)

To advance an accordion load using a shoulder carry, follow the steps in SKILL DRILL 16-29:

1 Find the end of the accordion load, whether it is a nozzle or a coupling.

2 Using two hands, grasp the end of the load and the number of folds it will take to make an adequate shoulder load. (**STEP 1**)

3 Pull the accordion load about one-third of the way off the apparatus. (**STEP 2**)

SKILL DRILL 16-27

Performing a Working Hose Drag
Fire Fighter I (NFPA 5.3.10)

1 Place the end of the hose over the shoulder.

2 Hold onto the coupling with one hand.

3 Walk in the direction you want to advance the hose.

4 As the next hose coupling is ready to come off the hose bed, a second fire fighter grasps the coupling and places the hose over his or her shoulder.

5 Continue this process until enough hose has been pulled out of the hose bed.

SKILL DRILL 16-28 Performing a Shoulder Carry
Fire Fighter I (NFPA 5.3.10)

1 Stand at the tailboard of the engine. Grasp the end of the hose and place it over your shoulder so that the coupling is at chest height. Have a second fire fighter place additional hose on your shoulder so that the ends of the folds reach about knee level. Continue to place folds on your shoulder—but only as much as you can safely carry.

2 Hold the hose to prevent it from falling off your shoulder. Continue to hold the hose and move forward about 4.5 m (15 ft.). A second fire fighter then stands at the tailboard to receive a load of hose and then move forward.

3 When enough fire fighters have received hose loads, the hose can be uncoupled from the hose bed.

4 All of the fire fighters start walking toward the fire. The last fire fighter should not start offloading hose from his or her shoulder until all of the hose is laid out. The next fire fighter then starts laying out hose from his or her shoulder. Each fire fighter lays out his or her supply of hose until the entire length is laid out.

© LiquidLibrary

4 Twist the folds so that they become flat, with the end of the accordion load (nozzle or coupling) on the bottom of the now-flat shoulder load. (**STEP 3**)

5 Transfer the hose to the opposite shoulder while turning so that you face in the direction you will walk. (**STEP 4**)

6 Place the shoulder load over your shoulder, and grasp it tightly with both hands. (**STEP 5**)

7 Walk away from the apparatus, pulling the shoulder load out of the hose bed. (**STEP 6**)

8 If additional fire fighters are needed, they may follow steps 1–6 to assist in removing the required amount of hose.

SKILL DRILL 16-29   Advancing an Accordion Load
Fire Fighter I (NFPA 5.3.10)

1   Using two hands, grasp the end of the load and the number of folds it will take to make an adequate shoulder load.

2   Pull the accordion load about one-third of the way off the apparatus.

3   Twist the folds so that they become flat, with the end of the accordion load (nozzle or coupling) on the bottom of the now-flat shoulder load.

4   Transfer the hose to the opposite shoulder while turning so that you face in the direction you will walk.

5   Place the shoulder load over your shoulder and grasp it tightly with both hands.

6   Walk away from the apparatus, pulling the shoulder load out of the hose bed.

■ Connecting Supply Hose Lines to Standpipe and Sprinkler Systems

Another water supply evolution is furnishing water to standpipe and sprinkler systems. Fire department connections on buildings are provided so that the fire department can pump water into standpipe and sprinkler systems. This setup is considered to be a supply line, because it supplies water to the standpipe and sprinkler systems. What is different is that this supply line is connected to the discharge side of the attack engine.

The function of the hose line in this case is to provide either a primary or secondary water supply for the sprinkler or standpipe system. The same basic techniques are used to connect the hose lines to either type of system.

Standpipe systems are used to provide a water supply for attack lines that will be operated inside the building. Outlets are provided inside the building, where fire fighters can connect attack lines. The fire fighters inside the building must then depend on fire fighters outside the building to supply the water to the fire department connection FIGURE 16-42 .

FIGURE 16-42 A standpipe connection.

Two types of standpipe systems exist:
- A dry standpipe system depends on the fire department to provide all of the water.
- A wet standpipe system has a built-in water supply, but the fire department connection is provided to deliver a higher flow or boost the pressure.

The pressure requirements for standpipe systems depend on how high up in the building the water will be used.

The fire department connection for a sprinkler system is also used to supplement the normal water supply. The required pressures and flows for different types of sprinkler systems can vary significantly. As a guideline, sprinkler systems should be fed at 1034 kPa (150 psi) unless more specific information is available.

To connect a hose line to supply a fire department connection, follow the steps in SKILL DRILL 16-30 :

❶ Locate the fire department connection to the standpipe or sprinkler system.

❷ Extend a hose line from the engine discharge to the fire department connection, using the size of hose required by the fire department's SOGs/SOPs. Some fire departments use a single hose line, whereas others call for two or more lines to be connected. (STEP ❶)

❸ Remove the caps on the standpipe inlet. Some caps are threaded into the connections and must be unscrewed. Other caps are designed to break away when struck with a tool such as a hydrant wrench or spanner wrench. (STEP ❷)

❹ Visually inspect the interior of the connection to ensure that it does not contain any debris that might obstruct the water flow. Do not stick your hand or fingers inside the connections; fire fighters have been injured by broken glass or needles inside these connections.

❺ Attach the hose line to the connection.

❻ Notify the pump driver/operator when the connection has been completed. (STEP ❸)

■ Replacing a Defective Section of Hose

With proper maintenance and testing, the risk of fire hose failure should be low, but even so, it is always a possibility. Fire fighters need to know what to do if a section of hose bursts or develops a major leak while it is being used. Every fire fighter should know how to quickly replace a length of defective hose and restore the flow.

A burst hose line should be shut down as soon as possible. If the line cannot be shut down at the pump, at the fire hydrant, or at a control valve, a hose clamp can be used to stop the flow. Place the hose clamp on an undamaged section of hose upstream from the problem. After the water flow has been shut off, quickly remove the damaged section of hose and replace it with two sections of hose. Using two sections of hose will ensure that the replacement hose is long enough to replace the damaged section.

To replace a hose section, follow the steps in SKILL DRILL 16-31 :

❶ Shut down or clamp off the damaged line.

❷ Remove the damaged section of hose.

Fire Fighter Safety Tips
© Photos.com

Some departments use LDH with Storz-type couplings to connect to fire department connections. Hose that is rated for use as attack hose should be used to connect to a standpipe system.

Large-diameter supply line hose is rated for a safe working pressure of 1275 kPa (185 psi). Standpipe systems generally require at least 1034 kPa (150 psi) to be provided at the fire department connection, but a water hammer can cause a spike of much higher pressure. The excess pressure could burst the hose and place fire fighters who are working inside the building in danger.

SKILL DRILL 16-30 Connecting a Hose Line to Supply a Fire Department Connection
Fire Fighter I (NFPA 5.3.15)

1 Locate the fire department connection to the standpipe or sprinkler system. Extend a hose line from the engine discharge to the fire department connection using the size of hose required by the fire department's SOGs/SOPs. Some fire departments use a single hose line, whereas others call for two or more lines to be connected.

2 Remove the caps on the standpipe inlet. Some caps are threaded into the connections and must be unscrewed. Other caps are designed to break away when struck with a tool such as a hydrant wrench or spanner wrench.

3 Visually inspect the interior of the connection to ensure that it does not contain any debris that might obstruct the water flow. Never stick your hand or fingers inside the connections; fire fighters have been injured from broken glass or needles inside these connections. Attach the hose line to the connection. Notify the pump driver/operator when the connection has been completed.

© LiquidLibrary

3 Replace the damaged section with two new sections to ensure that the replacement hose's length will be adequate.

4 Restore the water flow.

Many fire department SOGs/SOPs require two hose lines to be connected to a fire department connection. If one line breaks or if the flow is interrupted, water will still be delivered to the system through the other line.

■ Draining and Picking Up Hose

To put a hose back into service, it must be drained of water. To do so, lay the hose straight on a flat surface, and then lift one end of the hose to shoulder level. Gravity will allow the water to flow to the lower portion of the hose and eventually out of the hose. As you proceed down the length of hose, fold the hose back and forth over your shoulder. When you reach the end of the section, you will have the whole section of hose on your shoulder.

To drain a hose, follow the steps in **SKILL DRILL 16-32**:

1 Lay the section of hose straight on a flat surface. (**STEP 1**)

2 Starting at one end of the section, lift the hose to shoulder level. (**STEP 2**)

3 Move down the length of hose, folding it back and forth over the shoulder. (**STEP 3**)

4 Continue down the length until the entire section is on the shoulder. (**STEP 4**)

■ Unloading Hose

There are times other than a fire when fire fighters will need to unload the hose from an engine; for example, hose should be unloaded and reloaded on a regular basis to place the bends in different portions of the hose, because leaving bends in the same locations for long periods of time is likely to cause weakened areas. In addition, hose might have to be unloaded to change out apparatus. Sometimes all of the equipment from one engine must be transferred to another vehicle. It could also be necessary to offload the hose in order to conduct annual testing.

The following procedure should be used to unload a hose from a hose bed:

1. Use a large area such as a parking lot for this procedure. Be sure it is clean.

2. Disconnect any gate valves or nozzles from the hose before you begin.

3. Grasp the end of the hose, and pull it off the engine in a straight line.

4. When a coupling comes off the engine, disconnect the hose and pull off the next section of hose.

5. When all of the hose has been removed from the hose bed, use a broom to brush off any dirt or debris on both sides of the hose jacket.

6. Sweep out any debris or dirt from the hose bed.

7. Store hose rolls off the floor on a rack in a cool dry area.

SKILL DRILL 16-32   Draining a Hose
Fire Fighter I (NFPA 5.5.2)

1   Lay the section of hose straight on a flat surface.

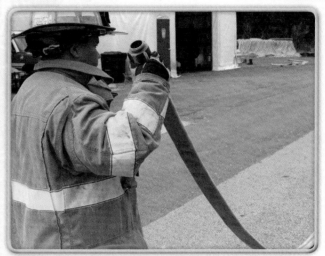

2   Starting at one end of the section, lift the hose to shoulder level.

3   Move down the length of hose, folding it back and forth over the shoulder.

4   Continue down the length until the entire section is on the shoulder.

© LiquidLibrary

Wrap-Up

Chief Concepts

- In rural areas, fire departments often depend on water from static sources to maintain their water supply. Water may be used from the static source directly or transported via a mobile water supply tanker/tender. These trucks are designed to carry large volumes of water—from 1892 to 18,925 L (500 to 5000 gal.).
  - Dry or drafting hydrants have a strainer on one end and connect to a fire apparatus to supply water from a static source.
  - Portable tanks are carried on fire apparatus and can hold between 2271 to 18,925 L (600 and 5000 gal.) of water. These tanks can be set up quickly and linked together to increase the water storage capacity.
  - A tanker shuttle can deliver water from a static-source fill site to the fire scene.
- Municipal water systems make clean water available to people in populated areas and provide water for fire protection. Hydrants make this water supply available to the fire department.
- Municipal water systems can draw water from wells, rivers, or lakes; they carry the water via pipelines or canals to a water treatment facility and then to the water distribution system, a complex network of underground pipes.
  - Most municipal water supply systems use both pumps and gravity to deliver water.
  - Underground water mains come in several sizes. Large mains, or primary feeders, carry large quantities of water to a section of a city. Smaller mains, or secondary feeders, distribute water to a smaller area. The smallest pipes, or distributors, carry water to the users and hydrants along individual streets.
  - Shut-off valves are located at the connection points where the underground mains meet the distributor pipes. These valves can be used to prevent water flow if the water system in the building or the fire hydrant is damaged.
- The two types of fire hydrants are wet barrel hydrants and dry barrel hydrants.
  - Wet barrel hydrants are used in areas where temperatures do not drop below freezing. These hydrants do not have to be drained after each use.
  - Dry barrel hydrants are used in areas where temperatures drop below freezing. When this type of hydrant is not in use, the barrel must be dry.
- Fire hydrants are located according to local standards and nationally recommended practices. In many communities, hydrants are located at every street intersection.
- Fire fighters must be proficient in operating fire hydrants, including the tasks of turning on the hydrant, shutting off the hydrant, and inspecting the hydrant.

- To understand how a water supply system works, you must understand some basic concepts:
  - The flow or quantity of water moving through a pipe, hose, or nozzle is described in terms of its volume, usually specified in units of litres per minute or gallons per minute.
  - Water pressure refers to an energy level and is measured in units of kPa (or psi). Volume and pressure are two different, but mathematically related, measurements.
  - Water that is not moving has potential (static) energy. When water is moving, it has a combination of potential energy and kinetic (in-motion) energy. Both the quantity of water flowing and the pressure under a specific set of conditions must be measured when testing any water system, including hydrants.
  - Static pressure is created by elevation pressure and/or pump pressure.
  - Normal operating pressure refers to the amount of pressure in a water distribution system during a period of normal consumption.
  - Residual pressure is the amount of pressure that remains in the system when water is flowing.
  - Flow pressure measures the quantity of water flowing through an opening. A Pitot gauge is used to measure flow pressure in kilopascals or psi and to calculate the flow in litres per minute or gallons per minute.
  - Knowing the static pressure, the flow in litres per minute or gallons per minute, and the residential pressure enables fire fighters to calculate the amount of water that can be obtained from a hydrant or a group of hydrants on the same water main.
- Fire hoses are used as supply hoses and as attack hoses. Supply hoses are used to deliver water from a static source or from a fire hydrant to an attack engine. Fire hoses range in size from 25 to 152 mm (1 to 6 in.).
  - Small-diameter hose (SDH) ranges in size from 25 to 51 mm (1 to 2 in.) in diameter. It is commonly used for attack lines.
  - Medium-diameter (MDH) hose has a diameter of 64 or 76 mm (2.5 or 3 in.). It is used for both attack and supply lines.
  - Large-diameter hose (LDH) has a diameter of 89 mm (3.5 in.) or more. It is used for supply lines.
- Fire hose is constructed with a waterproof liner surrounded by either one or two outer layers.
- Couplings are used to connect individual lengths of fire hose together. They are also used to connect a hose line to a hydrant, to an intake or discharge valve on an engine, or to a variety of nozzles, fittings, and appliances. A coupling is permanently attached to each end of a section of fire hose.

- The two types of couplings are threaded couplings and Storz-type couplings.
  - Threaded couplings are used on most hoses up to 76 mm (3 in.) in diameter and on both soft and hard suction hoses. Threaded couplings consist of a male coupling and a female coupling.
  - Storz-type couplings are the same on both ends.
- Supply hose can either be soft suction or hard suction.
  - Soft suction hose is a short section of LDH that is used to connect a fire department engine directly to the large steamer outlet on a hydrant. It has a female connection on each end, with one end matching the local hydrant threads and the other end matching the threads on a large-diameter inlet on the engine.
  - A hard suction hose is a special type of supply hose that is used to draft water from a static source, such as a river, lake, or portable drafting basin. It is designed to remain rigid and will not collapse when a vacuum is created in the hose to draft the water into the pump. It can also be used to carry water from a fire hydrant to the pumper.
- Fire hose should be inspected and tested following the procedures in NFPA 1962.
- The most common types of hose damage are mechanical damage, heat and cold damage, chemical damage, and mildew damage.
- Visual hose inspections should be performed at least quarterly. A visual inspection of hose should be performed after each use. If any defects are found, the length of hose should be marked and removed from service.
- Hose appliances include wyes, gated wyes, water thieves, Siamese connections, adaptors, reducers, hose jackets, hose rollers, and hose clamps.
- Rolled hose is compact and easy to manage. A fire hose can be rolled many different ways, depending on how it will be used. Follow the SOGs/SOPs of your department when rolling hose.
- The objective of laying a supply line is to deliver water from a hydrant or a static water source to an attack engine. In most cases, this operation involves laying a continuous hose line out of the bed of the fire apparatus as the vehicle drives forward. It can be done using either a forward lay or a reverse lay.
  - A forward lay starts at the hydrant and proceeds toward the fire. The hose is laid in the same direction as the water flows—from the hydrant to the fire.
  - A reverse lay involves laying the hose from the fire to the hydrant (i.e., opposite to the water flow).
  - A split hose lay is performed by two engine companies in situations where hose must be laid in two different directions to establish a water supply. The

attack engine drops the end of its supply hose at the corner of the street and performs a forward lay toward the fire. The supply engine stops at the same intersection, pulls off enough hose to connect to the end of the dropped supply line, and then performs a reverse hose lay to the hydrant or static water source. Once the two lines are connected, the supply engine can pump water to the attack engine.

- Hose can be loaded in several different ways, depending on how it will be laid out at the fire scene.
  - The hose must be easily removable from the hose bed, without kinks or twists, and without the possibility of becoming caught or tangled.
  - The ideal hose load would be easy to load, avoid wear and tear on the hose, have few sharp bends, and allow the hose to play out of the hose bed smoothly and easily.
- Several different techniques are used to carry and advance supply hose.
  - The best technique for a particular situation will depend on the size of the hose, the distance over which it must be moved, and the number of fire fighters available to perform the task.
  - The same techniques can be used for both supply lines and attack lines.
- Whenever possible, a hose line should be laid out and positioned as close as possible to the location where it will be operated before it is charged with water.
  - A charged line is much heavier and more difficult to maneuver than a dry hose line.
  - A suitable amount of extra hose should be available to allow for maneuvering after the line has been charged.

Hot Terms

Accordion hose load A method of loading hose onto a vehicle in accordion-like sections. It is achieved by standing the hose on its edge, then placing the next fold on its edge, and so on.

Adaptor A device that joins hose couplings of the same type, such as male to male or female to female.

Attack engine An engine from which attack lines have been pulled.

Attack hose Hose designed to be used by trained fire fighters and fire brigade members to combat fires beyond the incipient stage; also known as an attack line. (NFPA 1961)

Ball valve A valve used on nozzles, gated wyes, and engine discharge gates. They consist of a ball with a hole in its middle.

Booster hose A noncollapsible hose used under positive pressure with an elastomeric or thermoplastic tube, a braided or spiralled reinforcement, and a protective cover; also known as a booster line. (NFPA 1962)

Butterfly valve A valve found on the large pump intake where the suction hose connects to the inlet of the fire pump.

Distributors Relatively small-diameter underground pipes that deliver water to local users within a neighbourhood.

Double-female adaptor A hose adaptor used to join two male hose couplings.

Double-jacket hose A hose constructed with two layers of woven fibres.

Double-male adaptor A hose adaptor used to join two female hose couplings.

Dry barrel hydrant A type of hydrant used in areas subject to below-freezing weather. The valve that allows water to flow into the hydrant is located underground, and the barrel of the hydrant is normally dry.

Dry hydrant An arrangement of pipe permanently connected to a water source other than a piped, pressurized water supply system, which provides a ready water supply for firefighting purposes and utilizes the drafting (suction) capability of a fire department pump. (NFPA 1142)

Dump valve A large opening from the water tank of a mobile water supply apparatus, used for unloading purposes. (NFPA 1901)

Dutchman A short fold placed in a hose when loading it into the bed, which prevents the coupling from turning in the hose bed.

Elevated water-storage tower An above ground water-storage tank designed to maintain pressure in a water distribution system.

Elevation pressure The amount of pressure created by gravity; also known as head pressure.

Fire hydraulics The physical science of how water flows through a pipe or hose.

Flat hose load A method of putting a hose on a vehicle in which the hose is laid flat and stacked on top of the previous section.

Flow pressure The amount of pressure created by moving water.

Forward lay A method of laying a supply line where the line starts at the water source and ends at the attack engine.

Four-way hydrant valve A specialized type of valve that can be placed on a hydrant to allow another engine to increase the supply pressure without interrupting flow.

Friction loss The reduction in pressure caused by the water's contact with the side of the hose. This contact requires additional force to overcome the drag created by the hose wall.

Gate valve A valve found on hydrants and sprinkler systems.

Gated wye A valved device that splits a single hose into two separate hoses, so that each hose can be turned on and off independently.

Gravity-feed system A water distribution system that depends on gravity to provide the required pressure. The system storage is usually located at a higher elevation than the end-users.

Hard suction hose A hose used for drafting water from static supplies (e.g., lakes, rivers, and wells). It can also be used for supplying fire apparatus pumps from hydrants, if designed for that purpose. The hose contains a semirigid or rigid reinforcement. (NFPA 1963)

Higbee indicators Indicators on male and female threaded couplings that identify where the threads begin. These indicators should be aligned before fire fighters start to thread the couplings together.

Horseshoe hose load A method of loading hose in which the hose is laid into the bed along its three walls, so that it resembles a horseshoe.

Hose appliance A piece of hardware (other than a nozzle) that is generally intended to be connected to a fire hose to control or convey water. (NFPA 1962)

Hose clamp A device used to compress a fire hose so as to stop water flow.

Hose jacket A device used to stop a leak in a fire hose or to join hoses that have damaged couplings.

Hose liner The inner portion of a hose that is in direct contact with the flowing water; also known as the hose inner jacket.

Hose roller A device that is placed on the edge of a roof and used to protect hose as it is being hoisted up and over the roof's edge.

Large-diameter hose (LDH) A hose with a diameter of 89 mm (3.5 in.) or more that is designed to move large volumes of water to supply master stream appliances, portable hydrants, manifolds, standpipe and sprinkler systems, and fire department pumpers from hydrants and in relay. (NFPA 1410)

Master stream device A large-capacity nozzle that can be supplied by two or more hose lines of fixed piping. It can flow 1136 L/min. (300 gpm). These devices include deck guns and portable ground monitors.

Medium-diameter hose (MDH) Hose with a diameter of 64 or 76 mm (2.5 or 3 in.).

Mildew A fungus that can grow on hose if the hose is stored wet. Mildew can damage the jacket of a hose.

Mobile water supply apparatus A vehicle designed primarily for the transportation—pickup, transport, and delivery—of water to fire emergency scenes, where the water will then be applied by other vehicles or pumping equipment. (NFPA 1901)

Municipal water system A water distribution system that is designed to deliver potable water to end-users for domestic, commercial, industrial, and fire protection purposes.

Normal operating pressure The observed static pressure in a water distribution system during a period of normal demand.

Pitot gauge A type of gauge used to measure the velocity and pressure of water as it is discharged from an opening. In particular, it is used to determine the flow of water from a hydrant or nozzle.

Portable tanks Folding or collapsible tanks used at the fire scene to hold water for drafting.

Primary feeders The largest-diameter pipes in a water distribution system, which carry the greatest amounts of water.

Private water system A privately owned water system that operates separately from the municipal water system.

Reducer A fitting used to connect a small hose line or pipe to a larger hose line or pipe. (NFPA 1142)

Remote-controlled hydrant valve A valve attached to a fire hydrant to allow the operator to turn the hydrant on without flowing water into the hose line.

Reservoir A water storage facility.

Residual pressure The pressure that exists in the distribution system as measured at the residual hydrant at the time the flow readings are taken at the flow hydrants. (NFPA 24)

Reverse lay A method of laying a supply line where the supply line starts at the attack engine and ends at the water source.

Rocker lugs Fittings on threaded couplings that aid in coupling the hoses; also known as rocker pins.

Rubber-covered hose Hose whose outside covering is made of rubber, which is said to be more resistant to damage; also known as rubber-jacket hose.

Secondary feeders Smaller-diameter pipes that connect the primary feeders to the distributors.

Shut-off valve Any valve that can be used to shut down water flow to a water user or system.

Siamese connection A device that allows two hoses to be joined to flow into a single hose.

Small-diameter hose (SDH) Hose with a diameter of 25–51 mm (1–2 in.).

Soft suction hose A large-diameter hose designed to be connected to the large port on a hydrant (steamer connection) and into the engine.

Spanner wrench A type of tool used to couple or uncouple hoses by turning the rocker lugs on the connections.

Split hose bed A hose bed arranged so as to enable the engine to lay out either a single supply line or two supply lines simultaneously.

Split hose lay A scenario in which the attack engine forward lays a supply line from an intersection to the fire, and the supply engine reverse lays a supply line from the hose left by the attack engine to the water source.

Static pressure The pressure that exists in a distribution system at a given point under normal conditions as measured at the residual hydrant with no hydrants flowing. (NFPA 24)

Static water source A water source that is not under pressure, such as a pond, river, stream, or other body of water.

Steamer port The large-diameter port on a hydrant.

Storz-type hose coupling A hose coupling that has the properties of both a male and a female coupling. It is connected by engaging the lugs and turning the coupling clockwise one-third of a turn.

Supply hose Hose designed for the purpose of moving water between a pressurized water source and a pump that is supplying attack lines; also known as a supply line. (NFPA 1961)

Tender shuttle A method of transporting water from a source to a fire scene, using a number of mobile water supply apparatus; also known as a tanker shuttle.

Threaded hose coupling A type of coupling that requires a male fitting and a female fitting to be screwed together.

Water hammer The surge of pressure that occurs when a high-velocity flow of water is abruptly shut off. The pressure exerted by the flowing water against the closed system can be seven or more times that of the static pressure. (NFPA 1962)

Water main A generic term for any underground water pipe.

Water supply A source of water for firefighting activities. (NFPA 1144)

Water thief A device that has a 64-mm (2.5-in.) inlet and a 64-mm (2.5-in.) outlet, in addition to two 38-mm (1.5-in.) outlets. It is used to supply multiple hoses from one source.

Wet barrel hydrant A hydrant used in areas that are not susceptible to below-freezing temperatures. The barrel of the hydrant is normally filled with water.

Wye A device used to split a single hose into two separate lines.

You arrive as part of a third-alarm response at a fire at a large industrial complex. A huge column of black smoke is billowing into the sky; it is visible for kilometres. Your company is directed to catch a hydrant two blocks away and set up master streams to cool the fire. The engineer weaves the apparatus through the hose that is scattered everywhere like spaghetti on a plate. Once you reach your final destination, you jump out and begin to break out the LDH before setting up the master streams.

1. Large-diameter hose is _____ or larger and is designed to move large volumes of water to supply master stream appliances, portable hydrants, manifolds, standpipe and sprinkler systems, and fire department pumpers from hydrants and in relay.
   A. 64 mm (2.5 in.)
   B. 76 mm (3 in.)
   C. 89 mm (3.5 in.)
   D. 102 mm (4 in.)

2. The _____ hose load is achieved by standing the hose on its edge, then placing the next fold on its edge, and so on.
   A. flat
   B. forward
   C. reverse
   D. accordion

3. _____ is the pressure that exists in a distribution system at a given point under normal conditions as measured at the residual hydrant with no hydrants flowing.
   A. Static pressure
   B. Flow pressure
   C. Residual pressure
   D. Elevation pressure

4. The _____ are the largest-diameter pipes in a water distribution system, carrying the greatest amounts of water.
   A. distributors
   B. primary feeders
   C. secondary feeders
   D. large-diameter hose

5. A _____ is a device that has a 64-mm (2.5-in.) inlet and a 64-mm (2.5-in.) outlet, in addition to two 38-mm (1.5-in.) outlets. It is used to supply many hoses from one source.
   A. water thief
   B. gated wye
   C. four-way hydrant valve
   D. dump valve

6. A _____ is used to join two female hose couplings.
   A. hose jacket
   B. double-male
   C. double-female
   D. Siamese connection

You serve on a Volunteer Fire Department in your community. You annually perform hydrant test and inspection in your community. One of the new recruits indicates that he is not capable of direct fire response but would like to assist by joining, training, and performing peripheral tasks such as service tests and data entry that have become increasingly difficult for the other fire fighters to find the time to perform. The fire chief has asked you to work with him on how to properly conduct and document these service tests.

1. What equipment will be needed in order to conduct the tests?
2. What are the steps in the procedure?
3. How should the results of his tests be recorded?
4. Is this permitted by your insurance company? If so, to what level can he assist?

Fire Attack Lines and Foam

Fire Fighter I

Knowledge Objectives

After studying this chapter, you will be able to:

- List the standard sizes of attack hoses. (NFPA 5.3.10.A, 5.3.13.A , p 540–541)
- Describe the characteristics of booster hose. (NFPA 5.3.10.A , p 541)
- Describe the general procedures that are followed during attack line evolutions. (NFPA 5.3.10.A , p 541)
- Describe the types of loads used to organize attack hose. (p 541–547)
- Describe the procedures to follow when advancing attack hose. (NFPA 5.3.10.A , p 547–553)
- Describe how to extend an attack line. (NFPA 5.3.10.A , p 553–554)
- Describe how to advance an attack line from a standpipe. (NFPA 5.3.10.A , p 554)
- Describe how to replace a defective section of attack hose. (NFPA 5.5.2.A , p 554–555)
- List the three classifications of nozzles. (NFPA 5.3.10.A , p 556)
- Describe the characteristics of smooth-bore nozzles. (NFPA 5.3.10.A , p 556)
- Describe the characteristics of fog-stream nozzles. (NFPA 5.3.10.A , p 556–557)
- List the three types of fog-stream nozzles. (NFPA 5.3.10.A , p 557)
- Describe the specialized nozzles that may be used during fire-suppression operations. (NFPA 5.3.10.A , p 557–559)
- Describe how to maintain nozzles to ensure proper operation. (p 559)
- Describe how to inspect nozzles. (p 559)

Skills Objectives

After studying this chapter, you will be able to perform the following skills:

- Perform a minuteman hose load. (NFPA 5.5.2.B , p 542–543)
- Advance a minuteman hose load. (NFPA 5.3.10.B , p 542; 544)
- Perform a preconnected flat load. (NFPA 5.5.2.B , p 544–545)
- Advance a preconnected flat hose load. (NFPA 5.3.10.B , p 545–546)
- Perform a triple-layer hose load. (NFPA 5.5.2.B , p 546–548)
- Advance a triple-layer hose load. (NFPA 5.3.10.B , p 546–549)
- Unload and advance wyed lines. (NFPA 5.3.10.B , p 547–550)
- Advance a hose line up a stairway. (NFPA 5.3.10.B , p 550–551)
- Advance a hose line down a stairway. (NFPA 5.3.10.B , p 551–552)
- Advance an uncharged hose line up a ladder. (NFPA 5.3.10.B , p 551–553)
- Operate a hose stream from a ladder. (NFPA 5.3.10.B , p 553–554)
- Connect and advance an attack line from a standpipe outlet. (NFPA 5.3.15.B , p 554–555)
- Replace a defective section of hose. (NFPA 5.3.10.B , p 555)
- Operate a smooth-bore nozzle. (NFPA 5.3.10.B , p 556–558)
- Operate a fog-stream nozzle. (NFPA 5.3.10.B , p 557, 559)

Fire Fighter II — FFII

Knowledge Objectives

After studying this chapter, you will be able to:

- Describe how foam suppresses fire. (NFPA 6.3, 6.3.1.A , p 560)
- Describe the characteristics of Class A foam. (NFPA 6.3.1.A , p 560)
- Describe the characteristics of Class B foam. (NFPA 6.3.1.A , p 560–561)
- List the major categories of Class A foam concentrate. (NFPA 6.3.1.A , p 561)
- Describe the characteristics of compressed air foam. (NFPA 6.3.1.A , p 561)
- List the major categories of Class B foam concentrate. (NFPA 6.3.1.A , p 561–562)
- Describe the characteristics of protein foam. (NFPA 6.3.1.A , p 562)
- Describe the characteristics of fluoroprotein foam. (NFPA 6.3.1.A , p 562)
- Describe the characteristics of alcohol-resistant foam. (NFPA 6.3.1.A , p 562)
- Describe how foam proportioner equipment works with foam concentrate to produce foam. (NFPA 6.3.1.A , p 562)
- Describe how foam is applied to fires. (NFPA 6.3.1.A , p 563–564)

Skills Objectives

After studying this chapter, you will be able to perform the following skills:

- Place an eductor foam line in service. (NFPA 6.3.1.B , p 563)
- Apply foam using the sweep method. (NFPA 6.3.1.B , p 564)
- Apply foam using the bankshot method. (NFPA 6.3.1.B , p 564–565)
- Apply foam using the rain-down method. (NFPA 6.3.1.B , p 564)

Additional NFPA Standards

- NFPA 1962, *Standard for the Inspection, Care, and Use of Fire Hose, Couplings, and Nozzles and the Service Testing of Fire Hose*

You Are the Fire Fighter

"It's a worker!" is the first thought you have as you step out of the cab of the engine at a small ranch-style house that is set back from the street about 38.1 metres (125 feet). It appears to be a room-and-contents fire. You tighten your self-contained breathing apparatus (SCBA) waist strap as your captain directs you to pull a preconnect. He says you will be doing a direct attack and to make an entry from the rear of the structure. You step up to the engine and come face to face with three preconnected lines of various lengths from which to choose. Two are 44 millimetres (1.75 inch). One has a smooth-bore nozzle and the other has a combination nozzle. The other hose line is 64 mm (2.5 in.) with a combination nozzle on it. It is up to you to decide which hose is the proper one to pull.

1. What size and length of hose would you pull for this fire?
2. What type of nozzle is needed for this type of fire?
3. Why does it not matter about the type of hose load that is used?

Introduction

Fire hoses are used for two main purposes: as supply hoses and as attack hoses. This chapter deals with the use of attack lines, nozzles, and foam. Attack lines (attack hoses) are used to discharge water from an **attack engine** onto the fire. Most attack hoses carry water directly from the attack engine to a nozzle that is used to direct the water onto the fire. Attack hoses usually operate at higher pressures than do supply lines.

Nozzles are attached to the discharge end of attack lines to give fire streams shape and direction. Without a nozzle, the water discharged from the end of an attack line would reach only a short distance. The types of nozzles used with attack lines and how to use them are discussed in this chapter.

Not all fires can be suppressed with water. Foam can be a valuable tool in suppressing certain types of fires. This chapter discusses the types of foam used in the fire service and how to apply them.

Attack Hose

The hoses most commonly used to attack interior fires are either 38 mm (1.5 in.) or 44 mm (1.75 in.) in diameter **FIGURE 17-1**. Some fire departments also use 64-mm (2.5-in.) attack lines. Each section of attack hose is usually 15.2 m (50 ft.) long.

Medium-diameter hose (MDH) with a diameter of 64 mm (2.5 in.) or 76 mm (3 in.) can be used as either supply lines or attack lines. Sixty-four-millimetre (2.5-in.) attack handlines are used to extinguish larger fires. When used as an attack hose, the 76-mm (3-in.) size is more often used to deliver water to a master stream device or a fire department connection. Both 64-mm (2.5-in.) and 76-mm (3-in.) hoses usually come in 15.2-m (50-ft.) lengths.

Large-diameter hose (LDH) has a limited role as a fire attack tool. Large-diameter hose is used to supply portable monitors or is mounted on an aerial ladder. These commonly produce a water flow between 1591 litres per minute (350 gallons per minute [gpm]) and 6818 L/min. (1500 gpm).

FIGURE 17-1 Attack hoses are used during fire-suppression operations.

Large-diameter hose is used to supply **master stream devices**, such as portable monitors, or is mounted on an aerial ladder.

Attack hose must withstand high pressure and is designed to be used in a fire environment where it can be subjected to high temperatures. Attack hose must be tested annually at a pressure of at least 2068 kilopascals (300 pounds per square inch [psi]) and is intended to be used at pressures up to 1896 kPa (275 psi).

Because attack hose is designed to be used for fire suppression, it may be exposed to heat and flames, hot embers, broken glass, sharp objects, and many other potentially damaging conditions. For this reason, attack hose must be tough, yet flexible and light in weight. Attack lines can be either double-jacket or rubber-covered construction.

■ Sizes of Attack Lines

38-mm (1.5-Inch) and 44-mm (1.75-Inch) Attack Hose
Most fire departments use either 38-mm (1.5-in.) or 44-mm (1.75-in.) hose as the primary attack line for most fires. Both sizes of hose use the same 38-mm (1.5-in.) couplings.

This attack hose is the type used most often during basic fire training. Handlines of this size can usually be operated by one fire fighter, although a second person on the line makes it much easier to advance and control the hose. This hose is often stored on fire apparatus as a preconnected attack line in lengths ranging from 45 to 106 m (150 to 350 ft.), ready for immediate use.

The primary difference between 38-mm (1.5-in.) and 44-mm (1.75-in.) hose is the amount of water that can flow though the hose. Depending on the pressure in the hose and the type of nozzle used, a 38-mm (1.5-in.) hose can generally flow between 273 and 568 L/min (60 and 125 gpm). An equivalent 44-mm (1.75-in.) hose can flow between 545 and 818 L/min (120 and 180 gpm). This difference is important because the amount of fire that can be extinguished is directly related to the amount of water that is applied to it. A 44-mm (1.75-in.) hose can deliver much more water and is only slightly heavier and more difficult to advance than a 38-mm (1.5-in.) hose line.

64-mm (2.5-In.) Attack Hose

A 64-mm (2.5-in.) hose is used as an attack line for fires that are too large to be controlled by a 38-mm (1.5-in.) or 44-mm (1.75-in.) hose line. A 64-mm (2.5-in.) handline hose is generally considered to deliver a flow of approximately 1136 L/min. (250 gpm). It takes at least two fire fighters to safely control a 64-mm (2.5-in.) handline hose owing to the weight of the hose and the water and the nozzle reaction force. A 152-m (50-ft.) length of dry 64-mm (2.5-in.) hose weighs about 14 kilograms (30 pounds). When the hose is charged and filled with water, however, it can weigh as much as 91 k (200 lb.) per length. A 64-mm (2.5-in.) hose is most often used for interior attacks in large buildings and for defensive exterior attacks.

Higher flows—up to approximately 1591 L/min (350 gpm)—can be achieved with higher pressures and larger nozzles. It is difficult to operate a handline hose at these high flow rates, however. Such flows are more likely to be used to supply a master stream device.

Booster Hose

A **booster hose** (booster line) is usually carried on a hose reel that holds 45 to 61 m (150 or 200 ft.) of rubber hose. Booster hose contains a steel wire that gives it a rigid shape. This rigid shape allows the hose to flow water without pulling all of the hose off the reel. Booster hose is light in weight and can be advanced quickly by one person.

The disadvantage of booster hose is its limited flow. The normal flow from a 38-mm (1.5-in.) booster hose is in the range of 182 to 227 L/min (40 to 50 gpm), which is not an adequate flow for extinguishing structure fires. As a consequence, the use of booster hose is typically limited to small outdoor fires and trash-dumpster fires. This type of hose should not be used for structural firefighting.

Forestry Lines

Small-diameter hose lines, typically 25 mm (1 in.) or 38 mm (1.5 in.) in diameter, are often used for fighting wildland and ground fires. Large volumes of water are not often needed for these fires, and the small diameter provides much better maneuverability through brush and trees. These hose lines are sometimes extended for hundreds of feet.

Attack Line Evolutions

Attack lines are used to deliver water from an attack engine to a nozzle, which discharges the water onto the fire. Attack line evolutions are standard methods of working with attack lines to accomplish different objectives in a variety of situations. Most fire departments set up their equipment and conduct regular training so that fire fighters will be prepared to perform a set of standard hose evolutions. Hose evolutions involve specific actions that are assigned to specific members of a crew, depending on their riding positions on the apparatus. Every fire fighter should know how to perform all of the standard evolutions quickly and proficiently. When a fire officer calls for a particular evolution to be performed, each crew member should know exactly what to do.

Attack lines are usually stretched from an attack engine to the fire. The attack engine is usually positioned close to the fire, and attack lines are stretched manually by fire fighters. In some situations, an attack engine will drop an attack line at the fire and drive from the fire to a hydrant or water source. This procedure is similar to the reverse lay evolution for supply lines, except that the hose will be used as an attack line.

Most engines are equipped with preconnected attack lines, which provide a predetermined length of attack hose that is already equipped with a nozzle and connected to a pump discharge outlet. An additional supply of attack hose that is not preconnected is usually carried in another hose bed or compartment. To create an attack line with this hose, the desired length of hose is removed from the bed; a coupling is then disconnected and attached to a pump discharge outlet. This hose can also be used to extend a preconnected attack line or to attach to a wye or a water thief.

Attack hose is loaded in such a manner that it can be quickly and easily deployed. There are many ways to load attack lines into a hose bed, and this section presents only a few of the most common hose loads. Your department might use a variation of one of these techniques. It is important for you to master the hose loads used by your department.

FIRE FIGHTER II Tips   **FFII**

The optimal hose line crew is four personnel. Whether you have three or four, the officer of the crew should be the second or third person in the "stack." This allows him or her to monitor the progress of the fire attack as well as provide direction the fire fighter operating the nozzle. Being positioned at the front also allows the officer to direct the third and fourth fire fighters to help with hose movement, fix kinks, and so forth. The officer also remains mobile to continue to check fire and building conditions.

■ Preconnected Attack Lines

Preconnected hose lines are intended for immediate use as attack lines. A preconnected hose line has a predetermined length of hose with the nozzle already attached and is connected to a discharge outlet on the fire engine. The most commonly used attack lines are 44-mm (1.75-in.) hoses, generally ranging from 45.7 to 762 m (150 to 250 ft.) in length. Many engines are also equipped with preconnected 64-mm (2.5-in.) hose lines to enable them to make a quick attack on larger fires.

Attack lines should be loaded in the hose bed in a manner that ensures hat they can be quickly stretched from the attack engine to the fire. It should be possible for one or two fire fighters to remove the hose quickly from the hose bed and advance the hose to the fire. In many cases, an attack line is stretched in two stages: First, the hose is laid out from the attack engine to the building entrance or to a location close to the fire. Then, the hose is advanced into the building to reach the fire. Extra hose should be flaked out at the entrance to the fire building.

Whichever hose load technique is used, the hose should not become tangled as it is being removed from the bed and advanced. Laying out the hose should not require multiple trips between the engine and the fire building, and it should be easy to lay the hose around obstacles and corners. If additional fire fighters are available, they can remain behind to help move hose line around any obstacles. It should also be possible to repack the hose quickly and with minimal personnel. There is no perfect hose load that works well for every situation. The three most common hose loads for preconnected attack lines are the minuteman load, the flat load, and the triple-layer load. Variations in circumstances can make one type of hose load preferable for your community.

Because they face different types of fire situations, most departments load attack lines of different lengths; an engine could be equipped with a 45.7-m (150-ft.) preconnected line and a 76.2-m (250-ft.) preconnected line, for example. Fire fighters should pull an attack line that is long enough to reach the fire, but not so long that an excess of hose might slow down the operation and become tangled.

Preconnected hose lines can be placed in several different locations on a fire engine; for example, a section of a divided hose bed at the rear of the apparatus can be loaded with a preconnected attack line. Transverse hose beds are installed above the pump on many engines and loaded so that the hose can be pulled off from either side of the apparatus. Preconnected lines can be loaded into special trays that are mounted on the side of fire apparatus.

Many engines include a special compartment in the front bumper that can store a short preconnected hose line. This line is often used for vehicle fires and dumpster fires, where the apparatus can drive up close to the incident and a longer hose line is not needed.

Booster hose is another type of preconnected attack line. Booster reels holding 19-mm (0.75-in.) or 25-mm (1-in.) hose can be mounted in a variety of locations on fire apparatus. Booster line is not used for structure fires and has limited applications.

Minuteman Load

To prepare a 61-m (200-ft.) minuteman hose load, connect the female coupling to the preconnect discharge. Lay the first two sections in the hose bed using a flat load. After the first loop of hose, make short pulling handles at the end of the hose bed. Leave the male end of the second section of hose at the side of

FIRE FIGHTER Tips

When laying out an attack line with threaded couplings, you might find that the wrong end of the hose is on top of the hose bed. Double-male connectors and double-female connectors enable you to attach a male coupling to a male discharge or to attach a female coupling to a female coupling, respectively. A set of adaptors (one double-male and one double-female) should be easily accessible for these situations. Some departments place a set of adaptors on the end of the hose for this purpose.

the hose compartment. Attach the remaining two sections of hose together, and place the nozzle on the male end of the last section of hose. Place these sections of hose in the hose bed, starting with the nozzle end. Leave long pulling loops in the first loops attached to the nozzle. When you reach the female end of the last section of hose, attach it to the male end of the second section of hose placed into the hose bed.

To perform a minuteman load, follow the steps in **SKILL DRILL 17-1**:

❶ Connect the female end of the first length of hose to the preconnect discharge outlet. (**STEP** ❶)

❷ Flat load the hose even to the edges of the hose bed. At the second fold, make a loop or ear. This loop will create a handle to be used when advancing the hose load. (**STEP** ❷)

❸ Flat load the rest of the 30 m (100 ft.) of hose. Place the male hose coupling to the side of the hose bed. (**STEP** ❸)

❹ Couple the remaining two hose sections and attach the nozzle. (**STEP** ❹)

❺ Place the nozzle in the hose bed facing outward. Feed one layer of hose under the nozzle. Wrap the first layer of hose around the tip of the nozzle. (**STEP** ❺)

❻ Continue flat loading the remaining two sections of hose. (**STEP** ❻)

❼ Connect the female coupling from the last section of hose to the male coupling from the first section of hose. (**STEP** ❼)

To advance a minuteman hose load, follow the steps in **SKILL DRILL 17-2**:

❶ Grasp the nozzle and the folds on top of it.

❷ Pull this top part of the load approximately one-third out of the hose bed. (**STEP** ❶)

❸ Turn away from the hose bed, and place this top part of the hose load on your shoulder. Walk away from the apparatus until the rest of the top section of hose drops from the hose bed. (**STEP** ❷)

❹ Turn back around toward the hose bed and grasp the loop or ear from the first section of hose. (**STEP** ❸)

❺ Walk away from the apparatus until all hose is clear from the hose bed.

❻ Continue walking away, allowing the rest of the hose to deploy from the top of the load on your shoulder. (**STEP** ❹)

SKILL DRILL 17-1

Performing a Minuteman Hose Load
(Fire Fighter I, NFPA 5.5.2)

1 Connect the female end of the first length of hose to the preconnect discharge outlet.

2 Flat load the hose even to the edges of the hose bed. At the second fold, make a loop or ear. This loop will create a handle to be used when advancing the hose load.

3 Flat load the rest of the 30 m (100 ft.) of hose. Place the male hose coupling to the side of the hose bed.

4 Couple the remaining two hose sections and attach the nozzle.

5 Place the nozzle in the hose bed facing outward. Feed one layer of hose under the nozzle. Wrap the first layer of hose around the tip of the nozzle.

6 Continue flat loading the remaining two sections of hose.

7 Connect the female coupling from the last section of hose to the male coupling from the first section of hose.

SKILL DRILL 17-2   Advancing a Minuteman Hose Load
(Fire Fighter I, NFPA 5.3.10)

1 Grasp the nozzle and the folds on top of it. Pull this top part of the load approximately one-third out of the hose bed.

2 Turn away from the hose bed, and place this top part of the hose load on your shoulder. Walk away from the apparatus until the rest of the top section of hose drops from the hose bed.

3 Turn back around toward the hose bed and grasp the loop or ear from the first section of hose.

4 Walk away from the apparatus until all hose is clear from the hose bed. Continue walking away, allowing the rest of the hose to deploy from the top of the load on your shoulder.

© LiquidLibrary

Fire Fighter Safety Tips
© Photos.com

When loading hose on an apparatus, always use caution in climbing up and down on the apparatus. If you are loading hose at a fire scene, watch out for wet, slippery surfaces, ice, or other hazards. Also, wet hose can be heavy and you might need to reach, stretch, or lift the hose to get it into the hose bed. Use caution!

Preconnected Flat Load

To prepare the preconnected flat load, attach the female end of the hose to the preconnect discharge. Begin placing the hose

flat in the hose bed. When about one-third of the hose is in the bed, make an 20-centimetre (8-in.) loop at the end of the hose bed. This loop will be used as a pulling handle. When two-thirds of the hose is loaded, make a second pulling loop about twice the size of the first loop. Finish loading the hose, attach the nozzle, and place it on top of the hose bed. The preconnected flat load is now ready for use.

To perform a preconnected flat hose load, follow the steps in **SKILL DRILL 17-3**:

1 Attach the female end of the hose to the preconnect discharge. (**STEP 1**)

2 Begin flat loading the hose in the hose bed. (**STEP 2**)

SKILL DRILL 17-3   Performing a Preconnected Flat Load
(Fire Fighter I, NFPA 5.5.2)

1 Attach the female end of the hose to the preconnect discharge.

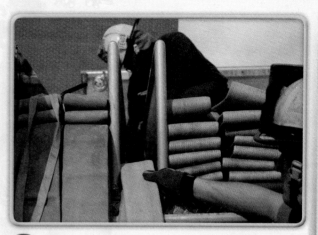

2 Begin flat loading the hose in the hose bed.

3 After the first layer is loaded, make the first fold even with the edge of the hose bed. On the second layer, make a loop or ear.

4 Flat load the remainder of the hose in the hose bed. Attach the nozzle to the male coupling at the end of the hose.

© LiquidLibrary

3 After the first layer is loaded, make the first fold even with the edge of the hose bed. On the second layer, make a loop or ear. **(STEP 3)**

4 Flat load the remainder of the hose in the hose bed. Attach the nozzle to the male coupling at the end of the hose. **(STEP 4)**

To advance a preconnected flat hose load, follow the steps in **SKILL DRILL 17-4**:

1 Grasp the top section of hose and place it over your shoulder. **(STEP 1)**

2 Walk away from the apparatus until the top load drops onto your shoulder. **(STEP 2)**

3 Turn back to the hose bed and grasp the lower loop or ear. Pull the remainder of the hose from the hose bed. **(STEP 3)**

4 Turn away from the apparatus and continue walking, letting the hose lay out off of your shoulder. **(STEP 4)**

FIRE FIGHTER Tips

When referring to the hose bed, the end closest to the cab is called the front of the hose bed. The end closest to the tailboard is called the rear of the hose bed.

Triple-Layer Load

To prepare the **triple-layer load**, attach the female end of the hose to the preconnect discharge. Connect the sections of hose together. Extend the hose directly from the hose bed. Pick up the hose two-thirds of the distance from the discharge to the

SKILL DRILL 17-4

Advancing a Preconnected Flat Hose Load
(Fire Fighter I, NFPA 5.3.10)

1 Grasp the top section of hose and place it over your shoulder.

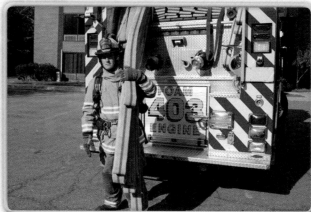

2 Walk away from the apparatus until the top load drops onto your shoulder.

3 Turn back to the hose bed and grasp the lower loop or ear. Pull the remainder of the hose from the hose bed.

4 Turn away from the apparatus and continue walking, letting the hose lay out off of your shoulder.

© LiquidLibrary

hose nozzle. Carry the hose back to the apparatus, forming a three-layer loop. Pick up the entire length of folded hose (this will take several people). Lay the tripled-folded hose in the hose bed in an S shape with the nozzle on top.

To perform a triple-layer hose load, follow the steps in **SKILL DRILL 17-5**:

1 Attach the female end of the hose to the preconnect discharge. (**STEP 1**)

2 Connect the sections of hose together. (**STEP 2**)

3 Extend the hose directly from the hose bed. Pick up the hose two-thirds of the distance from the discharge to the hose nozzle. (**STEP 3**)

4 Carry the hose back to the apparatus, forming a three-layer loop. (**STEP 4**)

5 Pick up the entire length of folded hose. (This will take several people.) (**STEP 5**)

FIRE FIGHTER Tips

The triple-layer load requires several fire fighters and must be practised often.

6 Lay the tripled-folded hose in the hose bed in an S shape with the nozzle on top. (**STEP 6**)

To advance a triple-layer hose load, follow the steps in **SKILL DRILL 17-6**:

1 Grasp the nozzle and the top fold. (**STEP 1**)

2 Turn away from the hose bed and place the hose on the shoulder. (**STEP 2**)

3 Walk away from the vehicle until the entire load is out of the bed. (**STEP 3**)

4 When the load is out of the bed, drop the fold. (**STEP 4**)

5 Extend the nozzle the remaining distance. (**STEP 5**)

■ Wyed Lines

To reach a fire that is some distance from the engine, it can be necessary first to advance a larger-diameter line, such as a 64-mm (2.5-in.) hose line, and then split it into two 44-mm (1.75-in.) attack lines. This is accomplished by attaching a gated wye or a water thief to the end of the 64-mm (2.5-in.) line, and then attaching the two attack lines to the gated outlets.

To unload and advance wyed lines, follow the steps in **SKILL DRILL 17-7**:

1 Grasp one of the attack lines and pull it from the bed. (**STEP 1**)

2 Pull the second attack line from the bed and place it far enough from the first line so that you can walk between the hose lines. This will keep the lines from becoming entangled. (**STEP 2**)

3 Grasp the gated wye and pull it from the bed.

4 The apparatus will deploy the remaining hose from the wye back to a water source. Place the gated wye so that one attack line is on the other side. (**STEP 3**)

5 The individual attack lines can now be extended to the desired positions. (**STEP 4**)

FIRE FIGHTER Tips

To prepare to advance the hose, form an upright loop about 1.2 m (4 ft.) in diameter outside the door to the fire building. If more hose is needed, one person can roll this loop toward the nozzle. This will yield several additional feet of hose at the nozzle.

■ Advancing Attack Lines

To attack an interior fire, an attack line is usually advanced in two stages: The first stage involves laying out the hose to the building entrance. The second stage is to advance the line into the building to the location where it will be operated.

When the attack line has been laid out to the entry point, the extra hose that will be advanced into the building should be flaked out in a serpentine pattern, ensuring that it will not become tangled when it is charged. Specifically, the hose should be flaked out with lengths of hose running parallel to the front of the fire building so that it can be easily advanced into the building **FIGURE 17-2**. It should also be set back from the doorway so that it does not obstruct the entry and exit path.

Make sure that you flake out the hose before it is charged with water. Once the line is charged, the hose becomes much more difficult to maneuver and advance. It is also important to verify that the hose length will reach the location where the line is needed inside the building.

While fire fighters are flaking out the hose and preparing to enter the building, the company officer will be completing the size-up. Other members of the team will be carrying out

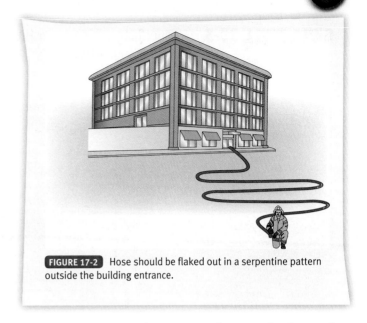

FIGURE 17-2 Hose should be flaked out in a serpentine pattern outside the building entrance.

other tasks to support the operation; for example, they might be forcing entry, getting into position for ventilation, establishing a water supply, and performing search and rescue. All of these tasks must be performed in sequence to maintain a safe environment and to extinguish the fire efficiently.

Once the hose is flaked out, signal the driver/operator to charge the line. Open the nozzle slowly to bleed out any trapped air and to make sure the hose is operating properly. If you are using an adjustable nozzle, make sure the nozzle is set to deliver the appropriate stream. For an interior attack, this setting will usually be a straight stream pattern. Once this is done, slowly close the nozzle.

Quickly recheck all parts of your personal protective equipment (PPE). Make sure your coat is fastened and your collar is turned up and fastened in front. Check your gloves and make sure you have a hand light. Check your partner's equipment and have your partner check your equipment. If you have time, catch your breath while you are breathing ambient air. Be ready to start breathing air from your SCBA and to advance the charged hose line as soon as your officer directs you to do so.

When you are given the command to advance the hose, keep safety as your number one priority. Make sure the other members of the nozzle team are ready. Do not stand in front of the door as it is opened: you do not know what might happen.

As you move inside the building, stay low to avoid the greatest amount of heat and smoke. If you cannot see because of the dense smoke, use your hands to feel the pathway in front of you. Feel in front of you so that you do not fall into a hole or other opening. Look for the glow of fire, and check for the sensation of heat coming through your face piece. Communicate with the other members of the nozzle team as you advance.

As you advance the hose line, you need to have enough hose to enable you to move forward. Ideally, a hose line crew consists of at least three members at the nozzle and a fourth member outside the door. As resistance is encountered in advancing the line, the fire fighter at the nozzle can help to

SKILL DRILL 17-5   Performing a Triple-Layer Hose Load
(Fire Fighter I, NFPA 5.5.2)

1   Attach the female end of the hose to the preconnect discharge.

2   Connect the sections of hose together.

3   Extend the hose directly from the hose bed. Pick up the hose two-thirds of the distance from the discharge to the hose nozzle.

4   Carry the hose back to the apparatus, forming a three-layer loop.

5   Pick up the entire length of folded hose.

6   Lay the triple-folded hose in the hose bed in an S shape with the nozzle on top.

SKILL DRILL 17-6

Advancing a Triple-Layer Hose Load
(Fire Fighter I, NFPA 5.3.10)

1 Grasp the nozzle and the top fold.

2 Turn away from the hose bed and place the hose on the shoulder.

3 Walk away from the vehicle until the entire load is out of the bed.

4 When the load is out of the bed, drop the fold.

5 Extend the nozzle the remaining distance.

© LiquidLibrary

Unloading and Advancing Wyed Lines
(Fire Fighter I, NFPA 5.3.10)

1 Grasp one of the attack lines and pull it from the bed.

2 Pull the second attack line from the bed, and place it far enough from the first line so that you can walk between the hose lines. This will keep the lines from becoming entangled.

3 Grasp the gated wye and pull it from the bed. The apparatus will deploy the remaining hose from the wye back to a water source. Place the gated wye so that one attack line is on the other side.

4 The individual attack lines can now be extended.

© LiquidLibrary

pull more hose, while the fire fighter at the door is responsible for feeding more hose into the building.

Charged hose lines are not easy to advance through a house or other building; for example, the hose line can become caught on a door frame. It is only with good teamwork that efficient hose line advancement can occur.

Advancing an Attack Line up a Stairway

When advancing a hose line up stairs, arrange to have an adequate amount of extra hose close to the bottom of the stairs. Utilizing a shoulder carry with the minuteman hose load allows for easy advancement. Make sure all members of the team are ready to move on command. It is difficult to move a charged hose line up a set of stairs while flowing water through the nozzle; shutting down the hose line while you are moving up the stairs will often allow you to get to the top of the stairs more quickly and safely. Follow the directions of your officer.

To advance an uncharged hose line up a stairway, follow the steps in **SKILL DRILL 17-8**:

1 Use a shoulder carry to advance up the stairs. (**STEP 1**)

②　When ascending the stairway, lay the hose against the outside of the stairs to avoid sharp bends and kinks and to reduce tripping hazards. (**STEP** ②)

③　Arrange excess hose so that it is available to fire fighters entering the fire floor. (**STEP** ③)

Advancing an Attack Line down a Stairway

Advancing a charged hose line down a stairway is also difficult. The smoke and flames from the fire tend to travel up the stairway of a residential structure, so you want to get down the stairway and position yourself below the heat and smoke as quickly as possible. Keep as low as possible to avoid the worst of the heat and smoke. When you advance a hose line down a stairway, you have a major advantage on your side: Gravity is working with you to bring the hose line down the stairs. You should never advance toward a fire unless your hose line is charged and ready to flow water. Wearing PPE and SCBA changes your centre of gravity. If you try to crawl down a stairway headfirst, you are likely to find yourself tumbling head over heels. Instead, move down the stairway feet first, using your feet to feel for the next step. Move carefully but as quickly as possible to get below the worst of the heat and smoke.

To advance a hose line down a stairway, follow the steps in **SKILL DRILL 17-9**:

①　Advance forward with the charged hose line. (**STEP** ①)

②　Descend stairs feet first. (**STEP** ②)

③　Position fire fighters at areas where hose lines could snag. (**STEP** ③)

Advancing an Attack Line up a Ladder

If a hose line has to be advanced up a ladder, it should be positioned correctly before the line is charged. To do so, place the hose on the left side of the ladder. Place the hose across your chest, with the nozzle draped over your right shoulder. Climb up the ladder with the uncharged hose line in this position. With the hose arranged in this way, if the line is mistakenly charged while you are climbing, it will not push you away from the ladder.

Additional fire fighters should pick up the hose about every 76 m (25 ft.) and help to advance it up the ladder. The nozzle is then passed over the top rung of the ladder and into the fire building. Additional hose should be fed up the ladder until sufficient hose is inside the building to reach the fire. The hose should be secured to the ladder with a hose strap or other rope or webbing to keep it from becoming dislodged.

To advance a hose line up a ladder, follow the steps in **SKILL DRILL 17-10**:

①　If a hose line needs to be advanced up a ladder, it should be advanced before it is charged.

②　Advance the hose line to the ladder. (**STEP** ①)

③　Pick up the nozzle; place the hose across the chest, with the nozzle draped over the shoulder.

④　Climb up the ladder with the uncharged hose line.

⑤　Once the first fire fighter reaches the first fly section of the ladder, a second fire fighter shoulders the hose to assist advancing the hose line up the ladder. To avoid overloading of the ladder, enforce a limit of one fire fighter per fly section.

SKILL DRILL 17-8　Advancing a Hose Line up a Stairway
(Fire Fighter I, NFPA 5.3.10)

1　Use a shoulder carry to advance up the stairs.

2　When ascending the stairway, lay the hose against the outside of the stairs to reduce tripping hazards. Avoid sharp bends.

3　Arrange excess hose so that it is available to fire fighters entering the fire floor.

Near-Miss REPORT

Report Number: 05-0000281

Synopsis: Early in the afternoon, a commercial structure fire call was sent out. Ten varying units were dispatched.

Event Description: Early in the afternoon, a commercial structure fire call was sent out. Ten varying units were dispatched. While en route, dispatch advised numerous calls. I was on a rescue unit that was second on the scene. The first engine reported nothing showing and entered the structure. With no following radio traffic, the first engine exited the building and reentered with a preconnect. They eventually reported fire on the top floor of the two-storey apartment. Rescue was assigned search. When we entered the structure, other units were receiving other assignments or arriving at the scene. The first engine was Fire Attack One. They had advanced no further than the top of the stairs. We (Search) entered and went upstairs to find high heat and blackout conditions. We eventually were trying to do fire attack duties and search. Thermal imaging showed no victims in the room of ignition, but high heat temps were obvious. Once a window was located, I knocked it out. A rush of heat was felt and flames became quickly visible and growing. When we made our way back to the door of the room, the flames were rolling over us. My partner started a combination pattern with the fog nozzle. The fire quickly subsided. At the time, we didn't feel as if we were in much danger. Once we exited the building other fire fighters stated a large fireball rushed out of the window we had broken out to vent. It was then we realized what the conditions were. It was a closer call than I realized.

Lessons Learned: Never underestimate fire conditions, whether they are improving or not. We probably should have worked together on venting and having a nozzle ready simultaneously. Communication before and after we made entry would have been helpful. We also took on too many tasks for a two-man crew.

SKILL DRILL 17-9   Advancing a Hose Line down a Stairway
(Fire Fighter I, NFPA 5.3.10)

1 Advance the charged hose line.

2 Descend stairs feet first. If there is smoke, position yourself underneath it.

3 Position fire fighters at areas where hose lines could snag.

© LiquidLibrary

SKILL DRILL 17-10   Advancing an Uncharged Hose Line up a Ladder
(Fire Fighter I, NFPA 5.3.10)

1 Advance the hose line to the ladder. Pick up the nozzle; place the hose across the chest, with the nozzle draped over the shoulder. Climb up the ladder with the uncharged hose line.

2 Once the first fire fighter reaches the first fly section of the ladder, a second fire fighter shoulders the hose to assist advancing the hose line up the ladder. To avoid overloading of the ladder, enforce a limit of one fire fighter per fly section. The nozzle is placed over the top rung of the ladder and advanced into the fire area.

3 Additional hose can be fed up the ladder until sufficient hose is in position. The hose can be secured to the ladder with a hose strap to support its weight and keep it from becoming dislodged.

© LiquidLibrary

6 The nozzle is placed over the top rung of the ladder and advanced into the fire area. (**STEP 2**)

7 Additional hose can be fed up the ladder until sufficient hose is in position.

8 The hose can be secured to the ladder with a hose strap to support its weight and keep it from becoming dislodged. (**STEP 3**)

Operating an Attack Line from a Ladder

A hose stream can be operated from a ladder and directed into a building through a window or other opening. To operate a fire hose from a ladder, follow the steps in **SKILL DRILL 17-11**:

1 Climb the ladder with a hose line to the height at which the line will be operated.

2 Apply a leg lock or use a ladder belt. (**STEP 1**)

3 Place the hose between two rungs and secure the hose to the ladder with a rope hose tool, rope, or piece of webbing. (**STEP 2**)

4 Carefully operate the hose stream from the ladder. Be careful when opening and closing nozzles and redirecting the stream because of the nozzle backpressure. This force could destabilize the ladder. (**STEP 3**)

■ Extending an Attack Line

When choosing a preconnected hose line or assembling an attack line, it is better to have too much hose than not enough. With a hose that is longer than necessary, you can flake out the excess hose. With a hose that is too short, you cannot advance it to the seat of the fire without shutting down the line and taking the time to extend it.

In some circumstances, it can be necessary to extend a hose line by adding additional lengths of hose—for example, if the fire is farther from the apparatus than initially estimated. More hose line could also be needed to reach the burning area or to complete extinguishment.

There are two basic ways to extend a hose line. First, fire fighters can disconnect the hose from the discharge gate on the attack engine and add the extra hose at that location. This method requires advancing the full length of the attack line to take advantage of the extra hose, which could take time and considerable effort.

Alternatively, fire fighters can add the hose to the discharge end of the hose. This addition can be achieved if the nozzle is a **breakaway type nozzle** that can be separated between the shut-off and the tip. With this type of nozzle, the valve is shut down, the nozzle tip is removed, and the extra hose is attached to the shut-off valve. The nozzle tip can then be installed on the male end of the added hose. This evolution allows fire fighters to lengthen the attack hose without shutting it down at the pump. A standpipe kit can be used to supply the added section of hose.

Advancing an Attack Line from a Standpipe Outlet

Standpipe outlets inside a building are provided for fire fighters to connect attack hose lines. Their availability eliminates the need to advance hose lines all the way from the attack engine on the street outside to an upper floor or a fire deep inside a large-area building. In tall buildings, it would be impossible to advance hose lines up the stairways in a reasonable time and to supply sufficient pressure to fight a fire on an upper floor.

Standpipe outlets are often located in stairways, and suggested operating guidelines (SOGs)/standard operating procedures (SOPs) generally require attack lines to be connected to an outlet one floor below the fire. The working space in the stairway and around the outlet valve is usually limited. Before opening the door, it is important to flake out the hose line properly so that it will be ready to advance into the fire floor. Before charging the hose line, the hose line should be flaked out on the stairs going up from the fire floor. When the hose line is charged and advanced into the fire floor, gravity will help to move the line forward. This technique is much easier than trying to pull the charged hose line up the stairs.

To connect and advance an attack line from a standpipe outlet, follow the steps in **SKILL DRILL 17-12**:

1. Carry a standpipe hose bundle to the standpipe connection below the fire. Remove the cap from the standpipe. (**STEP 1**)
2. Attach the proper adaptor or an appliance such as a gated wye. (**STEP 2**)
3. Flake the hose up the stairs to the floor above the fire. (**STEP 3**)
4. Extend the hose to the fire floor and prepare for the fire attack. (**STEP 4**)

Replacing a Defective Section of Hose

With proper maintenance and testing, the risk of fire hose failure should be low—but even so, it is always a possibility. Fire fighters need to know what to do if a section of hose bursts or develops a major leak while it is being used. Every fire fighter

---

SKILL DRILL 17-11 Operating a Hose Stream from a Ladder
(Fire Fighter I, NFPA 5.3.10)

1 Climb the ladder with a hose line to the height at which the line will be operated. Apply a leg lock or use a ladder belt.

2 Place the hose and secure the hose to the ladder with a rope hose tool, rope, or piece of webbing.

3 Carefully operate the hose stream from the ladder. Be careful when opening and closing nozzles and redirecting the stream because of the nozzle backpressure. This force could destabilize the ladder.

© LiquidLibrary

should know how to quickly replace a length of defective hose and restore the flow.

A burst hose line should be shut down as soon as possible. If the line cannot be shut down at the pump or at a control valve, a hose clamp can be used to stop the flow in an undamaged section of hose upstream from the problem. After the water flow has been shut off, quickly remove the damaged section of hose and replace it with two sections of hose. Using two sections of hose line ensures that the replacement hose is long enough to replace the damaged section.

To replace a hose section, follow the steps in **SKILL DRILL 17-13** (Fire Fighter I, NFPA 5.3.10):

1 Shut down or clamp off the damaged hose line.

2 Remove the damaged section of the hose line.

3 Replace the damaged section with two new sections to ensure that the hose line's length will be adequate. Restore the water flow.

Nozzles

<u>Nozzles</u> are attached to the discharge end of attack lines to give fire streams shape and direction. Without a nozzle, the water discharged from the end of a hose would reach for only a short distance. Nozzles are used on all sizes of handlines as well as on master stream devices.

SKILL DRILL 17-12 **Connecting and Advancing an Attack Line from a Standpipe Outlet**
(Fire Fighter I, NFPA 5.3.15)

1 Carry a standpipe hose bundle to the standpipe connection below the fire. Remove the cap from the standpipe.

2 Attach the proper adaptor or an appliance such as a gated wye.

3 Flake the hose up the stairs to the floor above the fire.

4 Extend the hose to the fire floor and prepare for the fire attack.

© LiquidLibrary

Nozzles can be classified into three groups:

- **Low-volume nozzles** flow 182 L/min. (40 gpm) or less. They are primarily used for booster hoses; their use is limited to small outside fires.
- **Handline nozzles** are used on hose lines ranging from 38 mm (1.5 in.) to 64 mm (2.5 in.) in diameter. Handline streams usually flow between 273 and 1591 L/min. (60 and 350 gpm).
- **Master stream nozzles** are used on deck guns, portable monitors, and ladder pipes that flow more than 1591 L/min. (350 gpm).

Low-volume and handline nozzles incorporate a shut-off valve that is used to control the flow of water. The control valve for a master stream is usually separate from the nozzle itself. All nozzles have some type of device or mechanism to direct the water stream into a certain shape. Some nozzles also incorporate a mechanism that can automatically adjust the flow based on the water's volume and pressure.

Nozzle Shut-Offs

The **nozzle shut-off** or bale enables the fire fighter at the nozzle to start or stop the flow of water. The handle that controls this valve is called a bale. Some nozzles incorporate a rotary control valve that is operated by rotating the nozzle in one direction to open it and in the opposite direction to shut off the flow of water.

Two different types of nozzles are manufactured for the fire service: **smooth-bore nozzles** and **fog-stream nozzles**. Smooth-bore nozzles produce a solid column of water, whereas fog-stream nozzles separate the water into droplets. The size of the water droplets and the discharge pattern can be varied by

FIGURE 17-3 Smooth-bore nozzle.

adjusting the nozzle setting. Nozzles must have an adequate volume of water and an adequate pressure to produce a good fire stream. The volume and pressure requirements vary according to the type and size of nozzle.

Smooth-Bore Nozzles

The simplest smooth-bore nozzle consists of a shut-off valve and a **smooth-bore tip** that gradually decreases the diameter of the stream to a size smaller than the hose diameter **FIGURE 17-3**. Smooth-bore nozzles are manufactured to fit both handlines and master stream devices. Those that are used for master streams and ladder pipes often consist of a set of stacked tips, where each successive tip in the stack has a smaller-diameter opening. Tips can be quickly added or removed to provide the desired stream size. This approach allows different sizes of streams to be produced under different conditions.

Using a smooth-bore nozzle offers several advantages; for example, a good smooth-bore nozzle has a longer reach than a combination fog nozzle operating at a straight-stream setting. In addition, a smooth-bore nozzle is capable of deeper penetration into burning materials, resulting in quicker fire knockdown and extinguishment. Smooth-bore nozzles also operate at lower pressures than adjustable-stream nozzles do. Most smooth-bore nozzles are designed to operate at 345 kPa (50 psi), whereas adjustable-stream nozzles generally require pressures of 517 to 689 kPa (75 to 100 psi). Lower nozzle pressure makes it easier for a fire fighter to handle the nozzle.

A straight stream extinguishes a fire with less air movement and less disturbance of the thermal layering than does a fog stream, which in turn makes the heat conditions less intense for fire fighters during an interior attack. It is also easier for the operator to see the pathway of a solid stream than a fog stream.

There are also some disadvantages associated with smooth-bore nozzles. Specifically, these nozzles do not absorb heat as readily as fog streams and are not as effective for hydraulic ventilation. A fire fighter cannot change the setting of a smooth-bore nozzle to produce a fog pattern; in contrast, a fog nozzle can be set to produce a straight stream but not to the same effect as a smooth-bore nozzle.

To operate a smooth-bore nozzle, follow the steps in **SKILL DRILL 17-14**.

1. Select the desired tip size and attach it to the nozzle shut-off valve. (**STEP 1**)
2. Attain a stable stance. (**STEP 2**)
3. Slowly open the valve, allowing water to flow. (**STEP 3**)
4. Open the valve completely to achieve maximum effectiveness. (Failure to open the valve fully results in reduced flow, depriving the line of the necessary L/min. or gpm rate.) (**STEP 4**)
5. Direct the stream to the desired location. (**STEP 5**)

Fog-Stream Nozzles

Fog-stream nozzles produce fine droplets of water **FIGURE 17-4**. The advantage of creating these droplets of water is that they absorb heat much more quickly and efficiently than does a solid column of water. This characteristic is important when

FIGURE 17-4 Fog-stream nozzle.

immediate reduction of room temperature is needed to avoid a flashover. Discharging 3.8 L (1 gal.) of water in 2.8 cubic metres (100 cubic feet) of involved interior space can extinguish a fire in 30 seconds. Fog nozzles can produce a variety of stream patterns, ranging from a straight stream to a narrow fog cone of less than 45 degrees to a wide-angle fog pattern that is close to 90 degrees.

The straight streams produced by fog nozzles have openings in the centre; therefore, a fog nozzle cannot produce a solid stream. The straight stream from a fog-stream nozzle breaks up faster and does not have the reach of a solid stream. A straight stream also is affected more dramatically by wind than a solid stream is.

The use of fog-stream nozzles offers several advantages. First, these nozzles can be used to produce a variety of stream patterns by rotating the tip of the nozzle. In addition, fog streams are effective at absorbing heat and can be used to create a water curtain to protect fire fighters from extreme heat.

Fog nozzles move large volumes of air along with the water. This can be an advantage or a disadvantage, depending on the situation. A fog stream can be used to exhaust smoke and gases through hydraulic ventilation. Unfortunately, this air movement can also result in sudden heat inversion in a room, which then pushes hot steam and gases down onto the fire fighters. If used incorrectly, a fog pattern can push the fire into unaffected areas of a building.

To produce an effective stream, nozzles must be operated at the pressure recommended by the manufacturer. For many years, the standard operating pressure for fog-stream nozzles was 689 kPa (100 psi). In recent years, some manufacturers have produced low-pressure nozzles that are designed to operate at 345 or 517 kPa (50 or 70 psi). The advantage of low-pressure nozzles is that they produce less reaction force, which makes them easier to control and advance. Lower nozzle pressure also decreases the risk that the nozzle will get out of control.

To operate a fog-stream nozzle, follow the steps in **SKILL DRILL 17-15**:

1. Select the desired nozzle.
2. Attain a stable stance (if standing). (**STEP 1**)
3. Slowly open the valve, allowing water to flow. (**STEP 2**)
4. Open the valve completely. (Failure to open the valve fully restricts water flow, reducing the necessary L/min. [gpm] rate). (**STEP 3**)
5. Select the desired water pattern by rotating the bezel of the nozzle.
6. Apply water where needed. (**STEP 4**)

Three types of fog-stream nozzles are available, with the difference among the types related to the water delivery capability:

- A **fixed-gallonage fog nozzle** delivers a preset flow at the rated discharge pressure. The nozzle could be designed to flow 136, 273, or 455 L/min. (30, 60, or 100 gpm).
- An **adjustable-gallonage fog nozzle** allows the operator to select a desired flow from several settings by rotating a selector bezel to adjust the size of the opening. For example, a nozzle could have the options of flowing 273, 432, or 568 L/min. (60, 95, or 125 gpm). Once the setting is chosen, the nozzle delivers the rated flow only as long as the rated pressure is provided at the nozzle.
- An **automatic-adjusting fog nozzle** can deliver a wide range of flows. As the pressure at the nozzle increases or decreases, its internal spring-loaded piston moves in or out to adjust the size of the opening. The amount of water flowing through the nozzle is adjusted to maintain the rated pressure and produce a good stream. A typical automatic nozzle could have an operating range of 409 to 1023 L/min. (90 to 225 gpm) while maintaining 689 kPa (100 psi) discharge pressure.

■ Other Types of Nozzles

Several other types of nozzles are used for special purposes. If your fire department has other types of specialty nozzles, you need to become proficient in their use and operation. **Piercing nozzles**, for example, are used to make a hole in automobile sheet metal, aircraft, or building walls so as to extinguish fires behind these surfaces **FIGURE 17-5**.

Cellar nozzles and **Bresnan distributor nozzles** are used to fight fires in cellars and other inaccessible places, such as attics and cocklofts **FIGURE 17-6**. These nozzles discharge water in a wide circular pattern as the nozzle is lowered

| **SKILL DRILL 17-14** | Operating a Smooth-Bore Nozzle
(Fire Fighter I, NFPA 5.3.10) |

1 Select the desired tip size and attach it to the nozzle shut-off valve.

2 Attain a stable stance (if standing).

3 Slowly open the valve, allowing water to flow.

4 Open the valve completely to achieve maximum effectiveness.

5 Direct the stream to the desired location.

SKILL DRILL 17-15 Operating a Fog-Stream Nozzle
(Fire Fighter I, NFPA 5.3.10)

1 Attain a stable stance (if standing).

2 Slowly open the valve, allow water to flow.

3 Open the valve completely.

4 Select the desired water pattern by rotating the bezel of the nozzle. Apply water where needed.

© LiquidLibrary

vertically through a hole into the cellar. They work like a large sprinkler head.

<u>Water curtain nozzles</u> are used to deliver a flat screen of water that then forms a protective sheet (curtain) of water on the surface of an exposed building **FIGURE 17-7**. The water curtains must be directed onto the exposed building because radiant heat can pass through the water curtain.

■ Nozzle Maintenance and Inspection

Nozzles should be inspected on a regular basis, along with all of the equipment on every fire department vehicle. In particular, nozzles should be checked after each use before they are placed back on the apparatus. They should be kept clean and clear of debris. Debris inside the nozzle will affect the performance of the nozzle, possibly reducing flow. Dirt and grit can also interfere with the valve operation and prevent the nozzle

from opening and closing fully. A light grease on the valve ball keeps it operating smoothly.

On fog nozzles, inspect the fingers on the face of the nozzle. Make sure all fingers are present and that the finger ring can spin freely. Any missing fingers or failure of the ring to spin drastically affects the fog pattern. Any problems noted should be referred to a competent technician for repair.

FIRE FIGHTER Tips

Some nozzles are made so that the tip of the nozzle can be separated from the shut-off valve. Such a breakaway nozzle allows fire fighters to shut off the flow, unscrew the nozzle tip, and then add more lengths of hose to extend the hose line without shutting off the valve at the engine.

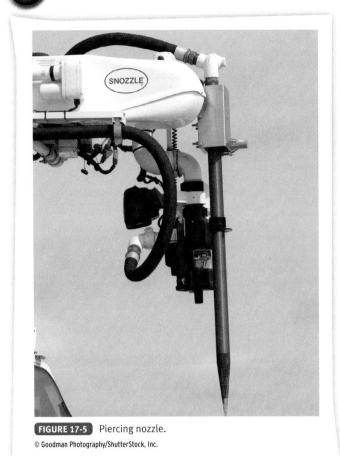

FIGURE 17-5 Piercing nozzle.
© Goodman Photography/ShutterStock, Inc.

FIGURE 17-6 Bresnan distributor nozzle.
Adapted from Akron Brass Company.

FIGURE 17-7 Water curtain nozzle.
Courtesy of POK of North America, Inc.

Foam

Firefighting foam can be used to fight multiple types of fires and to prevent the ignition of materials that could become involved in a fire. The use of foam is increasing as many new types of foam have become available and efficient systems for applying them have been developed. Foams have also been developed for use in neutralizing hazardous materials and decontamination. Many fire departments use several different types of foam for a variety of situations.

Firefighting foam is produced by mixing <u>**foam concentrate**</u> with water and air to produce a solution that can serve as an effective extinguishing agent. Several different types of foam are used for fires involving different types of fuels. Each type of foam requires the appropriate type of concentrate, the proper equipment to mix the concentrate with water in the required proportions, and the proper application equipment and techniques. Fire fighters must become familiar with the specific types of foam used by their fire department and the proper techniques for using them. It is particularly important to learn where and when to use each type of foam that is available.

■ Foam Classifications

The basic classifications of firefighting foams are either Class A or Class B. There are many types of foam within each classification.

<u>**Class A foam**</u> is used to fight fires involving ordinary combustible materials, such as wood, paper, and textiles. It is also effective on organic materials such as hay and straw. Class A foam is particularly useful for protecting buildings in rural areas during forest and brush fires when the supply of water is limited.

Class A foam increases the effectiveness of water as an extinguishing agent by reducing the surface tension of water. This allows the water to penetrate dense materials instead of running off the surface and allows more heat to be absorbed. The foam also keeps water in contact with unburned fuel to prevent its ignition. Class A foam can be added to water streams and applied with several types of nozzles.

<u>**Class B foam**</u> is used to fight Class B fires—that is, fires involving flammable and combustible liquids. The various types of Class B foam are formulated to be effective on different types of flammable liquids. Note that some liquids are incompatible with different foam formulations and will destroy the foam before the foam can control the fire.

In a flammable-liquid fire, the liquid itself does not burn. Only the flammable vapours that are evaporating from the surface of the liquid and mixing with air can burn. Depending on the temperature and the physical properties of the liquid, the

FIRE FIGHTER II

amount of vapours being released from the surface of a liquid varies; for example, gasoline produces flammable vapours down to a temperature of –42.7°C (–45°F). Some vapours are lighter than air and rise up into the atmosphere. Other vapours are heavier than air and flow across the surface and along the ground, collecting in low spots.

Foam extinguishes flammable-liquid fires by separating the fuel from the fire. When a blanket of foam completely covers the surface of the liquid, the release of flammable vapours stops. Preventing the production of additional vapours eliminates the fuel source for the fire, which extinguishes the fire.

Once a foam blanket has been applied, it must not be disturbed. The fuel under the foam blanket is still hot and capable of producing flammable vapours. Thus, if the foam blanket is disturbed by wind, by someone walking through the liquid, or by hose streams breaking up the foam blanket, flammable vapours will be released and could be reignited easily.

When using foam to extinguish a flammable-liquid fire, it is critically important to apply enough foam to cover the liquid surface fully. If not enough foam is available to completely cover the surface at one time, the fire will have the ability to continue burning and the suppression efforts will be ineffective. The rate of foam application must also be high enough to cover the surface and maintain a blanket on top of the liquid. If the application rate is too low, thermal updrafts caused by the fire will keep the foam from covering the surface, and the heat of the fire will destroy the foam that has already been applied.

Class B foam can also be applied to a flammable-liquid spill in an effort to prevent a fire. A foam blanket floating on the surface of the liquid will inhibit the production of vapours that might otherwise be ignited. In this situation, the rate of foam application is not as critical because there is no fire working to destroy the foam while it is being applied.

Fire fighters must ensure that they use the proper foam for the situation that is encountered. For example, they must use an alcohol-resistant foam if the incident involves a polar solvent. Polar solvents are chemicals (such as alcohols) that readily mix with water. An ordinary foam would be broken down quickly if it came in contact with this type of product. Make sure to use the proper foam given the product involved in the incident.

Class A foams are not designed to resist hydrocarbons or to self-seal on a liquid surface. If Class A foam is used on a flammable liquid, fire fighters would find it difficult to extinguish a fire and the foam blanket would not mend itself if it became disrupted. In this case, the risk of reignition could create a dangerous situation.

■ Foam Concentrates

Foam concentrate is the product that is mixed with water in different ratios to produce **foam solution**. The foam solution is the product that is actually applied to extinguish a fire or to cover a spill.

Class A Foams

Class A foams are usually formulated to be mixed with water in ratios from 0.1 percent 1 L of concentrate to 999 L of water or (1 gallon of concentrate to 999 gal. of water) to 1 percent (1 L of concentrate to 99 L of water, or 1 gal. of concentrate to 99 gal. of water). The end product can be tweaked to have different properties by varying the percentage of foam concentrate in the mixture and the application method. It is possible to produce "wet" foam that has good penetrating properties, for example, or "drier" foam that is more effective for applying a protective layer of foam onto a building.

Compressed Air Foam Systems

Compressed air foam systems (CAFS) are a relatively new method of making Class A foam.

Compressed air foam (CAF) is produced by injecting compressed air into a stream of water that has been mixed with 0.1 to 1 percent foam. This results in a foam with a small, highly compacted structure that provides a much larger surface area for heat absorption than aspirated foams. It achieves a rapid knockdown of a fire and a rapid cooling of the atmosphere.

CAF adheres to most surfaces and absorbs more heat than water. It decreases the amount of fuel available to ignite by isolating the fuel. CAF seems to decrease the amount of smoke in an enclosed fire, probably by absorbing some of the carbon particles from the smoke and depositing them on the surface of the foam. This has the effect of removing unburned fuel from the enclosed environment, thereby reducing the threat of a flashover. Because CAF coats the surface of a fuel, it reduces rekindling and smouldering of unburned fuel.

Hose lines containing CAF are lighter than those lines completely filled with water, because the CAF consists of a mixture of water and a significant amount of air. CAF is usually discharged at a rate of between 364 and 432 litres of water/min (80 and 90 gpm).

Class B Foams

Most Class B foam concentrates are designed to be used in strengths of either 3 or 6 percent. A 3 percent foam mixes 11 L (3 gal.) of foam concentrate with 367 L (97 gal.) of water to produce 378 L (100 gal.) of foam solution. A 6 percent foam mixes 22 L (6 gal.) of foam concentrate with 356 L (94 gal.) of water to produce 378 L (100 gal.) of foam solution. Some foams are designed to be used at 3 percent for ordinary hydrocarbons and at 6 percent for polar solvents. Fire fighters must determine which type of fuel is involved in the incident so that they can select the correct proportioning rate.

FIRE FIGHTER II Tips — **FFII**

The use of CAF requires special training, including practise with live burn situations.

The compatibility of foam agents with other extinguishing agents needs to be considered as well; for example, some combinations of dry chemical extinguishing agents and foam agents can cause an adverse reaction. The compatibility data needed are available from the manufacturers of the agents.

Fire fighters must not mix different types of foam concentrate or even different brands of the same type of foam concentrate unless they are known to be compatible. Some concentrates have been known to react with other types of foams, causing a congealing of the concentrate in the storage containers or in the proportioning system. This type of reaction can plug a foam system and render it useless.

Fire Fighter Safety Tips © Photos.com

Although most foam concentrates are somewhat corrosive, they pose few health risks to fire fighters if normal precautions are taken. In the past, some fire fighters have experienced minor skin irritations from coming into contact with foam concentrate. As always, before using any equipment, fire fighters must read and follow the manufacturer's directions when applying foam.

Some constituents in Class B foam concentrates that have been widely used in the past are being phased out now because of environmental concerns. Newer concentrates have been developed that are equally effective but lack the undesirable properties. The major categories of Class B foam concentrate are as follows:

- Protein foam
- Fluoroprotein foam
- Film-forming fluoroprotein (FFFP)
- Alcohol-resistant foam

Standard personal protective clothing provides appropriate protection when working with foam agents. Fire fighters are advised to rinse the skin after coming into contact with foam and to flush all equipment with clear water after foam use. Protective clothing should be rinsed with plain water.

Protein Foam

Protein foams are made from animal by-products. They are effective on Class B hydrocarbon fires and are applied in 3 percent or 6 percent delivery rates.

Fluoroprotein Foam

Fluoroprotein foams are made from the same base materials as protein foam but include additional fluorochemical surfactant additives. The additives allow this foam to produce a fast-spreading film.

Film-forming fluoroprotein (FFFP) agents are composed of protein plus film-forming fluorinated surface-active agents, which make them capable of forming water solution films on the surface of most flammable hydrocarbons and of conferring a fuel-shedding property to the foam generated. FFFP foams have a fast-spreading and levelling characteristic. Like other foams, they act as surface barriers that exclude air and prevent vapourization.

Alcohol-Resistant Foam

Alcohol-resistant foam it is formulated so that alcohols and other polar solvents do not dissolve the foam. Regular foams cannot be used on these products.

■ Foam Equipment

Foam equipment includes the proportioning equipment used to mix foam concentrate and water to produce foam solution, as well as the nozzles and other devices that are used to apply the foam. Many different types of proportioning and application systems are available. Most engine companies carry the necessary equipment to place at least one foam attack line into operation. Structural firefighting apparatus can also be designed with built-in foam-proportioning systems and on board tanks of foam concentrate to provide greater capabilities. Many fire departments specify that new apparatus include both Class A and Class B integrated foam systems or have special foam apparatus available for situations when large quantities of foam are needed.

Foam Proportioning Equipment

A **foam proportioner** is the device that mixes the foam concentrate into the fire stream in the proper percentage. The two types of proportioners—eductors and injectors—are available in a wide range of sizes and capacities. Foam solution can also be produced by batch mixing or premixing.

Foam Eductors

A **foam eductor** draws foam concentrate from a container or storage tank into a moving stream of water. An eductor can be built into the plumbing of an engine, or a portable eductor can be inserted in an attack hose line. A foam eductor is usually designed to work at a predetermined pressure and flow rate. A metering valve can be adjusted to set the percentage of foam concentrate that is educted into the stream.

The most common type of portable in-line eductor used by fire departments is sized to work with a 38-mm (1.5-in.) attack line. This type of eductor requires 1379 kPa (200 psi) of water pressure to draw foam concentrate from a portable container into the stream. An attack line with 45.7 m (150 ft.) of 38-mm (1.5-in.) hose can be connected on the discharge side of the eductor to deliver the foam to a nozzle.

Foam Injectors

Foam injectors add the foam concentrate to the water stream under pressure. Most injector-based proportioning systems can work across a range of flow rates and pressures. A metering system measures the flow rate and pressure of the water and adjusts the injector to add the proper amount of foam concentrate. This type of system is often installed on special foam apparatus.

To place an eductor foam line in service, follow the steps in **SKILL DRILL 17-16** (Fire Fighter II, NFPA 6.3.1):

1 Make sure all necessary equipment is available, including an in-line foam eductor and an air-aspirating nozzle. Ensure that enough foam concentrate is available to suppress the fire.

2 Don all PPE.

3 Procure an attack line, remove the nozzle, and replace it with the air-aspirating nozzle.

4 Place the in-line eductor in the hose line no more than 45.7 m (150 ft.) from the nozzle.

5 Place the foam concentrate container next to the eductor, check the percentage at which the foam concentrate should be used (found on container label), and set the metering device on the eductor accordingly.

6 Place the pickup tube from the eductor into the foam concentrate, keeping both items at similar elevations to ensure sufficient induction of foam concentrate.

7 Charge the hose line with water, ensuring there is a minimum of 1379 kPa (200 psi) at the eductor.

8 Flow water through the hose line until foam starts to come out of the nozzle. The hose line is now ready to be advanced onto burning flammable liquids.

9 Apply foam using one of the three application methods—sweep technique, bankshot technique, or rain-down technique—discussed later in this chapter, depending on the situation.

Batch Mixing

Foam concentrate can be poured directly into an apparatus booster tank to produce foam solution, a technique called **batch mixing**. If the booster tank has a capacity of 1892 2 L (500 gal.), 57 L (15 gal.) of 3 percent foam concentrate should be added. If 6 percent foam concentrate is used, 113 L (30 gal.) of foam concentrate should be added to the booster tank. It might be necessary to drain sufficient water from the tank first to make room for the foam concentrate. After the concentrate has been added, the solution should be mixed by circulating the water through the pump before it is discharged.

Premixing

Premixed foam is commonly used in 9.5 L (2.5 gal.) portable fire extinguishers. Foam fire extinguishers are filled with premixed foam solution and pressurized with compressed air or nitrogen. Some vehicles are equipped with large tanks holding 189 or 378 L (50 or 100 gal.) of premixed foam, which operate in the same manner.

■ Foam Application

Foam can be applied to a fire or spill through portable extinguishers, handlines, master stream devices, or a variety of fixed systems for special applications. Most fire departments apply foam through either handlines or master stream devices on fire apparatus. Several types of nozzles are used, with the goal of producing different types of foam. The major difference between these nozzles is the manner in which air is introduced into the stream of foam solution to produce the desired consistency of foam and air bubbles.

Foam can be applied with a wide range of expansion rates, depending on the amount of air that is mixed into the stream and the size of the bubbles that are produced. Low-expansion foam has little entrained air and a small bubble structure. This type of foam is often produced with standard adjustable fog nozzles, where the air is entrained by the flowing stream and mixed into the foam solution. This type of nozzle is often used to apply AFFF or Class A foam.

Medium-expansion foam is produced with special **aerating** nozzles that are designed to introduce more air into the stream and produce a consistent bubble structure (aeration). This type of nozzle is generally used with protein and fluoroprotein foams to produce a thicker blanket of foam. It is also recommended for use with alcohol-resistant foams to produce a thicker foam blanket.

High-expansion foam contains a much higher proportion of air and large bubbles. Such a system uses a high-expansion foam generator to introduce large quantities of air into the discharge stream. High-expansion foam is sometimes used in automatic systems that are designed to completely fill a large space with foam. These systems are most likely to be found in aircraft hangars or other large storage areas.

■ Foam Application Techniques

When applying foam from a handline, the correct application techniques must be used to produce the desired quality of foam and successfully blanket the surface of a burning liquid or spill. These techniques can be direct or indirect and include the sweep, bankshot, and rain-down methods of application.

Sweep Method

The **sweep (or roll-on) method** should be used only on a pool of flammable product that is on open ground. With this technique, the fire fighter sweeps the stream along the ground just in front of the target to produce a quantity of foam and then uses the energy of the stream to push the foam blanket across the surface. The stream is moved back and forth in a slow, steady horizontal motion to push the foam forward gently until the area is covered. It is important to push the foam gingerly so that the blanket is not broken. The fire fighter might need to move to different positions to be sure that the entire surface of the product is covered by the foam blanket.

To perform the sweep method of applying foam, follow the steps in **SKILL DRILL 17-17**:

1. Open the nozzle and test to ensure that foam is being produced.
2. Move within a safe range of the product and open the nozzle.
3. Direct the stream of foam onto the ground just in front of the pool of product.
4. Allow the foam to roll across the top of the pool of product until it is completely covered.
5. Be aware that the fire fighters might have to change positions along the spill so as to adequately cover the entire pool.

Bankshot Method

The **bankshot (or bank-down) method** is used at fires where the fire fighter can use an object to deflect the foam stream and let it flow down onto the burning surface. This method could be used to apply foam to an open-top storage tank or a rolled-over transport vehicle, for example. The fire fighter should sweep the foam back and forth against the object while the foam flows down and spreads back across the surface. As with all foam application methods, it is important to let the foam blanket flow gently on the surface of the flammable liquid to form a blanket.

To perform the bankshot method of applying foam, follow the steps in **SKILL DRILL 17-18**:

1. Open the nozzle and test to ensure that foam is being produced. (**STEP ①**)
2. Move within a safe range of the product and open the nozzle.
3. Direct the stream of foam onto a solid structure such as a wall or metal tank so that the foam is directed off the object and onto the pool of product.
4. Allow the foam to flow across the top of the pool of product until it is completely covered.
5. Be aware that the fire fighters might have to bank the foam off several areas of the solid object so as to extinguish the burning product. (**STEP ②**)

Rain-Down Method

The rain-down application method consists of lofting the foam stream into the air above the fire and letting it fall down gently onto the surface. The stream should be broken so that the amount of foam falling in the same area does not cause splashing or break up the blanket that has already been applied. Carefully observe how the foam blanket is building up and direct your stream so that the entire surface is covered.

To perform the rain-down method of applying foam, follow the steps in **SKILL DRILL 17-19** (Fire Fighter II, NFPA 6.3.1):

1. Open the nozzle and test to ensure that foam is being produced.

2. Move within a safe range of the product and open the nozzle.
3. Direct the stream of foam into the air so that the foam breaks apart in the air and falls onto the pool of product.
4. Allow the foam to flow across the top of the pool of product until it is completely covered.
5. Be aware that the fire fighters might have to move to several locations and shoot the foam into the air so as to extinguish the burning product.

FIRE FIGHTER II Tips   FFII

Utilizing either Class A or Class B foam on a brush fire after extinguishment can allow for additional penetration and reduce the likelihood of rekindle.

■ Backup Resources

When attempting to use foam to extinguish a flammable-liquid fire, it is critically important that enough foam concentrate be available to complete the job. If the flow of foam must be interrupted while additional foam supplies are obtained, the fire will destroy the foam that has already been applied. It is better to wait until an adequate supply of foam concentrate is on hand than to waste the limited supply that is immediately available.

Foam manufacturers provide specific formulas that enable fire fighters to calculate how much foam is required to extinguish fires of a certain size. Most fire departments have contingency plans to deliver quantities of foam to the scene of a major incident. This foam could be located on designated vehicles or in storage at fire stations, for example. Backup sources often include airport crash vehicles and petroleum facilities that have their own foam apparatus and often keep large quantities of foam in storage. Manufacturers of foam products also have emergency programs to deliver large quantities of foam to the scene of exceptionally large-scale incidents.

■ Foam Apparatus

Some fire departments operate apparatus that is specifically designed to produce and apply foam. The most common examples are found at airports, where these apparatus are used for aircraft rescue and firefighting. These large vehicles carry the foam concentrate and water on board and are designed to quickly apply large quantities of foam to a flammable-liquid fire. Remote-control monitors can be used to apply foam while the vehicles are in motion. If there is an airport close to your department, you should practise with them and know the best way to access them, before an emergency occurs.

SKILL DRILL 17-18 Apply Foam with the Bankshot Method
(Fire Fighter I, NFPA 6.3.1)

1 Open the nozzle and test to ensure that foam is being produced.

2 Move within a safe range of the product and open the nozzle. Direct the stream of foam onto a solid structure such as a wall or metal tank so that the foam is directed off the object and onto the pool of product. Allow the foam to flow across the top of the pool of product until it is completely covered. Be aware that the fire fighters might have to bank the foam off several areas of the solid object so as to extinguish the burning product.

© LiquidLibrary

Wrap-Up

Chief Concepts

- Most attack hoses carry water directly from the attack engine to a nozzle that is used to direct the water onto the fire.
- Attack hoses usually operate at higher pressures than supply lines do.
- The hoses most commonly used to attack interior fires are either 38 mm (1.5 in.) or 44 mm (1.75 in.) in diameter. Each section of attack hose is usually 15.2 m (50 ft.) long.
- Attack hose must be tested annually at a pressure of at least 2068 kPa (300 psi) and is intended to be used at pressures up to 1896 kPa (275 psi).
- Depending on the pressure in the hose and the type of nozzle used, a 38-mm (1.5-in.) hose can generally flow between 273 and 568 L/min. (60 and 100 gpm). An equivalent 44-mm (1.75-in.) hose can flow between 545 and 818 L/min. (120 and 180 gpm).
- A (64-mm (2.5-in.) hose is used as an attack line for large fires and generally delivers a flow of approximately 1136 L/min. (250 gpm) It takes at least two fire fighters to control a 64-mm (2.5-in.) handline safely because when the hose is charged with water it can weigh as much as 91 kg (200 lb.).
- Booster hose contains a steel wire that gives it a rigid shape so that the hose can flow water without pulling all of the hose off the reel. The normal flow from a 25-mm (1-in.) booster hose is in the range of 182 to 227 L/min. (40 to 50 gpm). This type of hose should not be used for structural firefighting.
- Attack lines are usually stretched from an attack engine to the fire. The attack engine is usually positioned close to the fire, and attack lines are stretched manually by fire fighters. In some situations, an engine will drop an attack line at the fire and drive from the fire to a hydrant or water source. This procedure is similar to the reverse lay evolution, except that the hose will be used as an attack line.
- Attack hose is loaded so that it can be quickly and easily deployed. The three most common hose loads used for preconnected attack lines are the minuteman load, the flat load, and the triple-layer load.
- Preconnected hose lines can be placed in several different locations on a fire engine:
  - A section of a divided hose bed at the rear of the apparatus can be loaded with a preconnected attack line.
  - Transverse hose beds are installed above the pump on many engines and loaded so that the hose can be pulled off from either side of the apparatus.
  - Preconnected lines can be loaded into special trays that are mounted on the side of fire apparatus.
  - Many engines include a special compartment in the front bumper that can store a short preconnected hose line.
- To reach a fire that is some distance from the engine, it might be necessary to first advance a larger-diameter line, and then split it into two 44-mm (1.75-in.) attack lines. This is accomplished by attaching a gated wye or a water thief to the end of the 64-mm (2.5-in.) line, and then attaching the two attack lines to the gated outlets.
- A basic attack evolution looks like this:
  - Attack line is stretched in two stages. First, the hose is laid out from the attack engine to the building entrance or to a location close to the fire. Then, the hose is advanced into the building to reach the fire. Extra hose should be deposited at the entrance to the fire building. Make sure you flake out the hose before it is charged with water.
  - Once the hose is flaked out, signal the pump driver/operator to charge the line.
  - Once the hose is flaked out, check your PPE and your partner's PPE.
  - When you are given the command to advance the hose, keep safety as your number one priority.
  - As you move inside the building, stay low to avoid the greatest amount of heat and smoke. Communicate with the other members of the nozzle team as you advance.
  - As you advance the hose line, you need to have enough hose to enable you to move forward.
  - Charged hose lines are not easy to advance. It is only with good teamwork that efficient hose line advancement can occur.
- Nozzles are attached to the discharge end of attack lines to give fire streams shape and direction. Nozzles are used on all sizes of handlines as well as on master stream devices.
- Nozzles can be classified into three groups:
  - Low-volume nozzles flow 1821 L/min. (40 gpm) or less. They are primarily used for booster hoses; their use is limited to small outside fires.
  - Handline nozzles are used on hose lines ranging from 38 mm (1.5 in.) to 64 mm (2.5 in.) in diameter. Handline streams usually flow between 273 and 1591 L/min. (60 and 350 gpm).
  - Master stream nozzles are used on deck guns, portable monitors, and ladder pipes that flow more than 1591 L/min. (350 gpm).
- Nozzle shut-off enables the fire fighter at the nozzle to start or stop the flow of water.

- Two different types of nozzles are manufactured for the fire service:
  - *Smooth-bore nozzles.* Produce a solid column of water.
  - *Fog-stream nozzles.* Separate the water into droplets. The size of the water droplets and the discharge pattern can be varied by adjusting the nozzle setting.
- Specialized nozzles include these:
  - *Piercing nozzles.* Used to make a hole in sheet metal or building walls to extinguish fires.
  - *Cellar nozzles and Bresnan distributor nozzles.* Used to fight fires in cellars and other inaccessible places, such as attics and cocklofts.
  - *Water curtain nozzles.* Used to deliver a flat screen of water that forms a protective sheet of water on the surface of an exposed building.
- Firefighting foam can be used to fight multiple types of fires and to prevent the ignition of additional fuels. Several different types of foam are used for fires involving different types of fuels.
- Foams are either Class A or Class B.
  - Class A foam is used to fight fires involving ordinary combustible materials, such as wood, paper, and textiles.
  - Class B foam is used to fight Class B fires, or fires involving flammable and combustible liquids.
- Foam extinguishes flammable-liquid fires by separating the fuel from the fire. When a blanket of foam completely covers the surface of the liquid, the release of flammable vapours stops. Preventing the production of additional vapours eliminates the fuel source for the fire, which extinguishes the fire.
- Foam concentrate is the product that is mixed with water in different ratios to produce foam solution. The foam solution is the product that is actually applied to extinguish a fire or to cover a spill.
- The major categories of Class B foam concentrate are protein foam, fluoroprotein foam, film-forming fluoroprotein, aqueous film-forming foam, and alcohol-resistant foams.
- Compressed air foam systems are a new method of making Class A foam. Compressed air foam is produced by injecting compressed air into a stream of water that has been mixed with 0.1 percent to 1.0 percent foam.
- A foam proportioner is the device that mixes the foam concentrate into the fire stream in the proper percentage. The two types of proportioners—eductors and injectors—are available in a wide range of sizes and capacities.
- Foam solution can also be produced by batch mixing or premixing.
  - Batch mixing is a technique where foam concentrate is poured directly into an apparatus booster tank to produce foam solution.
  - Premixed foam is commonly used in 9.5-L (2.5-gal.) portable fire extinguishers.

Wrap-Up, continued

Hot Terms

Adjustable-gallonage fog nozzle A nozzle that allows the operator to select a desired flow from several settings.

Aerating The process of inducing air into the foam solution, which expands and finishes the foam.

Alcohol-resistant foam Used for fighting fires involving water-soluble materials or fuels that are destructive to other types of foams. Some alcohol-resistant foams are capable of forming a vapour-suppressing aqueous film on the surface of hydrocarbon fuels. (NFPA 412)

Attack engine An engine from which attack lines have been pulled.

Automatic-adjusting fog nozzle A nozzle that can deliver a wide range of water stream flows. It operates by means of an internal spring-loaded piston.

Bankshot (or bank-down) method A method that applies the stream onto a nearby object, such as a wall, instead of directly aiming at the fire.

Batch mixing Pouring foam concentrate directly into the fire apparatus water tank, thereby mixing a large amount of foam at one time.

Booster hose (booster line) A noncollapsible hose used under positive pressure having an elastomeric or thermoplastic tube, a braided or spiralled reinforcement, and an outer protective cover. (NFPA 1962)

Breakaway type nozzle A nozzle with a tip that can be separated from the shut-off valve.

Bresnan distributor nozzle A nozzle that can be placed in confined spaces. The nozzle spins, spreading water over a large area.

Cellar nozzle A nozzle used to fight fires in cellars and other inaccessible places. The device works by spreading water in a wide pattern as the nozzle is lowered through a hole into the cellar.

Class A foam Foam intended for use on Class A fires. (NFPA 1150)

Class B foam Foam intended for use on Class B fires. (NFPA 1901)

Compressed air foam (CAF) Class A foam produced by injecting compressed air into a stream of water that has been mixed with 0.1 percent to 1.0 percent foam.

Compressed air foam system (CAFS) A foam system that combines air under pressure with foam solution to create foam. (NFPA 1901)

Film-forming fluoroprotein (FFFP) A protein-based foam concentrate incorporating fluorinated surfactants that forms a foam capable of producing a vapour-suppressing aqueous film on the surface of hydrocarbon fuels. This foam can show an acceptable level of compatibility with dry chemicals and might be suitable for use with those agents. (NFPA 412)

Fixed-gallonage fog nozzle A nozzle that delivers a set number of litres per minute (gallons per minute) as per the nozzle's design, no matter what pressure is applied to the nozzle.

Fluoroprotein foam A protein-based foam concentrate to which fluorochemical surfactants have been added. This has the effect of giving the foam a measurable degree of compatibility with dry chemical extinguishing agents and an increase in tolerance to contamination by fuel. (NFPA 402)

Foam concentrate A concentrated liquid foaming agent as received from the manufacturer. (NFPA 11)

Foam eductor A device placed in the hose line that draws foam concentrate from a container and introduces it into the fire stream.

Foam injector A device installed on a fire pump that meters out foam by pumping or injecting it into the fire stream.

Foam proportioner A device or method to add foam concentrate to water to make foam solution. (NFPA 1901)

Foam solution A homogeneous mixture of water and foam concentrate in the proper proportions. (NFPA 1901)

Fog-stream nozzle A nozzle that is placed at the end of a fire hose and separates water into fine droplets to aid in heat absorption.

Handline nozzle A nozzle with a rated discharge of less than 1591 L/min. (350 gpm). (NFPA 1964)

Low-volume nozzle A nozzle that flows 182 L/min. (40 gpm) or less.

Master stream device A large-capacity nozzle that can be supplied by two or more 64-mm (2.5-in.) hose lines or large-diameter hose or fixed piping. It can flow between 1591 and 6818 L/min and (350 and 1500 gpm). These devices include deck guns and portable ground monitors.

Master stream nozzle A nozzle with a rated discharge of 1591 L/min (350 gpm) or greater. (NFPA 1964)

Nozzle A constricting appliance attached to the end of a fire hose or monitor to increase the water velocity and form a stream. (NFPA 1965)

Nozzle shut-off A device that enables the fire fighter at the nozzle to start or stop the flow of water.

Piercing nozzle A nozzle that can be driven through sheet metal or other material to deliver a water stream to that area.

Premixed foam Solution produced by introducing a measured amount of foam concentrate into a given amount of water in a storage tank. (NFPA 11)

Protein foam A protein-based foam concentrate that is stabilized with metal salts to make a fire-resistant foam blanket. (NFPA 402)

Smooth-bore nozzle A nozzle that produces a straight stream that is a solid column of water.

Smooth-bore tip A nozzle device that is a smooth tube, which is used to deliver a solid column of water.

Sweep (or roll-on) method A method of applying foam that involves sweeping the stream just in front of the target.

Triple-layer load A hose loading method in which the hose is folded back onto itself to reduce the overall length to one-third before loading the hose in the bed. This load method reduces deployment distances.

Water curtain nozzle A nozzle used to deliver a flat screen of water that forms a protective sheet of water to protect exposures from fire.

You are on the scene of a fire involving a large, two-and-a-half-storey, wood-frame house. There is heavy fire coming from the second floor at the rear of the structure and thick, black, turbulent smoke pushing from the windows and eaves of the remainder of the second floor and the attic area. There are similar structures beside it, each only 4.6 m (15 ft.) away. The siding on one of them is already smoking. The resident has told you that everyone is out of the building. Your captain tells you to pull a preconnect to protect an exposure until additional units arrive.

1. What steps will you take to pull a minuteman load?

A. While facing the engine, pull the hose one-third of the way out, place the hose on your shoulder, turn, and then walk toward the proper location.

B. While facing away from the engine, pull the hose one-third of the way out, turn, place the hose on your shoulder, and then walk toward the proper location.

C. Place your arm though the larger loop, and then grasp the smaller loop with that hand. Pull the hose and walk toward the proper location.

D. Place your arm through the larger loop, and then grasp the smaller loop with the opposite hand. Pull the hose and walk toward the proper location.

2. Hose with a diameter of 64 or 76 mm (2.5 or 3 in.) is considered _____.

A. small diameter

B. medium diameter

C. intermediate diameter

D. large diameter

3. A nozzle that can deliver a wide range of water stream flows and that operates by means of an internal spring-loaded piston is called a/an _____.

A. adjustable-gallonage fog nozzle

B. Bresnan distributing nozzle

C. fog stream nozzle

D. automatic-adjusting fog nozzle

4. What is a device placed in the hose line that draws foam concentrate from a container and introduces it into the fire stream called?

A. Foam injector

B. Aerator

C. Foam eductor

D. Batch mixer

5. What is the method that applies the stream onto a nearby object, such as a wall, instead of directly aiming at the fire called?

A. Roll-on

B. Sweep

C. Rain-down

D. Bank-down

6. _____ foam increases the effectiveness of water as an extinguishing agent by reducing the surface tension of water.

A. Class A

B. Class B

C. Class C

D. Class D

You are dispatched to a motor vehicle accident in the downtown area involving a small passenger car and a tanker transporting alcohol. Passersby are reporting that everyone is out of the vehicles, but that the tanker rolled over the guardrail of the interstate, is on its side, and a liquid is running out of a hole in it. While you are en route, the dispatcher reports that the tanker is now on fire. Your engine has a Class A compressed air foam system as well as a Class B foam system with AFFF. Your captain says to pull a line to protect exposures.

1. How will you protect the exposed buildings?
2. Do you have the proper foam to extinguish the fire?
3. Will CAFs be of any value in this scenario?
4. What concerns do you have with this incident?

© MarcoCoda/iStockphoto

Fire Fighter Survival

Fire Fighter I

Knowledge Objectives

After studying this chapter, you will be able to:

- Describe how to apply a risk–benefit analysis to an emergency incident. (p 574–575)
- List the common hazard indicators that should alert fire fighters to a potentially life-threatening situation. (NFPA 5.3.5.A , p 575)
- List the 11 rules of engagement for fire fighter survival. (p 576)
- Explain how to maintain team integrity during emergency operations. (NFPA 5.3.5 , p 576–577)
- Define personnel accountability system. (NFPA 5.3.5.A , p 577–578)
- Describe the types of personnel accountability systems and how they function. (NFPA 5.3.5.A , p 578)
- Explain how a personnel accountability report (PAR) is taken. (p 578)
- Describe how to initiate emergency communications procedures. (NFPA 5.3.5.A , p 578–579)
- Describe the methods used for maintaining orientation. (p 581)
- Describe common self-rescue techniques. (NFPA 5.3.5.A , p 581–582)
- Describe how to find a safe location while awaiting rescue. (NFPA 5.3.5, 5.3.5.A , p 584–587)
- Describe air management procedures. (NFPA 5.3.1.B, 5.3.5.A, 5.3.5.B , p 587)

- Describe how rapid intervention crews (RICs) provide an air supply to a trapped fire fighter. (NFPA 5.3.9B , p 588)
- Explain the importance of the rehabilitation process. (p 589–590)
- List the common causes of critical incident stress. (p 592)
- Explain the goal of critical incident stress management. (p 590–592)

Skills Objectives

After studying this chapter, you will be able to:

- Initiate a mayday call. (NFPA 5.2.4, 5.2.4.A, 5.2.4.B , p 579–580)
- Perform a self-rescue using a hose line. (NFPA 5.3.5.B , p 582)
- Perform a self-rescue by locating a door or window for an emergency exit. (p 582–583)
- Perform a self-rescue by opening a wall to escape. (NFPA 5.3.1.B, 5.3.9B , p 584–585)
- Perform a self-rescue by removing an entanglement. (p 584, 586)
- Rescue a downed fire fighter using the fire fighter's self-contained breathing apparatus (SCBA) straps. (NFPA 5.3.9B , p 588–589)
- Rescue a downed fire fighter using a drag rescue harness. (NFPA 5.3.9B , p 588–590)
- Rescue a downed fire fighter as a two-person team. (NFPA 5.3.5B , p 588–589, 591)

Fire Fighter II — FFII

Knowledge Objectives

There are no knowledge objectives for Fire Fighter II candidates. NFPA 1001 contains no Fire Fighter II Job Performance Requirements for this chapter.

Skills Objectives

There are no skills objectives for Fire Fighter II candidates. NFPA 1001 contains no Fire Fighter II Job Performance Requirements for this chapter.

Additional NFPA Standards

- NFPA 1500, *Standard on Fire Department Occupational Safety and Health Program*

I t is the middle of August, and you are dispatched to a fire in a barn in a neighbouring jurisdiction. As you leave your station, you can see a large column of black smoke in the distance. Upon your arrival, you find that the barn is fully involved, the plastic on a recreational vehicle (RV) that is parked nearby is smoking heavily, and embers have ignited a ground fire across the road in grass dried out by the recent drought. The incident commander (IC) has assigned crews to protect the RV and other exposures, as well as crews to address the grass fire. A water shuttle has been established, so the IC directs your captain to take over personnel accountability and orders you to establish a rehabilitation (rehab) unit. You quickly search for the optimal location for the rehab unit and begin to pull the essential items together.

1. What should be considered when establishing a rehab unit?
2. Why is it important to establish rehab early in an incident such as this one?
3. Why should personnel be assigned to personnel accountability?

Introduction

Fire fighter survival is the most important outcome in any successful fire department operation. Fire fighters often perform their duties in environments that are inherently dangerous. Survival is a positive outcome that can be achieved only by working and training in a manner that recognizes risks and hazards and consistently applies safe operating policies and procedures. "Everybody goes home" should be the goal of every member of the team. This chapter introduces the actions, attitudes, and systems that are important in achieving that goal.

Fire fighters often encounter situations where their survival depends on making the right decisions and taking the appropriate actions. Many of the factors that cause fire fighter deaths and injuries appear again and again in postincident studies as risks that are often present at fires and other emergency incidents. Fire fighters must learn to recognize dangerous situations and to take actions to prevent more dangerous situations from occurring.

Risk–Benefit Analysis

A **risk–benefit analysis** approach to emergency operations can limit the risk of fire fighter deaths and injuries. This kind of analysis weighs the positive results that can be achieved against the probability and severity of potential negative consequences. A standard approach to risk–benefit analysis should be incorporated into the suggested operating guidelines (SOGs) or standard operating procedures (SOPs) for all fire departments. NFPA 1500, *Standard on Fire Department Occupational Safety and Health Program*, is an excellent reference guide to this approach as it applies to fire department operations.

In the fire service, risk–benefit analysis should be practised and implemented by everyone on the fire ground. The incident commander (IC) is responsible for the analysis of the overall scene, while company officers and fire fighters need to process the risks and benefits of the strategies implemented by command and be cognitive of changes on the fire ground.

The IC must always assess the risks and benefits before committing crews to the interior of a burning structure. If the potential benefits are marginal and the risks are too high, fire fighters should not be engaged to conduct an interior attack. The IC must also reassess the risks and benefits periodically during an operation. There should be no hesitation to terminate an interior attack and move to an exterior defensive operation at any time if the risks associated with an offensive operation change so that they outweigh the benefits of this strategy.

The risk–benefit method attempts to predict the potential risks that fire fighters might face when entering a burning structure and weighs them against the results that might be achieved by undertaking such a mission. If a building is known to be unoccupied and has no value, there is no justification for risking the lives of fire fighters to save it. Similarly, if a fire has already reached the stage where no occupants could survive and no property of value can be saved, there is no justification for risking the lives of fire fighters in an interior attack **FIGURE 18-1**.

FIGURE 18-1 If a fire has already reached the stage where no occupants could survive and no property of value can be saved, there is no justification for risking the lives of fire fighters in an interior attack.

© Condor 36/ShutterStock, Inc.

In a situation where no lives are at stake but there is a reasonable expectation that property could be saved, this policy allows for fire fighters to be committed to an interior attack. It recognizes that a standard set of hazards are anticipated in an interior fire attack situation, but that the combination of personal protective clothing and equipment, training, and SOGs/SOPs are designed to allow fire fighters to work safely in this type of environment. All standard approaches to operational safety must be followed without compromise. No property is ever worth the life of a fire fighter.

It is permissible to risk the life of a fire fighter only in situations where there is a reasonable and realistic possibility of saving a life. The determination that a risk is acceptable in a particular situation does not justify taking unsafe actions, however; it merely justifies taking actions that involve a higher level of risk.

Company officers and safety officers engage in risk–benefit analysis on an ongoing basis. If they perceive that the risk–benefit balance has shifted, it is their responsibility to report their observations and recommendations to the IC. The IC reevaluates the risk–benefit balance whenever conditions change and decides whether the strategy must be changed.

In addition, each individual fire fighter involved in the operation should conduct a risk–benefit analysis from his or her own perspective. A report from an observant fire fighter to his or her company officer can be crucial to the safety of an operation.

Hazard Indicators

As you progress in your career as a fire fighter, it will become evident that fire suppression and other types of emergency operations involve many different types of hazards. Fire fighters must be capable of working safely in an environment that includes a wide range of inherent hazards. While the danger of firefighting should never be taken for granted or perceived as routine, a fire fighter must learn to routinely follow safe SOGs/SOPs. Some hazards are encountered at almost every incident, whereas others are rare occurrences. A fire fighter should recognize the various types of hazardous conditions and react to them appropriately.

An example of a commonly encountered hazard would be the presence of smoke inside a structure. Fires produce smoke, and breathing smoke is well recognized as dangerous. The proper response to this hazard is to always wear appropriate personal protective equipment (PPE) for the hazard at hand. At the same time, you should recognize that turbulent smoke and black fire indicate that the atmosphere may soon prove deadly even for fire fighters who are in complete PPE, including self-contained breathing apparatus (SCBA). (The chapter "Fire Behaviour" has more information on smoke reading.) This is a situation when hazards are recognized and standard solutions are applied.

Hazardous conditions may or may not be evident by simple observation; for example, smoke and flames are usually visible, whereas hazards related to the construction details of a building might not be so easily detectable. You should be aware of the indicators of hazards that present themselves in a less obvious manner. Making this kind of mental connection requires knowledge of what can be hazardous, given a particular set of circumstances. Hazard recognition becomes easier through study and experience:

- Building construction—Knowledge of building construction can help in anticipating fire behaviour in different types of buildings and recognizing the potential for structural collapse. Is the building designed to withstand a fire for some period of time, or is it likely to weaken quickly? Has the building been modified from its original use? Does the structure incorporate void spaces that would allow a fire to spread quickly? Is it constructed using lightweight construction techniques? Does the building incorporate truss roofs or floors or lightweight construction that could fail in a short period of time?
- Weather conditions—Inclement weather can turn a routine incident into a hazardous one. Rain and snow make operations on a roof even more dangerous than usual. Raising a ladder close to power lines during windy weather involves an increased level of risk. Wind-driven fires can rapidly change direction, or wind can rapidly increase the intensity of a fire, often with deadly results.
- Occupancy—A name like *Central Plating Company* on a building should cause fire fighters to anticipate that hazardous materials could be present. A warning placard outside would be a more specific hazardous materials indicator **FIGURE 18-2**. Many common commercial buildings such as hardware stores and home improvement centres contain huge amounts of highly flammable and hazardous materials.

These are just a few examples of observable factors that might indicate a hazard. At each and every incident, fire fighters should look for what could be critical indications of a hazard.

FIGURE 18-2 The NFPA 704 diamond indicates that hazardous materials are present.

Safe Operating Procedures

SOGs/SOPs define the manner in which a fire department conducts operations at an emergency incident. Many of these procedures were developed to protect fire fighters' health and safety.

SOGs/SOPs must be learned and practised before they can be implemented. Training is the only way for individuals to become proficient at performing any set of skills. To be effective, the fire fighter must apply the same skills routinely and consistently at every training session and at every emergency incident. When under great pressure, most people will revert to habit; thus, when self-survival is at stake, well-practiced habits can be lifesavers.

■ Rules of Engagement for Fire Fighter Survival

To help define safe practices of engaging structural firefighting, the International Association of Fire Chiefs (IAFC) and National Volunteer Fire Council (NVFC) developed rules of engagement for fire fighter survival. These 11 rules are summarized here:

1. **Size up your tactical area of operation**. Fire fighters and company officers need to pause for a moment and look over their area of responsibility. During this size-up, you need to evaluate your individual risk exposure and determine a safe approach to complete your assigned objectives.
2. **Determine the occupant survival profile**. Fire fighters and company officers need to consider the fire conditions to determine where an occupant could be alive in the conditions that present themselves. They must continue an ongoing assessment of these conditions.
3. *Do not* **risk your life for lives or property that cannot be saved**. Do not engage in high-risk search and rescue and firefighting activities when the fire conditions prevent occupant survival or if the property destruction is inevitable.
4. **Extend** *limited* **risk to protect** *savable* **property**. Limit your risk to a reasonable, cautious, and conservative level when trying to save a building.
5. **Extend** *vigilant* **and** *measured* **risk to protect and rescue** *savable* **lives**. Engage in search and rescue and firefighting operations in a calculated, controlled, and safe manner, while remaining alert to changing conditions during high-risk search and rescue operations, only in conditions where lives can be saved.
6. **Go in together, stay together, and come out together**. Ensure that you and your fellow fire fighters always enter burning buildings as a team of two or more and that no fire fighter is allowed to be alone at any time while entering, operating in, or exiting a hazardous building.
7. **Maintain continuous awareness of your air supply, situation, location, and fire conditions**. Maintain constant situational awareness of your SCBA air supply, your location in the building, and all that is happening in your area of operation and elsewhere on the fire ground that may affect your risk and safety. Monitor your air supply and ensure that you are outside the hazard zone before your low-air alarm activates.
8. **Constantly monitor fire-ground communications for critical radio reports**. Maintain constant awareness of all fire-ground radio communications on their assigned channels for progress reports, critical messages, and other information that may affect your risk and safety.
9. **Report unsafe practices or conditions that could harm you. Stop, evaluate, and decide**. Report unsafe practices and exposure to unsafe conditions to your supervisor. There should be no penalty for this reporting.
10. **Abandon your position and retreat before deteriorating conditions can harm you**. Be aware of unsafe fire conditions and exit immediately to a safe area when you are exposed to deteriorating conditions, unacceptable risk, or a life-threatening condition.
11. **Declare a mayday as soon as you *think* you are in danger**. Declare a mayday as soon as you are faced with a life-threatening situation; declare a mayday as soon as you think you are in trouble. Failure to declare a mayday as soon as possible puts your life and the lives of fellow fire fighters in even more danger.

■ Team Integrity

Teamwork is an essential requirement in firefighting **FIGURE 18-3**. A team is defined as two or more individuals attempting to achieve a common goal. Given the intense physical labour necessary to pull hose or raise ladders, team efforts are imperative in firefighting. The most important reason for maintaining team integrity in an emergency operation is safety: Team members keep track of one another and provide immediate assistance to other team members when needed. The standard operational team in firefighting operations is a company, usually consisting of three to five fire fighters working under the direct supervision of a company officer. Career fire fighters may arrive as a company on the same apparatus, whereas on-call or volunteer fire fighters should assemble in companies upon arrival.

Individual companies are assigned as a group to perform tasks at the scene of an emergency incident. The IC keeps track of the companies involved in the response and communicates with company officers. Every company officer must have a

FIGURE 18-3 A team attempts to achieve a common goal.

Near-Miss REPORT

Report Number: 07-0000726

Synopsis: Truss roof collapse causes near miss at church fire.

Event Description: We were involved in fire-suppression activities on a working attic fire at a church. Company crew members were attempting to perform a trench cut to ventilate and confine the fire to an area adjacent to the fire wall. Upon laddering the roof, and while crew members were still operating on and from the aerial apparatus, a sudden collapse of the truss roof occurred, endangering the crew due to fire and radiant heat. The collapse involved approximately 12.2 metres (40 feet) of the roof line, and a fireball extended upward approximately 9 m (30 ft.) in the air. The ventilation crew had to make an immediate descent down the aerial apparatus to escape harm. The collapse occurred about 10 minutes after arrival of the first-in companies. All fire fighters involved were in full PPE and SCBA. Awareness of wood truss construction must always be considered during fire-suppression activities.

Lessons Learned: Fire-ground accountability, awareness of wood truss construction, the fire's impact due to direct impingement, and the duration of the fire must always be considered during fire-suppression activities. Had the crew actually been actively ventilating or on the roof, there would have been serious injuries and most likely fire fighter fatalities.

Awareness training for wood truss construction should be included in any ventilation and structural fire-suppression training. Command and control, fire-ground accountability, and fire-ground safety officer training should also be included as part of the training activities.

We must always keep in mind that the fire ground is a dynamic, ever-changing scene, and we must adapt to those conditions and stay aware. Remember: No structure is worth our lives. Everyone goes home.

---

portable radio to maintain contact with the IC. The company officer is responsible for knowing the location of every company member at all times; he or she should also know exactly what each individual fire fighter is doing at all times.

Team integrity means that a company arrives at a fire together, works together, and leaves together. The members of a company should always be oriented to one another's location, activities, and condition. In a hazardous atmosphere, such as inside a structure, a fire fighter should always be able to contact his or her company members by voice, sight, or touch. When air cylinders need to be replaced, all of the company members should leave the building together, change cylinders together, and return to work together. If any members require rehabilitation, the full company should go to rehabilitation together and return to action together.

In some cases, a company will be divided into teams of two to perform specific tasks; for example, a ladder company might be split into a roof team and an interior search team. The initial attack on a room-and-contents fire in a single-family dwelling might be made by two fire fighters, while two other fire fighters remain outside, following the two-in–two-out rule. For safety and survival reasons, fire fighters should always use a buddy system, working in teams of at least two. If one team member should become incapacitated, his or her partner can provide help immediately and call for assistance. The two team members must remain in direct contact with each other. At least one member of every team must have a portable radio to maintain contact with the IC; in many fire departments, every member carries a portable radio.

■ Personnel Accountability System

A personnel accountability system provides a systematic way to keep track of the locations and functions of all personnel operating at the scene of an incident **FIGURE 18-4**. The IC has the responsibility to account for each fire fighter involved in an emergency incident. A personnel accountability system SOG/SOP should

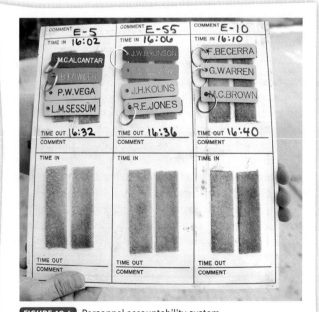

FIGURE 18-4 Personnel accountability system.

be designed to establish standard procedures for performing this function and managing the necessary information.

A personnel accountability system must keep track of every company or team of fire fighters working at the scene of the incident—from the time they arrive until the time they are released. In addition, it must be able to identify each individual fire fighter who is assigned to a particular company or team. The system assumes that all of the company members will be together—that is, working as a unit under the supervision of the company officer—unless the IC has been advised of a change. At any point in the operation, the IC should be able to identify the location and function of each company and, within seconds, determine which individual fire fighters are assigned to that company.

The personnel accountability systems used by fire departments take many different forms. Whatever type of system is used, the only way for it to be truly effective is if every fire fighter takes responsibility for its use at all incidents. This rule includes even the smallest of incidents. Some fire departments start with a written roster of the team members assigned to each company; others use a computer database tied to a daily staffing program. Lists, roll calls, or electronic tracking databases should be used to account for all personnel at the beginning of every tour of duty or when changes are made to assignments throughout the tour.

Each piece of apparatus should carry a list of assigned company members. One popular personnel accountability system uses a Velcro® or magnetic name tag for each company member. The tags for all members assigned to the apparatus are affixed to a special board carried in the cab. This board, which is sometimes called a passport, should be kept up to date throughout the tour of duty and should denote any change in the assigned team. In the case of a volunteer company, the tags can be affixed to the board at the time of an alarm.

At the scene of an incident, these boards are physically left with a designated person at the command post or at the entry point to a hazardous area. Depositing the board and the tags indicates that the company and those individual fire fighters have entered the hazardous area. Later, when the company leaves the hazardous area or is released from the scene, the company officer must retrieve the board—an action that indicates the entire company has safely left the hazardous area.

A **personnel accountability report (PAR)** is a roll call taken by each supervisor at an emergency incident. When the IC requests a PAR, each company officer physically verifies that all assigned members are present and confirms this information to the IC. The IC should also request a PAR at tactical benchmarks, such as when a change from an offensive strategy to a defensive strategy is made.

The company officer must be in visual or physical contact with all company members to verify their status. The company officer then communicates with the IC by radio to report a PAR. Whenever a fire fighter cannot be accounted for, he or she is considered missing until proven otherwise. A report of a missing fire fighter always becomes the highest priority at the incident scene.

If unusual or unplanned events occur at an incident, a PAR should always be performed; for example, if there is a report of an explosion, a structural collapse, or a fire fighter missing or in need of assistance, a PAR should immediately take place so that command can determine if any personnel are missing, identify any missing individuals, and establish their last known locations and assignments. These locations would then serve as the starting points for any search and rescue crews.

■ Emergency Communications Procedures

Communication is a critical part of any emergency operation, but it is especially important in relation to fire fighter safety and survival **FIGURE 18-5**. Breakdowns in the communication process are a major contributor to fire fighter injuries and deaths. Conversely, good communication can save the lives of fire fighters in a dangerous situation.

The chapter "Fire Service Communications" discussed the proper radio communications procedures. The following is a model of good radio communication:

"Central IC, this is ladder 203."
"Ladder 203, this is Central IC, go ahead."
"Ladder 203 has completed roof ventilation and is available for another assignment."
"IC copies. Roof ventilation completed, and Ladder 203 available for reassignment."

Such standard communications procedures ensure that the message is clearly stated and then repeated back as confirmation that it has been received and understood. Note that

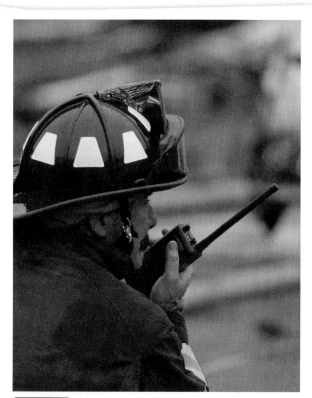

FIGURE 18-5 Communication is a critical part of any emergency operation.

© Mark C. Ide

some fire departments might open radio communications differently; for example, "Central IC, this is Ladder 203." Consult your department's SOGs/SOPs.

When a fire fighter is missing or in trouble, or when an imminent safety hazard is detected, it is even more important to communicate clearly and to ensure that the message is understood. For this reason, much of fire fighters' communications have been standardized. There are standard methods to transmit emergency information, standard information items to convey, and standard responses to emergency messages. Reserved phrases, sounds, and signals for emergency messages should be a part of every fire department's SOGs/SOPs. Likewise, a separate set of standard phrases should be used for conveying information about emergency situations and hazardous conditions. These phrases should be known and practised by everyone in the department. The same procedures and terminology should be used by all of the fire departments that can be expected to work together at emergency incidents. In many areas, these procedures are coordinated regionally.

The word *mayday* is used to indicate that a fire fighter is in trouble and requires immediate assistance. It could be used if a fire fighter is lost, is trapped, has an SCBA failure, or is running out of air. A fire fighter can transmit a mayday message to request assistance, or another fire fighter could transmit a mayday message to report that a fire fighter is missing or in trouble. A mayday call takes precedence over all other radio communications.

The term *emergency traffic* is used to indicate an imminent fire-ground hazard, such as a potential explosion or structural collapse. This kind of message would also be used to order fire fighters to immediately withdraw from interior offensive attack positions and switch to a defensive strategy. An emergency traffic message takes precedence over all other radio communications with the exception of a mayday call. Any use of the term *mayday* should bring an immediate response. The IC will interrupt any other communications and direct the person reporting the mayday to proceed. That person will then state the emergency, including all pertinent information, as in the following example:

> "This is Fire Fighter Jones with Engine 212. I'm running low on air and I'm separated from my crew. I'm on the third floor and I believe that I'm in the C side of the building."

The IC will repeat this information back and then initiate procedures to rescue Fire Fighter Jones. The next actions could include committing the rapid intervention crew (RIC) to the rescue effort, calling for additional resources, and redirecting other teams to support a search and rescue operation.

In many fire departments, the communications centre sounds a special tone over the radio to alert all members that an emergency situation is in progress. It then repeats the information to be certain that everyone at the incident scene heard the message correctly. Some departments sound a horn in such circumstances.

Similar procedures are followed for emergency traffic situations. All imminent hazards and emergency instructions should be announced in a manner that captures the attention of everyone at the incident scene and ensures that the message is clearly understood.

Initiating a Mayday

A mayday message, as described previously, is used to indicate that a fire fighter is in trouble and needs immediate assistance. Analysis of fire fighter fatalities and serious injuries has shown that fire fighters often wait too long before calling for assistance. Instead of initiating a mayday call when they first get into trouble, all too often fire fighters wait until the situation is absolutely critical before requesting help. This delay could reflect a fear of embarrassment if the situation turns out to be less severe than anticipated, combined with a hope that the fire fighter will be able to resolve the problem without assistance.

The failure to act promptly can be fatal in many situations. Systems have been designed to do everything possible to protect fire fighters and to rescue fire fighters from dangerous situations—but those systems are effective only when fire fighters act appropriately when they find themselves in trouble. Do not hesitate to call for help when you think you need it.

To initiate a mayday call, follow your fire department's SOGs/SOPs. In most cases, this will involve transmitting the message "Mayday, mayday, mayday" over the radio to initiate the process. When this transmission is acknowledged by the IC, give a LUNAR (location, unit number, name, assignment, and resources needed) report. Manually activate your personal alert safety system (PASS) device so that other fire fighters will hear the signal and come to your location. Also activate the emergency alert button on your portable radio, if it has this feature. All of these actions will increase the probability that the RIC or other fire fighters will be able to locate you. It is also important to remember that you should continue to work to free yourself from the dangerous situation.

To initiate a mayday call, follow the steps in **SKILL DRILL 18-1**:

1. Use your radio to call, "Mayday, mayday, mayday."
2. Give a LUNAR report: (**STEP ❶**)
   A. Location
   B. Unit number
   C. Name
   D. Assignment
   E. Resources needed
3. Activate your PASS device.
4. Attempt self-rescue.
5. If you are able to move, identify a safe location where you can await rescue. (**STEP ❷**)
6. Lie on your side in a fetal position with your PASS device pointing out so that it can be heard. (**STEP ❸**)
7. Point your flashlight toward the ceiling.
8. Slow your breathing as much as possible to conserve your air supply. (**STEP ❹**)

■ Rapid Intervention Crews

RICs are established for the sole purpose of rescuing fire fighters who are operating at emergency incidents. In some fire departments, the term *rapid intervention team* or *fire fighter assistance and safety team* is used to describe this assignment. An RIC is a company or crew that is assigned to stand by at the incident scene, fully dressed in PPE and SCBA and equipped for action,

SKILL DRILL 18-1 Initiating a Mayday Call
(Fire Fighter I, NFPA 5.2.4)

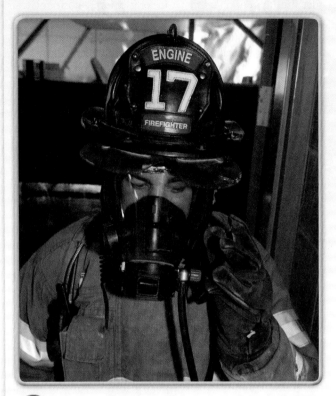

1 Use your radio to call, "Mayday, mayday, mayday."
Give a LUNAR report: your location, unit number,
name, assignment, and resources needed.

2 Activate your PASS device. Attempt self-rescue. If you
are able to move, identify a safe location where you
can await rescue.

3 Lie on your side in a fetal position with your PASS
device pointing out so it can be heard.

4 Point your flashlight toward the ceiling. Slow your
breathing as much as possible to conserve your air
supply.

© LiquidLibrary

and that is ready to deploy immediately when assigned to do
so by the IC.

An RIC is an extension of the two-in–two-out rule.
During the early stages of an interior fire attack, a minimum
of two fire fighters are required to establish an entry team,
and a minimum of two additional fire fighters are required
to remain outside the hazardous area. The two fire fighters
who remain outside the hazardous area can perform other

functions, but they must have donned full PPE and SCBA, so that they can immediately enter and assist the entry team if the interior fire fighters need to be rescued. The two fire fighters who remain outside the hazardous area fulfill the first stage of an RIC procedure. The second stage involves a dedicated RIC specifically assigned to stand by for fire fighter rescue assignments.

NFPA 1500, *Standard on Fire Department Occupational Safety and Health Program*, includes a complete description of RIC procedures and provides examples of their use at typical scenes. A number of options are available for establishing and organizing RIC crews or teams, including dispatching one or more additional companies to incidents with this specific assignment. Some small fire departments rely on an automatic response from an adjoining community to provide a dedicated RIC at working incidents.

SOGs/SOPs should dictate when a RIC is required, how it is assigned, and where it should be positioned at an incident scene. A RIC should be in place at any incident where fire fighters are operating in conditions that are immediately dangerous to life and health (IDLH), which includes any structure fire where fire fighters are operating inside a building and using SCBA. Many fire departments specify a special set of rescue equipment, including extra SCBA air cylinders, to be brought to the standby location. All RIC members must wear full PPE and SCBA and be ready for action at an instant.

The IC should quickly deploy the RIC to any situation where a fire fighter needs immediate assistance. Situations that call for the use of the RIC would include a lost or missing fire fighter, an injured fire fighter who has to be removed from a hazardous location, and a fire fighter trapped by a structural collapse.

Fire Fighter Survival Procedures

Some of the most important procedures you need to learn are directly related to your personal safety and to the safety of the fire fighters who will be working with you. These procedures are designed to keep you from entering into dangerous situations and to guide you in situations where you could be in immediate danger. Your personal safety will depend on learning, practicing, and consistently following these procedures; however, do not lose sight of the fact that preventing life-threatening situations is preferable to utilizing survival procedures.

■ Maintaining Orientation

As you practise working with SCBA, you will discover that it is very easy to become disoriented in a dark, smoke-filled building. Sometimes the visibility inside a burning building can be measured in centimetres. In these conditions, it is extremely important to stay oriented. If you become lost, you could run out of air before you find your way out. The atmosphere outside your face piece could be fatal in one or two breaths. Even if you call for assistance, rescuers might not be able to find you. For all these reasons, safety and survival inside a fire building are directly related to remaining oriented within the building.

Several methods can help fire fighters stay oriented inside a smoke-filled building. Before entering a building, look at it from the outside to get an idea of the size, shape, arrangement, and number of storeys. After entering the structure, paint a picture in your mind of your surroundings. Given that most of your senses will be disrupted by dense smoke, touch may become your only means of moving through a structure. By feeling out your surroundings, you will be able to tell which type of room you have entered. Feel for changes in floor coverings, and always look for alternative means of egress.

Preplanning helps with orientation. By reviewing the preincident plan before entering the structure you will have a general idea of the structure's floor plan. See the chapter "Preincident Planning" for more information.

One of the most basic methods for remaining oriented is to always stay in contact with a hose line—for example, by always keeping a hand or foot on the hose line. To find your way out, feel for the couplings. Place one hand on the male coupling and the other hand on the female coupling. To exit the building, travel in the direction of the hand holding the male coupling.

Team integrity is an important factor in maintaining orientation. Everyone works as a team to stay oriented. When the team members cannot see one another, they must stay in direct physical contact or within the limits of verbal contact. Standard procedures are employed to ensure that fire fighters work efficiently as a team, such as having one member remain at the doorway of the room while the other members perform a systematic search inside. The searchers then work their way around the room in one direction until they get back to the fire fighter at the door.

A **guide line** can be used for orientation when inside a structure. A guide line is a rope that is attached to an object on the exterior or a known fixed location; it is stretched out as a team enters the structure. The fire fighters can follow the guide line back out, or another team can follow the guide line in to find them. Some fire departments use a series of knots to mark distances along the rope to indicate how far it is stretched. Be aware that using the guide line technique requires intense practice.

Training sessions in conditions with no visibility will build confidence. This type of training can be conducted inside any building while using an SCBA face piece with the lens covered. More realistic conditions can be simulated by using smoke machines inside training facilities. Fire fighters should practise navigating around furniture, following charged hose lines, climbing and descending stairs, and fitting through narrow spaces. Distracting noises, such as operating fans, can add a realistic feel. Team integrity must always be maintained in these practise sessions.

Self-Rescue

If you become separated from your team, or if you become lost, disoriented, or trapped inside a structure, you can use several techniques for **self-rescue**. The first step is always to call for assistance. Do not wait until it is too late for anyone to find you before making it known that you are in trouble. As described in the section "Initiating a Mayday," time is critical when your safety is at risk. You need to initiate the process as soon as you *think* you are in trouble, not when you are absolutely *sure* you are in grave danger.

If you are simply separated from your team, the best option is to take a normal route of egress to the exterior. Follow a hose line back to an open doorway, descend a ladder, or locate a window or door—all of these options can provide a direct exit and require no special tools or techniques. You must immediately communicate with your company officer or the IC to inform them that you have exited safely.

To rescue yourself by locating and following a hose line to an exit, follow the steps in **SKILL DRILL 18-2** :

1 Initiate a mayday.

2 Stay calm and control your breathing. (**STEP 1**)

3 Systematically search the room to locate a hose line.

4 Follow the hose line to a hose coupling.

5 Identify the male and female ends of the coupling.

6 Move from the female coupling to the male coupling. (**STEP 2**)

7 Follow the hose out.

8 Exit the hazard area.

9 Notify command of your location. (**STEP 3**)

To locate a door or window that can be used for an exit, follow the steps in **SKILL DRILL 18-3** :

1 Initiate a mayday.

2 Stay calm and control your breathing. (**STEP 1**)

3 Systematically locate a wall. (**STEP 2**)

4 Use a sweeping motion on the wall to locate an alternative exit.

5 Identify the opening as a window, interior door, or external door. (**STEP 3**)

6 If the first opening identified is not adequate for an exit, continue to search.

7 Maintain your orientation and stay low.

8 Exit the room safely, if possible.

9 If unable to exit, assume the downed fire fighter position in a safe location or find refuge.

10 Keep command informed of your situation. (**STEP 4**)

In addition to these actions, fire fighters can use many more complicated techniques to escape from dangerous predicaments, such as breaching a wall. These methods must be learned and practised regularly so that you will be ready to

---

SKILL DRILL 18-2

Performing Self-Rescue
(Fire Fighter I, NFPA 5.3.5)

1 Initiate a mayday. Stay calm and control your breathing.

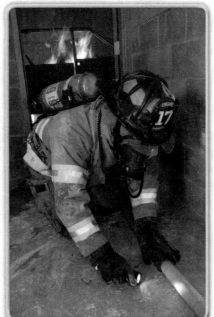

2 Systematically search the room to locate a hose line. Follow the hose line to a hose coupling. Identify the male and female ends of the coupling. Move from the female coupling to the male coupling.

3 Follow the hose out. Exit the hazard area. Notify command of your location.

© LiquidLibrary

SKILL DRILL 18-3   Locating a Door or Window for Emergency Exit
(Fire Fighter I, NFPA 5.3.5)

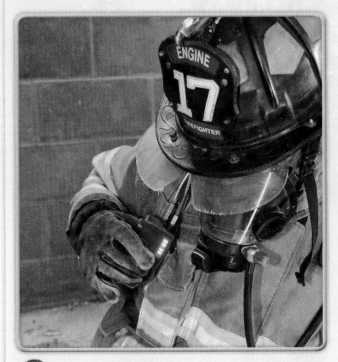

1 Initiate a mayday. Stay calm and control your breathing.

2 Systematically locate a wall.

3 Use a sweeping motion on the wall to locate an alternative exit. Identify the opening as a window, interior door, or external door.

4 If the first opening identified is not adequate for an exit, continue to search. Maintain your orientation and stay low. Exit the room safely, if possible. If unable to exit, assume the downed fire fighter position in a safe location or find refuge. Keep command informed of your situation.

© LiquidLibrary

perform them without hesitation if you are ever in a situation that requires them.

To escape through a wall that is constructed with wood or metal studs, you need to expose the studs and then reduce the size of your profile to get your body and your SCBA through the narrow opening. Sometimes you can get through a narrow opening by loosening your waist strap and one strap of your SCBA and then repositioning your SCBA as you twist through the limited opening. At other times you may have to remove your SCBA backpack while keeping your face piece on. You will then have to push your SCBA in front of you. Hold on to your backpack carefully to prevent it from becoming separated from your face piece. The technique you need to use will depend on your size. A small person may be able to squeeze through a limited-size opening without removing SCBA; a larger person trying to escape through the same small opening may have to remove the SCBA backpack to fit through it.

One method of opening a wall and using a reduced profile to escape through a wall is shown in **SKILL DRILL 18-4** :

1. Identify deteriorating conditions that require exiting through a wall.
2. Initiate a mayday.
3. Use a hand tool or your feet to open a hole in the wall between two studs.
4. If using a hand tool, drive the tool completely through the wall to check for obstacles on the other side. (**STEP 1**)
5. Enlarge the hole.
6. Enter the hole head first to check the floor and fire conditions on the other side of the wall. (**STEP 2**)
7. Loosen the SCBA waist strap and remove one shoulder strap.
8. Sling the SCBA to one side to reduce your profile. (**STEP 3**)
9. Escape through the opening in the wall. Adjust the SCBA straps to their normal position. (**STEP 4**)
10. Assist others through the opening in the wall.
11. Report your status to command. (**STEP 5**)

Disentanglement is an important skill that fire fighters need to learn and practise. When crawling in a smoke-filled building, a fire fighter can easily become entangled in wires, cables, and debris. Above most dropped ceilings is a huge and varied mass of electrical wires and conduit, computer and TV cables, plumbing and sewage pipes, telephone cables, and heating and air-conditioning tubing and ducts. Because many dropped ceilings are held up only with wires, it does not take much fire or water damage to bring this huge tangle of utilities crashing down. If these wires fall on your back while you are wearing an SCBA, you could be easily trapped.

Many fire fighters carry small tools in the pockets of their PPE that can be used to cut through wires or small cables. This procedure can be very difficult if poor-visibility conditions prevent the fire fighter from seeing and identifying the entangling material.

To escape from an entanglement, follow the steps in **SKILL DRILL 18-5** :

1. Initiate a mayday.
2. Stay calm and control your breathing. (**STEP 1**)
3. Change your position—back up and turn on your side to try to free yourself. (**STEP 2**)
4. Use the swimmer stroke to try to free yourself. (**STEP 3**)
5. Loosen the SCBA straps, remove one arm, and slide the air pack to the front of your body to try to free the SCBA. (**STEP 4**)
6. Cut the wires or cables causing the entanglement. Be aware of any possible electrocution risk.
7. If you are unable to disentangle yourself, notify command of your situation.
8. If you are able to exit, notify command that you are out of danger. (**STEP 5**)

Some additional self-rescue methods involve using tools and equipment in ways that they were not designed for. These so-called last-resort methods should be taught only by experienced and qualified instructors and must be practised with strict safety measures in place. Some of these methods are quite controversial. Some fire departments consider them unacceptable, owing to the high risk of injury when they are used or even practised; other fire departments provide training in them, because these measures could save the life of a fire fighter in an extreme situation. This policy decision must be made by each fire department.

Safe Locations

A **safe location** is a temporary location that provides refuge while awaiting rescue or finding a method of self-rescue from an extremely hazardous situation. The term *safe* in this case is relative: the location may not be particularly safe, but it is less dangerous than the alternatives. Wildland fire fighters carry lightweight reflective emergency shelters as a form of safe location. When overrun by a fire, taking refuge in the shelter might be the fire fighter's only chance for survival.

Safe locations are important when situations become critical and few alternatives are available. When a critical situation occurs, fire fighters should know where to look for a safe location and recognize one when it presents itself.

If fire fighters become trapped inside a burning building, a room with a door and a window could be a safe location. Closing the door could keep the fire out of the room for at least a few minutes, and breaking the window could provide fresh air and an escape route. In this case, the safe location ideally provides time for a rescue team to reach the fire fighters, a ladder to be raised to the window, or another group of fire fighters to control the fire.

A roof or floor collapse often leaves a void adjacent to an exterior wall. If trapped fire fighters can reach this void, it could be the best place to go. Breaching the wall might be the only way out of the void or the only way for a rescue team to reach the trapped personnel.

SKILL DRILL 18-4

Opening a Wall to Escape
(Fire Fighter I, NFPA 5.3.5)

1 Identify deteriorating conditions that require exiting through a wall. Initiate a mayday. Use a hand tool or your feet to open a hole in the wall between two studs. If using a hand tool, drive the tool completely through the wall to check for obstacles on the other side.

2 Enlarge the hole. Enter the hole head first to check the floor and fire conditions on the other side of the wall.

3 Loosen the SCBA waist strap and remove one shoulder strap. Sling the SCBA to one side to reduce your profile.

4 Escape through the opening in the wall. Adjust the SCBA straps to their normal position.

5 Assist others through the opening in the wall. Report your status to command.

SKILL DRILL 18-5

Escaping from an Entanglement
(Fire Fighter I, NFPA 5.3.5)

1 Initiate a mayday. Stay calm and control your breathing.

2 Change your position—back up and turn on your side to try to free yourself.

3 Use the swimmer stroke to try to free yourself.

4 Loosen the SCBA straps, remove one arm, and slide the air pack to the front of your body to try to free the SCBA.

5 Cut the wires or cables causing the entanglement. Be aware of any possible electrocution risk. If you are unable to disentangle yourself, notify command of your situation. If you are able to exit, notify command that you are out of danger.

Maintaining team integrity is important when escaping to a safe location. In some cases, fire fighters have survived in safe locations long enough to work together to create an opening. The ability to use a radio to call for help and to describe the location where fire fighters are trapped can be critical.

Like all of the emergency procedures described in this chapter, these activities require good instruction and practise. Always follow your department's operating guidelines when it comes to self-rescue techniques. Follow the SOGs/SOPs used by your department for these evolutions.

Air Management

Air management is important to all fire fighters when using SCBA. Air management reflects a basic fact: air equals time. How long a fire fighter can survive in a dangerous atmosphere depends on how much air is in the SCBA and how quickly the fire fighter uses it. You have to manage your air supply to manage your time in a hazardous area.

The time in the hazardous atmosphere must take into account the time it takes to enter and get to the location where the fire fighter plans to operate and the time it takes to exit safely. The work time is limited by the amount of air in the SCBA cylinder, after allowing for entry and exit time. If it will take 6 minutes to reach a safe atmosphere and the fire fighter has only 3 minutes of air left, the results could be disastrous.

Recall that the time rating for an SCBA is based on a standard rate of consumption for a typical adult under low-exertion conditions in a test laboratory. Fire fighters typically consume air much more rapidly than occurs in the standard test. In fact, fire fighters often consume a 45-minute-rated air supply in just 20–25 minutes.

The actual rate of air consumption varies significantly among individual fire fighters and depends on the activities that they are performing; chopping holes with an axe or pulling ceilings with a pike pole, for example, will cause a fire fighter to consume air more rapidly than will conducting a survey with a thermal image camera. Even given the same workload, no two individuals will use exactly the same amount of air from their SCBA cylinders. Differences in body size and fitness levels, as well as heredity factors, cause air consumption rates to be faster or slower than the average.

Air management must be a team effort as well as an individual responsibility. Fire fighters will always be working as part of a team when using SCBA, and team members inevitably use air at different rates. The member who uses the air supply most rapidly determines the working time for the team.

A good way to determine your personal air usage rate is to participate in an SCBA consumption exercise. This drill puts fire fighters through a series of exercises and obstacles to learn how quickly each fire fighter uses air when performing different types of activities. Repeated practise can help you learn to use air more efficiently by building your confidence and learning to control your breathing rate. Keeping physically fit results in a more efficient use of oxygen and allows you to work for longer periods with a limited air supply.

Practise sessions should take place in a nonhazardous atmosphere. A typical skill-building exercise might include the following components:

- While wearing full PPE and breathing air from SCBA, perform typical activities such as carrying hose packs up stairs, hitting an object with an axe or sledgehammer, or crawling along a hallway.
- Record the following information: how much air is in the cylinder when you start, how long you work, whether the physical labour is light or heavy, when the low-air alarm sounds, and when the cylinder is completely empty.

This type of exercise will allow you to determine your personal rate of consumption and work on improving your times. With practise, you will be able to predict how long your air supply should last under a given set of conditions.

Knowledge of your team members' physical conditions and their workloads can help you look out for their safety as well. Because many tasks are divided up in any given operation, one member may be working to his or her maximum level while another team member is acting only in a supportive role. This difference in exertion levels could cause the first team member to use up his or her air supply much faster without realizing it.

When using SCBA at an emergency incident, be aware of the limitations of the device. Do not enter a hazardous area unless your air cylinder is full. Keep track of your air supply by using the heads up display (HUD) to determine how much air you have left. Always follow the rule of air management: know how much air you have in your SCBA before you go in, and manage that amount so that you leave the hazardous environment *before* your SCBA low-air alarm activates. Do not wait until the low-pressure alarm sounds to start thinking about leaving the hazardous area; you should be out of the hazardous area before this alarm sounds.

Fire Fighter Safety Tips
© Photos.com

Follow the rule of air management: know how much air you have in your SCBA before you go in, and manage that amount so that you leave the hazardous environment *before* your SCBA low-air alarm activates.

Even when fire fighters employ the best possible air management practices, emergency situations can occur. An SCBA can malfunction or fire fighters may be trapped inside a building by a structural collapse or an unexpected situation. In these circumstances, remaining calm and knowing how to use all of the emergency features of the SCBA can be critically important. Remaining calm, controlling your breathing rate, and taking shallow breaths can slow air consumption. All of these techniques require practise to build confidence.

Fire Fighter Safety Tips

Before entering a hazard zone:
- Ensure that all crew members have portable radios and appropriate hand tools.
- Keep crews together.
- Work under the Incident Management System (IMS).
- Follow guidelines for managing your air supply.

Rescuing a Downed Fire Fighter

One of the most critical and demanding situations in the fire service arises when fire fighters have to rescue another fire fighter who is in trouble. Air management has critical implications in a fire fighter rescue situation. Specifically, air management must be considered for both the rescuers and the fire fighter who is in trouble.

When you reach a downed fire fighter, the first step is to assess the fire fighter's condition. Is the fire fighter conscious and breathing? Does he or she have a pulse? Is the fire fighter trapped or injured? Make a rapid assessment and notify the IC of your situation and location. Have the RIC deployed to your location and quickly determine the additional resources needed.

The most critical decision at this point is whether the fire fighter can be moved out of the hazardous area quickly and easily or whether considerable time and effort will be required to bring the fire fighter to safety. If it will take more than a minute or two to remove the fire fighter from the building, air supply will be an important consideration. You need to provide enough air to keep the fire fighter breathing for the duration of the rescue operation.

Fire fighters must know and understand all aspects of their SCBA in this type of situation. If the downed fire fighter has a working SCBA, determine how much air remains in the cylinder. A fire fighter who is breathing and has an adequate air supply is not in immediate, life-threatening danger. If the SCBA contains very little air or no air, however, air management is a critical priority. In this case, rescuers need to move the fire fighter out of the hazardous area immediately or provide an additional air supply. A spare SCBA air cylinder should be brought to the location and connected to the downed fire fighter's SCBA through the rapid intervention crew/company universal air connection (RIC-UAC). This procedure provides for rapid emergency refilling of breathing air to an SCBA for the downed or trapped fire fighter. Many RIC teams bring a complete replacement SCBA as part of their rescue effort, in case it is needed.

If the fire fighter's SCBA is not working, check whether a simple fix can make it operational. Did a hose or regulator come off? Was the cylinder valve accidentally closed? If regulator failure is a possibility, open the bypass valve.

When it is necessary to quickly remove a downed fire fighter, you can use the fire fighter's SCBA straps to create a rescue harness. To rescue a downed fire fighter by making a rescue harness from the fire fighter's SCBA straps, follow the steps in **SKILL DRILL 18-6**:

1. Locate the downed fire fighter.
2. Activate the mayday procedure, if that step has not already been taken.
3. Quickly assess the condition of the downed fire fighter and the situation.
4. Shut off the PASS device as needed to aid in communication. (**STEP 1**)
5. Position yourself at the legs of the downed fire fighter.
6. Lift one leg of the downed fire fighter up onto your shoulder. (**STEP 2**)
7. Locate and loosen the waist straps and shoulder straps (if needed) of the downed fire fighter's SCBA harness. (**STEP 3**)
8. Unbuckle the downed fire fighter's waist strap, and buckle the waist strap under the lifted leg.
9. Tighten the straps.
10. Remove the downed fire fighter from the hazard area to a safe area. (**STEP 4**)

To rescue a downed fire fighter who is wearing PPE and SCBA from the immediate hazard using a **drag rescue device**, follow the steps in **SKILL DRILL 18-7**:

1. Locate the downed fire fighter.
2. Activate the mayday procedure, if that step has not already been taken.
3. Shut off the PASS device to aid in communication.
4. Assess the situation and the condition of the downed fire fighter. (**STEP 1**)
5. Use the RIC-UAC to fill the downed fire fighter's air supply cylinder, if needed. (**STEP 2**)
6. Access the fire fighter's drag rescue device. (**STEP 3**)
7. Remove the downed fire fighter from the hazard area to a safe area. (**STEP 4**)

As a member of a two-person team, you can also remove a downed fire fighter with PPE and SCBA from a hazardous area by having one fire fighter pull on the downed fire fighter's SCBA straps and the second fire fighter support the legs of the downed fire fighter. To do so, follow the steps in **SKILL DRILL 18-8**:

1. Locate the downed fire fighter.
2. Activate the mayday procedure, if that step has not already been taken.
3. Shut off the PASS device to aid in communication.
4. Assess the situation and the condition of the downed fire fighter.
5. Use the RIC-UAC to fill the downed fire fighter's air supply cylinder, if needed. (**STEP 1**)
6. The second fire fighter converts the downed fire fighter's SCBA harness into a rescue harness. (**STEP 2**)

SKILL DRILL 18-6   Rescuing a Downed Fire Fighter Using SCBA Straps as a Rescue Harness
(Fire Fighter I, NFPA 5.3.5)

1 Locate the downed fire fighter. Activate the mayday procedure, if that step has not already been taken. Quickly assess the condition of the downed fire fighter and the situation. Shut off the PASS device as needed to aid in communication.

2 Position yourself at the legs of the downed fire fighter. Lift one leg of the downed fire fighter up onto your shoulder.

3 Locate and loosen the waist straps and shoulder straps (if needed) of the downed fire fighter's SCBA harness.

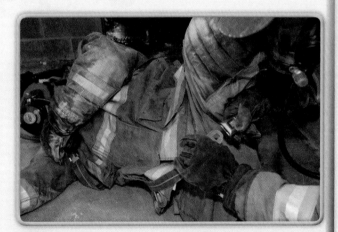

4 Unbuckle the downed fire fighter's waist strap, and buckle the waist strap under the lifted leg. Tighten the straps. Remove the downed fire fighter from the hazard area to a safe area.

© LiquidLibrary

7 The first fire fighter grabs the shoulder straps or uses the webbing to create a handle to pull the downed fire fighter. (**STEP 3**)

8 The second fire fighter stays at the downed fighter's legs and supports the legs of the downed fire fighter. (**STEP 4**)

9 Remove the downed fire fighter from the hazard area to a safe area. (**STEP 5**)

Rehabilitation

The chapter "Fire Fighter Rehabilitation" covers fire fighter rehabilitation in detail. The purpose of rehabilitation is to reduce the effects of fatigue during an emergency operation **FIGURE 18-6**. Firefighting involves very demanding physical labour. When combined with weather extremes and the mental stresses associated with emergency incidents, it

SKILL DRILL 18-7

Rescuing a Downed Fire Fighter Using a Drag Rescue Device
(Fire Fighter I, NFPA 5.3.5)

1 Locate the downed fire fighter. Activate the mayday procedure, if that step has not already been taken. Shut off the PASS device to aid in communication. Assess the situation and the condition of the downed fire fighter.

2 Use the RIC-UAC to fill the downed fire fighter's air supply cylinder, if needed.

3 Access the fire fighter's drag rescue device.

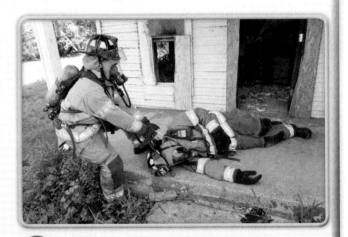

4 Remove the downed fire fighter from the hazard area to a safe area.

© LiquidLibrary

can challenge the strength and endurance of even the most physically-fit and well-prepared fire fighters. Rehabilitation helps fire fighters retain the ability to perform at the current incident and restores their capacity to work at later incidents.

At the simplest of incidents, a rehabilitation area can be set up on the tailboard of an apparatus with a water cooler. At larger incidents, a complete rehabilitation operation should be established, including personnel to monitor the vital signs of fire fighters and provide first aid. In whatever size or form it takes, rehabilitation is integral to fire fighter safety and survival.

The personnel accountability system must continue to track fire fighters who are assigned to report to rehabilitation and note when they are released from rehabilitation for another assignment.

Counselling and Critical Incident Stress

Fire fighters are often exposed to stressful situations. Sometimes they are involved in situations that are described as critical incidents; these incidents challenge the capacity of

SKILL DRILL 18-8

Rescuing a Downed Fire Fighter as a Two-Person Team
(Fire Fighter I, NFPA 5.3.5)

1 Locate the downed fire fighter. Activate the mayday procedure, if that step has not already been taken. Shut off the PASS device to aid in communication. Assess the situation and the condition of the downed fire fighter. Use the RIC-UAC to fill the downed fire fighter's air supply cylinder, if needed.

2 The second fire fighter converts the downed fire fighter's SCBA harness into a rescue harness.

3 The first fire fighter grabs the shoulder straps or uses the webbing to create a handle to pull the downed fire fighter.

4 The second fire fighter stays at the downed fire fighter's legs and supports the legs of the downed fire fighter.

5 Remove the downed fire fighter from the hazard area to a safe area.

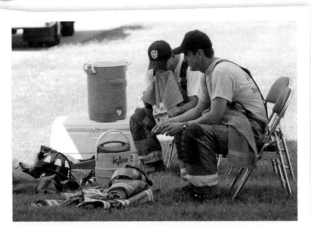

FIGURE 18-6 Rehabilitation reduces the effects of fatigue produced by an operation.

FIGURE 18-7 Group stress debriefings are sometimes used to alleviate the stress reactions generated by high-stress emergency situations.

most individuals to deal with stress. Examples of critical incidents include the following:

- Line-of-duty deaths (police, fire/rescue, Emergency Medical Services [EMS])
- Suicide of a colleague
- Serious injury to a colleague
- Situations that involve a high level of personal risk to fire fighters
- Events in which the victim is known to the fire fighters
- Multiple-casualty/disaster/terrorism incidents
- Events involving death or life-threatening injury or illness to a victim, especially a child
- Events that are prolonged or end with a negative or unexpected outcome

This list is not complete, nor is it necessarily a fact that any of these situations will seriously trouble every individual. Normal coping mechanisms help many fire fighters to handle many situations. Some individuals have a high capacity for dealing with stressful situations and/or use exercise, support from friends and family, or their religious beliefs to help them cope. These are healthy, nondestructive ways to deal with the pressures of being exposed to a critical incident.

Sometimes individuals react to critical incidents in ways that are not positive—alcohol or drug abuse, depression, the inability to function normally, or a negative attitude toward life and work. These symptoms can occur in anyone, even individuals who normally have healthy coping skills. Reactions will vary from one individual to the next, both in type and severity. Many times, fire fighters do not realize they are affected in a deeply negative manner. A somewhat routine incident, however, may trigger negative reactions from a critical incident that occurred in the past. Critical incident stress can also be cumulative, building up over time.

This condition, which is called burnout, cannot be traced to any one incident.

The aim of counselling and **critical incident stress management (CISM)** is to prevent these reactions from having a negative impact on the fire fighter's work and life, over both the short term and the long term **FIGURE 18-7**. Fire fighters should understand the counselling resources and CISM systems available to them and know how they can access them. Maintaining good emotional health is a simple but important part of fire fighter survival.

The recognized stages of emotional reaction experienced by fire fighters and other rescue personnel after a stressful incident can include the following:

1. Anxiety
2. Denial/disbelief
3. Frustration/anger
4. Inability to function logically
5. Remorse
6. Grief
7. Reconciliation/acceptance

These stages can occur within minutes or hours, or they can take several days or months to unfold. Not all of the steps will occur for every event, and they do not necessarily occur in the order given here.

Counselling aims to help fire fighters recognize and deal with these reactions in the most positive manner possible. Although the ways in which this outcome is accomplished may vary, the purpose remains the same.

Your fire department should have some type of counselling program in place. All department members should know how it is activated and be supportive of its intent and function. Fire fighters should realize that their emotional and mental health must be protected just as much as their physical health and safety.

Chief Concepts

- Risk–benefit analysis weighs the positive results that can be achieved against the probability and severity of potential negative consequences. It should be practised and implemented by everyone on the fire ground. While the IC is responsible for the analysis of the overall scene, company officers and fire fighters need to process the risks and benefits of the strategies implemented by command.

- No property is ever worth the life of a fire fighter.

- Smoke is an obvious hazard of fire. You should also be familiar with the less obvious hazard indicators, including inclement weather conditions and the likelihood that certain occupancies might hold hazardous materials.

- Safe operating procedures define the manner in which a fire department conducts operations at an emergency incident.

- The 11 rules of engagement identified by the IAFC and NVFC are as follows:
  1. Size up your tactical area of operation.
  2. Determine the occupant survival profile.
  3. Do not risk your life for lives or property that cannot be saved.
  4. Extend limited risk to protect savable property.
  5. Extend vigilant and measured risk to protect and rescue savable lives.
  6. Go in together, stay together, and come out together.
  7. Maintain continuous awareness of your air supply, situation, location, and fire conditions.
  8. Constantly monitor fire-ground communications for critical radio reports.
  9. Report unsafe practices or conditions that can harm you. Stop, evaluate, and decide.
  10. Abandon your position and retreat before deteriorating conditions can harm you.
  11. Declare a mayday as soon as you think you are in danger.

- Team integrity means that a company arrives at a fire together, works together, and leaves together. Personnel accountability systems help to ensure team integrity by tracking all of the locations and functions of the personnel on the scene.

- A personnel accountability report (PAR) is a roll call taken by each supervisor at an emergency incident. These headcounts are made at regular, predetermined intervals throughout the incident, generally every 10 minutes.

- Mayday is used if a fire fighter is lost, is trapped, has an SCBA failure, or is running out of air. A fire fighter can transmit a mayday message to request assistance, or another fire fighter can transmit a mayday message to report that a fire fighter is missing or in trouble.

- A mayday call takes precedence over all other radio communications.

- Emergency traffic is used to indicate an imminent fire-ground hazard. This kind of message would also be used to order fire fighters to immediately withdraw from interior offensive attack positions and switch to a defensive strategy. An emergency traffic message takes precedence over all other radio communications except a mayday call.

- The sole purpose of RICs is to rescue fire fighters operating at emergency incidents. An RIC is an extension of the two-in–two-out rule. This team should be in place at any incident where fire fighters are operating in conditions that are immediately dangerous to life and health (IDLH) and should be deployed to any situation where a fire fighter needs immediate assistance.

- One of the most basic methods for remaining oriented in a dark, smoke-filled environment is to always stay in contact with a hose line or guide line.

- The first step in attempting a self-rescue is to initiate a mayday.

- A safe location is a temporary location that provides refuge while awaiting rescue or finding a method of self-rescue. In this case, safe means less hazardous.

- Always follow the rule of air management: know how much air you have left in your SCBA before you enter a building, and manage that amount so that you leave the hazardous environment before your low-air alarm sounds.

- SCBA straps or a drag rescue harness may be used to help drag a downed fire fighter to a safe location.

- Critical incident stress is a known hazard that emergency personnel can learn to handle. This hazard can affect fire fighters' emotional and mental health. Numerous resources are available to help fire fighters maintain good emotional and mental health.

Hot Terms

Air management The use of a limited air supply in such a way as to ensure that it will last long enough to enter a hazardous area, accomplish needed tasks, and return safely.

Critical incident stress management (CISM) A program designed to reduce both acute and chronic effects of stress related to job functions. (NFPA 450)

Drag rescue device (DRD) A component integrated within the turnout coat to aid in the rescue of an incapacitated fire fighter. (NFPA 1851)

Guide line A rope used for orientation when fire fighters are inside a structure where there is low or no visibility. The line is attached to a fixed object outside the hazardous area.

Personnel accountability report (PAR) One of a series of periodic reports verifying the status of responders assigned to an incident or planned event. (NFPA 1026)

Risk–benefit analysis An assessment of the risk to rescuers versus the benefits that can be derived from their intended actions. (NFPA 1006)

Safe location A location remote or separated from the effects of a fire so that such effects no longer pose a threat. (NFPA 101)

Self-rescue Escaping or exiting a hazardous area under one's own power. (NFPA 1006)

FIRE FIGHTER
in action

It is 2:00 a.m., and you are on the scene of a fire in an apartment complex. The building is a five-storey, ordinary construction, with a centre-hallway design. The fire is on the second floor, and there are heavy smoke conditions on the second, third, fourth, and fifth floors. The IC assigns your captain to the search group; you and your crew are assigned to the primary search of the third floor.

1. Who is responsible for conducting a risk–benefit analysis during this fire?
- **A.** Incident commander
- **B.** Company officer
- **C.** Fire fighter
- **D.** All of the above

2. Which of the following is *not* part of the 11 rules of engagement?
- **A.** Constantly monitor fire-ground communications for critical radio reports.
- **B.** Report unsafe practices or conditions that can harm you.
- **C.** Do not abandon your position and retreat before deteriorating conditions can harm you unless directed to do so by the IC.
- **D.** Declare a mayday as soon as you *think* you are in danger.

3. If you become lost during your search, what should you do?
- **A.** Continue searching until you find something that is familiar.
- **B.** Switch the hand that is in contact with the wall and try to find your way out.
- **C.** Transmit a mayday communication over the radio.
- **D.** Transmit an emergency traffic communication over the radio.

4. The time rating for an SCBA is based on _____.
- **A.** a standard rate of consumption for a typical fire fighter under high exertion conditions in a test laboratory
- **B.** a standard rate of consumption for a typical adult under high exertion conditions in a test laboratory
- **C.** a standard rate of consumption for a typical adult under low exertion conditions in a test laboratory
- **D.** a standard rate of consumption for a typical fire fighter under low exertion conditions in a test laboratory

5. When is a PAR not necessary?

A. When changing from an offensive to a defensive strategy

B. When an emergency evacuation has been declared

C. Every 10 minutes

D. Upon transfer of command

6. What is the purpose of CISM?

A. To prevent the public's reaction to critical incidents from having a negative impact on their work and life

B. To prevent a fire fighter's reaction to critical incidents from having a negative impact on the fire fighter's work and life

C. To prevent fire fighters from having an emotional response to critical situations

D. To prevent the public from having an emotional response to critical situations

FIRE FIGHTER II
in action

The night before Christmas, faulty lights on a Christmas tree caused a fire in a multi-unit apartment building. When your crew arrived on scene, a mother in bare feet and a nightgown was screaming, "I can't find my daughter! I can't find my daughter!" Thankfully, you and your partner found a four-year-old girl underneath a bunk bed in the second-floor apartment. The paramedics transported her to the hospital, where she is now in stable condition.

Despite the positive outcome, you find yourself having recurring nightmares about being trapped in a pitch-black room, with the sound of a child crying ringing in your ears. After a week of the same nightmare, you ask your partner how he has been sleeping. "Not so great," he admits. "Why do you ask?"

1. What would your response be?

2. What would your next step be to resolve this issue?

3. Would you inform your captain about this issue?

4. If your partner is reluctant to seek help, how would you convince him of his best interests?

Salvage and Overhaul

Fire Fighter I

Knowledge Objectives

After studying this chapter, you will be able to:

- Describe the equipment used to illuminate an emergency scene. (NFPA 5.3.17 , p 598)
- Describe the safety precautions to take when working with lighting equipment. (NFPA 5.3.17, 5.3.17.A , p 598)
- Describe the types of lights used to illuminate exterior and interior scenes. (NFPA 5.3.17 , p 600)
- Describe how to operate lighting equipment to light exterior and interior scenes. (NFPA 5.3.17, 5.3.17.A , p 600–601)
- Explain the purpose of salvage operations. (NFPA 5.3.14.A , p 602–603)
- List the tasks involved in a salvage operation. (p 602–603)
- Describe the safety precautions that need to be considered when performing salvage. (p 603)
- List the tools used to perform salvage operations. (p 603)
- Describe the salvage techniques commonly used to prevent water damage. (p 603–604)
- Describe the general procedures for preventing excess water damage from fire sprinklers. (NFPA 5.3.14.A , p 604–609)
- List the equipment used to shut down fire sprinklers. (NFPA 5.3.14.A , p 604)
- Describe the identifying characteristics of a main control valve of a fire sprinkler system. (NFPA 5.3.14.A , p 605)
- Describe the general procedures and equipment used to remove excess water from a structure. (NFPA 5.3.14.B , p 608–611)
- Describe the general procedures and equipment used to limit smoke and heat damage. (p 611)
- Describe how to maintain salvage covers. (NFPA 5.5.1 , p 613)
- Explain when fire investigators should become involved in salvage operations. (NFPA 5.3.14.B , p 616)
- Describe the purpose of overhaul operations. (NFPA 5.3.13, 5.3.14 , p 617)
- List the concerns that must be addressed to ensure the safety of fire fighters who are performing overhaul. (NFPA 5.3.13.A , p 618)
- List the indicators of a possible structural collapse. (NFPA 5.3.13.A , p 618)
- Explain how to preserve structural integrity during overhaul. (NFPA 5.3.13 , p 618)
- Describe how to coordinate overhaul operations with fire investigators. (NFPA 5.3.13.A, 5.3.13.B , p 618)
- Explain how fire fighters determine overhaul locations. (NFPA 5.3.13 , p 619)
- List the tools that are used for overhaul operations. (NFPA 5.3.13, 5.3.13.A , p 620–621)
- Describe the general techniques used in overhaul operations. (NFPA 5.3.13, 5.3.14 , p 620)

Skills Objectives

After studying this chapter, you will be able to perform the following skills:

- Illuminate an emergency scene. (NFPA 5.3.17.B , p 600–601)
- Use a sprinkler stop to shut down a sprinkler head. (NFPA 5.3.14.B , p 604–605)
- Use a sprinkler wedge to shut down a sprinkler head. (NFPA 5.3.14.B , p 604, 606)
- Close and reopen a main control valve. (NFPA 5.3.14.B , p 605–607)
- Close and open a post indicator valve. (NFPA 5.3.14.B , p 606–608)
- Construct a water chute. (NFPA 5.3.14.B , p 609)
- Construct a water catch-all. (NFPA 5.3.14.B , p 610)
- Fold a salvage cover for one–fire fighter deployment. (NFPA 5.3.14.B , p 611–612)
- Fold a salvage cover for two–fire fighter deployment. (NFPA 5.3.14.B , p 611–613)
- Fold and roll a salvage cover. (NFPA 5.3.14.B , p 612–614)
- Perform a one–fire fighter salvage cover roll. (NFPA 5.3.14.B , p 612, 615)
- Perform a shoulder toss. (NFPA 5.3.14.B , p 613, 616)
- Perform a balloon toss. (NFPA 5.3.14.B , p 613, 617)
- Open a ceiling to check for fire using a pike pole. (NFPA 5.3.8.B, 5.3.13.B , p 621–622)
- Open an interior wall to check for fire. (NFPA 5.3.8.B, 5.3.13.B , p 621, 623)

© Photos.com

Fire Fighter II

FFII

Knowledge Objectives

After studying this chapter, you will be able to:

- Describe the types of generators used to power lighting equipment. (p 600)
- Describe how generators operate. (p 600)
- Describe how to clean and maintain lighting equipment. (NFPA 6.5.4.A , p 601)
- Describe how to maintain generators. (NFPA 6.5.4, 6.5.4.A , p 601)

Skills Objectives

After studying this chapter, you will be able to perform the following skill:

- Conduct a weekly/monthly generator test. (NFPA 6.5.4.B , p 601–602)

You Are the Fire Fighter

After a bitterly cold and snowy winter with several sub-zero days with strong winds, the sun is finally shining. Today the temperature is climbing to near record highs. The sun and unexpected warmth have brightened the collective mood of the fire station when the calls for activated fire alarms come rolling in.

Sprinkler pipes throughout the community are breaking due to the thaw. You arrive at a four-storey warehouse with water flowing out the front door. You find several broken pipes on various floors that are flowing onto the merchandise, as well as water dripping through the office ceiling.

1. How would you go about protecting the contents of the office?
2. Where would you begin in order to protect the merchandise?
3. Which equipment will you need to protect the office and merchandise from water damage?

Introduction

The basic goals of firefighting are first to save lives, second to control the fire, and third to protect property. **Salvage** and overhaul operations help meet that third goal by limiting and reducing property losses resulting from a fire. Salvage efforts protect property and belongings from damage, particularly from the effects of smoke and water. **Overhaul** ensures that a fire is completely extinguished by finding and exposing any smouldering or hidden pockets of fire in an area that has been burned.

Salvage and overhaul are usually conducted in close coordination with each other and have a lower priority than conducting search and rescue operations or controlling the fire. If a department has enough personnel on scene to address more than one priority, salvage can be performed concurrently with fire suppression. Overhaul, however, can begin only once the fire is under control.

Throughout the salvage and overhaul process, fire fighters must attempt to preserve evidence related to the cause of a fire, particularly when working near the suspected area of fire origin and when the fire's cause is not obvious. In general, disturb as little as possible in or near the area of origin until a fire investigator determines the cause of the fire and gives permission to clear the area. Knowing what caused a fire can help prevent future fires and enable victims to recoup losses under their insurance coverage. This subject is covered in greater depth in the chapter "Fire Cause Determination."

Fire fighters performing salvage and overhaul must be able to see where they are going, what they are doing, and which potential hazards, if any, are present. The smoke might have cleared, but electrical power might still be unavailable. Portable lights should be set up to fully illuminate areas during salvage and overhaul.

Lighting

Lighting is an important concern at many fires and other emergency incidents. Because many emergency incidents occur at night, lighting is required to illuminate the scene and enable safe, efficient operations. Inside fire buildings, heavy smoke can completely block out natural light. In addition, electrical service is often interrupted or disconnected for safety purposes. As a consequence, setting up interior lighting is essential for fire fighters to conduct search and rescue, ventilation, fire-suppression, salvage, and overhaul operations.

Most fire departments use different types of lighting equipment, depending on the particular situation. **Spotlights**, for example, project a narrow concentrated beam of light. **Floodlights** project a more diffuse light over a wide area. Lights can be portable or permanently mounted on fire apparatus.

Although the primary responsibility for fire-scene lighting often belongs to truck (ladder) companies or support units, any company can be assigned this duty. Lighting equipment can be mounted on any apparatus, and some fire departments equip special vehicles with high-powered lights to illuminate incident scenes.

■ Safety Principles and Practices

The lighting and power equipment used at a fire scene generally operates on 110-volt AC (alternating current), which is the same as standard household current. Some systems require higher voltage. All electrical cords, junction boxes, lights, and power tools must be maintained properly and handled carefully to avoid electrical shocks. In addition, all electrical equipment must be properly grounded. Electrical cords must be well insulated, without cuts or defects, and properly sized to handle the required amperage.

Generators should have ground-fault interrupters (GFI) to prevent a fire fighter from receiving a potentially fatal electric shock. A GFI senses when there is a problem with an electrical ground and interrupts the current in such a case, shutting down both the power source and the equipment it is feeding.

Some portable generators are equipped with a grounding rod that must be inserted into the ground. Always use a grounding device if one is provided. Avoid areas of standing or flowing water when placing power cords and junction boxes at a fire scene, and put electrical equipment on higher ground whenever possible.

■ Lighting Equipment

Portable lights can be taken into buildings to illuminate the interior. They can also be set up outside the structure to illuminate the fire or incident scene **FIGURE 19-1**. Portable lights usually range from 300 to 1500 watts and can use several types of bulbs, including quartz bulbs, halogen bulbs, and light-emitting diodes.

Portable light fixtures are connected to the generator with electrical cords. These electrical cords should be stored neatly coiled or on permanently mounted reels attached to the fire apparatus. The cord is then pulled from the reel to the place

FIGURE 19-2 An apparatus-mounted light tower can provide near-daylight conditions at an incident scene.
© Glen E. Ellman

FIGURE 19-1 Portable emergency lights come in various sizes and levels of brightness.
© Lusoimages/ShutterStock, Inc.

where the power is needed. Portable reels can be taken from the apparatus into the fire scene.

Junction boxes are used as mobile power outlets. They are placed in convenient locations so that cords for individual lights and electrical equipment can be attached to them. Junction boxes used by fire departments are protected by waterproof covers and are often equipped with small lights so that they can be easily located in the dark.

The connectors and plugs used for fire department lighting have special connectors that attach with a slight clockwise twist. This setup keeps the power cords from becoming unplugged during fire department operations.

Lights also can be permanently mounted on fire apparatus to illuminate incident scenes **FIGURE 19-2**. In such cases, the vehicle operator can immediately illuminate the incident scene simply by pressing the generator starter button. Some vehicle-mounted lights can be manually raised to illuminate a

larger area. Mechanically operated light towers, which can be raised and rotated by remote control, can create near-daylight conditions at an incident scene.

■ Battery-Powered Lights

Battery-powered lights are generally used by individual fire fighters to find their way in dark areas or to illuminate their immediate work area. These kinds of lights are most often used during the first few critical minutes of an incident, because they are lightweight, are easily transported, do not require power cords, and can be used immediately **FIGURE 19-3**. The lights are powered by either disposable or rechargeable batteries; thus they have a limited operating time before the batteries need to be recharged or replaced.

Powerful lights are needed to pierce the smoke inside a fire building. Large hand lights that project a powerful beam of light are preferred for search and rescue activities and interior

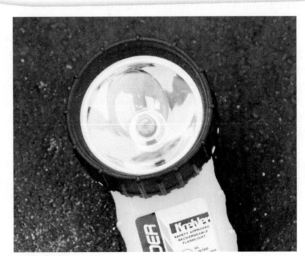

FIGURE 19-3 Hand lights manufactured for use by fire fighters have light outputs ranging from 3500 to 75,000 candlepower.
© 2003, Berta A. Daniels

fire-suppression operations when fire fighters must penetrate smoke-filled areas quickly. These lights can be equipped with a shoulder strap for easy transport. Every crew member entering a fire building should be equipped with a high-powered hand light.

A personal flashlight is another type of battery-operated light used by fire fighters. Always carry a flashlight as part of your personal protective equipment (PPE); it could be a lifesaver if your primary light source fails. Flashlights are not as powerful as the larger hand lights, however, and they will not operate for as long on one set of batteries. A fire fighter's flashlight should be rugged and project a strong light beam.

■ Electrical Generators

More powerful lighting equipment requires a separate source of electricity, generally a 110-V AC delivered through power cords. The electricity can be supplied by a **generator**, an **inverter**, or a building's electrical system, if the power has not been interrupted.

Power inverters convert 12-V DC (direct current) from a vehicle's electrical system to 110-V AC power. An inverter can provide a limited amount of AC current and is typically used to power one or two lights mounted directly on the apparatus. Some vehicles have additional outlets for operating portable lights or small electrically powered tools. Most power inverters do not produce enough power to operate high-intensity lighting equipment, large power tools, or ventilation fans. Be aware that connecting devices that draw too much current to an inverter can seriously damage it.

Electrical generators are powered by gasoline or diesel engines; they can be either portable or permanently mounted on fire apparatus. Portable generators are small enough to be removed from the apparatus and carried to the fire scene. These devices come in various sizes and can produce as much as 6 kilowatts of power. A portable generator has a small gasoline- or diesel-powered motor with its own fuel tank.

Apparatus-mounted generators have much larger power capacities, sometimes exceeding 20 kW. The smaller units, which are similar to portable generators, can be permanently mounted in a compartment on the apparatus. By contrast, larger generators are usually mounted directly on the vehicle, often have diesel engines, and draw fuel directly from the vehicle's fuel tank. Permanently mounted generators can also be powered by the apparatus engine through a power takeoff and a hydraulic pump.

Another potential power source for lighting is a building's normal power supply. It is usually not an option if a serious fire has occurred in the building, because the current will be disconnected for safety reasons. In contrast, if the fire is relatively minor, the power might not have to be interrupted and can be used for lighting. Alternatively, power can sometimes be obtained from a nearby building. When drawing power from a building, be aware that you might not know the capacity of the circuit from which you are obtaining electricity. Do not overload a household electrical circuit.

Fire Fighter Safety Tips
© Photos.com

Never run a portable generator in a confined space or an area without adequate ventilation unless you and everyone else in the area are wearing self-contained breathing apparatus (SCBA). The exhaust from the generator contains carbon monoxide, which can kill you.

■ Lighting Methods

Effective lighting improves the efficiency and safety of all crews at an emergency incident, but it requires practise to set up quickly and efficiently. Departmental suggested operating guidelines (SOGs), or standard operating procedures (SOPs), or the incident commander (IC) will dictate the responsibility for setting up lighting.

Exterior Lighting

Exterior lighting should be provided at all incident scenes during hours of darkness so that fire fighters can see what they are doing, recognize hazards, and find victims who need rescuing. Scene lighting also makes fire fighters more visible to drivers who are approaching the scene or maneuvering emergency vehicles. Powerful exterior lights can often be seen through the windows and doors of a dark, smoke-filled building and can guide disoriented fire fighters or victims to safety.

Apparatus operators should turn on their apparatus-mounted floodlights and position them to illuminate as much of the area as possible. Position vehicles with light towers to use their lights for maximum effectiveness. The entire area around a fire-struck building should be illuminated. If some areas cannot be covered with apparatus-mounted lights, use portable lights.

Interior Lighting

Portable lights can provide interior lighting at a fire scene. During interior operations, quickly extend a power cord and set up a portable light at the entry point to the structure. Fire fighters looking for the building exit can then use this light at the door as a beacon. Extend additional lights into the building to illuminate interior areas as needed and as time permits.

Interior operations often progress through the search and rescue phase and the fire-suppression stage quickly, with these steps sometimes being completed before effective interior lighting can be established. If interior operations continue for a lengthy period, portable lighting should be used to provide as much light as possible in the areas where fire fighters are working.

When operations reach the salvage and overhaul phase, take extra time to provide adequate interior lighting in all areas so that crews can work safely. Proper lighting enables fire fighters to see what they are doing and to observe any dangerous conditions that need to be addressed. Although exterior lighting is generally needed only at night, interior lighting may be needed even when it is a bright day outside.

Follow the steps in **SKILL DRILL 19-1** (Fire Fighter I, NFPA 5.3.17) to illuminate an emergency scene (**FIGURE 19-4**):

❶ Wear PPE. Depending on the conditions, you may or may not wear SCBA.

❷ Inspect all equipment while setting it up.

3 Start the power plant, or engage the inverter.

4 Connect cords, plug adaptors, GFIs, and lighting equipment.

5 Ensure proper grounding and GFI use.

6 Ensure that the scene is adequately and safely illuminated.

Fire Fighter Safety Tips

© Photos.com

A good rule to remember: light early, light often, and light safely.

FIGURE 19-4 Inspect all equipment while setting it up.

■ Cleaning and Maintenance

It is important to properly maintain lighting equipment and generators to keep them in good operating condition. To clean lighting equipment, follow the manufacturer's instructions for cleaning. In general, the directions specify how to clean the floodlights and search lights with a mild soap or detergent. Always use a soft cloth to clean and polish the lens. Avoid the use of strong solvents, as they may damage plastic or rubber components.

When cleaning and maintaining a generator, also follow the manufacturer's instructions. Avoid overfilling the fuel tank on portable generators. When a fuel tank is too full, it can overflow onto the hot engine and start a fire. An overfull tank can also leak fuel as the fuel expands, especially if the tank was filled in a cool environment and then brought into a warmer place. Spilled fuel can damage paint under certain circumstances.

Clean the external parts of the generator with a mild soap or detergent as recommended by the manufacturer. Remove any dirt, vegetation, or charred materials on the device. Avoid the use of strong solvents, as they can damage rubber and plastic parts of the generator and can remove paint.

If a solvent is recommended by the manufacturer for any cleaning operation, follow the directions for its use on the product label. Also read the material safety data sheet before starting to work with any solvent. Check whether the product is flammable. If it is, be sure to follow the safety recommendations for its use. Always use flammable solvents in a well-ventilated area that is not close to any flame or heat source.

Fire Fighter Safety Tips

© Photos.com

Solvents can irritate or burn the skin, eyes, and lungs. Some of these agents are known to cause cancer and should be avoided. Solvents can be absorbed through the lungs as well as through the skin. For solvents that can be absorbed through skin, check whether chemically resistant gloves are required during their use. Also check whether eye protection or respiratory protection is required.

Test and run generators on a weekly or monthly basis to confirm that they will start, run smoothly, and produce power. Run gasoline-powered generators for 15 to 30 minutes to reduce any deposit buildup that could foul the spark plugs and make the generator hard to start.

At the same time, inspect and test other electrical equipment on the apparatus. Take junction boxes out and check them for cracked covers or broken outlets. Plug them in to make sure the GFIs work properly. Check power cords for tears in the protective covering, mechanical damage, fraying, heat damage, or burns. Inspect plugs for loose or bent prongs. Inspect and test the operation of all power tools and equipment on the apparatus. Check lighting fixtures to confirm that they are in good repair and that the bulbs are working.

After shutting down the generator, refill the fuel tank. Follow the manufacturer's recommendations for generator engine maintenance, including regular oil changes. To conduct a weekly/monthly generator test, follow the steps in **SKILL DRILL 19-2**:

1 Remove the generator from the apparatus compartment, or open all doors as needed for ventilation. Install the grounding rod, if needed. **(STEP 1)**

2 Check the oil and fuel levels, and start the generator.

3 Connect the power cord or junction box to the generator, connect a load such as a fan or lights, and make sure the generator attains the proper speed. Check the voltage and amperage gauges to confirm efficient operation. **(STEP 2)**

4 Run the generator under load for 15–30 minutes.

5 Turn off the load and listen as the generator slows down to idle speed. Allow the generator to idle for approximately 2 minutes before turning it off.

6 Disconnect all power cords and junction boxes, and remove the grounding rod if present.

7 Clean all power cords, plugs, adaptors, GFIs, and tools, and replace them in proper storage areas. Allow the generator to cool for 5 minutes.

8 Refill the generator with fuel and oil as needed, and return the generator to its compartment. Fill out the appropriate paperwork. **(STEP 3)**

SKILL DRILL 19-2 | Conducting a Weekly/Monthly Generator Test | **FFII**
(Fire Fighter II, NFPA 6.5.4)

1 Remove the generator from the apparatus compartment, or open all doors as needed for ventilation. Install the grounding rod, if needed.

2 Check the oil and fuel levels, and start the generator. Connect the power cord or junction box to the generator, connect a load such as a fan or lights, and make sure the generator attains the proper speed. Check the voltage and amperage gauges to confirm efficient operation.

3 Run the generator under load for 15–30 minutes. Turn off the load and listen as the generator slows down to idle speed. Allow the generator to idle for approximately 2 minutes before turning it off. Disconnect all power cords and junction boxes; clean all power cords, plugs, adaptors, GFIs, and tools, and replace them in proper storage areas. Allow the generator to cool for 5 minutes. Refill the generator with fuel and oil as needed, and return the generator to its compartment. Fill out the appropriate paperwork.

© LiquidLibrary

Salvage Overview

Salvage operations are conducted to save property from a fire and to reduce the damage that results from a fire. Often, the damage caused by smoke and water can be more extensive and costly to repair or replace than the property that is burned. Salvage operations are also conducted in other situations that endanger property, such as floods or water leaking from burst pipes.

Salvage efforts usually are aimed at preventing or limiting <u>secondary losses</u> that result from smoke and water damage, fire-suppression efforts, and other causes. They must be performed promptly to preserve the building and protect the contents from avoidable damage. Salvage operations include ejecting smoke, removing heat, controlling water runoff, removing water from the building, securing a building after a fire, covering broken windows and doors, and providing temporary patches for ventilation openings in the roof to protect the structure and contents.

Fire fighters must help protect property from avoidable loss or damage as part of their professional responsibilities. At the scene of a fire or any other type of emergency, the victims' property is entrusted to the care of fire fighters, who are expected to protect that property to the best of their ability.

Salvage efforts at residential fires often focus on protecting personal property. These items often have personal implications for the victims' lives. Fire fighters may be able to identify and protect critical personal belongings, such as wallets, purses, car keys, or medications. The value of some items, such as photographs and family heirlooms, cannot be measured in dollars, but these items are certainly treasured by victims and are often irreplaceable.

FIRE FIGHTER Tips

One way many fire departments help homeowners after a fire is by having a list of board-up and fire-restoration companies that they can give to a homeowner. Board-up companies are prepared to respond quickly and secure a property to protect it from rain, snow, or damage from vandals. Fire-restoration companies are skilled in removing smoke and water from a building and its contents before more damage occurs. Because many building owners are not aware of these companies, your recommendations are often greatly appreciated by the owner or occupant of a damaged building at this highly stressful time.

At commercial or industrial fires, the salvage priorities are quite different from those at residential fires. The preincident plan should identify critical items and enable fire fighters to focus their salvage efforts appropriately; for example, a law firm might identify its files as the most valuable property, whereas a library might place more importance on books and computer files than on furniture. A preincident plan should note whether the contents of a warehouse consist of valuable electronic components or industrial steel pipe. Factory machinery could also be a higher priority than the finished products stored on premises.

Fire Fighter Safety Tips

At the emergency scene, the IC should know where every crew is working and what those crew members are doing at all times. The IC relies on your company officer to know your location and activities and to relay this information to the command post.

■ Safety Considerations During Salvage Operations

Safety is a primary concern during salvage operations. The PPE required will depend on the site of the salvage operations, the stage of the fire, and any hazards present. Salvage operations often begin while the fire is still burning and continue for several hours after it has been extinguished. During firefighting operations and immediately afterward, fire fighters should wear a full set of protective clothing and equipment, including SCBA. After the fire is out, the area must be ventilated, and the atmosphere tested and determined to be safe by the safety officer before fire fighters can work without SCBA.

If salvage is conducted in a different part of the building than where the fire occurred, several floors below the fire, or in a neighbouring building, fire fighters may have other protective clothing requirements. Hazards in these areas might be quite different from those in the fire area. Fire fighters conducting salvage operations in remote areas can wear partial PPE—boots, turnout pants, helmet, and gloves—with the approval of the safety officer.

Structural collapse is always a possibility during salvage operations because of fire damage to the building's structural components; for example, lightweight trusses can collapse quickly when damaged by a fire. Heavy objects, such as

air-conditioning units and appliances, can crash through roofs or ceilings weakened by fire damage.

Even the water used to douse the fire can contribute to structural collapse. Recall that each litre of water weighs 1 kilogram (2.2 pounds). Extinguishing a major fire can, therefore, load the structure with tons of extra water weight. If the structure is already weakened by the fire, the extra water weight can cause a catastrophic structural collapse, with the potential to injure or kill fire fighters; for example, a ladder pipe discharging water into a building at a rate of 3028 litres per minute (800 gallons per minute [gpm]) is adding 3 tonnes (3.3 tons) of weight to the building every minute.

Sometimes this danger may be masked by absorbent materials, such as bolts of cloth or bales of rags, stored in the building. Uneven absorption of water also presents a danger; for example, stacked rolls of newsprint in a printing factory can collapse if the bottom layer becomes waterlogged and, therefore, unable to support the stack.

Water can also create potential hazards for fire fighters conducting salvage operations on floors below the fire. Water used to extinguish a fire inevitably leaks down through the building and can cause major damage. If it becomes trapped in the void space above a ceiling, it may cause sections of the ceiling to collapse. Fire fighters should watch for warning signs such as water leaking through ceiling joints and sagging areas of ceilings. Water that accumulates in a basement can hide a stairway and present additional hazards if utility connections are also located in the basement.

Damage to the building's utility systems presents significant risks to fire fighters who are conducting salvage operations. Gas and electrical services should always be shut off to eliminate the potential hazards of electrocution or explosion stemming from gas leaks.

■ Salvage Tools

Fire fighters conducting salvage operations use a variety of special tools and equipment TABLE 19-1 . Salvage covers and other protective items shield and cover building contents. Other salvage tools remove smoke, water, and other products that can damage property FIGURE 19-5 .

TABLE 19-1	Salvage Tools

- Salvage covers (treated canvas or plastic)
- Box cutter for cutting plastic
- Floor runners
- Wet/dry vacuums
- Squeegees
- Submersible pumps and hose
- Sprinkler shut-off kit
- Ventilation fans, power blowers
- Small tool kit
- Pike poles to construct water chutes

Using Salvage Techniques to Prevent Water Damage

The best way to prevent water damage at a fire scene is to limit the amount of water used to fight the fire. In other words, use only enough water to knock down the fire quickly; do not use

FIGURE 19-5 Salvage tools.

FIGURE 19-6 Most fires are controlled by the activation of only one or two sprinkler heads.
Courtesy of Tyco Fire and Building Products

more water than is needed. Do not continue to douse a fire that is already out, because this action merely creates unnecessary water damage. Turn off hose nozzles when they are not in use. If a nozzle must be left partly open to prevent freezing during cold weather, direct the flow through a window. Take the time to tighten any leaky hose couplings after the fire has been knocked down, because this measure prevents additional water damage during overhaul. Look for leaking water pipes in the building; if you find any, shut off the building's water supply.

■ Deactivating Sprinklers

Buildings equipped with sprinkler systems require special steps to limit water damage. Sprinklers can control a fire efficiently, but will keep flowing until the activated sprinklers are shut off or the entire system is shut down. This ongoing flow can cause significant water damage.

Sprinklers should be shut down as soon as the IC declares that the fire is under control. Do not shut down sprinklers prematurely; the fire may rekindle, causing major damage. Between the time when sprinklers are shut down and when overhaul is completed, a fire fighter with a portable radio should stand by and be ready to reactivate the sprinklers if necessary.

Sprinkler systems are usually designed so that only those sprinklers directly over or close to the fire will be activated. Indeed, most fires can be contained or fully extinguished by the release of water from only one or two sprinklers **FIGURE 19-6**. Although movies may show all sprinklers activating at once, this does not occur in real life except in specially designed deluge sprinkler systems. More information on sprinkler systems is presented in the chapter "Fire Detection, Protection, and Suppression Systems."

If only one or two sprinklers have been activated, inserting a sprinkler wedge or a sprinkler stop can quickly stop the flow. This action keeps the rest of the system operational during overhaul.

A **sprinkler wedge** is a small triangular piece of wood. Inserting two wedges on opposite sides, between the orifice

and the deflector of the sprinkler, and pushing them together effectively plugs the opening, stopping the flow from standard upright and pendant sprinklers.

A **sprinkler stop** is a more sophisticated mechanical device with a rubber stopper that can be inserted into a sprinkler head. Several types of sprinkler stops are available, including some that work only with specific sprinklers. Some types of sprinkler stops, such as a Shutgun™, are manufactured with a fusible link, so they can shut off a sprinkler head and return that sprinkler to temporary service.

Not all sprinklers can be shut off with a wedge or a sprinkler stop. Recessed sprinklers, which are often installed in buildings with finished ceilings, are usually difficult to shut off. If the individual sprinklers cannot be shut off, stop the water flow by closing the sprinkler control valve.

To stop the flow of water from a sprinkler using a sprinkler stop, follow the steps in **SKILL DRILL 19-3**:

1. Have a sprinkler stop in hand. (**STEP 1**)
2. Place the flat-coated part of the sprinkler stop over the sprinkler orifice and between the frame of the sprinkler. (**STEP 2**)
3. Push the lever to expand the sprinkler stop until it snaps into position. (**STEP 3**)

To stop the flow of water from a sprinkler using a pair of sprinkler wedges, follow the steps in **SKILL DRILL 19-4**:

1. Hold one wedge in each hand. (**STEP 1**)
2. Insert the two wedges, one from each side, between the discharge orifice and the sprinkler-head deflector. (**STEP 2**)
3. Bump the wedges securely into place to stop the water flow. (**STEP 3**)

If individual sprinklers cannot be plugged or if several sprinklers have been activated, you can stop the flow by closing the sprinkler control valve. After the valve is closed, the water trapped in the piping system will drain out through the activated sprinklers for several minutes. If a drain valve is found near the control valve, open it to drain the system quickly. This step also directs the water flow to a location where it will not cause additional damage.

SKILL DRILL 19-3 Using a Sprinkler Stop
(Fire Fighter I, NFPA 5.3.14)

1 Have a sprinkler stop in hand.

2 Place the flat-coated part of the sprinkler stop over the sprinkler orifice and between the frame of the sprinkler.

3 Push the lever to expand the sprinkler stop until it snaps into position.

© LiquidLibrary

The main control valve for a sprinkler system is usually an **outside stem and yoke (OS&Y) valve** or a **post indicator valve (PIV)**. An OS&Y valve is typically found in a mechanical room in the basement or on the ground-floor level of a building. A PIV is located outside the building or on an exterior wall. Some sprinkler systems also have zone valves that control the flow of water to sprinklers in different areas of the building; high-rise buildings, for example, usually have a zone valve for each floor. Closing a zone valve stops the flow to sprinklers in the zone; in the rest of the building, however, the sprinklers remain operational. The locations of the main control valves and zone valves should be identified during preincident planning visits to a building.

Sprinkler control valves should always be locked in the open position, usually with a chain and padlock, and should be equipped with a tamper alarm that is tied into the fire alarm system. Fire fighters who are sent to shut off a sprinkler control valve should take a pair of bolt cutters to remove the lock if the key is not available.

To close and reopen the main control valve, follow the steps in **SKILL DRILL 19-5** :

1 Locate the OS&Y valve as indicated on the preincident plan. It may be either inside the building or nearby outside the building. The stem of an OS&Y valve protrudes from the valve handle in the open position.

2 Identify the valve that controls sprinklers in the fire area. Smaller systems will have only one OS&Y valve; large buildings may have either multiple sprinkler systems or one main control valve and additional zone valves controlling the flow to different areas.

SKILL DRILL 19-4   Using Sprinkler Wedges
(Fire Fighter I, NFPA 5.3.14)

1   Hold one wedge in each hand.

2   Insert the two wedges, one from each side, between the discharge orifice and the sprinkler deflector.

3   Bump the wedges securely into place to stop the water flow.

© LiquidLibrary

3 If the valve is locked in the open position with a chain and padlock and the key is readily available, unlock and remove the chain. If no key is available, cut the lock or the chain with a pair of bolt cutters. Cut a link close to the padlock so that the chain can be reused. (**STEP 1**)

4 Turn the valve handle clockwise to close the valve. Keep turning until resistance is strong and little of the stem is visible. (**STEP 2**)

5 To open the OS&Y valve, turn the handle counterclockwise until resistance is strong and the stem is visible again.

6 Lock the valve in the open position. (**STEP 3**)

To close and open a PIV, follow the steps in **SKILL DRILL 19-6** :

1 Locate the PIV (it is usually found outside the building). The location of the valve should be indicated on a preincident plan.

2 Most PIVs will be locked in the open position by a padlock. Unlock the padlock if the key is readily available. If no key is available, cut the lock with a pair of bolt cutters. (**STEP 1**)

3 Remove the handle from its storage position on the PIV, and place it on top of the valve as you would a hydrant wrench.

SKILL DRILL 19-5
Closing and Reopening a Main Control Valve (OS&Y)
(Fire Fighter I, NFPA 5.3.14)

1 Locate the OS&Y valve as indicated on the preincident plan. Identify the valve that controls sprinklers in the fire area. If the valve is locked in the open position with a chain and padlock and the key is readily available, unlock and remove the chain. If no key is available, cut the lock or the chain with a pair of bolt cutters.

2 Turn the valve handle clockwise to close the valve. Keep turning until resistance is strong and little of the stem is visible.

3 To open the OS&Y valve, turn the handle counter-clockwise until resistance is strong and the stem is visible again. Lock the valve in the open position.

© LiquidLibrary

4 Turn the valve stem in the direction indicated on top of the valve to close the valve. Keep turning until resistance is strong and the visual indicator changes from "open" to "shut." (**STEP 2**)

5 To reopen the PIV, turn the valve stem in the opposite direction until resistance is strong and the indicator changes back to "open."

6 Lock the valve in the open position. (**STEP 3**)

Before a sprinkler system can be restored to normal operation, the sprinklers that have been activated must be replaced. Every sprinkler system should have spare sprinklers stored somewhere, usually near the main control valves. An activated sprinkler must be replaced with another of the same design, size, and temperature rating.

The main sprinkler control valve or the appropriate zone valve must be closed, and the system must be drained, before a sprinkler can be changed. Special wrenches must be used when replacing sprinklers to prevent damage to the operating mechanism.

After the activated sprinklers are replaced, the sprinkler system can be restored to service. Restoring a sprinkler system to service takes special training and should be performed only by qualified individuals.

SKILL DRILL 19-6

Closing and Reopening a Main Post Indicator Control Valve
(Fire Fighter I, NFPA 5.3.14)

1 Locate the PIV. Unlock the padlock with a key or cut the lock with a pair of bolt cutters.

2 Remove the handle from its storage position on the PIV and place it on top of the valve as you would a hydrant wrench. Turn the valve stem in the direction indicated on top of the valve to close the valve. Keep turning until resistance is strong and the visual indicator changes from "open" to "shut."

3 To reopen the PIV, turn the valve stem in the opposite direction until resistance is strong and the indicator changes back to "open." Lock the valve in the open position.

© LiquidLibrary

■ Removing Water

Water that accumulates within a building or drips down from higher levels should be channelled to a drain or to the outside of the building to prevent or limit water damage to the structure. A salvage pump may be needed to help remove the water in some cases.

Some buildings have floor drains, which will funnel water into the below-ground sewer system. These floor drains should be kept free from debris so that the water can drain freely. Fire fighters can use squeegees to direct the water into the drain.

In buildings or houses without floor drains, it may be possible to shut off the water supply to a floor-mounted toilet, remove the nuts that hold the toilet bowl to the floor, and remove the toilet from the floor flange. This maneuver creates a large drain capable of handling large quantities of water, as long as someone keeps the opening from becoming clogged with debris.

Water on a floor at ground level can often be channelled to flow outside of the structure through a doorway or other opening. Water on a floor above ground level can sometimes be drained to the outside by making an opening at floor level in an exterior wall of the building.

Water chutes or water catch-alls are often used to collect water leaking down from firefighting operations on higher floor levels and to help protect property on calls involving burst pipes or leaking roofs. A **water chute** catches dripping water and directs it toward a drain or to the outside through a window or doorway. A **water catch-all** is a temporary pond that holds dripping water in one location. The accumulated water must then be drained to the outside of the building. Water chutes and catch-alls can be constructed quickly using salvage covers.

A water chute can be constructed with a single salvage cover or with a cover and two pike poles for support. To construct a water chute, follow the steps in **SKILL DRILL 19-7** :

1 Fully open a large salvage cover flat on the ground. (**STEP 1**)

2 Roll the cover tightly from one edge toward the middle. If using pike poles, lay one pole on the edge, and roll the cover around the handle. (**STEP 2**)

3 Roll the opposite edge tightly toward the middle in the same manner. Stop when the rolls are 30–91 centimetres (1–3 feet) apart. (**STEP 3**)

4 Turn the cover upside down. Position the chute so that it collects the dripping water and channels it toward a drain or outside opening. Place the chute on the floor with one end propped up by a chair or other object. (**STEP 4**)

5 Use a stepladder or other tall object to support chutes constructed with pike poles. (**STEP 5**)

SKILL DRILL 19-7 — Constructing a Water Chute
(Fire Fighter I, NFPA 5.3.14)

1 Fully open a large salvage cover flat on the ground.

2 Roll the cover tightly from one edge toward the middle. If using pike poles, lay one pole on the edge, and roll the cover around the handle.

3 Roll the opposite edge tightly toward the middle until the two rolls are 30–91 cm (1–3 ft.) apart.

4 Turn the cover upside down. Position the chute so that it collects dripping water and channels it toward a drain or outside opening. Place the chute on the floor with one end propped up by a chair or other object.

5 Use a stepladder or other tall object to support chutes constructed with pike poles.

To construct a water catch-all, follow the steps in **SKILL DRILL 19-8**:

1. Open a large salvage cover on the ground, and roll two edges inward from the opposite sides (approximately 90 cm [3 ft.] on each side). (**STEP 1**)
2. Fold each corner over at a 90-degree angle, starting each fold approximately 90 cm (3 ft.) in from the edge. (**STEP 2**)
3. Roll the remaining two edges inward approximately 60 cm (2 ft.). (**STEP 3**)
4. Lift the rolled edge over the corner flaps, and tuck it in under the flaps to lock the corners in place. (**STEP 4**)

Water Vacuum

Special vacuum cleaners that suck up water can also be used during salvage operations. Two types of **water vacuums** are available: a small-capacity backpack type and a larger wheeled unit. The backpack vacuum cannot be used by someone who is wearing SCBA. Wet/dry shop vacuums may also be used as a low-cost alternative to water vacuums.

Fire Fighter Safety Tips © Photos.com

You cannot operate a gasoline-powered pump safely inside a structure without wearing SCBA because the operation of the pump generates deadly carbon monoxide gas.

SKILL DRILL 19-8   Constructing a Water Catch-All
(Fire Fighter I, NFPA 5.3.14)

1 Open a large salvage cover on the ground, and roll each edge of the salvage cover towards the opposite side.

2 Fold each corner over at a 90-degree angle, starting each fold approximately 90 cm (3 ft.) in from the edge.

3 Roll the remaining two edges inward approximately 60 cm (2 ft.).

4 Lift the rolled edge over the corner flaps, and tuck it in under the flaps to lock the corners in place.

© LiquidLibrary

Drainage Pumps

Drainage pumps remove water that has accumulated in basements or below ground level. Portable electric submersible pumps can be lowered directly into the water to pump it out of a building. Gasoline-powered portable pumps must be placed outside a building because they exhaust poisonous carbon-monoxide gas. These pumps use a hard suction hose, which is lowered into the basement through a window and drafts water out of the building.

Using Salvage Techniques to Limit Smoke and Heat Damage

One way to reduce property loss is to keep heat and smoke out of areas that are not involved in the fire. A closed door can effectively keep smoke and heat out of a room or area. Given this fact, fire fighters doing search and rescue operations should remember to close doors after searching a room.

Properly timed and effective ventilation practices also limit smoke and heat damage. Rapid ventilation will often reduce smoke damage in an area already filled with smoke. The visible components of smoke are mostly soot particles and other products of combustion, including corrosive chemicals, which will settle on any horizontal surface when the smoke cools. Blowing the smoke out of the building is a better option than allowing contaminants to settle on the contents. All members of the fire-suppression team—not just the salvage crews—should recognize opportunities to prevent smoke and heat damage. Ventilation is discussed in detail in the chapter "Ventilation."

■ Salvage Covers

The most common method of protecting building contents is to cover them with salvage covers. <u>Salvage covers</u> are large square or rectangular sheets of heavy canvas or plastic material that are used to protect furniture and other items from water runoff, falling debris, soot, and particulate matter in smoke residue.

Salvage crews usually begin their work on the floor immediately below the fire, with the goal being to prevent water damage to furniture and other contents on lower floors. If the fire is in an attic, fire fighters may have enough time to spread covers over the furniture in the rooms directly below the fire before pulling the ceilings to attack the flames.

The most efficient way to protect a room's contents is to move all the furniture to the centre of the room, away from the walls, where water could damage the backs of the furniture. This approach reduces the total area that must be covered, enabling one or two fire fighters to cover the pile quickly and move on to the next room **FIGURE 19-7**.

Remove any pictures from the walls and place them with the furniture. Put smaller pictures and valuable objects in drawers or wherever they will be protected from breakage. If you have enough time, roll up any rugs and place them on the pile.

Some departments use rolls of construction-grade polyethylene film for protecting structures' contents instead of salvage covers. This material comes in rolls as long as

FIGURE 19-7 Move furniture to the centre of the room so that a single salvage cover can protect all of the contents.

36.6 metres (120 ft.), so it can be unrolled over the room's contents and cut to the correct length with a box cutter. Polyethylene film is particularly useful for covering long surfaces, such as retail display counters. This material is disposable and can be left behind after the fire; in contrast, traditional salvage covers must be picked up, washed, dried, and properly folded for storage in fire-apparatus compartments after each use. Occasionally, salvage covers may be left behind as protection for the building's contents, but they should be picked up when they are no longer needed.

Special folding and rolling techniques are used to store salvage covers so that they can be deployed quickly by one or two fire fighters. Be certain to follow your fire department's SOGs/SOPs when folding or rolling a salvage cover.

To fold a salvage cover to prepare it for one fire fighter to deploy, follow the steps in **SKILL DRILL 19-9**:

1. Spread the salvage cover flat on the ground with a partner facing you. (**STEP 1**)
2. On the right side of the salvage cover, make a fold at the quarter point of the cover. Next, make a second fold that ends in the middle of the cover. (**STEP 2**)
3. On the left side of the salvage cover, make a fold at the quarter point of the cover. Next, make a second fold that ends in the middle of the cover. (**STEP 3**)
4. Fold the two halves together, and flatten the salvage cover to remove any trapped air. (**STEP 4**)
5. Make 30-cm (1-ft.) folds from each end of the cover until you reach the centre of the salvage cover. (**STEP 5**)
6. Fold the two halves together (**STEP 6**).

To fold a salvage cover to prepare it for two fire fighters to deploy, follow the steps in **SKILL DRILL 19-10**:

1. Spread the salvage cover flat on the ground with a partner facing you. Together, fold the cover in half. (**STEP 1**)
2. Together, grasp the unfolded edge and fold the cover in half again. Flatten the salvage cover to remove any trapped air. (**STEP 2**)

SKILL DRILL 19-9 Folding a Salvage Cover for Use by One Fire Fighter
(Fire Fighter I, NFPA 5.3.14)

1 Spread the salvage cover flat on the ground with a partner facing you.

2 On the right side of the salvage cover, make a fold at the quarter point of the cover. Next, make a second fold that ends in the middle of the cover.

3 On the left side of the salvage cover, make a fold at the quarter point of the cover. Next, make a second fold that ends in the middle of the cover.

4 Fold the two halves together, and flatten the salvage cover to remove any trapped air.

5 Make 30-cm (1-ft.) folds from each end of the cover until you reach the centre of the salvage cover.

6 Fold the two halves together.

© LiquidLibrary

3 Move to the newly created narrow ends of the salvage cover, and fold the salvage cover in half lengthwise. (**STEP 3**)

4 Fold the salvage cover in half lengthwise again. Make certain that the open end is on top. (**STEP 4**)

5 Fold the cover in half a third time. (**STEP 5**)

To fold and roll a salvage cover, follow the steps in **SKILL DRILL 19-11**:

1 Spread the salvage cover flat on the ground with a partner facing you. (**STEP 1**)

2 Together, fold the outside edge in to the middle of the cover, creating a fold at the quarter point. (**STEP 2**)

3 Fold the outside fold in to the middle of the cover, creating a second fold. (**STEP 3**)

4 Repeat steps 2 and 3 from the opposite side of the cover, so the folded edges meet at the middle of the cover, with the folds touching but not overlapping. (**STEP 4**)

5 Tightly roll up the folded salvage cover from the end. (**STEP 5**)

Although it takes two people to fold and roll a salvage cover for storage, this technique enables one person to unroll the cover easily and quickly. To perform the one-person salvage cover roll, follow the steps in **SKILL DRILL 19-12**:

1 Stand in front of the end of the object that you are going to cover. (**STEP 1**)

2 Start to unroll the cover over one end of the object. (**STEP 2**)

3 Continue unrolling the cover over the top of the object. Allow the remainder of the roll to settle at the other side of the object. (**STEP 3**)

4 Spread the cover, unfolding each side outward over the object to the first fold. (**STEP 4**)

5 Unfold the second fold on each side to drape the cover completely over the object. (**STEP 5**)

6 Tuck in all loose edges of the cover around the object. (**STEP 6**)

SKILL DRILL 19-10   Performing a Salvage Cover Fold for Two–Fire Fighter Deployment
(Fire Fighter I, NFPA 5.3.14)

1 Spread the salvage cover flat on the ground with a partner facing you. Together, fold the cover in half.

2 Together, grasp the unfolded edge and fold the cover in half again. Flatten the salvage cover to remove any trapped air.

3 Move to the newly created narrow ends of the salvage cover, and fold the salvage cover in half lengthwise.

4 Fold the salvage cover in half lengthwise again. Make certain that the open end is on top.

5 Fold the cover in half a third time.

© LiquidLibrary

A single fire fighter can also use a shoulder toss to spread a salvage cover. To perform a shoulder toss, follow the steps in **SKILL DRILL 19-13**:

1 Place the folded salvage cover over one arm. (**STEP 1**)

2 Toss the cover over the salvaged object with a straight-arm movement. (**STEP 2**)

3 Unfold the cover until it completely drapes the object. (**STEP 3**)

Two fire fighters can use a balloon toss to cover a pile of building contents quickly. To perform a balloon toss, follow the steps shown in **SKILL DRILL 19-14**:

1 Place the cover on the ground beside the object. (**STEP 1**)

2 Unfold the cover so that it runs along the entire base of the object. Each fire fighter grabs one edge of the cover, and brings it up to waist height. (**STEP 2**)

3 Together, lift the cover quickly so that it fills with air like a balloon. (**STEP 3**)

4 Move quickly to the other side of the item, using the air to help support the cover, and spread the entire cover over the object. (**STEP 4**)

■ Salvage Cover Maintenance

Salvage covers must be adequately maintained to preserve their shelf life. Salvage cover maintenance depends on the type of cover used. A canvas cover can usually be cleaned with a scrub brush and clean water. If the cover becomes particularly dirty, however, cleaning with a mild detergent may be necessary. Covers should be adequately rinsed when a detergent is used.

Canvas covers must be properly dried before being returned to service. Effectively drying a canvas cover will reduce mildew. Vinyl-type covers are easily maintained by rinsing and do not mildew as easily as canvas covers.

Once dried, salvage covers should be inspected for tears and holes. Any damage found can be patched by using duct tape or a sewn-on patch.

SKILL DRILL 19-11   Folding and Rolling a Salvage Cover
(Fire Fighter I, NFPA 5.3.14)

1   Spread the salvage cover flat on the ground with a partner facing you.

2   Together, fold the outside edge in to the middle of the cover, creating a fold at the quarter point.

3   Fold the outside fold in to the middle of the cover, creating a second fold.

4   Repeat steps 2 and 3 from the opposite side of the cover, so the folded edges meet at the middle of the cover, with the folds touching but not overlapping.

5   Tightly roll up the folded salvage cover from the end.

SKILL DRILL 19-12 Performing a One-Person Salvage Cover Roll
(Fire Fighter I, NFPA 5.3.14)

1 Stand in front of the end of the object that you are going to cover.

2 Start to unroll the cover over one end of the object.

3 Continue unrolling the cover until you reach the top of the object. Allow the remainder of the cover to unroll and settle at the end of the object.

4 Spread the cover, unfolding each side outward over the object to the first fold.

5 Unfold the second fold on each side, and drape the cover completely over the object.

6 Tuck in all loose edges of the cover around the object.

© LiquidLibrary

SKILL DRILL 19-13 Performing a Shoulder Toss
(Fire Fighter I, NFPA 5.3.14)

1 Place the folded salvage cover over one arm.

2 Toss the cover over the salvaged object with a straight-arm movement.

3 Unfold the cover until it completely drapes the object.

© LiquidLibrary

■ Floor Runners

A **floor runner** is a long section of protective material used to cover a section of flooring or carpet. Floor runners protect carpets or hardwood floors from water, debris, fire fighters' boots, and firefighting equipment. Fire fighters entering an area for salvage operations should unroll the floor runner ahead of themselves and stay on the floor runner while working in the area.

■ Other Salvage Operations

The best way to protect the contents of a building may be to move them to a safe location. The IC will make this determination. Any items removed from the building should be placed in a dry, secure area—preferably a single location. In some cases, salvaged items can be moved to a suitable location within the same building. If items are moved outside, they should be protected from further damage caused by firefighting operations or the weather. Valuable items should be placed in the care of a law-enforcement officer if the property owner is not present.

Sometimes the building contents can provide clues about the cause or spread of the fire. In these situations, fire investigators should be consulted and supervise the removal process.

Salvage operations sometimes extend outside the building to include valuable property such as vehicles or machinery. Use a salvage cover to protect outside property, or move vehicles if this action can be done without compromising the fire-suppression efforts.

SKILL DRILL 19-14 | Performing a Balloon Toss
(Fire Fighter I, NFPA 5.3.14)

1 Place the cover on the ground beside the object.

2 Unfold the cover so that it runs along the entire base of the object. Each fire fighter grabs one edge of the cover and brings it up to waist height.

3 Together, lift the cover quickly so that it fills with air like a balloon.

4 Move quickly to the other side of the item, using the air to help support the cover, and spread the entire cover over the object.

© LiquidLibrary

Overhaul Overview

Overhaul is the process of searching for and extinguishing any pockets of fire that remain after a fire has been brought under control. A new fire remains a possibility even if 99 percent of a fire is out and just 1 percent is smouldering. Indeed, a single pocket of embers can rekindle after fire fighters leave the scene and cause even more damage and destruction than the original fire. A fire cannot be considered fully extinguished until the overhaul process is complete.

The process of overhaul begins after the fire is brought under control. Overhaul can be a time-consuming, physically demanding process. The greatest challenge during this phase of fire operations is to identify and open any void spaces in a building where the fire might be burning undetected. If the fire extends into any void spaces, fire fighters must open the walls and ceilings to expose the burned area. Any materials that are still burning must be soaked with water or physically removed from the building. This process must continue until all of the burned material is located and unburned areas are exposed. Overhaul is also required for nonstructure fires, such as those occurring in automobiles, bulk piles (e.g., tires or wood chips), vegetation, and even garbage.

■ Safety Considerations During Overhaul

Many injuries can occur during overhaul. As mentioned earlier, overhaul is strenuous work conducted in an area already damaged by fire. Fire fighters who were involved in the fire-suppression efforts may be physically fatigued and so overlook hazards. Fire departments should summon fresh crews to conduct overhaul, giving the crews that fought the fire time to recover from their efforts. Provide adequate breaks for rehabilitation during the overhaul process.

Several hazards may be present in the overhaul area. Notably, the structural safety of the building is often compromised. Catastrophic building collapses have occurred during overhaul. Heavy objects could lead to roof or ceiling collapse, debris could litter the area, and there could be holes in the floor. Visibility is often limited, so fire fighters may have to depend on portable lighting. The presence of wet or icy surfaces makes falls more likely. Smouldering areas may burst into flames, and the air is probably not safe to breathe. In addition, during overhaul operations, dangerous equipment—including axes, pike poles, and power tools—is used in close quarters.

Fire fighters must be aware of these hazards and proceed with caution. When necessary, take extra time to evaluate the hazards and determine the safest way to proceed. There is no need to rush during overhaul, and there is no excuse for risking injury during this phase of operations.

A charged hose line must always be ready for use during overhaul operations because flare-ups may occur. For example, a smouldering fire in a void space could reignite suddenly when fire fighters open the space and allow fresh air to enter; such a fire could quickly become intense and dangerous. In addition, fine-grain combustible materials like sawdust can smoulder for a long time and then ignite explosively when they are disturbed and oxygenated.

Fire fighters must wear full PPE during overhaul. SCBA use is mandatory until the air is tested and found safe to breathe. Because working for extended periods in full PPE creates heat stress and dehydration, overhaul crews should work for short periods and take frequent breaks. Fresh crews or crews that have been properly rehabilitated should replace fatigued crews.

A safety officer should always be present during overhaul operations to note any hazards and ensure that operations are conducted safely. Company officers should supervise operations, look for hazards, and make sure that all crew members work carefully. The work should proceed at a cautious pace, with an appropriate number of crew members; too many people working in a small area creates chaos.

Always evaluate the structural condition of a building before beginning overhaul. Look for the following indicators of possible structural collapse:

- Lightweight and/or truss construction
- Cracked walls, misaligned walls, or sagging floors
- Heavy mechanical equipment on the roof
- Overhanging cornices or heavy signs
- Water accumulation

During the overhaul work, do not compromise the structural integrity of the building. When opening walls and ceilings, remove only the outer coverings, and leave the structural members in place. If more invasive overhaul is required, be careful not to compromise the structure's load-bearing members. Avoid cutting lightweight wood trusses, load-bearing wall studs, and floor or ceiling joists, especially when using power tools.

If fire damages the structural integrity of a building, the IC may call for a hydraulic overhaul rather than a standard overhaul. In this situation, large-calibre hose streams are used to completely extinguish a fire from the exterior. This strategy is appropriate if the site poses excessive risks to fire fighters and damage to the property is so extensive that it has no salvage value. Heavy mechanical equipment may be used to demolish unsafe buildings and expose any remaining hot spots. A condemned or abandoned building that is going to be torn down is not a place to risk the life or health of a fire fighter.

If a complete overhaul cannot be conducted, the IC can establish a fire watch. The fire watch team remains at the fire scene and watches for signs of rekindling. This team can then request additional help if the fire reignites.

Fire Fighter Safety Tips © Photos.com

A charged hose line must always be available during overhaul.

■ Coordinating Overhaul with Fire Investigators

Overhaul crews must work with fire investigators to ensure that important evidence is not lost or destroyed as a result of their efforts to extinguish all remnants of the fire. Ideally, a fire investigator should examine the area before overhaul operations begin, identifying evidence and photographing the scene before it is disturbed.

If a fire investigator is not immediately available, the overhaul crews should make careful observations and report to the investigator later. When performing overhaul in or near the suspected area of fire origin, note burn patterns on the walls or ceilings that could indicate the exact site of origin or the path of fire travel. A piercing nozzle can be used to extinguish hidden fire in these void areas; its use limits the damage in and around the suspected area of origin. When moving appliances or other electrical items, note whether they were plugged in or turned on. Always look for evidence that the fire investigator can use to determine the cause of the fire.

If you observe anything suspicious—particularly indications of arson—you should delay the overhaul operations until an investigator can examine the scene. Ensure that the fire will not rekindle, but do not go any further until the investigator arrives. The chapter "Fire Cause Determination" provides additional information on preservation of evidence in case of fire.

■ Where to Overhaul

Determining when, where, and how much property needs to be overhauled requires good judgment. Overhaul must be thorough and extensive enough to ensure that the fire is completely out. At the same time, fire fighters should try not to destroy any more property than necessary.

Generally, it is better to make sure that the fire is definitely out than to be too careful about damaging property. If the fire rekindles, more property damage will occur, and the fire department could be held responsible for the additional losses.

The area that must be overhauled depends on the building's construction, its contents, and the size of the fire. All areas directly involved in the fire must be overhauled. If the fire was confined to a single room, all of the furniture in that room must be checked for smouldering fire. The overhaul process also must ensure that the fire did not extend into any void spaces in the walls, above the ceiling, or into the floor. If signs indicate that the fire spread into the structure itself or into the void spaces, all suspect areas must be opened to expose any hidden fire.

If the fire involved more than one room, overhaul must include all possible paths of fire extension. The paths available for a fire to spread within a building are directly related to the type of construction. As a fire fighter, you must learn to anticipate where and how a fire is likely to spread in different types of buildings to find any hidden pockets of fire.

Fire-resistant construction can help contain a fire and keep it from spreading within a building, although this outcome is not guaranteed. Look for any openings that would allow the fire to spread, including utility shafts, pipe chases, and fire doors or dampers that failed to close tightly.

Wood-frame and ordinary construction buildings may contain several areas where a hidden fire could be burning. In particular, these structures often have void spaces under the wall coverings. When a serious fire strikes a building of ordinary or wood-frame construction, fire fighters may need to open every wall, ceiling, and void space to check for fire extension. Neighbouring buildings may also need to be overhauled if the fire could have spread into them.

In balloon-frame construction, a fire can extend directly from the basement to the attic, without obvious signs of fire on any other floor **FIGURE 19-8**. For this reason, these buildings require a thorough floor-by-floor overhaul. The problems could be compounded if the attic insulation consists of blown-in cellulose materials; these materials can smoulder for a long time. Additional information on how building construction may affect fire spread is found in the chapter "Building Construction."

If a building has been extensively remodelled, overhaul presents special challenges. Fire can hide in the space between a dropped ceiling and the original ceiling, or it may extend into a different section of the building through doors and windows covered by new construction. Some buildings have two roofs—one original and one added later—with a void space

FIGURE 19-8 A fire in a balloon-frame construction building can spread directly from the basement to the attic.

between them. A fire in this void space presents especially difficult overhaul problems.

The cause of the fire can also indicate the extent of necessary overhaul. A kitchen stove fire will probably extend into the kitchen exhaust duct and could ignite combustible materials in the immediate vicinity. Follow the path of the duct, and open the areas around it to locate any residual fire. A lightning strike releases enormous energy through the wiring and piping systems, which can start multiple fires in different parts of the building. Overhaul after this incident must be quite extensive.

Using Your Senses

Efficient and effective overhaul requires the use of all your senses. Look, listen, and feel to detect signs of potential burning.

Look for these signs:
- Smoke seeping from cracks or from around doors and windows
- Fresh or new smoke
- Red, glowing embers in dark areas
- Burned areas
- Discoloured material
- Peeling paint or cracked plaster
- Hot spots on a thermal imagery

Listen for these sounds:

- Crackling sounds that indicate active fire
- Hissing sounds that indicate water has touched hot objects

Feel for this condition:

- Heat, using the back of your hand (but only if it is safe to remove your glove)

Experience is a valuable trait during overhaul situations. By using his or her senses in a systematic manner, an experienced fire fighter can often determine which parts of the structure need to be opened and which areas remain untouched by the fire.

Thermal Imaging

A thermal imaging device is a valuable high-tech tool, which is often used during overhaul. The same type of thermal imager used for search and rescue (as discussed in the chapter "Search and Rescue") can also be used to locate hidden hot spots or residual pockets of fire. It can quickly differentiate between unaffected areas and areas that need to be opened. Using a thermal imager can decrease the amount of time needed to overhaul a fire scene and reduce the amount of physical damage to the building.

The thermal imager distinguishes between objects or areas with different temperatures and displays hot areas and cold areas as different colours on the video screen. Because it is very sensitive, this device can "see" a hot spot even if the heat source is located behind a wall. It can also show the pattern of fire within a wall, even if there is only a few degrees' difference in the wall's surface temperature.

Interpreting the readings from a thermal imaging device requires practise and training **FIGURE 19-9** . This device displays relative temperature differences, so an object will appear to be warmer or cooler than its surroundings. If the room has been superheated by fire, all of the contents—including the walls, floor, and ceiling—will appear hot. In these circumstances, an extra-hot spot behind a wall could be difficult to distinguish.

FIGURE 19-9 Thermal imaging devices can help locate hot spots.

Before overhaul operations begin, the safety officer should confirm that electric power and gas service in the overhaul area are shut off. Be careful: even though the gas supply may be turned off, gas under pressure might still be present in the lines that needs to be bled off.

In such a case, it might be better to look at the wall from the side that was not exposed to the fire to identify any hot spots.

Do not rely solely on the information from a thermal imaging device to identify areas of persistent fire, because anything that acts as an insulator can hide hot spots; hot spots behind heavy padding and carpeting or in insulated walls, for example, might not show up as hot spots on the thermal imager screen.

The thermal imager, then, is just a tool to complement the rest of the overhaul operation; it cannot completely rule out the possibility of concealed fire. The only way to ensure that no hidden fire exists is to open a ceiling or wall and do a direct inspection.

Overhaul Techniques

The objective of overhaul is to find and extinguish any fires that could still be burning after a fire is brought under control. Any fire uncovered during overhaul must be thoroughly extinguished. Overhaul operations should continue until the IC is satisfied that no smouldering fires are left.

During overhaul operations, a charged hose line should be available to douse any sudden flare-ups or exposed pockets of fire. If necessary, use direct stream from the hose line, but avoid unnecessary water damage. Extinguish small pockets of fire or smouldering materials with the least possible amount of water by using a short burst from the hose line or simply drizzling water from the nozzle directly onto the fire.

Extinguish smouldering objects that can be safely picked up by dropping them into a bathtub or bucket filled with water. Remove materials prone to smouldering, such as mattresses and cushioned furniture, from the building and thoroughly soak them outside. Roll mattresses and secure them with rope or webbing to move them out of the building and decrease the possibility of rekindling.

Place debris that is moved outside far enough away from the building to prevent any additional damage if it reignites. Do not allow debris to block entrances or exits. In some cases, a window opening can be enlarged so that debris can be removed more easily. Heavy machinery, such as a front-end loader, may be used to remove large quantities of debris from commercial buildings.

■ Overhaul Tools

Tools used during overhaul are designed for cutting, prying, and pulling, thereby ensuring that fire fighters can access spaces that might contain hidden fires. Many of the tools used for overhaul are also used for ventilation and forcible entry. The following tools are frequently required in overhaul operations:

- Pike poles and ceiling hooks for pulling ceilings and removing gypsum wallboard
- Crowbars and Halligan-type tools for removing baseboard and window or door casings
- Axes for chopping through wood, such as floorboards and roofing materials
- Power tools, such as battery-powered saws, for opening up walls and ceilings
- Pitchforks and shovels for removing debris
- Rubbish hooks and rakes for pulling things apart
- Thermal imaging cameras for identifying hot spots

Because overhaul situations usually do not require high pressures or large volumes of water, a 38- or 44-millimetre (1.5- or 1.75-inch) hose line is usually sufficient to extinguish hot spots. Follow your department's SOGs/SOPs when choosing a hose line during overhaul.

Buckets, tubs, wheelbarrows, and carryalls (rubbish carriers) are used to remove debris from a building. A **carryall** is a 0.6-square-metre (6-square-foot) piece of heavy canvas material with rope handles in each corner **FIGURE 19-10**.

FIGURE 19-10 A carryall is used to remove debris during overhaul.
Courtesy of Cascade Fire Equipment Company

FIRE FIGHTER Tips

Overhaul is physically demanding work. Pace yourself and use proper technique to increase your efficiency. Ask experienced members of your team for tips on improving your technique.

■ Opening Walls and Ceilings

Pike poles are used to open ceilings and walls to expose hidden fire. To pull down a ceiling with a pike pole, follow the steps in **SKILL DRILL 19-15**:

1 Select the appropriate length of pike pole, based on the height of the ceiling.

2 Determine which area of the ceiling will be opened. Open the most heavily damaged areas first, followed by the surrounding areas.

3 Position yourself to begin work with your back toward a door, so the debris you pull down will not block your access to the exit. (**STEP 1**)

4 Using a strong, upward-thrusting motion, penetrate the ceiling with the tip of the pike pole. Face the hook side of the tip away from you. (**STEP 2**)

5 Pull down and away from your body, so the ceiling material falls away from you. (**STEP 3**)

6 Continue pulling down sections of the ceiling until the desired area is open. Pull down any insulation, such as rolled fibreglass, found in the ceiling. (**STEP 4**)

Use the pike pole to break through and pull down large sections of ceilings made with gypsum board. Pulling down laths and breaking through plaster ceilings require more force. Power saws may be required to cut through ceilings made with plywood or solid boards.

Pike poles, axes, power saws, and handsaws can all be used to open a hole in a wall. Use the same technique to open a wall with a pike pole as you do to open a ceiling. When using an axe, make vertical cuts with the blade of the axe, and then pull the wallboard away from the studs by hand or with the pick end of the axe. A power saw can also be used to make vertical cuts. Pull the wall section away with another tool or by hand.

To open an interior wall with a pick-head axe, follow the steps in **SKILL DRILL 19-16**:

1 Determine which area of the wall will be opened. Open those areas most heavily damaged by the fire first, followed by the surrounding areas. (**STEP 1**)

2 Use the axe blade to begin cutting near the top of the wall. Cut downward between wall studs. Be alert for electrical switches or receptacles, as they indicate the presence of electrical wires behind the wall. (**STEP 2**)

3 Make two vertical cuts, using the pick end of the axe to pull the wall material away from the studs and open the wall. Work from top to bottom. Remove items such as baseboards or window and door trim with a Halligan tool or axe. (**STEP 3**)

4 Continue opening additional sections of the wall until the desired area is open. Pull out any insulation, such as rolled fibreglass, found behind the wall. (**STEP 4**)

SKILL DRILL 19-15 | Pulling a Ceiling Using a Pike Pole
(Fire Fighter I, NFPA 5.3.13)

1 Select the appropriate length of pike pole based on the height of the ceiling. Determine which area of the ceiling will be opened. Typically, the most heavily damaged areas are opened first, followed by the surrounding areas. Position yourself to begin work with your back toward a door, so the debris you pull down will not block your access to the exit.

2 Using a strong, upward-thrusting motion, penetrate the ceiling with the tip of the pike pole. Face the hook side of the tip away from you.

3 Pull down and away from your body, so the ceiling material falls away from you.

4 Continue pulling down sections of the ceiling until the desired area is opened. Pull down any insulation, such as rolled fibreglass, found in the ceiling.

© LiquidLibrary

SKILL DRILL 19-16   Opening an Interior Wall
(Fire Fighter I, NFPA 5.3.13)

1 Determine which area of the wall will be opened. Open those areas most heavily damaged by the fire first, followed by the surrounding areas.

2 Use the axe blade to begin cutting near the top of the wall. Cut downward between wall studs. Be alert for electrical switches or receptacles, as they indicate the presence of electrical wires behind the wall.

3 Make two vertical cuts, using the pick end of the axe to pull the wall material away from the studs and open the wall. Work from top to bottom. Remove items such as baseboards or window and door trim with a Halligan tool or axe.

4 Continue opening additional sections of the wall until the desired area is open. Pull out any insulation, such as rolled fibreglass, found behind the wall.

© LiquidLibrary

Wrap-Up

Chief Concepts

- Salvage and overhaul limit and reduce property losses from a fire. Salvage efforts protect property and belongings from damage, particularly from the effects of smoke and water. Overhaul ensures that a fire is completely extinguished by finding and exposing any smouldering or hidden pockets of fire in an area that has been burned.
- Because many emergency incidents occur at night, lighting is required to illuminate the scene and enable safe, efficient operations. Inside fire buildings, heavy smoke can completely block out natural light, and electrical service is often interrupted or disconnected for safety purposes. As a consequence, interior lighting is essential for fire fighters to conduct search and rescue, ventilation, fire suppression, salvage, and overhaul operations.
- Spotlights project a narrow, concentrated beam of light. Floodlights project a more diffuse light over a wide area.
- All electrical equipment must be properly grounded. Electrical cords must be well insulated, without cuts or defects, and properly sized to handle the required amperage.
- Lights can be portable or mounted on apparatus. Large hand lights that project a powerful beam of light are preferred for search and rescue activities and interior fire-suppression operations when fire fighters must penetrate smoke-filled areas quickly.
- Electricity for lighting equipment is supplied by a generator, an inverter, or a building's electrical system. An inverter is powered by a vehicle, whereas a generator is powered by gasoline or diesel fuel.
- Portable electrical equipment should be cleaned and properly maintained to ensure that it will work when needed. Test and run generators on a weekly or monthly basis to confirm that they will start, run smoothly, and produce power. Run gasoline-powered generators for 15 to 30 minutes to reduce any deposit buildup that could foul the spark plugs and make the generator hard to start.
- Salvage efforts usually are aimed at preventing or limiting secondary losses that result from smoke and water damage, fire suppression efforts, and other causes. Salvage operations include ejecting smoke, removing heat, controlling water runoff, removing water from the building, securing a building after a fire, covering broken windows and doors, and providing temporary patches for ventilation openings in the roof to protect the structure and its contents.
- Safety is a primary concern during salvage operations. Structural collapse is always a possibility because of fire damage to the building's structural components. In addition, the water used to douse the fire can add extra weight to the damaged structural components.
- The best way to prevent water damage at a fire scene is to limit the amount of water used to fight the fire. One way to do so is to shut down the sprinkler system. Sprinklers should be shut down as soon as the IC declares that the fire is under control.
- Sprinklers can be shut down by using sprinkler wedges or sprinkler stops, or by shutting off the main control valve.
- Water may be removed from a structure with a salvage pump, water chutes, water catch-alls, water vacuums, or drainage pumps.
- Salvage techniques to limit smoke and heat damage include closed doors, ventilation, and salvage covers. Salvage covers can be used to protect floors and furniture from smoke and water damage.
- The best way to protect an object from water and smoke damage is to remove it from the structure and place it in a safe location.
- Overhaul is the process of searching for and extinguishing any pockets of fire that remain after a fire has been brought under control. A single pocket of embers can rekindle after fire fighters leave the scene and cause even more damage and destruction than the original fire.
- Many injuries can occur during overhaul. During this time, fire fighters may be physically fatigued and require rehabilitation. In addition, the structure may be compromised and visibility limited. Be aware of these hazards and proceed with caution.
- A charged hose line must always be ready for use during overhaul operations to suppress flare-ups and explosions.
- Always evaluate the structural condition of a building before beginning overhaul. Look for the following indicators of possible structural collapse:
  - Lightweight and/or truss construction
  - Cracked walls, misaligned walls, or sagging floors
  - Heavy mechanical equipment on the roof
  - Overhanging cornices or heavy signs
  - Water accumulation

- During overhaul operations, do not compromise the structural integrity of the building.
- The area that must be overhauled depends on the building's construction, its contents, and the size of the fire. All areas directly involved in the fire must be overhauled.
- Use your senses of sight, hearing, and touch to determine where overhaul is needed. You can also use a thermal imaging device to find hot spots.

Hot Terms

<u>Carryall</u> A piece of heavy canvas with handles, which can be used to tote debris, ash, embers, and burning materials out of a structure.

<u>Floodlight</u> A light that can illuminate a broad area.

<u>Floor runner</u> A piece of canvas or plastic material, usually 91–122 cm (3–4 ft.) wide and available in various lengths, that is used to protect flooring from dropped debris and dirt from shoes and boots.

<u>Generator</u> An electromechanical device for the production of electricity. (NFPA 1901)

<u>Inverter</u> Equipment used to change the voltage level or waveform, or both, of electrical energy. Commonly, an inverter is a device that changes DC input to AC output. Inverters may also function as battery chargers that use alternating current from another source and convert it into direct current for charging batteries. Also known as a power conditioning unit (PCU or power conversion system (PCS). (NFPA 70)

<u>Junction box</u> A device that attaches to an electrical cord to provide additional outlets.

<u>Outside stem and yoke (OS&Y) valve</u> A sprinkler control valve with a valve stem that moves in and out as the valve is opened or closed.

<u>Overhaul</u> The process of final extinguishment after the main body of a fire has been knocked down. All traces of fire must be extinguished at this time. (NFPA 402)

<u>Post indicator valve (PIV)</u> A sprinkler control valve with an indicator that reads either open or shut, depending on its position.

<u>Salvage</u> A firefighting procedure for protecting property from further loss following an aircraft accident or fire. (NFPA 402)

<u>Salvage cover</u> A large square or rectangular sheet made of heavy canvas or plastic material that is spread over furniture and other items to protect them from water runoff and falling debris.

<u>Secondary loss</u> Property damage due to smoke, water, or other measures taken to extinguish a fire.

<u>Spotlight</u> A light designed to project a narrow, concentrated beam of light.

<u>Sprinkler stop</u> A mechanical device inserted between the deflector and the orifice of a sprinkler head to stop the flow of water.

<u>Sprinkler wedge</u> A wedge-shaped piece of wood placed between the deflector and the orifice of a sprinkler head to stop the flow of water.

<u>Water catch-all</u> A salvage cover that has been folded to form a container to hold water until it can be removed.

<u>Water chute</u> A salvage cover that has been folded to direct water flow out of a building or away from sensitive items or areas.

<u>Water vacuum</u> A device similar to a wet/dry shop vacuum cleaner that can pick up liquids. It is used to remove water from buildings.

Dusk is rapidly turning to evening. You have responded to a fire in a three-storey wood-frame apartment building. The fire appears to have begun on the third floor balcony and spread into the attic through the soffit. Four units on one end of the building have been destroyed by the fire, and about two-thirds of the roof is burned away.

The fire is now extinguished, and there is a tremendous amount of water on the upper floors due to the use of master stream devices. The IC has determined that it is safe to reenter the building in order to knock down any remaining hot spots and to protect the personal property of the tenants. It is time to assemble the equipment that you will need to perform overhaul operations.

1. A _____ is a piece of canvas or plastic material, usually 91–122 cm (3–4 ft.) wide and available in various lengths, that is used to protect flooring from dropped debris and/or dirt from shoes and boots.
   - **A.** floor runner
   - **B.** water chute
   - **C.** water catch-all
   - **D.** carryall

2. A _____ is a salvage cover that has been folded to direct water flow out of a building or away from sensitive items or areas.
   - **A.** floor runner
   - **B.** water chute
   - **C.** water catch-all
   - **D.** carryall

3. Which is the first choice for stopping water from flowing through a sprinkler system that has activated?
   - **A.** Close the OS&Y valve.
   - **B.** Close the PIV.
   - **C.** Use a sprinkler wedge to deactivate the sprinklers.
   - **D.** Close the sprinkler system control valve.

4. Which of the following should you perform during overhaul operations?
   - **A.** Clear off all floors so that when the fire investigator arrives, he can check for pour patterns.
   - **B.** Open all interior walls in the structure.
   - **C.** Remove all furniture from the room to prevent rekindling.
   - **D.** Allow the fire investigator to inspect the structure before beginning any overhaul operations.

You are in rehabilitation recovering from stopping a fast-moving fire in a strip mall. After rehabilitation, your captain notifies the IC that you are available for assignment. The IC asks you and your crew to begin the overhaul of the scene. It is one o'clock in the morning, and the power was cut to the building, so your captain asks you to set up lighting for the overhaul operations.

1. What will you need to adequately light the interior of the 12.2- by 24.4-m (40- by 80-ft.) unit of the building that was damaged by the fire?
2. What can be done to avoid possible electrocution?
3. How would you determine how many lights you can run from a single generator?

Fire Fighter Rehabilitation

© Photos.com

Fire Fighter II — FFII

Knowledge Objectives

There are no knowledge objectives for Fire Fighter II candidates. NFPA 1001 contains no Fire Fighter II Job Performance Requirements for this chapter.

Skills Objectives

There are no skills objectives for Fire Fighter II candidates. NFPA 1001 contains no Fire Fighter II Job Performance Requirements for this chapter.

Additional NFPA Standards

- NFPA 1500, *Standard on Fire Department Occupational Safety and Health Program*
- NFPA 1561, *Standard on Emergency Services Incident Management System*
- NFPA 1584, *Standard on the Rehabilitation Process for Members During Emergency Operations and Training Exercises*

You Are the Fire Fighter

I t is shaping up to be a hot day, so you decide to get your workout in early. You have just completed 5 kilometres (3.3 miles) on the treadmill when you are dispatched to a house fire in an old section of town, which dates back to the late 1890s. Upon arrival, you are assigned to fire attack and are told that the fire is in the basement. You and your crew battle through the heat as you make your way down the stairs. You have trouble locating the seat of the fire in the dirty, brown smoke. When you finally reach it, you realize it is a large fire that is spreading quickly to the upper floors. Just as you are able to get a knockdown of the fire in the basement, you notice you are getting low on air. Your officer orders a withdraw and he notifies command of your status. You and your crew quickly change out their air cylinders and are re-assigned to the first floor to begin overhaul and assist with fire attack.

You begin pulling walls and notice that you are breathing much harder than usual. You begin getting mild cramps in your arms as you tire from swinging the axe. You strive to keep up with the others in your crew. You are relieved when your audible alarm goes off, knowing you will be sent to rehabilitation.

1. What is the purpose of rehabilitation?
2. Should an incident commander (IC) consider sending personnel to rehabilitation sooner than after two bottles of air have been consumed?
3. Which actions could you have taken to reduce the heat stress on your body?

Introduction

A common saying within the fire service is "Take care of yourself first, take care of the rest of your team second, and take care of the people involved in the incident third." At first glance, this statement might seem self-centred. After all, fire fighters have a deep commitment to help others. If you were to put your personal comfort first, you probably would not want to get up in the middle of a cold winter night to answer a call. But this statement actually has a deeper meaning: You must take care of yourself physically and mentally so you can continue to help others. You must place a high priority on the health and well-being of your colleagues for the good of the fire service as a whole. Only when you and your teammates are physically fit and mentally alert will you be able to perform the tasks necessary to save lives and protect property.

To **rehabilitate** means to restore someone or something to a condition of health or to a state of useful and constructive activity. Fighting fires is a job that requires excellent physical conditioning to combat the rigours of heat, cold, smoke, flames, physical exertion, and emotional stress that are routinely encountered by team members. Even a seasoned, well-conditioned fire fighter can quickly become fatigued when battling a tough fire **FIGURE 20-1**. New fire fighters are frequently amazed at the amount of energy expended during fire-suppression activities. This exertion takes its toll on your body, leaving you hot, dehydrated, hungry, and tired. The goal of rehabilitation (usually called simply "rehab") is to take time to recharge your body's batteries, so you can continue to be a productive member of the team. Rehabilitation is a critical factor in maintaining your health and well-being. It is so critical, in fact, that it is the focus of several National Fire Protection Association (NFPA) standards, including NFPA 1500, *Standard on Fire Department Occupational Safety and*

FIGURE 20-1 Firefighting often involves extreme physical exertion and results in rapid fatigue.

Health Program; NFPA 1561, *Standard on Emergency Services Incident Management System*; and NFPA 1584, *Standard on the Rehabilitation Process for Members During Emergency Operations and Training Exercises*.

Emergency incident rehabilitation is part of the overall emergency effort. This aspect of the emergency response ensures that fire fighters and other emergency workers who are exhausted, thirsty, hungry, ill, injured, or emotionally upset can take a break for rest, fluids, food, medical monitoring, and treatment of illnesses or injuries **FIGURE 20-2**. Without the opportunity to rest and recover, you may develop physical symptoms such as fatigue, headaches, gastrointestinal problems, or collapse. Rehabilitation gives you the opportunity to evaluate and address these issues.

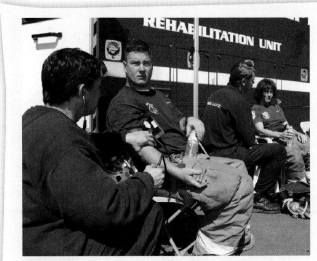

FIGURE 20-2 Rehabilitation provides fire fighters with an opportunity to take a break for rest, cooling or rewarming, rehydration, calorie replacement, medical monitoring, and ensuring accountability.

Rehabilitation enables fire fighters to perform more safely and effectively at an emergency scene. A tired or dehydrated fire fighter is not able to accomplish as much work, is more likely to be injured, and runs the risk of collapsing. The effort that is required to rescue a collapsed or injured fire fighter takes time and resources away from fire-suppression activities. For all these reasons, rehabilitation is essential to fight fires safely and effectively.

Factors, Causes, and Need for Rehabilitation

Many conditions come together during a firefighting operation to produce a stressful environment. Consider the stresses involved in a typical middle-of-the-night call for a working fire. The loud, jarring sound of the alarm jolts your sleeping body awake. Without hesitation, you must immediately get up, get dressed, and put on your personal protective equipment (PPE). You do not have time to eat or get something to drink. You may have to drive an emergency vehicle, position a ladder, and climb to the roof to cut a ventilation hole. All of these tasks require a significant amount of energy and concentration, and you must be able to move into action quickly with no time to warm up your muscles, as athletes do before an event.

FIRE FIGHTER Tips

The amount of rest needed to recover from physical exertion is directly related to the intensity of the work performed. Fire fighters who have expended a tremendous amount of energy will require a longer recovery period than those who have performed moderate work.

You may be called to a fire on the hottest day of the year, the coldest day of the year, and under all types of adverse circumstances. Because you know that lives and property are at risk, you may feel an added emotional stress, which in turn affects your body. Finally, fire fighters often work in unfamiliar, smoke-filled environments. All of these factors make firefighting very difficult and stressful.

Fire Fighter Safety Tips
© Photos.com

A tired or dehydrated fire fighter is more likely to be injured and runs the risk of collapsing. Rehabilitation is essential to correct imbalances in the body that, if left untreated, could endanger you, your crew, and others at the scene.

FIRE FIGHTER Tips

The SAID principle (Specific Adaptation to Imposed Demands) identifies a need for training that mimics the type of work to be performed. In other words, the best way to prepare for physical work is to do activities that match the type, intensity, and duration of the work. The optimal training program should include both high- and low-intensity cardiovascular training and muscular work that involves strength, endurance, and power.

■ Personal Protective Equipment

Wearing PPE—the protective clothing and breathing apparatus used by fire fighters to reduce and prevent injuries—adds heat stress on the body **FIGURE 20-3**. PPE can weigh 18.1 kilograms (40 pounds) or more, and the extra weight increases the amount of energy needed simply to move around. PPE creates a protective envelope around a fire fighter that protects his or her body from the smoke, flames, heat, and steam of a fire; at the same time, however, it traps almost all body heat inside that protective envelope.

Normally, evaporating perspiration helps cool your body so that it does not overheat. PPE contains a vapour barrier that keeps hot liquids and steam from getting in. Depending on your vapour barrier and thermal liner, evaporation may take place on a limited basis. When a fire fighter is wearing PPE, perspiration soaks the inner clothing, but evaporative cooling—such as might take place from bare skin—does not occur. Because evaporative cooling cannot occur, a fire fighter's core temperature quickly rises to dangerous levels.

Regardless of the temperature and relative humidity in the ambient air, the temperature and relative humidity inside your PPE will quickly rise to high levels. These high temperatures and high relative humidity produce heat stress. Heat stress reduces your ability to do work, and excessive amounts can lead to collapse if not treated. Because your body continues to perspire in an attempt to cool your body, you can quickly become dehydrated as well as overheated under conditions of heat stress.

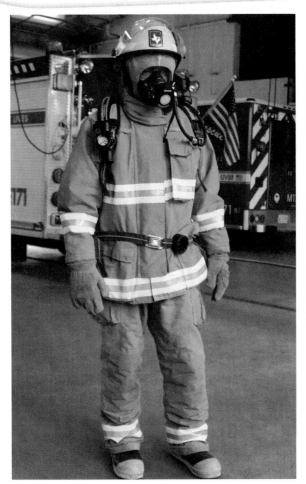

FIGURE 20-3 The PPE that is worn to protect a fire fighter contributes to heat stress.

FIGURE 20-4 When perspiring, fire fighters can lose as much as 1 litre (1 quart) of fluid in 20 minutes.

Fire Fighter Safety Tips

© Photos.com

The heat stress index combines temperature and relative humidity to measure the degree of heat stress. **TABLE 20-1** shows the heat stress index.

■ Dehydration

Dehydration is a state in which fluid losses are greater than fluid intake into the body. If left untreated, this imbalance can lead to shock and even death. Fighting fires is a very strenuous activity, and the large amounts of muscular energy required in firefighting activities produce a significant amount of heat. During this exertion, the body loses a substantial amount of water through perspiration. In fact, fire fighters in action can lose as much as 1 litres (1 quart.) of fluid in 20 minutes **FIGURE 20-4**.

Dehydration reduces strength, endurance, and mental judgment, as evidenced by a variety of signs and symptoms

TABLE 20-2. Extreme dehydration can lead to confusion and total collapse. It is important to prevent or correct dehydration as quickly as possible. To prevent dehydration, drink 0.5 L (16 ounces) of water 2 hours before physical exertion. It is simple to plan for and accomplish this fluid intake when you have a scheduled physical-conditioning session or in-service drill. Fires are unscheduled events, however, so you must try to keep your body prehydrated when you are on duty. In addition, replenishing fluids during rehabilitation is essential to correct any fluid imbalance in your body.

■ Energy Consumption

Food provides the fuel your muscles need to work properly. In times of strenuous activity, the body burns carbohydrates and fats for energy. These energy sources then need to be replenished. Without a sufficient supply of the right food for energy, your body cannot continue to perform at peak levels for extended periods. The number of calories you need depends on the duration of the activity, the time since your last meal, and your general physical condition; for example, a 190-centimetre (6 ft., 3 in.) muscular athlete who has not eaten in four hours will need more calories to fuel his body than a 157-cm (5 ft., 2 in.) video gamer who has been munching on

TABLE 20-1   Heat Index

Air Temperature (°C)										
21	24	27	29	32	35	38	41	43	46	49

Relative Humidity (%) / Apparent Temperature (°C):

RH (%)	21	24	27	29	32	35	38	41	43	46	49
0	18	21	24	26	28	31	33	35	37	39	42
10	18	21	24	27	29	32	35	38	41	44	43
20	19	22	25	28	31	34	37	41	44	49	54
30	19	24	26	29	32	36	40	45	51	57	64
40	20	23	26	30	34	38	43	51	58	66	
50	21	24	27	31	36	42	49	57	66		
60	21	44	28	32	38	46	56	65			
70	21	25	29	34	41	51	62				
80	22	26	30	36	45	58	69				
90	22	26	31	39	50	66	77				
100	22	27	33	42	56	74					

Apparent Temperature (°C)	Danger Category	Injury Threat
Below 27	None	Little or no danger under normal circumstances
27–32	Caution	Fatigue possible if exposure is prolonged and there is physical activity
33–38	Extreme Caution	Heat cramps and heat exhaustion possible if exposure is prolonged and there is physical activity
38–54	Danger	Heat cramps or exhaustion likely, heat stroke possible if exposure is prolonged and there is physical activity
Above 54	Extreme Danger	Heat stroke imminent

Courtesy of the International Association of Fire Chiefs

TABLE 20-2   Signs and Symptoms of Dehydration

Percentage of Body Weight (Lost)	Signs and Symptoms
1 percent	Increased thirst
2 percent	Loss of appetite Dry skin Dark urine Fatigue Dry mouth
3 percent	Increased heart rate
4–5 percent	Decreased work capacity by up to 30 percent
5 percent	Increased respiration Nausea Decreased sweating Decreased urine output Increased body temperature Markedly increased fatigue Muscle cramps Headache
10 percent	Muscle spasms Markedly elevated pulse rate Vomiting Dim vision Confusion Altered mental status

Source: International Association of Fire Chiefs, A Guide for Best Practices: An Introduction to NFPA 1584 (2008 Standards).

Fire Fighter Safety Tips

© Photos.com

You should always sanitize your hands before you eat.

potato chips. The type of calories that you consume is also important: it is important to refuel the body with a balanced diet consisting of nutritious food.

■ Tolerance for Stress

Each individual has a different tolerance level for the stresses encountered when fighting fires. For example, younger individuals tend to have greater endurance and can tolerate higher levels of stress. Likewise, a person who is well rested and in good condition will have a greater level of endurance than someone who is tired and in poor condition. In addition, carrying extra weight and performing strenuous tasks will strain the cardiovascular system and greatly increase the risk of heart attack.

Conditioning plays a significant role in a fire fighter's level of endurance. A well-conditioned person with good cardiovascular capacity, good flexibility, and well-developed muscles will be better able to tolerate the stresses of fighting fires than a person who is out of shape **FIGURE 20-5**. Nevertheless, even the most impressive conditioning will not keep a fire fighter from becoming exhausted under physically stressful situations.

■ The Body's Need for Rehabilitation

Rehabilitation provides fire fighters with periods of rest and time to recover from the fatigue and stresses of fighting fires and participating in emergency operations. Studies have shown that proper rehabilitation is one way to prevent fire fighters from collapsing or suffering injuries during fire-suppression activities and emergency operations. Taking short breaks, replacing fluids, ingesting healthy food, and cooling or rewarming are all measures that reduce the risks of injury and illness **FIGURE 20-6**. Rehabilitation even helps to improve the quality of decision making, because people who are tired tend to make poor decisions.

FIGURE 20-5 A well-conditioned fire fighter will have a greater tolerance for the stresses encountered when fighting fires.

FIGURE 20-6 Taking short breaks to rehabilitate lowers your risk of injury and illness.

Types of Incidents Affecting Fire Fighter Rehabilitation

The concept of rehabilitation needs to be addressed at all incidents, but it will not be necessary to implement all components of a rehabilitation centre for every incident; for example, fire departments should always have fluids and high-quality foods available for rehabilitation. Fire fighters who put out a small fire in a single room might require only water for rehydration, whereas those who are involved in extinguishing a

FIRE FIGHTER Tips

Engaging in a variety of activities can help develop optimal stair-climbing endurance. At the health club, the fire fighter may participate in step aerobics classes or use stairclimbers (with or without gear) to develop his or her leg muscles and lung capacity. Outside of the health club, a fire fighter can use step boxes or stairwells to enhance his or her climbing capacity.

FIRE FIGHTER Tips

The introduction of structured rehabilitation procedures has contributed to a reduction in the number of injuries caused by heat stress and exhaustion. The old philosophy was to fight hard until you drop, and it was not unusual to see exhausted fire fighters being carried out of burning buildings to waiting ambulances. Of course, this procedure disrupted effective firefighting operations because the fire fighters who could still function were busy rescuing their injured and exhausted teammates instead of attacking the fire.

major wildland fire may need a full-fledged rehabilitation station.

■ Extended Fire Incidents

Structure Fires

Large emergency incidents require full-scale rehabilitation efforts. Major structure fires that involve extended time on the scene will be hard on crews **FIGURE 20-7**. Crews working on the interior will become dehydrated and fatigued quickly because of the intense heat and stressful conditions brought on by working in full PPE. They will require rehabilitation so they can continue working at a healthy and safe level. Rotating

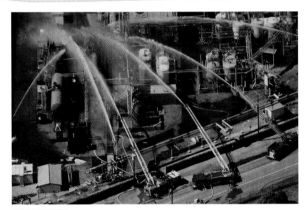

FIGURE 20-7 Major fires often require a strong effort that lasts for an extended period of time. Rehabilitation and crew rotation are important measures that limit the risk of exhaustion and injuries to fire fighters in such situations.
© Jerry S. Mendoza/AP Photos

crews off the fire ground and bringing in fresh crews also promotes health and safety and helps get the job done in an efficient manner.

High-Rise Fires

High-rise fires place increased stress on fire fighters at the same time that they drain crew members' energy resources. It takes considerable energy to walk up many flights of stairs and then work in PPE and self-contained breathing apparatus (SCBA). Hose packs, hand tools, and extra oxygen cylinders must be carried up staircases that may be crowded with people exiting the building. Just getting everyone and everything up to the location of the fire may require a truly marathon effort, even before the attack on the fire begins. It is no

FIRE FIGHTER Tips

Reducing the amount of clothing worn while at a safe distance from the fire can help to avoid heat stress before actually engaging in firefighting activities.

wonder that fatigue and dehydration can soon occur in these circumstances.

During a high-rise fire, the incident commander (IC) may assign three companies to do the work that is normally done by one company. This strategy enables the companies to rotate between attacking the fire, replenishing their air supplies, and meeting their needs for rest, fluids, and nutrition. At high-rise fires, the rehabilitation centre will often be located two or three floors below the level of the fire to reduce the time and energy expended walking up and down the stairs.

FIRE FIGHTER Tips

During active operations, fire fighters may be very reluctant to admit that they need a rest. Asking an obviously exhausted fire fighter if he or she needs to go to rehabilitation almost invariably brings the response, "No—I'm okay." All company members need to watch out for one another, and company officers must monitor their crews for indications of fatigue. It is better to go to rehabilitation a few minutes early than to wait too long and risk the consequences of injury or exhaustion. The IC should always plan ahead so that a fresh or rested crew is ready to rotate with a crew that needs rehabilitation.

Wildland Fires

Emergency incident rehabilitation was first used on a widespread basis during firefighting operations carried out in wildland areas such as grasslands and forests. Given the size and intensity of major wildland fires, crews need to work in shifts so that their bodies can recover from the heat, smoke, and stresses of the fire. Minimal rehabilitation efforts may be all that is needed for a small ground fire that is easily extinguished. By contrast, a large forest fire may require hundreds of fire fighters and take several weeks to control and extinguish; such a fire requires a significant and ongoing rehabilitation effort for all crews involved FIGURE 20-8 .

Proper nutrition and rehydration are essential in keeping the crews healthy and ready to resume duty. Some studies have shown that wildland fire fighters may require as much as 5000 to 6000 calories (21 to 25 joules) per day.

■ Other Types of Incidents Requiring Rehabilitation

Other incidents may also require extensive rehabilitation efforts to maintain the health and safety of fire fighters. For example, hazardous materials incidents that require responders to wear a **fully encapsulating suit** (a protective suit that fully covers the responder, including the breathing apparatus) are especially stressful FIGURE 20-9 . These kinds of incidents can expose responders to strenuous conditions for extended periods of time. They may also require that

FIGURE 20-8 Wildland fires often involve hundreds of fire fighters who work for days or weeks at the site.
Courtesy of Michael Rieger/FEMA News Photo

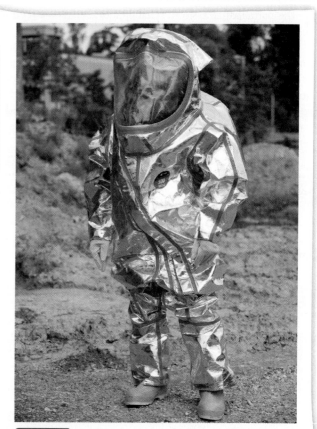

FIGURE 20-9 Responders wearing fully encapsulating suits must be carefully monitored for symptoms of heat stress.

the staging area be located at significant distance from the hazardous materials, such that responders have to walk for long distances while wearing the heavy PPE. As a result, hazardous materials incidents require adequate rehabilitation for responders.

On other occasions, the fire department may be involved in long-duration search and rescue activities or other incidents that require the presence of public safety agencies for extended periods of time. These situations can be both mentally and physically taxing. The establishment of a rehabilitation centre where fire fighters can recuperate during these stressful incidents is essential.

The need for rehabilitation is not limited to emergency situations. Training exercises, athletic events, and even standby assignments may also require rehabilitation. Whenever fire fighters are required to be ready for action for an extended period of time, some provision should be made for providing replenishing fluids, nourishing foods, and rewarming or cooling for crew members.

Large-scale training activities, including live fire exercises, generate the same concerns as major fire incidents, and rehabilitation should likewise be incorporated in the planning for these activities. Training exercises may be conducted over a full day and involve a series of activities as well as time to set up each exercise. As part of this process, time should be set aside for the participants to go to rehabilitation between strenuous activities.

Weather conditions can significantly affect the need for rehabilitation. Fire fighters should always dress appropriately for the weather, and plans for rehabilitation procedures should take into account the anticipated environmental conditions. Whether it is hot or cold, returning the body's temperature back to normal is one of the primary goals of rehabilitation.

Emergencies that occur when the temperature is very hot will increase the need to rotate crews and allow extensive rehabilitation. Crews working on the interior fire attack may not notice much difference, because their environment during the attack will be hot regardless of the external temperature. However, hot weather will definitely affect crews working on the outside. These fire fighters will tend to become dehydrated and fatigued much faster than they would if the ambient temperature was in a more comfortable range.

Another factor that must be considered is the humidity of the air—humidity plays an important role in evaporative cooling. High humidity reduces evaporative cooling, making it more difficult for the body to regulate its internal temperature.

Cold weather also increases the need for rehabilitation and crew rotation. Indeed, cold temperatures can be just as dangerous as heat. **Hypothermia**, a condition in which the internal body temperature falls below 35°C (95°F), can lead to loss of coordination, muscle stiffness, coma, and death

FIGURE 20-10 Winter weather presents a different set of problems for fire fighters than summer weather. Rehabilitation procedures must be conducted in a warm space, and fire fighters should be given hot liquids and food, as well as dry clothing.

© Jack Dagley Photography/ShutterStock, Inc.

FIGURE 20-10 . Even in cold weather, the weight of PPE and the physical exertion of fighting a fire will cause the body to sweat inside the protective clothing. The combination of damp clothing and cold temperatures can quickly lead to hypothermia.

FIRE FIGHTER Tips

At most fire department incidents, only one rehabilitation centre will be established. In contrast, at large-scale incidents, it may be necessary to set up multiple rehabilitation centres. In some communities, rehabilitation is initiated entirely by fire service personnel. In other communities, establishing the rehabilitation facility may be a combined effort between the Salvation Army, the Canadian Red Cross, the fire department auxiliary unit, or a separate Emergency Medical Services (EMS) agency.

How Does Rehabilitation Work?

One way to understand how an emergency incident rehabilitation centre works is to look at the functions it is designed to perform. A widely used model of rehabilitation consists of seven parts:

- Relief from climatic conditions
- Rest and recovery
- Active or passive cooling or warming
- Rehydration and calorie replacement
- Medical monitoring
- Member accountability
- Release and reassignment

Fire Fighter Safety Tips

© Photos.com

PPE is designed to protect fire fighters from hazards encountered in emergency operations. Fire fighters who are overheated should remove their PPE as soon as possible to permit evaporative cooling. PPE must be removed only in a safe place, outside the hazard area.

Although these seven activities are not necessarily completed in the order listed, it is important that each step be addressed.

Relief from Climatic Conditions

Few days provide the perfect climatic conditions for rehabilitation. When the temperature is between approximately 20°C and 22°C (68°F and 72°F), the humidity is low, and plenty of shade is available, indoor rehabilitation may not be needed. Conversely, on days when the temperature is warmer or colder than 20°C–22°C (68°F–72°F), or when there is high humidity or wind, rehabilitation needs to take place in a climate-controlled location. In that case, a rehabilitation centre may be set up in a nearby building such as a church, school, business, or shopping centre, or rehabilitation services may be provided in a transit bus that is brought to the emergency scene.

When entering the rehabilitation centre, it is important to take off your PPE to help your body warm up or cool down as needed **FIGURE 20-11**. If your inner clothing is wet, it may be necessary to change clothing. Many fire fighters carry an extra pair of socks and a second pair of gloves to handle such emergencies. Use a towel to dry off if needed.

Rest and Recovery

Rest should begin as soon as you arrive for rehabilitation. The rehabilitation centre should be located away from the central activity of the emergency, so you can disengage from all other stressful activities and remove personal protective clothing. Rest continues as you go through the other parts of revitalization. Rehabilitation centres are equipped with chairs or cots so fire fighters can sit or lie down and relax in a comfortable position.

FIRE FIGHTER Tips

The rehabilitation centre will provide some type of restroom facility; for example, the existing restrooms in the building may be used for the rehabilitation centre, or portable restrooms may be brought to the emergency scene.

FIGURE 20-11 In the rehabilitation centre, fire fighters should be able to remove personal protective clothing and get some rest before returning to action.

© Tom Carter/PhotoEdit, Inc.

Active or Passive Cooling or Warming

The third step of rehabilitation is stabilizing the body's internal temperature. Interior firefighting requires wearing full PPE including SCBA. This encapsulating gear and the rigorous exercise required during firefighting generate huge amounts of body heat that cannot be dissipated by normal means. Performing strenuous external firefighting, especially in times of hot temperatures and high relative humidity, but even without the use of SCBA, also generates huge amounts of body heat. At other times, fire fighters may suffer from cold temperatures. Mounting a defensive attack on a large outdoor lumber yard in –15°C (5°F) temperatures, for example, can lead to hypothermia. Rehabilitation centres must be ready to provide means to warm cold fire fighters and cool overheated fire fighters.

Active and Passive Cooling

For the overheated fire fighter, PPE should be removed as soon as possible to allow the body to cool. Two types of cooling are possible: active and passive.

Passive cooling is performed by removing PPE and other clothing and moving to a cooler environment. These measures allow the body to cool through effective sweating. Passive cooling works best when both the ambient temperature and the relative humidity are low, which is why stabilizing a fire fighter's body temperature through passive cooling becomes even more complicated during periods of hot weather or high humidity. During these periods, rehabilitation centres need to be climate controlled so that fire fighters can achieve a normal body temperature before resuming active duty. Some fire departments have air-conditioned vehicles available for rehabilitation. Others use buses or establish rehabilitation centres in an already-cooled area such as a shopping centre.

If passive cooling is not adequate to produce a reduction in body temperature, active cooling may be required. Active cooling is performed by placing something cool on the fire fighter. This can be done by engaging a misting fan, by placing cool, wet towels around the head and neck, or by dipping both arms in cool water. Active cooling can produce a more rapid reduction in the body temperature.

Passive and Active Warming

Fighting fires in cold temperatures can cause problems as well. Fire fighters responding to incidents during cold weather are subject to both hypothermia and **frostbite** (damage to tissues resulting from prolonged exposure to cold) **FIGURE 20-12**. In these cases, the rehabilitation centre needs to be heated so that fire fighters can warm up before returning to the cold environment. Passive warming occurs when the fire fighter gets out of the cold, and the body restores itself to an equilibrium temperature. Active heating may be needed when a fire fighter's body is unable to restore itself to the normal temperature balance. Active warming is achieved by medical professionals through the use of heat packs, warming blankets, or warmed IV fluids.

Fire fighters who are wet or severely chilled should be wrapped in warming blankets and moved into a well-heated

FIGURE 20-12 Prolonged exposures to freezing weather can result in severe frostbite.

Courtesy of Neil Malcom Winkelmann

area before they remove their PPE. As soon as possible, all wet clothing should be removed and replaced with warm, dry clothing.

■ Rehydration and Calorie Replacement

Rehydration

When the body becomes overheated, it sweats so that evaporative cooling can reduce body temperature. Perspiration is composed of water and other dissolved substances such as salt. During fire-suppression activities, a fire fighter can perspire enough to lose 2 L (2 qt.) of water in the time it takes to go through two cylinders of air. Because 1 L (1 qt.) of water weighs 1 kg (2.2 lb.), losing 2 L (2 qt.) of water is equivalent to a 2 percent loss in body weight for a 91 kg (201 lb.) fire fighter. The loss of that much bodily fluid can result in impaired body temperature regulation. Further dehydration can result in reduced muscular endurance, reduced strength, and heat cramps. In addition, severe cases of dehydration can contribute to heatstroke and death. Because a fire fighter can lose so much water during fire-suppression activities, it is important to replace fluids before severe dehydration takes place **FIGURE 20-13**.

When the body perspires, it loses **electrolytes** (certain salts and other chemicals that are dissolved in bodily fluids and cells) as well as water. If a fire fighter loses large amounts of water through perspiration, his or her electrolytes must be replenished in addition to the water. Sports-type drinks are often used in rehabilitation for this purpose; they contain water, balanced electrolytes, and some sugars.

Rehydrating the body is not as simple as just drinking lots of fluids quickly. Drinks such as colas, coffee, and tea should be avoided because they contain caffeine; caffeine acts as a diuretic that causes the body to excrete more water. Sugar-rich carbonated beverages are not tolerated or absorbed as well as straight water or sports drinks, because too much sugar is difficult to digest and causes swings in the body's energy level. In addition, drinks that are too cold or too hot may be difficult to consume and may prevent the fire fighter from ingesting enough liquids.

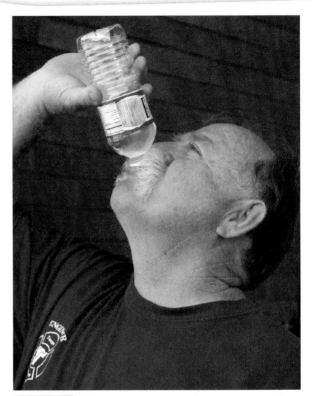

FIGURE 20-13 The rehabilitation centre should have plenty of fluids available to rehydrate fire fighters. Plain water or diluted sports drinks are preferred.

An additional concern is the rate at which fluids are absorbed from the stomach. Drinking too much too quickly can cause bloating, a condition in which air fills the stomach. It causes a feeling of fullness and can lead to discomfort, nausea, and even vomiting.

Studies have demonstrated that the stomach can absorb approximately 1 L (1 qt.) of fluid per hour but, as previously noted, the body can lose up to 2 L (2 qt.) of fluid per hour during intense activity; thus, the body may lose fluids much more rapidly than they can be replaced. Once the initial 2 L (2 qt.) of water are lost, the body will require 1–2 hours to recover. One key to remaining hydrated while fighting fires is to make sure you are hydrated before you reach the fire ground. To do so, keep your body hydrated whenever you are on duty. This will reduce the need for playing catch-up with fluids during rehabilitation.

Try to drink enough fluids to keep your body properly hydrated at all times, particularly during hot weather.

FIRE FIGHTER Tips

The sensation of thirst is not a reliable indicator of the amount of water the body has lost. Thirst develops only after the body is already dehydrated. Do not wait until you feel thirsty to focus on your hydration. Instead, start to drink before you get thirsty, and remember to drink early and often. Fire fighters should strive to drink a litre of water per hour during periods of work.

Near-Miss REPORT

Report Number: 10-0000880

Synopsis: Heat stress overcomes fire fighter.

Event Description: The morning of my shift, the weather conditions were as follows: temperature 38°C (101°F), heat index 43, humidity 80 percent or more, clear, sunny, and dry, and no rain in the past 12–24 hrs.

I was assigned to the driver/operator position of my engine company, which is staffed only by a driver and officer. Approaching [time deleted], my company was dispatched for a working structure fire in a subdivision. Homes in this subdivision are approximately 165- to 205-square-metre (1800- to 2200-square-foot) wood-frame construction built within the past 15 years.

Upon arrival to the scene we found a 18.6-sq. m (2000-sq.-ft.) home with 25 percent involvement with exposures on the two sides. I hand-jacked 121.9 metres (400 feet) of 127-millimetre (5-inch) LDH to the hydrant and then proceeded to charge two 44-mm (1.75-in.) cross lays. There was one engine (staffed with two), one truck (staffed with two), and one heavy rescue (staffed with three).

I assisted with pulling a 64-mm (2.5-in.) preconnected portable master stream to the rear of the building where the fire was located. My pump was taken over by the next arriving apparatus, and I was reassigned to interior operations. I performed interior operations for approximately 12 minutes and went to rehab with the rest of my crew. I was in rehab for about 10 minutes before I began to feel chest palpitations, weakness, and dizziness. According to medical crews on scene, I was also lethargic, disoriented, and hypotensive (BP 80 palp). I had hydrated prior to and on shift and while in rehab. At this point, I do not know the actual events that happened as this is all from secondhand accounts of people on the scene. I was moved to the medic unit, and an IV was established with normal saline. The cardiac monitor was applied, which showed sinus tachycardia. All my gear was removed, and active cooling took place. After several minutes of cooling and IV fluids, my core temperature was 37°C (99°F). I was in the hospital for several hours and was administered three 1-L (1-qt.) bags of saline. I was sent home from the ER at approximately 1600 and placed on sick leave for the remainder of my shift.

Lessons Learned: The entire shift (two on the engine and two on the ambulance) were hydrating prior to and during the shift to prevent this very problem. Nobody else had this happen during this call, so does this mean it is simply the way one person's body reacted? I believe so. Hydration is important and must be constant during periods of high heat and humidity; however, hydration is not the only way to ensure our safety and health in this weather. We must monitor ourselves and not be afraid to say, "I need a break," earlier than a normal day. I was not pushing myself any harder than normal—but then again, who can say since I had this problem. Initial staffing levels would have allowed me to perform less work than I did. I delivered hose to the hydrant by hand (since the hydrant was not on our direction of travel), and I also pulled the master stream line because the fire fighters operating in the house were not in a position to perform this task. Proper staffing would have meant some of the tasks I did would have been performed by somebody else.

This habit will decrease your risk of dehydration when an emergency incident occurs. Adequate hydration also ensures peak physical and mental performance.

Calorie Replacement

A fire fighter performs intense physical work and burns many calories during active firefighting, and it is necessary to replace those calories during rehabilitation. In much the same way that an engine runs on diesel fuel, the human body runs on **glucose**. Glucose (also known as blood sugar) is carried throughout the body by the bloodstream and is needed to burn fat efficiently and release energy.

For the body to work properly, its glucose levels need to be in balance: If blood sugar drops too low, the body becomes weak and shaky. If blood sugar is too high, the body becomes sluggish. Blood sugar levels can be balanced by eating a proper diet of carbohydrates, proteins, and fats.

Carbohydrates—a major source of fuel for the body—can be found in grains, vegetables, and fruits. The body converts carbohydrates into glucose, so carbs are an excellent energy source. A common dietary myth is that carbohydrates are fattening and should be avoided. In fact, carbohydrates should account for 55–65 percent of calories in a balanced diet. Carbohydrates have the same number of calories per gram as proteins and fewer calories than fat. Additionally, carbohydrates are the only fuel that the body can readily use during high-intensity physical activities, such as fighting fires.

Proteins perform many vital functions within the body. Most protein comes from meats and dairy products, but smaller amounts are found in grains, nuts, legumes, and vegetables. The body uses proteins (amino acids) to grow and repair tissues; proteins are used as a primary fuel source only in extreme conditions, such as during starvation. Like other nutrients, excess proteins are converted and stored in the body as fat. A balanced diet will get 10–12 percent of its calories from proteins.

Fats are also essential for life. These nutrients are used as a source of energy, for insulating and protecting organs, and for breaking down certain vitamins. Some fats, such as those found in fish and nuts, are healthier and more beneficial to the body than others, such as the fats found in margarine. No more than 25–30 percent of the diet should come from fats, however. Excess fat consumption—particularly of saturated fats (which come mostly from animal products)—is linked to high cholesterol, high blood pressure, and cardiovascular disease.

Both candy and soft drinks contain sugar (unless they are specifically labelled as "sugar-free"), so the body can quickly absorb and convert these foods to fuel. Unfortunately, simple sugars also stimulate the production of insulin, which reduces blood glucose levels. As a consequence, eating a lot of sugar can actually result in lower energy levels.

For a fire fighter to sustain peak performance levels, it is necessary for him or her to refuel during rehabilitation. During short-duration incidents, low-sugar, high-protein sports bars can be used to keep the glucose balance steady. During extended-duration incidents, however, a fire fighter should eat a more complete meal **FIGURE 20-14**. The proper balance of carbohydrates, proteins, and fats will maintain energy levels throughout the emergency. To ensure peak performance, the meal should include complex carbohydrates, such as whole-grain breads, whole-grain pasta, rice, and vegetables. It is also better to eat a series of smaller meals over time, rather than one or two large meals, because larger meals can increase glucose levels and slow down the body.

Healthy, balanced eating is the recommended lifestyle for all fire fighters. Proper nutrition reduces stress, improves health, and provides more energy. It is just as important to keep blood sugar levels balanced throughout the day as it is to remain hydrated. If you follow this regimen, you will be ready to react to an emergency at any time. In Canada, the federal government has created the Canada Food Guide. It is suggested that you review the Canada Food Guide in order to understand what is recommended for a balanced diet. The Canada Food Guide is regularly reviewed and amended to ensure that it remains current.

■ Medical Monitoring

Medical monitoring during rehabilitation is important to identify potentially dangerous health problems. The checking of vital signs is one way to identify potential problems. Many

FIGURE 20-14 Plans for rehabilitation at extended-duration incidents should include complete balanced meals.
© Photodisc/Creatas

FIRE FIGHTER Tips

During rehabilitation, it is important to eat foods that contain carbohydrates. Foods such as energy bars or meal-replacement bars are readily available and can be stored for extended periods of time, as opposed to perishable foods such as fruit. Fruit, however, provides carbohydrates and contains a lot of water, which helps maintain a good degree of hydration.

rehabilitation protocols specify the following types of medical monitoring:

- Checking the fire fighter's heart rate, respiratory rate, blood pressure, and temperature
- Determining how much oxygen is in the fire fighter's bloodstream by taking a pulse oximetry reading
- Performing a carbon-monoxide assessment

If the fire fighter's vital signs or test results are abnormal upon entering the rehabilitation centre, the fire fighter should be reexamined before being permitted to return to firefighting.

In addition, signs of illness or injury should be assessed in the rehabilitation centre before the fire fighter resumes active duty. As a fire fighter, you should be alert for any of the following signs and symptoms:

- Chest pain
- Dizziness
- Shortness of breath

- Weakness
- Headache
- Cramps
- Mental status changes
- Behavioural changes
- Changes in speech
- Changes in gait

Report these signs to medical personnel at the rehabilitation centre if you notice any of them in yourself or in one of your crew members.

FIRE FIGHTER Tips

Depending on their levels of exertion, the type of work that they have performed, and how they feel, fire fighters entering rehabilitation should be quickly checked by medical personnel before returning to active firefighting. It is important to note that Canadian law prohibits forced medical procedures except under exceptional circumstances, such as diminished mental ability to understand the consequences and specific circumstances involving minors. Therefore, one cannot be forced to submit to medical evaluation in a rehabilitation unit.

Transportation to a Hospital

Another function of the rehabilitation centre is to arrange for transportation of ill or injured fire fighters. An ambulance should be available to ensure that ill or injured fire fighters receive prompt transport.

FIRE FIGHTER Tips

Many fire services have found that assigning a fire officer to be in charge of the rehabilitation centre ensures that fire fighters stay in rehabilitation as long as needed and that those fire fighters who need to be transported to a hospital for further evaluation are transported rapidly. Some fire services involve EMS in their rehabilitation centre.

■ Member Accountability

Crew integrity should be maintained during rehabilitation. Crew members should be released to rehabilitation together and should be released from rehabilitation together. This practice makes accountability on the fire ground much easier for the IC to control. It also ensures that all members of the crew report to rehabilitation to receive the rest, cooling or warming, hydration, food, and medical monitoring that they require. In the event that one member of a crew is not able to return to the active fire scene, the IC must determine how to handle the situation.

■ Release and Reassignment

Once they are rested, rehydrated, refuelled, and rechecked to make certain that they are fit for duty, fire fighters can be released from rehabilitation and reassigned to active duty. Reassignments may be to the same job performed before the trip to rehabilitation or to a different task, depending on the decision of the IC.

Personal Responsibility in Rehabilitation

The goal of every firefighting team is to save lives first and property second. To achieve this goal, you need to take care of yourself first, take care of the rest of your crew second, and then take care of the people involved in the incident. Another way of looking at this issue is to remember that safety begins with you. The other members of your crew depend on you to bear your share of the load. To meet their expectations, you need to maintain your body in peak condition.

Part of your responsibility is to know your own limits. No one else can know what you ate, whether you are light-headed or dehydrated, whether you are feeling ill, or whether you need a breather. You are the only person who knows these things. Therefore, you may be the only person who knows when you need to request rehabilitation. It may be difficult to say, "I need a break," while the rest of the crew is still hard at work. Even so, it is better to take a break when you need it than to push yourself too far and have to be rescued by other members of your crew.

Regular rehabilitation enables fire fighters to accomplish more work during a major incident, and it decreases the risk of stress-related injuries, illness, and death. Remember: safety begins and ends with you.

Fire Fighter Safety Tips   © Photos.com

Stress-management skills are crucial for high-quality performance, regardless of the type of stressor involved and the role that the person plays in emergency assistance. A fire fighter needs to implement basic stress-management skills to deliver high-quality performance. Exercise, proper nutrition, and emotional support are all helpful in managing stress.

FIRE FIGHTER Tips

You must take care of yourself first if you are to be able to help others.

FIRE FIGHTER Tips

It is important to consider when and how often rehabilitation is necessary. NFPA 1584 indicates fire fighters should undergo rehabilitation after exhausting a single 45- or 60-minute SCBA cylinder, after every second 30-minute SCBA cylinder, or after 40 minutes of intense work without SCBA. Another recommendation is that a team should have at least 20 minutes of rest during its period of rehabilitation. Additional rehabilitation time may be needed after completing very stressful activities.

Wrap-Up

Chief Concepts

- Rehabilitation is an important part of firefighting. It is designed to ensure that emergency personnel can rest, cool off or warm up, receive fluids and nourishment, and be evaluated for potential medical problems.
- Conditions such as stress, physical exertion, and weather extremes all take a toll on the body.
- Rehabilitation is essential to correct physical imbalances, such as overheating and dehydration, that could endanger members of the firefighting crew.
- Due to their protective qualities and weight, PPE and SCBA add heat stress on the body.
- Dehydration is a state in which fluid losses are greater than fluid intake into the body. If left untreated, it can lead to shock and even death. To avoid dehydration, frequently drink small amounts of nonsugary and noncaffeinated beverages throughout the shift.
- To work at peak capacity, the body needs to be fuelled with the right amount of nutrient-rich foods. A physically active fire fighter needs more calories than an office worker.
- Rehabilitation provides fire fighters with periods of rest and time to recover from the physical fatigue and mental stresses of fighting fires. Taking short breaks, replacing fluids, ingesting healthy food, and cooling or rewarming are all measures that can reduce the risk of injury and illness. They also improve mental focus.
- Rehabilitation needs to be addressed at all incidents, but establishing a rehabilitation centre may not be necessary for every incident. Types of incidents that require a dedicated rehabilitation centre include structure fires, high-rise fires, and wildland fires. Rehabilitation is also needed for hazardous materials incidents, search and rescue operations, and training activities.
- Emergencies that occur during weather extremes require a rehabilitation centre for cooling or rewarming.
- Rehabilitation includes seven components:
  - Relief from climatic conditions
  - Rest and recovery
  - Active or passive cooling or warming
  - Rehydration and calorie replacement
  - Medical monitoring
  - Member accountability
  - Release and reassignment
- Report the following signs to medical personnel at the rehabilitation centre if you notice any of them in yourself or in one of your crew members:
  - Chest pain
  - Dizziness
  - Shortness of breath
  - Weakness
  - Headache
  - Cramps
  - Mental status changes
  - Behavioural changes
  - Changes in speech
  - Changes in gait
- Your responsibilities in rehabilitation are to know your limits, to listen to your body, and to use the rehabilitation facilities when needed.

Hot Terms

<u>Dehydration</u> A state in which fluid losses are greater than fluid intake into the body. If left untreated, dehydration may lead to shock and even death.

<u>Electrolytes</u> Certain salts and other chemicals that are dissolved in bodily fluids and cells. Proper levels of electrolytes need to be maintained for good health and strength.

<u>Emergency incident rehabilitation</u> A function on the emergency scene that cares for the well-being of the fire fighters. It includes relief from climatic conditions, rest, cooling or warming, rehydration, calorie replacement, medical monitoring, member accountability, and release.

<u>Frostbite</u> A localized condition that occurs when the layers of the skin and deeper tissue freeze. (NFPA 704)

<u>Fully encapsulating suit</u> A protective suit that completely covers the fire fighter, including the breathing apparatus, and does not let any vapour or fluids enter the suit. It is commonly used in hazardous materials emergencies.

<u>Glucose</u> The source of energy for the body. One of the basic sugars, it is the body's primary fuel, along with oxygen.

<u>Hypothermia</u> A condition in which the internal body temperature falls below 35°C (95°F), usually a result of prolonged exposure to cold or freezing temperatures.

<u>Rehabilitate</u> To restore someone or something to a condition of health or to a state of useful and constructive activity.

FIRE FIGHTER
in action

You arrive at your station for your tour of duty, but are told by your captain that you need to report to station 8 and will be assigned to the rehabilitation unit for the shift. As you drive to station 8, you dread the thought of being assigned to rehab. Your first thought is that you will not get to do any of the "fun" stuff like fighting fires, doing extrications, or providing patient care. Your second thought is even more concerning: "What do I do if I have to set up rehab?" For the remainder of your drive, you have a nagging feeling that you are not prepared for this assignment. When you finally arrive, you pull the captain aside and express your concerns. He begins to go over the basics of rehabilitation.

1. What is a state in which fluid losses are greater than fluid intake into the body?
   - **A.** Homeostasis
   - **B.** Dehydration
   - **C.** Hypothermia
   - **D.** Hyperthermia

2. Where should the rehabilitation centre be located?
   - **A.** Near the SCBA bottle-changing station
   - **B.** Away from the central activity of the emergency
   - **C.** Close to the ancillary activities of the emergency
   - **D.** Close to the central activity of the emergency

3. Many rehabilitation medical monitoring protocols recommend checking all of the following except _____.
   - **A.** vital signs
   - **B.** glucometry
   - **C.** pulse oximetry
   - **D.** carbon-monoxide assessment

4. Which of the following is *not* one of the seven parts of rehabilitation?
   - **A.** Active or passive heating or cooling
   - **B.** Member accountability
   - **C.** Rest and recovery
   - **D.** PPE replacement

5. Removing PPE and other clothing and moving to a cooler environment is considered _____.
   - **A.** active cooling
   - **B.** passive cooling
   - **C.** active warming
   - **D.** passive warming

Your crew takes great pride in its fitness level and is well known for its stamina on the fire ground. One day, you and your crew are aggressively fighting a large fire. You have already gone through your first cylinder when your partner's low-pressure alarm begins to sound. You both exit and make your way to the engine to swap out the cylinder. He mentions how he dreads having to go to rehab because he feels great. Before you can say that rehab is for your own good, your partner radios the IC that you are ready for another assignment.

1. What will you do now?
2. What do you say when you get back to the station?
3. What do you say when your captain says that he noticed you did not go through rehabilitation?
4. If you had it to do over again, what would you do differently?

Extreme Cold Weather: Cautions and Operations

Fire Fighter I

Knowledge Objectives

After studying this chapter, you will be able to:

- Describe the tasks to perform to prepare a fire apparatus for an extreme-cold-weather response. (p 648–649)
- Describe how extreme cold weather can obstruct the water supply. (p 649)
- Describe the measures the fire department must be prepared to take to ensure a reliable water supply. (p 649)
- Describe the tasks to perform to prepare energized equipment for an extreme-cold-weather response. (p 650–651)
- Describe how extreme winter conditions impact a structure. (p 650–651)
- Describe how fire fighters can prepare themselves for an extreme-cold-weather response. (p 651)
- List the five ways the body loses heat. (p 651)
- Define *hypothermia* and describe how it occurs. (p 651–652)
- Define *frostnip* and describe how it occurs. (p 652)
- Define *frostbite* and describe how it occurs. (p 652)
- Describe the precautions the fire apparatus driver must take when responding to an incident in wintry conditions. (p 653)
- Describe the precautions a fire fighter should take when dismounting a fire apparatus and arriving on scene. (p 653)
- Describe the specialized fire-ground tactics used when suppressing a fire in extreme cold weather. (p 653–654)
- Describe the precautions a fire fighter should take when using a ladder in extreme cold weather. (p 655)
- Describe the precautions a fire fighter should take to ensure the safety and success of rescue operations. (p 655–656)
- Describe how extreme cold can impair self-contained breathing apparatus (SCBA). (p 656)
- Explain why rehabilitation is critical for fire fighters during extreme cold weather. (p 656–657)
- List the signs that a fire fighter should report to rehabilitation due to a suspected cold injury. (p 657–658)

Skills Objectives

There are no skills objectives for Fire Fighter I candidates. NFPA 1001 contains no Fire Fighter I Job Performance Requirements for this chapter.

© Photos.com

Fire Fighter II

FFII

Knowledge Objectives

There are no knowledge objectives for Fire Fighter II candidates. NFPA 1001 contains no Fire Fighter II Job Performance Requirements for this chapter.

Skills Objectives

There are no skills objectives for Fire Fighter II candidates. NFPA 1001 contains no Fire Fighter II Job Performance Requirements for this chapter.

You Are the Fire Fighter

On a late-January night, your phone rings, waking you from a sound sleep. Dispatch reports that there is a chimney fire that has spread to the roof structure of a home. You quickly dress, jump into your vehicle, and drive to the fire hall to prepare for response. While en route, you notice a fresh snowfall, but the clouds have cleared, it's windy and the thermometer in your vehicle says -27 degrees C. You realize that given the weather conditions, this will be a challenging situation.

1. What precautions do you need to consider when responding to an incident like this one?
2. What special fire ground tactics will you need to employ to suppress the fire under these extreme weather conditions?
3. How do you prepare yourself for extreme weather fire response?

Introduction

Responding to incidents in extreme cold temperatures is a challenging reality that the Canadian fire service faces every winter. Responding to emergency incidents—fires, rescues, hazardous materials spills, and emergency medical calls—is both dangerous and difficult in normal weather conditions; adding extreme cold temperatures, ice-covered roads, and deep snow to the mix can make response, arrival, and operations particularly treacherous **FIGURE 21-1**.

In this chapter, we will discuss how to prepare fire apparatus, the water supply, and personnel for the special challenges that an incident in extreme cold weather can bring. We will discuss the precautions to take when responding to and arriving at an incident. Finally, we will discuss the fire-ground tactics used when suppressing a fire in extreme cold weather.

FIGURE 21-1 Winter presents a different set of problems for fire fighters.

© Keith Muratori/ShutterStock, Inc.

Preplanning and Preparing

■ Prepping the Fire Apparatus

When preparing fire apparatus for extreme-cold-weather operations, it is imperative to perform preventive methods to reduce the risk of the pumps, valves, and piping freezing. Unfortunately, a frozen pump cannot deliver an adequate supply of water. There have been incidents where fire apparatus have been dispatched to a scene and discovered at the scene that they were unable to supply water to the hoses due to ice in the piping, valves, or the actual pump itself. Preplanning and preparing the fire apparatus for extreme cold weather would have prevented such dangerous situations; for example, a fire department may plan to have a backup water supply during certain extreme weather conditions.

To prepare fire apparatus for extreme winter weather, several preventive measures are available, including:

- Ensure that all drains are in working condition on fire apparatus with pumps.
- On fire apparatus with pumps, ensure that the pumps are drained and stored dry, depending on the manufacturers' guidelines. Some manufacturers may require a pump to be stored wet.
- On fire apparatus with pumps, keep a spray bottle on the fire apparatus that contains alcohol or antifreeze, which can be applied to threads on pump outlets and intakes in case of an ice buildup.
- Before leaving the warmth of the fire station, close all drains on fire apparatus with pumps. If left open, the drains may freeze in the open position and compromise pump performance at the scene.
- Ensure that the seating compartments are dry. Compartment doors may have to be left open at the fire station to allow for complete drying. If they are left closed with moisture still inside, the doors and windows may freeze during an incident.

- Ensure that the cab and compartment heaters are in working order.
- On fire apparatus with pumps, ensure that the heat pans under the pump are in place and that the heaters are working.
- Keep sandbags, salt, or other materials that can be used to spread over ice to provide traction for feet and wheels. Covering icy or snowy spots with such materials will prevent slips and falls and can also help to prevent the fire apparatus from becoming stuck in the snow.
- Keep shovels on the fire apparatus to shovel snow away from fire hydrants and walking paths.

Prepping Tankers/Tenders

When using tankers/tenders to supply water in extreme-cold-weather conditions, there is a tendency for the valves to freeze or for ice buildup to occur on the tank vents. This can create pressure in the tank and possibly cause a collapse. On fire apparatus with pumps, it is important to always drain water from the pump, valves, and piping after use to prevent freezing and damage.

The turnaround time for the tankers/tenders may also be affected by the wintry conditions. As discussed in the chapter "Water Supply," when tankers/tenders are used, fire departments establish a dump area where tarps are placed on the ground and portable tanks are placed on top of the tarps. In extreme-cold-weather conditions, the tarps may freeze to the ground and become difficult to remove after the incident. In addition, the portable tanks may become frozen to the tarps and the ground if enough overflowing water creates an icy seal. To prevent this, dumping areas for the tankers/tenders should be well thought out during the preplanning stage to minimize the risk of frozen tarps and tanks.

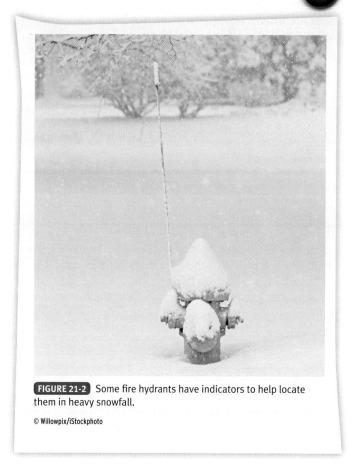

FIGURE 21-2 Some fire hydrants have indicators to help locate them in heavy snowfall.
© Willowpix/iStockphoto

Fire Fighter Safety Tips
© Photos.com

It is important to ensure that when the incident is over, the water is drained away from the established dump area. This makes it easier for the incident cleanup and reduces the risk of slip-and-fall injuries on any newly formed ice.

■ Prepping the Water Supply

Prepping Fire Hydrants

In extreme-cold-weather operations, supplying water using a fire hydrant can become a difficult task. A major issue in extreme cold weather is the fact that if the water is not moving, it will freeze very quickly. Fire hydrants may become frozen due to lack of maintenance or from being drained improperly from a previous use. The outlet caps may be frozen and be very difficult to remove without first applying heat to thaw the ice.

Fire hydrants may be buried from heavy snowfall and may not be quickly locatable. Some areas have started to identify fire hydrant locations with indicators that are 1.5 metres (5 feet) high and not easily buried in snowfall **FIGURE 21-2**. Fire departments in some areas have encouraged citizens to remove snow from the fire hydrants on their properties through public education programs.

Prepping for Alternative Water Sources

Fire hydrants may not even be a water supply source outside of major urban centres. In rural areas, **alternative water sources** that can act as reliable water supply sources are identified during the preplanning stage. In extremely cold conditions, static sources of water, such as ponds and lakes, will develop a layer of ice, making access to the water supply challenging. The static source of water may have a layer of ice ranging from 2 centimetres to 2 metres (0.75 inches to 6.6 feet) thick. Because of this, a common practice in many rural areas is to use ice augers or a chainsaw to open a hole in the ice to reach the water supply. When crews are working on or near the ice, their safety is an added concern. Only properly trained personnel well versed on the fire department's written operating guidelines should be permitted to work on or near the ice. Personnel working on or near the ice should be supplied with personal floatation devices and life lines.

Given the access issues with static water sources caused by extreme cold, some fire departments may install and use dry hydrants for drafting from static sources in their area. Unfortunately, because of the costs involved, dry hydrants may not be readily available in all rural fire departments.

■ Prepping Energized Equipment

When operating battery-powered equipment in extreme-cold-weather conditions, battery life is dramatically reduced. When flashlights, radios, atmospheric monitoring devices, self-contained breathing apparatus (SCBA) heads-up displays (HUDs), and personal alert safety system (PASS) devices are used, the batteries can drain very quickly and require replacement often. Ensure that there are extra batteries available for each item of battery-powered equipment. If the batteries are rechargeable, charge them as soon as possible. Equipment must always be in a ready state; in extreme cold conditions, this is critical for incident success.

■ Prepping for Structures under Extreme Conditions

Building construction becomes important when dealing with extreme-cold-weather conditions. Structures are built to protect occupants from the outside elements. During the planning and building phase, architects and engineers consider the different types of loads the building must endure: buildings must be able to endure stresses such as wind, rain, snow, extreme temperatures, and gravity.

One critical factor that needs to be considered is the load or weight that the more vulnerable structures of the building will have to support. Architects and engineers in many parts of Canada must take into consideration the weight of the heavy, wet snow that will accumulate on the roof areas; however, wet snow, when combined with the additional water added during firefighting operations, can lead to a roof collapse and an early collapse of the structure **FIGURE 21-3**.

FIGURE 21-3 Buildings are vulnerable to overloading during extreme weather conditions due to the extra weight provided by the fire-suppression effort.

Courtesy of Bernard Dallaire, Fire Chief of Alma Fire Department, Quebec, Canada

Near-Miss REPORT

Report Number: 08-0000106

Synopsis: Building collapse leads to a near miss.

Event Description: It was a six-alarm fire. Trucks were being moved because of the imminent collapse of a three-storey structure. A captain and two-person crew entered the adjacent building via a staircase. The captain was still outside the door when the building fully collapsed. Debris and bricks hit and covered the captain, who fell on the ground face first. The crew inside was shaken because of building movement. Smoke covered the ground, and we lost sight of the captain. We found him and cleared the debris, and he walked away okay! There was no clear reason for the crew to enter the structure, as all occupants had been evacuated.

It was very cold out, with temperatures at −20 degrees Celcius (−4 degrees Fahrenheit) and icy. With 150 members on scene, the fire was beginning to come under some control. In addition to dozens of apartments, eight stores were involved as well.

Lessons Learned: I am not sure why a crew would enter a building with a huge collapse problem. It appeared they acted on their own. It serves as a great lesson on the dangers of freelancing.

When fire fighters suppress structure fires during extreme-cold-weather operations, the water will freeze very quickly and will not drain out of the structure. The water may freeze to the floor, walls, and ceilings of the structure and add many kilograms of weight into the structure. If you have fire-suppression crews applying water at a rate of 1020 litres per minute (270 gallons per minute), this may add 1000 kilograms (2200 pounds) of weight onto the building every minute, which may be more than the structure was designed to carry. The risk of collapse occurs when the building cannot withstand the weight and force generated by the water and ice buildup. When working in these conditions, be aware of any signs that the structure may be compromised and evacuate immediately.

Preplanning for the Effects of Extreme Cold on Personnel

■ Factors Affecting Exposure

A number of factors will affect how a person deals with a cold environment. This information can be used when personnel are preparing for extreme environmental temperatures.

- Physical condition—People who are in poor physical condition will not be able to tolerate extreme temperatures as well as those with a strong cardiovascular system. A well-conditioned athlete performs much better and is less likely to experience injury due to the cold than the "weekend warrior" who is not in peak physical condition. In addition, it is easier for a well-conditioned athlete to pick up the pace, even in cold weather, and thereby generate more heat.
- Nutrition and hydration—Your body needs calories in order for your metabolism to function. Staying well hydrated provides water to catalyze this metabolism. A decrease in either will aggravate cold stress. Calories provide fuel to burn, which creates heat during the cold.

Fire Fighter Safety Tips   © Photos.com

The temperature does not need to be extreme for a cold injury to occur. Many cases of hypothermia occur at temperatures between −1°C and 10°C (30°F and 50°F).

■ Cold Exposure

Normal body temperature must be maintained within a very narrow range for the body's chemistry to work efficiently. If the body, or any part of it, is exposed to a cold environment, these mechanisms may be overwhelmed. Cold exposure may cause injury to individual parts of the body, such as the feet, hands, ears, or nose, or to the body as a whole. When the entire body temperature falls, the condition is called __hypothermia__.

Because heat always travels from a warmer place to a cooler place, the body tends to lose heat to the environment. The body can lose heat in the following five ways:

1. __Conduction__ is the direct transfer of heat from a part of the body to a colder object by direct contact, such as when a warm hand touches cold metal or ice, or is immersed in water with a temperature of less than 2.8°C (37°F). Heat passes directly from the body to the colder object.
2. __Convection__ occurs when heat is transferred to circulating air, such as when cool air moves across the body's surface. A person standing outside in windy winter weather, wearing lightweight clothing, is losing heat to the environment mostly by convection.
3. __Evaporation__ is the conversion of any liquid to a gas, a process that requires energy or heat. Evaporation is the natural mechanism by which sweating cools the body. This is why swimmers coming out of the water feel a sensation of cold as the water evaporates from their skin. Individuals who exercise vigorously in a cool environment may sweat and feel warm at first, but later, as their sweat evaporates, they can become exceedingly cooler. This is why it is critical to keep moisture away from the skin in extreme cold conditions.
4. __Radiation__ is the transfer of heat by radiant energy. Radiant energy is a type of invisible light that transfers heat. The body can lose heat by radiation, such as when a person stands in a cold room.
5. __Respiration__ causes body heat to be lost as the warm air in the lungs is exhaled into the atmosphere and cooler air is inhaled.

The rate and amount of heat loss by the body can be modified in three ways:

1. **Increase heat production.** One way for the body to increase its heat production is to increase the rate of metabolism, which the body can accomplish by shivering. Also, people often have a natural urge to move around when they are cold.
2. **Move to an area where heat loss is decreased.** The most obvious way to decrease heat loss from radiation and convection is to move out of a cold environment and seek shelter from the wind. Just covering the head will minimize radiation heat loss by up to 70 percent.
3. **Wear insulated clothing.** This helps to decrease heat loss in several ways. Insulators, such as specific materials or dry, still air, do not conduct heat; thus, layers of clothing that trap air provide good insulation, as do wool, down, and synthetic fabrics that contain small pockets of trapped air. Protective clothing also traps perspiration and prevents evaporation. Sweating without evaporation will not result in cooling.

■ Hypothermia

Hypothermia literally means low temperature. It is diagnosed when the __core temperature__ of the body—the temperature of the heart, lungs, and vital organs—falls below 35°C (95°F). The body can usually tolerate a drop in core temperature of a few degrees; however, below this critical point, the body loses the ability to regulate its temperature and to generate body heat. Progressive loss of body heat then begins.

To protect itself against heat loss, the body normally constricts blood vessels in the skin; this results in the characteristic appearance of blue lips and/or fingertips. As a secondary precaution against heat loss, the body tends to create additional heat by shivering, which is the active movement of many muscles to generate heat. As cold exposure worsens and these mechanisms are overwhelmed, many body functions begin to slow down. Eventually, the functioning of key organs, such as the heart, begins to slow, and if no emergency care is given, this can lead to death.

Hypothermia can develop either quickly, as when someone is immersed in cold water, or more gradually, as when a lost person is exposed to a cold environment for several hours or more. The temperature does not have to be below freezing for hypothermia to occur. In winter, homeless people and those whose homes lack heating may develop hypothermia at higher temperatures.

In the "Rehabilitation" section, we will discuss the signs and symptoms of hypothermia and the steps to take if a fire fighter or victim at the incident presents with these signs and symptoms.

■ Localized Cold Injuries

Most injuries from cold are confined to exposed parts of the body. The extremities, particularly the feet, and the exposed ears, nose, and face, are especially vulnerable to cold injuries. When exposed parts of the body become very cold but not frozen, that condition is called frostnip, chilblain, or immersion foot (trench foot). When these parts become frozen, the injury is called **frostbite**.

In the "Rehabilitation" section, we will discuss the signs and symptoms of local cold injuries and the steps to take if a fire fighter or victim at an incident presents with these signs and symptoms.

Frostnip and Immersion Foot

After prolonged exposure to the cold, the skin may be freezing, whereas the deeper tissues are unaffected. This condition, which often affects the ears, nose, and fingers, is called frostnip. Because frostnip is usually not painful, the victim often is unaware that a cold injury has occurred. Immersion foot, or trench foot, occurs after prolonged exposure to cold water. It is particularly common in hikers or hunters who stand for a long time in a river or lake. Due to the ever-flowing hose lines at an extreme-cold-weather fire ground, fire fighters may be at risk if their feet get wet during an incident. With both frostnip and immersion foot, the skin is pale (blanched) and cold to the touch; normal colour does not return after palpation of the skin. In some cases, the skin of the foot will be wrinkled, but it can also remain soft. The victim experiences a loss of feeling and sensation in the injured area.

Frostbite

Frostbite happens only in below-freezing temperatures. Tissue is not composed of water alone, so it will not freeze until it has been cooled to about −2°C (28.4°F). Tissue is damaged in two

ways: (1) actual tissue freezing, which results in the formation of ice crystals within the tissues (the ice crystals expand as they freeze, damaging cells), and (2) the obstruction of the blood supply to the tissue, which causes sludgy blood clots and further prevents blood from flowing to the tissues. The second type of tissue damage is more extensive than the first. In severely cold temperatures, flesh can freeze in less than a minute.

Frostbite can be identified by the hard, frozen feel of the affected tissues. Most frostbitten parts are hard and waxy **FIGURE 21-4**. The injured part feels firm to frozen as you gently touch it. If the frostbite is only skin deep, it will feel leathery or thick, not hard and frozen through. Blisters and swelling may also be present. In light-skinned individuals with a deep injury that has thawed or partially thawed, the skin may appear red with purple and white, or it may be spotty and blue.

■ Preplanning Actions for Extreme Cold

While preparing for the next call, fire fighters must ensure that their personal protective equipment (PPE) is dry, because this will decrease their likelihood of developing hypothermia. Fire fighters must ensure that they have extra clothes to change into during rehabilitation after they get wet. Fire fighters should keep extra gloves, balaclavas, T-shirts, sweatshirts, and extra socks in their PPE cache so they can replace their wet clothes as soon as possible. Wool is one of the best materials a fire fighter can use, because it can keep them warm even when wet.

Working in adverse cold conditions also compromises the integrity of firefighting boots. The rubber in firefighting boots may crack in certain situations. In some extreme cold temperatures, the rubber sole of the boot can become very hard, with decreased ability for traction, which creates a higher risk for slip-and-fall injuries in ice and snow conditions.

FIGURE 21-4 Frostbitten parts are hard and usually waxy to touch.

Courtesy of AAOS

Response

■ Driving in Winter Conditions

The temperature is −30°C (−22°F) with the wind at 60 kilometres per hour (37.3 miles per hour) and a wind chill of −48°C (−54.4°F). It is snowing heavily, and you are dispatched to a structure fire 8 km (5 mi.) away. This is a common scenario in many areas of Canada, and fire personnel must ensure that their vehicles and fire apparatus are prepared for the icy conditions and have alternative preplanned routes. Road conditions may be so challenging that some fire departments keep front-end loaders on standby in order to plow their way to the incident or to ensure that the incident area can be cleared of snow during blizzard conditions. Occasionally, snowmobiles have been used to transport medical personnel to an incident, even when the incident is located in an urban area.

Driving on snowy or icy roads requires the driver to reduce his or her speed **FIGURE 21-5**. Even when the plows can keep the pavement bare, the roads will be wet, requiring drivers to slow down. Under these wet conditions, oil slicks often occur and can lead to slipping and skidding. Wet roads can cause a loss of tire contact with the road known as

FIGURE 21-5 Driving on snow or icy roads requires the driver to reduce his or her speed.

© Zoran Milich/age fotostock

hydroplaning, which results in loss of control. Depending on road conditions, the stopping distance of the fire apparatus may be reduced, and visibility may also be affected by these conditions.

Fire apparatus that carry water will have increased stopping distances due to liquid surge. **Liquid surge** is the movement of liquid inside a container as the container is moved. As the fire apparatus decelerates or accelerates, the water inside the tank will slosh around. This movement can be very hazardous due to the large amount of water in the tank. If the fire apparatus has to brake quickly, then the liquid surge will try and force the fire apparatus forward, putting an additional strain on the fire apparatus' braking system and increasing the total stopping distance. To reduce the effects of liquid surge, the water tanks are designed with baffles. These plates inside the tank slow down the movement of water and displace some of the energy transferred during the movement of the fire apparatus.

During a snowstorm, visibility may be decreased. In addition to limiting the fire apparatus driver's field of vision, all other drivers on the road will also have a limited field of vision, thus increasing the likelihood of a collision. This is one reason why other drivers on the road have to be taken into consideration during extreme weather conditions, because they can be unpredictable.

Extreme caution is always needed when driving in adverse weather conditions. Extreme weather conditions will affect the response time of departments. Fire personnel should preplan accordingly to ensure that they make it to the incident safely and efficiently.

Arrival

After you have arrived safely at the scene, be careful when dismounting apparatus, because the steps on fire apparatus are often high and can be slippery. Use caution when walking on snow- and ice-covered ground. You cannot help your crew or any victims if you are injured. In order to prevent slips, sand or salt stored on the fire apparatus may be spread underfoot.

Incident Operations

■ Fire-Ground Tactics

There are many things to consider when dealing with fire-ground tactics in extreme-cold-weather operations. Accessibility to the fire apparatus along with frozen hoses and nozzles all add time and effort to fire-ground tactics. Accessibility to the actual incident site may be an issue if snow drifts are 2 m (6.6 ft.) high. Fire apparatus may have to stage and set up at a distance from the scene, which requires hose lays to be stretched farther to reach the fire **FIGURE 21-6**. Because of this, fire apparatus should keep extra lengths of hose to quickly attach to the hose lays. In addition, fire fighters may want to add extra lengths to their preconnected hose lays during the winter months to ensure that the hose will be long enough to cover extended distances.

Near-Miss REPORT

Report Number: 07-0001168

Synopsis: Fire fighter slips on packed snow.

Event Description: It was −20°C (−4°F) and overcast, with compact snow on the ground. The fire fighter was wearing leather structural firefighting boots compliant with NFPA 1971, *Standard on Protective Ensembles for Structural Fire Fighting and Proximity Fire Fighting*. He was also wearing full PPE in accordance with department suggested operating guidelines (SOGs).

At the end of the incident, while walking back to the rescue apparatus, he slipped on the compact snow and fell, landing on his right hip.

Lessons Learned: Leather boots have very little traction in freezing weather. Departments should review their use in freezing weather when ice and snow are present.

FIGURE 21-6 Fire apparatus may have to stage and set up at a distance from the scene, due to heavy snow.

© Joe McDaniel/iStockphoto

FIRE FIGHTER Tips

If the placement of the fire apparatus is located far from the actual incident scene, fire fighters may not have ready access to their tools and equipment, which could delay efforts in fire suppression.

In extreme-cold-weather conditions, fire fighters need to keep the water flowing through the nozzles at all times to prevent them from freezing. This is often called "cracking" the nozzle—that is, opening the fire nozzle slightly so that a small amount of water will flow through the tip in order to prevent freezing. Cracking the nozzle also prevents freezing because it allows water to keep moving through the entire system—from hydrant to hose. If water becomes static in an extreme cold environment, it will freeze and detrimentally impact the suppression efforts.

The pump operator should recirculate the water in the tank on the fire apparatus to keep the water above a freezing temperature. The pump operator also needs to keep an eye on the water temperature and ensure that it does not become too warm. If left to recirculate too long, the water in the tank can actually boil in extreme cold temperatures, which can cause extensive damage to the pump on the fire apparatus.

If unused hose lines are left in place on the ground for a long period of time, they will freeze to the ground. When it is time to clean up the incident scene, fire fighters may cause damage to the hose if they have to use tools to pry the hose from the ground. To minimize this, place unused hose lines in a warmer environment as soon as it is tactically possible.

■ Use of Ladders

During extreme-cold-weather incidents, using ladders can be a challenge. There may be a large accumulation of snow or ice in the area where the ladders are needed. Fire fighters must ensure that the ladders are stable, even on icy or snow-covered surfaces, in order to perform their duties safely. When roof ladders are required for operations, fire fighters must ensure that the area of placement is free of snow and that the ladder will remain in place with the hooks. This can be very challenging when there is a large amount of snow on the roof and the snow is close to the roof's peak.

Near-Miss REPORT

Report Number: 08-0000040

Synopsis: Extreme-cold-weather conditions contribute to near miss.

Event Description: While recovering hose, a fire fighter fell on the ice.

Lessons Learned: In −28°C (−18.4°F) weather, hoses can freeze into the ice, and the fire ground will likely be very icy.

FIRE FIGHTER Tips

Spraying the nozzles with alcohol or antifreeze will aid in freezing prevention.

Additional threats to safety occur when discharged water accumulates on the ladder rungs and become a slip-and-fall hazard. If a fire fighter's gloves are wet, then they may become frozen to the ladder, effectively leaving the fire fighter without the use of his or her hands or arms.

■ Rescue Operations

Trying to keep power generators and rescue tools running can be a challenge in extreme cold weather, as the fuel can start to solidify (gel). Hydraulic tools will respond slower than usual in extreme cold temperatures, which can affect the time it takes to perform certain tasks, such as vehicle extrication. Keeping hand tools readily available is a common practice to combat this. It is important to have fuel additives in both the fire apparatus and equipment fuel tanks to minimize the risk of fuel gelling and ensure that all fuel is indicated for winter use.

When responding to incidents on roads and highways, all responders must ensure that the apparatus are parked appropriately and safe zones are created. Wintry road conditions cause the stopping distances for traffic on roadways to increase, which may put responders at greater risk. There have been many occasions when the traffic driving close to the incident have been involved in secondary incidents that then need to be addressed **FIGURE 21-7**.

Emergency Personnel

Emergency personnel may be dispatched to rescue situations in extreme cold weather. The extreme cold may cause delays in patient removal, creating a risk of cold exposure and injury to responders and to the victim. Responders are typically dressed for the weather, but the victim may not be. To help a patient retain warmth on scene while awaiting removal,

FIGURE 21-7 Winter weather can cause treacherous driving conditions.

© Boris Roessler/EPA/Landov

FIRE FIGHTER Tips

In extreme-cold-weather conditions, any incident—vehicle extrication, hazardous materials, water/ice rescue, rope rescue, confined-space rescue, trench rescue, and structural collapse—will have unique challenges that the fire department must be prepared to respond to. As a fire fighter, you can assist in this effort by ensuring that you are prepared for extreme cold with an extra set of clothing, gloves, and a hat.

responders may deploy extra blankets and cover the victim. Responders may also utilize a tarp as a windbreak to protect themselves and the victim against the wind chill and to help prevent hypothermia. When there is a wind chill of −50°C (−58°F), it becomes critical to create protection against the wind.

Portable heat packs can ensure that the victim's body temperature remains stable. Portable heaters can also be

Near-Miss REPORT

Report Number: 08-0000038

Synopsis: Icy gloves make a hose line hard to control.

Event Description: Nearly lost control of 64-millimetre (2.5-inch) hose due to wet, icy gloves.

Lessons Learned: Hold on tightly to hose lines, and keep a supply of dry gloves available in extreme cold weather.

implemented if the victim is in a position where he or she cannot be removed quickly. In certain circumstances, responders may have to perform rapid removal techniques to get the victim to a warm area as soon as possible.

Self-Contained Breathing Apparatus

In extreme cold weather, SCBA can experience several issues that will complicate its use. For example, when the face piece is donned outside during extreme cold temperatures, it will quickly fog up, and visibility will be reduced. This may lead to trip hazards prior to the fire fighter's entry into the structure. Once the fire fighter enters the structure and the heat from the fire reaches the face piece, condensation will form on the outside of the face piece. This can be alleviated by using a gloved hand to wipe the moisture from the lens on the face piece. This same condition can occur when the fire fighter's helmet visor is lowered.

After the SCBA is donned in the extreme cold, fire fighters must be cognizant of the fact that any condensation in the regulator could lead to its freezing before entry into the structure is even attempted. To reduce the risk of this occurrence, personnel should ensure that the regulator is dry before donning the SCBA. Often, when personnel exit the structure for a full oxygen cylinder, the regulator may freeze.

Another issue that can arise is ice forming on the SCBA while the company is working outside **FIGURE 21-8**. This ice can affect the low-pressure alarm (PASS alarm) on the breathing apparatus, effectively muffling it, so it is not as loud as it should be. The ability to use the PASS alarm may also be compromised by a large formation of ice on the PASS alarm. There is also the risk of a fire fighter's gloves freezing to the point where the fire fighter has little dexterity and cannot activate the device.

Rehabilitation

At the incident, command should establish a warm area for the rehabilitation (rehab) of all responders and to ensure that

FIGURE 21-8 If ice forms on the SCBA, it can affect the low-pressure alarm on the breathing apparatus, making it not as loud as it should be.

© Mark Sauer/Mesabi Daily News/AP Photo

Emergency Medical Services (EMS) are addressing the signs and symptoms of working in the extreme cold. Fire fighters should be trained to identify the signs and symptoms of frostbite or hypothermia not only for themselves but also for all personnel working at an incident during extreme cold temperatures and conditions. Fire fighters need to know when it is time to request rehab, because time is critical in treating cold injuries.

Frostnip

It is difficult to tell the difference between frostnip and frostbite. Signs of frostnip include:

- Skin appears red and sometimes swollen.
- Pain, but usually no further damage after rewarming
- Repeated frostnip in the same spot can dry the skin, causing it to crack and become sensitive.

If you or a crew member have any of these signs, go to rehab and see an EMS professional. Do not rub the affected area.

Frostbite

The severity and extent of frostbite are difficult to judge until hours after thawing. Frostbite can be classified as superficial or deep before thawing.

The signs and symptoms of superficial frostbite are as follows:

- The skin is white, waxy, or greyish yellow.
- The affected part feels very cold and numb. There might be a tingling, stinging, or aching sensation.
- The skin's surface feels stiff or crusty and the underlying tissue soft when depressed gently and firmly.

The following signs and symptoms indicate deep frostbite:

- The affected part feels cold, hard, and solid and cannot be depressed; it feels like a piece of wood or frozen meat.
- Blisters might appear after rewarming.
- The affected part is cold, with pale, waxy skin.
- A painfully cold part suddenly stops hurting.

After a part has thawed, frostbite can be categorized by degrees:

- First-degree frostbite—The affected part is warm, swollen, and tender.
- Second-degree frostbite—Blisters form minutes to hours after thawing and enlarge over several days **FIGURE 21-9** and **FIGURE 21-10**.
- Third-degree frostbite—Blisters are small and contain reddish-blue or purplish fluid. The surrounding skin can be red or blue and might not turn white when pressure is applied.
- Fourth-degree frostbite—No blisters or swelling occur. The part remains numb, cold, and white to dark purple.

If you or a crew member have any of these signs, get out of the cold, go to rehab, and see an EMS professional. Frostbite can cause serious damage, especially to fingers and toes.

Hypothermia

The signs and symptoms of hypothermia are as follows:

- Change in mental status—This is one of the first symptoms of developing hypothermia. Examples include disorientation, apathy, and changes in personality, such as unusual aggressiveness.
- Shivering—Shivering is the first, and most important, of the body's defences against a falling body temperature. Shivering can produce more heat than many rewarming methods. As the core temperature continues

FIGURE 21-9 Second-degree frostbite.
Courtesy of AAOS

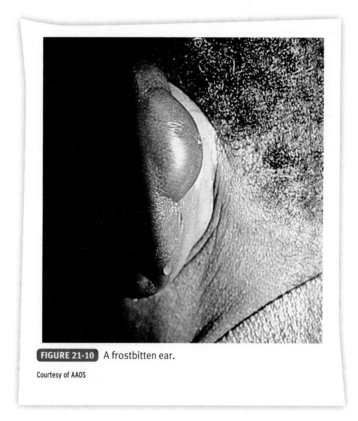

FIGURE 21-10 A frostbitten ear.
Courtesy of AAOS

to fall, shivering decreases and usually stops at about 35°C (95°F).

- Cool abdomen—Place the back of your hand between the clothing and the victim's abdomen to assess the victim's temperature. When the victim's abdominal skin under clothing is cooler than your hand, consider the victim hypothermic until proven otherwise.

There are two types of hypothermia: mild and severe. With severe hypothermia, the victim becomes so cold that shivering stops, which means the victim's body cannot rewarm itself internally and requires external body heat for recovery. In fact, 50–80 percent of all victims of severe hypothermia die.

Signs of mild hypothermia include the following:

- Vigorous, uncontrollable shivering
- Victim has the "umbles."
  - Grumbles—Decreased mental skills
  - Mumbles—Slurred speech
  - Fumbles—Difficulty using fingers or hands
  - Stumbles—Staggers while walking
- Has cool or cold skin on the abdomen, chest, or back.

If you or a crew member have any of these signs, get out of the cold, go to rehab, and see an EMS professional.

You may find a victim at an emergency incident with severe hypothermia. The signs of severe hypothermia are:

- No shivering
- Skin feels ice cold and appears blue.
- Muscles can be stiff and rigid, similar to rigor mortis.
- Altered mental status, not alert
- Breathing and pulse are slow.
- Victim might appear to be dead.

Remove the victim from further exposure to the cold and into a warm area. Handle the victim gently. Do not allow the victim to exert himself or herself. Do not rub the victim's arms or legs. Contact EMS to provide emergency medical care to the victim.

FIRE FIGHTER Tips

In certain regions of Canada during the winter, it is a challenge to train and learn when the mind is focused on just keeping warm. It is a great benefit to these fire departments if they have access to indoor training facilities. In Manitoba, all facets of firefighting training (with the exception of live fires) can be performed in an indoor facility.

Chief Concepts

- Responding to incidents in extreme cold temperatures is a challenging reality that the Canadian fire service faces every winter.
- When preparing the apparatus for cold-weather operations, fire fighters must take preventive action to reduce the risk of the pumps, valves, and piping from freezing.
- Battery life on energized equipment is greatly diminished in extreme cold weather. Ensure that there are extra batteries available for each item of battery-powered equipment.
- In extreme cold weather, building structures have a greater load or weight to support when dealing with wet, heavy snow, or the added weight of firefighting operations, which can lead to roof collapse or collapse of the structure.
- A number of factors affect how a person deals with a cold environment, including:
  - Physical condition
  - Nutrition
  - Hydration
- Cold exposure may cause injury to individual parts of the body, such as hands, feet, ears or nose, or to the body as a whole.
- Extreme caution is always needed when driving in adverse weather conditions. Extreme weather conditions will affect the response time of departments. Fire personnel should preplan accordingly to ensure that they make it to the incident safely and efficiently.
- After arrival on the scene, fire fighters should use caution when dismounting the apparatus in order to prevent slip-and-fall injuries.
- Accessibility to the apparatus and frozen hoses and nozzles all add time and effort to fire-ground tactics. Accessibility to the actual incident may become an issue if snow drifts are 2 m (6.6 ft.) high or greater. Consideration should be taken to keep extra hose lengths for quick attachment to hose lays to ensure the hose will be long enough to cover extended distances.
- Rescue operations become more difficult during extreme weather due to tools responding slower than normal because of fuel "gelling," so it is necessary to keep hand tools readily available.
- SCBA use can be complicated by extreme cold weather. Fire fighters should be cognizant of any moisture gathering on their SCBA. Moisture on the SCBA can lead to fogging, freezing, or muffling of the PASS alarm.
- Fire fighters should be trained to identify the signs and symptoms of frostbite and/or hypothermia for themselves and for all personnel working at an incident during extreme cold weather.

Wrap-Up, continued

Hot Terms

Alternative water sources Reliable water supply sources found in rural areas, such as ponds or lakes.

Conduction The loss of heat by direct contact (e.g., when a body part comes into contact with a colder object).

Convection The loss of body heat caused by air movement (e.g., a breeze blowing across the body).

Core temperature The temperature of the central parts of the body (i.e., heart, lungs, and vital organs).

Evaporation The loss of body heat caused by the conversion of water or another fluid (often sweat) from a liquid to a gas.

Frostbite Damage to tissues as the result of exposure to cold; frozen body parts.

Hydroplaning The loss of steering or braking ability (and thus, control) because of a layer of water on the road that prevents sufficient contact between the vehicle's tires and the road's surface.

Hypothermia A condition in which the internal body temperature falls below 35°C (95°F), usually as a result of prolonged exposure to cold or freezing temperatures.

Liquid surge The force imposed upon a fire apparatus by the contents of a partially filled water or foam-concentrate tank when the vehicle accelerates, decelerates, or turns. (NFPA 1002)

Radiation The transfer of heat to colder objects in an environment by radiant energy—for example, heat gain from a fire.

Respiration The loss of body heat as warm air in the lungs is exhaled into the atmosphere and cooler air is inhaled.

It is a frigid January night, and you are called to a structure fire. The homes in the neighbourhood are built close together. When your crew arrives on scene, the house looks abandoned. A call to the utilities company during the scene size-up confirms that the electricity and heat are off. To your surprise, a woman yells out to you, "Fred's still inside!" After donning your gear and receiving your assignment, you and your partner begin to search the house.

You find a young man lying on his back in the living room with an empty bottle of alcohol. Despite the cold temperature, he is dressed in a T-shirt and jeans.

1. Hypothermia occurs when the core body temperature falls below what temperature?
   - **A.** 37°C (98.6°F)
   - **B.** 35°C (95°F)
   - **C.** 33°C (91.4°F)
   - **D.** 32°C (89.6°F)

2. From his outward appearance, what type of hypothermia does the victim seem to have?
   - **A.** Mild
   - **B.** Severe
   - **C.** Life threatening
   - **D.** Critical

3. To maintain a reliable water supply, what action should the fire fighters on the hose line take?
   - **A.** Only use a fog nozzle.
   - **B.** Only use a spray nozzle.
   - **C.** Keep the water running at all times.
   - **D.** All of the above

4. What precautions should the fire fighters on the ladder take?
   - **A.** Ensure that the ladders are stable before climbing.
   - **B.** Replace wet gloves.
   - **C.** Be aware of any accumulated water.
   - **D.** All of the above

Wildland and Ground Fires

Fire Fighter I

Knowledge Objectives

After studying this chapter, you will be able to:

- Define the terms *wildland fires* and *ground fires*. (NFPA 5.3.19.A , p 664)
- Explain how the three elements of the wildland fire triangle affect each side of the fire triangle. (p 664–667)
- Describe light fuels, heavy fuels, subsurface fuels, surface fuels, and aerial fuels. (NFPA 5.3.19.A , p 665)
- Explain the relationship between a fuel's properties and the speed at which the fuel ignites and the ensuing fire spreads. (NFPA 5.3.19.B , p 665–666)
- Describe how weather factors and topography influence the growth of wildland fires. (NFPA 5.3.19.A , p 667)
- Label the parts of a wildland fire. (NFPA 5.3.19.A , p 668)
- Describe the methods and tools used to cool a fuel with water. (NFPA 5.3.19.A , p 668)
- Describe the methods and tools used to remove a fuel from wildland fires. (NFPA 5.3.19.A , p 668–669)
- Describe the methods and tools used to smother wildland fires. (NFPA 5.3.19.A , p 669)
- Describe how a direct attack is mounted on a wildland fire. (NFPA 5.3.19.A , p 669–670)
- Describe how an indirect attack is mounted on a wildland fire. (NFPA 5.3.19.A , p 670)
- Itemize the characteristics of the fire apparatus used to suppress wildland fires. (NFPA 5.3.19.A , p 670–671)
- Describe the hazards associated with wildland and ground firefighting. (NFPA 5.3.19 , p 672)
- Describe the personal protective equipment (PPE) needed for wildland firefighting. (NFPA 5.3.19 , p 672–673)
- Explain the problems created by the wildland–urban interface. (p 674)

Skills Objectives

After studying this chapter, you will be able to perform the following skill:

- Suppress a ground cover fire. (NFPA 5.3.19.B , p 673–674)

© Photos.com

Fire Fighter II — FFII

Knowledge Objectives

There are no knowledge objectives for Fire Fighter II candidates. NFPA 1001 contains no Fire Fighter II Job Performance Requirements for this chapter.

Skills Objectives

There are no skills objectives for Fire Fighter II candidates. NFPA 1001 contains no Fire Fighter II Job Performance Requirements for this chapter.

Additional NFPA Standards

- NFPA 1051, *Standard for Wildland Fire Fighter Professional Qualifications*
- NFPA 1901, *Standard for Automotive Fire Apparatus*
- NFPA 1906, *Standard for Wildland Fire Apparatus*
- NFPA 1977, *Standard on Protective Clothing and Equipment for Wildland Fire Fighting*

You Are the Fire Fighter

Y ou have been watching the national news with awe as you see homes with only their foundation remaining as another major wildland fire sweeps through the formerly pristine forest. Lately it seems that these types of fires have become a regular nightly news item. You think about your own community and whether it is possible that you would ever face a wildland fire since you are part of a large municipal fire department that is not surrounded by forests.

1. How is wildland firefighting similar to structural firefighting?

2. Are urban/suburban fire fighters likely to face wildland fires? Why or why not?

3. How are wildland vehicles and equipment similar to the equipment used in structural firefighting?

Introduction

Wildland fires are defined by the National Fire Protection Association (NFPA) as unplanned and uncontrolled fires burning in vegetative fuel—grass, leaves, crop fields, and trees—that sometimes includes structures. Wildland fires can consume grasslands, brush, and trees of all sizes. Some of these fires burn vegetation that is located many metres above the ground. Their incidence varies from season to season and from one year to the next. The threat and intensity of wildland fires are closely tied to the moisture content and the combustibility of the vegetative fuels, as well as to weather conditions and topography.

Wildland fires are sometimes referred to by different terminology, including brush fires, forest fires, grass fires, ground cover fires, ground fires, natural cover fires, and wildfires. While these names provide descriptions of specific types of wildland fires, all are properly classified as wildland fires.

Ground cover fires burn loose debris on the surface of the ground. This debris includes vegetation such as grass, as well as dead leaves, needles, and branches that have fallen from shrubs and trees. In most fire departments, ground cover fires are the most common type of wildland fire encountered.

Some fire departments respond to more wildland and ground fires than to structure fires. Although many calls for wildland and ground fires involve small incidents, some calls will escalate into major incidents. Most fire departments are responsible for at least some areas that can be classified as wildland. **Wildland** is land in an uncultivated natural state that is covered by timber, woodland, brush, or grass. Most large cities have some parklands that are covered with natural vegetation.

Most structural fire fighters will be called on to extinguish wildland and ground fires at some point in their careers. The information in this chapter is intended to teach you how to extinguish wildland and ground fires safely; it will not train you to be a wildland fire fighter, however. Fire fighters whose primary responsibility is suppressing wildland fires need to complete separate wildland firefighting training.

Large wildland fires are handled by agencies that specialize in this type of firefighting. Each province/territory has an agency designated to coordinate wildland firefighting. Federal government agencies such as the National Resources Canada, Canadian Forest Service, Parks Canada, Canadian Council of Forest Ministers (CCFM), and Canadian Interagency Forest Fire Centre are also responsible for coordinating firefighting activities at large incidents and incidents that occur on federal lands. The Incident Management System (IMS) is also activated at these incidents. See the chapter "Incident Management System" for more information on working within the IMS during a wildland fire.

Wildland and Ground Fires and the Fire Triangle

To safely and effectively extinguish wildland and ground fires, fire fighters need to understand the various factors that cause fire ignition and affect the growth and spread of wildland fires. The fire triangle provides a good model for explaining the behaviour of wildland and ground fires **FIGURE 22-1**.

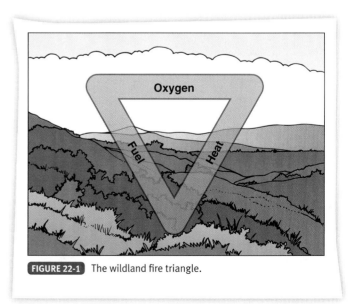

FIGURE 22-1 The wildland fire triangle.

The fire triangle, as discussed in the chapter "Fire Behaviour," consists of three elements: fuel, oxygen, and heat. Wildland and ground fires require the same three elements as structural fires to enable them to start and continue, but the difference between these types of fires is the conditions under which the fuel, oxygen, and heat come together to produce an uncontrolled fire. Because weather conditions have a major impact on the behaviour of wildland fires, we will also consider how each side of the fire triangle is influenced by certain weather conditions.

■ Fuel

The primary fuel for wildland and ground fires is the vegetation —grasslands, brush, and trees—in the area. The amount of fuel in an area may range from sparse grass to heavy underbrush and large trees. Some fuels ignite readily and burn rapidly when dry, whereas others are more difficult to ignite and burn more slowly. Vegetative fuels can be located under the ground (roots), on the surface (grass and fallen leaves), or above the ground (tree branches). These fuels are further classified as fine fuels or heavy fuels. TABLE 22-1 summarizes the types of fuels that are typically associated with wildland and ground fires.

Fine Fuels

Fine fuels and light fuels include dried vegetation, such as twigs, leaves, needles, grass, moss, and light brush. **Ground duff**—the partly decomposed organic material on a forest floor—is another type of fine fuel. Fine fuels have a large surface area relative to their volume, which causes them to ignite easily, burn quickly, and produce a lot of heat. Such fuels are usually the main type of fuel present in ground cover fires. They facilitate the ignition of heavier fuels and cause fire to burn and spread with great intensity. In addition, they burn rapidly and unpredictably. As a consequence of these characteristics, fires spread more quickly in fine fuels than in heavy timber and brush.

Heavy Fuels

Heavy fuels are of a larger diameter than fine fuels. They include large brush, heavy timber, stumps, branches, and dead

TABLE 22-1	Fuel Classification
Fine fuels	1-hour fuels; are 0.6 cm (0.25 in.) in diameter or less; consist of grass, leaves, tree moss, and loose surface litter. Fine fuels can absorb or lose moisture very rapidly, and their flammability can change very quickly from no flammability to very high flammability in just a short period of time.
Light fuels	10-hour fuels; range from 0.6 to 2.5 cm (0.25 to 1 in.) in diameter; consist of small twigs and stems, both living and dead. Light fuels dry out very quickly and can become highly flammable.
Medium fuels	100-hour fuels; range from 2.5 to 7.6 cm (1 to 3 in.) in diameter; consist of sticks, branch wood, and medium duff. Medium fuels gain and lose moisture relatively slowly and cumulatively. They also add to wildfire intensity.
Coarse and heavily compacted fuels	1000-hour fuels; are 7.6 cm (3 in.) or more in diameter; include large logs, deep duff, and peaty material. The moisture content of coarse and heavily compacted fuels changes very slowly and cumulatively over long periods of time.

timber on the ground. Another type of heavy fuel is **slash**—the leftovers of logging and land-clearing operations. Slash includes both large and small pieces of logs, branches, bark, stumps, and other vegetative debris. Although fires involving heavy fuels do not spread as rapidly as fires involving fine fuels do, heavy fuels can burn with a high intensity.

Subsurface, Surface, and Aerial Fuels

Subsurface fuels are those fuels located under the ground. Roots, moss, duff, and decomposed stumps are examples of subsurface fuels. Fires involving subsurface fuels are difficult to locate and challenging to extinguish.

Surface fuels are located close to the surface of the ground. They include grass, leaves, twigs, needles, small trees, and logging slash. Brush less than 1.8 metres (6 feet) above the ground is also classified as a surface fuel. Surface fuels, which are sometimes called ground fuels, are involved in ground cover fires.

Aerial fuels (also called canopy fuels) are located more than 1.8 m (6 ft.) above the ground. They usually consist of trees, including tree limbs, leaves and needles on limbs, and moss attached to the tree limbs.

Other Fuel Characteristics

Other characteristics of wildland fuels may determine how quickly the fuel ignites, how rapidly it burns, and how readily it spreads to other areas. These properties include the fuel's size and shape, its compactness, its continuity, its volume, and its moisture level.

The size and shape of a fuel influence how it burns; for example, fine fuels burn more quickly than heavy fuels and require less heat to reach their ignition temperature. This difference arises because fine fuels have a greater surface area relative to the total volume of the fuel.

Fuel compactness influences the rate at which a wildland fire burns. When fuels are tightly compressed, adequate oxygen and heat cannot get to large surfaces of the fuel in a short period of time; therefore compact fuels often burn more slowly and over a long period of time. For example, a compact fuel such as a large tree trunk burns more slowly than a less compact fuel such as dry grass.

Fuel continuity refers to wildland fuels that have uninterrupted connections; for example, intertwined tree branches provide for fuel continuity. Fuels with continuity touch one another or are located close enough together that a fire can spread from one location to another—a characteristic that contributes to a rapidly spreading wildland fire. In situations where there is no fuel continuity, a fire may burn out at the spot where it started.

Fuel volume refers to the quantity of fuel available in a specific area. The amount of fuel in a given area influences the growth and intensity of the fire.

Fuel moisture refers to the amount of moisture contained in a fuel. This characteristic influences the speed of ignition, the rate of spread, and the intensity of the fire. Fuels with high moisture content will not ignite and burn as readily as fuels with low moisture content. Some types of vegetation naturally contain more moisture than others. The amount of moisture in a fuel is also related to the season of the year, the amount of rain that has fallen, and the relative humidity in the air. In a year with decreased rainfall, less moisture will be present in vegetative fuels compared with the same vegetation during a rainy year. The capacity of fuels to vary in fuel moisture content depends on the type and condition of the fuels.

■ Oxygen

The second side of the fire triangle is oxygen. Oxygen is needed to initiate and support the process of combustion. In a structure fire, a limited supply of oxygen results in a slow-burning fire. If additional oxygen becomes available to a fire, it will begin to burn more rapidly.

In wildland and ground fires, which are usually free burning, oxygen is rarely an important variable in the ignition or spread of the fire. Unlike in structure fires, oxygen is available in unlimited quantities in these open-air fires. Nevertheless, air movement around wildland and ground fires will influence the speed with which the fire moves. Wind blowing on a wildland and ground fire brings more oxygen to the fire, accelerates the process of combustion, and influences the direction in which the fire travels.

■ Heat

The third side of the fire triangle is heat. Sufficient heat must be applied to fuel in the presence of adequate oxygen to produce a fire. Three types of factors may ignite wildland and ground fires: natural causes, accidental causes, and intentional causes.

Lightning is the source of almost all naturally caused wildland and ground fires. Some storms produce dry lightning, in which lightning strikes occur without rain. A single lightning storm can produce many lightning strikes and start many fires.

Near-Miss REPORT

Report Number: 11-0000349

Synopsis: Wildland crew saved by wind shift.

Event Description: We were dispatched to a brush/wildland fire in a neighbouring district with 183.2 hectares (458 acres) of woods on fire. The fire was moving higher and faster than a person can move. We arrived in brush units with about 378 litres (100 gallons) of water at most. From the time we arrived, we were teamed up with other departments. We never once made contact with command concerning which radio channels we would be communicating on. We stopped at the top of the hill and saw several brush units flying down the hill to the right flank. We heard what sounded like a freight train coming over the hill. There were no trains nearby. No communications were made to tell us that the fire was coming right at us. We were just able to back up, turn around, and start to fly down the hill.

The brush truck engine died. The fire was moving faster than we could run. Our only saving grace was the wind changing direction and blowing the fire away from us. The wind blew hard enough to stop the fire and stand it straight up. The wind saved 19 lives that day. The fire was further than you could see with nowhere to go.

Lessons Learned:

We need to establish good communications prior to entering a fire scene.

Equipment needs to be checked every day.

Communication is key. We need to make sure we have communication with all mutual aid companies.

A variety of accidental causes may also result in wildland and ground fires, such as discarded smoking materials and improperly extinguished campfires. In addition, downed electric wires may provide a heat source that can ignite vegetative fuels.

Finally, some wildfires and ground fires are the result of intentional acts of arson.

Other Factors That Affect Wildland Fires

Once a wildland and ground fire has started, the growth of the fire is influenced by weather factors and by the topography of the land involved in the fire. Weather conditions and topography have a greater effect on wildland and ground fires than they do on structure fires.

■ Weather

Weather conditions have a major impact on the course of a wildland fire. The two most critical aspects of weather that influence a wildland fire are moisture and wind.

Moisture

Moisture can be present in the form of either relative humidity or precipitation. **Relative humidity** is the ratio of the amount of water vapour present in the air compared to the maximum amount of water vapour that the air can hold at a given temperature. Warm air has a higher capacity for moisture content than cool air does. If the air contains the maximum amount of water vapour possible, the relative humidity is said to be 100 percent. On a rainy day, the relative humidity is usually 100 percent. In dry climates, the relative humidity may be 20 percent or less, which means the water vapour contained in the air is only 20 percent of the maximum amount that the air is capable of holding. If the relative humidity drops and stays at 10 percent, for example, eventually all fuels will equalize at that level—but light fuels will reach that level much more quickly than heavy fuels do.

Relative humidity is a major factor in the behaviour of wildland and ground fires. When relative humidity is low, vegetative fuels dry out, making them more susceptible to ignition. Conversely, when relative humidity is high, the moisture from the air is absorbed by the vegetative fuels, making them less susceptible to ignition. Changes in relative humidity affect light fuels more dramatically than they affect heavy fuels.

Relative humidity typically varies with the time of day. As the temperature increases throughout the day, the relative humidity generally drops. Conversely, as the temperature decreases during the evening and night hours, the relative humidity rises. A rise in the relative humidity can help to slow the spread of a wildland fire and aid the work of fire crews who are trying to control the fire.

In some parts of Canada, the average relative humidity also varies with the season of the year. These changes contribute to variations in the moisture content of vegetative fuels—a factor that partly explains why many regions of Canada demonstrate seasonal patterns of wildland and ground fires. The seasons with the most wildland and ground fires are often the seasons with the lowest relative humidity.

The second type of moisture that influences the behaviour of wildland and ground fires is precipitation. Moisture falling from the sky increases the relative humidity. In addition, much precipitation is absorbed by plants, which increases their moisture content and renders the plants less susceptible to combustion. In seasons with adequate precipitation, the incidence of wildland and ground fires is typically much lower than in years with below-average precipitation.

Wind

Air movement also has an effect on the spread and direction of wildland and ground fires. On a still day, a wildland or ground fire can burn and grow in size; a fire with similar conditions will spread more rapidly, however, if wind pushes the fire so that it comes in contact with additional fuels. Wind has the ability to move a fire at great speed. The effect of wind on a wildland and ground fire is similar to fanning a flame to help it burn more rapidly.

■ Topography

The movement of wildland and ground fires also reflects the topography of the land. **Topography**, which refers to the changes of elevation in the land as well as the positions of natural and human-made features, has a major effect on the behaviour of wildland and ground fires. Heat from these fires rises just like the heat in a structure fire. When a wildland or ground fire burns on flat land, much of the heat from the fire will rise into the air, thereby preheating the fuels above the main body of fire; for example, when a fire is burning on a hillside, convective and radiant heat from the fire will preheat the fuels on the slope above, causing them to dry out and be more flammable. This factor may result in more rapid rates of spread over a larger area as the fire moves up the slope.

Other features of the topography influence the spread of a fire as well. Natural barriers, such as streams and lakes, may help contain fires. Human-built barriers, such as highways, also make it easier to contain a fire.

Fire Fighter Safety Tips   © Photos.com

Certain topographical features, such as valleys, steep hillsides, and saddles, can cause sudden and rapid fire progression. Many fire fighter fatalities and near misses have occurred in these types of terrain. Lookouts, communications, escape routes, and safety zones (LCES) must be rigidly enforced when operating in these areas.

Extinguishing Wildland Fires

To safely attack and effectively extinguish wildland and ground fires, fire fighters must consider how the factors of weather, fuels, and topography come together to influence the behaviour of the fire. They also need to know the terms that are commonly used to describe different parts of the fire so that they can understand directions from command and report information back to fire officers.

■ Anatomy of a Wildland Fire

The location where a wildland or ground fire begins is called the **area of origin**. As the fire grows and moves into new fuel, the most rapidly moving area—the travelling edge of the fire—is called the **head of the fire** FIGURE 22-2. As the fire gets bigger, the area close to the area of origin may be referred to as the **heel of the fire** or the **rear of the fire**.

As the fire grows, a change in wind or topography may cause it to move in such a way that a long, narrow extension of fire projects out from the fire. This type of expansion is called a **finger**. A finger can grow because of a change in topography, fuel, or wind direction and produce a secondary direction of travel for the fire. The unburned area between a finger and the main body of the fire is called a **pocket**. A pocket is a dangerous place for fire fighters because this area of unburned fuel is surrounded on three sides by fire. An area of land that is left untouched by the fire, but is surrounded by burned land, is called an **island**.

A **spot fire** is a new fire that starts outside the perimeter of the main fire. Spot fires usually begin when flaming vegetation, in the form of an ember or spark, is picked up by the convection currents generated by the fire and dropped some distance away from the original fire.

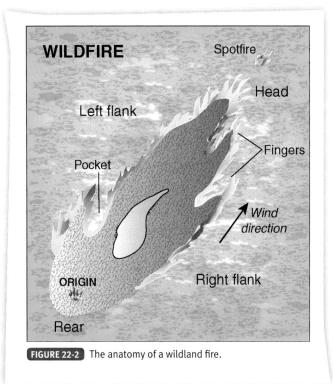

FIGURE 22-2 The anatomy of a wildland fire.

Green and black are terms often used by wildland fire fighters. *Green* describes areas containing unburned fuels; these areas are more likely to become involved in the fire than are black areas. *Black* describes areas that have already been burned. Because black areas contain few unburned fuels, they are often relatively safe for fire fighters.

■ Methods of Extinguishment

The methods used to extinguish wildland and ground fires focus on disrupting the conditions represented by the fire triangle. Wildland and ground fires can be controlled and extinguished by cooling the fuel, by removing fuel from the fire, or by smothering the fire.

Cooling the Fuel

Water is used to cool the fuels that drive wildland and ground fires. Although the principles involved in cooling wildland and ground fires are the same as those used for structural fires, the methods used to apply the water may be different.

For small fires with a light fuel load, **backpack pump extinguishers** may be effective. These devices can be transported to the fire and quickly put into operation before the fire has an opportunity to grow. Portable backpack pumps hold between 15 and 30 L (4 and 8 gal.) of water. Fire fighters expel the water by using a hand pump.

Water from booster tanks carried on structural fire apparatus or on special wildland fire trucks is another means of cooling a wildland fire's fuel. The hoses used for wildland fires may be smaller than the hoses required for structural firefighting. Unfortunately, it is not always possible to get trucks into a position where they can attack a wildland fire. The water supply may also be limited to the amount of water carried by the apparatus, unless static water sources can be tapped by fire pumpers or portable pumps. Backpack pumps and small hose lines are used to extinguish most ground cover fires.

A third means of applying water to a wildland fire is from aircraft. Fixed-wing aircraft can take on a load of water from a lake and apply it to the fire. Rotary-wing aircraft (helicopters) can carry a container of water and drop it on the fire. The use of aircraft can be an effective means to control wildland fires, but it requires close communication between different agencies to coordinate this activity. Aircraft are most commonly used for fires involving heavy fuels.

Removing the Fuel

The second method of controlling and extinguishing wildland and ground fires is to remove the fuel from the fire. This can be done by fire fighters using hand tools or mechanized equipment.

In terms of hand tools, removal of fine fuels can be accomplished with a fire broom to sweep away light grass or weeds. Steel fire rakes can be used to create a fire line in light brush. A combination hoe and rake called a **McLeod tool** is also used to create a fire line. When working in heavier brush, fire fighters can use an **adz** to grub out brush to create a fire break. The **Pulaski axe** combines an adz and an axe for brush removal. A **council rake** is a long-handled rake constructed with hardened triangular-shaped steel teeth that is used for raking a fire line down to soil with no subsurface fuel (mineral soil), for digging, for rolling burning logs, and for cutting grass and small brush.

In some situations, saws are used to remove heavy brush and trees from the fire. The types of saws employed for this purpose range from handsaws to gasoline-powered chainsaws. Other powered equipment includes tractors, plows, and bulldozers.

Backfiring is another technique used to remove the fuel from a wildland fire. When properly set, a backfire can burn an area of vegetation in front of the fire, thereby creating a wide area devoid of vegetation. Backfires must be set at the right time and at the proper place to work correctly. When improperly set, a backfire may worsen a wildland fire or create a new one. These fires should be set only by fire fighters who have had special training in the control of wildland and ground fires.

FIRE FIGHTER II Tips	FFII

Only specially trained personnel should operate chainsaws. When operating a chainsaw, you must wear the following protective gear: helmet, eye protection, hearing protection, gloves, and chainsaw chaps to protect the legs. In addition, always have a lookout and a preplanned escape route when trees are being cut.

FIRE FIGHTER II Tips	FFII

When operating around motorized apparatus or bulldozers, stay at least 15.2 m (50 ft.) away from the equipment. Wear a light on your helmet and a reflective vest at night. Avoid standing downhill from motorized apparatus. Finally, make sure that the equipment operator can see you.

Smothering the Fire (Removing Oxygen)

The third method that can be used to control and extinguish wildland fires is to remove oxygen from the fire. The smothering technique is most commonly used when fire fighters are overhauling the last remnants of a wildland and ground fire. In these circumstances, mineral soil (dirt with no organic material) and water are often thrown on smouldering vegetation until it is cool to the touch to prevent flare-ups. Subsurface fuels are also removed to prevent flare-ups. Smothering is less useful during the more active phases of a fire, although some vegetative fires in grasses can be smothered using a special fire-beater sheet constructed of heavy rubber.

Compressed-air foam systems (CAFS) combine foam concentrate, water, and compressed air to produce a foam that will stick to vegetation and structures that are in a wildland and ground fire's path. When the heat of the fire reaches the foam, the foam absorbs the fire's heat and breaks down, thereby cooling the fuel. CAFS can also extinguish a fire by coating the fuel and preventing oxygen from combining with the fuel. CAFS can be delivered from a backpack extinguisher or mounted on apparatus. Such systems help to extinguish fires with less water because the compressed air foam achieves better penetration of fuels and clings to the fuel, thereby increasing the effectiveness of the water used. The use of CAFS reduces the incidence of fires rekindling.

Fire Fighter Safety Tips	© Photos.com

Before attacking a wildland fire, ensure that lookouts, communication systems, escape routes, and safety zones (LCES) are in place.

■ Types of Attacks

Wildland and ground fires can advance quickly and can change direction quickly. Given their unpredictability, they present a very hazardous environment for fire fighters. For these reasons, it is important for the incident commander (IC) to match the type of firefighting attack to the conditions present at the fire. Two general types of attacks are used to contain and extinguish wildland fires: direct attacks and indirect attacks.

An attack on a wildland fire needs to ensure the safety of fire fighters and citizens, and to be as effective as possible in minimizing the property losses from the fire. Fire attacks should start from an anchor point close to the place where the fire started. The anchor point provides an area of safety where the vegetation has already burned. Ideally, it should be located close to a highway or road. This strategy helps prevent a situation in which the fire might change direction, circle back on fire fighters, and trap them in a point of danger.

FIRE FIGHTER Tips

Two primary methods of attack are used for fighting wildland and ground fires: the pincer attack and the flanking attack.

- **Pincer attack:** Two crews of fire fighters mount a two-pronged attack, with one crew suppressing one side of the fire and the other crew suppressing the other side of the fire. The two crews work together to "pinch out" the fire.
- **Flanking attack:** Only one crew is required. The IC positions the crew on one side of the fire and the crew suppresses it.

Direct Attack

A **direct attack** on a wildland and ground fire is mounted by containing and extinguishing the fire at its burning edge. A direct attack can be made with a fire crew and hand tools, a structural engine company, a wildland engine company, a plow, or aircraft. Fire fighters might smother the fire with dirt, use hoses to apply water or compressed air foam to cool the fire, or remove the fuel. A direct attack is typically used on a small wildland fire before it has grown to a large size.

Any direct attack needs to be coordinated through the IMS. Before a direct attack is started, the IC will assess the resources available and determine the best way to attack the fire.

A direct attack is dangerous to fire fighters because they must work in smoke and heat close to the fire. Such an attack has the advantage of accomplishing quick containment of a fire, however. By extinguishing the fire while it is small, fire fighters reduce the risk posed by the fire. Most wildland and ground fires are kept small because the initial crews used an aggressive direct attack on the fire.

Two major types of direct attacks are used on wildland and ground cover fires: the pincer attack and the flanking attack.

Pincer Attack

The pincer attack requires two teams of fire fighters. One team establishes an anchor point on the left side of the fire near the point of origin and mounts a direct attack along the left flank of the fire, working toward the head of the fire. A second team of fire fighters establishes an anchor point on the right side of the fire near the point of origin and mounts a direct attack along the right flank of the fire, also working toward the head of the fire. As the teams advance, the fire gets "pinched" between them, which reduces its growth. By starting at a safe anchor point, the risk to fire fighters is minimized. A successful pincer attack requires the availability of sufficient personnel to be able to mount a two-pronged attack **FIGURE 22-3** .

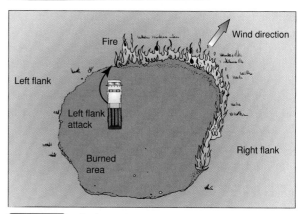

FIGURE 22-4 A flank attack is made on the left flank or right flank of a wildland fire.

FIGURE 22-3 A pincer attack is made on both flanks of a wildland fire.

Flanking Attack

The **flanking attack** requires only one team of fire fighters. This team starts at an anchor point and attacks along one flank of the fire. Such an attack is used when sufficient resources are not available to mount two simultaneous attacks or when the risks posed by the fire are very high on one flank of the fire but lower on the other flank of the fire. The decision regarding which flank of the fire to attack is based on the determination of which side of the fire poses the greatest risk **FIGURE 22-4** .

Indirect Attack

An **indirect attack** is most often used for large wildland and ground fires that are too dangerous to approach through a direct attack. An indirect attack is mounted by building a fire line along natural fuel breaks, at favourable breaks in the topography, or at considerable distance from the fire and then burning out the intervening fuel. This strategy is similar to using a defensive attack on a structure fire. The fire line may be constructed fairly close to the fire, or it may be constructed several kilometres away. An indirect attack can be mounted using either hand tools or mechanized machinery.

Indirect attacks are employed when not enough equipment or personnel are available to mount a direct attack. They are also appropriate when the topography is so rough that a direct attack is dangerous or when the weather conditions pose a high risk for fire fighter safety.

Priorities of Attack

Before determining how to attack a wildland fire, the IC must assess and evaluate the priorities for preserving lives and property. The top priority, of course, is to ensure the safety of both fire fighters and civilians. All fire fighters should make sure that they know the escape route and escape plan before attacking the fire. The second priority is to preserve any structures threatened by the blaze.

■ Fire Apparatus Used for Wildland Fires

The fire apparatus used for wildland fire suppression include both structural fire apparatus and specialized wildland fire apparatus. Structural fire apparatus are commonly used because most small wildland fires are fought by departments whose primary mission is structural firefighting. Structural fire pumpers carry enough water to mount an attack on many small wildland fires. Booster lines provide enough water for attacking small wildland fires. For larger fires, preconnected 44-mm (1.75-in.) hose lines may be needed.

Using structural fire apparatus for wildland fires has several drawbacks. In particular, most structural fire apparatus are not designed to pump water at the same time that the truck is moving forward. The lack of a pump-and-roll capacity means that the apparatus must stop and be shifted into pump gear before a stream of water can be pumped. It also makes the apparatus less efficient when a continuous stream of water is needed while the truck moves along the flank of a fire. In addition, most structural fire apparatus is designed for on-the-road use. As a consequence, trucks that lack all-wheel drive and underside protection may become stuck or damaged when used in off-road situations. Structural fire apparatus that meet the requirements of NFPA 1901, *Standard for Automotive Fire Apparatus*, are not primarily designed for fighting wildland fires.

By contrast, wildland fire apparatus are especially designed for fighting wildland fires. Because the terrain and conditions vary from one geographic location to another, many variations in wildland fire apparatus exist. The design of any such apparatus needs to be suited to the local wildland conditions. Wildland apparatus range from small pickup trucks or Jeep-type vehicles to large trucks. Most of this apparatus is designed with all-wheel drive, underside protection, and high road clearance, which enables the vehicles to travel off the road. Wildland apparatus normally pump between 38 and 2082 litres per minute (10 and 550 gallons per minute [gpm]), are equipped with a water tank, and have pump-and-roll capabilities. This pump-and-roll capacity is achieved by powering the pump with a separate engine or by powering the pump from a **power take-off shaft**. Wildland fire apparatus should meet the requirements of NFPA 1906, *Standard for Wildland Fire Apparatus*.

Small apparatus for fighting wildland fires typically carry 757–1135 L (200–300 gal.) of water. Most such small apparatus have a pump that is powered by a small gasoline or diesel engine, an arrangement that gives the vehicle a pump-and-roll capacity FIGURE 22-5 . Small brush trucks, for example, can respond quickly and contain many fires before they grow into larger blazes FIGURE 22-6 .

FIGURE 22-5 This apparatus has a pump-and-roll capacity.
© Steven Townsend/Code 3 Images

FIGURE 22-6 A smaller wildland or brush truck.
© Michael Routh/Alamy Images

Larger wildland fire apparatus are often built on a midsized or large truck chassis. These trucks carry more water than small apparatus, which enables them to operate for a longer period of time without being refilled. Many of these vehicles have crew cabs that enable an entire crew of fire fighters to travel to the fire scene in the truck.

A third type of wildland fire apparatus is a tanker or water tender FIGURE 22-7 . These trucks are designed to transport water to the scene of a wildland fire. They are usually equipped with a pump that enables them to fill from a drafting site and to offload water when needed. Many tenders are equipped with a dump valve for rapidly unloading water into a portable tank. Some tenders are suited for off-road use, whereas others must remain on a road.

FIGURE 22-7 A water tender (tanker) is used to transport water to wildland fires.
© Steven Townsend/Code 3 Images

Wildland fire apparatus also include plows and bulldozers FIGURE 22-8 . These vehicles are used to create fire breaks.

As mentioned earlier, fixed-wing and rotary-wing aircraft are valuable resources for dropping water where the terrain makes it dangerous or impossible to fight a fire with ground-based fire fighters FIGURE 22-9 . Aircraft can be used to attack the head of a fire, thereby slowing the growth and spread of the fire.

FIGURE 22-8 A wildland bulldozer is used to create fire breaks.
Courtesy of Bob McMillan/FEMA

FIGURE 22-9 Aircraft can be used for making water drops on wildland fires.

Courtesy of Andrea Booher/FEMA

Safety in Wildland Firefighting

Fighting wildland and ground fires is a hazardous duty. The wildland and ground fire environment shares many of the hazards of structural firefighting but also includes other hazards associated with driving, falls, smoke, fire, electricity, and falling trees.

Hazards of Wildland Firefighting

Responses for wildland and ground fires involve all of the hazards encountered by personnel responding to structural fires. In addition, they may involve some unique hazards associated with driving fire apparatus on poorly maintained roads or trails and driving apparatus off-road. Driving on unimproved roads and steep terrain greatly increases the chance of fire apparatus rollovers. Given these dangers, drivers must thoroughly understand the operating characteristics of their fire apparatus and operate the apparatus within the safe limits specific to its design. All fire fighters must keep their seat belts fastened whenever the apparatus is moving.

When working in rough terrain, wildland and ground fire fighters are at increased risk for falls. Rough ground often contains holes that are difficult to see in smoky conditions. Steep terrain also increases the likelihood of falls. Because wildland and ground firefighting involves working with sharp tools, it is important to prevent injuries caused by these tools.

Other hazards of fighting wildland and ground fires include burns, smoke inhalation, dehydration, and heat stroke. Because the personal protective equipment (PPE) worn by wildland and ground fire fighters provides less protection than the PPE worn by structural fire fighters, fire fighters must keep far enough from the heat of the fire to prevent burns. Also, because much wildland and ground firefighting is done without self-contained breathing apparatus (SCBA), fire fighters must avoid inhaling poisonous gases and suspended smoke particles. Use SCBA in any conditions where it is needed.

When engaged in wildland and ground firefighting, be alert for the hazards posed by falling trees. During a fire, the lower parts of trees may burn away and weaken the support for the rest of the tree. As a result, trees of all sizes can fall with little warning.

Also be alert for the presence of electrical hazards. Electrical transmission lines and other electrical wires may be present in the location of a wildland fire. Wires that drop on vegetation may ignite a wildland and ground fire and pose an electrical hazard to fire fighters. Many of these safety hazards can be difficult to see at night and in smoky conditions. Electricity from a downed electrical wire, however, may spark a ground fire. Be alert for electrical hazards even in the middle of an area covered with vegetation with no buildings around.

Because many wildland fires occur during hot weather and because combating them involves strenuous physical activity, all fire fighters must be alert to the possibility of dehydration and heat stroke. Make sure you are drinking enough water, and take breaks when needed to give your body time to recover.

Fire Fighter Safety Tips   © Photos.com

An ash-covered pit with hot embers can be left behind when stumps and large logs burn out. Watch your step!

Personal Protective Equipment

It is important for wildland fire fighters to wear appropriate PPE **FIGURE 22-10**. Specifically, they should be equipped with a one-piece jumpsuit, or a coat, shirt, and trousers that

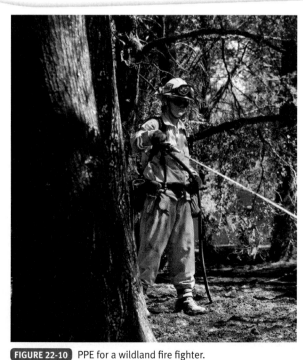

FIGURE 22-10 PPE for a wildland fire fighter.

© Michael Routh/Alamy Images

meet the requirements of NFPA 1977, *Standard on Protective Clothing and Equipment for Wildland Fire Fighting*. These garments should be constructed of a fire-resistant material, such as Nomex®. Such construction ensures that the clothing will stop burning as soon as the heat and flame are removed, which reduces the chance of fire fighters being burned. Wildland fire fighters should also wear an approved helmet with a protective shroud, eye protection, gloves, and protective footwear. All of this equipment should meet the provisions of NFPA 1977.

Respiratory protection for wildland and ground firefighting is usually limited to a filter mask that filters out small particles of smoke. Such masks provide a very limited degree of protection and should not be relied upon beyond the limits for which they are approved. SCBA should be used in conditions of heavy smoke and poisonous gases.

The PPE that is provided for wildland and ground firefighting varies from one location to another, depending on the types of wildland fires encountered and on the climate in a given location. Follow the local fire department's standards regarding PPE. The use of structural PPE for wildland fires can quickly result in rapid body overheating and dehydration. Given this risk, fighting wildland fires in structural PPE for long periods of time is not recommended.

Fire Fighter Safety Tips

© Photos.com

Avoid the use of structural PPE for wildland firefighting. Wearing structural PPE for long periods of time will result in the retention of body heat and will greatly increase the chance of dehydration, heat exhaustion, and heatstroke.

Fire Fighter Safety Tips

© Photos.com

Safety zones are areas of refuge you can retreat to. The size of the zone will be based on the behaviour of the fire and your location relative to the fire. A basic guideline is to choose an area that provides a separation of four times the flame height. This guideline is based on a three-person engine crew on flat ground with no wind. The size of the safety zone would need to be adjusted to take into account wind, location on a slope, and the number of resources.

Suppressing a Ground Cover Fire

To suppress a ground cover fire, follow the steps in **SKILL DRILL 22-1** :

1. Don appropriate PPE.
2. Identify safety and exposure risks.
3. Protect exposures if necessary. (**STEP ❶**)
4. Construct a fire line by removing fuel with hand tools. (**STEP ❷**) OR
5. Extinguish the fire with a backpack pump extinguisher or a hand line. (**STEP ❸**)
6. Overhaul the area completely to ensure complete extinguishment of the ground cover. (**STEP ❹**)

The Challenge of the Wildland–Urban Interface

When people think of wildland fires, they tend to think of burning grass, brush, and forests. In fact, wildland fires are defined by the NFPA as "unplanned and uncontrolled fires burning in vegetation, including structures and other improvements." In other words, many wildland fires threaten not only vegetative fuels, but also buildings and other improvements that have been made on the land.

Today many people want a house that is surrounded by the beauty of nature; that is, they want a house that has modern conveniences but is set in the middle of the woods. The mixing of wildlands and developed areas in this way has created massive problems for fire departments in many parts of Canada. Wildland fires regularly ignite buildings and become structure fires. Structure fires, conversely, regularly ignite vegetative fuels and become wildland fires. In many cases, it is not possible to clearly separate one type of fire from the other. This phenomenon explains why most fire departments are involved in fighting some types of wildland fires.

The term **wildland–urban interface** is used to describe the mixing of wildland with developed areas. It is defined as an area where undeveloped land with vegetative fuels is mixed with human-made structures. The wildland–urban interface presents a major challenge for fire fighters. Because many people live in the wildland–urban interface, fires in this zone present a significant life-safety hazard. Also, because so many structures are built in this zone, there is a huge potential for property loss whenever a fire occurs. Finally, many parts of the wildland–urban interface do not have adequate municipal water systems.

The wildland–urban interface represents a unique challenge for fire fighters and emphasizes the importance of having structural fire fighters learn the basic principles of fighting wildland fires. While the material presented in this chapter will help fire fighters understand the basic principles of fighting wildland fires, it is not intended to provide all of the skills and knowledge needed by wildland fire fighters. These types of fires require special training on what to look for during a size-up and which tactical and strategic considerations to take into account. If your department is regularly involved in fighting wildland fires, you should complete a separate course on wildland firefighting based on NFPA 1051, *Standard for Wildland Fire Fighter Professional Qualifications*.

Much of the effort geared toward reducing the loss from wildland fires needs to be directed at prevention. Many communities work to involve homeowners by educating them about the risks their homes pose and the steps that they can take to reduce this risk. One important measure that has been implemented in many jurisdictions is to outlaw the use of flammable cedar shake roofs in houses in the wildland–urban interface. Another important step is to encourage or require homeowners to maintain a **defensible space** around their structures. A defensible space is an area around a structure where combustible vegetation that can spread fire has been cleared, reduced, or replaced. It acts as a barrier between a structure and an advancing fire. Clearly, our best offence against the huge problem of fire in the wildland–urban interface is often vigorous prevention.

Fire Fighter Safety Tips

© Photos.com

Many areas that are prone to wildland fires have limited numbers of roads leading into and out of them. Be aware of the routes by which you can escape a rapidly moving fire or a fire that changes direction quickly. Do not let yourself get caught in a wildland fire with no escape route.

Wildland Fire Safety

The 10 standard firefighting orders were developed by the U.S. Department of Agriculture Forest Service Chief after a task force investigated 16 tragic wildland fires. These orders were developed in an effort to prevent further tragedies while fighting these fires. The ten orders are as follows:

1. Keep informed on fire weather conditions and forecasts.
2. Know what your fire is doing at all times.
3. Base all actions on the current and expected behaviour of the fire.
4. Identify escape routes and safety zones and make them known.
5. Post lookouts when there is possible danger.
6. Be alert. Keep calm. Think clearly. Act decisively.
7. Maintain prompt communications with your forces, your supervisor, and adjoining forces.
8. Give clear instructions and ensure that they are understood.
9. Maintain control of your forces at all times.
10. Fight fire aggressively, having provided for safety first.

SKILL DRILL 22-1

Suppressing a Ground Cover Fire
(Fire Fighter I, NFPA 5.3.19)

1 Don appropriate PPE. Identify safety and exposure risks. Protect exposures if necessary.

2 Construct a fire line by removing fuel with hand tools. **OR**

3 Extinguish the fire with a backpack pump extinguisher or a hand line.

4 Overhaul the area completely to ensure complete extinguishment of the ground cover.

© LiquidLibrary

Wrap-Up

Chief Concepts

- Wildland fires are unplanned and uncontrolled fires burning in vegetative fuel that sometimes includes structures.
- Wildland fires are sometimes referred to as brush fires, forest fires, grass fires, ground cover fires, ground fires, natural cover fires, and wildfires.
- Wildland is land in an uncultivated natural state that is covered by timber, woodland, brush, or grass.
- Large wildland fires are handled by specialized agencies, such as Natural Resources Canada.
- The wildland fire triangle consists of fuel, oxygen, and heat. These three elements affect the intensity of the fire.
- The primary fuel for wildland fires is vegetation, such as grasslands, brush, and trees. Vegetative fuels range from fine fuels—dried twigs, leaves, or grass—to heavy fuels —large brush, dead timber, or large stumps.
- The size and shape of a fuel influence how it burns. Fine fuels burn more quickly than heavy fuels and require less heat to reach their ignition temperature.
- In wildland fires, air movement influences the speed and direction which the fire moves. Wind blowing on a wildland fire brings more oxygen to the fire and speeds the process of combustion.
- Lightning is the source of almost all naturally caused wildland fires. A single storm can produce many lightning strikes and many fires. Discarded smoking materials and campfires are common human-made causes of wildland fires.
- Weather conditions and topography have a greater effect on wildland fires than they do on structure fires.
- The two most critical weather conditions that influence a wildland fire are moisture and wind. When the relative humidity is low, vegetative fuels dry out, making them more susceptible to ignition. Wind can then push the fire into additional fuels.
- Topography can influence the spread of the fire and the speed of the fire. A wildland fire will spread more quickly over dry, flat land than over a river valley.
- The location where the wildland fire begins is the area of origin.
- The most rapidly moving area of a wildland fire is the head of the fire.

- A long, narrow extension of fire that projects from the head of the fire is called a finger. A finger can grow and produce a secondary direction of travel for the fire.
- A pocket is an unburned area between a finger and the head of the fire. It is a dangerous place for fire fighters because it is surrounded by three sides of fire, leaving no options for escape.
- Green areas contain unburned fuels, whereas black areas have already been burned. Green areas are more likely to become involved in a wildland fire.
- Wildland fires can be controlled and extinguished by cooling the fuel, by removing fuel from the fire, or by smothering the fire.
- Backpack pump extinguishers may be effective against small wildland fires with light fuel loads. Water may also be applied by hoses on special wildland fire trucks or by aircraft.
- Hand tools such as rakes and power tools such as chainsaws may be used to remove fuel from the fire.
- CAFS foam may be used to smother wildland fires.
- The two general types of attacks used to contain and extinguish wildland fires are direct attacks and indirect attacks. A direct attack is mounted by containing and extinguishing the fire at its burning edge. An indirect attack is mounted by building a fire line along natural fuel breaks, at favourable breaks in the topography, or at considerable distance from the fire and then burning out the intervening fuel.
- All fire fighters should ensure that they know the escape route and escape plan before attacking a wildland fire.
- The hazards of wildland fires include fire apparatus rollovers, falls, burns, smoke inhalation, dehydration, and heat stroke.
- The fire shelter is one of the most important pieces of PPE for wildland fire fighters. Fire shelters are designed to reflect 95 percent of a fire's radiant heat for a short period of time.
- The wildland–urban interface is the mixing of wildland and developed areas. Because many people live in the wildland–urban interface, fires in this zone present a significant risk to both lives and structures.

Wrap-Up, continued

Hot Terms

Adz A hand tool constructed of a thin, arched blade set at right angles to the handle. It is used to chop brush for clearing a fire line or to mop up a wildland fire.

Aerial fuels Fuels located more than 1.8 m (6 ft.) off the ground, usually part of or attached to trees.

Area of origin The room or area where a fire began. (NFPA 921)

Backfiring A planned operation to remove fuel from a wildland fire by burning out large selected areas. This operation should be carried out only under the close supervision of experienced, authorized fire officers.

Backpack pump extinguisher A portable fire extinguisher consisting of a 15- to 30-L (4- to 8-gal.) water tank that is worn on the user's back and features a hand-powered piston pump for discharging the water.

Black An area that has already been burned.

Compressed-air foam system (CAFS) A system that combines foam concentrate, water, and compressed air to produce foam that can stick to both vegetation and structures.

Council rake A long-handled rake constructed with hardened triangular-shaped steel teeth that is used for raking a fire line down to soil with no subsurface fuel, for digging, rolling burning logs, and cutting grass and small brush.

Defensible space An area as defined by the authority having jurisdiction (typically 9.1 m [30 ft.] wide or more) between an improved property and a potential wildland fire where combustible materials and vegetation have been removed or modified to reduce the potential for fire on improved property spreading to wildland fuels or to provide a safe working area for fire fighters protecting life and improved property from wildland fire. (NFPA 1051)

Direct attack (wildland and ground fire) A method of fire attack in which fire fighters focus on containing and extinguishing the fire at its burning edge.

Fine fuels Fuels that ignite and burn easily, such as dried twigs, leaves, needles, grass, moss, and light brush.

Finger A narrow point of fire whose extension is created by a shift in wind or a change in topography.

Fire shelter An item of protective equipment configured as an aluminized tent utilized for protection, by means of reflecting radiant heat, in a fire-entrapment situation. (NFPA 1500)

Flanking attack A direct method of suppressing a wildland or ground fire that involves placing a suppression crew on one flank of a fire.

Fuel compactness The extent to which fuels are tightly packed together.

Fuel continuity The relative closeness of wildland fuels, which affects a fire's ability to spread from one area of fuel to another.

Fuel moisture The amount of moisture present in a fuel, which affects how readily the fuel will ignite and burn.

Fuel volume The amount of fuel present in a given area.

Green An area of unburned fuels.

Ground cover fire A fire that burns loose debris on the surface of the ground.

Ground duff Partly decomposed organic material on a forest floor; a type of light fuel.

Head of the fire The main or running edge of a fire; the part of the fire that spreads with the greatest speed.

Heavy fuels Fuels of a large diameter, such as large brush, heavy timber, snags, stumps, branches, and dead timber on the ground. These fuels ignite and are consumed more slowly than light fuels.

Heel of the fire The side opposite the head of the fire, which is often close to the area of origin.

Indirect attack A method of attack in which the control line is located along natural fuel breaks, at favourable breaks in the topography, or at considerable distance from the fire, and the intervening fuel is burned out.

Island An unburned area surrounded by fire.

McLeod tool A hand tool used for constructing fire lines and overhauling wildland fires. One side of the head consists of a five- to seven-toothed fire rake; the other side is a hoe.

Pincer attack A direct method of suppressing a wildland or ground fire, in which one suppression crew attacks the right flank of the fire and a second suppression crew attacks the left flank of the fire.

Pocket A deep indentation of unburned fuel along the fire's perimeter, often found between a finger and the head of the fire.

Power take-off shaft A supplemental mechanism that enables a fire engine to operate a pump while the engine is still moving.

Pulaski axe A hand tool that combines an adz and an axe for brush removal.

Rear of the fire The side opposite the head of the fire. Also called the heel of the fire.

Relative humidity The ratio between the amount of water vapour in the air (gas) at the time of measurement and the amount of water vapour that could be in the air (gas) when condensation begins, at a given temperature. (NFPA 79)

Slash Debris resulting from natural events such as wind, fire, snow, or ice breakage, or from human activities such as building or road construction, logging, pruning, thinning, or brush cutting. (NFPA 1144)

Spot fire A new fire that starts outside areas of the main fire, usually caused by flying embers and sparks.

Subsurface fuels Partially decomposed matter that lies beneath the ground, such as roots, moss, duff, and decomposed stumps.

Surface fuels Fuels that are close to the surface of the ground, such as grass, leaves, twigs, needles, small trees, logging slash, and low brush. Also called ground fuels.

Topography The land surface configuration. (NFPA 1051)

Wildland Land in an uncultivated, more or less natural state and covered by timber, woodland, brush, and/or grass. (NFPA 901)

Wildland fire An unplanned fire burning in vegetative fuels. (NFPA 1051)

Wildland–urban interface An area where undeveloped land with vegetative fuels mixes with human-made structures.

FIRE FIGHTER
in action

The alarm squawks out, "Grass fire. A little more than three klicks east of Highway 125." You jump to your feet and scramble to the engine. Your captain directs you to the brush unit.

Highway 125 is a hilly road that winds through the countryside. The road is lined with pastures that are interspersed with groves of trees. You notice a large plume of billowing white smoke in the distance. This is your first real wildland fire and your mind begins to race. To calm yourself, you quickly cycle through what you learned in your basic training,

1. A grass fire is characterized by what type of fuel?
   A. Fine
   B. Light
   C. Medium
   D. Coarse and heavily compacted

2. Which of the following would be considered surface fuels?
   A. Roots
   B. Duff
   C. Brush
   D. Trees

3. Which of the following is a leg of the fire triangle?
   A. Weather
   B. Fuel
   C. Humidity
   D. Slash

4. A pocket is _____.
   A. a safe place for fire fighters
   B. a dangerous place for fire fighters

   C. at the heel of the fire
   D. at the head of the fire

5. A McLeod tool is a(n) _____.
   A. axe with an adz
   B. tool that combines a hoe and an adz
   C. tool that combines a hoe and a rake
   D. long-handled rake with hardened, triangular-shaped steel teeth

6. A backfire is a(n) _____.
   A. pincer attack
   B. flanking attack
   C. direct attack
   D. indirect attack

7. The basic guideline for a safety zone should include a minimum separation area of at least _____ times the height of the fire.
   A. two
   B. three
   C. four
   D. five

You are normally assigned to a wildland engine and have gained some valuable experience during your first fire season. As the season draws to a close, many of the fire fighters have returned to school and their off-season jobs. To fill the void, a senior fire fighter who normally is assigned to structural firefighting is assigned to your crew. Upon first getting a chance to visit with him, you realize he has little training and no experience in wildland firefighting.

1. How would you build trust with the newly assigned fire fighter?
2. How will you approach the subject and provide guidance?
3. How will you approach your supervisor about the issue of this fire fighter's lack of training?
4. What would be your first priority to teach him?

Fire Suppression

Fire Fighter I

Knowledge Objectives

After studying this chapter, you will be able to:

- Describe the objectives of an offensive attack. (NFPA 5.3.10 , p 682–683)
- Describe the operations performed during an offensive attack. (NFPA 5.3.10 , p 682–683)
- Describe the objectives of a defensive attack. (NFPA 5.3.10 , p 682–683)
- Describe the operations performed during a defensive attack. (NFPA 5.3.10 , p 682–683)
- Describe the characteristics of a fog stream. (NFPA 5.3.8 , p 684)
- Describe the characteristics of a straight stream. (NFPA 5.3.8 , p 684)
- Describe the characteristics of a solid stream. (NFPA 5.3.8 , p 684)
- Describe the objectives of an interior fire attack. (NFPA 5.3.10 , p 684–685)
- Describe the objectives of a direct attack. (NFPA 5.3.10 , p 685)
- Describe the objectives of an indirect attack. (NFPA 5.3.10 , p 685–686)
- Describe the objectives of a combination attack. (NFPA 5.3.10 , p 686)
- Describe the techniques used to advance a large handline. (NFPA 5.3.1.A , p 686–689)
- Describe the characteristics of a master stream device. (NFPA 5.3.8.A , p 689–690)
- Describe the characteristics of a deck gun. (NFPA 5.3.8.A , p 690–691)
- Describe the characteristics of a portable monitor. (NFPA 5.3.8.A , p 691)
- Describe the characteristics of elevated master streams. (NFPA 5.3.8.A , p 691–692)
- Describe the tactics used to protect exposures. (NFPA 5.3.10.A , p 692)
- Describe the characteristics of concealed-space fires. (NFPA 5.3.10 , p 692–693)
- Describe the characteristics of basement fires. (NFPA 5.3.10 , p 693)
- Describe the tactics used to suppress basement fires. (NFPA 5.3.10, 5.3.10.A, 5.3.10.B , p 693)
- Describe the tactics used to suppress fires above ground level. (NFPA 5.3.10, 5.3.10.A, 5.3.10.B , p 693–694)
- Describe the tactics used to suppress fires in large buildings. (NFPA 5.3.10 , p 695)
- Describe the tactics used to suppress fires in buildings under construction, renovation, or demolition. (NFPA 5.3.10 , p 695)
- Describe the tactics used to suppress fires in lumberyards. (NFPA 5.3.8A , p 695)
- Describe the tactics used to suppress fires in stacked or piled materials. (NFPA 5.3.8, 5.3.8.A , p 695–696)
- Describe the tactics used to suppress fires in trash containers. (NFPA 5.3.8 , p 696)
- Describe the tactics used to suppress fires in confined spaces. (NFPA 5.3.8.A , p 696–697)
- Describe the tactics used to suppress fires in vehicles. (NPFA 5.3.7 , p 697–699)
- Describe the tactics used to suppress fires in alternative-fuel vehicles. (NFPA 5.3.7.A , p 698)
- Describe the tactics used to suppress fires in the passenger compartment of a vehicle. (NPFA 5.3.7 , p 698)
- Describe the tactics used to suppress fires in the engine compartment of a vehicle. (NPFA 5.3.7.A , p 699)
- Describe the tactics used to suppress fires in the trunk of a vehicle. (NPFA 5.3.7.A , p 699)
- Describe how to overhaul a vehicle fire. (NPFA 5.3.7.A , p 699)
- Describe the hazards presented by flammable-liquid fires. (NFPA 5.3.7 , p 699–700)
- Describe the tactics used to suppress flammable-liquid fires. (NFPA 5.3.7 , p 700)
- Discuss when gas service should be shut off. (NFPA 5.3.18, 5.3.18A , p 702)
- Describe when the electrical system should be shut off. (NFPA 5.3.18, 5.3.18A , p 702)
- Describe the hazards posed by electrical fires. (NFPA 5.3.18.A , p 702–703)
- Describe the tactics used to suppress an electrical fire. (p 702–703)

© Photos.com

Skills Objectives

After studying this chapter, you will be able to perform the following skills:

- Perform a direct attack. (NFPA 5.3.10.B , p 685)
- Perform an indirect attack. (NFPA 5.3.10.B , p 686)
- Perform a combination attack. (NFPA 5.3.10.B , p 686)
- Perform the one–fire fighter method for operating a large handline. (NFPA 5.3.8.B , p 687–688)
- Perform the two–fire fighter method for operating a large handline. (NFPA 5.3.8.B , p 687–689)
- Operate a deck gun. (NFPA 5.3.8.B , p 690–691)
- Deploy and operate a portable monitor. (NFPA 5.3.8.B , p 691)
- Locate and suppress concealed-space fires. (NFPA 5.3.8.B , p 693)
- Extinguish an outdoor trash fire or other outdoor Class A fire. (NFPA 5.3.8.B , p 696)
- Extinguish a vehicle fire. (NPFA 5.3.7.B , p 698)
- Shut off gas utilities. (NFPA 5.3.18.B , p 702)
- Control the electric utility system. (NFPA 5.3.18.B , p 702)

Fire Fighter II   FFII

Knowledge Objectives

After studying this chapter, you will be able to:

- List the factors that the incident commander (IC) evaluates when determining whether to perform a defensive attack or an offensive attack. (NFPA 6.3.2 , p 683)
- Describe the characteristics of a fog stream. (NFPA 6.3.2A , p 684)
- Describe the characteristics of a straight stream. (NFPA 6.3.2A , p 684)

- Describe the characteristics of a solid stream. (NFPA 6.3.2A , p 684)
- Describe the objectives of an interior fire attack. (NFPA 6.3.2B , p 684–685)
- Describe the objectives of a direct attack. (NFPA 6.3.2B , p 685)
- Describe the objectives of an indirect attack. (NFPA 6.3.2B , p 685–686)
- Describe the objectives of a combination attack. (NFPA 6.3.2B , p 686)
- Explain how ventilation is coordinated with fire-suppression operations. (NFPA 6.3.2 , p 692)
- Describe the characteristics of concealed-space fires. (NFPA 6.3.2 , p 692–693)
- Describe the characteristics of basement fires. (NFPA 6.3.2 , p 693)
- Describe the tactics used to suppress basement fires. (NFPA 6.3.2B , p 693)
- Describe the tactics used to suppress fires above ground level. (NFPA 6.3.2B , p 693–694)
- Describe the characteristics of flammable-gas cylinders. (NFPA 6.3.3A , p 700–702)
- Describe the hazards presented by flammable-gas fires. (NFPA 6.3.3A , p 700–702)
- Describe a boiling liquid/expanding vapour explosion (BLEVE). (NFPA 6.3.3A , p 701)
- Describe the tactics used to suppress flammable-gas fires. (NFPA 6.3.3, 6.3.3A , p 701–702)

Skills Objectives

After studying this chapter, you will be able to perform the following skills:

- Coordinate an interior attack. (NFPA 6.3.2.B , p 685)
- Suppress a flammable-gas cylinder fire. (NFPA 6.3.3.B , p 701–702)

You Are the Fire Fighter

This is it: your first night at the fire station. You lie in your bed wide awake, just hoping for a call. Just past midnight, the alarm sounds for a fire in a house. You don your PPE quickly before mounting the apparatus. Your heart is racing as the other crew members mount the apparatus as if it were just another day at the office. As you get close to the scene, you can see the orange glow: the incident is a working fire.

1. What is the difference between an offensive attack and a defensive attack?
2. When would you use an indirect attack rather than a direct attack?
3. Which type of nozzle would you use for an indirect attack?

Introduction

The term *fire suppression* refers to all of the tactics and tasks that are performed on the fire scene to achieve the final goal of extinguishing the fire. Fire suppression can be accomplished through a variety of methods that will stop the combustion process. All of these methods involve removal of one of the four components of the fire tetrahedron; that is, a fire can be extinguished by removing the oxygen, the fuel, or the heat from the combustion process. Interrupting the chemical chain reactions will also stop the combustion process and extinguish the fire.

This chapter presents the methods that fire fighters most frequently use to extinguish fires. The apparatus and equipment employed by most fire departments are designed to apply large volumes of water to a fire in an attempt to cool the fuel below its ignition temperature. Although fire departments typically deploy a variety of extinguishing agents for different situations, water is used more often than any other agent. Water is a very effective extinguishing agent for many different types of fires, because tremendous quantities of heat energy are required to convert water into steam. When water is applied to a fire, all of the heat that is used to create steam is removed from the combustion process. If a sufficient quantity of water is applied, the fuel is cooled below its ignition temperature and the fire is extinguished.

FIRE FIGHTER Tips

While fire-suppression operations are ongoing, a variety of other activities will be occurring simultaneously. All firefighting activities are important because successful firefighting requires a team effort and coordinated attack.

Offensive versus Defensive Operations

All fire-suppression operations are classified as either offensive or defensive. When fire fighters advance hose lines into a building to attack a fire, the strategy is offensive. By contrast, defensive operations are conducted from the exterior, by directing water streams toward the fire from a safe distance.

Offensive operations expose fire fighters to the heat and smoke of the fire inside the building as well as to several other risk factors, such as the possibility of being trapped by a structural collapse. The objective in an offensive operation is for fire fighters to get close enough to the fire to apply extinguishing agents at close range. This allows the extinguishing agent to be applied directly where it is needed to overpower the fire. When an offensive attack is successful, the fire can be controlled with the least amount of property damage.

An offensive strategy is used in situations where the fire is not too large or too dangerous to be extinguished by applying water from interior handlines. Small handlines (38- or 44-millimetre [1.5- or 1.75-inch] lines) are used for most interior fire attack operations. Large handlines (64-mm [2.5-in.] lines) are often used to conduct an interior attack on a large fire. Although they are much heavier and more difficult to maneuver inside a building than their smaller-diameter counterparts, the volume of fire may require the use of these larger-diameter hoses.

Large handlines and master streams are more often used in defensive operations. A defensive strategy is used in situations where the fire is too large to be controlled by an offensive attack and in situations where the level of risk to fire fighters conducting an interior attack would be unacceptable. The primary objective in a defensive operation is to prevent the fire from spreading; thus water is directed into the building through doorways, windows, and openings in roofs, or onto exposures to keep the fire from spreading. During this type of fire-suppression operation, fire fighters remain outside the building and operate from safe positions.

The decision whether to conduct offensive or defensive operations must be made by the incident commander (IC) at the beginning of each fire-suppression operation and is periodically reevaluated throughout the incident. This decision must be made before operations begin and must be clearly communicated and understood by everyone who is involved in the operation. There is no room for confusion: it would be an extremely dangerous situation if one group of fire fighters initiated defensive operations while another group of fire fighters was conducting an offensive attack inside the building.

If the decision is made to switch from offensive to defensive strategy, or from defensive to offensive strategy, at any point during an operation, this change must be clearly communicated and understood by all fire fighters. A change from offensive to defensive operations could be warranted if an interior attack is unsuccessful or if the risk factors are determined to be too great to justify having fire fighters work inside the building. Sometimes the strategy switches from defensive to offensive after an exterior attack has reduced the volume of fire inside a building to the point at which fire fighters can enter and complete extinguishments with handlines. An offensive fire attack requires well-planned coordination among crews performing different tasks, such as ventilating, operating hose lines, and conducting aggressive search and rescue.

Fire Fighter Safety Tips
© Photos.com

One of the most crucial decisions made by the IC is whether to initiate a defensive or offensive fire attack. Mounting an offensive attack on an unsafe building can result in fire fighter deaths or injuries.

A defensive strategy should be implemented when the IC determines that it would be impossible to enter the burning building and control the fire with handlines, as well as in situations where the risk of injury to or death of a fire fighter is excessive. A defensive operation involves the use of large-diameter hoses and master stream devices from the exterior of the structure in an attempt to confine the fire. Exposure protection should be a high priority during a defensive operation.

Fire Fighter Safety Tips
© Photos.com

Offensive (interior attack) operations and defensive (exterior attack) operations must never be performed simultaneously.

Fire Fighter Safety Tips
© Photos.com

An IC should never risk the lives of fire fighters when there are no lives to save.

■ Command Considerations

The IC must evaluate a whole range of factors to decide whether an offensive strategy (interior attack) or a defensive strategy (exterior attack) should be used at a particular fire. If the risk factors are too great, an exterior attack is the only acceptable option. If the decision is made to launch an interior attack, the IC must determine where and how to attack, after considering both safety issues and the potential effectiveness of the operation. The factors to be evaluated when considering whether to enter the structure to mount an attack include the following:

- What are the risks versus the potential benefits?
- Is it safe to send fire fighters into the building? Do not risk fire fighters' lives to retrieve the dead or save a building about to be demolished.
- What are the structural concerns?
- Is this building made of lightweight construction?

- Are there any lives at risk?
- Does the size of the fire prohibit entry?
- Are enough fire fighters on the scene to mount an interior attack? (Remember the two-in–two-out rule.)
- Is an adequate water supply available?
- Can proper ventilation be carried out to support offensive operations?

After sizing up the situation, the IC must determine which type of attack is appropriate. As a new fire fighter, you are not responsible for determining the type of fire attack that will be used. Nevertheless, you should understand the various factors that go into making these decisions and recognize why the IC orders different types of fire attacks for different types of fires.

Operating Hose Lines

Some of the most basic skills that must be mastered by every fire fighter involve the use of hose lines to apply water onto a fire. Put simply, a fire fighter must be able to advance and operate a hose line effectively to extinguish a fire. The proper operation of a hose line is also essential to protect yourself, your crew, and any trapped victims from the fire. Fire attack operations are often conducted under extremely stressful conditions, including high heat conditions, limited or zero visibility, and unfamiliar surroundings. Care should be taken not to have opposing hose lines, such that two crews are working against each other.

Fire fighters must learn how to operate both large and small handlines, as well as master stream appliances. Small handlines can be as large as 51 mm (2 in.) in diameter. The most frequently used size for interior fire attack is 44-mm (1.75-in.) handlines. Although one fire fighter can usually operate the nozzle on a small hose line, a second fire fighter will provide valuable assistance when a small hose line must be advanced and maneuvered.

Large handlines are defined as hoses that are at least 64 mm (2.5 in.) in diameter. Because water can flow through these hoses at a rate of more than 1136 litres per minute (250 gallons per minute [gpm]), large handlines are heavier and less maneuverable than smaller lines. At least two fire fighters are required to advance and control a large handline, although one fire fighter can control a large handline if it is firmly anchored. This task can be accomplished by utilizing a webbing strap or by looping the line and sitting down on it **FIGURE 23-1**.

Master streams are used when large quantities of water are needed to control a large fire. Such a stream can deliver water at a rate of at least 1591 L/min. (350 gpm), and some master stream devices can flow more than 9092 L/min. (2000 gpm). The most commonly used master stream devices deliver flows between 1591 and 6819 L/min. (350 and 1500 gpm). Master stream devices are operated from a fixed position—either on the ground, on top of a piece of fire apparatus, or on an aerial ladder or elevated platform. They are typically used for defensive operations, although master streams can also be used to "blitz" a fire before beginning an offensive attack. This fire-suppression method knocks the main body of fire down with a heavy stream; crews can then stretch handlines into the site and extinguish the remaining fire.

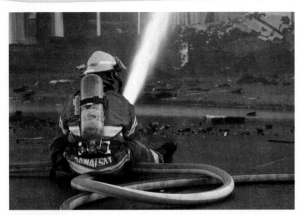

FIGURE 23-1 One fire fighter can control a large handline if it is firmly anchored. This task can be accomplished by utilizing a webbing strap or by looping the line and sitting down on it.

© Jack Dagley Photography/ShutterStock, Inc.

Fire Streams

Different types of fire streams are produced by using different types of nozzles. As described in the chapter "Fire Attack Lines and Foam," the nozzle defines the pattern and form of the water that is discharged onto the fire. A fire stream can be produced with a smooth-bore nozzle or an adjustable nozzle. Fire department policies and suggested operating guidelines (SOGs), or standard operating procedures (SOPs), usually dictate the types of nozzles that are used with different types of hose lines. The nozzle operator must know which type of nozzle should be used in the specific situation at hand. When an adjustable nozzle is used, the nozzle operator must know how to set the discharge pattern to produce different kinds of streams.

The first major distinction in nozzle discharge patterns is between a fog stream and a straight stream **FIGURE 23-2**. A fog stream divides water into droplets, which have a very large surface area and can absorb heat efficiently. When heat levels in a building need to be lowered quickly, a combination of ventilation and a fog stream may be the fire-suppression method of choice. A fog stream can also be used to protect fire fighters from the heat of a large fire. Most adjustable nozzles can be adjusted from a straight stream to a narrow fog pattern to a

FIGURE 23-2 A straight stream and a fog stream are both produced using a fog nozzle.

wide fog pattern, depending on the reach that is required and how the stream will be used.

A **straight stream** has a greater reach than a fog stream, so it can hit the fire from farther away. A straight stream also keeps the water concentrated in a small area, so it can penetrate through a hot atmosphere to reach and cool the burning materials. To produce a straight stream, the fire fighter sets the adjustable nozzle to the narrowest pattern it can discharge. This type of stream is made up of a highly concentrated pattern of droplets that are all discharged in the same direction.

A **solid stream** is produced by a smooth-bore nozzle **FIGURE 23-3**. A solid stream has a greater reach and penetrating power than a straight stream, because it is discharged as a continuous column of water.

Solid Stream

FIGURE 23-3 A solid stream is produced by a smooth-bore nozzle.

One consideration when selecting and operating nozzles is the amount of air that is moved along with the water. A fog stream naturally moves a large quantity of air along with the mass of water droplets. This air flows into the fire area along with the water. When this air movement is combined with steam production as the water droplets encounter a heated atmosphere, the thermal balance is likely to be disrupted quickly. In such a case, the hot fire gases and steam may be displaced back toward the nozzle operator. Straight and solid streams move little air in comparison to a fog stream, so fewer concerns with displacement and disruption of the thermal balance arise when these types of streams are used for fire suppression.

When applied correctly, the air movement created by a fog stream can be used for ventilation; discharging a fog stream out through a window or doorway, for example, will draw smoke and heat out in the same manner as an exhaust fan. This operation must be performed carefully to prevent accidentally drawing hidden fire toward the nozzle operator. The use of a water stream to provide ventilation is called hydraulic ventilation and is discussed further in the chapter "Ventilation."

Interior Fire Attack

An interior fire attack is an offensive operation that requires fire fighters to enter a building and discharge an extinguishing agent (usually water) onto the fire. An interior structure fire is a fire that occurs inside a building or structure. Its fuel could be the contents of the building, or the structure itself might be burning. The larger the fire, the greater the challenge in suppressing it, and the more ominous the risks that are involved in interior fire suppression.

Interior fire attack can be conducted on many different scales. In many cases, such an attack is geared toward a fire that is burning in only one room; this kind of fire may be controlled quickly by one attack hose line. Larger fires require more water, which could be provided by two or more small handlines working together or by one or more larger handlines. Fires that involve multiple rooms, large spaces, or concealed spaces are more complicated and require more extensive coordination; nevertheless, the basic techniques for attacking these fires are similar to the techniques used when extinguishing smaller fires.

As a trained Fire Fighter II, you should be able to understand and coordinate an interior fire attack. To coordinate an interior attack, follow the steps in **SKILL DRILL 23-1** (Fire Fighter II, NFPA 6.3.2):

1. Don full personal protective equipment (PPE), including self-contained breathing apparatus (SCBA). Enter the personnel accountability system, and proceed to work as a team.
2. Perform size-up and give an arrival report. Call for additional resources if needed.
3. Ensure that an adequate water supply and backup resources are available.
4. Select the appropriate attack technique.
5. Communicate the attack technique to the team.
6. Maintain constant team coordination.
7. Evaluate conditions on an ongoing basis.
8. Communicate and manage search, rescue, and ventilation requirements.
9. Report hazards.
10. Inform incident command of changing conditions.
11. Assess burn patterns to determine the fire's origin.
12. Preserve signs of the fire's origin, cause, and arson.
13. Ensure complete extinguishment of fire during overhaul.
14. Exit the hazard area, account for all members of the team, and report to incident command.

Direct attack and indirect attack are two different methods of discharging water onto a fire. A combination attack is performed in two stages, beginning with an indirect attack and then continuing with a direct attack.

Fire Fighter Safety Tips

© Photos.com

It is imperative to extinguish all fire as you proceed to the seat of the fire. Failure to do so may allow the fire to grow behind you, entrapping you by cutting off your escape route.

Direct Attack

The most effective means of fire suppression in most situations is a **direct attack**. This kind of attack uses a straight or solid hose stream to deliver water directly onto the base of the fire **FIGURE 23-4**. The water cools the fuel until it is below its ignition temperature. Water should be applied until flame is no

longer visible. To perform a direct attack, follow the steps in **SKILL DRILL 23-2** (Fire Fighter I, NFPA 5.3.10):

1. Exit the fire apparatus wearing full PPE, including SCBA.
2. Select the proper hose line to fight the fire based on the fire's size, location, and type.
3. Advance the hose line from the apparatus to the entry point of the structure. Flake out excess hose in front of the door.
4. Don a face piece and activate the SCBA and personal alert safety system (PASS) device prior to entering the building.
5. Signal the pump operator/driver that you are ready for water.
6. Open the nozzle to purge air from the system and make sure water is flowing.
7. Make sure that ventilation is completed or in progress.
8. Enter into the structure and locate the seat of the fire.
9. Apply water in either a straight or solid stream onto the base of the fire until all visible flame has been extinguished.
10. Watch for changes in fire conditions.
11. Shut down the nozzle and listen.
12. Locate and extinguish hot spots.

FIGURE 23-4 In a direct attack, a straight or solid hose stream is used to deliver water directly onto the base of the fire.

Indirect Attack

Indirect application of water is used in situations where the temperature is increasing and it appears that the room or space is ready to flash over. With this fire-suppression method, the fire fighter aims a short burst of water at the ceiling to cool the superheated gases in the upper levels of the room or space. This action can prevent or delay flashover long enough for fire fighters to apply water directly to the seat of the fire or to make a safe exit. Follow your department's SOGs/SOPs regarding the application of water.

The objective of an **indirect attack** is to quickly remove as much heat as possible from the fire atmosphere. An indirect attack is particularly effective at preventing flashover from occurring. This method of fire suppression should be used when a fire has produced a layer of hot gases at the ceiling level. When water is injected into the hot fire gases, it is converted to steam, absorbing tremendous quantities of heat in the process. The atmosphere cools quickly down to 100 degrees Celcius (212 degrees Fahrenheit), the boiling point of water. In this way, heat—which is used to convert the water to steam—is removed from the combustion process.

Fire fighters can make an indirect attack by using a straight stream, a solid stream, or a narrow fog stream. With this strategy, they direct their water toward the ceiling of the intensely heated area where the hot gases are layered, so as to create steam **FIGURE 23-5**. This practice is often referred to as "painting the ceiling" with the water stream. The water is distributed over a large surface area so that it will absorb heat as quickly as possible. Once the temperature has been reduced and the area

FIGURE 23-5   In an indirect attack, fire fighters direct a fog stream at the ceiling of the intensely heated area so as to create steam.

has been properly ventilated, fire fighters can switch to a direct attack to complete extinguishment.

As soon as enough steam has been produced to reduce the fire, the fire stream should be shut down so that the thermal layering of the superheated gases is disturbed as little as possible. When water is converted to gaseous steam, it expands to occupy a volume 1700 times greater than the volume of an equivalent amount of liquid water. This expansion tends to displace the hot gases that were near the ceiling and push them down toward the floor. The resulting mixture of steam and hot gases is capable of causing serious steam burns to fire fighters—even those who are wearing PPE. Serious injuries can occur if fire fighters put too much water into the upper atmosphere and the hot gases are forced down on top of them.

To perform an indirect attack, follow the steps in **SKILL DRILL 23-3** (Fire Fighter I, NFPA 5.3.10):

1. Exit the fire apparatus wearing full PPE, including SCBA.
2. Select the correct hose line to be used to attack the fire depending on the type of fire, its location, and its size.
3. Advance the hose line from the apparatus to the opening in the structure where the indirect attack will be made.
4. Don a face piece, and activate the SCBA and PASS device.

5. Notify the operator/driver that you are ready for water.
6. Open the nozzle and make sure that air is purged from the hose line and that water is flowing. If using a fog nozzle, ensure that it is set to the proper nozzle pattern for entry. Shut down the nozzle until you are in a position to apply water.
7. Advance with a charged hose line to the location where you will apply water.
8. Direct the water stream toward the upper levels of the room and ceiling into the heated area overhead, and move the stream back and forth. Flow water until the room begins to darken. Shut the nozzle off, and reassess the fire conditions.
9. Watch for changes and a reduction in the amount of fire. Once the fire is reduced, shut down the nozzle.
10. Confirm that ventilation has been completed.
11. Attack any remaining fire and hot spots until the fire is completely extinguished.

Combination Attack

A **combination attack** employs both indirect attack and direct attack methods in a sequential manner. This strategy should be used when a room's interior has been heated to the point that it is nearing a flashover condition. Fire fighters should first use an indirect attack method to cool the fire gases down to safer temperatures and prevent flashover from occurring. This operation is followed with a direct attack on the main body of fire.

In a combination attack, the fire fighter who is operating the nozzle should be given plenty of space to maneuver. Only enough water as is needed to control the fire should be used, so as to avoid unnecessary water damage.

To perform a combination attack, follow the steps in **SKILL DRILL 23-4** (Fire Fighter I, NFPA 5.3.10):

1. Don full PPE and SCBA. Select the correct hose line to accomplish the suppression task at hand.
2. Stretch the hose line to the entry point of the structure, and signal the operator/driver that you are ready to receive water.
3. Open the nozzle to get the air out, and make sure that water is flowing.
4. Enter the structure, and locate the room or area where the fire originated.
5. Aim the nozzle at the upper-left corner of the fire and make either a T, O, or Z pattern with the nozzle. Start high and then work the pattern down to the fire level.
6. Use only enough water to darken down the fire without upsetting the thermal layering.
7. Once the fire has been reduced, find the remaining hot spots and complete fire extinguishment using a direct attack.

■ Large Handlines

Large handlines can be used either for offensive fire attacks or for defensive operations. In an offensive attack situation, a 64-mm (2.5-in.) attack line can be advanced into a building to apply a heavy stream of water onto a large volume of fire. The same direct and indirect attack techniques that were described

Fire Fighter Safety Tips

Indicators of possible building collapse include the following signs:

- Leaning walls
- Free-standing walls
- Smoke emitting from cracks
- Creaking sounds
- Sagging roofs or floors

Considerations for possible building collapse include the following factors:

- Exposure to serious fire
- Water loading from use of master streams
- Lightweight construction or trusses

FIRE FIGHTER Tips

When using an indirect or combination attack, watch for droplets of water raining down; they indicate lowered ceiling temperatures. If no droplets fall, the ceiling is still too hot.

for small hose lines can also be used with large handlines. A 64-mm (2.5-in.) handline can overwhelm a substantial interior fire if it can be discharged directly into the involved area. The extra reach of the stream can also prove valuable when making an interior attack in a large building.

It is more difficult for fire fighters to advance and maneuver a large handline inside a building, particularly in tight quarters or around corners. At least three team members are usually needed to advance and maneuver a 64-mm (2.5-in.) handline inside a building. These fire fighters must contend with the nozzle reaction force as well as the combined weight of the hose and the water. In situations where the hose line must be advanced over a considerable distance into a building, additional fire fighters will be required to move the line; the extra effort required to deploy such a line, however, is balanced by the powerful fire-suppression capabilities of a large handline.

Large handlines are often used in defensive situations to direct a heavy stream of water onto a fire from an exterior position. In these cases, the nozzle is usually positioned so that it can be operated from a single location by one or two fire fighters. The stream can be used to attack a large exterior fire or to protect exposures. It can also be directed into a building through a doorway or window opening to knock down a large volume of fire inside. If the exterior attack is successful in reducing the volume of fire, the IC might decide to switch to an offensive (interior) attack to complete extinguishment of the fire.

One–Fire Fighter Method

One fire fighter can control a large attack hose by forming a large loop of hose about 0.6 m (2 ft.) behind the nozzle. When the loop is placed over the top of the nozzle, the weight of the hose stabilizes the nozzle and reduces the nozzle reaction. To add more stability, lash the hose loop to the hose behind the nozzle where they cross. This technique reduces the energy needed to control the line if it is necessary to maintain the

water stream for a long period of time. Although this method does not allow the hose to be moved while water is flowing, it is a good choice for protecting exposures when fire fighters are operating in a defensive attack mode.

To perform the one–fire fighter method for operating a large handline, follow the steps in **SKILL DRILL 23-5**:

1. Select the correct size of fire hose for the task to be performed.
2. While wearing full PPE and SCBA, advance the hose into the position from which you plan to attack the fire.
3. Signal the pump operator that you are ready for water.
4. Open the nozzle to allow air to escape the hose and to ensure that water is flowing.
5. Close the nozzle and then make a loop with the hose, ensuring that the nozzle is under the hose line that is coming from the fire apparatus. (STEP **1**)
6. Lash the hose sections together where they cross, or use your body weight to kneel or sit on the hose line at the point where the hose crosses itself. (STEP **2**)
7. Allow enough hose to extend past the section where the line crosses itself for maneuverability. (STEP **3**)
8. Open the nozzle and direct water onto the designated area. (STEP **4**)

Two–Fire Fighter Method

When two fire fighters are available to operate a large handline, one should act as the nozzle operator, while the other serves as a backup. The nozzle operator grasps the nozzle with one hand and holds the hose behind the nozzle with the other hand. The hose should be cradled across the fire fighter's hip for added stability. The backup fire fighter should be positioned approximately 1 m (3 ft.) behind the nozzle operator. This person grasps the hose with both hands and holds the hose against a leg or hip. The backup fire fighter can also use a hose strap to maintain a better hand grip on a large handline. When the line is operated from a fixed position, the second fire fighter can kneel on the hose with one knee to stabilize it against the ground.

To perform the two–fire fighter method for operating a large handline, follow the steps in **SKILL DRILL 23-6**:

1. Don all PPE and SCBA.
2. Select the correct hose line for the task at hand.
3. Stretch the hose line from the fire apparatus into position. (STEP **1**)
4. Signal the pump operator that you are ready for water.
5. Open the nozzle a small amount to allow air to escape and to ensure that water is flowing.
6. Advance the hose line as needed. (STEP **2**)
7. Before attacking the fire, the fire fighter on the nozzle should cradle the hose on his or her hip while grasping the nozzle with one hand and supporting the hose with the other hand.
8. The second fire fighter should stay approximately 1 m (3 ft.) behind the fire fighter who is on the nozzle. The second fire fighter should grasp the hose with two hands and, if necessary, use a knee to stabilize the hose against the ground.

SKILL DRILL 23-5 Performing the One–Fire Fighter Method for Operating a Large Handline
(Fire Fighter I, NFPA 5.3.8)

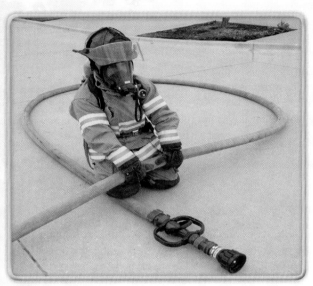

1 Select the correct size of fire hose. Advance the hose into position. Signal that you are ready for water, and open the nozzle to allow air to escape and to ensure that water is flowing. Close the nozzle and then make a loop with the hose, ensuring that the nozzle is under the hose line that is coming from the fire apparatus.

2 Lash the hose sections together where they cross, or use your body weight to kneel or sit on the hose line at the point where the hose crosses itself.

3 Allow enough hose to extend past the section where the line crosses itself for maneuverability.

4 Open the nozzle and direct water onto the designated area.

© LiquidLibrary

9 Use a hose strap to maintain a better grip on the hose. (**STEP 3**)

10 Open the nozzle in a controlled fashion, and direct water onto the fire or designated exposure. (**STEP 4**)

If it is necessary to advance a flowing 64-mm (2.5-in.) handline over a short distance and only two fire fighters are available, be aware of the large reaction force exerted by the flowing water. It is much easier to shut down the nozzle momentarily and move it to the new position than to relocate a flowing line. If the line must be moved while water is flowing, both fire fighters must brace the hose against their bodies to keep it under control. Three fire fighters can stabilize and advance a large handline more comfortably and safely than can two fire fighters.

SKILL DRILL 23-6 — Performing the Two–Fire Fighter Method for Operating a Large Handline
(Fire Fighter I, NFPA 5.3.8)

1 Stretch the hose line from the fire apparatus into position.

2 Signal that you are ready for water, and open the nozzle to allow air to escape and to ensure water is flowing. Advance the hose line as needed.

3 Before attacking the fire, the fire fighter on the nozzle should cradle the hose on his or her hip while grasping the nozzle with one hand and supporting the hose with the other hand. The second fire fighter should stay approximately 1 m (3 ft.) behind the fire fighter who is on the nozzle. The second fire fighter should grasp the hose with two hands and may use a knee to stabilize the hose against the ground, if necessary.

4 Open the nozzle in a controlled fashion and direct water onto the fire or designated exposure.

© LiquidLibrary

■ Master Stream Devices

<u>Master stream devices</u> are used to produce high-volume water streams for large fires. Several types of master stream devices exist, including portable monitors, deck guns, ladder pipes, and other elevated stream devices. Most master streams discharge water at a rate of between 1591 and 6819 L/min. (350 and 1500 gpm), although much larger capacities are available for special applications. In addition, the stream that is discharged from a master stream device has a greater range than the stream from a handline, so it can be effective from a greater distance.

A master stream device can be either manually operated or directed by remote control. Many of these devices can be set up and then left to operate unattended. This capability may prove extremely valuable in a high-risk situation, because it eliminates the need to leave a fire fighter in an unsafe location or a hazardous environment to operate the device.

Master streams are used mainly during defensive operations. They should never be directed into a building while fire fighters are operating inside the structure, because these streams can push heat, smoke, or fire onto the fire fighters.

Fire Fighter Safety Tips

© Photos.com

A nozzle flowing at 4546 L/min. (1000 gpm) is adding 3.8 tonnes (4.2 tons) of weight to the building every minute. Be aware of the increasing risk of building collapse in such a scenario.

Fire Fighter Safety Tips

© Photos.com

Master streams should never be directed into a building while fire fighters are operating inside the structure. The force of the stream can push heat, smoke, or fire onto the fire fighters.

The force and impact of the stream can also dislodge loose materials or cause a structural collapse.

Deck Guns

A **deck gun** is permanently mounted on a vehicle and equipped with a piping system that delivers water to the device. These devices are sometimes called turret pipes or wagon pipes **FIGURE 23-6** . If the vehicle is equipped with a pump, the pump operator can usually open a valve to start the flow of water.

Sometimes, however, a hose must be connected to a special inlet to deliver water to the deck gun. If your fire apparatus is equipped with a deck gun, you need to learn your role when placing it in operation.

To operate a deck gun, follow the steps in **SKILL DRILL 23-7** :

1. Make sure that all firefighting personnel are out of a structure before using a deck gun.

SKILL DRILL 23-7   Operating a Deck Gun
(Fire Fighter I, NFPA 5.3.8)

1 Make sure that all firefighting personnel are out of a structure before using a deck gun. Place the deck gun in the correct position.

2 Aim the deck gun at the fire or at the target exposure. Signal the pump operator that you are ready for water.

3 Once water is flowing, adjust the angle, aim, or water flow as necessary.

© LiquidLibrary

FIGURE 23-6 A deck gun is permanently mounted on a vehicle and equipped with a piping system that delivers water to the device.

FIGURE 23-7 A portable monitor is placed on the ground and supplied with water from one or more hose lines.

2. Place the deck gun in the correct position. (**STEP ①**)
3. Aim the deck gun at the fire or at the target exposure.
4. Signal the pump operator that you are ready for water. (**STEP ②**)
5. Once water is flowing, adjust the angle, aim, or water flow as necessary. (**STEP ③**)

FIRE FIGHTER Tips

Remote-controlled wildland monitors can be mounted on the front bumper of wildland apparatus. These devices operate through an electrically operated remote control and are built to flow anywhere between 60 and 1320 L/min. (15 and 350 gpm). They can be used to mount an attack while the operator remains in the vehicle.

Portable Monitor

A **portable monitor** is a master stream device that can be positioned wherever a master stream is needed **FIGURE 23-7**. It is placed on the ground, and hose lines are then connected to the portable monitor to supply the water. Most of these devices come equipped with either one, two, or three inlets. Portable monitors can be built with 64-mm (2.5-in.) inlets or with a large-diameter hose inlet. Smaller portable monitors are sometimes set up attached to a preconnected hose. This arrangement enables one fire fighter to step away from the engine with the portable monitor and attached hose, walk to an assigned area, and quickly place the portable monitor in operation. Some master stream devices can be used as deck guns or taken off the fire apparatus and used as portable monitors.

To deploy a portable monitor, remove it from the apparatus and carry it to the location where it will be used. Advance an adequate number of hose lines from the engine to the monitor. The number of hose lines needed depends on the volume of water to be delivered and the size of the hose lines. Form a large loop in the end of each hose line in front of the monitor, and then attach the male coupling to the inlets of the monitor. The loops serve to counteract the force created by the flow of the water through the nozzle.

To set up and operate a portable monitor, follow the steps in **SKILL DRILL 23-8** (Fire Fighter I, 5.3.8):

1. Remove the portable monitor from the fire apparatus and move it into the desired position.
2. Attach the necessary hose lines to the monitor as per SOGs/SOPs or the manufacturer's instructions.
3. Loop the hose lines in front of the monitor to counteract the force created by water flowing out of the nozzle.
4. Signal the pump operator that you are ready for water.
5. Aim the water stream at the fire or onto the designated exposure, and adjust the stream as necessary.

If the portable monitor is not adequately secured, the nozzle reaction force can cause it to move from the position where it was originally placed. A moving portable monitor poses a danger to anyone in its path. Many of these master stream devices are equipped with a strap or chain that may be secured to a fixed object to prevent the monitor from moving. Pointed feet on the base also help to keep a portable monitor from moving. If the stream is operated at a low angle, the reaction force will tend to make the monitor unstable. For this reason, a safety lock is usually provided to prevent the monitor from being lowered beyond a safe angle of 35 degrees. When setting up any portable monitor, always follow the manufacturer's instructions and your department's SOGs/SOPs to ensure its safe and effective operation.

Elevated Master Streams

Elevated master stream devices are mounted on aerial ladders, aerial platforms, or special hydraulically operated booms **FIGURE 23-8**. A **ladder pipe** is an elevated master stream device that is mounted at the tip of an aerial ladder or tower ladder. On many aerial ladders, the ladder pipe is attached to the top of the ladder only when it is actually needed, and a hose is run up the ladder to deliver water to the device. Most newer aerial ladders and tower ladders are equipped with a fixed piping system to deliver water to a permanently mounted master stream device at the top. This arrangement saves valuable

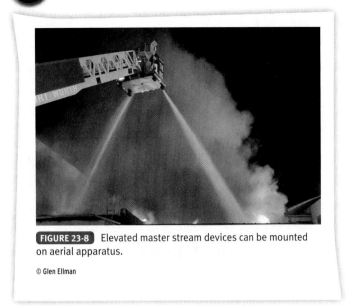

FIGURE 23-8 Elevated master stream devices can be mounted on aerial apparatus.

© Glen Ellman

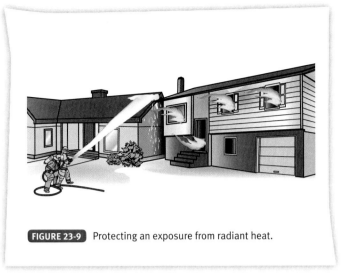

FIGURE 23-9 Protecting an exposure from radiant heat.

setup time at a fire scene. If your apparatus is equipped with a ladder pipe, you need to learn how to assist in its setup.

Protecting Exposures

Protecting exposures refers to actions that are taken to prevent the spread of a fire to areas that are not already burning. Exposure protection is a consideration at every fire; it becomes even more important with a large fire. If the fire is relatively small and contained within a limited area, the best way to protect exposures is usually to extinguish the fire; when the fire is extinguished, the exposure problem ceases to exist. In cases where the fire is too large to be controlled by an initial attack, exposure protection becomes a priority. In some cases, the best outcome that fire fighters can hope to obtain is to stop the fire from spreading.

The IC must consider the size of the fire and the risk to exposures in relation to how much firefighting capability is available and how quickly those resources can be assembled. In some cases, the IC will direct the first-arriving companies to protect exposures while a second group of companies prepares to attack the fire. At other times, the IC must identify a point where the progress of the fire can be stopped and direct all firefighting efforts toward that objective.

Protecting exposures involves very different tactics from offensive fire attacks. At a large free-burning fire, the first priority is to protect exposed buildings and property from a combination of radiant heat, convective heat, and burning embers **FIGURE 23-9**. The best option in these circumstances is usually to direct the first hose streams at the exposures rather than at the fire itself. Wetting the exposures will keep the fuel from reaching its ignition temperature. Because radiant heat can travel through a water stream, directing water onto the exposed surface is more effective than aiming a stream between the fire and the exposure. Master stream devices, such as deck guns, portable monitors, and elevated master streams, are excellent tools for protecting exposures. They also ensure that large volumes of water can be directed onto the exposures from a safe distance without putting fire fighters in the path of excess heat

or in danger of building collapse. Fog streams can sometimes be used to absorb some of the heat coming from the main body of fire.

Ventilation

Ventilation must be coordinated with the suppression efforts to ensure that both events occur simultaneously and in a manner that supports the attack plan. Proper ventilation is designed to allow hot gases and smoke to be removed from the building, thereby improving visibility and tenability in the building for any trapped victims and fire fighters. Conversely, improper ventilation can create conditions that allow a fire to burn more aggressively and make it more difficult for fire fighters to enter the structure and attack the fire.

Coordination is essential to ensure that the hose lines will be ready to attack when the ventilation openings are made. These openings must be located so that the hot smoke and gases will be drawn away from the attack crews.

Fire Fighter Safety Tips © Photos.com

Due to the lightweight construction and highly flammable contents of modern homes, many fires today are likely to be underventilated when you arrive. Remember that even the act of opening the front door can introduce additional oxygen to the fire and contribute to a violent flashover in as little as 100 seconds!

Specific Fire-Ground Operations

■ Concealed-Space Fires

Fires in ordinary and wood-frame construction can burn in combustible void spaces behind walls and under subfloors and ceilings. To prevent the fire from spreading, these fires must be found and suppressed **FIGURE 23-10**. To locate and suppress

FIRE FIGHTER II

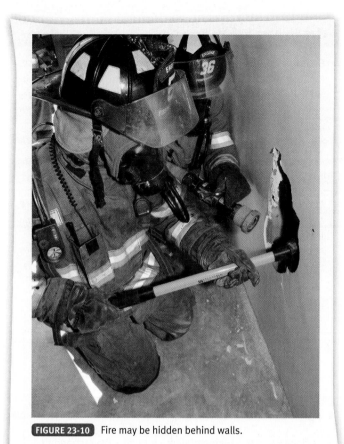

FIGURE 23-10 Fire may be hidden behind walls.

Floors that are built using lightweight materials, such as prefabricated wood I-joists or trusses, may fail faster than floors built with more solid lumber.

difficult to ventilate, which means that an interior attack must often be made in conditions of high heat and low visibility. Likewise, it will be difficult to remove the fire gases and steam produced by the attack lines. As a consequence, fire fighters may find it hard to see in a basement even after ventilation has been performed. Basements are often used for storage, so fire fighters may find it challenging to keep their sense of orientation in the narrow, disorganized, cluttered spaces.

Fire fighters should identify the safest means of entry and exit into the area where firefighting operations will be conducted. An exterior access point allows them to enter a basement without passing through the layers of hot gas at the basement ceiling level. If the only point of entry is an interior stairway, fire fighters must protect that opening to keep the fire from extending to the upper floors of the building **FIGURE 23-11**. Ventilation must be planned and conducted early. If this operation is not managed properly, the interior stairwell will act as a chimney and bring heat and smoke up from the basement.

fires behind walls and under subfloors, follow the steps in **SKILL DRILL 23-9** (Fire Fighter I, 5.3.8):

1. Locate the area of the building where a hidden fire is believed to exist.
2. Look for signs of fire, such as smoke coming from cracks or openings in walls, charred areas with no outward evidence of fire, and peeling or bubbled paint or wallpaper. Listen for cracks and pops or hissing steam.
3. Use a thermal imaging camera to look for areas of heat that may indicate a hidden fire.
4. Use the back of your hand to feel for heat coming from a wall or floor.
5. If a hidden fire is suspected, use a tool such as an axe or Halligan tool to remove the building material over the area.
6. If fire is found, expose the area as much as possible without causing unnecessary damage and extinguish the fire using conventional firefighting methods.

■ Basement Fires

Fires in basements or below grade level present several different challenges. First and foremost, they are difficult to recognize. Basement fires can damage the floor above the fire. If fire fighters do not identify a basement fire and enter the building above the basement fire, they are at risk of falling through the damaged floor and ending up in the burning basement.

Basements are difficult and dangerous spaces to enter, and they have limited routes of egress. They are also usually

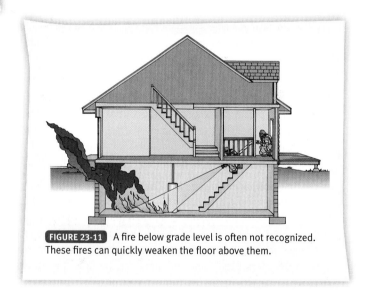

FIGURE 23-11 A fire below grade level is often not recognized. These fires can quickly weaken the floor above them.

Cellar fires can also spread to upper floors in houses with balloon-frame construction. Thermal imaging cameras are especially useful in identifying fires below the floor.

■ Fires above Ground Level

Advancing charged hose lines up stairs and along narrow hallways requires much more physical effort than advancing a charged hose line on a level surface. It is important to protect stairways and other vertical openings between floors when fighting a fire in a multiple-level structure. Specifically, hose lines must be placed to keep the fire from extending vertically and to ensure that exit paths remain available.

When working with a hose above the ground floor, fire fighters should advance the line uncharged until they reach the fire floor and have extra hose available. This approach allows for easier advancement of attack lines.

Interior fire crews must always look for a secondary exit path in case their entry route becomes blocked by the fire or by a structural collapse. This secondary exit could be a second interior stairway, an outside fire escape, a ground ladder placed to a window, or an aerial device.

In high-rise buildings, the standpipe system is typically used to supply water for hose lines. Fire fighters must practise connecting hose lines to standpipe outlets and extending lines from stairways into remote floor areas. Additional hose lines, tools, air cylinders, and EMS equipment should be staged one or two floors below the fire.

Fire Fighter Safety Tips
© Photos.com

Be alert to the risk of structural instability and collapse. Remember that modern lightweight construction does not provide clear signs of weakness before collapse. Modern construction fails quickly, suddenly, and without warning signs, and there are no effective ways to check the floor for stability.

FIRE FIGHTER Tips

Do not use more water than is needed to extinguish the fire, because retained water adds weight to the structure and may lead to structural collapse.

Near-Miss REPORT

Report Number: 06-0000533

Synopsis: Aerial master stream hits second-floor attack crew.

Event Description: On the afternoon of July 4, 2006, our department was dispatched to a residential structure fire. We responded with three engine companies, one truck company, and two squads. The responding units were staffed with two people, and there were two battalion chiefs responding. One of the battalion chiefs was on duty, and the other responded from his house less than a block away. All other units were staffed with three people, and another engine was requested later in the incident.

The first engine arrived on the scene of an approximately 325-square-metre (3500-square-foot), two-storey, wood-frame dwelling. Heavy smoke and fire were showing from the first floor, and heavy smoke was showing from the second floor. This engine company began setting up for fire attack using one 44-mm (1.75-in.) handline, and the second engine company laid a supply line and assumed rapid intervention crew (RIC) duties. The third engine and truck company responded from the same station and arrived at the same time. The truck company was assigned ventilation duties, the third engine company was assigned to the second fire attack, and one of the squad companies was assigned search and rescue responsibility. The on-duty battalion chief (A) established command, and the other battalion chief (B) was assigned to operations.

The first-arriving engine company attacked the fire in division 1, while the third-arriving engine company went to division 2. They encountered heavy fire and high-heat conditions. The fire was growing in size and intensity, so command ordered a defensive attack. No evacuation warnings (e.g., air horns) were given. I saw division 1 fire attack search crews coming out of the house. As they exited, the battalion chief in charge of operations called for an aerial master stream. The stream was directed to the second storey through a side window. The master stream hit the division 2 fire attack team, knocking them down the stairs and injuring two firefighters. After command contacted the crew, the RIC was sent in to assist them with egress.

During the RIC operation, all other operations were stopped. This allowed the fire to continue to grow.

Lessons Learned:

- Accountability is priceless.
- Know the crew assignments and number of personnel in the crew.
- When evacuating a structure, use evacuation tones (e.g., air horns).
- Big fire calls for big water. Do not be afraid to use something bigger than 44-mm (1.75-in.) handlines.
- Engine companies and truck companies should be staffed with more than three people.
- When going to a defensive attack, slow down, take time to make time, and make sure your people are where they are supposed to be.

If the structure is a residential building, fire personnel should conduct a thorough primary search as soon as possible, unless the fire conditions are incompatible with life. Do not risk your life to save a building that will be torn down next week.

Fires in Large Buildings

Large buildings, such as big box stores or office buildings, contain one very large open space surrounded by smaller rooms and storage areas. When these structures experience fires, the fire load varies greatly depending on the building's contents. Big box stores and home-improvement centres contain large amounts of flammable materials (contents), ranging from lumber to flammable fuels.

Many large buildings have floor plans that can cause fire fighters to become lost or disoriented while working inside, particularly in low-visibility or zero-visibility conditions. In such a setting, guide lines may be necessary to keep fire fighters from becoming lost and running out of air. A well-organized preincident plan of the structure can be essential when fighting this type of fire. Knowing the occupancy and the other hazards beforehand will help in determining the best strategy and tactics.

Consider the need for long hose lays when fighting a fire in a large building if an adequate number of standpipe connections are not available.

In large buildings that are equipped with automatic sprinkler systems, it is important to augment the water supply by connecting to the fire department connection (FDC) to increase the water volume and pressure in the sprinkler system. The chapter "Fire Detection, Protection, and Suppression Systems" discusses this procedure in detail.

Fires in Buildings During Construction, Renovation, or Demolition

Buildings that are under construction, renovation, or demolition are all at increased risk for destruction by fire. These buildings often have large quantities of combustible materials exposed, while lacking the fire-resistant features of a finished building. If the building lacks windows and doors, an almost unlimited supply of oxygen is available to fuel a fire. Fire detection, fire alarm, and automatic fire-suppression systems are often not installed or are inoperable. In addition, construction workers using torches and other flame-producing devices pose a notable fire risk. Moreover, these buildings are often unoccupied and can be easy targets for arsonists.

Under these conditions, a fire in a large building can be impossible to extinguish. If no life-safety hazards are involved, fire fighters may need to use a defensive strategy for this type of fire. In such a case, no fire fighters should enter the building, and a collapse zone should be established. A defensive exterior operation should be conducted using master streams, aerial streams, and large handlines to protect exposures.

It is important to maintain safety precautions during overhaul. Large, unsupported, or fire-damaged walls may collapse hours after the main body of fire has been knocked down.

Fires in Lumberyards

Lumberyard fires are often prime candidates for a defensive firefighting strategy. A typical lumberyard contains large quantities of highly combustible materials that are stored in the open or in sheds where plenty of air is available to support combustion. Given this rich fuel supply, a lumberyard fire will usually produce tremendous quantities of radiant heat and release burning embers that will cause the fire to spread quickly from stack to stack. Lumberyards may be sited in locations where there is an inadequate water supply. Such fires may quickly extend to nearby buildings and other structures. Large buildings at lumberyards are often constructed using trusses and other lightweight building materials, so you must be aware of the potential for rapid building collapse at these incidents.

Protecting exposures is often the primary objective at lumberyard fires, as there may be little life-safety risk. Exposure protection should be dealt with early in the operation by placing large handlines and master stream devices where they can be most effective. A collapse zone must be established around any stacks of burning material and buildings to keep fire fighters out of dangerous positions.

Fires in Stacked or Piled Materials

Fires occurring in stacked or piled materials can present a variety of hazards. The greatest danger to fire fighters is the possibility that a stack of heavy material, such as rolled paper or baled rags, will collapse without warning. Such an event might occur, for example, if a fire has damaged the stacked materials or if water has soaked into them. Absorbed water can increase the weight of many materials and weaken cardboard and paper products. In fact, the water discharged by automatic sprinklers alone can be sufficient to make some stacked materials unstable. A tall stack of material that falls on top of a team of fire fighters can cause injury or death.

Given these risks, fires in stacked materials should be approached cautiously. All fire fighters must remain outside potential collapse zones. Mechanical equipment should be used to move material that has been partially burned or water soaked.

Conventional methods of fire attack can often be used to gain control of the fire; however, water must penetrate into the stacked material to fully extinguish the residual combustion. Class A foam and wetting agents can be applied to extinguish smouldering fires in tightly packed combustible materials. Overhaul will require fire fighters to separate the materials to expose any remaining deep-seated fire. This operation can be a labour-intensive process unless mechanical equipment can be used to dig through the material.

■ Trash-Container and Rubbish Fires

Trash-container (i.e., dumpster) fires usually occur outside of any structure and appear to present fewer challenges than fires inside buildings. Even so, fire fighters must be vigilant in wearing full PPE and using SCBA when fighting trash-container fires or trash-pile fires, because there is no way of knowing what might be included in a collection of trash. Some trash containers may contain hazardous materials or materials that are highly flammable or explosive.

If the fire is deep-seated, fire fighters will have to overhaul the trash to make sure that the fire is completely extinguished. Manual overhaul involves pulling the contents of a trash container apart with pike poles and other hand tools so that water can reach the burning material. This process can be labour-intensive and involves considerable risk to fire fighters. The fire fighters are exposed to any contaminants in the container as well as to the risks of injury from burns, smoke, or other causes. Considering the low value of the contents of a trash container, it is difficult to justify any risk to fire fighters' safety.

Class A foam is useful for extinguishing many trash-container fires, because it allows water to soak into the materials and, therefore, can eliminate the need for manual overhaul. Some fire departments use the deck gun on the top of an engine to extinguish large trash-container fires and then complete extinguishment by filling the container with water. This is done by pointing the deck gun at the dumpster and slowly opening the discharge gate to lob water into the container **FIGURE 23-12**.

Trash containers are often placed behind large buildings and businesses. If the container is close to the structure, be sure to check for fire extension. Also look around the container for the presence of telephone, cable, and power lines that might have been damaged by the fire.

To extinguish an outdoor trash fire or other outdoor Class A fire, follow the steps in **SKILL DRILL 23-10** (Fire Fighter I, 5.3.8):

1. Don full PPE, including SCBA, enter the accountability system, and work as a team.
2. Perform size-up and give an arrival report. Call for additional resources if needed.
3. Ensure that apparatus is positioned uphill and upwind of the fire and that it protects the scene from traffic.
4. Develop and implement a fire-suppression strategy.
5. Protect the crew from hazards.
6. Deploy an appropriate attack line (at least 38 mm [1.5 in.]) in diameter).
7. Direct the crew to attack the fire in a safe manner—specifically, uphill and upwind from the fire.
8. Break up compact materials with hand tools or hose streams.
9. Evaluate the water application technique and modify if necessary.
10. Maintain good body mechanics during the fire attack.
11. Notify command when the fire is under control.
12. Investigate the origin and cause of the fire. Preserve any evidence of arson.
13. Return the equipment and crew to service.

■ Confined Spaces

Both fires and other types of emergencies can occur in confined spaces. Fires in underground vaults and utility rooms, such as transformer vaults, are too dangerous to enter. In such cases, fire fighters should summon the utility company and keep the area around manhole covers and other openings clear while awaiting the arrival of utility company personnel. In the event of an emergency in these areas, Canada Occupational Health and Safety Regulations stipulate that a qualified person specially trained in entry and emergency procedures must be in attendance outside the confined space. Learn your fire department's operational procedures for handling such incidents.

It is also important that fire fighters know about confined spaces that exist in the industries within their individual response areas. Fire fighters should visit these areas with plant personnel so that preincident plans can be developed.

Unique hazards may arise in confined spaces, including oxygen deficiencies, toxic gases, and standing water. Although conditions might appear to be safe, the confined space might not have enough oxygen to sustain life. It is common for a victim to enter these spaces and pass out, only to be followed by another victim—an individual who comes in to assist the first person and also passes out. When entering confined spaces to attempt rescue, fire fighters should wear full SCBA gear and be attached to a life line. An additional life line should be lowered or brought with the fire fighter to tie the victim in a bowline on a bight (rescue knot), which allows surface personnel to raise the victim out of the hole.

FIGURE 23-12 Some fire departments use a deck gun on top to extinguish large trash-container fires.

Owing to the lack of ventilation in most confined spaces, fire fighters may notice an intense amount of heat once they have entered the space. Fire fighters will tire quickly in this environment, so they must recognize the signs of heat exhaustion and heatstroke.

Because confined spaces commonly have low oxygen levels and high levels of combustible gases, such as methane, fire fighters without breathing apparatus can quickly be overcome in confined spaces with these conditions. For this reason, fire fighters who must enter a confined space should carry a monitoring device. Air quality must be checked constantly, looking for the buildup of explosive gases as well as any decline in oxygen levels.

Fire fighters must adhere to a strict accountability system when they enter into a confined space. This procedure ensures that only those personnel with proper training and equipment enter the space. It is important for a safety officer to track the movement of personnel and the amount of time that they remain in the confined space.

Fire suppression in confined spaces must not begin until all utilities and industrial processes have been turned off. Potential suppression agents include hose streams, high- and low-expansion foams, carbon-dioxide (CO_2) flooding systems, and built-in sprinkler systems.

Vehicle Fires

Vehicle fires are one of the most common types of fires handled by fire departments. In 2012, the NFPA reported that approximately 184,000 highway vehicle fires occurred in the United States. These fires may result from a variety of causes; for example, discarded smoking materials can cause fires in upholstery. Electrical short-circuits may cause fires in many different parts of a vehicle. Friction caused by dragging brakes or defective wheel bearings may cause fires. Collisions may lead to ruptured fuel lines, resulting in fires.

Fire fighters save few cars. In fact, when active flames are seen at a vehicle fire, the vehicle will likely be classified as totalled by the insurance company. Given this fact, it is important to understand the hazards involved in fighting vehicle fires and to mount a safe attack on the fire. Only when a viable victim is trapped in a burning vehicle does the scenario become a life-or-death situation.

Many hazards are associated with vehicle fires—traffic hazards, fuel, pressurized cylinders and containers that can explode, fire, and toxic smoke. Because vehicle fires usually occur on streets and highways, one of the biggest hazards faced by fire fighters is the danger posed by traffic. Drivers are easily distracted by the sight of a burning vehicle, which may lead to subsequent collisions. To counteract this risk, you should use your apparatus to block traffic. Do not be afraid to shut down traffic flow if necessary to ensure safety for fire fighters. Fire apparatus operators often place their vehicles 30.5 m (100 ft.) behind the burning vehicle to stop traffic and to position the apparatus a safe distance from the burning vehicle. Follow the steps outlined by your department for guarding against traffic hazards.

Modern vehicles contain a variety of gas-filled, pressurized cylinders and containers containing explosive materials. Hydraulic pistons are used to support hatchbacks, trunks, tailgates, and automobile hoods. Energy-absorbing bumper systems also contain hydraulic pistons. In addition, many modern vehicles use a MacPherson strut suspension system to absorb road shocks. When these gas-filled components are quickly heated to high temperatures in a fire, they can release pressure explosively, sending metal parts hurtling away from the vehicle. In addition, the supplemental restraint systems (SRS) found in most vehicles consist of airbags and air curtains containing chemicals that can ignite explosively during a fire.

Modern automobiles are constructed from hundreds of kilograms of plastics, which give off large quantities of toxic smoke and heat when they burn. They also contain a variety of petroleum products, including gasoline or diesel fuel, motor oil, brake fluid, and automatic transmission fluid. These products ignite easily, burn with high intensity, and produce large quantities of toxic gases. For this reason, it is important to always wear full PPE, including SCBA, when fighting a vehicle fire.

If the owner or driver of the burning vehicle is present, ask about any specific hazards that may be present in the vehicle, such as portable propane cylinders, propane torches, medical oxygen equipment, cans of spray paint, and other hazardous materials. If no driver or occupant is present, do not assume that the vehicle is safe; always be cautious as you approach a vehicle fire. Perform a risk–analysis assessment; that is, look at the big picture, and weigh the options. Do not risk injuring fire fighters in a vehicle fire.

■ Attacking Vehicle Fires

As you prepare to approach a vehicle fire, make sure that you have created a safe area around the vehicle. The only people closer than 15.2 m (50 ft.) to the vehicle should be fire fighters in full PPE and SCBA who are extinguishing the fire. Use a hose line at least 38 mm (1.5 in.) in diameter; such a hose will provide sufficient cooling power to overwhelm the fire and provide protection from a sudden flare-up. The use of compressed air foam or Class B foam aids in fire suppression. Charge the hose line while you are at least 15.2 m (50 ft.) from the fire. Bleed all the air from the hose line. Set the nozzle to initially deliver a straight stream or a fog pattern that is no wider than a 30-degree angle. Approach the vehicle from an uphill and upwind position, moving in from the side and at a 45-degree angle **FIGURE 23-13**. This path will help you avoid debris in case of an exploding bumper.

From a point approximately 9 m (30 ft.) away from the vehicle, open the nozzle and sweep the bottom part of the vehicle using a horizontal motion. Extinguish all visible fire while advancing toward the vehicle. Sweeping along the undercarriage helps to cool the bumper pistons, shock absorbers, and hydraulic struts; cool the tires before they explode; and cool the fuel tank before it fails. By applying a sufficient quantity of water to the lower part of the vehicle, you reduce the chance of an explosive event. Observe the area under the car during the approach for any sign of leaking flammable liquids. If burning flammable liquids are present, widen the spray pattern on the nozzle. Foam can be used to extinguish the burning liquid and provide a vapour barrier to prevent reignition. Once this sweep is complete, fire fighters can begin to attack any fire in the passenger compartment, the engine compartment, and the cargo compartment.

FIGURE 23-13 When possible, approach vehicle fires in the hood or engine area from the upwind side.

To extinguish a vehicle fire, follow the steps in **SKILL DRILL 23-11** (Fire Fighter I, 5.3.7):

1. Don full PPE, including SCBA, enter the accountability system, and work as a team.
2. Perform size-up and give an arrival report. Call for additional resources if needed.
3. Ensure that the fire apparatus is positioned uphill and upwind and that it protects the scene from traffic.
4. Ensure that the crew is protected from any hazards.
5. Identify the type of fuel used in the burning vehicle and look for fuel leaks.
6. Advance a fire attack line of at least 38-mm (1.5-in.) diameter using water or foam.
7. Attack from uphill and upwind of the fire, and at a 45-degree angle to the side of the vehicle.
8. Suppress the fire.
9. Overhaul all areas of the vehicle, passenger compartment, engine compartment, and trunk.
10. Notify command when the fire is under control.
11. Investigate the origin and cause of the fire. Preserve any evidence of arson.
12. Return the equipment and crew to service.

■ Alternative-Fuel Vehicles

Always be alert for signs that a burning vehicle could be powered by an alternative fuel, such as compressed natural gas (CNG) or liquefied petroleum gas (LPG). Fully involved fires in vehicles powered by either type of fuel should be fought with an unmanned master stream to prevent injuries from exploding gas cylinders.

Automobiles powered by CNG contain storage cylinders that are very similar to SCBA cylinders. These cylinders are usually located in the trunk and contain CNG at high pressures. When a fire occurs, they must be cooled and protected just like any gas cylinder. CNG is a nontoxic, lighter-than-air gas that will rise and dissipate if it is released into the atmosphere.

Propane is stored in the same types of cylinders as are used for heating or cooking purposes. Propane is heavier than air, so its vapours will pool or collect in low areas.

Hybrid automobiles have small gasoline-powered engines and large battery banks. The batteries power electric motors that drive the wheels, much like a train locomotive. Two noteworthy hazards are associated with these vehicles:

1. The nickel metal hydride batteries are hazardous when burning.
2. High-voltage, direct-current cables connect the batteries to the electric motors that power the wheels. The power to these cables can usually be disconnected by activating the power shut-off switch that is built into these vehicles. Cutting these orange cables can be dangerous. These cables usually run from the battery bank to the front of the car via the undercarriage; they typically pass directly under the centre of the driver's seat. Given this setup, fire fighters must use extra care when using hydraulic metal cutters or spreaders on these types of vehicles.

Follow the manufacturer's instructions for disconnecting the power to alternative-powered vehicles. More information on alternative-powered vehicles is presented in the chapter "Vehicle Rescue and Extrication."

■ Fire in the Passenger Area

Fires in the passenger area of a vehicle are more visible and accessible than fires in the engine compartment or the cargo area. Usually it is logical to extinguish the fire in the passenger compartment before moving on to extinguish any fire in the engine and cargo compartments. Often the windows are broken or a door is ajar, presenting an opening through which fire fighters can direct a stream of water. If the doors will not open, stand upwind from the window and use a striking tool to break out one or more windows. Be cautious, because a backdraft may occur.

As you get closer to the vehicle, change the nozzle to produce a wider pattern that will cool a wider area and give you some protection from the heat of the fire. Pay special attention to cooling areas such as the steering column and the dashboard on the passenger side. Cool areas that contain side-curtain air bags; cooling will greatly reduce the chance for accidental deployment of the SRS. Once the fire in the passenger compartment has been knocked down, attack any fire in the engine compartment or in the cargo compartment.

Fire in the Engine Compartment

The engine compartment of a vehicle is filled with a variety of devices that use petroleum products to power or lubricate them. They also contain components made of plastics and rubber. As a consequence, these devices produce a large amount of smoke when they burn. Other hazards present in the engine compartment include suspension struts and hydraulic-lift cylinders for the hood. Vehicle batteries contain sulfuric acid, which can cause serious burns. Some vehicles have magnesium parts within the engine compartment; when magnesium is present, you must use a Class D extinguishing agent.

A challenging part of extinguishing a fire in the engine compartment is gaining access to the fire. An initial attack can be made through the wheel well. The plastic liner between the engine compartment and the wheel well is often consumed during such a fire, so it may be possible to spray water into the engine compartment through this opening. An alternative initial approach is to spray water through the grille after thoroughly cooling the area around the bumper. When using this tactic, avoid standing directly in front of the bumper until this area is thoroughly cooled. Although neither of these two methods is effective at totally extinguishing the fire, they will help to diminish the volume of fire while fire fighters are gaining access to the engine compartment.

The fastest way to gain access to the engine compartment is to pull the hood-release lever inside the passenger compartment to open the hood. Unfortunately, this method rarely works during a vehicle fire, so other methods usually need to be tried. A quick means of gaining access to the engine compartment is to insert a pry bar along the side of the hood between the edge of the hood and the fender. Pry the side of the hood away from the fender to produce an opening big enough to apply a stream of water into the engine compartment.

A second method of gaining access to the engine compartment is to open the hood of the vehicle. In most fires, the cable that normally opens the hood is damaged and will not release from the inside of the vehicle. To open the hood of a burning vehicle, break out the plastic grille and find the hood-release cable. With a gloved hand, pull on the cable to release the hood or use the forked end of a Halligan tool to twist the cable until the hood releases. Fire fighters can also use vise-grip pliers to grasp the cable and pull it.

Once the hood is raised, fire fighters should have good access to any remaining fire and hotspots. Use plenty of water or foam to cool this area. Once the vehicle's hood has been opened and any fire in the engine compartment extinguished, disconnect the power to the vehicle by cutting the battery cables; if the battery is not under the hood, check the wheel wells or trunk. First, remove a section of the negative cable at least 152 mm (6 in.) long by making two cuts with wire cutters. Then, remove a similar section of the positive cable. This excision will disrupt the flow of power to the vehicle and ensure that nothing is powered accidentally.

Fire in the Trunk

A fire in the trunk of a vehicle may present unknown hazards because it is not possible to know what the trunk contains. In addition, fires in this area are challenging to access. Initial access can sometimes be made by knocking out the tail light assembly on one side, which enables fire fighters to direct a hose stream into the trunk to cool down and partially extinguish fire in this area. A fire in the trunk area of an automobile can also be accessed by first using the pike of a Halligan tool to force the lock into the trunk, and then using a screwdriver or key tool (K tool lock puller) to turn the lock cylinder in a clockwise direction. A charged hose line must be ready when the trunk lid is raised.

Fires in the rear of light trucks and vans must always be approached cautiously. In addition, vehicles using alternative fuels (discussed earlier in this chapter) may contain CNG or propane cylinders in the trunk. Vans are often used by couriers and could contain medical waste, laboratory specimens, and radioactive material.

Overhauling Vehicle Fires

Overhaul of vehicle fires is just as important as overhaul of structure fires. As soon as it is safe to approach the vehicle, chock the wheels to prevent the vehicle from moving. If a vehicle fire erupts quickly, the driver may not have time to set the brake and parking gear. Also, fire can damage cables and wires that control the operation of parking and braking mechanisms. After all visible fire has been knocked down, allow a few minutes for the steam and smoke to dissipate before starting overhaul. This delay will allow visibility to improve so that overhaul can be completed safely.

During overhaul of interior fires, remember that airbags can deploy without warning in a burning automobile. Never place any part of your body in the path of a front or side airbag.

As you overhaul the vehicle, be systematic and thorough. Do not miss areas that may contain lingering sparks. Direct the hose stream under the dashboard, and soak and remove smouldering upholstery. Apply water over and under all parts of the engine compartment. Confirm that no fluids are leaking from the vehicle. If contents in the car might potentially be salvaged, treat them with respect. Continue to use your SCBA as long as smoke or fumes are present.

> **Fire Fighter Safety Tips** © Photos.com
>
> Some vehicle components (in the engine or body) may be constructed from magnesium or other flammable or explosive metals that can react violently when water is applied during fire suppression. A Class D extinguishing agent should be used in these cases instead of water.

Flammable-Liquid Fires

Flammable-liquid fires can be encountered in almost any type of occupancy. Most fires involving a vehicle (e.g., airplane, train, ship, car, truck) will involve a combustible or flammable liquid. Special tactics must be used when attempting to extinguish a flammable-liquid fire, and special extinguishing agents such as foam or dry chemicals may be needed.

Hazards

Fires involving flammable liquids, such as gasoline, require special extinguishing agents. Most flammable liquids can be extinguished using either foam or dry chemicals. Class B extinguishing agents are approved for use on Class B (flammable-liquid) fires. Flammable-liquid fires can be classified as either two-dimensional or three-dimensional. A two-dimensional fire refers to a spill, pool, or open container of liquid that is burning only on the top surface. A three-dimensional fire refers to a situation in which the burning liquid is dripping, spraying, or flowing over the edges of a container.

A two-dimensional flammable-liquid fire can usually be controlled by applying the appropriate Class B foam to the burning surface. Several different formulations of Class B foam are suitable for a variety of liquids and situations. The foam will flow across the surface of the liquid and create a seal that stops the fuel from vapourizing; this separates the fuel from the oxygen and extinguishes the fire. Foam will also cool the liquid and reduce the possibility of reignition.

When dealing with flammable-liquid fires, fire fighters should look for hot surfaces or open flames that could cause the vapours to reignite after a fire has been extinguished. It is important to determine the identity of the liquid that is involved so as to select the appropriate extinguishing agent and to determine whether the vapours are lighter or heavier than air.

A three-dimensional flammable-liquid fire is much more difficult to extinguish with foam, because the foam cannot establish an effective seal between the fuel and the oxygen. Either dry chemical or gaseous extinguishing agents are usually more effective than foam in controlling these kinds of fires. These agents can also be used to extinguish two-dimensional fires, although they do not provide a long-lasting seal between the fuel and the oxygen. In some cases, a fire can be extinguished with a dry chemical, and then the surface can be covered with foam to prevent reignition.

Fire fighters should avoid standing in pools of flammable liquids or contaminated runoff from them, because their PPE will absorb the flammable product and become contaminated. In cases of serious contamination, the PPE itself can become flammable.

Suppression

The skills used in suppressing small flammable-liquid fires are presented in the chapter "Portable Fire Extinguishers." Larger flammable liquid fires may require the use of Class B foam. The equipment and methods used to apply Class B foam are described in the chapter "Fire Attack Lines and Foam."

FIRE FIGHTER II

Flammable-Gas Cylinders

Flammable-gas cylinders can be found in many places. Many types of flammable gases are stored in many different types and sizes of containers. A variety of flammable gases can be found in industrial occupancies.

Propane Gas

The popularity of propane gas for heating and cooking has meant that these cylinders have become commonplace in residential areas and many industrial and commercial locations. In addition, propane is used as an alternative fuel for vehicles and is often stored to power emergency electrical generators. Fire fighters should be familiar with the basic hazards and characteristics of propane, as well as with procedures for fighting propane fires.

Propane (LPG) exists as a gas in its natural state at temperatures higher than −42.2°C (−44°F). When the gas is placed into a storage cylinder under pressure, it is changed into a liquid. Storing propane as a liquid is very efficient, because it has an expansion ratio of 270:1 (i.e., 1 cubic metre [1.3 cubic yards] of liquid propane is converted to 270 cu. m [353.1 cu. yd.] of gaseous propane when it is released into the atmosphere). Put simply, a large quantity of propane fuel can be stored in a small container.

Inside a propane container, there is a space filled with propane gas above the level of the liquid propane. As the contents of the cylinder are used, the liquid level becomes lower and the vapour space increases. The internal piping is arranged so as to draw product from the vapour space.

Propane gas containers come in a variety of sizes and shapes, with capacities ranging from a few millilitres to thousands of litres. The cylinder itself is usually made of steel or aluminum. A discharge valve keeps the gas inside the cylinder from escaping into the atmosphere and controls the flow of gas into the system where it is used. This valve should be easily visible and accessible. In the event of a fire, closing the valve should stop the flow of the product and extinguish the fire. The valve should be clearly marked to indicate the direction in which it should be turned or moved to reach the closed position.

A connection to a hose, tubing, or piping allows the propane gas to flow from the cylinder to its destination. In the case of portable tanks, this connection is often the most likely place for a leak to occur. If the gas is ignited, this area could become involved in fire.

A propane cylinder is always equipped with a relief valve to allow excess pressure to escape, thereby preventing an explosion if the tank becomes overheated. Propane cylinders must be stored in an upright position so that the relief valve remains within the vapour space. If the cylinder is placed on its side, the relief valve could fall below the liquid level. If a fire were then to heat the tank and cause an increase in pressure, the relief valve would release liquid propane, which would expand by the 270:1 ratio and create a huge cloud of potentially explosive propane gas.

Propane Hazards

Propane is highly flammable. Although it is nontoxic, this gas can displace oxygen and cause asphyxiation. By itself, propane is odourless, so leaks of pure propane cannot be detected by a human sense of smell. For this reason, mercaptan is added to propane to give it a distinctive odour.

Propane gas is heavier than air, so it will flow along the ground and accumulate in low areas. When using meters to check for LPG, be sure to check storm drains, basements, and other low-lying areas for concentrations of the gas.

When responding to a reported LPG leak, fire fighters and their apparatus should be staged uphill and upwind of

FIRE FIGHTER II

the scene. Because an explosion can happen at any time, fire fighters should wear full PPE and SCBA at this type of incident. Life safety should be the highest priority; depending on the type and size of the leak, an evacuation might be necessary.

The greatest danger with propane and similar products is a boiling liquid/expanding vapour explosion (BLEVE). If an LPG tank is exposed to heat from a fire, the temperature of the liquid inside the container will increase. The fire could then be fuelled by propane escaping from the tank or from an external source. As the temperature of the product increases, the vapour pressure will also increase. The increasing pressure creates added stress on the container. If this pressure exceeds the strength of the cylinder, the cylinder can rupture catastrophically. An exploding LPG cylinder can produce the same explosive power as dynamite.

To protect the container from rupture, the relief valve will open to release some of the pressure. This relief valve is designed to exhaust the vapour until the pressure drops to a preset level. When the pressure returns to a safe level, the valve will close. If the heating continues, however, the liquid will begin to boil within the container. If the flame impinges directly on the tank, the container can weaken and fail somewhere above the liquid line. When this happens, the container will rupture and release its contents with explosive speed. The boiling liquid will expand, vapourize, and ignite in a giant fireball, accompanied by flying fragments of the ruptured container. Fire fighters have been killed in these explosions.

The best method to prevent a BLEVE is to direct heavy streams of water onto the tank from a safe distance. The water should be directed at the area where the tank is being heated. Cool the upper part of the tank to cool the gas vapours. The fire fighters operating these streams should work from shielded positions or use remote-controlled or unmanned monitors. Horizontal tanks are designed to fail at the ends if a catastrophic failure occurs, so fire fighters should operate only from the sides of the tank.

Flammable-Gas Fire Suppression

Fighting fires involving LPG or other flammable-gas cylinders requires careful analysis and logical procedures. If the gas itself is burning because of a pipe or regulator failure, the best way to extinguish the fire is to shut off the main discharge valve at the cylinder. If the fire is extinguished and the fuel continues to leak, there is a high probability that it will reignite explosively. Do not attempt to extinguish the flames unless the source of the fuel has been shut off or all of the fuel has been consumed. If the fire is heating the storage tank, use hose streams to cool the cylinder, being careful not to extinguish the fire.

Unless a remote shut-off valve is available, the flow of propane can be stopped only if it is safe to approach the cylinder. Fire fighters should inspect the integrity of the cylinder from a distance before they make any attempt to approach and shut off the valve. If the container is damaged or the valve is missing, the fuel should be allowed to burn off, while hose streams continue to cool the tank from a safe distance.

Approach a flammable-gas fire with two 44-mm (1.75-in.) hose lines working together. When approaching a horizontal LPG tank, always approach it from the sides.

The nozzles should be set on a wide fog pattern, with the discharge streams interlocked to create a protective curtain. The team leader should be located between the two nozzle operators. On the command of the leader, the crew should move forward, remaining together and never turning their backs to the burning product. Upon reaching the valve, the fire fighter in the centre can turn off the valve, stopping the flow of gas. Any remaining fire may then be extinguished by normal means. Continue the flow of water as a protective curtain and to reduce sources of ignition.

If the fire is extinguished prematurely, the valve should still be turned off as soon as the team reaches it. Always approach and retreat from these types of fires while facing the objective with water flowing, in case of reignition.

Unmanned master streams should be used to protect flammable-gas containers that are exposed to a severe fire. Direct the stream so that it is one-third of the way down the container. This technique will allow half of the water to roll up and over the container, while the remainder projects downward FIGURE 23-14 . The objective is to cover as much of the exposed tank as possible.

FIGURE 23-14 Using a master stream to protect a flammable-gas container that is exposed to fire.

If the LPG container is located next to a fully involved building or a fire that is too large to control, evacuate the area and do not fight the fire. If there is nothing to save, risk nothing.

Keep in mind that if the relief valve is open, the flammable-gas container is under stress. Exercise extreme caution in this scenario. As the gas pressure is relieved, it will sound like the whistle on a teakettle; if the sound is rising in frequency, an explosion could be imminent and evacuation should be ordered.

To suppress a flammable-gas cylinder fire, follow the steps in SKILL DRILL 23-12 (Fire Fighter II, 6.3.3):

1. Cool the tank from a distance until the relief valve resets.

2. Wearing full PPE, two teams of fire fighters using a minimum of two 44-mm (1.75-in.) hose lines advance using an interlocking 90-degree-wide fog pattern for protection. Do not approach the cylinder from the ends. The team leader should be located between the two nozzle persons. The leader coordinates the advance toward the cylinder.

❸ When the cylinder is reached, the two nozzle teams isolate the shut-off valve from the fire with their fog streams, while the leader closes the tank valve, eliminating the fuel source.

❹ After the burning gas is extinguished, the fire fighters continue to apply water to the cylinder to cool the metal, with the goal of preventing tank failure and a subsequent BLEVE.

❺ As cooling continues, fire fighters slowly back away from the cylinder, never turning their backs to it.

■ Shutting Off Gas Service

Many structures use either natural gas or propane gas for heat or cooking. In addition, these two energy sources have many industrial applications. If a gas line inside a structure becomes compromised during a fire, the escaping gas can add fuel to the fire. The means by which the gas is supplied to the structure must be located to stop the flow.

Most residential gas supplies are delivered through a gas meter connected to an underground utility network or from a storage tank located outside the building. If the gas is supplied by an underground distribution system, the flow can be stopped by closing a quarter-turn valve on the gas meter; if the gas is supplied from an outside LPG storage cylinder, closing the cylinder valve will stop the flow.

After the gas service has been shut off, use a lockout tag to ensure that it is not turned back on. Only a professional can reestablish the flow of gas to a structure.

To shut off gas utilities, follow the steps in **SKILL DRILL 23-13** (Fire Fighter I, NFPA 5.3.18) **FIGURE 23-15**:

❶ Don PPE, including SCBA; enter the personnel accountability system.

❷ Acknowledge the assignment.

❸ Locate the exterior gas shut-off valve.

❹ Shut off the gas valve.

❺ Attach a shut-off tag and lock if required.

❻ Notify command that the gas is shut off.

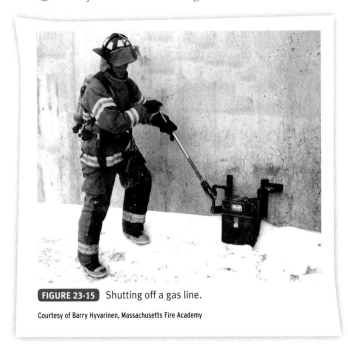

FIGURE 23-15 Shutting off a gas line.

Courtesy of Barry Hyvarinen, Massachusetts Fire Academy

Fires Involving Electricity

The greatest danger with most fires involving electrical equipment is electrocution. For this reason, only Class C extinguishing agents should be used when energized equipment is involved in a fire. All electrical equipment should be considered as potentially energized until the power company or a qualified electrical professional confirms that the power is off. Once the electrical service has been disconnected, most fires in electrical equipment can be controlled using the same tactics and procedures as are used for a Class A fire.

When a fire occurs in a building, the electrical service should be turned off quickly to reduce the risks of injury or death to fire fighters, even if there is no involvement of electrical equipment in the fire. If possible, this shut-off should take place at the main circuit breaker box, with a lockout tag being used to prevent someone from accidentally turning the electrical current back on.

If it is not possible to turn the electricity off at the breaker box, fire fighters must notify the electrical utility company to send a representative to the fire scene to disconnect the service from a location outside the building. In many cases, the local utility company is automatically notified of any working fire.

To control the electric utility system, follow the steps in **SKILL DRILL 23-14** (Fire Fighter I, 5.3.18):

❶ Don PPE, including SCBA; enter the personnel accountability system.

❷ Acknowledge the assignment

❸ Locate the exterior shut-off.

❹ Shut off the power.

❺ Attach a lockout tag and lock.

❻ Notify command that the electricity is shut off.

■ Electrical Fire Suppression

Fire-suppression methods for fires involving electrical equipment vary according to the type of equipment and the power supply. In many cases, the best approach is to wait until the power is disconnected and then use the appropriate extinguishing agents to control the fire. If the power cannot be disconnected or the situation requires immediate action, only Class C extinguishing agents—such as halon agents, CO_2, or dry chemicals—should be used.

When delicate electronic equipment is involved in a fire, either halons or CO_2 should be used to limit the damage as much as possible. These agents cause less damage to computers and sensitive equipment than do water or dry chemical agents.

When power distribution lines or transformers are involved in a fire, special care must be taken to ensure the safety of both emergency personnel and the public. No attempt should be made to attack these fires until the power has been disconnected. In some cases, fire fighters may need to protect exposures or extinguish a fire that has spread to other combustible materials, if this can be accomplished without coming in contact with the electrically energized equipment. If a hose stream comes in contact with the energized equipment, the current can flow back through the water to the nozzle and electrocute fire fighters who are in contact with the hose line.

Many electrical transformers contain a cooling liquid that includes polychlorinated biphenyls, a cancer-causing material. Do not apply water to a burning transformer. Water can cause the transformer's cooling liquid to spill or splash, contaminating both fire fighters and the environment. If the transformer is located on a pole, it should be allowed to burn until electrical utility professionals arrive and disconnect the power. Dry chemical extinguishers can then be used to control the fire. Fires in ground-mounted transformers can also be extinguished with dry chemical agents after the power has been disconnected. Fire fighters should stay out of the smoke and away from any liquids that are discharged from a transformer, and they must wear full PPE and SCBA to attack the fire.

Some very large transformers contain large quantities of cooling oil and require foam for fire extinguishment. This foam can be applied only after the power has been disconnected. Until the power is off, fire fighters can still protect exposures while taking care to avoid contamination.

Underground power lines and transformers are often located in vaults beneath the ground's surface. Explosive gases can build up within these vaults. If a spark then ignites the gases, the resulting explosion can lift a manhole cover from a vault and hurl it for a considerable distance. Products of combustion can also leak into buildings through the underground conduits. Fire fighters should never enter an underground electrical vault while the equipment is energized. Even after the power has been disconnected, these vaults should be considered to be confined spaces containing potentially toxic gases, explosive atmospheres, or oxygen-deficient atmospheres. Special precautions are required to enter this type of confined space.

Large commercial and residential structures often have high-voltage electrical service connections and interior rooms containing transformers and distribution equipment. These areas should be clearly marked with electrical hazard signs, and fire fighters should not enter them unless there is a rescue to be made. Until the power has been disconnected, fire-suppression efforts should be limited to protecting exposures. Fire fighters must wear full PPE and SCBA owing to the inhalation hazards presented by the burning of the plastics and cooling liquids that are often used with equipment of this size.

Wrap-Up

Chief Concepts

- Fire suppression refers to all of the tactics and tasks performed on the fire scene to extinguish a fire.
- All fire-suppression operations are either offensive or defensive:
  - Offensive—Tasks performed inside. Used when the fire is not too large or dangerous to be extinguished using interior handlines.
  - Defensive—Tasks performed outside. Used when the fire is too large to be controlled by an offensive attack or when the level of risk is too high.
- The IC evaluates conditions constantly to determine the type of attack that should be used. An interior attack may be switched to a defensive attack, if necessary, and vice versa.
- A fire fighter must be able to advance and operate a hose line effectively to extinguish a fire.
- Care should be taken not to have opposing hose lines, such that two crews work against each other.
- The most frequently used hose line size for interior fire attack is 44-mm (1.75 in.). Although one fire fighter can usually operate the nozzle on a small hose line, a second fire fighter will provide valuable assistance when a small hose line must be advanced and maneuvered.
- Different types of fire streams are produced by different types of nozzles. A fog stream divides water into droplets, which can absorb heat efficiently. Fog streams can be used to protect fire fighters from the intense heat of the fire and to assist in ventilation. A straight stream can hit a fire from a longer distance away; it also keeps the water concentrated in a small area. A solid stream has an even greater reach and penetrating power than a straight stream and is charged as a continuous column of water.
- An interior fire attack is an offensive operation that requires fire fighters to enter a building and discharge an extinguishing agent onto the fire. Interior fire attack can be conducted on many different scales. In many cases, an interior attack is conducted on a fire that is burning in only one room and may be controlled quickly by one attack hose line. Larger fires require more water, which could be provided by two or more small handlines working together or by one or more larger handlines.
- The most effective means of fire suppression in most situations is a direct attack. This kind of attack uses a straight or solid hose stream to deliver water directly onto the base of the fire. The water cools the fuel until it is below its ignition temperature.
- The objective of an indirect attack is to quickly remove as much heat as possible from the fire atmosphere. Such an approach is particularly effective at preventing flashover from occurring. This method of fire suppression should be used when a fire has produced a layer of hot gases at the ceiling level.
- A combination attack employs both indirect attack and direct attack methods in a sequential manner. This strategy should be used when a room's interior has been heated to the point that it is nearing a flashover condition. Fire fighters first use an indirect attack method to cool the fire gases, then make a direct attack on the main body of fire.
- Large handlines can be used either for offensive fire attacks or for defensive operations.
- Master stream devices are used to produce high-volume water streams for large fires. Several types of master stream devices exist, including portable monitors, deck guns, ladder pipes, and other elevated stream devices. Most master streams discharge water at between 1591 and 6819 L/min. (350 and 1500 gpm).
- Fires in ordinary and wood-frame construction can burn in combustible void spaces behind walls and under subfloors and ceilings.
- Fires in basements or below grade level may cause the floor on the ground level to collapse.
- Protect stairways and other vertical openings between floors when fighting a fire in a multiple-level structure. Hose lines must be placed to keep the fire from extending vertically and to ensure that exit paths remain available to fire fighters and victims.
- Many large buildings have floor plans that can cause fire fighters to become lost or disoriented while working inside, particularly in low-visibility or zero-visibility conditions.
- Buildings that are under construction, renovation, or demolition are all at increased risk for destruction by fire. These buildings often have large quantities of combustible materials exposed, while lacking the fire-resistant features of a finished building.
- A lumberyard fire will usually produce tremendous quantities of radiant heat and release burning embers that can cause the fire to spread quickly from stack to stack.
- Fires in stacked materials should be approached cautiously. All fire fighters must remain outside potential collapse zones. Mechanical equipment should be used to move material that has been partially burned or water soaked.
- Fires in trash containers may be fought with foam or a deck gun.
- Some unique hazards exist with fires in confined spaces, including oxygen deficiencies, toxic gases, and standing water.

- Vehicle fires are one of the most common types of fires handled by fire departments. These fires may result from a variety of causes, including discarded smoking materials, electrical short-circuits, friction caused by dragging brakes or defective wheel bearings, and ruptured fuel lines due to a collision.
- Many hazards related to vehicle fires are possible—traffic hazards, fuel, pressurized cylinders and containers that can explode, fire, and toxic smoke.
- Overhaul of vehicle fires is just as important as overhaul of structure fires. As soon as it is safe to approach the vehicle, chock the wheels to prevent the vehicle from moving. After all visible fire has been knocked down, allow a few minutes for the steam and smoke to dissipate before starting overhaul. As you overhaul the vehicle, be systematic and thorough.
- Always be alert for signs that a burning vehicle could be powered by an alternative fuel, such as CNG or LPG. Fully involved fires in vehicles powered by either type of fuel should be fought with an unmanned master stream to prevent injuries due to exploding gas cylinders.
- Flammable-liquid fires can be encountered in almost any type of occupancy. Most fires involving a vehicle (e.g., airplane, train, ship, car, truck) will involve a combustible or flammable liquid. Special tactics must be used when attempting to extinguish a flammable-liquid fire, and special extinguishing agents such as foam or dry chemicals may be needed.
- Flammable-gas cylinders can be found in many places. Many types of flammable gases are stored in many different types and sizes of containers. A variety of flammable gases can be found in industrial occupancies.
- The greatest danger with propane and similar flammable-gas products is a BLEVE. If an LPG tank is exposed to heat from a fire, the temperature of the liquid inside the container will increase. The fire itself could be fuelled by propane escaping from the tank or from an external source. As the temperature of the product increases, the vapour pressure will increase in tandem. The increasing pressure creates added stress on the container. If this pressure exceeds the strength of the cylinder, the cylinder can rupture catastrophically.
- The greatest danger with most fires involving electrical equipment is electrocution. Only Class C extinguishing agents should be used when energized equipment is involved in a fire. All electrical equipment should be considered as potentially energized until the power

company or a qualified electrical professional confirms that the power is off.
- Once the electrical service has been disconnected, most fires in electrical equipment can be controlled using the same tactics and procedures that are used for Class A fires.

Hot Terms

Combination attack A type of attack employing both direct attack and indirect attack methods.

Deck gun An apparatus-mounted master stream device that is intended to flow large amounts of water directly onto a fire or exposed building.

Direct attack (structure fire) Firefighting operations involving the application of extinguishing agents directly onto the burning fuel. (NFPA 1145)

Elevated master stream device A nozzle mounted on the end of an aerial device that is capable of delivering large amounts of water onto a fire or exposed building from an elevated position.

Indirect application of water The use of a solid object such as a wall or ceiling to break apart a stream of water, creating more surface area on the water droplets and thereby causing the water to absorb more heat.

Indirect attack (structure fire) Firefighting operations involving the application of extinguishing agents to reduce the buildup of heat released from a fire without applying the agent directly onto the burning fuel. (NFPA 1145)

Ladder pipe A monitor that is fed by a hose and that holds and directs a nozzle while attached to the rungs of a vehicle-mounted aerial ladder. (NFPA 1965)

Master stream device A large-capacity nozzle that can be supplied by two or more hose lines or fixed piping. Such devices include deck guns, portable ground monitors, and elevated streams, and commonly flow between 1591 and 6819 L/min. (350 and 1500 gpm).

Portable monitor A monitor that can be lifted from a vehicle-mounted bracket and moved to an operating position on the ground by not more than two people. (NFPA 1965)

Solid stream A stream made by using a smooth-bore nozzle to produce a penetrating stream of water.

Straight stream A stream made by using an adjustable nozzle to provide a straight stream of water.

You are just returning to the station from a medical emergency when your engine is dispatched to the report of a car on fire in a garage. The battalion chief arrives first and establishes command. The IC gives a size-up indicating that the structure is a single-storey, wood-frame house with dark, turbulent smoke coming from the garage area. He then assigns your crew to fire attack.

1. A(n) _____ uses a solid object such as a wall or ceiling to break apart a stream of water, creating more surface area on the water droplets and thereby causing the water to absorb more heat.

   A. interior attack

   B. indirect attack

   C. direct attack

   D. combination attack

2. Which of the following is not a master stream device?

   A. Deck gun

   B. Portable ground monitor

   C. Elevated stream

   D. Solid stream

3. A _____ is made by using a smooth-bore nozzle to produce a penetrating stream of water.

   A. solid stream

   B. fog stream

   C. straight stream

   D. master stream

4. What is the expansion ratio of propane?

   A. 5:1

   B. 17:1

   C. 70:1

   D. 270:1

5. What should be done once the gas has been shut off to a building?

   A. Attach a shut-off tag.

   B. Report to staging.

   C. Report to rehabilitation.

   D. Notify the gas company.

6. Large handlines will flow at least _____ L/min. (gpm).

   A. 681 (150)

   B. 909 (200)

   C. 1136 (250)

   D. 1363 (300)

7. Which challenge do fire fighters frequently face when extinguishing an engine compartment fire?

   A. Confining the fire

   B. Vehicle movement

   C. Gaining access

   D. Locating the seat of the fire

It is a beautiful Saturday morning, and you are out washing the engine on the front apron of your fire station. A car pulls in, and the driver walks over and begins to chat with you. He says he is a fire fighter from a neighbouring jurisdiction, so he decided to stop and visit when he saw you out washing the engine. It is evident that he has been a fire fighter for only a short period of time, but he is very enthusiastic about learning and seems to be a very likeable individual. You know that his department has very limited financial resources, so you engage him in an informative discussion about fire attack.

1. Which factors would you tell him affect the IC's decision on whether to make an interior attack rather than an exterior attack?
2. How would you describe to him the characteristics of the various types of nozzles?
3. Which other activities would you tell him need to be coordinated with fire attack?

Preincident Planning

Fire Fighter I

Knowledge Objectives

There are no knowledge objectives for Fire Fighter I candidates. NFPA 1001 contains no Fire Fighter I Job Performance Requirements for this chapter.

Skills Objectives

There are no skills objectives for Fire Fighter I candidates. NFPA 1001 contains no Fire Fighter I Job Performance Requirements for this chapter.

Fire Fighter II   FFII

Knowledge Objectives

After studying this chapter, you will be able to:

- Describe why and for which types of properties a preincident plan is created. (p 710–712)
- List the typical target hazards that may be found in a community. (p 712)
- Describe how a preincident survey is performed. (NFPA 6.5 , p 712–713)
- List the information that is gathered during a preincident survey. (NFPA 6.5.3 , p 713)
- Describe the information included in any sketches or drawings created during the preincident survey. (NFPA 6.5.3, 6.5.3A, 6.5.3B , p 713–715)
- Describe the symbols commonly used in preincident plans. (NFPA 6.5.3A , p 714)
- Describe how preincident planning for safe and rapid response is performed. (p 713–715)
- Describe the information that needs to be gathered to assist the incident commander (IC) in making a rapid and correct size-up during an emergency incident. (p 715–719)

© Photos.com

- Explain how to identify built-in fire detection and suppression systems during a preincident survey. (NFPA 6.5.3B), p 718–719)
- Describe the tactical information that is collected during a preincident survey. (NFPA 6.5.3 , p 719–723)
- Describe how the sources of water supply for fire-suppression operations are identified. (NFPA 6.5.3A), p 719–721)
- Explain why the locations of utilities are noted on the preincident plan. (p 721–722)
- Describe how preincident planning for an efficient search and rescue is performed. (p 722)
- Describe how preincident planning for rapid forcible entry is performed. (p 722–723)
- Describe how preincident planning for safe ladder placement is performed. (p 723)
- Describe how preincident planning for effective ventilation is performed. (p 723)
- List the occupancy considerations to take into account when conducting a preincident survey. (NFPA 6.5.3 , p 724)
- List the types of locations that require special considerations in preplanning. (NFPA 6.5.3 , p 725–726)

Additional NFPA Standards

- NFPA 1, *Fire Code,* 2009 ed.
- NFPA 170, *Standard for Fire Safety and Emergency Symbols*
- NFPA 220, *Standard on Types of Building Construction*
- NFPA 704, *Standard System for the Identification of the Hazards of Materials for Emergency Response*
- NFPA 1620, *Standard for Pre-incident Planning*

Skills Objectives

After studying this chapter, you will be able to perform the following skills:

- Conduct a preincident survey. (NFPA 6.5.3B , p 719)

You Are the Fire Fighter

Your chief advises that the Insurance Services Office (ISO) will be evaluating your department during the fall, so the chief wants to ensure that all of the commercial buildings in your community have an up-to-date preincident plan. He has laid out a plan in which the town is divided into smaller sections, and each section is assigned to a different fire fighter. You are given a section with a convenience store, a group of storage units, and an old shoe factory, which is now vacant.

1. Why do fire departments conduct preincident planning?
2. What should you do to prepare?
3. Which process will you use when you arrive at a building?

Introduction

Preincident planning gives you the tools and knowledge that you need to become a much more effective fire fighter. Without a preincident plan, you will go into an emergency situation blind. You will not be familiar with the structure, the location of hydrants, or the potential hazards. By contrast, with a preincident plan, you will know where the hydrants and exits are and which hazards to anticipate. A preincident plan puts all of this information at your fingertips, either on paper, in a computer file, or online and available to you at the fire scene.

Preincident planning helps your fire department to make better command decisions because important information is assembled before the emergency occurs. At the emergency scene, the incident commander (IC) can use the information gathered prior to the incident to direct the emergency operations much more effectively **FIGURE 24-1**. Because a preincident plan identifies potentially hazardous situations before an emergency occurs, fire fighters can be made aware of hidden dangers and prepare for them.

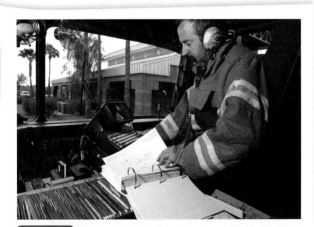

FIGURE 24-1 The preincident information is supplied to the IC at the emergency scene.

Preincident Plan

Preincident planning is the process of obtaining information about a building or a property and storing that information in a system so that it can be retrieved quickly for future reference. It is performed under the direction of a fire officer. The completed **preincident plan** should be made available to all units that might respond to an incident at that location.

The preincident plan is intended to help the IC make informed decisions when an emergency incident occurs at the location **FIGURE 24-2**. Some of this information would be useful to any fire company or unit responding to an incident at that location. In addition, a preincident plan can be used in training activities to help fire fighters become familiar with the properties within their jurisdiction.

The objective of a preincident plan is to make valuable information available quickly during an emergency incident that otherwise would not be readily evident or easily determined. The amount and the nature of the information provided for specific properties will depend on the size and complexity of the property, the types of risks present, and the particular hazards or challenges likely to be encountered.

The use of modern information technology has greatly enhanced the ability of fire departments to capture, store, organize, update, and retrieve preincident planning information. Accurate and current information such as drawings, maps, satellite and aerial imagery, photographs, descriptive text, lists of hazardous materials, and Safety Data Sheets (SDS) can be made available instantly to the fire fighters who need it during an emergency incident. **Geographic information systems (GIS)** collect data on a geographic environment, such as population characteristics, types of occupancies, water, other utilities, and road maps; organize those data; and present the information digitally to the entire emergency response system, including the fire department.

A preincident plan makes information immediately available to fire fighters who need it during an emergency incident. It is particularly important to inform fire fighters of potential safety hazards. The most critical information, such as hydrant locations and life hazards, should be instantly available,

Tactical Priorities

Address:

Occupancy Name:

Preplan #:	Number Drawings:	Revised Date:

District:	Subzone:	By:

Rescue Considerations: Yes (   ) No (   )

Occupancy Load Day:	Occupancy Load Night:

Building Size:	Best Access:

Knox Box:	Knox Switch:	Opticom:

Roof Type:	Attic Space: Yes ( ) No ( )	Attic Height:

Ventilation Horizontal:	Ventilation Vertical:

Sprinklers: Yes (   )   No (   )   Full (   )   Partial (   )

Standpipes: Yes (   )   No (   )   Wet (   )   Dry (   )

Gas: Yes (   )   No (   )   Lpg (   )

Hazardous Materials: Yes (   ) No (   )

Firefighter Safety Considerations:

Property Conservation And Special Considerations:

FIGURE 24-2 An example of a preincident plan.

Courtesy of Tempe Fire Department

with additional data being accessible as needed. All of the information must be presented in an uniform and understandable format.

A preincident plan usually includes one or more diagrams that shows details such as the building location and layout, access routes, entry points, exposures, and hydrant locations or alternative water supplies. In addition, the location and nature of any special hazards to the public or to fire fighters should be highlighted on the diagrams. Information about the actual building should include its height and overall dimensions, its type of construction, the nature of the occupancy, and the types of contents in different areas of the building. Additional information should include interior floor plans, stairway and elevator locations, utility shut-off locations, and information about built-in fire detection and suppression systems. For more information, see NFPA 1620, *Standard for Pre-incident Planning*.

■ Target Hazards

Most fire departments are not able to create a preincident plan for every individual property in their jurisdiction. Instead, they identify properties that are particularly large or that present unusual risks. These properties are identified as **target hazards** **TABLE 24-1** . Target hazard properties pose an increased risk to fire fighters.

A preincident plan should be prepared for every property that poses a high life-safety hazard to its occupants or presents safety risks for responding fire fighters. Preincident plans should also be prepared for properties that have the potential to create a large fire or **conflagration** (a large fire involving multiple structures).

Properties that have an increased life-safety hazard include the following structures:

- Hospitals
- Nursing facilities
- Assisted-living facilities
- Large apartment buildings
- Hotels and rooming houses
- Schools
- Public-assembly occupancies

■ Developing a Preincident Plan

The information that goes into a preincident plan is gathered during a **preincident survey**. This survey is usually completed by one of the crews that would respond to an emergency incident at the location. Its performance enables the crew to visit the property and become familiar with the location as they collect the preincident information.

Survey information can be collected and updated automatically in real time, through mobile devices. It can be

Preincident planning is not the same thing as fire prevention. With fire prevention, the goal is to identify hazards and minimize or correct them so that fires do not occur or have limited consequences. With preincident planning, the person creating the plan assumes that a fire will occur and compiles information that responding fire fighters would need.

A preincident survey can often identify the need for a preincident plan or a fire prevention inspection. For this reason, many departments conduct preincident surveys simultaneously with fire safety inspections. If this method is used, different members of the department should be assigned to conduct the inspection and to complete the preincident survey. This separation of duties allows each fire fighter to concentrate on the objectives of one specific assignment.

uploaded to a central system or submitted manually in a hard-copy preincident plan format. Drawings and other graphics are usually prepared at the fire station, where the fire officer takes the time to organize the information properly and to generate reports and forms that can be used effectively during an emergency.

If the preincident plan is stored on paper, copies should be made and distributed to all of the companies and command officers who might respond to the location on an initial alarm. The plans should be kept in binders or in a filing system on each vehicle **FIGURE 24-3** . In some jurisdictions, the communications centre can send a copy of the preincident plan to a portable fax machine in a command unit at the scene of an incident.

Fire departments using an integrated information technology system to organize data and create preincident plans can manage access to these plans through a network. This network can be accessed in multiple ways—directly, through computer terminals at fire stations, and through mobile computer devices **FIGURE 24-4** . When this approach is used, the preincident plan for a specific property can be automatically transmitted to the

TABLE 24-1	Typical Target Hazard Properties

- Bulk oil facilities and refineries
- High-rise buildings
- Hospitals
- Hotels and rooming houses
- Large apartment buildings
- Lumberyards
- Manufacturing plants
- Nursing homes and assisted-living facilities
- Public-assembly occupancies
- Schools
- Shopping centres
- Storage structures for hazardous materials
- Warehouses

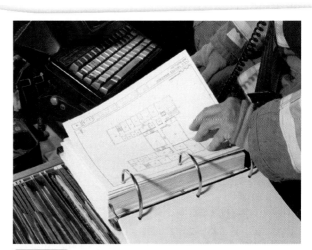

FIGURE 24-3 Copies of preincident plans can be kept in three-ring binders.

responding fire stations and to responding vehicles. Updating preincident plans is also easier when they are stored on a computer network. As soon as the database is updated, everyone has access to the new information, which eliminates the need to make and distribute new copies to everyone. No matter which system your department uses, every effort should be made to keep preincident plans current, consistent, and readily available to personnel in the communications centre, in responding vehicles, and to command staff.

FIGURE 24-4 Computers allow access to preincident plans from any location.

Conducting a Preincident Survey

A preincident survey should be conducted with the knowledge and cooperation of the property owner or occupant. The property owner or a representative should be contacted before the preincident survey is conducted. Making this initial contact enables the fire department to schedule an acceptable time, to explain the purpose and the importance of the preincident survey, and to clarify that the information is needed to prepare fire fighters in the event that an emergency occurs at the location.

The team members who conduct the survey should dress and conduct themselves in a manner appropriate to the department's mission. A representative of the property should accompany the survey team to answer questions and provide access to different areas. Every effort should be made to obtain accurate, useful information.

The preincident survey is conducted in a systematic fashion, following a uniform format. Begin with the outside of the building, gathering all of the necessary information about the building's geographic location, external features, and access points. Then survey the inside of the building to collect information about every interior area. A good, systematic approach starts at the roof and works down through the building, covering every level of the structure, including the basement. If the property is large and complicated, it may be necessary to make more than one visit to ensure that all the required information is obtained and recorded accurately.

The same set of basic information must be collected for each property that is surveyed **TABLE 24-2**. Additional information should be gathered for properties that are unusually large, are complicated, or might be the site of a particularly hazardous situation. Most fire departments use standard forms to record the survey information. After the team returns to the fire station, they can use the information to develop the preincident plan.

The fire fighters conducting the survey should prepare sketches or drawings to show the building layout and the location of important features such as exits. It takes practise and experience to learn how to sketch the required information while doing the survey, and then to convert the information to a final drawing. In some cases, the building owner can provide the survey team with a copy of a plot plan or a floor plan. Some departments use computer-assisted software to create and store these diagrams. These systems have the advantage of creating graphics more quickly and accurately than is possible with hand drawings. Many fire departments also use digital cameras to record information during the survey.

The completed drawing should use standard, easily understood map symbols **FIGURE 24-5** and **FIGURE 24-6**.

Preincident Planning for Response and Access

Building layout and access information is particularly important during the response phase of an emergency incident. The preincident plan should provide information that would be valuable to units en route to an incident; for instance, it might identify the most efficient route for the apparatus to take to the fire building and note an alternate route if the time of the day and local traffic patterns might affect the primary route. An alternate route

TABLE 24-2	Information Gathered During the Preincident Survey

- Building location
- Apparatus access to the exterior of the building
- Access points to the interior of the building
- Hydrant locations and alternative water supplies
- Size of the building (height, number of storeys, length, width)
- Exposures to the fire building and separation distances
- Type of building construction
- Building use
- Type of occupancy (assembly, institutional, residential, commercial, industrial)
- Floor plan
- Life hazards
- Building exit plan and exit locations
- Stairway locations (note whether the stairways are enclosed or unenclosed)
- Elevator locations and emergency controls
- Built-in fire protection systems (sprinklers, sprinkler control valves, standpipes, standpipe connections)
- Fire detection and alarm systems and location of the **fire alarm annunciator panel** (part of the fire alarm system that indicates the location of an alarm within the building)
- Utility shut-off locations
- Ventilation locations
- Presence of hazardous materials
- Presence of unusual contents or hazards
- Type of incident expected
- Sources of potential damage
- Special resources required
- General firefighting concerns

Preincident Plan Symbols

 Emergency Exit

 Emergency Exit Use of Arrows

 Manual Station— Pull Station/ Fire Alarm Box

 Area of Refuge

 Automated External Defibrillator (AED)

 Fire Extinguisher

 Fire Hose or Standpipe

 Ordinary Combustibles

 Flammable Liquid

 Electrical Equipment

 Combustible Metals

 Fire Department Automatic Sprinkler Connection—Siamese

 Fire Department Automatic Sprinkler Connection—Single

 Fire Department Standpipe Connection

 Fire Department Combined Automatic Sprinkler/Standpipe Connection

 Fire Hydrant (All Types)

 Automatic Sprinkler Control Valve

 Electric Panel or Electric Shutoff

 Gas Shutoff Valve

 Firefighting Hose or Standpipe Outlet

 Fire Extinguisher

Self-Contained Breathing Apparatus (SCBA)

 Opening in wall

Rated fire door in wall (less than 3 hours)

Fire door in wall (3-hour rated)

 Elevator in combustible shaft

 Elevator in noncombustible shaft

 Open hoistway

 Escalator

 Stairs in combustible shaft

 Stairs in fire-rated shaft

 Stairs in open shaft

 Skylight

FIGURE 24-5 Preincident plan symbols.

should also be identified if the primary route requires crossing railroad tracks, drawbridges, or other potentially blocked routes.

When conducting a preincident survey, fire fighters should ensure that the building address is clearly visible to save time in locating the property during an emergency incident. If the building is part of a complex, the best route to each individual building, section, or apartment should be indicated on a map. The locations of hydrants and fire department connections (FDCs) for sprinkler and standpipe systems should always be identified.

FIGURE 24-6 The completed drawing should use standard, easily understood map symbols.

Points of access into the building must also be noted, particularly if the building has multiple entrances or if access points are not easily seen from the street. In some cases, different entrances should be used, depending on the location of the emergency within the building. In other cases, the first-due unit should always respond to a designated entrance where the fire alarm annunciator panel is located or to meet a security guard who can act as a guide.

Diagrams should clearly indicate points at which gates, fences, or other barriers block access to parts of a building. The preincident survey should also note whether the topography makes one or more sides of a building inaccessible to fire apparatus or if an underground parking area will not support the weight of fire apparatus.

Access to the Exterior of the Building

The preincident plan should address all possible issues related to access to the exterior of the building; for example, you should ask the following questions during your survey:

- Do several roads lead to the building, or just a few?
- Where are the hydrants located?
- Where are the FDCs for automatic sprinkler and stand-pipe systems located?
- Do any security barriers limit access to the site?
- Are there fire lanes to provide access to specific areas?
- Are any barricades, gates, or other obstructions so narrow or low that they would prevent passage of fire apparatus?

- Are there bridges or underground structures that will not support the weight of apparatus?
- Do any gates require keys or a code to gain entry?
- Will it be necessary to cut fences to gain access to the site?
- Does the site include natural barriers, such as streams, lakes, or rivers, that might limit access?
- Does the topography limit access to any parts of the building?
- Might the landscaping or the presence of snow prevent access to certain parts of the building?

Access to the Interior of the Building

The preincident survey should also consider access to the interior of the building. The following questions are helpful in this regard:

- Is there a lockbox containing keys to the building? Do the keys work?
- Where is the lockbox located? Is it clearly visible?
- Are key codes needed to gain access to the building? Who has them? Are they in the lockbox?
- Does the building have security guards? Is a guard always on duty? Does the guard have access to all areas of the building?
- Is a key holder available to respond to the alarm within a reasonable amount of time? How can this person be reached?
- Where is the fire alarm annunciator panel located?
- Is the fire alarm annunciator panel properly programmed so you can quickly determine the exact location of an alarm?

■ Preincident Planning for Scene Size-Up

The preincident survey must also obtain essential information about the building that is important for size-up—that is, the ongoing observation and evaluation of factors that influence the objectives, strategies, and tactics for fire suppression. This information should include the building's construction, height, area, use, and occupancy, as well as the presence of hazardous materials or other risk factors. The locations of other buildings or structures that might be jeopardized by a fire in the building should be recorded as well.

Fire protection system information is also important for size-up. The preincident survey should identify areas that are protected by automatic sprinklers or other types of fire-extinguishing systems, the locations of standpipes, and the locations of fire walls and other features designed to limit the spread of a fire. Likewise, the survey should note any areas where fire protection is lacking, such as an area without sprinklers in a building that otherwise has a sprinkler system.

Features inside a building that would allow a fire to spread but are not readily visible from the outside should be noted as well. A common attic space over several occupancies, unprotected openings between floors, or buildings connected by overhead passages or conveyor systems are examples of these features.

Construction

Building construction is an important factor to identify in the preincident survey. Although they might have similar appearances, two buildings might nevertheless be constructed differently and, therefore, behave differently during a fire. Five types of construction are defined in NFPA 220, *Standard on Types of Building Construction* TABLE 24-3 . (Building construction is discussed in detail in this book in the chapter

TABLE 24-3	Types of Building Construction	
Type I: fire resistive	Buildings where the structural members are made of noncombustible materials that have a specified fire resistance FIGURE 24-7 . Materials include concrete, protected steel beams, and masonry block walls, among others.	 FIGURE 24-7 A Type I building. © John Foxx/Alamy Images
Type II: noncombustible	Buildings where the structural members are made of noncombustible materials but may not have fire resistance protection FIGURE 24-8 . Includes unprotected steel beams.	 FIGURE 24-8 A Type II building. © Rick's Photography/ShutterStock, Inc.
Type III: ordinary	Buildings where the exterior walls are made of noncombustible or limited-combustible materials, but the interior floors and walls are made of combustible materials FIGURE 24-9 .	 FIGURE 24-9 A Type III building. © Brandon Bourdages/ShutterStock, Inc.
Type IV: heavy timber	Buildings where the exterior walls are made of noncombustible or limited-combustible materials, but the interior walls and floors are made of combustible materials FIGURE 24-10 . The dimensions of the interior materials are greater than those of ordinary construction (typically with minimum dimensions of 20 centimetres by 20 centimetres [8 inches by 8 inches]).	 FIGURE 24-10 A Type IV building. © Helen Filatova/ShutterStock, Inc.
Type V: wood frame	Buildings where the exterior wall, interior walls, floors, and roof are made of combustible wood materials FIGURE 24-11 .	 FIGURE 24-11 A Type V building.

"Building Construction.") Preincident planning enables you to take the time that is needed to determine the exact type of building construction and identify any problem areas. Then, if an emergency occurs, the IC can check the preincident plan for information on construction, instead of hazarding a guess.

Lightweight Construction

It is important to identify whether the building contains <u>**lightweight construction**</u>. Lightweight construction uses assemblies of small components, such as trusses or fabricated beams, as structural support materials. This type of construction can be found in newer buildings as well as in older buildings with extensive remodelling. In many cases, the lightweight components are located in void spaces or concealed above ceilings and are not readily visible.

> ### Fire Fighter Safety Tips
> © Photos.com
>
> *Lightweight construction* is a relative term. A structure may be lightweight in comparison to other methods of construction, but its components are still heavy enough to cause serious injury or death if they collapse on a fire fighter.

FIGURE 24-12 A lightweight wood-truss roof assembly.

Courtesy of Underwriters Laboratories

> ### Fire Fighter Safety Tips
> © Photos.com
>
> Because lightweight construction uses truss supports or I-beams, a floor or roof in this type of structure may appear sturdier than it actually is. Two fire fighters in Houston, Texas, died while battling an early-morning blaze in a fast-food restaurant. The fire burned through the lightweight wood-truss roof supports, and the roof-mounted air conditioner dropped into the building, trapping and killing the fire fighters. The fire was later determined to be arson.

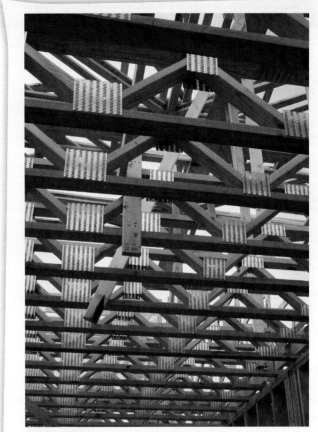

FIGURE 24-13 A lightweight wood truss with gusset plates.

© David Huntley/ShutterStock, Inc.

A wood truss constructed from 5- by 10-cm (2- by 4-in.), or 5- by 15-cm (2- by 6-in.), pieces of wood, for example, can be used to span a wide area and support a floor or roof. A truss uses the principle of a triangle to build a structure that can support a great deal of weight with much less supporting material than conventional construction methods **FIGURE 24-12** and **FIGURE 24-13**.

Remodelled Buildings

Buildings that have been remodelled or renovated may present unique hazards. Remodelling can remove some of the original, built-in fire protection and create new hazards. Examples of remodelling-derived hazards include multiple ceilings with void spaces between them, new construction of concrete topping over a wooden floor, and new openings between floors or through walls that originally were fire resistant.

If you can conduct the preincident survey during construction or remodelling, you will be able to see the construction from the inside out, before everything is covered over. It is also important to realize that all buildings under construction—including those that are being remodelled or demolished—are especially vulnerable to fire. Unfinished

Fire Fighter Safety Tips

© Photos.com

Buildings that are being remodelled or demolished are at higher risk for major fires. You should consider doing a special preincident survey for buildings undergoing substantial remodelling during the construction phase. This visit will enable you to gather more thorough information for a preincident plan of the completed building.

construction is open, so it lacks many of the fire-resistant features and fire detection/suppression systems that will be part of the finished structure.

FIRE FIGHTER II Tips

FFII

A new component in lightweight building construction is the truss joist I-beam, or TGI beam. This framing material is made up of three wooden components. Channels may be drilled through TGI beams, thus providing another avenue for fire to spread.

Building Use

The second major consideration during size-up—and hence during a preincident survey—is the building's use and occupancy. Buildings are used for many purposes, such as residences, offices, stores, restaurants, warehouses, factories, schools, and churches. For each use, different types of problems, concerns, and hazards may be present.

For example, knowing the building's use can help fire fighters determine the number of occupants and predict their ability to escape if a fire occurs. It is also a key factor in determining the probable contents of the building. These factors can have a major effect on the problems and hazards encountered by fire fighters responding to an emergency incident at the location.

A building is usually classified by its major use, which identifies the basic characteristics of the building **TABLE 24-4**. Within the major use classification, various occupancy subclassifications may provide a more specific description of the possible uses of sections of the building and their associated characteristics.

Many large building complexes contain multiple occupancy subcategories under one roof. For example, a shopping mall usually includes both retail stores and restaurants. The

TABLE 24-4	Classifications of Buildings
Major Use Classification	**Occupancy Subcategories**
Public assembly	Theatres, auditoriums, and churches Arenas and stadiums Convention centres and meeting halls Bars and restaurants
Institutional	Hospitals and nursing homes Schools
Commercial	Retail stores Industrial factories Warehouses Parking garages Offices

mall might also be part of an even larger complex that includes offices, residential condominiums, a hotel, a movie theatre, an underground parking garage, and a subway station. Each area may have very different characteristics and, therefore, different risk factors.

FIRE FIGHTER II Tips

FFII

A building can have multiple uses during its lifetime or can incorporate more than one use within the same structure.

Occupancy Changes

Building use may change over time. An outdated factory may be transformed into a residential building, or an unused school may be converted into an office building. A warehouse that once stored concrete blocks may now be filled with flammable foam plastic insulation or hazardous materials such as swimming pool chemicals. Given this propensity for change, preincident plans should be checked and updated on a regular schedule. A building's current occupancy information must always be determined during a preincident survey.

Exposures

An **exposure** is any other building or item that might be in danger if an incident occurs in another building or area; for example, an exposure could be an attached building, separated from the other structure by a common wall, or it could be a building across an alley or street. Exposures can include other buildings, vehicles, outside storage, or anything else that could be damaged by or involved in a fire; for example, heat from a burning warehouse could ignite adjacent buildings and spread the fire.

A preincident survey should identify any potential exposures for the property that is being evaluated. It should take into account the size, construction, and **fire load** (the amount of combustible material and the rate of heat release) of the property being evaluated, the distance to the exposure, and the ease of ignition of the exposure by radiation, convection, or conduction of heat.

Built-In Fire Protection Systems

The preincident survey should identify built-in fire detection and suppression systems on the property. These systems may include automatic sprinklers, standpipes, fire alarms, and fire detection systems, as well as systems designed to control or extinguish particular types of fires. Some buildings have automatic smoke control or exhaust systems. Most high-rise buildings have systems that control elevator function during an emergency. Each of these systems is covered in detail in the chapter "Fire Detection, Protection, and Suppression Systems."

Automatic Sprinkler Systems

A properly designed and maintained automatic **sprinkler system** can help control or extinguish a fire before the arrival of the fire services unit. When properly designed and maintained, such systems are extremely effective and can play a major role in reducing the loss of life and property at an incident. They also create a safer situation for fire fighters.

The preincident survey should determine whether the building has a sprinkler system and which parts of the building that system covers. Note the locations of the valves that control water flow to different sections of the system. Normally, these valves are found in the open position.

In addition to the control valves, sprinkler systems should have an FDC outside the building. The FDC is used to supplement the water supply to the sprinklers by pumping water from a hydrant or other water supply through fire hoses into this connection. The location of the FDC for the sprinkler system must be marked on the preincident plan.

Standpipe Systems

Standpipe systems are installed to deliver water to fire hose outlets on each floor of a building. Their use eliminates the need to extend hose lines from an engine at the street level up to the fire level. Standpipes may also be used in low-rise buildings, such as convention centres. In such structures, fire fighters can bring attack hose lines inside the building and connect them to a standpipe outlet close to the fire, while the engine delivers water to a FDC outside the building.

The location of the FDC, as well as the locations of the outlets on each floor, should be clearly marked on the preincident survey. The locations of nearby hydrants that will be used to supply water to the pumper that feeds the FDC must also be recorded. In a large building, multiple FDCs may be available to deliver water to different parts of the building. The area or floor levels served by each connection should be carefully noted on the preincident plan and labelled at the connection.

Fire Alarm and Fire Detection Systems

The primary role of a fire alarm system is to alert the occupants of a building so that they can evacuate or take appropriate action when an incident occurs. Some fire alarm systems are connected directly to the fire department; others are monitored by a service that calls the fire department when the system is activated.

In some cases, the fire alarm system must be manually activated. In other cases, it is activated automatically. A smoke or heat detection system can trigger the alarm, for example, or it may respond to a device that indicates when water is being discharged from the sprinkler system.

The annunciator panel, which is usually located close to a building entrance, indicates the location and type of device that activated the alarm system. In most cases, fire fighters who respond to the alarm must check the annunciator panel to determine the actual source of the alarm within the building or complex. The preincident survey should identify which type of system is installed, where the annunciator panel is located, and whether the system is remotely monitored.

Special Fire-Extinguishing Systems

Several types of fixed fire-extinguishing systems can be installed to protect areas where automatic sprinklers are not suitable. Most commercial kitchens and computer rooms are required to have installed fire-suppression systems, for example. Special extinguishing systems are also found in many industrial buildings. The type of system and the area that is protected should be identified in the preincident survey.

Areas where flammable liquids are stored or used are often protected by sophisticated foam or dry chemical fire-suppression systems. The locations of these systems should be noted on the preincident plan. In addition, details about the method of operation, location of equipment, and foam

supplies should be recorded during the preincident survey. More information about built-in systems is presented in the chapter "Fire Detection, Protection, and Suppression Systems."

Conducting a Preincident Survey

To conduct a preincident survey, follow the steps in **SKILL DRILL 24-1** (Fire Fighter II, NFPA 6.5.3):

1. Schedule the survey in advance.
2. Make contact with a responsible person.
3. Present a neat and professional image.
4. Identify yourself by name, title, and department.
5. Ensure that a representative from the facility accompanies you during the survey.
6. Take notes and pictures as needed, and start outside.
7. Note the building location.
8. Identify the building construction.
9. Identify the building use and occupancy.
10. Note any life hazards.
11. Note the access points to the interior of the building.
12. Note the utility shut-off locations.
13. Assess the apparatus access locations to the building.
14. Note hydrant locations and alternative water supplies.
15. Note ventilation concerns.
16. Record information about built-in fire detection and suppression systems.
17. Sketch floor plans.
18. Note the elevator and stairway locations.
19. Review exit plans and exit locations.
20. Identify any special hazards and hazardous materials.
21. Note the building exposures.
22. Anticipate the type of incident expected.
23. Identify any special resources needed.
24. Complete and file the preincident survey form.

Tactical Information

Many different types of tactical information can be obtained during a preincident survey and documented in a preincident plan. Such information would be of particular value to an IC in directing operations during an emergency incident or to fire fighters who are assigned to perform specific functions. When conducting a survey, all of the factors that could be significant when conducting emergency operations at that location should be considered.

■ Considerations for Water Supply

During the preincident survey, the amount of water needed to fight a fire in the building should be determined. The water supply source should be identified as well. The required flow rate, measured in litres per minute or gallons per minute, can be calculated based on the building's size, construction, contents, and exposures. In Canada, the formula to calculate required water flow is: Length × Width × Height × 2 = water flow required in litres per minute (L × W × H = L/min. flow).

In most urban areas, the water supply will come from municipal hydrants. It is important to locate the hydrants closest to the building. In addition, for large buildings, it will be necessary to locate enough hydrants to supply the volume of water required to control a fire. The ability of the municipal water system to provide the required flow must be determined as well. In some cases, it may be necessary to use hydrants that are supplied by different water mains of sufficient size to achieve the needed water flow **FIGURE 24-14** .

In areas without municipal water systems, water may have to be obtained from a static water supply, such as a lake or stream,

FIRE DEPARTMENT TACTICAL RESPONSE GUIDE

BUSINESS		ADDRESS		
STATION **5C**	DATE **08/30/00**	PAGE **1** OF **5**	MAPSCO**656Y**	BOX **3415**

TO PRESTON

N

TO MAIN

FAIRFEILD INN

16.8 m
PH

K
OFFICE

30.5 m
SH

BANK

ALLIED DRIVE

TO PRESTON   ALLIANCE BLVD

FIGURE 24-14 A diagram showing hydrant locations around a building. PH indicates the primary hydrant for connecting to the FDC. The number 16.8 indicates that the primary hydrant is located 16.8 metres (55 feet) from the FDC. SH indicates the secondary hydrant for connecting to the FDC. The number 30.5 indicates that the secondary hydrant is located 30.5 m (100 ft.) from the FDC.

or be delivered by fire department tanker (or tender). When static water sources are used, the preincident plan must identify drafting sites (locations where an engine can draft water directly from the static source) or the locations of the nearest dry hydrants (an arrangement of pipes that is permanently connected to a static water supply). It is also important to measure the distance from the water source to the fire building to determine whether a large-diameter hose (LDH) can be used and whether additional engines will be needed. Finally, the preincident plan should outline the operation that would be required to deliver water to the fire.

If a tanker (or tender) shuttle (a system of tankers [or tenders] transporting water from a water supply to the fire scene) will be used to deliver water, the preincident plan must include several additional details; for instance, sites for filling the tankers (or tenders) and for discharging their loads must be identified (FIGURE 24-15). The preincident plan should also identify how many tankers (or tenders) will be needed based on the total distance they must travel, the quantity of water

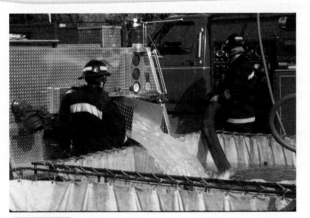

FIGURE 24-15 The points where tankers (or tenders) can be filled and where they can discharge their loads must be identified in the preincident plan.

© Jim Smalley

each vehicle can transport, and the total time it takes to empty and refill the vehicle.

A large industrial or commercial complex may have its own water supply system that provides water for automatic sprinkler systems, standpipes, and private hydrants. These systems usually include storage tanks or reservoirs as well as fixed fire pumps to deliver the water under pressure. The details and arrangement of the private water supply system must be determined during the preincident survey.

The survey should indicate how much water is stored on the property and where the tanks are located. If a fixed fire pump is available, its location, capacity (in litres per minute), and power source (electricity or a diesel engine) should be noted. It is also important to confirm that these private water systems are being maintained in good operating condition; for example, the private water supply system on an abandoned property may be useless if it is no longer properly maintained.

In many cases, the same private water main provides water for both the sprinkler system and the private hydrants. If the fire department uses the private hydrants, the water supply to the sprinklers may be compromised. The preincident plan for these sites should note whether public, off-site hydrants should be used instead of the private hydrants.

■ Utilities

During an emergency incident, it may be necessary to turn off utilities, such as electricity or natural gas, as a safety measure. The preincident survey should note the locations of shut-offs for electricity, natural gas, propane gas, fuel oil, and other energy sources. These sites, as well as contact information for the appropriate utility companies, should be included in the preincident plan. In some cases, special knowledge or equipment may be required to disconnect utilities. This information must be noted in the preincident plan, along with the procedure for contacting the appropriate individual or organization.

Electrical wires pose particular problems during firefighting operations. Electricity may be supplied by overhead wires or through underground cables. Overhead wires can be deadly if a ground ladder or aerial ladder comes in contact with them. The preincident plan should show the locations of high-voltage electrical lines and equipment that could prove dangerous to fire fighters. If electricity is supplied by underground cables, the shut-off may be located inside the basement of a building or in an underground vault. These locations should also be noted on the preincident plan.

If propane gas or fuel oil is stored on the property, the preincident plan should show the location and note the capacity of each tank. The presence of an emergency generator should

Near-Miss REPORT

Report Number: 06-0000508

Synopsis: Fire fighter falls through skylight.

Event Description: My engine company (three-man crew) responded to an apartment fire. We had received multiple calls reporting the possibility of victims trapped.

We were the first unit on the scene and found heavy smoke and fire from the second storey of an eight-unit apartment building. Upon exiting the apparatus, I was advised by sheriff's deputies that there were possibly two victims trapped. Another fire fighter and I pulled a 51-millimetre (2-inch) attack line, while the engine operator worked with the truck company to establish a supply line. The centre of the building was fully involved and contained the entrances to all of the apartments.

My plan was to knock down the fire by the entrances to allow a rapid search for victims. I knew there was plenty of help on the way, because I had called for a second alarm while responding to the scene. We knocked down the fire by the doors to the lower units and began moving up the stairs to the second floor. While advancing the attack line to the balcony/entrance area, I fell through the floor and landed on the enclosed concrete patio that was below me. This area was still fully involved with fire. The fire fighter who was with me attempted to call a mayday but could not get through, owing to heavy radio traffic. I was able to exit the building on my own and sustained only a bruised shoulder and hip; my injuries could have been much worse.

I gathered my crew and exited out of the building. Command decided to switch to a defensive mode using master streams to knock down the majority of the fire. This incident eventually escalated to three alarms. There were no victims found.

After reviewing the incident and looking at some of the other buildings in the complex, it appeared that I actually fell through a 0.6- by 1.2-m (2- by 4-ft.) skylight/atrium located on the second floor. The railing had burned away, and I could not see it, owing to the smoke, fire, and steam. If we had not received reports of victims trapped, I would not have been as aggressive with our initial attack. In our jurisdiction (population of more than 45,000), we rarely find victims in the fire building. We have experienced only four fire fatalities in 30-plus years and have only a few "saves."

Lessons Learned: I fell through a hole that I did not know was there and could not see. This problem might have been prevented with better knowledge of preincident plans. I have been responding to this complex for 10 years but had never noticed the small atriums. My station has 11 large apartment complexes in our first-due area and more than 75 large apartment complexes in our first-alarm area. This makes it difficult to remember small details about each one of them.

FIRE FIGHTER II Tips　　　　　　　　**FFII**

The locations of utility shut-offs are as varied as the layout of structures. Whereas some structures may have their utility shut-offs outside on the top of a utility pole, other structures may have their utility shut-offs inside in a basement closet. It is important to note these locations on the preincident plan to avoid lengthy delays in shutting down utilities during an emergency.

be noted as well, along with the fuel source and a list of equipment powered by the generator.

■ Preincident Planning for Search and Rescue

Fire fighters who are conducting search and rescue operations will need to know the locations of the occupants of a building as well as the locations of exits. The preincident survey should identify all entrances and exits to the building, including fire escapes and roof exits.

In addition to conducting a search for occupants who are unable to escape on their own, search and rescue teams may need to assist occupants who are trying to use the exits **FIGURE 24-16** ; for example, fire fighters may need to assist occupants who are descending fire escapes or need to place ladders near windows or other locations for occupants to use.

During the preincident survey, team members should obtain the interior floor plan for the building. Information about the floor plan of a building can be life-saving knowledge when the structure is filled with smoke. It is also much easier to understand a floor plan when you can tour a building under nonemergency conditions. In large buildings, it may be necessary to plan for the use of ropes during search and rescue to prevent disorientation in conditions of limited visibility.

■ Preincident Planning for Forcible Entry

As previously noted, the preincident survey should consider both exterior and interior access problems. Locations where forcible entry may be required should be identified and marked

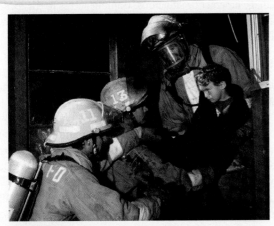

FIGURE 24-16 Fire fighters should plan to assist occupants who are trying to use the building exits.

© Keith D. Cullom

FIGURE 24-17 Considerations for the use of ladders should include identifying the best locations to place ground ladders and use aerial apparatus.

© Colin Archer/AP Photos

on the site diagrams and building floor plan. Noting which tools would be needed to gain entry can save time during the actual emergency. The location of a lockbox and instructions on obtaining keys should be identified as well.

Preincident Planning for Ladder Placement

The preincident survey is an excellent time to identify the best locations for placing ground ladders or using aerial apparatus **FIGURE 24-17**. The length of ladder needed to reach a roof or entry point should be noted. When planning ladder placement, pay careful attention to any electrical wires and other obstructions that might not be visible at night or in a smoky atmosphere **FIGURE 24-18**. Plan to place ladders in locations that will not disturb other vital functions; for example, you should not place ladders in front of a door that will be used for entrance or exit from the fire building.

Preincident Planning for Ventilation

While performing a preincident survey, fire fighters should consider which information would be valuable to the members of a ventilation team during a fire. For example, what would be the best means to provide ventilation? How useful are the existing openings for ventilation? Are there windows and doors that would be suitable for horizontal ventilation? Where could fans be placed? Could the roof be opened to provide vertical ventilation? What is the best way to reach the roof? Is the roof construction safe? Are there ventilators or skylights that could be easily removed or bulkhead doors that could be opened easily? Will fire fighters need to use saws and axes to cut through the roof? Would multiple ceilings have to be punctured to allow smoke and heat to escape?

It is also important to determine whether the **heating, ventilation, and air conditioning (HVAC) system** can be used to remove smoke without circulating it throughout the building. Many buildings with sealed windows have controls that enable the fire department to set the HVAC system to deliver outside air to some areas and exhaust smoke from other areas.

FIGURE 24-18 The preincident survey should note overhead obstructions—in particular, electrical wires that might not be visible at night or in a smoky environment.

© Photos.com

The instructions for controlling the HVAC system should be included in the preincident plan.

Roof construction must also be evaluated to determine whether it would be safe for fire fighters to work on the roof when there is a fire below. If the roof is constructed with lightweight trusses, the risk of collapse is great. Many fire departments do not permit fire fighters on these roofs. The presence of an attic that might allow a fire to spread quickly under the roof should also be noted. In addition, common attics in townhouses and shopping malls can spread fire from one occupancy to another.

Occupancy Considerations

Each type of occupancy involves unique considerations that should be taken into account when preparing a preincident plan. Fire fighters should keep these factors in mind when conducting a preincident survey.

High-Rise Buildings

A high-rise building is generally defined as a structure that is more than 22.9 m (75 ft.) high, which is usually six or seven storeys. High-rise buildings present special problems during an emergency because of the difficulty in gaining access and because of the large numbers of occupants. As a consequence, major fires in high-rise buildings often result in injuries, fatalities, and millions of dollars in property losses.

A preincident survey should identify both the building construction and any special features that have been installed. It should note all of the systems that are present in a particular building and identify how they are designed to function. This information, in turn, should be incorporated into the preincident plan.

The fire protection features of a high-rise building will vary depending on the age of the building and the specific building and fire code requirements in different jurisdictions. Older high-rise buildings are generally constructed of noncombustible materials and designed to confine a fire within a limited area. Newer buildings generally include automatic sprinklers, smoke detection and alarm systems, emergency generators, elevator control systems, smoke control systems, and building control stations where all these systems are monitored and controlled. The details of each system must be documented in the preincident plan. In many cases, expert consultants may be contacted to develop an emergency plan for such a building. These plans are intended to be used by the building tenants and management as well as the fire department.

Assembly Occupancies

Public-assembly venues—such as theatres, nightclubs, stadiums, churches, hotel meeting rooms, and arenas—present the possibility that large numbers of people could become involved in an emergency incident. These structures are often very large and complicated, and they may be equipped with complex systems intended to manage emergency situations. Gaining access to the location of the fire or emergency situation may prove difficult, however, when all of the occupants are trying to evacuate at the same time.

Health Care Facilities

Health care facilities—such as hospitals, nursing homes, assisted-living facilities, surgery centres, and ambulatory health care centres—require special preincident planning. Hospitals are often very large and include many different areas, ranging from operating rooms to walk-in clinics. The most challenging problem during an emergency incident at a health care facility is protecting **nonambulatory** patients.

A **defend-in-place** philosophy is used when designing fire protection for these facilities. This approach presumes that patients will not be able to escape from a fire without assistance and that there may not be enough staff present to move all of the patients. Therefore, the facility itself is designed to protect the patients from the fire. Most health care facilities utilize fire-resistant construction, have fire detection systems and sprinkler systems, and are compartmentalized.

If it becomes necessary to move patients from a dangerous area, the preferred approach is often **horizontal evacuation** (moving patients from a dangerous area to a safe area on the same floor). It is much easier to wheel a bed to another area on the same floor level than to carry patients down stairways.

Detention and Correctional Facilities

By their very nature, detention and correctional facilities are designed to ensure that the occupants cannot leave the building, even under normal conditions. The preincident plan must consider the problems involved in removing the inmates from a dangerous situation while protecting the fire fighters from angry or frightened inmates. Additionally, security concerns can make it difficult for fire fighters to gain rapid access to the building or for occupants and rescuers to exit the facility.

Residential Occupancies

Preincident plans are usually prepared for multifamily residential properties, such as apartment complexes and condominiums. Most fire departments do not prepare individual preincident plans for one- and two-family residential buildings, unless there are unusual risk factors such as a dwelling with handicapped occupants. In addition to the standard preincident plan information, the following information will be helpful when developing a preincident plan for a private residence:

- Detailed floor plan
- Location of sleeping areas
- Information about any handicapped occupants
- Escape routes and an outside rendezvous area
- Fire hydrant or water source location

Even if individual homes are not surveyed, many fire departments will survey different neighbourhoods making special note of addresses, access routes, and hydrant locations. Knowing which styles and types of private residences are present in your jurisdiction will help you prepare for emergency incidents.

Some homeowners may request an individual fire safety survey of their property. These surveys are intended to identify fire hazards and to educate the homeowner about steps that can be taken to reduce the risk of fire.

FIRE FIGHTER II Tips **FFII**

Modern high-rise buildings are often equipped with multiple fire protection systems—but these systems must be maintained properly to be effective. One Meridian Plaza (a modern high-rise office building in Philadelphia, Pennsylvania) had both a standpipe system and a partial sprinkler system on a few floors. When a fire broke out in this building in 1991, fire fighters attached their hose lines to the standpipe system; because the pressure-reducing valves on the standpipe outlets had been set incorrectly, however, the fire fighters were unable to obtain sufficient pressure to control the fire. Three fire fighters died, and the fire burned several floors of the building, which eventually had to be demolished. The fire was stopped only when it reached one of the floors that was equipped with a sprinkler system. An NFPA Fire Investigation Report is available on this fire.

Several fires have occurred in home-improvement stores. In one case, pool chemicals ignited, creating a special challenge for the fire department because of the hazardous materials involved. In another case, a fire started in a seat cushion display and spread rapidly through several aisles. The responding fire fighters were faced with a large fire, no visibility, and products dropping onto them from the upper shelves. Reports on these fires are available from NFPA Fire Investigations.

Locations Requiring Special Considerations

Preincident planning should extend beyond planning for fires and other types of emergency situations that could occur in a building. Specifically, it should anticipate the types of incidents that could occur at locations such as airports, bridges, and tunnels, as well as incidents along highways or railroad lines, or at construction sites **FIGURE 24-19** .

Detailed preincident plans are prepared for airports, for example; these plans should note the various types of aircraft that use the airport. In addition, the airport facilities themselves often require extensive preincident planning because of their size, the number of people who may be present, and security issues. Similar planning should be done for bridges, tunnels, and any other locations where complicated situations could occur.

The following special locations would also require preincident planning:

- Gas or liquid fuel transmission pipelines
- Electrical transmission lines **FIGURE 24-20**
- Ships and waterways **FIGURE 24-21**
- Subways **FIGURE 24-22**
- Railroads **FIGURE 24-23**

■ Special Hazards

One of the most important reasons for developing a preincident plan is to identify any special hazards and to provide information that would be valuable during an emergency incident. This information includes chemicals or hazardous materials that are stored or used on the premises, structural conditions that could result in a building collapse, industrial processes that pose special hazards, high-voltage electrical equipment, and confined spaces. Preincident plans warn fire fighters of potentially dangerous situations and could include detailed instructions about what to do in those circumstances—information that could ultimately save the lives of fire fighters.

Fire fighters who are conducting a preincident survey should always look for special hazards and obtain as much information as possible at that time **FIGURE 24-24** . If the information is not readily available, it may be necessary to contact specialists for advice or conduct further research before completing the preincident plan.

Hazardous Materials

A preincident survey should obtain a complete description of all hazardous materials that are stored, used, or produced at the property. This assessment includes an inventory of the types and quantities of hazardous materials that are on the

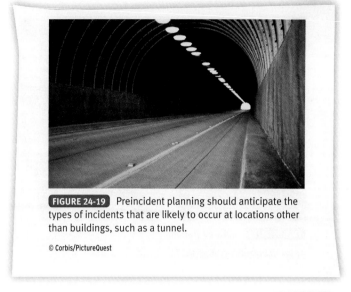

FIGURE 24-19 Preincident planning should anticipate the types of incidents that are likely to occur at locations other than buildings, such as a tunnel.

© Corbis/PictureQuest

FIGURE 24-20 Electrical transmission lines.

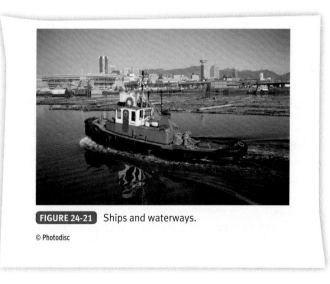

FIGURE 24-21 Ships and waterways.

© Photodisc

premises, as well as information about where they are located, how they are used, and how they are stored. The appropriate actions and precautions for fire fighters in the event of a spill, leak, fire, or other emergency incident should also be listed.

Many jurisdictions require businesses that store or use hazardous materials to obtain a fire department permit. They may also require that hazardous materials specialists conduct special

FIGURE 24-22 Subway station.

© Arthur S. Aubry/Photodisc/Getty Images

FIGURE 24-24 During the preincident survey, look for special hazards and obtain as much information as possible.

FIGURE 24-23 Railroad.

© JJJ/ShutterStock, Inc.

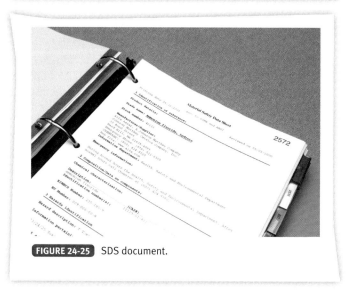

FIGURE 24-25 SDS document.

inspections. If the quantity of hazardous materials on hand exceeds a specified limit, federal and state regulations require a property owner to provide the local fire department with current inventories and **Safety Data Sheet (SDS) documents** **FIGURE 24-25** . Fire fighters conducting preincident surveys should learn where the SDSs are stored, ensure that the required information has been provided, and verify that it is up to date.

Fire fighters should expect to encounter hazardous materials at certain types of occupancies, such as chemical companies, garden centres, swimming pool supply stores, hardware stores, and laboratories. Of course, hazardous materials can also be found in unexpected locations, so fire fighters should always be alert for their presence.

Some communities put placards on the outside of any building that contains hazardous materials. These placards should use the marking system specified in NFPA 704, *Standard System for the Identification of the Hazards of Materials for Emergency Response.*

In case of fire, some hazardous materials require use of special suppression techniques or extinguishing agents. This information should be noted in the preincident plan. The plan should also contain contact information for individuals or organizations that can provide advice if an incident occurs. Hazardous materials are discussed in detail in the Hazardous Materials chapters.

FIRE FIGHTER II Tips   **FFII**

Once the preincident plan is completed, train on it using a tabletop scenario. Invite neighbouring fire departments to train with you and become familiar with the preincident plan.

FIRE FIGHTER II Tips   **FFII**

All information in preincident plans, including SDS, is confidential. The general public is not at liberty to see the information collected in your fire department's preincident plans. Keep all information that you learn about an occupancy during a preincident plan survey confidential.

Chief Concepts

- Preincident planning helps your department to make better command decisions, because important information is assembled before the emergency occurs.

- Preincident planning is the process of obtaining information about a building or a property and storing the information in an accessible system. It is intended to help the IC make informed decisions when an emergency incident occurs at a specific location.

- The use of modern information technology has greatly enhanced the ability of fire departments to capture, store, organize, update, and retrieve preincident plans. Accurate and current information such as drawings, maps, satellite and aerial imagery, photographs, descriptive text, lists of hazardous materials, and MSDSs can be made available instantly to the fire fighters who need it during an emergency incident.

- A preincident plan usually includes one or more diagrams that shows details such as the building location and layout, access routes, entry points, exposures, and hydrant locations and alternative water supplies.

- The location and nature of any special hazards to the public or to fire fighters should be highlighted on the diagrams. Information about the actual building should include its height and overall dimensions, its type of construction, the nature of the occupancy, and the types of contents in different areas of the building. Additional information should include interior floor plans, stairway and elevator locations, utility shut-off locations, and information about built-in fire detection and suppression systems.

- A preincident plan should be prepared for every property that poses a high life-safety hazard to its occupants or presents safety risks for responding fire fighters. Preincident plans should also be prepared for properties that have the potential to create a large fire or conflagration.

- The information that goes into a preincident plan is gathered during a preincident survey.

- A preincident survey should be conducted with the knowledge and cooperation of the property owner or occupant.

- The preincident survey is conducted in a systematic fashion, beginning with the outside of the building and moving inside. A good, systematic approach starts at the roof and works down through the building, covering every level of the structure, including the basement.

- The fire fighters conducting the preincident survey should prepare sketches or drawings to show the building layout and the location of important features such as exits.

- Building layout and access information is particularly important during the response phase of an emergency incident. The preincident plan should provide information that would be valuable to units en route to an incident. Points of access into the building must also be noted.

- The preincident survey must obtain essential information about the building that is important for size-up, including the building's construction, height, area, use, and occupancy; the presence of hazardous materials; locations of exposed structures; and fire protection system information.

- To conduct a preincident survey:
  - Schedule the survey in advance.
  - Make contact with a responsible person.
  - Present a neat and professional image.
  - Identify yourself by name, title, and department.
  - Ensure that a representative accompanies you during the survey.
  - Take notes and pictures as needed, and start outside.
  - Note the building location.
  - Identify the building construction.
  - Identify the building use and occupancy.
  - Note any life hazards.
  - Note the access points to the interior of the building.
  - Note the utility shut-off locations.
  - Assess the apparatus access to the building.
  - Note hydrant locations and alternative water supplies.
  - Note ventilation concerns.
  - Record information about built-in fire detection and suppression systems.
  - Sketch floor plans.
  - Note the elevator and stairway locations.
  - Review exit plans and exit locations.
  - Identify any special hazards and hazardous materials.
  - Note the building exposures.
  - Anticipate the type of incident expected.
  - Identify any special resources needed.
  - Complete and file the preincident survey form.

- During the preincident survey, collect tactical information about water supply considerations and locations of shut-offs for utilities.

- To facilitate a safe search and rescue operation at a site, the preincident survey should identify all entrances and exits to a building, including fire escapes and roof exits.

- To ensure a safe forcible entry operation, the preincident survey should identify any exterior and interior access issues and the locations where forcible entry may be required.

- To support safe ladder operations, the preincident survey should identify the best locations for placing ground ladders or aerial apparatus.

- To ensure safe ventilation operations, the preincident survey should identify the best locations for ventilation and determine whether the HVAC system can be used for ventilation.

- The following occupancies involve unique considerations:
  - High-rise buildings—Special issues include difficulty in gaining access and the large number of occupants to evacuate.
  - Public-assembly venues—These structures are often very large and contain large numbers of people to be evacuated.
  - Health care facilities—These structures are very large and contain nonambulatory occupants who need assistance in evacuating.
  - Detention and correctional facilities—Security concerns may make it difficult for fire fighters to gain rapid access to the building or for occupants to exit the facility.
  - Residential occupancies—Include apartment complexes and condominiums. The preincident plan should identify the locations of sleeping areas and the water supplies.
- Preincident planning should anticipate the types of incidents that could occur at locations such as airports, bridges, and tunnels, as well as incidents along highways or railroad lines, or at construction sites.

Hot Terms

Conflagration A large fire, often involving multiple structures.

Defend-in-place A strategy in which the victims are protected from the fire without relocating them.

Exposure Any person or property that could be endangered by fire, smoke, gases, runoff, or other hazardous conditions. (NFPA 402)

Fire alarm annunciator panel Part of the fire alarm system that indicates the source of an alarm within a building.

Fire load The weight of combustibles in a fire area (per sq. m or sq. ft.) or on a floor in buildings and structures, including either contents or building parts, or both. (NFPA 914)

Geographic information system (GIS) A system of computer software, hardware, data, and personnel to describe information tied to a spatial location. (NFPA 450)

Heating, ventilation, and air conditioning (HVAC) system A system to manage the internal environment that is often found in large buildings.

Horizontal evacuation Moving occupants from a dangerous area to a safe area on the same floor level.

Lightweight construction Lightweight materials or advanced engineering, or both practices, which result in a weight savings without any sacrifice of strength or efficiency. (NFPA 414)

Safety Data Sheet (SDS) document A form, provided by manufacturers and compounders (blenders) of chemicals, containing information about chemical composition, physical and chemical properties, health and safety hazards, emergency response, and waste disposal of the material. (NFPA 472)

Nonambulatory A term describing individuals who cannot move themselves to an area of safety owing to their physical condition, medical treatment, or other factors.

Preincident plan A written document resulting from the gathering of general and detailed information to be used by public emergency response agencies and private industry for determining the response to reasonably anticipated emergency incidents at a specific facility.

Preincident survey The process used to gather information to develop a preincident plan.

Sprinkler system For fire protection purposes, an integrated system of underground and overhead piping designed in accordance with fire protection engineering standards. The installation includes at least one automatic water supply that supplies one or more systems. The portion of the sprinkler system above ground is a network of specially sized or hydraulically designed piping installed in a building, structure, or area, generally overhead, and to which sprinklers are attached in a systematic pattern. Each system has a control valve located in the system riser or its supply piping. Each sprinkler system includes a device for actuating an alarm when the system is in operation. The system is usually activated by heat from a fire and discharges water over the fire area. (NFPA 13, 2007)

Standpipe system An arrangement of piping, valves, hose connections, and allied equipment installed in a building or structure, with the hose connections located in such a manner that water can be discharged in streams or spray patterns through attached hose and nozzles, for the purpose of extinguishing a fire, thereby protecting a building or structure and its contents in addition to protecting the occupants. (NFPA 14, 2007)

Target hazard Any occupancy type or facility that presents a high potential for loss of life or serious impact to the community resulting from fire, explosion, or chemical release.

FIRE FIGHTER *in action*

It is a beautiful spring day, and your captain lays out the plan for the day: vehicle checks, trip to the store, ladder training, lunch, preincident planning, and then supper. Your captain goes on to say that he would like to knock out several buildings that are all in a row during preincident planning, so he will break the crew into two groups to speed up the process. You feel good when he looks at you and says that he wants you to act as the lead for the second group.

1. Which of the following would normally have a pre-incident plan because it is typically considered a target hazard?

A. Single-family residence

B. Lumberyard

C. Office building

D. Hardware store

2. Buildings where the exterior walls are made of non-combustible or limited-combustible materials, but the interior floors and walls are made of combustible materials, are termed _____.

A. fire resistive

B. noncombustible

C. ordinary

D. wood frame

3. A(n) _____ is a form, provided by manufacturers and compounders (blenders) of chemicals, containing information about chemical composition, physical and chemical properties, health and safety hazards, emergency response, and waste disposal of the material.

A. Safety Data Sheet (SDS)

B. preincident survey

C. preincident plan

D. geographic information system (GIS)

4. In which type of occupancy would you most likely consider a defend-in-place strategy?

A. Residential

B. Retail

C. Hospital

D. Factory

5. Over which height is a building normally considered to be a "high-rise"?

A. 9.7 m (32 ft.)

B. 15.2 m (50 ft.)

C. 22.9 m (75 ft.)

D. 30.5 m (100 ft.)

6. When preincident planning for ladder placement, which of the following is not normally a consideration?

A. Length of ladders required

B. Type of ladder construction

C. Locations of entrances and exits

D. Overhead obstructions

You are out conducting a preincident survey at a local strip shopping centre. As you walk into one of the suites, the tenant becomes visibly upset about having the fire department in his business. He immediately begins complaining about the fire department trying to run businesses out of town through all of its rules. The tenant continues to raise his voice even more loudly as he goes on to complain about the fire department wasting fuel by driving the big apparatus for no good reason. He then goes on to state that he pays your "overpriced" salary, and he is barely getting by.

1. How would you respond to this situation?
2. Which techniques can be used to defuse the situation?
3. Which comments would you choose to address?
4. How would you explain the difference between an inspection and a preincident plan?

Vehicle Rescue and Extrication

Fire Fighter I

Knowledge Objectives

After studying this chapter, you will be able to:

- Describe the types of motor vehicles. (NFPA 5.3.7.A , p 734–735)
- Describe the four different types of alternative fuels that power motor vehicles. (NFPA 5.3.7.A , p 735–738)
- Describe the extrication tools that are used for accessing locked compartments in a motor vehicle. (NFPA 5.3.7.A , p 745–746)
- Describe how to gain access to a victim of a motor vehicle collision. (NFPA 5.3.7.A , p 746–750)

Skills Objectives

There are no skills objectives for Fire Fighter I candidates. NFPA 1001 contains no Fire Fighter I Job Performance Requirements for this chapter.

Fire Fighter II   FFII

Knowledge Objectives

After studying this chapter, you will be able to:

- Describe a vehicle's anatomy. (NFPA 6.4.1.A , p 735)
- List the hazards involved in responding to an emergency scene. (NFPA 6.4.1.A , p 738–741)
- List the hazards to look for when arriving on the scene of a vehicle extrication situation. (NFPA 6.4.1.A , p 740–741)
- Describe cribbing. (NFPA 6.4.1.A , p 741–742)
- Describe the extrication tools that are used for stabilizing, bending, cutting, and disassembling. (NFPA 6.4.1, 6.4.1.A , p 745–746)
- Describe how to gain access to a victim of a motor vehicle collision. (NFPA 6.4.1, 6.4.1.A , p 746–750)
- Describe how to disentangle a victim of a motor vehicle collision. (NFPA 6.4.1 , p 750–756)
- Describe how to remove and transport victims of a motor vehicle collision. (NFPA 6.4.1 , p 756)

© Photos.com

Skills Objectives

After studying this chapter, you will be able to perform the following skills:

- Disable the electrical system of a hybrid vehicle.
  (NFPA 6.4, 6.4.1.B , p 737)
- Perform scene size-up at a motor vehicle crash.
  (NFPA 6.4, 6.4.1.B , p 739–740)
- Mitigate the hazards at a motor vehicle crash.
  (NFPA 6.4, 6.4.1.B , p 741–742)
- Stabilize a vehicle following a motor vehicle crash.
  (NFPA 6.4, 6.4.1.B , p 743–744)
- Break tempered glass. (NFPA 6.4, 6.4.1.B , p 747–748)
- Gain access to a vehicle following a motor vehicle crash.
  (NFPA 6.4, 6.4.1.B , p 747, 749)
- Open a vehicle door. (NFPA 6.4, 6.4.1.B , p 750–751)
- Gain access and provide medical care to a victim in a vehicle.
  (NFPA 6.4, 6.4.1.B , p 750, 752)
- Displace the dashboard of a vehicle. (NFPA 6.4, 6.4.1.B , p 754–755)
- Remove the roof of a vehicle. (NFPA 6.4, 6.4.1.B , p 754, 756–757)

Additional NFPA Standards

- NFPA 1006, *Standard for Technical Rescuer Professional Qualifications*
- NFPA 1670, *Standard on Operations and Training for Technical Search and Rescue Incidents*

You Are the Fire Fighter

You have been assigned to the vehicle extrication company for the day. It begins in the usual manner with a detailed check of the vehicle extrication tools and equipment. As you look at all of the cribbing, the hydraulic spreaders, cutters, and rams, it dawns on you the gravity of your responsibility. You think of the past motor vehicle collisions that you have been to and how the vehicle extrication crew always seemed to know just the right tools and techniques to use to swiftly extricate the victims.

Then the station alerting system taps out the response for a person trapped. Bystanders are reporting that a semi has crossed the median at kilometre 78 on Highway 44, hitting a passenger van head-on. The driver and passenger of the van are trapped in the van.

1. What type of equipment will you most likely need in this situation?
2. What extrication techniques will be required?
3. What safety concerns do you have?

Introduction

Beginning fire fighters must understand the process of extrication and have some proficiency in extrication skills. Most fire departments respond to certain types of vehicle collisions. This chapter presents the knowledge and skills needed to assist rescuers in extricating victims from various types of motor vehicles involved in crashes.

While this chapter provides the knowledge and skills a beginning fire fighter needs, it does not cover all of the knowledge and skills needed by members of special extrication teams. Fire fighters who are members of special rescue teams or who will be expected to perform vehicle extrication should complete a course in rescue techniques that meets the requirements of NFPA 1006, *Standard for Technical Rescuer Professional Qualifications*, and NFPA 1670, *Standard on Operations and Training for Technical Search and Rescue Incidents*.

Types of Vehicles

Most of the vehicles on the road today are **conventional vehicles**, which use internal combustion engines for power. Internal combustion engines burn gasoline or diesel fuel to produce power. Fuel creates a hazard if it leaks after a crash. Other hazards associated with these vehicles include short-circuits and battery-acid leaks after a collision—not to mention the unknown of what people are carrying in their vehicles while they travel. The 12-volt electrical systems in these vehicles pose a minimal threat to rescuers, however.

Alternative-powered vehicles are powered by compressed natural gas (CNG), liquefied natural gas (LNG), or liquefied petroleum gas (LPG). These fuels are carried in cylinders that may be mounted in the trunk of smaller vehicles or on the roofs of buses. In particular, buses, delivery vans, and fleet sedans may be powered by CNG. CNG-powered vehicles should be identified by "CNG" stickers mounted on the front and back of the vehicle **FIGURE 25-1**. After a crash, damage to the natural gas cylinders or fuel lines could result in the escape of flammable natural gas. In addition, a fire in a natural gas–powered vehicle poses the threat of a boiling liquid/expanding vapour explosion (BLEVE). Any sign of fire in a vehicle powered by CNG should be treated with extreme caution.

Electric vehicles are propelled by an electric motor that is powered by batteries. These vehicles contain a large number of batteries that provide the needed power. Hazards posed by these vehicles include the large amount of energy stored in the batteries, the potential for electrical shock or arcing and vehicle movement. Additional hazards include off-gassing and delayed-reaction electrical fire from damaged batteries. Be aware that these vehicles use much higher voltage systems in addition to the 12-V system found in conventional vehicles.

FIGURE 25-1 The CNG sticker identifies a vehicle that is powered by compressed natural gas.

Courtesy of DOE/ NREL, Credit - Warren Gretz

Hybrid vehicles use both a battery-powered electric motor and a gasoline-powered engine. Hybrid vehicles contain all of the same hazards associated with conventional gasoline-powered vehicles, plus additional hazards linked to the electrical system. It is important for fire fighters to become familiar with the features of these vehicles, because their numbers are increasing. Vehicle manufacturers maintain websites that you can access to learn more about the operation and hazards of hybrid vehicles. More information on alternative-powered vehicles is found later in this chapter.

FIGURE 25-2 The anatomy of a vehicle.

Fire Fighter Safety Tips

© Photos.com

Some vehicles—for example, some mass-transit buses, forklifts, and school buses—run on liquefied petroleum gas, compressed natural gas, or liquefied natural gas. These vehicles, such as gasoline- and diesel-powered vehicles, can produce flammable vapours. Given this potential threat, it is important to stretch a charged hose line at the scene of all motorized vehicle/equipment extrications.

Vehicle Anatomy

Parts of a Motor Vehicle

To reduce confusion and minimize the potential for mistakes at extrication scenes, it is important to use standardized terminology when referring to specific parts of vehicles. The front of a vehicle normally travels down the road first. The hood is located on the front of the vehicle. The trunk is usually located at the rear of the vehicle.

The sides of the vehicle are referred to as either the driver's side or the passenger side. This helps avoid the confusion of determining whether it is the left side or the right side of the car (depending on position). The steering wheel is always located on the driver's side.

Vehicles contain "A", "B", and "C" **posts**, which are the vertical support members of a vehicle that hold up the roof and form the upright columns of the passenger compartment. The **"A" posts** are located closest to the front of the vehicle; they form the sides of the windshield. In four-door vehicles, the **"B" posts** are located between the front and rear doors of a vehicle. In some vehicles, these posts do not reach all of the way to the roof of the vehicle.

In four-door vehicles, the **"C" posts** are located behind the rear doors—specifically, behind the rear passenger windows. In two-door vehicles, the C post is the rear post.

The hood covers the engine compartment. The structure that divides the engine compartment from the passenger compartment is called the **bulkhead** or firewall. The passenger compartment includes the front and back seats. This part of a vehicle is also sometimes called the occupant cage, or the occupant compartment **FIGURE 25-2**. The trunk may be exposed to the back seats, as in a station wagon, or enclosed, as in a sedan.

Motor Vehicle Frames

Two types of vehicle frames are commonly found in today's motor vehicles: conventional-frame construction and **unibody (unit-body)** construction.

Conventional-frame construction uses beams to fabricate the load-bearing frame of a vehicle. The engine, transmission, and body components are then attached to this basic conventional frame. This type of frame construction is found primarily in trucks and larger sport-utility vehicles (SUVs), but is rarely present in smaller passenger cars. Platform-frame construction provides a structurally sound place for stabilizing the vehicle and an anchor point for attaching cables or extrication tools.

Unibody construction, which is used for most modern passenger cars, combines the vehicle body and the frame into a single component. By folding multiple thicknesses of metal together, a column can be formed that is strong enough to serve as the frame for a lightweight vehicle. Unibody construction has the advantage of enabling auto manufacturers to produce lighter-weight vehicles. When extricating a person from a unibody vehicle, remember that these vehicles do not have the frame rails that are present with conventional frame–constructed vehicles.

Alternative-Powered Vehicles

Several types of alternative-powered vehicles are currently either in production or in the planning stages. Alternative-powered vehicles may be powered by any of four different types of fuels: blended liquid fuels, compressed gases, electric and gasoline hybrid combination, and fuel cells. Following a crash or fire, these vehicles present hazards that are not encountered in incidents involving conventionally powered vehicles. It is important for rescuers to recognize the hazards these vehicles pose both to rescuers and to victims, and to be familiar with the additional steps needed to mitigate these hazards.

Blended Liquid Fuel–Powered Vehicles

The most common blended liquid fuel is a combination of gasoline and another flammable liquid. Two types of alcohol—methanol and ethanol—are commonly mixed with gasoline to produce a blended fuel. The mixture of 10 percent alcohol and 90 percent gasoline is sold in many cities in Canada as a measure to reduce pollution. Diesel fuel may also be blended with other combustibles to produce a blended diesel fuel.

Biofuels are derived from agriculturally produced products such as grains, grasses, or recycled animal oils. They are processed to produce a type of alcohol that is added to gasoline to produce a blended fuel. Biofuels can also be produced to power diesel vehicles.

Vehicles operating on these blended fuels are identified by special placards or designations. They are singled out for attention primarily because of their environmental benefits. Currently, biofuel-powered vehicles account for less than 3 percent of all vehicles in Canada.

Crashes involving vehicles that rely on blended fuels are handled in the same manner as crashes involving vehicles powered by conventional fuels. In the event of a fire involving blended fuel, it is necessary to use an alcohol-resistant foam as a fire-suppressing agent. Other types of foam will break down when exposed to alcohol.

■ Compressed Gas–Powered Vehicles

Three forms of compressed gas are used to power vehicles. CNG is commonly stored in steel, aluminum, or carbon fibre–wound tanks, under pressures ranging from 20,685 to 24,822 kilopascals (3000 to 3600 pounds per square inch [psi]). Many cities power some of their municipal bus fleets with CNG. These vehicles can be identified by the pregnant shape of the rooftop.

LNG is stored at very low temperatures. Storage tanks for LNG are double walled and must be insulated to keep the liquefied gas cold. The use of LNG is not as common as the use of CNG.

LPG (also called propane) is the third type of compressed gas used to power vehicles. It is stored in a liquid state at a pressure of 345 kPa (50 psi). LPG-powered vehicles are often found in industrial settings; for example, LPG is used to power forklifts in factories.

When responding to crashes involving vehicles powered by compressed gases, rescuers need to able to identify and mitigate the special hazards posed by these vehicles. Look for signage or placards on the vehicle identifying it as being powered by a compressed gas. If a compressed gas is present, check for the presence of leaking cylinders or hoses. If you suspect that a leak is present, keep all sources of ignition away from the vehicle and take measures to stop the leak, if possible. If LNG is leaking, avoid spraying a stream of water on the escaping liquid, as this action heats the very cold liquid; it may then convert to a gaseous state with explosive force. It is acceptable to spray a fog stream on an escaping cloud of vapour, however. Follow the procedures described in the hazardous materials chapters for mitigating hazardous materials. Ensure that a fire fighter in full protective equipment is standing by with a charged hose line that is least 38 millimetres (1.5 inches) in diameter. In the event of a fire, keep the gas cylinders cool to avoid a BLEVE.

■ Hybrid and Electric Vehicles

A hybrid vehicle has two separate systems to power the drive train **FIGURE 25-3**. One part of the power system is a fuel-driven power train. This part of the drive train contains the same parts as a conventional vehicle. The second part of the power system is powered by an electrical motor. The electrical power system consists of an electrical motor, a high-voltage battery pack, a DC/DC converter (and possibly an AC/DC inverter), and electrical cables to connect these components. In some hybrid vehicles, the high-voltage batteries are charged by the onboard gasoline or diesel engine. In others, the batteries can also be charged from an external electrical power source. Usually hybrid vehicles are identified by placards such as the one shown in Figure 25-3.

The components of these vehicles continue to change, and the design and location of each component vary from one manufacturer to another. It is important to keep up with changes in vehicle architecture, as they affect the response to incidents involving the vehicles. Most manufacturers offer a wealth of information that can be viewed or downloaded from the manufacturer's website.

When responding to crashes involving hybrid vehicles, you will encounter the hazards normally associated with gasoline- or diesel-powered vehicles, as these hazards are still present in hybrid vehicles. The additional hazards posed by the electrical drive train are similar to those you would encounter around other high-voltage electrical hazards.

The high-voltage battery pack in a hybrid vehicle is usually located under or behind the rear seats or the trunk. High-voltage batteries are protected by a strong structural cage and located out of the way for most extrication evolutions. Hybrid vehicles have automatic shut-offs that are designed to disable the high-voltage electrical system in the event of a crash. According to the current manufacturers' recommendations, water remains the extinguishing agent of choice in case of a fire involving a hybrid vehicle.

FIGURE 25-3 A hybrid sticker indicates that this vehicle is powered by a gasoline engine and an electric motor, and it contains a high-voltage battery or batteries.

As part of the electrical power system, brightly coloured high-voltage cables are routed from the high-voltage battery pack to the inverter and electric motor, which are located in the engine compartment. These cables are usually located beneath the passenger compartment toward the centre of the vehicle, so they do not interfere with most extrication procedures. Because these cables carry high voltage, however, they pose a risk to rescuers if responders try to cut them without first disabling the electrical system. Look for placards and high voltage labels that will identify a potential shock hazard.

Although the procedures for disabling the electrical system of a hybrid vehicle vary from one manufacturer to another, generally they follow similar steps. To disable the electrical system of a hybrid vehicle, follow the steps in **SKILL DRILL 25-1**:

1. Identify the vehicle's propulsion system. Is it conventional, hybrid, or CNG?
2. Immobilize the vehicle by chocking the wheels. Set the parking brake. Place the vehicle in park. (**STEP 1**)
3. Disable both the high-voltage and low-voltage systems by shutting off the vehicle's ignition (power button or conventional key) and disconnecting or cutting the 12-V battery cable (usually the negative). For vehicles equipped with a proximity key, remove the key at least 4.9 metres (16 feet) from the vehicle to prevent the possibility of unintentional restart. (**STEP 2**)

Alternate disabling methods involve removing fuses or the high-voltage manual service disconnect. Be aware that some manufacturers require the use of electrical personal protective equipment (PPE) for the removal of manual service disconnects, while other manufacturers do not recommend the use of service disconnects at all for emergency responders. Check with the manufacturer's emergency response guide or the NFPA's *Electric Vehicle Emergency Field Guide* for alternate disabling methods or before attempting to use the service disconnect.

Electric vehicles are powered by an electric motor and batteries, but do not contain a gasoline- or diesel-powered engine. Instead, these vehicles rely solely on batteries, which must be recharged by plugging them into an external power source. The procedures used to deal with electric vehicles involved in collisions are nearly the same as those used in dealing with a hybrid vehicle.

Review the manufacturers' fact sheets regularly to keep up-to-date with the steps needed to disable the electrical systems in hybrid vehicles. The National Fire Protection Association (NFPA) also offers its *Electric Vehicle Emergency Field Guide* to assist fire fighters in responding safely to motor vehicle accidents involving electric vehicles.

■ Fuel Cell–Powered Vehicles

A fuel cell is a device that generates electricity through a chemical reaction between hydrogen gas and oxygen gas to

---

SKILL DRILL 25-1 | Disable the Electrical System of a Hybrid Vehicle | **FFII**
Fire Fighter II (NFPA 6.4.1)

1 Identify the vehicle's propulsion system. Is it conventional, hybrid, or CNG? Immobilize the vehicle by chocking the wheels. Set the parking brake. Place the vehicle in park.

2 Disable both the high-voltage and low-voltage systems by shutting off the vehicle's ignition (power button or conventional key) and disconnecting or cutting the 12-V battery cable (usually the negative). For vehicles equipped with a proximity key, remove the key at least 4.9 m (16 ft.) from the vehicle to prevent the possibility of unintentional restart.

© LiquidLibrary

produce water. The purpose of fuel cells is not to produce water, however, but rather to produce electricity that can then be used to drive an electric motor. Fuel cell–powered vehicles are essentially electric vehicles with a different device for generating the electricity. The major components needed for a fuel cell–powered vehicle are a hydrogen-gas storage tank, a high-output battery, a group of <u>fuel cells</u> called a <u>fuel cell stack</u>, and an electric motor. The use of fuel cells is being advocated because they produce less pollution than internal combustion engines.

Several issues remain to be worked out before fuel cell–powered vehicles are likely to be widely produced. Because hydrogen gas contains one-third the amount of energy per volume that gasoline does, it is difficult to store enough hydrogen gas to propel the vehicle as far as gasoline vehicle can travel on a full tank. Several ways to store hydrogen gas are currently being explored.

Fire Fighter Safety Tips

© Photos.com

Hybrid vehicles and electric vehicles contain powerful batteries, and their electrical systems produce a higher voltage than would be encountered in a conventional-powered vehicle. Be alert for special electrical hazards when you respond to incidents involving these vehicles. Also be aware that batteries may be placed anywhere inside the vehicle—from under the hood to the trunk to behind the backseat. See the section on alternative-powered vehicles.

Responding to the Scene

The first step in the vehicle extrication process is response. To perform extrication, you need to safely and efficiently arrive on the scene. Safe response includes picking the best route for the time of the day, driving in a safe manner, and keeping in mind the limitations of your vehicle and its warning devices. Always fasten your seat belt. Follow the same safety practices while responding to a motor vehicle crash that you would practise when responding to a structural fire call.

When responding to an emergency, evaluate the dispatch information that you have received, because it may help you to anticipate the necessary types of equipment and the procedures that you may need to perform. Use this time to mentally review the scenarios that you might encounter. Listen carefully for any directions that your officer may give during the response phase.

Arrival and Size-Up of the Scene

The next step in the extrication process is arrival and size-up. After arriving at the scene of a motor vehicle crash, it is important to assess the hazards present and to determine the scope of the incident and the need for any additional resources.

FIRE FIGHTER II Tips   **FFII**

Before doing any type of work on a vehicle that has been involved in an accident, scan the entire area for hazards. Did the vehicle hit an electric box or rupture a gas line? Hazards are everywhere, and a quick scan might save everyone on the scene from further injury.

■ Traffic

The decision of where to position emergency vehicles should take into account the safety of emergency workers, the victims, and the motorists travelling along the road. Whenever possible, place emergency vehicles in a manner that will ensure safety and avoid disrupting traffic any more than necessary; however, do not hesitate to request that the road be closed, if necessary. Sometimes the most important action you need to take at the scene of a vehicle crash is to slow, stop, or divert the flow of traffic before proceeding with additional actions. Remember: safety first!

Position large emergency vehicles so that they provide a barrier against motorists who fail to recognize or heed emergency warning lights. Many departments place apparatus at an angle to the crash, pointing away from the rescue or crash scene **FIGURE 25-4**. This position helps to push the apparatus to the side of a crash in the event that the emergency apparatus is struck from behind. Pylons or flares can be placed to direct motorists away from the crash **FIGURE 25-5**. Call for law enforcement to assist in traffic control whenever needed.

Fire fighters need to be readily visible at a crash scene. PPE should be bright to help ensure fire fighters' visibility during daylight hours; PPE that is used at night should be equipped with reflective material to increase visibility in the darkness. PPE must be worn at all motor vehicle crashes. Any time you are operating in a roadway or are close to traffic, you are required to wear a bright-coloured reflective safety vest that meets the provincial or territorial Workplace Safety and

FIGURE 25-4 Many fire departments place an apparatus at an angle to the crash.

FIRE FIGHTER II

FIGURE 25-5 Traffic cones or flares can be placed to direct motorists away from the crash.

Courtesy of David Jackson, Saginaw Township Fire Department

Health Regulations. This standard requires the vest to contain 0.3 square metres (450 square inches) of high-visibility fabric and 0.1 sq. m (201 sq. in.) of retro-reflective material.

Before exiting fire apparatus at an emergency scene, be alert for any vehicles that might cause injury to fire fighters. Do not assume that motorists will always heed the warning lights. Let law-enforcement personnel coordinate traffic control.

The incident commander (IC) will usually perform a size-up of the scene by conducting a 360-degree walk-around of the scene. During this size-up, the IC evaluates the hazards present and determines the number of crash victims. Using the information obtained from the scene size-up, the IC can create an action plan and call for additional resources, if needed.

■ Fire Hazards

Look for spilled fuel and other flammable substances. Motor vehicles carry a variety of fuels and lubricants that might pose fire hazards. Also look for the presence of fire, because a short in the electrical system or a damaged battery may cause a postcrash fire by releasing sparks and igniting spilled fuel. These fires may trap the occupants of the vehicle and require fire suppression.

Fire Fighter Safety Tips
© Photos.com

The locations of vehicle batteries can vary widely. If you open the hood to cut the battery cables and do not find a battery, look in the not-so-obvious locations. Batteries may be behind the back seat, in the trunk, and inside front wheel wells.

■ Electrical Hazards

Downed and low-hanging power lines represent an electrical hazard. Look closely to determine whether the crash has damaged any electrical power poles. Downed and low-hanging power lines may be either clearly visible or difficult to see, but they inevitably create a deadly hazard for rescuers, the victims of the crash, and bystanders.

As noted earlier, hybrid vehicles also contain high-voltage batteries and electrical cables that require special handling because they may pose an electrical hazard.

Fire Fighter Safety Tips
© Photos.com

High-intensity discharge (HID) headlights are becoming more common on vehicles as standard equipment or as after-market modifications. HID headlights discharge several thousand volts of electricity. If the HID bulb is broken or removed, you may be exposed to an electrical hazard.

■ Other Hazards

Environmental conditions can lead to unique hazards at a crash scene. Crashes that occur in rain, sleet, or snow, for example, present an added hazard for rescuers and the victims of the crash.

Crashes that occur on hills are harder to handle than those that occur on level ground.

Assume all vehicles are carrying hazardous materials until proven otherwise. Assess the scene for hazardous material placards, unusual odours, or leaking liquids. More information about hazardous materials is found in the Hazardous Materials chapters.

Be especially alert for the presence of infectious bodily substances. Be prepared for the presence of blood and exercise universal precautions. Specifically, do not let blood or other bodily fluids come in contact with your skin, and wear gloves that will protect you from both contaminated fluids and sharp objects that are present at a crash site. If you or your clothes become contaminated, report the contamination, document it, and then clean and wash the affected clothes and equipment.

Threats of violence may be present at some crash scenes. In particular, intoxicated people or those who are upset with other motorists may pose a threat to you or to other people present at the scene. Be alert for weapons that are carried in vehicles.

Occasionally, animals become a hazard at crash scenes. Dogs and other family pets may be protective of their owners and threaten rescuers. Farm animals or horses that have been involved in a crash may need care. You may need to call in specialized resources to assist with this type of incident.

To perform the scene size-up at a motor vehicle crash, follow the steps in **SKILL DRILL 25-2** :

1. Position emergency vehicles to protect the crash scene and the rescuers. Take any additional actions needed to prevent further crashes. (**STEP ❶**)

2. Perform a quick initial assessment from the first-arriving vehicle, establish command, and give a brief initial radio report. (**STEP ❷**)

3. Perform a 360-degree walk-around looking for potential hazards. Look for overhead hazards and hazards under the vehicles, and determine the stabilization needed to prevent further movement of the vehicles involved in the incident. (**STEP ❸**)

FIRE FIGHTER II

SKILL DRILL 25-2   Performing a Scene Size-up at a Motor Vehicle Crash   FFII

Fire Fighter II (NFPA 6.4.1)

1 Position emergency vehicles to protect the crash scene and the rescuers. Take any additional actions needed to prevent further crashes.

2 Perform a quick initial assessment from the first-arriving vehicle, establish command, and give a brief initial radio report.

3 Perform a 360-degree walk-around looking for potential hazards. Look for overhead hazards and hazards under the vehicles, and determine the stabilization needed to prevent further movement of the vehicles involved in the incident.

4 Determine the number of patients, the severity of their injuries, and the amount of entrapment. Give an updated report and call for additional resources, if needed.

5 Establish a secure working area and an equipment staging area.

6 Direct personnel to perform initial tasks.

© LiquidLibrary

FIRE FIGHTER II

4 Determine the number of patients, the severity of their injuries, and the amount of entrapment. Give an updated report and call for additional resources, if needed. (**STEP 4**)

5 Establish a secure working area and an equipment staging area. (**STEP 5**)

6 Direct personnel to perform initial tasks. (**STEP 6**)

Stabilization of the Scene

The next step of the extrication process is scene stabilization. This step consists of reducing, removing, or mitigating the hazards at the scene. These hazards should have been identified during arrival and scene size-up. The order in which these hazards are addressed will depend on the specific conditions at a scene and the amount of risk that each hazard poses.

■ Traffic Hazards

The hazards posed by traffic need to be handled quickly before they lead to additional crashes. In some cases, the most critical issue is slowing or stopping traffic before further damage and injuries can occur. As mentioned previously, emergency vehicles can be used to block traffic from the crash scene. Pylons and flares can slow motorists and direct them in a safe pattern around the crash scene. Avoid positioning emergency vehicles in a manner that confuses oncoming motorists.

Traffic hazards are best handled by the appropriate law-enforcement agency. This is their area of expertise, and giving this duty to law-enforcement personnel leaves fire department personnel free to handle other extrication tasks. All fire fighters need to work together with law-enforcement officials to control traffic in a manner that is safe for emergency workers, victims of the crash, and motorists. If law-enforcement officials are not on the scene when you arrive, verify that they are aware of the incident and that they have been dispatched.

■ Fire Hazards

Because there is a significant risk of spilled fuel in many motor vehicle crashes, the suggested operating guideline (SOG), or standard operating procedure (SOP), should be to advance a charged hose line close to the damaged vehicles to protect them from the threat of fire. This line, which should be at least 38 mm (1.5 in.) in diameter, should be staffed by a trained fire fighter in full turnout gear. It protects both rescuers and victims of the crash. Crashes that pose large fire hazards or actual fires may require additional fire-suppression resources, which should be requested as soon as possible. Small fuel spills can be handled by using an absorbent material to remove the fuel from the area around the damaged vehicle. See the chapter "Fire Attack Lines and Foam" for the steps required to advance an attack line. See the chapter "Fire Suppression" for more information on extinguishing vehicle fires.

To mitigate fire hazards at a motor vehicle crash, follow the steps in **SKILL DRILL 25-3**:

1. Don PPE, including self-contained breathing apparatus (SCBA). Attach a regulator if indicated. (**STEP ❶**)
2. Assess the incident scene for hazards.
3. Communicate with other crews and command.
4. Advance a charged hose line to the proximity of the vehicle. Extinguish any fires. (**STEP ❷**)
5. Open the vehicle's hood. (**STEP ❸**)
6. Mitigate electrical sources of ignition by cutting the 12-V battery cables. In hybrid vehicles, it is also necessary to disable the high-voltage electrical system. Follow the manufacturer's advice for disabling the high-voltage electrical system. (**STEP ❹**)

■ Electrical Hazards

Disconnecting the vehicle's electrical system can mitigate electrical hazards posed by a damaged vehicle. Follow your department's SOG/SOP for disconnecting the electrical system. When deciding whether to pursue this path, you must weigh the need to disconnect the electrical system against the advantage of being able to operate electrically powered components of the vehicle. Additional information on handling the electrical hazards posed by hybrid and electric vehicles is presented later in this chapter.

Stabilizing electrical hazards posed by downed electrical wires is essential before fire fighters attempt to approach the crash site. At times it may be necessary to instruct victims of a crash to remain in their vehicles until the power can be turned off. Disconnecting the power to damaged electrical lines is a job for properly trained members of the power company. Do not attempt to approach downed electrical lines until the utility company personnel have turned off the electricity. Although

it is frustrating to delay extrication and treatment because of downed power lines, this step is essential to avoid potentially life-threatening injuries to both rescuers and victims of the crash.

■ Other Hazards

Stabilize the environmental hazards present at a crash scene. At scenes where the temperature is hot, fire fighters may need to provide some shade for a victim during a prolonged extrication. In addition, rescue crews require rehabilitation more quickly in a hot environment. At scenes where the temperature is cold, provide covering to help the victim maintain body warmth.

Many crashes occur at night. At such crashes, it is important to provide adequate lighting, so that rescuers can work quickly and safely. The skill drill "Illuminating an Emergency Scene" in the chapter "Salvage and Overhaul" describes the steps required to light an emergency scene.

Be careful when conditions are wet or icy, because it is important to prevent slips and falls. Sprinkling sand, kitty litter, or the like may help to give solid footing to rescuers.

Vehicle crashes are often characterized by a variety of sharp objects; for example, mangled pieces of sheet metal, plastic, and glass all have sharp edges that pose significant hazards for rescuers and victims of the crash. To reduce the chance of injury, rescuers should wear proper personal protective clothing. Another way to reduce the threat of injury from sharp edges is to cover them with some type of padding— for example, with short sections of old fire hose. Be especially careful when moving victims, making sure that they are protected from contact with any sharp objects.

In incidents involving animals, it may be necessary to deal with the animals before proceeding with other activities. In particular, dogs can be very possessive of their owners. If the dog's owner is injured, remove the dog from the crash site, and place it with an uninjured family member or other responsible person. If a pet is injured, call the agency that is responsible for transporting animals to a veterinary clinic. As part of preincident planning, you should learn who in your community is responsible for caring for large animals at rescue scenes.

Cribbing

Unstable objects pose a threat to both rescuers and victims of a crash. Such objects need to be stabilized before fire fighters approach the scene. Most often, the objects that need to be stabilized are the damaged vehicles.

Vehicles that end up on their wheels need to be stabilized vertically with **cribbing**. Cribbing consists of short lengths of sturdy timber (10 by 10 centimetres [4 by 4 inches], 10 by 46 cm [4 by 18 in.], and 10 by 61 cm [4 by 24 in.], as well as other lengths and sizes) **FIGURE 25-6**. After cribbing has been placed against the wheels, a vehicle may still be able to move because of the give and motion that the suspension system causes when rescuers get into the vehicle and victims are extricated from the vehicle. This type of instability can cause further injuries to the victims of the crash.

FIRE FIGHTER II

SKILL DRILL 25-3　Mitigating the Hazards at a Motor Vehicle Crash　FFII

Fire Fighter II (NFPA 6.4.1)

1 Don PPE, including SCBA. Attach a regulator if indicated.

2 Assess the incident scene for hazards. Communicate with other crews and command. Advance a charged hose line to the proximity of the vehicle. Extinguish any fires.

3 Open the vehicle's hood.

4 Mitigate electrical sources of ignition by cutting the 12-V battery cables. Recognize and disable electrical systems in hybrid vehicles.

FIRE FIGHTER II

The suspension system of most vehicles can be stabilized with **step chocks**, which are shaped like stair steps and are placed under both sides of the vehicle. Place one step chock toward the front of the vehicle and a second step chock toward the rear of the vehicle **FIGURE 25-7**. Repeat this process on the other side of the vehicle. Once the step chocks are in place, the tires can be deflated by pulling out the valve stems to create a stable vehicle.

If step chocks are not available or are not the right size, you can build a box crib. To do so, place cribbing at right angles to the preceding layer of cribbing **FIGURE 25-8**.

After a crash, some vehicles might come to rest on their roofs or sides. These vehicles are very unstable, and placing the slightest amount of weight on them can cause them to move. Overturned vehicles can be stabilized using box cribs or step chocks on each end of the vehicle.

Besides cribbing, **wedges** are used to snug loose cribbing under the load or, when using rescue-lift air bags, to fill the void between the crib and the object as it is raised **FIGURE 25-9**. Only lift or raise the vehicle enough to place additional cribbing under it; remember the saying, "Lift an inch, crib an inch." Wedges should be the same width as the cribbing, with the tapered end no less than 6 mm (0.25 in.) thick. When the ends are feathered to less than 6 mm (0.25 in.), the end will commonly fracture under a load.

FIGURE 25-6 The application of cribbing in the front and back of the vehicle wheels stabilizes the vehicle.

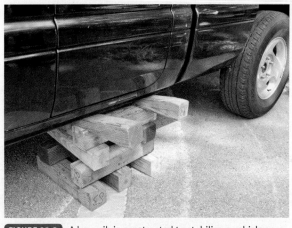

FIGURE 25-8 A box crib is constructed to stabilize a vehicle.

FIGURE 25-7 Step chocks are used to stabilize a vehicle.

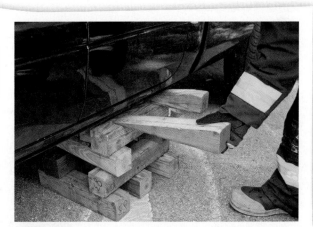

FIGURE 25-9 Wedges are used to snug loose cribbing when the void space is not big enough to accommodate a regular-sized crib.

To stabilize a vehicle following a motor vehicle crash, follow the steps in **SKILL DRILL 25-4** :

1. Don PPE, including face protection.
2. Minimize the hazards to rescuers and victims.
3. Chock both sides of one tire to prevent the vehicle from rolling by placing one chock in front of a wheel and a second chock in back of the wheel. (**STEP 1**)
4. Consider deflating tires for added stability. Assess the need for step chocks for additional stability. (**STEP 2**)
5. Turn off the ignition, and remove the key or fob. (**STEP 3**)
6. Place the gear shift in park, and apply the parking brake. (**STEP 4**)

Rescue-Lift Air Bags

Rescue-lift air bags are pneumatic-filled bladders made out of rubber or synthetic material **FIGURE 25-10**. They are used to lift an object or to spread one or more objects away from each other to assist in freeing a victim. Rescue-lift air bags are never

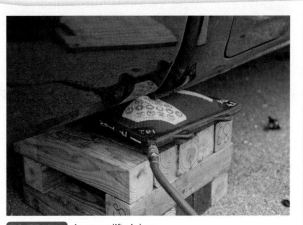

FIGURE 25-10 A rescue-lift air bag.

used to shore or stabilize a vehicle by themselves; cribbing is always required in conjunction with rescue-lift air bags.

Rescue-lift air bags are often used to lift a vehicle or object off a victim. Fire fighters should use extreme caution when

SKILL DRILL 25-4 Stabilizing a Vehicle Following a Motor Vehicle Crash **FFII**

Fire Fighter II (NFPA 6.4.1)

1 Don PPE, including face protection. Minimize hazards to rescuers and victims. Chock both sides of one tire to prevent the vehicle from rolling by placing one chock in front of a wheel and a second chock in back of the wheel.

2 Consider deflating tires for added stability. Assess the need for step chocks for additional stability.

3 Turn off the ignition and remove the key or fob.

4 Place the gear shift in park, and apply the parking brake.

applying this technique and should adhere to all safety precautions outlined in the manufacturer's instructions for the rescue-lift air bag. Cribbing must be used whenever fire fighters are lifting a load, because instability can occur from weight shifts or the rescue-lift air bags may fail under a load. Given these risks, it is a must—not an option—that cribbing be used in conjunction with rescue-lift air bags.

Several types of pneumatic rescue-lift air bags are in common use today: low pressure, medium pressure, and high pressure. None of these devices should be used without first properly blocking the wheels opposite of the lift air bag and cribbing the vehicle or object as it is lifted. These safety

precautions will prove invaluable in case the rescue-lift air bag suffers a catastrophic failure.

As they age, rescue-lift air bags become more prone to failure. Given this fact, they should be tested regularly according to the manufacturer's recommendations. Rescue-lift air bags that are 10 years of age or older have generally reached the end of their useful life.

Low-Pressure Lift Air Bags

Low-pressure lift air bags are commonly used for recovery operations and are sometimes used for vehicle rescue operations. These devices come in many sizes and shapes; square air bags, however, offer greater stability. Because of their lightweight construction, low-pressure lift air bags can be less stable (at least until they are fully inflated) than high-pressure air bags, which have a lower height lift and stronger construction.

Medium-Pressure Lift Air Bags

Medium-pressure lift air bags are designed so that they contain either two or three cells. For general vehicle rescue, these devices are not appropriate. Medium-pressure lift air bags are more suitable for aircraft, medium or heavy truck, or bus rescue, and for recovery work.

High-Pressure Lift Air Bags

The rescue-lift air bags most frequently used by fire fighters are high-pressure lift air bags. These devices have a very sturdy construction and are generally made of vulcanized rubber mats that are reinforced by steel or other material woven into a fibre mat and then covered with rubber. When using this type of air bag, cribbing remains essential to provide a safe working environment.

Fire Fighter Safety Tips   © Photos.com

- Never stack high-pressure lift air bags more than two units high.
- Do not use a rescue-lift air bag to pull a steering column.
- Do not use a rescue-lift air bag as the sole means to stabilize a vehicle; cribbing must be the primary stabilizer.
- Never operate the rescue-lift air bag system without having been properly trained and fully understanding how the system works.
- When stacking rescue-lift air bags, the largest size should be on the bottom and the smallest on top. The bottom rescue-lift air bag should be inflated first.
- Place a sheet of plywood on the ground under the air bag to protect it.
- Do not use boards or plywood between or above rescue-lift air bags.
- Clean rescue-lift air bags by following the manufacturer's recommendations.
- Test rescue-lift air bags regularly.
- Never store a rescue-lift air bag near gasoline.

Principles of Gaining Access and Disentangling the Victim

Before discussing the steps of gaining access and disentanglement, it is important to understand the principles involved with these steps of extrication. Although many different techniques and methods are used to gain access and disentangle victims from crashed vehicles, the objective of these techniques is accomplished by performing four functions:

- Stabilize or hold an object or vehicle. Vehicles are stabilized when cribbing is used to keep them from moving.
- Bend, distort, or displace—for example, bending a vehicle door back to get it out of the way.

- Cut or sever—for example, cutting a roof.
- Disassemble—for example, removing a vehicle door by unbolting the door hinges.

FIRE FIGHTER II Tips   **FFII**

Extrications are dynamic events, especially on roadways, and fire fighters must be constantly aware of what is going on around them. Maintain situational awareness at all times.

Tools Used for Extrication

Many different tools can be used for extrication. It is important to understand the purpose of each tool, because a certain tool will be more appropriate in certain circumstances. Fire fighters need to know which tools are available in their department and how to operate each of these tools.

Cribbing, rescue-lift air bags, step clocks, high-lift jacks, and stabilization jacks are all types of stabilization devices. When using these tools, be certain that they are placed firmly under the vehicle and that the vehicle is stable before attempting to enter it.

Hand tools and power tools are used to displace, bend, and distort metal parts of a vehicle to gain access to a victim and to disentangle a victim from the vehicle. Hand tools include pry axes, short pry bars, long pry bars, hacksaws, screwdrivers, hammers, and Halligan tools; these tools can be used to bend sheet metal, so that it can be pulled away from a victim. In addition, manual hydraulic rams and manual hydraulic spreaders can exert a large amount of force onto the vehicle. Powered hydraulic rams and powered hydraulic spreaders use a hydraulic pump powered by an electric motor or a gasoline engine. Because newer vehicles are designed to tear and rip (as a safety feature), sometimes hydraulic power is not the best option for extrication purposes.

Hand tools that can be used for cutting include axes, bolt cutters, cable cutters, hacksaws, and manual hydraulic cutters. Each of these cutting tools is designed to cut certain materials. If you try to use a tool on an inappropriate material, you will not only fail to cut the material but also may damage the tool and injure the victim or yourself. In case of powered equipment failure, hand tools are an appropriate backup. Training sessions should include hand tools use and practise to show advantages and limitations.

A variety of power tools for cutting or severing are available: some are air powered, some are electrically powered, and others are hydraulically powered. Air chisels are powered by pressurized air from portable SCBA air cylinders or from a cascade system mounted on a vehicle **FIGURE 25-11**. Air chisels are efficient at cutting sheet metal and some types of plastic; they can also be used to cut metal posts.

A variety of powered saws are used for extrication. Some electric saws require power cords, whereas others operate on battery packs. Powered saws produce different types of motion;

FIGURE 25-11 An air chisel is useful for cutting sheet metal.

for example, reciprocating saws move back and forth very quickly, whereas rotary saws move in a circular motion. One of the most commonly used extrication tools is the powered hydraulic cutter, which can quickly cut through the roof posts of a vehicle. Whenever fire fighters use power and hydraulic tools, they should cover the victim to protect him or her.

Specialized equipment, such as the hand-operated come along, is also used during extrication. You can learn more about the proper use of this and other specialized equipment through special rescue courses or in-service training.

Sometimes the most effective way to remove a part of a vehicle is to disassemble it. This method of extrication has the advantage of being controlled; it may also provide an added degree of safety for the victim and rescuers. A disadvantage is that it is very slow and may not always be possible. The types of tools needed for disassembly are the same tools that are used for construction and repair of vehicles. A mechanic's tool kit that contains an assortment of wrenches and screwdrivers will enable a qualified rescuer to disassemble some parts of a vehicle efficiently and safely.

Gaining Access to the Victim

■ Open the Door

After stabilizing the vehicle, the simplest way to access a victim of a crash is to open a door. Try all of the doors first—even if they appear to be badly damaged. Remember the first rule of forcible entry: "Try before you pry." It is an embarrassing waste of time and energy to open a jammed door with heavy rescue equipment, only to discover that another door can be opened easily and without any special equipment.

Attempt to unlock and open the least damaged door first. Make sure the locking mechanism is released. Then try the outside and inside handles at the same time, if possible.

Near-Miss REPORT

Report Number: 10-0000452

Synopsis: Unsecured vehicle rolls down hill during extrication.

Event Description: We responded to a motor vehicle accident with injury. The victim in the driver's seat was restrained. A small amount of extrication was required. The fire crew was already on the scene and stated they needed a medic in the vehicle. The vehicle was off the road on top of a hill, with a pond on both sides of the hill. I arrived on the scene and got in the backseat to hold the victim's c-spine and do patient care during extrication.

The vehicle had been chocked, but no one had checked whether it was in park or whether the emergency brake was on. As the extrication started, the vehicle started moving down the hill toward the pond. The patient and I were in the vehicle. It stopped prior to making it down the hill, and we finished the extrication and transported the patient to the hospital.

Lessons Learned

- Scene safety—Always be aware of surroundings.
- Teamwork—Work together and remind one another of possible safety issues.
- Command—Just because the scene has been established and units are on scene, it does not mean that all safety aspects have been taken care of.
- Training—Always chock vehicles and make sure they are in park, with the emergency brake set.

FIRE FIGHTER II

■ Break Tempered Glass

If a victim's condition is serious enough to require immediate care and you cannot enter through a door, consider breaking a window. You cannot easily enter through the windshield, however; it is made of **laminated glass**, which is more difficult to break. In contrast, the side and rear windows are made of **tempered glass**, which will break easily into small pieces when hit with a sharp, pointed object such as a spring-loaded centre punch, a multipurpose tool, or the point of an axe **FIGURE 25-12**. Because these windows do not pose as great of a safety threat, they can be fire fighters' primary access route to the victim.

If you must break a window to unlock a door or gain access, cover the victim and try to break a window that is farthest away from the victim; if the victim's condition warrants your immediate entry, however, do not hesitate to break the closest window. Small pieces of tempered glass fragments are sharper than regular glass. Use caution with your victim around glass fragments. Advise Emergency Medical Services (EMS) personnel if a victim is covered with broken glass, so that they can notify the hospital emergency department.

After breaking the window, use a tool to remove broken glass, or as a last resort, use your gloved hands to pull the remaining glass out of the window frame, so that it does not fall onto any victims or injure any rescuers. If you use something other than a spring-loaded centre punch to break the window, always aim for a low corner.

To break tempered glass, follow the steps in **SKILL DRILL 25-5**:

1. After donning complete PPE, including face and eye protection, select a tool. (**STEP 1**)
2. Ensure that the victim and other fire fighters are as properly protected as possible. (**STEP 2**)
3. Place the tool in the lower corner of the window and sharply apply pressure until the spring is activated. (**STEP 3**)
4. Remove loose glass around the window opening. (**STEP 4**)

Once you have broken the glass and removed the pieces from the frame, try to unlock the door again. Release the locking mechanism, and then use both the inside and outside door handles at the same time. This technique will often force a jammed locking mechanism to open the door, even when the door appears to be badly damaged.

Breaking the rear window will sometimes provide an opening large enough to enable a rescuer to gain access to the victim, if there is no other rapid means for doing so. The simple techniques of opening a door and breaking the rear window will enable fire fighters to gain access to most vehicle crash victims, even those in an upside-down vehicle.

Follow the steps in **SKILL DRILL 25-6** to gain access to a vehicle, following a motor vehicle crash. The steps in this skill drill may have to be modified depending on the situation.

1. Don PPE, including face protection.
2. Minimize hazards to rescuers and victims.
3. Ensure immobilization of the vehicle: chock wheels, turn off the ignition, remove smart keys away from the

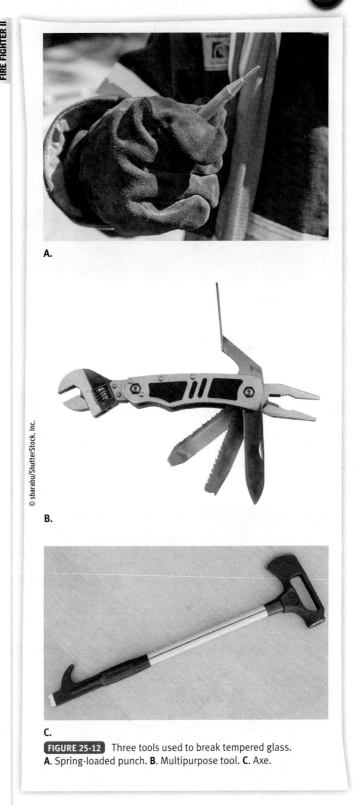

A.

© sbarabu/ShutterStock, Inc.

B.

C.

FIGURE 25-12 Three tools used to break tempered glass. **A.** Spring-loaded punch. **B.** Multipurpose tool. **C.** Axe.

vehicle, place the gearshift in park, and apply the parking brake. Isolate the power by cutting the battery cables.

4. Determine the best access point, and then open a door, break needed glass, or distort metal. (**STEP 1**)
5. Enter the vehicle and approach the victim.
6. Communicate with the victim. (**STEP 2**)

FIRE FIGHTER II

SKILL DRILL 25-5 — Breaking Tempered Glass

Fire Fighter II (NFPA 6.4.1)

FFII

1 After donning complete PPE, including face and eye protection, select a tool.

2 Ensure that the victim and other fire fighters are as properly protected as possible.

3 Place the tool in the lower corner of the window and sharply apply pressure until the spring is activated.

4 Remove loose glass around the window opening.

© LiquidLibrary

Fire Fighter Safety Tips

© Photos.com

Some newer vehicles may soon be equipped with side and rear windows made out of a new type of glass that cannot be broken with a sharp object. This glass promises to provide more protection during a crash and offers an increased level of protection against theft.

■ Force the Door

If fire fighters cannot gain access by the previously mentioned methods, they must use heavier extrication tools to gain access to the victim. The most commonly used technique for gaining access to a vehicle following a collision is door displacement **FIGURE 25-13**. The opening and displacement of vehicle doors may be difficult and somewhat unpredictable, however. As

FIRE FIGHTER II

mentioned earlier, doors on newer vehicles will sometimes be hard to force, and several attempts may be necessary.

When it is necessary to force a door to gain access to the victim, choose a door that will not endanger the safety of the victim; for example, do not try to force a door open if the victim is leaning against it.

To force a door, fire fighters typically use a prying tool to bend the sheet metal at the edge of the door near the locking mechanism. Once this metal has been exposed, a hydraulic spreading tool can be inserted into the space between the door and the pillar of the vehicle and can pull the locking mechanism apart. Once a door has been removed, it should be placed away from the vehicle where it will not be a safety hazard to rescuers.

Powered hydraulic tools are the most efficient and widely used tools for opening jammed doors, which can be opened by releasing the door from the latch side or the hinge side of the

SKILL DRILL 25-6 Gaining Access to a Vehicle Following a Motor Vehicle Crash **FFII**

Fire Fighter II (NFPA 6.4.1)

1 Don PPE, including face protection. Minimize hazards to rescuers and victims. Ensure immobilization of the vehicle: chock wheels, turn off the ignition, place the gear shift in park, and apply the parking brake. Isolate the power by cutting the battery cables. Determine the best access point, and then open a door, break needed glass, or distort metal.

2 Enter the vehicle and communicate with the victim.

FIGURE 25-13 The most commonly used technique for gaining access to a vehicle following a collision is door displacement.

door. The decision regarding which method to use will depend on the structure of the vehicle and the type of damage the door has sustained.

The first step in opening a vehicle door is to use a pry tool to bend the sheet metal away from the edge of the door to create an opening for the tips of the hydraulic tool. This action provides a place where the spreader of the hydraulic tool can be inserted.

Place the spreader in a position so that it is not in the pathway that the door will take when the hinges or latch

break. Do not stand in a position that might put you in danger. To ensure that the door does not swing open violently, you can use a long pry bar to limit the movement of the door when it opens. Activate the hydraulic tool to create a larger opening to the hinges or the door latch. Once the outer sheet metal has been exposed, close the tips of the spreader and remove them.

Insert the closed tips onto the inner skin of the door and the doorjamb just above the latch or above the hinges. Activate the spreader to spread the tips until the latch or the hinge separates. When separating a door at the latch side, you can place 10- by 10-cm (4- by 4-in.) cribbing under the bottom of the door to hold it up and then start to separate the hinges of the door. Some hydraulic tools are capable of cutting door hinges. Check the manufacturer's recommendations for the proper operation of the hydraulic tool.

When separating a door from the hinge side, place the spreader on the top of the bottom hinge and then use the hydraulic spreader to separate the door from the hinge. Once the hinges have separated, place 10- by 10-cm (4- by 4-in.) cribbing underneath the door to hold it in place while you work on the latch side of the door.

Another way to expose the hinges of the door is to vertically pinch the door in the centre of the door, from the bottom of the window down, using the hydraulic spreader. This will pull the sheet metal away from the edges of the door, exposing the latch and hinges.

FIRE FIGHTER II Tips   **FFII**

As vehicle manufacturers strive to make new vehicles stronger and lighter, they have begun using increasing quantities of high-strength steel. These stronger steels, in turn, require more robust rescue tools. Tool manufacturers continue to increase the capacity of rescue tools to cut and distort the metals in newer vehicles. By studying individual vehicle manufacturers' specifications, you can determine the easiest places to cut or distort specific vehicle parts.

To open a door following a motor vehicle collision, follow the steps in **SKILL DRILL 25-7**:

1. Don PPE, including face protection.
2. Retrieve and set up the required tools. Check the equipment for readiness.
3. Assess the vehicle for hazards, including passive restraint devices. (**STEP 1**)
4. Communicate with the victim; minimize hazards to both the rescuers and the victim. (**STEP 2**)
5. Engage hand tools or a power unit to force the door, using good body mechanics. (**STEP 3**)
6. Ensure that the forced door has been stabilized. (**STEP 4**)

FIRE FIGHTER II Tips   **FFII**

Having one leader coordinate the extrication and rescue effort will increase efficiency and help prevent the actions or one team from interfering with another.

■ Provide Initial Medical Care

As soon as you have secured access to the victim, begin to provide emergency medical care. A qualified emergency medical provider should perform this function. Being trapped in a damaged vehicle is a frightening experience, so one caregiver should remain with the victim and provide both emotional comfort and physical care. Victim care should be provided at the same time that extrication is ongoing. Although it might be necessary to delay one part of these processes for a short period, all personnel should work toward the goal of getting the victim stabilized and removed from the vehicle as quickly and safely as possible.

To gain access and provide medical care to a victim in a vehicle, following a motor vehicle collision, follow the steps in **SKILL DRILL 25-8**:

1. Don PPE, including eye protection.
2. Communicate your actions with the victim.
3. Enter the vehicle through the best access point. Minimize the hazards present.
4. Perform an initial assessment of the victim, begin initial medical care, and communicate the victim's condition to the outside rescuer. (**STEP 1**)

5. Perform as complete an assessment as possible. (**STEP 2**)
6. Provide immediate care, and begin packaging the victim for extrication. (**STEP 3**)

Disentangling the Victim

The next step in the extrication sequence is disentangling the victim. The purpose of this step is to remove those parts of the vehicle that are trapping the victim. Note that the goal of this step is to remove the sheet metal and plastic from around the victim.

Before beginning disentanglement, study the situation. What is trapping the victim in the vehicle? Perform only those disentanglement procedures that are necessary to remove the victim safely from the vehicle. The order in which these procedures are carried out will be dictated by the conditions at the scene. Many times it will be necessary to perform one procedure before you can access the parts of the vehicle to perform another procedure.

As you work to disentangle the victim from the vehicle, be sure to protect the victim by covering him or her with a blanket or by using a backboard. Be sure that the victim understands what is being done. In addition, provide emotional support, because the sounds made by extrication procedures are frightening for victims.

To learn all of the methods of disentangling a victim, fire fighters need to take an approved extrication course. The five procedures presented here are the ones that are most commonly performed.

■ Displace the Seat

In frontal and rear-end crashes, the vehicle may become compressed **FIGURE 25-14**. That is, as the front of the vehicle is pushed back, the space between the steering wheel and the seat becomes smaller. In some cases, the driver may be trapped between the steering wheel and the back of the front seat.

FIGURE 25-14 In frontal crashes, the vehicle may become compressed.

© EuroStyle Graphics/Alamy Images

FIRE FIGHTER II

SKILL DRILL 25-7 Opening a Vehicle Door FFII

Fire Fighter II (NFPA 6.4.1)

1 Don PPE, including face protection. Retrieve and set up the required tools. Check the equipment for readiness. Assess the vehicle for hazards, including passive restraint devices.

2 Communicate with the victim; minimize hazards to both the rescuers and the victim.

3 Engage hand tools or a power unit to force the door, using good body mechanics.

4 Ensure that the forced door is stable.

© LiquidLibrary

Displacing the seat can relieve pressure on the driver and give rescuers more space for removal.

If it is necessary to displace a seat backward, start with the simplest steps. Many times you can gain some room by moving the seat backward on its track. This maneuver works well with short drivers who have their seat forward. To move a seat back, first make sure that the victim is supported. With manually operated seats, release the seat adjustment lever and carefully slide the seat back as far as it will go. Attempt to use these simple methods first.

With seats that have adjustable backs and adjustable heights, use these features to move the seat back to give the victim more room or lower the seat to help disentangle the victim. Make sure that the victim is adequately supported before attempting these maneuvers.

If this technique is unsuccessful, perform a dash displacement or roll-up. As a last resort, use a manual hydraulic spreader or a powered hydraulic tool to move the seat back. Place one tip of the hydraulic tool on the bottom of the seat, but avoid pushing on the seat channel that is attached to the floor of the vehicle. Place the other tip of the spreader at the bottom of the A-post door jamb. Support the victim carefully. Engage the seat adjustment lever on manually operated seats, and open the spreader in a careful and controlled fashion.

In some cases, it may be helpful to remove the back of the seat. To accomplish this task, cut the upholstery away from the bottom of the seat back where it joins the main part of the seat. A reciprocating saw or a hydraulic cutter can then be used to cut the supports for the seat back. Be certain

FIRE FIGHTER II

SKILL DRILL 25-8 Gaining Access and Providing Medical Care to a Victim in a Vehicle   **FFII**
Fire Fighter II (NFPA 6.4.1)

1 Don PPE, including eye protection. Communicate your actions with the victim. Enter the vehicle through the best access point. Minimize the hazards present. Perform an initial assessment of the victim, begin initial medical care, and communicate the victim's condition to the outside rescuer.

2 Perform as complete an assessment as possible.

3 Provide immediate care, and begin packaging the victim for extrication.

© LiquidLibrary

that the victim is supported and protected throughout this procedure.

Displacing the seat or removing the seat back will give fire fighters more room to administer treatment and remove the victim from the vehicle.

■ Remove the Windshield

A second technique that is often part of the disentanglement process involves the removal of glass. Removing the rear window, the side glass, or the windshield may improve communication between rescue personnel inside the vehicle and personnel outside the vehicle. Sometimes all that rescuers need to do is to roll a window down to gain this advantage. Open windows provide a good route for passing medical care supplies to the inside caregiver. When the roof of a vehicle must be removed, it is necessary to first remove all of the glass from the vehicle.

The side windows and rear window on most vehicles are made of tempered glass, which can be removed by striking the glass in a lower corner with a sharp object, such as a spring-loaded centre punch. Whenever this measure becomes necessary, fire fighters should protect the victim as much as possible by covering him or her to prevent the victim from being cut by the small pieces of glass.

Windshields, however, cannot be broken with a spring-loaded centre punch. Windshields consist of plastic laminated glass, which has a sandwich type of construction: the two pieces of "bread" consist of thin sheets of specially constructed glass, and the "filling" of the sandwich is a thin layer of a flexible plastic. When laminated glass is struck by a sharp stone or by a spring-loaded centre punch, a small mark appears, but the structure of the glass remains intact. Because of this special construction, it is necessary to remove the windshield in one large piece.

The windshields of most passenger vehicles are glued in place with urethane adhesive glue that is very strong. Nevertheless, there are several ways to remove a windshield after a collision, including cutting the windshield with a special glass saw or reciprocating saw, or carefully using an axe.

To remove a windshield, first make sure that the victim is protected from flying glass. One rescuer begins to cut the top of the windshield at the middle of the windshield. The rescuer continues cutting along the top of the windshield, and then cuts cut down the side of the windshield close to the A post **FIGURE 25-15**. Finally, the rescuer cuts along the bottom of the windshield. At this point, the first rescuer stabilizes the half of the windshield that has been cut free.

Once this operation is complete, a second rescuer starts at the top of the windshield where the initial cuts were made by the first rescuer and cuts the second half of the windshield, following the same sequence of cuts used by the first rescuer. When the second rescuer has completed cutting the windshield, the entire windshield is lifted out of its frame and placed under the vehicle or in another safe place, where it will not present a safety hazard.

Removing a windshield is an essential step prior to removing the roof of a vehicle. It also provides extra space when administering emergency medical care to an injured victim.

FIGURE 25-15 Removing the windshield by cutting close to the A post.

Fire Fighter Safety Tips

© Photos.com

- On most recently manufactured vehicles, the steering wheel contains the driver's-side airbag, which is a life-saving safety feature for the driver of the vehicle. Front-passenger airbags may also be present.
- Airbags may present a hazard after deployment because some airbags deploy in two stages.
- If the airbag did not deploy during the crash, it presents a hazard both for the occupant of the vehicle and for rescue personnel. An undeployed airbag could deploy if wires are cut or if it becomes activated during the rescue operation.
- If the airbag did not deploy, disconnect the battery and allow the airbag capacitor to discharge. The time required to discharge the capacitor varies for different models of airbags.
- Do not place an object between the victim and an undeployed airbag.
- For your safety, never get in front of an undeployed airbag. You could suffer serious injury if it activates unexpectedly.
- Some vehicles contain side-mounted airbags or curtains that provide lateral protection for occupants. Check vehicles for the presence of these devices.

■ Remove the Steering Wheel

During normal driving, the steering wheel must be located close to the driver; during a crash, however, any compression of the front part of the vehicle may push the steering wheel back into the victim's abdomen or chest. Removing the steering wheel can be a major help in disentangling a victim from the vehicle, in such a case.

Steering wheels can be cut with hand tools, such as a hacksaw or bolt cutters. Hydraulic cutters and reciprocating saws will also cut steering wheel spokes and the steering wheel hoop. One method of removing a steering wheel is to cut the

spokes as close to the hub of the steering wheel as possible. A second method is to cut the hoop or ring of the steering wheel. The steering wheel ring can be removed completely, or one section can be cut and removed.

Note that cutting the steering wheel hoop or spokes leaves sharp edges that present a safety hazard. These sharp edges must be covered to prevent injury both to the victim and to rescuers.

■ Displace the Dashboard

During frontal crashes, the dashboard is often pushed down or backward. When a victim is trapped by the dash, it is necessary to remove this component **FIGURE 25-16**. The technique used to remove the dash is called dash displacement or dash roll-up. The objective with this procedure is to lift the dashboard up and move it forward. This evolution is also sometimes used to lift the steering column up and away from a victim.

Dash displacement requires a cutting tool, such as a hacksaw, a reciprocating saw, an air chisel, or a hydraulic cutter. A cutting tool is needed to make a cut on the A post. A mechanical high-lift jack or a hydraulic ram is needed to push the dash forward, and cribbing is needed to maintain the opening made with these tools.

The first step of the dash displacement is to open both front doors. Tie them in the open position so that they will not move as the dash is being displaced. Alternatively, the doors can be removed, as discussed earlier in this chapter in the section on gaining access. Additionally, the roof could be removed.

Place a backboard or other protective device between the victim and the bottom part of the A post where the relief cut will be made. Do not place a backboard in this position if the driver's-side airbag has not deployed. Next, cut the bottom of the A post where it meets the sill or floor of the vehicle. It is critical to make the cut perpendicular to the A post; failure to do so may cut the fuel or power line located in the rocker panels.

Place the base of a high-lift mechanical jack or a hydraulically powered ram at the base of the B post where the sill and the B post meet. Place the tip of the jack or ram at the bend

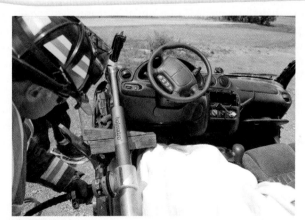

FIGURE 25-16 When a victim is trapped by the dash, it is necessary to displace this component.

in the A post, which is located toward the top of the A post. In a controlled manner, extend the jack or ram to push the dash up and off of the victim. This evolution may be difficult if the crashed vehicle contains high-strength steel.

Once the dash has been removed from the victim, build a crib to hold the sill in the proper position and to prevent the dash from moving. Only then can the jack or ram be removed.

Performing dash displacement requires careful monitoring of the dashboard's movement to be certain that it is moving away from the victim and is not causing any additional harm to the victim. Throughout this evolution, make sure that the victim is protected and that someone keeps the victim informed about what is happening. Communication is key in this procedure.

To displace the dashboard of a vehicle following a motor vehicle collision, follow the steps in **SKILL DRILL 25-9**:

1. Don PPE, including face protection.
2. Retrieve and set up the required tools. Check the equipment for readiness.
3. Assess the vehicle for hazards, including passive restraint devices. (**STEP 1**)
4. Communicate with the victim, and ensure that both the victim and rescuers are protected from hazards. (**STEP 2**)
5. Make relief cuts at the bottom of the A post. (**STEP 3**)
6. Insert the hydraulic spreader in the cut, and lift the dashboard and firewall. Place cribbing under the dash as it is being lifted. (**STEP 4**)

FIRE FIGHTER II Tips   **FFII**

Cribbing should be placed under the vehicle whenever downward pressure is expected. This will maximize the displacement of the dash.

■ Displace the Roof

Removing the roof of a vehicle has several advantages: First, it enables equipment to be easily passed to the emergency medical provider. In addition, it increases the amount of space available to perform medical care, and it increases the visibility and space for performing disentanglement. Likewise, visibility and fresh air supply are improved. The increased space also helps to reduce the feeling of panic caused by the confined space of the crashed vehicle. Finally, removing the roof provides a large exit route for the victim **FIGURE 25-17**.

One method of displacing the roof is to cut the A posts and fold the roof back toward the rear of the vehicle. This technique provides limited space. Given this fact, it is often preferable to remove the entire roof.

FIRE FIGHTER II Tips   **FFII**

When removing the roof, remember to cut all of the seat belts, so they will not interfere with the operation.

FIGURE 25-17 Removing the roof provides a large exit route for the victim.

Roof displacement can be accomplished with hand tools, such as hacksaws, air chisels, and manual hydraulic cutters. Appropriate power tools include reciprocating saws and powered hydraulic cutters, especially the latter tools. Be aware that some A posts are made of high-strength steel and may be difficult to cut, especially with older cutting tools. In addition, most newer vehicles include additional reinforcement beams for the seat belts. Older hydraulic cutters may not be able to sever these beams, so cuts must be made above the seat belts.

The first step in displacing the roof is to ensure the safety of both rescuers and the victim inside of the vehicle. In particular, as rescuers cut the posts that support the roof, they must support the roof to keep it from falling on the victim.

To remove the roof, first remove the glass in the vehicle's windows and windshields to prevent it from falling on the victim. Cut the vehicle posts farthest away from the victim. Cut each post at a level that will ensure that the least amount of post remains in place after roof removal. When cutting the wider rear posts, cut them at the narrowest point of the post. As each post is cut, a rescuer needs to support that post. Cut the post closest to the victim last, because it provides some protection for the victim.

Working together, several rescuers should remove the roof and place it away from the vehicle, where it does not pose a safety hazard. They should then cover the sharp ends of the cut posts with a protective device, thereby ensuring the safety of both rescuers and the victim.

To remove the roof of a vehicle following a motor vehicle collision, follow the steps in **SKILL DRILL 25-10**:

1. Don PPE, including face protection.
2. Retrieve and set up the required tools. Check the equipment for readiness.
3. Assess the vehicle for hazards, including passive restraint devices.
4. Communicate with the victim, and minimize hazards to both the rescuers and the victim.
5. Remove any remaining glass. (**STEP 1**)

SKILL DRILL 25-9 Displacing the Dashboard of a Vehicle **FFII**

Fire Fighter II (NFPA 6.4.1)

1 Don PPE, including face protection. Retrieve and set up the required tools. Check the equipment for readiness. Assess the vehicle for hazards, including passive restraint devices.

2 Communicate with the victim, and ensure that both the victim and rescuers are protected from hazards.

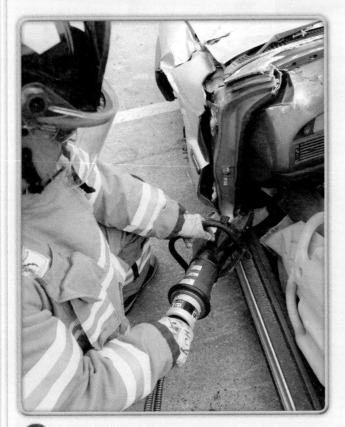

3 Make relief cuts at the bottom of the "A" post.

4 Insert the hydraulic spreader in the cut, and lift the dashboard and firewall. Place cribbing under the dash as it is being lifted.

© LiquidLibrary

6 Engage appropriate tools to remove the roof, while using good body mechanics. (**STEP 2**)

7 Control the vehicle roof and protect against sharp edges at all times. (**STEP 3**)

8 Remove the roof to a safe location.

9 Return the tools to the staging area upon completion of the tasks. (**STEP 4**)

Removing and Transporting the Victim

The final phase of the extrication process consists of removing the victim from the crashed vehicle. During this phase, the victim needs to be stabilized and packaged in preparation for removal. Follow the direction of the medical care provider who is in charge of victim care. The definitive treatment of trauma victims will occur at a hospital; therefore, the amount of stabilization that should be attempted while the victim is in the vehicle should be limited to only those steps that are needed to prevent further injury to the victim. See the skill drill "Performing a Long Backboard Rescue" in the chapter "Search and Rescue" for more information on using a long backboard to remove a victim from a vehicle.

Develop a plan for the victim's removal, and make sure that a clear exit pathway is available. The actual victim removal should be directed by a designated person—usually the rescuer controlling c-spine immobilization—and conducted with clear commands. Ensure that an adequate number of rescuers are able to assist with the victim removal, and confirm that everyone involved in this process understands the commands that will be given.

Try to make the removal process as seamless as possible. Ideally, the ambulance cot should be positioned close by, so that the victim can be placed on the cot without any delay. In any event, the victim should be moved to an ambulance as soon as possible.

Transport the victim to an appropriate medical facility as soon as the initial stabilization has been completed. If your department uses a helicopter service to transport victims, you also need to be familiar with the procedures used to load victims into helicopters.

FIRE FIGHTER II Tips   FFII

Always have a separate set of eyes, or a lookout, observing the extrication operation for safety.

Terminating an Incident

Terminating an incident includes securing the scene by removing the damaged vehicle and equipment from the scene and ensuring the scene is left in a safe condition. After you have secured the scene and packed your equipment, it is important to return to the station and fully inventory, clean, service, and maintain all the equipment (per the manufacturer's instructions) to prepare it for the next call. Some items will inherently

need repair, but most will need simple maintenance before they are placed back on the fire apparatus and considered in service.

■ Securing the Scene

Just because the victim has been removed from the wreckage does not mean the scene can be abandoned. Law enforcement will secure the scene by conducting their on-scene investigation. A potential crime scene, such as a homicide, should be managed by law enforcement, who will preserve and secure any evidence and close off the scene or roadway. Once a scene has been released by law enforcement, the removal of vehicles will be coordinated with a tow agency.

In some cases, a vehicle may need to be turned right side up. This may cause sparks or a fire. In addition, the high-voltage batteries in hybrid and electric vehicles may be a delayed-reaction fire risk. Be aware of smoke, sparks, irritating fumes, or a gurgling sound from the battery. These are warning signs of a damaged high-voltage battery that is undergoing a thermal event and is at risk for fire hours or even days later. There is also a risk of toxic off-gassing. Being proactive by standing by with a charged hose line is always a good practice during one of these procedures.

Any fluid hazards that have spilled from the vehicle, such as gasoline, motor oil, transmission fluid, or radiator fluid that was either captured with an absorbent or not, will be removed by the towing agency. Towing agencies must be licenced to transport and dispose of any hazardous substances; fire rescue agencies are typically not licenced to do so. To assist the towing agency, place any vehicle parts, such as doors, roofs, and fenders, back in a heavily damaged vehicle. A word of advice is to always ask the tow-agency representative if they need assistance before you start throwing items back into a vehicle. The tow agency may have procedures or a policy of their own for stowing and transporting loose objects. Being considerate of their preferences will help to maintain a good working relationship with the tow agency for future endeavours such as acquiring vehicles for use in an extrication class. It is also advisable to carry a contact list of various resources for any private and public organization or business that can offer a particular resource that can be utilized on the incident, such as public works/utilities, Transport Canada, or a heavy-equipment company.

FIRE FIGHTER II Tips   FFII

All electric and hybrid vehicles should be towed using a flatbed. Towing with the drive wheels on the ground could result in an electrical fire.

The exception to these procedures will be an incident involving a fatality. Law enforcement will acquire the scene immediately to conduct a fatality investigation. In the event that law-enforcement investigators take possession of the vehicle, brief them on the hazards and warning signs of a damaged high-voltage battery.

SKILL DRILL 25-10 | Removing the Roof of a Vehicle | **FFII**

Fire Fighter II (NFPA 6.4.1)

1 Don PPE, including face protection. Retrieve and set up the required tools. Check the equipment for readiness. Assess the vehicle for hazards, including passive restraint devices. Communicate with the victim, and minimize hazards to both the rescuers and the victim. Remove any remaining glass.

2 Engage appropriate tools to remove the roof, while using good body mechanics.

3 Control the vehicle roof and protect against sharp edges at all times.

4 Remove the roof to a safe location. Return the tools to the staging area upon completion of the tasks.

© LiquidLibrary

FIRE FIGHTER II Tips | **FFII**

To ensure your safety and the safety of every member of your crew:
- Follow the manufacturers' recommendations when using power tools.
- Inspect all power tools before and after use.

FIRE FIGHTER II Tips | **FFII**

After the incident, conduct a brief meeting with the crew to discuss what went right during the incident and what could be done better.

Wrap-Up

Chief Concepts

- The design of motor vehicles is constantly evolving. As a fire fighter, you need to keep up to date on the design of motor vehicles.
- The most common types of vehicles on the road are conventional vehicles, which use internal combustion engines that burn gasoline for power. Alternative-powered vehicles may be powered by CNG, LNG, or LPG (i.e., propane). The hazards posed by these vehicles include fuel fires.
- Electric-powered vehicles are propelled by an electric motor that is powered by batteries. These vehicles pose electrical hazards due to the large number of batteries needed to power the engine.
- Hybrid vehicles use both a battery-powered electric motor and a gasoline-powered engine. As a consequence, these vehicles pose both electrical hazards and fuel fire hazards.
- The main parts of a motor vehicle include the following components:
  - The A posts form the sides of the windshield.
  - The B posts are located between the front and rear doors of a vehicle.
  - The C posts are located behind the rear passenger windows.
  - The bulkhead divides the engine compartment from the passenger compartment.
  - The passenger compartment (also called the occupant cage) includes the front and back seats.
  - The two common types of vehicle frames are the platform frame and the unibody. The platform frame is used for SUVs and trucks; it uses beams to form the load-bearing frame of a vehicle. The unibody is used in most passenger vehicles; it combines the vehicle body and the frame into a single component.
- Extrication should follow a series of logical steps:
  - Safely respond to the scene—Pick the best route, drive in a safe manner, and wear your safety belt.
  - Arrival and scene size-up—Assess the hazards and determine whether additional resources are needed. Be aware of the traffic and any spilled fuel.
  - Stabilize the scene—Reduce, remove, or mitigate the hazards at the scene.
  - Stabilize the vehicle—Use cribbing and/or rescue-lift air bags to stabilize the vehicle before the victim can be removed.
  - Access the victim—The victim may be accessed (and assessed) through a door, window, or a removed roof.

- Disentangle the victim—Using hand and power tools, remove the tangle of metal from around the trapped victim.
- The following tools are used for stabilizing a vehicle:
  - Cribbing
  - Step chocks
  - Wedges
  - Rescue-lift air bags
- Four techniques may be used for gaining access to and disentangling the victim of a motor vehicle collision:
  1. Stabilize or hold an object or vehicle.
  2. Bend, distort, or displace.
  3. Cut or sever.
  4. Disassemble.
- The following classes of tools are used for extrication:
  - Tools for stabilization—Cribbing, rescue-lift air bags, step chocks, high-lift jacks, and stabilization jacks
  - Tools to displace, bend, and distort metal—Pry axes, short pry bars, long pry bars, hacksaws, screwdrivers, hammers, and Halligan tools
  - Tools for cutting—Axes, bolt cutters, cable cutters, hacksaws, manual hydraulic cutters, air chisels, and power saws
- To gain access to the victim, the following techniques are used:
  - Open the door.
  - Break a window.
  - Force open the door.
- To disentangle the victim from the vehicle, the following techniques are used:
  - Displace the seat.
  - Remove the windshield.
  - Remove the steering wheel.
  - Remove the steering wheel: loops or spokes.
  - Displace the dashboard.
  - Displace the roof.
- The final phase of the extrication process consists of removing and transporting the victim to the hospital. During this phase, the victim is stabilized, packaged, removed from the vehicle, and transported to a medical facility.

Hot Terms

Alternative-powered vehicle A vehicle that is powered by compressed natural gas or other fuel.

"A" posts Vertical support members that form the sides of the windshield of a motor vehicle.

"B" posts Vertical support members located between the front and rear doors of a motor vehicle.

Bulkhead The wall that separates the engine compartment from the passenger compartment in a motor vehicle; also known as a firewall.

Conventional vehicle A vehicle that uses an internal combustion engine and is fuelled by either gasoline or diesel fuel.

"C" posts Vertical support members located behind the rear doors of a motor vehicle.

Cribbing Short lengths of timber/composite materials, usually 10 by 10 cm (4 by 4 in.) and 46 by 61 cm (18 by 24 in.) long that are used in various configurations to stabilize loads in place or while a load is moving. (NFPA 1006)

Electric vehicle An automotive-type vehicle for on-road use, such as passenger automobiles, buses, trucks, vans, electric vehicles, and the like, primarily powered by an electric motor that draws current from a rechargeable storage battery, fuel cell, photovoltaic array, or other source of electric current. Electric motorcycles and similar-type vehicles and off-road, self-propelled electric vehicles, such as industrial trucks, hoists, lifts, transports, golf carts, airline ground-support equipment, tractors, boats, and the like, are not included in this definition. (NFPA 70)

Fuel cell An electrochemical system that consumes fuel to produce an electric current. The main chemical reaction used in a fuel cell for producing electric power is not combustion, but sources of combustion may be used within the overall fuel cell system, such as reformers/fuel processors. (NFPA 70)

Fuel cell stack Many fuel cells assembled into a group to produce larger quantities of electricity.

Hybrid vehicle A vehicle that uses a battery-powered electric motor and an internal combustion engine.

Laminated windshield glass A type of window glazing that incorporates a sheeting material that stops the glass from breaking into individual shards.

Platform frame A type of vehicle frame resembling a ladder, which is made up of two parallel rails joined by a series of cross members. This kind of construction is typically used for luxury vehicles, SUVs, and all types of trucks.

Post One of the vertical support members, or pillars, of a vehicle that holds up the roof and forms the upright columns of the passenger compartment.

Rescue-lift air bags Extrication tools consisting of air bags, filler hoses, air regulators, control valves, and a supply of compressed air. These tools are used to lift vehicles that have been involved in crashes.

Step chocks Specialized cribbing assemblies made out of wood or plastic blocks in a step configuration. They are typically used to stabilize vehicles.

Tempered glass A type of safety glass that is heat treated so that it will break into small, less dangerous pieces.

Unibody (unit-body) The frame construction most commonly used in vehicles. The base unit is made of formed sheet metal; structural components are then added to the base to form the passenger compartment. Subframes are attached to each end. This type of construction does away with the rail beams used in platform-frame vehicles; also known as unit-body.

Wedges Material used to tighten or adjust cribbing and shoring systems. (NFPA 1006)

It is midnight, and you find yourself arriving at the scene of a serious motor vehicle collision where a pickup has T-boned the driver's door of a 2012 Honda Civic. The driver of the pickup, who appears highly intoxicated, is wandering around with minor cuts and scrapes. His passenger has her legs pinned between the seat and the dash. She has open fractures to her femurs.

The 19-year-old driver of the car is deceased, and her body is trapped between the door and the centre console. The front-seat passenger is unconscious, and her back is resting against the passenger door. The back-seat passenger is semi-conscious and is unable to get out of the back seat. Your captain looks at you and says, "I'll check out the pickup, you check the car."

1. You decide to remove the roof and begin with the vertical support member that runs along the windshield and the front of the passenger door. What is the term for this member?
   A. "A" post
   B. "B" post
   C. "C" post
   D. "D" post

2. Prior to making your first cut, you know you must first stabilize the vehicle. What will you use?
   A. Rescue-lift air bag
   B. Struts
   C. Cribbing
   D. Spreaders

3. What type of glass is generally used in the side and rear windows of automobiles?
   A. Laminated
   B. Tempered
   C. Heat-resistant
   D. Flat

4. Which of the following would precede the others in the extrication process?
   A. Disentangle the patient
   B. Stabilize the vehicle
   C. Assess the patient
   D. Transport the patient

5. Which of the following tools is used to displace, bend, or distort metal?
   A. Hacksaw
   B. Axe
   C. Step chock
   D. Air chisel

6. After covering the patient, the next step in removing a windshield with an axe is to cut _____.
   A. along the "A" post starting at the top
   B. in the centre, starting at the top
   C. along the top
   D. along the bottom of the windshield

FIRE FIGHTER II in action

You are the first to arrive at a motor vehicle collision in your privately owned vehicle (POV). It involves two vehicles, one of which has rolled down an embankment and is now resting on its side. You find two people in the overturned vehicle. One is pinned and has moderate injuries, while the other is in her seat belt, but unharmed. In the other vehicle, one has a serious head injury, another has an amputated arm, while the last is yelling and screaming at the occupants of the overturned vehicle for causing the accident.

You recognize the need for more help than your small, volunteer department can provide. Your department has a single engine with a basic set of extrication tools. In addition, there is an ambulance responding. You quickly begin to process the information and then key up your radio.

1. What additional resources should you request?
2. How will you assign other fire fighters that arrive by POV?
3. How will you handle the irate occupant?
4. How will you balance your need to manage the incident with your need to provide care?

Assisting Special Rescue Teams

Fire Fighter I

Knowledge Objectives

There are no knowledge objectives for Fire Fighter I candidates. NFPA 1001 contains no Fire Fighter I Job Performance Requirements for this chapter.

Skills Objectives

There are no skills objectives for Fire Fighter I candidates. NFPA 1001 contains no Fire Fighter I Job Performance Requirements for this chapter.

Fire Fighter II   FFII

Knowledge Objectives

After studying this chapter, you will be able to:

- Define the types of special rescues encountered by fire fighters. (NFPA 6.4.2.A), p 764)
- Describe the steps of a special rescue. (NFPA 6.4.2, 6.4.2.A), p 766–770)
- Describe the general procedures at a special rescue scene. (NFPA 6.4.2, 6.4.2.A), p 770–772)
- Describe how to safely approach and assist at a vehicle or machinery rescue incident. (NFPA 6.4.2, 6.4.2.A), p 773–774)
- Describe how to safely approach and assist at a confined-space rescue incident. (NFPA 6.4.2, 6.4.2.A), p 774–775)
- Describe how to safely approach and assist at a rope rescue incident. (NFPA 6.4.2, 6.4.2.A), p 775–776)
- Describe how to safely approach and assist at a trench and excavation rescue incident. (NFPA 6.4.2, 6.4.2.A), p 776–777)

26

- Describe how to safely approach and assist at a structural-collapse rescue incident. (NFPA 6.4.2, 6.4.2.A , p 777–778)
- Describe how to safely approach and assist at a water or ice rescue incident. (NFPA 6.4.2, 6.4.2.A , p 778–780)
- Describe how to safely approach and assist at a wilderness search and rescue incident. (NFPA 6.4.2, 6.4.2.A , p 780–781)
- Describe how to safely approach and assist at a hazardous materials rescue incident. (NFPA 6.4.2, 6.4.2.A , p 781–782)
- Describe how to safely respond to an elevator or escalator rescue. (NFPA 6.4.2, 6.4.2.A , p 782–783)

Skills Objectives

After studying this chapter, you will be able to perform the following skills:

- Establish a barrier. (NFPA 6.4, 6.4.2.B , p 771–772)
- Identify and retrieve rescue tools. (NFPA 6.4, 6.4.2.B , p 773)

Additional NFPA Standards

- NFPA 472, *Standard for Competence of Responders to Hazardous Materials/Weapons of Mass Destruction Incidents*
- NFPA 1006, *Standard for Technical Rescuer Professional Qualifications*
- NFPA 1500, *Standard on Fire Department Occupational Safety and Health Program*
- NFPA 1670, *Standard on Operations and Training for Technical Search and Rescue Incidents*
- NFPA 1951, *Standard on Protective Ensembles for Technical Rescue Incidents*

© Photos.com

Y ou hear a call come in that is only two blocks from your home. A construction worker has been buried in a trench collapse. You travel with your crew to the scene. On arrival, you find an idling backhoe at the end of the trench and two workers in the trench digging feverishly with hand shovels to find their co-worker.

1. What is the first priority?
2. What do you say to the workers?
3. What is your greatest concern?

Introduction

In recent years, the number of fires and the associated human and property losses from fires in Canada have generally decreased. This decrease is attributable to better fire prevention measures, better fire safety education, better building codes, and better fire safety design. At the same time, the mission of fire departments has expanded. As fire departments take on added roles, such as Emergency Medical Services (EMS), hazardous materials response, and technical rescue responses, they are generally seen as the agency that can handle any and all types of emergencies.

A **technical rescue incident (TRI)** is a complex rescue incident involving vehicles or machinery, water or ice, rope techniques, trench or excavation collapse, confined spaces, structural collapse, wilderness search and rescue, or hazardous materials that requires specially trained personnel and special equipment. This chapter shows you how to assist specially trained rescue personnel in carrying out the tasks for which they have been trained, but it will not make you an expert in the skills needed to handle these eight types of rescue situations. Specialized training is required to acquire true proficiency in TRIs. The more training you receive in these areas, the better you will understand how to conduct yourself at these scenes.

Training in technical rescue areas is conducted at three levels: awareness, operations, and technician.

- Awareness level—This training serves as an introduction to the topic, with an emphasis on recognizing hazards, securing the scene, and calling for appropriate assistance. There is no actual use of rescue skills at the awareness level.
- Operations level—This level is geared toward working in the warm zone of an incident (the area directly around the hazard). Such training will allow you to directly assist those conducting the rescue operation and use certain skills and procedures to conduct the rescue.
- Technician level—This level allows you to be directly involved in the rescue operation itself. Such training includes use of specialized equipment, care of victims during the rescue, and management of the incident and of all personnel at the scene.

Most of the training you receive as a beginning fire fighter aims at preparing you at the awareness level, so that you can identify hazards and secure the scene to prevent further injuries.

Types of Rescues Encountered by Fire Fighters

Most fire departments respond to a variety of special rescue situations **FIGURE 26-1** :

- Vehicle and machinery rescue
- Confined-space rescue
- Rope rescue
- Trench and excavation rescue
- Structural-collapse rescue
- Water and ice rescue
- Wilderness rescue
- Hazardous materials incidents

FIGURE 26-1 Most fire departments respond to a variety of special rescue situations.

To become proficient in handling these situations, you must take a formal course to gain specialized knowledge and skills. NFPA 1670, *Standard on Operations and Training for Technical Search and Rescue Incidents*, identifies operational benchmarks and training needed to establish operational capacity for technical search and rescue events.

It is important for awareness-level responders to have an understanding of these special types of rescues. Often, the first emergency unit to arrive at a rescue incident is a fire department engine or truck company. As a consequence, the initial actions taken by that company may determine the safety of both the victims and the rescuers. These initial actions may also determine how efficiently the rescue is completed.

It is also important for you to understand and recognize the tools needed for rescue situations. You may not have to know how to use all of them, but you may be called upon to assist with any tool at any time.

Guidelines for Operations

When assisting rescue team members, keep in mind the five guidelines that you follow during other firefighting operations:

1. Be safe.
2. Follow orders.
3. Work as a team.
4. Think.
5. Follow the golden rule of public service, which emphasizes the ethic of reciprocity: treat others as you would like to be treated.

FIRE FIGHTER II Tips	FFII

Emergency incidents are unpredictable. Don't just have one backup plan; have two or three.

■ Be Safe

Rescue situations may be full of hidden hazards, including oxygen-deficient atmospheres, weakened floors, electrical hazards, and strong water currents. Knowledge and training are required to recognize signs indicating that a hazardous rescue situation exists. Once the hazards are recognized, determine which actions are necessary to ensure your own safety as well as the safety of your team members, the victims of the incident, and bystanders. It requires experience and skill to determine that a rescue scene is not safe to enter, but that determination could ultimately save lives.

■ Follow Orders

When you begin your career as a fire fighter, you will have limited training and experience. By contrast, your officers and the rescue teams with whom you will work on special rescue incidents have received extensive specialized training. They have been chosen for their duties precisely because they have experience and skills in a particular area of rescue. It is critical to follow the orders of those who understand exactly what needs to be done to ensure safety and to mitigate the dangers involved in the rescue situation. Moreover, orders should be followed exactly as given. If you do not understand what is expected of you, ask for more information. Have the orders clarified so you will be able to complete your assigned task safely.

To do well as a fire fighter, you must have a strong appreciation for the basic philosophy of many firefighting organizations. Most fire departments are run like military organizations. Grasping the paramilitary structure will enable you to understand the command and control concept of fire departments.

- The fire officer's knowledge base and experience are greater than yours.
- Orders come from superiors. Legitimate orders are those given by a fire officer or other designated person.
- Follow rules and procedures. A fire fighter is required to follow rules, procedures, and guidelines regardless of his or her personal opinions.
- You must do your own job.
- Get the job done. In emergency situations, time is critical. Nevertheless, you must not act beyond your own skill and training level, and you must not violate any rules, procedures, suggested operating guidelines (SOGs), or standard operating procedures (SOPs), orders of a superior in an effort to get the job done quickly.

Acting with these ideas clear in your head will enable you to deal with the orders of your commanding fire officers quickly and confidently.

■ Work as a Team

Rescue efforts often require many people to complete a wide variety of tasks. Some personnel have been trained in specific tasks, such as rope rescue or swift water rescue **FIGURE 26-2**. To do their jobs, however, they need the support and assistance of other fire fighters.

Firefighting requires members of the team to work together to complete the goal of fire suppression. Rescue is a different goal, but it requires the same kind of team effort. Your role in the rescue effort is an essential part of this team effort.

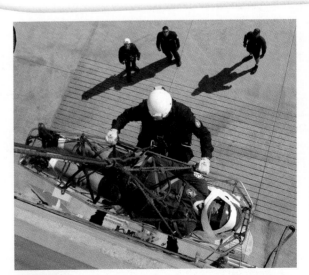

FIGURE 26-2 Some personnel are trained in specific rescue skills, such as rope rescue.

Think

As you are working on a rescue situation, you must constantly assess and reassess the scene. You might see something that your company officer does not see. If you think your assigned task may be unsafe, bring this matter to the attention of your superior officer. Do not try to reorganize the total rescue effort, as it is being directed by people who are highly trained and experienced—but do not ignore what is going on around you, either.

Observations that a fire fighter should bring to the attention of a superior fire officer include unsafe structures, changing weather conditions, suspicious packages, and broken equipment. In a hazardous materials rescue incident, the wind direction and speed are important to know; for example, an increase in wind speed can cause hazardous vapours to spread more quickly, which in turn can affect the response plan.

Follow the Golden Rule of Public Service

When you are involved in carrying out a rescue effort, it is easy to concentrate on the technical parts of the rescue and forget that there is a frightened person who needs your emotional support and encouragement. It is helpful to have a rescuer stay with the victim whenever possible. He or she can tell the victim which actions will be performed during the rescue process. This rescuer is often the one whom the victim remembers and the one who helps to fulfill the golden rule.

FIRE FIGHTER II Tips | **FFII**

In any TRI, it is important to know why rescuers fail. The reasons for rescue failures are most commonly referred to by the acronym FAILURE:

- F—Failure to understand the environment, or underestimating it
- A—Additional medical problems not considered
- I—Inadequate rescue skills
- L—Lack of teamwork or experience
- U—Underestimating the logistics of the incident
- R—Rescue versus recovery mode not considered
- E—Equipment not mastered

Steps of Special Rescue

Although special rescue situations may take many different forms, all rescuers take certain basic steps to perform special rescues in a safe, effective, and efficient manner:

1. Preparation
2. Response
3. Arrival and size-up
4. Stabilization
5. Access
6. Disentanglement
7. Removal
8. Transport
9. Security of the scene and preparation for the next call
10. Postincident analysis

Preparation

You can prepare for responses to emergency rescue incidents by training with the specialized rescue teams with whom you will be working. This kind of experience will enable you to learn about their techniques and equipment. It will help you to understand what you can do to assist them as well as what you should *not* do. It is also helpful to train with other fire departments in your area, as such training will better enable you to respond to a mutual aid call. In addition, you will be able to learn about the types of rescue equipment to which other departments have access as well as the training levels of their personnel.

Pay attention to terminology; knowing the terminology used in the field will make communicating with other rescuers easier and more effective. Also, learn to recognize the different types of rescue situations that you could encounter. Prior to a technical rescue call, your department must consider these issues:

- Does the department have the personnel and equipment to handle a technical rescue incident from start to finish?
- Does the department meet National Fire Protection Association (NFPA) and Public Safety Canada standards for technical rescue calls?
- Which equipment and personnel will the department send on a technical rescue call?
- Do members of the department know the hazards located within the department's response area, and have they visited these areas with local representatives?

Response

A TRI should have its own dispatch protocol. If your agency has its own **technical rescue team**, it will usually respond with a rescue unit supported by other equipment and personnel, such as a medic unit, an engine company, a truck company, and a chief. The makeup of the response is based on the ability of the agency or department to provide resources for the response. Even if a technical rescue team is available, it may not be trained or equipped to handle all incidents that could be encountered. Your department should make provisions for additional resources and trained personnel to assist in handling incidents when the TRI is beyond the local technical rescue team's capabilities.

Often, it is necessary to notify power and utility companies during a TRI for possible assistance. Many technical rescues involve electricity, sewer pipes, or other factors that may create the need for additional heavy equipment, to which utility companies have ready access.

Arrival and Size-Up

The first company officer to arrive on the scene should assume command. A rapid and accurate size-up is needed to avoid placing would-be rescuers in danger and to identify additional resources that may be needed. Determine how many victims are involved, and note the extent of their injuries. This information will help to decide how many EMS units and other resources are needed.

When responding to a work site or industrial facility, the incident commander (IC) should make contact with the foreman or supervisor, also known as the responsible party. This individual can often provide valuable information about the work site and the emergency situation. An important part of any rescue is the identification of all hazards. It is also important to determine whether a TRI is a rescue event or a body-recovery operation. After these determinations have been made, the IC can decide whether to call for additional resources and order other actions intended to stabilize the incident.

Do *not* rush into the incident scene until you have adequately assessed the situation, because you may inadvertently make the situation worse. For example, a fire fighter approaching a trench collapse may cause further collapse. A fire fighter entering a swiftly flowing river might be quickly knocked down and carried downstream. A fire fighter climbing down into a well to evaluate an unconscious victim may be overcome by an oxygen-deficient atmosphere. Stop and think about the dangers that may be present; do not make yourself part of the problem. Always be alert for invisible dangers such as electrical hazards and oxygen-deficient or poisonous atmospheres.

> ### Fire Fighter Safety Tips
> © Photos.com
>
> At many fire and emergency scenes, fire line tape is set up to keep bystanders back. Emergency responders operating at such a scene routinely ignore this barrier by simply ducking under or climbing over the tape. This complacency toward fire line tape as a warning device creates a potential hazard for responders at TRIs and other emergency scenes where the tape is placed to warn not just bystanders but responders as well. All fire line tape should be considered a hazard warning to fire fighters unless and until it is proved to be for bystanders only.

■ Stabilization

Once needed resources are on the way and the scene is safe to enter, begin to stabilize the incident. Establish an outer perimeter to keep the public and media out of the staging area, and establish a smaller perimeter directly around the rescue site. A rescue area is an area that surrounds the incident site (e.g., a collapsed structure, a collapsed trench, or a hazardous spill area) and whose size is proportional to the hazards that exist. Three controlled zones should be established:

1. <u>Hot zone</u>—This area is for entry teams and rescue teams only. The hot zone immediately surrounds the dangers of the site (e.g., hazardous materials releases) and is established to protect personnel outside the zone.
2. <u>Warm zone</u>—This area is for properly trained and equipped personnel only. The warm zone is where personnel and equipment decontamination and hot zone support take place.
3. <u>Cold zone</u>—This area is for staging vehicles and equipment. The cold zone contains the command post. The public and the media should be kept clear of the cold zone at all times.

To figure out the scope of each of these zones, fire fighters should identify and evaluate the hazards that are discovered at the scene, observe the geographical area, note the routes of access and exit, observe weather and wind conditions, and consider evacuation problems and transport distances.

The most common method of establishing the control zones for an emergency incident site is to use police- or fire-line tape. Police- or fire-line tape is available in a variety of colours. The most commonly used colours are red, for the hot zone; orange, for the warm zone; and yellow, for the cold zone. For other methods of securing the perimeters of the controlled zones, see the chapter "Hazardous Materials: Operations and Specialties."

Once the fire department has established the three controlled zones, responders should ensure that the zones of the emergency scene are enforced. Because of lack of firefighting personnel, scene-control activities are sometimes assigned to law-enforcement personnel.

<u>Lockout and tagout systems</u> should be used at this time to ensure an electrically safe environment **FIGURE 26-3**. Lockout and tagout systems are intended to ensure that the electricity has been shut off and that electrically powered equipment is not unintentionally turned on.

In an emergency situation, all power must be turned off and locked out. As part of this effort, switches or valves must be tagged with labels to protect personnel and emergency response workers from accidental machine start-up. Lockout and tagout procedures are used to warn personnel and ensure that the electrical power is disconnected. Authorized and

FIGURE 26-3 Lockout and tagout systems are methods of ensuring that the electricity has been shut down.

trained personnel can disconnect the source of power, lock it out, and tag it. Locks and tags are used for everyone's protection against electrical dangers. For your safety and that of others, never remove or ignore a lock or tag. Always be alert for electrical hazards, because they are not necessarily easy to recognize. Vehicle crashes, building collapses, entrapments, and natural disasters, for example, can all create electrical hazards for rescuers and patients. These scenes require a careful overview of the scene and continual monitoring of the scene.

During stabilization, atmospheric monitoring should be started to identify any immediately dangerous to life or health (IDLH) environments for rescuers and victims. The next steps involve looking at the type of incident and planning how to rescue victims safely; for example, in a trench rescue, these measures might include setting up ventilation fans for air flow, setting up lights for visibility, or protecting a trench from further collapse.

Access

Once the scene is stabilized, rescuers need to consider how to gain access to the victim. How is the victim trapped? In a trench situation, a dirt pile may be to blame. In a rope rescue, scaffolding may have collapsed. In a confined space, a hazardous atmosphere may have caused the victim to collapse. To reach a victim who is buried or trapped beneath debris, it is sometimes necessary to dig a tunnel as a means of rescue and escape. Identify the actual reason for the rescue, and work toward freeing the victim safely.

Communicate with the victims at all times during the rescue to make sure they are not injured further by the rescue operation. Even if they are not injured, all victims need to be reassured that the team is working as quickly as possible to free them.

Emergency medical care should be initiated as soon as safe access is made to the victim. Technical rescue paramedics are vital resources at TRIs; not only can these responders start intravenous lines and treat medical conditions, but they also know how to deal with the equipment and procedures that are going on around them. It is vitally important that EMS personnel be effectively integrated into ongoing operations during rescue incidents. Their main functions are to treat victims and to stand by in case a rescue team member needs medical assistance. As soon as a rescue area or scene is secured and stabilized, EMS personnel must be allowed access to the victims for medical assessment and stabilization. Some fire departments have trained paramedics to the technical rescue level, so that they can enter hazardous areas and provide direct assistance to the victim. Throughout the course of the rescue operation (which can sometimes span many hours), EMS personnel must continually monitor and ensure the stability of the victims and, therefore, must be allowed access to the victims.

The strategy used to gain access to victims depends on the type of incident. For example, in an incident involving a motor vehicle, the location and position of the vehicle, the damage to the vehicle, and the position of the victim are important considerations. The means of gaining access to the victim must also take into account the victim's injuries and their severity.

The chosen means of access may have to be changed during the course of the rescue as the nature or severity of the victim's injuries becomes apparent.

Disentanglement

Once precautions have been taken and the reason for entrapment has been identified, the victim needs to be freed as quickly and as safely as possible. A team member should remain with the victim to direct the rescuers who are performing the disentanglement. In a trench incident, this step would include digging either with a shovel or by hand to free the victim.

In a vehicle accident, the most important point to remember is that the vehicle is to be removed from around the victim, rather than trying to remove the victim through the wreckage. Many different parts of the vehicle may trap the occupants, including the steering wheel, seats, pedals, and dashboard. In this scenario, disentanglement entails the cutting of a vehicle (and or machinery) away from trapped or injured victims, which is accomplished by using extrication and rescue tools with a variety of extrication methods.

FIRE FIGHTER II Tips — FFII

Technical rescue incidents may involve longer operational periods than structure fires. Pay attention to your physical state and that of your team members. If you are tired, go to the rehabilitation area.

Removal

Once the victim has been disentangled, efforts will be redirected toward removing the victim **FIGURE 26-4**. In some instances, this may simply amount to having someone assist the victim up a ladder. In other cases, it may require removal with spinal immobilization due to possible injuries. Stokes baskets, backboards, stretchers, and other immobilization devices are all commonly used to remove an injured victim from a trench, a confined space, or an elevated point.

FIGURE 26-4 Victim removal.

Preparing the victim for removal involves maintaining continued control of all life-threatening problems, dressing wounds, and immobilizing suspected fractures and spinal injuries. The application of standard splints in confined areas may be difficult or sometimes even impossible, but stabilization of the arms to the victim's trunk and of the legs to each other will often be adequate until the victim is positioned on a long spine board, which may serve as a splint for the whole body. A short spine immobilization device is sometimes used to stabilize a sitting victim.

Sometimes a victim has to be removed quickly (rapid extrication) because his or her general condition is deteriorating, and time will not permit meticulous splinting and dressing procedures. Quick removal may also be necessary if hazards, such as spilled gas or hazardous materials, pose a danger to the victim or the rescue personnel. The only time the victim should be moved prior to completion of initial care, assessment, stabilization, and treatment is when the victim's life or the emergency responder's life is in immediate danger.

Packaging is the process of preparing the victim for movement as a unit. It is often accomplished by means of a long spine board or similar device. Such boards are essential for moving victims with potential or actual spine injuries.

Rough-terrain rescues may require passing a stretcher up and over rough terrain, fording streams, or rock climbing. These kinds of rescues often require ingenuity to suspend or pad a stretcher so that the victim is provided with a reasonably comfortable ride. Padding such as inflated inner tubes, 10- to 13-centimetre (4- to 5-inch) foam padding, or loosely rolled blankets may be used in such circumstances. Try to achieve smooth movement of the stretcher to prevent excess swaying and bouncing of the patient.

The overriding objective for each rescue, transfer, and removal is to complete the process as safely and efficiently as possible. It is important that the rescuer utilize good body mechanics as well as victim-packaging, removal, and transportation skills.

■ Transport

Once the victim has been removed from the hazard area, transport to an appropriate medical facility is accomplished by EMS personnel. The type of transport will vary depending on the severity of the victim's injuries and the distance to the medical facility; for example, if a victim is critically injured or if the rescue is taking place some distance from the hospital, air transportation may be more appropriate than use of a ground ambulance.

In rough-terrain rescues, four-wheel drive, high-clearance vehicles may be required to transport victims on stretchers to an awaiting ambulance. Snowmobiles with attached sleds can be used to transport victims down a snowy mountain. Helicopters are increasingly used for quick evacuation from remote areas, but weather conditions may sometimes limit their use.

Transporting victims who have been injured in a hazardous materials incident may pose many problems. Prior to transport, it is extremely important that adequate decontamination occur. Placing a poorly decontaminated victim inside an ambulance or helicopter and then closing the doors puts both the victim and the rescue personnel in danger. Any toxic fumes given off by the victim or by the victim's clothing can contaminate the inside of the transport vehicle, causing injury to the EMS personnel. In such an incident, any bags containing contaminated clothing or other personal effects must be properly sealed if they are going with the victim. Receiving hospitals should be notified of the impending arrival of victims who have been involved in a hazardous materials–exposure incident. This will alert the hospital's triage teams, so that they can take the appropriate precautions once the victims begin to arrive.

Throughout any extended-duration technical rescue event, it is important to address the rehabilitation needs of rescue personnel. During many such situations, it will be necessary to rotate personnel. This course of action reduces the chance for injury or sickness to the rescuers and also provides a safer rescue for the victim. Rehabilitation is especially important when temperatures are very high or very low, and when rescuers are subjected to wet conditions. It is better to plan for rehab and not need it than to not have rehab available when it is needed. For more information on rehabilitation, see the chapter "Fire Fighter Rehabilitation."

Postincident Duties

■ Security of the Scene and Preparation for the Next Call

Once the rescue is complete, the scene must be stabilized by the rescue crew to ensure that no one else becomes injured. In a trench incident, this includes filling the trench with dirt and roping off the surrounding area.

In a hazardous materials incident, the cleanup of equipment and personnel takes place after the hazardous materials incident has been completely controlled and all victims have been treated and transported. Trained disposal crews should be available to clean the site. All equipment, protective gear, and clothing, as well as the rescue personnel themselves, must be decontaminated.

In a vehicle rescue incident, the vehicle is removed; usually, it is transported by a tow truck to a storage lot. The street or roadway is cleaned of all debris from the vehicle. This includes the clean-up of any spilled fluids, including gasoline, from the vehicle that was involved in the incident.

In an industrial setting, the supervisor of the facility is responsible for securing the scene. Nevertheless, the technical rescue team must follow up with the supervisor or the safety coordinator to ensure that further problems are prevented.

Once you have secured the scene and packed up your equipment, it is important to return to the station and fully inventory, clean, service, and maintain all of the equipment to prepare it for the next call. Although some items may need repair, most will require only simple maintenance before being placed back on the apparatus and considered in service.

Back at the station, as you prepare for the next call, complete all paperwork and document the rescue incident. Record keeping serves several important purposes: Adequate reporting and the keeping of accurate records ensure the continuity of quality care, guarantee proper transfer of responsibility, and fulfill the administrative needs of the fire department for local, provincial/territorial, and federal reporting requirements. In addition, the reports can be used to evaluate response times, equipment usage, and other areas of administrative interest.

Postincident Analysis

As with any type of call, the best way to prepare for the next rescue call is to review the last one and identify the strengths and weaknesses in your response: What did you do well? What could have been done better? Would any other equipment have made the rescue safer or easier? If a death or serious injury occurred during the call, a critical incident stress debriefing (CISD) may occur to assist fire fighters. Reviewing a TRI with everyone involved will allow everyone to learn from the call and make the next call even more successful.

General Rescue Scene Procedures

At any scene you respond to—whether it is a fire, EMS, or technical rescue call—the safety of you, your company, and the public is paramount. At a TRI, many issues need to be considered during the response. While the temptation may be to immediately approach the victim or the accident area, it is critically important to slow down and properly evaluate the situation.

In confined-space rescue incidents, for example, potential hazards may include deep or isolated spaces, multiple complicating hazards (e.g., water or chemicals), failure of essential equipment or service, and environmental conditions (e.g., snow or rain). In all rescue incidents, fire fighters should consider the potential general hazards and risks posed by utilities, hazardous materials, confined spaces, and environmental conditions, as well as IDLH hazards.

Approaching the Scene

As you approach the scene of a TRI, you will not always know which kind of scene you are entering. Is it a construction scene? Do you see piles of dirt that would indicate a trench? Has a structural collapse occurred in a building? Which actions are the civilians taking? Are they attempting to rescue trapped people, possibly placing themselves at great danger?

Beginning with the initial dispatch of the rescue call, fire fighters should be compiling facts and factors about the call. Size-up begins with the information gained first from the person reporting the incident and then from the bystanders at the scene upon arrival.

The information received when an emergency call is received is important to the success of the rescue operation. It should include the following details:

- The location of the incident
- The nature of the incident (kind, cause, what is involved, etc.)
- The conditions and positions of victims
- The conditions and positions of vehicles, building, structure, terrain, etc.
- The number of people trapped or injured, and the types of injuries

- Any specific or special hazard information
- The name of the person calling and a number where that person can be reached

Many times the information received on the emergency call will be incomplete. You will have to evaluate the incident with the information available. Once on the scene, you can better identify life-threatening hazards and take the appropriate corrective measures to mitigate those dangers. If additional resources are needed, they should be ordered by the IC.

A size-up should include the initial and continuous evaluation of the following issues:

- Scope and magnitude of the incident
- Risk–benefit analysis
- Number of known and potential victims
- Hazards
- Access to the scene
- Environmental factors
- Available and necessary resources
- Establishment of control perimeters

Careful size-up and coordination of rescue efforts are a must to avoid further injuries to the victims and to provide for the safety of the fire fighters.

Dealing with Utility Hazards

Are there any downed electrical wires near the scene **FIGURE 26-5**? Is the equipment or machinery present electrically charged so that it might present a hazard to the victim or the rescuers? The IC should ensure that the proper procedures have been taken to shut off the utilities in the area where the rescuers will be working.

Utility hazards require the assistance of trained personnel. For electrical hazards, such as downed lines, park at least one utility pole span away. Watch for falling utility poles: a damaged pole may bring other poles down with it. Do not touch any wires, power lines, or other electrical sources until they have been deactivated by a power company representative. It is not just the wires that are hazardous; any metal that they touch is also energized. Metal fences that become energized,

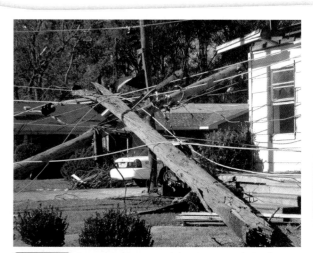

FIGURE 26-5 Downed electrical wires present a hazard at many types of rescue scenes.

Courtesy of Lara Shane/FEMA Photo News

for example, are energized for their entire unbroken length. Be careful around running or standing water, because it is an excellent conductor of electricity.

Both natural gas and liquefied petroleum gas (LPG) are nontoxic but are classified as asphyxiants because they displace breathing air. In addition, both gases are explosive. If a call involves leaking gas, call the gas company immediately. If a victim has been overcome by leaking gas, wear positive-pressure self-contained breathing apparatus (SCBA) when making the rescue. Remove the victim from the hazardous atmosphere before beginning treatment.

■ Providing Scene Security

Has the area been secured to prevent people from entering the area? Often co-workers, family members, and sometimes other rescuers will enter an unsafe scene and become additional victims. The IC should coordinate with law enforcement to help secure the scene and control access. A strict personnel accountability system should be implemented by the fire department to control access to the rescue scene.

One task you will often be called upon to perform is to establish barriers to provide scene security at a rescue incident. When emergency incidents occur on streets or highways, vehicle warning lights and pylons are often used to ward off oncoming traffic. When crowd control is an issue, plastic barrier tape marked with the words Caution or Fire Line is often used to keep out nonessential personnel. The steps for establishing a barrier are listed in **SKILL DRILL 26-1** :

1 Respond safely to the emergency scene.

2 Place the emergency vehicle in a safe position that protects the scene. (**STEP 1**)

3 Perform a size-up to assess for hazards.

SKILL DRILL 26-1 | Establishing a Barrier | **FFII**

Fire Fighter II (NFPA 6.4.2)

1 Place the emergency vehicle in a safe position that protects the scene.

2 Perform a size-up to assess for hazards. Secure the scene.

3 Use appropriate devices to establish a barrier, following the IC's orders.

Note: These are artistic renderings. The fire apparatus should be positioned away from the building collapse field.

© LiquidLibrary

④ Secure the scene. (**STEP** ②)

⑤ Call for needed assistance.

⑥ Don the appropriate personal protective equipment (PPE).

⑦ Use appropriate devices to establish a barrier, following the IC's orders. (**STEP** ③)

■ Using Protective Equipment

Firefighting gear is designed to protect the body from the high temperatures of fire; it does, however, restrict movement. Technical rescues generally require the ability to move around freely, so most firefighting gear does not work well in a TRI. For this reason, most specialized teams also carry items such as harnesses; smaller, lighter helmets; and jumpsuits that are easier to move in than turnout gear **FIGURE 26-6**.

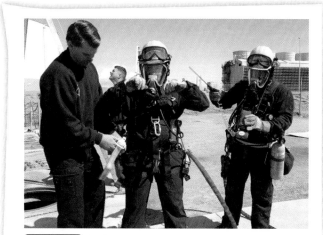

FIGURE 26-6 Technical rescue technicians may need to be able to move freely in their gear, depending on the type of incident.

For personnel operating in a water rescue hazard zone, the minimum PPE includes a Canadian Coast Guard–approved **personal flotation device (PFD)**, thermal protection, a helmet appropriate for water rescue, a cutting device, a whistle that works in water, and contamination protection, if necessary.

Other equipment items that are easily carried by fire fighters include binoculars, chalk or spray paint for marking searched areas, a compass for wilderness rescues, first-aid kits, a whistle, a hand-held global positioning system (GPS), and Cyalume®-type light sticks. The IC and the technical rescue team will help determine which PPE you will need to wear while assisting in the rescue operation. Also see NFPA 1951, *Standard on Protective Ensembles for Technical Rescue Incidents*.

■ Using the Incident Management System

The first-arriving officer immediately assumes command and starts using the Incident Management System (IMS). The IMS's activation is critically important, because many TRIs will eventually become very complex and require a large number of assisting units. Without the IMS in place, it will be difficult—if not impossible—to ensure the safety of the rescuers. For more information on IMS, see the chapter "Incident Management System."

■ Ensuring Accountability

Accountability should be practised at all emergencies, no matter how small. A **personnel accountability system** is important to ensure rescuers' safety. It tracks the personnel on the scene, including their identities, assignments, and locations. This system ensures that only rescuers who have been given specific assignments are operating within the area where the rescue is taking place. By using a personnel accountability system and working within the IMS, an IC can track the resources at the scene, make appropriate assignments, and ensure that every person at the scene operates safely. For more information on the personnel accountability system, see the chapter "Fire Fighter Survival."

■ Making Victim Contact

At any rescue scene, every effort should be made to communicate with the victim, if at all possible. Technical rescue situations often last for hours, with the victim being left alone for long periods of time. Sometimes making contact with the victim is just not possible, but do try to communicate via a radio, by cell phone, or by yelling. Reassure the victim that everything possible is being done to ensure his or her safety.

For the fire fighter, it is important to stay in constant communication with the victim. If possible, assign someone to talk to the victim, while other fire fighters carry out the rescue. Realize that the victim could be sick or injured and is probably frightened. If the fire fighter is calm, his or her reassuring demeanor will calm the victim. To help keep a victim calm, do the following:

- Make and keep eye contact with the victim.
- Tell the truth. Lying destroys trust and confidence. You may not always tell the victim everything, but if the victim asks a specific question, answer it truthfully.
- Communicate at a level that the victim can understand.
- Be aware of your own body language.
- Always speak slowly, clearly, and distinctly.
- Use the victim's name.
- If a victim is hard of hearing, speak clearly and directly to the person, so that the person can read your lips.
- Allow time for the victim to answer or respond to your questions.
- Try to make the victim comfortable and relaxed whenever possible.

Many victims at TRIs require medical care, but this care should be given only if it can be done safely. Do not become a victim yourself during a rescue attempt. Many fire fighters have been killed while entering a TRI in an attempt to help a dead victim.

Assisting Rescue Crews

Every TRI is different. If you have the role of assisting a technical rescue team, training with the team is probably the most important thing you can do to prepare for your participation in a TRI. By training with the team you will respond with, you will get a feel for how the team members operate and they will get an idea of what they can trust you with. The more knowledge you have, the more you will be able to do.

At any TRI, follow the orders of the company officer who receives direction from the IC. Many of the tasks will involve moving equipment and objects from one place to another. Other tasks will involve protecting the team and victims. Do not take these tasks lightly; they might not seem important but they are essential to the effort. Without scene security, traffic control, protective hose lines, and people to maintain equipment, the rescue cannot and will not happen.

One task you will often be called upon to perform at any type of special rescue incident is to retrieve needed rescue tools. To do so, you must become familiar with the tools that are used by the specialized rescue teams in your department. The steps for identifying and retrieving rescue tools are listed in **SKILL DRILL 26-2** :

1. Respond safely to the emergency scene.
2. Place the emergency vehicle in a safe position that protects the scene.
3. Perform a size-up to assess for hazards.
4. Secure the scene.
5. Call for needed assistance.
6. Don the appropriate PPE.
7. Receive the request from the special rescue team to retrieve rescue tools.
8. Identify and retrieve the rescue tools **FIGURE 26-7** .

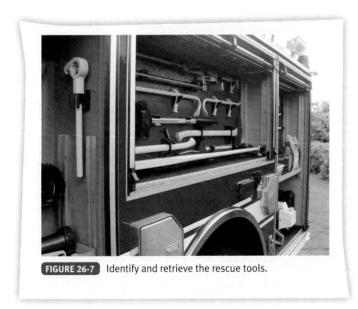

FIGURE 26-7 Identify and retrieve the rescue tools.

As you read the information that follows about assisting at specific types of rescue scenes, bear in mind the three factors about safely approaching the scene, which apply to all TRIs:

- Approach the scene cautiously.
- Position the apparatus properly.
- Assist the specialized team members as needed.

■ Vehicles and Machinery

A wide variety of rescue situations involving vehicles and machinery are possible. By far, the most common type that fire fighters encounter is the motor vehicle crash. The actions

commonly taken at the scene of motor vehicle crashes are covered in the chapter "Vehicle Rescue and Extrication." This section focuses on other types of rescues involving vehicles and machinery, including motorcycles, trucks, buses, trains, mass-transit vehicles, aircraft, watercraft, farm machinery, agricultural implements, construction machinery, and industrial machinery. Rescues involving mass transit often involve many victims, whereas most other types of rescues involve single victims.

Safe Approach

Vehicle and machinery rescues present a wide variety of problems to rescuers and victims. Hazards present at these scenes may, for example, include flammable liquids, electrical hazards, and unstable machines or vehicles.

Electricity is an invisible hazard. Consider any machine you encounter to be electrically charged until proven otherwise. Because of this risk, it is important to use lockout and tagout systems with any potential electrical source before approaching the scene. The operator of the machine or a maintenance person who is familiar with the machine can be a valuable resource in this regard.

With vehicle-related calls, traffic control is a crucial issue. Every year, fire fighters and EMS workers are struck and killed at accident scenes by passing drivers who became distracted by the commotion. Placement of the apparatus is key to the protection of you and those around you. In addition, whenever you are in a traffic-related situation, you must wear an approved vest to increase your visibility and prevent injuries.

Stabilization of the rescue scene includes making sure that the machine cannot move. Gaining access to some victims will be easy; in other situations, access to the victim can occur only after extensive stabilization of the scene.

Unusual and special rescue situations include extrication and disentanglement operations at incidents involving vehicles on their tops or sides, trucks and large commercial vehicles, and vehicles on top of other vehicles. To ensure responder safety, be wary of fuel spilled at these scenes.

How You Can Assist

At a vehicle or machinery rescue call, you may be called upon to assist in the extrication and treatment of the victim. Adequate protection of the victim (e.g., with c-spine immobilization or holding of blankets) will allow the technical rescue team to operate hydraulic tools and cut the vehicle or machine apart without further hurting the victim. Many hydraulic tools are heavy, so you may need to help support the tools while technicians operate these pieces of equipment.

One skill with which you can assist is the removal of a patient from a vehicle using a long backboard. This skill is covered in the chapter "Search and Rescue." You may also be called upon to do the following:

- Assist with controlling site security and the perimeter of the rescue incident.
- Obtain information from witnesses.
- Retrieve more rescue tools and equipment from the fire apparatus or rescue trucks.
- Console a family member of a victim.

- Keep bystanders out of the way.
- Assist in moving items that are in the way of the rescue team.
- Service equipment.
- Set up power tools.
- Stand by with a hose line.
- Extinguish a fire.

Tools Used

Many tools—including some inexpensive items—are required for a successful vehicle or machinery rescue situation. Simple tools such as a spring-loaded punch, for example, can safely break out a vehicle window to access a victim. Tools and items that will be required for a vehicle or machinery rescue include the following:

- PPE
- Hydraulic tools (spreaders, cutters, rams)
- Halligan tool
- Cutting torch
- Air chisels
- Cribbing
- Saber saw
- Windshield cutter
- Spring-loaded punch
- Chains
- Air bags (high and low pressure)
- SCBA air cylinders
- Basic hand tools
- Come along
- Portable generator
- Seat belt cutters
- Hand lights and other scene lighting
- Hose lines (protection)
- Blanket

Confined Spaces

A **confined space** is an enclosed area that is not designed for people to occupy. Confined spaces have limited openings for entrance and exit and may have limited ventilation to provide air circulation and exchange. They can occur in natural, farm, commercial, and industrial settings. For example, grain silos, industrial pits, tanks, and below-ground structures are all confined spaces. Likewise, cisterns, well casings, sewers, and septic tanks are confined spaces that are found in residential settings.

Confined spaces present special hazards, as they may be oxygen deficient, contain poisonous gases, or have combustible atmospheres. Entering a confined space without testing the atmosphere for safety and without the proper breathing apparatus can result in death. Confined spaces may also contain hazardous chemicals as well as harmful dusts and moulds. Other serious safety or health hazards present at these sites may include slips, trips, falls, noise, dust, poor lighting, and sensitizing agents. A confined-space call is sometimes dispatched as a heart attack or medical-illness call, because the caller assumes that the person who entered a confined space and became unresponsive because of a heart attack or medical illness rather than because of a hazardous atmosphere.

All confined spaces should be considered to contain, or have the potential to contain, a hazardous atmosphere until proven otherwise.

Fire Fighter Safety Tips © Photos.com

The two greatest hazards in a confined space are a lack of oxygen and the presence of poisonous gases or asphyxiants. If rescuers fail to identify a confined space and to ensure that it contains a safe atmosphere, injury and death can occur to rescuers as well as to the original victim. In several cases, well-meaning rescuers have gone into a confined space, become incapacitated, and died because they did not properly evaluate the situation.

Safe Approach

As you approach a rescue scene, look for a bystander who might have witnessed the emergency. Information gathered prior to the technical rescue team's arrival will save valuable time during the actual rescue. Do not assume that a person in a pit has simply suffered a heart attack; instead, assume that an IDLH atmosphere is present at any confined-space call. An IDLH atmosphere can immediately incapacitate anyone who enters the area without breathing protection. Toxic or explosive gases may be present, or the confined space may not have enough oxygen to support life.

When a rescue involves a confined space, remember that it will take some time for qualified rescuers to arrive on the scene and prepare for a safe entry into the confined space. Given this delay, the victim of the original incident may have died before your arrival. Do not put your life in danger on behalf of a body recovery. The majority of deaths in confined spaces are would-be rescuers.

How You Can Assist

Your main role in confined-space rescues is to secure the scene, preventing other people from entering the confined space until additional rescue resources arrive. If the incident has occurred at an industrial site, ascertain whether a confined-space entry permit was prepared. If so, locate the entry supervisor and attendant listed on the permit and have them stand by at the space. As additional highly trained personnel arrive, your company may provide help by giving the rescuers a situation report.

The first-responding company must share whatever information it discovered at the rescue scene with other arriving crews. Anything that might be important to the response should be noted by the first-arriving unit. Observed conditions should be compared to reported conditions, and a determination should be made as to the relative change over the time period. Whether an incident appears to be stable or has changed greatly since the first report will affect the operation strategy for the rescue. A size-up should be completed quickly and immediately upon arrival, and this information should be relayed to the special rescue team members upon their arrival at the scene. Other items of importance that should be included in a situation report are a description of any rescue attempts that have been made, the exposures, any hazards present, the extinguishment of fires, the facts and probabilities of the scene, the situation and resources of the fire company, the identities of any hazardous materials present, and an evaluation of the progress made so far.

Many engine and ladder companies carry atmospheric or gas detection devices; if you have this capability, obtain readings to determine whether the situation presents a hazard. If you do encounter an oxygen-deficient atmosphere, set up a ventilation fan to help remove toxic gases and improve air flow to the victim. Sometimes you can help a victim without entering the confined space by passing down SCBA, oxygen, first-aid supplies, or even a ladder that the victim can use to climb out.

Confined-space rescues can be complex and take a long time to complete. You may be asked to assist by bringing rescue equipment to the scene, maintaining a charged hose line, or providing crowd control. By understanding the hazards associated with confined spaces, you will be better prepared to assist a specialized team that is dealing with an emergency involving a confined space.

Tools Used

One of the key tools used in any confined-space rescue is a **supplied-air respirator** system **FIGURE 26-8**. These devices are similar to SCBA, except that instead of carrying the air supply in a cylinder on your back, you are connected by a hose line to an air supply located outside the confined space. This kind of system provides the rescuer with a continuous supply of air that is not limited by the capacity of a back-mounted air cylinder.

A number of other tools and pieces of equipment can also be used during a confined-space rescue operation:

- Air lines
- Air carts
- Extra SCBA air cylinders
- Tripod and winch for raising and lowering rescuers and victims
- Pry bar or manhole-cover remover
- Rescue rope
- Personnel harnesses
- Gas monitors
- Ventilation fans
- Explosion-proof lights

- Radios
- Hard-wired communications equipment
- Protective gear
- Lockout, tagout, and blankout kits
- Accountability boards
- Victim-removal devices such as short backboard devices, Stokes baskets, backboards, and other commercially made devices
- Medical equipment
- Carabiners and locking D rings
- Descenders
- Safety goggles
- Hand light (battery-operated)
- Hearing-protection headsets

■ Rope Rescue

Rope rescue skills are used in many types of TRIs. For example, an injured victim might be lifted from inside a tank using a raising system, or rescuers might need to lower themselves down to the area where victims are trapped in a structural collapse. Rope rescue skills are the most versatile and widely used technical rescue skills.

Rope rescue incidents are divided into low-angle and high-angle operations. **Low-angle operations** are situations where the slope of the ground over which the fire fighters are working is less than 45 degrees, fire fighters are dependent on the ground for their primary support, and the rope system becomes the secondary means of support **FIGURE 26-9**. An example of a low-angle system is a rope stretched from the top of an embankment and used for support by rescuers who are carrying a victim up an incline. In this scenario, the rope is the secondary means of support for the rescuers; the primary support is the rescuers' contact with the ground.

Low-angle operations are used when the scene requires ropes to be used only as assistance to pull or haul up a victim or rescuer. This step usually becomes necessary when adequate footing is not present in areas such as a dirt or rock embankment. In such an incident, a rope will be tied to the rescuer's harness, and the rescuer will climb the embankment on his or

FIGURE 26-8 Supplied-air respirator system.

FIGURE 26-9 Low-angle operations.
© Keith D. Cullom

her own, using the rope only to make sure he or she does not fall. Low-angle operations also include life lines deployed during ice or water rescues.

Ropes can also be used to assist in carrying a Stokes basket. This use of ropes aids the rescuers and frees them from having to carry all of the weight over rough terrain. Rescuers at the top of the embankment can help to pull up or lower the basket using a rope system.

<u>High-angle operations</u> are situations where the slope of the ground is greater than 45 degrees, and rescuers or victims depend on life-safety rope for support, rather than on a fixed surface, such as the ground. High-angle rescue techniques are used to raise or lower a person when other means of raising or lowering are not readily available. Sometimes, a rope rescue is performed to remove a person from a position of peril; at other times, it is needed to remove an ill or injured victim.

The chapter "Ropes and Knots" discusses how to tie some basic knots and to use ropes to lift and lower selected tools. Rope rescue training courses will build on these skills. There is much more to learn before you are ready to perform high-angle rescues, however, because you need to be able to perform complex tasks quickly and safely in these incidents. Successful rope rescue operations also require a cohesive team effort.

Safe Approach

If you respond to an incident that may require a rope rescue operation, consider your safety and the safety of those around you. Rope rescues are among the most time-consuming calls that you will encounter. Extensive setup is necessary, and a significant body of equipment must be assembled prior to initiating any rescue. Protect your safety by remaining away from the area under the victim and away from any loose materials that may fall or slide. Likewise, when you are above the victim, stay back from the edge unless you are securely tied off to keep from falling. Use extreme caution so you do not cause any loose debris to fall on the victim. Work to control the scene so that the bystanders and friends of the victim move to an area where they will not be injured.

You can do a lot to stabilize the scene and prevent further injuries by remaining calm and putting your skills to work. A rope can, for example, be used to secure a rescuer when making a rescue attempt, haul a stretcher or litter up an embankment, or lower a rescuer into a trench or cave.

How You Can Assist

Rope rescues are labour-intensive operations. By remembering the introductory rope skills you have learned, you may be able to assist with many phases of the rescue operation; for instance, a technical rescue team member might ask you to tie knots and get anchors ready. Do not be offended if the team members go back and check your work—it is protocol to check a system two or three times to ensure its safety.

Avoid stepping on ropes, because any damage or friction to the outer layers of a rope can decrease its tensile strength and possibly cause a catastrophic failure. Keep everyone clear of areas where falling objects could cause injuries. Keep in mind the importance of following the IMS. Work to complete your assigned tasks.

Tools Used

Rope rescues rely almost completely on the equipment used. When you are rappelling down to rescue a victim who is 90 metres (300 feet) off the ground, you need to have equipment that has been properly designed and maintained to keep you alive. Some of the tools and equipment that you will need to recognize and use are listed here:

- PPE (helmet, rescue gloves)
- Personnel harnesses
- Stokes basket
- Harness for Stokes basket
- Rescue ropes
- Carabiners
- Webbing and Prussic cord
- Miscellaneous hardware (racks, pulleys)

■ Trench and Excavation Collapse

Trench and excavation rescues become necessary when earth has been removed for placement of a utility line or for other construction and the sides of the excavation collapse, trapping a worker **FIGURE 26-10**. <u>Entrapments</u> can occur when children play around a pile of sand or earth that collapses. Many entrapments occur because the required safety precautions were not taken.

Whenever a collapse has occurred, you need to understand that the collapsed product is unstable and prone to further collapse. Earth and sand are heavy, and a person who is partly entrapped in these materials cannot simply be pulled out. Instead, the victim must be carefully dug out. This step can be taken only after <u>shoring</u> has stabilized the sides of the excavation.

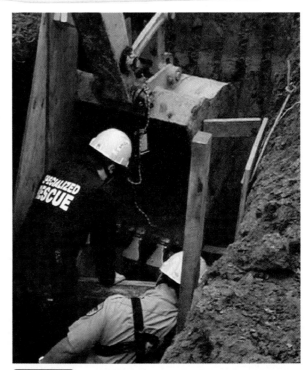

FIGURE 26-10 Trench rescue.

Courtesy of Captain David Jackson, Saginaw Township Fire Department

Vibration or additional weight on top of displaced earth will increase the probability of a **secondary collapse**—that is, a collapse that occurs after the initial collapse. A secondary collapse can be caused by equipment vibration, personnel standing at the edge of the trench, or water eroding away the soil. Safe removal of trapped persons requires a special rescue team who are trained and equipped to erect shoring to protect the rescuers and the entrapped person from secondary collapse.

Safe Approach

Safety is of paramount importance when approaching a trench or excavation collapse. Walking close to the edge of a collapsed area can trigger a secondary collapse, so stay away from the edge of the collapse and keep all workers and bystanders at a distance from the site. Vibration from equipment and machinery can also cause secondary collapses, so shut off all heavy equipment. Likewise, vibrations caused by nearby traffic can cause collapse, so it may be necessary to stop or divert traffic near the scene.

Soil that has been removed from the excavation and placed in a pile is called the **spoil pile**. This material is unstable and may collapse if placed too close to the excavation. Avoid disturbing the spoil pile.

Make verbal contact with the trapped person, if possible, but do not place yourself in danger while doing so. If you plan to approach the trench at all, do so from the narrow end, where the soil will be more stable; however, it is best not to approach the trench unless absolutely necessary. Stay out of a trench unless you have been properly trained in trench rescue techniques.

Provide reassurance by letting the trapped person know that a trained rescue team is on the way. By moving people away from the edges of the excavation, shutting down machinery, and establishing contact with the victim, you start the rescue process.

You can also size up the scene by looking for evidence that would indicate where the trapped victims may be located. Hand tools are an indicator of where victims may have been working, as are hardhats. By questioning the bystanders, you can also determine where the victims were last seen.

How You Can Assist

As the rescue team starts to work, your company will be assigned to carry out certain tasks, ranging from unloading lumber for shoring to assisting with cutting timbers at a safe distance away from the entrapment. This type of rescue can take a long time. If it is hot or cold, a rehabilitation sector may need to be set up. Early implementation of the IMS will help make this type of rescue go smoothly. If someone in your company has a specialty such as carpentry, he or she should make this fact known, because this skill is valuable at a trench rescue. Cutting and measuring timber and shores for a trench operation is an important aspect of the rescue.

Like confined spaces, trenches may have an IDLH atmosphere inside, owing to the presence of poisonous gases, such as methane or other sewer gases. Another hazard may be gasoline fumes from gas-powered equipment or gas cans in the trench. Buried power lines may also be present. Setting up ventilation fans can often make a difference in victim and rescuer survivability in these scenarios.

It may also be necessary to pump water out of a trench. If the collapse occurred because of a ruptured pipe or during a rainstorm, for example, rising water in the trench may endanger the trapped victim and cause additional soil to collapse into the trench; removing this water can, therefore, stabilize the situation.

When extricating victims from the trench, it may be necessary to lift them out using a raising system, just as in a high-angle rope rescue operation. This effort will require that the rescuers have rescue ropes, harnesses, pulleys, carabiners, and all of the other associated equipment available. Personnel must be trained in how to use and apply these systems in this type of rescue scenario.

Tools Used

Trench and excavation rescue uses equipment, rescue techniques, and skills similar to those used with both confined-space rescue and structural-collapse rescue. Tools and equipment used in trench and excavation rescue include the following:

- PPE—helmet, gloves, personal protective clothing, harness, flashlight, work boots, knee pads, elbow pads, eye protection, SCBA, and supplied-air breathing apparatus (SABA)
- Hydraulic, pneumatic, and wood shores
- Lumber and plywood for shoring
- Cribbing
- Power cutting tools and saws
- Carpentry hand tools
- Shovels
- Buckets for moving soil
- Rescue rope, harnesses, webbing, and associated hardware
- Utility rope
- Ventilation fans
- Water pumps
- Lighting
- Ladders
- Shovels
- Extrication equipment, such as Stokes baskets or backboards
- Harness sets for Stokes baskets
- Medical equipment

■ Structural Collapse

Structural collapse is the sudden and unplanned fall of part or all of a building **FIGURE 26-11**. Such collapses may occur because of fires, removal of supports during construction or renovation, vehicle crashes, explosions, rain, wind, snowstorms, earthquakes, and tornados.

The chapter "Building Construction" emphasizes the importance of all building components working together to provide a stable building. Consider the type of building construction when determining the potential for collapse. When any part of a building becomes compromised, the dynamics of the building will change. As a consequence, fire fighters should always be alert for signs of a possible building collapse.

FIGURE 26-11 Structural collapse is the sudden and unplanned fall of part or all of a building.

Courtesy of Dave Gatley/FEMA

A partial building collapse may also be hazardous to rescuers because of the potential for secondary collapse.

Safe Approach

Many different factors can lead to building collapses, so you must approach the scene carefully. The causes of some collapses—for example, a vehicle crashing through the wall of a house—will be readily evident. Other causes, such as a natural gas explosion, may require extensive investigation before they are definitively identified.

> ### Fire Fighter Safety Tips
> © Photos.com
>
> Fires often lead to structural collapses. Be alert for the possibility of a building collapse anytime a fire has damaged a building.

As you approach any building collapse, consider the need to shut off utilities. Entering a structure with escaping natural gas or propane is extremely hazardous. Electricity from damaged wiring can also present a deadly hazard.

A prime safety consideration is the stability of the building. Even a well-trained engineer cannot always determine the stability of a building simply by looking at its exterior; therefore, you must operate as though the building might experience a secondary collapse at any time. The IC must make the decision regarding whether the building is safe to enter. This is

> ### FIRE FIGHTER II Tips   FFII
>
> New fire fighters must learn when to offer helpful suggestions and when to remain quiet. Sometimes you might see something that the company officer does not see, or you might have an idea that the company officer has not considered. In such a case, the need to provide helpful information must be balanced against the risk of overwhelming a company or chief officer with "helpful" suggestions. No hard-and-fast rule covers all situations, of course, but stay mindful of the fact that remaining quiet is sometimes the most helpful thing you can do.

a difficult decision, especially given that bystanders cannot always tell you whether people were present in the building before the collapse.

How You Can Assist

Rescue operations at a structure collapse vary depending on the size of the building and the amount of damage to the building. In cases of large building collapses, the rescue operation will be sizable. Urban search and rescue teams and structural-collapse teams have received special training in dealing with these types of situations. Specifically, they are trained in shoring and specialized techniques for gaining access and extricating victims that enable them to systematically search the affected building. These types of rescue situations take a lot of time and require a lot of personnel.

Personnel who do not have this kind of special training will usually be assigned to support operations; for example, you might be assigned to be part of a bucket brigade that works to remove debris from the building. A substantial amount of manual digging and searching may be necessary, all of which is physically taxing. You may be able to assist with this activity if the work is not in an unstable location.

For a rescue effort of this type to work, there must be teamwork and a well-organized and well-implemented IMS. Without a solid command structure, most large-scale rescue efforts are doomed to fail.

Tools Used

Fire fighters should know the tools and equipment designed for structural-collapse emergency rescue incidents:

- PPE—helmet, gloves, work boots, harness, elbow pads, knee pads, eye protection, SCBA, SABA, and dust masks
- Shoring equipment
- Lumber for shoring and cribbing
- Power tools
- Hand tools
- Lighting
- Rescue ropes, harnesses, webbing, and associated hardware
- Utility ropes
- Buckets
- Shovels

■ Water and Ice Rescue

Almost all fire departments have the potential for being called to perform a water rescue **FIGURE 26-12**. Water is present in small streams and large rivers, and it fills lakes, oceans, reservoirs, irrigation ditches, canals, and swimming pools. A static source such as a lake may have no current, whereas a whitewater stream or flooded river may have a swift current. During a flood, a dry wash in the desert can temporarily become a huge raging, swiftly flowing river.

> ### Fire Fighter Safety Tips
> © Photos.com
>
> Wearing turnout gear while conducting water rescue is dangerous and can be deadly. Do not wear turnout gear within 3 m (10 ft.) of water, especially around swift-moving water.

Near-Miss REPORT

Report Number: 08-0000557

Synopsis: Embankment rescue could have ended tragically.

Event Description: We were called on a technical rescue response for a report of two dogs stuck on a cliff. The response consisted of an engine, a rescue company, a truck company, a chief, and a safety officer. The reported location was a rugged area of suburban/wildland at the mouth of a steep canyon. The first-due engine responded from the station approximately 1.6 kilometre 1 mile away. The rest of the units came from substantially farther away and would have a delayed arrival.

Upon arrival, the first-due engine company met with the reporting person and were escorted up to the area where the dogs were. They were on small ledge approach, 22.9 m (75 ft.) down from a small promontory, which served as their access point. The terrain below the access point was a series of steep, loose rock and grass slopes ending in small 1.5- to 3-m (5- to 10-ft.) drop-offs to a ledge. The dogs were on the third ledge down at the end of a 3-m (10-ft.) drop-off. Below that ledge was an approximately 15-m (50-ft.) drop to a dirt access road.

The two fire fighters had grabbed a utility rope bag and a rope rescue pack before heading up. Upon reaching the access point, the first fire fighter began to scramble down to the dogs. As the dogs were initially agitated and growling, the second fire fighter felt that he should proceed down to assist his partner, who had now reached the dogs.

About halfway down the slope, the second fire fighter realized the severity of the terrain and asked the bystander to throw down a rope. He threw down a utility rope while holding on to one end. The closest fire fighter tied a loop around his waist and tossed the end down to his partner, who did the same. The fire fighters asked the bystander to belay them.

The engine officer observed this part of the operation from a vantage point below the access point. He gave a size-up that said they had reached the dogs and it was a simple low-angle rescue.

When the rescue company arrived on scene, the rescue officer was assigned to rescue operations and proceeded up to the access point. He updated the size-up to include the fact that the rescue was a high-angle rescue and that the first priority was to remove the two fire fighters to safety. A lowering system with safety was established, and a technical rescue team member was lowered to the stuck fire fighter. He was put in a harness and safety and was hauled up. The technical rescue team member proceeded down to the first fire fighter and the dogs. He was placed in a harness and hauled up. The dogs were then assisted up to the access point and walked down to animal-control officers, who checked them over and figured out their ownership. The technical rescue team member came up safely, and the operation was broken down.

Lessons Learned: The main lesson learned was that you should not let emotion and adrenaline override judgment. The dogs had been on the cliff for at least several days and were in no immediate danger. There was plenty of time to assess the situation and plan for a proper, safe operation.

The first-in officer needed to be more of a presence and control his crew to prevent them from acting hastily.

- Awareness—The first fire fighters began to scramble down and were already in a dangerous situation before they realized that fact and asked a bystander for assistance.
- Training—Frequent training with the rescue equipment carried on the first-due engine would have allowed the team members to set up a safe lowering system in a timely manner and still achieve their objectives.

In North America, the most common swiftwater rescue scenario involves a car that has tried to drive through a pool of water created by a storm. When the vehicle stalls because of the depth of the water, it leaves the vehicle occupants stranded in a rising stream with a swift current. If the water is high enough, the vehicle can be swept away. These incidents are especially dangerous for rescue personnel, because it is difficult to determine the depth of the water around the vehicle.

Fire departments should be prepared for all types of water rescue incidents that may occur in their communities. Fixed bodies of water in your community are easy to identify. Identifying the areas in your community where flooding is likely to occur will require preplanning. Other common water rescue scenarios include cars rolling into lakes or streams, cars going off bridges, people falling into bodies of water, and swimmers getting into trouble.

FIGURE 26-12 Water rescue.

© Nate Rawner

All fire fighters should be trained in the use of water rescue throw bags. This equipment contains a rope that is thrown to a struggling victim in water or on ice. The rescuer secures one end of the rope before throwing the water rescue throw bag. Once the victim has grasped the water rescue throw bag, the rescuer can pull the victim to shore.

Safe Approach

When responding to water rescue incidents, your safety and the safety of other teammates is your primary concern. Your turnout gear is designed for structural firefighting, and it provides amazing protection for that purpose. Unfortunately, it is not designed for water rescue activities. Fire fighters who fall into the water while wearing turnout gear quickly find that their movements are severely limited by the bulky gear. For this reason, when working at a water rescue scene, you should use PPE designed for water rescue, not structural firefighting. Anytime you are within 3 m (10 ft.) of the water, you should be wearing a Canadian Coast Guard–approved PFD. Shoes that provide solid traction are preferred to fire boots in this situation.

If you are part of the first-arriving engine or truck company, and the endangered people are in a vehicle or holding on to a tree or other solid object, try to communicate with them. Let them know that additional help is on the way.

Do not exceed your level of training. If you cannot swim, operating around or near water is not recommended. A person who is trained as a lifeguard for still water is not prepared to enter flowing water with a strong current, such as is found in a river, stream, or ocean. Trained swiftwater rescuers may consider using a throw rope, a pike pole, or a ladder to reach a person. Many people who are in trouble in the water are close to shore. Do not use a boat unless you are trained in its use. Ensure that bystanders do not try to rescue the victim and place themselves in a situation where they need to be rescued as well.

Ice rescues are common in colder climates, and many departments in colder regions have developed specialized equipment for these scenarios. Throwing a rope or flotation device may be helpful initially. In addition, ladders can be extended to the victim or used to distribute the weight of the rescuer on ice-covered water. A fire hose that has been capped at each end and then inflated with an SCBA air cylinder can be used to create to create a flotation buoy. In addition, many other devices can be used to keep a victim above the ice and pull him or her to safety. Rescuers in this situation need special ice rescue suits to prevent hypothermia and provide flotation. If your department is involved in ice rescues, you need to receive training in these specialized procedures.

How You Can Assist

Water and ice rescues are dangerous for rescuers. PFDs should be worn by everyone operating at the scene. Good scene control is needed to prevent additional people from becoming victims.

At water and ice rescues, you may be called upon to assist by keeping victims in sight, retrieving equipment, assisting with rescue ropes, and changing rescuers' air supplies. You can also assist special rescue teams by relaying communications.

Be alert for changing weather conditions, such as an increase in wind speed or a change in temperature that could affect the rescue operations.

Tools Used

Fire fighters should become familiar with the tools used in water rescue operations:

- PFD
- Self-contained underwater breathing apparatus (SCUBA)
- Helmet
- Waterproof whistle
- Throw bag with rope
- Boat
- Rescue rope, webbing, and associated hardware
- Lighting and flashlights
- Gloves
- Goggles or other eye protection
- Suitable footwear for the environmental and rescue conditions

■ Wilderness Search and Rescue

Wilderness search and rescue (SAR) is conducted by a limited number of fire departments. It is included in NFPA 1670 as part of technical rescue training. SAR missions consist of two parts: search and rescue. Search is defined as looking for a lost or overdue person. Rescue, in the SAR context, is defined as removing a victim from a hostile environment.

Several types of situations may result in the initiation of SAR missions. For example, small children may wander off and be unable to find their way back to a known place. Older adults who are suffering from disorders such as Alzheimer's disease may fail to remember where they are going and become lost. People who are hiking, hunting, or participating in other wilderness activities may become lost because they do not have the proper training or equipment for that activity, because the weather changes unexpectedly, or because they become sick or injured.

Safe Approach

As a beginning fire fighter, you may respond to a possible SAR mission as part of your fire company. In cases of a lost person, some small fire departments will request all available personnel to respond to assist with a search. It is important to respond to such a call as part of your department and to work together as a team. The IC for such a mission should be a person who is well trained in directing SAR missions. In some communities, SAR is the responsibility of law-enforcement personnel; in others, it becomes the responsibility of volunteer search and rescue groups.

The term *wilderness* may be used to describe a variety of environments, such as forests, mountains, deserts, parks, and animal refuges. Depending on the terrain and environmental factors, the wilderness can be as little as a few minutes into the backcountry or a couple metres off the roadway. An incident with a short access time could require an extended evacuation and, therefore, qualify as a wilderness incident. Examples of terrains found in wilderness areas include cliffs, steep slopes, rivers, streams, valleys, mountainsides, and beaches. Terrain hazards may include cliffs, caves, wells, mines, avalanches, and rock slides.

When you participate in SAR missions, prepare for the weather conditions by bringing suitable clothing. Make sure that you do not exceed your physical limitations, and do not get in situations that are beyond your ability to handle yourself in the wilderness. Call for a special wilderness rescue team, depending on the needs of the situation and your local protocols.

How You Can Assist

The fire service teaches its members to work in a buddy system; wilderness rescues are another situation in which you never go out alone. By working in teams of at least two and having a radio, teams can be methodically deployed and assigned to a search. The more knowledge you have about the search area, the more vital your role is during the search. Knowing your response area and the places where a child or lost person might hide is always beneficial to the search. A well-coordinated team can be an effective SAR force; conversely, an unorganized group of people will produce increasing chaos. Do not enter the search area before the search team arrives. If the team will be using dogs, your scent will distract them.

Tools Used

Each specialized rescue team will bring its own equipment to the scene. Many of the same tools, equipment, and personal protective clothing that you would use for other specialized rescue situations will be used in SAR emergencies:

- Personal clothing appropriate for the environment
- Water
- Food
- Lighting and flashlights
- Communications equipment, including radios
- Medical equipment
- Maps, compass, and GPS
- Shelter
- Rescue ropes, harnesses, webbing, and associated hardware

- Extrication equipment, such as a Stokes basket and harness
- Flare gun and flares, whistles, or other signalling devices

■ Hazardous Materials Incidents

Hazardous materials are defined as any materials or substances that pose a significant risk to the health and safety of persons or to the environment if they are not properly handled during manufacture, processing, packaging, transportation, storage, use, or disposal **FIGURE 26-13** .

FIGURE 26-13 Hazardous materials incident.

Courtesy of Photographer's Mate 2nd Class Daniel R. Mennuto/U.S. Navy

Although hazardous materials incidents often involve petroleum products, many other chemicals in our society have toxic effects when they are not handled properly. We tend to think of hazardous materials incidents as occurring during transportation or at a large industrial setting, but it is important to consider that many retail businesses contain significant quantities of hazardous materials. Home and garden stores, farm co-operatives, hardware stores, and pool supply stores, for example, are places where hazardous materials can be found in significant quantities.

Most fire departments are trained and equipped to recognize these incidents, contain the hazards, and evacuate people from the area, if necessary. See the hazardous materials chapters for complete coverage of hazardous materials incidents.

Safe Approach

Hazardous materials incidents are not always dispatched as hazardous materials incidents. Given this fact, you must be able to recognize the signs indicating that hazardous materials may be present as you approach the scene. Preincident plans may note that a fixed facility stores chemicals, or the use of a specific type of transport container may tip you off to the presence of hazardous materials. You might also see an escaping chemical or smell a suspicious odour. Warning **placards** are required for most hazardous materials for storage or in transit.

Once you have recognized the presence of a hazardous material, you must protect yourself by staying out of the area exposed to the material. If you have happened upon a hazardous materials incident, it is important to have the hazardous materials team dispatched as soon as possible. In addition to standard action and precautions, call for a special hazardous materials rescue team immediately upon your arrival at such an incident, implement site control and scene management, and assist specialized personnel after their arrival according to your training level.

How You Can Assist

To assist in operations at a hazardous materials incident, you need formal training in techniques to deal with this kind of threat. Training at the awareness level will provide you with the knowledge and skills you need to be able to recognize the presence of a hazardous material, protect yourself, call for appropriate assistance, and evacuate or secure the affected area. As part of your training, you will learn how to assist other responders to the hazardous materials incident.

The four major objectives of training at the operational level are to analyze the magnitude of the hazardous materials incident, plan an initial response, implement the planned response, and evaluate the progress of the actions taken to mitigate the incident. Being trained and equipped to perform emergency decontamination of victims may help to minimize the effects of chemical exposures, while hazardous materials technicians are en route.

Tools Used

The following tools are used in a hazardous materials incident:

- PPE appropriate to the level of the hazard
- Two-way radios
- Lighting
- Gas monitors
- Sensing and monitoring tools to identify the materials involved
- Extensive research material
- Decontamination equipment
- A wide variety of hand tools (e.g., hammers, screwdrivers, axes, wrenches, pliers)
- Devices for sealing breached containers
- Leak-control devices
- Binoculars
- Fire-line tape
- Control agents

Elevator and Escalator Rescue

Elevator and escalator rescues are technical responses that require specialized knowledge. Rescue crews should review the American Society of Mechanical Engineers' *A17.4 Guide for Emergency Personnel* and the Canadian Centre for Occupational Health and Safety's (CCOHS) lockout and tagout procedures.

All elevator and escalator responses begin with knowledge of the machinery found in your department's jurisdiction— information that can be gathered only through a preincident analysis of these structures. Specialized training classes are also an essential component of safe and successful operations.

Elevator Rescue

As metropolitan areas continue to grow, building heights are also increasing. Because of our aging population, more buildings are being equipped with elevators. In turn, elevator emergencies are increasing in frequency **FIGURE 26-14**.

Fire fighters should never attempt to move or relocate an elevator under any circumstances; only professional elevator technicians who are thoroughly trained and authorized to do so should consider this step. When responding to elevator incidents, consider these recommendations:

1. Always cut power to a malfunctioning elevator, and secure the power supply through the appropriate lockout and tagout procedures. Remember that elevator technicians and building maintenance personnel will have been summoned to the same elevator incident; they may try to turn on the elevator equipment. To eliminate any possibility of this happening, use the lockout and tagout procedures

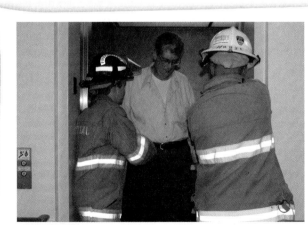

FIGURE 26-14 Elevator emergencies are a common emergency call for many fire departments.

© Dennis Wetherhold, Jr

Elevator and Escalator Rescue (*Continued*)

at every incident. Turn off the power supply in the elevator machine room, and identify the location of the stalled elevator within the hoist way or elevator shaft. Are there victims in the stalled elevator? If so, communicate with them, and reassure them that the situation is under control, they are not in danger, and the situation will soon be corrected.

2. Implement incident risk management and risk assessment. Can the victims in the disabled elevator be removed without increasing the risk of injury or death to fire fighters as well as to the victims? Elevator cabs are passenger compartments that are securely locked to ensure that the victims are contained within the cab for their own safety. Fire fighters must carefully evaluate the situation and the inherent risks involved in a rescue and extrication operation, and understand that the victims are in a much safer environment inside the elevator than outside it. The best option may be to reassure the victims and wait for the elevator technician to arrive and correct the situation.

3. Are there enough trained personnel and the proper equipment on scene to perform the rescue and extrication operation safely? Elevator hall door release keys, safety harnesses, rope bags, ladders, forcible entry tools, portable lighting, and an assortment of hand tools will be necessary in such a case. Always have additional trained personnel on hand, and anticipate obstacles and problems. You need to know the procedure your department uses to call for assistance when confronted with an elevator rescue.

4. Once the incident has been resolved, leave the power supply off. Always leave the building with the elevator power supply turned off.

Escalator Rescue

Escalator emergencies can be challenging and complex. Fire fighters should never attempt to move an escalator under any circumstances; only professional escalator technicians who are thoroughly trained and authorized to do so should consider this step. This is particularly true if a victim (usually a child) is caught or entangled in the machinery.

Good incident risk-management and risk-assessment skills are crucial in this type of incident, because an error in judgment or a poorly executed reaction on the part of rescue personnel can easily turn an abrasion, laceration, or fractured bone into an amputated limb. Victims who are caught in machinery must be reassured, calmed, and stabilized medically by the rescuers.

Consider these recommendations:

1. The escalator machinery must be stopped. Stop switches are located at the top and bottom of most escalators. One member of your company must maintain constant pressure on the stop switch until the power supply can be shut down. If the escalator machinery is stopped upon your arrival, which is generally the case, secure the power source. Escalator power supplies are usually located under the top landing or step plate of the escalator. Always turn off the power supply to the escalator and secure the deenergized (power off) power source with the lockout and tagout procedures.

2. Are personnel competently trained for this type of incident? Is the proper equipment on the scene to perform the rescue and extrication operation safely? These are reasonable questions, as you might not always have the experienced, adequately trained personnel or the equipment needed to handle the situation. Again, exercise good judgment through risk management and risk assessment; simply reacting to the situation could have a very negative outcome. Unnecessary or excessive damage to the escalator equipment is also undesirable.

3. The best course may be to take no action until a trained professional arrives on the scene to apply special expertise regarding this type of incident. The machinery may have to be dismantled to reduce the possibility of causing more damage to the entangled victim (usually the victim's hands or feet) or the equipment involved.

4. Leave the power supply to the escalator turned off after the escalator incident has been resolved. Assume nothing. Remember: safety is the number one priority.

Wrap-Up

Chief Concepts

- A TRI is a complex rescue incident involving vehicles or machinery, water or ice, rope techniques, trench or excavation collapse, confined spaces, structural collapse, SAR, or hazardous materials that requires specially trained personnel and special equipment.
- Training in technical rescue areas is conducted at three levels: awareness, operations, and technician. Awareness training serves as an introduction to the topic with an emphasis on recognizing hazards, securing the scene, and calling for appropriate assistance. There is no actual use of rescue skills at the awareness level. Most of the training that beginning fire fighters receive is at the awareness level.
- The following types of TRIs are often encountered by fire fighters:
  - Vehicle and machinery rescue
  - Confined-space rescue
  - Rope rescue
  - Trench and excavation rescue
  - Structural-collapse rescue
  - Water and ice rescue
  - Wilderness rescue
  - Hazardous materials incidents
- To become proficient in handling these types of situations, you must take a formal course to gain specialized knowledge and skills.
- When assisting rescue team members, keep five guidelines in mind:
  - Be safe.
  - Follow orders.
  - Work as a team.
  - Think.
  - Follow the golden rule of public service: treat others as you would like to be treated.
- The basic steps of special rescue operations are outlined here:
  - Preparation—Train with the specialized rescue teams in your area, and become familiar with the terminology used in the field.
  - Response—A technical rescue team will usually respond with a rescue unit supported by other units such as a medic unit, an engine company, a truck company, and a chief.
  - Arrival and size-up—The first company officer to arrive on the scene should assume command and rapidly assess the situation. Do not rush into the incident scene until the situation has been assessed and the hazards identified.

- Stabilization—Once the needed resources are on the way and the scene is safe to enter, begin to stabilize the scene by establishing a perimeter around the rescue site.
- Access—Identify the actual reason for the rescue, and work toward freeing the victim safely. Communicate with the victims at all times, and initiate emergency medical care as soon as safe access is made to the victim.
- Disentanglement—Once precautions have been taken and the reason for entrapment has been identified, the victim needs to be freed as safely as possible. Disentanglement removes what is confining the victim from around the victim.
- Removal—This step could be as simple as assisting a victim up a ladder or as complex as packaging the victim in a Stokes basket and lifting him or her out of a trench.
- Transport—Once the victim has been removed from the hazard area, transport to an appropriate medical facility is accomplished by EMS personnel.
- Security of the scene and preparation for the next call—Once the victim has been transported, the scene must be stabilized by the rescue crew to ensure that no one else becomes injured.
- Postincident analysis—Identify what worked well and which procedures could work better.
- Beginning with the initial dispatch of the rescue call, begin compiling facts and identifying factors pertinent to the call. The information received when an emergency call is received is important to the success of the rescue operation. The information should include the following details:
  - The location of the incident
  - The nature of the incident (kinds and number of vehicles)
  - The conditions and positions of victims
  - The conditions and positions of vehicles
  - The number of people trapped or injured, and the types of their injuries
  - Any specific or special hazard information
  - The name of the person calling and a number where that person can be reached
- A size-up should include the initial and continuous evaluation of the following issues:
  - Scope and magnitude of the incident
  - Risk–benefit analysis
  - Number of known and potential victims
  - Hazards
  - Access to the scene

www.FireFighter.jbpub.com/Canadian3e

- Environmental factors
- Available and necessary resources
- Establishment of control perimeters
- After sizing up the scene, take the following precautions to avoid further injuries to the victims and to provide for the safety of other fire fighters:
  - Secure utility hazards.
  - Provide scene security.
  - Use the proper protective equipment for the emergency.
  - Activate the IMS.
  - Activate the personnel accountability system.
  - Make contact with the victim, and keep the victim calm.
- At any type of TRI, follow the orders of the company officer who receives direction from the IC. Many of the tasks assigned to fire fighters will involve moving equipment and objects from one place to another; others will involve protecting the team and victims.
- Vehicle and machinery rescues occur in many settings. These situations require responders to stabilize the machinery and ensure that the electricity is off.
- With a confined space, rescuers must ensure that a pure and adequate air supply is available before entering confined spaces.
- Low-angle and high-angle rope rescues require safe equipment, adequate training, and much caution.
- Trench and excavation rescues are hazardous and require responders to minimize the chance for secondary collapses.
- A damaged building is prone to structural collapse. Whenever a building has been damaged, assume that it may collapse. Consider fire scenes to be potential building collapses.
- Water rescue training is needed in almost all communities. Training, proper PFDs, and appropriate clothing are important to ensure rescuer safety.
- Wilderness rescue may be called for even when initial access to a lost or stranded individual occurs quickly.
- Hazardous materials incidents are not always dispatched as hazardous materials incidents, so be cautious when approaching any incident. Once the presence of a hazardous material is identified, protect yourself by staying out of the area exposed to the material.
- Never attempt to move or relocate an elevator under any circumstances. Always cut the power to a malfunctioning elevator and secure the power supply. Implement incident risk management and risk assessment. Always have additional trained personnel on hand, and anticipate obstacles and problems. Once the incident has been resolved, leave the power supply off.

Hot Terms

Cold zone (technical rescue incident) The control zone of an incident that contains the command post and such other support functions as are deemed necessary to control the incident. (NFPA 1500)

Confined space An area large enough and so configured that a member can bodily enter and perform assigned work, but which has limited or restricted means for entry and exit and is not designed for continuous human occupancy. (NFPA 1500)

Entrapment A condition in which a victim is trapped by debris, soil, or other material and is unable to extricate himself or herself.

Hazardous material A substance that, when released, is capable of creating harm to people, the environment, and property. (NFPA 472)

High-angle operation A rope rescue operation where the angle of the slope is greater than 45 degrees. In this scenario, rescuers depend on life-safety rope rather than a fixed support surface, such as the ground.

Hot zone (technical rescue incident) The area immediately surrounding a hazardous materials spill/incident site that is directly dangerous to life and health. All personnel working in the hot zone must wear complete, appropriate protective clothing and equipment. Entry requires approval by the IC or a designated sector officer. Complete backup, rescue, and decontamination teams must be in place at the perimeter before operations begin.

Lockout and tagout systems Methods of ensuring that electricity and other utilities have been shut down and switches are locked so that they cannot be switched on, so as to prevent flow of power or gases into the area where rescue is being conducted.

Low-angle operation A rope rescue operation on a mildly sloping surface (less than 45 degrees) or flat land. In this scenario, fire fighters depend on the ground for their primary support, and the rope system is a secondary means of support.

Packaging The process of securing a victim in a transfer device, with regard to existing and potential injuries or illness, so as to prevent further harm during movement. (NFPA 1006)

Personal flotation device (PFD) A displacement device worn to keep the wearer afloat in water. (NFPA 1925)

Personnel accountability system A system that readily identifies both the locations and the functions of all members operating at an incident scene. (NFPA 1500)

Placards Signage required to be placed on all four sides of highway transport vehicles, railroad tank cars, and other forms of hazardous materials transportation that identifies the hazardous contents of the vehicle, using a standardized system—namely, 250- by 250-mm (10.75- by 10.75-in.) diamond-shaped indicators.

Secondary collapse A subsequent collapse in a building or excavation. (NFPA 1006)

Shoring A structure such as a metal hydraulic, pneumatic/mechanical, or timber system that supports the sides of an excavation and is designed to prevent cave-ins. (NFPA 1670)

Spoil pile A pile of excavated soil next to the excavation or trench. (NFPA 1006)

Supplied-air respirator An atmosphere-supplying respirator for which the source of breathing air is not designed to be carried by the user. (NFPA 1404)

Technical rescue incident (TRI) A complex rescue incident requiring specially trained personnel and special equipment to complete the mission. (NFPA 1670)

Technical rescue team A group of rescuers specially trained in the various disciplines of technical rescue.

Warm zone (technical rescue incident) The area located between the hot zone and the cold zone at an incident. The decontamination corridor is located in the warm zone.

Wilderness search and rescue (SAR) The process of locating and removing a victim from the wilderness.

FIRE FIGHTER *in action*

You find yourself having to travel from your normal station assignment to a downtown station where the technical rescue team is located. You report to your captain who assigns you to the engine company; however, he advises you that the engine crew will be training on technical rescue with his rescue company. You are thrilled by the prospect of getting to learn more about these highly-specialized duties. As soon as the morning briefing is over, the captain asks you to follow him to the bays. He wants to get a feel for your current level of knowledge.

1. What level of responder takes offensive actions at technical rescue incidents?

   A. Technician

   B. First responder

   C. Operations

   D. Awareness

2. In what zone is the contamination corridor usually located?

   A. Cold

   B. Tepid

   C. Warm

   D. Hot

3. A rope rescue operation of greater than 45 degrees is considered a _____ rescue.

   A. low-angle

   B. high-angle

   C. obtuse-angle

   D. acute-angle

4. What must responders ensure is available before entering into a confined-space incident?

   A. Ropes and rope hardware

   B. Confined-space permit

   C. Operations-level responders

   D. Pure and adequate air supply

5. Which of the following is true about elevator incidents?

A. Do not move an elevator to rescue victims.

B. Only move the elevator in an upward direction to rescue victims.

C. Only move the elevator in a downward direction to rescue victims.

D. Elevators can be moved in either direction to rescue victims.

6. When at the scene of a swiftwater incident should a PFD be worn?

A. Whenever a rescuer is in a boat

B. Whenever a rescuer is in the water

C. Whenever a rescuer is within 3 m (10 ft.) of the water

D. Whenever a rescuer is on scene

7. What task will you likely be called upon to perform at a technical rescue incident?

A. Set up safety zones.

B. Lockout, tagout utilities.

C. Carry tools and equipment.

D. Assist with the extrication of the victim.

FIRE FIGHTER II
in action

Since you first began your career, you have dreamed of getting on the rescue company so you can perform the technical rescue duties. While not exactly the same, you are thrilled when you are assigned to the engine company that is co-located with the heavy rescue company. You have heard that the rescue company crew is highly seasoned and is very competent at their jobs. However, they are also known to be a bit arrogant. At times, their behaviour is downright demeaning to others.

During your first shift at the station, the rescue company was going over their equipment with the rookie fire fighter. When asked about the K tool, his mind drew a blank and he froze. They immediately began riding him about being incompetent. You noticed that everything the rookie did, from cleaning the toilets to washing the battalion vehicle, was not good enough for them, and they all made fun of him. You could tell that it really bothered him.

1. How would you handle the situation?

2. With whom would you discuss the issue?

3. What would you do if your captain was present when they were teasing him?

4. What responsibility do you have in addressing this?

Fire Prevention and Public Education

Fire Fighter I

Knowledge Objectives

There are no knowledge objectives for Fire Fighter I candidates. NFPA 1001 contains no Fire Fighter I Job Performance Requirements for this chapter.

Skills Objectives

There are no skills objectives for Fire Fighter I candidates. NFPA 1001 contains no Fire Fighter I Job Performance Requirements for this chapter.

Fire Fighter II | FFII

Knowledge Objectives

After studying this chapter, you will be able to:

- Describe the activities that prevent fires and limit their consequences if fire occurs. (NFPA 6.5.2 , p 790)
- Identify elements of the public fire-safety education programs covering how and when to stop, drop, and roll; exit drills in the home; the selection and use of portable fire extinguishers; and the importance of smoke alarms and residential sprinkler systems in preventing fire deaths. (NFPA 6.5.1.A , p 791–796)
- Stress the importance of having portable fire extinguishers. (NFPA 6.5.1.B , p 796)
- Recognize hazards during a fire safety survey of an occupied structure. (NFPA 6.5, 6.5.1 , p 797–800)
- Describe the steps in conducting a fire station tour. (NFPA 6.5.2.A , p 800–803)

Skills Objectives

After studying this chapter, you will be able to perform the following skills:

- Complete an occupied fire safety survey. (**NFPA 6.5.1.B**, p 800–801)
- Perform a public fire-safety education presentation on the Stop, Drop, and Roll program. (**NFPA 6.5.2**, p 793–794)
- Execute a public fire-safety education presentation on exit drills in the home. (**NFPA 6.5.2**, p 794)
- Install and maintain a smoke alarm. (**NFPA 6.5.2**, p 795–796)
- Give a public education tour of a fire station. (**NFPA 6.5.2.B**, p 801–802)

Additional NFPA Standards

- NFPA 1, *Fire Code*
- NFPA 101, *Life Safety Code*
- NFPA 1452, *Guide for Training Fire Service Personnel to Conduct Dwelling Fire Safety Surveys*

© Photos.com

You Are the Fire Fighter

You are attending a family reunion, and the conversation turns to your new job as a fire fighter. You beam with pride as you tell your family about your training and your new assignment. The discussion turns to your role in fire prevention. Your cousin and his wife who recently had a new baby boy begin to question you on what they can do to ensure that their home is safe for their new addition. You think back to what you learned in your Fire Fighter I and II courses.

1. How often should you change the batteries in your smoke alarm?
2. Does your family have an evacuation plan in case of a fire?
3. What actions would you recommend to your cousin?

Introduction

Fire prevention is an extremely important facet of fire service provision. As a fire fighter, you will be expected to understand the essence of fire prevention service delivery within your respective community. Members of the public will look to you for fire safety information and advice on matters such as smoke alarms, fire extinguishers, barbecue safety, and procedures for preventing fires. For most fire fighters, the exercise of teaching fire prevention skills is not as exciting as fighting fires; however, what fire prevention lacks in excitement, it more than makes up for in job satisfaction.

This introductory chapter on fire prevention is designed to ignite your interest in becoming a responsible supporter of fire prevention initiatives. It also provides valuable information on how to communicate the information in an interesting manner. Regardless of whether or not you are a career professional or a part-time volunteer, your commitment to fire prevention initiatives is a vitally important function of your responsibilities. Fire prevention initiatives are not only a means to reduce fire loss and injury but also a way to reduce other societal costs associated with fires.

FIRE FIGHTER II Tips   FFII

Fires that are prevented do not cause any deaths, injuries, or property loss. Most fires are caused by actions or situations that could have been avoided. A key element in reducing the number of fire fighter line-of-duty deaths and civilian deaths is proactive fire prevention.

What Is Fire Prevention?

Fire prevention includes a range of activities that are intended to prevent the outbreak of fires and to limit the consequences if a fire does occur. Fire prevention includes, but is not limited to, provincial/territorial regulations and legislation, municipal bylaws, condominium rules and regulations, code enforcement, inspections, public education, fire prevention advertising, fire prevention programming, fire investigations, and fire-frequency analysis. All of these activities share a common objective—limiting life loss, injuries, and property damage. These activities increase fire fighter safety as well. Your highest priority as a fire fighter should always be to prevent fires. It is important for fire personnel to recognize and understand the important role that government regulations play in ensuring that fire safety is incorporated into building design and fire prevention programming.

■ Building and Fire Codes

Provinces/territories within Canada have either adopted the National Building Code (NBC) or their own building codes for establishing construction standards. In either case, all cities, towns, villages, and counties within a given jurisdiction are responsible for ensuring that buildings are built to a recognized code. Some departments use on-duty suppression personnel to perform basic fire inspections within their response area. These regulations ensure that public safety is maintained and that buildings are constructed in a manner that resists collapse, ensures rapid evacuation, and, as is the case for high-rise construction, limits fire spread.

Fire codes are companion documents to building codes. An important function of these codes is to ensure that all structures are constructed with fire safety in mind. The building code dictates the thickness of fire separations, the nature of closures, the number and size of means of egress, and the type and nature of all fire protection and early warning systems. Fire codes do not come into effect during construction; rather, fire code enforcement comes into force and effect post-construction and occupancy. This differentiation is important for fire personnel to understand. Fire service enforcement cannot take place until after a structure is constructed and occupied. During construction, the building inspector is the authority who has jurisdiction over all code enforcement initiatives, including those aspects of the building code that refer to fire safety. Some provinces, such as Ontario, have enacted their own building codes. Other provinces, such as Manitoba and Prince Edward Island, have adopted the NBC to regulate building construction.

■ Inspection and Code Enforcement

Fire personnel should be able to differentiate between standards and legally enforceable codes. Standards are designed on widely accepted or recognized norms; standards establish a means to do a specific thing. There are many recognized organizations that produce standards for a number of agencies and government. The National Fire Protection Association (NFPA) is an organization that establishes a wide variety of standards, which are widely referenced by governments and industry. Standards are usually designed by experts who have specific knowledge with regards to what is being standardized.

Standards are not codes, because, unless specifically outlined in legislation or regulation, they are not able to be legally enforced. Codes are legally legitimized laws or regulations, which are enforceable by the authority having jurisdiction over a specific area. After a jurisdiction adopts a fire code, it must be able to enforce it. Citizens, building owners, companies, and corporations have a legal obligation to comply with all rules and regulations related to building and fire codes. In Canada, every jurisdiction has a mechanism for conducting inspections, issuing compliance orders, and code enforcement through judiciary means; however, not every jurisdiction performs these activities in the same manner. In large municipalities such as Calgary, Vancouver, Ottawa, Toronto, and Montreal, the fire service has a fire prevention division, which is responsible for public education, code enforcement, and fire investigation. In smaller towns and municipalities (less than 50,000 inhabitants), the responsibility to conduct inspections typically falls on the chief fire official or a person whom they designate.

Fire codes usually specify the types of occupancies that can be inspected and the frequency of the inspections; for example, some fire codes require all schools to be inspected twice each year. In other cases, the schedule of inspections is discretionary. The agency responsible for performing inspections and enforcing codes is usually named in the fire code; it may vary in different communities. Sometimes the responsibility belongs to specific individuals or positions within the fire department, such as fire inspectors or fire marshals. Other fire departments train all fire fighters to conduct inspections and take code enforcement actions. In some communities, building inspectors or special code-enforcement personnel are given this responsibility. Generally speaking, rookie fire fighters are not directly involved in code enforcement activities; however, this does mitigate the importance of understanding the legislation and code enforcement practices within their jurisdiction.

Every fire fighter should know how to conduct a home fire-safety survey in a private dwelling—including knowing what to look for during the survey, how to contact the occupant, and how to communicate the survey findings. This information is presented in detail later in this chapter.

■ Public Fire and Life Safety Education in Canada

In order for fire prevention programs to be effective they must be based on factual analysis. Unfortunately, the Canadian federal government does not have a department whose sole purpose is to collect fire loss statistics. In lieu of this information, vacuum organizations such as the Council of Canadian Fire Marshals and Fire Commissioners collect and publish data on fire loss.

The goals of public fire and life safety education are to help people understand how to prevent fires from occurring and to teach them how to react in an appropriate manner if a fire occurs. Making people aware of common fire risks and hazards and providing information about reducing or eliminating those dangers can prevent many fires. Public education teaches techniques to reduce the risks of death or injury in the event of a fire. The programs are also designed to prevent other types of accidents and injury.

Public fire-safety education programs include, but are not limited to the following topics.

For business owners, companies, and corporations:

- Fire code inspections
- Employee fire-safety awareness
- Fire safety programs, which encourage profit maximization by limiting risk
- Code enforcement

For homeowners and occupants of rented residential units:

- Explain how to react to a fire alarm activation
- Kitchen and cooking safety
- Proper disposal of smoking materials
- Chimney and wood-burning fireplace maintenance
- Stop, Drop, and Roll program **FIGURE 27-1**
- Exit Drills in the Home (EDITH) program
- Installation and maintenance of smoke alarms
- Advantages of residential sprinkler systems
- Selection and use of portable fire extinguishers
- Learn Not to Burn®
- Change Your Clock—Change Your Battery campaign
- Fire safety for special populations
- Bike safety
- Fall prevention
- Wildland fire prevention programs

Many fire and life safety programs are presented to groups such as school classes, scout troops, church groups, senior citizens' groups, civic organizations, hospital staff, and business employees. Presentations are often made at community events and celebrations, as well as during Fire Prevention Week activities. Seven popular subjects for fire safety presentations include:

1. Cooking safety
2. Electrical safety
3. Child-proofing your home for fire safety
4. Stop, drop, and roll
5. EDITH
6. Installation and maintenance of smoke alarms
7. Placement and use of portable fire extinguishers

FIGURE 27-1 The Stop, Drop, and Roll program teaches young children what to do if their clothing catches fire.

Another popular fire-safety education activity is a fire station tour. Both children and adults enjoy the opportunity to tour their local fire station, and a fire station tour is an excellent opportunity to promote fire prevention. Every fire fighter must understand how to conduct an effective fire station tour. The skills and knowledge for conducting a fire station tour are presented later in this chapter.

Cooking Safety

The improper use and handling of cooking materials is one of the leading causes of fires in Canada. Fire personnel must be able to advise residents about simple steps to follow when cooking, including:

1. Never leave the stove unattended while you are cooking, especially if you have children.
2. Never place cooking oil over high heat as the oil will eventually overheat and create a fire hazard.
3. When cooking with oil, always have the pan lid close at hand. Simply placing a lid on a pot that is boiling over or has ignited can greatly reduce the possibility that the fire can spread.
4. Never put water on hot oil, as it will react violently. This violent reaction can cause burns or create a fire as the heated oil lands on combustible materials, such as countertops and shelving.
5. Never walk with a burning pot of food in order to place it outside or dump it into the sink. People often cover a heated pot with a lid and then immediately proceed to dispose of it; as they walk, there is the potential to splash its contents and burn themselves or to drop the pot onto the floor, which creates an opportunity for a fire to occur. Teach people to simply place a lid on a burning pot, turn off the heat source, and let the pot cool in place until it is safe to handle without the possibility of creating a fire.
6. Never cook while intoxicated or under the influence of drugs.
7. Avoid cooking while wearing loose clothing or sleepwear; most adult sleepwear is not required to have a

fire rating. Always have a well-maintained pair of baking gloves nearby when cooking or baking to protect your hands from the heat.
8. Always use a timer when baking as a reminder to remove cooked food.
9. If at all possible, have an ABC extinguisher close at hand and near an exit.

Electrical Safety

Remind homeowners to avoid making any changes to their residential electrical system on their own; all electrical work should be performed by a qualified electrical professional. People should never try to wire a stove, dishwasher, or any other large appliance without first consulting an electrician.

Child-Proofing for Fire Safety

Fire personnel should be able to offer advice on how parents can make their homes safer for their children in order to avoid fire occurrence and burns. The following is a small list of simple child-safety measures:

- Avoid having children play in the kitchen area.
- Teach young children how hot a stove can become and to avoid the stove area while you are cooking.
- Place child-safety covers over all electrical sockets not in use.
- Teach your child to avoid sticking anything into electrical sockets and not to remove anything that is plugged in.
- Teach your children how to react when they hear the smoke alarm.
- Store all toxic and combustible/flammable materials in a safe place where children cannot access them.
- Teach your children not to play with any type of ignition source, such as matches or lighters.
- Never leave candles unattended. To ensure safety, candles should be enclosed in a glass candle holder.
- Establish an emergency plan for all occupants that includes a rapid evacuation plan and a process for accounting for everyone once they have exited the building.

- As soon as your child is old enough to understand, have them memorize their address, and teach them how to use 911 in a responsible manner.

Stop, Drop, and Roll

The Stop, Drop, and Roll public fire-safety education program can be taught to children as young as preschoolers, but it is also a valuable educational program for adults. This program is designed to instruct people what to do if their clothing catches fire. When clothing catches fire, the normal reaction is to run for help; unfortunately, running tends to fan the flames and increases the chance for severe burns. Teaching the mnemonic "stop, drop, and roll" and demonstrating each step helps to reduce burn injuries.

When teaching the Stop, Drop, and Roll program to children, ask them to discuss situations that might result in their clothing catching fire. These circumstances might include being too close to a kitchen stove or fireplace or playing with matches

or lighters. The fire fighter should talk about ways to reduce these hazards, such as keeping matches and lighters out of the reach of children. He or she should then stress each step of the sequence:

1. **Stop** means just that: Stay in place; do not run. Stress the fact that running will fan the flames and spread the fire.
2. **Drop** means getting down on the ground or on the floor. The person should cover his or her face with the hands to help protect the airway and the eyes.
3. **Roll** involves encouraging children to tuck in their elbows, keep their legs together, and roll like a bug. They should roll over and over and then roll back in the other direction. This helps to smother the flames.

To teach the Stop, Drop, and Roll program, follow the steps in **SKILL DRILL 27-1**:

1. Explain the purpose of Stop, Drop, and Roll.
2. Instruct students to **stop**: stay in place. (**STEP 1**)

SKILL DRILL 27-1   Teaching Stop, Drop, and Roll   **FFII**
(Fire Fighter II, NFPA 6.5.2)

1 Instruct students to STOP: Stay in place.

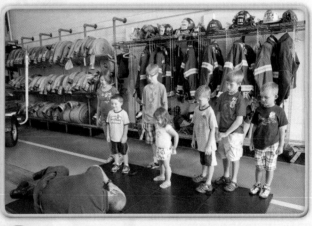

2 Instruct students to DROP to the floor or ground and cover their faces with their hands.

3 Instruct students to tuck in their elbows, keep their legs together, and ROLL over and then back to extinguish the flames.

4 Have students practise the procedure.

© LiquidLibrary

3 Instruct students to **drop** to the floor or ground and cover their faces with their hands. (**STEP 2**)

4 Instruct students to tuck in their elbows, keep their legs together, and **roll** over and then back to extinguish the flames. (**STEP 3**)

5 Demonstrate the Stop, Drop, and Roll procedure.

6 Have students practise the procedure. (**STEP 4**)

FIRE FIGHTER II Tips   FFII

Fire personnel should understand how to design a fire safety plan in order to guide members of the public in designing their own. A written plan offers the following benefits:

- Guides the presentation
- Outlines the main points the presentation will cover
- Ensures consistency when multiple presenters are involved

FIRE FIGHTER II Tips   FFII

Whenever you are dealing with the public, pay attention to the language that you use. It is vitally important for personnel to act and speak in a professional tone. Avoid making jokes or using questionable language when speaking to, or near, members of the public. Under no circumstance should the interaction become personal or social. Members of the public look up to fire personnel and maintain the expectation that they will not take advantage of their goodwill for personal gain. Avoid using firefighting terms and jargon. Speak at a slower pace for young children or the elderly. Small children can be intimidated by adults, so whenever feasible, personnel should bend down or crouch in order to be able to look at the child at his or her height level. Regardless of one's age, gender, race, religion, or socioeconomic circumstance, all members of the public are to be treated with the utmost respect at all times.

■ Exit Drills in the Home

A second public fire-safety education program called **Exit Drills in the Home (EDITH)** teaches residents how to safely get out of their homes in the event of a fire or other emergency. Many fire deaths could be prevented if everyone learned a few simple rules about what to do in the event of a fire in the home. As part of this exercise, residents need to understand the importance of having properly installed and working smoke alarms. Working smoke alarms alert residents to a fire and give them more time to escape from a burning building. A public fire-safety program about smoke alarms is described later in this chapter. More information about the operation of smoke alarms is also found in the chapter "Fire Detection, Protection, and Suppression Systems."

As part of your EDITH presentation, stress the importance of keeping bedroom doors closed during sleeping hours, as this step prevents smoke and heat from reaching bedrooms. Emphasize the need to have two escape routes from each bedroom. If smoke, heat, or flames are present in the primary escape route, use the secondary escape route. The secondary escape route may be through a window or through a different door. Stress the importance of alerting other occupants to the presence of a fire.

When alerted to the presence of a fire, all occupants should roll out of bed, stay low, and crawl toward the designated exit. By staying low, individuals decrease the amount of smoke and gases they inhale.

Before opening any door, touch the closed door to see if it is hot. Teach students to use the back of a bare hand to sense the temperature of the door. If the door is hot, do not open it; instead, use a window or another door for escape.

Once outside the house, occupants should assemble at a preestablished meeting spot. Stress the importance that no one should reenter any structure that is involved in a fire. Also, make sure that the fire service has been called. If any doubt arises about whether the fire service has been called, use a neighbour's phone or a cell phone to call 911.

Once the steps for emergency drills have been taught, families should practise EDITH regularly. Some drills should be held in the daytime, and some should take place at night.

To teach EDITH, follow the steps in **SKILL DRILL 27-2**:

1 **Plan.** Describe the importance of making a quick exit when a fire occurs.

2 Stress the importance of having a sufficient number of working smoke alarms on each level of the house.

3 Explain why you should sleep with bedroom doors closed.

4 Draw a diagram of the house.

5 Plan two escape routes from each room. (**STEP 1**)

6 **Practise.** When you hear a smoke alarm, roll out of bed and crawl to the exit. (**STEP 2**).

7 Check exit doors for heat. Do not open a door if it is hot. (**STEP 3**)

8 Follow the escape route to the preestablished meeting place.

9 Ensure that the fire department has been called. (**STEP 4**)

10 Do *not* reenter a burning building.

■ Smoke Alarms

A third type of public fire-safety presentation covers the proper installation and maintenance of smoke alarms. Smoke alarms can save lives—but only if they are working properly. It is important to note that many Canadian jurisdictions have created legislation for the use of smoke detectors in residential housing. The goal of this presentation is to help people properly install and maintain approved smoke alarms in their homes. More information about this topic is presented in the chapter "Fire Detection, Protection, and Suppression Systems."

During your presentation on this topic, stress the importance of having a working smoke alarm on every floor of a house. The area outside each sleeping area should contain a smoke alarm. Remind residents to follow the manufacturer's instructions for installation. Stress the importance of keeping smoke alarms in working order. Alarms should be tested once a month by pressing the test button. Identify the type of battery in the smoke alarm. Smoke alarms powered by alkaline batteries should have their batteries changed every 6 months. Many communities participate in the Change Your Clock–Change

SKILL DRILL 27-2 Teaching EDITH FFII
(Fire Fighter II, NFPA 6.5.2)

1 Plan. Draw a diagram of your house. Plan two escape routes from each room.

2 Practise. When you hear a smoke alarm or smell smoke, roll out of bed and crawl to the exit.

3 Check exit doors for heat. Do not open a door if it is hot.

4 Follow the escape route to the preestablished meeting place. Ensure that the fire department has been called.

© LiquidLibrary

Your Battery campaign, which encourages regular battery replacement. Many new smoke alarms contain a 10-year battery. Once the smoke alarm signals a low battery alarm, replace the smoke alarm, not the battery. Remind homeowners to keep circuit breakers supplying current to smoke alarms on to ensure that these alarms remain operational. Vacuum the detector chamber regularly to prevent the buildup of dust or insects, which can cause false alarms. Some alarms are manufactured to detect the presence of smoke as well as carbon monoxide; it is important to explain the function of each alarm when these dual-purpose alarms are present.

FIRE FIGHTER II Tips FFII

When advising members of the public with regards to the installation, maintenance, and use of any smoke or carbon-monoxide alarms, refer to the manufacturer's instructions for use.

Teaching these simple facts about the installation and maintenance of smoke alarms will help families keep their homes safe in the event of a fire. To present a public fire-safety education program on installation and maintenance of smoke alarms, follow the steps listed in **SKILL DRILL 27-3**:

1 Explain the importance of properly installing and maintaining smoke alarms.

2 Stress the importance of installing a smoke alarm on each floor and outside each sleeping area.

3 Remind students to avoid installing smoke alarms in kitchens and garages, or near fireplaces, windows, and exit doors.

4 Mount smoke alarms on the ceiling or as high as possible on walls. (**STEP 1**)

5 Test smoke alarms once a month using the test button. (**STEP 2**)

SKILL DRILL 27-3 | Installing and Maintaining Smoke Alarms | FFII

(Fire Fighter II, NFPA 6.5.1)

1 Ensure that smoke alarms are mounted on the ceiling or as high as possible on walls.

2 Test smoke alarms once a month using the test button.

3 Change alkaline batteries in the smoke detector every 6 months if not a 10-year battery.

4 Clean alarms regularly to prevent false alarms.

© LiquidLibrary

6 Change alkaline batteries in smoke alarms every 6 months if not 10-year batteries. (**STEP 3**)

7 Clean alarms regularly to prevent false alarms. (**STEP 4**)

■ Selection and Use of Portable Fire Extinguishers

Many groups visiting your fire station will have questions about which types of fire extinguishers they should have in their home, kitchen, garage, car, or boat. You should be familiar with the information you learned in the chapter "Portable Fire Extinguishers," so you can answer any questions that arise.

Fire Cause Determination

Fire cause determination is the process of establishing the cause of a fire through careful investigation and analysis of available evidence. It is considered part of fire prevention, because fire cause determination can assist in focusing education programs and enforcement. Education programs can be tailored with alignment to specific areas of concern in an effort to prevent similar occurrences. Fire cause determination is extremely important for both accidental and intentional fires. The fire fighter's role in fire cause determination is covered in the chapter "Fire Cause Determination."

Conducting a Fire Safety Survey in a Private Dwelling

Some fire services may offer to conduct fire safety surveys for private dwellings. Fire service personnel, unless responding to an emergency call or complaint, have no right of entry to conduct fire safety surveys. Residential fire safety surveys cannot be conducted without the occupant's permission. These programs are usually based upon a voluntary request. Home fire-safety surveys help identify fire and life-safety hazards and provide the occupants with recommendations on making their home safer; in Ottawa however, the city's fire service has created the Wake Up fire prevention program, which is part of a smoke alarm bylaw. As part of this program, fire personnel visit every residential dwelling and check to ensure that each dwelling has at least one working smoke alarm. If a residence does not have an alarm, then the fire service provides one with a new battery to that residence. This program involved teaming up with smoke alarm manufacturers and battery companies to supply the needs of the program.

Other municipalities in Canada have similar programs. These programs are effective because they combine public education, legislation, and code enforcement in an effort to reduce fire loss and injury.

Many fire services provide educational materials that address safety issues such as installation and maintenance of smoke alarms, the role of residential fire sprinklers in preventing fire deaths, and the selection and use of portable fire extinguishers. These materials can be left with the occupant.

■ Getting Started

During a home inspection, you should present a neat, professional image FIGURE 27-2 . Identify yourself by name, title, and organization. You should be wearing an identification badge or card if you have been provided with one. Inform the occupant of the purpose of your visit and the survey format. Always remember that you are a guest in a private home, and respect the privacy of the people who live there. You are there to help teach this person how to protect his or her family and property from fire.

During the survey, you will need to consider several types of hazards and issues. Concentrate on the hazard categories that most often cause residential fires—for example, cooking equipment, heating equipment, electrical wiring and appliances, smoking materials, candles and other open flames, and children playing with fire. Some of the hazards you encounter will depend on the local climate; for example, wood-burning stoves and fireplaces are more common in colder climates. Other hazards may be related to the age of the occupants; the fire hazards in a residence with young children, occupants with a physical or mental disability, or senior citizens would be quite different, for example.

Always look for fire protection equipment, such as smoke alarms, fire extinguishers, and residential fire sprinkler systems. Make sure that any fire alarm or suppression equipment in the home is properly installed, maintained, and fully operational FIGURE 27-3 . Reference your particular service's policies and procedures before making recommendations on fire protection equipment. Most jurisdictions do not allow recommending a specific manufacturer's product.

Conduct the survey in a systematic fashion for both the inside and the outside of the home. Some fire fighters begin by inspecting the outside of the house, whereas others prefer to start with the interior. Many fire inspectors will start at the top of a building and systematically work their way down to ensure that all areas are covered. Inside the house, systematically inspect each room for fire hazards. Each section will have its own hazards.

FIGURE 27-2 Present a neat, professional image during a home inspection.

FIGURE 27-3 Ensure that smoke alarms are properly installed, maintained, and fully operational during a home inspection.

Courtesy of Nampa Fire Department - Nampa, Idaho

■ Outside Hazards

On the exterior of a residential occupancy, you need to check several items. Make sure that the house number, or address, is clearly visible from the street, so emergency responders can identify the correct house quickly **FIGURE 27-4**. Look for accumulated trash that could present a fire hazard or obstruct an exit. Note any flammable materials that are located close to the house or cars that are parked so close that a vehicle fire could spread to the structure. Point out shrubs and vegetation that need to be trimmed or removed; for example, in an urban–wildland interface zone, no dead vegetation should be found within 9.1 metres (30 feet) of a house.

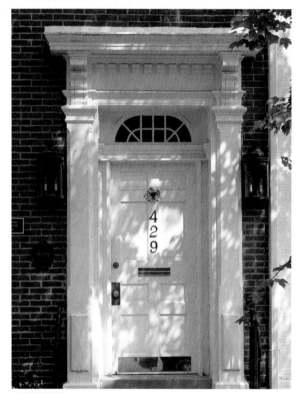

FIGURE 27-4 The home's address should be clearly visible from the street.

© David Papazian/Corbis

Locate any chimneys and try to determine the condition of the mortar. Ask the occupant whether the chimney has been cleaned within the past year, and recommend a cleaning if it has not.

Fire Fighter Safety Tips

© Photos.com

Security bars and double-cylinder door locks are great deterrents to burglars but can prove deadly for occupants trying to escape from a house fire. These measures can also waste precious time if fire fighters must enter to rescue the occupants or control a fire. Occupants should always be able to release security bars and open locked doors from the inside.

■ Inside Hazards

During the interior inspection, explain why different situations are considered potential fire risks and hazards. A person who understands why an overloaded electrical circuit is a fire hazard will be more likely to avoid overloading circuits in the future. Your goal should be to educate the occupant not simply to point out fire hazards.

As you walk through the house, help the occupant identify alternate escape routes that can be used in case of fire. Look for a primary exit and an alternate exit from each room, and discuss these points of egress with the occupant. Be sure to mention the importance of EDITH involving all family members.

Smoke Alarms

Studies show that the presence of working smoke alarms on each level of a house reduces the risk of death from fire by 50 percent. Take the time to verify and test all smoke alarms. Some fire departments keep donated smoke alarms available and install one in any dwelling that lacks this important protection. Give the residents a copy of the NFPA fact sheet on smoke alarms, and request that they read it. More information about the proper maintenance of smoke alarms is presented in the chapter "Fire Detection, Protection, and Suppression Systems."

The following important smoke alarm tips should be shared with the home's occupants:

- Smoke alarms should be installed on every floor and in or near every sleeping room of the home.
- Smoke alarms should be mounted on the ceiling at least 10 cm (4 in.) from a wall or high on the wall, 10–30 cm (4–12 in.) below the ceiling.
- Smoke alarms should not be located near windows, exterior doors, or duct vents that could interfere with their operation.
- Only qualified electricians should install or replace AC-powered smoke alarms.
- Smoke alarms should be tested at least once a month by using the test button on the device **FIGURE 27-5**.
- Smoke alarms should be dusted and vacuumed regularly.
- Alkaline batteries in smoke alarms should be replaced twice each year or as soon as the intermittent "chirp" warning that the battery is low sounds. Remind homeowners to "Change the battery when you change your clock." If the smoke alarm is equipped with lithium batteries, the smoke alarm batteries need to be replaced only once every 10 years.

Bedrooms

The most common causes of fires in bedrooms are defective wiring, improper use of heating devices, improper use of candles, children playing with matches, and smoking in bed. Although smoking in bed causes only 5.4 percent of all fires in the United States, these fires are responsible for 23 percent of civilian fire deaths. In recent years, Canada and almost half of U.S. states, have passed legislation requiring that cigarettes sold should be "fire safe," with greatly reduced ignition strength (ability to start a fire), as determined by ASTM Standard E2187-04. Fire fighters conducting home safety surveys should stress the fire

FIGURE 27-6 Encourage homeowners to purchase an approved ABC-rated fire extinguisher.

hazard that is associated with smoking in bed. In houses with children, it is equally important to discuss the risk of children playing with lighters and matches.

Kitchens

Kitchen fires are one of the more common fires in residential occupancies. In particular, these fires are often caused by leaving cooking food on the stove unattended and by faulty electric appliances. In the kitchen, look for signs of frayed wires on appliances, flammable towels or pot holders stored too close to the stove, and flammable cooking oils stored near the stove. Nothing that can be ignited—for example, a plastic food container or paper packaging—should ever be placed on a cooking surface, even if the burners are turned off and the surface is cold. Many fires have resulted from someone inadvertently turning on the wrong burner and igniting something that was left on the stovetop.

Every kitchen should be equipped with an approved ABC-rated fire extinguisher **FIGURE 27-6**. Kitchens that are equipped with a deep fryer should also have a Class K fire extinguisher. Check whether this extinguisher is visible and properly charged. Additional recommendations for kitchen safety include the following measures:

- Do not leave anything unattended on the stove while cooking.
- Keep all flammable materials, cleaning supplies, cooking oils, and aerosols away from the stove.
- Do not place anything that could ignite on a cooking surface—even when it is turned off and cold.
- Do not place towel racks near the stove.

- Store smoking materials out of the reach of children.
- Do not overload electrical outlets or extension cords.
- Keep electric cords properly maintained; replace any frayed cords.

Living Rooms

The "living room" category includes formal living rooms, family rooms, and entertainment rooms. The primary causes of fires in these areas are smoking and electrical equipment. Look for indications of careless smoking or improperly installed home entertainment equipment, such as multiple devices plugged into a light-duty extension cord. Also look for overloaded electrical outlets and extension cords in home offices with computers, printers, scanners, fax machines, and other devices. Stress the importance of using power strips that contain circuit breakers to prevent overloading of extension cords.

If the room contains a fireplace, a wood stove, or a portable heater, ensure that no flammable or easily ignited materials are stored nearby. A fireplace should have a screen to keep sparks and hot embers from escaping. If solid fuels are used, the chimney or flue pipe should be professionally cleaned at least once each year.

Garages, Basements, and Storage Areas

Explain the importance of good housekeeping and the need to clear junk out of garages, basements, and storage areas. These areas are often filled with large quantities of unneeded materials that can easily ignite, such as old newspapers and trash. Furnaces and water heaters that operate with an open flame are often located in a basement or garage as well; they can become ignition sources if flammable materials are stored too close to them. Excessive storage can result in inhibiting proper airflow to gas-burning appliances and create the conditions for carbon-monoxide poisoning.

Flammable liquids should not be stored inside homes. Gasoline and other flammable liquids should be stored only in approved containers and in outside storage areas or outbuildings. Propane tanks, such as those used in gas grills, also should only be stored outside or in outbuildings. Small quantities of flammable and combustible liquids—paint, thinners, varnishes, and cleaning fluids—should be stored in closed metal containers away from heat sources. Oily or greasy rags should also be stored in closed metal containers. Recommend that homeowners maintain fully charged fire extinguishers (2-A:10-B:C classification) for basements and garages.

Closing Review

During the home fire-safety survey, listen carefully to any questions from the occupants, because they will enable you to address any special concerns or correct inaccurate information. At the end of the survey, complete the inspection form, and give a copy to the occupants. Carefully review the findings, and describe the steps that need to be taken to eliminate any identified hazards. Talk to the occupants, and take the time to answer their questions fully. Emphasize the importance of smoke alarms, home exit plans, and fire drills.

After you have completed a home fire-safety survey or any other type of inspection, you must file your report according to the suggested operating guidelines (SOGs), or standard operating procedures (SOPs), of your department. If you identified hazards that require further action or follow-up, discuss them with your company officer or a designated representative from the code-enforcement office.

Conducting a home fire-safety survey is a complex and important task. Initially, you may accompany your company officer, a more experienced fire fighter, or an inspector to observe how it is done. You will work with a more experienced partner until you are ready to conduct home surveys on your own. Remember: a home fire-safety survey is important part of helping occupants prevent fires to reduce loss of lives and property.

To conduct a home fire-safety survey, follow the steps in **SKILL DRILL 27-4**:

1. Explain that the purpose of the home fire-safety survey is to make a home safer for the occupants.
2. Check for a plainly visible house number.
3. Look outside the house for accumulated trash, overgrown shrubs, and blocked exits. (**STEP 1**)
4. Check inside the house for properly working smoke alarms, fire extinguishers, and fire sprinkler systems.
5. Help building occupants identify primary and alternate escape routes.
6. Look for improper wiring, use of candles, and evidence of smoking in bedrooms.
7. Explain how improper cooking procedures can start kitchen fires. Stress the safe storage of cooking oils and flammable objects away from the stove.
8. Check living rooms for improper wiring and hazards from smoking.
9. Explain the safe use of fireplaces, heating stoves, and portable heaters. (**STEP 2**)
10. Stress the importance of good housekeeping practices and safe storage of flammable liquids.
11. Review the results of the fire safety survey with the building's occupants. (**STEP 3**)
12. Leave fire- and life-safety brochures with the building's occupants. (**STEP 4**)

Conducting Fire Station Tours

Fire station tours present a unique opportunity to help people learn how the fire department operates and how they can help prevent fires. Most fire services have SOGs/SOPs for conducting fire station tours. Fire personnel should be familiar with their procedures whenever they are responsible for conducting a tour.

Some tours start in the day room or in a meeting room with a general overview of the fire station, which is then followed by a tour of the fire apparatus. A tour may also start with an inspection of the apparatus and end in a meeting room with a question-and-answer session and information about the department and fire prevention techniques. Showing a fire prevention video or a video about the operation of your department is a good way to start or finish a tour and to stimulate questions from your audience.

In preparing for a fire station tour, remember that you are representing your department to the visitors. Your professional appearance and presentation project a positive image of the department to the community. One of the first things you should cover in your welcoming remarks is to tell the visitors what to expect and what they should do if the station receives an alarm during the tour. Visitors should not wander around the station while fire fighters respond to an emergency call. This is particularly important if the tour guide also might have to leave in response to a call.

The tour format will vary depending on the age and interests of the group, which could consist of small children, teenagers, adults, or senior citizens. The tour plan, including your presentation and fire prevention messages, should be designed to meet the needs of each group. The weather may also be an important factor if you are planning any outdoor demonstrations.

Young children like to see action, so squirting some water or raising an aerial ladder can be a real crowd-pleaser **FIGURE 27-7**. Fire prevention messages for young people include use of the Stop, Drop, and Roll procedure; learning how to call

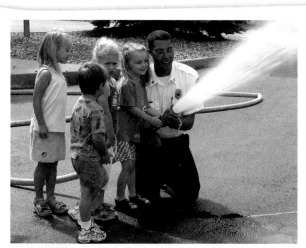

FIGURE 27-7 Young children always like to see fire fighters.
Courtesy of Captain David Jackson, Saginaw Township Fire Department

SKILL DRILL 27-4   Home Safety Survey   **FFII**

(Fire Fighter II, NFPA 6.5.1)

1 Look outside the house for accumulated trash, overgrown shrubs, and blocked exits.

2 Check inside the house for properly working smoke alarms, fire extinguishers, and fire sprinkler systems. Explain how improper cooking procedures can start kitchen fires. Explain the safe use of fireplaces, heating stoves, and portable heaters.

3 Review the results of the fire survey with the building occupants.

4 Leave fire- and life-safety brochures with the building occupants.

© LiquidLibrary

FIRE FIGHTER II Tips   **FFII**

Fire stations can be potentially dangerous for the general public. Before conducting a tour, ask yourself:

- Is the area safe?
- Am I showing the fire service in a positive light?
- Will there be enough fire fighters to supervise the tour group?
- Are the planned activities age-appropriate?

911; the importance of EDITH; selection and use of portable fire extinguishers; the importance of residential fire sprinklers; and the dangers of playing with lighters and matches.

Teenagers are ready for lessons that they can apply in everyday life. Many teenagers work as babysitters, so prevention messages can be geared toward cooking safety and evacuating children in the event of a fire. Teenagers are also a high-risk group for motor vehicle collisions, so messages about seat belts, proper cellphone use, safe driving, the dangers of drinking and driving, and actions to take if they see or hear an emergency vehicle while driving are appropriate. Some teenagers might be interested in becoming fire fighters or joining a fire fighter explorer group. This age group can be a good source of recruits for both career and volunteer departments.

Adults are probably more interested in home fire safety. Remind smokers not to smoke in bed or throw burning cigarettes into wastepaper baskets. Some senior citizens may forget to turn off burners or may leave cooking food unattended. They also may be worried about their ability to move quickly if a fire starts in their home. Tailor your prevention messages to the specific concerns of the age group you are addressing.

FIRE FIGHTER II Tips   FFII

Adults often ask about funding sources for the fire department. If you give station tours, you should be aware of how your fire department is funded. Be prepared to answer questions on how taxpayer money is used to provide the myriad services the fire department provides.

FIGURE 27-8 Always leave visitors with printed fire prevention materials.

Try to leave every tour group with both a message and materials. First, reinforce a simple prevention or safety message that expands their awareness of fire safety and helps them to deal with potential emergency situations. Second, give them printed fire prevention materials **FIGURE 27-8**. These materials could range from colouring books for young children to a home fire prevention checklist for adults.

The steps for conducting a fire station tour are listed in **SKILL DRILL 27-5**:

1 Introduce yourself to the tour participants, and explain the mission of the fire department.

2 Explain the layout and purpose of the fire station.

3 Consider the use of an educational video presentation. (**STEP 1**)

4 Tour the living areas of the station. (**STEP 2**)

5 Explain the purpose of each piece of fire apparatus. (**STEP 3**)

6 Conclude the tour in the meeting room.

7 Encourage questions from participants.

8 Hand out appropriate fire safety materials. (**STEP 4**)

Near-Miss REPORT

Report Number: 06-0000531

Synopsis: Smoke from peanut oil causes severe reaction.

Event Description: We were holding our annual fire prevention night activities, which included the smoke house trailer. The kids and adults were having a good time. We were teaching the children about safety in the home. The trailer then filled with smoke and everything seemed to be fine. After our first few tours a woman screamed for help and stated that her child was having trouble breathing. We did not realize at first that it was an allergic reaction until we were en route to the hospital. We asked the mother if the child had any allergies. She stated yes, he was severely allergic to peanuts. We did not put two and two together until after we were returning from the hospital. We kept thinking of how his allergic reaction could have occurred at the station. We tried to think where the child could have gotten peanuts. Then one of us thought about the smoke trailer. The smoke is made from peanut oil, which started his reaction. This is one of those things you just don't think of until something goes wrong.

Lessons Learned: We learned that we need to have signs and announcements made regarding the smoke in the trailer. We now ask prior to all people entering if they or their children are allergic to peanuts as the smoke is made from peanut oil.

SKILL DRILL 27-5 | Conducting a Fire Station Tour | **FFII**

(Fire Fighter II, NFPA 6.5.2)

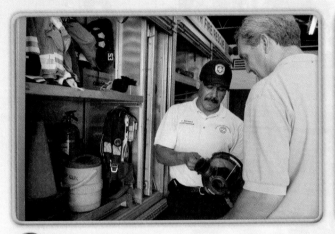

1 Introduce yourself to the tour participants and explain the mission of the fire department. Explain the layout and purpose of the fire station. Consider the use of an educational video presentation.

2 Tour the living areas of the station.

3 Explain the purpose of each piece of fire apparatus.

4 Conclude the tour in the meeting room. Encourage questions from participants. Hand out appropriate fire safety materials.

© LiquidLibrary

Wrap-Up

Chief Concepts

- Fire prevention is as important as fire suppression. Countless fires could be prevented by teaching safe behaviours and eliminating unsafe conditions. This kind of education ultimately saves lives and reduces the losses caused by fires.
- Fire prevention includes a range of activities that are intended to prevent the outbreak of fires and to limit the consequences if a fire does occur. These activities include enacting and enforcing fire codes, conducting property inspections, and presenting public fire-safety education programs.
- Your highest priority as a fire fighter should always be to prevent fires. Although Fire Fighter I and Fire Fighter II personnel may have limited responsibilities for formal fire prevention activities, it is imperative that you understand the objectives of fire prevention, the delineation of fire prevention responsibilities within your department, and the importance of your role in fire prevention.
- Fire codes are enacted to ensure a minimum level of fire safety in the home and workplace environments. After a jurisdiction adopts a fire code, it must be able to enforce it. Regular inspections ensure compliance, note any violations, and require their correction.
- Public education teaches techniques to reduce the risks of death or injury in the event of fire.
- Public fire and life safety education programs include those focused on the following topics:
  - Stop, Drop, and Roll program
  - Exit Drills in the Home (EDITH) program
  - Installation and maintenance of smoke alarms
  - Advantages of residential sprinkler systems
  - Selection and use of portable fire extinguishers
  - Learn Not to Burn®
  - Change Your Clock–Change Your Battery
  - Fire safety for special populations
  - Bike safety
  - Fall prevention
  - Wildland fire prevention programs

- The Stop, Drop, and Roll program is designed to instruct people what to do if their clothing catches fire.
  - **Stop** means just that: stay in place; do not run. Stress the fact that running will fan the flames and spread the fire.
  - **Drop** means getting down on the ground or on the floor. Cover the face with the hands to help protect the airway and the eyes.
  - **Roll** involves tucking in the elbows and keeping the legs together while rolling like a bug. Individuals should then roll over and over, and then roll back in the other direction; this helps to smother the flames.
- EDITH teaches residents how to safely get out of their homes in the event of a fire or other emergency.
- Another type of public fire-safety presentation covers the proper installation and maintenance of smoke alarms. In discussing this topic, stress the importance of having a working smoke alarm on every floor of a house. Also stress the importance of keeping smoke alarms in good working order.
- Many fire departments offer to conduct fire safety surveys in private dwellings. A home fire-safety survey helps identify fire and life-safety hazards and provides occupants with recommendations on making their home safer.
- During a home fire-safety survey, consider several types of hazards and issues. Concentrate on the hazard categories that most often cause residential fires, such as cooking equipment, heating equipment, electrical wiring, smoking materials, candles and other open flames, and children playing with fire.
- Most fire departments have a set format for conducting fire station tours; for example, a tour might start with an inspection of the apparatus and end in a meeting room with a question-and-answer session and information about the department and fire prevention techniques.
- Try to leave every tour group with both a fire prevention message and materials.

Hot Terms

<u>Exit Drills in the Home (EDITH)</u> A public safety program designed to teach occupants how to safely exit a house in the event of a fire or other emergency. It stresses the importance of having a preestablished meeting place for all family members.

<u>Fire code</u> A set of legally adopted rules and regulations designed to prevent fires and protect lives and property in the event of a fire.

<u>Fire prevention</u> Activities conducted to prevent fires and protect lives and property in the event of a fire. Fire prevention activities include the enactment and enforcement of fire codes, the inspection of properties, the presentation of fire safety education programs, and the investigation of the causes of fires.

You are a volunteer fire fighter in a small rural community. One night after training, the Chief asks if you would be willing to serve as the department's public educator. You know there is a desperate need and you enjoy talking about the fire department, but you just aren't sure where you would begin. You think back to the programs and concepts you learned in Fire Fighter I and II training.

1. Regulations designed to prevent fires from occurring, eliminate fire hazards, protect lives, and limit fire losses are called _____ codes.
   - **A.** building
   - **B.** electrical
   - **C.** fire
   - **D.** model

2. Where do fire codes not generally apply?
   - **A.** Interior of residential dwellings
   - **B.** Abandoned property
   - **C.** New construction
   - **D.** Commercial property

3. Which program focuses on instructing children what to do when their clothes catch fire?
   - **A.** Learn Not to Burn®
   - **B.** Stop, drop, and roll
   - **C.** Get low and go
   - **D.** EDITH

4. Which of the following is accurate about fire safety in the home?
   - **A.** Preidentify an exit.
   - **B.** Keep bedroom doors closed.
   - **C.** Keep bedroom doors open.
   - **D.** Keep a heavy object next to the window.

5. Where should smoke alarms be installed?
   - **A.** In accordance with the manufacturer's instructions
   - **B.** Kitchen
   - **C.** Near an exterior door
   - **D.** All of the above

You are in your recruit academy, and several of the other recruits are talking as they clean the room after a day of live fire burns. The typical banter is occurring about how much better one group did over another. Eventually, a fire fighter who had previously worked for another fire service makes the comment that his group will be the ones fighting the fires, while the other group will be the ones assigned to public education. Others quickly jump in with similar comments. Having spent a day during the previous week being trained on the importance of public education, including a personal visit by the chief of the fire department, the comments surprise you.

1. What should you say?
2. Should you mention this discussion to the Chief?
3. How would you go about changing the attitudes of the group?
4. How do you maintain your attitude when others around you do not share your views?

Fire Detection, Protection, and Suppression Systems

Fire Fighter I

Knowledge Objectives

After studying this chapter, you will be able to:

- Explain why all fire fighters should have a basic understanding of fire protection systems. (**NFPA 5.3.14.A**, p 810)
- Describe when and how water is shut off to a building's sprinkler system and how to stop water at a single sprinklers. (**NFPA 5.3.14.A**, p 834)

Skills Objectives

There are no skills objectives for Fire Fighter I candidates. NFPA 1001 contains no Fire Fighter I Job Performance Requirements for this chapter.

Fire Fighter II   FFII

Knowledge Objectives

After studying this chapter, you will be able to:

- Describe the basic components and functions of a fire alarm system. (**NFPA 6.5.3.A**, p 810–814)
- Describe the basic types of fire alarm initiation devices and indicate where each type is most suitable. (**NFPA 6.5.3.A**, p 814–820)
- Describe the fire department's role in resetting fire alarms. (**NFPA 6.5.3.A**, p 814–820)
- Describe the basic types of alarm notification appliances. (p 820)
- Describe the basic types of fire alarm annunciation systems. (p 821)
- Explain the different ways that fire alarms may be transmitted to the fire department. (**NFPA 6.5.3.A**) (p 821–823)
- Identify the four types of sprinklers. (**NFPA 6.5.3.A**) (p 823–826)
- Identify the different styles of indicating valves. (**NFPA 6.5.3.A**) (p 827–828, 830)

- Describe the operation and application of the following types of automatic sprinkler systems:
  - Wet pipe system
  - Dry-pipe system
  - Preaction system
  - Deluge system (NFPA 6.5.3.A) (p 833–834)
- Describe the differences between commercial and residential sprinkler systems. (NFPA 6.5.3.A) (p 834–835)
- Identify the three types of standpipes, and point out the differences among them. (NFPA 6.5.3.A) (p 835–836)
- Describe two problems that fire fighters could encounter when using a standpipe in a high-rise building. (NFPA 6.5.3.A) (p 836–837)
- Identify the hazards that specialized extinguishing systems can pose to responding fire fighters. (NFPA 6.5.3.A) (p 837–840)

Skills Objectives

There are no skills objectives for Fire Fighter II candidates. NFPA 1001 contains no Fire Fighter II Job Performance Requirements for this chapter.

You are riding in the back seat as you arrive at your first fire in a high-rise structure. As you step out of the cab, you see fire coming out of the 18th floor window. Over the radio, you hear the IC direct your crew to connect to the standpipe. Your heart rate quickens as you think about the task ahead and worry about getting things right.

1. What advantages do fire protection systems offer when fighting high-rise fires?
2. What is the difference between a standpipe system and a sprinkler system?
3. How would you know what systems the fire department connection serves?

Introduction

Today, fire prevention and building codes require that almost all new structures have some sort of fire protection system installed. This requirement makes it more important than ever for the line fire fighter to have a working knowledge of these systems. Fire protection systems include fire alarm, automatic fire detection, and fire-suppression systems. Understanding how these systems operate is both important for the fire fighter's safety and necessary to provide effective customer service to the public.

From a safety standpoint, fire fighters need to understand the operations and limitations of fire detection and suppression systems. A building with a fire protection system will have very different working conditions during a fire than an unprotected building. Fire fighters need to know how to fight a fire in a building with a working fire-suppression system and how to shut down the system after the fire is extinguished.

From a customer service standpoint, fire fighters who understand how fire protection systems work can help dispel misconceptions about these systems and advise building owners and occupants appropriately after an alarm is sounded. Most people have no idea of how the fire protection or detection systems in their building work, and there are often more **false alarms** in buildings with fire protection systems than actual fires. Fire fighters can help the owners and occupants determine what activated the system, how they can prevent future false alarms, and what needs to be done to restore a system to service.

The term *false alarm* is not always completely accurate. Often the sensors within the fire alarm system record an error somewhere within the system. In the modern construction of multifamily apartments, smoke detectors may be placed too close to the kitchen, which results in numerous false alarms. In these situations, the alarm is not raised in error; rather, smoke or heat has truly activated the system, though no crisis is in process. The major problem with these nuisance alarms is that people become used to hearing them and no longer respond to them with any urgency. Such was the case in January 2000, when three students died in their dormitory rooms at Seton Hall University. The students ignored the alarm, which signalled a real fire emergency, because they had become attuned to hearing the alarm on a regular basis for nonemergencies.

Fire alarms should be treated as true emergencies until someone from the fire department arrives on the scene and dictates otherwise. This chapter discusses the basic design and operation of fire alarm, fire detection, and fire-suppression systems, each of which could be the subject of an entire book. It is not necessary to understand all of the design criteria for a fire protection system, but simply to appreciate how the system is intended to function. Fire protection systems have fairly standardized design requirements across North America: most areas follow the applicable National Fire Protection Association (NFPA) standards. Local fire and building codes may require different types of systems for different buildings.

Fire Alarm and Detection Systems

Practically all new construction includes some sort of fire alarm and detection system. In most cases, both fire detection and fire alarm components are integrated into a single system. A fire detection system recognizes when a fire is occurring and activates the fire alarm system, which alerts the building's occupants and, in some cases, the fire department. Some fire detection systems also automatically activate fire-suppression systems to control or extinguish the fire.

Fire alarm and detection systems range from simple, single-station smoke alarms for private homes to complex fire detection and control systems for high-rise buildings. Many fire alarm and detection systems in large buildings also control other systems to help protect occupants and control the spread of fire and smoke. Although these systems can be complex, they generally include the same basic components as the simpler versions.

■ Fire Alarm System Components

A fire alarm system has three basic components: an alarm initiation device, an alarm notification device, and a control panel. The **alarm initiation device** is either an automatic or manually operated device that, when activated, causes the system to indicate an alarm. The **alarm notification device** is generally an audible device—often accompanied by a visual device—that alerts the building's occupants when the system is activated. The control panel links the initiation device to the notification device and performs other essential functions.

A single-station smoke alarm in a residential dwelling has the same basic components as a fire alarm system in a multistorey office building; that is, a single-station smoke alarm includes a **smoke detector**, an audible notification device, and a control unit. See the chapter "Fire Prevention and Public Education" for the correct methods of testing a smoke alarm.

Fire Alarm System Control Panels

Most fire alarm systems in buildings other than single residential units have several alarm initiation devices in different areas and use both audible and visible devices to notify the occupants of an alarm. The **fire alarm control panel** serves as the brain of the system, linking the initiation devices to the notification devices.

The control panel manages and monitors the proper operation of the system. It can indicate the source of an alarm, so that responding fire personnel will know what activated the alarm and where the initial activation occurred. The control panel also manages the primary power supply and provides a backup power supply for the system. It may perform additional functions, such as notifying the fire department or central station monitoring company when the alarm system is activated, and may interface with other systems and facilities.

Control panels vary greatly depending on the age of the system and the manufacturer; for example, an older system might simply indicate that an alarm has been activated, whereas a newer system might indicate that an alarm occurred in a specific zone within the building **FIGURE 28-1**. The most modern panels actually specify the exact location of the activated initiation device.

Fire alarm control panels are used to silence the alarm and reset the system. These panels should always be locked. Alarms should not be silenced or reset until the activation source has been found and checked by fire fighters to ensure that the situation is under control. In some systems, each initiating device must be reset before the entire system can be reset; others reset themselves after the problem is resolved.

Many buildings have an additional display panel in a separate location, usually near the front door of the building. This panel, which is called a **remote annunciator**, enables fire fighters to ascertain the type and location of the activated alarm device as they enter the building **FIGURE 28-2**.

The fire alarm control panel should also monitor the condition of the entire alarm system to detect any faults. Faults within the system are indicated by a trouble alarm, which shows that a component of the system is not operating properly and requires attention. Trouble alarms do not activate the building's fire alarm, but should make an audible sound and illuminate a special light on the alarm control panel. They may also transmit a notification to a remote service location.

A fire alarm system is usually powered by a 110-volt line, even though the system's appliances may use a lower voltage.

In addition to the normal power supply, fire codes call for a backup power supply for all alarm systems **FIGURE 28-3**. In some systems, a battery in the fire alarm control panel becomes activated automatically when the external power

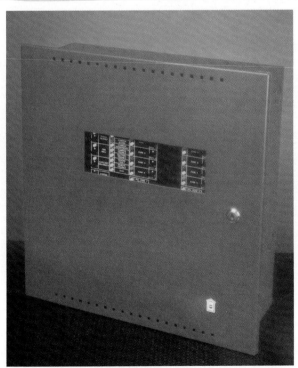

FIGURE 28-1 Most modern fire alarm control panels indicate the zone where an alarm was initiated.

Courtesy of Honeywell/Gamewell

FIGURE 28-2 A remote annunciator allows fire fighters to quickly determine the type and location of the activated alarm device.

Courtesy of Aaron Fire & Safety and Honeywell/Gamewell

is interrupted. Fire codes will specify how long the system must be able to function on the battery backup. In large buildings, the backup power supply could consist of an emergency generator or a secondary power source. If either the main power supply or the backup power source fails, the trouble alarm should sound.

Depending on the building's size and floor plan, an activated alarm may sound throughout the building or only in particular areas. In high-rise buildings, fire alarm systems are often programmed to alert only the occupants on the same floor as the activated alarm and those on the floors immediately above and below the affected floor. Alarms on the remaining floors can be manually activated from the system control panel. Some systems include a public address feature, which enables a fire department officer to provide specific instructions or information for occupants in different areas.

The control panel in a large building may be programmed to perform several additional functions; for example, it may automatically shut down or change the operation of air-handling systems, recall elevators to the ground floor, and unlock stairwell doors so that a person in an exit stairway can reenter an occupied floor.

■ Residential Fire Alarm Systems

Current building and fire codes usually require the installation of a fire detection and alarm system in all residential units. The most common type of residential fire alarm system is a **single-station smoke alarm** **FIGURE 28-4**. This life-safety device includes a smoke detection device, an automatic control unit, and an audible alarm, within a single unit. It alerts occupants quickly when a fire occurs. Millions of single-station smoke alarms have been installed in private dwellings and apartments.

Smoke alarms can be either battery-powered or hard-wired to a 110-V electrical system or both. Most building codes require hard-wired, AC-powered smoke alarms in all newly constructed dwellings; battery-powered units are popular in existing residences. The major concern with a battery-powered smoke alarm is ensuring that the battery is replaced on a regular basis. Newer battery-powered smoke alarms are available with lithium batteries that will last for 10 years.

The most up-to-date codes require new homes to have a smoke alarm in every bedroom and on every floor level. They also require a battery backup in the event of a power failure. In newer installations, all smoke alarms in the dwelling are interconnected; as a consequence, all of them will sound if one is activated. For example, if the basement smoke alarm is activated, alarms will sound on all levels to alert the occupants.

Studies have shown that not all people are able to hear and respond to a smoke alarm sounding. This may be due to deep sleep, hearing impairment or consumption of drugs and/or alcohol. For sound sleepers, particularly children, a parent may purchase an alarm that permits them to record a personalized message warning the child to wake up. The hearing impaired may purchase a smoke alarm that activates a mini–strobe light with the alarm or one that can be powered to a vibration device, which can be placed under their pillow to alert them. Either of the devices previously described could also potentially work for people impaired from substance consumption.

Many home smoke alarms are part of security systems. Like commercial systems, these home fire alarm/security systems have an alarm control panel and require a passcode to set or reset the system. They may also be monitored by a central station.

FIGURE 28-3 Fire prevention codes require that alarm systems have a backup power supply, which activates automatically when the normal electrical power is interrupted.

Courtesy of Honeywell/Gamewell

FIGURE 28-4 A single-station smoke alarm.

Courtesy of Kidde Residential and Commercial Division

■ Ionization versus Photoelectric Smoke Detectors

Two types of fire detection devices may be used in a smoke alarm to detect combustion:

- Ionization detectors are triggered by the invisible products of combustion.
- Photoelectric detectors are triggered by the visible products of combustion.

Ionization smoke detectors work on the principle that burning materials release many different products of combustion, including electrically charged microscopic particles. An ionization detector senses the presence of these invisible charged particles (ions).

An ionization smoke detector contains a very small amount of radioactive material inside its inner chamber. This radioactive material releases charged particles into the chamber, and a small electric current flows between two plates FIGURE 28-5. When smoke particles enter the chamber, they neutralize the charged particles and interrupt the current flow. The detector then senses this interruption and activates the alarm.

Photoelectric smoke detectors use a light beam and a photocell to detect larger visible particles of smoke FIGURE 28-6. They operate by reflecting the light beam either away from or onto the photocell, depending on the design of the device. When visible particles of smoke pass through the light beam, they reflect the beam onto or prevent it from striking the photocell, thereby activating the alarm.

Ionization smoke detectors are more common and less expensive than photoelectric smoke detectors. They react more quickly to fast-burning fires, such as a fire in a wastebasket, which may produce little visible smoke FIGURE 28-7. On the downside, fumes and steam from common activities such as cooking and showering may trigger unwanted alarms.

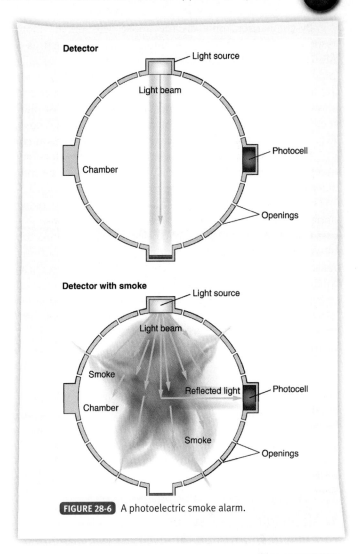

FIGURE 28-6 A photoelectric smoke alarm.

FIGURE 28-5 An ionization smoke alarm.

© AbleStock

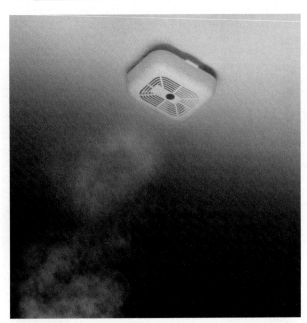

FIGURE 28-7 Ionization smoke detectors react more quickly than photoelectric smoke detectors to a fast-burning fire, such as a fire in a wastepaper basket, which may produce little visible smoke.

© Brendan Byrne/age fotostock

By comparison, photoelectric smoke detectors are more responsive to slow-burning or smouldering fires, such as a fire caused by a cigarette caught in a couch, which usually produces a large quantity of visible smoke **FIGURE 28-8**. They are less prone to false alarms from regular cooking fumes and shower steam than are ionization smoke detectors. Most photoelectric smoke alarms require more current to operate, however, so they are typically connected to a 110-V power source.

Both ionization and photoelectric smoke detectors are acceptable life-safety devices. Indeed, in most cases, they are considered interchangeable. Some fire codes require that photoelectric smoke detectors be used within a certain distance of cooking appliances or bathrooms to help prevent unwanted alarms.

Combination ionization–photoelectric smoke alarms are also available. These alarms react quickly to both fast-burning and smouldering fires. They are not suitable for use near kitchens or bathrooms, because they are prone to the same unwanted alarms as regular ionization smoke detectors.

Most ionization and photoelectric smoke alarms look very similar to each other. The only sure way to identify the type of alarm is to read the label, which is often found on the back of the case. An ionization alarm must have a label or engraving stating that it contains radioactive material.

There are now combined carbon monoxide (CO) and smoke alarms available for residential use. The manufacturer's instructions should be followed for the correct installation location and maintenance. If a person wishes to install a separate CO detector, they should be encouraged to do so but also provided similar education about how to respond when the alarm sounds. (See page 818 of this chapter for more information.)

■ Alarm Initiation Devices

Alarm initiation devices begin the fire alarm process either manually or automatically. Manual initiation devices require human activation; automatic devices function without human intervention. Manual fire alarm boxes are the most common type of alarm initiation devices that require human activation. Automatic initiation devices include various types of heat and smoke detectors and other devices that automatically recognize the evidence of a fire.

Manual Initiation Devices

Manual initiation devices are designed so that building occupants can activate the fire alarm system on their own if they discover a fire in the building. Many older alarm systems could only be activated manually. The primary manual initiation device is the manual fire alarm box, or **manual pull-station** **FIGURE 28-9**. Such a station has a switch that either opens or closes an electrical circuit in order to activate the alarm.

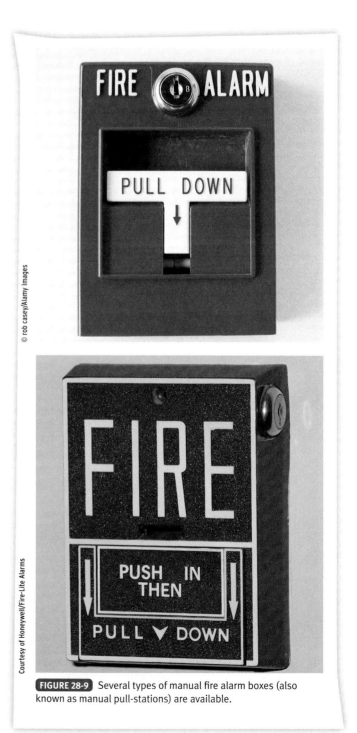

© rob casey/Alamy Images

Courtesy of Honeywell/Fire-Lite Alarms

FIGURE 28-9 Several types of manual fire alarm boxes (also known as manual pull-stations) are available.

FIGURE 28-8 Photoelectric smoke alarms react more quickly than ionization smoke detectors to a slow-burning or smouldering fire, such as a cigarette in furniture, which usually produces a large quantity of visible smoke.

Courtesy of Kevin Reed

Troubleshooting Smoke Alarms

Most problems with smoke alarms are caused by a lack of power, dirt in the sensor chamber, or an alarm defect.

Power Problems

Smoke alarms require power from a battery or from a hard-wired 110-V power source. Some smoke alarms that are hard-wired to a 110-V power source are equipped with a battery that serves as a backup power source.

The biggest problem with battery-powered alarms is that they may contain a dead or missing battery. Some people do not change batteries when recommended, resulting in inoperable smoke alarms. People may also remove a smoke alarm battery to use it for another purpose or to prevent nuisance alarm activations. The solution in this situation is to install a new battery in the alarm. New lithium batteries will last for 10 years in a smoke alarm.

Hard-wired smoke detectors become inoperable if someone turns the power off at the circuit breaker. The solution in this situation is to turn the circuit breaker back on.

Most smoke alarms containing batteries will signal a low-battery condition by emitting a chirp every few seconds. This chirp indicates that the battery needs to be replaced to keep the smoke alarm functional. Many hard-wired smoke alarms that are equipped with battery backups will also chirp to indicate that the backup battery is low, even when the 110-V power source is operational. This feature ensures that the smoke alarm always has an operational backup system.

Dirt Problems

A second problem with smoke alarms is the increased sensitivity that results from dust or an insect becoming lodged in the photoelectric or ionization chamber. Such an obstruction causes the alarm to become activated when a small quantity of water vapour or smoke enters the chamber, leading to unnecessary alarms. The solution in this situation is to remove the cover of the chamber and gently vacuum out the chamber. Alternatively, you can use a puff of air from a can of compressed air to remove the dust or insect. Follow the manufacturer's instructions for this process.

Alarm Problems

The third common problem you may encounter with smoke alarms is a worn-out detector. It is recommended that smoke alarms be replaced every 10 years because the sensitivity of the alarm can change. Some worn out alarms may become overly sensitive and emit false alarms; others may develop decreased sensitivity and fail to emit an alarm in the event of a fire. The date of manufacture should be stamped on each smoke alarm. If a detector is more than 10 years old, it should be replaced with a new alarm.

Understanding the basic functions of simple smoke alarms and how to troubleshoot them enables you to respond to citizens who call your fire department when they encounter problems with their smoke alarms. Most importantly, it enables you to keep smoke alarms operational. Remember: only operational smoke alarms can save lives. All battery-operated alarms should be tested if the resident has been gone for an extended period of time, in addition to annual testing.

Pull-stations come in a variety of sizes and designs, depending on the manufacturer. They can be either single-action or double-action devices. **Single-action pull-stations** require a person to pull down a lever, toggle, or handle to activate the alarm; the alarm sounds as soon as the pull-station is activated. **Double-action pull-stations** require a person to perform two steps before the alarm will activate. They are designed to reduce the number of false alarms caused by accidental or intentional pulling of the alarm. The person must move a flap, lift a cover, or break a piece of glass to reach the alarm activation device. Designs that use glass are no longer in favour, both because the glass must be replaced each time the alarm is activated and because the broken glass poses a risk of injury.

Once activated, a manual pull-station should stay in the "activated" position until it is reset. This status enables responding fire fighters to determine which pull-station initiated the alarm. Resetting a pull-station requires a special key, screwdriver, or Allen key. In many systems, the pull-station must be reset before the building alarm can be reset at the alarm control panel.

A variation on the double-action pull-station designed to prevent malicious false alarms is covered with a piece of clear plastic **FIGURE 28-10**. These covers are often used in areas where malicious false alarms occur frequently, such as high schools and university dormitories. The plastic cover must be opened before the pull-station can be activated. Lifting the cover triggers a loud tamper alarm at that specific location, but does not activate the fire alarm system. Snapping the cover back into place resets the **tamper alarm**. The intent is that a person planning to initiate a false alarm will drop the cover and run when the tamper alarm sounds. The pull-station tamper alarm is not connected to the fire alarm system in any way.

Even automatic fire alarm systems can be manually activated by pressing a button or flipping a switch on the alarm

FIGURE 28-10 A variation on the double-action pull-station, designed to prevent malicious false alarms, has a clear plastic cover and a separate tamper alarm.

Courtesy of STI-USA

system control panel. In large buildings (particularly high-rise buildings), the control panel has separate manual activation switches for different areas or on particular floors.

Automatic Initiation Devices

Automatic initiation devices are designed to function without human intervention and will activate the alarm system when they detect evidence of a fire. These systems can be programmed to transmit the alarm to the fire department, even if the building is unoccupied, and to perform other functions when a detector is activated.

Automatic initiation devices can use any of several types of detectors. Some detectors are activated by smoke or by the invisible products of combustion; others react to heat, the light produced by an open flame, or specific gases. Automatic initiation devices sense fire in the same manner as a single-station alarm does in a residence; that is, they use either an ionization or a photoelectric chamber to detect products of combustion.

Smoke Detectors

A **smoke detector** is designed to sense the presence of smoke. Among fire fighters, the term *smoke detector* generally refers to a sensor that is part of a fire alarm system. This type of device is commonly found in schools, hospitals, businesses, and commercial occupancies that are equipped with fire alarm systems **FIGURE 28-11**.

Smoke detectors come in a variety of designs and styles geared toward different applications. The most commonly used smoke detectors are ionization and photoelectric models, which operate in the same way that residential smoke alarms do; however, the smoke detectors used in commercial fire alarm systems are usually more sophisticated and more expensive than residential smoke alarms.

Each detection device is rated to protect a certain floor area, so in large areas, the detectors are often placed in a grid pattern. Newer smoke detectors also have a visual indicator, such as a steady or flashing light, that signals when the device has been activated.

A **beam detector** is a type of photoelectric smoke detector used to protect large, open areas such as churches, auditoriums, airport terminals, and indoor sports arenas. In these facilities, it would be difficult or costly to install large numbers of individual smoke detectors, but a single beam detector could be used for the entire area **FIGURE 28-12**.

A beam detector has two components: a sending unit, which projects a narrow beam of light across the open area,

FIGURE 28-11 Commercial ionization smoke detector.

FIGURE 28-12 Beam detectors are used in large, open spaces.

Courtesy of Fire Fighting Enterprises Ltd.

and a receiving unit, which measures the intensity of the light when the beam strikes the receiver. When smoke interrupts the light beam, the receiver detects a drop in the light intensity and activates the fire alarm system. Most photoelectric beam detectors are set to respond to a certain **obscuration rate**, or percentage of light blocked. If the light is completely blocked, such as when a solid object is moved across the beam, the trouble alarm will sound, but the fire alarm will not be activated.

Smoke detectors are usually powered by a low-voltage circuit and send a signal to the fire alarm control panel when they are activated. Both ionization and photoelectric smoke detectors are self-restoring. After the smoke condition clears, the alarm system can be reset at the control panel.

Heat Detectors

Heat detectors are also commonly used as automatic alarm initiation devices. These devices can provide property protection but cannot provide reliable life-safety protection, because they do not react quickly enough to incipient fires. They are generally used in situations where smoke alarms cannot be used, such as dusty environments and areas that experience extreme cold or heat. Heat detectors are often installed in unheated areas, such as attics and storage rooms, as well as boiler rooms and manufacturing areas.

Heat detectors are generally very reliable and less prone to false alarms than are smoke alarms. You may come across heat detectors that were installed 30 or more years ago and are still in service. Unfortunately, older units lack a visual trigger that indicates which device was activated, so tracking down the cause of an alarm may be very difficult. Newer models have an indicator light that shows which device was activated.

Several types of heat detectors are available, each of which is designed for specific situations and applications. Single-station heat alarms are sometimes installed in unoccupied areas of buildings that do not have fire alarm systems, such as attics or storage rooms. **Spot detectors** are individual units that can be spaced throughout an occupied space, so that each detector covers a specific floor area. Spot detectors are usually installed in light commercial and residential settings; the units may be in individual rooms or spaced at intervals along the ceiling in larger areas.

Heat detectors can be designed to operate at a fixed temperature or to react to a rapid increase in temperature. Either fixed-temperature or rate-of-rise devices can be configured as spot or line detectors.

Fixed-Temperature Heat Detectors **Fixed-temperature heat detectors**, as the name implies, are designed to operate at a preset temperature FIGURE 28-13 . A typical temperature for a light-hazard occupancy, such as an office building, would be 57 degrees Celcius (135 degrees Fahrenheit). Fixed-temperature detectors typically include a metal alloy that will melt at the preset temperature. This melting alloy releases a lever-and-spring mechanism, which either opens or closes a switch. Most fixed-temperature heat detectors must be replaced after they have been activated, even if the activation was accidental.

Rate-of-Rise Heat Detectors **Rate-of-rise heat detectors** become activated when the temperature of the surrounding air increases by more than a set amount in a given period of time. A typical rating might be "greater than 7°C (12°F)

in 1 minute." If the temperature increase occurs more slowly than this rate, the rate-of-rise heat detector will not activate. By contrast, a temperature increase at a pace greater than this rate will activate the detector and set off the fire alarm. Rate-of-rise heat detectors should not be located in areas that normally experience rapid changes in temperature, such as near garage doors in heated parking areas.

Some rate-of-rise heat detectors have a **bimetallic strip** made of two metals that respond differently to heat; a rapid increase in temperature causes the strip to bend unevenly, which opens or closes a switch. Another type of rate-of-rise heat detector uses an air chamber and diaphragm mechanism: As air in the chamber heats up, the pressure increases. Gradual increases in pressure are released through a small hole, but a rapid increase in pressure will press upon the diaphragm and activate the alarm. Most rate-of-rise heat detectors are self-restoring, so they do not need to be replaced after activation, unless they were directly exposed to a fire.

Rate-of-rise heat detectors generally respond more rapidly to most fires than do fixed-temperature heat detectors; however, a slow-burning fire, such as a smoldering couch, might not activate a rate-of-rise heat detector until the fire is well established.

Combination rate-of-rise–fixed-temperature heat detectors are available. These devices balance the faster response of the rate-of-rise detector with the reliability of the fixed-temperature heat detector.

Line-Type Detectors

Line detectors use wire or tubing strung along the ceiling of large, open areas to detect an increase in heat. An increase in temperature anywhere along the line will activate the detector. Line detectors are often found in churches, warehouses, and industrial or manufacturing settings.

Line detectors use wires or a sealed tube to sense heat. One wire-type model has two wires inside, separated by an insulating material. When heat melts the insulation, the wires short and activate the alarm. The damaged section of insulation must be replaced with a new piece after activation of the detector.

Another wire-type model measures changes in the electrical resistance of a single wire as it heats up. This device is self-restoring and does not need to be replaced after activation, unless it was directly exposed to a fire.

The tube-type line detector contains a sealed metal tube filled with air or a nonflammable gas. When the tube is heated, the internal pressure increases and activates the alarm. Like the single-wire line-type detector, this device is self-restoring and does not need to be replaced after activation, unless it was directly exposed to a fire.

Flame Detectors

Flame detectors are specialized devices that detect the electromagnetic light waves produced by a flame **FIGURE 28-14**. These devices can quickly recognize even a very small fire.

Typically, flame detectors are found in places such as aircraft hangars or specialized industrial settings in which early detection and rapid reaction to a fire are critical. Such detectors are also used in explosion-suppression systems, where they detect and suppress an explosion as it is occurring.

Flame detectors are both complicated and expensive. A key disadvantage of these models is that other infrared or ultraviolet sources, such as the sun or a welding operation, can set off an unwanted alarm. Flame detectors that combine infrared and ultraviolet sensors are sometimes used to lessen the chance of a false alarm.

Gas Detectors

Gas detectors are calibrated to detect the presence of a specific gas that is created by combustion or that is used in the facility. Depending on the system, a gas detector may be programmed to activate either the building's fire alarm system or a separate alarm. Some of these specialized instruments need regular calibration if they are to operate properly. Gas detectors for specific gases are often found in commercial or industrial applications. The most common type of gas detector you will encounter in residential settings is a CO detector.

Air-Sampling Detectors

Air-sampling detectors continuously capture air samples and measure the concentrations of specific gases or products of combustion. These devices draw in air samples through a sampling unit and analyze them using an ionization or photoelectric smoke detector **FIGURE 28-15**. Duct detectors (air-sampling detectors) are often installed in the return air ducts of large buildings. They will sound an alarm and shut down the air-handling system if they detect smoke.

More complex systems are sometimes installed in special hazard areas to draw air samples from rooms, enclosed spaces, or equipment cabinets. These samples pass through gas analyzers, which can identify smoke particles, products of combustion, and concentrations of other gases associated with a dangerous condition. Air-sampling detectors are most often used in areas that hold valuable contents or sensitive equipment, where it is important to detect problems early.

Residential Carbon Monoxide Detectors

Residential CO detectors are designed to sound an audible or visual alarm when the concentration of CO becomes high enough to pose a health risk to the occupants of the building. Such an alarm is triggered by a very high concentration of CO that accumulates in a short period of time or by a lower concentration of CO that occurs over a longer period of time. To most people, many CO alarms look like smoke alarms. Some of these units are combined with smoke alarms to warn against fire or elevated levels of CO **FIGURE 28-16**. These combination alarms have one audible or visible alarm for fire and a different audible or visible alarm for elevated levels of CO.

CO alarm activations need to be investigated by personnel who can use proper gas detection devices to determine the level of the gas and the cause of the alarm activation. Anytime you respond to a CO alarm, you need to check all building occupants to determine whether they are suffering from the following signs or symptoms of CO poisoning:

- Headache
- Nausea
- Disorientation
- Unconsciousness
- Flulike symptoms (present in multiple people in the same location)

Be especially alert for multiple people who are all exhibiting flulike symptoms, as they may be suffering from CO poisoning.

FIGURE 28-14 Flame detectors are specialized devices that detect the electromagnetic light waves produced by a flame.

FIGURE 28-15 Air-sampling detector.

FIGURE 28-16 One type of combination CO–smoke alarm.
© kvisel/iStockphoto

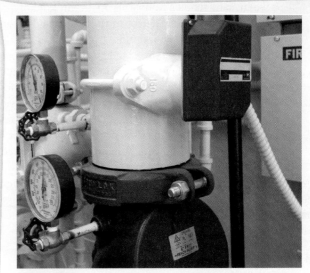

FIGURE 28-17 Automatic sprinkler systems use an electric flow switch to activate the building's fire alarm system.

Alarm Initiation by Fire-Suppression Systems

Other fire protection systems in a building may be used to activate the fire alarm system. **Automatic sprinkler systems** are usually connected to the fire alarm system and will activate the alarm if a water flow occurs **FIGURE 28-17**. Such a system not only alerts the building occupants and the fire department to a possible fire but also ensures that someone is made aware that water is flowing, in case of an accidental discharge. Any other fire extinguishing systems in a building, such as those found in kitchens or containing halogenated agents, should also be tied into the building's fire alarm.

False, Unwanted, and Nuisance Alarms

As the number of fire detection and alarm systems increases, so does the number of false, unwanted, or nuisance alarms. Knowing how to handle these alarms is just as important as knowing how to deal with an actual fire. Fire department personnel who are informed about fire alarm and detection systems can advise building owners and operators when these systems experience problems.

The term *false alarm* is generally used by the public to describe all fire alarm activations that are not associated with a true emergency. In reality, three distinct types of false alarms are possible: malicious false alarms, unwanted alarms, and nuisance alarms. It is important to distinguish among the three types in determining the root cause of the fire alarm activation, because the problem must be recognized before it can be rectified.

Regardless of the underlying cause, all three types of false alarms have the same results: they waste fire department resources and may delay legitimate responses. In addition, frequent false alarms at the same site can desensitize building occupants to the alarm system so that they might not respond appropriately to a real emergency.

Malicious False Alarms

Malicious false alarms occur when individuals deliberately activate a fire alarm when there is no fire, causing a disturbance.

Manual fire alarm boxes are popular targets for pranksters. A malicious false alarm is an illegal act.

Unwanted Alarms

An **unwanted alarm** occurs when an alarm system is activated by a condition that is not really an emergency; for example, a smoke alarm placed too close to a kitchen may be triggered by normal cooking activities. An unwanted alarm could also occur if a smoker unknowingly stands under a smoke detector. The smoke detector does its job—recognizing smoke and activating the alarm system—but there is no real emergency.

Nuisance Alarms

Nuisance alarms are caused by the improper functioning of an alarm system or one of its components. Alarm systems must be properly maintained on a continual basis. A mechanical failure or a lack of maintenance that causes an alarm system to be activated when no emergency is actually occurring means that the system might not be activated when there is a real fire.

Preventing Unwanted and Nuisance Alarms

Systems that experience unwanted and nuisance alarms should be examined to determine the cause of the alarms and correct the problem. Many unwanted alarms could be avoided by relocating a detector or using a different type of detector in a particular location. Proper design, installation, and system maintenance are essential to prevent nuisance alarms.

If an alarm system is activated whenever it rains, water is probably leaking into the wiring. Detectors may also respond to a buildup of dust, dirt, or other debris by becoming more sensitive and needlessly going off. That risk explains why smoke alarms and smoke detectors must be cleaned periodically.

Several methods can be used to reduce unwanted and nuisance alarms caused by smoke detection systems:

- In a **cross-zoned system**, the activation of a single smoke detector will not sound the fire alarm, although it will usually set off a trouble alarm. In this kind of system, a second smoke detector must be activated before the actual fire alarm will sound.

- In a **verification system**, a delay of 30–60 seconds separates activation and notification. During this time, the system may show a trouble or pre-alarm condition at the system control panel. After the preset interval, the system rechecks the detector. If the condition has cleared, the system returns to normal; if the detector is still sensing smoke, however, the fire alarm is activated.

Both cross-zoned and verification systems are designed to prevent brief exposures to common occurrences such as cigarette smoke from activating the alarm system. For more information, see the NFPA's *Fire Service Guide to Reducing Unwanted Fire Alarms.*

■ Alarm Notification Appliances

Audible alarm notification devices, such as bells, horns, and electronic speakers, produce an audible signal when the fire alarm is activated. Newer systems also incorporate visual alerts. These audible and visual alarms alert occupants of a building to a fire.

Older systems used a variety of sounds as notification devices, but this inconsistency often led to confusion over whether the sound was actually an alarm. More-recent fire codes have adopted a standardized audio pattern called the **temporal-3 pattern** that must be produced by any audio device used as a fire alarm. The temporal-3 pattern consists of three tones followed by a short pause. Even single-station smoke alarms designed for residential occupancies now use this sound pattern. As a consequence, people can recognize a fire alarm immediately.

Some public buildings also play a recorded evacuation announcement in conjunction with the temporal-3 pattern. This recorded message is played through the fire alarm speakers and provides instructions for safe evacuation from the site **FIGURE 28-18**. In facilities such as airport terminals, this announcement is recorded in multiple languages. Such a system may also include a public address feature that fire department or building security personnel can use to provide specific instructions, information about the situation, or notice when the alarm condition is terminated. Research labs are now reviewing and studying human reaction to fire alarm systems. As children, we are conditioned to the sound of a fire alarm bell or smoke alarm buzzer; it has been shown that any other sound or any different form of notification may not receive immediate recognition as indicative of an emergency situation.

Many new fire alarm systems incorporate visual notification devices such as high-intensity strobe lights or other types of flashing lights, as well as audio devices **FIGURE 28-19**. Visual devices alert hearing-impaired occupants to a fire alarm and are very useful in noisy environments where an audible alarm might not be heard.

■ Other Fire Alarm Functions

In addition to alerting occupants and summoning the fire department, fire alarm systems may control other building functions, such as air-handling systems, fire doors, and elevators. To control smoke movement through the building, the system may either shut down or start up air-handling systems.

FIGURE 28-18 An alarm notification appliance featuring a speaker.
Courtesy of Honeywell/Fire-Lite Alarms

FIGURE 28-19 This alarm notification appliance has both a loud horn and a high-intensity strobe light.

Fire doors that are normally held open by electromagnets may be released to compartmentalize the building and confine the fire to a specific area. Doors allowing reentry from exit stairways into occupied areas may be unlocked. Elevators may be summoned to a predetermined floor, usually the main lobby, so they can be used by fire crews.

Responding fire personnel must understand which building functions are being controlled by the fire alarm, for both safety and fire-suppression reasons. This information should be gathered during preincident planning surveys and should be available in printed form or on a graphic display at the control panel location.

Fire Alarm Annunciation Systems

Some fire alarm systems give little information at the alarm control panel; others specify exactly which initiation device activated the fire alarm. The systems can be further subdivided based on whether they are zoned or coded systems.

In a zoned system, the alarm control panel indicates where in the building the alarm was activated. Today, almost all alarm systems are zoned to some extent; only the most rudimentary alarm systems give no information at the alarm control panel about where the alarm was initiated. In a coded system, the zone is identified not only at the alarm control panel but also through the audio notification device. Using these two variables, systems can be broken down into four categories: noncoded alarm, zoned noncoded alarm, zoned coded alarm, and master-coded alarm TABLE 28-1 .

Noncoded Alarm System

In a noncoded alarm system, the control panel does not provide any information as to where in the building the fire alarm was activated. The alarm typically sounds a bell or horn. Fire department personnel must then search the entire building to find which initiation device was activated. This type of system is generally found only in older small buildings.

Zoned Noncoded Alarm System

The zoned noncoded alarm system is the most common type of system encountered by fire fighters, particularly in newer buildings. With this model, the building is divided into multiple zones, often by floor or by wing. The alarm control panel indicates in which zone the activated device is located and, sometimes, which type of device was activated. Responding personnel can go directly to that part of the building to search for the problem and check the activated device.

Many zoned noncoded alarm systems maintain an individual indicator light for each zone. When a device in that area is activated, the indicator light goes on. Computerized alarm systems also may use "addressable devices." In these systems, each individual initiation device—whether it is a smoke detector, heat detector, or pull-station—has its own unique identifier. When the device is activated, the identifier is indicated on a display or printout at the control panel. Responding personnel can then determine exactly which device or devices have been activated.

Zoned Coded Alarm

In addition to having all the features of a zoned alarm system, a zoned coded alarm system indicates which zone has been activated over the announcement system. Hospitals often use this type of system, because it is not possible to evacuate all staff and patients for every fire alarm. The audible notification devices give a numbered code that indicates which zone was activated. A code list tells building personnel which zone is in an alarm condition and which areas must be evacuated.

More modern systems in these occupancies use speakers as their alarm notification devices. As a consequence, a voice message indicating the nature and location of the alarm accompanies the audible alarm signal.

Master-Coded Alarm

In a master-coded alarm system, the audible notification devices for fire alarms also are used for other purposes; for example, a school may use the same bell to announce a change in classes, to signal a fire alarm, to summon the janitor, or to make other notifications. Most of these systems have been replaced by modern speaker systems that use the temporal-3 pattern fire alarm signal and have public address capabilities. This type of system is rarely installed in new buildings.

Fire Department Notification

The fire department should always be notified when a fire alarm system is activated. In some cases, a person must make a telephone call to the fire department. In other cases, the fire alarm system may be connected directly to the fire department or to a remote location where someone on duty calls the fire department. Fire alarm systems can be classified into five categories, based on how the fire department is notified of an alarm TABLE 28-2 .

Local Alarm Systems

A local alarm system does not notify the fire department. Instead, the alarm sounds only in the building to notify its occupants. Buildings with this type of system should have notices posted requesting occupants to call the fire department and report the alarm after they exit FIGURE 28-20 .

Remote Station Systems

A remote station system sends a signal directly to the fire department or to another monitoring location via a telephone line or a radio signal. This type of direct notification can be installed only in jurisdictions where the fire department is equipped to handle direct alarms. If the signal goes to a monitoring location, that site must be continually staffed by someone who will call the fire department.

TABLE 28-1	Categories of Alarm Annunciation Systems
Category	**Description**
Noncoded alarm	No information is given on which device was activated or where it is located.
Zoned noncoded alarm	The alarm system control panel indicates the zone in the building that was the source of the alarm. It may also indicate the specific device that was activated.
Zoned coded alarm	The system indicates which zone has been activated over the audible warning device. This type of system is often used in hospitals, where it is not feasible to evacuate the entire facility.
Master-coded alarm	The system is zoned and coded. The audible warning devices are also used for other emergency-related functions.

TABLE 28-2	Fire Department Notification Systems
Type of System	**Description**
Local alarm	The fire alarm system sounds an alarm only in the building where it was activated. No signal is sent out of the building. Someone must call the fire department to respond.
Remote station	The fire alarm system sounds an alarm in the building and transmits a signal to a remote location. The signal may go directly to the fire department or to another location where someone is responsible for calling the fire department.
Auxiliary	The fire alarm system sounds an alarm in the building and transmits a signal to the fire department via a public alarm-box system.
Proprietary	The fire alarm system sounds an alarm in the building and transmits a signal to a monitoring location owned and operated by the facility's owner. Depending on the nature of the alarm and arrangements made with the local fire department, facility personnel may respond and investigate, or the alarm may be immediately retransmitted to the fire department. These facilities are monitored 24 hours a day.
Central station	The fire alarm system sounds an alarm in the building and transmits a signal to an off-premises alarm monitoring facility. The off-premises monitoring facility is then responsible for notifying the fire department to respond.

FIGURE 28-20 Buildings with local alarm systems should post notices requesting occupants to call the fire department and report the alarm after they exit.

A.

B.

FIGURE 28-21 In a proprietary system, fire alarms from several buildings are connected to a single monitoring site owned and operated by the buildings' owner.

Auxiliary Systems

<u>Auxiliary systems</u> can be used in jurisdictions with a public fire alarm–box system. With this approach, the building's fire alarm system is tied into a master alarm box located outside the building. When the alarm activates, it trips the master box, which transmits the alarm directly to the fire department's communications centre. Most cities have phased out public fire alarm boxes, because they are expensive to maintain. They have been largely replaced by more modern communication devices.

Proprietary Systems

In a <u>proprietary system</u>, the building's fire alarms are connected directly to a monitoring site that is owned and operated by the building's owner. Proprietary systems are often installed at facilities where multiple buildings belong to the same owner, such as universities or industrial complexes. Each building is connected to a monitoring site on the premises (usually the security centre), which is staffed at all times **FIGURE 28-21**. When an alarm sounds, the staff at the monitoring site reports

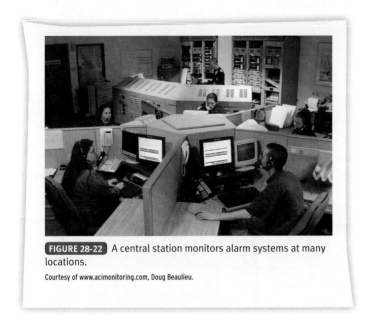

FIGURE 28-22 A central station monitors alarm systems at many locations.
Courtesy of www.acimonitoring.com, Doug Beaulieu.

the alarm to the fire department, usually by telephone or a direct line.

Central Stations

A **central station** is a third-party, off-site monitoring facility that monitors multiple alarm systems. Individual building owners contract with and pay the central station to monitor their facilities **FIGURE 28-22**. When an activated alarm at a covered building transmits a signal to the central station by telephone or radio, personnel at the central station notify the appropriate fire department of the fire alarm. The central station facility may be located in the same city as the facility or in a different part of the country.

Usually, building alarms are connected to the central station through leased or standard telephone lines; the use of either cellular telephone frequencies or radio frequencies, however, is becoming more common. Cellular or radio connections may be used to back up regular telephone lines in case they fail; in remote areas without telephone lines, they may be the primary transmission method.

Fire-Suppression Systems

Fire-suppression systems include automatic sprinkler systems, standpipe systems, and specialized extinguishing systems, such as dry chemical systems. Most newly constructed commercial buildings incorporate at least one of these systems, and increasing numbers of single-family dwellings are being built with residential sprinkler systems as well.

Understanding how these systems work is important because they can affect fire behaviour. In addition, fire fighters should know how to interface with the fire-suppression system and how to shut down a system to prevent unnecessary damage.

■ Automatic Fire Sprinkler Systems

The most common type of fire-suppression system is the automatic sprinkler system. Automatic sprinklers are reliable and effective, with a history of more than 100 years of successfully

controlling fires. When properly installed and maintained, these systems can help control fires and protect lives and property.

Unfortunately, few members of the general public have an accurate understanding of how automatic sprinklers work. In movies and on television, when one sprinkler is activated, the entire system begins to discharge water. This inaccurate portrayal of how automatic sprinkler systems operate has made people hesitant to install automatic sprinklers for fear that they will cause unnecessary water damage.

The reality is quite different: in most automatic sprinkler systems, the sprinklers open one at a time as they are heated to their operating temperature. Usually, only one or two sprinklers open before the fire is controlled.

The basic operating principles of an automatic sprinkler system are simple: A system of water pipes is installed throughout a building to deliver water to every area. Depending on the design and occupancy of the building, these pipes may be placed above or below the ceiling. Automatic sprinklers are located along this system of pipes, such that each sprinkler covers a particular floor area. A fire in that area will activate the sprinklers, which then discharges water onto the fire. This kind of system is analogous to having a fire fighter in every room with a charged hose line, just waiting for a fire.

One of the major advantages of a sprinkler system is that it can function as both a fire detection system and a fire-suppression system. An activated sprinkler not only discharges water on the fire but also triggers a **water-motor gong**, a flow alarm that signifies water is flowing in the system. In addition, the system prompts electric flow or pressure switches to activate the building's fire alarm system, notifying the fire department and the building's occupants. This kind of system is so effective that by the time fire fighters arrive, the sprinklers have often extinguished the fire.

Automatic Sprinkler System Components

The overall design of automatic sprinkler systems can be complex, especially in large buildings. This complexity is somewhat deceiving, however, because even the largest systems include just four major components: the automatic sprinklers, piping, control valves, and a water supply, which may or may not include a fire pump **FIGURE 28-23**. Most sprinkler systems are directly connected to a public water supply system and do not need electricity to function.

Automatic Sprinklers

Automatic sprinklers, which are commonly referred to as sprinklers, are the working ends of a sprinkler system. In most systems, the sprinklers serve two functions: they activate the sprinkler system and apply water to the fire. Sprinklers are composed of a body, which includes the orifice (opening); a release mechanism, which holds a cap in place over the orifice; and a deflector, which directs the water in a spray pattern **FIGURE 28-24**. Standard sprinklers have a 13-millimetre (0.5-inch) orifice, but several other sizes are available for special applications.

Although sprinklers come in several styles, all of them are categorized according to the type of release mechanism used and the intended mounting position—upright, pendant, or horizontal. They are also rated according to their release temperature. The release mechanisms hold the cap in

FIGURE 28-23 The basic components of an automatic sprinkler system include sprinklers, piping, control valves, and a water supply.

FIGURE 28-24 Automatic sprinklers have a body with an opening, a release mechanism, and a water deflector.

FIGURE 28-25 Automatic sprinklers are designed to activate at a wide range of temperatures. They may react slowly or quickly, depending on their settings.

FIGURE 28-26 In fusible-link sprinklers, two pieces of metal are linked together by an alloy such as solder.

place until the release temperature is reached. At that point, the mechanism is released, and the water pushes the cap out of the way as it discharges onto the fire **FIGURE 28-25**.

<u>**Fusible-link sprinklers**</u> use a metal alloy, such as solder, that melts at a specific temperature **FIGURE 28-26**. This alloy links two pieces of metal that keep the cap in place. When

the designated operating temperature is reached, the solder melts and the link breaks, releasing the cap. Fusible-link sprinklers come in a wide range of styles and temperature ratings.

<u>**Frangible-bulb sprinklers**</u> use a glass bulb filled with glycerin or alcohol to hold the cap in place **FIGURE 28-27**. This bulb also contains a small air bubble. As the bulb is heated, the liquid absorbs the air bubble and expands until it breaks the glass, releasing the cap. The volume and composition of the liquid and the size of the air bubble determine the temperature at which the sprinkler activates and the speed with which it responds.

<u>**Chemical-pellet sprinklers**</u> use a plunger and a small chemical pellet to hold the cap in place. The chemical pellet liquefies when the temperature reaches a preset point. When the pellet melts, the liquid compresses the plunger, releasing the cap and allowing water to flow.

FIGURE 28-27 Frangible-bulb sprinklers activate when the liquid in the bulb expands and breaks the glass.

FIRE FIGHTER Tips

Here are some popular misconceptions about sprinklers:

- The entire building will be drowned when the sprinklers go off.
- The manual alarm will set off every sprinkler in the building.
- Water does more damage than fire.
- The activation of smoke detectors sets off all of the sprinklers in the building.

These beliefs are all wrong!

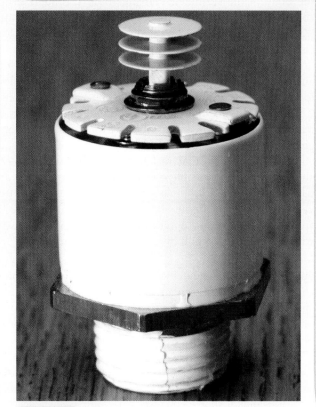

FIGURE 28-28 An ESFR sprinkler.

Specialized Sprinklers

Sprinklers can also be designed for special applications, such as covering large areas or discharging the water in extra-large droplets or as a fine mist. Some sprinklers have protective coatings to help prevent corrosion. Builders and installers should consider these characteristics when designing a system and selecting appropriate sprinklers. It is important to ensure that the proper sprinklers are installed and that any replacement sprinklers inserted in subsequent years are of the same type.

Sprinklers that are intended for residential occupancies are manufactured with a release mechanism that provides for a fast response.

By contrast, **early suppression fast-response (ESFR) sprinklers** have improved heat collectors that speed up the response and ensure the rapid release of water **FIGURE 28-28**. They are used in large warehouses and distribution facilities, where early fire suppression is important. These sprinklers often have large orifices to discharge large volumes of water onto a fire.

Deluge Sprinklers

Deluge sprinklers are easily identifiable by the fact that they have no cap or release mechanism. The orifice is always open **FIGURE 28-29**. Deluge sprinklers are used only in deluge sprinkler systems, which are covered later in this chapter.

Temperature Ratings

Sprinklers are rated according to their release temperature. A typical rating for sprinklers in a light-hazard occupancy, such as an office building, would be 74°C (165°F). Sprinklers that are used in areas with warmer ambient air would have higher ratings. The rating should be stamped on

FIGURE 28-29 A deluge sprinkler has no release mechanism.
Courtesy of Tyco Fire & Building Products

the body of the sprinkler. Frangible-bulb sprinklers use a colour-coding system to identify the temperature rating TABLE 28-3. Some fusible-link and chemical-pellet sprinklers also use this system.

The temperature rating on a sprinkler must match the anticipated ambient air temperature. If the rating is too low for the ambient air temperature, accidental alarms may occur. Conversely, if the rating is too high, the system will be slow to react to a fire, and the fire may be able to establish itself and grow before the system becomes activated.

Spare sprinklers that match those used in the system should always be available on site. The spare sprinklers are usually kept in a clearly marked box near the main control valve FIGURE 28-30. Having spare sprinklers handy enables sprinkler systems to be returned to full service quickly, whether they were set off by a fire or by accident.

Mounting Position

Sprinklers with different mounting positions are not interchangeable, because each mounting position has deflectors specifically designed to produce an effective water stream down or out toward the fire. Each automatic sprinkler is designed to be mounted in one of three positions FIGURE 28-31:

- **Pendant sprinklers** are designed to be mounted on the underside of the sprinkler piping, hanging down into the room. These sprinklers are commonly marked SSP (standard spray pendant).
- **Upright sprinklers** are designed to be mounted on top of the supply piping, as their name suggests. Upright sprinklers are usually marked SSU (standard spray upright).
- **Sidewall sprinklers** are designed for horizontal mounting and project out from a wall.

Old-Style versus New-Style Sprinklers

In automatic sprinklers manufactured before the mid-1950s, deflectors in both pendant and upright sprinklers directed part of the water stream up toward the ceiling. At the time, it was believed that this action helped cool the area and extinguish the fire. Sprinklers with this design are called old-style sprinklers, and many of them remain in service today.

FIGURE 28-30 Spare sprinklers should be kept in a special box near the main sprinkler system valve.

Automatic sprinklers manufactured after the mid-1950s deflect the entire water stream down toward the fire. These types of sprinklers are referred to as new-style sprinklers or standard spray sprinklers. New-style sprinklers can replace old-style sprinklers, but the reverse is not true. Due to their differing coverage patterns, old-style sprinklers should not be used to replace any new-style sprinklers.

Sprinkler Piping

Sprinkler piping, the network of pipes that delivers water to the sprinklers, includes the main water supply lines, risers, feeder lines, and branch lines. Although sprinkler pipes are usually made of steel, other metals can be used as well FIGURE 28-32; for example, plastic pipe is sometimes used in residential sprinkler systems.

Sprinkler system designers use hydraulic calculations to determine the size of pipe and the layout of the grid. Older systems were designed using piping schedules. Most new systems are designed using computer software. Near the main control valve, pipes have a large diameter; as the pipes approach the sprinklers, the diameter generally decreases.

TABLE 28-3	Temperature Rating Determined by Colour of Sprinklers		
Maximum Ceiling Temperature	**Temperature Rating**	**Colour Code**	**Glass Bulb Colours**
38°C (100°F)	57–77°C (135–170°F)	Uncoloured or black	Orange or red
66°C (150°F)	79–107°C (175–225°F)	White	Yellow or green
107°C (225°F)	121–149°C (250–300°F)	Blue	Blue
149°C (300°F)	163–191°C (325–375°F)	Red	Purple
191°C (375°F)	204–246°C (400–475°F)	Green	Black
246°C (475°F)	260–302°C (500–575°F)	Orange	Black
329°C (625°F)	343°C (650°F)	Orange	Black

Source: Originally published in a slightly different form in NFPA 13, *Standard for the Installation of Sprinkler Systems*, 2013 ed. (Quincy, MA: National Fire Protection Association, Inc., 2012), 28, table 6.2.5.1. Reprinted with permission. This reprinted material is not the complete and official position of the NFPA on the referenced subject, which is represented only by the standard in its entirety.

A.

B.

C.

FIGURE 28-31 Sprinklers mounting positions. **A.** Upright.
B. Pendant. **C.** Sidewall.

Courtesy of Tyco Fire & Building Products

FIGURE 28-32 Most new sprinkler systems use steel pipes.
© Foto Factory/Alamy.

Valves

A sprinkler system includes several different valves, such as the main water supply control valve, the alarm valve, and other smaller valves used for testing and service. Many large systems have zone valves, which enable the water supply to different areas to be shut down without turning off the entire system. All of these valves play a critical role in the design and function of the system.

Water Supply Control Valves

Every sprinkler system must have at least one main control valve that allows water to enter the system. This water supply control valve must be of the indicating type, meaning that the position of the valve itself indicates whether it is open or closed. Two examples are the <u>**outside stem and yoke (OS&Y) valve**</u> and the <u>**post indicator valve (PIV)**</u>.

An OS&Y valve has a stem that moves in and out as the valve is opened or closed **FIGURE 28-33**. If the stem is out, the valve is open; if the stem is in, the valve is closed. OS&Y valves are often found in a mechanical room in the building, where the water supply to the sprinkler system enters the building. In warmer climates, they may be found outside.

The PIV has an indicator that reads either open or shut depending on its position **FIGURE 28-34**. A PIV is usually located in an open area outside the building and controls an underground valve. Opening or closing a PIV requires a wrench, which is usually attached to the side of the valve. The valve must always be either in the fully open or closed position.

A <u>**wall post indicator valve (WPIV)**</u> is similar to a PIV but is designed to be mounted on an outside wall of a building **FIGURE 28-35**.

The main control valve—be it an OS&Y valve or a PIV—should be locked in the open position. This practice ensures that the water supply to the sprinkler system is never shut off unless the proper people are notified that the system is out of service. It is critically important that the sprinkler system always remain charged with water and ready to operate, if needed.

As an alternative to locking the valves open, the valves may be equipped with <u>**tamper switches**</u> **FIGURE 28-36**. These devices monitor the position of the valve. If someone opens

FIGURE 28-33 An OS&Y valve is often used to control the flow of water into a sprinkler system.

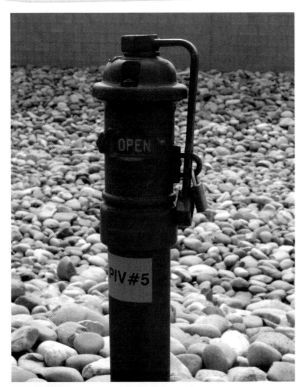

FIGURE 28-34 A PIV is used to open or close an underground valve.

FIGURE 28-35 A WPIV controls the flow of water from an underground pipe into a sprinkler system.

FIGURE 28-36 A tamper switch activates an alarm if someone attempts to close a valve that should remain open.

or closes the valve, the tamper switch sends a signal to the fire alarm control panel, indicating a change in the valve position. If the change has not been authorized, the cause of the signal can be investigated, and the problem can be corrected.

Main Sprinkler System Valves

The type of main sprinkler system valve used depends on the type of sprinkler system installed. Options include an **alarm valve**, a **dry-pipe valve**, or a **deluge valve**. These valves are usually installed on the main riser, above the water supply control valve.

The primary functions of an alarm valve are to signal an alarm when a sprinklers is activated and to prevent nuisance alarms caused by pressure variations and surges in the water supply to the system. This valve has a **clapper mechanism** that remains in the closed position until a sprinklers opens. The closed clapper prevents water from flowing out of the system and back into the public water mains when water pressure drops.

When a sprinklers is activated, the clapper opens fully and allows water to flow freely through the system. The open

Near-Miss REPORT

Report Number: 05-0000472

Synopsis: Responded to an automatic fire alarm. Fire found on second floor. Officer trapped.

Event Description: Our company responded to an automatic fire alarm. The occupancy was a renovated grade/middle school building that now housed the local park district offices and the area headquarters for a local representative. The building was of brick-and-block construction with large windows. The still-alarm assignment consisted of two engines, one tower ladder with the shift commander aboard, and one ambulance.

The first-arriving engine's lieutenant and number 3 fire fighter entered the building, leaving the engineer with the rig. The ambulance crew finished donning self-contained breathing apparatus (SCBA), gathered tools, and staged. Both the lieutenant and fire fighter had portable radios, and the number 3 fire fighter had irons. The crew located the annunciator and found it indicating a smoke alarm on the second floor. The lieutenant radioed command and advised him that it smelled like paper burning, "like a trash can."

Command acknowledged this information and requested to be advised of developments/needs. The crew proceeded to the second floor and, upon gaining the landing, reported a "light haze of smoke" at the ceiling. Again, command acknowledged the message. The lieutenant and fire fighter began searching numerous offices and cubicles in the local representative's office area.

The second-due engine arrived and requested orders. Command ordered the lieutenant and number 3 fire fighter to enter the building from the C side and meet up with the additional crew to assist in the search. Hearing this order via the radio, the first-due lieutenant sent his number 3 fire fighter down to the rear entrance to open the door for the second crew, as the Knox-Box® was on the front of the building and had already been accessed to enter the building. Not long after the number 3 fire fighter left, the lieutenant, alone on the second floor, radioed command stating that conditions had deteriorated greatly, with thick, black, hot smoke. Command acknowledged the message. This was then followed by another transmission from the lieutenant: "It's getting hot up here, and I can't find the door."

Command then exited the tower ladder and, dressed only in his bunker pants and coat, entered the building to attempt to locate the lieutenant. He did not remain inside long. When he did emerge, he finally radioed our dispatch centre to upgrade to a full still and then a general alarm. In the meantime, the second-due engine crew and the number 3 fire fighter were attempting to retrace the fire fighter's steps back to the lieutenant's last location; the doors in this government office, however, automatically locked upon closing. About this time the lieutenant made a third radio transmission repeating that it was "hot up here." Command acknowledged the message and reported that a crew was on their way up to get him.

The second-due crew never reported the trouble they were encountering with the locked doors to command. The lost lieutenant eventually made his way out of the office on his own after locating a receptionist's desk and remembering that the door was just in front of it. He emerged into the hallway as the second-due crew and his number 3 fire fighter were coming down the hall. The lieutenant suffered no injuries as a result of this incident. The fire eventually went to a second alarm.

Lessons Learned:

1. An automatic fire alarm can still be a fire. Treat every one like it is.
2. There are very few things on the fire ground that can safely be done alone.
3. Be aware of building construction and features (e.g., doors that lock behind you).
4. Locate second and even third exits when you enter a space. If there is only one way in and out, ensure that you know where you are in relation to that exit at all times.
5. Call the mayday early, and give as much information as you can regarding your whereabouts.
6. Command is just that—command. The incident commander (IC) is not a line participant.
7. Communicate your needs/difficulties/assessments to command so that other plans can be made if needed.
8. Complacency kills. This incident was the exception.

clapper also allows water to flow to the water-motor gong, sounding an alarm. Electrical flow and pressure switches then activate connections to external alarm systems.

Without a properly functioning alarm valve, sprinkler system flow alarms would occur frequently; the normal pressure changes and surges in a public water system will not open the clapper, however, which prevents water from flowing to the water-motor gong or tripping the electrical water flow or pressure switches unnecessarily.

In dry-pipe and deluge systems, the main valve functions both as an alarm valve and as a dam, holding back the water until the sprinkler system is activated. When the system is activated, the valve opens fully so that water can enter the sprinkler piping. Both dry-pipe and deluge systems are described later in this chapter.

Additional Valves

Sprinkler systems are equipped with a variety of other control valves. Several smaller valves are usually located near the main control valve, and still others are located elsewhere in the building. These smaller valves include drain valves, test valves, and connections to alarm devices. All of these valves should be properly labelled.

In larger facilities, the sprinkler system may be divided into zones, with a specific valve controlling the flow of water to each zone. This design makes maintenance easy and can prove extremely valuable when a fire occurs. After the fire is extinguished, water flow to the affected area can be shut off so that the activated heads can be replaced. Fire protection in the rest of the building, however, is unaffected by this shutdown.

Water Supplies

The water used in an automatic sprinkler system may come from a municipal water system, from on-site storage tanks, or from static water sources, such as storage ponds or rivers. Whatever its source, the water supply must be able to satisfy the demands of the sprinkler system as well as meet the needs of the fire department in the event of a fire.

The preferred water source for a sprinkler system is a municipal water supply, if one is available. If the municipal supply cannot meet the water pressure and volume requirements of the sprinkler system, alternative supplies must be established.

Fire pumps are used when the water comes from a static source. They may also be deployed to boost the pressure in some sprinkler systems, particularly those of tall buildings **FIGURE 28-37**. Because most municipal water supply systems do not provide enough pressure to control a fire on the upper floors of a high-rise building, fire pumps will turn on automatically when the sprinkler system activates or when the pressure drops to a preset level. In high-rise buildings, a series of fire pumps on upper floors may be needed to provide adequate pressure.

A large industrial complex could have more than one water source—for example, a municipal system and a backup storage tank **FIGURE 28-38**. Multiple fire pumps can provide water to the sprinkler and standpipe systems in different areas through underground pipes. Private hydrants may also be connected to the same underground system.

Each sprinkler system should also have a **fire department connection (FDC)**. This connection allows the department's

FIGURE 28-37 In tall buildings, a fire pump may be needed to maintain appropriate pressure in the sprinkler system.

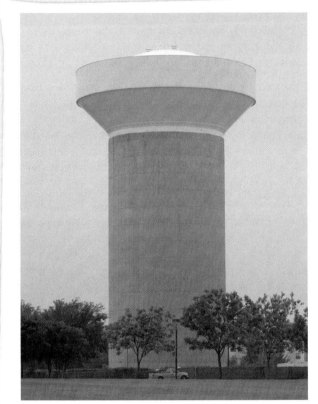

FIGURE 28-38 An elevated storage tank ensures that sufficient water and adequate pressure will be available to fight a fire.

engine to pump water into the sprinkler system **FIGURE 28-39**. The FDC may be used as either a supplement or the main source of water to the sprinkler system if the regular supply is interrupted or a fire pump fails.

The FDC usually has two or more 64-mm (2.5-in.) female couplings or one large-diameter hose (LDH) coupling mounted on an outside wall or placed near the building. It ties directly into the sprinkler system after the main control valve or alarm valve. Each fire department should establish a suggested operating guideline (SOG), or standard operating procedure (SOP), for first-arriving companies that specifies how to connect to the FDC and when to charge the system. For fire fighter safety reasons, the pump operator should never exceed 862 kilopascals (125 pounds per square inch [psi]) at the FDC. When hooking up the pumper to a supply hydrant, it is important to confirm that the supplying hydrant to the truck is not the same one feeding the operating standpipe or sprinkler system. Otherwise, the water supply to those systems may be reduced, thus causing the fire to grow in size and possibly limiting the available water for hose line suppression tactics.

In large facilities, a single FDC may be used to deliver water to all fire protection systems in the complex. The water from this connection flows into the private underground water mains instead of into each system. Water pumped into this type of FDC should come from a source that does not service the complex, such as a public hydrant on a different grid.

Water Flow Alarms

All sprinkler systems should be equipped with a method for sounding an alarm whenever water begins flowing in the pipes. This type of warning is important both in an actual fire and in an accidental activation. Without these alarms, the occupants or the fire department might not be aware of the sprinkler activation. If a building is unoccupied, the sprinkler system could continue to discharge water long after a fire is extinguished, leading to extensive water damage.

Most systems incorporate a mechanical flow alarm called a water-motor gong **FIGURE 28-40**. When the sprinkler system is activated and the main alarm valve opens, some water is fed through a pipe to a water-powered gong located on the outside of the building. This gong alerts people outside the building that there is water flowing. This type of alarm will function even if there is no electricity.

Accidental soundings of water-motor gongs are rare. If a water-motor gong is sounding, water is probably flowing from the sprinkler system somewhere in the building. Fire companies that arrive and hear the distinctive sound of a water-motor gong know that there is a fire or that something else is causing the sprinkler system to flow water.

FIGURE 28-39 A fire department connection delivers additional water and boosts the pressure in a sprinkler system.

FIGURE 28-40 A water-motor gong sounds when water is flowing in a sprinkler system.

Sprinkler systems also are connected to the building's fire alarm system by either an electric **flow switch** or a pressure switch. This connection triggers the alarm to alert the building's occupants of its activation; a monitored system will notify the fire department as well. Unlike water-motor gongs, flow and pressure switches can be accidentally triggered by water pressure surges in the system. To reduce the risk of accidental activation, these devices usually have a time delay before they will sound an alarm.

Types of Automatic Sprinkler Systems

Automatic sprinkler systems are classified into four categories, depending on which type of sprinklers is used and how the system is designed to activate:

- Wet sprinkler systems
- Dry sprinkler systems
- Preaction sprinkler systems
- Deluge sprinkler systems

Although many buildings may use the same type of system to protect the entire facility, it is not uncommon to see two or three systems combined in one building. Some facilities use a wet sprinkler system to protect most of the structure but implement a dry sprinkler or preaction system in a specific area. In many cases, a dry sprinkler or preaction system will branch off from the wet sprinkler system.

Wet Sprinkler Systems

A **wet sprinkler system** is the most common and least expensive type of automatic sprinkler system. As its name implies, the piping in a wet system is always filled with water. When a sprinklers activates, water is immediately discharged onto the fire. The major drawback to a wet sprinkler system is that it cannot be used in areas within the building where temperatures drop below freezing. Water will also begin to flow in such a system if a sprinklers is accidentally opened or a leak occurs in the piping.

If only a small, unheated area needs to be protected, two options are available. A dry-pendant sprinklers can be used in very small areas, such as walk-in freezers FIGURE 28-41. The bottom part of a dry-pendant head, which resembles a standard sprinklers, is mounted inside the freezer. The head has an elongated neck, usually 15–46 centimetres (6–18 in.) long, that extends up and connects to the wet sprinkler piping in the heated area above the freezer. The vertical neck section is filled with air and capped at each end. The top cap prevents water from entering the lower section, where it would freeze; the bottom cap acts just like the cap on a standard sprinklers. When the head is activated and the lower cap drops out, a device inside the neck releases the upper cap, so water can flow down. The entire dry-pendant head assembly must be replaced after it has been activated.

Larger unheated areas, such as a loading dock, can be protected with an antifreeze loop. This small section of the wet sprinkler system is filled with glycol or glycerin instead of water. A check valve separates the antifreeze loop from the rest of the sprinkler system. When a sprinklers in the unheated area is activated, the antifreeze sprays out first, followed by water. These areas can also be protected by dry sprinkler systems.

Dry Sprinkler Systems

A **dry sprinkler system** operates much like a wet sprinkler system, except that the pipes are filled with pressurized air instead of water. A dry-pipe valve keeps water from entering the pipes until the air pressure is released. Dry systems are used in facilities that may experience below-freezing temperatures, such as unheated warehouses, garages, or attics.

The air pressure in this kind of sprinkler system is set high enough to hold a clapper inside the dry-pipe valve in the closed position FIGURE 28-42. When a sprinklers opens, the air escapes. As the air pressure drops, the water pressure on the other side of the clapper forces it open and water begins to flow into the pipes. When the water reaches the open sprinklers, it is discharged onto the fire.

Dry sprinkler systems do not eliminate the risk of water damage from accidental activation. If a sprinklers breaks, the air pressure will drop, and water will flow, just as in a wet sprinkler system.

The action of the clapper assembly located inside most dry-pipe valves relies on a pressure differential. The system (air) side of the clapper has a larger surface area than the supply (wet) side. As a consequence, a lower air pressure can hold back a higher water pressure. A small compressor is used to maintain the air pressure in the system.

Dry sprinkler systems should have an air pressure alarm to alert building management personnel if the air pressure drops. The activation of this alarm could mean one of two things: the compressor is not working, or there is an air leak in the system. If the air pressure in the system is too low, the clapper will open and the system will fill with water. At that point, the system would essentially function as a wet sprinkler system, which could cause it to freeze in low temperatures. In this

FIGURE 28-41 A dry-pendant sprinklers can be used to protect a freezer box.

FIGURE 28-42 Water pressure on one side of the dry-pipe valve is held down by air pressure on the other side of the valve.

scenario, the system would have to be drained and reset to prevent the pipes from freezing.

Dry sprinkler systems must be drained after every activation, so the dry-pipe valve can be reset. The clapper also must be reset, and the air pressure must be restored, before the water is turned back on.

Accelerators and Exhausters

One problem encountered in dry sprinkler systems is the delay between the activation of a sprinklers and the actual flow of water out of the head. The pressurized air that fills the system must escape through the open head before the water can flow. For personal-safety and property-protection reasons, any delay longer than 90 seconds is unacceptable; large systems, however, can take several minutes to empty out the air and refill the pipes with water. To compensate for this problem, two additional devices are used: accelerators and exhausters.

An **accelerator** is installed at the dry-pipe valve. The rapid drop in air pressure caused by an open sprinklers triggers the accelerator, which allows air pressure to flow to the supply side of the clapper valve. This quickly eliminates the pressure differential, opening the dry-pipe valve and allowing the water pressure to force the remaining air out of the piping.

An **exhauster** is installed on the system side of the dry-pipe valve, often at a remote location in the building. Like an accelerator, the exhauster monitors the air pressure in the piping. If it detects a drop in pressure, it opens a large-diameter portal, allowing the air in the pipes to escape. The exhauster closes when it detects water, diverting the flow to the open sprinklers. Large systems may have multiple exhausters located in different sections of the piping.

Preaction Sprinkler Systems

A **preaction sprinkler system** is similar to a dry sprinkler system with one key difference: in a preaction sprinkler system, a secondary device, such as a smoke detector or a manual-pull alarm, must be activated before water is released into the sprinkler piping. When the system is filled with water, it functions as a wet sprinkler system.

A preaction system uses a deluge valve instead of a dry-pipe valve. The deluge valve will not open until it receives a signal that the secondary device has been activated. Because a detection system usually becomes activated more quickly than a sprinkler system does, water in a preaction system will generally reach the sprinklers before a head is activated.

The primary advantage of a preaction sprinkler system is its ability to prevent accidental water discharge. If a sprinklers is accidentally broken or the pipe is damaged, the deluge valve will prevent water from entering the system. This feature makes preaction sprinkler systems well suited for locations where water damage is a major concern, such as computer rooms and electrical rooms.

Deluge Sprinkler Systems

A **deluge sprinkler system** is a type of dry sprinkler system in which water flows from all of the sprinklers as soon as the system is activated **FIGURE 28-43**. A deluge system does

FIGURE 28-43 Water flows from all of the heads in a deluge system as soon as the system is activated.

Fire Fighter Tips

Some dry-pipe valves will not open if there is water above the clapper valve (a mechanical device that allows the water to flow in only one direction). Unless the main valve is designed to operate as either a wet- or dry-pipe valve, a dry system that has been filled with water should be immediately drained and reset for dry-pipe operation.

not have closed heads that open individually at the activation temperature; instead, all of the heads in a deluge system are always open.

Deluge systems can be activated in three ways:

1. A detection system can release the deluge valve when a detector is activated.
2. The deluge system can be connected to a separate pilot system of air-filled pipes with closed sprinklers. When a head on the pilot line is activated, the air pressure drops, opening the deluge valve.
3. Most deluge valves can be released manually.

Deluge systems are used in special applications such as aircraft hangars or industrial settings, where rapid fire suppression is critical. In some cases, foam concentrate is added to the water, so that the system will discharge a foam blanket over the hazard. Deluge systems are also used for special hazard applications, such as liquid-propane-gas loading stations. In these situations, a heavy deluge of water is needed to protect exposures from a large fire that ignites very rapidly.

Shutting Down Sprinkler Systems

Responding fire companies should know where and how to shut down any automatic sprinkler system, if needed. If the system is accidentally activated, it should be shut down as soon as possible to avoid excessive water damage. In an actual fire, the order to shut off the sprinkler system should only come from the IC. Generally, the sprinkler system should not be shut down until the fire is completely extinguished.

When the system is shut down, a fire fighter with a portable radio should stand by at the control valve, in case it has to be reopened quickly.

In most cases, the entire sprinkler system can be shut down by closing the main control valve (OS&Y or PIV). In a zoned system, the affected zone can be shut down by closing the appropriate valve; the rest of the system will then remain operational while the activated heads are being replaced.

Placing a wooden wedge or a commercial sprinkler stopper into the sprinklers can quickly stop the flow of water, although this measure will not work with all types of heads FIGURE 28-44. Blocking the sprinklers shuts off the flow of water until the control valve can be located and shut down. Some types of commercial sprinkler stoppers incorporate a fusible link, which allows the sprinklers to function until it can be permanently replaced FIGURE 28-45. After the main valve is closed, opening the drain valve drains the system so that the activated head can be replaced.

FIGURE 28-44 A sprinkler stopper can be used to stop the flow of water from a sprinklers that has been activated.

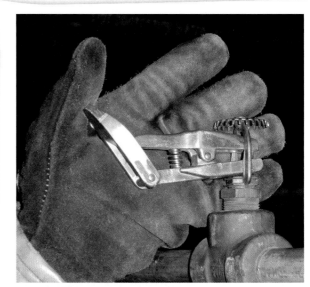

FIGURE 28-45 This shut–off device shuts off the sprinkler and provides a temporary fusible link to keep the sprinklers in service.
Courtesy of Scott Dornan, ConocoPhillips Alaska

Residential Sprinkler Systems

<u>Residential sprinkler systems</u> are becoming more accepted, with many recently constructed homes including them. Some communities require residential sprinkler systems in every newly constructed home.

The design of and theory underlying residential sprinkler systems are similar to those of commercial systems with some significant variations. The primary objective of a residential sprinkler system is life safety, so the sprinklers are designed to respond quickly FIGURE 28-46. A fast response helps protect the occupants and control incipient-stage fires.

Residential systems typically use smaller piping and sprinklers with smaller orifices and less water discharge. To control costs, plastic pipe may be used instead of metal pipe.

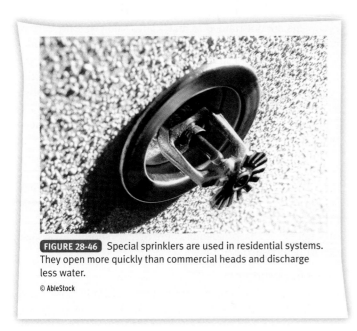

FIGURE 28-46 Special sprinklers are used in residential systems. They open more quickly than commercial heads and discharge less water.

© AbleStock

These systems typically protect high-risk areas such as rooms and corridors; small areas such as closets and bathrooms may not be covered.

Residential systems are usually wet sprinkler systems. They are typically connected to the domestic water supply of a public water main. Residential sprinkler systems do not have an FDC or have to be tied into a separate fire alarm system unless required locally. But they usually do require the system to include a water flow alarm.

The responding fire companies in the jurisdiction may not be familiar with every home that has a sprinkler system, but fire fighters should know both how these systems work and how to shut down a system that has been activated. Usually the main shut-off valve for the sprinklers can be found near the location where water enters the house. If the house has a basement, the shut-off valve may be located near other utility controls. If a separate shut-off valve for the sprinkler system cannot be found, the main water shut-off for the house can be used to deactivate the sprinkler system. Installing working smoke alarms and a residential sprinkler system reduces the chance of death from a home fire by 82 percent.

Fire Fighter Safety Tips

© Photos.com

Do not shut down the water to a sprinkler system until ordered to do so by the IC. The fire must be completely extinguished, and hose lines must be available should the fire reignite. A fire fighter should be stationed at the valve with a portable radio, ready to reopen it, if necessary.

■ Standpipe Systems

A **standpipe system** consists of a network of inlets, pipes, and outlets for fire hoses that is built into a structure to provide water for firefighting purposes. It includes one or more inlets supplied by a municipal water supply for fire department hoses, piping to carry the water closer to the fire, and one or more outlets equipped with valves to which fire hoses can be

connected. Standpipe systems are required in many high-rise buildings, and they are found in many other structures as well. Standpipes are also installed to carry water to large bridges and to supply water to limited-access highways that are not equipped with fire hydrants **FIGURE 28-47** .

Standpipes are found in buildings both with and without sprinkler systems. In many newer buildings, sprinklers and standpipes are combined into a single system. In older buildings, the sprinkler and standpipe systems are typically separate entities. Three categories of standpipes—Class I, Class II, and Class III—are distinguished based on their intended use.

Class I Standpipe

A **Class I standpipe** is designed for use by fire department personnel only. Each outlet has a 64-mm (2.5-in.) male coupling and a valve to open the water supply after the attack line is connected **FIGURE 28-48** . Often the connection is located inside a cabinet, which may or may not be locked. Responding fire personnel carry the hose into the building with them, usually in some sort of roll, bag, or backpack. A Class I standpipe system must be able to supply an adequate volume of water with sufficient pressure to operate fire department attack lines.

FIGURE 28-47 Standpipe outlets allow fire hoses to be connected inside a building.

Class II Standpipes

A **Class II standpipe** is designed for use by the building's occupants. The outlets are generally equipped with a length of 38-mm (1.5-in.) single-jacket hose preconnected to the system . These systems are intended to enable occupants to attack a fire before the fire department arrives, but their safety and effectiveness are questionable. After all, most occupants are not trained to attack fires safely. If a fire cannot be controlled with a regular fire extinguisher, it is usually safer for the occupants to simply evacuate the building and call the fire department. Class II standpipes may be useful at facilities such as refineries and military bases, where workers are trained as an in-house fire brigade.

Responding fire personnel should not use Class II standpipes for fighting a fire, because the hoses and nozzles may be of inferior quality and lack maintenance and testing. The water flow may not be adequate to control a fire. Class II standpipe outlets are frequently connected to the domestic water piping system in the building rather than an outside main or a separate system. Instead of using equipment (hose and nozzle) that may not be reliable or adequate, fire fighters should always use department-issued equipment.

Class III Standpipes

A **Class III standpipe** has the features of both Class I and Class II standpipes in a single system. This kind of system has 64-mm (2.5-in.) outlets for fire department use, as well as smaller outlets with attached hoses for occupant use. The occupant hoses may have been removed—either intentionally or by vandalism—in many facilities, so the system basically becomes a Class I system. Fire fighters should use only the 64-mm (2.5-in.) outlets, even if they employ an adapter to connect a smaller hose. The 38-mm (1.5-in.) outlets may have flow-restriction devices to limit the flow and pressure for use by untrained civilians.

Water Flow in Standpipe Systems

Standpipes are designed to deliver a minimum amount of water at a particular pressure to each floor. The design requirements depend on the code requirements in effect when the building was constructed.

Flow-restriction devices **FIGURE 28-50** or pressure-reducing valves **FIGURE 28-51** are often installed at the outlets to limit the pressure and flow. A vertical column of water, such as the water in a standpipe riser, exerts a back pressure (also called head pressure). In a tall building, this back pressure can amount to thousands of kilopascals at lower floor levels. If a hose line is connected to an outlet without a flow restrictor or a pressure-reducing valve, the water pressure could rupture the hose, and the excessive nozzle pressure could make the line difficult or dangerous to handle.

If flow restrictors or pressure-reducing valves are not properly installed and maintained, these devices can cause problems for fire fighters. An improperly adjusted pressure-reducing valve, for example, could severely restrict the flow to a hose line. Likewise, a flow-restriction device could limit the flow of water to fight a fire.

The flow and pressure capabilities of a standpipe system should be determined during preincident planning. Many standpipe systems deliver water at a pressure of only 448 kPa (65 psi) at the top of the building. The combination fog–straight-stream nozzles used by many fire departments are designed to operate at 689 kPa (100 psi). For this reason, many fire departments use low-pressure combination nozzles for fighting fires in high-rise buildings or require the use of only smooth-bore nozzles when operating from a standpipe system.

FIGURE 28-48 A Class I standpipe provides water for fire department hose lines.

© Stephen Coburn/ShutterStock, Inc.

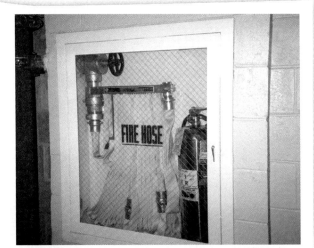

FIGURE 28-49 A Class II standpipe is intended to be used by building occupants to attack incipient-stage fires.

Courtesy of Ralph G. Johnson (nyail.com/fsd)

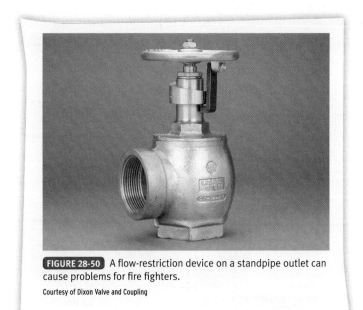

FIGURE 28-50 A flow-restriction device on a standpipe outlet can cause problems for fire fighters.

Courtesy of Dixon Valve and Coupling

FIGURE 28-51 A pressure-reducing valve on a standpipe outlet may be necessary on lower floors to avoid problems caused by back pressure.

Courtesy of Dixon Valve and Coupling

Preincident planning for high-rise buildings should include an evaluation of the building's standpipe system and a determination of the anticipated flows and pressures. This information should be used to make decisions about the appropriate nozzles and tactics for those buildings.

Engine companies who respond to buildings equipped with standpipes should carry a kit that includes the appropriate hose and nozzle, a spanner wrench, and any required adapters. This kit should also include tools to adjust the settings of pressure-reducing valves or to remove restrictors that are obstructing flows.

Water Supplies

Both standpipe systems and sprinkler systems are supplied with water in essentially the same way. That is, many wet standpipe systems in modern buildings are connected to a public water supply and equipped with an electric or diesel fire pump to provide additional pressure. Some also have a water storage tank that serves as a backup supply. In these systems, the FDC on the outside of the building can be used to increase the flow, boost the pressure, or obtain water from an alternative source.

Dry standpipe systems are found in many older buildings. If freezing weather is a problem, such as in open parking structures, bridges, and tunnels, dry standpipe systems are still acceptable. Most of these systems do not have a permanent connection to a water supply, so the FDC must be used to pump water into the system. If a fire occurs in a building with dry standpipes, connecting the hose lines to the FDC and charging the system with water are high priorities.

Like sprinkler systems, some dry standpipe systems are connected to a water supply through a dry-pipe or deluge valve. In such systems, opening an outlet valve or tripping a switch next to the outlet releases water into the standpipes.

High-rise buildings often incorporate complex systems of risers, storage tanks, and fire pumps to deliver the needed flows to upper floors. The details of these systems should be obtained during preincident planning surveys. Department procedures should dictate how responding units will supply the standpipes with water as well as how crews should use the standpipes inside the building.

In some jurisdictions, there is now a provision for a midrise combustible structure up to 6 storeys in height. These new buildings must be equipped with working standpipes erected to aid in any suppression activities during construction. In newer codes, these buildings will also have sprinkler systems, including a wet system, that are operational prior to occupancy.

■ Specialized Extinguishing Systems

Automatic sprinkler systems are used to protect whole buildings—or at least major sections of buildings. Nevertheless, in certain situations, more specialized extinguishing systems are needed. These kinds of systems are often used in areas where water would not be an acceptable extinguishing agent **FIGURE 28-52**; for example, water is not the agent of choice for areas containing sensitive electronic equipment or contents such as computers, valuable books, or documents. Water is also incompatible with materials such as flammable liquids or water-reactive chemicals. Given these facts, areas where these materials are stored or used may have a separate extinguishing system.

■ Dry Chemical and Wet Chemical Extinguishing Systems

Dry chemical and wet chemical extinguishing systems are the most common specialized agent systems. In commercial kitchens, they are used to protect the cooking areas and exhaust systems. In addition, some gas stations have dry chemical systems that protect the dispensing areas. These systems are also installed inside buildings to protect areas where flammable liquids are stored or used. Dry chemical and wet chemical extinguishing systems are similar in their basic design and arrangement.

Dry chemical extinguishing systems use the same types of finely powdered agents as dry chemical fire extinguishers **FIGURE 28-53**. The agent is kept in self-pressurized tanks or in tanks with an external cartridge of carbon dioxide or nitrogen that provides pressure when the system is activated.

Wet chemical extinguishing systems (also known as Type K) are used in most new commercial kitchens **FIGURE 28-54**. They use a liquid extinguishing agent, which is much more effective on vegetable oils than are the dry chemicals used in older kitchen systems. Wet chemical systems are also easier to clean up after a discharge, so the kitchen can resume operations more quickly after the system has discharged and been recharged by a qualified technician.

Wet chemical extinguishing agents are not compatible with normal all-purpose dry chemical extinguishing agents, however. Only Type K wet extinguishers should be used where these systems are installed and only in a mop-up-type operation, never in a first-response mode. Fire codes require a sign be posted over this type of extinguisher to make users aware of this fact.

With both dry chemical and wet chemical extinguishing agent systems, fusible-link or other automatic initiation devices are placed above the target hazard to activate the

FIGURE 28-53 Dry chemical extinguishing systems are installed at many self-service gasoline filling stations.

FIGURE 28-52 Specialized extinguishing systems are used in areas where water would not be effective or desirable.

FIGURE 28-54 Wet chemical extinguishing systems are used in most new commercial kitchens.

system **FIGURE 28-55**. In the event that the fusible link fails, a manual discharge button is also provided in the path of egress next to an exit door, so that workers can activate the system if they discover a fire **FIGURE 28-56**. Open nozzles are located over the target areas to discharge the agent directly onto a fire. When the system is activated, the extinguishing agent flows out of all the nozzles.

Many kitchen systems discharge the extinguishing agent into the ductwork above the exhaust hood, as well as onto the cooking surface. This approach helps prevent a fire from igniting any grease buildup inside the ductwork and spreading throughout the system. Although the ductwork should be cleaned regularly, it is not unusual for a kitchen fire to extend into the exhaust system.

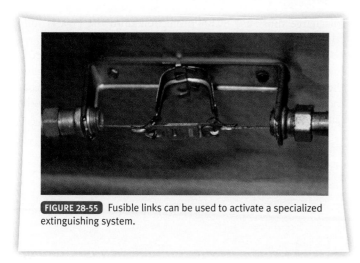

FIGURE 28-55 Fusible links can be used to activate a specialized extinguishing system.

FIGURE 28-56 Most specialized extinguishing systems can also be manually activated.

Most systems are tied into the building's fire alarm system. Kitchen extinguishing systems should also shut down the gas or electricity to the cooking appliances and exhaust fans.

■ Clean-Agent Extinguishing Systems

Clean-agent extinguishing systems are often installed in areas where computers or sensitive electronic equipment are used or where valuable documents are stored. The agents used in these systems are nonconductive and leave no residue. Notably, halogenated agents or carbon dioxide can be used for this purpose, because they will extinguish a fire without causing significant damage to the contents of a room.

Clean-agent systems operate by discharging a gaseous agent into the atmosphere at a concentration that will extinguish a fire. Smoke detectors or heat detectors installed in these areas activate the system, although a manual discharge button is also provided with most installations. Discharge is usually delayed 30–60 seconds after the detector is activated to allow workers to evacuate the area.

During this delay (also known as the pre-alarm period), an abort switch can be used to stop the discharge. In some systems, the abort button must be pressed until the detection system is reset; releasing the abort button too soon causes the

system to discharge. If there is a fire, the clean-agent system should be completely discharged before fire fighters arrive. Whether the fire was successfully extinguished or not, fire fighters entering the area should use SCBA until the area has been properly ventilated. Although the gaseous agents used in clean-agent extinguishing systems are not considered immediately dangerous to life and health, it is better to avoid any unnecessary exposure to them. Toxic products or by-products of combustion could be present in the atmosphere as well, or the oxygen level could be reduced.

Clean-agent extinguishing systems should be tied to the building's fire alarm system and indicated as a zone on the control panel. This notification scheme alerts fire fighters that they are responding to a situation where a clean agent has discharged. If the system has a preprogrammed delay, the pre-alarm should activate the building's fire alarm system.

Until the 1990s, <u>Halon 1301</u> was the agent of choice for protecting areas such as computer rooms, telecommunications rooms, and other sensitive areas. Halon 1301 is a nontoxic, odourless, colourless gas that leaves behind no residue. It is very effective at extinguishing fires because it interrupts the chemical reaction of combustion. Unfortunately, Halon 1301 is also damaging to the environment—a fact that led to its discontinuation as a firefighting agent. Alternative agents have since been developed for use in new systems; they are also replacing Halon 1301 in many existing systems. A stockpile of Halon 1301 is still available to recharge existing systems when no suitable substitute agent exists.

■ Carbon-Dioxide Extinguishing Systems

<u>Carbon-dioxide extinguishing systems</u> are similar in design to clean-agent systems. The primary difference is that carbon dioxide extinguishes a fire by displacing the oxygen in the room and smothering the fire. Large quantities of carbon dioxide are required for this purpose, because the area must be totally flooded to extinguish a fire **FIGURE 28-57** .

Carbon-dioxide systems may be designed to protect either a single room or a series of rooms. They usually have

FIGURE 28-57 Carbon dioxide extinguishes a fire by displacing the oxygen in the room and smothering the fire.
Courtesy of Chemetron Fire Systems

the same series of pre-alarms and abort buttons as are found in Halon 1301 systems. Because the carbon-dioxide discharge creates an oxygen-deficient atmosphere in the room, the activation of such a system is immediately dangerous to life. Any occupant who is still in the room when the agent is discharged is likely to be rendered unconscious and asphyxiated. Fire fighters responding to a carbon-dioxide extinguishing system discharge must use SCBA protection until the area is fully vented.

Carbon-dioxide extinguishing systems should be connected to the building's fire alarm system. Responding fire fighters should see that a carbon-dioxide system discharge has been activated. Using this knowledge, they can deal with the situation safely.

Chief Concepts

- Fire protection systems include fire alarms, automatic fire detection, and fire-suppression systems.
- Fire alarm and detection systems range from simple, single-station smoke alarms for private homes to complex fire detection and control systems for high-rise buildings. Many fire alarm and detection systems in large buildings also control other systems to help protect occupants and control the spread of fire and smoke.
- A fire alarm system has three basic components: an alarm initiation device, an alarm notification device, and a control panel. The alarm initiation device is either an automatic or manually operated device that, when activated, causes the system to indicate an alarm. The alarm notification device is generally an audible device but is often accompanied by a visual device, which alerts the building's occupants. The control panel links the initiation device to the notification device and performs other essential functions.
- The most common type of residential fire alarm system is a single-station smoke alarm. This life-safety device includes a smoke detection device, an automatic control unit, and an audible alarm within a single unit. It alerts occupants quickly when a fire occurs.
- The most up-to-date codes require new homes to have a smoke alarm in every bedroom and on every floor level. They also require a battery backup for smoke alarms in the event of a power failure.
- Two types of fire detection devices may be used in a smoke alarm to detect combustion:
  1. Ionization detectors use radioactive material within the device to detect invisible products of combustion.
  2. Photoelectric detectors use a light beam to detect the presence of visible particles of smoke.
- Alarm initiation devices begin the fire alarm process either manually or automatically. Manual alarm initiation devices include single-action pull-stations and double-action pull-stations. Automatic initiation devices sense fire using an ionization detector or a photoelectric chamber to detect products of combustion and trigger an audible alarm.
- Various types of automatic initiation devices are available:
  - Smoke detectors—Designed to sense the presence of smoke.
  - Beam detectors—Photoelectric smoke detectors used to protect large open areas.
  - Heat detectors—Designed to sense the presence of too much heat.
  - Fixed-temperature heat detectors—Designed to activate at a preset temperature.
  - Rate-of-rise heat detectors—Designed to activate if the temperature of the surrounding air increases by more than a set amount in a given period of time.
  - Line detectors—Use wire or tubing strung along the ceiling of large, open areas to detect an increase in heat.
  - Flame detectors—Specialized devices that detect the electromagnetic light waves produced by a flame.
  - Gas detectors—Designed to detect the presence of a specific gas that is created by combustion or that is used in the facility.
  - Air-sampling detectors—Designed to continuously capture air samples and measure the concentrations of specific gases or products of combustion.
- Residential carbon-monoxide detectors are designed to sound an audible or visual alarm when the concentration of carbon monoxide is high enough to pose a health risk to the occupants of the building.
- Knowing how to handle false, unwanted, and nuisance alarms is just as important as knowing how to deal with an actual fire.
- Three types of false alarms are possible:
  1. Malicious false alarm—Occurs when individuals deliberately activate a fire alarm when there is no fire, causing a disturbance.
  2. Unwanted alarm—Occurs when an alarm system is activated by a condition that is not really an emergency, such as smoke from normal cooking activities.
  3. Nuisance alarm—Occurs when an alarm system malfunctions.
- Several methods can be used to reduce unwanted and nuisance alarms caused by smoke detection systems:
  - Cross-zoned system—The activation of a single smoke detector will not sound the fire alarm, although it will usually set off a trouble alarm. In this kind of system, a second smoke detector must be activated before the actual fire alarm will sound.
  - Verification system—A delay of 30–60 seconds separates activation and notification. During this time, the system may show a trouble or pre-alarm condition at the system control panel. After the preset interval, the system rechecks the detector. If the condition has cleared, the system returns to normal; if the detector is still sensing smoke, the fire alarm is activated.
- Some fire alarm systems give little information at the alarm control panel, while others specify exactly which initiation device activated the fire alarm. Alarm annunciation systems are classified into four categories:
  1. Noncoded alarm—No information is given on the control panel.
  2. Zoned noncoded alarm—The control panel indicates which zone in the building was the source of the alarm.

3. Zoned coded alarm—The system indicates over the audible warning device which zone has been activated.
4. Master-coded alarm—The audible notification device is also used for other purposes, such as a public address system.

- There are five categories of fire department notification systems:
  1. Local alarm—Sounds an alarm only in the building where it was activated.
  2. Remote station—Sounds an alarm in the building, and transmits a signal to a remote location.
  3. Auxiliary system—Sounds an alarm in the building, and transmits a signal to the fire department via a public alarm-box system.
  4. Proprietary system—Sounds an alarm in the building, and transmits a signal to a monitoring location owned and operated by the facility's owner.
  5. Central station—Sounds an alarm in the building and transmits a signal to an off-premises alarm monitoring facility.

- Fire-suppression systems include sprinkler systems, standpipe systems, and specialized extinguishing systems.
- The most common type of fire-suppression system is the automatic sprinkler system.
- The basic operating principles of an automatic sprinkler system are simple. A system of water pipes is installed throughout a building to deliver water to every area where a fire might occur. Depending on the design and occupancy of the building, these pipes may be placed above or below the ceiling. Automatic sprinklers are located along this system of pipes, such that each sprinkler covers a particular floor area. A fire in that area will activate the sprinklers, which then discharges water on the fire.
- Every automatic sprinkler system includes four major components:
  1. Automatic sprinklers—Commonly called sprinklers and available in several styles.
  2. Piping—Carries the water or extinguishing agent to the sprinklers.
  3. Control valves—Control the flow of water to the piping.
  4. Water supply—May come from a municipal water system, from on-site storage tanks, or from static water sources.
- Automatic sprinkler systems are divided into four categories:
  1. Wet sprinkler systems—The most common type of automatic sprinkler system in which the pipes are always filled with water.
  2. Dry sprinkler systems—The pipes are filled with pressurized air instead of water until the system is activated; used when the pipes may be exposed to freezing temperatures.

3. Preaction sprinkler systems—A type of dry sprinkler system in which a secondary device must be activated before water is released into the sprinkler piping.
4. Deluge sprinkler systems—A type of dry sprinkler system in which water flows from all of the sprinklers as soon as the system is activated.

- A standpipe system consists of a network of inlets, pipes, and outlets for fire hoses that are built into a structure to provide water for firefighting purposes. Three types of standpipes are distinguished based on their intended use:
  - Class I—Designed for use by fire department personnel only.
  - Class II—Designed for use by the building's occupants.
  - Class III—Has the features of both Class I and Class II standpipes in a single system.
- Specialized extinguishing systems may be installed to protect areas where water may not be used, such as computer rooms.

Hot Terms

Accelerator A device that speeds up the removal of the air from a dry-pipe or preaction sprinkler system. An accelerator reduces the time it takes for water to start flowing from sprinklers.

Air-sampling detector A system that captures a sample of air from a room or enclosed space and passes it through a smoke detection or gas analysis device.

Alarm initiation device An automatic or manually operated device in a fire alarm system that, when activated, causes the system to indicate an alarm condition.

Alarm notification device An audible and/or visual device in a fire alarm system that makes occupants or other persons aware of an alarm condition.

Alarm valve A valve that signals an alarm when a sprinklers is activated and prevents nuisance alarms caused by pressure variations.

Automatic sprinklers The working ends of a sprinkler system, which serve to activate the system and apply water to the fire.

Automatic sprinkler system A system of pipes filled with water under pressure that discharge water immediately when a sprinklers opens.

Auxiliary system A fire alarm system that sounds an alarm in the building and transmits a signal to the fire department via a public alarm-box system.

Beam detector A smoke detection device that projects a narrow beam of light across a large, open area from a sending unit to a receiving unit. When the beam is interrupted by smoke, the receiver detects a reduction in light transmission and activates the fire alarm.

Bimetallic strip A thermostatic component made from two distinct metals that respond differently to heat, often found on rate-of-rise heat detectors; when heated, the metals will bend or change shape.

Carbon-dioxide extinguishing system A fire-suppression system designed to protect either a single room or series of rooms by flooding the area with carbon dioxide.

Central station An off-premises facility that monitors alarm systems and is responsible for notifying the fire department of an alarm. These facilities may be geographically located some distance from the protected building(s).

Chemical-pellet sprinklers A sprinklers activated by a chemical pellet that liquefies at a preset temperature.

Clapper mechanism A mechanical device installed within a piping system that allows water to flow in only one direction.

Class I standpipe A standpipe system designed for use by fire department personnel only. Each outlet should have a valve to control the flow of water and a 64-mm (2.5-in.) male coupling for fire hose.

Class II standpipe A standpipe system designed for use by occupants of a building only. Each outlet is generally equipped with a length of 38-mm (1.5-in.) single-jacket hose and a nozzle, which are preconnected to the system.

Class III standpipe A combination system that has features of both Class I and Class II standpipes.

Coded system A fire alarm system that divides a building or facility into zones and utilizes audible notification devices to identify the zone in which an alarm originated.

Cross-zoned system A fire alarm system that requires activation of two separate detection devices before initiating an alarm condition. If a single detection device is activated, the alarm control panel will usually indicate a problem or trouble condition.

Deluge sprinkler A sprinkler with no release mechanism; the orifice is always open.

Deluge sprinkler system A sprinkler system in which all sprinklers are open. When an initiation device, such as a smoke or heat detector, is activated, the deluge valve opens and water discharges from all of the open sprinklers simultaneously.

Deluge valve A valve assembly designed to release water into a sprinkler system when an external initiation device is activated.

Double-action pull-station A manual fire alarm activation device that requires two steps to activate the alarm: the user must push in a flap, lift a cover, or break a piece of glass before activating the alarm.

Dry chemical extinguishing system An automatic fire extinguishing system that discharges a dry chemical agent.

Dry-pipe valve The valve assembly on a dry sprinkler system that prevents water from entering the system until the air pressure is released.

Dry sprinkler system A sprinkler system in which the pipes are normally filled with compressed air. When a sprinklers is activated, it releases the air from the system, which opens a valve so that pipes fill with water.

Early suppression fast-response (ESFR) sprinkler A sprinkler designed to react quickly and suppress a fire in its early stages.

Exhauster A device that accelerates the removal of the air from a dry-pipe or preaction sprinkler system.

False alarm The activation of a fire alarm system when no fire or emergency condition exists.

Fire alarm control panel The component in a fire alarm system that controls the functions of the entire system.

Fire department connection (FDC) A fire hose connection through which the fire department can pump water into a sprinkler system or standpipe system.

Fixed-temperature heat detector A sensor that responds when its operating element is heated to a predetermined temperature.

Flame detector A sensor that detects the radiant energy emitted by a flame.

Flow switch An electrical switch that is activated by water moving through a pipe in a sprinkler system.

Frangible-bulb sprinkler A sprinkler with a liquid-filled bulb. The sprinklers becomes activated when the liquid is heated and the glass bulb breaks.

Fusible-link sprinkler A sprinkler with an activation mechanism that incorporates two pieces of metal held together by low-melting-point solder. When the solder melts, it releases the link and water begins to flow.

Gas detector A device that detects and/or measures the concentration of dangerous gases.

Halon 1301 A liquefied-gas extinguishing agent that puts out a fire by chemically interrupting the combustion reaction between fuel and oxygen. Halon agents leave no residue.

Heat detector A fire alarm device that detects an abnormally high temperature, an abnormally high rate of rise in temperature, or both.

Ionization smoke detector A device containing a small amount of radioactive material that ionizes the air between two charged electrodes to sense the presence of smoke particles.

Line detector Wire or tubing that can be strung along the ceiling of large, open areas to detect an increase in heat.

Local alarm system A fire alarm system that sounds an alarm only in the building where it was activated; that is, no signal is sent out of the building.

Malicious false alarm A fire alarm signal sent when there is no fire, usually initiated by an individual who wishes to cause a disturbance.

Manual pull-station A device with a switch that either opens or closes a circuit, activating the fire alarm; also known as a manual fire alarm box.

Master-coded alarm An alarm system in which audible notification devices can be used for multiple purposes beyond a fire alarm.

Noncoded alarm An alarm system that provides no information at the alarm control panel to indicate where the activated alarm is located.

Nuisance alarm A fire alarm signal caused by the malfunction or improper operation of a fire alarm system or component.

Obscuration rate A measure of the percentage of light transmission blocked between the sender and receiver unit of a beam detector.

Outside stem and yoke (OS&Y) valve A sprinkler control valve with a valve stem that moves in and out as the valve is opened or closed.

Pendant sprinkler A sprinkler designed to be mounted on the underside of sprinkler piping so that the water stream is directed downward; usually marked SSP (standard spray pendant).

Photoelectric smoke detector A device that uses a light source and a photosensitive sensor to detect visible products of combustion.

Post indicator valve (PIV) A sprinkler control valve with an indicator that reads either open or shut, depending on its position.

Preaction sprinkler system A dry sprinkler system that uses a deluge valve instead of a dry-pipe valve and requires the activation of a secondary device before the pipes will fill with water.

Proprietary system A fire alarm system that transmits a signal to a monitoring location owned and operated by the facility's owner.

Rate-of-rise heat detector A fire detection device that responds when the temperature increases at a rate that exceeds a predetermined value.

Remote annunciator A secondary fire alarm control panel in a different location than the main alarm panel; it is usually located near the front door of a building.

Remote station system A fire alarm system that sounds an alarm in the building and transmits a signal to the fire department or an off-premises monitoring location.

Residential sprinkler system A sprinkler system designed to protect residential dwellings.

Sidewall sprinkler A wall-mounted sprinkler that discharges water horizontally into a room.

Single-action pull-station A manual fire alarm activation device in which the user takes a single step—moving a lever, toggle, or handle—to activate the alarm.

Single-station smoke alarm A single device that both detects visible and invisible products of combustion (namely, smoke) and sounds an alarm.

Smoke detector A device that detects visible and invisible products of combustion (namely, smoke) and sends a signal to a fire alarm control panel.

Spot detector A single-station heat alarm. These devices are often spaced throughout an area.

Sprinkler piping The network of piping in a sprinkler system that delivers water to the sprinklers.

Standpipe system An arrangement of piping, valves, and hose connections installed in a structure to deliver water for fire hoses.

Tamper alarm An alarm that sounds when the clear plastic cover of a pull-station is lifted; it is intended to prevent malicious false alarms.

Tamper switch A switch on a sprinkler valve that transmits a signal to the fire alarm control panel if the normal position of the valve is changed.

Temporal-3 pattern The standard audible signal made by a fire alarm to alert occupants of a building.

Unwanted alarm A fire alarm signal caused by a device reacting properly to a condition that is not a true fire emergency (e.g., cigarette smoke).

Upright sprinkler A sprinkler designed to be installed on top of the supply piping; usually marked SSU (standard spray upright).

Verification system A fire alarm system that does not immediately initiate an alarm condition when a smoke detector activates. The system will wait a preset interval (generally 30–60 seconds) before checking the detector again. If the condition is clear, the system returns to normal status; if the detector still senses smoke, the system activates the fire alarm.

Wall post indicator valve A sprinkler control valve that is mounted on an outside wall of a building. The position of the indicator tells whether the valve is open or shut.

Water-motor gong An audible alarm notification device that is powered by water moving through the sprinkler system.

Wet chemical extinguishing system An extinguishing system that discharges a proprietary liquid extinguishing agent. It is often installed over stoves and deep fryers in commercial kitchens.

Wet sprinkler system A sprinkler system in which the pipes are normally filled with water.

Zoned coded alarm A fire alarm system that indicates which zone was activated both on the alarm control panel and through a coded audio signal.

Zoned noncoded alarm A fire alarm system that indicates the activated zone on the alarm control panel.

Zoned system A fire alarm system that divides a building or facility into zones so that the area where an alarm originated can be identified.

You are dispatched to the report of an automatic alarm for a 10-storey, low-income housing facility. The alarm company reports a trip on a single alarm on the fourth floor. Immediately after the alarm was received, calls began coming into the 911 centre advising there is light smoke on the fourth floor and that the occupants are evacuating. You arrive on scene and prepare to go to work.

1. An _____ is an audible and/or visual device in a fire alarm system that makes occupants or other persons aware of an alarm condition.
   A. alarm initiating device
   B. accelerator
   C. auxiliary system
   D. alarm notification device

2. A type _____ standpipe system is designed for use by occupants of a building only.
   A. I
   B. II
   C. III
   D. IV

3. A(n) _____ fire alarm system transmits a signal to a monitoring location owned and operated by the facility's owner.
   A. proprietary
   B. auxiliary
   C. central station
   D. local

4. A _____ is a fire hose connection through which the fire department can pump water into a sprinkler system or standpipe system.
   A. fire department connection
   B. post indicator valve
   C. dry-pipe valve
   D. outside stem and yoke

5. A(n) _____ is a sprinklers designed to be mounted on the underside of sprinkler piping so that the water stream is directed down.
   A. frangible-bulb sprinklers
   B. fusible-link sprinklers
   C. pendant sprinklers
   D. upright sprinklers

6. A(n) _____ alarm occurs when an alarm system is activated by a condition that is not really an emergency.
   A. nuisance
   B. malicious false
   C. undesired
   D. unwanted

You are at your son's baseball game when you strike up a conversation with the couple sitting next to you. As you get to know them, you find out that he is a local contractor who builds houses in your city. When asked what you do, you tell him that you are a fire fighter with the local fire department. He mentions that trying to make homes safer has been on his mind ever since his mother's apartment was destroyed by a fire caused by a neighbour's cigar. He is committed to installing fire sprinklers in his new homes, but has some questions that he hopes you can answer.

1. How will you respond when he asks if the sprinklers will accidently flood the house?
2. How will you respond if he asks how he can justify the cost to his customers?
3. How will you respond if he asks if you have sprinklers in your home?
4. How will you encourage this contractor to continue to promote fire sprinklers in residential structures?

Fire Cause Determination

Fire Fighter I

Knowledge Objectives

After studying this chapter, you will be able to perform the following skills:

- Differentiate accidental fires from incendiary fires. (NFPA 5.3.8, p 850)
- Describe the point of origin. (NFPA 5.3.8, 5.3.8.A, 5.3.8.B, p 849–850)
- Define the chain of custody. (NFPA 5.3.13, 5.3.13.A, 5.3.13.B, p 853–854)
- Describe techniques for preserving fire-cause evidence. (NFPA 5.3.14.B, p 853–854)
- Describe the observations that fire fighters should make during fire-ground operations. (NFPA 5.3.13.B, p 856–859)

Skills Objectives

There are no skills objectives for Fire Fighter I candidates. NFPA 1001 contains no Fire Fighter I Job Performance Requirements for this chapter.

Fire Fighter II   FFII

Knowledge Objectives

After studying this chapter, you will be able to perform the following skills:

- Describe the role and relationship of the Fire Fighter II to criminal investigators and insurance investigators. (NFPA 6.3.4A, p 849)
- Describe how the origin and cause of a fire are determined. (NFPA 6.3.4.A, p 850)
- Describe how to assist the fire investigators in the process of digging out the fire scene. (NFPA 6.3.4.B, p 851–852)
- Describe the types of evidence that may be uncovered at a fire scene. (NFPA 6.3.4.A, p 852)
- Describe techniques for preserving fire-cause evidence. (NFPA 6.3.4.B, p 852–853)
- Describe the steps needed to secure a property. (NFPA 6.3.4, 6.3.4A, 6.3.4B, p 859–860)
- Explain the importance of protecting a fire scene to aid in cause determination. (NFPA 6.3.4, 6.3.4A, 6.3.4B, p 859–860)
- Describe the common motives of an arsonist. (p 860–862)

© Photos.com

Skills Objectives

There are no skills objectives for Fire Fighter II candidates. NFPA 1001 contains no Fire Fighter II Job Performance Requirements for this chapter.

Additional NFPA Standards

- NFPA 921, *Guide for Fire and Explosion Investigations*

You Are the Fire Fighter

t is 2:00 a.m. on Saturday morning, and you have just put out a fire in a church. After going through rehab, the chief of your small department asks you to assist him as he begins an investigation. As he begins taking photos of the exterior of the building and making notes, he asks you to set up a fire line and, once you are done with that, to get some tools to dig it out and meet him at the area of origin.

1. How do you properly secure a scene?

2. What tools would you bring to dig it out?

3. How would you know where the area of origin is located?

© Duncan Noakes/Dreamstime.com

Introduction

Fire fighters usually reach a fire scene before a trained fire investigator arrives. As a consequence, fire fighters are able to observe important signs and patterns that the fire investigators can use in determining how and where the fire started. By identifying and preserving possible evidence, as well as recalling and reporting objective findings, fire fighters provide essential assistance to fire investigators. In some cases, the observations and actions of fire fighters could be significant in apprehending and convicting arsonists.

Determining the causes of fires allows fire departments to take steps to prevent future fires. For example, a fire department might develop an education program to reduce accidental fires, such as those caused when food is left unattended on a hot stove. A series of fires could point to a product defect, such as a design error in a chimney flue or the improper installation of chimney flues. Identifying fires that were intentionally set could lead to the arrest of the person responsible for the crimes.

Fire fighters must understand the basic principles of fire investigation and participate in taking care of this important departmental responsibility. Fire cause determination is often difficult, because important evidence may be consumed by the fire or destroyed during fire suppression operations or during salvage and overhaul. Fire investigators must rely on fire fighters to observe, capture, and retain information, as well as to preserve evidence until it can be examined. Evidence that is lost can never be replaced.

FIRE FIGHTER Tips

The preliminary investigation report will include names, dates, and witnesses. Make a mental note of what you witnessed—that is, who you saw, who opened or closed doors, and so on. Although the investigation report will be written by a trained fire investigator, your information is vital to its creation. A report is only as good as the information in it.

FIRE FIGHTER II Tips   FFII

Fire departments can use fire cause data to support ordinances or create policies designed to eliminate major fires from occurring; for example, such data can lead to development of ordinances related to removing combustible landscape materials such as pine needles, removing construction materials from the outside of multiple-family buildings, or requiring the use of residential sprinklers.

Who Conducts Fire Investigations?

In most jurisdictions, the chief of the fire department has a legal responsibility to determine and report the causes of fires. In large fire departments, the fire chief usually delegates this responsibility to a special unit staffed with trained fire investigators. In smaller jurisdictions, one or two fire department members may be trained to conduct these investigations. Sometimes a special unit conducts fire investigations for several area fire departments, under an established policy that determines when those unit members should be called to conduct an investigation. In very small communities, the fire chief may personally examine the scene to determine the cause.

Many fire departments automatically dispatch a fire investigator to all working structure fires as well as to any other fire where the incident commander (IC) suspects something unusual. Other jurisdictions have policies that require the use of a fire investigator only when the damage exceeds a predetermined level or when the incident involves injuries or fatalities. In other departments, the IC may be expected to conduct a preliminary investigation and decide whether a fire investigator is needed. If the cause of the fire is evident and accidental, the IC would be responsible for gathering the information and filing the necessary reports. If the cause cannot be determined or appears to be intentional, the IC will summon a fire investigator.

When a fire investigator is not on the scene while the fire is being extinguished or overhauled, fire fighters must be particularly careful to preserve evidence. In such a situation, they serve as the eyes and ears of the fire investigator and must pass on their observations after extinguishing the fire. In some cases, fire fighters may need to be interviewed or to prepare a written report documenting their observations. These tasks are part of the fire fighter's role and responsibility in the fire investigation process.

Fire cause investigation should not be confused with a criminal investigation of **arson**, which is the malicious burning of property with a criminal intent. Arson is defined in the Canadian Criminal Code Sections 433 to 436. Any person convicted of the crime of arson can face a jail term of 5 to 14 years, depending on the type of arson. The objective of a fire cause investigation is to identify the cause of the fire; the fire investigator will never have a preconceived idea about what started the fire. The evidence and the facts will indicate whether the fire was started deliberately or resulted from an accidental cause. Only if the fire investigator determines that arson is the cause will a criminal investigation begin to find the person responsible.

■ Law-Enforcement Authority

An arson investigation must determine not only the origin and cause of the fire but also the person who was responsible for starting the fire and the sequence of events that led up to it. Determination of origin and cause is a small, yet important, step in the overall investigation. To prove that a crime took place, fire investigators must establish a link from the fire to a particular origin and cause, eliminating any other cause as a possibility.

In Canada, once there is clear evidence of arson, control of the fire investigation is turned over to the local police. In larger communities, there may be an arson squad consisting of experienced police and fire investigators. In smaller communities, most often the police investigate the fire. They may invite the local fire department to assist them in processing the scene and providing scene security, but the fire department's role as investigator is over.

■ Investigation Assistance

Most fire departments and government agencies have limited time, personnel, and resources available for fire investigations. To provide for more extensive investigations, a provincial/territorial fire marshal or commissioner's office may establish an investigations unit that concentrates on major incidents and supports local investigators on larger fires. The same office will normally investigate all fatal fires and undetermined fire causes if requested by the local fire department. In some jurisdictions, qualified fire investigators, often retired fire department personnel, provide this service. Because these investigators cannot always reach the fire scene quickly, the local fire department must be prepared to a) conduct a thorough preliminary investigation, b) to protect the scene, and c) preserve evidence.

Federal resources are also available for major investigations. The U.S. Bureau of Alcohol, Tobacco, Firearms and Explosives (ATF), which has field offices in Canada, has individual agents who can assist local jurisdictions with fire investigations. The ATF also maintains response teams spread across the country that respond to large-scale incidents in the United States. Each of these teams includes approximately 15 agents and uses equipment that ranges from simple tools for digging out a fire scene to laser-surveying devices for fully documenting the scene.

Insurance companies often investigate fires to determine the validity of a claim for damages or to identify factors that might help prevent future fires. The cost of an investigation is more than offset by the savings the company would realize by identifying a fraudulent claim. Some insurance companies employ their own fire investigators, whereas others retain the services of independent fire investigators. These outside fire investigators often have valuable experience and can provide critical technical support to determine the cause of a fire.

Causes of Fires

Every fire has a cause, which the fire investigator tries to uncover. An origin and cause investigation determines where, why, and how the fire originated. Some fires have simple causes that are easily identified and understood; others result from a complex set of circumstances that must be examined carefully to determine what actually happened. In some cases, the cause of a fire will never be determined with absolute certainty.

Every fire has a starting point where ignition occurs and fuel begins to burn. At this **point of origin**, an ignition source comes into contact with a fuel supply, such as a lighted match touching a piece of paper. To find the point of origin, fire investigators first search for the general area of origin and then attempt to locate the exact point of origin. The cause of the fire is the particular set of circumstances that brought the ignition source into contact with the fuel.

Fires result when a **competent ignition source** and a fuel come together for long enough to ignite. A competent ignition source must have enough heat energy to ignite the fuel and must remain in contact with the fuel until the fuel reaches its ignition temperature—a period that could range from a fraction of a second to hours, days, or even weeks.

A fire can be caused either by an act or by an omission. Igniting a piece of paper with a match is an act; leaving a pot of grease unattended on a hot stove is an omission. The cause of a fire can also be classified as either incendiary or accidental. Arson is an **incendiary** cause. According to NFPA 921, *Guide for Fire and Explosion Investigations*, arson is a crime of maliciously and intentionally, or recklessly, starting a fire or causing an explosion. By contrast, **accidental** fires do not involve a criminal or malicious intent, even if they are caused by human error or carelessness. Falling asleep with a lit cigarette, for example, would be considered an accidental cause.

Fire investigators always consider a fire to have an **undetermined** cause until the specific cause is established. The evidence from both incendiary and accidental fires can be very similar, so untrained individuals should not attempt to categorize fires as one or the other of these types. Instead, fire fighters should focus on helping to identify and preserve possible evidence for the fire investigator to examine. This may be

FIRE FIGHTER Tips

NFPA 921, *Guide for Fire and Explosion Investigations*, contains information and guidance for fire investigators.

done by securing a suspicious area/point that shows possible evidence that is found during overhaul operations. Another useful method is for the fire department to use a photographer from start to finish during suppression operations. It is often said, "A picture is worth a thousand words," and this is true when an investigator has to determine point of origin and cause from a pile of ash and rubble.

Fire Cause Statistics

The latest national statistics (2007–2008) for Canada indicate that there were a total of 42,753 fires resulting in 224 civilians losing their lives. The total property damage was 1.5 billion dollars. On average, home fires accounted for 30% of all fires, and 73% of all fire deaths. Cooking fires are the leading cause of home fires and injuries. However, smoker's material is still the leading cause of fatal fires with the point of origin typically occurring in the living room. These statistics are similar to losses in the United States.

Accidental Fire Causes

Accidental fires have hundreds of possible causes and involve multiple factors and circumstances. The most important reason for investigating and determining the causes of accidental fires is to prevent future fires. To reduce the number of fires, efforts must concentrate on the most frequent causes and those involving the greatest risks of death, injury, and property damage.

Most fires, fire deaths, and injuries occur in residential occupancies. The most commonly reported accidental causes of fire in these occupancies involve cooking, heating equipment, electrical equipment, and smoking. These statistics provide a foundation for fire prevention and public education efforts. Additional analysis can provide more specific information; for instance, fires caused by electrical equipment can be divided into four groups: those caused by worn-out or defective equipment, those caused by improper use of approved equipment, those caused by defective installations, and those caused by other accidents. A proper, thorough investigation of an electrical fire would identify and classify the specific cause within one of these groups.

For example, worn-out or defective electrical equipment that could cause fires would include deteriorating 50-year-old wiring circuits or a computer circuit board with an internal defect. Other worn-out equipment that might still be in use and cause fires includes electrical motors, switches, appliances, and extension cords. Properly used and maintained equipment that has been tested and listed by a recognized laboratory rarely causes a fire, but it should be replaced when it wears out.

Placing a portable heater too close to combustible materials or using a toaster oven to heat a container of flammable glue are examples of the improper use of electrical equipment. Defective installations are those not acceptable under electrical codes or printed instructions, such as using a light-duty extension cord to connect a heavy-duty appliance to a wall outlet. It is important to stress to the residents of a house the importance of reading and following the manufacturer's instructions when using any electrical device.

Some electrical fires result from accidental misuse or oversight, such as unintentionally leaving a cooking appliance turned on. Natural events can also cause accidental electrical fires, such as when a tree falls on a wire. A fire caused by an electrical overload or a short-circuit can start wherever electricity is present, such as in the electrical panel, fuses, fuse boxes, circuit breakers, wiring, and appliances.

Sometimes it may be difficult or impossible to identify the specific source of ignition because the fire destroyed every trace of evidence. Fire investigators will classify these fires as having an undetermined cause, rather than making a best guess. For example, low-temperature ignition can occur when wood is subjected to low heat, such as that generated by steam pipes or incandescent light bulbs, over a long period of time. In this scenario, the wood can gradually deteriorate and eventually ignite through a process called pyrolysis. If the resulting fire destroys the building, its point of origin would be difficult to identify.

Incendiary Fire Causes

People may set fires for several reasons and in many different ways. The same type of origin and cause investigation that is employed for an accidental fire is needed to identify where and how an incendiary fire started. With incendiary fires, it is particularly important to rule out possible accidental causes to prove beyond doubt that the fire was deliberately set.

A fire caused by arson requires a second phase of investigation to identify the person responsible. All of the evidence relating to the cause of the fire must be handled in a way that ensures that it will be admissible as evidence in a criminal trial. A trained, qualified fire investigator should always be called to determine the cause of any fire that may have been deliberately set. Arson and the factors that could indicate an incendiary fire cause are discussed in more detail later in this chapter.

Determining the Origin and Cause of a Fire

According to NFPA 921, a scientific method and a systematic analysis are needed to determine the origin and cause of a fire. As part of his or her responsibilities, the investigator must determine where the fire started and how it was ignited. He or she must look at the situation objectively to be sure that the evidence is convincing and fully explains the situation. If more than one explanation for the observations exists, each possibility must be considered. The cause cannot be determined with absolute certainty until all alternative explanations have been ruled out.

Identifying the Point of Origin

One of the first steps in a fire investigation is identifying the point of origin. At this location, the fire investigator can look for clues indicating the specific cause of the fire.

The investigation process usually begins with an examination of the building's exterior. The fire investigator will look for indications that the fire originated outside the building before looking inside, and he or she will size up the building to identify important information. The overall size, construction, layout, and occupancy of the building will be noted, as well as the extent of damage that is visible from the exterior. The fire investigator will look for any openings that might have created drafts that influenced the fire spread and will examine the condition of outside utilities, such as the electrical power connection and gas meter.

The search for indications of the fire's point of origin continues inside the building, beginning with the area of lightest damage and moving systematically to the area of heaviest damage. The area of heaviest damage is probably the area that was burning for the longest time; areas with less damage were probably not as heavily involved in the fire or were involved for shorter periods.

In the past, the **depth of char** has been used to help determine how intensely and how long a fire burned in a particular location. Post-fire patterns found on combustible material (wood) are generally more pronounced and extensive than those found on non-combustible materials. Due to a change in building construction material from natural to engineered wood, as detailed in the chapter "Building Construction," using the depth of char is no longer considered a reliable method to determine the intensity of a fire. In fact, some well-recognised investigation experts are now saying that charring rates are meaningless with respect to fire investigations. NFPA 921 states that, "the investigator is cautioned that no specific time of burning can be determined based solely on the depth of char." Although charring may be deepest at the point of origin, the presence of ignitable liquids, combustible materials, and the binding material used in the combustible products, as well as the effects of ventilation, can also influence charring FIGURE 29-1.

Burn patterns and smoke residue can be helpful in identifying the point of origin, but—like depth of char—are not conclusive evidence. Burn patterns and damage will often spread outward from the room or area where the damage is most severe. Because heat rises, the flow of heated gases from a fire will often be up and out from the point of origin. This upward, outward flow can usually be recognized, even when all of the building's contents were involved in the fire. Often the point of origin is found directly below the most damaged area on the ceiling, where the heat of the fire was most intensely concentrated.

A charred V-, hourglass-, or U-shaped burn pattern on a wall indicates that fire spread up and out from something at the base. The best place to start looking for a specific fire cause in a room with a burn pattern on the wall, a pile of charred debris at the base, and minimal damage to the other room contents is in the pile of debris FIGURE 29-2.

FIGURE 29-2 Often the point of a burn pattern is near or at the point of origin.

Courtesy of Charles B. Hughes/Unified Investigations & Sciences, Inc.

An experienced fire investigator knows that many factors can influence burn patterns, including ventilation, fire-suppression efforts, and the burning materials themselves. A burn pattern on a wall could indicate that an easily ignited and intensely burning fuel source was present at that location. It might also indicate that something fell from a higher level and burned on the floor. Such patterns may be caused by direct flame contact, heat, smoke, soot, or a combination of these elements. The patterns produced by the fire may be apparent on any surface that displays damage indicating a movement from the point of origin outward. Examples include wall and ceiling surfaces, appliances and furnishings, doors and windows, and even fire victims.

An inverted cone may be present when short-duration fires do not spread horizontally from the point of origin. An inverted cone on a wall may also indicate that an ignitable liquid was spread along the base of the wall as someone set the fire intentionally.

Once the fire investigator identifies the exact or approximate location of the point of origin, the search for indications of a specific cause can begin. The fire investigator must determine what happened at that location to cause the fire. This effort involves identifying both the source of ignition and the fuels that were involved. The fire investigator must ultimately determine how the source of ignition and fuel came together, whether accidentally or intentionally.

■ Digging Out

After the fire investigator identifies the area of origin, fire fighters could be asked to assist in digging out the fire scene. *Digging out* is a term used to describe the process of carefully looking for evidence within the debris. This may typically take place at the same time as overhaul or post fire salvage. Sometimes the entire fire scene must be closely examined to determine the cause of the fire and gather evidence. During

FIGURE 29-1 Depth of char.

Courtesy of Charles B. Hughes/Unified Investigations & Sciences, Inc.

this process, scene safety is paramount. All fire fighters and investigators should wear proper PPE, including airway protection. This is due to the high levels of HCN, CO, and CO_2 that can be released when material is being moved. The level of protection supplied can often be determined through the constant use of a gas meter.

The fire investigator will take extensive photographs of the fire scene as it first appears. He or she will then begin to remove and inspect the debris, layer by layer, working from the top of the pile down to the bottom. The type of evidence uncovered and its location within the layers of debris can provide important indications of how the fire originated and progressed.

Removing and inspecting the layers of debris enables the fire investigator to determine the sequence in which items burned, whether an item burned from the top down or from the bottom up, and how long it burned. Did the fire start at a low point and burn up, or did burning items fall down from above and ignite combustible materials below? Did the fire spread along the ceiling or along the floor? Is a residue of rags containing an ignitable liquid found under the furniture? Why are papers that should have been in a metal filing cabinet stacked on the floor and partially burned?

Systematically digging out through the debris often can uncover the exact point of origin and point to the cause of both accidental and deliberate fires. If circumstances or eyewitness accounts indicate that a deliberate fire is a possibility, the fire investigator may request that fire fighters help examine the entire area. Common search methods include the grid search (also known as the double-strip search) and the strip search. The investigator will explain generally what to look for, how to search, and what to do with any potential evidence.

When you find possible evidence, stop and inform the fire investigator so that he or she can examine it in place. It is the fire investigator's job to document, photograph, and remove any potential evidence, whether or not it supports the suspected cause.

To determine the origin and cause of a fire, the fire investigator must evaluate all potential causes of the fire. The process of eliminating alternative causes and documenting the reasons for rejecting them is as important as properly documenting the ultimate cause of the fire. For example, if the fire started in the kitchen, potential causes could include:

- The stove
- Any of the electrical appliances
- Light fixtures
- Smoking materials
- Cleaning supplies
- Other factors, such as candles

■ Collecting and Processing Evidence

Evidence refers to all of the information gathered and used by a fire investigator in determining the cause of a fire. Evidence can be used in a legal process to establish a fact or prove a point. To be admissible in court, it must be gathered and processed under strict procedures.

Types of Evidence

<u>Physical evidence</u> consists of items that can be observed, photographed, measured, collected, examined in a laboratory, and presented in court to prove or demonstrate a point **FIGURE 29-3**.

FIGURE 29-3 Physical evidence at a fire scene.
Courtesy of Charles B. Hughes/Unified Investigations & Sciences, Inc.

Fire investigators can gather physical evidence at the fire scene, such as a burn pattern on a wall or an empty gasoline can, to explain how the fire started or to document how it burned.

<u>Trace (transfer) evidence</u> consists of a minute quantity of physical evidence that is conveyed from one place to another; for example, a suspect's clothing may contain the residue of the same ignitable liquid found at the scene of a fire.

<u>Demonstrative evidence</u> is anything that can be used to validate a theory or to show how something could have occurred; to demonstrate how a fire could spread, for example, a fire investigator might use a computer model of the burned building.

Evidence used in court can be classified as either direct or circumstantial. <u>Direct evidence</u> includes facts that can be observed or reported first-hand. Testimony from an eyewitness who saw a person actually ignite a fire and a videotape from a security camera showing the person starting the fire are examples of direct evidence. Whereas direct evidence is rare in arson cases, eyewitnesses often can describe the circumstances that led to an accidental fire.

<u>Circumstantial evidence</u> is information that can be used to prove a theory, based on facts that were observed directly. For example, an investigation might show that the gasoline from a container found at the scene was used to start a fire. Two different witnesses might testify that the suspect purchased a container of gasoline before the fire and walked away from the fire scene without the container a few minutes before the fire department arrived. Such circumstantial evidence clearly places the suspect at the fire site with an ignition source at the time the fire started. Fire investigators must often work with circumstantial evidence to connect an arson suspect to a fire.

Preservation of Evidence

Fire fighters have a responsibility to preserve evidence that could indicate the cause or point of origin of a fire. Fire fighters who discover something that could be evidence while digging out a fire scene or performing other activities should leave it in place, make sure that no one interferes with the item or the surrounding area, and notify a fire officer or fire investigator immediately.

Evidence is most often found during the salvage and overhaul phases of a fire. Salvage and overhaul should always be performed carefully and can often be delayed until a fire investigator has examined the scene. Do not move debris any more than is absolutely necessary, and never discard debris until the fire investigator gives his or her approval to do so. It is the fire investigator's job to decide whether the evidence is relevant, not the fire fighter's.

Fire fighters at the scene are not in the position to decide whether the evidence they find will be admissible in court and, therefore, is worthy of preservation; that is also the fire investigator's decision. As a fire fighter, your responsibility is to make sure that potential evidence is not destroyed or lost. Too much evidence is better than too little, so no piece of potential evidence should be considered insignificant.

If evidence could be damaged or destroyed during fire-suppression activities, cover it with a salvage cover or some other type of protection, such as a garbage can. Use barrier tape to keep others from accidentally walking through evidence. These methods may prove to be ineffective, however, if no indication of their purpose is given. For this reason, if evidence is at risk of being damaged or altered by fire-suppression activities, it should be attended to by personnel or recovered. Before moving an object to protect it from damage, be sure that witnesses are present, a location sketch is drawn, and a photograph is taken.

Evidence should not be **contaminated** (i.e., altered from its original state) in any way. Fire investigators use special containers to store evidence and prevent contamination from any other products.

FIRE FIGHTER Tips

Gasoline-powered tools and equipment, including saws and generators, are often used during firefighting operations. Gasoline from such equipment could contaminate the area and lead to an erroneous assumption that accelerants (materials used to initiate or increase the spread of fire) were used in the fire. Limit the use of gasoline-powered equipment inside the structure, particularly in the area of the fire's origin. At a minimum, refuel such equipment completely outside the investigation area and the structure itself to prevent contamination of the site by spilled fuel.

FIRE FIGHTER Tips

To avoid contaminating evidence, fire investigators always wash their tools between taking samples. This step ensures that material from one piece of evidence will not be unintentionally transferred to another piece. Fire investigators also change their gloves each time they take a sample of evidence and place evidence only in absolutely clean containers. Before entering the fire scene, fire investigators often wash their boots to keep from transporting any contamination into the fire scene.

FIRE FIGHTER Tips

Ensure that your personal protective equipment (PPE) does not retain flammable liquids. Wash your PPE anytime it has come in contact with a flammable liquid or smoke.

Chain of Custody

To be admissible in a court of law, physical evidence must be handled according to certain prescribed standards. Because the cause of the fire may not be known when evidence is first collected, all evidence should be handled according to the same procedure. In an incendiary or arson fire, evidence relating to the cause of the fire will probably have to be presented in court. The same is true for accidental fires, because lawsuits might be filed to claim or recover damages.

Chain of custody (also known as *chain of evidence* or *chain of possession*) is a legal term that describes the process of maintaining continuous possession and control of the evidence from the time it is discovered until it is presented in court. Every step in the capture, movement, storage, and examination of the evidence must be properly documented; for example, if a gasoline can is found in the debris of a fire suspected to involve arson, documentation must record the person who found the can, along with where and when the can was found. Photographs should be taken to show the site where the gasoline can was found and its condition. In court, the fire investigator must be able to show that the gas can presented is the same can that was found at the fire site.

The person who takes initial possession of the evidence must keep it under his or her personal control until the item is turned over to another official. Each successive transfer of possession must be properly recorded. If evidence is stored, documentation must indicate where and when it was placed in storage, whether the storage location was secure, and when the evidence was removed from that location. Often, evidence is maintained in a secured evidence locker to ensure that only authorized personnel have access to it.

Everyone who had possession of the evidence must be able to attest that it has not been contaminated, damaged, or changed in any way. If evidence is examined in a laboratory, the laboratory tests must be documented. The documentation for chain of custody must establish that the evidence was never out of the control of the responsible agency and that no one could have tampered with it.

Fire fighters are frequently the first link in the chain of custody. The fire fighter's responsibility in protecting the integrity of this chain is relatively simple: Report everything to a supervisor and disturb nothing needlessly. The individual who finds the evidence should remain with it until the material is turned over to a company officer or to the fire investigator **FIGURE 29-4** . Remember: you could be called as a witness to state that you are the fire fighter who discovered this particular piece of evidence.

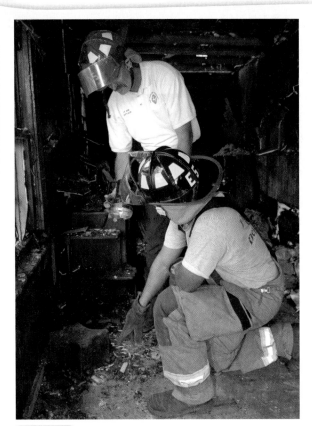

FIGURE 29-4 Evidence should remain where you find it until you can turn it over to your fire officer or the fire investigator.

The fire investigator's suggested operating guidelines (SOGs), or standard operating procedures (SOPs), for collecting and processing evidence generally include the following steps:

- Take photographs of each piece of evidence as it is found and collected. If possible, photograph the item exactly as it was found, before it is moved or disturbed.
- On the fire scene, sketch, mark, and label the location of the evidence. Sketch the scene as near to scale as possible.
- Place evidence in appropriate containers to ensure its safety and prevent contamination. Unused paint cans with lids that automatically seal when closed are good containers for transporting evidence. Glass Mason jars sealed with a sturdy sealing tape are appropriate for transporting smaller quantities of materials. Plastic containers and plastic bags should not be used to hold evidence containing petroleum products, as these chemicals may lead to deterioration of the plastic. Paper bags can be used for collecting dry clothing or metal articles, matches, or papers. Soak up small quantities of liquids with either a sterile cellulose sponge or sterile cotton batting. Protect partially burned paper and ash by placing them between layers of glass.
- Tag all evidence at the fire scene. The container for evidence being transported to the laboratory should be labelled with the date, time, location, discoverer's name, and witnesses' names.

- Record the time when the evidence was found, the location where it was found, and the name of the person who found it. Keep a record of each person who handled the evidence.
- Keep a constant watch on the evidence until it can be stored in a secure location. Evidence that must be moved temporarily should be put in a secure place that is accessible only to authorized personnel.
- Preserve the chain of custody in handling all the evidence. A broken chain of custody may result in a court ruling that the evidence is inadmissible.

Only one person should be responsible for collecting and taking custody of all evidence at a fire scene, no matter who discovers it. If someone other than the assigned evidence collector must seize the evidence, that person must photograph, mark, and contain the evidence properly and turn it over to the evidence collector as soon as possible. The evidence collector must also document all evidence that is collected, including the date, time, and location of discovery; the name of the finder; the known or suspected nature of the evidence; and the evidence number from the evidence tag or label. A log of all photographs taken should be recorded at the scene as well.

■ Identifying Witnesses

Although fire fighters may not interview witnesses, you can identify potential witnesses to the fire investigator. People who were on the scene when fire fighters arrived could have invaluable information about the fire. If a fire fighter learns something that might be related to the cause of the fire, he or she should pass this information on to a supervisor or to the fire investigator.

Interviews with witnesses should be conducted by the fire investigator or by a police officer. If the fire investigator is not on the scene or does not have the opportunity to interview the witness, the fire fighter should obtain the witness's name, address, and telephone number and give it to the fire investigator. A witness who leaves the scene without providing this information could be difficult or impossible to locate later.

Fire fighters have a primary responsibility to save lives and property. Until the fire is under control, they must concentrate on fighting the fire, not investigating its cause. Even so, fire fighters should pay attention to the situation and make mental notes about any observations. They must tell the fire investigator about any odd or unusual happenings. Information, suspicions, or theories about the fire should be shared only with the fire investigator and only in private.

Do not make any statements of accusation, personal opinion, or probable cause to anyone other than the fire investigator. Comments that are overheard by the property owner, the occupant, a news reporter, or a bystander can impede the efforts of the fire investigator to obtain complete and accurate information. A witness who is trying to be helpful might report an overheard comment as a personal observation. In this way, inaccurate information can generate a rumour, which then becomes a theory, which then turns into a reported "fact" as it passes from person to person.

Never make jesting remarks or jokes at the scene . Careless, unauthorized, or premature remarks could embarrass the fire department. Statements to news reporters about the fire's cause should be made only by an official spokesperson after the fire investigator and ranking fire officer have agreed on their accuracy and validity. Until then, "The fire is under investigation," is a sufficient reply to any questions concerning the cause of the fire. Any further questions should be directed to the scene incident commander, scene investigator, public information officer, or fire chief (if they are on the scene).

FIGURE 29-5 Jesting remarks and jokes should never be made at the scene, because your comments could be overheard by the media and other bystanders.

© Mark C. Ide

Observations During Fire-Ground Operations

Although a fire fighter's primary concern is saving lives and property, you will inevitably make observations and gather information as you perform your duties. What you observe could prove significant in the subsequent investigation of the incident. Remember, the fire investigator is the only person who can determine if something is suspicious and needs to be investigated further.

■ Dispatch and Response

During dispatch and response, form a mental image of the scene you expect to encounter. Note the time of day, the weather conditions, and any route obstructions. As the incident progresses, pay attention to things that do not match your expectations; they could help the fire investigator determine the origin and cause of the fire.

Time of Day

Time of day and type of occupancy can indicate the number and type of people at an incident. Offices, stores, and other business places are filled with people during the day but are mostly empty at night. Conversely, restaurants and nightclubs are likely to be crowded after dark. Residential occupancies are more likely to be occupied at night and on weekends and holidays.

As victims evacuate the building, some may stand out, either because they are fully dressed when everyone else is in pajamas, or because their behaviour and demeanour are quite different from those of the other victims.

Weather Conditions

Note whether the day is hot, cold, cloudy, or clear, and whether conditions in the burning structure match the weather. On a cold day, windows should be closed; on a hot day, the furnace should be off.

Lightning, heavy snow, ice, flooding, fog, or other hazardous conditions can help cover an arsonist's activities, because they delay the fire department's arrival and make a fire fighter's job more difficult. Because wind direction and velocity help determine the natural path of fire spread, being aware of these conditions will help to determine if the fire behaved in an unnatural way.

Route Obstructions

Unusual traffic patterns or barriers blocking the route to the scene may be early indications of an incendiary fire. Be sure to note these and any other obstructions, such as barricades, felled trees, downed cables, or trash containers, that cause delays.

■ Arrival and Size-Up

Size-up operations can provide valuable information for fire investigators. Pay attention to the fire conditions, the building characteristics, positions of doors and windows, and any vehicles and people at or leaving the scene.

Description of the Fire

The fire investigator will compare the dispatcher's description with the actual fire conditions. If the fire has intensified dramatically in a short time, an accelerant could have been used. During fire-suppression operations, note whether flames are visible or whether only smoke is apparent. Also observe the quantity, colour, and source of the smoke. Does the fire appear to be burning in one place or in multiple locations?

People on the Scene

The appearance and behaviour of people and vehicles at the scene of a fire can provide valuable clues. Note the attitude and dress of the owner and/or the occupants of the building, as well as any other individuals at the scene of a fire. Normally, the owner or occupants of a building will be distressed and wearing clothing appropriate to the time of day; for example, at 4 a.m., you would expect to see residents in pajamas.

Anyone who seems out of place at a fire or someone who has been observed at several fires in various locations should be reported to the fire investigator. Some arsonists are emotionally disturbed individuals who receive personal satisfaction in watching a working fire. Sometimes fire investigators will photograph the crowd at a fire scene, particularly if a series of similar fires has occurred recently, and look for the same faces at different incidents.

Most people at a fire scene are serious and intent on watching the drama unfold **FIGURE 29-6**. Someone who is talking loudly, laughing, or making light of the situation should be considered suspicious. Fire investigators may also want to know about someone who eagerly volunteered multiple theories or too much information.

FIGURE 29-6 Most persons at the fire scene are intent on watching the fire fighters at work.

© Louis Brems, The Herald-Dispatch/AP Photos

Unusual Items or Conditions

Always note any unusual items or conditions about the property, such as whether windows and doors are open or closed. Arsonists commonly draw the shades or cover windows and doors with blankets to delay the discovery of a fire. Gasoline cans, forcible entry tools, and a damaged hydrant or sprinkler connection might all suggest an intentionally set fire.

■ Entry

As you prepare to enter the burning structure, note if you observe any evidence of prior entry, such as footprints leading into or out of the structure or tracks from vehicle tires. Note whether the windows and doors are intact, whether they are locked or unlocked, and whether any unusual barriers limit access to the structure. Also note any signs of forced entry by others.

Forced entry could leave impressions of tools on the windows and doors. Likewise, cut or torn edges of wood, metal, or glass may indicate forced entry. Fire investigators might be able to determine whether the glass had been broken by heat or by mechanical means. Note if you observe any signs of a burglary, too; the fire may have been set to destroy evidence of another crime.

FIRE FIGHTER Tips

Photographs—whether taken by film or digital cameras—can document conditions before the arrival of fire investigators, but specific procedures must be followed for these photographs to be admissible in court as evidence. Follow your department's SOGs/SOPs regarding picture taking.

■ Search and Rescue

As you enter the building to perform search and rescue or interior fire-suppression activities, consider the location and extent of the fire. Be especially alert for signs of multiple fires in a building. Arsonists often set more than one fire in an attempt

to ensure the destruction of a building. The presence of multiple fires increases the risk of injury and death both to the building's occupants and to responding fire fighters.

The location and condition of the building's contents may provide clues for the fire investigator as well. Unusual building contents or conditions—such as barriers in doors and windows, the absence of inventory from a warehouse, or insurance papers or deeds left out on a desk—should be noted to the fire investigator.

The team responsible for shutting off electric power to the building should identify whether the circuit breakers were on or off when they arrived. The location of any people found in the building should be noted as well. When cutting off electrical panels, try to cut off main breakers and do not disturb breakers that are tripped.

■ Ventilation

The ventilation crew should note whether the windows and doors of the building were open or closed, locked or unlocked. They should also note the colour and quantity of the smoke as well as the presence of any unusual odours.

Colour of Smoke

The colour of smoke sometimes indicates which kind of material is burning. Unusual smoke might indicate that more fuel was added to the mix.

Unusual Odours

Self-contained breathing apparatus (SCBA) protects fire fighters from hazardous fumes and toxic odours. Nevertheless, sometimes an odour is so strong that it can be detected en route or lingers after the fire has been extinguished. Cooking fires and overheated light ballasts produce distinctive, identifiable odours that are highly familiar to experienced fire fighters. Other common odours familiar to fire fighters include gasoline, kerosene, paint thinner, lacquers, turpentine, linseed oil, furniture polish, or natural gas.

Odours often linger in the soil under a building without a basement, particularly if an accelerant has been used and if the ground is wet. Concrete, brick, and plaster will all retain vapours after a fire has been extinguished. These observations should be passed along to the fire investigator.

Effects of Ventilation

Fire department ventilation operations can dramatically influence the behaviour of a fire and alter burn patterns. Certain ventilation situations can cause a fire to become more involved. Relay information about these operations to the fire investigator so that he or she can correctly interpret burn patterns.

■ Suppression

Fire behaviour, the presence of **incendiary devices** (materials used to start a fire), obstacles encountered during fire-suppression operations, and charring and burn patterns are among the factors that might help the fire investigator determine the origin and cause of the blaze. You may encounter weakened floors, booby traps, holes cut in floors, and reinforced doors and walls at sites for the production of illegal drugs and grow operations. These may pose a post fire hazard to you and the fire investigator.

Behaviour of Fire

During the fire attack, observe the behaviour of the fire and note how it reacts when an extinguishing agent is applied. Look for unusual flame colours, sounds, or reactions. For example, most ignitable liquids will float, continue to burn, and spread the fire when water is applied to them. Rekindled fire in the same area or a flare-up when water is applied could indicate the presence of an ignitable liquid. Fires set in multiple locations and accelerants present an increased risk to you and your fellow fire fighters during suppression activities.

Incendiary Devices, Trailers, and Accelerants

While fighting the fire, be aware of streamers or **trailers** (combustible materials positioned to spread the fire). Note if there are any combustible materials such as wood, paper, or rags in unusual locations. Note containers of ignitable liquids that are normally not found in the type of occupancy. Very intense heat or rapid fire spread might indicate the use of an accelerant to increase the fire spread.

Incendiary devices can include unusual items in unlikely places, such as a packet of matches tied to a bundle of combustible fibres or attached to a mechanical device. Often, incendiary devices fail to ignite or burn out without igniting other materials.

Condition of Fire Alarm or Suppression Systems

If the building is equipped with a fire alarm or fire-suppression system, fire fighters should note whether the system operated properly or was disabled. Note if the outside stem and yoke (OS&Y) valve was open or closed in an occupied structure. If the fire-suppression system failed to work, the culprit could be poor maintenance or deliberate tampering. Notify the fire investigator of any findings. Removal of the batteries from a smoke alarm is a common way that an arsonist attempts to delay discovery of a fire.

Obstacles

An arsonist may place obstacles to hinder the efforts of fire fighters. Note whether any furniture was obviously moved to block entry. Arsonists also may prop open fire doors, pull down plaster to expose the wood structure, or punch holes in walls and ceilings to increase the rate of fire spread. In such circumstances, be alert for unexpected openings in floors.

Fire Fighter Safety Tips
© Photos.com

If you encounter an incendiary device that has not ignited, notify your sector officer or incident commander in the area immediately! Do not make any attempt to move or disable the device.

Contents

Fire fighters who are involved in interior fire suppression like the members of the search and rescue team should make note of anything unusual about the contents of the building.

The absence of personal items in a residence may indicate that they were removed before the incident and that the fire was intentionally set. Empty boxes in a warehouse may belong there, or they may indicate that valuable contents were removed prior to the fire. If only one item or a particular stock of products burns, and there is no reasonable explanation for the fire, arson could be suspected. Sometimes the contents will be removed and replaced by less expensive or nonfunctioning items. Neighbours may report that they observed items being removed or replaced prior to the fire. Check whether appliances are plugged in at the time of the fire.

In some cases of arson, clothing or other valuables may be placed in areas of a structure considered to be protected from the fire. In addition, the arsonist may place clothing in vinyl or plastic bags to reduce its absorption of smoke. Valuable documents or items may be placed under floors, in crawl spaces, or in detached garages. The fire investigator will consider these factors with other evidence before reaching any conclusion.

Charring and Burn Patterns

Charring in unusual places—such as open floor space away from any likely accidental ignition source—could indicate that the fire was deliberately set. Char on the underside of doors or on the underside of a low horizontal surface, such as a tabletop, could indicate that a pool of an ignitable liquid was placed below the item.

■ Overhaul

During overhaul, the smoke and steam should begin to dissipate, enabling both fire fighters and fire investigators to get a better look at the surroundings. Fire fighters should continue to look for the signs, patterns, and evidence previously discussed while conducting overhaul in a way that allows evidence to be identified and preserved. The overhaul process, if not done carefully, can quickly destroy valuable evidence.

If possible, the fire investigator will take a good look at an area before overhaul begins. He or she can often quickly identify potential evidence and can help direct or guide the overhaul operation so that this evidence is properly preserved. Evidence located during overhaul should be left where it is found, untouched and undisturbed, until the fire investigator examines it. Evidence that must be removed from the scene should be properly identified, documented, photographed, packaged, and placed in a secure location.

Fire-suppression personnel and fire investigators must work as a team to ensure that the fire is completely extinguished and properly overhauled while continually searching for and preserving signs, patterns, and evidence. Taking photographs during this phase of the fire operation is a good idea.

In short, be careful not to destroy evidence during overhaul. Avoid throwing materials into a pile. Use low-velocity hose streams to avoid breaking up and scattering debris needlessly. Thermal imaging devices can be used to find hotspots without tearing apart the interior structure.

Watch for evidence that was shielded from the fire and is lying beneath burned debris; for example, a wall clock may have fallen during the fire and been covered with debris. If the clock is near the area of origin and stopped approximately when the fire broke out, it could be an important piece of evidence.

A fire investigator will always try to determine whether the building contents were changed or removed prior to a fire. If all of a business's computers are missing; for example, the fire might have been set to conceal a theft. In other cases, missing portraits or other sentimental items might have been removed by the building's owner before setting the fire.

The cause of wildland and vehicle fires should be investigated as well, although the evidence collected at these fires may be somewhat different from the evidence gathered at a building fire. An accidental wildland fire might have started from an untended campfire, for example, whereas intentional wildland fires can be started from a book of matches or other incendiary device.

Automobiles, recreational vehicles, and boats are common targets for arson. Some people believe that starting a fire in a vehicle or boat will destroy evidence of an incendiary fire. Fire investigators, however, will gather much evidence about the area of origin and seek to establish whether the fire was accidental or intentional. Take care at the scene of a vehicle fire to preserve the scene as much as possible.

A vehicle fire may also start from an accidental electrical short or from a combination of gasoline and matches placed in the passenger compartment. Signs of spilled fuel on the ground around a vehicle may be evidence of an intentional fire. Do not move the car without documenting this spilled fuel; otherwise, the evidence will be lost.

FIRE FIGHTER Tips

Some of the best friends of fire investigators have four legs. Specially trained dogs are invaluable in searching for accelerants.

FIRE FIGHTER II Tips　　FFII

While looking for hidden fires, leave as much of the room of origin intact by breeching walls on the opposite side of the burned areas from the room of origin. This approach will leave valuable burn patterns intact and aid the fire investigator in his or her scene examination.

■ Injuries and Fatalities

Any fire that results in an injury or fatality must be thoroughly investigated and the fire scene documented. A search and rescue operation should never be compromised, but it is important to document the location and position of any victims, especially in relation to the fire and the exits.

Clothing removed from any victim should be preserved as evidence. It may contain traces of ignitable liquids, and burn patterns on the clothing can indicate the fire flow. If the clothing is removed in the ambulance or at the hospital, personnel should be assigned to collect it and keep this evidence as intact as possible.

Document anything lying under a victim's body after it is removed, because this is often a protected area and may reveal important evidence. Victims who do not escape the fire should not be disturbed, and no attempts should be made to try to identify them. Fire investigators will work with other investigators, including the coroner or medical examiner, to examine the victims and the surrounding area for items of evidentiary value.

Securing and Transferring the Property

Fire departments have the legal authority to maintain control of a building after a fire to determine the cause of the fire. Maintaining site integrity is critical to the fire investigation. To meet this requirement, the building and premises must be properly secured and guarded until the fire investigator has finished gathering evidence and documenting the fire scene. This could be accomplished by posting fire fighters or police personnel. Otherwise, any efforts to determine the cause of a malicious or incendiary fire, no matter how efficient or complete, may be wasted.

Fire fighters must be aware of any provincial/territorial or local laws pertaining to right of access by the owners or occupants while the property is under the control of the fire department. The fire department should keep a log of the name of anyone who enters the property, the times of his or her entry and departure, and a description of any items taken from the scene. After receiving permission from the fire investigator, a fire officer or a fire fighter should accompany anyone who enters the premises for any reason until the scene is released.

If a fire investigator is not immediately available, the premises should be guarded and maintained under the control of the fire department until the investigation takes place and all evidence is collected. In the interim, take the following steps:

- Suspend salvage and overhaul, and secure the scene. Keep nonessential personnel out of the area. Deny entry to all unauthorized and unnecessary persons.
- Photograph the fire scene extensively. Start with the area with the least amount of damage, and work your way toward the area of possible origin. Take several pictures of the point of origin from various angles. Photograph any incendiary devices on the premises exactly where they were found.
- If weather, traffic, or other factors could destroy evidence, take steps to preserve these items in the best way possible. Protect tire tracks or footprints by placing boxes over them to prevent dust accumulation. Use barricades to block off the area to further traffic. Rope off areas surrounding plants, trailers, and devices, and post a guard at the site.

To secure the property, cordon off the area with fire- or police-line tape. A member of the fire department or law-enforcement agency should remain at the scene to ensure that no unauthorized persons cross the line. As previously mentioned, the fire department has the authority to deny access to a building for as long as necessary to conduct a thorough investigation and ensure the safety of the public.

Before leaving the scene, make sure that the building is properly secured and that no hazards to public safety exist. Shut off all utilities and seal any openings in the roof to prevent additional water damage. Board up and secure windows and doors to prevent unauthorized entry.

Fire departments can secure and protect the premises in several ways. Lock and guard gates if necessary. Rope off dangerous areas and mark them with signs. Most fire departments have contracts with local companies that provide 24-hour board-up services, and they sometimes hire private security guards to secure a property after a fire.

Eventually, when fire department operations are over, the property will be returned to the owner. This handover should not occur until the investigation is complete and all evidence is collected. The fire department's authority ends when the property is formally released to the property owner. Afterward, the fire investigator might need a search warrant or written consent to search the site, and unguarded evidence could be contaminated by subsequent visitors to the property.

Incendiary Fires

The term *incendiary fires* refers to all fires that are deliberately started with malicious or criminal intent. In many jurisdictions, arson has a narrower, specific legal definition. This chapter uses the term *arson* to mean the malicious burning of property with criminal intent.

Fire fighters must be aware of factors that could indicate an intentionally set fire, report any observations to a supervisor or fire investigator, and help protect evidence.

■ Indications of Arson

Arson fires have several distinct, recognizable patterns or indications. For example, a deliberate fire might have multiple points of origin or include multiple simultaneous fires. An arsonist may ignite fires in many different areas so as to create a large fire as quickly as possible and involve the entire building. Arsonists may also use trailers made from combustible materials, such as paper, rags, clothing, curtains, kerosene-soaked rope, or other fuels, to enhance the fire's spread. Such trailers often leave distinctive char and burn patterns that fire investigators can use to trace the spread of a fire.

An incendiary device is a device or mechanism, such as a candle, timer, or electrical heater, that is used to start a fire or explosion. Many incendiary devices leave behind valuable evidence—for example, metal parts, electrical components, or mechanical devices—at the point of origin. Often an arsonist will use more than one incendiary device; sometimes, a failed incendiary device is found at the fire scene. If trailers were used, investigators may be able to trace their burn patterns back to an incendiary device.

Evidence consisting of an ignitable liquid often indicates the presence of an incendiary fire but does not necessarily

establish arson as the cause: an accident could also explain the presence of an ignitable liquid spill. To prove arson, the fire investigator must determine that an ignitable liquid was used and that no explanation other than arson justifies its presence.

Extensive burn damage on a floor's surface could indicate that an ignitable liquid was poured and ignited; for example, a wood floor might demonstrate a pool-shaped burn pattern, or floor tiles could be discoloured or blistered with an irregular burn pattern. On concrete and masonry floors, an ignitable liquid pattern is usually irregular and consists of various shades of grey to black. Liquids flow to the lowest level possible, so pooled liquids could be visible in corners and along the base of walls, where low levels of charring might also be apparent.

Sometimes the first indications of arson are entirely circumstantial. For example, a particular fire might fit an existing pattern, such as a series of fires in the same area, at about the same time, and on the same days of the week. Alternatively, a series of fires might occur in the same type of business or in properties with the same owner. A fire fighter who notices a pattern or a set of similar circumstances should immediately report that observation to a supervisor or a fire investigator.

Cause Determination

Fire investigations involve much more than simply determining what caused the fire, where it started, and whether it was accidental or incendiary (intentionally set). Many fire investigations also examine many other aspects of a fire incident. If the fire results in injuries or fatalities, the investigation often will study every factor that could have contributed to or prevented those losses. In a fire that causes millions of dollars in damage, a whole range of factors could contribute to the large loss. Fire investigators typically examine the construction and contents of the building to determine whether fire and building codes were followed, if the built-in fire protection systems functioned properly, or if lessons from the fire could be applied by amending codes or inspection procedures. A fire investigation can teach many lessons, even if the specific cause and origin of the fire are never determined.

Arsonists

An **arsonist** is a person who deliberately sets a fire with criminal intent. Arsonists fall into several categories, with various explanations for their behaviour. The fire service has identified two groups who are responsible for a large number of fires: pyromaniacs and juvenile fire-setters. Many other arsonists start fires for a wide range of motives.

■ Pyromaniacs

A **pyromaniac** is a pathological fire-setter. Most are adult males, and they are often loners. These individuals are usually introverted, are polite but timid, and have difficulty relating to other people. The fires set by pyromaniacs have the following characteristics:

- Fires are set in easily accessible locations, such as immediately inside entrances, on basement stairs, in trash bins, or on porches.

- Fires are set in structures such as occupied residences of all types, barns, and vacant buildings.
- Accelerants are rarely used. The pyromaniac is impulsive, so materials readily at hand are used.
- Each pyromaniac follows a unique pattern—for example, setting fires at the same time of day or night, using the same method, and in similar locations.

Juvenile Fire-Setters

Juvenile fire-setters are usually divided into three groups according to age: 8 years old and younger, 9–12 years old (preadolescent), and 13–17 years old (adolescent). Preadolescent and adolescent fire-setters often exhibit the same personality traits as adult pyromaniacs: they are introverted, they have difficulty with interpersonal relationships, and they are extremely polite when questioned.

Children younger than 8 years old are seldom criminally motivated when they set fires; they usually are just curious and experimenting. Children of this age do not really understand the danger of fire. They typically set fires in or near their homes—in attics, basements, bathrooms, garages, closets, or nearby fields or vacant lots. They start fires with matches or by sticking combustible material into equipment such as electric heaters that provide an ignition source. The remains of matches, matchboxes, or matchbooks are often found at the point of origin when these young fire-setters are responsible for a fire incident.

Preadolescent fire-setters do not venture far from home, yet members of this group set most of the fires that involve schools and churches. Preadolescents have motivations other than idle curiosity. The motivations of boys range from spite to revenge and disruptive behaviour. They typically set fires in closets, classrooms, and wastebaskets, and they may try burning hallway bulletin boards. Girls usually set less aggressive fires and are motivated by a need for attention or as a response to a particular stress, such as a test they do not want to take.

Preadolescent fire-setters rarely use elaborate trailers or incendiary devices. Instead, they use common, readily available accelerants, such as gasoline, kerosene, or lighter fluid, to fuel the fire. Preadolescents often take advantage of whatever materials are at hand and on site, including trash-container contents, loose papers, and rags. The fire set by a preadolescent will show a lack of planning, preparation, and sophistication, particularly if it was a group effort.

In some cases, preadolescents set fires in an attempt to cover vandalism and theft. The items taken from the scene are typically personal, rather than especially valuable; for example, a child may take athletic equipment, which is later displayed to classmates, or may reclaim an item that had been confiscated by a teacher.

The fires set by adolescents are similar to those set by adults. Adolescents have better access to transportation and can travel farther, so they have access to a wider variety of buildings. They also have many of the same motivations of adult fire-setters, such as revenge or attempts to hide larceny, and they often use accelerants. Two-thirds of all fires set in

vacant buildings are initiated by adolescents, but no class of property is exempt from harm. Often a large amount of vandalism at the scene serves as a clue that the fire was set by an adolescent.

FIRE FIGHTER II Tips **FFII**

Arson Facts
- In 2007, 10% of all fires in Canada were indicated as arson or set fires.
- In Alberta there were 1429 arson-related fires, which equal 27% of all fires in Alberta. These resulted in a financial loss of 64 million dollars, 6 deaths, and 10 injuries.
- In Ontario, there were 2112 arson-related fires, which equals 15% of total fires in Ontario. These resulted in a financial loss of 37 million dollars, 6 deaths and 63 injuries.

FIRE FIGHTER II Tips **FFII**

Serial Arson
NFPA 921 identifies three types of serial (repetitive) arson:
- **Serial arson** involves an offender who sets three or more fires, with a cooling-off period between fires.
- **Spree arson** involves an arsonist who sets three or more fires at separate locations, with no emotional cooling-off period between fires.
- **Mass arson** involves an offender who sets three or more fires at the same site or location during a limited period of time.

Source: NFPA 921, *Guide for Fire and Explosion Investigations*, section 22.4.9.3.1.2.

Arsonist Motives

Arsonists set all types of fires and have a range of motives. Male arsonists are usually motivated by profit, revenge, or vanity, and they often use accelerants. In the past, female fire-setters were seldom motivated by profit and rarely used accelerants, but this pattern seems to be changing.

Six common motives of arsonists are listed in NFPA 921:
1. Vandalism
2. Excitement
3. Revenge
4. Crime concealment
5. Profit
6. Extremism

Arsonists who are motivated by excitement are relatively easy to apprehend, because they usually have some identifiable connection with their targets, such as a failing student who sets fire to a school or a disgruntled employee who sets fire to a business. One category of excitement fires includes fires started by a would-be hero, who can be either a discoverer or an assister. The discoverer sets the fire, "discovers" it, and initiates the alarm. The assister sets the fire, sometimes discovers it

Near-Miss REPORT

Report-Number: 09-0000909

Event Description: Units responded to a report of a natural-gas odour. Upon arrival, the deputy chief located the caller who stated that the vacant house next door has a strong odour of gas coming from it. After a complete 360-degree walk-around, the deputy chief confirmed the dispatch.

The house had a For Sale sign in the front yard and was reported to be vacant. The gas line going into the structure was shut off by the fire department. The gas utility company was notified and responded. Once the gas company arrived, the fire department gained access to the structure but did not enter. After venting the structure for approximately 20 minutes, the fire department entered. Once in the house, they entered the utility room to find a gas line for a gas clothes dryer was left in the "on" position. The dryer had been removed. Right next to the gas dryer was a wash sink with water slowly running from the faucet. The water was hot water. In the same room and in close proximity to the gas line was a hot water heater. This created the perfect recipe to blow up a house. Upon further investigation, we discovered that the house had just gone into foreclosure.

We consider this a close call because, if we had just entered this house, we potentially could have lost an entire company. The decision to not enter this vacant building was important and communication was a key factor.

Lessons Learned:

- Always conduct a 360-degree walk-around.
- Notice For Sale signs, identify vacant properties, and talk to neighbours.
- Step back and proceed cautiously when responding to situations like this.
- Good decision-making and communications skills are always key factors.
- Maintain situational awareness when planning what to do.

Remember to conduct a complete size-up and use caution when responding to reported gas odours. Good communication and command structure protected the responders from a potential disaster. When approaching a structure with a suspected gas leak inside, use a meter and check explosive levels before entering. To help mitigate this situation, we utilized a gas-powered fan to ventilate; however, for safety purposes, we started the fan at a remote location and then walked it to the position just outside the front door. Remember, if a scene seems suspicious, do not tamper with evidence. In this case, we notified the police department and the fire investigation unit for further investigation.

FIRE FIGHTER II

and reports it, and is always on hand to help fire fighters pull hose, raise ladders, and rescue people.

Another category of excitement arsonists is closely related to the would-be heroes: fire fighters and would-be fire fighters (persons who have tried to become fire fighters and failed) who intentionally start fires. Some just want the opportunity to operate a particular piece of apparatus or equipment; others want to get some practise. Some of these individuals have even started fires in nearby jurisdictions in an attempt to harm another fire department's reputation.

Revenge arsonists are often very careful and make detailed plans for setting the fire and escaping detection afterward. Such persons motivated by revenge seldom consider the extent of damage that fire can inflict; instead, their intent is to harm a particular person or group so as to get revenge for a perceived injustice. This type of fire is dangerous for fire fighters and occupants because the revenge arsonist will often set fires near

FIRE FIGHTER II

doorways, thereby blocking an exit point. Revenge is often a factor for arsonists in the other motive categories as well.

Arsonists also may set fires to conceal a crime. In such a case, the fire is intended to destroy evidence of another crime, such as burglary, vandalism, fraud, or murder. Fires have also been set to divert attention while another crime is being committed or during attempted escapes from jails, hospitals, and other institutions.

Arson for profit occurs because the arsonist hopes to benefit from the fire either directly or indirectly. Fires that directly benefit the arsonist often involve businesses; for example, a business owner might set a fire or arrange to have one set, with the goal being to collect an insurance settlement that can be used to cover other financial needs. An increasing problem connected with this kind of arson-for-profit is fraud; for example, a person may buy and insure a property, only to burn it for the sole purpose of defrauding an insurance company.

During times of economic hardship, the incidence of arson may increase. People who owe more on a house than the house is worth may set fire to try to get insurance money. When large numbers of houses are unoccupied in a neighbourhood, the incidence of arson often increases. Given this relationship, keeping vacant houses secure is often a priority for fire departments. Fire fighters must be aware of the added risks that occur in vacant houses, especially when houses have been vandalized.

The other type of profit-motivated arsonist sets a fire to benefit from another person's loss. Included in this motive category are the following individuals:

- Insurance agents who want to sell more insurance to the victim's neighbours

- Contractors who want to secure a contract for rebuilding or wrecking the burned building
- Competitors who want to drive the victim out of business
- Owners of adjoining property who would like to buy the property and expand their own holdings
- Owners of nearby property who want to prevent an occupancy change, such as the conversion of a building to a drug rehabilitation centre

Some individuals with extreme beliefs set fires in an attempt to aid their cause. These fires are prompted by political, religious, and racial differences. Churches, abortion clinics, and government buildings have been targets of arsonists who belong to the extremism category.

Chief Concepts

- Fire fighters usually reach a fire scene before a trained fire investigator and are able to observe important signs and patterns that the fire investigators can use in determining how and where the fire started.
- Determining the causes of fires allows fire departments to take steps to prevent future fires.
- The size of the fire department and the size of the fire determine who performs the fire investigation and when a fire is investigated.
- When a fire investigator is not on the scene while the fire is being extinguished or overhauled, fire fighters must be particularly careful to preserve evidence.
- Basic fire investigation includes locating the point of origin, determining the fuel used, and identifying the ignition source for the fire.
- Fire investigation should be performed by one of the following: trained fire department investigators, the fire marshal's office, insurance-company investigators, or a law-enforcement agency.
- An origin and cause investigation determines where, why, and how the fire originated.
- At the point of origin, an ignition source comes into contact with a fuel supply. As part of their examination of a fire scene, fire investigators first search for the general area of origin and then attempt to locate the exact point of origin.
- The cause of a fire can be classified as either incendiary or accidental. Arson is an incendiary cause, whereas accidental fires do not involve a criminal or malicious intent.
- Accidental fires have hundreds of possible causes and involve multiple factors and circumstances. The most important reason for investigating and determining the causes of accidental fires is to prevent future fires. To reduce the number of fires, efforts must concentrate on the most frequent causes and those involving the greatest risks of death, injury, and property damage.
- Most fires, fire deaths, and injuries occur in residential occupancies. The most commonly reported accidental causes of fire in these occupancies involve cooking, heating equipment, electrical equipment, lightning strikes, and smoking.
- Sometimes it may difficult or impossible to identify the specific source of ignition because the fire destroyed every trace of evidence. Fire investigators will classify these fires as having an undetermined cause.
- According to NFPA 921, a scientific method and a systematic analysis are needed to determine the origin and cause of a fire. The fire investigator must determine where the fire started and how it was ignited.
- At the location of the fire, the fire investigator will look for clues indicating the specific cause of the fire. This fire investigation will begin on the exterior of the structure and move to the interior.
- Depth of char, burn patterns, and smoke residue can be helpful in identifying the point of origin but are not conclusive evidence.
- After the fire investigator identifies the area of origin, fire fighters may be asked to assist in digging out the fire scene to help look for evidence within the debris.
- Evidence refers to all of the information gathered and used by fire investigators in determining the cause of a fire:
  - Physical evidence—Items that can be observed, photographed, measured, collected, examined in a laboratory, and presented in court to prove or demonstrate a point
  - Trace or transfer evidence—A minute quantity of physical evidence that is conveyed from one place to another
  - Demonstrative evidence—Anything that can be used to validate a theory or to show how something could have occurred
  - Direct evidence—Facts that can be observed or reported first-hand
  - Circumstantial evidence—Information that can be used to prove a theory, based on facts that were observed directly
- Fire fighters have a responsibility to preserve evidence that could indicate the cause or origin of a fire. If fire fighters discover something that could be evidence, they should leave it in place and notify a supervisor or fire investigator immediately. Evidence is most often found during the salvage and overhaul phases.
- Physical evidence must be handled in a manner that protects the integrity of the chain of custody. Every step in the capture, movement, storage, and examination of the evidence must be properly documented.
- The fire investigator's SOGs/SOPs (for collecting and processing evidence generally include the following steps:
  - Take photographs of each piece of evidence as it is found and collected.

- On the fire scene, sketch, mark, and label the location of the evidence.
- Place evidence in appropriate containers to ensure its safety and prevent contamination. Tag all evidence at the fire scene.
- Keep a constant watch on the evidence until it can be stored in a secure location.
- Preserve the chain of custody in handling all the evidence.

■ Although fire fighters may not interview witnesses, they can identify potential witnesses to the fire investigators.

■ The fire fighter's role in identifying and preserving evidence continues throughout the fire-suppression sequence, and takes into account the following factors:
- The time of day and the weather conditions
- People leaving the scene
- The extent of the fire and the number of locations in which fire is found
- The security of the building
- Any signs of property break-in
- Vehicles or people in the area
- Indications of unusual fire situations
- Unusual colour of smoke
- The positions of the windows and roof
- The reaction of the fire during initial attack
- Any signs of abnormal behaviour of fire
- The condition of the building contents
- The need to coordinate overhaul and evidence-preservation activities

■ The building and premises must be properly secured and guarded until the fire investigator has finished gathering evidence and documenting the fire scene—for example, by posting fire fighters or police personnel at the site. Fire fighters must be aware of any provincial/territorial or local laws pertaining to right of access by the owners or occupants while the property is under the control of the fire department.

■ If a fire investigator is not immediately available, the premises should be guarded and maintained under the control of the fire department until the investigation takes place and all evidence is collected. In the interim, take the following steps:
- Suspend salvage and overhaul, and secure the scene.
- Keep nonessential personnel out of the area. Deny entry to all unauthorized and unnecessary persons.
- Photograph the fire scene extensively.
- If weather, traffic, or other factors could destroy the evidence, take steps to preserve it in the best way possible.

■ Incendiary fires refer to all fires that are deliberately started for malicious or criminal intent. You must be aware of factors that could indicate an intentionally set fire, report any relevant observations to a supervisor or fire investigator, and help protect evidence in such a scenario.

■ Arson fires have several distinct, recognizable patterns or indications; for example, a deliberately set fire might have multiple points of origin or include multiple simultaneous fires.

■ Fire investigations involve much more than simply determining what caused the fire, where it started, and whether it was accidental or incendiary (intentionally set). Many fire investigations examine other aspects of a fire incident; for example, if the fire results in injuries or fatalities, the investigation often will study every factor that could have contributed to or prevented those losses.

■ Fire investigators typically examine the construction and contents of the building to determine whether fire and building codes were followed, if the built-in fire protection systems functioned properly, or if lessons from the fire could be applied by amending codes or inspection procedures.

■ An arsonist is someone who deliberately sets a fire with criminal intent. Arsonists fall into several categories, with various explanations for their behaviour. The fire service has identified two groups who are responsible for a large number of fires: pyromaniacs and juvenile fire-setters.

■ The six common motives of arsonists are listed in NFPA 921:
  1. Vandalism
  2. Excitement
  3. Revenge
  4. Crime concealment
  5. Profit
  6. Extremism

Hot Terms

<u>Accidental</u> Fire-cause classification that includes fires with a proven cause that does not involve a deliberate human act.

<u>Arson</u> The crime of maliciously and intentionally, or recklessly, starting a fire or causing an explosion. (NFPA 921)

<u>Arsonist</u> A person who deliberately sets a fire to destroy property with criminal intent.

<u>Chain of custody</u> A legal term used to describe the paperwork or documentation describing the movement, storage, and custody of evidence, such as the record of possession of a gas can from a fire scene.

<u>Circumstantial evidence</u> The means by which alleged facts are proven by deduction or inference from other facts that were observed directly.

<u>Competent ignition source</u> An ignition source that can ignite a fuel under the existing conditions at the time of the fire. It must have sufficient heat and be in close enough proximity to the fuel for a sufficient amount of time to ignite the fuel.

<u>Contaminated</u> A term used to describe evidence that may have been altered from its original state.

<u>Demonstrative evidence</u> Materials that are used to demonstrate a theory or explain an event.

<u>Depth of char</u> The thickness of the layer of a material that has been consumed by a fire. The depth of char on wood may be used to help determine the intensity of a fire at a specific location.

<u>Direct evidence</u> Evidence that is reported first-hand, such as statements from an eyewitness who saw or heard something.

<u>Ignitable liquid</u> Classification for liquid fuels including both flammable and combustible liquids.

<u>Incendiary</u> A fire that is intentionally ignited under circumstances in which the person knows that the fire should not be ignited.

<u>Incendiary device</u> A device or mechanism used to start a fire or explosion.

<u>Mass arson</u> Arson in which an offender sets three or more fires at the same site or location during a limited period of time.

<u>Physical evidence</u> Items that can be observed, photographed, measured, collected, examined in a laboratory, and presented in court to prove or demonstrate a point.

<u>Point of origin</u> The exact location where a heat source and a fuel come in contact with each other and a fire begins. (NFPA 921)

<u>Pyromaniac</u> A pathological fire-setter.

<u>Serial arson</u> A series of fires set by the same offender, with a cooling-off period between fires.

<u>Spree arson</u> A series of fires started by an arsonist who sets three or more fires at separate locations, with no emotional cooling-off period between fires.

<u>Trace (transfer) evidence</u> A minute quantity of physical evidence that is conveyed from one place to another.

<u>Trailers</u> Combustible materials, such as rolled rags, blankets, and newspapers or ignitable liquid, that are used to spread fire from one point or area to other points or areas, often used in conjunction with an incendiary device.

<u>Undetermined</u> Fire-cause classification that includes fires for which the cause has not or cannot be proven.

You are excited to get to job shadow the fire marshal during your shift. At 7:00 in the evening, the two of you get called out for a fire in a vacant house. The fire marshal says that he has had a rash of fires in this area and is concerned that this will be another one caused by an arsonist. He said that he suspects a serial arsonist because each previous one has been set in exactly the same manner: forced entry into the back door of a vacant structure and then a trailer from the back door leading to ignitable liquids that were poured in every room. Being along for this definitely peaks your curiosity.

1. What is a trailer?

   A. A device or mechanism used to start a fire or explosion.

   B. Evidence of a minute quantity of physical evidence that is conveyed from one place to another.

   C. Combustible materials that are used to spread fire from one point or area to other points or areas.

   D. The exact location where a heat source and a fuel come in contact with each other and a fire begins.

2. Which of the following is a fire-cause classification that includes fires that are deliberately set?

   A. Arson

   B. Natural

   C. Incendiary

   D. Accidental

3. Arson in which an offender sets three or more fires at the same site or location during a limited period of time is considered _____.

   A. mass arson

   B. spree arson

   C. serial arson

   D. pyromania

4. In which type of evidence are alleged facts proven by deduction or inference from other facts that were observed directly?

   A. Demonstrative

   B. Circumstantial

   C. Direct

   D. Indirect

5. Which pattern might indicate a short-duration fire that does not spread horizontally from the point of origin?

   A. U-pattern

   B. Y-pattern

   C. V-pattern

   D. Inverted V-pattern

6. When securing a scene, a fire department should _____.

   A. allow anyone to enter the premises for any reason

   B. not allow anyone to enter the premises for any reason

   C. keep a log of the name of nonfirefighting personnel who enters the property and the times of his or her entry and departure

   D. keep a log of the name of anyone who enters the property and the times of his or her entry and departure

7. After the property is officially turned over to the owner, the fire department _____ to search the house.

   A. will not have to obtain a search warrant

   B. may have to obtain a search warrant

   C. must have a search warrant

   D. may not return under any condition

FIRE FIGHTER II
in action

Having just returned to the station from an overnight fire in a vacant structure, several of you sit around the table discussing the incident. The crews made a great stop; particularly considering the time of night and that there was heavy fire involvement upon your arrival. When the discussion naturally turns to how the fire started, you notice that one fire fighter doesn't say much, but sits there with a smirk on his face as it is discussed. At one point, he mumbles that this is only the beginning, but no one else seems to notice the comment. You become suspicious that he may have something to do with this and other recent fires.

1. How would you handle the situation?
2. To whom would you report your suspicions?
3. Is there a moral obligation to report this? Why or why not?
4. How would you handle it if you reported it and you were wrong?

Hazardous Materials: Introduction

Fire Fighter I

Knowledge Objectives

After studying this chapter, you will be able to:

- Define a hazardous material.
  (NFPA 472, 5.1, 5.1.1, 5.1.1.1, 6.6, 6.6.1, 6.6.1.1, 6.6.1.1.1 , p 870)
- Define hazardous waste. (NFPA 5.1.1.1 , p 871)
- List the common locations that may contain hazardous materials. (NFPA 5.1.1.2 , p 871)
- Distinguish between a regulation and a standard. (p 872)
- Describe which regulations and standards govern how fire departments respond to hazardous materials incidents. (p 872)
- Describe the roles and responsibilities of awareness-level personnel. (NFPA 5.1.1.1, 5.1.1.2, 5.1.1.3, 6.6.1.1.2, 6.6.1.1.3, 6.6.1.1.4 , p 872)
- Describe the roles and responsibilities of operations-level personnel.
  (NFPA 5.1.1.1, 5.1.1.2, 5.1.1.3, 6.6.1.1.2, 6.6.1.1.3, 6.6.1.1.4, 6.6.1.2, 6.6.1.2.1 ,
  p 872–873)
- Describe the roles and responsibilities of technician-level personnel. (NFPA 6.6.1.1.3 , p 873)
- Describe the roles and responsibilities of specialist-level personnel. (p 873)
- Describe the roles and responsibilities of a hazardous materials officer. (p 873)
- Describe the roles and responsibilities of hazardous materials safety officers. (p 873)
- Describe the roles and responsibilities of hazardous material technicians with specialties. (p 873)
- List the laws that govern hazardous material response activities. (p 873–874)
- Explain the differences between hazardous materials incidents and other emergencies. (p 874)
- Explain the need for a planned response to a hazardous materials incident. (p 874)
- Describe how to identify a substance's state of matter.
  (NFPA 472, 5.2.3, 6.6.1.1.2 , p 874–875)
- Describe the process of chemical change.
  (NFPA 472, 5.2.3, 6.6.1.1.2 , p 875–876)
- Describe the process of physical change.
  (NFPA 472, 5.2.3, 6.6.1.1.2 , p 875)
- Define *boiling point* and explain how this principle affects hazardous materials. (NFPA 472, 5.2.3, 6.6.1.1.2 , p 876)
- Define *flash point* and explain how this principle affects hazardous materials. (NFPA 472, 5.2.3, 6.6.1.1.2 , p 876–877)
- Define *fire point* and explain how this principle affects hazardous materials. (NFPA 472, 5.2.3, 6.6.1.1.2 , p 877)
- Define *ignition temperature* and explain how this principle affects hazardous materials. (NFPA 472, 5.2.3, 6.6.1.1.2 , p 877)
- Define *flammable range* and explain how this principle affects hazardous materials. (NFPA 472, 5.2.3, 6.6.1.1.2 , p 877)
- Define *vapour density* and explain how this principle affects hazardous materials. (NFPA 472, 5.2.3, 6.6.1.1.2 , p 877)
- Define *vapour pressure* and explain how this principle affects hazardous materials. (NFPA 472, 5.2.3, 6.6.1.1.2 , p 878–879)
- Define *specific gravity* and explain how this principle affects hazardous materials. (NFPA 472, 5.2.3, 6.6.1.1.2 , p 879)
- Define *water miscibility* and explain how this principle affects hazardous materials. (NFPA 472, 5.2.3, 6.6.1.1.2 , p 879–880)
- Define *corrosivity* and explain how this principle affects hazardous materials. (NFPA 472, 5.2.3, 6.6.1.1.2 , p 880)
- Define *pH* and explain how this principle affects hazardous materials. (NFPA 472, 5.2.3, 6.6.1.1.2 , p 880)
- Describe how to determine a substance's pH in the field. (NFPA 472, 5.2.3, 6.6.1.1.2 , p 880)
- Describe the physical hazards posed by the toxic products of combustion. (NFPA 472, 5.2.3, 6.6.1.1.2 , p 880)
- Describe the differences between non-ionizing and ionizing radiation. (NFPA 472, 5.2.3, 6.6.1.1.2 , p 881)
- Describe how radiation is detected in the field. (p 881)
- Define *alpha particles* and describe how to avoid exposure.
  (NFPA 472, 5.2.2, 5.2.3, 5.2.4, 6.6.1.1.2 , p 881)
- Define *beta particles* and describe their potential effects on the human body. (NFPA 472, 5.2.3, 6.6.1.1.2 , p 881–882)
- Define *gamma radiation*. (NFPA 472, 5.2.3, 6.6.1.1.2 , p 882)
- Describe the differences between contamination and secondary contamination. (NFPA 472, 5.2.3, 6.6.1.1.2 , p 882)
- List the types of weapons of mass destruction (WMDs).
  (NFPA 472, 5.2.3, 6.6.1.1.2 , p 882–883)
- Describe how nerve agents damage the human body.
  (NFPA 472, 5.2.3, 6.6.1.1.2 , p 883)
- List the signs and symptoms of nerve agent exposure.
  (NFPA 472, 5.2.3, 6.6.1.1.2 , p 883)
- Describe how blister agents damage the human body.
  (NFPA 472, 5.2.3, 6.6.1.1.2 , p 883–884)
- List the signs and symptoms of blister agent exposure.
  (NFPA 472, 5.2.3, 6.6.1.1.2 , p 883–884)

- Explain the route of exposure for cyanide. (**NFPA 472, 5.2.3, 6.6.1.1.2**, p 884)
- List the signs and symptoms of cyanide exposure. (**NFPA 472, 5.2.3, 6.6.1.1.2**, p 884)
- Describe how choking agents damage the human body. (**NFPA 472, 5.2.3, 6.6.1.1.2**, p 884)
- Describe how irritants damage the human body. (**NFPA 472, 5.2.3, 6.6.1.1.2**, p 884)
- List common convulsant chemicals, and describe how they damage the human body. (**NFPA 472, 5.2.3, 6.6.1.1.2**, p 884)
- List the four ways chemicals can enter the human body. (**NFPA 472, 5.2.3, 6.6.1.1.2**, p 884)
- Describe the precautions fire fighters take to avoid chemical exposure through inhalation. (**NFPA 472, 6.6.1.1.2**, p 884–885)
- Describe the precautions fire fighters take to avoid chemical exposure through absorption. (**NFPA 472, 6.6.1.1.2**, p 886–887)
- Explain the differences between chronic and acute health effects. (**NFPA 472, 5.2.3, 6.6.1.1.2**, p 887)
- List the four major sections of the *Emergency Response Guidebook* (*ERG*). (**NFPA 472, 6.6.1.1.2**, p 887–888)

Skills Objectives

After studying this chapter, you will be able to perform the following skill:

- Demonstrate the ability to properly utilize the current edition of the *ERG*. (p 888)

Fire Fighter II | FFII

Knowledge Objectives

There are no knowledge objectives for Fire Fighter II candidates. NFPA 1001 contains no Fire Fighter II Job Performance Requirements for this chapter.

Skills Objectives

There are no skills objectives for Fire Fighter II candidates. NFPA 1001 contains no Fire Fighter II Job Performance Requirements for this chapter.

Additional NFPA Standards

- NFPA 1072, *Hazardous Materials/Weapons of Mass Destruction Emergency Response Personnel Professional Qualifications*

You and your crew are dispatched to a report of an unusual odour in a multi-storey residential structure. Initial dispatch information is sketchy but indicates the residents have smelled the odour for several days, and it is causing irritation to their eyes and making them sick to their stomachs. As you arrive on scene, you and your crew observe several of the residents outside of the building: they have skin irritation, are vomiting, and complain of trouble breathing. You and your crew are asked to conduct an initial size-up and report back to the incident commander (IC).

1. Which initial dispatch information would lead you to believe that this incident might involve some type of hazardous material?
2. Which other indicators or direct observations will help identify the presence of hazardous materials?
3. Which actions should be taken to immediately to protect other people in the area, assist in identification of the problem, and make appropriate notifications?

Introduction

The modern fire service must respond to both existing and emerging threats to lives and property. In addition to fighting fires, today's fire fighters may be called to incidents involving chemical spills, gas leaks, emergencies at industrial plants, or railroad or truck crashes resulting in material leaks or spills. These incidents frequently involve hazardous materials that threaten lives, property, and the environment.

This chapter designed to give you the basic information required to recognize, respond to, and take initial action at a hazardous materials incident. The information is required for Fire Fighter I and II certifications and is designed to help you to build a foundation for safe and efficient response practices.

The following material describes the competency levels for first responders at the awareness and operations levels as found in NFPA 472, *Standard for Competence of Responders to Hazardous Materials/Weapons of Mass Destruction Incidents*, depending on your fire department's policies and procedures as they relate to hazardous materials response, and you likely will receive additional training in this area once you complete your basic training.

What Is a Hazardous Material?

A **hazardous material**, as defined by the **National Fire Protection Association (NFPA)**, is a substance—either matter (solid, liquid, or gas) or energy—that, when released, is capable of causing harm to people, the environment, and property, including weapons of mass destruction (WMDs). It has been noted in previous chapters that the governmental organization of Canada commands a multi-jurisdictional approach to major emergencies; however, for the vast majority of hazardous materials emergencies, it is municipal emergency personnel who handle the initial phases of the incident.

It is important for fire personnel to understand basic hazardous material response from a fire perspective. Most major

(census metropolitan area [CMA]) Canadian fire services that protect more than 500,000 people have hazardous materials units. In Ontario, the province is divided into three response areas that cover the entire province during a hazardous materials technical response: Ottawa, Toronto, and Windsor. These three areas have agreements to provide services beyond their jurisidiction to other areas within the province. Similar agreements may be in force within your province/territory. Hazardous materials response is a low-frequency, high-risk response that is expensive to provide; therefore, many jurisdictions have mutual aid agreements in place to ensure coverage.

Transport Canada is responsible for overseeing the regulations for the transportation of hazardous materials throughout the country, regardless of provincial/territorial boundaries. Historically speaking, the close trading relationship between the United States and Canada has influenced regulations in both countries, particularly with regards to the transportation industry; as a result, the respective regulations pertaining to the safe transportation of hazardous materials are similar with regards to identification and labelling.

The placard number is used to identify the nature of the substance being transported and provide instructions with regards to fire control, containment, personal protective clothing, safe distances, and actions to be taken and avoided. The chapter "Hazardous Materials: Awareness" will explain this in greater detail. The more accurate the information you obtain, both prior to and upon your arrival at the scene of a hazardous materials call, the more control (power) you will have to ensure your own safety and that of the public.

The ability to recognize a potential hazardous materials incident is critical to ensuring your own safety **FIGURE 30-1**. To avoid entering a dangerous area, you must be alert to the possibility that hazardous materials are present at the scene. Next, you must attempt to identify and isolate the released material and then prevent and address any injuries or exposures encountered. After all immediate life-safety hazards have been addressed, you must decide which level of action is

FIGURE 30-1 The ability to recognize a potential hazardous materials incident is critical to ensuring your safety.
© FIORENTINI MASSIMO/ShutterStock, Inc.

A.

B. © Richard Goldberg/ShutterStock, Inc.

C. © Harris Shiffman/ShutterStock, Inc.

D. © michael rubin/ShutterStock, Inc.

FIGURE 30-2 Hazardous materials can be found anywhere.
A. Residential. **B.** Industrial. **C.** Business. **D.** Transportation.

required and request any additional resources (i.e., appropriate equipment and expertise) necessary to mitigate the incident.

A hazardous material can be almost anything. Milk, for example, is not regarded as a dangerous chemical substance, but 18,930 litres (5000 gallons) of milk leaking into a creek does pose a risk to the environment. A large chlorine cloud also fits the NFPA's definition, as does a sustained leak from a 208-L (55-gal) drum of sulphuric acid. A fire fighter must be able to identify the substances or hazards involved and place them in the proper context. In other words, ask yourself this question: What is the chemical or hazard, and what are the conditions under which it was released?

Hazardous materials can be found anywhere. Pure chemicals and chemical mixtures are used to create millions of consumer products in sites ranging from hospitals to petrochemical plants. More than 80,000 chemicals are registered for commercial use in the United States, and it is estimated that 2,000 new ones are introduced annually **FIGURE 30-2**. The bulk of the new chemical substances are industrial chemicals, household cleaners, and lawn-care products. The integration of the Canadian economy with that of the United States means that many of these chemicals are present in Canadian workplaces, homes, and institutions as well.

In addition, manufacturing processes sometimes generate **hazardous waste**. Hazardous waste is a potentially harmful by-product or residue that remains after a manufacturing process. This type of waste can be just as dangerous as pure chemicals. Hazardous waste can consist of mixtures of several chemicals, resulting in a hybrid substance. It may be difficult to determine how such a substance will react when it is released or comes into contact with other chemicals.

The good news is that compared to the volumes of hazardous materials used and the amounts of hazardous waste generated in North America, very few hazardous materials incidents occur that require intervention. When accidents and incidents do happen, the fire department is usually the agency that handles the initial emergency response phase.

Levels of Training: Regulations and Standards

To better understand hazardous materials service provision, you must first understand some of the regulatory processes that govern hazardous materials response, beginning with the difference between a regulation and a standard. Regulations are issued and enforced by governmental bodies—for example, federal and provincial/territorial acts or regulations. In contrast, standards are issued by nongovernmental entities and are generally consensus based. Essentially, consensus organizations, such as the NFPA, issue standards that the public and other interested parties can comment on before committee members agree to adopt them. It is important to note that, while the NFPA is based in the United States, NFPA standards are recognized internationally. In Canada, many fire and building codes reference NFPA standards, as do the regulations governing manufacturing processes and laboratory activities.

NFPA standards governing hazardous materials response come from the Technical Committee on Hazardous Materials Response Personnel. This group includes members from private industry, the fire service, and state and federal agencies, as well as other stakeholders from outside of the fire service who may be called on to respond to emergencies that involve hazardous materials.

NFPA has two current hazardous materials–specific standards that relate directly to how fire departments respond to hazardous materials incidents: the aforementioned NFPA 472 and NFPA 473, *Standard for Competencies for EMS Personnel Responding to Hazardous Materials/Weapons of Mass Destruction Incidents*. The NFPA has recently withdrawn NFPA 471, *Recommended Practice for Responding to Hazardous Materials Incidents*.

The NFPA 472 levels of competencies are awareness, operations, technician, technician with specialty, and incident commander (IC). In 2008, the operations level was divided into the basic core competencies that every responder should have when responding to a hazardous materials incident and mission-specific competencies. These core competencies include understanding basic chemical and physical properties, concepts of chemical-protective clothing, and decontamination. The mission-specific competencies allow the authority having jurisdiction (AHJ) to decide which additional individual competencies responders should have based on their expected duties. For example, if a fire service has air-monitoring equipment and expects to use it, then the responders must

Fire Fighter Safety Tips
© Photos.com

In Canada, the various provincial/territorial governments have enacted regulations that require workplace training and awareness with regard to the identification, storage, and use of hazardous materials. These requirements form part of each province's individual occupational health and safety regulations or other relevant legislation.

receive additional training and demonstrate competency in its use; air monitoring becomes a mission-specific skill that is added to the list of core competencies in which a fire fighter is expected to show proficiency.

A concerted effort is being made by many of the larger fire services in Canada to ensure that fire fighters are trained in the operations core competencies.

■ Awareness Level

According to the NFPA, <u>awareness-level</u> personnel are individuals who, in the course of their normal duties, could encounter an emergency involving hazardous materials or WMDs and who are expected to recognize the presence of the hazardous materials or WMDs, protect themselves, call for trained personnel, and secure the scene.

According to the United States <u>Occupational Safety and Health Administration (OSHA)</u>, awareness-level personnel are first responders who are likely to witness or discover a hazardous substance release and who have been trained to initiate an emergency response sequence by notifying the proper authorities of the release. They do not take any further action beyond notifying the authorities of the release. First responders at the awareness level should have sufficient training or sufficient experience to objectively demonstrate competency.

In other words, any person who comes upon an incident and has been trained to recognize, identify, and notify is considered to be operating at the awareness level. This definition applies to personnel during their regular course of work; for example, the public works employee who is picking up road debris and comes upon a drum leaking a liquid, surrounded by lots of dead foliage and animals should be able to protect the area and make a notification.

■ Operations Level

According to the NFPA, <u>operations-level</u> personnel respond to hazardous materials or WMD incidents for the purpose of protecting nearby persons, the environment, or property from the effects of the release. A responder should be trained to meet all competencies at the awareness level and the core competencies of the operational level. The operations-level responder receives additional training to meet applicable governmental occupational health and safety regulations.

According to OSHA, first responders at the operations level are individuals who respond to releases or potential releases of hazardous substances as part of the initial response to the site for the purpose of protecting nearby persons, property, or the environment from the effects of the release. They are trained to respond in a defensive fashion without actually trying to stop the release. Their function is to contain the release from a safe distance, keep it from spreading, and prevent exposures. First responders at the operational level must have received at least 8 hours of training or have sufficient experience to objectively demonstrate competency in the following areas, in addition to those listed for the awareness level: basic hazard and risk-assessment techniques; selection and use of personal protective equipment (PPE); basic hazardous materials terms; and control, containment, and/or confinement operations.

In other words, operations-level personnel respond to hazardous materials and WMD incidents for the purpose of protecting nearby persons, the environment, or property from the effects of the release. Fire fighters in modern society are usually trained to the operations level, because they should be able to recognize potential hazardous materials incidents, isolate and deny entry to the affected area by other responders and the public, evacuate persons in danger, and take defensive actions, such as remotely shutting off valves and protecting drains, without having contact with the product. Operations-level responders act in a defensive fashion **FIGURE 30-3**.

Many hazardous materials incidents require some form of product control—hence the need for product-control training for the operations-level responder. These responders may be called upon to attempt to confine, contain, or otherwise control a released substance. The ultimate goal is to control the release and reduce or eliminate the threat to people, property, or the environment. Operations-level responders should work under the guidance of a hazardous material technician, an allied professional, or suggested operating guidelines (SOGs), or standard operating procedures (SOPs), and should be familiar with the policies and procedures of the local jurisdiction.

FIGURE 30-4 Hazardous materials technicians.
Courtesy of Staff Sgt. Brandie Session/U.S. Air Force

FIGURE 30-3 Operations-level responders.

■ Additional Levels

Additional hazardous materials responders include:

- **Technician level**—At this level, fire fighters are trained to enter heavily contaminated areas using the highest levels of personal protection. Hazardous materials technicians take offensive actions using a variety of specialized tools, equipment, and training in an effort to stop the release **FIGURE 30-4**.
- **Specialist level**—This level receives more specialized training than a hazardous materials technician; practically speaking, however, the two levels are not very different. The majority of the specialized training relates to a specific product, such as chlorine, or to a specific mode of transportation, such as rail emergencies.
- **Hazardous materials branch director/group supervisor**—This level of training is intended for those

persons assuming command of a hazardous materials incident beyond the operations level. Individuals trained as **hazardous materials officers** receive operations-level training as well as additional training specific to commanding a hazardous materials incident. The hazardous materials officer is trained to act as a branch director or group supervisor for the hazardous materials component of the incident.

- **Incident commander (IC)**—The IC's primary responsibilities during an incident include the development of strategies and tactics and acquisition of the resources necessary for the response.
- **Hazardous materials safety officer**—This person is assigned to the hazardous materials branch and is responsible for ensuring the safety of hazardous materials personnel and recognizing that appropriate hazardous materials/WMD practices are followed.
- **Hazardous materials technician with specialties**—This individual is a hazardous materials technician with training in specialized areas such as railway tank cars, highway cargo vehicles, intermodal containers, and others as listed in NFPA 472. Three levels of specialist employees are distinguished: A, B, and C. These designations are primarily used in the industrial world and describe the employee's capabilities with regard to his or her response during incidents on or off site. Such personnel may have specialties with regard to specialized chemicals, containers, and their uses.

Difference Between Hazardous Materials Incidents and Other Types of Emergencies

A critical distinction must be made between a hazardous materials event and other emergencies, such as a structure fire. Fire fighters cannot approach a hazardous materials incident with the same mindset they would a structure fire. Structural firefighting employs a fairly basic strategy determined during the size-up phase, with the ultimate goal being to perform rescue and extinguish the fire. By comparison, hazardous materials incidents are more complex. Consequently, the strategies used in addressing these incidents must be based on a reasoned approach that is both methodical and process based.

It is also important to make a distinction between a hazardous materials call and a chemical, biological, radiological, nuclear, and explosives (CBRNE) call. Hazardous materials calls generally involve a substance(s) or chemical(s) that has leaked by accident or by a neglectful act, whereas it is commonly viewed that CBRNE incidents are events that are created with the intent to inflict mass casualties and disrupt societal order. Fire personnel responding to a hazardous materials call must always look for indications that it might be categorized as a CBRNE occurrence and make every effort to preserve evidence and work with local, provincial/territorial, and federal authorities, as required.

Due to the cost associated with providing these programs and their legal nature, it is normal for CBRNE teams to be comprised of members from various emergency service providers and government agencies. It is important to note that Canada has a multi-jurisdictional approach to CBRNE incidents. It is important to always follow your individual fire service protocols.

Most of the time, a hazardous materials emergency moves more slowly than a structure fire. If a rescue is required, or if the situation is imminently dangerous and requires rapid intervention, events may move quickly, but a hazardous materials emergency requires time, forethought, and planning to ensure that the incident is handled safely and effectively.

The actions taken at hazardous materials incidents are dictated largely by the chemicals involved and the available resources. In many cases, it takes considerable effort to identify them and assemble the appropriately trained personnel and equipment. It becomes much more difficult and time-consuming to handle an incident involving an unknown substance. Response objectives, the choice of PPE, and the type of decontamination needed are all complex decisions that depend on the chemical properties of the hazardous substance.

In the case study at the beginning of this chapter, there are some clues to the nature of the material involved but not enough to indicate all of the hazards involved. Ask yourself these important questions:

- Is the material flammable, toxic, or corrosive?
- What is the chemical name of the material involved?
- How will it respond when exposed to air, water, or other materials?

When responding to hazardous materials calls, you must make a conscious effort to change your attitude and shift your thinking from that of a fire/rescue response to one that is more reflective. Decisions made at the outset of any emergency usually dictate the outcome; hazardous materials response is no different. Hazardous materials incidents demand an extremely disciplined, methodical approach. The most important function of the first-arriving fire personnel is to properly identify the problem and make every effort to prevent further injury by isolating the area involved. Once the threats have been identified, a strategy and approach must be undertaken. This approach must be communicated to all first responders. Some agencies establish SOGs/SOPs that guide their response to particular hazardous

materials incidents. Those parameters outline incident severity based on the nature of the chemical, the amount released, or the type of occupancy involved in the incident.

Planning a Response

It is a mistake to assume that the response to a hazardous materials incident begins when the alarm sounds. The response really begins with training, learning about the regulations and agencies involved, and finding out about the potential hazards in your area. Response agencies also should conduct incident planning activities at **target hazards** and other potential problem areas throughout the jurisdiction. Preplanning activities enable agencies to develop logical and appropriate response procedures for anticipated incidents. Planning should focus on the real threats that exist in your community or in adjacent communities where you could be assisting other departments at an incident **FIGURE 30-5**.

Once the threats have been identified, agencies and fire departments must determine how they will build their response. Some agencies establish SOGs/SOPs that guide their response to particular hazardous materials incidents. Those parameters outline incident severity based on the nature of the chemical, the amount released, or the type of occupancy involved in the incident.

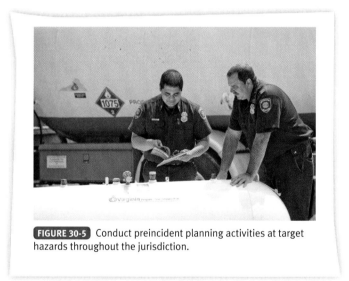

FIGURE 30-5 Conduct preincident planning activities at target hazards throughout the jurisdiction.

Hazardous Materials Fundamentals

To understand hazardous materials incidents, fire fighters must understand the chemical and physical properties of the materials and substances involved. Chemical and physical properties are the characteristics of a substance that are measurable, such as vapour density, flammability, corrosivity, and water reactivity. This requirement does not mean that you need to be a chemist, however; in most cases, being an astute observer and correctly interpreting the visual clues presented provide enough information. The basis for interpreting those visual clues lies in having a solid grasp of some basic hazardous materials concepts.

The first step in understanding the hazard posed by any substance involves identifying its state of matter. The **state of matter**

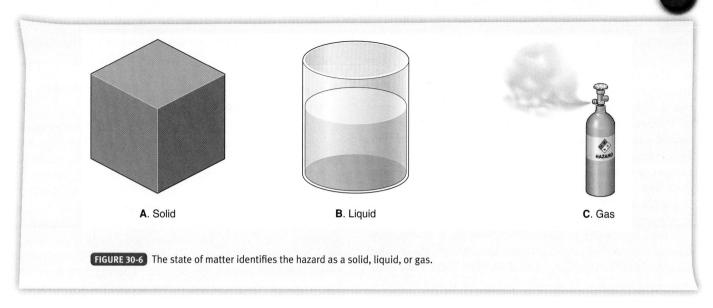

A. Solid   **B.** Liquid   **C.** Gas

FIGURE 30-6 The state of matter identifies the hazard as a solid, liquid, or gas.

identifies the hazard as a solid, liquid, or gas **FIGURE 30-6**. Why is this information important? If you know the state of matter and physical properties of the chemical, you can predict what the substance will do if it escapes from its container. Remember that the working definition for a hazardous material is potential energy that could be harmful when released from its container; hence, the importance of being able to predict the behaviour of a material when it escapes its container is obvious.

For example, it is vital to know whether a released gas is heavier or lighter than air. In a trench rescue, a gas that is heavier than air will sink to the lowest point, posing a significant problem for the victim. It is also critical to determine whether the released gas is flammable, is potentially **toxic**, or could displace oxygen.

Based on the various potential hazards associated with a material, you need to see the big picture before committing to a course of action. The bottom line is this: think before you act. This simple statement could save your life and/or the lives of your team members.

This chapter identifies concepts and key terms to help you understand the hazards a chemical poses as well as the information contained in **Safety Data Sheets (SDSs)** and other chemical reference sources.

Characteristics of Hazardous Materials

■ Physical and Chemical Changes

As previously mentioned, chemicals exist in three physical states: solid, liquid, or gas. The state that they are in at any given moment is a matter of the environment that they are in or how they are stored in a given container. Chemicals can undergo a **physical change** when subjected to different forms of energy, such as changes in temperature or pressure. When water is frozen, for example, it undergoes a physical change from a liquid to a solid. The actual chemical makeup (H_2O) is still the same, but the state in which it exists is different.

Propane gas has the potential to undergo physical change based on how it is stored inside a vessel. Liquefied propane has a large expansion ratio. The **expansion ratio** is a description of the volume increase that occurs when a liquid material changes to a gas. For example, propane has an expansion ratio of 270:1. This means that for every 1 volume unit of liquid propane, there would be 270 times that amount of propane **vapour**; in other words, 1 litre of liquid propane has the ability to vaporize to 270 litres of propane vapour when released.

If the propane cylinder is exposed to heat, the liquid propane inside changes into gaseous propane, which increases the pressure inside the vessel. If this uncontrolled expansion takes place faster than the relief valve can vent that excess pressure, a catastrophic container failure could occur. Such a catastrophic failure in a pressurized cylinder of liquid is referred to as a boiling liquid/expanding vapour explosion (BLEVE). The pressure in the container exceeds the container's ability to hold the pressure, and the container fails in a catastrophic fashion. The chemical makeup of propane has not changed; the substance has just physically changed state from liquid to gas.

The term **chemical change** describes the ability of a chemical to undergo a change in its chemical makeup, usually with the release of some form of energy. The reactive nature of a chemical may be influenced in many ways, such as by mixing it with another chemical or by applying heat to it.

Chemical change is different from physical change: physical change is essentially a change in state, whereas chemical change results in an alteration of the chemical nature of the material. Steel rusting and wood burning are examples of chemical changes. When the chemical reaction is complete, the substance itself is not the same as it was.

To understand this distinction between a physical and chemical change, one could use the example of your personal car: when you first purchased it, the paint was shiny, and the door panels were solid; however, after several winters of being exposed to the Canadian climate and road salt, rust begins to show on the lower part of the door panels (chemical change). You decide to take the vehicle to a repair shop to have the rusted door replaced (physical change). Understanding how a given chemical reacts or can affect its environment when it is in the three different states of matter can be a matter of life and death for a fire fighter.

The concepts of physical and chemical change can be applied to nearly any incident that involves chemicals. Assume that the owner of a landscape company left some rags soaked with linseed oil wadded up in a corner of a shed. The rags ignited spontaneously and started a fire. The heat from that fire ultimately causes the catastrophic failure of a 9.1-kilogram (20-pound) propane cylinder. The fire also caused the valve on a Type A 45.5-kg (100-pound) chlorine cylinder to fail (frangible disc functions as designed), allowing chlorine gas to be released into the atmosphere.

Looking closely at each of these events, it is apparent how the various physical and chemical changes altered the course of the fire incident. Linseed oil is an organic material that generates heat as it decomposes (chemical change). That heat, in the presence of the oxygen in the surrounding air, ignited the rags, which in turn ignited other combustibles (a common occurrence). Propane is gas liquefied by pressure and temperature and stored in steel cylinders. The surrounding heat from the fire caused the propane to expand (physical change) until it overwhelmed the ability of the container and the relief valve to handle the buildup of pressure. The cylinder ultimately exploded, and damaged the chlorine cylinder. That cylinder then leaked chlorine gas, which mixed with the moisture in the air and formed an acidic compound called hypochlorous acid (chemical change). The breeze carried that acid mist toward an apartment complex. Because chlorine vapour is heavier than air, most of the residents who complained about eye, nose, and throat irritation would likely be located on the bottom floor of the apartment complex.

Boiling Point

Boiling point is the temperature at which a liquid will continually give off vapours in sustained amounts and, if held at that temperature long enough, will eventually turn completely into a gas. The boiling point of water is 100 degrees Celsius (212 degrees Fahrenheit). At this temperature, the water molecules have enough kinetic energy (energy in motion) to break free from the liquid and overcome the force of the surrounding atmospheric pressure. Standard atmospheric pressure at sea level is 101.4 kilopascals (kPa), or 14.7 pounds per square inch (psi). In other words, at sea level, 101.4 kPa (14.7 psi) of pressure is exerted on every surface of every object.

An example of boiling point can be observed in the preparation of popcorn. Popcorn pops because of water inside the kernel of corn. Water has an expansion ratio of 1700:1 and, as previously referenced, a boiling point of 100°C (212°F). A kernel of corn has no relief valve, so a rapid buildup of pressure causes it to rupture. When popcorn kernels are heated, the small amount of water inside exceeds its boiling point, expands to 1700 times its original water volume, and the kernel pops. Unpopped kernels are those with insufficient water inside the kernel to allow for this change.

Flammable liquids with low boiling points are dangerous because of their potential to produce large volumes of **flammable vapour** when exposed to relatively low temperatures.

Flash Point, Fire Point, Ignition Temperature, and Flammable Range

When looking at the fire potential of a chemical, there are three important aspects to consider: flash point, ignition temperature, and flammable range. When it comes to combustion, materials must be in a gaseous or vaporous state to burn. Neither solids nor liquids actually burn; rather, they must give off a gas or a vapour that must be ignited to sustain combustion.

Think of a log in a fireplace: if you look closely, the fire is not directly on the log, but rather slightly above the surface. The reason for this is that the wood must be heated to the point where it produces enough "wood gas." The log, then, does not burn; instead, the gas produced by heating the log starts the process of combustion. Liquids behave in the same way as a log in a fire; that is, one way or another, there must be vapour production before there can be fire.

Flash point is an expression of the temperature at which a liquid fuel gives off sufficient vapour that, when an ignition source is present, will result in a flash fire. The flash fire involves only the vapour of the liquid, so it goes out once that fuel has been consumed.

− 41.7° C (− 43° F)
Flash Point
(Gasoline)

FIGURE 30-7 Gasoline at −41.7 °C (−43 °F) gives off sufficient vapour to result in a flash point when an ignition source is present.

The flash point of gasoline is −41.7°C (−43°F); in other words, when the temperature of gasoline is above −41.7°C (−43°F), the gasoline gives off sufficient flammable vapours to support combustion. If those vapours reach an ignition source, they will ignite in a ball of fire **FIGURE 30-7**. Diesel fuel has a much higher flash point—approximately 48.9°C (120°F). In either case, once the temperature of the liquid surpasses its flash point, the fuel will give off sufficient flammable vapours to support combustion.

Flash point is merely one aspect of how flammable and combustible liquids behave in a fire situation. The **fire point** is the temperature at which sustained combustion occurs. It is usually only slightly above the flash point for most materials.

The **ignition temperature** is another important landmark in the burning process. When a liquid fuel is heated beyond its ignition temperature, the substance ignites without an external ignition source. Think of a pan full of cooking oil being heated on the stove. For illustration purposes, assume that the ignition temperature of the oil is 148.9°C (300°F). What would happen if the oil were heated past that point? Once the temperature exceeds the ignition, or auto-ignition, temperature, the oil ignites; there is no need for an external ignition source. This is a common cause of stove fires.

Flammable range is another important concept. In broad terms, a substance's flammable range reflects an amount of vapour mixed with a volume of air, defined by upper and lower flammable limits. To illustrate this concept, think about gasoline. The flammable range for gasoline vapours is 1.4–7.6 percent. These two percentages, called the lower flammable limit (UFL) and the upper flammable limit (LFL), respectively, define the boundaries of the fuel–air mixture necessary for gasoline to burn properly. These terms are also known as the lower explosive limit (LEL) and upper explosive limit (UEL). The **upper flammable limit** defines the amount of flammable vapour needed to keep a fire burning; the **lower flammable limit** defines the amount of vapour needed to cause ignition. If a given gasoline–air mixture falls between the UFL and LFL and that mixture is exposed to an ignition source, there will likely be a flash fire. This fire will appear to be explosive in nature, even though it is not technically an explosion, owing to the speed of the flame front/pressure wave generated. Of course, this is a moot point should the fire fighter be near the event as it occurs: the result, though not an explosion by strict definition, will nevertheless be catastrophic for the fire fighter.

The concept underlying the functioning of an older-style automobile carburetor is based on flammable range. The carburetor is the place where gasoline and air are mixed. With the right mixture of gasoline vapours and air, the car runs smoothly. When there is too much fuel and not enough air, the mixture is considered to be "too rich"; in the case of too much air and not enough fuel, the mix is called "too lean." In either case, optimal combustion is not achieved and the motor does not run correctly.

Flammable range in the hazardous materials environment works in much the same way. If gasoline is released from an overturned tanker, gasoline vapours and air will mix. If the mixture of vapours and air falls between the UFL and LEL and those vapours encounter an ignition source, a flash fire will occur. As with carburetion, if there is too much or too little fuel, the mixture will not adequately support combustion.

To accurately assess the flammable range of a released vapour or gas, specialized combustible gas indicators must be used. These indicators are generally used by hazardous materials technicians and help determine the presence of a dangerous flammable atmosphere.

Generally speaking, the wider the flammable range, the more dangerous the material **TABLE 30-1**. This relationship exists because the wider the flammable range, the more opportunity there is for an explosive mixture to find an ignition source. The one major concern in the case of an unconfined

TABLE 30-1	Flammable Ranges of Common Gases
Hydrogen	4–75 %
Natural gas	5–15 %
Propane	2.5–9 %

vapour cloud is that as the cloud expands outward from the point of release, the likelihood of the edges of the vapour cloud having a flammable mixture at one or more points becomes much greater.

When operating in any environment where a flammable or combustible vapour cloud is suspected to be present, always err on the side of safety and assume that at some point in the suspected vapour cloud, the mixture is of such a concentration that the danger of an ignition exists. Always assume that any ignition source could ignite the vapour cloud and that, once ignited, that vapour cloud will continue to burn, possibly with explosive forces. For this reason, it is essential to always operate in a very defensive mode in such situations.

■ Vapour Density

In addition to addressing the flammability of a particular vapour or gas, fire fighters must determine whether the vapour or gas is heavier or lighter than air; in essence, you must know the vapour density of the substance in question. **Vapour density** is the weight of an airborne concentration (vapour or gas) as compared to an equal volume of dry air **FIGURE 30-8**.

Basically, vapour density compares the weight of a given vapour with the weight of air. What will the gas or vapour do when it is released into the air? Will it collect in low spots in the topography or somewhere inside a building, or will it float upward into the ventilation system? These are important response considerations for hazardous materials incidents, and the answers to these questions can influence the severity of the incident and directly affect fire fighters' ability to operate safely in the environment.

To find a chemical's vapour density, consult a reference source such as an SDS. The vapour density is expressed in numerical fashion: 1.2, 0.59, 4.0, etc.; the vapour density of propane, for example, is 2.4. Air has an index value of 1. Gases such as propane or chlorine are heavier than air and, therefore, have

FIGURE 30-8 Vapour density.

FIGURE 30-9 Vapour pressure.

a vapour density value greater than 1; by contrast, substances such as acetylene, natural gas, and hydrogen are lighter than air and, therefore, have a vapour density value less than 1.

In the absence of reliable reference sources in the field, you can use the HA HA MICEN mnemonic to remember a number of lighter-than-air gases:

H—Hydrogen
A—Acetylene
H—Helium
A—Ammonia
M—Methane
I—Illuminating gas (neon and hydrogen cyanide)
C—Carbon monoxide
E—Ethylene
N—Nitrogen

Although this list is not all inclusive, most materials not on this list are heavier than air (i.e., vapour density > 1). If you encounter a leaking propane cylinder, for example, just think of the HA HA MICEN mnemonic. Propane is not on that list, so by default, you can assume that propane is heavier than air.

■ Vapour Pressure

For fire fighters' purposes, the definition of **vapour pressure** can be assumed to pertain to liquids **FIGURE 30-9**. When liquids are held in a closed container, such as a 208-L (55-gal.) drum, pressure will develop inside that container. All liquids—even water—develop a certain amount of pressure in the air space between the top of the liquid and the container.

The key point to understanding vapour pressure is this: the vapours released from the surface of the liquid must be contained if they are to exert any pressure. Essentially, the liquid inside the container will vaporize until the molecules given off by the liquid reach equilibrium with the liquid itself. It is a balancing act between the liquid and the vapours in which some

liquid molecules turn into vapour, even as other molecules leave the vapour phase and return to the liquid phase.

Carbonated beverages illustrate this concept very clearly. Inside the basic carbonated soft drink is the formula for the soda and a certain amount of carbon-dioxide (CO_2) molecules. The CO_2 makes the bubbles in the soda. When the soda sits in an unopened plastic bottle on the shelf, the CO_2 balancing act is taking place: CO_2 from the liquid is turning into gaseous CO_2 above the liquid, and vice versa. The pressure inside the bottle can be indirectly observed by feeling the rigidity of the unopened plastic bottle. When the CO_2 is not in a closed vessel (i.e., the bottle is opened), however, it escapes into the atmosphere. With the pressure released, you can easily dent the sides of the bottle with your fingers. Ultimately, the opened bottle of soda goes flat, because all of the CO_2 has escaped. If you put the cap back on the bottle tightly before all of the CO_2 has escaped, after a short time, the bottle will feel rigid again. If you shake the closed bottle, you will release more of the CO_2 and the bottle (container) could fail as a result of the buildup of vapour pressure.

Armed with that knowledge, the technical definition of vapour pressure makes ready sense. The vapour pressure of a liquid is the pressure exerted by its vapour, until the liquid and the vapour are in equilibrium. Again, this process occurs inside a closed container, and temperature has a direct influence on the vapour pressure. For example, if heat impinges on a drum of acetone, the pressure above the liquid will increase and perhaps cause the drum to fail; if the temperature is reduced dramatically, the vapour pressure will drop. This is true for any liquid held inside a closed container.

What happens if the container is opened or spilled onto the ground to form a puddle? The liquid still has a vapour pressure, but it is no longer confined to a container. In such a case, liquids with high vapour pressures evaporate much more quickly than do liquids with low vapour pressures. Vapour pressure directly correlates to the speed at which a material evaporates, once it is released from its container.

For example, motor oil has a low vapour pressure. When it is released, it will stay on the ground for a long time. By contrast, chemicals such as isopropyl alcohol and diethyl ether have high vapour pressures. When either of these materials is released and collects on the ground, it evaporates rapidly. If the ambient air temperature or pavement temperature it elevated, the evaporation rate will increase. Wind speed, shade, humidity, and the surface area of the spill also influence how fast the chemical evaporates.

In reference sources, vapour pressure might be expressed in several different units—kilopascals, atmospheres (atm), torr, or millimetres of mercury (mm Hg)—and is usually shown at a temperature of 20°C (68°F). For a frame of reference, use the following comparison:

101.4 kPa (14.7 psi) = 1 atm = 760 torr = 760 mm Hg

The vapour pressure of water at room temperature is approximately 15 torr. Standard 40-weight motor oil, by comparison, has an approximate vapour pressure of less than 0.1 mm Hg at 20°C (68°F); it is practically vapourless at room temperature. Isopropyl alcohol has a very high vapour pressure of 30 mm Hg at room temperature. The temperature of the materials and the ambient temperature both have a direct correlation to vapour pressure.

Specific Gravity

<u>Specific gravity</u> is to liquids what vapour density is to gases and vapours—that is, a comparison value. In this case, the comparison is between the weight of a liquid and the weight of water, where water is assigned an index value of 1 as its specific gravity. Any material with a specific gravity less than 1 floats on water, whereas any material with a specific gravity greater than 1 sinks and remains below the surface of water. Consult a comprehensive reference source to find the specific gravity of the chemical in question.

Most flammable liquids float on water **FIGURE 30-10**. Gasoline, diesel fuel, motor oil, and benzene are examples of liquids that float on water. Carbon disulphide, however, has a specific gravity of approximately 2.6; consequently, if water is applied to a puddle of carbon disulphide, the water will cover it completely.

Water Miscibility

When discussing the concept of specific gravity, it is also necessary to determine whether a chemical will mix with water. Water is the predominant agent used to extinguish fires; it has often been referred to as the "universal solvent and decontamination agent." Both statements are general in nature, and there are certainly exceptions to every rule; nevertheless, both are also relatively correct. Always remember that when dealing with chemical emergencies, water might not always be the best or safest choice to mitigate the situation. It is a tremendously aggressive solvent and has the ability to react violently with certain other chemicals.

Concentrated sulphuric acid, metallic sodium, and magnesium are just a few examples of substances that react adversely with water. If water is applied to burning magnesium, for example, the heat of the fire will break apart the water molecules, creating an explosive reaction. Adding water to sulphuric acid would be like throwing water on a pan full of hot oil, in that popping and spattering might occur.

Conversely, water can be a friendly ally when fire fighters are attempting to mitigate a chemical emergency. Depending on the chemical involved, fog streams operated from handlines can knock down vapour clouds. Additionally, fires involving heavier-than-water flammable liquids can be extinguished by gently applying water to the surface of the liquid. The term

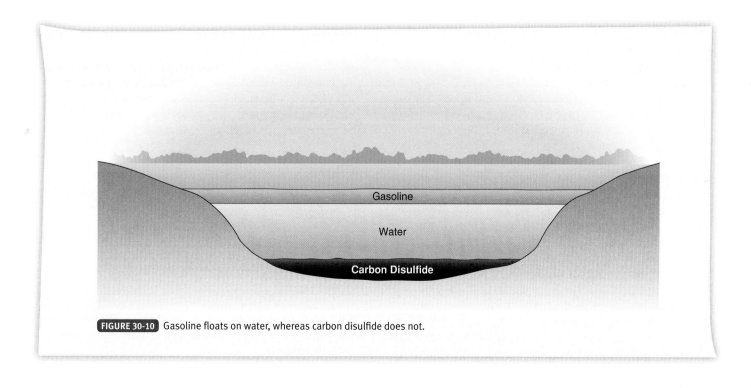

FIGURE 30-10 Gasoline floats on water, whereas carbon disulfide does not.

water solubility describes a substance's ability to dissolve in water.

It is important to consider the application of water carefully in all cases involving spilled or released chemicals. Do not approach any problem with the "it is just water" mindset. The results could be something worse than the initial problem.

Corrosivity (pH)

Corrosivity is the ability of a material to cause damage (on contact) to skin, eyes, or other parts on the body. The technical definition, as written by the United States Department of Transportation (DOT), includes the following language: "destruction or irreversible damage to living tissue at the site of contact." Corrosive materials are often also damaging to clothing, rescue equipment, and other physical objects in the environment.

Corrosives constitute a complex group of chemicals and should not be taken lightly. Tens of thousands of corrosive chemicals are used in general industry, semiconductor manufacturing, and biotechnology, all of which can be further categorized into two classes: **acid** and **base**.

A number of technical definitions for *acid* and *base* have been put forth, but the most common way to define them is in terms of **pH**. Think of pH as the potential of hydrogen. Essentially, pH is an expression of the concentration of hydrogen ions in a given substance. For fire fighters' purposes, pH serves as a measurement of corrosive strength, which loosely translates to the substance's degree of hazard. In simple terms, fire fighters can use pH to judge how aggressive a particular corrosive might be. To anticipate the seriousness of a particular hazard, they must understand the pH scale.

Acids such as sulphuric acid, hydrochloric acid, phosphoric acid, nitric acid, and acetic acid (vinegar) have pH values less than 7; chemicals that are considered to be bases, such as sodium hydroxide, potassium hydroxide, sodium carbonate, and ammonium hydroxide, have pH values greater than 7. The midpoint of the pH scale (pH = 7) is where a chemical is considered to be neutral—that is, neither acidic nor basic. At this value, a chemical does not harm human tissue.

Generally, substances with pH values of 2.5 and lower or 12.5 and higher are considered to be strong. Strong corrosives (both acids and bases) react more aggressively with metallic substances, such as steel and iron; they cause more damage to unprotected skin; they react more adversely when contacting other chemicals; and they might react violently with water.

To determine pH in the field, fire fighters can use specialized pH test paper. They can also obtain this information from the SDS or call for specialized hazardous materials teams to determine the pH level of a chemical. Hazardous materials technicians have more specialized ways of determining pH and can be a useful resource when handling corrosive incidents.

At the operations level, it is critical to understand that incidents involving corrosive substances might not be as straightforward as other types of chemical emergencies. The materials themselves are more complicated, and therefore, the tactics employed to deal with them could, in fact, be outside the fire fighter's scope of practice.

Toxic Products of Combustion

Toxic products of combustion are the hazardous chemical compounds that are released when a material decomposes under heat. The process of combustion is a chemical reaction and, like other reactions, is certain to generate a given amount of by-products. The smoke produced by a structure fire is obvious, but its composition and the toxic gases that are liberated during a residential structure fire might be less well known.

An easy way to think about the process of combustion is to apply the adage "Garbage in, garbage out." In other words, whatever serves as fuel for the fire makes up whatever is given off as a by-product of the reaction: the combustion reaction produces the toxic gases and other chemical substances found in the smoke.

Burning wood might seem like simple combustion reaction, but consider that the incineration of a piece of Douglas fir (a type of wood commonly used in residential construction) gives off more than 70 harmful chemical compounds. Other substances found in most fire smoke include soot, carbon monoxide, carbon dioxide, water vapour, formaldehyde, cyanide compounds, and many oxides of nitrogen. Each of these substances is unique in its chemical makeup, and most are toxic to humans in small doses.

For example, carbon monoxide affects the ability of the human body to transport oxygen. When the body inhales too much of this gas, the red blood cells cannot get oxygen to the cells, and the person dies from tissue asphyxiation. Cyanide compounds also affect oxygen uptake in the body and are a common cause of civilian deaths in structure fires. Formaldehyde is found in many plastics and resins and is one of the many components of smoke that causes eye and lung irritation. The oxides of nitrogen—including nitric oxide, nitrous oxide, and nitrogen dioxide—are deep-lung irritants, which can cause a serious medical condition called **pulmonary edema**, or fluid buildup in the lungs.

Fire fighters should always consider the origin of any fire they encounter. The conclusions reached might require evacuating affected citizens, dictate unique tactics (e.g., no water), or help them to better understand and treat injuries to civilians or fire-operations personnel.

Radiation

Most fire fighters have not been trained on the finer points of handling incidents involving radioactive materials and, therefore, have many misconceptions about radiation. Oddly enough, in many ways, radiation incidents are some of the easiest hazardous materials events to deal with—assuming that fire fighters have the right training and the right equipment to respond to them.

Radiation is energy transmitted through space in the form of electromagnetic waves or energetic particles. Electromagnetic radiation (light or radio waves) has no mass or charge, whereas energetic particles do have mass.

Radiation can be described as either ionizing or non-ionizing. Ionizing radiation consists of energetic particles or waves that have the potential to ionize an atom or molecule through atomic interactions. Ionization is a function of the energy of the individual particles or waves rather than a function of the number of particles or waves present. If enough of these ionizations occur, they can prove destructive to biological organisms by damaging the DNA in individual cells. In contrast, non-ionizing radiation refers to any type of electromagnetic radiation that does not carry enough energy to ionize atoms or molecules—that is, to completely remove an electron from an atom or molecule. Microwaves and radio waves are examples of non-ionizing radiation. Although it is not as harmful to humans as ionizing radiation, non-ionizing radiation can sometimes cause negative biological effects. Nevertheless, the form of radiation that is most dangerous to humans in general terms is ionizing radiation.

No one can completely escape radiation, because it is all around. During the course of his or her life, each person receives a certain amount of radiation from the sun, from the soil, or while having an X-ray. Radiation is everywhere and has been around since the beginning of time. The focus here, however, is not background radiation, but rather the occupational exposures encountered in the field. For the most part, the health hazards posed by radiation are a function of two factors:

- The amount of radiation absorbed by the body has a direct relationship to the degree of damage done.
- The amount of exposure time ultimately determines the extent of the injury.

The periodic table lists all of the known elements that make up every known chemical compound. Those elements, in turn, are made up of atoms. In the nucleus of those atoms are protons, which have a positive (+) electrical charge, and neutrons, which do not have any electrical charge (neutral). Orbiting the nucleus are electrons, which have a negative (−) electrical charge.

All atoms of any given element have the same number of protons in their nucleus; some elements, however, have varying numbers of neutrons in their nucleus. Each variation in the number of neutrons results in a different isotope of the element. For most purposes, different isotopes of the same element can be considered to behave identically in chemical reactions. A **radioactive isotope** has an unstable configuration of protons and neutrons in the nucleus of the atom. Carbon-14, for example, is a radioactive isotope of a carbon atom. Other examples include sulphur-35 and phosphorus-32, which are radioactive isotopes of sulphur and phosphorus atoms, respectively. **Radioactivity** is the natural and spontaneous process by which unstable atoms (isotopes) of an element decay to a different state and emit or radiate excess energy in the form of particles or waves.

Small radiation detectors are available that can be worn on turnout gear. These detectors sound an alarm when dangerous levels of radioactivity are encountered and alert fire fighters to leave the scene and call for more specialized assistance. The detection of radiation is usually the responsibility of hazardous materials technicians or specialists.

If the presence of radiation is suspected, steps must be taken to identify the type of radiation being emitted. In simple terms, radioactive isotopes give off energy from the nucleus of

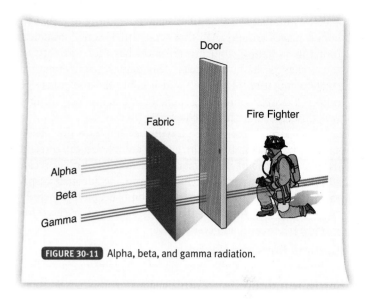

FIGURE 30-11 Alpha, beta, and gamma radiation.

an unstable atom in an attempt to reach a stable state. Typically, a combination of alpha, beta, and gamma radiation is present. Each of these forms of energy given off by a radioactive isotope varies in intensity and, consequently, determines how fire fighters attempt to reduce the exposure potential **FIGURE 30-11**.

Alpha Particles

As previously mentioned, radiation stems from an imbalance in the numbers of protons and neutrons contained in the nucleus of an atom. Alpha radiation is a reflection of that instability. This form of radiation energy occurs when an electrically charged particle is given off by the nucleus of an unstable atom. **Alpha particles** have weight and mass and, therefore, cannot travel very far (typically less than a few centimetres) from the nucleus of the atom. For the purpose of comparison, alpha particles are like dust particles.

Typical alpha emitters include americium (found in smoke detectors), polonium (identified in cigarette smoke), radium, radon, thorium, and uranium. Fire fighters can protect themselves from alpha emitters by staying a couple of metres away from the radioactive material and by protecting their respiratory tract with a **high-efficiency particulate air (HEPA) filter** on a simple respirator or by using self-contained breathing apparatus (SCBA). HEPA filters catch particles as small as 0.3 micron—a size much smaller than a typical dust particle. Although fire fighters can protect themselves from exposure to alpha particles by wearing HEPA filters and proper personal protective equipment (PPE), this type of radiation still poses a serious health risk if it enters into the body, especially if it is inhaled.

Beta Particles

Beta particles are more energetic than alpha particles and, therefore, pose a greater health hazard. Essentially, beta particles are like electrons, except that a beta particle is ejected from the nucleus of an unstable atom. Depending on the strength of the source, beta particles can travel one metre in the open air. Beta particles themselves are not radioactive; rather, the radiation energy is generated by the speed at which the particles are emitted from the nucleus. As a consequence, beta radiation can break chemical bonds at the molecular level and cause damage to living tissue. This breaking of chemical bonds creates an ion, such that beta particles are considered **ionizing radiation**. Ionizing radiation has the capability to cause changes in human cells. This damage can ultimately create a mutation of the cell and become the root cause of cancer. Examples of ionizing radiation are X-rays and gamma rays (discussed later in this chapter).

Beta particles can redden (erythema) and burn skin. They can also be inhaled. Beta particles that are inhaled can directly damage the cells of the human body. Most solid objects can stop beta particles, and SCBA should provide adequate respiratory protection against them. Common beta emitters include tritium (found in luminous dials on gauges), iodine (used for medical treatment), and cesium.

Gamma Rays and Neutrons

Gamma radiation is the most energetic type of radiation that fire fighters might encounter. Gamma rays differ from alpha and beta particles in that they are not particles ejected from the nucleus but, rather, pure electromagnetic energy. Gamma radiation has no mass, has no electrical charge, and travels at the speed of light. Gamma rays can pass through thick, solid objects (including the human body) very easily and generally follow the emission of a beta particle. If the nucleus still has too much energy after ejecting beta particles, a photon (a packet of pure energy) is released.

Like beta particles, gamma radiation is ionizing radiation and can be deadly. Structural firefighting gear with SCBA does not protect fire fighters from gamma rays; if fire fighters are near the source, they will be exposed to gamma radiation. Typical sources of gamma radiation are cesium (used for cancer treatment and in soil-density testing at construction sites) and cobalt (used for medical-instrument sterilization).

Neutrons are also penetrating particles found in the nucleus of an atom. They can be removed from the nucleus through nuclear fusion or fission. Although neutrons themselves are not radioactive, exposure to neutrons can create radiation, such as gamma radiation. Most fire fighters will never encounter neutrons in any form.

Hazard Exposure and Contamination

■ Hazard and Exposure

According to NFPA 472, a **hazard** is defined as a material capable of posing an unreasonable risk to health, safety, or the environment—that is, a material capable of causing harm. The same source defines **exposure** as the process by which people, animals, the environment, and equipment are subjected to or come into contact with a hazardous material.

For fire fighters' purposes, the two definitions are intertwined. When fire fighters respond to hazardous materials incidents, it is vital that they determine the risk to reduce their potential for exposure. This is done by taking the time to understand the problem at hand, including the physical properties of the chemical and the release scenario, and determining whether fire fighters have the right training, equipment, and protective gear to positively influence the outcome of the incident. All of the information presented to this point has been directed at understanding the nature of the chemicals involved.

■ Contamination

When a chemical has been released and physically contacts people, the environment, animals, tools, or other items—either intentionally or unintentionally—the residue is called **contamination**. In some cases, the process of removing those contaminants is complex and requires a significant effort to complete. In other cases, the contamination is easily removed and does not pose much of a threat to the overall success in handling the incident. For now, simply understand that when a dangerous chemical escapes its container, fire fighters might need to organize a system to remove it safely and efficiently or reduce its physical hazard.

■ Secondary Contamination

Secondary contamination occurs when a person or an object transfers the contamination or the source of contamination to another person or object by direct contact. In such cases, fire fighters might become contaminated and subsequently handle tools and equipment, grasp door handles, or touch other responders, thereby spreading the contamination. The possibility of spreading contamination brings up a point that all fire fighters should understand: the cleaner fire fighters stay during the response, the less decontamination they will have to do later. Many fire fighters have the misperception that PPE is worn to enable them to contact the product. In fact, quite the opposite is true: PPE, including specialized chemical-protective gear, is worn to protect fire fighters in the event that they cannot avoid product contact. The less contamination encountered or spread, the easier the decontamination will be.

Types of Hazardous Materials: Weapons of Mass Destruction

In recent years, **weapons of mass destruction (WMDs)** have become a highly relevant topic for fire fighters. WMDs represent a real threat in Canada, and each fire fighter should have a basic knowledge of those potential threats. To that end, this section briefly addresses the threats posed by nerve, blister, blood, and choking agents, as well as other potential hazards associated with various irritant materials.

WMD events can be complex, and the harm inflicted can be the result of several factors—nerve agents dispersed by explosions, radioactive contamination spread by the detonation of conventional explosives (so-called dirty bombs), and so on. Generally speaking, however, the damages caused by a terrorist attack or any other hazardous materials incident can be broadly classified into seven categories, best remembered by the mnemonic **TRACEMP**:

1. Thermal—Heat created from intentional explosions or fires, or cold generated by cryogenic liquids
2. Radiologic—Radioactive contamination from dirty bombs; alpha, beta, and gamma radiation

3. Asphyxiation—Oxygen deprivation caused by materials such as nitrogen; tissue asphyxiation from **blood agents**

4. Chemical—Injury and death caused by the intentional release of toxic industrial chemicals, nerve agents, vesicants, poisons, or other chemicals

5. Etiologic—Illness and death resulting from biohazards, such as anthrax, plague, and smallpox; hazards posed by bloodborne pathogens

6. Mechanical—Property damage and injury caused by explosion, falling debris, shrapnel, firearms, explosives, and slips, trips, and falls.

7. Psychogenic—Mental harm from being potentially exposed to, contaminated by, and even just being in close proximity to an incident of this nature.

■ Nerve Agents

Nerve agents pose a significant threat to civilians and fire fighters alike. The main threat stems from the ability of a nerve agent to enter the body through the lungs or skin and then systemically affect the function of the human body. Nerve agents, including many common pesticides, disrupt the central nervous system, possibly causing death or serious impairment.

The human central nervous system is laid out much like the electrical wiring in a house: The brain serves as the main panel. The wires—that is, the nerves—go to muscles, organs, glands, and other critical areas of the body. When a nerve agent, such as **sarin** or **VX**, gets into the body, it affects the ability of the body to transact nerve impulses at the junction points between nerves and muscles, nerves and vital organs, and nerves and glands. Subsequently, those critical areas do not function properly and associated bodily functions are affected.

Signs and symptoms of nerve agent exposure include pinpoint pupils, tearing of the eyes, twitching muscles, loss of bowel and bladder control, and a slow or rapid heartbeat. Mild exposures can result in mild cases of some or all of these symptoms. Significant exposures can result in rapid death, serious impairment, or a more pronounced manifestation of symptoms. The bottom line is that an exposure to a nerve agent is not an automatic death sentence. The dose absorbed directly correlates to the degree of damage done and the symptoms presented by the victim.

Recognition of the signs and symptoms of nerve agent exposure is vital in determining the presence of such a threat. Nerve agents in general are reported to have fruity odours. In most cases, fire fighters do not have the specialized detection devices necessary to identify the presence of a nerve agent; the symptoms presented by the victims should tip them off, however. To that end, the mnemonic SLUDGEM briefly summarizes some of the more common signs and symptoms of nerve agent exposure:

S—Salivation
L—Lachrymation (tearing)
U—Urination
D—Defecation
G—Gastric disturbance
E—Emesis (vomiting)
M—Miosis (constriction of the pupils)

Many times, nerve agents are incorrectly referred to as nerve gases. Nerve agents are liquids, not gases. Each evaporates at a different rate. Sarin, for example, is a waterlike liquid that evaporates more easily than VX, which is a thick, oily liquid. VX is the most toxic of the weapons-grade nerve agents. Other nerve agents include soman and tabun. The organophosphate and carbamate classes of pesticides are similar to the weapons-grade agents but are less toxic in general terms.

■ Blister Agents

Blister agents (also known as vesicants) are classified as such because of their ability to cause blistering of the skin. Common blister agents include **sulphur mustard** and **lewisite**, which interact in unique ways with the human body.

Sulphur mustard was first used as a weapon in World War I. Chemical casualties suffered horrific injuries to the skin and lungs; in some cases, those exposures were fatal. Although sulphur mustard is commonly referred to as mustard gas, it is typically found in its liquid state. This chemical is colourless and odourless when pure; when mixed with other substances or otherwise impure, however, it can appear brownish in colour and smell like garlic or onions.

Exposures to sulphur mustard might not be immediately apparent. A victim might not experience pain or any other sign indicating current contact with this agent. Indeed, redness and blistering might not appear for 2–24 hours after exposure. Once the blistering does occur, skin decontamination will not reduce its effects **FIGURE 30-12**.

If sulphur-mustard exposure is suspected, however, immediate skin decontamination is indicated. Early skin decontamination can reduce the effects of exposure.

Sulphur mustard produces injuries similar to second- and third-degree thermal burns. In cases involving mild skin exposures, victims typically make a complete recovery. Inhalation exposures are much more serious, with the damage to lung tissues sometimes resulting in permanent respiratory impairment. Additionally, sulphur mustard is very persistent in the

FIGURE 30-12 Blister agent exposure.
Courtesy of Dr. Saeed Keshavarz/RCCI (Research Center of Chemical Injuries)/IRAN

environment once released. It can remain intact, sticking to the ground or other surfaces, for several days.

Lewisite has many of the same characteristics of sulphur mustard. Lewisite, however, contains arsenic; thus, when it is absorbed by the skin, this blister agent can produce some symptoms specific to arsenic poisoning, such as vomiting and low blood pressure. Unlike sulphur mustard, lewisite exposures to the skin produce immediate pain. If a lewisite exposure is suspected, immediate and aggressive skin decontamination is indicated. Like sulphur mustard, lewisite does not occur naturally in the environment and is considered to be a chemical warfare agent.

■ Cyanide

Cyanide compounds, including hydrogen cyanide and cyanogen chloride, prevent the body from using oxygen. Victims exposed to high levels of cyanide, for example, could be placed on supplemental oxygen but still not survive the exposure. The cause of death in this case is not a lack of oxygen but, rather, the inability of the body to get oxygen to the cells.

The main route of exposure to cyanide compounds is through the lungs, but many of these agents can also be absorbed through the skin. Hydrogen cyanide has an odour of bitter almonds. Sixty percent of the general population cannot detect the presence of hydrogen cyanide by smell; thus, because the odour threshold and the amount of cyanide that would be lethal are very close, many times odour alone cannot be used to detect dangerous amounts of hydrogen cyanide. By contrast, cyanogen chloride, because of the chlorine atom within the chemical compound, has a much more irritating and pungent odour.

Some of the typical signs and symptoms of cyanide exposure include vomiting, dizziness, watery eyes, and deep and rapid breathing. High concentrations lead to convulsions, inability to breathe, loss of consciousness, and death.

■ Choking Agents

<u>Choking agents</u> are predominantly designed to incapacitate rather than to kill; nevertheless, death and serious injuries from these agents are certainly possible. The nature of choking agents—namely, the extremely irritating odour—alerts potential victims of their presence and allows for escape from the environment, when possible. Some commonly recognized choking agents include <u>**chlorine**</u>, phosgene, and chloropicrin.

Each of these substances is associated with a specific odour. Chlorine smells like a swimming pool, whereas phosgene and chloropicrin have been reported to smell like freshly mown grass or hay. In either case, the odours are noticeable and, in many cases, strong enough and irritating enough to alert victims to the presence of an unusual situation.

In the event of a significant exposure, the main threat from these materials is pulmonary edema (also known as dry drowning, because the victim's lungs fill with fluid). When inhaled, choking agents damage sensitive lung tissue and cause fluid to be released by the injury. That fluid results in chemically induced pneumonia, which is fatal (over time) in many cases. Additionally, these agents act as skin irritants and can cause mild to moderate skin irritation and significant burning when victims are exposed to them in high concentrations.

■ Irritants (Riot-Control Agents)

<u>Irritants</u> (also known as riot-control agents) include substances such as mace that can be dispersed to briefly incapacitate a person or groups of people. Irritants chiefly cause pain and a burning sensation in exposed skin, eyes, and mucous membranes. The onset of symptoms occurs within seconds after contact and lasts several minutes to several hours, usually with no lasting effects. Irritants can be dispersed from canisters, hand-held sprayers, and grenades.

From a terrorism perspective, irritants can be employed to incapacitate rescuers or to drive a group of people into another area where a more dangerous substance can be released. Of all the groups of WMDs, irritants pose the least amount of danger in terms of toxicity. Exposed patients can be decontaminated with clean water, and the residual effects of the exposure are rarely significant. Fire fighters are also likely to encounter these agents when they respond to calls in which law-enforcement personnel have used these agents for riot control or suspect apprehension. It is important for fire fighters to know which specific agents are used by local law-enforcement agencies and any specific decontamination techniques recommended for use with those agents.

■ Convulsants

Chemicals classified as <u>**convulsants**</u> are capable of causing convulsions or seizures when absorbed by the body. Chemicals that fall into this classification include nerve agents such as sarin, soman, tabun, VX, and the organophosphate and carbamate classes of pesticides. These substances interfere with the central nervous system and disrupt normal transmission of neuromuscular impulses throughout the body. Pesticides capable of functioning as convulsants include parathion, aldicarb, diazinon, and fonofos.

When dealing with these materials, it is important to identify their presence and to avoid breathing vapours or allowing liquid to contact skin. In some cases, even a small exposure can be fatal.

Harmful Substances' Routes of Entry into the Human Body

For a chemical or other harmful substance to injure a person, it must first get into his or her body. Throughout the course of his or her fire service career, a fire fighter will inevitably be bombarded by harmful substances, such as diesel exhaust, bloodborne pathogens, smoke, and accidentally and intentionally released chemicals. Fortunately, most of these exposures are not immediately deadly. Unfortunately, repetitive exposures to these materials can have a negative health effect over a 20-year career. To protect yourself now and give yourself the best shot at a healthy retirement and long life, it is important to understand some basic concepts about toxicology and the four ways that chemicals enter the body **FIGURE 30-13** :

1. Inhalation—Through the lungs
2. Absorption—By permeating the skin
3. Ingestion—Via the gastrointestinal tract
4. Injection—Through cuts or other breaches in the skin

A.

B.

C.

D.

FIGURE 30-13 The four ways a chemical substance can enter the body. **A.** Inhalation. **B.** Absorption. **C.** Ingestion. **D.** Injection.

Toxicology is the study of the adverse effects of chemical or physical agents on living organisms. The following sections discuss the potential routes of entry and methods fire fighters can employ to protect themselves against these agents.

■ Inhalation

Inhalation exposures occur when harmful substances are brought into the body through the respiratory system. The lungs are a direct point of access to the bloodstream, so they can quickly transfer an airborne substance into the bloodstream.

The entire respiratory system is vulnerable to attack from most hazardous materials and WMDs; corrosive materials, such as chlorine and ammonia; solvent vapours, such as gasoline and acetone; superheated air; and any other material finding its way into the air. In addition to gases and vapours, small particles of dust and smoke can lodge in sensitive lung tissue, causing substantial irritation.

Given these facts, it is imperative that fire fighters always consider using respiratory protection when they are operating in the presence of airborne contamination. Fortunately, fire fighters have ready access to an excellent form of respiratory protection—namely, the positive pressure open-circuit SCBA. This equipment is by far the single most important piece of PPE that fire fighters have at their disposal—so use it! Sometimes fire fighters might need to use other forms of respiratory protection, depending on the specific respiratory hazard they are facing while performing a given task—for example, full-face and half-face air-purifying respirators (APRs) or battery-powered APRs that offer specific degrees of protection if the chemical hazard is known and the appropriate filter canister is used **FIGURE 30-14**.

Respirators do not provide oxygen or breathing air, however; thus, if the oxygen content of the work area is low,

FIGURE 30-14 Air-purifying respirators offer some degree of protection against airborne chemical hazards.
Courtesy of Survivair

respirators are not a viable option. According to the Canadian Centre for Occupational Health and Safety (CCOHS), any work environment containing less than 18 percent oxygen (or 19.5 percent in some jurisdictions) is considered to be oxygen deficient and requires the use of an air-supplying respirator. Another drawback to APRs is that filters can be costly and are often needed in large quantities, depending on the nature of the response.

Respirators are lighter than SCBA, are more comfortable to wear, and usually allow for longer work periods because they do not depend on a limited source of breathing air. SCBA certainly offer a higher level of respiratory protection, but in the right circumstances, APRs represent a viable alternative.

TABLE 30-2	Particle Sizes of Common Types of Respiratory Hazards
Fume	< 1 micron
Smoke	Up to 1 micron
Dust	1 micron and up
Fog	< 40 microns
Mist	> 40 microns

When considering protection against airborne contamination, it is important to understand the origin of the contamination, its concentration, and the impact that the contamination might potentially have on oxygen levels in the area. In short, determining the appropriate type of respiratory protection must be accomplished by looking at the overall situation, including the nature of the contaminant. In some cases, the anticipated particle size of the contamination determines the level of respiratory protection employed **TABLE 30-2** .

Anthrax spores offer an excellent example to illustrate this point. Weaponized anthrax spores typically vary in size from 0.5 to 1 micron. Given that range, a typical full-face APR with a nuisance dust filter would not offer sufficient protection against this hazard. Anyone operating in an area contaminated with anthrax should wear SCBA or, at minimum, an APR with HEPA filtration, which offers protection from 0.3-micron or larger particles.

Now consider this scenario: a container of cryogenic nitrogen is leaking inside a poorly ventilated storage room. The main threat is the cold vapour produced and the possibility of oxygen deficiency. In this case, SCBA is the appropriate type of respiratory protection, based on the anticipated hazard. In many cases, hazardous materials technicians might need to respond; they will use air-monitoring devices to characterize the work area prior to entry. The choices made regarding respiratory protection will then be based on definitive and known information.

Particle size also determines where the inspired contamination ultimately ends up; bigger particles that make up visible mists are captured in the nose and upper airway, for example, whereas smaller particles work their way deeper into the lungs

TABLE 30-3	Location of Respiratory Trapping by Particle Size
> 7 microns	Nose
5–7 microns	Larynx
3–5 microns	Trachea and bronchi
2–3 microns	Bronchi
1–2.5 microns	Respiratory bronchioles
0.5–1 micron	Alveoli

TABLE 30-3 . Gaseous substances, such as carbon monoxide and hydrogen cyanide, can diffuse into the bloodstream and affect entire body systems.

Respiratory protection is one of the most important components of any fire fighter's PPE. In all cases where airborne contamination is encountered, take the time to understand the nature of the threat, evaluate the respiratory protection available, and decide whether it provides adequate protection. When protecting the lungs, "good enough" is not an option.

■ Absorption

The largest organ in the body, the skin, is susceptible to damage by a number of substances. In addition to serving as the body's protective shield against heat, light, and infection, the skin helps to regulate body temperature, stores water and fat, and serves as a sensory centre for painful and pleasant stimulation. Without this important organ, humans would not be able to survive.

When discussing chemical exposures, however, absorption is not limited only to the skin. **Absorption** is the process by which hazardous materials travel through bodily tissues until they reach the bloodstream. The eyes, nose, and mouth—that is, any mucous membrane—and, to a certain degree, the intestinal tract are all part of this equation; the eyes, for example, absorb a high amount of liquid and vapour that comes into contact with the sensitive tissues. This absorption is particularly problematic, because the eyes connect directly to the optic nerve, which provides a direct route to the brain and the central nervous system.

Although the skin functions as a shield for the body, that shield can be pierced by a number of chemicals; aggressive solvents, such as methylene chloride (found in paint stripper), for example, can be readily absorbed through the skin. A secondary hazard associated with this chemical occurs when the body attempts to metabolize the methylene chloride after it has been absorbed. A by-product of that metabolic reaction is carbon monoxide, a cellular asphyxiant. **Asphyxiants** are substances that prevent the body from using oxygen, thus causing suffocation. In this scenario, the initial chemical is broken down into another substance that poses a greater health hazard than the original chemical. Methylene chloride is also suspected to be a human **carcinogen**.

Many other hydrocarbon-based substances can be absorbed through the skin and ultimately enter the bloodstream. A hydrocarbon is a substance consisting of carbon and hydrogen atoms, among other components. Examples include gasoline, diesel fuel, benzene, and other solvents.

Absorption hazards are not limited to hydrocarbons, however; hydrofluoric acid, for example, poses a significant threat to life when it is absorbed through the skin. This unique corrosive has the ability to bind with certain substances in the body (predominantly calcium). Secondary health effects occurring after exposure can include muscular pain and potentially lethal cardiac arrhythmias.

In the field, fire fighters must constantly evaluate the possibility of chemical contact with the skin and eyes. In many cases, structural firefighting turnout gear provides little or no protection against liquid chemicals. Consult the *ERG* or an SDS for response guidance when deciding whether turnout gear is appropriate, given the hazards at hand.

In the event that turnout gear does not offer adequate protection, fire fighters might have to increase their level of protection to include specialized chemical-protective clothing. Keep in mind that the operations level of training is primarily intended to allow fire fighters to take defensive actions while avoiding contact with the released product. If they have to wear something other than turnout gear, they might be exceeding their level of training. The use of SCBA and proper protective firefighting gear can reduce many of the standard routes of exposure—but not always! Refer to your fire department's SOGs/SOPs for guidance in such situations.

■ Ingestion

In addition to absorption through the skin, chemicals can be brought into the body through the gastrointestinal tract by the process of **ingestion**. The water, nutrients, and vitamins the body needs are predominantly absorbed in this manner. This is an uncommon route of significant chemical exposure for most people—except for fire fighters. For example, at a structure fire, fire fighters generally have an opportunity to rotate out of the building for rest and refreshment. If they do not take the time to wash up prior to eating or drinking, they have a high probability of spreading contamination from the hands to the food and, subsequently, to the intestinal tract. If you do not think about every situation where you might become exposed, you put yourself in harm's way.

■ Injection

Chemicals brought into the body through open cuts and abrasions qualify as **injection** exposures. Protecting yourself from this route of exposure starts with realizing that you are going to work in a compromised state. Any cuts or open wounds should be addressed before reporting for duty. If they are significant, you might be excluded from operating in contaminated environments. Open wounds are a direct portal to the bloodstream and, subsequently, to muscles, organs, and other bodily systems. If a chemical substance contacts this open portal, the health effects could be immediate and pronounced. Remember: Intact skin is a good protective shield. Do not go into battle if your shield is not up to the task.

Chronic versus Acute Health Effects

A **chronic health hazard**, or chronic health effect, is an adverse health effect to a substance that occurs gradually over time. Chronic health effects can occur either after long-term exposures or after multiple short exposures over a shorter period. The exposures can occur by all routes of entry into the body and can result in health issues such as cancer, permanent loss of lung function, or repetitive skin rashes; for example, inhaling asbestos fibres for years without respiratory protection can result in a form of lung cancer called asbestosis.

Chemicals that pose a hazard to health after only relatively short exposure periods are said to cause **acute health effects**. Such an exposure period is considered to be subchronic, with effects occurring either immediately after a single exposure or up to several days or weeks after an exposure. Essentially, acute exposures are "right now" exposures that produce some observable condition such as eye irritation, coughing, dizziness, or skin burns.

Skin irritation and burning after a dermal exposure to sulphuric acid are classified as acute health effects. Other chemicals, such as formaldehyde, are also capable of causing acute health effects. Among other things, formaldehyde is a sensitizer that is capable of causing an immune system response when it is inhaled or absorbed through the skin. This reaction is similar to other allergic reactions, in that an initial exposure produces mild health effects but subsequent exposures cause much more severe reactions such as breathing difficulties and skin irritation. Health Canada's Workplace Hazardous Materials Information System (WHMIS) defines **sensitizer** as a material that can cause severe skin and/or respiratory responses in a sensitized person after exposure to a very small amount of the material. Sensitization develops over time: when a person is first exposed, there may be no obvious reaction; however, repeated exposure can lead to increasingly severe reactions.

Using the *Emergency Response Guidebook*

Probably one of the most frequently consulted hazardous materials references used by emergency response agencies is the *Emergency Response Guidebook* (*ERG*). The *ERG* is jointly developed by Transport Canada, the DOT, and Mexico's Secretaria de Comunicaciones y Transportes, as well as the Centro de Información Química para Emergencias (CIQUIME) of Argentina. It is designed to provide basic information to keep both responders and the general public safe during the initial stages of a hazardous materials incident.

The *ERG* is divided into four major sections, plus some additional reference materials:

- Yellow section—Lists hazardous materials in numerical order by the chemical identification number. This identification number is followed by a guide number and the chemical name.
- Blue section—Lists hazardous chemicals in alphabetical order. The chemical name is followed by the guide number and the four-digit chemical identification number.

- Orange section—The most important section of the *ERG*. Its 62 guides provide basic response information for specific groups of chemicals—namely, potential hazards (health and fire/explosion), public safety (general information regarding chemical hazards, protective clothing requirements, and evacuation needs), and emergency response (fire situations, spill or leak situations, and first-aid requirements).

- Green section—Provides information on the evacuation/isolation distance for specific chemicals. This section lists the chemical identification number, the chemical name, and isolation and protection distances for both small and large spill situations.

- White pages—List reference resources, including who to call for assistance, placarding information, rail and highway vehicle identification charts, and a glossary of key terms.

Emergency responders also use numerous other chemical reference manuals. For example, the *Fire Fighter's Handbook of Hazardous Materials*, 7th edition (Jones and Bartlett Publishers, Inc., 2006), is similar to the *ERG*. This handbook is divided into three major sections: The first section lists chemicals in alphabetical order and provides a listing of the basic properties for each chemical in a spreadsheet format. The second section provides emergency guides and is very similar to the guides section of the *ERG*. The third section lists chemicals by the DOT identification number.

To use the *ERG*, follow the steps in **SKILL DRILL 30-1**:

1 Identify the chemical name and/or the chemical identification number for the suspect material.

2 Look up the material name in the appropriate section. Use the yellow section to obtain information by the chemical identification number or the blue section to obtain information by the chemical name.

3 Determine the correct emergency action guide to use for the chemical identified.

4 Identify the isolation distance and the protective actions required for the chemical identified.

5 Identify the potential hazards of the chemical identified.

6 Identify the emergency response actions of the chemical identified.

Chief Concepts

- A hazardous material is a material that poses an unreasonable risk to the health and safety of operating emergency personnel, the public, and/or the environment if it is not properly controlled during handling, storage, manufacture, processing, packaging, use and disposal, or transportation.
- All fire fighters must be able to recognize the presence of hazardous materials, analyze information, and take appropriate action.
- The ability to recognize a potential hazardous material incident is critical to ensuring your own safety.
- To avoid entering a dangerous area, you must be alert to the possibility that hazardous materials are present at the scene.
- At a hazardous materials incident, you must attempt to identify and isolate the released material, and then prevent and address any injuries or exposures encountered. After any life-safety hazards are addressed, you must decide which level of action and protection is required to alleviate or resolve the incident.
- Hazardous materials can be found anywhere. At sites ranging from hospitals to petrochemical plants, pure chemicals and chemical mixtures are used to create millions of consumer products. In addition, manufacturing processes sometimes generate hazardous waste.
- Standards are issued by nongovernmental entities, such as the NFPA, and are generally consensus based. The NFPA has two current standards that pertain to hazardous materials incidents: NFPA 472 and NFPA 473.
- Any person who comes upon a hazardous materials incident and has been trained to recognize it, identify it, and notify the appropriate agencies is considered to be operating at the awareness level.
- Operations-level responders respond to hazardous materials and WMD incidents for the purpose of protecting nearby persons, the environment, or property from the effects of the release.
- Technician-level personnel are trained to enter heavily contaminated areas using the highest levels of personal protection.
- Specialist-level personnel have a similar level of training as technician-level personnel, but have additional training related to specific chemicals, container types, or transportation modes.
- Hazardous materials officers are trained to assume command of a hazardous materials incident beyond the operations level.

- Hazardous materials safety officers are responsible for ensuring the safety of hazardous materials personnel and the use of appropriate hazardous materials and WMD practices.
- Hazardous materials incidents cannot be approached in the same manner as a structural fire. These incidents are more complex and structural firefighting gear does not provide adequate protection from hazardous materials.
- The actions taken at hazardous materials incidents are dictated largely by the chemicals involved and the resources available.
- A hazardous materials response begins with preincident planning.
- The first step in understanding the hazard posed by a substance is to identify its state of matter—solid, liquid, or gas.
- A physical change occurs when a substance is subjected to outside influences such as heat, cold, or pressure; for example, when water is exposed to cold, it becomes ice. A chemical change is a change in the substance's chemical makeup with a release of some form of energy; for example, if oxygen is removed from water (H_2O), the chemical makeup of the substance changes (H).
- The boiling point is the temperature at which a liquid eventually turns to gas. Flammable liquids with low boiling points are dangerous because of their potential to produce large volumes of flammable vapour when exposed to relatively low temperatures.
- The three aspects to consider when evaluating the fire potential of a chemical are as follows:
    1. Flash point—An expression of the temperature at which a liquid fuel gives off sufficient vapour that, when an ignition source is present, results in a flash fire.
    2. Ignition temperature—When a liquid fuel is heated beyond its ignition temperature, the substance ignites without an external ignition source.
    3. Flammable range—Reflects an amount of vapour mixed with a volume of air, defined by UFL and LFL; the wider the flammable range, the more dangerous the material.
- When operating in any environment where a flammable or combustible vapour cloud is suspected to be present, always err on the side of safety and assume that the danger of ignition exists.
- In addition to addressing the flammability of a particular vapour or gas, fire fighters must determine the vapour density (or weight) of the substance. To find a chemical's vapour density, consult a reference source, such as the SDS.

- In the absence of reliable reference sources in the field, fire fighters can use the HA HA MICEN mnemonic to remember a number of lighter-than-air gases:

  H—Hydrogen
  A—Acetylene
  H—Helium
  A—Ammonia
  M—Methane
  I—Illuminating gas (neon and hydrogen cyanide)
  C—Carbon monoxide
  E—Ethylene
  N—Nitrogen

- The key point to understanding vapour pressure is this: the vapours released from the surface of the liquid must be contained if they are to exert any pressure. Essentially, the liquid inside a container will vaporize until the molecules given off by the liquid reach equilibrium with the liquid itself. It is a balancing act between the liquid and the vapours, in which some liquid molecules turn into vapour, even as other molecules leave the vapour phase and return to the liquid phase.

- Always remember that when dealing with chemical emergencies, water might not always be the best or safest choice to mitigate the situation. It is a tremendously aggressive solvent and has the ability to react violently with certain other chemicals.

- Corrosivity is the ability of a material to cause damage to the body on contact. Corrosives constitute a complex group of chemicals and fall into two classes: acids and bases. Acids have pH values less than 7; bases have pH values greater than 7. Generally, substances with pH values of 2.5 and lower and 12.5 and higher are considered to be strong.

- Radiation is the energy transmitted through space in the form of electromagnetic waves or energetic particles. It can be described as ionizing or non-ionizing. Ionizing radiation has the potential to ionize an atom or molecule through atomic interactions. Non-ionizing radiation does not carry enough energy to remove an electron from an atom or molecule. Radiation is everywhere and has been around since the beginning of time. For the most part, the health hazards posed by radiation are a function of two factors:

  1. The amount of radiation absorbed by the body has a direct relationship to the degree of damage done.
  2. The amount of exposure time ultimately determines the extent of the injury.

- Fire fighters can protect themselves from alpha-particle exposure by wearing HEPA filters. Fire fighters can protect themselves from beta-particle exposure by wearing SCBA. Structural firefighting gear and SCBA do not protect fire fighters from gamma rays.

- The residue from a chemical is contamination. Secondary contamination occurs when a person or object transfers the contamination to another person or object by direct contact. To prevent this, decontamination is necessary.

- The seven categories of WMDs are as follows:

  1. Thermal—Heat created from intentional explosions or fires, or cold generated by cryogenic liquids
  2. Radiological—Radioactive contamination from dirty bombs; alpha, beta, and gamma radiation
  3. Asphyxiation—Oxygen deprivation caused by materials such as nitrogen; tissue asphyxiation from blood agents
  4. Chemical—Injury and death caused by the intentional release of toxic industrial chemicals, nerve agents, vesicants, poisons, or other chemicals
  5. Etiological—Illness and death resulting from biohazards, such as anthrax, plague, and smallpox; hazards posed by bloodborne pathogens
  6. Mechanical—Property damage and injury caused by explosion, falling debris, shrapnel, firearms, and explosives, as well as slips, trips, and falls
  7. Psychogenic—The mental harm from being potentially exposed to, contaminated by, and even just being in close proximity to an incident of this nature

- Nerve agents attack the body's central nervous system. The signs and symptoms of nerve agent exposure can be remembered through the mnemonic SLUDGEM:

  S—Salivation
  L—Lachrymation (tearing)
  U—Urination
  D—Defecation
  G—Gastric disturbance
  E—Emesis (vomiting)
  M—Miosis (constriction of the pupils)

- Blister agents include sulphur mustard and lewisite. Immediate skin decontamination is required.

- Cyanide compounds prevent the body from using oxygen. The typical signs and symptoms include vomiting, dizziness, watery eyes, and deep and rapid breathing.

- Choking agents are mostly designed to incapacitate rather than to kill. Irritants can be dispersed to briefly incapacitate a person or groups of people.

- Convulsants are capable of causing seizures. They include nerve agents, such as sarin, soman, and VX. It is critical to identify their presence and avoid exposure.

- The four ways that chemicals enter the body are as follows:

  1. Inhalation—Through the lungs
  2. Absorption—By permeating the skin

3. Ingestion—Via the gastrointestinal tract
4. Injection—Through cuts or other breaches in the skin

■ A chronic health hazard is an adverse health effect that occurs gradually over time. Acute health effects occur after only relatively short exposure periods.

■ One of the most frequently consulted hazardous materials references used by emergency response agencies is the *ERG*. It is designed to provide basic information to keep both responders and the general public safe during the initial stages of a hazardous materials incident. The *ERG* is divided into four major sections, plus some additional reference materials:

- Yellow section—Lists hazardous materials in numerical order by the chemical identification number
- Blue section—Lists hazardous chemicals in alphabetical order
- Orange section—Provides basic response information for specific groups of chemicals
- Green section—Provides information on the evacuation/isolation distance for specific chemicals
- White pages—List reference resources, including who to call for assistance, placarding information, rail and highway vehicle identification charts, and a glossary of key terms

Wrap-Up, continued

Hot Terms

Absorption The process by which hazardous materials travel through bodily tissues until they reach the bloodstream.

Acid A material with a pH value of less than 7.

Acute health effect A health problem caused by relatively short exposure periods to a harmful substance, which produces observable conditions such as eye irritation, coughing, dizziness, and skin burns.

Alpha particle A positively charged particle emitted by certain radioactive materials, identical to the nucleus of a helium atom. (NFPA 801)

Asphyxiant A material that causes a victim to suffocate.

Awareness level: (29 CFR 1910.12 First Responder at the Awareness Level) Personnel who, in the course of their normal duties, could encounter an emergency involving hazardous materials and weapons of mass destruction (WMDs) and who are expected to recognize the presence of the hazardous materials and WMDs, protect themselves, call for trained personnel, and secure the scene. (NFPA 472)

Base A material with a pH value greater than 7.

Beta particle An elementary particle emitted from a nucleus during radioactive decay, with a single electrical charge and a mass equal to 1/1837 that of a proton. (NFPA 801)

Blister agent A chemical that causes the skin to blister.

Blood agent A chemical that, when absorbed by the body, interferes with the transfer of oxygen from the blood to the cells.

Boiling point The temperature at which the vapour pressure of a liquid equals the surrounding atmospheric pressure. (NFPA 30)

Carcinogen A cancer-causing substance identified in one of several published lists, including, but not limited to, the *NIOSH Pocket Guide to Chemical Hazards* (Centers for Disease Control and Prevention 2007), *Hazardous Chemicals Desk Reference*, 6th edition (Wiley-Blackwell, 2008), and *2013 TLVs® and BEIs®* (American Conference of Governmental Industrial Hygeienists 2013). (NFPA 1851)

Census metropolitan area (CMA) An area consisting of one or more adjacent municipalities positioned around a major urban hub.

Chemical change The ability of a chemical to undergo an alteration in its chemical makeup, usually accompanied by the release of some form of energy.

Chlorine A yellowish gas that is about two-and-a-half times heavier than air and slightly water soluble. Chlorine has many industrial uses but also damages the lungs when inhaled; it is a choking agent.

Choking agents A chemical designed to inhibit breathing. It is typically intended to incapacitate rather than kill.

Chronic health hazard A health problem that occurs as a result of long-term exposure to a substance; also known as a chronic health effect.

Contamination The process of transferring a hazardous material from its source to people, animals, the environment, or equipment, all of which can act as carriers for the material.

Convulsant A chemical capable of causing convulsions or seizures when absorbed by the body.

Corrosivity The ability of a material to cause damage (on contact) to skin, eyes, or other parts of the body.

Expansion ratio A description of the volume increase that occurs when a liquid material changes to a gas.

Exposure The process by which people, animals, the environment, and equipment are subjected to or come in contact with a hazardous material. (NFPA 472)

Fire point The lowest temperature at which a liquid will ignite and achieve sustained burning when exposed to a test flame in accordance with ASTM D92, *Standard Test Method for Flash and Fire Points by Cleveland Open Cup Tester*. (NFPA 704)

Flammable range The range of concentrations between the UFL and LEL. (NFPA 68)

Flammable vapour A concentration of constituents in air that exceeds 10 percent of its LFL. (NFPA 115)

Flash point The minimum temperature of a liquid at which sufficient vapour is given off to form an ignitable mixture with the air, near the surface of the liquid, or within the vessel used. (NFPA 30)

Gamma radiation A type of radiation that can travel significant distances, penetrating most materials and passing through the body. Gamma rays are the most destructive type of radiation to the human body.

Hazard The potential for harm or damage to people, property, or the environment. (NFPA 1710)

Hazardous material A substance that, when released, is capable of causing harm to people, the environment, and property. (NFPA 472)

Hazardous materials branch director/group supervisor Commanders of hazardous materials incidents beyond the operations level.

Hazardous materials officer The person who is responsible for directing and coordinating all operations involving hazardous materials and weapons of mass destruction as assigned by the incident commander. (NFPA 472)

Hazardous materials safety officer The person who is responsible for ensuring the safety of hazardous materials personnel and ensuring that appropriate hazardous materials/weapons of mass destruction practices are followed.

Hazardous materials technician with specialties A hazardous materials technician with training in areas such as specialized chemicals, containers, and their uses.

Hazardous waste Waste that is potentially damaging to the environment or human health, due to its toxicity, ignitability, corrosivity, or chemical reactivity, or another cause. (NFPA 80)

High-efficiency particulate air (HEPA) filter A filter capable of catching particles as small as 0.3 micron—much smaller than a typical dust of alpha radiation particle.

Ignition temperature The minimum temperature at which a fuel, when heated, will ignite in air and continue to burn; also known as auto-ignition temperature.

Incident commander (IC) The person who is responsible for all decisions relating to the management of an incident and is in charge of the incident site. (NFPA 1500)

Ingestion Exposure to a hazardous material by swallowing it.

Inhalation Exposure to a hazardous material by breathing it into the lungs.

Injection Exposure to a hazardous material through cuts or other breaches in the skin.

Ionizing radiation Radiation of sufficient energy to alter the atomic structure of materials or cells with which it interacts, including electromagnetic radiation, such as X-rays, gamma rays, and microwaves, and particulate radiation, such as alpha and beta particles. (NFPA 1991)

Irritant A substance such as mace that can be dispersed to briefly incapacitate a person or groups of people.

Lewisite An oily, colourless to dark brown, liquid blister agent, with an odour of geraniums.

Lower flammable limit (LFL) The minimum concentration of combustible vapour or gas in a mixture of the vapour or gas and gaseous oxidant above which propagation of flame will occur on contact with an ignition source. (NFPA 115)

National Fire Protection Association (NFPA) A private organization that develops and maintains nationally recognized minimum consensus standards on many areas of fire safety as well as specific standards on hazardous materials.

Nerve agent A toxic substance that attacks the central nervous system in humans.

Neutron A penetrating particle found in the nucleus of an atom that is removed through nuclear fusion or fission. Although neutrons are not radioactive, exposure to neutrons can create radiation.

Operations level Persons who respond to hazardous materials/weapons of mass destruction incidents for the purpose of implementing or supporting actions to protect nearby persons, the environment, or property from the effects of the release. (NFPA 472)

Occupational Safety and Health Administration (OSHA) The federal agency that regulates worker safety and, in some cases, responder safety. OSHA is a part of the United States Department of Labor.

pH A measure of the acidity or basic nature of a material; more technically, an expression of the concentration of hydrogen ions in the substance.

Physical change A transformation in which a material changes its state of matter—for instance, from a liquid to a solid.

Pulmonary edema Fluid buildup in the lungs.

Radiation The emission and propagation of energy through matter or space by means of electromagnetic disturbances that display both wave- and particle-like behaviour. (NFPA 801)

Radioactive isotope An atom that has unequal numbers of protons and neutrons in its nucleus and that emits radioactivity.

Radioactivity The spontaneous decay or disintegration of an unstable atomic nucleus, accompanied by the emission of radiation. (NFPA 801)

Safety Data Sheet (SDS) A form, provided by manufacturers and compounders (blenders) of chemicals, containing information about the chemical composition, physical and chemical properties, health and safety hazards, emergency response, and waste disposal of the material.

Sarin A nerve agent that, when dispersed, sends droplets into the air that, when inhaled, harm intended victims.

Secondary contamination The process by which a contaminant is carried out of the hot zone and contaminates people, animals, the environment, or equipment.

Sensitizer A chemical that causes a large portion of people (or animals) to develop an allergic reaction after repeated exposure to the substance.

Specific gravity The ratio of the weight of a given volume of gas to that of the same volume of air, both measured under the same conditions. (NFPA 54)

Specialist level Persons who respond to hazardous materials/weapons of mass destruction incidents who have received more specialized training than hazardous materials technicians. Most of the training that specialist employees receive is either product or transportation mode specific.

State of matter The physical state of a material—solid, liquid, or gas.

Sulphur mustard A clear, yellow, or amber oily liquid blister agent with a faint, sweet odour of mustard or garlic. Sulphur mustard can be dispersed in an aerosol form.

<u>Technician level</u> Persons who respond to hazardous materials/weapons of mass destruction incidents and are allowed to enter heavily contaminated areas using the highest levels of protection. This is the most aggressive level of training as identified in NFPA 472.

<u>Toxic</u> Any material whose properties contain the inherent capacity to produce injury to a biological system, which is dependent on concentration, rate, method, and site of absorption. (NFPA 306)

<u>Toxicology</u> The study of the adverse effects of chemical or physical agents on living organisms.

<u>Toxic product of combustion</u> A hazardous chemical compound released when a material decomposes under heat.

<u>TRACEMP</u> An mnemonic to help remember the effects and potential exposures to a hazardous materials incident—that is, thermal, radiation, asphyxiant, chemical, etiologic, mechanical, and psychogenic.

<u>Upper flammable limit</u> The highest concentration of a combustible substance in a gaseous oxidizer that will propagate a flame. (NFPA 68)

<u>Vapour</u> The gaseous state of a substance, particularly of those that are normally liquids or solids at ordinary temperatures. (NFPA 92)

<u>Vapour density</u> The weight of an airborne concentration (vapour or gas) as compared to an equal volume of dry air.

<u>Vapour pressure</u> The pressure, measured in kilopascals or psi absolute (psia), exerted by a liquid, as determined by ASTM D323, *Standard Test Method for Vapor Pressure of Petroleum Products (Reid Method)*. (NFPA 385)

<u>VX</u> A nerve agent.

<u>Weapons of mass destruction (WMDs)</u> A. Any destructive device, such as any explosive, incendiary, or poison-gas bomb; grenade; rocket having a propellant charge of more than 113 grams (4 ounces); missile having an explosive or incendiary charge of more than 7 g (0.25 oz.); mine; or device similar to the above. B. Any weapon involving toxic or poisonous chemicals. C. Any weapon involving a disease organism. D. Any weapon that is designed to release radiation or radioactivity at a level dangerous to human life. (NFPA 472)

It is 1300 hours, and your engine company is dispatched to an incident that involves several leaking drums at a local self-storage facility near the elementary school. En route to the incident, you start thinking about the potential scene: Are children outside the school playing and running around? Is there pedestrian traffic on the street? What does this facility look like?

As your apparatus approaches the dispatched address, you see a white cloud appearing from the doorway of one of the storage lockers. Your officer immediately stops the apparatus and notifies dispatch with a report of what he has observed. He starts requesting additional resources.

1. Based on the type of incident, which level of hazardous materials training will be needed to mitigate it?
   A. Awareness
   B. Operations
   C. Technician
   D. Specialist

2. Which actions should the first-arriving engine company consider as first priority?
   A. Life safety of responders
   B. Evacuation
   C. Property conservation
   D. Environmental considerations

3. Which actions, if any, should awareness-level personnel take in the incident described?
   A. None
   B. Identification of product
   C. Notification of emergency/trained personnel
   D. Entry to assist in evacuation

4. Which actions, if any, should operations-level responders take in the incident described?
   A. Evacuation
   B. Entry to identify product
   C. Entry to stop leaking drum
   D. Defensive actions

Your engine company has been asked to give a presentation on a recent incident involving a suspected hazardous material. Your captain has assigned you the task of reporting on the activities and responsibilities of the hazardous materials technician during the incident. As you interview the rest of the company, which includes both Fire Fighter I and II members, you receive some suggestions on how to improve the safety of the operation.

1. Which safety considerations should be the primary concern of each responder on the scene of a hazardous materials incident?
2. As a Fire Fighter II, what are your responsibilities to ensure the safety of other responders?
3. Who should be responsible for the accountability of personnel in the hot zone?

Hazardous Materials: Awareness and Specialties

© Huntstock, Inc./Alamy Images

Fire Fighter I

Knowledge Objectives

- Describe how your senses can be used to detect the presence of hazardous materials safely. (NFPA 472, 5.1.2.2, 5.2.1, 6.6.1.1.2 , p 898)
- Describe how to identify a placard and label. (NFPA 472, 5.1.2.2, 6.6.1.1.2 , p 899)
- List the nine chemical families as outlined by the *Emergency Response Guidebook* (*ERG*). (NFPA 472, 5.2.2, 6.6.1.1.2 , p 899)
- Explain how to use the *ERG*. (NFPA 472, 5.2.2, 6.6.1.1.2 , p 899–900)
- Describe the NFPA 704, *Standard System for the Identification of the Hazards of Materials for Emergency Response*, hazard identification system. (NFPA 472, 5.2.2, 6.6.1.1.2 , p 900–901)
- List the information about a hazardous material found on Safety Data Sheets (SDSs). (NFPA 472, 5.2.1.5, 5.2.2, 6.6.1.1.2 , p 902)
- Explain when shipping papers are utilized. (NFPA 472, 5.2.1.5, 5.2.2, 6.6.1.1.2 , p 902)
- Describe the signs of a potential terrorist incident. (NFPA 472, 5.2.1.6, 6.6.1.1.2 , p 902–903)
- Identify the potential skin-contact hazards encountered at hazardous materials incidents. (NFPA 472, 5.3.3, 6.6.1.1.2, 6.6.3.2 , p 904)
- Explain how a hazardous material's threshold limit value (TLV) determines the level of protection required for responders. (NFPA 472, 5.3.3, 6.6.1.1.2, 6.6.3.2 , p 904)
- List the three categories of TLVs. (NFPA 472, 5.3.3, 6.6.1.1.2, 6.6.3.2 , p 904)
- List the subcategories of immediately-dangerous-to-life-or-health (IDLH) atmospheres. (NFPA 472, 5.3.3, 6.6.1.1.2, 6.6.3.2 , p 904)
- List and define the three basic types of atmospheres at a hazardous materials incident. (NFPA 472, 5.3.3, 6.6.1.1.2, 6.6.3.2 , p 904)
- List the categories of personal protective equipment (PPE). (NFPA 472, 5.3.3, 6.6.1.1.2, 6.6.3.2 , p 904–907)
- Describe the purpose and components of street clothing and work uniforms. (NFPA 472, 5.3.3, 6.6.1.1.2, 6.6.3.2 , p 905)
- Describe the purpose and components of structural firefighting protective clothing. (NFPA 472, 5.3.3, 6.6.1.1.2, 6.6.3.2 , p 905)
- Describe the purpose and components of high temperature–protective equipment. (NFPA 472, 5.3.3, 6.6.1.1.2, 6.6.3.2 , p 905–906)
- Describe the purpose and components of chemical-protective clothing and equipment. (NFPA 472, 5.3.3, 6.6.1.1.2, 6.6.3.2 , p 906)

- Describe the purpose and components of liquid splash–protective clothing. (NFPA 472, 5.3.3, 6.6.1.1.2, 6.6.3.2 , p 906)
- Describe the purpose and components of vapour-protective clothing. (NFPA 472, 5.3.3, 6.6.1.1.2, 6.6.3.2 , p 906–907)
- Discuss respiratory protection in a hazardous materials incident. (NFPA 472, 5.3.3, 6.6.1.1.2, 6.6.3.2 , p 907–908)
- Describe the levels of hazardous materials PPE. (NFPA 472, 5.3.3, 6.6.1.1.2, 6.6.3.2 , p 908–916)
- List the ratings of chemical-protective clothing. (NFPA 472, 5.3.3, 6.6.1.1.2, 6.6.3.2 , p 908–916)
- Describe the purpose and components of Level A protection. (NFPA 472, 5.3.3, 6.6.1.1.2, 6.6.3.2 , p 908–909)
- Describe the purpose and components of Level B protection. (NFPA 472, 5.3.3, 6.6.1.1.2, 6.6.3.2 , p 909)
- Describe the purpose and components of Level C protection. (NFPA 472, 5.3.3, 6.6.1.1.2, 6.6.3.2 , p 913–915)
- Describe the purpose and components of Level D protection. (NFPA 472, 5.3.3, 6.6.1.1.2, 6.6.3.2 , p 915–916)
- Define exposures in regard to a hazardous materials incident. (NFPA 472, 5.2.4, 6.6.1.1.2 , p 916–917)
- List two protective actions that may be taken in a hazardous materials incident. (NFPA 472, 5.4, 5.4.1, 6.6.1.1.2 , p 916–917)
- Describe the role of fire fighters during an evacuation. (NFPA 472, 5.4.1, 6.6.1.1.2 , p 917)
- Describe how and when a fire department conducts shelter-in-place operations with a local population. (NFPA 472, 5.4.1, 6.6.1.1.2 , p 917)

Skills Objectives

- Don a Level B encapsulating chemical-protective clothing ensemble (CPE). (NFPA 472, 5.3.3, 6.6.1.1.2, 6.6.3.2 , p 909–911)
- Don a Level B nonencapsulating CPE. (NFPA 472, 5.3.3, 6.6.1.1.2, 6.6.3.2 , p 912–913)
- Doff a Level B encapsulating CPE. (NFPA 472, 5.3.3, 6.6.1.1.2, 6.6.3.2 , p 909–912)
- Doff a Level B nonencapsulating CPE. (NFPA 472, 5.3.3, 6.6.1.1.2, 6.6.3.2 , p 912, 914)
- Don a Level C CPE. (NFPA 472, 5.3.3, 6.6.1.1.2, 6.6.3.2 , p 914–915)
- Doff a Level C CPE. (NFPA 472, 5.3.3, 6.6.1.1.2, 6.6.3.2 , p 915)
- Don a Level D CPE. (NFPA 472, 5.3.3, 6.6.1.1.2, 6.6.3.2 , p 916)
- Doff a Level D CPE. (NFPA 472, 5.3.3, 6.6.1.1.2, 6.6.3.2 , p 916)

© Photos.com

Fire Fighter II — FFII

Knowledge Objectives

There are no knowledge objectives for Fire Fighter II candidates. NFPA 1001 contains no Fire Fighter II Job Performance Requirements for this chapter.

Skills Objectives

There are no skills objectives for Fire Fighter II candidates. NFPA 1001 contains no Fire Fighter II Job Performance Requirements for this chapter.

Additional NFPA Standards

- NFPA 1951, *Standard on Protective Ensembles for Technical Rescue Incidents*
- NFPA 1981, *Standard on Open-Circuit Self-Contained Breathing Apparatus (SCBA) for Emergency Services*
- NFPA 1991, *Standard on Vapor-Protective Ensembles for Hazardous Materials Emergencies*
- NFPA 1992, *Standard on Liquid Splash–Protective Ensembles and Clothing for Hazardous Materials Emergencies*
- NFPA 1994, *Standard on Protective Ensembles for First Responders to CBRN Terrorism Incidents*

You Are the Fire Fighter

The pump and ladder from your station have been dispatched for reports of several visible leaks from railcars and highway cargo vehicles at an industrial park in your district. En route to the site, your officer receives an update that a strong chemical odour has been detected in the air near the park's main intersection. On one corner of the intersection is the entrance to the rail yard, which includes several tracks and nonpressurized tank cars; on the other corner is a trucking company, which has several different highway cargo tanker trucks; on the other side of the street is an industrial facility that stores large amounts of chemicals. Upon your company's arrival, your officer requests that you recognize and identify the area for potential indicators of the source of the chemical odour.

1. Which markings would you look for on the outside of the facility to indicate whether any potential hazardous materials are present at the site?
2. What placard information would you look for on the transportation vehicles, railcars, and highway cargo tanker trucks that would indicate the products they might carry and the hazards they might present?
3. Which other markings would you look for to indicate other potential hazards on site?

Introduction

Scene size-up is important in any emergency but is especially so in hazardous materials incidents. Situational awareness, or the ability to "read" the scene, is a critical skill. Fire fighters must be able to identify sensory clues to be useful and safe.

In hazardous materials incidents, it is not always possible to know the whole story before action is required. You may be able to identify the incident as involving hazardous materials based on information from the dispatcher, knowledge of the response area, and visual, auditory, or olfactory (odourous) clues. A leaking cargo trailer, coloured placards on abandoned drums, and certain odours are all indications that hazardous materials may be present. Departmental standard operating procedures (SOPs) in concert with suggested operating guidelines (SOGs) and your level of training should guide any initial actions. Following departmental SOGs/SOPs will help you take a consistent approach to the size-up of hazardous materials incidents as well as the material identification process. SOGs/SOPs serve as a road map to lead properly trained personnel to an end point.

The proper use of personal protective equipment (PPE) and proper scene control is also important at all emergencies; at a hazardous materials incident, however, PPE, scene control, site management, and personnel accountability are critical issues that have a direct bearing on the life safety of all responders charged with handling such incidents. The safe handling of the incident is often determined in the first 5 to 15 minutes, based on the actions of the fire fighters who are the initial responders. Mistakes made by those responders can lead to disaster.

This chapter focuses on how to identify and obtain necessary information on hazardous materials incidents in order to maintain the safety of everyone involved. Recognizing the potential for hazardous materials based on the labels or placards or on your knowledge of the response will help you take proper action at the scene.

Recognizing a Hazardous Materials Incident

A **hazardous material** is any material that poses an unreasonable risk of damage or injury to persons, property, or the environment if it is not properly controlled during handling, storage, manufacture, processing, packaging, use and disposal, or transportation. Recognizing a hazardous materials event, and properly identifying the materials involved, requires detective work. Fire personnel must take the time to look at the enitre scene, in an effort to identify critical visual indicators, and fit them into what is known about the problem.

If the dispatch information indicates that an incident involves hazardous materials, it may be a good idea to use binoculars and view the scene, upwind from a safe distance. Be sure to question anyone involved in the incident, such as the truck driver or the plant owner. Take time to scan the scene and interpret visual clues such as dead animals near the release, discoloured pavement, dead grass, visible vapours or puddles, or labels that might help identify the presence of a hazardous material.

■ Senses as a Hazardous Materials Detector

Another way to detect the presence of hazardous materials is to use your senses, although this must be done carefully to avoid exposure. Many hazardous materials can be recognized by smell, taste, or touch. Getting close enough to exercise these senses, however, may expose you to the hazard. The farther you are from the incident when you notice a problem, the safer you will be.

Clues that are seen or heard provide warning information from a distance, enabling you to take precautionary steps. Vapour clouds at the scene, for example, are a signal to move away to safety; the sound of an alarm from a toxic-gas cabinet is a warning to retreat. This chapter provides you with information on visual clues such as containers, placards, labels, and shipping information that signal possible hazardous materials incidents.

Facility and Transportation Markings and Colourings

Labels, placards, and other markings on buildings, packages, boxes, and containers often enable fire fighters to identify a spilled or released chemical. When used correctly, marking systems indicate the presence of a hazardous material and provide clues about the substance. This section provides an introduction to various marking systems in use. It does not cover all of the intricacies and requirements for every marking system; it will, however, acquaint you with the most common systems.

■ Dangerous Goods Safety Marks

Transport Canada's Transportation of Dangerous Goods (TDG) regulations employ a system of coded labels and placards similar to the United States **Department of Transportation (DOT) marking system** to identify dangerous goods in transit. The *Emergency Response Guidebook (ERG)*—developed jointly by Transport Canada, the DOT and Mexico's Secretaria de Comunicaciones y Transportes—is also part of this system and offers a certain amount of guidance for fire fighters operating at a hazardous materials incident.

Labels and Placards

Placards are diamond-shaped indicators (250 millimetres [10.75 inches] on each side) that must be placed on all four sides of highway transport vehicles, railroad tanker cars, and other forms of transportation carrying hazardous materials **FIGURE 31-1**. **Labels** are smaller versions (100-mm [4-in.] diamond-shaped indicators) of placards that are used on any side (except top or bottom) of individual boxes and smaller packages being transported. (A package containing radioactive materials must display a label on two opposite sides.)

Placards and labels are intended to give fire personnel a general idea of the hazard inside a particular container. A placard may identify the broad hazard class (e.g., explosive, flammable, corrosive) that a tanker contains, while the label on a box inside a delivery truck relates only to the potential hazard inside that package **FIGURE 31-2**.

The *ERG* can be used to identify types of threats during the initial phase of a hazardous materials incident. The book organizes chemicals into nine basic hazard classes, or families, the members of which exhibit similar properties:

- Class 1—Explosives
- Class 2—Gases
- Class 3—Flammable liquids
- Class 4—Flammable solids, spontaneously combustible materials, and water-reactive substances
- Class 5—Oxidizers
- Class 6—Poisons (including blood agents and choking agents)
- Class 7—Radioactive materials
- Class 8—Corrosives
- Class 9—Other regulated materials

There also is a Danger placard, which indicates that more than one hazard class is contained in the same load. The TDG marking system is a broad spectrum look at chemical hazards and a valuable resource to fire fighters.

Other Considerations

TDG regulations do not require that all chemical shipments be marked with placards or labels. In most cases, the package or tank must contain a certain amount of hazardous material before a placard is required; for example, the "150-kg Gross Mass Exemption" allows for the transport of up to 150 kilograms of a variety of potentially hazardous materials, including propane, gasoline, oxygen and fireworks, without safety marks.

In contrast, some chemicals are so hazardous that shipping any amount of them requires the use of labels or placards. These include most explosives and all poisonous gases, water-reactive solids, and high-level radioactive substances. A four-digit United Nations (UN) number may be required on some placards. This number identifies the specific material being shipped; a list of UN numbers is included in the *ERG*.

Using the *Emergency Response Guidebook*

The *ERG* is a preliminary action guide. It is useful during the initial 30 minutes of an incident, but it should not be used to create a comprehensive incident action plan (IAP).

When the *ERG* refers to a small spill, it means a leak from one small package, a small leak in a large container (up to a 208-litre [55-gallon] drum), a small cylinder leak, or any small leak, even one in a large package. A large spill is a large leak or spill from a larger container or package, a spill from a number of small packages, or anything from a 0.9-metric-ton (1-ton) container, tanker truck, or railcar. The *ERG* is divided into four coloured sections: yellow, blue, orange, and green.

- Yellow section—Chemicals in this section are listed numerically by their four-digit UN identification number; entry number 1017, for example, identifies chlorine. Use the yellow section when the UN number is known or can be identified. The entries include the name of the chemical and the emergency action guide number.

FIGURE 31-1 A dangerous goods safety mark placard.

Class 1: Explosives

Class 2: Gases

Class 3: Flammable Liquids

Class 4: Flammable Solids, Substances liable to spontaneous combustion; Substances that on contact with water emit flammable gases (water-reactive substances)

Class 5: Oxidizing Substances and Organic Peroxides

Class 6: Toxic Substances and Infectious Substances

Class 7: Radioactive Materials

Class 8: Corrosives

Class 9: Miscellaneous Products, Substances, or Organisms

FIGURE 31-2 TC hazardous materials warning labels.

- Blue section—Chemicals in the blue section are listed alphabetically by name. The entry will include the emergency action guide number and the identification number. The same information, organized differently, appears in both the blue and yellow sections.
- Orange section—This section contains the emergency action guides. Guide numbers are organized by general hazard class and indicate which basic emergency actions should be taken, based on the hazard class.
- Green section—This section is organized numerically by UN identification number and provides the initial isolation distances for specific materials. Chemicals included in this section are highlighted in the blue and yellow sections. Any materials listed in the green section are always extremely hazardous. This section also directs the reader to consult the tables listing **toxic inhalation hazard (TIH) materials**—gases and volatile liquids that are extremely toxic to humans and pose a hazard to health during transportation. In addition, the green section includes a table of water-reactive materials that produce toxic gases and the associated protective actions to be taken.

■ NFPA 704 Marking System

The TDG marking system is used when materials are being transported from one location to another. The National Fire Protection Association (NFPA) system—more specifically, the **NFPA 704 hazard identification system**—is designed for fixed-facility use. The use of the NFPA 704 marking system is purely voluntary in most jurisdictions, but local fire and building codes may require its use in some instances. Check with your code enforcement agencies for your local situation.

NFPA 704 diamonds are found on the outsides of buildings, on doorways to chemical storage areas, and on fixed storage tanks. Fire fighters can use the NFPA diamonds to determine a course of action at a hazardous material incident **FIGURE 31-3**.

The NFPA 704 hazard identification system uses a diamond-shaped symbol of any size, which is itself broken into four smaller diamonds, each representing a particular property or characteristic: The blue diamond at the nine o'clock position indicates the health hazard posed by the material. The top red diamond indicates flammability. The yellow diamond at the three o'clock position indicates reactivity. The bottom white diamond is used for special symbols and handling instructions.

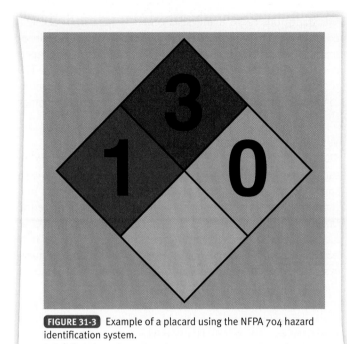

FIGURE 31-3 Example of a placard using the NFPA 704 hazard identification system.

TABLE 31-1	Hazard Levels in the NFPA 704 Hazard Identification System

Flammability Hazards

4	Materials that will rapidly or completely vapourize at atmospheric pressure and normal ambient temperature or that are readily dispersed in air and will burn readily. Liquids with a flash point below 22.8 degrees Celcius (73 degrees Fahrenheit) and a boiling point below 37.8°C (100°F).
3	Liquids and solids that can be ignited under almost all ambient temperature conditions. Liquids with a flash point below 22.8°C (73°F) and a boiling point above 37.8°C (100°F) or liquids with a flash point above 22.8°C (73°F), but not exceeding 37.8°C (100°F), and a boiling point below 37.8°C (100°F).
2	Materials that must be moderately heated or exposed to relatively high ambient temperatures before ignition can occur. Liquids with flash point above 37.8°C (100°F) but not exceeding 200°F (93°C).
1	Materials that must be preheated before ignition can occur. Liquids that have a flash point above 93.3°C (200°F).
0	Materials that will not burn.

Health Hazards

4	Materials that on very short exposure could cause death or major residual injury.
3	Materials that on short exposure could cause serious temporary or residual injury.
2	Materials that on intense or continued, but not chronic, exposure could cause incapacitation or possible residual injury.
1	Materials that on exposure would cause irritation but only minor residual injury.
0	Materials that on exposure under fire conditions would offer no hazard beyond that of an ordinary combustible material.

Reactivity Hazards

4	Materials that are readily capable of detonation, explosive decomposition, or reaction at normal temperatures and pressures.
3	Materials that are capable of detonation, explosive decomposition, or reactions but require a strong initiating source; that must be heated under confinement before initiation; or that react explosively with water.
2	Materials that readily undergo violent chemical change at elevated temperatures and pressures, that react violently with water, or that may form explosive mixtures with water.
1	Materials that are normally stable but can become unstable at elevated temperatures and pressures.
0	Materials that are normally stable, even under fire exposure conditions, and are not reactive with water.

Special Hazards

ACID	Acid
ALK	Alkali
COR	Corrosive
OX	Oxidizer
W̶	Reacts with water

The blue, red, and yellow diamonds will each contain a numerical rating of 0–4, with 0 being the least hazardous and 4 being the most hazardous for that type of hazard. The white quadrant will not hold a number but may contain special symbols. Among the symbols used are a burning *O* (oxidizing capability), a three-bladed fan (radioactivity), and a *W* with a slash through it (water reactive). **TABLE 31-1** provides a description of the numerical ratings in each category.

For more complete information on the NFPA 704 system, consult NFPA 704, *Standard System for the Identification of the Hazards of Materials for Emergency Response.*

■ Canadian Legislative Jurisdiction

Unlike OSHA, in the United States, there are constitutional jurisdictional divisions with regards to occupational health and safety. However, it is important to note that much of the legislation, regardless of whether it is federal, provincial, or territorial is based on similar, if not the same source information, such as NFPA standards and commonly used transportation markings and identification placarding.

Not only are firefighters required to follow occupational health and safety rules and regulations pertaining to their particular workplace, such as a fire hall, fire ground, or training ground, but they must be aware of the rules and regulations as they pertain to the transportation of hazardous materials. This knowledge is necessary in order to be able to contain and control any hazardous materials incident.

Other Reference Sources

Labels and placards may be helpful in identifying hazardous materials, but other sources of information are also available. Among these reference sources are Safety Data Sheets (SDSs), shipping papers, and staffed national resources such as the Canadian Transport Emergency Centre (CANUTEC) or CHEMTREC® and the National Response Center (NRC) in the United States. First-responding fire fighters should place a high priority on identifying the released material and finding a reliable source of information about the chemical and physical properties of the released substance. The most appropriate source will depend on the situation; use your best judgment in making your selection.

■ Safety Data Sheets

A commonly used source of information about a particular chemical is the SDS specific to that substance. Essentially, an SDS provides basic information about the chemical makeup of a substance, the potential hazards it presents, appropriate first aid in the event of an exposure, and other pertinent data for safe handling of the material. Generally, an SDS will include the following data:

- Physical and chemical characteristics
- Physical hazards of the material
- Health hazards of the material
- Signs and symptoms of exposure
- Routes of entry
- Permissible exposure limits
- Responsible party contact
- Precautions for safe handling, including hygiene practices, protective measures, and procedures for cleaning up spills or leaks
- Applicable control measures, including PPE
- Emergency and first aid procedures
- Appropriate waste disposal

When responding to a hazardous materials incident at a fixed facility such as a factory or plant, fire fighters should ask the site manager for an SDS for the spilled material. All facilities that use or store chemicals are required by law to have an SDS on file for each chemical used or stored in the facility. Although the SDS is not a definitive response tool, it is a piece of the puzzle. Fire fighters should investigate as many sources as possible (preferably at least three) to obtain emergency information about a released substance.

Shipping papers are required whenever materials are transported from one place to another. They include the names and addresses of the shipper and the receiver, identify the material being shipped, and specify the quantity and weight of each part of the shipment. Shipping papers for road and highway transportation are called **bills of lading** or **freight bills** and are located in the cab of the vehicle. Drivers transporting chemicals are required by law to have a set of shipping papers on their person or within easy reach inside the cab at all times.

A bill of lading may include additional information about a hazardous substance, such as its packaging-group designation. The packaging-group designation is also used by shippers to identify special handling requirements or hazards. Some hazard classes require shippers to assign packaging groups based on the material's flash point and toxicity. A packaging-group designation might, for example, signal that the material poses a greater hazard than similar materials in a hazard class. Three packaging group designations are used:

- Packaging group I—high danger
- Packaging group II—medium danger
- Packaging group III—minor danger

Shipping papers for railroad transportation are called **waybills**; the list identifying every car on the train is called a **consist**. The conductor, engineer, or a designated member of the train crew will have a copy of the consist when the train is travelling in main line service. They may have the waybill as well, but it is more often used in the rail yards. On a marine vessel, shipping papers are called the **dangerous cargo manifest**.

The dangerous cargo manifest is generally kept in a tube-like container in the wheelhouse, in the custody of the captain or master. For air transport, the **air bill** is the shipping paper; it is kept in the cockpit and is the pilot's responsibility. The responsible person in each situation should maintain the information about hazardous cargo and provide it in an emergency situation.

Where to Find a Safety Data Sheet

SDSs are stored in a several different formats, depending on the number of chemicals stored within a facility. Facilities with small chemical inventories typically store the SDS forms in a three-ring binder. In some cases, fire departments require the binder to be stored in a lockbox located near an entrance. In facilities with large chemical inventories, the SDS forms may be stored in filing cabinets or in multiple binders in a library-system format. In some cases, large SDS libraries may be stored electronically on a computer system with a paper backup.

Potential Terrorist Incidents

■ Chemical versus Biological Incidents

The biggest difference between a chemical incident and a biological incident is the speed of onset of health effects from the involved agents. Chemical incidents are typically characterized by a rapid onset of symptoms (usually minutes to hours after exposure to the chemical agent). Such incidents also usually have some type of easily identifiable signature, such as an odour, liquid or solid residue, or dead insects or foliage. Biological incidents, by comparison, feature a delayed onset of symptoms (usually days to weeks after the initial exposure). Because most biological agents are odourless and colourless, there are typically no outward indicators that the agents have been spread.

■ Chemical Agents

Indicators of possible criminal or terrorist activity involving chemical agents may vary depending on the complexity of the operation. The perpetrators may have left behind protective equipment such as rubber gloves, chemical suits, and respirators. Chemical containers of various shapes and sizes may be present, especially glass containers. The chemicals that are involved in the incident may provide unexplained odours that are out of character for the surroundings. Residual chemicals—in liquid, powder, or gas form—might also be observed in the area. Chemistry books or other reference materials may be present as well, along with equipment that is used to manufacture chemical weapons, such as scales, thermometers, or torches. Finally, personnel working around the materials may exhibit symptoms of chemical exposure—for example, irritation to the eyes, nose, and throat; difficulty breathing; tightness in the chest; nausea and vomiting; dizziness; headache; blurred vision; blisters or rashes; disorientation; or even convulsions.

■ Biological Agents

Indicators of incidents that potentially involve biological agents are somewhat similar to those that involve chemical agents. For example, production or containment equipment may be present, such as petri dishes, vented hoods, Bunsen burners, pipettes, microscopes, and incubators. Microbiology or biology textbooks or reference manuals may be found. Containers used to transport biological agents may be found at the site, including metal cylindrical cans or red plastic boxes or bags with biohazard labels. PPE (e.g., respirators, chemical or biological suits, and latex gloves) may be observed as well. Excessive amounts of antibiotics may be present as a means to protect personnel working with the agents. Other potential indicators may include abandoned spray devices and unscheduled or unusual sprays being disseminated (especially if outdoors at night). Finally, personnel working in the area of the lab may exhibit symptoms consistent with contact with the biological weapons with which they are working.

■ Radiological Agents

Because radiological agents are difficult to access, you are less likely to see incidents involving them. Indicators of radiological agents will typically include production or containment equipment, such as lead or stainless-steel containers (with nuclear or radiological labels), and equipment that can be used to detonate the radioactive source, such as pipes, caps, fuses, gunpowder, timers, wire, and detonators. PPE present may include radiation-protective suits and respirators. Radiation monitoring equipment such as Geiger counters may be present. Personnel working around the radiological agents may exhibit exposure symptoms such as burns or difficulty breathing.

■ Illicit Laboratories

There are many indicators of possible criminal or terrorist activity involving illicit labs. One set of indicators may be terrorist paraphernalia such as terrorist training manuals, ideological propaganda, and documents indicating affiliation with known terrorist groups. Certain locations are also commonly associated with clandestine labs—for example, basements with unusual or multiple vents, buildings with heavy security, buildings with windows obscured, and buildings with odd or unusual odours.

Fire personnel who suspect that they are responding to a suspected drug lab must exercise extreme caution. Illicit drug manufacturing is a very lucrative and competitive business and those who engage in this activity tend to be very protective of their investment. It is not uncommon to find weapons, guard dogs, and or booby traps in these buildings. In Canada, many drug labs are hidden in plain site as in a single family home on a quiet residential street. Single family homes are commonly used as grow operations. As clandestine labs require inordinate electricity supply the operators will usually circumvent the electrical meter and tap into the electrical grid directly. This can create a significant electrical hazard for fire personnel.

People working in illegal lab settings. may be nervous and have a high level of anxiety. They may be very protective of the lab area and not want to grant access to anyone, for any reason. They may rush people to leave the lab area as soon as possible.

Other equipment that may be present in these types of labs include surveillance materials, such as videotapes, photographs, maps, blueprints, or time logs of the target locations;

nonweaponized supplies, such as identification badges, uniforms, and decals that might be used to allow the terrorist to access target hazards; and weaponized supplies, such as timers, switches, fuses, container, wires, projectiles, and gun powder or fuel. Drug labs are by far the most common type of clandestine labs encountered. Most are very primitive. Many materials used to manufacture the drugs will be common, everyday—jars, bottles, glass cookware, coolers, and tubing—modified to produce the illicit drugs. Specific chemicals and materials present will include large quantities of cold-medicine tablets (ephedrine or pseudoephedrine), hydrochloric or sulphuric acid, paint thinner, drain cleaners, iodine crystals, table salt, aluminum foil, and camera batteries. The strong smell of urine or unusual chemical smells like that of ether, ammonia, or acetone are also commonly encountered in drug labs. Clandestine drug labs should be considered to represent significant hazardous materials scenes. The inexperienced chemists who run these labs take many shortcuts and disregard typical safety protocols in an effort to increase production.

■ Possible Criminal or Terrorist Activity Involving Explosives

Indicators of possible criminal or terrorist activity involving explosives will typically include materials that fit into four major categories: protective equipment, production and containment materials, explosive materials, and support materials. Protective equipment may include rubber gloves, goggles and face shields, and fire extinguishers. Production and containment equipment may include funnels, spoons, threaded pipe, caps, fuses, timers, wires, and detonators, as well as concealment containers such as briefcases, backpacks, or other common packages. Explosive materials may include gunpowder, gasoline, fertilizer, and similar materials. Support materials may include reference manuals on explosives, Internet reference materials, and military information.

■ Indicators of Secondary Devices

Indicators of potential secondary devices may include trip devices, such as timers, wires, or switches, as well as common concealment containers, such as briefcases, backpacks, boxes, or other common packages, and uncommon concealment containers, such as pressurized vessels (e.g., propane tanks) or industrial chemical containers (e.g., chlorine storage containers). Personnel may be watching the site at which the primary devices are placed, waiting to manually activate the secondary devices when first responders arrive at the scene.

Levels of Damage Caused to Humans by Chemicals

The damage that a hazardous material inflicts on a human depends on the material's threshold limit value (TLV). The TLV, which is the point at which the material begins to affect a person, can be defined in several ways. The various definitions indicate the levels of protection that are required. Most provincial/territorial governments have adopted TLV exposure guidelines as their legal limits.

The <u>threshold limit value–time-weighted average (TLV–TWA)</u> is the maximum concentration of a material to which a worker could be exposed for 8 hours a day, 40 hours a week, with no ill effects; the lower the TLV–TWA, the more toxic the substance.

The <u>threshold limit value–short-term exposure limit (TLV–STEL)</u> is the maximum concentration of a material to which a person can be exposed for 15-minute intervals, up to four times a day, without experiencing irritation or chronic or irreversible tissue damage. There should be a minimum one-hour rest period between any exposures to this concentration of the material; the lower the TLV–STEL, the more toxic the substance.

The <u>threshold limit value–ceiling (TLV–C)</u> is the maximum concentration of a material to which a worker should not be exposed, even for an instant; again, the lower the TLV–C, the more toxic the substance.

The <u>threshold limit value–skin</u> indicates the concentration at which direct or airborne contact with a material could result in possible and significant exposure from absorption through the skin, mucous membranes, and eyes. This designation is intended to suggest that appropriate measures be taken to minimize skin absorption so that the TLV–TWA is not exceeded.

- Immediately Dangerous to Life or Health (IDLH) means that an atmospheric concentration of a toxic, corrosive, or asphyxiating substance poses an immediate threat to life or could cause irreversible or delayed adverse health effects. Three general IDLH atmospheres are distinguished: toxic, flammable, and oxygen-deficient.

The last term that is used is the <u>recommended exposure level (REL)</u>. The National Institute for Occupational Safety and Health (NIOSH) sets RELs for other chemicals; these limits do not have the force of United States federal law, however. In Canada this is often referred to as "Occupational Exposure Limits" and is defined in various provincial and territorial legislation. An example, in the province of Ontario there are OELs for over 725 substances set out in R.R.O. 1990, Regulation 833 (Control of Exposure to Biological or Chemical Agents) and Ontario Regulation 490/09 (Designated Substances) under Ontario's Occupational Health and Safety Act.

Depending on the substance, individuals exposed to atmospheric concentrations below the IDLH value could—in theory—escape from the atmosphere without experiencing irreversible damage to their health, even if their respiratory protection fails. Individuals who might be exposed to atmospheric concentrations at or above the IDLH value must use positive pressure self-contained breathing apparatus (SCBA) or equivalent respiratory protection to guard against harm from the substance.

At most hazardous materials emergencies, fire fighters are not able to measure concentrations of specific chemicals. Measuring concentrations of chemicals requires specific monitoring instruments and trained response personnel to interpret the results.

Exposure guidelines are intended to minimize the possibility that either emergency responders or the public will be exposed to hazardous materials or atmospheres that will lead to harm. Once these exposure guidelines are understood, they can be applied at the scene of a hazardous materials emergency. For example, exposure guidelines can be used to define the three basic atmospheres at a hazardous materials emergency:

1. Safe atmosphere—No harmful hazardous materials effects exist, so personnel can handle routine emergencies without donning specialized PPE.
2. Unsafe atmosphere—A hazardous material that is no longer contained has created an unsafe condition or atmosphere. A person who is exposed to the material for long enough will probably experience some form of acute or chronic injury.
3. Dangerous atmosphere—Serious, irreversible injury or death can occur in the environment.

All exposure guidelines share a common goal: to ensure the safety and health of people exposed to a hazardous material. As long as the exposure remains below these values, it is considered safe to the average healthy adult, based on current information. As additional research and studies on the toxic effects of various chemicals are completed, these values may be adjusted.

Personal Protective Equipment

PPE is the clothing and protection provided to shield or insulate a person from chemical, physical, and thermal hazards. It is essential that a fire fighter's PPE meet both NFPA standards and Canadian Standards Association (CSA) regulations and that it be properly maintained and used.

Selecting the appropriate PPE for a hazardous materials incident is an important, even potentially life-saving decision. Without the proper protective clothing and respiratory protection, fire fighters might be injured. Adequate PPE should protect the fire fighter's respiratory system, as well as the skin, eyes, face, hands, feet, head, body, and hearing. Each member of the responding team must know the shielding capabilities and limitations of his or her PPE. When product-specific chemical suits are unavailable, fire fighters must know how much and what kind of protection their regular firefighting clothing provides. Remember: protective clothing is your last line of defence, not your first.

Which type of PPE is appropriate for the incident depends on the activities that a fire fighter is expected to perform—specifically, the tasks that the fire fighter is expected to complete. Standard structural firefighting gear might be suitable for support operations at many hazardous materials incidents, but personnel who are involved in decontamination activities may require chemical-protective clothing because of the possibility of cross-contamination. The incident commander (IC) has the ultimate authority and responsibility to approve the level of PPE required for a given activity.

Fire fighters should operate only at the incident level that matches their knowledge, training, and equipment. If conditions indicate a need for a higher response level, additional personnel with the appropriate training and PPE should be summoned.

Hazardous Materials–Specific Personal Protective Equipment

Different levels of PPE are required at a hazardous materials incident. This section reviews the protective qualities of various outfits—from those offering the least protection to those providing the greatest protection.

Street Clothing and Work Uniforms

Street clothing and work uniforms offer the least amount of protection in a hazardous materials emergency. Normal clothing can prevent a non-caustic powder from coming into direct contact with the skin, but it offers no more protection. A one-piece flame-resistant or lightly flame-resistant coverall can enhance protection slightly. For this very reason, such PPE is often used in industrial applications such as oil refineries as a general work uniform **FIGURE 31-4**.

Structural Firefighting Protective Clothing

The next level of protection is provided by structural firefighting protective equipment, also known as turnout or bunker gear. Such an outfit includes a helmet, a bunker coat, bunker pants, boots, gloves, a hood, and a personal alert safety system (PASS) device. Standard firefighting turnout gear offers no chemical protection, although it does have some abrasion resistance and prevents direct skin contact; however, it easily absorbs materials, breaks down when exposed to chemicals, and does not provide adequate protection from the harmful gases, vapours, liquids, or dusts that are encountered during hazardous materials incidents **FIGURE 31-5**.

High Temperature–Protective Equipment

High temperature–protective equipment protects the wearer during short-term exposures to high temperatures **FIGURE 31-6**. It allows the properly trained fire fighter to work in extreme fire conditions (i.e., it provides protection against high temperatures) and is not designed to protect the fire fighter from hazardous materials.

FIGURE 31-5 Standard turnout gear.

FIGURE 31-6 High temperature–protective equipment protects the wearer from high temperatures during a short exposure.
© Photodisc

FIGURE 31-4 Nomex® jumpsuit.
© KellyBoreson/ShutterStock, Inc.

Depending on the radioactive isotope, this same gear and SCBA can be a good choice for alpha emitters and perhaps beta emitters. Remember that this equipment is not intended to be radiation protection; it is designed to shield the wearer against radioactive particles and perhaps from a radiological dispersal device.

■ Chemical-Protective Clothing and Equipment

Chemical-protective clothing is designed to prevent chemicals from coming in contact with the body and can offer varying degrees of resistance. It comprises a material, or combination of materials, used in an item of clothing to isolate parts of the wearer's body from contact with a hazardous chemical. Other performance requirements that must be considered when selecting chemical-protective materials include chemical resistance, flexibility, abrasion, temperature resistance, shelf life, and sizing criteria.

Chemical resistance is the ability to resist chemical attack. Time, temperature, stress, and reagent are all factors that affect the chemical resistance of materials. **Chemical-resistant materials** are specifically designed to inhibit or resist the passage of chemicals into and through the material by the processes of penetration, permeation, or degradation:

- **Penetration** is the flow or movement of a hazardous chemical through closures (e.g., zippers), seams, porous materials, pinholes, or other imperfections in the material. Although liquids are the most likely to penetrate materials, solids (e.g., asbestos) can also penetrate protective clothing.
- **Permeation** is the process by which a hazardous chemical moves through a given material on the molecular level. It differs from penetration in that permeation occurs through the material itself rather than through openings in the material.
- **Degradation** is the physical destruction or decomposition of a clothing material, owing to chemical exposure, general use, or ambient conditions (e.g., storage in sunlight). It might be evidenced by visible signs such as charring, shrinking, swelling, colour changes, or dissolving. Materials can also be tested for weight changes, loss of tensile strength, and other properties that measure degradation.

Chemical-protective clothing can be constructed as a single- or multipiece garment. A single-piece garment completely encloses the wearer and is known as an encapsulating suit. Encapsulating protective clothing offers full body protection from hostile environments and requires the use of supplied-air respiratory protection devices, such as SCBA; some materials are so caustic or corrosive that they could destroy SCBA after only a short exposure, however. A multipiece garment works with the wearer's respiratory protection, an attached or detachable hood, gloves, and boots to protect against a specific hazard.

A variety of materials is used in both encapsulating and nonencapsulating protective clothing. The most common include butyl rubber, Tyvek®, Saranex™, polyvinyl chloride, and Viton®, as well as many other materials that are protected by the manufacturer, either singly or in multiple layers of several materials. Specialized chemical-protective clothing is adequate for some chemicals yet useless for other chemicals; no single material provides satisfactory protection from all chemicals. Protective clothing materials must be compatible with the chemical substances involved in the incident and use must be consistent with the manufacturer's instructions. The manufacturer's guidelines and recommendations should be consulted for material compatibility information.

Personnel trained at the operations level should not be operating in encapsulating suits. Operational-level personnel can wear some nonencapsulating chemical-protective clothing while providing support for those entering the hazardous materials area.

Liquid splash–protective clothing consists of several pieces of clothing and equipment that protect the skin and eyes from chemical splashes **FIGURE 31-7**. It does not provide total body protection from gases or vapours. For this reason, liquid splash–protective clothing should not be used for incidents involving liquids that emit vapours known to affect or be absorbed through the skin.

Vapour-protective clothing must be used when hazardous vapours are present **FIGURE 31-8**. This type of clothing traps the fire fighter's body heat, so fire fighters and support personnel must be aware of the possibility of heat-related emergencies. SCBA or air lines must also be used for respiratory protection.

Fire fighters can wear liquid splash–protective clothing over or under structural firefighting clothing in some situations. This multiple-PPE approach provides limited chemical-splash and thermal protection. Fire fighters trained at the operational level can wear liquid splash–protective clothing when they are assigned to enter the initial site, protect decontamination

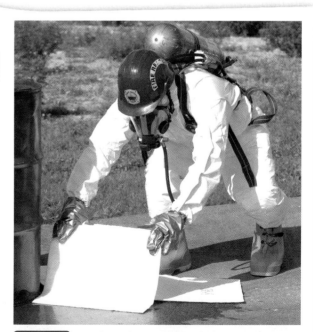

FIGURE 31-7 Liquid splash–protective clothing must be worn when there is the danger of chemical splashes.

FIGURE 31-8 Vapour-protective clothing retains body heat and increases the possibility of heat-related emergencies.

Courtesy of Lakeland Industries, Inc.

personnel, or construct isolation barriers such as dikes, diversions, retention areas, or dams.

■ Respiratory Protection

Positive Pressure Self-Contained Breathing Apparatus

Respiratory protection, which is usually provided by SCBA, is an important—and separate—level of protection at a hazardous materials incident. Using positive pressure SCBA prevents both inhalation and ingestion (two primary routes of exposure) and should be mandatory for fire service personnel. Although it is an excellent barrier in many hazardous environments, positive pressure SCBA is not designed for—and has never been tested or approved for use in—atmospheres that might contain unknown hazards. Fire fighters must be aware of the limitations of SCBA in a hazardous atmosphere.

New standards are now in effect for the testing of SCBA in environments that are contaminated with certain concentrations of chemical, biological, radiological, and nuclear (CBRN) materials. Check with your fire department to see whether it has adopted any of these specialized SCBA units for day-to-day or special use.

Other Respiratory Protection

Supplied-air respirators (SARs) have an external air source such as a compressor or storage cylinder. A hose connects the user to the air source and provides air to the face piece. SARs

are useful during extended operations such as decontamination, clean-up, and remedial work. These devices might be less bulky and weigh less than SCBA, but the length of the air hose can limit the fire fighter's movement **FIGURE 31-9**.

Air-purifying respirators (APRs) and powered air-purifying respirators (PAPRs) are filtering devices that remove particulates and contaminants from the air. The difference between the two devices is that the PAPR has a built-in fan that assists in drawing air into a hose to the face piece. This action provides the wearer with a positive flow of air while it is operating. They should be worn only in atmospheres where the type and quantity of the contaminants are known and where sufficient oxygen for breathing is available. APRs/PAPRs might be appropriate for operations involving volatile solids and for remedial clean-up and recovery operations where the type and concentration of contaminants are verifiable **FIGURE 31-10**.

APRs/PAPRs use filtration devices to remove particulate matter, gases, and vapours from the atmosphere. PAPRs include a fan motor and pack where the filters mount and a battery pack for operation, as well as a hose from the fan to a full face piece. APRs range from full-face-piece, dual-cartridge masks with eye protection to half-mask, face piece–mounted cartridges with no eye protection. APRs/PAPRs do not have a separate source of air but rather filter and purify ambient air before it is inhaled. The models used in environments containing hazardous gases or vapours are commonly equipped with an absorbent material that soaks up or reacts with the gas. Consequently, cartridge selection is based on the expected contaminants. Particle-removing respirators use a mechanical filter to separate the contaminants from the air. Both types of devices require that the ambient atmosphere contain a minimum of 19.5 percent oxygen.

FIGURE 31-9 A supplied-air respirator is less bulky than an SCBA but is limited by the length and structural integrity of the air hose.

© Jim Ruymen/Reuters/Landov

A. **B.**

FIGURE 31-10 Air-purifying respirators and powered air-purifying respirators can be used only by trained personnel. **A.** Air-purifying respirator. **B.** Powered air-purifying respirator.

Courtesy of Survivair

APRs are easy to wear and offer no hose line restriction, but they do have some drawbacks. Because filtering cartridges are specific to expected contaminants, they will be ineffective if the contaminant changes suddenly, endangering the lives of fire fighters. The air must be continually monitored for both the known substance and the ambient oxygen level throughout the incident. For these reasons, APRs should not be employed at hazardous materials incidents until qualified personnel have tested the ambient atmosphere and determined that such devices can be used safely.

Chemical-Protective Clothing Ratings

Chemical-protective clothing is rated for its effectiveness in several different ways. Testing measures whether chemicals will permeate the suit and, if so, how quickly and to what degree. This information can be obtained from the manufacturer of the clothing. The U.S. Environmental Protection Agency (EPA) defines levels of protection using an alphabetical system. Also consult NFPA 1992, *Standard on Liquid Splash–Protective Ensembles and Clothing for Hazardous Materials Emergencies*, and NFPA 1994, *Standard on Protective Ensembles for First Responders to CBRN Terrorism Incidents*.

■ Level A

<u>Level A protection</u> consists of a heavy, encapsulating suit that completely envelops both the wearer and his or her SCBA. It should be used when the hazardous material identified requires the highest level of protection for skin, eyes, and respiration. Level A protection is effective against vapours, gases, mists, and even dusts **FIGURE 31-11**. To warrant this rating, chemical-protective clothing must meet the requirements for vapour–gas protection outlined in NFPA 1991, *Standard on Vapor-Protective Ensembles for Hazardous Materials Emergencies*. Level A protection also requires open-circuit, positive pressure SCBA or an SAR for respiratory protection.

FIGURE 31-11 Level A protection envelops the wearer in a totally encapsulating suit.

Recommended PPE for Level A protection includes the following components:

- SCBA or SAR
- Fully encapsulating chemical-resistant suit

- Inner chemical-resistant gloves
- Chemical-resistant safety boots/shoes
- Two-way radio

Optional PPE for Level A protection includes the following components:

- Coveralls
- Cotton long underwear
- Hard hat
- Disposable glove and boot covers

Fire Fighter Safety Tips

© Photos.com

Mechanical hazards are best addressed by understanding the NFPA performance requirements for chemical-protective clothing ensembles (CPEs). Typical performance requirements include tests for durability, barrier integrity after flex and abrasion challenges, cold-temperature flex, and flammability. Essentially, each part of the suit must pass a particular set of challenges prior to receiving certification based on the NFPA testing standards. Users of any garment that meets the performance requirements set forth by the NFPA can rest assured that the garment will withstand reasonable insults from most mechanical-type hazards encountered on the scene. This does not mean the suit is "bulletproof" and cannot fail; it just means that it will hold up under normal conditions. It is up to the user to be aware of the hazards and avoid situations that might cause the garment to fail.

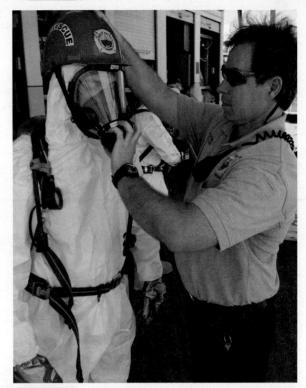

FIGURE 31-12 Level B protection provides a high level of respiratory protection but less skin protection.

■ Level B

Level B protection consists of chemical-protective clothing, boots, gloves, and SCBA **FIGURE 31-12**. This type of PPE should be used when the type and atmospheric concentration of identified substances require a high level of respiratory protection but less skin protection. The type of gloves and boots worn depends on the identified chemical; however, wrists and ankles must be properly sealed to prevent splashed liquids from contacting the skin.

Recommended PPE for Level B protection includes the following components:

- SCBA or SAR
- Chemical-resistant clothing
- Inner and outer chemical-resistant gloves
- Chemical-resistant safety boots/shoes
- Hard hat
- Two-way radio

Optional PPE for Level B protection includes the following components:

- Coveralls
- Cotton long underwear
- Disposable glove and boot covers

To don a Level B encapsulating CPE, follow the steps in **SKILL DRILL 31-1**:

1. Conduct a pre-entry briefing, medical monitoring, and equipment inspection.
2. While seated, pull on the suit to waist level; pull on the chemical boots over the top of the chemical suit. Pull the suit boot covers over the tops of the boots. (**STEP ①**)

3. Stand up and don SCBA and the SCBA face piece, but do not connect the regulator to the face piece. (**STEP ②**)
4. Place the helmet on the head if required. (**STEP ③**)
5. Don the inner gloves. (**STEP ④**)
6. With assistance, complete donning the suit by placing both arms in the suit, pulling the expanded back piece over the SCBA, and placing the chemical suit over the head.
7. Instruct the assistant to connect the regulator to the SCBA face piece, and ensure that air flow is working. Note when you begin to use the SCBA. (**STEP ⑤**)
8. Instruct the assistant to close the chemical suit with the zipper and seal the splash flap. (**STEP ⑥**)
9. Review hand signals, and indicate that you are okay. (**STEP ⑦**)

To doff a Level B encapsulating CPE, follow the steps in **SKILL DRILL 31-2**:

1. After completing decontamination, proceed to the clean area for suit doffing.
2. Pull the hands and arms out of the suit gloves and sleeves, and cross the arms in front, inside the suit. (**STEP ①**)
3. Instruct the assistant to open the chemical splash flap and suit zipper. (**STEP ②**)
4. Instruct the assistant to begin at the head and roll the suit down and away, until the suit is below waist level. (**STEP ③**)

5 Sit and instruct the assistant to complete rolling down the suit and remove the outer boots and suit. Rotate on the bench to the direction that will allow you to place the feet on a dry, clean area. (**STEP 4**)

6 Stand and doff the SCBA using the quick-release method. The face piece should be kept in place, while the SCBA frame is placed on the ground. (**STEP 5**)

7 Take a deep breath, doff the SCBA mask, and walk away from the clean area.

8 Go to the rehabilitation area for medical monitoring, rehydration, and a personal decontamination shower. (**STEP 6**)

SKILL DRILL 31-1   Donning a Level B Encapsulating Chemical-Protective Clothing Ensemble

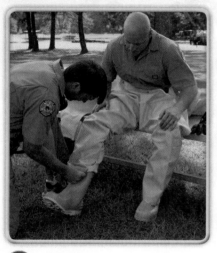

1 While seated, pull on the suit to waist level; put on the chemical boots over the chemical suit. Pull the suit boot covers over the tops of the boots.

2 Stand up and don SCBA and the SCBA face piece, but do not connect the regulator to the face piece.

3 Place the helmet on the head if required.

4 Don the inner gloves.

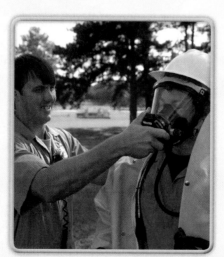

5 With assistance, complete donning the suit. Instruct the assistant to connect the regulator to the SCBA face piece, and ensure air flow.

© LiquidLibrary

SKILL DRILL 31-1 | Donning a Level B Encapsulated Chemical-Protective Clothing Ensemble *(Continued)*

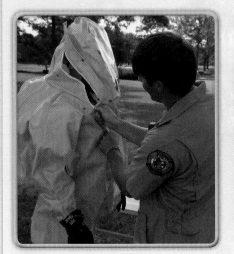

6 Instruct the assistant to close the chemical suit with the zipper and seal the splash flap.

7 Review hand signals, and indicate that you are ready to operate.

© LiquidLibrary

SKILL DRILL 31-2 | Doffing a Level B Encapsulating Chemical-Protective Clothing Ensemble

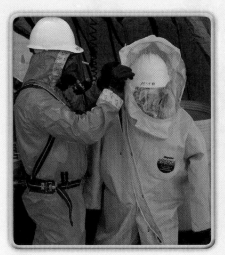

1 After completing decontamination, proceed to the clean area. Remove the hands and arms from the suit gloves and sleeves, and cross the arms in front, inside the suit.

2 Instruct the assistant to open the chemical splash flap and suit zipper.

3 Instruct the assistant to begin at the head and roll the suit down and away from you, until the suit is below waist level.

© LiquidLibrary

(Continues)

SKILL DRILL 31-2   Doffing a Level B Encapsulating Chemical-Protective Clothing Ensemble (*Continued*)

4   Sit and instruct the assistant to complete rolling down the suit and remove the outer boots and suit. Rotate on the bench to the direction that will allow you to place the feet on a dry, clean area.

5   Stand and doff the SCBA using the quick-release method. Keep the face piece in place, while the SCBA frame is placed on the ground.

6   Take a deep breath, doff the SCBA mask, and walk away from the clean area. Go to rehabilitation area for medical monitoring, rehydration, and a personal decontamination shower.

© LiquidLibrary

To don a Level B nonencapsulating CPE, follow the steps in **SKILL DRILL 31-3**:

1   Conduct a pre-entry briefing, medical monitoring, and equipment inspection.

2   Sit down and pull on the suit to waist level; pull on the chemical boots over the top of the chemical suit. Pull the suit boot covers over the tops of the boots, or tape if needed. (**STEP 1**)

3   Don the inner gloves. (**STEP 2**)

4   With assistance, complete donning the suit by placing both arms in the suit and pulling the suit over the shoulders.

5   Instruct the assistant to close the chemical suit with the zipper and seal the splash flap. (**STEP 3**)

6   Stand up and don SCBA and the SCBA face piece, but do not connect the regulator to the face piece.

7   With assistance, pull the hood over the head and SCBA face piece. (**STEP 4**)

8   Place the helmet on the head.

9   Pull the gloves over or under the sleeves, depending on the situation.

10   Instruct the assistant to connect the regulator to the SCBA face piece, and ensure that air flow is working. (**STEP 5**)

11   Review hand signals, and indicate that you are okay.

To doff a Level B nonencapsulating CPE, follow the steps in **SKILL DRILL 31-4**:

1   After completing decontamination, proceed to the clean area for suit doffing.

2   Stand and doff the SCBA. Keep the face piece in place, while the SCBA frame is placed on the ground. (**STEP 1**)

3   Instruct the assistant to open the chemical splash flap and suit zipper. (**STEP 2**)

4   Remove your hands and arms from the suit gloves (except the inner gloves) and sleeves, and cross your arms in front, inside the suit.

5   Instruct the assistant to begin at the head and roll the suit down and away, until the suit is below waist level. (**STEP 3**)

6   Sit down and instruct the assistant to complete rolling down the suit and remove the outer boots and suit. Rotate on the bench to the direction that will allow you to place the feet on a dry, clean area. The assistant helps remove the inner gloves. (**STEP 4**)

7   Remove the SCBA face piece.

8   Go to the rehabilitation area for medical monitoring, rehydration, and a personal decontamination shower. (**STEP 5**)

SKILL DRILL 31-3 Donning a Level B Nonencapsulating Chemical-Protective Clothing Ensemble

1 While seated, pull on the suit to waist level; pull on the chemical boots over the top of the chemical suit. Pull the suit boot covers over the tops of the boots, or tape if needed.

2 Don the inner gloves.

3 With assistance, complete donning the suit by placing both arms in the suit and pulling the suit over the shoulders. Instruct the assistant to close the chemical suit with the zipper and seal the splash flap.

4 Stand up and don SCBA and SCBA face piece, but do not connect regulator to the face piece. With assistance, pull the hood over the head and SCBA face piece.

5 Place the helmet on the head. Instruct the assistant to connect the regulator to the SCBA face piece, and ensure you have air flow.

© LiquidLibrary

■ Level C

<u>Level C protection</u> consists of standard work clothing plus chemical-protective clothing, chemical-resistant gloves, and a form of respiratory protection. Level C protection is appropriate when the type of airborne substance is known, its concentration is measured, the criteria for using APRs are met, and skin and eye exposure is unlikely.

Level C respiratory protection can be provided with a filter or cartridge mask or an APR. The APR could be a half-face or full-face mask that includes eye protection. Head protection and goggles are also required.

Recommended PPE for Level C protection includes the following components:

- Full-face APR
- Chemical-resistant clothing
- Inner and outer chemical-resistant gloves
- Chemical-resistant safety boots/shoes

SKILL DRILL 31-4   Doffing a Level B Nonencapsulating Chemical-Protective Clothing Ensemble

1 After completing decontamination, proceed to the clean area for suit doffing. Stand and doff the SCBA. Keep the face piece in place, while the SCBA frame is placed on the ground.

2 Instruct the assistant to open the chemical splash flap and suit zipper.

3 Remove your hands and arms from the suit gloves (except the inner gloves) and sleeves, and cross your arms in front, inside the suit. Instruct the assistant to begin at the head and roll the suit down and away, until the suit is below waist level.

4 Sit down and instruct the assistant to complete rolling down the suit and remove the outer boots and suit. Rotate on the bench to the direction that will allow you to place the feet on a dry, clean area. The assistant helps remove the inner gloves.

5 Remove the SCBA face piece. Go to the rehabilitation area for medical monitoring, rehydration, and a personal decontamination shower.

© LiquidLibrary

- Two-way radio
- Hard hat

Optional PPE for Level C protection includes the following components:
- Coveralls
- Cotton long underwear
- Disposable glove and boot covers

To don a Level C CPE, follow the steps in **SKILL DRILL 31-5**:

1 Conduct a pre-entry briefing, medical monitoring, and equipment inspection.

2 While seated, pull on the suit to waist level; pull on the chemical boots over the top of the chemical suit. Pull the suit boot covers over the tops of the boots. (**STEP 1**)

3 With assistance, complete donning the suit by placing both arms in the suit and pulling the suit over the shoulders.

4 The assistant closes the chemical suit with the zipper and seals the splash flap. (**STEP 2**)

5 Don the inner gloves. (**STEP 3**)

6 Stand up and don the APR/PAPR and APR/PAPR face piece.

7 With assistance, pull the hood over the head and the APR/PAPR face piece.

8 Place the helmet on the head. (**STEP 4**)

9 Review hand signals, and indicate that you are okay.

SKILL DRILL 31-5 — Donning a Level C Chemical-Protective Clothing Ensemble

1 Conduct a pre-entry briefing, medical monitoring, and equipment inspection. While seated, pull on the suit to waist level; pull on the chemical boots over the top of the chemical suit. Pull the suit boot covers over the tops of the boots.

2 With assistance, complete donning the suit by placing both arms in the suit and pulling the suit over the shoulders. The assistant closes the chemical suit with the zipper and seals the splash flap.

3 Don the inner gloves.

4 Stand up and don the APR/PAPR and APR/PAPR face piece. With assistance, pull the hood over the head and the APR/PAPR face piece. Place the helmet on the head.

© LiquidLibrary

To doff a Level C CPE, follow the steps in **SKILL DRILL 31-6** **FIGURE 31-13** :

1 After completing decontamination, proceed to the clean area for suit doffing.

2 Instruct the assistant to open the chemical splash flap and suit zipper.

3 Remove the hands and arms from the suit gloves (except the inner gloves) and sleeves, and cross the arms in front, inside the suit.

4 Instruct the assistant to begin at the head and roll the suit down and away, until the suit is below waist level.

5 Sit down. Instruct the assistant to complete rolling down the suit and take the outer boots and suit away.

6 Instruct the assistant to help remove the inner gloves.

7 Rotate on the bench to the direction that allows the feet to be placed on a dry, clean area.

8 Remove the APR/PAPR.

9 Go to the rehabilitation area for medical monitoring, rehydration, and a personal decontamination shower.

■ Level D

Level D protection is the lowest level of protection and includes coveralls, work shoes, hard hats, gloves, and standard work clothing. It should be used when the atmosphere contains no known hazard, and when work functions

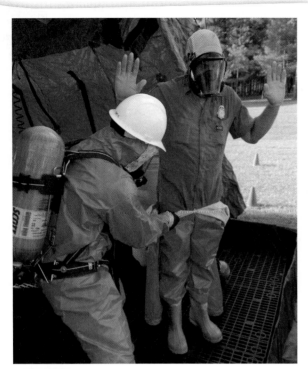

FIGURE 31-13 An assistant helps in doffing a Level C CPE.

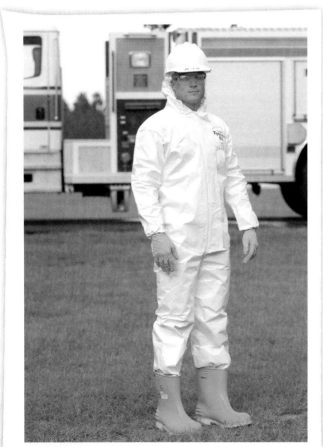

FIGURE 31-14 Level D Protection is primarily a work uniform that includes coveralls and provides minimal protection.

preclude splashes, immersion, or the potential for unexpected inhalation of or contact with hazardous levels of chemicals. Level D PPE should be used for nuisance contamination (e.g., dust) only; it should not be worn at any site where respiratory or skin hazards exist.

Recommended PPE for Level D protection includes the following components:

- Coveralls
- Safety boots/shoes
- Safety glasses or chemical-splash goggles
- Hard hat
- Gloves

Optional PPE for Level D protection includes the following components:

- Face shield
- Escape mask

To don a Level D CPE, follow the steps in **SKILL DRILL 31-7** **FIGURE 31-14**:

1. Conduct a pre-entry briefing, medical monitoring, and equipment inspection.
2. Don the Level D suit.
3. Don the boots.
4. Don safety glasses or chemical goggles.
5. Don a hard hat.
6. Don gloves, a face shield, and any other required equipment.

To doff a Level D CPE, follow the steps in **SKILL DRILL 31-8**:

1. After completing decontamination, proceed to the clean area for suit doffing.
2. Remove the gloves and hard hat.
3. Begin at the head and roll the suit down and away, until the suit is below waist level.
4. Sit down. The assistant completes rolling down the suit and takes the boots and suit away.
5. Instruct the assistant to help remove the inner gloves.
6. Rotate on the bench to the direction that allows the feet to be placed on a dry, clean area.
7. Go to the rehabilitation area for medical monitoring and rehydration.

Exposures

In case of a hazardous materials incident, exposures can consist of people, property, structures, or the environment that are subject to influence, damage, or injury as a result of contact with the hazardous material. The amount of exposures that remain is determined both by the location of the incident and by the amount of progress that has been made in protecting those exposures via isolation and other indirect responses. Incidents in urban areas will likely have more exposures and therefore will likely need more resources to protect those exposures from the hazardous materials.

■ Protective Actions

The IC must consider many factors before deciding on protective actions for responders and the general public, such as evacuation or sheltering-in-place. The IC must first evaluate the risk assessment of the hazardous material, the container holding the hazardous material, and the environment surrounding the incident, as well as the resources available to handle the incident.

Evacuating the exposed area is one of the first response priorities at a hazardous materials incident **FIGURE 31-15**. Evacuation involves some significant risks even when it is properly planned. Once the evacuation order is given, fire fighters are charged with assisting in the physical evacuation of residents to a safe area. This duty might include travelling to homes and informing residents that they must leave their homes for a shelter.

Before an evacuation order is given, a safe area and suitable shelter should be established. In many cases, schools are used for shelters.

Depending on the time of day and the season, it may take a considerable amount of time to evacuate even a small area. The evacuation area should be located close enough to the exposure for the evacuation to be practical, yet far enough away from the incident to be safe. Temporary evacuation areas may be needed to shelter residents until longer-term evacuation sites or structures indicated in the community's emergency plan can be accessed and opened.

Transportation to temporary evacuation areas must be arranged for all populations. The needs of handicapped and special-needs persons should be considered as well. The initial areas that need to be evacuated can be determined from the

ERG. The guidebook provides the minimum basic guidelines for the initial response, but it does not give complete detailed information for all conditions that responders might potentially encounter, such as weather extremes, road conditions, or actual conditions at the scene.

The best way for the IC to determine the appropriate area of evacuation and decide how far to extend actual evacuation distances is to have hazardous materials technicians with detection devices monitor both the concentration of hazardous material and the direction and speed that the material is moving due to wind currents and terrain. Access to local weather conditions is invaluable in this decision-making process.

Another concern when considering evacuation is the potential for exposure to the hazardous material during the evacuation attempt. Personnel may have to go into or through the hazardous material to remove residents. Also, when the evacuated residents are removed from protected areas, they may pass into or through the hazardous material. These individuals must be adequately protected from the hazardous material during their egress, or they will become victims.

If the area to be evacuated is more than a few blocks or if it involves any special hazards such as jails, hospitals, or nursing homes, the IC will need to activate local emergency management plans and involve agencies other than the fire department for the safe and orderly evacuation of these types of facilities.

Shelter-in-place is a method of safeguarding people located near or in a hazardous area by keeping them in a safe atmosphere, usually inside structures. Local emergency planning should address local facilities that have large populations to evacuate, such as schools, churches, hospitals, nursing homes, and apartment complexes. When residents are sheltered-in-place, they remain indoors with windows and doors closed. All ventilation systems are turned off to prevent outside air from being drawn into the structure.

The toxicity of the hazardous material is a major factor in the decision whether to evacuate or to shelter-in-place. The more toxic the substance, the less likely a shelter-in-place strategy is going to be appropriate. The expected duration of the incident is another factor in determining whether shelter-in-place is a viable option. The longer the incident, the more time the material has to enter or permeate into protected areas.

Flammability is another consideration in the decision whether to evacuate or to shelter-in-place. If the risk of flammability is high, then sheltering-in-place may no longer be the safest option.

In areas that are home to many chemical facilities, it is not uncommon for fire departments and emergency management agencies to provide preincident information on shelter-in-place actions. The dissemination of this type of information may be a function of the fire department's fire prevention bureau.

The ability to evacuate an area and to staff an adequate emergency shelter will be determined by the local resources available. There is no mathematical formula to calculate how many residents can be evacuated at any given time. In severe weather, evacuation can be challenging at best. Even under perfect circumstances, the evacuation of residents will be a process that is measured in hours, not minutes. Local emergency planning for evacuations is part of a safe and successful emergency evacuation.

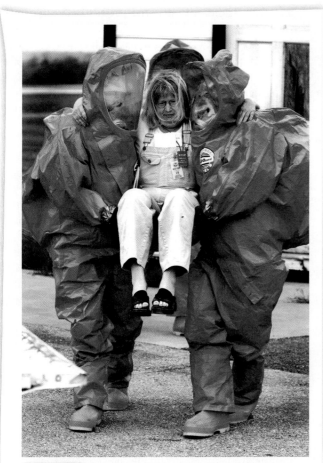

FIGURE 31-15 Evacuating the exposed area is one of the first response priorities at a hazardous materials incident.

© Ric Dugan, Herald-Mail/AP Photos

Wrap-Up

Chief Concepts

- To ensure your safety and the safety of your crew, you must be able to recognize a potential hazardous material as soon as possible.
- Many hazardous materials can be recognized by smell, taste, or touch; getting close enough to exercise these senses, however, may expose you to the hazard. The farther you are from the incident, the safer you will be.
- Transport Canada's TDG regulations employ a system of labels and placards to identify dangerous goods in transit. The *ERG* is also part of this system and offers a certain amount of guidance for fire fighters operating at a hazardous materials incident.
- Labels, placards, and other markings on buildings, packages, boxes, and containers often enable fire fighters to identify a spilled or released chemical.
- Placards are diamond-shaped indicators (250 mm [10.75 in.] on each side) that must be placed on all four sides of highway transport vehicles, railroad tank cars, and other forms of transportation carrying hazardous materials.
- Labels are smaller versions of placards (100 mm [4-in.] diamond-shaped indicators) and are used on any side of individual boxes and smaller packages being transported.
- Nine chemical families are recognized by the *ERG*:
  1. Class 1—Explosives
  2. Class 2—Gases
  3. Class 3—Flammable liquids
  4. Class 4—Flammable solids, spontaneously combustible materials, and water-reactive substances
  5. Class 5—Oxidizers
  6. Class 6—Poisons (including blood agents and choking agents)
  7. Class 7—Radioactive materials
  8. Class 8—Corrosives
  9. Class 9—Other regulated materials
- The *ERG* is a preliminary action guide for hazardous materials incidents. It is divided into four coloured sections:
  1. Yellow section—Used when the UN number is known or can be identified.
  2. Blue section—Lists chemicals alphabetically by name.
  3. Orange section—Contains the emergency action guides.
  4. Green section—Organized numerically by UN identification number and provides the initial isolation distances for specific materials.

- The NFPA 704 hazard identification system is designed for fixed-facility use. It uses a diamond-shaped symbol of any size, which is itself broken into four smaller diamonds, each representing a particular property or characteristic: The blue diamond at the nine o'clock position indicates the health hazard posed by the material; the top red diamond indicates flammability; the yellow diamond at the three o'clock position indicates reactivity; and the bottom white diamond is used for special symbols and handling instructions.
- Additional reference sources for hazardous materials include SDSs, shipping papers, and staffed national resources such as CANUTEC. An SDS provides basic information about the chemical makeup of a substance, the potential hazards it presents, appropriate first aid in the event of an exposure, and other pertinent data for safe handling of the material. SDSs may be kept in a three-ring binder or a file cabinet, or a computer system.
- Industries that routinely use radioactive materials include food testing labs, hospitals, medical research centres, biotechnology facilities, construction sites, and medical laboratories. Radioactive materials are not detectable by sight, smell, taste, or other senses. If you have any suspicion that an incident involves radiation, it is necessary to call a hazardous materials team or some other resource with radiation detection capabilities.
- The proper use of PPE and scene control are critical to ensuring the safety of fire fighters, victims, and bystanders.
- TLV exposure guidelines are as follows:
  - TLV–TWA is the maximum concentration of a material to which a worker could be exposed for 8 hours a day, 40 hours a week, with no ill effects.
  - TLV–STEL is the maximum concentration of a material to which a person can be exposed for 15-minute intervals, up to four times a day, without experiencing irritation or chronic or irreversible tissue damage.
  - TLV–C is the maximum concentration of a material to which a worker should not be exposed, even for an instant.
  - Threshold limit value–skin is the concentration at which direct or airborne contact with a material could result in possible and significant exposure from absorption through the skin, mucous membranes, and eyes.
  - IDLH—An atmospheric concentration of a toxic, corrosive, or asphyxiating substance that poses an immediate threat to life or could cause irreversible or delayed adverse health effects.

- Exposure guidelines are intended to minimize the possibility that either responders or the public will be exposed to hazardous materials or atmospheres that will lead to harm. Exposure guidelines can be used to define the three basic atmospheres at a hazardous materials emergency:
  1. Safe atmosphere—No harmful hazardous materials effects exist, so personnel can handle routine emergencies without donning specialized PPE.
  2. Unsafe atmosphere—A hazardous material that is no longer contained has created an unsafe condition or atmosphere. A person who is exposed to the material for long enough will probably experience some form of acute or chronic injury.
  3. Dangerous atmosphere—Serious, irreversible injury or death can occur in the environment.
- Selecting the appropriate PPE for a hazardous materials incident is a potentially life-saving decision. Adequate PPE should protect the fire fighter's respiratory system, as well as his or her skin, eyes, face, hands, feet, head, body, and hearing.
- Which type of PPE is required for the incident depends on the activities that the fire fighter is expected to perform.
- The categories of hazardous materials–specific PPE include the following:
  - Street clothing and work uniforms (least amount of protection)
  - Structural firefighting protective clothing (no chemical protection but protects against abrasion and direct skin contact)
  - High temperature–protective equipment (high-temperature protection during short-term exposures)
  - Chemical-protective clothing and equipment (e.g., liquid splash–protective clothing and vapour-protective clothing)
- Respiratory protection includes the following:
  - SCBA
  - SARs—Have an external air source, such as a compressor or storage cylinder
  - APRs/PAPRs—Filtering devices
- CPE ratings are as follows:
  - Level A protection—Provides the highest level of protection for skin, eyes, and respiration.
  - Level B protection—Used when the type and atmospheric concentration of identified substances require a high level of respiratory protection but less skin protection.
  - Level C protection—Used when the type of airborne substance is known, its concentration is measured, the criteria for using APRs are met, and skin and eye exposure is unlikely.
  - Level D protection—Used when the atmosphere contains no known hazard and when work functions preclude splashes, immersion, or the potential for unexpected inhalation of or contact with hazardous levels of chemicals.
- Exposures can consist of people, property, structures, or the environment. The number of exposures is determined by the location and the amount of progress made in protecting the exposures.
- The two main protective actions utilized to protect human exposures are evacuation and sheltering-in-place.

Hot Terms

Air bill Shipping document for materials sent via air transport.

Air-purifying respirator (APR) Filtering device that removes particulates and contaminants from the air.

Bill of lading Shipping document for materials sent via roads and highways.

Chemical-resistant material A material specifically designed to inhibit or resist the passage of chemicals into and through the material by the processes of penetration, permeation, or degradation.

Consist A list of every car on a train.

Dangerous cargo manifest Shipping document for materials sent via a marine vessel, typically located in a tube-like container.

Degradation The physical destruction or decomposition of a clothing material, owing to chemical exposure, general use, or ambient conditions (e.g., storage in sunlight).

Department of Transportation (DOT) marking system A unique system of labels and placards that, in combination with the *ERG*, offers guidance for first responders operating at a hazardous materials incident in the United States.

Emergency Response Guidebook (ERG) A reference book, written in plain language, to guide emergency responders in their initial actions at the incident scene. (NFPA 472)

Freight bill Shipping document for materials sent via roads and highways.

Hazardous material A substance that, when released, is capable of causing harm to people, the environment, and property. (NFPA 472)

Hazardous Materials Information System (HMIS) A colour-coded marking system used in the United States by which employers give their personnel the necessary information to work safely around chemicals.

High temperature–protective equipment Protective equipment designed to safeguard the wearer from high temperatures during short-term exposures.

Label A smaller version (100-mm [4-in.] diamond-shaped marking) of a placard, placed on any side of an individual box or smaller package.

Level A protection A heavy encapsulating suit that provides the highest level of chemical protection for skin, eyes, and respiration.

Level B protection PPE used when the type and atmospheric concentration of identified substances require a high level of respiratory protection but less skin protection.

Level C protection PPE used when the type of airborne substance is known, its concentration is measured, the criteria for using APRs are met, and skin and eye exposure is unlikely.

Level D protection PPE used when the atmosphere contains no known hazard, and when work functions preclude splashes, immersion, or the potential for unexpected inhalation of or contact with hazardous levels of chemicals.

Liquid splash–protective clothing The garment portion of a CPE that is designed and configured to protect the wearer against chemical liquid splashes but not against chemical vapours or gases. (NFPA 472)

NFPA 704 hazard identification system A hazardous materials marking system designed for fixed-facility use.

Penetration The movement of matter through a suit's closures, including zippers, buttonholes, seams, flaps, or other design features of chemical-protective clothing, as well as through punctures, cuts, and tears. (NFPA 472)

Permeation A chemical action involving the movement of chemicals on a molecular level through intact material. (NFPA 472)

Permissible exposure limit (PEL) The maximum permitted 8-hour, time-weighted average concentration of an airborne contaminant per United States regulations. (NFPA 5000, *Building Construction and Safety Code*®)

Placard Signage required to be placed on all four sides of highway transport vehicles, railroad tanker cars, and other forms of hazardous materials transportation that identifies the hazardous contents of the vehicle using a standardized system with 250-mm (10.75-in.) diamond-shaped indicators.

Powered air-purifying respirator (PAPR) An air-purifying respirator with a hood or helmet, breathing tube, canister, cartridge, filter, and a blower that passes ambient air through the purifying element. The blower can be stationary or portable. It can also be designed and equipped with a full face piece, in which case it is referred to as a full-face-piece powered air-purifying respirator (FFPAPR). (NFPA 1404)

Recommended exposure level (REL) The established standard limit of exposure to a hazardous material. It is based on the maximum time-weighted concentration at which 95 percent of exposed, healthy adults suffer no adverse effects over a 40-hour workweek. RELs are set by the NIOSH but do not have the force of United States federal law.

Shelter-in-place A method of safeguarding people in a hazardous area by keeping them in a safe atmosphere, usually inside structures.

Shipping paper A shipping order, bill of lading, manifest, or other shipping document serving a similar purpose and containing the information required by regulations of Transport Canada. (NFPA 498)

Supplied-air respirator (SAR) An atmosphere-supplying respirator for which the source of breathing air is not designed to be carried by the user. (NFPA 1404)

Threshold limit value–ceiling (TLV–C) The concentration of a hazardous material to which a worker should not be exposed, even for an instant.

Threshold limit value–short-term exposure limit (TLV–STEL) The maximum concentration of a hazardous material to which a worker can sustain a 15-minute exposure, not more than four times daily, with a 60-minute rest period required between each exposure; the lower the value, the more toxic the substance.

Threshold limit value–skin A designation indicating that direct or airborne contact with a material could result in significant exposure from absorption through the skin, mucous membranes, and eyes.

Threshold limit value–time-weighted average (TLV–TWA) The airborne concentration of a material to which a worker can be exposed for 8 hours a day, 40 hours a week, and not suffer any ill effects.

Toxic inhalation hazard (TIH) material A gas or volatile liquid that is extremely toxic to humans and poses a hazard to health during transportation.

Vapour-protective clothing The garment portion of a CPE that is designed and configured to protect the wearer against chemical vapours or gases. (NFPA 472)

Waybill Shipping document for railroad transportation.

While you are conducting an commercial inspection in the industrial part of your first-due territory, your engine company is dispatched for a multiple-vehicle accident just down the street. Upon arrival on the scene, you observe a box truck and a highway cargo tank overturned in the middle of the street. Product is leaking out the cargo tank top. Your officer assigns you the task of sizing up the situation and determining the type of materials involved.

1. Given the task of scene assessment, what should your first action be?
   A. Look for SDS labels.
   B. Try to locate the driver.
   C. Use your initial senses (sight, smell) from a safe distance to determine the spilled product.
   D. Call CANUTEC.

2. What should you look for to determine if the spilled product is a hazardous material?
   A. The name of the trucking company on the door of vehicle
   B. Placards
   C. Writing on the tank
   D. A supervisor

3. An ERG is _____.
   A. a HazMat information manual that will identify products by assigned designations
   B. a PPE system for HazMat
   C. a foam system for HazMat
   D. a HazMat response company

4. What is the next step once you have identified the product that has spilled?
   A. Direct your crew to get attack line prepared for action.
   B. Inform any bystanders of the hazard.
   C. Set up a staging area.
   D. Inform your IC and contact CANUTEC.

It is the start of the Canada Day weekend. You and your company are sitting in the station discussing your plans for the weekend when the tones go off. There is a Hazardous Materials Incident in progress on the major interstate that cuts through your district.

You are assigned to the engine company along with two other fire fighters and a fire officer. As the engine turns onto the entrance ramp to the interstate, you notice that traffic is already backed up. As you approach the scene, you see three semi-tractor trailers involved in a multiple vehicle accident. Two of the drivers are walking around, but the third is trapped in his truck. You see that the last truck has several placards and is leaking a large amount of liquid from its trailer.

1. What key decision-making tools would you use to identify the hazards to personnel on the scene?
2. To whom would you report your observations?
3. How much detail should you provide when reporting your observations?
4. Which resources would you be able to use to assist your fire officer in identifying the product(s), shipper, and emergency contacts for the truck that is leaking product?

Hazardous Materials: Operations and Specialties

© Steve Morgan/Alamy Images

Fire Fighter I

Knowledge Objectives

After studying this chapter, you will be able to:

- Describe occupancies that may contain hazardous materials. (NFPA 472, 5.1, 5.1.2, 5.1.2.2, 5.2, 5.2.1, 6.6.1.1.2 , p 925)
- Describe the general characteristics of a bulk storage container. (NFPA 472, 5.2.1.1, 5.2.1.2.2, 6.6.1.1.2 , p 926)
- Describe the common types of bulk storage containers.
- List the hazardous materials commonly stored in each type of bulk storage container. (NFPA 472, 5.2.1.3, 6.6.1.1.2 , p 926)
- Describe the general characteristics of nonbulk storage vessels. (NFPA 472 5.2.1.1, 5.2.1.1.5, 5.2.1.2, 6.6.1.1.2 , p 927)
- Describe the general characteristics of drums. (NFPA 472, 5.2.1.1, 5.2.1.1.5, 6.6.1.1.2 , p 927–928)
- Describe the general appearance of intermediate bulk containers and ton containers. (NFPA 472, 5.2.1.1, 5.2.1.1.6, 6.6.1.1.2 , p 926–927)
- Describe the general characteristics of bags. (NFPA 472, 5.2.1.1, 5.2.1.1.5, 6.6.1.1.2 , p 928)
- List the information found on a pesticide label. (NFPA 472, 5.2.1.1, 5.2.1.3.2, 6.6.1.1.2 , p 928)
- Describe the general characteristics of carboys. (NFPA 472, 5.2.1.1, 5.2.1.1.5, 6.6.1.1.2 , p 928)
- Describe the general characteristics of cylinders. (NFPA 472, 5.2.1.1, 5.2.1.1.5, 6.6.1.1.2 , p 928–929)
- Describe the common types of cylinders. (NFPA 472, 5.2.1.1, 5.2.1.1.5, 6.6.1.1.2 , p 928–929)
- Describe the types of chemical tankers that transport hazardous materials and their common characteristics. (NFPA 472, 5.2.1.1, 5.2.1.1.3, 5.2.1.1.4, 6.6.1.1.2 , p 930)
- Describe the types of railcars that transport hazardous materials and their common characteristics. (NFPA 472, 5.2.1.1, 5.2.1.1.1, 5.2.1.1.2, 5.2.1.2.1, 6.6.1.1.2 , p 930–932)
- Describe how to identify the product, owner, and emergency telephone number on a pipeline marker. (NFPA 472, 5.2.1.1, 5.2.1.3.1, 6.6.1.1.2 , p 932)
- Describe the role of the Canadian Transport Emergency Centre (CANUTEC)/CHEMTREC® during a hazardous materials incident. (NFPA 472, 5.1.2.2, 5.2.2, 6.6.1.1.2 , p 932–933)
- List the information CANUTEC/CHEMTREC needs in order to assist with a hazardous materials incident. (NFPA, 472, 5.1.2.2, 5.2.2, 6.6.1.1.2 , p 932–933)

- Describe the common containers used to hold radioactive materials. (NFPA 472 5.2.1.1, 5.2.1.1.7, 5.2.1.3.3, 6.6.1.1.2 , p 933–934)
- List and describe the hazardous materials incident levels. (NFPA 472, 5.4.3, 6.6.1.1.2 , p 934)
- List the information to provide when reporting a hazardous materials incident. (NFPA 472, 5.4.3, 5.5.2, 6.6.1.1.2 , p 934–935)
- Describe how to plan an initial response. (NFPA 472, 5.1.2.2, 5.2.1.4, 5.3, 5.3.2 , p 935)
- List and describe defensive objectives. (NFPA 472, 5.1.2.2 , p 936)
- List and describe defensive actions. (NFPA 472, 5.1.2.2 , p 936)
- Describe how to estimate the size and scope of a hazardous materials/weapons of mass destruction (WMD) incident. (NFPA 472, 5.2.4 , p 937)
- Describe the resources available for determining the concentrations of a released hazardous material. (p 937)
- Describe a secondary attack and its impact on responders. (NFPA 472, 5.3.1 , p 937)
- List the signs that indicate the possible presence of a secondary device. (p 937)
- Explain how the Incident Management System (IMS) is adapted for hazardous materials incidents. (NFPA 472, 5.1.2.2 , p 937, 939)
- Identify the considerations for determining the location of the incident command post (ICP). (NFPA 472, 5.4.3 , p 939)
- Describe the safety precautions to be observed, including those for both heat and cold stress, when approaching and working at hazardous materials incidents. (NFPA 472, 5.3.3, 6.6.1.1.2, 6.6.3.2 , p 939–941)
- Describe the signs and symptoms of heat cramps. (NFPA 472, 5.4.4, 6.6.1.1.2 , p 939–940)
- Describe the signs and symptoms of heat exhaustion. (NFPA 472, 5.4.4, 6.6.1.1.2 , p 939–940)
- Describe how to prevent cold injuries. (NFPA 472, 5.4.4, 6.6.1.1.2 , p 940–941)
- Describe the physical capabilities required and limitations of personnel working in personal protective equipment (PPE). (NFPA 472, 5.4.4, 6.6.1.1.2 , p 941–942)
- List the three control zones. (NFPA 472, 5.4.1, 6.6.1.1.2 , p 942–943)
- Define *hot zone*, and describe the tasks performed in the zone. (NFPA 472, 5.4.1, 6.6.1.1.2 , p 942)

CHAPTER

32

© Photos.com

Skills Objectives

After studying this chapter, you will be able to perform the following skills:

- Use a multi-gas meter to provide atmospheric monitoring. (p 943–944)
- Utilize absorption/adsorption to manage a hazardous materials incident. (NFPA 472, 6.6.3, 6.6.3.1, 6.6.4, 6.6.4.1, 6.6.5 , p 946–947)
- Construct a dike. (NFPA 472, 6.6.3, 6.6.3.1, 6.6.4, 6.6.4.1, 6.6.5 , p 946)
- Construct an overflow dam. (NFPA 472, 6.6.3, 6.6.3.1, 6.6.4, 6.6.4.1, 6.6.5 , p 947)
- Construct an underflow dam. (NFPA 472, 6.6.3, 6.6.3.1, 6.6.4, 6.6.4.1, 6.6.5 , p 947–948)
- Construct a diversion. (NFPA 472, 6.6.3, 6.6.3.1, 6.6.4, 6.6.4.1, 6.6.5 , p 948)
- Construct a retention system. (NFPA 472, 6.6.3, 6.6.3.1, 6.6.4, 6.6.4.1, 6.6.5 , p 948)
- Perform vapour dispersion to manage a hazardous materials incident. (NFPA 472, 6.6.3, 6.6.3.1, 6.6.4, 6.6.4.1, 6.6.5 , p 949)
- Perform vapour suppression to manage a hazardous materials incident. (NFPA 472, 6.6.3, 6.6.3.1, 6.6.4, 6.6.4.1, 6.6.5 , p 949)
- Perform emergency decontamination. (NFPA 472, 5.3.4 , p 952–953)
- Perform mass decontamination. (NFPA 472, 5.3.4 , p 954)
- Perform technical decontamination. (NFPA 472, 5.3.4, 6.6.4.2 , p. 957–958)

Additional NFPA Standards

- NFPA 1072, *Standard for Hazardous Materials/Weapons of Mass Destruction Emergency Response Personnel Professional Qualifications*
- NFPA 1500, *Standard on Fire Department Occupational Safety and Health Program*
- NFPA 1561, *Standard on Emergency Services Incident Management System*
- NFPA 1584, *Standard on the Rehabilitation Process for Members During Emergency Operations and Training Exercises*

A 208-litre (55-gallon) steel drum has fallen off a loading dock at an abandoned warehouse in your first-due response area. The warehouse borders a small residential area with garden apartments. Your engine company is dispatched as part of the initial response and is the first unit to arrive on the scene. A police officer meets you at the front gate and points out the location of the drum on the ground. From the gate, you can see the drum, and it is obviously leaking from the top closure. The police officer has already checked the area, and there are no occupants present, except for at the playground in the residential complex, where children are playing. You and your crew must manage the incident until the department's hazardous materials team arrives.

1. Which immediate actions can you and your crew safely take, based on the level of training that you currently have?
2. What additional information should you obtain while waiting for the hazardous materials team to arrive?
3. Which resources do you currently have available to use, and which resources will you request to respond to the incident?

Introduction

The modern fire service must respond to both existing and emerging threats to lives and property. In addition to fighting fires, today's fire fighters may be called to incidents involving chemical spills, gas leaks, emergencies at industrial plants, or railroad or truck crashes resulting in material leaks or spills. In hazardous materials incidents, it is not always possible to know the whole story before action is required. Even so, you may be able to identify the incident as involving hazardous materials based on information from the dispatcher, knowledge of the response area, and visual, auditory, or olfactory (odourous) clues. A leaking cargo trailer, coloured placards on abandoned drums, and certain odours are all indications that hazardous materials may be present.

The primary functions of fire personnel at the scene of a hazardous materials incident include, but are not limited to:

- Scene size-up
- Accurately identifying the substance
- Establishing control zones
- Identifying safety hazards
- Identifying the need for search and rescue
- Identifying the container or transport vehicle
- Establishing an incident action plan (IAP) and achievable objectives
- Understanding the limitations of firefighting personal protective equipment (PPE) in relation to chemical, biological, and radiological exposure
- Assisting in establishing accountability and decontamination areas as is required
- Ensuring that the scene is controlled at all times and that other agencies are involved in the process

This chapter is designed to give you the basic information required to recognize, respond to, and take action at a hazardous materials incident. Fire personnel must be able to identify the nature and severity of the incident before creating a plan of action. Fire fighters must follow their local suggested operating guidelines (SOGs), or standard operating procedures (SOPs).

Occupancy and Location

Hazardous materials incidents are not confined to places such as chemical facilities or nuclear power plants; they can occur almost anywhere. Hazardous materials are stored in warehouses, hospitals, laboratories, industrial occupancies, residential garages, bowling alleys, home-improvement centres, garden supply stores, restaurants, and many other facilities and businesses in your response area. So many different chemicals are used in daily life that you could encounter several types of hazardous materials as you respond to emergencies. For example, a fire in a pesticide-storage facility could release dangerous chemicals; a person exposed to a solvent while cleaning automotive parts could require emergency medical assistance; a rollover on the highway could involve leaking diesel fuel. Nearly any emergency—or even nonemergency—response might potentially include a hazardous materials component.

The location and type of occupancy are two indicators of the presence of hazardous materials. The chapter "Hazardous Materials: Introduction" outlines the basic reporting requirements for occupancies that use hazardous materials. By reporting the kinds and quantities of hazardous materials present to the local fire department, facilities that pose a significant threat to the community become part of a comprehensive community plan. These reports also alert fire departments to the need for a preincident plan for the site, so that fire fighters are aware of the location and types of hazardous materials involved.

Fire Fighter Safety Tips

© Photos.com

Package delivery services such as UPS, DHL, and FedEx carry shipments that could contain hazardous materials. Although small amounts do not require placards, the package may still require labels. If a package is damaged, it could become a hazardous material. If a delivery transport vehicle is involved in an accident, multiple packages could open, allowing their contents to mix and produce a hazardous material.

Containers

A **container** is any vessel or receptacle that holds material, including storage vessels, pipelines, and packaging. Often, container type, size, and material provide important clues about the nature of the substance inside; fire fighters should not rely solely on the type of container, however, when making a determination about hazardous materials.

Red phosphorus from a drug lab, for example, might be found in an unmarked plastic jug. Acetone or other solvents might be stored in 208-L (55-gal.) steel drums with two capped openings on the top. Sulphuric acid at 97 percent concentration is typically found in a polyethylene drum, but that drum could be black, red, white, or blue. The same sulphuric acid might also be found in a 3.8-L (1-gal.) amber glass container. Hydrofluoric acid, in contrast, is incompatible with silica (glass) and would be stored in a plastic container.

■ Container Type

Hazardous materials can be found in many different types of containers—from 3.8-L (1-gal.) glass containers to 18,925-L (5,000-gal.) steel storage tanks. Steel or polyethylene plastic drums, bags, high-pressure gas cylinders, tank cars, plastic buckets, above ground and underground storage tanks, truck tankers, and pipelines are all types of containers used with hazardous materials.

Some very recognizable chemical containers, such as 208-L (55-gal.) drums and compressed gas cylinders, can be found in almost every commercial building. Materials in cardboard drums are usually in solid form. Stainless-steel containers are used to hold particularly dangerous chemicals, and cold liquids are kept in Thermos-like **dewar containers** designed to maintain the appropriate temperature.

Containers may be large or small. Sometimes, smaller containers may be stored and shipped in larger containers. One way to distinguish containers is to divide them into two separate categories: bulk storage containers and nonbulk storage vessels.

■ Container Volume

Bulk storage containers, or large-volume containers, are defined by their internal capacity, based on the following measures:

- Liquids—More than 450 L (119 gal.)
- Solids—More than 400 kilograms (882 pounds)
- Gases—More than 400 kg (882 lb.)

Bulk storage containers include fixed tanks, large transportation tankers, totes, and intermodal tanks. In general, these kinds of containers are found in occupancies that rely on and need to store large quantities of a particular chemical. Most manufacturing facilities have at least one bulk storage container. Often, these bulk storage containers are surrounded by a supplementary containment system to help control an accidental release. **Secondary containment** is an engineered method to control spilled or released product, if the main containment vessel fails. A 18,925-L (5000-gal.) vertical storage tank, for example, might be surrounded by a series of short walls that form a catch basin around the tank. This basin

typically can hold the entire volume of the tank and accommodate water from hose lines or sprinkler systems in the event of fire. Many storage vessels, including 208-L (55-gal.) drums, may have secondary containment systems. If bulk storage containers in your response area have secondary containment systems, large leaks in bulk storage containers will be easier to deal with.

Large-volume horizontal tanks are also common. When stored above ground, these tanks are referred to as above ground storage tanks (ASTs); if they are placed underground, they are known as underground storage tanks (USTs). These tanks can hold amounts ranging from 1,000 (3785 gal.) to more than 10,000 L (37,850 gal.) of product and are usually made of aluminum, steel, or plastic. USTs and ASTs can be pressurized or nonpressurized. Nonpressurized horizontal tanks are usually constructed from steel or aluminum. Because it is difficult to relieve internal pressure in these tanks, they are dangerous when exposed to fire. Typically, they hold flammable or combustible materials such as gasoline, oil, or diesel fuel.

Pressurized horizontal tanks have rounded ends and large vents or pressure-relief stacks. The most common above ground pressurized tanks are liquid-propane and liquid-ammonia tanks, which can hold volumes ranging from 1,000 L (3785 gal.) to more than 10,000 L (37,850 gal.) of product. These tanks also contain a small vapour space; anywhere from 10 to 15 percent of the tank's total capacity can be vapour. Refer to the discussion of liquefied gases in the chapter "Hazardous Materials: Introduction" for more information on these materials.

Another common bulk storage vessel is the tote. **Totes** are portable plastic tanks surrounded by a stainless-steel web, which adds both structural stability and protection. They can hold approximately 1000 L (3785 gal.) of product and may contain any type of chemical, including flammable liquids, corrosives, food-grade liquids, and oxidizers.

Shipping and storage totes can be hazardous. These containers are often stacked atop one another and moved with a forklift. A mishap during either the loading or moving process can compromise the container. Because totes have no secondary containment mechanism, any leak will create a large puddle. Additionally, the steel webbing around the tote makes leaks difficult to patch.

Intermodal tanks (IM or IMO tanks) are both shipping and storage vehicles TABLE 32-1. They hold between 18,925 and 22,710 L (5000 and 6000 gal.) of product and can be either pressurized or nonpressurized. In most cases, an intermodal tank is shipped to a facility, where it is stored and used, and then returned to the shipper for refilling. Intermodal tanks can be shipped by any mode of transportation—air, sea, or land. These horizontal, round tanks are surrounded by, or are part of, a boxlike steel framework for shipping. Three basic types of intermodal tanks are routinely encountered:

- IM101 containers have a 22,710-L (6000-gal.) capacity with internal working pressures between 172 and 689 kilopascals (25 and 100 pounds per square inch [psi]). These containers typically carry mild corrosives, food-grade products, or flammable liquids.
- IM102 containers have a 22,710-L (6000-gal.) capacity with internal working pressures between 96 and

TABLE 32-1	Common Bulk Storage Vessels, Locations, and Contents	
Tank Shape	**Common Locations**	**Hazardous Materials Commonly Stored**
Underground tanks	Residential, commercial	Fuel oil and combustible liquids
Covered floating-roof tanks	Bulk terminal and storage	Highly volatile flammable liquids (especially ethanol-blended fuels)
Cone-roof tanks	Bulk terminal and storage	Combustible liquids
Open floating-roof tanks	Bulk terminal and storage	Flammable and combustible liquids
Dome-roof tanks	Bulk terminal and storage	Combustible liquids
High-pressure horizontal tanks	Industrial storage and terminal	Flammable gases, chlorine, ammonia
High-pressure spherical tanks	Industrial storage and terminal	Liquid-propane gas, liquid-nitrogen gas
Cryogenic liquid storage tanks	Industrial and hospital storage	Oxygen, liquid-nitrogen gas

207 KPa (14 and 30 psi). They carry primarily flammable liquids and corrosives.

■ Spec 51, or IMO Type 5 (internationally), containers are high-pressure vessels with internal pressures of 1000 kilopascals that carry liquefied gases, such as propane and butane.

■ Nonbulk Storage Vessels

Essentially, **nonbulk storage vessels** are all other types of containers. Nonbulk storage vessels can hold volumes ranging from a few litres to many litres; they include drums, bags, carboys, compressed gas cylinders, cryogenic containers, and more. Nonbulk storage vessels hold commonly used commercial and industrial chemicals, such as solvents, industrial cleaners, and compounds. This section describes the most common nonbulk storage vessels.

Drums

Drums are easily recognizable, barrel-like containers. They are used to store a wide variety of substances, including food-grade materials, corrosives, flammable liquids, and grease. Drums may be constructed of low-carbon steel, polyethylene, cardboard, stainless steel, nickel, or other hybrid materials **FIGURE 32-1**. Generally, the nature of the material dictates the construction of the storage drum. Steel utility drums, for example, hold flammable liquids, cleaning fluids, oil, and other noncorrosive chemicals. Polyethylene drums are used for corrosives such as acids, bases, oxidizers, and other materials that cannot be stored in steel containers. Cardboard drums hold solid materials such as soap flakes, sodium-hydroxide pellets, and food-grade materials. Stainless-steel or other heavy-duty drums generally hold materials too aggressive for either plain steel or polyethylene.

Closed-head drums have a permanently attached lid with one or more small openings called **bungs**. Typically, these openings are threaded holes sealed by caps, which can be removed only by using a special wrench called a bung wrench. Closed-head drums usually have one 5-centimetre (2-inch) bung and one 1.9-cm (0.75-in.) bung. The larger hole is used to pump product from the drum, while the smaller bung functions as a vent.

An open-head drum has a removable lid fastened to the drum with a ring. The ring is tightened with a clasp or

A. © Joe Potato Photo/iStockphoto

B. © Natalia7/ShutterStock, Inc.

C. © sherwood/ShutterStock, Inc.

FIGURE 32-1 Drums may be constructed of **A.** cardboard, **B.** polyethylene, or **C.** stainless steel.

a threaded nut-and-bolt assembly. The lid of an open-head drum may or may not have bung-type openings.

Bags

Bags are commonly used to store solids and powders such as cement powder, sand, pesticides, soda ash, and slaked lime. Storage bags may be constructed of plastic, paper, or plastic-lined paper. Bags come in different sizes and weights, depending on their contents.

Pesticide bags must be labelled with specific information **FIGURE 32-2**. Fire fighters can learn a great deal from the label:

- Name of the product
- Statement of ingredients
- Total amount of product in the container
- Manufacturer's name and address
- Materials originating in Canada carry a Pest Control Products Act registration number. The Canadian Transport Emergency Centre (CANUTEC) operated by Transport Canada provides information about these materials when provided the registration number. Canadian products also have the same signal words and required information as they would in the United States (e.g., Environmental Protection Agency [EPA] registration number, which provides proof that the product was registered with the EPA, and EPA establishment number, which shows where the product was manufactured).
- **Signal words** to indicate the relative toxicity of the material:
  - Danger: Poison—Highly toxic by all routes of entry
  - Danger—Severe eye damage or skin irritation
  - Warning—Moderately toxic
  - Caution—Minor toxicity and minor eye damage or skin irritation

- Practical first-aid treatment description
- Directions for use
- Agricultural use requirements
- Precautionary statements, such as mixing directions or potential environmental hazards
- Storage and disposal information
- Classification statement on who may use the product

In addition, every pesticide label must include the following statement: "Keep out of reach of children."

Carboys

Some corrosives and other types of chemicals are transported and stored in vessels called **carboys** **FIGURE 32-3**. A carboy is a glass, plastic, or steel container that holds 19–57 L (5–15 gal.) of product. Glass carboys often have a protective wood or fibreglass box to help prevent breakage.

Nitric, sulphuric, and other strong acids are transported and stored in thick glass carboys protected by a wooden or Styrofoam crate to shield the glass container from damage during normal shipping.

Cylinders

Several types of **cylinders** are used to hold liquids and gases **FIGURE 32-4**. Uninsulated compressed gas cylinders are used to store gases such as nitrogen, argon, helium, and oxygen. They have a range of sizes and pressure readings. A typical oxygen cylinder used for medical purposes, for example, has a gas pressure reading of approximately 13,790 kPa (2000 psi). Very large compressed gas cylinders found at a fixed facility may have pressure readings of 34,475 kPa (5000 psi) or greater.

The high pressures exerted by these cylinders are potentially dangerous to fire fighters and others. If the cylinder is punctured or the valve assembly fails, the rapid release of compressed gas will turn the cylinder into a deadly missile. Additionally, if the cylinder is heated rapidly, it could explode with tremendous force, spewing product and metal fragments

FIGURE 32-2 A pesticide container must be labelled with specific information.

FIGURE 32-3 A carboy is used to transport and store corrosive chemicals.

Courtesy of EMD Chemicals, Inc.

over long distances. Compressed gas cylinders do include pressure-relief valves, but those valves might not be sufficient to relieve the pressure created by a fast-growing fire; the result would be a catastrophic explosion.

A propane cylinder is another type of compressed gas cylinder. Propane cylinders have lower pressures (1379–2068 kPa [200–300 psi]) and contain a liquefied gas. Liquefied gases, such as propane, are subject to the phenomenon boiling liquid/expanding vapour explosion (BLEVE). For more information on BLEVEs, refer to the chapter "Fire Suppression."

The low-pressure dewar is another common cylinder type. Dewars are Thermos-like vessels designed to hold **cryogenic liquids** (also known as cryogens)—that is, gaseous substances that have been chilled until they liquefy **FIGURE 32-5**. Typical cryogens are oxygen, helium, hydrogen, argon, and nitrogen. Under normal conditions, each of these substances is a gas. A complex process turns them into liquids, which can be stored in dewar containers; nitrogen, for example, becomes a liquid at –195.6°C (–320°F) and must be kept that cold to remain a liquid.

Cryogens pose a substantial threat if the dewar fails to maintain the low temperature. Cryogens have large expansion ratios—even larger than that of propane. Cryogenic helium, for example, has an expansion ratio of approximately 750–1; thus, if one volume of liquid helium is warmed to room temperature and vapourized in a totally enclosed container, it can generate a pressure of more than 99,977 kPa (14,500 psi). To offset this risk, cryogenic containers usually have two pressure-relief devices: a pressure-relief valve and a frangible (easily broken) disc.

A. Courtesy of Glen Rudner

B. Courtesy of Glen Rudner

FIGURE 32-4 Cylinders. **A.** 45-kg (100-lb.) cylinder. **B.** 68-kg (150-lb.) cylinder.

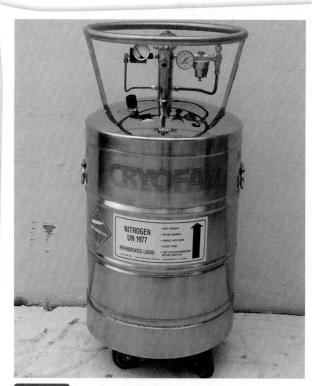

FIGURE 32-5 A small cryogenic dewar.

Courtesy of Cryofab, Inc.

Transporting Hazardous Materials

Although rail, air, and sea transport are used to deliver chemicals and other hazardous materials to their destinations, the most common method of transport is over land, by highway transportation vehicles. Even when another method of transport is used, motor vehicles often carry the shipments from the station, airport, or dock to the factory or plant. For this reason, fire fighters must become familiar with the types of chemical transport vehicles, or tankers, they might encounter during a traffic emergency.

One of the most common chemical tankers is the **TC306/ MC306/DOT406 flammable-liquid tanker** FIGURE 32-6A, which typically carries gasoline or other flammable and combustible materials. The oval-shaped tank is pulled by a diesel tractor and can carry between 22,710 and 37,850 L (6000 and 10,000 gal.). The TC306/MC306/DOT406 is nonpressurized, usually made of aluminum, and off-loaded through valves at the bottom of the tank. It is a common highway sight and a reliable way to transport chemicals.

A similar vehicle is the **TC307/MC307/DOT407 chemical hauler**, which has a round or horseshoe-shaped tank and typically carries 22,710–26,495 L (6000–7000 gallons) FIGURE 32-6B. The TC307/MC307/DOT407 is used to transport flammable liquids, mild corrosives, and poisons. Tanks that transport corrosives may have a rubber lining.

The **TC312/MC312/DOT412 corrosives tanker** is used for concentrated sulphuric and nitric acids and other corrosive substances FIGURE 32-6C. This tanker has a smaller diameter than either the TC306/MC306/DOT406 or the TC307/ MC307/DOT407 and is often, though not always, characterized by several reinforcing rings around the tank. These rings provide structural stability during transportation and in the event of a rollover. The inside of an TC312/MC312/DOT412 tanker operates at a pressure of approximately 517 kPa (75 psi) and holds approximately 22,710 L (6000 gal.).

The **TC331/MC331 pressure cargo tanker** carries materials such as ammonia, propane, and butane FIGURE 32-6D. The tank has rounded ends, typical of a pressurized vessel, and is commonly constructed of steel with a single tank compartment. The TC331/MC331 operates at a pressure of approximately 2068 kPa (300 psi) and could be a significant explosion hazard if it accidentally rolled over or is threatened by fire. Fire fighters must use great care when dealing with this type of highway emergency.

The **TC341/CGA341/MC338 cryogenic tanker** operates much like the dewar containers described earlier and carries many of the same substances FIGURE 32-6E. This low-pressure tanker relies on tank insulation to maintain the low temperatures required by the cryogens it carries. A boxlike structure containing the tank control valves is typically attached to the rear of the tanker. Special training is required to operate valves on this and any other tanker. An untrained individual who attempts to operate the valves may disrupt the normal operation of the tank, thereby compromising its ability to keep the liquefied gas cold and creating a potential explosion hazard.

Tube trailers carry compressed gases, such as hydrogen, oxygen, helium, and methane FIGURE 32-6F. Essentially, they are high-volume transportation vehicles comprising several individual cylinders banded together and affixed to a trailer. The individual cylinders on the tube trailer work much like the smaller compressed gas cylinders discussed earlier. These large-volume cylinders operate at a pressure of 20,685–34,475 kPa (3000–5000 psi). One trailer may carry several different gases in individual tubes. Typically, a valve control box can be found toward the rear of the trailer, and each individual cylinder has its own relief valve. These trailers can frequently be seen at construction sites or at facilities that use great quantities of these materials.

Dry bulk cargo tanks also are commonly seen on the road and carry dry bulk goods, such as powders, pellets, fertilizers, and grain FIGURE 32-6G. These tanks are not pressurized, but they may use pressure to off-load product. Dry bulk cargo tanks are generally V shaped with rounded sides that funnel the contents to bottom-mounted valves. Intermodal tanks can be laid out in a wide variety of configurations and are found in all modes of transportation FIGURE 32-6H.

■ Railroad Transportation

Railroads move almost two million carloads of freight per year yet are involved in relatively few hazardous materials incidents. Railway tank cars may carry volumes up to 113,550 L (30,000 gal.) and have the potential to create large leaks or vapour clouds if their contents leak. Hazardous materials incidents involving railroad transportation, although relatively rare, are extremely dangerous.

Fortunately, there are only three basic railcar configurations that fire fighters should recognize: nonpressurized, pressurized, and special use. Each type of railcar has a distinct profile that can be recognized from a long distance. Additionally, railcars are usually labelled on both sides with the volume and maximum working pressure inside the tank. Dedicated haulers often have the chemical name clearly visible.

DOT-111 nonpressurized (general service) railcars typically carry general industrial chemicals, consumer products such as corn syrup, flammable and combustible liquids, and mild corrosives. Nonpressurized railcars are easily identified by looking at the top of the car: Nonpressurized railcars will have visible valves and piping without a dome cover.

DOT-105 and DOT-112 pressurized railcars feature an enclosed dome on the top of the railcar. These cars transport materials such as propane, ammonia, ethylene oxide, and chlorine. Pressurized cars have internal working pressures ranging from 689 to 3392 kPa (100 to 500 psi) and are equipped with relief valves similar to those on bulk storage tankers. Unfortunately, the high volumes carried in these cars can generate long-duration, high-pressure leaks, which may be impossible to stop.

Special-use railcars include boxcars, flatcars, corrosive tank cars, and high-pressure compressed gas tube cars. DOT-113 cryogenic railway tank cars carry products that have varying temperatures that are less than 704.4°C (1300°F) FIGURE 32-7. Products such as oxygen, helium, hydrogen, argon, and others are shipped via this method to larger industrial facilities that use these materials in everyday industrial processes. In each case, the hazard will be unique to the particular railcar and its contents. Do not assume that only the chemical tank cars pose

A.

Courtesy of Polar Tank Trailer L.L.C.

B.

Courtesy of Polar Tank Trailer L.L.C.

C.

D.

Courtesy of Rob Schnepp

E.

Courtesy of Jack B. Kelley, Inc.

F.

Courtesy of Jack B. Kelley, Inc.

G.

Courtesy of Polar Tank Trailer L.L.C.

H.

Courtesy of Eurotainer

FIGURE 32-6 **A.** An MC306 flammable-liquid tanker. **B.** An MC307 chemical hauler. **C.** An MC312 corrosives tanker. **D.** An MC331 pressure cargo tanker. **E.** An MC338 cryogenic tanker. **F.** A tube trailer. **G.** A dry bulk cargo tank. **H.** Intermodal tanks.

FIGURE 32-7 DOT-113 cryogenic tank car.

© Mark Gibson/Index Stock/Alamy Images

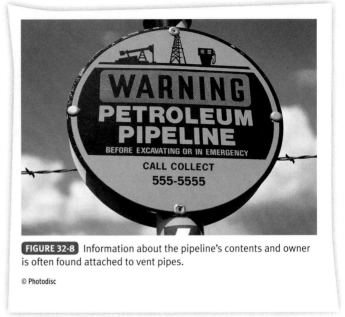

FIGURE 32-8 Information about the pipeline's contents and owner is often found attached to vent pipes.

© Photodisc

a threat; until you know what is in a particular car, assume the incident is a hazardous situation.

The train's engineer or conductor will have information about all the cars on that train and should be consulted during a railway emergency. Details about the shipping papers unique to rail transportation are presented later in this chapter.

■ Pipelines

Of the various methods used to transport hazardous materials, the high-volume **pipeline** is the least often involved in emergencies; however, every fire fighter should be aware of these structures and know where they run in their communities. Although the number of incidents involving pipelines is low, the incidents that have occurred have resulted in large losses of property and mass casualties and have affected many communities. In many areas, large-diameter pipelines transport natural gas, gasoline, diesel fuel, and other products from delivery terminals to distribution facilities. Pipelines are often buried underground, but they may be above ground in remote areas. The **pipeline right-of-way** is an area, patch, or roadway that extends a certain number of metres on either side of the pipe itself. This area is maintained by the company that owns the pipeline. The company is also responsible for placing warning signs at regular intervals along the length of the pipeline.

Pipeline warning signs include a warning symbol, the pipeline owner's name, and an emergency contact number **FIGURE 32-8** . Pipeline emergencies are complex events that require specially trained responders. If you suspect an emergency involving a pipeline, contact the owner of the line immediately. The company pipeline will dispatch a crew to assist with the incident.

Information about the pipe's contents and owner is also often found at the **vent pipes**. These inverted J-shaped tubes provide pressure relief or natural venting during maintenance and repairs. Vent pipes are clearly marked and are approximately 1 metre (3 feet) above the ground.

■ CANUTEC

Located in Ottawa, the **Canadian Transport Emergency Centre (CANUTEC)** is operated by Transport Canada. This national bilingual (English and French) advisory centre is part

of the Transportation of Dangerous Goods (TDG) Directorate. CANUTEC has a scientific data bank on chemicals manufactured, stored, and transported in Canada and is staffed by professional scientists who specialize in emergency response and are experienced in interpreting technical information and providing advice.

■ CHEMTREC®

Located in Falls Church, Virginia, established by the Chemical Manufacturers Association, and now operated by the American Chemical Council (ACC), the **Chemical Transportation Emergency Center (CHEMTREC®)** is a clearing house of emergency response information. The Canadian equivalent of CHEMTREC® is CANUTEC. Both centres provide a free service that connects fire fighters with chemical manufacturers, chemists, and other product specialists who can help during a chemical incident. The telephone number for CANUTEC is 613-996-6666; for cellular use (in Canada only), you can dial *666. The toll-free number for CHEMTREC® is 1-800-424-9300. Many fire fighters write the appropriate number inside their helmets so that they always have it in case of need. When calling CANUTEC/CHEMTREC®, be sure to provide the following information:

- Name and callback number
- Location of the actual incident or problem
- Name of the chemical involved in the incident, if known
- Shipper or manufacturer of the chemical, if known
- Container type
- Railcar or vehicle markings or numbers
- Shipping carrier's name
- Recipient of the material
- Local conditions and an exact description of the situation

When speaking with CANUTEC/CHEMTREC®, spell out all chemical names; if using a third party, such as a dispatcher, it is vital that you confirm all spellings so that there is no chance of a misunderstanding of any spelling or numbers. One number or letter could be vital; when in doubt, be sure to obtain clarification.

When CANUTEC is notified, Transport Canada may be notified; however, it will be necessary to contact municipal, regional, and provincial/territorial agencies directly to assist with any particular situation. Phone numbers for all CANUTEC and provincial/territorial resources can be found in the *Emergency Response Guidebook* (*ERG*). Fire fighters who operate at or near the United States border may need to be able to access CHEMTREC® and can do so both directly and through CANUTEC itself.

Radiation

Fire fighters must be able to recognize the situations in which radioactive materials could be encountered. Industries that routinely use radioactive materials include food testing labs, hospitals, medical research centres, biotechnology facilities, construction sites, and medical laboratories. Usually, some visual indicators (signs or placards) will indicate the presence of radioactive substances in a structure, but this is not always the case.

If you suspect a radiation incident at a fixed facility, you should initially ask for the facility's radiation safety officer. This person is responsible for the use, handling, and storage procedures for all the radioactive material at the site. He or she will likely be a tremendous resource to fire fighters and will know exactly which materials are being used at the facility.

If the incident occurs somewhere other than a fixed site, the presence of radiation might never be apparent. Radioactive materials cannot be detected by sight, smell, taste, nor the other senses; therefore, if you have any suspicion that the incident involves radiation, it is necessary to call a hazardous materials team or some other resource with radiation-detection capabilities.

Significant incidents involving radiation are few and far between. This is not to say accidents cannot or will not happen, but regulations for using, storing, and transporting significant radioactive sources are so comprehensive that the entire process has become quite safe. Most of the incidents you may encounter will involve low-level radioactive sources and can be handled safely. These low-level sources are typically found in radioactive packaging, industrial radioactive packaging, and Type A packaging FIGURE 32-9. The Type A packaging method is unique to radioactive substances and contains materials such as radiopharmaceuticals and other low-level emitters.

Type A packaging is characterized by an inner containment vessel of glass, plastic, or metal and packaging materials made of polyethylene, rubber, or vermiculite. This type of packaging is designed to protect the contents from damage during normal shipping and handling. The key is to be able to suspect, recognize, and understand when and where you may encounter radioactive sources involving Type A packaging.

More dangerous radioactive sources might be found in Type B packaging FIGURE 32-10. This type of containment vessel contains materials such as spent radioactive waste and other high-level emitters. Type B packages are designed to prevent greater exposure from the contents. The amount of protection is based on the potential severity of the hazard. Type B packages include small drums and heavily shielded casks weighing more than 100 metric tons.

Type C packages are authorized for international shipments of radioactive material by air. They are designed to withstand severe accident conditions associated with air transport. The Type C package performance requirements are significantly more stringent than those for Type B packages.

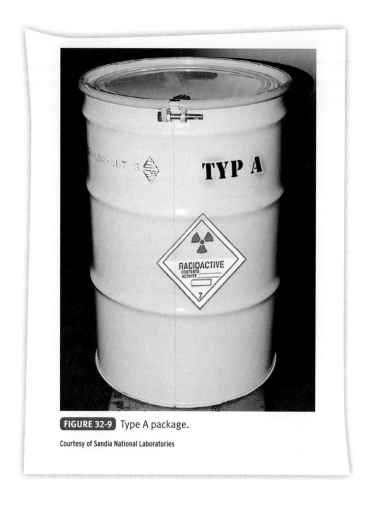

FIGURE 32-9 Type A package.
Courtesy of Sandia National Laboratories

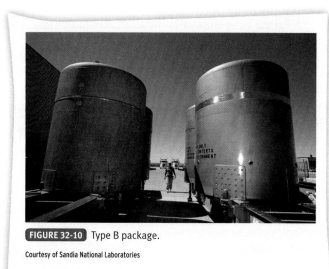

FIGURE 32-10 Type B package.
Courtesy of Sandia National Laboratories

Because Type C packages are not authorized for domestic use, it is unlikely that you would encounter them in Canada.

Determining a Hazardous Material

The first step after determining that a hazardous material is present is to identify those resources that can assist you in the process of reporting the hazardous materials incident. Fire fighters must follow their local SOGs/SOPs.

Who to Contact

Fire fighters should make a series of contacts after determining that they are responding to a hazardous materials incident. Some of these notifications are related to operational concerns; others are made for legal reasons TABLE 32-2 .

First and foremost, when a hazardous materials incident is detected, there should be an initial call for additional resources FIGURE 32-11 . The resources sought should include support personnel, trained hazardous materials technicians, and technical specialists who will help identify the hazardous material and control the incident. Additional calls should then request decontamination personnel and equipment. Other notifications could include the CANUTEC/CHEMTREC®, the National Response Center (United States), local and provincial/territorial environmental agencies, and the local emergency planning commission. A predetermined list of contact names, agencies, and numbers should be established and maintained at the dispatch centre.

No offensive action should be implemented until the identity of the hazardous material involved is confirmed. It makes no sense to try to deal with a hazardous material until it has been properly identified. This identification should be confirmed in a minimum of three references: For example, the hazardous material could first be identified by a placard, which may be used to look up information in the *ERG*. Then the shipping papers or Safety Data Sheets (SDSs) could be checked to ensure that they list the same hazardous material as the placard. Finally, the National Institute for Occupational Safety and Health (NIOSH) references and CANUTEC/CHEMTREC® could be consulted for a final confirmation of the identity of the hazardous material.

A variety of sources of information should be compared for consistency. If the information presented in the reference books is not consistent, the information that reflects a more

FIGURE 32-11 When a hazardous materials incident is detected, there should be an initial call for additional resources.

conservative course of action should be used. The protection and safety of fire fighters is always the first priority in any hazardous materials incident and cannot be compromised for any reason. Therefore, if two references indicate that the use of water is acceptable, but one reference cautions against it, the use of water should not be considered until further information is obtained. On-scene research should continue throughout the course of the incident.

After the material is identified, an operations-level responder should perform only actions that do not involve contact with the material. The responder must use full PPE during any activity and must complete decontamination procedures prior to leaving any area where the hazardous material is present.

FIRE FIGHTER Tips

If the hazardous material or the hazards of the product cannot be clearly identified, then the most conservative response strategy and tactics must be employed.

What to Report

When additional agencies are contacted regarding a potential hazardous materials incident, it is important that the

Table 32-2	Hazardous Materials Incident Levels	
Level I	**Level II**	**Level III**
Lowest level of threat. Small amount of hazardous material involved. Can usually be handled by the local fire department. Fire fighters must wear turnout gear and self-contained breathing apparatus (SCBA). Example: a small gasoline spill from a motor vehicle accident	A hazardous materials team is needed at this level. Fire fighters simply support the hazardous materials team. Additional PPE required will be specialized and carried only by the hazardous materials team. Civilian evacuations may be required. Decontamination may need to be performed. Example: a gasoline tanker overturns in a tunnel and begins spilling gasoline onto the highway.	The highest level of threat. Large-scale evacuations may be needed. Federal agencies will be called. Example: a ship in a highly populated harbour catches fire and begins to release chlorine vapours from its cargo area.

information given be as clear, concise, and accurate as possible. An error in spelling, an incorrect measurement, a mispronunciation of a chemical name, or incorrect identification of a hazardous substance may prove disastrous. The change or omission of just one letter in a chemical name, for example, could lead to incorrect identification of the material at hand. If a chemical is incorrectly identified, the proper response and safety procedures will not be carried out. This type of error can be deadly to both responders and the public. The fire fighter should keep information as simple as possible, spell out names that are complex or potentially confusing, and confirm that the receiver of the message has heard the correct information by repeating back what was heard.

If these agencies are to plan, prepare, and begin assisting fire fighters, they must have as much information as can be obtained. This will help to ensure that every possible situation can be taken into account and planned for, so as to prevent any further injury, property loss, or environmental damage. If possible, provide all of the following information:

- Exact address and specific location of the leak or spill
- Identification of indicators and markers of hazardous materials
- All colour or class information obtained from placards
- Four-digit United Nations (UN)/North American (NA) identification numbers for the hazardous materials
- Hazardous materials identification obtained from shipping papers or SDSs and the potential quantities of hazardous materials involved
- Description of the container, including its size, capacity, type, and shape
- The amount of chemical that could leak and the amount that has already leaked
- Exposures of people and the presence of special populations (e.g., children or elderly)
- Description of the environment in the immediate area
- Current weather conditions, including wind direction and speed
- A contact or callback number and two-way radio frequency or channel

Planning an Initial Response

When planning an initial hazardous materials incident response, the first priority is to consider the safety of the responding personnel. Responders are there to isolate, contain, and remedy the problem—not to become part of it. Proper incident planning will keep responders safe and provide a means to control the incident effectively, thereby preventing further harm to persons or property.

Planning the response begins with the initial call for help. Information obtained during this call is used to determine the safest, most effective, and fastest route to the hazardous materials scene. Choose a route that approaches the scene from an upwind and upgrade direction, so that natural wind currents blow the hazardous material vapours away from arriving responders. A route that places the responders uphill as well as upwind of the site is also desirable, so that any liquid or vapour flows away from responders.

Responders need to know the type of material involved. Is the material a solid, a liquid, or a gas? Is it contained in a drum, a barrel, or a pressurized tank? Response to a spill of a solid hazardous material will differ from response to a liquid release or vapour release. A solid is local and can be easily contained; whereas a released gas can be widespread and constantly moving, depending on the gas characteristics and weather conditions FIGURE 32-12 .

The characteristics of the affected area near the location of the spill or leak are also important factors in planning the response to an incident. If an area is heavily populated, evacuation procedures and a decontamination process must be established very early in the course of the incident; if the area is sparsely populated and rural, isolating the area from anyone trying to enter the location may be the top priority. A high-traffic area such as a major highway would necessitate immediate rerouting of traffic, especially during rush hours.

When responding to an incident, the more information that can be obtained, the better. If information is unknown or is unconfirmed, then assume the worst-case scenario. When planning for hazardous materials incidents, always provide the largest margin of safety possible.

A number of solutions to a given problem may exist. The challenge in dealing with hazardous materials incidents is arriving at a solution that can be employed quickly and safely, while minimizing the potential negative effects to people, property, and the environment. Handling a hazardous materials incident is a bit like a chess game: one move sets up and then positively or negatively influences the next move or series of moves.

Solid Spill

Liquid Spill

Vapor Release

FIGURE 32-12 The response to a spill of a solid hazardous material will differ from the response to a liquid release or vapour release.

For example, in the case of an anhydrous ammonia release, the incident commander (IC) might decide to use a water fog to knock down the vapour cloud. If the cloud is not immediately threatening lives or property and the water runoff is heading toward a drain that goes to a waste water treatment facility, this might not be a sound plan; conversely, if the ammonia cloud is immediately threatening to envelop an elementary school, the water runoff may be an acceptable risk. It is vital to use a thought process that takes incorporates the risk–reward trade-off when choosing to employ any product-control option. An initial response objective should be realistic and well planned, taking into account all potential positive and negative effects.

Sometimes no action is the safest course of action to choose; that is, on some occasions, the incident is so extreme that it may be prudent to create a safe perimeter and let the problem stabilize on its own. In firefighting, structures with inaccessible fires may be encountered. In these cases, an IC may choose to pull back, go defensive, and protect exposures. A similar mentality should be considered when dealing with extreme or unusual hazardous materials incidents. A fire in a gasoline tanker may have to burn itself out if not enough foam is available to fight the fire. When incompatible chemicals are mixed, it might be wise to let the reaction run its course before intervening. Highly volatile flammable liquids may be left to evaporate rather than being the subject of an offensive attack.

It is important that you understand the concepts behind the various product-control options available to you. It is equally important that you take the time to do enough hands-on training to become proficient with these skills. You should never learn a new skill under the pressure of a real incident.

It is also vitally important to identify the physical and chemical properties of the released substance. Such knowledge facilitates the selection of the proper PPE. It also provides operations-level responders with enough information to make a wise decision when selecting the appropriate control option. Lastly, as with any other facet of hazardous materials response, decontamination should be performed when and where appropriate.

Response Objectives

At the operational level of training, all response objectives should be primarily defensive in purpose; that is, personnel do not actually come in contact with the hazardous material. Some effective defence actions can be taken safely at a distance. Defensive objectives are those that do not involve stopping the leak or release of a hazardous material:

- Isolate the area affected by the leak or spill, and evacuate victims who could become exposed to the hazardous material if the leak or spill were to progress.
- Control where the spill or release is spreading.
- Contain the spill to a specific area.

■ Defensive Actions

Defensive actions that can be taken include diking and damming, absorbing or adsorbing the hazardous material, stopping the flow remotely from a valve or shut-off, diluting or diverting the material, and suppressing or dispersing vapour.

■ Proper Personal Protective Equipment

The determination of which PPE is needed is based on the hazardous material involved, specific hazards, and the physical state of the material, along with a consideration of the tasks to be performed by the operations-level responder. At a minimum, fire fighters should wear full protective gear with no skin exposed and use SCBA. Although this gear provides some respiratory protection, note that standard structural firefighting PPE offers only limited hazardous materials protection, depending on the type of hazardous material involved. The chapter "Hazardous Materials: Awareness and Specialties" discusses proper PPE for a hazardous materials incident in detail.

■ Emergency Decontamination Procedures

Even though there should be no intentional contact with the hazardous material involved, there must be a procedure or a plan in place to decontaminate any responder who accidentally becomes contaminated **FIGURE 32-13**. Victims who are removed from a contaminated zone must be decontaminated by personnel who have donned the appropriate protective gear and who have been trained in the proper methods of victim decontamination.

The decontamination methods employed for a specific incident are dictated by the hazardous material, the physical state of the material, and the hazards involved. Decontamination can be as simple as removing clothing and flushing material away with water or as complex as using drug therapy. It is

FIGURE 32-13 There must be a plan for decontamination at every hazardous materials incident.

© Bill Karrow, Dunkirk Observer/AP Photos

covered in detail later in this chapter in the sections "Types of Decontamination" and "Methods of Decontamination."

Gauging the Potential Harm or Severity of an Incident

When attempting to evaluate a hazardous materials incident's potential to cause injury to people or destruction of property and the environment, responders need to consider factors such as the size of the container, the nature of the hazardous material involved, the amount released, and the area exposed. Based on the toxicity and the concentration of the hazardous material, they may be able to gauge how the incident might progress; for example, the difference between a 9.1-kg (20-lb.) propane cylinder and a 68,130-L (18,000-gal.) propane tank will determine how extensive and lengthy an incident will become.

■ Resources for Determining the Size of the Incident

To determine the size of the incident, two resources should be used: The first is the *ERG*, which is published jointly by Transport Canada and the United States Department of Transportation (DOT). This reference book identifies and outlines predetermined protective-action zones and basic action plans based on spill size estimates for thousands of chemicals. The second resource is a computerized or hard-copy preincident plan, which should include both reports submitted by the fire department and topographical mapping information.

Monitoring devices such as wind-direction and weather-forecasting equipment are critical resources for the IC in formulating response plans. Computer modelling programs such as Areal Locations of Hazardous Atmospheres (ALOHA®) can plot the predicted movements of vapour clouds and plumes. The availability of monitoring and portable detection devices will allow the IC to determine the hot, warm, and cold zones and the required evacuation distances.

■ Reporting the Size and Scope of an Incident

Reporting the estimated physical size of the area affected by a hazardous materials incident is accomplished by using information available at the scene. If a vehicle is transporting a known amount of a material, an estimate of the size of the release can be made by subtracting the amount remaining in the container. For example, suppose a gasoline tanker containing 34,065 L (9000 gal.) overturns, and 17,032 L (4500 gal.) remains in the tanker. In this case, an estimated 17,032 L (4500 gal.) of gasoline has spilled. The actual spill area can then be estimated in square metres. Depending on its size, the affected area may be expressed in units as small as square metres or as large as square kilometres.

When it is initially unsafe to approach a vehicle and shipping papers are not immediately available, other methods must be used to determine the original amount of hazardous material and estimate the remaining contained amount. Remember that the safety of responders is paramount to maintaining an effective response to any hazardous materials incident.

■ Determining the Concentration of a Released Hazardous Material

The SDS usually states the concentration of the hazardous material and can be used to estimate the concentration of a chemical release **FIGURE 32-14**. Litmus paper (pH strips) can be used to determine the hazardous material's pH. To determine the concentration of a gaseous hazardous material, monitors can be used to analyze the atmosphere from a safe distance.

Once the hazardous material's concentration is known, the IC can evaluate the incident response plan; a high concentration of an acid, for example, would call for a higher level of personal protection. The revised plan might also require the evacuation of civilians, owing to the concentration of the hazardous material in vapourized form.

Secondary Attacks

Implementing a secondary attack on responders is a tactic that many terrorist organizations use to make their attacks even more dramatic. It is essential that responders be able to assess the potential for secondary attacks and devices; this may be much easier said than done, however, as the indicators of secondary attacks and devices can be difficult to locate.

Terrorists who want to injure responding personnel with a secondary device or attack will typically make the initial attack very dramatic to draw responders into very close proximity to the scene. Indeed, the primary attack may purposely injure members of the public to draw responders into the scene. As they begin to treat victims, the secondary attack then takes place. Signs of secondary devices may include tripwires, timers, and ordinary, everyday containers (e.g., luggage, briefcases, boxes) found in very close proximity to the initial incident site. The smell of chemicals or the sighting of chemical dispersion devices or other containers that may hold chemical, biological, or even radioactive agents could be an indicator that a secondary attack is possible.

Incident Management System

As described in the chapter "Incident Management System," an IMS can be expanded to handle an incident of any size and complexity. Hazardous materials incidents can be complex, and local, provincial/territorial, and federal responders and agencies may all become involved in many cases of long duration. The basic IMS consists of five functions: command, operations, planning, logistics, and finance/administration.

In an incident involving hazardous materials, the application of an IMS takes on two unique characteristics: First, Canadian federal law requires that an IMS be used in the response to all hazardous materials incidents, including the use of a written incident action plan (IAP). Second, a special technical group may be developed under the operations section known as the hazardous materials branch. The **hazardous materials branch** consists of some or all of the following positions, as needed for the safe control of the incident:

- A second safety officer reporting directly to the **hazardous materials officer** as well as to the incident

SAFETY DATA SHEET

Syngenta Crop Protection, Inc.
Post Office Box 18300
Greensboro, NC 27419

In Case of Emergency, Call
1-800-888-8372

1. PRODUCT IDENTIFICATION

Product Name:	**ENVOKE**	Product No.:	A9842A
EPA Signal Word:	Caution		
Active Ingredient(%):	Trifloxysulfuron-Sodium (75.0%)	CAS No.:	199119-58-9
Chemical Name:	N-[(4,6-Dimethoxy-2-pyrimidinyl)amino]carbonyl-3-(2,2,2-trifluoro-ethoxy)-pyridin-2-sulfonamide sodium salt		
Chemical Class:	Sulfonylurea Herbicide		

EPA Registration Number(s): 100-1132   **Section(s) Revised: 14**

2. HAZARDS IDENTIFICATION

Health and Environmental

Presents a low hazard during normal handling.

Hazardous Decomposition Products

May decompose at high temperatures forming toxic gases.

Physical Properties

Appearance: Light beige to brown granules

Odor: Not determined

Unusual Fire, Explosion and Reactivity Hazards

During a fire, irritating and possibly toxic gases may be generated by thermal decomposition or combustion.

3. COMPOSITION/INFORMATION ON INGREDIENTS

Material	OSHA PEL	ACGIH TLV	Other	NTP/IARC/OSHA Carcinogen
Diatomaceous Earth	80 mg/m³/%SiO2 (20 mppcf) TWA	Not Established	6 mg/m³ TWA **	IARC 3
Crystalline Silica, Quartz	10 mg/m³/(%SiO2+2) (respirable dust)	0.025 mg/m³ (respirable silica)	0.05 mg/m³ (respirable dust) **	IARC 1; ACGIH A2
Surfactant	Not Established	Not Established	15 mg/m³ TWA (total dust) *	No
Sodium Sulfite	Not Established	Not Established	Not Established	IARC Group 3
Trifloxysulfuron-Sodium (75.0%)	Not Established	Not Established	10 mg/m³ TWA ***	No

* recommended by manufacturer

** recommended by NIOSH

*** Syngenta Occupational Exposure Limit (OEL)

Ingredients not precisely identified are proprietary or non-hazardous. Values are not product specifications.
Syngenta Hazard Category: B

Product Name: **ENVOKE**   Page: 1

FIGURE 32-14 The SDS contains critical information.

Courtesy of Syngenta Crop Protection, Inc.

safety officer. This **hazardous materials safety officer** (the National Incident Management System [NIMS] calls this position assistant safety officer–hazardous materials) is responsible for the hazardous materials team's safety only.

- A **hot-zone entry team**
- A **decontamination team**
- A **backup entry team**
- A **hazardous materials information research team**

■ The Incident Command Post

The main hub of the IMS is the incident command post (ICP), which serves as the collection point for all information and resources. The ICP must be located upwind and upgrade from the spill or leak, in the cold zone, to keep it from becoming contaminated or unsafe. If the ICP, including those personnel housed there, were to become contaminated, the personnel inside the ICP would no longer be able to control the operation. The officers and members in that location would become victims, the ICP would become a part of the hot zone, and all operations would have to be reestablished elsewhere, using a completely different pool of personnel, equipment, and supplies. In this scenario, the overall efficiency of command would be negatively affected, and control of the scene would be temporarily lost during the transfer of command.

Based on the potential threat or severity of the hazard, the ICP could be as close as one block from the incident site or as far away as several kilometres from the hot zone. When choosing a site for the ICP, the maximum margin of safety must be used.

Skin Contact Hazards

The principal dangers of hazardous materials are toxicity, flammability, and reactivity; however, many hazardous materials have more than one dangerous characteristic. These materials can be extremely dangerous and produce harmful effects on the unprotected or inadequately protected human body. The best action to take when dealing with an unknown chemical or with a chemical that has more than one dangerous characteristic is to assume the worst and to provide the largest safety margin possible.

The skin can absorb harmful toxins without any sensation to the skin itself. Given this fact, fire fighters should not rely on pain or irritation as a warning sign of absorption. Some poisons are so concentrated that just a few drops placed on the skin can result in death.

Skin absorption is enhanced by abrasions, cuts, heat, and moisture. This relationship can create critical problems for fire fighters who are working at incidents that involve any form of chemical or biological agents. Fire fighters with large, open cuts, rashes, or abrasions should be prohibited from working in areas where they might be exposed to hazardous materials. Smaller cuts or abrasions should be covered with nonporous dressings.

The rate of absorption can vary depending on the body part that is exposed; for example, chemicals can be absorbed through the skin on the scalp much faster than they can through the skin on the forearm.

The high absorbency rate of the eyes makes them one of the fastest means of exposure. This type of exposure can occur when a chemical is splashed directly into the eye, when a chemical is carried from a fire by toxic smoke particles, or when gases or vapours are absorbed through the eyes. Absorption of a chemical through the eyes also can be seen as an early warning signal for either PPE or SCBA failure. This type of exposure is also a major problem because the eyes are directly connected to the optic nerve, which serves as a direct route to the brain.

Chemicals such as corrosives immediately damage skin or body tissues upon contact. Acids, for example, have a strong affinity for moisture and can create significant skin and respiratory tract burns. In addition, acid injuries create a clotlike barrier that blocks deep skin penetration of the chemical. In contrast, alkaline materials dissolve the fats and lipids that make up skin tissue and change solid tissue into a soapy liquid. This process is similar to the way caustic chemical drain cleaners dissolve grease and other materials in sinks and drains. As a result, alkaline burns are often much deeper and more destructive than acid burns are.

Safety Precautions

Fire fighters need to take several types of safety precautions during a hazardous materials incident. In addition to taking the standard safety precautions of using PPE and respirator protection, they must guard against the dangers posed by temperature-related conditions and stress.

■ Excessive-Heat Disorders

Fire fighters operating in protective clothing should be aware of the signs and symptoms of heat rash, heat cramps, heat exhaustion, heatstroke, and dehydration. If the body is unable to disperse heat because it is covered by a closed, impermeable garment, serious short- and long-term medical situations could result.

Heat rash can occur for a variety of reasons: In fire fighters, it occurs as a result of PPE rubbing against the skin when it is hot. Heat rash can also occur if there are folds or creases in the skin where sweat can collect and create clear, raised bumps on the skin. These bumps are usually not itchy; however, repeated or prolonged outbreaks can lead to infection.

Heat cramps are painful muscle spasms that can occur during physical exertion in high temperatures. The cause of heat cramps is electrolyte imbalance caused by excess sweating. Consuming a sports drink can rebalance the body.

Heat exhaustion is a mild form of shock that occurs when the circulatory system begins to fail because the body cannot dissipate excessive heat and becomes overheated. The body's core temperature rises, an event that is followed by weakness and profuse sweating. The fire fighter might become weak or dizzy and have episodes of blurred vision. Other signs and symptoms of heat exhaustion include rapid, shallow breathing; a weak pulse; cold, clammy skin; and sometimes, loss of consciousness. Although heat exhaustion is not an immediately life-threatening condition, the affected individual should be removed at once from the source of heat, rehydrated with electrolyte solutions, and kept cool. If not properly treated, heat exhaustion can progress to heatstroke.

<u>Heat stroke</u> is a severe and potentially fatal condition resulting from the failure of the temperature-regulating capacity of the body. It is caused by exposure to the sun or high temperatures. Reduction or cessation of sweating is an early symptom. The body temperature can rise to 40.5°C (105°F) or higher, with the victim experiencing a rapid pulse, hot skin, a headache, confusion, unconsciousness, and (possibly) convulsions. Heatstroke is a true medical emergency that requires immediate transport to a medical facility, preferably with advanced life support (ALS) in place.

The issue of dehydration should also be addressed. Fire fighters should be encouraged to drink 240 to 475 millilitres (8 to 16 ounces) of water before donning any protective clothing. During rehabilitation, they should be required to rehydrate with 475 mL (16 oz.) of water for each SCBA tank used or consumed.

Cooling Technologies

In an effort to combat heat stress while wearing PPE, many response agencies employ some form of cooling technology under the garment. These technologies include, but are not limited to, air, ice, and water-cooled vests, along with phase-change cooling technology. Many studies have been conducted on each form of cooling technology. Each is designed to accomplish the same goal: to reduce the impact of heat stress on the human body.

The human body is constantly seeking to reach an acceptable equilibrium with the external environment. To that end, heat generated by the body's core is carried through the circulatory system to the skin, where it is wicked away, as cooler ambient air passes over the skin. This process, called convective cooling, is how the body cools itself under normal conditions.

As the outside temperature rises, the difference between normal body heat and the surrounding environment narrows, and the normal process of convective cooling is ineffective. To compensate, the body switches to a process called evaporative cooling. Perspiration is generated during this process, and the ensuing evaporation of sweat produces a tremendous cooling effect. This method of cooling is only effective when the surrounding air is dry. When the moisture content in the surrounding environment rises (humidity), perspiration cannot evaporate, because the air is equally moist. Unfortunately, the human body has no other alternative when these natural mechanisms fail to work efficiently. The same suit that seals you up against the hazardous materials also seals in the heat, defeating the body's natural cooling mechanisms.

Forced-air cooling systems operate by forcing prechilled air through a system of hoses worn close to the body. As the cool air passes by the skin, heat is drawn away from the body (by convection) and released into the atmosphere. Forced-air systems are designed to function as the first level of cooling the body would naturally employ. Typically, these systems are lightweight and provide long-term cooling benefits, but mobility is limited because the hose is attached to an external fixed compressor.

Fluid-chilled systems operate by pumping ice-chilled liquids from a reservoir through a series of tubes held within a vestlike garment and back to the reservoir. Mobility may be limited with some varieties of this system, because the pump might be located away from the garment. Some units have a battery-operated unit worn on the hip, but the additional weight increases the workload, which generates more heat, thereby defeating the purpose of the cooling vest.

Passive systems such as ice- or gel-packed vests are commonly used because of their low cost, unlimited portability, and unlimited recharging capability (by refreezing the packs). These garments are vestlike in their design and intended to be worn around the torso. The principle is that the ice-chilled vest absorbs the heat generated by the body. This technology is bulkier and heavier than the aforementioned systems and can cause discomfort to the wearer because of the nature of an ice-cold vest near the skin. Additionally, the cold temperature near the skin can actually fool the body into thinking it is cold instead of hot, thereby causing it to retain even more heat.

Phase-change cooling technology operates in a similar fashion to the passive ice- or gel-packed vests. The main difference is that the temperature of the material in the packs is chilled to approximately 15.6°C (60°F) and the fabric of the vest is designed to wick perspiration away from the body. The packs typically recharge more quickly than ice- or gel-packed vests.

Cold-Temperature Exposures

Fire fighters at hazardous materials incidents can be exposed to two types of cold temperatures: those caused by materials and those caused by weather. Hazardous materials such as liquefied gases and cryogenic liquids expose fire fighters to the same low-temperature hazards as those created by cold-weather environments. Exposure to severe cold for even a short period can cause severe injury to body surfaces, particularly the ears, nose, hands, and feet.

Two factors in particular influence the development of cold-related injuries: temperature and wind speed. Because still air is a poor heat conductor, fire fighters working in low temperatures with little wind can endure these conditions for longer periods, assuming that their clothing remains dry. When low temperatures are combined with faster winds, however, wind chill occurs; for example, if the temperature with no wind is –6.6°C (20°F), it will feel like –15°C (5°F) when the wind speed is 24 kilometres per hour (15 miles per hour).

Regardless of the temperature, fire fighters perspire heavily when they are wearing impermeable chemical-protective clothing. When this wet inner clothing is removed during the decontamination process, particularly in a cold environment, the body can cool rapidly. Wet clothing extracts heat from the body as much as 240 times faster than dry clothing does. This phenomenon can also lead to hypothermia, a true medical emergency in which body temperature falls below 35°C (95°F).

Fire fighters must be aware of the dangers of frostbite, hypothermia, and the impaired ability to work when temperatures are low, when the wind chill factor is low, or when they are working in wet clothing. All personnel should wear appropriate, layered clothing and should be able to warm themselves in heated shelters or vehicles such as buses.

The layer of clothing next to the skin—especially the socks—should always be kept dry. Trench foot can result when wet socks are worn during long-term emergencies in cool (not cold) environments **FIGURE 32-15**. Fire fighters should carefully schedule their work and rest periods and monitor their physical working conditions. Warm shelters should be available for donning and doffing protective clothing.

■ Physical-Capability Requirements

Hazardous materials response operations put a great deal of both physiological and psychological stress on responders. During the incident, these personnel can be exposed to both chemical and physical hazards. They might face life-threatening emergencies, such as fire and explosions, or they might develop heat stress while wearing protective clothing or working under extreme temperatures.

For these reasons, every emergency response organization should have a health and safety management program. The Canadian Standards Association (CSA) outlines the components of a health and safety management system for hazardous materials responders in CAN/CGSB/CSA-Z1610-11, *Protection of First Responders from Chemical, Biological, Radiological, and Nuclear (CBRN) Events*, and CSA Z1000, *Occupational Health and Safety Management*. Among the components of such a program are medical surveillance, PPE, and site safety practices and procedures.

A medical surveillance program is the cornerstone of an effective health and safety management system for responders. The two primary objectives of a medical surveillance program are to determine whether an individual can perform his or her assigned duties, including the use of personal protective clothing and equipment, and to detect any changes in body system functions caused by physical and/or chemical exposures.

Medical monitoring is the ongoing, systematic evaluation of response personnel who might experience adverse effects because of exposure to heat, cold, stress, or hazardous materials. Regular medical monitoring can quickly identify problems

so that they can be treated in a timely fashion, thereby preventing severe adverse effects and maintaining the optimal health and safety of on-scene personnel.

Fire fighters should be examined on a regular basis, as determined by the fire department's medical director. For hazardous materials personnel, this might be on an annual basis. During this examination, the physician performs a respiratory-function test to measure lung capacity and lung function and to evaluate an individual's fitness for wearing SCBA and other respiratory protection devices. The specific requirements for any fire fighter who is assigned to any duty where any form of respiratory protection will be used can be found in the CSA's respiratory protection standards.

Whenever possible and immediately before fire fighters don PPE to enter a hazardous materials environment, they should have their vital signs taken as part of pre-entrance medical monitoring. This examination should also include an inspection of the individual's skin for rashes and open sores or wounds and an evaluation of the individual's mental status. Fire fighters at a hazardous materials incident should be alert and oriented to time and place, with clear speech, a normal gait, and the ability to respond appropriately. A recent medical history should be obtained, including current weight, medications taken within the past 72 hours, alcohol consumption within the past 24 hours, any new medical treatment or diagnosis made within the past two weeks, and any symptoms of fever, nausea, vomiting, diarrhea, or cough within the past 72 hours.

Preentrance medical monitoring should be performed on all individuals who will wear liquid splash or vapour-protective clothing and execute hazardous materials operations. It should be completed as soon as possible and within the hour before the fire fighter enters the hazardous environment.

Hazardous materials PPE puts significant stress on the body, particularly on the cardiovascular system. If a member of the hazardous materials team exhibits changes in gait, speech, or behaviour, he or she should leave the environment for immediate decontamination, doffing of protective clothing, and assessment. Anyone who complains of chest pain, dizziness, shortness of breath, weakness, nausea, or headache should also undergo immediate decontamination, doffing of protective clothing, and assessment by Emergency Medical Services (EMS) personnel on the scene, preferably at the advanced life-support (ALS) level.

After the hazardous materials team members undergo decontamination, they should once again have their vital signs checked and monitored. Components of postevent medical monitoring should include history, vital signs, weight, skin evaluation, and mental status. Normal baseline values should be attained within 15–20 minutes after a person leaves the environment. Anyone who does not return to normal baseline entry levels should be treated and transported to a definitive care facility for follow-up monitoring.

During this postevent medical monitoring, vital signs should be checked every 5–10 minutes. The monitoring team should obtain information from medical control on latent reactions or symptoms and communicate this information to response personnel. If any of these symptoms appear, medical control should be contacted for direction, and the affected

FIGURE 32-15 Trench foot can result when wet socks are worn during long-term emergencies in cool environments.

individual should be prepared for possible transport to a medical facility. Some hazardous materials teams maintain their own hazardous materials–specific medical control.

Response Safety Procedures

One of the first things that first responders at the awareness level should do at a hazardous materials incident scene is to isolate the area and prevent anyone from entering the scene. These basic protective actions should be taken immediately. For the first several minutes, not even the first responders themselves should enter the area. During this time, first responders can gather necessary scene information and identify the materials involved. Any actions taken should follow the predetermined local response plan or guideline; for example, local and state officials, the local emergency planning committee, and other federal response agencies should be notified as required.

Any possible ignition sources should be eliminated, particularly if the incident involves the release or probable release of flammable materials. Whenever possible, electrical devices used in the immediate area of the hazard should be certified as safe by recognized organizations. Note that some flammable vapours are heavier than air and, therefore, can travel along the ground to an ignition source.

The first step in gaining control of a hazardous materials incident is to isolate the problem and keep people away from it **FIGURE 32-16**. The IC cannot begin extended operations until the hazard area is identified and the perimeter is secured. Fire fighters can help to isolate the area by stretching banner tape across roads to divert traffic away from the spill and by coordinating their efforts with those of police, EMS, and other agencies.

■ Control Zones

Isolating the incident helps to limit the number of civilian and public service personnel who might be exposed to the hazardous materials. This can be done through a series of control zones around the incident site. **Control zones** are designated areas at a hazardous materials incident, based on safety and the degree of hazard. These zones should not be defined too narrowly **FIGURE 32-17**. Indeed, as the IC gets more information about the specifics of the chemical or material involved, the control zones can be reevaluated. If the incident is not as severe as first reported, the IC can readjust the control zones. It is always easier and safer to reduce the size of the control zones than to discover that they need to be larger. Control zones are labelled as hot, warm, or cold. They might also be colour-coded as red, yellow, and blue.

The **hot zone** is the area immediately around and adjacent to the incident. Although the size of this zone depends on the incident and the nature of the product, it should be large enough that adverse effects from the hazardous materials will not affect people outside the zone; an incident that involves a gas leak, for example, requires a larger hot zone than one involving a liquid leak. The hot zone should cover the incident site as well as the personnel and equipment necessary to control it. All personnel and equipment in this zone must be decontaminated when they leave the area, and personnel who work within it should be closely monitored. Access to the hot zone must be limited to only those persons necessary to control the incident. Individuals must check in at the access control point and record their entry and exit times. Only individuals trained to the hazardous materials–technician level should be permitted in the hot zone.

The **warm zone** (also referred to as the contamination-reduction zone) is the area where personnel and equipment are staged, before they enter and after they leave the hot zone. It contains control points for access corridors as well as the decontamination corridor (also referred to as the contamination-reduction corridor). Designating a warm zone helps to reduce the spread of contamination. Only the minimal amount of personnel and equipment necessary to perform a task or support those personnel who are operating in the hot zone should be permitted in the warm zone.

Personnel trained to the operations level can work with moderate supervision in the warm zone. Generally, personnel working in the warm zone can use PPE that is one level below the PPE used in the hot zone; if personnel in the hot zone are using Level A encapsulated chemical-protective suits, for example, personnel in the warm zone might dress in Level B protection. Ultimately, only the IC can determine which level of protection is needed for each responder.

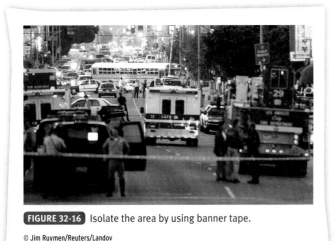

FIGURE 32-16 Isolate the area by using banner tape.

© Jim Ruymen/Reuters/Landov

FIGURE 32-17 Control zones spread outward from the centre of a hazardous materials incident.

Beyond the warm zone is the **cold zone**. In this "safe area," personnel do not need to wear any special protective clothing for safe operation. Personnel staging, the ICP, EMS providers, and an area for medical treatment after decontamination are all located in the cold zone.

Isolation Techniques

Whenever possible, approach a hazardous materials incident cautiously from upwind and uphill **FIGURE 32-18**. Resist the urge to rush in: do not attempt to help others until the situation has been fully assessed. Without entering the immediate hazard area, or hot zone, isolate the scene so that the safety of both people and the environment are ensured. Keep people away from the scene and outside the safety perimeter. Only those directly involved in the emergency response operations should be permitted in the area, and even they must wait until control has been established and the situation has been assessed.

There are several ways to isolate the hazard area and create control zones. The first step that a first responder should take is to identify the area verbally over the radio. Providing this information helps to ensure that those who arrive later will remain in a safe area and will not accidentally enter the hot zone. Cordon off the area using standard pylons and/or barrier tape to control access.

If the incident takes place inside a structure, the best place to control access is at the normal points of entry—that is, the doors. Once the doors are secured so that no unauthorized personnel can enter the facility, appropriately trained emergency response crews can begin to isolate areas above and below the hazard. The same concept applies to outdoor incidents. Secure the entry points and the area around the hazard. Begin by controlling intersections, on/off-ramps, service roads, or other access routes to the scene.

Police officers should assist fire fighters in this operation by diverting traffic at a safe distance outside the hazard area. They should block off streets, close intersections, or use their vehicles to redirect traffic. During a long-term incident, the highway department can set up traffic barriers. Crowd-control devices used by the police department can also be used to identify the hazard zones and restrict access. No matter which devices are used to restrict access, they should not limit or inhibit rapid withdrawal from the area by personnel working inside the hot zone.

The Buddy System and Backup Personnel

The personal alert safety system (PASS) device and the use of personal accountability and buddy systems are essential for fire fighter safety. On a hazardous materials incident,

backup personnel are also used to ensure the safety of emergency crews. They stand by at the scene, ready to remove personnel who are working in the hot zone if an emergency occurs. Backup personnel must be in place and prepared whenever personnel are operating in the hot zone.

All personnel must be fully briefed before they approach the hazard area or enter the hot zone. In addition, the decontamination team must be in place before anyone enters the hot zone. No one should enter the hot zone alone; instead, at least two team members should suit up and enter the hot zone together. At least two additional team members should suit up and stand by to assist those who entered first. The IC should ensure that backup personnel are prepared for immediate entry into the hot zone.

Team members should always remain within sight, sound, or touch of each other. Unfortunately, electronic communication devices sometimes fail, or the individual needing rescue might not be physically able to summon assistance. Radios should be used for communications at hazardous materials incident sites; they cannot, however, replace visual or voice contact or be the only way to contact an individual working in the hot zone.

Detecting the Hazard

When responding to hazardous material incidents, responders will be required to take appropriate steps to monitor the atmosphere for potential hazards. In general, atmospheric monitoring should be conducted in a prioritized manner. Monitoring priorities should include radiation, corrosive vapours, oxygen levels, combustibility (flammability), and final toxicity.

Radiation monitoring is conducted using a direct-reading radiation monitor, such as a Geiger counter. Corrosive vapour–monitoring can be conducted using either pH paper or a pH meter; for example, a stick with a dry piece of pH paper may be used to test liquids, and a wet piece of pH paper may be used to test vapours. Detector/colorimetric tubes are another method that can be used to monitor for a wide variety of chemicals.

Oxygen levels and combustible levels are typically monitored with the use of a two-, three-, four-, or five-gas monitor. These direct-reading monitors provide a direct indication of the oxygen level (as a percentage), combustible gas levels (typically expressed as a percentage of the lower flammability level [LFL]), and the levels of specific gases such as carbon monoxide, hydrogen sulphide, and sulphur dioxide. Other specialized chemical monitors can be used to detect individual chemicals. Photo- and flame-ionization detectors can be used to detect an even wider range of organic and inorganic gases. Finally, specialized instruments are available to detect chemical agents.

To use a multi-gas meter to provide atmospheric monitoring, follow the steps in **SKILL DRILL 32-1** **FIGURE 32-19**:

1. Prior to use of the multi-gas meter, make sure you understand the manufacturer's recommendations and local SOGs/SOPs for multi-gas analyzer use. The instruments should be calibrated and tested prior to use to verify that they are working properly.

2. Turn on the analyzer, allow the instrument to warm up, and zero it in a clean atmospheric environment.

Wind direction

FIGURE 32-18 Approach a hazardous materials incident cautiously.

FIGURE 32-19 A multi-gas meter can provide atmospheric monitoring.

③ Make sure you understand which hazards and conditions to avoid.

④ Approach the hazardous material and monitor the atmosphere.

⑤ Note the readings and interpret the meaning of the following conditions:

    a. Oxygen increase or decrease

    b. Lower explosive limit (LEL)/LFL increase

    c. Carbon-monoxide increase

    d. Hydrogen-sulphide increase

⑥ When the mission is completed, allow the meter to run in fresh air. Follow the appropriate procedures to turn the meter off and return it to service.

Search and Rescue

The protection of life is the first priority in any emergency response situation, and a hazardous materials incident is no exception. The search for and rescue of people involved in a hazardous materials event is the primary concern. However, this process becomes more complicated in a hazardous materials incident than in a fire situation.

In a structure fire, search and rescue operations are done quickly to remove victims from the smoke-filled environment. By contrast, in a hazardous materials incident, all emergency response personnel (e.g., fire, law enforcement, EMS) must first recognize and identify which hazardous materials may be present. After this task is accomplished, the IC must slowly

and cautiously evaluate the conditions before sending in personnel to attempt a search and rescue.

After identifying the hazards present, the IC must determine whether hazardous materials technicians can safely enter the hot zone to perform search and rescue. If the hazardous material is highly toxic, victims may not survive exposure to the material; in this case, no search would be necessary. If it is determined that a search can be safely performed, however, rescue teams wearing proper PPE may enter the hot zone to retrieve victims. These victims would then be removed to the warm zone, where they can be decontaminated and given emergency medical attention in preparation for transport to a medical facility.

Exposure Protection

Exposures can be protected in different ways. For example, the exposure from the threat can be eliminated by evacuating residents. In addition, a barrier might be constructed to prevent the hazardous materials from reaching the exposure. Alternatively, the hazardous material could be neutralized chemically, such as by treating an acid with a basic material to weaken the strength of the acid. Only a responder at the hazardous materials technician level should perform this kind of neutralization technique. The IC is ultimately responsible for determining which offensive and or defensive actions are to be taken and who will take those actions on the scene of a hazardous materials incident.

■ Confinement and Containment

<u>Confinement</u> is the process of attempting to keep the hazardous material on the site or within the immediate area of the release. It can usually be accomplished by damming or diking a material or by confining vapours to a specific area.

    <u>Containment</u> refers to actions that stop the hazardous material from leaking or escaping from its container. Examples of containment include patching or plugging a breached container or righting an overturned container. Only a responder at the hazardous materials technician level should patch or plug a container, given that direct product contamination is likely to occur while the responder is undertaking such an operation. A responder operating at the hazardous materials operations level can perform several different actions where no direct contact with the product would be likely to occur.

    When considering a containment response, a number of factors must be evaluated, including the maximum quantity of material that can be released, the level of risk posed by the hazard, and the likely duration of the incident if no intervention is attempted. Ultimately, the IC—and only the IC—is responsible for directing any offensive or defensive operations and for managing who is doing what at the scene.

■ Flammable-Liquid Vapour Control and Fire Extinguishment

Fires associated with a hazardous materials incident should be addressed with great caution, and the principles of extinguishment should be dealt with in a systematic manner. First and foremost is the need to recognize and identify the hazardous material or materials involved. Some hazardous materials react violently if water is applied—important knowledge for fire fighters, given that water is the most commonly used firefighting agent. After identification of the material is made and confirmed, its physical and chemical properties, such as its specific gravity, flash point, and flammable range, should be researched.

Most flammable- and combustible-liquid fires can be extinguished by the use of foam. A complete discussion of foam appears in the chapter "Fire Attack Lines and Foam."

Fluoroprotein foams contain protein products mixed with a synthetic fluorinated surfactant. Fluoroprotein foams can be used on fires or spills involving gasoline, oil, or similar products. Fluoroprotein foams rapidly spread over the fuel, causing fire knockdown and vapour suppression. These foams, like many of the synthetic foam concentrates, are resistant to polar solvents such as ketones and ethers.

Protein foam concentrates are made from hydrolyzed proteins along with stabilizers and preservatives. These types of foams are very stable with good expansion properties. They are quite durable and resistant to reignition when used on Class B fires or spills involving nonpolar substances, such as gasoline, toluene, oil, and kerosene.

High-expansion foams are used when large volumes of foam are required for spills or fires in warehouses, tank farms, and hazardous waste facilities. With these products, it is not uncommon to have the ability to produce more than 3780 L (1000 gal.) of finished foam for every gallon of foam concentrate used. This is accomplished by pumping large volumes of air through a small screen coated with a foam solution. Due to the large amount of air in the foam, high-expansion foam is referred to as dry foam. By excluding oxygen from the fire environment, high-expansion foam smothers the fire yet leaves very little water residue.

Remember that as foam is applied, more volume will be added to the spill.

Foam should be gently applied or bounced off another adjacent object so that it flows down across the liquid and does not directly upset the burning surface. Foam can also be applied in a rain-down method by directing the stream into the air over the material and letting the foam fall gently, much as rain would fall to the ground.

This technique is less effective when the fire is creating an intense thermal column. When the heat from the thermal column is high, the water content of the foam will turn to steam before it contacts the liquid's surface.

Foam can also be applied using the roll-on method, where you bounce the stream directly into the front of the spill area and allow it to gently push forward into the pool rather than directly splashing it in.

■ Pressurized-Gas-Cylinder Vapour Control and Fire Extinguishment

Incidents involving pressurized gas cylinders may consist of fires or releases of the cylinders' contents. Some materials stored in pressurized cylinders may be flammable, others may be flammable and toxic, and still other gases may pose asphyxiation hazards to personnel. The precise nature of the threat depends on the contents of the cylinder, the area in which the cylinder is stored at the time of the emergency, and the design and integrity of the cylinder itself.

Response to incidents involving cylinders may be very routine, extremely hazardous, or anywhere in between the two extremes. Owing to the unique nature of the hazards presented by cylinders involved in hazardous materials incidents, responders need to err on the side of caution at all times. Evacuation to a safe distance may well be the best response.

By definition, an emergency involving a cylinder or compressed gas comprises the actual or potential release of a hazardous material that cannot be stopped by closing the product's cylinder or container valve. Leaks involving compressed gases require immediate evacuation of the contaminated area. Isolate the cylinder in a well-ventilated and secure area, but only move it if the transport can be done in a safe manner. It is important to recognize all hazards associated with such a material, so that proper action can be taken without risk to anyone:

- Evacuation of all nonessential personnel
- Rescue of injured people by crews equipped with the necessary PPE
- Corrective action to minimize the leak or at least minimize exposure to people and equipment
- Assurance that all necessary resources are available for the final resolution of the situation
- Firefighting action
- Decontamination
- Written documentation and critique

These procedures describe the type of actions to take when the leak is of a minimal size and corrective measures can be taken without risk to personnel and/or direct product exposure; leaks of a large nature, however, require more sophisticated response efforts. Emergency action plans must be based on the nature of the product and the exact situation.

On the one hand, fires are less of a problem with pressurized gas cylinder incidents because the escaping gas is consumed by the fire. No fire should be extinguished unless fire fighters are certain that they will be able to control the escaping gas once the fire is out. Otherwise, allowing the fire to burn itself out is likely the best course of action.

Some fire departments may believe that the previously described procedures are too offensive in nature and prefer to take a defensive approach to all incidents that involve compressed gases. Always consult your fire department's SOGs/SOPs and follow them as well as the advice of the local hazardous materials team.

Hazardous Materials Control Activities

As fire fighters approach a hazardous materials incident, they should look for natural control points—that is, areas in the terrain or places in a structure where materials such as liquids, powders or solids can be contained. Hazardous materials technicians should use monitoring equipment to determine where a material has travelled. Some natural barriers that might be used to contain the material include doors to a room, doors to a building, and curb areas of roadways; these natural barriers make for an effective way to establish easily identifiable control zones. In addition, operational personnel can use a number of defensive control measures to manage the incident effectively.

■ Absorption/Adsorption

<u>Absorption</u> is the process of applying a material (an absorbent) that will soak up and hold the hazardous material in much the same way that a sponge gathers up water. This technique makes collection and disposal of a hazardous material in liquid form more manageable; for example, a dry, granular, clay-based material or dry sand may be added to a spill to help contain the spilled product. To apply the absorbent material, fire fighters should distribute it from a distance using shovels. Any tools used for dispersal and pickup of the absorbent material (e.g., shovels) need to be nonsparking to prevent ignition of the spilled material.

The technique of absorption is often difficult for operational personnel because it generally requires them to be in close proximity to the spill. It also involves the application of material to a spilled product, which adds volume to the spill. The absorbent material may also react with certain hazardous substances, so it is important to determine whether an absorbent substance is compatible with the hazardous material before it is applied to the spill. The addition of the absorbent material does not change the hazard of the spilled material, and appropriate PPE must still be worn.

<u>Adsorption</u> is a process that occurs when a liquid hazardous material accumulates on the surface of a solid, forming a molecular or atomic film. It is different from absorption, in which a substance passes into a material to form a solid.

Disposal is generally not the responsibility of a fire department but, rather, is carried out by the responsible party; if a responsible party cannot be quickly located, however, fire departments need to plan for the proper disposal of any contaminated materials collected from these types of incidents. Notify the appropriate local or provincial/territorial health or environmental regulatory agency to coordinate the removal and disposal of the spilled material.

To use absorption/adsorption to manage a hazardous materials incident, follow the steps in **SKILL DRILL 32-2** :

1 Collect the basic materials:

    a. Absorbents

    b. Adsorbents

    c. Absorbent pads

    d. Absorbent booms

2 Decide which material is best suited for use with the spilled product.

3 Assess the location of the spill and stay clear of any spilled product. (**STEP** **1**)

4 Apply the appropriate material to control and contain the spilled material. (**STEP** **2**)

5 Maintain materials and take appropriate steps for their disposal. (**STEP** **3**)

■ Diking, Damming, Diversion, and Retention

Controlling liquid spills can be difficult, because liquids tend to take the shape of their container regardless of where that container is located. Several techniques can be effective in managing these types of incidents.

Diking

<u>Diking</u> is the placement of impervious materials to form a barrier that will keep a hazardous material in liquid form from entering an area or that will hold the material in a specific area. To construct a dike, follow the steps in **SKILL DRILL 32-3** **FIGURE 32-20** :

1 Based on the spilled material and the topography conditions, determine the best location for the dike.

2 Dig a depression in the channel 15–20 cm (6–8 in.) deep, if needed.

3 Use sufficient plastic to line the bottom of the depression and cover the dike wall.

4 Build a dam with sandbags or other available materials.

5 Install plastic to the top of the dike, and secure it with additional sandbags.

6 Complete the dike installation, and ensure the size will contain all of the spilled product.

Damming

<u>Damming</u> is used when a liquid is flowing in a natural channel or depression and its progress can be stopped by blocking the channel. Three kinds of dams are typically used in such circumstances: a complete dam, an overflow dam, or an underflow dam.

FIGURE 32-20 Dikes keep a hazardous material from entering an area.

SKILL DRILL 32-2   Using Absorption/Adsorption to Manage a Hazardous Materials Incident

1 Collect the basic materials: absorbents, adsorbents, absorbent pads, absorbent booms. Decide which material is best suited for use with the spilled product. Assess the location of the spill and stay clear of any spilled product.

2 Apply the appropriate material to control and contain the spilled material.

3 Maintain materials and take appropriate steps for their disposal.

© LiquidLibrary

A complete dam is placed across a small stream or ditch to completely stop the flow of materials through the channel. This type of dam is used only in areas where the stream or ditch is basically dry and the amount of material that needs to be controlled is relatively small.

For streams or ditches with a continuous flow of water, an overflow or underflow dam will need to be constructed. An overflow dam is used to contain materials that are not water soluble and have a specific gravity greater than 1. It is constructed by building the dam base up to a level that holds back the flow and then installing polyvinyl chloride (PVC) pipe or hard suction hose at an angle that allows the water to flow over the dam, trapping the heavier material at a low level at the base of the dam. An underflow dam is used to contain materials that are not water soluble and have a specific gravity less than 1. It is basically constructed in the opposite manner of the overflow dam: The piping on the underflow dam is installed near the bottom of the dam, so that water flows under the dam and the materials floating on the water accumulate at the top of the dam. It is critical to have enough pipe or hose to allow sufficient water to flow past the dam without overflowing over the top of the dam.

To construct an overflow dam, follow the steps in **SKILL DRILL 32-4** **FIGURE 32-21**:

1 Determine the need for an overflow dam, based on whether the spilled material has a specific gravity greater than 1.

2 Dig a 15–20 cm (6–8 in.) depression in the channel.

3 Use plastic to line the bottom of the depression.

4 Build a dam with sandbags or other available materials.

5 Install 8- to 10-cm (3- to 4-in.) plastic pipe horizontally on top of the dam. Add more sandbags on top of the dam, if available.

6 Complete the dam installation, and ensure that its size will allow the proper flow.

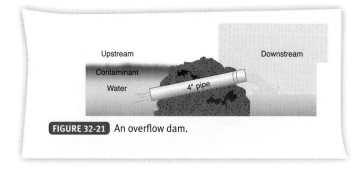

FIGURE 32-21 An overflow dam.

To construct an underflow dam, follow the steps in **SKILL DRILL 32-5** **FIGURE 32-22**:

1 Determine the need for an underflow dam, based on whether the spilled material has a specific gravity less than 1.

2 Dig a 15–20 cm (6–8 in.) depression in the channel.

③ Use plastic to line the bottom of the depression.

④ Build a dam with sandbags or other available materials.

⑤ Install 8- to 10-cm (3- to 4-in.) plastic pipe at a 20- to 30-degree angle at a level suitable to ensure that clean water passes through the pipes.

⑥ Add more sandbags on top of the dam, if available.

⑦ Complete the dam installation, and ensure its size will allow the proper flow.

FIGURE 32-22 An underflow dam.

Diversion

A **diversion** technique is used to redirect the flow of a liquid away from an area. Liquid hazardous materials spills can be controlled by using existing barriers, such as curbs, and channelling liquids away from storm drains, which would otherwise allow them to leave the site. Diversion can also be used to protect environmentally sensitive areas.

To use diversion to manage a hazardous materials incident, follow the steps in **SKILL DRILL 32-6** **FIGURE 32-23** :

① Determine the best location for the diversion, based on the spilled material and the topography conditions.

② Use sandbags or other materials to divert the product flow to an area of fewer hazards.

③ Stay clear of the product flow.

④ Monitor the diversion channel to ensure the integrity of the system.

Retention

Retention is the process of creating an area to hold hazardous materials. For instance, fire fighters might dig a depression and allow material to pool in the depression. The hazardous material would then be held there until a clean-up contractor could recover it. A hazardous material leak in an industrial facility might be safe to approach after its proper recognition and identification by operational personnel wearing the proper level of PPE, who could then close a door and contain the spill within one room. As another example, the public works department or an independent contractor might use a backhoe to create a retention area for runoff at a safe downhill distance from the release.

To use retention, follow the steps in **SKILL DRILL 32-7** **FIGURE 32-24** :

① Based on the spilled material and the topography conditions, determine the best location for the retention system.

② Dig a depression in the channel to serve as a retention area.

③ Use plastic to line the bottom of the depression.

④ Use sandbags or other materials to hold the plastic in place.

⑤ Stay clear of the product flow.

⑥ Monitor the retention system to ensure its integrity.

■ Dilution

Another protective action that can be taken from a safe distance is **dilution**. Dilution entails the addition of water or another liquid to weaken the strength or concentration of a hazardous material. This tactic can be used only when the identity and properties of the hazardous material are known with certainty. One concern with dilution relates to the fact that water applied to dilute a hazardous material increases the total volume; if the volume increases too much, it may overwhelm containment measures. Dilution is considered a risky operation because of these potential pitfalls and the possibility that it might exacerbate the contamination. As a consequence, it should be used with extreme caution and only on the advice of those persons who are knowledgeable regarding the nature of the chemicals involved in the incident.

FIGURE 32-23 A diversion.

FIGURE 32-24 A retention system.

To use dilution to manage a hazardous materials incident, follow the steps in **SKILL DRILL 32-8** **FIGURE 32-25** :

1. Determine the viability of a dilution operation, based on the spilled material's characteristics and the size of the spill. Obtain guidance from a hazardous materials technician.

2. Ensure that the water used will not overflow and affect other product-control activities.

3. From a distance, add small amounts of water to dilute the product.

4. Contact the hazardous materials technician if any additional issues arise.

◼ Vapour Dispersion and Suppression

When a hazardous material produces a vapour that collects in an area or increases in concentration, vapour dispersion or suppression is required.

__Vapour dispersion__ is the process of lowering the concentration of vapours by spreading them out. Vapours can be dispersed with hose streams set on fog patterns, large displacement fans, or heating and cooling systems found in fixed hazardous materials handling systems.

Before dispersing vapours, the consequences of this action should be considered. If the vapours are highly flammable, an attempt at dispersing the gases might ignite them. If the vapours are corrosive, the runoff will be corrosive and must be confined. The dispersed vapours might also contaminate areas outside the hot zone. Vapour dispersion with fans or a fog stream should be attempted only after the hazardous material is safely identified and only if the weather conditions will promote the diversion of vapours away from any populated areas. Conditions must be constantly monitored during the entire operation.

__Vapour suppression__ is the technique of controlling fumes or vapours that are given off by certain materials, such as flammable-liquid fuels. For example, gasoline gives off vapours, which are dangerous because they are flammable. To control these vapours, a blanket of firefighting or vapour-suppressing foam is layered on the surface of the liquid. The foam seals the surface of the liquid, prevents vapours from escaping, and reduces the danger of vapour ignition.

Reducing the temperature of some hazardous materials will also suppress vapour formation; however, there is no really effective way to accomplish this goal except in cases of the smallest spills.

To use vapour dispersion to manage a hazardous materials incident, follow the steps in **SKILL DRILL 32-9** **FIGURE 32-26** :

1. Based on the characteristics of the released material, the size of the release, and the topography, determine the viability of a dispersion operation. Stay upwind and uphill.

2. Use a four-gas monitor to identify a safe work area.

3. Ensure that ignition sources in the area have been removed if the material is flammable.

4. From a distance, use water to disperse vapours.

5. Monitor the environment until the vapours have been adequately dispersed.

To use vapour suppression to manage a hazardous materials incident, follow the steps in **SKILL DRILL 32-10** **FIGURE 32-27** :

1. Based on the characteristics of the released material, the size of the leak, and the topography, determine the viability of a vapour-suppression operation.

2. Use a four-gas monitor to identify a safe work area.

3. Ensure that ignition sources in the area have been removed.

4. Use a foam stream to suppress vapours from a distance.

5. Monitor the environment until the vapours have been adequately dispersed.

FIGURE 32-25 Using dilution.

FIGURE 32-26 Using vapour dispersion to manage a hazardous materials incident.

FIGURE 32-27 Using vapour suppression to manage a hazardous materials incident.

■ Remote Shut-off

A protective action that should always be considered, especially with transportation emergencies or incidents at fixed facilities, is the identification and isolation of **remote shut-off valves**. Many chemical processes or piped systems that carry chemicals have a way to remotely shut down a system or isolate a valve. This may prove to be a much safer way to mitigate the problem. Fixed ammonia systems provide good examples of the effectiveness of using remote shut-off valves. In the event of a large-scale ammonia release, most current systems have a main dump valve that will immediately redirect anhydrous ammonia into a water deluge system. This deluge system is designed to knock down the vapours and collect the resulting liquid in a designated holding system.

Many types of cargo tanks have emergency remote shut-off valves. TC306/MC306/DOT406 cargo tanks, for example, are certified to carry flammable and combustible liquids and Class B poisons. These single-shell aluminum tanks can hold 34,065 L (9000 gal.) or more of product at atmospheric pressures as high as 28 kPa (4 psi). The tanks have an oval/elliptical cross-section and overturn protection that serves to protect the top-mounted fittings and valves in the event of a rollover. In the event of fire, a fusible vent is designed to melt at a predetermined temperature, which in turn causes the closure of the internal product discharge valve. A baffle system within the cargo tank compartment reduces the surge of the liquid

contents. With these containers, remote shut-off valves are often found near the front (adjacent to the driver's door) or located at the rear of the cargo tank.

TC307/MC307/DOT407 cargo tanks are certified to carry chemicals that are transported at low pressure (less than 172 kPa [25 psi]), such as flammable and combustible liquids and other mild corrosives. These vehicles can carry as much as 26,495 L (7000 gal.) of product and are outfitted with similar safety features to those found on the TC306/MC306/DOT406 cargo tanks.

TC331/MC331 cargo tanks carry substances such as anhydrous ammonia, propane, butane, and liquefied petroleum gas (LPG). The tanks are not insulated and feature rounded ends and a circular cross-section. TC331/MC331 cargo tanks have a carrying capacity of between 35,768 and 43,527 L (9450 and 11,500 gal.). These containers have emergency shut-off valves either at both ends of the tank (tractor-trailer type) or at one location with possible radio control (bobtail or straight-truck types); internal shut-off valves; internal volume and pressure gauges; and two top-mounted vents.

Decision to Withdraw

After all factors have been considered, the IC may decide that the hazardous materials incident cannot be handled without incurring unnecessary risks to all personnel. Perhaps the resources are not available to mitigate the incident, or there are no exposures to be saved. In such a case, the IC may decide to withdraw to a safe distance, set a defensive perimeter, and wait for additional resources to arrive. He or she may also simply allow the hazardous materials incident to run its course. This situation arises with certain chemicals, such as pesticides; they are consumed in a hot fire and, therefore, may be allowed to burn. The decision to extinguish a fire involving pesticides could result in runoff contamination of surrounding areas, worsening the impact of the incident.

Recovery

The **recovery phase** of a hazardous materials incident occurs when the imminent danger to people, property, and the environment has passed or is controlled, and cleanup begins. During the recovery phase, both provincial/territorial and federal agencies may become involved in cleaning up the site, determining the responsible party, and implementing cost-recovery methods. In addition, a transition is made between on-scene public safety responders and private-sector commercial clean-up companies. The recovery phase also includes completion of the records necessary for documenting the incident.

It must be stressed that the hazardous materials incident is not over at this point. As long as public safety personnel and responders remain on the scene, they should maintain their vigilance for safety to ensure that no new hazards are created and no injuries are incurred during this final phase. Always remember that fire fighters' primary responsibility is to ensure the safety of themselves and other responders.

When to Terminate the Incident

The decision to terminate a hazardous materials incident is made by the IC. The recovery phase in large-scale incidents can go on for days, weeks, or even months. The recovery phase and cleanup will likely require amounts of resources and equipment that go far beyond the capabilities of local responders. The ultimate goal of incident termination is to return the property or site of the incident to its preincident condition, and then return the facility or mode of transportation to the responsible party. This hand-off often takes place long after fire fighters are released from the hazardous materials scene.

Crime or Terrorist Incident

When an incident involves potential criminal or terrorist activity, responders must take extra precautions and implement appropriate actions to ensure their own safety as well as the safety of the general public. One of the critical actions that must be taken is notifying the appropriate law-enforcement agencies of the situation. These notifications may include not only local law enforcement but also appropriate federal agencies, depending on the circumstances. Other actions that go hand in hand with law-enforcement notification are the preservation of the potential crime scene and good documentation of the initial scene activities by responders. Proper crime-scene preservation and strong documentation may prove invaluable to law enforcement during subsequent legal proceedings related to the incident.

In addition, specific activities should be implemented to ensure safety; for example, proper measures should be implemented to limit the potential for cross-contamination. These steps include isolating the scene from the general public and unneeded responders by establishing control zones, isolating people or animals that were potentially contaminated during the initial incident, and ensuring that potentially contaminated materials are not removed from the scene. Responders should also exercise extreme caution during incidents that involve potential criminal or terrorist activity and be aware of the potential risk of secondary devices that may be intended to injure arriving personnel.

As noted earlier, during hazardous materials incidents, especially those involving criminal or terrorist activity, responders should take appropriate measures to preserve incident evidence. Upon arriving on the scene, responders should note key information regarding the scene itself. This step will include documenting information such as who was present, where specific pieces of evidence were located, when time-specific elements of the incident took place, and any other information pertinent to the incident.

The use of photography to document the scene is encouraged if it can be done safely and will not create additional hazards. Document the date and time when the photographs were taken as well as the name of the person who took the pictures. Do not move potential evidence if at all possible. If evidence must be moved to mitigate the incident, document and photograph it if possible.

Limit the number of responders working within the area. All responders who work in the immediate incident area should be asked to document their specific actions in the area as well as their findings regarding evidence in the area. Work with law-enforcement agencies to coordinate the interviewing of witnesses, victims, responders, and other individuals who may have important information about the incident.

Decontamination

Decontamination is the physical or chemical process of reducing and preventing the spread of hazardous materials by persons and equipment. Decontamination makes personnel, equipment, and supplies safe by removing or eliminating hazardous materials or weapons of mass destruction (WMDs). Proper decontamination is essential at every hazardous material/WMD incident to ensure the safety of both personnel and property. In some ways, proper decontamination is the most vital function performed by the fire department at the scene of such an incident.

Contamination

Contamination is the process of transferring a hazardous material from its source to people, animals, the environment, or equipment, any of which may act as a carrier of the contaminant. Although contamination can occur though a variety of processes, at hazardous materials incidents, it often results from the release of a hazardous material from its intended container. Once a person, animal, or surface is contaminated, that person, animal, or surface can cause other people, animals, and surfaces to become contaminated. This type of contamination spread is known as cross-contamination or secondary contamination.

■ Cross-contamination

Cross-contamination is the process by which a contaminant is carried out of the hot zone and contaminates people, animals, the environment, or equipment. Cross-contamination may occur in several ways:

- A contaminated victim comes into physical contact with an emergency responder.
- A bystander or other emergency responder comes into contact with a contaminated object from the hot zone.
- A decontaminated responder comes into contact with another contaminated responder or object.

Because cross-contamination can occur in so many ways, control zones should be established, clearly marked, and enforced at all hazardous materials incidents.

Types of Decontamination

As noted earlier, the decontamination process makes personnel, equipment, and supplies safe by removing or eliminating the hazardous materials. Fire fighters are often responsible for establishing a decontamination corridor at a hazardous materials incident for the initial emergency response crews, for the victims of the incident, and in support of the hazardous materials response team. The **<u>decontamination corridor</u>** is a controlled area, usually located within the warm zone of the incident, where decontamination procedures take place. Everyone who leaves the hot zone is required to pass through the decontamination corridor as he or she exits through the warm zone and, eventually, into the cold zone.

The major categories of decontamination are emergency decontamination, gross decontamination, technical decontamination, and mass decontamination. Each category provides for a different level of decontamination.

■ Emergency Decontamination

<u>Emergency decontamination</u> is used in potentially life-threatening situations to rapidly remove the bulk of the contamination from an individual, regardless of the presence or absence of a technical decontamination corridor. A more formal and detailed decontamination process usually follows later. Emergency decontamination usually involves dousing the victim with adequate quantities of water via a hand-held hose line from an engine company, using a fog pattern; removal of contaminated clothing; and a second round dousing. This process is known as "flush, strip, and flush." If a decontamination corridor has not yet been established, fire fighters can isolate the exposed victims in a contained area and establish an emergency decontamination area.

Do not allow the water runoff from emergency decontamination to flow into drains, streams, or ponds; instead, try to divert it into an area where it can be treated or disposed of later. Nevertheless, do not delay decontamination; human life always comes first. If any extension of the contaminated area occurs as a result of this emergency decontamination, be sure to notify command as soon as possible to ensure an appropriate response to the newly contaminated area.

Whenever possible, identify the hazardous material before beginning decontamination. Some hazardous materials are water reactive and will require a special decontamination process. When in doubt, proceed with water-based decontamination, albeit with caution.

The best method of decontamination for fire fighters and first responders is to avoid contact with the hazardous material. If you do not become contaminated, there is no need to decontaminate!

To perform emergency decontamination, follow the steps in **SKILL DRILL 32-11**:

① Ensure that you have the appropriate PPE to protect against the chemical threat.

② Stay clear of the product, and do not make physical contact with it.

SKILL DRILL 32-11 Performing Emergency Decontamination

1 Ensure that you have the appropriate PPE to protect against the chemical threat. Stay clear of the product, and do not make physical contact with it. Make an effort to contain runoff by directing victims out of the hazard zone and into a suitable location for decontamination. Flush the victim to remove the product from the victim's clothing.

2 Instruct the victim in removing contaminated clothing.

3 Flush the contaminated victim. Assist or obtain medical treatment for the victim, and arrange for the victim's transport.

3 Make an effort to contain runoff by directing victims out of the hazard zone and into a suitable location for decontamination.

4 Flush the victim to remove the product from the victim's clothing. (**STEP 1**)

5 Instruct the victim in removing contaminated clothing. (**STEP 2**)

6 Flush the contaminated victim.

7 Assist or obtain medical treatment for the victim, and arrange for the victim's transport. (**STEP 3**)

■ Gross Decontamination

Gross decontamination, like emergency decontamination, aims to significantly reduce the amount of surface contaminant by delivery of a continuous shower of water and removal of outer clothing or PPE; it differs from emergency decontamination, however, in that gross decontamination is controlled through the decontamination corridor. Hazardous materials teams use a large shower system for gross decontamination as the first part of a multi-step decontamination process or line.

The shower system works well if water is appropriate for removing the hazardous material. With a shower-type arrangement, gross decontamination begins when a responder, still fully clothed in PPE and using the appropriate level of respiratory protection, steps into a portable shower where high-pressure, low-volume water flow (or spray) is used to rinse off the contaminants. This step will remove most solid or liquid contaminants from the PPE. Runoff water should be diverted to a safe area for treatment or disposal; it should not be allowed to flow into drains, streams, and ponds.

Do not confuse emergency decontamination with gross decontamination: the two are completely separate concepts. Emergency decontamination is performed for life-safety reasons. Gross decontamination is intended to carry out a brief and rapid contamination reduction—a sort of prewash before technical decontamination.

■ Technical Decontamination

Technical decontamination is performed after gross decontamination and is a more thorough cleaning process. This type of decontamination may involve several stations or steps. During technical decontamination, multiple personnel, known as the decontamination team, typically use brushes to scrub and wash the person or object so as to remove contaminants.

To clean a responder wearing chemical-protective clothing, for example, the decontamination team might use a water spray, long-handled scrub brushes, and a special cleaning solution. The type of cleaning solution will depend on the hazardous material involved. In some situations, special chemicals may be needed for decontamination.

In some cases, law enforcement or search canines may require decontamination after a mission. In these instances, it is best to check with the canine handler to better understand the process. Search dogs used by Canadian Urban Search and Rescue (USAR) teams have particular ways to perform canine decontamination; make sure you understand the process, or check with your local jurisdiction for guidance, before undertaking this specialized process.

Additionally, it may be necessary to perform decontamination on a criminal subject, firearms, or other specialized pieces of equipment. Unless life safety is a priority, it is wise to consult the owner of the piece of equipment or an otherwise knowledgeable person before doing anything that might have an adverse effect. Unusual items might require unusual methods of decontamination.

It may also be necessary to decontaminate a piece of evidence prior to turning it over to law-enforcement officials. In these cases, maintaining the chain of custody, even through the decontamination process, is important. The item should arrive at decontamination at least double bagged. The outer bag can then be washed, rendering it safe for transportation, laboratory analysis, or other further handling. Care should be taken not to compromise the integrity of any of the containers holding evidence. Additionally, the method used for decontamination should not alter the evidence in any way.

■ Mass Decontamination

Mass decontamination is often used in incidents involving unknown agents or in the case of a contamination of a large group of people. It takes place in the field and is a way of quickly performing gross decontamination on a large number of victims who has escaped from a hazardous materials incident. Because water is a generally effective solvent, washing off as much of the contaminant as possible with a massive spray fo water is the best and quickest way to decontaminate a large group of people.

The process itself relies on the *strip, flush, and cover* principle. The victims must first be herded into an area where a corridor has been set up and, if possible, excluded from public view. Next, the victims should be instructed to remove the outermost layer of clothing necessary to remove the contaminant from their person. They should then enter the shower area. If fire apparatus is used for this purpose, it should be set up to distribute large quantities of water at a low pressure so as not to injure victims or embed the contaminant in the skin. Once victims exit the shower area, they should be given a cover for protection. The cover could be a blanket, gown, Tyvek® suit, or even a black trash bag.

In mass decontamination, the primary concern is to remove a hazardous material from a large number of victims. This process can take place on any street, parking lot, or area where fire apparatus can be deployed with a continuous, uncontaminated water supply. Environmental concerns are secondary; there is rarely enough time to build structures to contain runoff. Although notifying command about the situation remains a priority, mass decontamination should not be delayed to allay environmental concerns.

Victims who go through mass decontamination will need further decontamination and medical monitoring after they emerge from the shower system. When more responders arrive, they can establish a decontamination corridor to clean and thoroughly decontaminate the victims before transporting them to a medical facility.

Mass decontamination can be set up in a variety of ways, and your fire department may have a specific procedure for it. Generally, this operation requires a minimum of two fire apparatus placed side by side. Place fog nozzles on the side discharge outlets, and use a spray pattern to douse the victims. An aerial ladder device can provide a complete overhead spray. Check with your department regarding its specific policies and procedures for mass decontamination.

To set up and use mass decontamination, follow the steps in **SKILL DRILL 32-12** :

1. Ensure that you have the appropriate PPE to protect against the chemical threat.
2. Stay clear of the product, and do not make physical contact with it.
3. Make an effort to contain runoff by directing victims out of the hazard zone and into a suitable location for decontamination. (**STEP ①**)
4. Set up the appropriate type of mass-decontamination system, based on the type of apparatus available.
5. Instruct victims to walk through the decontamination area and remove their contaminated clothing. (**STEP ②**)
6. Flush the contaminated victims.
7. Direct the decontaminated victims to the triage area. (**STEP ③**)

Methods of Decontamination

The previous descriptions of decontamination focused on using water to dilute or wash away the hazardous material. Some hazardous materials have chemical properties that require different methods of decontamination, or circumstances may dictate that fire fighters use an alternative to water as a decontamination measure. Although most of the methods described here can be carried out by the operations-level responder, you should consult a hazardous materials authority (e.g., your local hazardous materials response team) before attempting to implement any of them.

Alternative decontamination procedures include the following techniques:

- Absorption
- Adsorption
- Dilution
- Disinfection
- Disposal
- Solidification
- Emulsification
- Vapour dispersion
- Removal
- Vacuuming

■ Absorption/Adsorption for Decontamination

In absorption, a spongy material is mixed with a liquid hazardous material, and the contaminated mixture is then collected and disposed of. This technique is primarily used for

---

SKILL DRILL 32-12 Performing Mass Decontamination

1 Ensure that you have the appropriate PPE to protect against the chemical threat. Stay clear of the product, and do not make physical contact with it. Make an effort to contain runoff by directing victims out of the hazard zone and into a suitable location for decontamination.

2 Set up the appropriate type of mass-decontamination system, based on the type of apparatus available. Instruct victims to walk through the decontamination area and remove their contaminated clothing.

3 Flush the contaminated victims. Direct the decontaminated victims to the triage area.

decontaminating equipment and property; it has limited application for decontaminating personnel. Absorption may be used, for example, in a shuffle pit for the boots of responders, so that they are cleaned prior to entering the rest of the decontamination line.

Absorption minimizes the surface areas of liquid spills but is effective only on flat surfaces. Although absorbent materials, such as soil and sawdust, are inexpensive and readily available, they become hazardous themselves after coming in contact with the product and must be disposed of properly after the incident. One problem with this process is that it creates a hazardous waste that must be disposed of by someone (usually the generator), and the fire department may inadvertently become the generator of waste. Both the federal government and most provinces/territories have laws and regulations that dictate how to dispose of used absorbent materials. Consult a hazardous materials authority before disposing of any materials suspected of being contaminated.

The opposite of absorption is adsorption. In adsorption, the contaminant adheres to the surface of an added material, including sand and activated carbon, rather than combining with it, as in absorption. Similar issues related to the disposal of the contaminated materials arise whether you use an absorbent or an adsorbent.

■ Dilution

Dilution uses plain water or a soap–water mixture to lower the concentration of a hazardous material while flushing it off a contaminated person or object. Dilution is both fast and economical: water is a readily available solvent that usually generates no toxic fumes. Gross decontamination, technical decontamination, and mass decontamination processes all use dilution.

Fire fighters tend to use dilution as the first decontamination method, because water is readily available from the tank on an engine or from a hydrant. Before using water, consider whether the contaminant will react with water, be soluble in water, or spread to a larger area. The more water used, the more hazardous waste generated, which may spread the contamination and later must be disposed of safely. Dilution also can be effective in removing some solids, such as dusts and fibres; however, care must be used to prevent the spread of these contaminants to larger areas.

■ Disinfection

Disinfection is a process used to destroy recognized disease-carrying (pathogenic) micro-organisms. Commercial disinfectants are packaged with a detailed brochure that describes the limitations and capabilities of each product. Fire fighters who have medical research laboratories, hospitals, clinics, mortuaries, medical waste disposal facilities, blood banks, or universities in their response areas should be familiar with the specific types of biological hazards present and the best disinfectants to use for each hazard.

Take special precautions when responding to an emergency at a facility where blood-borne pathogens may be involved in an accident or spill. Also, be aware that the overuse of disinfectants contributes to the development of drug-resistant organisms, so always follow the manufacturer's instructions when such products are used.

■ Disposal

Disposal is a two-step removal process for items that cannot be properly decontaminated. First, the contaminated article is removed and isolated in a designated area; then, the item is packaged in a suitable container and transported to an approved waste facility, where it is either incinerated or buried in a hazardous waste landfill. Contaminated items such as disposable coveralls should be collected, bagged, and tagged. Contaminated tools and equipment should be placed in bags, barrels, or buckets.

■ Solidification

Solidification is the chemical process of treating a hazardous liquid so as to turn it into a solid material. This transformation makes the material easier to handle but does not change its inherent chemical properties.

■ Emulsification

Emulsification is the process of neutralizing or reducing the harmful effects of a hazardous material by changing its chemical properties. This process can be used with some petroleum-based products. Although emulsification may change the chemical nature of the hazardous material, the resultant product still needs to be disposed of properly. Federal, provincial/territorial, and local regulations may apply to the use of emulsifier products.

The principal advantage of emulsification is that it renders the hazardous material less harmful, thereby decreasing risks to emergency responders. It can also help limit clean-up costs. Unfortunately, the chemicals required for emulsification may be harmful to fire fighters. In addition, it may take some time to determine which chemicals should be used and whether they are available.

■ Vapour Dispersion

Vapour dispersion is the process of separating and diminishing harmful vapours. A water spray is commonly used to disperse or move vapours away from certain areas or materials. Vapour dispersion should be used with extreme caution because the rapid introduction of a large volume of air can create dynamic results.

■ Removal

Removal as a mode of decontamination applies specifically to contaminated soil that can be taken away from the scene and disposed of properly. Removal may be necessary when highly toxic materials that cannot be rendered harmless are involved, when on-site treatment presents unacceptable risks to fire fighters, or when on-site treatment costs exceed disposal costs.

For example, it may be more cost-effective and less risky to pump the hazardous material into a drum and transport it to a hazardous waste dump site than to neutralize a spill and decontaminate emergency equipment. Removing the contaminants reduces clean-up time and limits the exposure risk to fire fighters; it is not always the most economical decision, however, and often requires the purchase of hazardous waste and transportation permits.

■ Vacuuming

Vacuuming is the removal of dusts, particles, and some liquids by sucking them into a container. A filtering system prevents the contaminated material from recirculating and reentering the atmosphere. A special high-efficiency particulate air (HEPA) vacuum cleaner is used to remove hazardous dusts, powders, or fibres that are 0.3 micron or larger. HEPA filters allow air to pass through but capture the particulate matter in the air. The filters must be replaced regularly if they are to maintain their effectiveness.

The Decontamination Process

The decontamination process takes place in the decontamination corridor between the hot zone and the warm zone. Within this corridor, fire fighters pass through several stations to complete the decontamination process.

A clearly marked, easily seen, readily accessible entry point to the decontamination corridor should be established. If the decontamination operation occurs at night, the area should be well lit. All personnel leaving the hot zone must pass through the decontamination corridor.

The decontamination team must wear SCBA and a level of PPE equal to those being decontaminated. All decontamination team members must undergo decontamination before they leave the area.

■ Steps in Decontamination

Personnel leaving the hot zone should place any tools that they used in a tool drop area located near the decontamination corridor. This tool drop area can consist of a container, a recovery drum, or a special tarp. The tools can then be either used by another team entering the hot zone or collected for decontamination.

Next, the fire fighter, still wearing full PPE and SCBA, steps into the decontamination corridor for gross decontamination. The first step usually involves a portable shower, where medium-pressure, low-volume water flow rinses off and dilutes the contaminants on the PPE.

The decontamination team is responsible not only for scrubbing and swabbing the PPE worn by personnel but also for containing runoff of the water used for washing the equipment. This step generally involves one to three wash-and-rinse stations. At each station, one decontamination team member handles the scrubbing, and another does the rinsing. Only one fire fighter is allowed in any given station at a time. The decontamination team member who is scrubbing should

pay special attention to the gloves, the kneecaps, behind the knees, under the arms, under the boot bottoms, and any areas where significant folds in the personal protective gear are noted, because these are the places with the highest contamination levels.

After outerwear is thoroughly scrubbed and rinsed, it can be removed along with chemical-protective clothing; SCBA, however, should remain in place. The decontamination team member then unzips the PPE and peels back the suit, exposing the SCBA. The team should fold the PPE back, so that the contaminated side contacts only itself. If the procedure is done properly, the contaminated side of the suit will not touch either the interior of the suit or the person wearing it. If the fire fighter is wearing outer gloves, they can be removed at this point; the inner gloves will be removed later. The fire fighter then moves down the decontamination corridor to remove the remainder of his or her PPE, SCBA, and other support equipment.

When removing protective clothing, turn as much of the PPE inside out as possible. This tactic keeps any remaining trace contaminants inside the suit or glove. The fire fighter should be seated and allow the decontamination team members to remove PPE. Removing outside clothing properly is especially important when a fire fighter is wearing structural PPE, because these clothing items are porous and can absorb the hazardous material.

Remove the helmet, boots, and other support equipment along with the SCBA. The SCBA mask is usually the last item removed, because it serves as an additional safeguard against trace contaminants. Place equipment on the contaminated side of the decontamination corridor. Deposit the SCBA in a plastic bag, or place it on a tarp for bagging later. Highly contaminated SCBA should be removed and isolated until it can undergo complete decontamination. Remove the inner gloves, turn them inside out, and sort them into individual containers for cleanup or disposal. Heavy plastic bags can be used as containers for these items, because they usually provide sufficient temporary protection from most materials. The bags should be sealed with tape and transported elsewhere for cleanup or disposal. Place the bags in a properly marked recovery drum when disposing of them.

After removing personal clothing, the fire fighter should wash his or her entire body. An overhead shower produces much better results than a hose line. Apply ample soap to all areas of the body but especially to the head and the groin. Liquid surgical soaps in plastic squeeze bottles give the best results; local hospitals may be able to provide guidance about obtaining prepackaged cleaning kits. Use small brushes and sponges for scrubbing the body's surface. Afterwards, bag and mark all cleaning items for disposal. The water runoff from this step should be controlled and contained, just as in previous decontamination steps.

After going through the decontamination corridor, the fire fighter should dry off using towels or sheets. Each towel can be used only once; it is then placed in a bag on the contaminated side for disposal. Plastic garbage cans lined with plastic bags are good storage containers for this purpose.

Contact lenses may trap contaminants under the lens. Anyone wearing contact lenses should remove them and flush the eyes thoroughly. If any discomfort or irritation remains, the eye should be evaluated.

To perform technical decontamination at a hazardous materials incident, follow the steps in **SKILL DRILL 32-13**:

1. Drop any tools and equipment into a tool drum or a designated tarp. (**STEP** ❶)
2. Perform gross decontamination. (**STEP** ❷)
3. Wash and rinse the entry team member. (**STEP** ❸)
4. Remove outer hazardous materials–protective clothing and PPE. (**STEP** ❹)
5. Remove SCBA and the face piece. (**STEP** ❺)
6. Remove personal clothing. Bag and tag all personal clothing. (**STEP** ❻)
7. Shower and wash the body. Dry off the body and put on clean clothing.
8. The entry team member should proceed to medical monitoring.

9. At the station, the entry members should fill out any record-keeping information about the incident to include in exposure reports. (**STEP** ❼)

Decontaminated personnel can now don clean clothes. Disposable cotton coveralls, hospital gowns, hospital booties, slippers, and flip-flops are inexpensive and easy-to-use options for this step. They can be prepackaged according to size and stored for easy access. Provisions should be made to bag clothing in the event that uncontaminated clothing can be salvaged. During mass-decontamination operations, agencies have also used large, opaque plastic bags with holes cut out for the head and arms as a means to cover personnel after their contaminated clothes have been removed. The use of the bags provides a very cost-effective solution to providing coverage to personnel of many different sizes.

■ Medical Follow-up

After personnel are thoroughly decontaminated, they should proceed to an EMS station for a medical evaluation. EMS personnel will take vital signs for each person leaving decontamination and compare those data to baseline data (taken prior to the person's entry into the hot or warm zone). Any open wounds or breaks in skin surfaces should be reported immediately to medical control. Unless advised otherwise by the medical command physician, all open wounds should be cleaned on the scene.

SKILL DRILL 32-13   Performing Technical Decontamination

1 Drop any tools and equipment into a tool drum or onto a designated tarp.

2 Perform gross decontamination.

3 Wash and rinse the entry team member.

© LiquidLibrary

(Continues)

SKILL DRILL 32-13   Performing Technical Decontamination (*Continued*)

4   Remove outer hazardous materials–protective clothing and isolate PPE.

5   Remove SCBA and the face piece.

6   Remove personal clothing. Bag and tag all personal clothing.

7   Shower and wash the body. Dry off the body and put on clean clothing. The entry team member should proceed to medical monitoring. At the station, the entry members should fill out any record-keeping information about the incident to include exposure reporting.

© LiquidLibrary

Chief Concepts

- To ensure your safety and that of your crew, you must be able to recognize a potential hazardous material as soon as possible.
- Hazardous materials incidents can occur almost anywhere. Hazardous materials are stored in warehouses, hospitals, laboratories, industrial occupancies, residential garages, bowling alleys, home-improvement centres, garden supply stores, restaurants, and many other businesses.
- A container is any vessel or receptacle that holds material, including storage vessels, pipelines, and packaging. Often the container type, size, and material provide important clues about the nature of the substance inside.
- Bulk storage containers include the following types:
  - Fixed tanks—Include vertical storage tanks, above ground storage tanks, and underground storage tanks. Often these containers are surrounded by a supplementary containment system to help control an accidental release.
  - Large transportation tankers—Used to transport hazardous materials.
  - Totes—Portable plastic tanks surrounded by a stainless-steel web that adds both structural stability and protection. Both shipping and storing totes can be hazardous operations.
  - Intermodal tanks—Used as both shipping and storage vehicles. They hold between 18,925 and 22,710 L (5000 and 6000 gal.) of product and can be either pressurized or nonpressurized.
- Bulk storage containers, or large-volume containers, are defined by their internal capacity based on the following measures:
  - Liquids—More than 450 L (119 gal.)
  - Solids—More than 400 kg (882 lb.)
  - Gases—More than 400 kg (882 lb.)
- Nonbulk storage vessels include the following types:
  - Drums—Barrel-like containers used to store a wide variety of substances.
  - Bags—Commonly used to store solids and powders. Pesticide bags must be labelled with specific information.
  - Carboys—A glass, plastic, or steel container that holds 19–57 L (5–15 gal.) of product, such as some corrosives and other types of chemicals.
  - Compressed gas cylinders—Used to hold liquids and gases.
  - Cryogenic containers—Include dewar containers that hold gaseous substances that have been chilled until they liquefy.

- Chemical transport vehicles or tankers include the following types:
  - TC306/MC306/DOT406 flammable liquid tanker—Typically carries gasoline or other flammable and combustible materials.
  - TC307/MC307/DOT407 chemical hauler—Round or horseshoe-shaped tank that typically carries 22,712 to 26,498 litres (6000 to 7000 gallons) of flammable liquids, mild corrosives, and poisons.
  - MC312/DOT412 corrosives tanker—Used to transport concentrated sulfuric and nitric acids and other corrosive substances.
  - TC331/MC331 pressure cargo tanker—Carries materials such as ammonia, propane, and butane.
  - TC341/CGA341/MC338 cryogenic tanker—Low-pressure tanker that relies on tank insulation to maintain low temperatures to transport cryogenic gases.
  - Tube trailers—Carry compressed gases, such as hydrogen, oxygen, helium, and methane.
  - Dry bulk cargo tanks—Commonly seen on the road and carry dry bulk goods, such as powders, pellets, fertilizers, and grain.
- Railway tank cars can carry volumes up to 113,550 L (30,000 gal.) and have the potential to create large leaks or vapour clouds. There are three basic railcar configurations:
  - DOT-111 nonpressurized railcars—Carry general industrial chemicals, consumer products, flammable and combustible liquids, and mild corrosives.
  - DOT-105 and DOT 112 pressurized railcars—Transport materials such as propane, ammonia, ethylene oxide, and chlorine.
  - Special-use railcars—Include boxcars, flatcars, corrosive tank cars, and high-pressure compressed gas tube cars.
- Pipelines often transport hazardous materials. They may be located either underground or above ground.
- Additional reference sources for hazardous materials include SDSs, shipping papers, and staffed national resources such as CANUTEC and CHEMTREC®.
- Industries that routinely use radioactive materials include food testing labs, hospitals, medical research centres, biotechnology facilities, construction sites, and medical laboratories. Radioactive materials are not detected by sight, smell, taste, or other senses. If you have any suspicion that the incident involves radiation, it will be necessary to call a hazardous materials team or some other resource with radiation-detection capabilities.
- After identifying that a hazardous material is present, you must identify the resources needed to properly respond to the incident.

- The resources sought for a hazardous materials incident should include support personnel, trained hazardous materials technicians, and technical specialists who will help identify the hazardous material and control the incident. Additional calls should then request decontamination personnel and equipment. Other notifications could include the CANUTEC (or CHEMTREC®, where applicable), local and provincial/territorial environmental agencies, and the local emergency planning commission.
- A predetermined list of contact names, agencies, and numbers should be established and maintained at the dispatch centre.
- No offensive action should be taken until the hazardous material has been properly identified. After the material is identified, an operations-level responder should perform only actions that do not involve contact with the material. The responder must use full PPE during any activity and must complete decontamination procedures prior to leaving any area where the hazardous material is present.
- When reporting the hazardous material incident, if possible, provide all of the following information:
  - The exact address and specific location of the leak or spill
  - Identification of indicators and markers of hazardous materials
  - All colour or class information obtained from placards
  - Four-digit United Nations (UN)/North American (NA) identification numbers for the hazardous materials
  - Hazardous material identification obtained from shipping papers or SDS and the potential quantities of hazardous materials involved
  - Description of the container, including its size, capacity, type, and shape
  - The amount of chemical that could leak and the amount that has already leaked
  - Exposures of people and the presence of special populations (e.g., children or elderly)
  - The environment in the immediate area
  - Current weather conditions, including wind direction and speed
  - A contact or callback number and two-way radio frequency or channel
- The highest priority of an IAP is to consider the safety of the responding personnel. Responders are there to isolate, contain, and remedy the problem—not to become a part of it.
- Sometimes no action is the safest course of action.
- At the operational level, all response objectives should be primarily defensive (i.e., personnel do not come in contact with the hazardous material).
- Defensive objectives do not involve stopping the leak or release of a hazardous material and include the following:

- Isolating the area affected by the leak or spill and evacuating victims who could become exposed to the hazardous material if the leak or spill were to progress
- Controlling where the spill or release is spreading
- Containing the spill to a specific area
- Diking and damming
- Absorbing/adsorbing the hazardous material
- Stopping the flow remotely
- Diluting or diverting the material
- Suppressing or dispersing vapour
- Terrorists who want to injure responding personnel with a secondary device or attack will typically make the initial attack very dramatic to draw responders into very close proximity to the scene. As responders begin to treat victims, the secondary attack then takes place.
  - Signs of secondary devices may include tripwires, timers, and ordinary, everyday containers (e.g., luggage, briefcases, boxes) found in very close proximity to the initial incident site.
  - The smell of chemicals or the sighting of chemical dispersion devices or other containers that may hold chemical, biological, or even radioactive agents could be an indicator that a secondary attack is possible.
- The IMS can be expanded to handle a hazardous materials incident. A special technical group may be developed under the operations section known as the hazardous materials branch.
- Temperature-related conditions and stress pose dangers. If the body cannot disperse heat because it is covered by a closed, impermeable garment, serious short- and long-term medical situations could result. These medical conditions include the following:
  - Heat exhaustion—: Mild form of shock that occurs when the body overheats.
  - Heatstroke—Potentially fatal condition of which the reduction or cessation of sweating is the first symptom. Immediate medical attention is required.
- To avoid cold-related injuries, such as frostbite, fire fighters should dress in layers and try to keep dry.
- The control zones at a hazardous materials incident are as follows:
  - Hot zone—Area immediately around the incident. Only properly trained and equipped personnel should enter this zone.
  - Warm zone—Contains the decontamination corridor.
  - Cold zone—No special PPE is required. This zone contains the ICP, personnel staging area, and rehabilitation area.
- Team members should always remain within sight, sound, or touch of each other.
- Responders are required to take appropriate steps to monitor the atmosphere for potential hazards.

Monitoring priorities should include radiation, corrosive vapours, oxygen levels, combustibility, and final toxicity. Radiation monitoring is conducted using a Geiger counter. Oxygen levels and combustible levels are typically monitored with the use of a two-, three-, four-, or five-gas monitor. Other specialized chemical monitors can be used to detect individual chemicals.

■ The search and rescue process is more complicated in a hazardous materials incident than in a structure fire. Before beginning search and rescue operations, all emergency response personnel must first recognize and identify when hazardous materials may be present. After this is accomplished, the IC must evaluate the conditions. If hazardous materials are present, the IC must determine whether hazardous materials technicians can safely enter the hot zone to perform search and rescue operations.

■ Decontamination makes personnel, equipment, and supplies safe by removing or reducing hazardous materials or WMDs. Proper decontamination is essential at every hazardous material/WMD incident to ensure the safety of both personnel and property.

■ Cross-contamination is the process by which a contaminant is carried out of the hot zone and contaminates people, animals, the environment, or equipment. Cross-contamination may occur in several ways:
  • A contaminated victim comes into physical contact with an emergency responder.
  • A bystander or other emergency responder comes into contact with a contaminated object from the hot zone.
  • A decontaminated responder comes into contact with another contaminated responder or object.

■ Fire fighters are often responsible for establishing a decontamination corridor.

■ There are four major categories of decontamination:
  • Emergency decontamination—Aims to reduce the amount of surface contaminant. Involves dousing the victim with water via a hose line outside of a decontamination corridor. Formal decontamination is performed later.
  • Gross decontamination—Uses a large shower system in a decontamination corridor. This type of decontamination is used for a brief and rapid contamination reduction.
  • Technical decontamination—Performed after gross decontamination and may involve several steps.
  • Mass decontamination—Used to decontaminate large groups of people in the field. Victims will need to undergo further and more thorough decontamination.

■ Alternative decontamination procedures include the following options:
  • Absorption—Use of a spongy material to soak up the hazardous material. This technique is used to decontaminate equipment and property.
  • Adsorption—Binding of the contaminant to the surface of an added material, which is then disposed of.
  • Dilution—Use of soap and water to flush the hazardous material off a person or object.
  • Disinfection—Process used to destroy disease-carrying organisms.
  • Disposal—Two-step removal process for items that cannot be decontaminated properly.
  • Solidification—Chemical process of treating a hazardous liquid to become a solid.
  • Emulsification—Process of neutralizing or reducing the harmful effects of a hazardous material by changing its chemical properties.
  • Vapour dispersion—Process of separating and diminishing harmful vapours, often with a water spray.
  • Removal—Process of removing contaminated soil from a hazardous materials site.
  • Vacuuming—Removal of dusts, particles, and liquids by sucking them into a container.
  • Decontamination process—The decontamination process takes place in the decontamination corridor between the hot zone and the warm zone. Within this corridor, fire fighters pass through several stations to complete the decontamination process.

■ After all of the factors have been considered, the IC may decide that the hazardous materials incident cannot be handled without unnecessary risks and that all personnel should withdraw.

■ The recovery phase of a hazardous materials incident occurs when the imminent danger to people, property, and the environment has passed or is controlled and clean-up begins. During this phase, state and federal agencies may become involved in cleaning up the site, determining the responsible party, and implementing cost-recovery methods.

■ The recovery phase in large-scale invidents can last for days, weeks, or even months and may require resources and equipment beyond those available to local responders.

■ If a hazardous materials incident involves potential criminal or terrorist activity, responders must notify law enforcement and appropriate federal agencies. Specific activities should be implemented to ensure safety, such as isolating the scene.

Hot Terms

Absorption The process of applying a material (an absorbent) that will soak up and hold a hazardous material in a spongelike manner, for collection and disposal.

Adsorption A process that occurs when a liquid hazardous material accumulates on the surface of a solid.

Backup entry team A dedicated team of fully qualified and equipped responders who are ready to enter the hot zone at a moment's notice to rescue any member of the hot-zone entry team.

Backup personnel Individuals who remove or rescue those working in the hot zone in the event of an emergency.

Bulk storage container A large-volume container that has an internal volume greater than 450 L (119 gal.) for liquids or 400 kg (882 lb.) for solids, or a capacity of greater than 400 kg (882 lb.) for gases.

Bung A small opening in a closed-head drum.

Carboy A glass, plastic, or steel container ranging in volume from 19 to 57 L (5 to 15 gallons).

Canadian Transport Emergency Centre (CANUTEC) A national advisory centre operated by the TDG Directorate of Transport Canada. CANUTEC maintains a scientific data bank on chemicals manufactured, stored, and transported in Canada and is staffed by professional scientists who specialize in emergency response and are experienced in interpreting technical information and providing advice.

Chemical Transportation Emergency Center (CHEMTREC®) A national call centre for basic chemical information, established by the Chemical Manufacturers Association and now operated by the ACC.

Cold zone The control zone of hazardous materials/WMD incidents that contains the ICP and other such support functions as are deemed necessary to control the incident. (NFPA 472)

Confinement Actions taken to keep a material, once released, in a defined or local area. (NFPA 472)

Containment Actions taken to keep a material in its container (e.g., stop the release of the material or reduce the amount being released). (NFPA 472)

Container Any vessel or receptacle that holds material, including storage vessels, pipelines, and packaging.

Control zone A designated area at a hazardous materials incident, based on safety and the degree of hazard, established to help limit the number of civilian and public service personnel who might be exposed to the hazardous materials.

Cryogenic liquid A gaseous substance that has been chilled to the point at which it liquefies; a liquid having a boiling point lower than −101.1°C (−150°F) at an absolute pressure of 101 kPa (14.7 psia); also known as a cryogen.

Cylinder A portable compressed gas container.

Damming A process used when liquid is flowing in a natural channel or depression and its progress can be stopped by blocking the channel.

Decontamination The physical and/or chemical process of reducing and preventing the spread of contaminants from people, animals, the environment, or equipment

involved at hazardous materials/WMD incidents. (NFPA 472)

Decontamination corridor The area where decontamination is performed, usually located within the warm zone. (NFPA 472)

Decontamination team The team responsible for reducing and preventing the spread of contaminants from persons and equipment used at a hazardous materials incident. Members of this team establish the decontamination corridor and conduct all phases of decontamination.

Defensive objective A response objective that is primarily defensive in purpose and does not involve stopping the leak or release of a hazardous material.

Dewar container A container designed to preserve the temperature of a cold liquid held inside.

Diking The placement of materials to form a barrier that will keep a hazardous material in liquid form from entering an area or that will hold the material in an area.

Dilution The process of adding some substance (usually water) to a product to weaken its concentration.

Disinfection The process used to inactivate virtually all recognized pathogenic micro-organisms but not necessarily all microbial forms (e.g., bacterial endospores). (NFPA 1581)

Disposal A two-step removal process for contaminated items that cannot be properly decontaminated. Items are bagged and placed in appropriate containers for transport to a hazardous waste facility.

Diversion Redirecting spilled or leaking material to an area where it will have less impact.

Drum A barrel-like container used to store a wide variety of substances, including food-grade materials, corrosives, flammable liquids, and grease. Drums may be constructed of low-carbon steel, polyethylene, cardboard, stainless steel, nickel, or other hybrid materials.

Dry bulk cargo tank A tanks designed to carry dry bulk goods, such as powders, pellets, fertilizers, or grain; they are generally V-shaped with rounded sides that funnel toward the bottom.

Emergency decontamination The physical process of immediately reducing contamination of individuals in potentially life-threatening situations with or without the formal establishment of a decontamination corridor. (NFPA 472)

Emulsification The process of neutralizing or reducing the harmful effects of a hazardous material by changing its chemical properties.

Gross decontamination The phase of the decontamination process during which the amount of surface contaminants is significantly reduced. (NFPA 472)

Hazardous materials branch The function within an overall IMS that deals with the mitigation and control of the hazardous materials/WMD portion of an incident. (NFPA 472)

Hazardous materials information research team A dedicated team of responders who serve as an information gathering and referral point for the IC as well as the hazardous materials officer.

Hazardous materials officer The person who is responsible for directing and coordinating all operations involving hazardous materials/weapons of mass destruction as assigned by the incident commander; also known as hazardous materials branch director/group supervisor. (NFPA 472)

Hazardous materials safety officer The person who works within the IMS (specifically the hazardous materials branch) to ensure that recognized hazardous materials/WMD safe practices are followed at hazardous materials/WMD incidents; also known as assistant safety officer–hazardous material. (NFPA 472)

Heat exhaustion A mild form of shock caused when the circulatory system begins to fail as a result of the body's inadequate effort to give off excessive heat.

Heat stroke A severe and sometimes fatal condition resulting from the failure of the temperature-regulating capacity of the body. It is caused by prolonged exposure to the sun or high temperatures. Reduction or cessation of sweating is an early symptom; a body temperature of 40.5°C (105°F) or higher, a rapid pulse, hot and dry skin, a headache, confusion, unconsciousness, and convulsions can also occur. Heatstroke is a true medical emergency requiring immediate transport to a medical facility.

Hot zone The control zone immediately surrounding hazardous materials/WMD incidents, which extends far enough to prevent adverse effects of hazards to personnel outside the zone. (NFPA 472)

Hot-zone entry team The team of fire fighters assigned to enter into the designated hot zone.

Intermodal tank A bulk container that can be shipped by all modes of transportation—air, sea, and land.

Mass decontamination The physical process of reducing or removing surface contaminants from large numbers of victims in potentially life-threatening situations, in the fastest time possible. (NFPA 472)

National Response Center (NRC) An agency maintained and staffed by the United States Coast Guard; it should always be notified if any spilled material could possibly enter a navigable waterway.

Nonbulk storage vessel Any container other than a bulk storage container.

Pipeline A length of pipe including pumps, valves, flanges, control devices, strainers, and/or similar equipment for conveying fluids. (NFPA 70)

Pipeline right-of-way An area, patch, or roadway that extends about a metre on either side of the pipe itself and that may contain warning and informational signs about hazardous materials carried in the pipeline.

Recovery phase The stage of a hazardous materials incident after the imminent danger has passed, when cleanup and the return to normalcy have begun.

Remote shut-off valve A device that allows for the shut-off of a system or the isolation a valve from a remote location.

Removal A mode of decontamination that applies specifically to contaminated soil that can be taken away from the scene.

Retention The process of creating an area to hold hazardous materials.

Secondary containment Any device or structure that prevents environmental contamination when the primary container or its appurtenances fail. Secondary containment shall be designed and constructed to intercept and contain pesticide spills and leaks and to prevent runoff or leaching of pesticides into the environment. Examples of secondary containment include dikes, curbing, and double-walled tanks. (NFPA 434)

Signal words Information on a pesticide label that indicates the relative toxicity of the material.

Solidification The process of chemically treating a hazardous liquid so as to turn it into a solid material, thereby making the material easier to handle.

Special-use railcars Boxcars, flatcars, cryogenic and corrosive tank cars, and high-pressure compressed gas tube cars.

TC306/MC306/DOT406 flammable-liquid tanker Commonly known as a gasoline tanker; a tanker that typically carries gasoline or other flammable and combustible materials.

TC307/MC307/DOT407 chemical hauler A tanker with a rounded or horseshoe-shaped tank.

TC312/MC312/DOT412 corrosives tanker A tanker that often carries aggressive acids, such as concentrated sulphuric or nitric acid, and features reinforcing rings along the side of the tank.

TC331/MC331 pressure cargo tanker A tank commonly constructed of steel with rounded ends and a single open compartment inside; there are no baffles or other separations inside the tank.

TC341/CGA341/MC338 cryogenics tanker A low-pressure tanker designed to maintain the low temperature required by the cryogens it carries.

Technical decontamination The planned and systematic process of reducing contamination to a level that is as low as reasonably achievable.

Tote A portable tank characterized by its unique style of construction, which usually holds around 1000 L (264 gal.) of product.

Tube trailers High-volume transportation devices made up of several individual compressed gas cylinders banded together and affixed to a trailer.

Vacuuming The process of cleaning up dusts, particles, and some liquids using a vacuum with HEPA filtration to prevent recontamination of an environment.

Vapour dispersion The process of lowering the concentration of vapours by spreading them out.

Vapour suppression The process of controlling vapours given off by hazardous materials by covering the product with foam or other material or by reducing the temperature of the material, thereby preventing vapour ignition.

Vent pipes Inverted J-shaped tubes that allow for pressure relief or natural venting of a pipeline for maintenance and repairs.

Warm zone The zone where the decontamination process takes place in the decontamination corridor, which is set up in between the cold and hot zones.

FIRE FIGHTER *in action*

Your department is requested, as part of a mutual aid agreement, to send additional apparatus to assist at a large nursery/garden centre fire just before Christmas. The nursery/garden centre is at the peak of its growing season for holiday plants, which require the facility to construct large greenhouses made of plastic sheeting. The ensuing fire has enveloped the entire area in flames and has impinged on the chemical and fuel storage areas of the complex. Your engine company has been assigned to set up a decontamination area for all personnel operating at the scene. Hazardous materials personnel direct you to set up both a technical decontamination and an emergency decontamination. While setting up the decontamination area, a report comes in stating that a fire fighter has fallen near the chemical building and landed in several bags of open fertilizer.

1. Based on the reports, the fire fighter should go through emergency decontamination because the process would allow the victim to receive treatment as soon as possible.
   - **A.** True
   - **B.** False

2. As fire fighters exit the hot zone, what is the first thing they should do as they begin to enter the decontamination corridor?
   - **A.** Report to emergency decontamination
   - **B.** Drop their tools in the tool drop area prior to entering the decontamination area
   - **C.** Report to the mass decontamination wash site with their tools
   - **D.** Drop their tools, gloves, and boots in the drop area before entering the decontamination area

3. Because the likelihood of contamination is very small, decontamination team members need not pass through the decontamination process after they carry out the decontamination process for victims and other responders.
   - **A.** True
   - **B.** False

4. When conducting decontamination operations, decontamination team members should pay close attention to which of the following areas when scrubbing?
   - **A.** Gloves
   - **B.** Knee caps
   - **C.** Boot bottoms
   - **D.** All of the above

5. What is the purpose of the gross decontamination in a decontamination system?

   A. To remove small amounts of contaminants using small brushes

   B. To decontaminate multiple personnel at the same time

   C. To conduct final cleaning of the contaminated person after formal decontamination

   D. To remove large amounts of contaminants in a controlled manner

6. What is the next step in the decontamination process after all protective clothing and SCBA have been removed?

   A. Remove personal clothing and shower

   B. Report to staging for reassignment

   C. Assist the decontamination team in bagging the contaminated clothing

   D. Have a medical evaluation completed

FIRE FIGHTER II
in action

You arrive on the scene of a motor vehicle accident where a tanker has overturned. The tank is round and it has reinforcing ribs visible on the outside. There is a large amount of liquid that is pooling around the vehicle. Bystanders are screaming that the driver is trapped in the cab.

1. How would you manage the bystanders?
2. How would you provide emergency decontamination?
3. What are your greatest risks with this incident?

Terrorism Awareness

Fire Fighter I

Knowledge Objectives

After studying this chapter, you will be able to:

- Describe the threat posed by terrorism. (p. 968–969)
- Identify potential terrorist targets in your jurisdiction. (p. 969–970)
- Describe the dangers posed by explosive devices. (p. 971–975)
- Explain the difference between chemical and biological agents. (p. 975–980)
- Describe the dangers posed by radiological incidents. (p. 980–981)
- Describe the need for decontamination of exposed victims and response personnel. (p. 983)

Skills Objectives

There are no skills objectives for Fire Fighter I candidates. NFPA 1001 contains no Fire Fighter I Job Performance Requirements for this chapter.

© Photos.com

Fire Fighter II   FFII

Knowledge Objectives

There are no knowledge objectives for Fire Fighter II candidates. NFPA 1001 contains no Fire Fighter II Job Performance Requirements for this chapter.

Skills Objectives

There are no skills objectives for Fire Fighter II candidates. NFPA 1001 contains no Fire Fighter II Job Performance Requirements for this chapter.

Additional NFPA Standards

- NFPA 472, *Standard for Competence of Responders to Hazardous Materials/Weapons of Mass Destruction Incidents*

I t is 10:12 a.m. and you are dispatched to a water flow alarm in the basement of a 22-storey office building. The basement consists of four floors of parking as well as the heating, ventilating, and air conditioning (HVAC) system. The dispatcher advises you en route that there are reports of smoke coming from the basement and that the entire building shook just before the alarm sounded. Currently, the building is being evacuated.

1. What has most likely occurred, and how would you approach this incident?
2. How would your thinking change if you found out that the Royal Canadian Mounted Police (RCMP) has an office in the building?
3. How would you approach this incident if you received news reports that a bomb had gone off in the basement?

© Duncan Noakes/Dreamstime.com

What Is Terrorism?

Terrorism can be described as the unlawful use of violence or threats of violence to intimidate or coerce a government, the civilian population, or any segment thereof to further political or social objectives. This broad definition encompasses a wide range of acts committed by different groups for different purposes.

The goal of terrorism is to produce feelings of terror or fear in a population or a group. Terrorism is not limited to certain tools or weapons, however; almost anything can potentially be used as a tool by terrorists. For example, a gasoline tanker could be intentionally wrecked and burned to destroy property or lives, and such an event would be classified as a terrorist incident; yet if an identical gasoline tanker wrecked and caught fire because of driver error, we would call that event an accident. The steps the fire department needs to take to mitigate each of these incidents might be the same. Many times, first responders will not know whether an incident is caused by accidental circumstances or by intentional actions until a thorough investigation has been completed.

The Royal Canadian Mounted Police (RCMP) classifies terrorism as espionage or sabotage against Canada; foreign-influenced activities detrimental to the interests of Canada; activities directed toward or in support of the threat or use of acts of serious violence against Canadians for political, religious, or ideological objectives; and, activities leading to the destruction or overthrow by violence of the government of Canada.

The Canadian Security Intelligence Service (CSIS) notes that there are more terrorist groups active in Canada today than in any other country in the world. Their presence can be attributed to Canada's proximity to the United States, one of the world's preeminent terrorist targets, but also to the fact that Canada is an attractive place to live and do business because of its openness and its respect for human rights and freedoms.

Within the United States, 24 terrorist incidents occurred between 1994 and 1999; three other crimes that occurred during this period were also suspected of being terrorist in nature. Most of these incidents were classified as domestic terrorism. On September 11, 2001, the largest terrorist event

in the history of the United States—the successful attacks on the World Trade Center and the Pentagon and the crash of a hijacked plane in Pennsylvania, which aborted another attack—resulted in almost 3000 deaths **FIGURE 33-1**. Although terrorism can occur at any time, it is, thankfully, not a common

FIGURE 33-1 The September 11 attacks on the World Trade Center and the Pentagon, along with the crash of a hijacked plane in Pennsylvania, was the largest terrorist attack in the history of the United States.

© Todd Hollis/AP Photos

event in the United States. In 2010, the U.S. Department of State reported no deaths in the United States from terrorist events. In Canada, all terrorist groups are banned, and while activity does exist, the number of terrorist events are dramatically lower than those in other countries.

FIRE FIGHTER Tips

Emergency responders at a terrorist event may later be called to testify in court. All personnel must make appropriate and accurate notes. It is important that the responder who moves evidence document it and retain a copy of his or her report, so that they can directly speak to it if called to attend court.

FIRE FIGHTER Tips

The United States *Code of Federal Regulations* (CFR) defines terrorism as "the unlawful use of force and violence against persons or property to intimidate or coerce a government, the civilian population, or any segment thereof in furtherance of political or social objectives" (CFR, title 28, section 0.85).

The *Criminal Code of Canada* has a broad definition of terrorism as it relates to aviation, protected persons, hostages, nuclear material, maritime navigation, bombings, violence, and the financing of terrorist groups. It further notes that terrorist activity is usually directed at an entity with political, religious, or ideological purposes in mind.

■ Fire Service Response to Terrorist Incidents

The role of the fire service in handling terrorist events consists of the same functions that fire fighters perform on a day-to-day basis. The fire service's role includes supplying emergency medical services (in concert with emergency medicine providers), mitigation of hazardous materials, technical rescue, and fire suppression. All of these functions may be needed when a terrorist incident occurs. Terrorism also presents new challenges for the fire service, because it requires fire fighters to both perform the tasks they have been doing successfully and to learn to handle some new tasks. In this sense, responding to terrorist incidents represents an extension of the threats fire fighters encounter on a routine basis.

The terrorist threat requires fire fighters to work closely with local, provincial, and federal law-enforcement agencies; emergency response agencies (ERAs); allied health agencies; and the military. It is critical that all of these agencies work together in a coordinated and cooperative manner. Emergency responders and law-enforcement agencies must be prepared to face a wide range of potential situations.

One threat posed by terrorists is the use of **weapons of mass destruction (WMD)**—devices that are designed to cause maximum damage to property or people. These weapons could be made from chemical, biological, or radiological agents, or they could involve conventional weapons and explosives. They pose a grave danger to life, such that mass casualties could result from a WMD attack in an urban area. An incident involving a WMD could quickly overwhelm ERAs and the local healthcare system.

To prepare for the threat of terrorist attacks, fire fighters need to understand the technologies used to manage WMD incidents. In addition, first responders need to know their role in handling incidents involving the release of chemical, biological, and radiological agents.

Potential Targets and Tactics

Terrorists are motivated by a cause and choose targets that they believe will help them achieve their goals and objectives. The aim of a terrorist incident is to instill fear and panic in the general population and to disrupt their daily way of life. Given their goal, terrorists tend to choose symbolic targets, such as a place of worship, an embassy, a monument, or a prominent government building. Sometimes the objective is sabotage—that is, to destroy or disable a facility that is significant to the terrorist cause. The ultimate goal could be to cause economic turmoil by interfering with transportation, trade, or commerce.

Terrorists choose a method of attack they think will make a desired statement or achieve the maximum results. They may vary their methods or change them over time; explosive devices have remained popular weapons for terrorists, however, and have been used in thousands of terrorist attacks. In recent years, the number of suicide bombings has increased dramatically.

The type and location of terrorist incidents also change over time. Political instability in a country or region may result in an increase in terrorist incidents; for example, in 2011, the lack of a stable government in Somalia resulted in 25 incidents of ships being hijacked off its coast. The purpose of these terrorist incidents appeared to be financial gain rather than political change.

Terrorism can occur in any community and in connection with many different issues. There are many causes whose supporters range from peaceful, nonviolent organizations to fanatical fringe groups; the latter may be the wellspring of terrorism; for example, a rural ski lodge tucked away in the mountains might be attacked by an environmental group that is upset with its plans for expansion. Or a small retailer in an upscale suburban community could become the target of an animal-rights group that objects to the sale of fur coats. An antiabortion group might even plant a bomb at a local community health clinic.

It is often possible to anticipate likely targets and potential attacks. Law-enforcement agencies routinely gather intelligence about terrorist groups, threats, and potential targets. The fire service is increasing its capacity to engage in intelligence sharing so that preincident plans can be developed for possible targets and scenarios. Even if no specific threats have been made, certain types of occupancies are known to be potential targets; preincident planning for those locations should include the possibility of both a terrorist attack and an accidental fire.

News accounts of terrorist incidents abroad can help keep fire fighters up to date with trends related to terrorist tactics. They can provide useful information about situations that could occur in your jurisdiction in the future. A variety of Internet resources can also help you to keep abreast of current threats. Periodic e-mails sent by listservs, which are available through federal response agencies, can provide useful information. By becoming familiar with potential targets and current tactics, emergency responders can plan appropriate strategies and tactics for potential attacks.

FIRE FIGHTER Tips

Potential Terrorist Targets

Infrastructure Targets
- Bridges and tunnels
- Emergency facilities
- Hospitals
- Oil refineries
- Pipelines
- Power plants
- Hydropower dams
- Railroads
- Telecommunications systems
- Water reservoirs and treatment plants

Symbolic Targets
- Embassies
- Government buildings
- Military bases
- National monuments
- Places of worship

Civilian Targets
- Arenas and stadiums
- Airports and railroad stations
- Mass transit systems
- Schools and universities
- Shopping malls
- Theme parks

Eco-terrorism Targets
- Controversial development projects
- Environmentally sensitive areas
- Research facilities

Agro-terrorism Targets
- Crops
- Feed storage
- Grain elevators
- Livestock and poultry

Cyberterrorism Targets
- Banking and finance computer systems
- Business computer systems
- Court computer systems
- Government computer systems
- Law-enforcement computer systems
- Military computer systems

Potential targets include both natural landmarks and man-made structures. These sites can be classified into three broad categories: infrastructure targets, symbolic targets, and civilian targets.

Terrorism can also be classified according to the part of society it targets—for example, agro-terrorism, eco-terrorism, and cyberterrorism. These types of terrorism will probably not have as direct an impact on fire departments as incidents involving natural landmarks and man-made structures, however. Incidents involving cyberterrorism, for instance, may have an indirect impact on the fire department and could significantly disrupt its operations but would probably not require a fire department response.

■ Infrastructure Targets

Terrorists might select bridges, tunnels, or subways as targets in an attempt or disrupt transportation and inflict a large number of casualties **FIGURE 33-2**. They might also plan to attack the public water supply or try to disable the electrical power distribution system, telephones, or the Internet. Disruption of a community's 911 system or public-safety radio network would directly affect ERAs.

■ Symbolic Targets

National landmarks, such as Parliament Hill in Ottawa, provincial legislative assembly buildings, or the CN Tower, could be targeted by groups who want to attack symbols of Canadian pride **FIGURE 33-3**. Foreign embassies and institutions might be attacked by groups promoting revolution within those countries or protesting their international policies. Religious institutions are potential targets of hate groups. By attacking these symbols, terrorists seek to make people aware of their demands and to create a sense of fear in the public.

■ Civilian Targets

Civilian targets include places where large numbers of people gather, such as shopping malls, schools, or stadiums. In incidents involving these sites, the goal of terrorists is to indiscriminately kill or injure large numbers of people, thereby creating chaos in the society as a whole **FIGURE 33-4**. Attacking civilian targets tends to produce a high degree of fear in the general population.

■ Eco-terrorism

Eco-terrorism refers to illegal acts committed by groups supporting environmental or related causes. Examples include spiking trees to sabotage logging operations, vandalizing a university research laboratory that is conducting experiments on animals, and firebombing a store that sells fur coats. Unless such incidents involve arson or explosions,

A.

© Arthur S. Aubry/Photodisc/Getty Images

C.

© AbleStock

B.

© Steve Allen/Brand X Pictures/Alamy Images

D.

© Michael Dwyer/Alamy Images

FIGURE 33-2 **A.** Subways, **B.** airports, **C.** bridges, and **D.** hospitals are all vulnerable to attack by terrorists who seek to interrupt a country's infrastructure.

fire departments are rarely directly involved in responding to eco-terrorism.

■ Agro-terrorism

Agro-terrorism includes the use of chemical or biological agents to attack the agricultural industry or food supply **FIGURE 33-5**. The deliberate introduction of an animal disease, such as foot-and-mouth disease, into the livestock population could result in major losses to the food industry and produce fear among members of the general population.

■ Cyberterrorism

Groups of terrorists may engage in **cyberterrorism** by electronically attacking government or private computer systems.

In the past, several attempts have been made to disrupt the Internet or to attack government computer systems and other critical networks. This type of terrorism would disrupt many day-to-day activities in our society, and disrupt the notification, dispatch, response, and operation of emergency services.

Agents and Devices

Terrorists can turn the most ordinary objects into powerful weapons. We tend to think of gasoline tankers and commercial airliners as valuable parts of our society, but in the hands of determined terrorists, they can become WMD. Acts of terrorism reflect the instigator's intentions more than the

FIGURE 33-3 Terrorists might attempt to destroy visible national icons.
© Tania Thomson/ShutterStock, Inc.

FIGURE 33-4 By attacking a civilian target, such as a crowded stadium, terrorists might make citizens feel vulnerable in their everyday lives.
© Photodisc/Getty Images

FIGURE 33-5 Agro-terrorism might affect the food supply.
© Larry Rana/USDA

FIRE MARKS

The Earth Liberation Front (ELF) claimed credit for arson fires that destroyed several mountaintop buildings at a ski resort in Vail, Colorado, in 1998. The group was trying to prevent the development of additional ski trails on the mountain.

FIRE MARKS

MAJOR TERRORIST INCIDENTS IN THE UNITED STATES

September 1984: The Dalles, Oregon
In an effort to influence local elections, a religious sect spread *Salmonella* on salad bars in four restaurants, resulting in 750 cases of *Salmonella* poisoning.

February 1993: New York City
A large explosive device was detonated in a van parked in the underground garage of the World Trade Center. Six workers were killed and more than 1000 people were injured.

April 1995: Oklahoma City
The Alfred P. Murrah Federal Building was demolished by a truck bomb, which also killed 168 people. (A National Fire Protection Association [NFPA] fire investigation report is available on this incident.)

1978–1995: United States
Over a period of 17 years, the so-called Unabomber (né Theodore Kaczynski) mailed at least 16 packages containing explosives to university professors, corporate executives, and other targeted individuals. These attacks killed 3 people and injured 23 others.

July 1996: Atlanta
A pipe bomb exploded in Centennial Olympic Park, killing 1 person and injuring 111 others.

January 1997: Atlanta
Following the bombing of an abortion clinic in suburban Atlanta, a secondary device exploded, wounding several emergency responders. A month later, another secondary device was found and disarmed at the scene of a bombing at a gay nightclub in Atlanta.

use of specific devices. For example, demolition companies use explosives to bring down unnecessary structures quickly and safely; terrorists may use the same agents to bring down an occupied building, thereby killing many people.

Terrorists use several different kinds of weapons. Although bombings are the most frequent international terrorist acts,

January 1998: Birmingham, Alabama

A bomb killed a police officer who was providing security at an abortion clinic.

October 1998: Vail, Colorado

Arson destroyed eight buildings at a ski resort. An extremist environmental group opposed to expansion of the resort claimed responsibility.

September 2001: New York City/Washington/Pennsylvania

Terrorists hijacked four commercial jets: two were flown into the World Trade Center, one struck the Pentagon, and the fourth crashed into a field in rural Pennsylvania. More than 3000 people died in the various incidents.

Fall 2001: United States

Five people died after letters containing anthrax were sent to various locations in the eastern United States.

FIRE MARKS

MAJOR TERRORIST INCIDENTS IN CANADA

June 1985: Montreal

Air India Flight 182 operating the Montreal–London–Delhi route was blown up on June 23, 1985, in Irish air space and crashed into the Atlantic Ocean. All of the 329 people on board were killed.

October 2000: Montreal

The *Front de Libération de Québec* (*FLQ*)/*La Brigade d'autodéfense du français* (*BAF*) firebombed three Second Cup locations in Montreal. The locations were thought to be targeted because of the company's use of its incorporated English name, Second Cup. Rhéal Mathieu, a previously convicted *FLQ* terrorist was convicted of all three bombings.

April 2004: Montreal

A Canadian of Lebanese–Muslim heritage firebombed the library of the United Talmud Torahs Jewish day school.

June 2006: Toronto

Canadian counterterrorism forces arrested 18 people (the "Toronto 18") with ties to al Qaeda. They were accused of planning to detonate truck bombs; open fire in a crowded area; storm the Canadian Broadcasting Centre (CBC), Parliament Hill, and CSIS headquarters; take hostages; and decapitate the prime minister and other leaders.

October 2008–July 2009: Dawson Creek, British Columbia

Six natural gas pipelines owned by Encana Corporation were bombed after editorials opposing the gas industry were sent to a local newspaper.

May 2010: Ottawa

A Royal Bank of Canada was firebombed by an anarchist group, causing $300,000 in damages.

September 2012: Montreal

During the Quebec provincial elections, Richard Henry Bain attempted to assassinate Parti Québécois (PQ) leader and Premier-elect Pauline Marois at a victory gathering in Montreal. He also set fire to the Métropolis concert hall, where the event was being held. One man was killed and another was injured in the terrorist act. It is said that Bain's ultimate goal was to kill Marois following the PQ victory. Bain, who was arrested shortly after the attack, is currently incarcerated and awaiting trial.

---

fire fighters must also be aware of the threats posed by other potential weapons: Discharging an automatic weapon into a crowd at a shopping mall or train station could cause devastating carnage. The release of a microscopic biological agent into a subway system could cause numerous people to become ill and die, thereby creating a public panic response that might overwhelm the resources of the fire service and Emergency Medical Services (EMS). A computer virus that attacks the banking industry could lead to tremendous economic losses. Given the breadth of these threats, planning should consider the full range of possibilities.

Fire Fighter Safety Tips
© Photos.com

A terrorist event may be designed to target emergency responders as well as civilians. Explosives do not discriminate: anyone nearby could be severely injured or killed when they detonate.

■ Explosives and Incendiary Devices

Explosives and incendiary devices are often utilized by terrorists. Incendiary devices account for 20–25 percent of all bombings in the United States. In Canada, these statistics are gathered on a voluntary basis from police services and logged in the RCMP's Canadian Bomb Data Centre (CBDC). In 2007 alone, there were 5 reported attempted bombings, 36 recovered improvised explosive devices (IEDs), and 35 recovered hoax devices. The incendiary weapons are of special concern to fire fighters because they are capable of causing large fires in highly flammable settings.

Both groups and individuals have used explosives for many different purposes: to further a cause, to intimidate a co-worker or former spouse, to take revenge, or simply to experiment with a recipe found in a book or on the Internet.

Each year, thousands of kilograms of explosives are stolen from construction sites, mines, and military facilities. How much of this material makes its way to criminals and terrorists is not known **FIGURE 33-6** . Terrorists can also use commonly available materials, such as **ammonium-nitrate fertilizer and fuel oil (ANFO)**, to create their own explosives.

An **improvised explosive device (IED)** is any explosive device that is fabricated from readily available materials. An IED could be contained in almost any type of package—from a letter bomb to a truckload of explosives. The Unabomber, for example, constructed at least 16 bombs that were delivered in small packages through the United States Postal Service (USPS). By contrast, the bombings of the World Trade Center in 1993 and the Alfred P. Murrah Federal Building in 1995 both involved delivery vehicles loaded with ANFO, which were detonated via a simple timer **FIGURE 33-7** .

Pipe Bombs

The most common IED is the **pipe bomb**. A pipe bomb is simply a length of pipe filled with an explosive substance and rigged with some type of detonator **FIGURE 33-8** . Most pipe bombs are simple devices made with black powder or smokeless powder and ignited by a hobby fuse. More sophisticated

FIGURE 33-6 Every year, thousands of kilograms of explosives are stolen from their rightful owners.

From When Violence Erupts: A Survival Guide for Emergency Responders, courtesy of Dennis R. Krebs

FIGURE 33-8 Pipe bombs come in many shapes and sizes.

Courtesy of Captain David Jackson, Saginaw Township Fire Department

![Figure 33-7](The Alfred P. Murrah Federal Building)

FIGURE 33-7 The Alfred P. Murrah Federal Building in Oklahoma City was destroyed by a truck bomb.

Courtesy of FEMA News Photo

FIRE FIGHTER Tips

A radio-free area, spreading out for at least 300 metres (1000 feet), should be established at any incident for which a suspected explosion area is identified. The establishment of such an area should be noted in an operating procedure. Within the radio-free area, all personnel should be ordered to cease any and all transmissions through any radio-wave-transmitting device, including mobile radios, mobile data terminals, and cell phones.

pipe bombs may use a variety of chemicals and incorporate electronic timers, mercury switches, vibration switches, photocells, or remote-controlled detonators as triggers.

Pipe bombs are sometimes packed with nails or other objects, which act as shrapnel, reflecting the bomber's goal of inflicting as much injury as possible on anyone in the vicinity.

A chemical or biological agent or radiological material could also be added to a pipe bomb to create a much more complicated and dangerous incident.

Secondary Devices

Emergency responders must realize that terrorists sometimes place **secondary devices** in the area where an initial event has occurred. These devices, which are intended to explode some time after the initial device detonates, are designed to kill or injure emergency responders, law-enforcement personnel, spectators, and members of the media. Terrorists use this tactic to attack the best-trained and most-experienced investigators and emergency responders or simply to increase the levels of fear and chaos following an attack.

The use of secondary devices is a common tactic in incidents abroad and has occurred at a few incidents in North America as well. In 1998, for example, a bomb exploded outside a Georgia abortion clinic; approximately one hour later, a second explosion injured seven people, including two emergency responders. A similar secondary device was discovered nearly a month later at the scene of a bombing in a nearby community, although responders were able to disable that device before it detonated.

Although the possibility of a secondary device exists at every terrorist incident, do not forget that the hazards of a secondary explosion or collapse exist at any incident involving an explosion. As a consequence, fire fighters and other first responders must be alert to all hazards at the scene of *any* explosion. The hazards at an accidental explosion can be just as deadly as secondary devices at an incident caused by terrorists. In Canada, the fire department response to the scene occurs when the police give the all clear. The fire response typically stages at a safe distance until cleared by the police to approach.

Potentially Explosive Devices

All fire fighters should know their department's suggested operating guidelines (SOGs), or standard operating procedures (SOPs), for handling incidents involving explosives. Fire department units responding to an incident that involves a potentially explosive device—that is, a device that has not yet exploded—should move all civilians away from the area and establish a perimeter at a safe distance. At no time should fire fighters handle a potential explosive device unless they have received special training in doing so. Instead, trained explosive ordnance disposal (EOD) personnel should assess the device and render it inoperative.

While waiting for the properly trained EOD personnel to arrive, the fire department should establish a command post and a staging area **FIGURE 33-9**. Because secondary devices may be present, the staging area should be located at least 914.4 metres (3000 feet) from the incident site. In this type of incident, a unified command must be established with multiagency coordination, including all agencies involved in the incident.

If the bomb disposal team decides to disarm the device, an emergency action plan must be developed in case of accidental detonation. The incident commander (IC) should work in concert with law enforcement and EOD personnel to determine a safe perimeter where fire fighters and emergency medical personnel will be staged.

In some cases, a <u>forward staging area</u> will be established. A rapid intervention crew (RIC) then stands by in this area, ready to provide immediate assistance to the bomb disposal team if something goes wrong. Other fire fighters and emergency medical personnel would remain in a remote staging area in this scenario.

Actions Following an Explosion

Following an explosion, as with any type of emergency incident, your first priority should be to ensure the safety of the scene. During the initial stages of an incident, you will

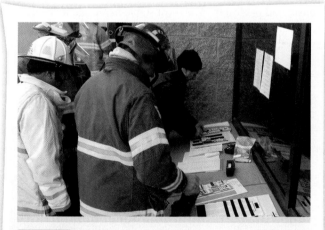

FIGURE 33-9 If the fire department is first to arrive at the scene of an explosion, it should establish a unified command post.
Courtesy of Captain David Jackson, Saginaw Township Fire Department

Fire Fighter Safety Tips   © Photos.com

Radio waves can trigger electric blasting caps, initiating an explosion. For this reason, radio transmitters should not be used near a suspicious device.

not know whether the event was caused by an intentional act or by accidental circumstances. In any incident involving an explosion, follow departmental procedures to ensure the safety of rescuers, victims, and bystanders. Consider the possibility that a secondary device and additional hazards might still be present in the vicinity. Quickly survey the area for any suspicious bags, packages, or other items.

It is also possible that chemical, biological, or radiological agents might be involved in a terrorist bombing. Qualified personnel with monitoring instruments should be assigned to check the area for potential contaminants. The initial size-up should also include an assessment of hazards and dangerous situations. The stability of any building involved in the explosion must be evaluated before anyone is permitted to enter, because entering an unstable area without proper training and equipment may complicate rescue and recovery efforts.

Working with Other Agencies

Joint training with the local, provincial, and federal agencies charged with handling incidents involving explosive devices should occur on a routine basis. Among these agencies are local and provincial police, RCMP, CSIS, and military EOD units. In many communities, specially trained law-enforcement personnel are responsible for handling explosive devices.

■ Chemical Agents

Chemical agents have the potential to kill or injure great numbers of people. Many chemical agents are readily available because they are produced in the United States and Canada and used in a wide variety of industrial and commercial processes. For instance, ammonia is used as a fertilizer for crops; chlorine is used for many purposes, including the purification of drinking water and swimming pools. These same chemicals, if released in a terrorist event, could cause injury or death to many people. Mitigating an accidental release of chlorine gas, however, requires the same safety precautions as handling an intentional release of chlorine gas by terrorists.

Chemical agents are not new; they have been used as weapons for at least 100 years; for example, phosgene, chlorine, and mustard agents were used as weapons in World War I, resulting in thousands of battlefield deaths and permanent injuries. Chemical weapons were also used during the Iran–Iraq War (1980–1988). During the same period, the Iraqi government used these weapons against the minority Kurdish people. Although international agreements prohibit the use of chemical and biological agents on the battlefield, concern remains that chemical weapons could be used by terrorists.

In 1995, the religious cult Aum Shinrikyo released a nerve gas, <u>sarin</u>, into the Tokyo subway system. Although the attack resulted in 12 deaths and more than 1000 injuries, many experts believe it to have been ineffective, because there were relatively few casualties, considering the potential exposure of thousands of subway riders to the nerve agent. Even so, this attack achieved one of the major goals of terrorism: it instilled fear in a large population.

The basic instructions for making chemical weapons can be readily found through the Internet, books, and other publications, and the necessary chemicals can be obtained fairly easily. Many of these chemicals are routinely used in legitimate industrial processes and therefore are widely available; for example, <u>chlorine</u> gas is used in swimming pools, at water treatment facilities, and in industry **FIGURE 33-10**. Despite its legitimate uses, chlorine is classified as a pulmonary irritant (a <u>choking agent</u>), because its inhalation causes severe pulmonary damage.

Like chlorine, <u>cyanide</u> has many legitimate uses, but it kills quickly once it enters the body. Cyanide compounds, such as hydrogen cyanide (HCN), are used in the production of paper and synthetic textiles, as well as in photography and printing. These chemicals are stored and transported in containers of various sizes, ranging from small containers and pressurized cylinders to railway tank cars.

Chemical weapons can be disseminated in several ways: Simply releasing chlorine gas from a storage tank in an unguarded rail yard, for example, might cause many injuries and deaths. To ensure a broader distribution of the chemical agent, it might be added to an explosive device. Crop-dusting aircraft, truck-mounted spraying units, or hand-operated-pump tanks could also be used to disperse an agent over a wide area **FIGURE 33-11**.

Dissemination of a toxic gas or suspended particles depends on wind direction, wind speed, air temperature, and humidity. Because these factors can change quickly, it is difficult to predict the exact direction that might be taken by a chemical-release cloud. Hazardous materials teams use computer models to predict the path of a toxic cloud. They also have the training and equipment to safely handle these situations.

Protection from Chemical Agents

Air flow is the most commonly used dispersal method for chemical agents. Agents released outdoors will follow wind currents; agents released inside will be dispersed through a building's air-circulation system. Some chemical agents have distinctive odours; hence, if any unusual odour is noticed at an emergency scene, fire fighters must use full personal protective equipment (PPE), including self-contained breathing apparatus (SCBA) **FIGURE 33-12**. The presence of some chemicals requires the use of fully encapsulated suits. In addition, the area should be monitored with the appropriate detection devices.

It is never sufficient to rely strictly on odour to determine the presence of a chemical agent. Instead, definitive detection and identification of a chemical agent require the activation of a well-trained hazardous materials team. As a fire fighter, your job is to prevent further injuries to rescuers and citizens—not to neutralize the chemical agent.

Nerve Agents

<u>Nerve agents</u> are toxic substances that attack the central nervous system. These weapons were first developed in Germany before World War II. Today, several countries maintain stockpiles of these agents. Nerve agents are similar to some pesticides but are much more toxic—in some cases, 100 to 1000 times more toxic than their pesticide counterparts. Exposure to these agents can result in injury or death within minutes.

In their normal states, most nerve agents are liquids **FIGURE 33-13**. As liquids, these agents are unlikely to contaminate large numbers of people, because direct contact with the agent is required for the chemical to inflict its health-damaging effects. To be an effective weapon, the liquid must be dispersed in aerosol form or broken down into fine droplets that can then be inhaled or absorbed by the skin.

Pouring a liquid nerve agent onto the floor of a crowded building would probably not immediately contaminate large numbers of people, but it would produce fear and panic. In this scenario, the effectiveness of the agent would depend on how long it remained in its liquid state and how widely it became dispersed throughout the building. Sarin, the most volatile nerve agent, evaporates at the same rate as water and is not considered persistent. The most stable nerve agent, <u>V-agent</u>, is considered persistent because it takes several days or weeks to evaporate. Common nerve agents, their methods of contamination, and specific characteristics are listed in **TABLE 33-1**.

FIGURE 33-10 Although chlorine is regularly used in swimming pools and water treatment facilities, it is also classified as a choking agent and can be used in a terrorist attack.

© TFoxFoto/ShutterStock, Inc.

FIGURE 33-11 Crop-dusting equipment could be used to distribute chemical agents.

© Tim McCabe/USDA

When a person is exposed to a nerve agent, symptoms will become evident within minutes. These symptoms include pinpoint pupils, a runny nose, drooling, difficulty breathing, tearing, twitching, diarrhea, convulsions or seizures, and loss of consciousness. The same symptoms are seen in victims who have been exposed to pesticides. Several mnemonics may be used to help you remember these symptoms, including the one used most often by the fire service: S.L.U.D.G.E.M. **TABLE 33-2**.

The United States military has developed a nerve agent antidote kit, which contains two medications: atropine and pralidoxime chloride. These antidotes can be quickly injected into a person who has been contaminated by a nerve agent. The first antidote kit to be widely distributed was the Mark I kit, which was issued to many fire departments' hazardous materials teams. The **DuoDote® kit** has since been introduced to replace expired Mark I kits. One cautionary note about use of the DuoDote® kit: the DuoDote® auto-injector should be administered by EMS personnel who have received adequate training in the recognition and treatment of nerve agent or insecticide intoxication. The DuoDote® auto-injector is intended as an initial treatment of the symptoms of organophosphorus nerve agent or insecticide poisoning. Definitive medical care should be sought immediately.

Blister Agents

Two chemical agents are generally used as **blistering agents**, so called because contact with these chemicals causes the skin to blister:

- **Sulphur mustard** is a translucent, yellow or amber, oily liquid marked by the faint, sweet odour of mustard or garlic. It vaporizes slowly at temperate climates and can be dispersed in an aerosol form.
- **Lewisite** is an oily, colourless to dark-brown liquid with an odour of geraniums.

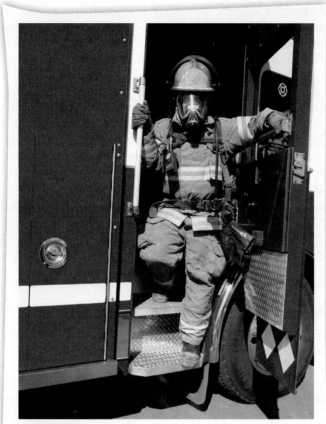

FIGURE 33-12 If an unusual odour is reported at an emergency scene, a fire fighter must don full PPE, including SCBA.

FIGURE 33-13 In their normal states, nerve agents are liquids. They must be dispersed in aerosol form if they are to be inhaled or absorbed by the skin.

© AbleStock

TABLE 33-1	Common Nerve Agents	
Nerve Agent	**Method of Contamination**	**Characteristics**
Tabun (GA)	Skin contact Inhalation	Disables the chemical connections between nerves and target organs
Soman (GD)	Skin contact Inhalation	Odour of camphor
Sarin (GB)	Inhalation	Evaporates quickly
V-agent (VX)	Skin contact	Oily liquid that can persist for weeks

TABLE 33-2	Symptoms of Nerve Agent Exposure
S—salivation (drooling)	
L—lacrimation (tearing)	
U—urination	
D—defecation	
G—gastric upset (upset stomach)	
E—emesis (vomiting)	
M—miosis (constriction of the pupils)	

Blistering agents produce painful burns and blisters with even minimal exposure **FIGURE 33-14**. The major difference between these two agents is that lewisite causes pain immediately upon contact with the skin, whereas the signs and symptoms of exposure to mustard gas may not appear for several hours. The victim will complain of burning at the site of the exposure, the skin will redden, and blisters will appear. His or her eyes may also itch, burn, and turn red. Inhalation of mustard gas produces significant respiratory damage.

If you suspect exposure to a chemical agent, treat victims as if they are contaminated. Use full PPE and prevent other people from being exposed to the chemical agent.

Pulmonary Agents

Pulmonary agents (also referred to as choking agents) cause severe damage to the lungs and lead to asphyxia. Two chemicals that might be used as pulmonary agents in terrorist attacks are **phosgene** and chlorine. Both of these agents were used extensively as weapons in World War I, and both also have several industrial uses.

Phosgene and chlorine are heavier than air, so they tend to settle in low areas; thus, subways, basements, and sewers are prime areas for these agents to accumulate. Fire fighters wearing appropriate PPE should quickly evacuate people from these areas when the possibility of phosgene or chlorine exposure exists.

Exposure to high concentrations of phosgene or chlorine will immediately irritate the eyes, nose, and upper airway. Within hours, the exposed individual will begin to develop pulmonary edema (fluid in the lungs). Individuals who are exposed to lower concentrations may not exhibit any initial symptoms but can still experience respiratory damage.

Although neither pulmonary agent is absorbed through the skin, both phosgene and chlorine can cause skin burns on contact. **Decontamination** consists of removing exposed individuals from the contaminated area and flushing the skin with water.

FIGURE 33-14 Typical effects of a blistering agent.
Courtesy of Dr. Saeed Keshavarz/RCCI (Research Center of Chemical Injuries) - Iran

Metabolic Agents

Metabolic agents, such as cyanide, interfere with the utilization of oxygen by the cells of the body. As mentioned earlier, cyanide compounds are highly toxic poisons that can cause death within minutes of exposure. The two most common cyanide compounds are hydrogen cyanide and cyanogen chloride, both of which have legitimate uses in industry.

Cyanide can be inhaled or ingested. Historically, the gaseous form was used as a means of carrying out the death penalty (hence the term *gas chamber*). Liquid cyanide mixed with fruit punch was used in the mass suicide/murder of 913 members of a religious cult in Guyana in 1978.

The symptoms associated with cyanide exposure appear quickly. A person who is exposed to the gas will begin gasping for air. In addition, if enough of the agent is inhaled, the skin may begin to appear red. Seizures are also possible.

In the face of cyanide exposure, a complete and rapid evacuation of the area is the most appropriate course of action.

■ Biological Agents

Biological agents are organisms that cause diseases and attack the body. They include bacteria, viruses, and toxins. Some of these organisms, such as anthrax, can live in the ground for years; others are rendered harmless after being exposed to sunlight for only a short period of time. The highest potential for infection is through inhalation, although some biological agents can be absorbed or injected through the skin or ingested. The effects of a biological agent depend on the specific organism or toxin, the dose, and the route of entry. Most experts believe that a biological weapon is most likely to be spread by a device similar to a garden sprayer or crop-dusting plane. After the release of a biological weapon, residents would begin experiencing symptoms in anywhere from 1 to 17 days.

Some of the diseases caused by biological agents, such as smallpox and pneumonic plague, are contagious and can be passed from person to person. Physicians are concerned about the use of contagious diseases as weapons, because the resultant epidemic could overwhelm the health care system. Experts have differing opinions about how difficult it would be to infect large numbers of people with one of these naturally occurring organisms. Because of their incubation period, it would take between 2 and 17 days for people to start showing signs of infection after exposure to one of these organisms.

Anthrax

Anthrax is an infectious disease caused by the bacterium *Bacillus anthracis*, which is typically found around farms and infects livestock. When anthrax is to be used as a weapon, the bacteria are cultured to develop anthrax spores. The spores (in powdered form) can then be dispersed in a variety of ways **FIGURE 33-15**. Approximately 8000 to 10,000 spores are typically required to cause an anthrax infection. Spores infecting the skin cause cutaneous anthrax, ingested spores cause gastrointestinal anthrax, and inhaled spores cause respiratory anthrax. Experts are not sure whether it would be possible to infect a large number of people with anthrax.

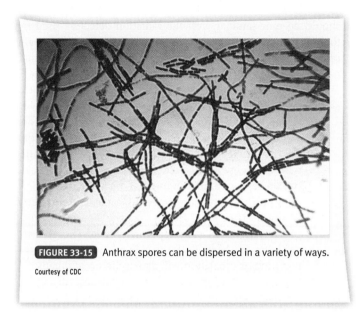

FIGURE 33-15 Anthrax spores can be dispersed in a variety of ways.
Courtesy of CDC

FIGURE 33-16 A bubo—one of the symptoms of plague—is a swollen, painful lymph node.
Courtesy of CDC

Later in 2001, four letters containing anthrax were mailed to locations in New York City; Boca Raton, Florida; and Washington. Five people died after being exposed to the contents of these letters, including two postal workers who came in contact with the anthrax as the letters passed through a USPS distribution centre. Additional exposures resulted in more than 20 injuries that have had lasting effects. Several major government buildings had to be shut down for months to be decontaminated. These incidents followed shortly after the terrorist attacks of September 11, 2001, and caused tremendous public concern. In their wake, emergency personnel had to respond to many incidents involving suspicious packages, and many citizens believed that they might have been exposed to anthrax. Today, there are simple tests that hazardous materials teams can perform in the field to determine whether the threat of anthrax is real.

On January 30, 2011, an anthrax hoax letter was received at the Ottawa Citizenship and Immigration office. It is suspected that the letter was sent by a militant group in protest of the detention of Mohammad Mahjoub, who was being held without charge as a threat to national security with alleged links to al Qaeeda.

Anthrax has an incubation period of 2–6 days. The disease can be successfully treated with a variety of antibiotics if it is diagnosed early enough.

Plague

<u>Plague</u> is caused by *Yersinia pestis*, a bacterium that is commonly found on rodents. It is most often transmitted to humans by fleas that feed on infected animals and then bite humans.

Three forms of plague are distinguished: bubonic, septicemic, and pneumonic. Individuals who are bitten generally develop bubonic plague, which attacks the lymph nodes **FIGURE 33-16**. Pneumonic plague can be contracted by inhaling the bacterium.

Y. pestis can survive for weeks in water, moist soil, or grains. These bacteria have been cultured for distribution as a weapon in aerosol form. Inhalation of the aerosol form would put the target population at risk for pneumonic plague.

The incubation period for the plague ranges from 2 to 6 days. This disease can be treated with antibiotics.

Smallpox

<u>Smallpox</u> is a highly infectious and often fatal disease caused by *Variola*, a virus; it kills approximately 30 percent of infected persons. Although smallpox was once routinely encountered throughout the world, by 1980, it had been successfully eradicated as a public health threat through the use of an extremely effective vaccine. Two countries maintained cultures of the disease for research purposes, however, and international terrorist groups may have acquired the virus.

The smallpox virus could potentially be dispersed over a wide area in an aerosol form. Infection of only a small number of people might lead to the rapid spread of the disease throughout a targeted population, given smallpox's highly contagious nature: the disease is easily spread by direct contact, droplet, and airborne transmission. Patients are considered highly infectious and should be quarantined until the last scab has fallen off **FIGURE 33-17**. The incubation period for smallpox is between 4 and 17 days (12 days on average).

There is no treatment for smallpox except for supportive care. Although an effective vaccine against this disease exists, it has not been widely used in recent decades, because smallpox was deemed to be eliminated in 1980. Today, millions of people have never been vaccinated against this infection, and millions more have reduced immunity as decades have passed since their last immunization.

Dealing with Biological Agents

It is unlikely that fire fighters or emergency responders would immediately recognize that a biological weapon had been released in their communities, in large part because of these agents' <u>incubation period</u>—that is, the lag between the actual infection and the appearance of symptoms. The symptoms of a biological attack would typically manifest over a period of days after the exposure. If left untreated, many of these diseases could kill a large proportion of the infected population.

FIGURE 33-17 Smallpox is a highly contagious disease with a mortality rate of approximately 30 percent.
Courtesy of CDC

Area hospitals would typically be the first to recognize the situation, after identifying significant numbers of people arriving at their facilities with similar symptoms. In Canada, preventing and controlling the spread of infectious disease is under the purview of the Public Health Agency of Canada (PHAC). It is the Canadian equivalent to the CDC in the United States. Multiple medical calls about patients exhibiting similar symptoms could provide a clue about the use of a biological weapon, especially if their location is considered a potential target.

Fire fighters or emergency responders treating people who may have been exposed to a biological agent should follow the recommendations of their departments regarding **universal precautions** (protective measures for use in dealing with objects, blood, bodily fluids, or other items that pose a potential risk of exposure to communicable diseases). These measures include wearing gloves, masks, eye protection, and surgical gowns when treating patients.

Emergency responders who exhibit flulike symptoms after a potential terrorist incident should seek medical care immediately. It is important that fire fighters report any possible exposure to department officials. Postincident actions may include medical screening, testing, or vaccinations.

■ Radiological Agents

A third type of agent that might potentially be used by terrorists is radiation. The most probable scenarios involve the use of various approaches to disperse **radiological agents** and contaminate an area with radioactive materials. This threat is very different from that posed by a nuclear detonation.

Radiation

Radioactive materials release energy in the form of electromagnetic waves or energy particles. The resulting radiation cannot be detected by the normal senses of smell and taste, although several instruments have been developed that can detect the presence of radiation and measure dose rates. Fire fighters

should become familiar with the radiation detectors used by their departments. Do not enter any place that might contain radiation unless you have been trained in measuring radiation, have a working radiation detector, and are properly protected.

Three types of radiation exist: alpha particles, beta particles, and gamma rays.

- **Alpha particles** quickly lose their energy and therefore can travel only 2.5–5.1 centimetres (1–2 inches) from their source. Clothing or a sheet of paper can stop this type of energy. If ingested or inhaled, alpha particles can damage a number of internal organs.
- **Beta particles** are more powerful, capable of travelling 3–4.6 m (10–15 ft.). Heavier materials such as metal, plastic, and glass can stop this type of energy. Beta radiation can harm both the skin and the eyes, and its ingestion or inhalation will damage internal organs.
- **Gamma radiation** can travel significant distances, penetrate most materials, and pass through the body. This type of radiation is the most destructive to the human body. The only materials with sufficient mass to stop gamma radiation are concrete, earth, and dense metals, such as lead.

Effects of Radiation

Symptoms of low-level radioactive exposure might include nausea and vomiting. Exposure to high levels of radiation can cause vomiting and damage to the digestive system within a short time. Other symptoms of high-level radiation exposure include the destruction of bone marrow, damage to the nervous system, and radioactive skin burns. An extreme exposure may rapidly cause death; more typically, it takes a considerable amount of time before the signs and symptoms of radiation sickness become obvious. As the effects progress over the years, a prolonged spiral into death caused by leukemia or carcinoma is likely.

A contaminated person will have radioactive materials on his or her exposed skin and clothing. In severe cases, this contamination could penetrate to the internal organs of the body. For this reason, decontamination procedures are required to remove the radioactive materials. Decontamination must be complete and thorough, because the situation could escalate rapidly if this procedure is inadequate. Specifically, the contamination could quickly spread to ambulances, hospitals, and other locations where contaminated individuals or items are transported.

Radioactive materials both expose rescuers to radiation injury and continue to affect the exposed person. Anyone who comes in contact with a contaminated person must be protected by appropriate PPE and shielding, depending on the particular contaminant. Rescuers and medical personnel must all be decontaminated after interacting with a contaminated person; the only way to know whether radiation is present on their bodies or clothing is to use radiation detectors.

A person can be exposed to radiation without coming into direct contact with a radioactive material through inhalation, skin absorption, or ingestion. A person can also be injured by exposure to radiation without being contaminated by the radioactive material itself. Someone who has been exposed to radiation but has not been contaminated by radioactive materials requires no special handling.

There are three ways to limit exposure to radioactivity:

- Keep the time of the exposure as short as possible.
- Stay as far away from the source of the radiation as possible.
- Use shielding to limit the amount of radiation absorbed by the body.

In addition, do not enter any area that might contain radiation until trained personnel have assessed the radiation level using approved detection devices.

If radioactive contamination is suspected, everyone who enters the area should be equipped with a **personal dosimeter** to measure the amount of radioactive exposure FIGURE 33-18 . Rescue attempts should be made only by properly trained and equipped personnel, each of whom must carry an approved radiation monitor to measure the amount of radiation present.

The Dirty Bomb

In recent years, the **radiation dispersal device**, or "dirty bomb," has emerged as a source of concern in terms of terrorism. Packing radioactive material around a conventional explosive device could contaminate a wide area, with the size of the affected area ultimately dependent on the amount of radioactive material and the power of the explosive device.

To limit this threat, radioactive materials, even in small amounts, are kept secure and protected. Such materials are widely used in industry and health care, particularly in conjunction with X-ray machines. A terrorist could potentially construct a dirty bomb with just a small quantity of stolen radioactive material.

Security experts are also concerned that a large explosive device, such as a truck bomb, could potentially be detonated near a nuclear power plant. Such an explosion could damage the reactor containment vessel and release radioactive material into the atmosphere, allowing it to spread over a large area.

FIGURE 33-18 A personal dosimeter measures the amount of radioactive exposure received by an individual.

Courtesy of Fluke Biomedical

Fire Fighter Safety Tips © Photos.com

Casualty Management After Detonation of a Radiological Weapon

- Do not enter an area that is suspected to be contaminated until radiation levels have been measured and found to be safe.
- Enter the area only to save lives; radiation levels will be very high.
- Wear PPE, including SCBA.
- All personnel and equipment must be decontaminated after leaving the contaminated area. Because radioactive dust collects on clothing, all clothing should be removed and discarded after leaving the area. Failure to do so will result in continued radiation exposure to yourself and others.
- Wash your entire body thoroughly with lukewarm water as soon as possible after leaving the scene, even if you have been through the decontamination process.
- Do not eat, drink, or smoke while exposed to potential radioactive dust or smoke.
- Use radiation-monitoring devices to map the locations of areas with high radioactivity. These monitoring devices should be wrapped in plastic bags to prevent their contamination.
- Vehicles should be washed before they leave the scene. The only exception is emergency units transporting critical victims.
- Life-threatening injuries are more serious than radioactive contamination. Treat such injuries before decontaminating the individual.
- Alert local hospitals that they may receive radioactive-contamination victims.

Operations

Responding to a terrorist incident puts fire fighters and other emergency personnel at risk. Although responders must ensure their own safety at every incident, a terrorist incident may carry an extra level of risk. Because the terrorist's objective is to cause as much harm as possible, emergency responders are just as likely to be targets as ordinary civilians.

In most cases, the first emergency units will not be dispatched for a known WMD or terrorist incident; rather, the initial dispatch might be for an explosion, for a possible hazardous materials incident, for a single person with difficulty breathing, or for multiple victims with similar symptoms. Emergency responders will usually not know that a terrorist incident has occurred until personnel on the scene begin to piece together information gained from their own observations and from witnesses.

If appropriate precautions are not taken, initial responders may find themselves in the middle of a dangerous situation before they realize what has happened. Initial responders should be aware of any factors that suggest the possibility of a terrorist incident and immediately implement appropriate procedures. The possibility of a terrorist incident should be considered when responding to any location that has been identified as a potential terrorist target. It could be difficult to determine the true nature of the situation until a size-up is conducted at the emergency scene.

■ Initial Actions

Fire fighters should approach a known or potential terrorist incident just as they would a hazardous materials incident. The possibility that chemical, biological, or radiological agents are involved cannot be ruled out until a hazardous materials team checks the area with detection and monitoring instruments. Apparatus and personnel should, if possible, approach the scene from a position that is uphill and upwind. Emergency responders should don PPE, including SCBA. Additional arriving units should be staged an appropriate distance away from the incident.

The first units to arrive should establish an outer perimeter to control access to and from the scene. They should deny access to all but emergency responders, and they should prevent potentially contaminated individuals from leaving the area before they have been decontaminated. The perimeter must completely surround the affected area, with the goal of keeping people who were not initially involved from becoming additional victims.

The incident command post should be established in a safe location, which could be as far as 914.4 m (3000 ft.) from the actual incident scene. It must be set up outside the area of possible contamination and beyond the distance where a secondary device might be planted. The initial priority should be to determine the nature of the situation, the types of hazards that might be encountered, and the magnitude of the problems faced by the responders.

An initial reconnaissance, or recon, team should be sent out to quickly examine the involved area and to determine how many people are involved. Proper use of PPE, including SCBA, is essential for the recon team, and the initial survey must be conducted very cautiously—albeit as rapidly as possible. The possibility that chemical, biological, or radiological agents are involved cannot be ruled out until qualified personnel with appropriate instruments and detection devices have surveyed the area. Responders should take special care not to touch any liquids or solids or to walk through pools of liquid. Any emergency responders who become contaminated must not leave the area until they have been decontaminated.

A process of elimination may be required to determine the nature of the situation. Occupants and witnesses should be asked if they observed any unusual packages or detected any strange odours, mists, or sprays. The presence of a large number of casualties who have no outward signs of trauma might indicate a chemical-agent exposure. In such a case, victims' symptoms might include difficulty breathing, skin irritations, or seizures.

A visible vapour cloud would be a strong indicator of a chemical release, which may have occurred as the result of either an accident or a terrorist attack. The presence of dead or dying animals, insects, or plant life might also signal a chemical-agent release.

When approaching the scene of an explosion, responders should consider the possibility of a terrorist bombing and remain vigilant for secondary explosive devices. If they note any suspicious packages, responders must notify the IC immediately.

Fire fighters who have not been specially trained should never approach a suspicious object. Instead, EOD personnel should examine any suspicious articles and disable them.

FIRE FIGHTER Tips

In 1993, the first units to respond to a call at the World Trade Center thought a transformer explosion had occurred somewhere beneath the huge complex. As responders explored the smoke-filled interior, they realized that a large explosion had occurred in the underground parking area. As they pieced together the information, they determined that terrorists had detonated a truck bomb, causing tremendous destruction and filling the twin towers with smoke.

At the same location in 2001, many of the first responders saw a plane crash into one of the towers and believed that they were responding to a terrible accident. They realized the true nature of the situation only when a second plane crashed into the other tower.

■ Interagency Coordination

If a terrorist incident is suspected, the IC should immediately consult with local law-enforcement officials. If a mass-casualty situation is evident, the IC should notify area hospitals and activate the local medical response system. Local technical rescue teams should be requested to evaluate structural damage and initiate rescue operations.

Provincial emergency management officials should be notified as soon as possible. This step will help to ensure a quick response by both provincial and federal resources to a major incident. Large-scale search and rescue incidents could require the response of urban search and rescue task forces (as activated through the local emergency management plan), which may escalate the response through provincial and federal agencies (e.g., Emergency Management Ontario (EMO), Emergency Management British Columbia (EMBC), and Public Safety Canada). Medical response teams, such as disaster-medical-assistance teams, may be needed for incidents involving large numbers of people.

An emergency operations centre (EOC) can help to coordinate the actions of all involved agencies in a large-scale incident, particularly if terrorism is involved. The EOC is usually set up in a predetermined remote location and is staffed by experienced command and staff personnel **FIGURE 33-19**. The IC, who remains at the scene of the incident, should provide detailed situation reports to the EOC and request additional resources as needed.

Fire fighters must remember that a terrorist incident is also a crime scene. To avoid destroying important evidence that could lead to the conviction of those parties responsible for the attack, responders should not disturb the scene any more than is necessary. Where possible, law-enforcement personnel should be consulted prior to overhaul and before any material is removed from the scene. Fire fighters should also realize that one or more terrorists could be among the injured victims. Be alert for threatening behaviour, and make note of anyone who seems determined to leave the scene.

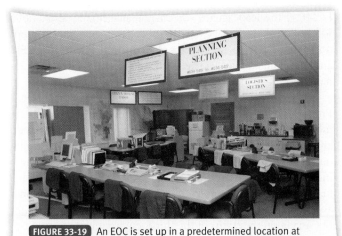

FIGURE 33-19 An EOC is set up in a predetermined location at large-scale incidents.

FIGURE 33-20 Mass decontamination procedures may be required to handle a large group of contaminated victims.

© Jim Mone/AP Photos

TEAMWORK Tips

A hazardous materials team should respond to any potential or suspected terrorist incident. These personnel will test for chemical, biological, or radiological contaminants, using appropriate test instruments.

■ Decontamination

Those persons who have been directly contaminated by chemical, biological, or radiological agents must be decontaminated to remove or neutralize any chemicals or substances on their bodies or clothing. Those who have been exposed to these materials should be evaluated first, before decontamination takes place, to avoid unnecessary exposure and waste of resources. If in doubt, decontaminate. Decontamination should occur as soon as possible to prevent further absorption of a contaminant and to reduce the possibility of spreading the contamination. Equipment must also be decontaminated before it leaves the scene. The chapter "Hazardous Materials: Operations and Specialties" explains these decontamination procedures in detail.

The perimeter must fully surround the area of known or suspected contamination. Every effort must be made to avoid contaminating any additional areas, particularly hospitals and medical facilities. Qualified personnel must monitor the perimeter with instruments or detection devices to ensure that contaminants are not spread.

Standard decontamination procedures usually involve a series of stations. At each station, clothing and protective equipment are removed, and the individual is cleaned. Although some contaminants require only soap and water to be removed, chemical and biological agents may require special neutralizing solutions.

A terrorist incident that results in a large number of casualties may require a somewhat different decontamination process. Special procedures for **mass decontamination** have been devised for incidents involving hundreds or thousands of people. These procedures use master stream devices from pumper units (engines) and aerial apparatus to create high-volume, low-pressure showers. This approach allows large numbers of people to be decontaminated quickly **FIGURE 33-20** .

FIRE FIGHTER Tips

ADDITIONAL RESOURCES FOR TERRORIST INCIDENTS

Resources	Action
Air evacuation	Air transportation of injured
Air-supply unit	Refill SCBA cylinders
Bomb squad	Render explosive devices safe
EMS	Emergency medical care
Hazardous materials unit	Neutralize hazardous substances, including chemical agents
Health department	Technical; advice on chemical and biological agents
Local law enforcement	Law enforcement, perimeter control, and evidence recovery
Structural collapse teams	Search and stabilize collapsed structures

FIRE FIGHTER Tips

Emergency responders at a terrorist event may later be called to testify in court. Take notes and document information if you must disturb or move any evidence to rescue someone in immediate danger.

■ Mass Casualties

A terrorist or WMD incident may result in a large number of casualties. As part of the overall planning process, mass-casualty plans are essential to manage this type of situation, which would quickly overwhelm the normal capabilities of most emergency response systems. Mass-casualty plans typically involve using the resources of multiple agencies in an effort to handle large numbers of victims efficiently.

The basic principles of mass-casualty operations are described in the chapter "Emergency Medical Care." A terrorist incident, however, might potentially involve several additional complications and considerations, including the possibility of contamination by chemical, biological, or radiological agents. The mass-casualty plan must be expanded to address these problems.

If contamination is suspected, the plan must ensure that it does not spread beyond a defined perimeter. In some cases, victims may need to be decontaminated as they are moved to the **triage** and treatment areas **FIGURE 33-21**. This step will keep these areas free of contaminants. In other cases, the triage and treatment areas may be considered contaminated zones. In this scenario, victims would be decontaminated before they are transported from the scene.

During the early stages of an incident, it may be difficult to determine which agent was used and therefore which treatment and decontamination procedures are appropriate. In such a case, it may be necessary to quarantine exposed individuals within an area that is assumed to be contaminated, until laboratory test results are evaluated. The strategic plan for this situation must assume that any personnel who provide assistance or treatment to victims during this period will also become contaminated, as will any equipment or vehicles used in the contaminated area.

■ Additional Resources

A terrorist incident would likely result in a massive response by local, provincial, and federal government agencies. The

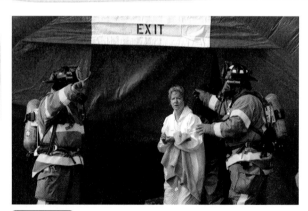

FIGURE 33-21 Victims should be decontaminated before they are placed in triage and treatment areas.
© Scott Fisher/MCT/Landov

RCMP is the lead agency for crisis management, and Public Safety Canada is the lead agency for consequence management during a terrorist incident. Other federal agencies, such as CSIS and the military, may also be involved. Several federal agencies have specific responsibilities regarding terrorist incidents.

The National Security Criminal Investigations (NSCI) Program integrates Canada by continuously improving its information-sharing practices with other federal agencies and partners, consistent with current policies; by developing outreach and education strategies that will lead to the engagement of stakeholders across the country and assist Canadians in avoiding being victimized; and by ensuring the situational awareness of the RCMP senior executive and the Government of Canada on the status of the threat environment.

In the United States, **National Terrorism Advisory System (NTAS)** alerts are issued when credible information is received that indicates a terrorist threat has been discovered that poses a significant danger. This system replaced the colour-coded system that had been put in place in 2002. The NTAS issues two types of alerts: the imminent threat alert and the elevated threat alert. An elevated threat alert warns of a credible terrorist threat against the United States; an imminent threat alert warns of a credible, specific, and impending terrorist threat against the United States. Both types of alerts provide a concise summary of the potential threat. In some cases, alerts are sent directly to law enforcement or specific areas of the private sector; in other situations, alerts are issued more broadly to the American people through the media. Each threat alert is issued for a specific time and then expires. It may be extended if new information becomes available or if the threat evolves.

Public Safety Canada develops national policy, response systems, and standards. Its goal is to issue timely alerts and similar products to help protect Canada's critical infrastructure. Public Safety Canada also works closely with emergency management organizations across Canada and supports regional partners and first responders with funds, tools, and training.

For the Government of Canada, Public Safety Canada houses the Government Operations Centre (GOC) at the hub of the national emergency management system. It's an advanced centre for monitoring and coordinating the federal response to an emergency. Its network of 11 regional offices and 2 satellite offices across Canada are the department's representatives who respond to any emergency or event of national interest. They provide regional operational support to the GOC.

Chief Concepts

- Terrorism can be described as the unlawful use of violence or threats of violence to intimidate or coerce a government, the civilian population, or any segment thereof in order to further political or social objectives.
- This broad definition can include a wide range of acts committed by various groups for various purposes.
- The goal of terrorism is to instill fear in a population or group.
- Terrorism can occur in any community. Be aware of potential targets in your area.
- Terrorists can turn ordinary objects into weapons.
- Responding to a terrorist incident puts fire fighters and other emergency personnel at risk.
- A call for a terrorist incident may at first appear to be a routine incident.
- The possibility of a terrorist incident should be considered when responding to any location that has been identified as a potential terrorist target.
- When responding to a potential terror-related incident, fire fighters should establish a staging area at a safe distance from the scene and follow the direction of the IC.

Hot Terms

Agro-terrorism The intentional use of chemical or biological agents against the agricultural industry or food supply.

Alpha particles A type of radiation that quickly loses energy and can travel only 2.5–5.1 cm (1–2 in.) from its source. Clothing or a sheet of paper can stop this type of energy. Alpha particles are not dangerous to plants, animals, or people unless the alpha-emitting substance enters the body.

Ammonium-nitrate fertilizer and fuel oil (ANFO) An explosive made of commonly available materials.

Anthrax An infectious disease spread by the bacterium *B. anthracis*; it is typically found around farms, infecting livestock.

Beta particles A type of radiation that is capable of travelling 3–4.6 m (10–15 ft.). Heavier materials, such as metal and glass, can stop this type of energy.

Biological agents Disease-causing bacteria, viruses, and other organisms that attack the human body.

Blistering agent A chemical that causes the skin to blister.

Chlorine A yellowish gas that is approximately two-and-a-half times heavier than air and slightly water-soluble. It has many industrial uses but is also a choking agent and can damage the lungs when inhaled.

Choking agent A chemical designed to inhibit breathing. It is typically intended to incapacitate rather than kill.

Cyanide A highly toxic chemical agent that attacks the circulatory system.

Cyberterrorism An intentional electronic attack against government or private computer systems.

Decontamination The physical or chemical process of removing any form of contaminant from a person, an animal, an object, or an environment.

DuoDote® kit A prefilled auto-injector that provides two medications—atropine and pralidoxime chloride—for use in treating the effects of exposure to nerve agents and insecticide poisoning.

Eco-terrorism Terrorism directed against causes that radical environmentalists think would damage the earth or its creatures.

Forward staging area A strategically placed area, close to the incident site, where personnel and equipment can be held at the ready for rapid response to an emergency event.

Gamma radiation A type of radiation that can travel significant distances, penetrate most materials, and pass through the body. Gamma rays are the most destructive type of radiation to the human body.

Improvised explosive device (IED) An explosive or incendiary device fabricated in an improvised manner.

Incubation period The time period between a victim's initial infection by an organism and the appearance of symptoms.

Lewisite An oily, colourless to dark-brown liquid blistering agent with an odour of geraniums.

Mass decontamination The process of showering a large number of victims with water for rapid decontamination in the field when the contaminating agent is unknown.

National Terrorism Advisory System (NTAS) A national system to communicate information about terrorist threats in the United States by providing timely information to the public, government agencies, and first responders. It issues imminent threat alerts and elevated threat alerts.

Nerve agent A toxic substance that attacks the central nervous system in humans.

Personal dosimeter A device that measures the dose of radiation to an individual.

Phosgene A chemical agent that causes severe pulmonary damage. It is a by-product of incomplete combustion.

Pipe bomb A device created by filling a section of pipe with an explosive material.

Wrap-Up, continued

Plague An infectious disease caused by the bacterium *Y. pestis*, which is commonly found on rodents.

Radiation dispersal device Any device that causes the purposeful dissemination of radioactive material without a nuclear detonation; also known as a dirty bomb.

Radiological agents Materials that emit radiation.

Sarin A nerve agent that is primarily a vapour hazard.

Secondary device An explosive device designed to injure emergency responders who have responded to an initial event.

Smallpox A highly infectious disease caused by the virus *Variola*.

Soman A nerve gas that is both a contact hazard and a vapour hazard; it has the odour of camphor.

Sulphur mustard A translucent, yellow or amber, oily liquid marked by the faint, sweet odour of mustard or garlic, which can be dispersed in an aerosol form. It causes exposed skin to blister.

Tabun A nerve gas that is both a contact hazard and a vapour hazard; it operates by disabling the chemical connection between the nerves and their target organs.

Triage The process of sorting victims based on the severity of their injuries and medical needs to establish treatment and transportation priorities.

Universal precautions Procedures for infection control that treat blood and certain body fluids as capable of transmitting infectious diseases.

V-agent An oily liquid nerve agent that can persist for several weeks; primarily a contact hazard.

Weapons of mass destruction (WMD) Weapons whose use is intended to cause mass casualties, damage, and chaos.

FIRE FIGHTER *in action*

You are dispatched along with the police to the local high school basketball game. The dispatch reports a call that a masked man ran out onto the court and sprayed some type of fog into the air. Some of the spectators began to chase after him, but he pulled out a handgun, injuring two of them, and escaped the area. People began coughing and experienced a burning sensation from the spray, which had a sweet, garlicky smell. The exits were quickly jammed, with people being crushed as they tried to escape to the gym.

1. The agent was most likely _____.
   A. sulphur mustard
   B. a radiological agent
   C. lewisite
   D. anthrax

2. Which of the following is an infectious disease that can be used in a WMD attack?
   A. Anthrax
   B. Soman
   C. V-agent
   D. Sarin

3. A type of radiation that quickly loses energy and can travel only 2.5–5.1 cm (1–2 in.) from its source is

_____.
   A. alpha
   B. beta
   C. gamma
   D. neutron

4. Which of the following is *not* a nerve agent?
   A. Tabun
   B. Phosgene
   C. Soman
   D. Sarin

5. The intentional act of using chemical or biological agents against the agricultural industry or food supply is termed _____.

A. eco-terrorism

B. cyberterrorism

C. nutri-terrorism

D. agro-terrorism

6. Because of the risk of a secondary device at incidents involving explosive devices, the staging area should be at least _____ away.

A. 91.4 m (300 ft.)

B. 304.8 m (1000 ft.)

C. 457.2 m (1500 ft.)

D. 914.4 m (3000 ft.)

FIRE FIGHTER II
in action

You are dispatched to a local convenience store for a man with shortness of breath. The owner says that he was waiting on a customer when he noticed some white powder on the counter. The owner says that the customer became short of breath right after the discovery of the powder. The owner says that the customer also became nauseated and broke out in a cold sweat.

1. How would you handle this situation?

2. What is the most likely cause of the customer's shortness of breath?

3. How will you calm and reassure the owner?

4. What concerns do you have about this incident?

An Extract from NFPA 1001, Standard for Fire Fighter Professional Qualifications, 2013 Edition

■ Chapter 5 Fire Fighter I

5.1 General. For qualification at Level I, the fire fighter candidate shall meet the general knowledge requirements in 5.1.1; the general skill requirements in 5.1.2; the JPRs defined in Sections 5.2 through 5.5 of this standard; and the requirements defined in Chapter 5, Core Competencies for Operations Level Responders, and Section 6.6, Mission-Specific Competencies: Product Control, of NFPA 472, *Standard for Competence of Responders to Hazardous Materials/Weapons of Mass Destruction Incidents.*

5.1.1 General Knowledge Requirements. The organization of the fire department; the role of the Fire Fighter I in the organization; the mission of fire service; the fire department's standard operating procedures (SOPs) and rules and regulations as they apply to the Fire Fighter I; the value of fire and life safety initiatives in support of the fire department mission and to reduce fire fighter line-of-duty injuries and fatalities; the role of other agencies as they relate to the fire department; aspects of the fire department's member assistance program; the importance of physical fitness and a healthy lifestyle to the performance of the duties of a fire fighter; the critical aspects of NFPA 1500, *Standard on Fire Department Occupational Safety and Health Program.*

5.1.2 General Skill Requirements. The ability to don personal protective clothing, doff personal protective clothing and prepare for reuse, hoist tools and equipment using ropes and the correct knot, and locate information in departmental documents and standard or code materials.

5.2 Fire Department Communications. This duty shall involve initiating responses, receiving telephone calls, and using fire department communications equipment to correctly relay verbal or written information, according to the JPRs in 5.2.1 through 5.2.4.

5.2.1 Initiate the response to a reported emergency, given the report of an emergency, fire department SOPs, and communications equipment, so that all necessary information is obtained, communications equipment is operated correctly, and the information is relayed promptly and accurately to the dispatch center.

(A) Requisite Knowledge. Procedures for reporting an emergency; departmental SOPs for taking and receiving alarms, radio codes, or procedures; and information needs of dispatch center.

(B) Requisite Skills. The ability to operate fire department communications equipment, relay information, and record information.

5.2.2 Receive a telephone call, given a fire department phone, so that procedures for answering the phone are used and the caller's information is relayed.

(A) Requisite Knowledge. Fire department procedures for answering nonemergency telephone calls.

(B) Requisite Skills. The ability to operate fire station telephone and intercom equipment.

5.2.3 Transmit and receive messages via the fire department radio, given a fire department radio and operating procedures, so that the information is accurate, complete, clear, and relayed within the time established by the AHJ.

(A) Requisite Knowledge. Departmental radio procedures and etiquette for routine traffic, emergency traffic, and emergency evacuation signals.

(B) Requisite Skills. The ability to operate radio equipment and discriminate between routine and emergency traffic.

5.2.4 Activate an emergency call for assistance, given vision obscured conditions, PPE, and department SOPs, so that the fire fighter can be located and rescued.

(A) Requisite Knowledge. Personnel accountability systems, emergency communication procedures, and emergency evacuation methods.

(B) Requisite Skills. The ability to initiate an emergency call for assistance in accordance with the AHJ's procedures, the ability to use other methods of emergency calls for assistance.

5.3 Fireground Operations. This duty shall involve performing activities necessary to ensure life safety, fire control, and property conservation, according to the JPRs in 5.3.1 through 5.3.19.

5.3.1 Use self-contained breathing apparatus (SCBA) during emergency operations, given SCBA and other personal protective equipment, so that the SCBA is correctly donned, the SCBA is correctly worn, controlled breathing techniques are used, emergency procedures are enacted if the SCBA fails, all low-air warnings are recognized, respiratory protection is not intentionally compromised, and hazardous areas are exited prior to air depletion.

(A) Requisite Knowledge. Conditions that require respiratory protection, uses and limitations of SCBA, components of SCBA, donning procedures, breathing techniques, indications for and emergency procedures used with SCBA, and physical requirements of the SCBA wearer.

(B) Requisite Skills. The ability to control breathing, replace SCBA air cylinders, use SCBA to exit through restricted passages, initiate and complete emergency procedures in the event of SCBA failure or air depletion, and complete donning procedures.

5.3.2* Respond on apparatus to an emergency scene, given personal protective clothing and other necessary personal protective equipment, so that the apparatus is correctly mounted and dismounted, seat belts are used while the vehicle is in motion, and other personal protective equipment is correctly used.

(A) Requisite Knowledge. Mounting and dismounting procedures for riding fire apparatus, hazards and ways to avoid hazards associated with riding apparatus, prohibited practices, and types of department personal protective equipment and the means for usage.

(B) Requisite Skills. The ability to use each piece of provided safety equipment.

5.3.3 Establish and operate in work areas at emergency scenes, given protective equipment, traffic and scene control devices, structure fire and roadway emergency scenes, traffic hazards and downed electrical wires, an assignment, and SOPs, so that procedures are followed, protective equipment is worn, protected work areas are established as directed using traffic and scene control devices, and the fire fighter performs assigned tasks only in established, protected work areas.

(A) Requisite Knowledge. Potential hazards involved in operating on emergency scenes including vehicle traffic, utilities, and environmental conditions; proper procedures for dismounting apparatus in traffic; procedures for safe operation at emergency scenes; and the protective equipment available for members' safety on emergency scenes and work zone designations.

(B) Requisite Skills. The ability to use personal protective clothing, deploy traffic and scene control devices, dismount apparatus, and operate in the protected work areas as directed.

5.3.4 Force entry into a structure, given personal protective equipment, tools, and an assignment, so that the tools are used as designed, the barrier is removed, and the opening is in a safe condition and ready for entry.

(A) Requisite Knowledge. Basic construction of typical doors, windows, and walls within the department's community or service area; operation of doors, windows, and locks; and the dangers associated with forcing entry through doors, windows, and walls.

(B) Requisite Skills. The ability to transport and operate hand and power tools and to force entry through doors, windows, and walls using assorted methods and tools.

5.3.5 Exit a hazardous area as a team, given vision-obscured conditions, so that a safe haven is found before exhausting the air supply, others are not endangered, and the team integrity is maintained.

(A) Requisite Knowledge. Personnel accountability systems, communication procedures, emergency evacuation methods, what constitutes a safe haven, elements that create or indicate a hazard, and emergency procedures for loss of air supply.

(B) Requisite Skills. The ability to operate as a team member in vision-obscured conditions, locate and follow a guideline, conserve air supply, and evaluate areas for hazards and identify a safe haven.

5.3.6 Set up ground ladders, given single and extension ladders, an assignment, and team members if needed, so that hazards are assessed, the ladder is stable, the angle is correct for climbing, extension ladders are extended to the necessary height with the fly locked, the top is placed against a reliable structural component, and the assignment is accomplished.

(A) Requisite Knowledge. Parts of a ladder, hazards associated with setting up ladders, what constitutes a stable foundation for ladder placement, different angles for various tasks, safety limits to the degree of angulation, and what constitutes a reliable structural component for top placement.

(B) Requisite Skills. The ability to carry ladders, raise ladders, extend ladders and lock flies, determine that a wall and roof will support the ladder, judge extension ladder height requirements, and place the ladder to avoid obvious hazards.

5.3.7 Attack a passenger vehicle fire operating as a member of a team, given personal protective equipment, attack line, and hand tools, so that hazards are avoided, leaking flammable liquids are identified and controlled, protection from flash fires is maintained, all vehicle compartments are overhauled, and the fire is extinguished.

(A) Requisite Knowledge. Principles of fire streams as they relate to fighting automobile fires; precautions to be followed when advancing hose lines toward an automobile; observable results that a fire stream has been properly applied; identifying alternative fuels and the hazards associated with them; dangerous conditions created during an automobile fire; common types of accidents or injuries related to fighting automobile fires and how to avoid them; how to access locked passenger, trunk, and engine compartments; and methods for overhauling an automobile.

(B) Requisite Skills. The ability to identify automobile fuel type; assess and control fuel leaks; open, close, and adjust the flow and pattern on nozzles; apply water for maximum effectiveness while maintaining flash fire protection; advance 1½ in. (38 mm) or larger diameter attack lines; and expose hidden fires by opening all automobile compartments.

5.3.8 Extinguish fires in exterior Class A materials, given fires in stacked or piled and small unattached structures or storage containers that can be fought from the exterior, attack lines, hand tools and master stream devices, and an assignment, so that exposures are protected, the spread of fire is stopped, collapse hazards are avoided, water application is effective, the fire is extinguished, and signs of the origin area(s) and arson are preserved.

(A) Requisite Knowledge. Types of attack lines and water streams appropriate for attacking stacked, piled materials and outdoor fires; dangers—such as collapse—associated with stacked and piled materials; various extinguishing agents and their effect on different material configurations; tools and methods to use in breaking up various types of materials; the difficulties related to complete extinguishment of stacked and piled materials; water application methods for exposure protection and fire extinguishment; dangers such as exposure to toxic or hazardous materials associated with storage building and container fires; obvious signs of origin and cause; and techniques for the preservation of fire cause evidence.

(B) Requisite Skills. The ability to recognize inherent hazards related to the material's configuration, operate handlines or master streams, break up material using hand tools and water streams, evaluate for complete extinguishment, operate hose lines and other water application devices, evaluate and modify water application for maximum penetration, search for and expose hidden fires, assess patterns for origin determination, and evaluate for complete extinguishment.

5.3.9 Conduct a search and rescue in a structure operating as a member of a team, given an assignment, obscured vision conditions, personal protective equipment, a flashlight, forcible entry tools, hose lines, and ladders when necessary, so that ladders are correctly placed when used, all assigned areas are searched, all victims are located and removed, team integrity is maintained, and team members' safety—including respiratory protection—is not compromised.

(A) Requisite Knowledge. Use of forcible entry tools during rescue operations, ladder operations for rescue, psychological effects of operating in obscured conditions and ways to manage them, methods to determine if an area is tenable, primary and secondary search techniques, team members' roles and goals, methods to use and indicators of finding victims, victim removal methods (including various carries), and considerations related to respiratory protection.

(B) Requisite Skills. The ability to use SCBA to exit through restricted passages, set up and use different types of ladders for various types of rescue operations, rescue a fire fighter with functioning respiratory protection, rescue a fire fighter whose respiratory protection is not functioning, rescue a person who has no respiratory protection, and assess areas to determine tenability.

5.3.10 Attack an interior structure fire operating as a member of a team, given an attack line, ladders when needed, personal protective equipment, tools, and an assignment, so that team integrity is maintained, the attack line is deployed for advancement, ladders are correctly placed when used, access is gained into the fire area, effective water application practices are used, the fire is approached correctly, attack techniques facilitate suppression given the level of the fire, hidden fires are located and controlled, the correct body posture is maintained, hazards are recognized and managed, and the fire is brought under control.

(A) Requisite Knowledge. Principles of fire streams; types, design, operation, nozzle pressure effects, and flow capabilities of nozzles; precautions to be followed when advancing hose lines to a fire; observable results that a fire stream has been properly applied; dangerous building conditions created by fire; principles of exposure protection; potential long-term consequences of exposure to products of combustion; physical states of matter in which fuels are found; common types of accidents or injuries and their causes; and the application of each size and type of attack line, the role of the backup team in fire attack situations, attack and control techniques for grade level and above and below grade levels, and exposing hidden fires.

(B) Requisite Skills. The ability to prevent water hammers when shutting down nozzles; open, close, and adjust nozzle flow and patterns; apply water using direct, indirect, and combination attacks; advance charged and uncharged 1½ in. (38 mm) diameter or larger hose lines up ladders and up and down interior and exterior stairways; extend hose lines; replace burst hose sections;

operate charged hose lines of 1½ in. (38 mm) diameter or larger while secured to a ground ladder; couple and uncouple various handline connections; carry hose; attack fires at grade level and above and below grade levels; and locate and suppress interior wall and subfloor fires.

5.3.11 Perform horizontal ventilation on a structure operating as part of a team, given an assignment, personal protective equipment, ventilation tools, equipment, and ladders, so that the ventilation openings are free of obstructions, tools are used as designed, ladders are correctly placed, ventilation devices are correctly placed, and the structure is cleared of smoke.

(A) Requisite Knowledge. The principles, advantages, limitations, and effects of horizontal, mechanical, and hydraulic ventilation; safety considerations when venting a structure; fire behavior in a structure; the products of combustion found in a structure fire; the signs, causes, effects, and prevention of backdrafts; and the relationship of oxygen concentration to life safety and fire growth.

(B) Requisite Skills. The ability to transport and operate ventilation tools and equipment and ladders, and to use safe procedures for breaking window and door glass and removing obstructions.

5.3.12 Perform vertical ventilation on a structure as part of a team, given an assignment, personal protective equipment, ground and roof ladders, and tools, so that ladders are positioned for ventilation, a specified opening is created, all ventilation barriers are removed, structural integrity is not compromised, products of combustion are released from the structure, and the team retreats from the area when ventilation is accomplished.

(A) Requisite Knowledge. The methods of heat transfer; the principles of thermal layering within a structure on fire; the techniques and safety precautions for venting flat roofs, pitched roofs, and basements; basic indicators of potential collapse or roof failure; the effects of construction type and elapsed time under fire conditions on structural integrity; and the advantages and disadvantages of vertical and trench/strip ventilation.

(B) Requisite Skills. The ability to transport and operate ventilation tools and equipment; hoist ventilation tools to a roof; cut roofing and flooring materials to vent flat roofs, pitched roofs, and basements; sound a roof for integrity; clear an opening with hand tools; select, carry, deploy, and secure ground ladders for ventilation activities; deploy roof ladders on pitched roofs while secured to a ground ladder; and carry ventilation-related tools and equipment while ascending and descending ladders.

5.3.13 Overhaul a fire scene, given personal protective equipment, attack line, hand tools, a flashlight, and an assignment, so that structural integrity is not compromised, all hidden fires are discovered, fire cause evidence is preserved, and the fire is extinguished.

(A) Requisite Knowledge. Types of fire attack lines and water application devices most effective for overhaul, water application methods for extinguishment that limit water damage, types of tools and methods used to expose hidden fire, dangers associated with overhaul, obvious signs of area of origin or signs of arson, and reasons for protection of fire scene.

(B) Requisite Skills. The ability to deploy and operate an attack line; remove flooring, ceiling, and wall components to expose void spaces without compromising structural integrity; apply water for maximum effectiveness; expose and extinguish hidden fires in walls, ceilings, and subfloor spaces; recognize and preserve obvious signs of area of origin and arson; and evaluate for complete extinguishment.

5.3.14 Conserve property as a member of a team, given salvage tools and equipment and an assignment, so that the building and its contents are protected from further damage.

(A) Requisite Knowledge. The purpose of property conservation and its value to the public, methods used to protect property, types of and uses for salvage covers, operations at properties protected with automatic sprinklers, how to stop the flow of water from an automatic sprinkler head, identification of the main control valve on an automatic sprinkler system, forcible entry issues related to salvage, and procedures for protecting possible areas of origin and potential evidence.

(B) Requisite Skills. The ability to cluster furniture; deploy covering materials; roll and fold salvage covers for reuse; construct water chutes and catch-alls; remove water; cover building openings, including doors, windows, floor openings, and roof openings; separate, remove, and relocate charred material to a safe location while protecting the area of origin for cause determination; stop the flow of water from a sprinkler with sprinkler wedges or stoppers; and operate a main control valve on an automatic sprinkler system.

5.3.15 Connect a fire department pumper to a water supply as a member of a team, given supply or intake hose, hose tools, and a fire hydrant or static water source, so that connections are tight and water flow is unobstructed.

(A) Requisite Knowledge. Loading and off-loading procedures for mobile water supply apparatus; fire hydrant operation; and suitable static water supply sources, procedures, and protocol for connecting to various water sources.

(B) Requisite Skills. The ability to hand lay a supply hose, connect and place hard suction hose for drafting operations, deploy portable water tanks as well as the equipment necessary to transfer water between and draft from them, make hydrant-to-pumper hose connections for forward and reverse lays, connect supply hose to a hydrant, and fully open and close the hydrant.

5.3.16 Extinguish incipient Class A, Class B, and Class C fires, given a selection of portable fire extinguishers, so that the correct extinguisher is chosen, the fire is completely extinguished, and correct extinguisher-handling techniques are followed.

(A) Requisite Knowledge. The classifications of fire; the types of, rating systems for, and risks associated with each class of fire; and the operating methods of and limitations of portable extinguishers.

(B) Requisite Skills. The ability to operate portable fire extinguishers, approach fire with portable fire extinguishers, select an appropriate extinguisher based on the size and type of fire, and safely carry portable fire extinguishers.

5.3.17 Illuminate the emergency scene, given fire service electrical equipment and an assignment, so that designated areas are illuminated and all equipment is operated within the manufacturer's listed safety precautions.

(A) Requisite Knowledge. Safety principles and practices, power supply capacity and limitations, and light deployment methods.

(B) Requisite Skills. The ability to operate department power supply and lighting equipment, deploy cords and connectors, reset ground-fault interrupter (GFI) devices, and locate lights for best effect.

5.3.18 Turn off building utilities, given tools and an assignment, so that the assignment is safely completed.

(A) Requisite Knowledge. Properties, principles, and safety concerns for electricity, gas, and water systems; utility disconnect methods and associated dangers; and use of required safety equipment.

(B) Requisite Skills. The ability to identify utility control devices, operate control valves or switches, and assess for related hazards.

5.3.19 Combat a ground cover fire operating as a member of a team, given protective clothing, SCBA (if needed), hose lines, extinguishers or hand tools, and an assignment, so that threats to property are reported, threats to personal safety are recognized, retreat is quickly accomplished when warranted, and the assignment is completed.

(A) Requisite Knowledge. Types of ground cover fires, parts of ground cover fires, methods to contain or suppress, and safety principles and practices.

(B) Requisite Skills. The ability to determine exposure threats based on fire spread potential, protect exposures, construct a fire line or extinguish with hand tools, maintain integrity of established fire lines, and suppress ground cover fires using water.

5.3.20 Tie a knot appropriate for hoisting tool, given personnel protective equipment, tools, ropes, and an assignment, so that the knots used are appropriate for hoisting tools securely and as directed.

(A) Requisite Knowledge. Knot types and usage; the difference between life safety and utility rope; reasons for placing rope out of service; the types of knots to use for given tools, ropes, or situations; hoisting methods for tools and equipment; and using rope to support response activities.

(B) Requisite Skills. The ability to hoist tools using specific knots based on the type of tool.

5.4 Rescue Operations. This duty shall involve no requirements for Fire Fighter I.

5.5 Preparedness and Maintenance. This duty shall involve performing activities that reduce the loss of life and property due to fire through response readiness, according to the JPRs in 5.5.1 and 5.5.2.

5.5.1 Clean and check ladders, ventilation equipment, SCBA, ropes, salvage equipment, and hand tools, given cleaning tools, cleaning supplies, and an assignment, so that equipment is clean and maintained according to manufacturer's or departmental guidelines, maintenance is recorded, and equipment is placed in a ready state or reported otherwise.

(A) Requisite Knowledge. Types of cleaning methods for various tools and equipment, correct use of cleaning solvents, and manufacturer's or departmental guidelines for cleaning equipment and tools.

(B) Requisite Skills. The ability to select correct tools for various parts and pieces of equipment, follow guidelines, and complete recording and reporting procedures.

5.5.2 Clean, inspect, and return fire hose to service, given washing equipment, water, detergent, tools, and replacement gaskets, so that damage is noted and corrected, the hose is clean, and the equipment is placed in a ready state for service.

(A) Requisite Knowledge. Departmental procedures for noting a defective hose and removing it from service, cleaning methods, and hose rolls and loads.

(B) Requisite Skills. The ability to clean different types of hose; operate hose washing and drying equipment; mark defective hose; and replace coupling gaskets, roll hose, and reload hose.

■ Chapter 6 Fire Fighter II

6.1 General. For qualification at Level II, the Fire Fighter I shall meet the general knowledge requirements in 6.1.1, the general skill requirements in 6.1.2, the JPRs defined in Sections 6.2 through 6.5 of this standard, and the requirements defined in Chapter 5.

6.1.1 General Knowledge Requirements. Responsibilities of the Fire Fighter II in assuming and transferring command within an incident management system, performing assigned duties in conformance with applicable NFPA and other safety regulations and AHJ procedures, and the role of a Fire Fighter II within the organization.

6.1.2 General Skill Requirements. The ability to determine the need for command, organize and coordinate an incident management system until command is transferred, and function within an assigned role in an incident management system.

6.2 Fire Department Communications. This duty shall involve performing activities related to initiating and reporting responses, according to the JPRs in 6.2.1 and 6.2.2.

6.2.1 Complete a basic incident report, given the report forms, guidelines, and information, so that all pertinent information is recorded, the information is accurate, and the report is complete.

(A) Requisite Knowledge. Content requirements for basic incident reports, the purpose and usefulness of accurate reports, consequences of inaccurate reports, how to obtain necessary information, and required coding procedures.

(B) Requisite Skills. The ability to determine necessary codes, proof reports, and operate fire department computers or other equipment necessary to complete reports.

6.2.2 Communicate the need for team assistance, given fire department communications equipment, SOPs, and a team, so that the supervisor is consistently informed of team needs, departmental SOPs are followed, and the assignment is accomplished safely.

(A) Requisite Knowledge. SOPs for alarm assignments and fire department radio communication procedures.

(B) Requisite Skills. The ability to operate fire department communications equipment.

6.3 Fireground Operations. This duty shall involve performing activities necessary to ensure life safety, fire control, and property conservation, according to the JPRs in 6.3.1 through 6.3.4.

6.3.1 Extinguish an ignitible liquid fire, operating as a member of a team, given an assignment, an attack line, personal protective equipment, a foam proportioning device, a nozzle, foam concentrates, and a water supply, so that the correct type of foam concentrate is selected for the given fuel and conditions, a properly proportioned foam stream is applied to the surface of the fuel to create and maintain a foam blanket, fire is extinguished, reignition is prevented, team protection is maintained with a foam stream, and the hazard is faced until retreat to safe haven is reached.

(A) Requisite Knowledge. Methods by which foam prevents or controls a hazard; principles by which foam is generated; causes for poor foam generation and corrective measures; difference between hydrocarbon and polar solvent fuels and the concentrates that work on each; the characteristics, uses, and limitations of fire-fighting foams; the advantages and disadvantages of using fog nozzles versus foam nozzles for foam application; foam stream application techniques; hazards associated with foam usage; and methods to reduce or avoid hazards.

(B) Requisite Skills. The ability to prepare a foam concentrate supply for use, assemble foam stream components, master various foam application techniques, and approach and retreat from spills as part of a coordinated team.

6.3.2 Coordinate an interior attack line for a team's accomplishment of an assignment in a structure fire, given attack lines, personnel, personal protective equipment, and tools, so that crew integrity is established; attack techniques are selected for the given level of the fire (e.g., attic, grade level, upper levels, or basement); attack techniques are communicated to the attack teams; constant team coordination is maintained; fire growth and development is continuously evaluated; search, rescue, and ventilation

requirements are communicated or managed; hazards are reported to the attack teams; and incident command is apprised of changing conditions.

(A) Requisite Knowledge. Selection of the nozzle and hose for fire attack, given different fire situations; selection of adapters and appliances to be used for specific fireground situations; dangerous building conditions created by fire and fire suppression activities; indicators of building collapse; the effects of fire and fire suppression activities on wood, masonry (brick, block, stone), cast iron, steel, reinforced concrete, gypsum wallboard, glass, and plaster on lath; search and rescue and ventilation procedures; indicators of structural instability; suppression approaches and practices for various types of structural fires; and the association between specific tools and special forcible entry needs.

(B) Requisite Skills. The ability to assemble a team, choose attack techniques for various levels of a fire (e.g., attic, grade level, upper levels, or basement), evaluate and forecast a fire's growth and development, select tools for forcible entry, incorporate search and rescue procedures and ventilation procedures in the completion of the attack team efforts, and determine developing hazardous building or fire conditions.

6.3.3 Control a flammable gas cylinder fire, operating as a member of a team, given an assignment, a cylinder outside of a structure, an attack line, personal protective equipment, and tools, so that crew integrity is maintained, contents are identified, safe havens are identified prior to advancing, open valves are closed, flames are not extinguished unless the leaking gas is eliminated, the cylinder is cooled, cylinder integrity is evaluated, hazardous conditions are recognized and acted upon, and the cylinder is faced during approach and retreat.

(A) Requisite Knowledge. Characteristics of pressurized flammable gases, elements of a gas cylinder, effects of heat and pressure on closed cylinders, boiling liquid expanding vapor explosion (BLEVE) signs and effects, methods for identifying contents, how to identify safe havens before approaching flammable gas cylinder fires, water stream usage and demands for pressurized cylinder fires, what to do if the fire is prematurely extinguished, valve types and their operation, alternative actions related to various hazards, and when to retreat.

(B) Requisite Skills. The ability to execute effective advances and retreats, apply various techniques for water application, assess cylinder integrity and changing cylinder conditions, operate control valves, and choose effective procedures when conditions change.

6.3.4 Protect evidence of fire cause and origin, given a flashlight and overhaul tools, so that the evidence is noted and protected from further disturbance until investigators can arrive on the scene.

(A) Requisite Knowledge. Methods to assess origin and cause; types of evidence; means to protect various types of evidence; the role and relationship of Fire Fighter IIs, criminal investigators, and insurance investigators in fire investigations; and the effects and problems associated with removing property or evidence from the scene.

(B) Requisite Skills. The ability to locate the fire's origin area, recognize possible causes, and protect the evidence.

6.4 Rescue Operations. This duty shall involve performing activities related to accessing and disentangling victims from motor vehicle accidents and helping special rescue teams, according to the JPRs in 6.4.1 and 6.4.2.

6.4.1 Extricate a victim entrapped in a motor vehicle as part of a team, given stabilization and extrication tools, so that the vehicle is stabilized, the victim is disentangled without further injury, and hazards are managed.

(A) Requisite Knowledge. The fire department's role at a vehicle accident, points of strength and weakness in auto body construction, dangers associated with vehicle components and systems, the uses and limitations of hand and power extrication equipment, and safety procedures when using various types of extrication equipment.

(B) Requisite Skills. The ability to operate hand and power tools used for forcible entry and rescue as designed; use cribbing and shoring material; and choose and apply appropriate techniques for moving or removing vehicle roofs, doors, windshields, windows, steering wheels or columns, and the dashboard.

6.4.2 Assist rescue operation teams, given standard operating procedures, necessary rescue equipment, and an assignment, so that procedures are followed, rescue items are recognized and retrieved in the time as prescribed by the AHJ, and the assignment is completed.

(A) Requisite Knowledge. The fire fighter's role at a technical rescue operation, the hazards associated with technical rescue operations, types and uses for rescue tools, and rescue practices and goals.

(B) Requisite Skills. The ability to identify and retrieve various types of rescue tools, establish public barriers, and assist rescue teams as a member of the team when assigned.

6.5 Fire and Life Safety Initiatives, Preparedness, and Maintenance. This duty shall involve performing activities related to reducing the loss of life and property due to fire through hazard identification, inspection, and response readiness, according to the JPRs in 6.5.1 through 6.5.5.

6.5.1 Perform a fire safety survey in an occupied structure, given survey forms and procedures, so that fire and life safety hazards are identified, recommendations for their correction are made to the occupant, and unresolved issues are referred to the proper authority.

(A) Requisite Knowledge. Organizational policy and procedures, common causes of fire and their prevention, the importance of a fire safety survey and public fire education programs to fire department public relations and the community, and referral procedures.

(B) Requisite Skills. The ability to complete forms, recognize hazards, match findings to preapproved recommendations, and effectively communicate findings to occupants or referrals.

6.5.2 Present fire safety information to station visitors or small groups, given prepared materials, so that all information is presented, the information is accurate, and questions are answered or referred.

(A) Requisite Knowledge. Parts of informational materials and how to use them, basic presentation skills, and departmental standard operating procedures for giving fire station tours.

(B) Requisite Skills. The ability to document presentations and to use prepared materials.

6.5.3 Prepare a preincident survey, given forms, necessary tools, and an assignment, so that all required occupancy information is recorded, items of concern are noted, and accurate sketches or diagrams are prepared.

(A) Requisite Knowledge. The sources of water supply for fire protection; the fundamentals of fire suppression and detection systems; common symbols used in diagramming construction features, utilities, hazards, and fire protection systems; departmental requirements for a preincident survey and form completion; and the importance of accurate diagrams.

(B) Requisite Skills. The ability to identify the components of fire suppression and detection systems; sketch the site, buildings, and special features; detect hazards and special considerations to include in the preincident sketch; and complete all related departmental forms.

6.5.4 Maintain power plants, power tools, and lighting equipment, given tools and manufacturers' instructions, so that equipment is clean and maintained according to manufacturer and departmental guidelines, maintenance is recorded, and equipment is placed in a ready state or reported otherwise.

(A) Requisite Knowledge. Types of cleaning methods, correct use of cleaning solvents, manufacturer and departmental guidelines for maintaining equipment and its documentation, and problem-reporting practices.

(B) Requisite Skills. The ability to select correct tools; follow guidelines; complete recording and reporting procedures; and operate power plants, power tools, and lighting equipment.

6.5.5 Perform an annual service test on fire hose, given a pump, a marking device, pressure gauges, a timer, record sheets, and related equipment, so that procedures are followed, the condition of the hose is evaluated, any damaged hose is removed from service, and the results are recorded.

(A) Requisite Knowledge. Procedures for safely conducting hose service testing, indicators that dictate any hose be removed from service, and recording procedures for hose test results.

(B) Requisite Skills. The ability to operate hose testing equipment and nozzles and to record results.

An Extract from NFPA 472, *Standard for Competence of Responders to Hazardous Materials/Weapons of Mass Destruction Incidents*, 2013 Edition

■ Chapter 4 Competencies for Awareness Level Personnel

4.1 General.

4.1.1 Introduction.

4.1.1.1 Awareness level personnel shall be persons who, in the course of their normal duties, could encounter an emergency involving hazardous materials/weapons of mass destruction (WMD) and who are expected to recognize the presence of the hazardous materials/WMD, protect themselves, call for trained personnel, and secure the area.

4.1.1.2 Awareness level personnel shall be trained to meet all competencies of this chapter.

4.1.1.3 Awareness level personnel shall receive additional training to meet applicable governmental occupational health and safety regulations.

4.1.2 Goal.

4.1.2.1 The goal of the competencies in this chapter shall be to provide personnel already on the scene of a hazardous materials/WMD incident with the knowledge and skills to perform the tasks in 4.1.2.2 safely and effectively.

4.1.2.2 When already on the scene of a hazardous materials/WMD incident, the awareness level personnel shall be able to perform the following tasks:

(1) Analyze the incident to determine both the hazardous materials/WMD present and the basic hazard and response information for each hazardous materials/WMD agent by completing the following tasks:

(a) Detect the presence of hazardous materials/WMD.

(b) Survey a hazardous materials/WMD incident from a safe location to identify the name, UN/NA identification number, type of placard, or other distinctive marking applied for the hazardous materials/WMD involved.

(c) Collect hazard information from the current edition of the DOT *Emergency Response Guidebook*.

(2) Implement actions consistent with the authority having jurisdiction (AHJ), and the current edition of the DOT *Emergency Response Guidebook* by completing the following tasks:

(a) Initiate protective actions

(b) Initiate the notification process

4.2 Competencies —Analyzing the Incident.

4.2.1 Detecting the Presence of Hazardous Materials/WMD. Given examples of various situations, awareness level personnel shall identify those situations where hazardous materials/WMD are present by completing the following requirements:

(1) Identify the definitions of both hazardous material (or dangerous goods, in Canada) and WMD

(2) Identify the UN/DOT hazard classes and divisions of hazardous materials/WMD and identify common examples of materials in each hazard class or division

(3) Identify the primary hazards associated with each UN/DOT hazard class and division

(4) Identify the difference between hazardous materials/WMD incidents and other emergencies

(5) Identify typical occupancies and locations in the community where hazardous materials/WMD are manufactured, transported, stored, used, or disposed of

(6) Identify typical container shapes that can indicate the presence of hazardous materials/WMD

(7) Identify facility and transportation markings and colors that indicate hazardous materials/WMD, including the following:

(a) Transportation markings, including UN/NA identification number marks, marine pollutant mark, elevated temperature (HOT) mark, commodity marking, and inhalation hazard mark

(b) NFPA 704, *Standard System for the Identification of the Hazards of Materials for Emergency Response*, markings

(c) Military hazardous materials/WMD markings

(d) Special hazard communication markings for each hazard class

(e) Pipeline markings

(f) Container markings

(8) Given an NFPA704 marking, describe the significance of the colors, numbers, and special symbols

(9) Identify U.S. and Canadian placards and labels that indicate hazardous materials/WMD

(10) Identify the following basic information on material safety data sheets (MSDS) and shipping papers for hazardous materials:

(a) Identify where to find MSDS

(b) Identify major sections of an MSDS

(c) Identify the entries on shipping papers that indicate the presence of hazardous materials

(d) Match the name of the shipping papers found in transportation (air, highway, rail, and water) with the mode of transportation

(e) Identify the person responsible for having the shipping papers in each mode of transportation

(f) Identify where the shipping papers are found in each mode of transportation

(g) Identify where the papers can be found in an emergency in each mode of transportation

(11) Identify examples of clues (other than occupancy/location, container shape, markings/color, placards/labels, MSDS, and shipping papers) to include sight, sound, and odor of which indicate hazardous materials/WMD

(12) Describe the limitations of using the senses in determining the presence or absence of hazardous materials/WMD

(13) Identify at least four types of locations that could be targets for criminal or terrorist activity using hazardous materials/WMD

(14) Describe the difference between a chemical and a biological incident

(15) Identify at least four indicators of possible criminal or terrorist activity involving chemical agents

(16) Identify at least four indicators of possible criminal or terrorist activity involving biological agents

(17) Identify at least four indicators of possible criminal or terrorist activity involving radiological agents

(18) Identify at least four indicators of possible criminal or terrorist activity involving illicit laboratories (clandestine laboratories, weapons lab, ricin lab)

(19) Identify at least four indicators of possible criminal or terrorist activity involving explosives

(20) Identify at least four indicators of secondary devices

4.2.2 Surveying Hazardous Materials/WMD Incidents. Given examples of hazardous materials/WMD incidents, awareness level personnel shall, from a safe location, identify the hazardous material(s)/WMD involved in each situation by name, UN/NA identification number, or type placard applied by completing the following requirements:

(1) Identify difficulties encountered in determining the specific names of hazardous materials/WMD at facilities and in transportation

(2) Identify sources for obtaining the names of, UN/NA identification numbers for, or types of placard associated with hazardous materials/WMD in transportation

(3) Identify sources for obtaining the names of hazardous materials/WMD at a facility

4.2.3 Collecting Hazard Information. Given the identity of various hazardous materials/WMD (name, UN/NA identification number, or type placard), awareness level personnel shall identify the fire, explosion, and health hazard information for each material by using the current edition of the DOT *Emergency Response Guidebook* by completing the following requirements:

(1) Identify the three methods for determining the guidebook page for a hazardous material/WMD

(2) Identify the two general types of hazards found on each guidebook page

4.3 Competencies — Planning the Response. (Reserved)

4.4 Competencies — Implementing the Planned Response.

4.4.1 Initiating Protective Actions. Given examples of hazardous materials/WMD incidents, the emergency response plan, the standard operating procedures, and the current edition of the DOT *Emergency Response Guidebook*, awareness level personnel shall be able to identify the actions to be taken to protect themselves and others and to control access to the scene by completing the following requirements:

(1) Identify the location of both the emergency response plan and/or standard operating procedures

(2) Identify the role of the awareness level personnel during hazardous materials/WMD incidents

(3) Identify the following basic precautions to be taken to protect themselves and others in hazardous materials/WMD incidents:

  (a) Identify the precautions necessary when providing emergency medical care to victims of hazardous materials/WMD incidents

  (b) Identify typical ignition sources found at the scene of hazardous materials/WMD incidents

  (c) Identify the ways hazardous materials/WMD are harmful to people, the environment, and property

  (d) Identify the general routes of entry for human exposure to hazardous materials/WMD

(4) Given examples of hazardous materials/WMD and the identity of each hazardous material/WMD (name, UN/NA identification number, or type placard), identify the following response information:

  (a) Emergency action (fire, spill, or leak and first aid)

  (b) Personal protective equipment necessary

  (c) Initial isolation and protective action distances

(5) Given the name of a hazardous material, identify the recommended personal protective equipment from the following list:

  (a) Street clothing and work uniforms

  (b) Structural fire-fighting protective clothing

  (c) Positive pressure self-contained breathing apparatus

  (d) Chemical-protective clothing and equipment

(6) Identify the definitions for each of the following protective actions:

  (a) Isolation of the hazard area and denial of entry

  (b) Evacuation

  (c) Shelter-in-place

(7) Identify the size and shape of recommended initial isolation and protective action zones

(8) Describe the difference between small and large spills as found in the Table of Initial Isolation and Protective Action Distances in the DOT *Emergency Response Guidebook*

(9) Identify the circumstances under which the following distances are used at hazardous materials/WMD incidents:

  (a) Table of Initial Isolation and Protective Action Distances

  (b) Isolation distances in the numbered guides

(10) Describe the difference between the isolation distances on the orange-bordered guidebook pages and the protective action distances on the green-bordered ERG (*Emergency Response Guidebook*) pages

(11) Identify the techniques used to isolate the hazard area and deny entry to unauthorized persons at hazardous materials/WMD incidents

(12) Identify at least four specific actions necessary when an incident is suspected to involve criminal or terrorist activity

4.4.2 Initiating the Notification Process. Given scenarios involving hazardous materials/WMD incidents, awareness level personnel shall identify the initial notifications to be made and how to make them, consistent with the AHJ.

4.5 Competencies — Evaluating Progress. (Reserved)

4.6 Competencies — Terminating the Incident. (Reserved)

■ Chapter 5 Core Competencies for Operations Level Responders

5.1 General.

5.1.1 Introduction.

5.1.1.1 The operations level responder shall be that person who responds to hazardous materials/weapons of mass destruction (WMD) incidents for the purpose of protecting nearby persons, the environment, or property from the effects of the release.

5.1.1.2 The operations level responder shall be trained to meet all competencies at the awareness level (*see Chapter 4*) and the competencies of this chapter.

5.1.1.3 The operations level responder shall receive additional training to meet applicable governmental occupational health and safety regulations.

5.1.2 Goal.

5.1.2.1 The goal of the competencies in this chapter shall be to provide operations level responders with the knowledge and skills to perform the core competencies in 5.1.2.2 safely.

5.1.2.2 When responding to hazardous materials/WMD incidents, operations level responders shall be able to perform the following tasks:

(1) Analyze a hazardous materials/WMD incident to determine the scope of the problem and potential outcomes by completing the following tasks:

  (a) Survey a hazardous materials/WMD incident to identify the containers and materials involved, determine whether hazardous materials/WMD have been released, and evaluate the surrounding conditions

  (b) Collect hazard and response information from MSDS; CHEMTREC/CANUTEC/SETIQ; local, state, and federal authorities; and shipper/manufacturer contacts

  (c) Predict the likely behavior of a hazardous materials/WMD and its container

  (d) Estimate the potential harm at a hazardous materials/WMD incident

(2) Plan an initial response to a hazardous materials/WMD incident within the capabilities and competencies of available personnel and personal protective equipment by completing the following tasks:

  (a) Describe the response objectives for the hazardous materials/WMD incident

  (b) Describe the response options available for each objective

  (c) Determine whether the personal protective equipment provided is appropriate for implementing each option

  (d) Describe emergency decontamination procedures

  (e) Develop a plan of action, including safety considerations

(3) Implement the planned response for a hazardous materials/WMD incident to favorably change the outcomes consistent with the emergency response plan and/or standard operating procedures by completing the following tasks:

  (a) Establish and enforce scene control procedures, including control zones, emergency decontamination, and communications

  (b) Where criminal or terrorist acts are suspected, establish means of evidence preservation

  (c) Initiate an incident command system (ICS) for hazardous materials/WMD incidents

(d) Perform tasks assigned as identified in the incident action plan

(e) Demonstrate emergency decontamination

(4) Evaluate the progress of the actions taken at a hazardous materials/WMD incident to ensure that the response objectives are being met safely, effectively, and efficiently by completing the following tasks:

(a) Evaluate the status of the actions taken in accomplishing the response objectives

(b) Communicate the status of the planned response

5.2 Core Competencies—Analyzing the Incident.

5.2.1 Surveying Hazardous Materials/WMD Incidents. Given scenarios involving hazardous materials/WMD incidents, the operations level responder shall collect information about the incident to identify the containers, the materials involved, the surrounding conditions, and whether hazardous materials/WMD have been released by completing the requirements of 5.2.1.1 through 5.2.1.6.

5.2.1.1 Given three examples each of liquid, gas, and solid hazardous material or WMD, including various hazard classes, operations level personnel shall identify the general shapes of containers in which the hazardous materials/WMD are typically found.

5.2.1.1.1 Given examples of the following tank cars, the operations level responder shall identify each tank car by type, as follows:

(1) Cryogenic liquid tank cars

(2) Nonpressure tank cars (general service or low pressure cars)

(3) Pressure tank cars

5.2.1.1.2 Given examples of the following intermodal tanks, the operations level responder shall identify each intermodal tank by type, as follows:

(1) Nonpressure intermodal tanks

(2) Pressure intermodal tanks

(3) Specialized intermodal tanks, including the following:

(a) Cryogenic intermodal tanks

(b) Tube modules

5.2.1.1.3 Given examples of the following cargo tanks, the operations level responder shall identify each cargo tank by type, as follows:

(1) Compressed gas tube trailers

(2) Corrosive liquid tanks

(3) Cryogenic liquid tanks

(4) Dry bulk cargo tanks

(5) High pressure tanks

(6) Low pressure chemical tanks

(7) Nonpressure liquid tanks

5.2.1.1.4 Given examples of the following storage tanks, the operations level responder shall identify each tank by type, as follows:

(1) Cryogenic liquid tank

(2) Nonpressure tank

(3) Pressure tank

5.2.1.1.5 Given examples of the following nonbulk packaging, the operations level responder shall identify each package by type, as follows:

(1) Bags

(2) Carboys

(3) Cylinders

(4) Drums

(5) Dewar flask (cryogenic liquids)

5.2.1.1.6 Given examples of the following packaging, the operations level responder shall identify the characteristics of each container or package by type as follows:

(1) Intermediate bulk container (IBC)

(2) Ton container

5.2.1.1.7 Given examples of the following radioactive material packages, the operations level responder shall identify the characteristics of each container or package by type, as follows:

(1) Excepted

(2) Industrial

(3) Type A

(4) Type B

(5) Type C

5.2.1.2 Given examples of containers, the operations level responder shall identify the markings that differentiate one container from another.

5.2.1.2.1 Given examples of the following marked transport vehicles and their corresponding shipping papers, the operations level responder shall identify the following vehicle or tank identification marking:

(1) Highway transport vehicles, including cargo tanks

(2) Intermodal equipment, including tank containers

(3) Rail transport vehicles, including tank cars

5.2.1.2.2 Given examples of facility containers, the operations level responder shall identify the markings indicating container size, product contained, and/or site identification numbers.

5.2.1.3 Given examples of hazardous materials incidents, the operations level responder shall identify the name(s) of the hazardous material(s) in 5.2.1.3.1 through 5.2.1.3.3.

5.2.1.3.1 The operations level responder shall identify the following information on a pipeline marker:

(1) Emergency telephone number

(2) Owner

(3) Product

5.2.1.3.2 Given a pesticide label, the operations level responder shall identify each of the following pieces of information, then match the piece of information to its significance in surveying hazardous materials incidents:

(1) Active ingredient

(2) Hazard statement

(3) Name of pesticide

(4) Pest control product (PCP) number (in Canada)

(5) Precautionary statement

(6) Signal word

5.2.1.3.3 Given a label for a radioactive material, the operations level responder shall identify the type or category of label, contents, activity, transport index, and criticality safety index as applicable.

5.2.1.4 The operations level responder shall identify and list the surrounding conditions that should be noted when a hazardous materials/WMD incident is surveyed.

5.2.1.5 The operations level responder shall describe ways to verify information obtained from the survey of a hazardous materials/WMD incident.

5.2.1.6 The operations level responder shall identify at least three additional hazards that could be associated with an incident involving terrorist or criminal activities.

5.2.2 Collecting Hazard and Response Information. Given scenarios involving known hazardous materials/WMD, the operations level responder shall collect hazard and response information using MSDS, CHEMTREC/CANUTEC/SETIQ, governmental authorities, and shippers and manufacturers by completing the following requirements:

(1) Match the definitions associated with the UN/DOT hazard classes and divisions of hazardous materials/WMD, including refrigerated liquefied gases and cryogenic liquids, with the class or division

(2) Identify two ways to obtain an MSDS in an emergency

(3) Using an MSDS for a specified material, identify the following hazard and response information:

    (a) Physical and chemical characteristics

    (b) Physical hazards of the material

    (c) Health hazards of the material

    (d) Signs and symptoms of exposure

    (e) Routes of entry

    (f) Permissible exposure limits

    (g) Responsible party contact

    (h) Precautions for safe handling (including hygiene practices, protective measures, and procedures for cleanup of spills and leaks)

    (i) Applicable control measures, including personal protective equipment

    (j) Emergency and first-aid procedures

(4) Identify the following:

    (a) Type of assistance provided by CHEMTREC/CANUTEC/SETIQ and governmental authorities

    (b) Procedure for contacting CHEMTREC/CANUTEC/SETIQ and governmental authorities

    (c) Information to be furnished to CHEMTREC/CANUTEC/SETIQ and governmental authorities

(5) Identify two methods of contacting the manufacturer or shipper to obtain hazard and response information

(6) Identify the type of assistance provided by governmental authorities with respect to criminal or terrorist activities involving the release or potential release of hazardous materials/WMD

(7) Identify the procedure for contacting local, state, and federal authorities as specified in the emergency response plan and/or standard operating procedures

(8) Describe the properties and characteristics of the following:

    (a) Alpha radiation

    (b) Beta radiation

    (c) Gamma radiation

    (d) Neutron radiation

5.2.3 Predicting the Likely Behavior of a Material and Its Container. Given scenarios involving hazardous materials/WMD incidents, each with a single hazardous material/WMD, the operations level responder shall describe the likely behavior of the material or agent and its container by completing the following requirements:

(1) Use the hazard and response information obtained from the current edition of the DOT Emergency Response Guidebook, MSDS, CHEMTREC/CANUTEC/SETIQ, governmental authorities, and shipper and manufacturer contacts, as follows:

    (a) Match the following chemical and physical properties with their significance and impact on the behavior of the container and its contents:

        i. Boiling point

        ii. Chemical reactivity

        iii. Corrosivity (pH)

        iv. Flammable (explosive) range [lower explosive limit (LEL) and upper explosive limit (UEL)]

        v. Flash point

        vi. Ignition (autoignition) temperature

        vii. Particle size

        viii. Persistence

        ix. Physical state (solid, liquid, gas)

        x. Radiation (ionizing and non-ionizing)

        xi. Specific gravity

        xii. Toxic products of combustion

        xiii. Vapor density

        xiv. Vapor pressure

        xv. Water solubility

    (b) Identify the differences between the following terms:

        i. Contamination and secondary contamination

        ii. Exposure and contamination

        iii. Exposure and hazard

        iv. Infectious and contagious

        v. Acute effects and chronic effects

        vi. Acute exposures and chronic exposures

(2) Identify three types of stress that can cause a container system to release its contents

(3) Identify five ways in which containers can breach

(4) Identify four ways in which containers can release their contents

(5) Identify at least four dispersion patterns that can be created upon release of a hazardous material

(6) Identify the time frames for estimating the duration that hazardous materials/WMD will present an exposure risk

(7) Identify the health and physical hazards that could cause harm

(8) Identify the health hazards associated with the following terms:

    (a) Alpha, beta, gamma, and neutron radiation

    (b) Asphyxiant

    (c) Carcinogen

    (d) Convulsant

    (e) Corrosive

    (f) Highly toxic

    (g) Irritant

    (h) Sensitizer, allergen

    (i) Target organ effects

    (j) Toxic

(9) Given the following, identify the corresponding UN/DOT hazard class and division:

    (a) Blood agents

    (b) Biological agents and biological toxins

    (c) Choking agents

    (d) Irritants (riot control agents)

    (e) Nerve agents

(f) Radiological materials

(g) Vesicants (blister agents)

5.2.4 Estimating Potential Harm. Given scenarios involving hazardous materials/WMD incidents, the operations level responder shall describe the potential harm within the endangered area at each incident by completing the following requirements:

(1) Identify a resource for determining the size of an endangered area of a hazardous materials/WMD incident

(2) Given the dimensions of the endangered area and the surrounding conditions at a hazardous materials/WMD incident, describe the number and type of exposures within that endangered area

(3) Identify resources available for determining the concentrations of a released hazardous materials/WMD within an endangered area

(4) Given the concentrations of the released material, describe the factors for determining the extent of physical, health, and safety hazards within the endangered area of a hazardous materials/WMD incident

(5) Describe the impact that time, distance, and shielding have on exposure to radioactive materials specific to the expected dose rate

5.3 Core Competencies—Planning the Response.

5.3.1 Describing Response Objectives. Given at least two scenarios involving hazardous materials/WMD incidents, the operations level responder shall describe the response objectives for each example by completing the following requirements:

(1) Given an analysis of a hazardous materials/WMD incident and the exposures, describe the number of exposures that could be saved with the resources provided by the AHJ

(2) Given an analysis of a hazardous materials/WMD incident, describe the steps for determining response objectives

(3) Describe how to assess the risk to a responder for each hazard class in rescuing injured persons at a hazardous materials/WMD incident

(4) Describe the potential for secondary attacks and devices at criminal or terrorist events

5.3.2 Identifying Action Options. Given examples of hazardous materials/WMD incidents (facility and transportation), the operations level responder shall identify the options for each response objective and shall meet the following requirements:

(1) Identify the options to accomplish a given response objective

(2) Describe the prioritization of emergency medical care and removal of victims from the hazard area relative to exposure and contamination concerns

5.3.3 Determining Suitability of Personal Protective Equipment.

Given examples of hazardous materials/WMD incidents, including the names of the hazardous materials/WMD involved and the anticipated type of exposure, the operations level responder shall determine whether available personal protective equipment is applicable to performing assigned tasks by completing the following requirements:

(1) Identify the respiratory protection required for a given response option and the following:

(a) Describe the advantages, limitations, uses, and operational components of the following types of respiratory protection at hazardous materials/WMD incidents:

i. Positive pressure self-contained breathing apparatus (SCBA)

ii. Positive pressure air-line respirator with required escape unit

iii. Closed-circuit SCBA

iv. Powered air-purifying respirator (PAPR)

v. Air-purifying respirator (APR)

vi. Particulate respirator

(b) Identify the required physical capabilities and limitations of personnel working in respiratory protection

(2) Identify the personal protective clothing required for a given option and the following:

(a) Identify skin contact hazards encountered at hazardous materials/WMD incidents

(b) Identify the purpose, advantages, and limitations of the following types of protective clothing at hazardous materials/WMD incidents:

i. Chemical-protective clothing such as liquid splash–protective clothing and vapor-protective clothing

ii. High temperature–protective clothing such as proximity suit and entry suits

iii. Structural fire-fighting protective clothing

5.3.4 Identifying Decontamination Issues. Given scenarios involving hazardous materials/WMD incidents, the operations level responder shall identify when decontamination is needed by completing the following requirements:

(1) Identify ways that people, personal protective equipment, apparatus, tools, and equipment become contaminated

(2) Describe how the potential for secondary contamination determines the need for decontamination

(3) Explain the importance and limitations of decontamination procedures at hazardous materials incidents

(4) Identify the purpose of emergency decontamination procedures at hazardous materials incidents

(5) Identify the methods, advantages, and limitations of emergency decontamination procedures

5.4 Core Competencies—Implementing the Planned Response.

5.4.1 Establishing Scene Control. Given two scenarios involving hazardous materials/WMD incidents, the operations level responder shall explain how to establish and maintain scene control, including control zones and emergency decontamination, and communications between responders and to the public by completing the following requirements:

(1) Identify the procedures for establishing scene control through control zones

(2) Identify the criteria for determining the locations of the control zones at hazardous materials/WMD incidents

(3) Identify the basic techniques for the following protective actions at hazardous materials/WMD incidents:

(a) Evacuation

(b) Shelter-in-place

(4) Demonstrate the ability to perform emergency decontamination

(5) Identify the items to be considered in a safety briefing prior to allowing personnel to work at the following:

(a) Hazardous material incidents

(b) Hazardous materials/WMD incidents involving criminal activities

(6) Identify the procedures for ensuring coordinated communication between responders and to the public

5.4.2 Preserving Evidence. Given two scenarios involving hazardous materials/WMD incidents, the operations level responder shall describe the process to preserve evidence as listed in the emergency response plan and/or standard operating procedures.

5.4.3 Initiating the Incident Command System. Given scenarios involving hazardous materials/WMD incidents, the operations level responder shall implement the incident command system as required by the AHJ by completing the following requirements:

(1) Identify the role of the operations level responder during hazardous materials/WMD incidents as specified in the emergency response plan and/or standard operating procedures

(2) Identify the levels of hazardous materials/WMD incidents as defined in the emergency response plan

(3) Identify the purpose, need, benefits, and elements of the incident command system for hazardous materials/WMD incidents

(4) Identify the duties and responsibilities of the following functions within the incident management system:

    (a) Incident safety officer

    (b) Hazardous materials branch or group

(5) Identify the considerations for determining the location of the incident command post for a hazardous materials/WMD incident

(6) Identify the procedures for requesting additional resources at a hazardous materials/WMD incident

(7) Describe the role and response objectives of other agencies that respond to hazardous materials/WMD incidents

5.4.4 Using Personal Protective Equipment. Given the personal protective equipment provided by the AHJ, the operations level responder shall describe considerations for the use of personal protective equipment provided by the AHJ by completing the following requirements:

(1) Identify the importance of the buddy system

(2) Identify the importance of the backup personnel

(3) Identify the safety precautions to be observed when approaching and working at hazardous materials/WMD incidents

(4) Identify the signs and symptoms of heat and cold stress and procedures for their control

(5) Identify the capabilities and limitations of personnel working in the personal protective equipment provided by the AHJ

(6) Identify the procedures for cleaning, disinfecting, and inspecting personal protective equipment provided by the AHJ

(7) Describe the maintenance, testing, inspection, and storage procedures for personal protective equipment provided by the AHJ according to the manufacturer's specifications and recommendations

5.5 Core Competencies—Evaluating Progress.

5.5.1 Evaluating the Status of Planned Response. Given two scenarios involving hazardous materials/WMD incidents, including the incident action plan, the operations level responder shall determine the effectiveness of the actions taken in accomplishing the response objectives and shall meet the following requirements:

(1) Identify the considerations for evaluating whether actions taken were effective in accomplishing the objectives

(2) Describe the circumstances under which it would be prudent to withdraw from a hazardous materials/WMD incident

5.5.2 Communicating the Status of Planned Response. Given two scenarios involving hazardous materials/WMD incidents, including the incident action plan, the operations level responder shall report the status of the planned response through the normal chain of command by completing the following requirements:

(1) Identify the procedures for reporting the status of the planned response through the normal chain of command

(2) Identify the methods for immediate notification of the incident commander and other response personnel about critical emergency conditions at the incident.

5.6 Competencies—Terminating the Incident. (Reserved)

■ Chapter 6 Competencies for Operations Level Responders Assigned Mission-Specific Responsibilities

6.2 Mission-Specific Competencies: Personal Protective Equipment.

6.2.1 General.

6.2.1.1 Introduction.

6.2.1.1.1 The operations level responder assigned to use personal protective equipment shall be that person, competent at the operations level, who is assigned to use personal protective equipment at hazardous materials/WMD incidents.

6.2.1.1.2 The operations level responder assigned to use personal protective equipment at hazardous materials/WMD incidents shall be trained to meet all competencies at the awareness level (*see Chapter 4*), all core competencies at the operations level (*see Chapter 5*), and all competencies in this section.

6.2.1.1.3 The operations level responder assigned to use personal protective equipment at hazardous materials/WMD incidents shall operate under the guidance of a hazardous materials technician, an allied professional, or standard operating procedures.

6.2.1.1.4 The operations level responder assigned to use personal protective equipment shall receive the additional training necessary to meet specific needs of the jurisdiction.

6.2.1.2 Goal. The goal of the competencies in this section shall be to provide the operations level responder assigned to use personal protective equipment with the knowledge and skills to perform the following tasks safely and effectively:

(1) Plan a response within the capabilities of personal protective equipment provided by the AHJ in order to perform mission-specific tasks assigned

(2) Implement the planned response consistent with the standard operating procedures and site safety and control plan by donning, working in, and doffing personal protective equipment provided by the AHJ

(3) Terminate the incident by completing the reports and documentation pertaining to personal protective equipment

6.2.2 Competencies —Analyzing the Incident. (Reserved)

6.2.3 Competencies — Planning the Response.

6.2.3.1 Selecting Personal Protective Equipment. Given scenarios involving hazardous materials/WMD incidents with known and unknown hazardous materials/WMD and the personal protective equipment provided by the AHJ, the operations level responder assigned to use personal protective equipment shall select the personal protective equipment required to support mission-specific tasks at hazardous materials/WMD incidents based on local procedures by completing the following requirements:

(1) Describe the types of personal protective equipment that are available for response based on NFPA standards and how these items relate to EPA levels of protection

(2) Describe personal protective equipment options for the following hazards:

    (a) Thermal

    (b) Radiological

(c) Asphyxiating

(d) Chemical

(e) Etiological/biological

(f) Mechanical

(3) Select personal protective equipment for mission-specific tasks at hazardous materials/WMD incidents based on local procedures

   (a) Describe the following terms and explain their impact and significance on the selection of chemical protective clothing:

      i. Degradation

      ii. Penetration

      iii. Permeation

   (b) Identify at least three indications of material degradation of chemical-protective clothing

   (c) Identify the different designs of vapor-protective and splash-protective clothing and describe the advantages and disadvantages of each type

   (d) Identify the relative advantages and disadvantages of the following heat exchange units used for the cooling of personnel operating in personal protective equipment:

      i. Air cooled

      ii. Ice cooled

      iii. Water cooled

      iv. Phase change cooling technology

   (e) Identify the physiological and psychological stresses that can affect users of personal protective equipment

   (f) Describe local procedures for going through the technical decontamination process

6.2.4 Competencies — Implementing the Planned Response.

6.2.4.1 Using Protective Clothing and Respiratory Protection.

Given the personal protective equipment provided by the AHJ, the operations level responder assigned to use personal protective equipment shall demonstrate the ability to don, work in, and doff the equipment provided to support mission specific tasks by completing the following requirements:

(1) Describe at least three safety procedures for personnel wearing protective clothing

(2) Describe at least three emergency procedures for personnel wearing protective clothing

(3) Demonstrate the ability to don, work in, and doff personal protective equipment provided by the AHJ

(4) Demonstrate local procedures for responders undergoing the technical decontamination process

(5) Describe the maintenance, testing, inspection, storage, and documentation procedures for personal protective equipment provided by the AHJ according to the manufacturer's specifications and recommendations

6.2.5 Competencies — Terminating the Incident.

6.2.5.1 Reporting and Documenting the Incident. Given a scenario involving a hazardous materials/WMD incident, the operations level responder assigned to use personal protective equipment shall document use of the personal protective equipment by completing the documentation requirements of the emergency response plan or standard operating procedures regarding personal protective equipment.

6.6 Mission-Specific Competencies: Product Control.

6.6.1 General.

6.6.1.1 Introduction.

6.6.1.1.1 The operations level responder assigned to perform product control shall be that person, competent at the operations level, who is assigned to implement product control measures at hazardous materials/WMD incidents.

6.6.1.1.2 The operations level responder assigned to perform product control at hazardous materials/WMD incidents shall be trained to meet all competencies at the awareness level (*see Chapter 4*), all core competencies at the operations level (*see Chapter 5*), all mission-specific competencies for personal protective equipment (*see Section 6.2*), and all competencies in this section.

6.6.1.1.3 The operations level responder assigned to perform product control at hazardous materials/WMD incidents shall operate under the guidance of a hazardous materials technician, an allied professional, or standard operating procedures.

6.6.1.1.4 The operations level responder assigned to perform product control at hazardous materials/WMD incidents shall receive the additional training necessary to meet specific needs of the jurisdiction.

6.6.1.2 Goal.

6.6.1.2.1 The goal of the competencies in this section shall be to provide the operations level responder assigned to product control at hazardous materials/WMD incidents with the knowledge and skills to perform the tasks in 6.6.1.2.2 safely and effectively.

6.6.1.2.2 When responding to hazardous materials/WMD incidents, the operations level responder assigned to perform product control shall be able to perform the following tasks:

(1) Plan an initial response within the capabilities and competencies of available personnel, personal protective equipment, and control equipment and in accordance with the emergency response plan or standard operating procedures by completing the following tasks:

   (a) Describe the control options available to the operations level responder

   (b) Describe the control options available for flammable liquid and flammable gas incidents

(3) Implement the planned response to a hazardous materials/WMD incident

6.6.2 Competencies—Analyzing the Incident. (Reserved)

6.6.3 Competencies—Planning the Response.

6.6.3.1 Identifying Control Options. Given examples of hazardous materials/WMD incidents, the operations level responder assigned to perform product control shall identify the options for each response objective by completing the following requirements as prescribed by the AHJ:

(1) Identify the options to accomplish a given response objective

(2) Identify the purpose for and the procedures, equipment, and safety precautions associated with each of the following control techniques:

   (a) Absorption

   (b) Adsorption

   (c) Damming

   (d) Diking

   (e) Dilution

   (f) Diversion

   (g) Remote valve shutoff

   (h) Retention

(i)   Vapor dispersion

(j)   Vapor suppression

6.6.3.2 Selecting Personal Protective Equipment. Given the personal protective equipment provided by the AHJ, the operations level responder assigned to perform product control shall select the personal protective equipment required to support product control at hazardous materials/WMD incidents based on local procedures (*see Section 6.2*).

6.6.4 Competencies—Implementing the Planned Response.

6.6.4.1 Performing Control Options. Given an incident action plan for a hazardous materials/WMD incident, within the capabilities and equipment provided by the AHJ, the operations level responder assigned to perform product control shall demonstrate control functions set out in the plan by completing the following requirements as prescribed by the AHJ:

(1)   Using the type of special purpose or hazard suppressing foams or agents and foam equipment furnished by the AHJ, demonstrate the application of the foam(s) or agent(s) on a spill or fire involving hazardous materials/WMD

(2)   Identify the characteristics and applicability of the following Class B foams if supplied by the AHJ:

(a)   Aqueous film-forming foam (AFFF)

(b)   Alcohol-resistant concentrates

(c)   Fluoroprotein

(d)   High-expansion foam

(3)   Given the required tools and equipment, demonstrate how to perform the following control activities:

(a)   Absorption

(b)   Adsorption

(c)   Damming

(d)   Diking

(e)   Dilution

(f)   Diversion

(g)   Retention

(h)   Remote valve shutoff

(i)   Vapor dispersion

(j)   Vapor suppression

(4)   Identify the location and describe the use of emergency remote shutoff devices on MC/DOT-306/406, MC/DOT-307/407, and MC-331 cargo tanks containing flammable liquids or gases

(5)   Describe the use of emergency remote shutoff devices at fixed facilities

6.6.4.2 The operations level responder assigned to perform product control shall describe local procedures for going through the technical decontamination process.

6.6.5 Competencies—Evaluating Progress. (Reserved)

6.6.6 Competencies—Terminating the Incident. (Reserved)

NFPA 1001 and 1072 Correlation Guide

Chapter 5, Fire Fighter I

NFPA 1001, *STANDARD FOR FIRE FIGHTER PROFESSIONAL QUALIFICATIONS*, 2013 EDITION	CORRESPONDING CHAPTER(S)	CORRESPONDING PAGE(S)
5.1	1, 3, 10, 30, 31, 32	5–6, 44–54, 256–286, 870–888, 898–904, 925–939
5.1.1	1, 2	6–8, 10, 29–30
5.1.2	1, 3, 10	9, 44–54, 73, 267–269, 271–288
5.2	4	90–91
5.2.1	4	91–97
5.2.1(A)	4	94
5.2.1(B)	4	99, 106–107
5.2.2	4	94–95, 106
5.2.2(A)	4	106
5.2.2(B)	4	106–107
5.2.3	4	100–105
5.2.3(A)	4	104–105
5.2.3(B)	4	103–104
5.2.4	18	579–580
5.2.4(A)	18	579–580
5.2.4(B)	18	579–580
5.3	3	49, 51, 54
5.3.1	3	62, 73
5.3.1(A)	3, 23	54–56, 58–64, 73, 686–689
5.3.1(B)	3, 16, 18	64–72, 75–76, 78–80, 491–492, 584–585, 587
5.3.2	11	295–296
5.3.2(A)	11	296–297
5.3.2(B)	11	296–297
5.3.3	11	297–300
5.3.3(A)	11	297–298
5.3.3(B)	11	297
5.3.4	9, 12	244–246, 314–315
5.3.4(A)	12	319–333, 337
5.3.4(B)	12	315, 321–328, 330–331, 334–340
5.3.5	18	576–577, 584–587
5.3.5(A)	18	575, 577–579, 581–582, 584–587
5.3.5(B)	18	582, 587
5.3.6	13	360–362
5.3.6(A)	13	351–356, 363–364

NFPA 1001, *STANDARD FOR FIRE FIGHTER PROFESSIONAL QUALIFICATIONS*, 2013 EDITION	CORRESPONDING CHAPTER(S)	CORRESPONDING PAGE(S)
5.3.6(B)	13	365–381, 383, 386–387, 389
5.3.7	23	697–699, 700
5.3.7(A)	23, 25	698–699, 734–738, 745–750
5.3.7(B)	23	698–702
5.3.8	9, 23, 29	245–246, 684, 695–697, 849–850
5.3.8(A)	23, 29	689–692, 695–697, 849–850
5.3.8(B)	19, 23, 29	621, 623, 687–691, 693, 696, 849–850
5.3.9	14	396–399, 405–406
5.3.9(A)	14	397–408, 412–413, 420
5.3.9(B)	14, 18	403–405, 408–424, 584–585, 587
5.3.10	23	682–686, 692–695
5.3.10(A)	6, 17, 23	139, 147–149, 540–541, 547–559, 692–694
5.3.10(B)	16, 17, 23	498–500, 542, 544–559, 685–686, 693–694
5.3.11	9, 15	245–246, 441–448
5.3.11(A)	6, 15	140–141, 146–151, 432–433, 436–437, 441, 445–449
5.3.11(B)	15	441–442
5.3.12	15	449–453, 456–457
5.3.12(A)	6, 7, 15	141–143, 150, 171–175, 436, 453, 459–464
5.3.12(B)	13, 15	384–386, 388, 452
5.3.13	9, 19, 29	246, 617–621, 853–854
5.3.13(A)	17, 19, 29	540–541, 618, 620–621, 853–854
5.3.13(B)	17, 19, 29	540–541, 618, 621–623, 685–686, 853–854, 856–859
5.3.14	19	617
5.3.14(A)	12, 19, 28	342, 602–609, 810, 834
5.3.14(B)	19, 29	604–617, 853–854
5.3.15	16	493, 496–499, 503–506, 509–523, 525–529, 531
5.3.15(A)	16	479–481, 484–487, 492–493
5.3.15(B)	16, 17	480–481, 486–488, 495, 512–529, 554–555
5.3.16	8	211–213, 220–222
5.3.16(A)	8	195–198, 202–211
5.3.16(B)	8	211, 213–221
5.3.17	19	598, 600–601
5.3.17(A)	19	598, 600–601
5.3.17(B)	19	600–601

NFPA 1001, *STANDARD FOR FIRE FIGHTER PROFESSIONAL QUALIFICATIONS*, 2013 EDITION	CORRESPONDING CHAPTER(S)	CORRESPONDING PAGE(S)
5.3.18	11, 23	299–300, 702
5.3.18(A)	11, 22	299–300, 702–703
5.3.18(B)	11, 22	299–300, 702
5.3.19	22	672–673
5.3.19(A)	22	664–665, 667–671
5.3.19(B)	22	665–666, 673–674
5.3.20	10	267–272, 281
5.3.20(A)	10	256–260, 263, 266–267
5.3.20(B)	10	267–269, 271–288
5.4	General	General
5.5	General	General
5.5.1	3, 9, 10, 13, 19	51, 54, 62, 73–82, 247–248, 263–266, 356–360, 613
5.5.1(A)	9, 10	247–248, 263–264
5.5.1(B)	9, 10	232–233, 247–248, 263
5.5.2	16	502
5.5.2(A)	16, 17	500–501, 503, 530–531
5.5.2(B)	16, 17	502, 542–548

Chapter 6, Fire Fighter II

NFPA 1001, *STANDARD FOR FIRE FIGHTER PROFESSIONAL QUALIFICATIONS*, 2013 EDITION	CORRESPONDING CHAPTER(S)	CORRESPONDING PAGE(S)
6.1	1	6
6.1.1	1, 5	6–7, 116–130
6.1.2	5	127–130
6.2	4	105–106
6.2.1	4, 5	105–107, 128
6.2.1(A)	4, 5	106–107, 128
6.2.1(B)	5	128
6.2.2	4	105
6.2.2(A)	4	105
6.2.2(B)	4	103
6.3	17	560
6.3.1	17	560
6.3.1(A)	17	560–564
6.3.1(B)	17	563–565
6.3.2	15, 23	449, 683, 692–693
6.3.2(A)	12, 15, 23	316–319, 453, 684
6.3.2(B)	12, 15, 23	316–319, 439, 684–686, 693–694
6.3.3	23	701–702
6.3.3(A)	23	700–702
6.3.3(B)	23	701–702
6.3.4	29	859–860
6.3.4(A)	29	849, 850, 852, 859–860
6.3.4(B)	29	851–853, 859–860
6.4	25, 26	737, 739–744, 747–752, 754–757

NFPA 1001, *STANDARD FOR FIRE FIGHTER PROFESSIONAL QUALIFICATIONS*, 2013 EDITION	CORRESPONDING CHAPTER(S)	CORRESPONDING PAGE(S)
6.4.1	25	745–746
6.4.1(A)	25	735, 738–742, 745–750
6.4.1(B)	25	737, 739–744, 747–752, 754–757
6.4.2	10, 26	260–262, 764, 771–773, 780–783
6.4.2(A)	9, 26	245, 766–783
6.4.2(B)	9, 26	245, 771–773
6.5	24, 27	712–713, 797–800
6.5.1	27	797–800
6.5.1(A)	27	791–796
6.5.1(B)	27	796, 800–801
6.5.2	27	790, 793–796
6.5.2(A)	27	800–803
6.5.2(B)	27	801–802
6.5.3	24	713–715, 719–726
6.5.3(A)	24, 28	713–715, 719–721, 810–828, 830, 833–840
6.5.3(B)	24, 28	713–715, 718–719, 810–840
6.5.4	9, 19	247–248, 601
6.5.4(A)	9, 19	247–248, 601
6.5.4(B)	9, 19	247, 601–602
6.5.5	16	502
6.5.5(A)	16	503
6.5.5(B)	16	503

NFPA 472, *STANDARD FOR COMPETENCE OF RESPONDERS TO HAZARDOUS MATERIALS/ WEAPONS OF MASS DESTRUCTION INCIDENTS*, 2013 EDITION	CORRESPONDING CHAPTER(S)	CORRESPONDING PAGE(S)
4.1	General	General
4.1.1	General	General
4.1.1.1	30	872–873
4.1.1.2	30	872–873
4.1.1.3	30	872–873
4.1.2	General	General
4.1.2.1	30, 31, 32	870–888, 898–904, 925–940
4.1.2.2	30, 31, 32	870–888, 898–904, 925–940
4.2	General	General
4.2.1	30, 31, 32	870, 873–874, 882–888, 899–903, 925–934
4.2.2	31	899–902
4.2.3	31, 32	899–902, 934
4.3	Reserved	Reserved
4.4	General	General
4.4.1	30, 31, 32	872–873, 903–916, 934–939, 942–943
4.4.2	31	934–936
4.5	General	General
4.65	General	General

NFPA 472, *STANDARD FOR COMPETENCE OF RESPONDERS TO HAZARDOUS MATERIALS/WEAPONS OF MASS DESTRUCTION INCIDENTS*, 2013 EDITION	CORRESPONDING CHAPTER(S)	CORRESPONDING PAGE(S)
5.1	30, 31	870, 925
5.1.1	30	870
5.1.1.1	30	870–873
5.1.1.2	30	871–873
5.1.1.3	30	873
5.1.2	32	925
5.1.2.1	General	General
5.1.2.2	31, 32	898–899, 925, 932–933, 935–937, 939
5.2	31	925
5.2.1	31	925
5.2.1.1	32	927–934
5.2.1.1.1	32	930–932
5.2.1.1.2	32	930–932
5.2.1.1.3	32	930, 933–934
5.2.1.1.4	32	930
5.2.1.1.5	32	927–929
5.2.1.1.6	32	926–928
5.2.1.1.7	32	932–933
5.2.1.2	32	927
5.2.1.2.1	32	930–932
5.2.1.2.2	32	926
5.2.1.3	32	926
5.2.1.3.1	32	928, 932
5.2.1.3.2	32	928
5.2.1.3.3	32	933–934
5.2.1.4	32	935
5.2.1.5	31	902
5.2.1.6	31	902–903
5.2.2	30, 31, 32	881, 899–902, 932–933
5.2.3	30	874–884, 887
5.2.4	30, 31, 32	881, 937
5.3	31, 32	916–917, 935
5.3.1	32	937
5.3.2	32	935
5.3.3	31, 32	904–916, 939–941
5.3.4	32	951–957
5.4	31, 32	916–917, 945–950
5.4.1	31, 32	916–917, 942–943, 945–950
5.4.2	32	951
5.4.3	32	934–935, 939
5.4.4	32	939–943
5.5	32	950
5.5.1	32	950

NFPA 472, *STANDARD FOR COMPETENCE OF RESPONDERS TO HAZARDOUS MATERIALS/WEAPONS OF MASS DESTRUCTION INCIDENTS,* 2013 EDITION	CORRESPONDING CHAPTER(S)	CORRESPONDING PAGE(S)
5.5.2	30, 31	934–935
5.6	32	950

NFPA 472, *STANDARD FOR COMPETENCE OF RESPONDERS TO HAZARDOUS MATERIALS/WEAPONS OF MASS DESTRUCTION INCIDENTS,* 2013 EDITION	CORRESPONDING CHAPTER(S)	CORRESPONDING PAGE(S)
6.2	General	General
6.2.1	General	General
6.2.1.1	General	General
6.2.1.1.1	31, 32	903–916, 939–943
6.2.1.1.2	30, 31, 32	870–871, 880–894, 898–914, 903–917, 934–935, 939–943
6.2.1.1.3	31, 32	903–916, 939–943
6.2.1.1.4	31, 32	903–916, 939–943
6.2.1.2	31, 32	903–916, 939–943
6.2.2	General	General
6.2.3	General	General
6.2.3.1	31, 32	903–916, 939–943
6.2.4	General	General
6.2.4.1	31, 32	903–916, 939–943
6.2.5	General	General
6.2.5.1	30, 31	903–908
6.6	30	870
6.6.1	30	870
6.6.1.1	30	870
6.6.1.1.1	30	870
6.6.1.1.2	30, 31, 32	872–888, 898–917, 925–945, 962–963
6.6.1.1.3	30	872–873
6.6.1.1.4	30	872–873
6.6.1.2	30	872–873
6.6.1.2.1	30	872–873
6.6.1.2.2	32	944–945
6.6.2	32	943–945
6.6.3	31, 32	910–916, 944–950
6.6.3.1	31, 32	910–916, 944–950
6.6.3.2	31, 32	904–916, 939–941
6.6.4	31, 32	910–916, 944–950
6.6.4.1	31, 32	910–916, 944–950
6.6.4.2	32	956–957
6.6.5	31, 32	910–916, 944–950
6.6.6	31, 32	916, 951

Pro Board Assessment Methodology Matrices for NFPA 1001

NFPA 1001 - Fire Fighter I - 2013 Edition

INSTRUCTIONS: In the column titled "Cognitive/Written Test" place the number of questions from the Test Bank that are used to evaluate the applicable JPR, RK, RS, or objective. In the column titled "Manipulative/Skill Station" identify the skill sheets that are used to evaluate the applicable JPR, RS, or objective. When the Portfolio or Project method is used to evaluate a particular JPR, RK, RS, or objective, identify the applicable section in the appropriate column and provide the procedures to be used as outlined in the NBFSPQ Operational Procedures, COA-5. Evaluation methods that are not cognitive, manipulative, portfolio, or project based should be identified in the "Other" column.

OBJECTIVE / JPR, RK, RS		COGNITIVE	MANIPULATIVE			
SECTION	ABBREVIATED TEXT	WRITTEN TEST	SKILLS STATION	PORTFOLIO	PROJECTS	OTHER
5.1	Requirements of NFPA 472, Operations Level; Chapter 5, Sec 6.6 (to include Sec 6.2)					Chapter 1 (p 5–10), Chapter 2 (29–30), Chapter 3 (p 44–54), Chapter 10 (p 265–286), Chapters 30, 31, 32 (p 870–888, 898–904, 925–939)
5.2.1	Initiate the response to a reported emergency					Chapter 4 (p 90–94, 96–97)
5.2.1(A)	RK: reporting an emergency					Chapter 4 (p 94)
5.2.1(B)	RS: operate communications equipment					Chapter 4 (p 99–107)
5.2.2	Receive a business or personal telephone call					Chapter 4 (p 94–95, 106)
5.2.2(A)	RK: Fire department procedures for answering nonemergency telephone calls					Chapter 4 (p 106)
5.2.2(B)	RS: fire station telephone and intercom equipment					Chapter 4, Skill Drill 4–4 (p 106–107)
5.2.3	Transmit and receive messages via the fire department radio					Chapter 4 (p 100–105)
5.2.3(A)	RK: Departmental radio procedures					Chapter 4 (p 104–105)
5.2.3(B)	RS: radio equipment and discriminate between routine and emergency traffic					Chapter 4, Skill Drill 4–3 (p 103)
5.2.4	Activate an emergency call for assistance					Chapter 18 (p 579–580)
5.2.4(A)	RK: Personal Accountability, emergency comms					Chapter 18 (p 579–580)
5.2.4(B)	RS: ability to initiate an emergency call for assistance					Chapter 18, Skill Drill 18–1 (p 579–580)
5.3.1	Use SCBA during emergency operations					Chapter 3 (p 62, 73)
5.3.1(A)	RK: respiratory protection					Chapter 3 (p 54–73)
5.3.1(B)	RS: use SCBA					Chapter 3 (p 64–80)
5.3.2	Respond on apparatus to an emergency scene					Chapter 11 (p 295–296)
5.3.2(A)	RK: riding fire apparatus					Chapter 11 (p 295–297)

OBJECTIVE / JPR, RK, RS		COGNITIVE	MANIPULATIVE			
SECTION	ABBREVIATED TEXT	WRITTEN TEST	SKILLS STATION	PORTFOLIO	PROJECTS	OTHER
5.3.2(B)	RS: use each piece of provided safety equipment					Chapter 11 (p 296–297)
5.3.3	Operate in established work areas at emergency					Chapter 11 (p 297–300)
5.3.3(A)	RK: Potential hazards at emergency scene					Chapter 11 (p 297–298)
5.3.3(B)	RS: Operate in protected area					Chapter 11 (p 297)
5.3.4	Force entry into a structure					Chapter 9, 12 (p 244–246, 314–315)
5.3.4(A)	RK: Basic construction					Chapter 12 (p 319–337)
5.3.4(B)	RS: transport and operate hand and power tools and to force entry					Chapter 12, Skill Drill 12–1, 12–2, 12–3, 12–4, 12–5, 12–6, 12–7, 12–8, 12–9, 12–10, 12–11, 12–12, 12–13 (p 321–328, 330–331, 334–340)
5.3.5	Exit a hazardous area as a team					Chapter 18 (p 576–577, 584–587)
5.3.5(A)	RK: Personnel accountability systems					Chapter 18 (p 577–578)
5.3.5(B)	RS: operate as a team					Chapter 18 (p 581–582)
5.3.6	Set up ground ladders					Chapter 13 (p 360–362)
5.3.6(A)	RK: parts and hazards of ladders					Chapter 13 (p 350–356)
5.3.6(B)	RS: carry raise and extend ladders					Chapter 13, Skill Drill 13–2, 13–3, 13–4, 13–5, 13–6, 13–7, 13–8, 13–9, 13–10, 13–11, 13–12, 13–13, 13–14, 13–15, 13–16, 13–17, 13–18, 13–19, 13–20 (p 365–388)
5.3.7	Attack a passenger vehicle fire operating as a member of a team					Chapter 23 (p 696–697)
5.3.7(A)	RK: fire streams					Chapter 23 (p 697–699)
5.3.7(B)	RS: automobile fire					Chapter 23 (p 697)
5.3.8	Extinguish fires in exterior Class A materials					Chapter 9, 23, 29 (p 245–246, 682, 695, 849–850)
5.3.8(A)	RK: attack lines and water streams					Chapter 23 (687–690, 694–696)
5.3.8(B)	RS: handlines or master streams					Chapter 23 (p 685–689, 692, 695)
5.3.9	Conduct a search and rescue in a structure operating as a member of a team					Chapter 14 (p 396–399, 405–406)

OBJECTIVE / JPR, RK, RS		COGNITIVE	MANIPULATIVE			
SECTION	ABBREVIATED TEXT	WRITTEN TEST	SKILLS STATION	PORTFOLIO	PROJECTS	OTHER
5.3.9(A)	RK: forcible entry tools					Chapter 9, 14 (p 248–249, 397–398, 401–408, 412–413, 420)
5.3.9(B)	RS: SCBA, set up ladders, rescue, and assess areas to determine tenability					Chapter 14, Skill Drill 14–3, 14–4, 14–5, 14–6, 14–7, 14–8, 14–9, 14–10, 14–11, 14–12, 14–13, 14–14, 14–15, 14–16, 14–17, 14–18, 14–19 (p 408–426)
5.3.10	Attack an interior structure fire operating as a member of a team					Chapter 23 (p 682–686, 692–696)
5.3.10(A)	RK: fire streams					Chapter 17, 23 (p 540–541, 547–550, 551–559, 692, 693–694)
5.3.10(B)	RS: water hammers; flow and patterns; apply water; and suppress interior wall and subfloor fires					Chapter 17, 23, Skill Drill 23–2, 23–3, 23–4, 23–9 (p 547–559, 685–686, 693–694)
5.3.11	Perform horizontal ventilation on a structure operating as part of a team					Chapter 9, 15 (p 245–246, 441–448)
5.3.11(A)	RK: ventilation;					Chapter 15 (p 432–441, 445–449)
5.3.11(B)	RS: ventilation tools and equipment					Chapter 15 (p 441–449)
5.3.12	Perform vertical ventilation on a structure operating as part of a team					Chapter 15 (p 449–453, 457–458)
5.3.12(A)	RK: heat transfer;					Chapter 6, 12, 15 (p 141–143, 150, 171–175, 436)
5.3.12(B)	RS: ventilation tools and equipment					Chapter 15 (p 458–470)
5.3.13	Overhaul a fire scene					Chapter 9, 19 (p 246, 617–621)
5.3.13(A)	RK: attack lines and water application devices					Chapter 17 (p 540–541)
5.3.13(B)	RS: attack line; expose void spaces; apply water; signs of area of origin and arson; and complete extinguishment					Chapter 17, 19, 29 (p 540–541, 621–623, 685–686, 853–858)
5.3.14	Conserve property as a member of a team					Chapter 19 (p 632, 636)
5.3.14(A)	RK: property conservation					Chapter 19, 28 (p 617–621, 810, 834)
5.3.14(B)	RS: covering materials;					Chapter 19, Skill Drill 19–11, 19–12, 19–13, 19–14 (p 613–619)
5.3.15	Connect a fire department pumper to a water supply as a member of a team					Chapter 16 (p 492–494, 496–501)
5.3.15(A)	RK: mobile water supply apparatus; fire hydrant operation; and suitable static water supply					Chapter 16 (p 479–481, 484–487, 491–492)
5.3.15(B)	RS: lay a supply hose,					Chapter 16 (p 514–526)
5.3.16	Extinguish incipient Class A, Class B, and Class C fires					Chapter 8 (p 211–213, 220–222)
5.3.16(A)	RK: classifications of fire;					Chapter 8 (p 195–211)
5.3.16(B)	RS: operate portable fire extinguishers					Chapter 8, Skill Drill 8–1, 8–2, 8–3, 8–4, 8–5, 8–6, 8–7, 8–8, 8–9 (p 211, 213–221)
5.3.17	Illuminate the emergency scene					Chapter 19 (p 598–601)
5.3.17(A)	RK: Safety principles					Chapter 19 (p 598, 600–601)

OBJECTIVE / JPR, RK, RS		COGNITIVE	MANIPULATIVE			
SECTION	ABBREVIATED TEXT	WRITTEN TEST	SKILLS STATION	PORTFOLIO	PROJECTS	OTHER
5.3.17(B)	RS: department power supply					Chapter 19, Skill Drill 19–1 (p 600–601)
5.3.18	Turn off building utilities					Chapter 11, 23 (p 299–300 701)
5.3.18(A)	RK: electricity, gas, and water systems;					Chapter 11, 23 (p 299–300, 702–703)
5.3.18(B)	RS: utility control devices					Chapter 11, 23 Skill Drill 23–13, 23–14 (p 299–300, 700)
5.3.19	Combat a ground cover fire operating as a member of a team					Chapter 22 (p 672–673)
5.3.19(A)	RK: ground cover fires					Chapter 22 (p 661–662, 664–669)
5.3.19(B)	RS: exposure threats					Chapter 22 (p 671–672)
5.3.20	tie a knot appropriate for hoisting tool					Chapter 10 (p 267–272, 279)
5.3.20(A)	RK: knot types and usage, diff between life safety and utility rope					Chapter 10 (p 255–260, 263–268)
5.3.20(B)	RS: ability to hoist tools using specific knots					Chapter 10, Skill Drill 10–5, 10–6, 10–7, 10–8, 10–9, 10–10, 10–11, 10–12, 10–13, 10–14, 10–15, 10–16, 10–17, 10–18, 10–19, 10–20, 10–21 (p 267–288)
5.5.1	Clean and check ladders, ventilation equipment, self-contained breathing apparatus (SCBA), ropes, salvage equipment, and hand tools					Chapter 3, 9, 10, 13, 19 (p 51, 54, 62, 72–82, 247–248, 262–265, 357–360, 613)
5.5.1(A)	RK: cleaning methods					Chapter 9, 10 (p 247–248, 263)
5.5.1(B)	RS: correct tools					Chapter 9, 10 (p 232–233, 247–248, 262)
5.5.2	Clean, inspect, and return fire hose to service					Chapter 16 (p 502–509, 510–521, 531)
5.5.2(A)	RK: noting a defective hose					Chapter 16 (p 501–502)
5.5.2(B)	RS: clean hose;					Chapter 16, Skill Drill 16–12 (p 502)

NFPA 1001 - Fire Fighter II - 2013 Edition

INSTRUCTIONS: In the column titled "Cognitive/Written Test" place the number of questions from the Test Bank that are used to evaluate the applicable JPR, RK, RS, or objective. In the column titled "Manipulative/Skill Station" identify the skill sheets that are used to evaluate the applicable JPR, RS, or objective. When the Portfolio or Project method is used to evaluate a particular JPR, RK, RS, or objective, identify the applicable section in the appropriate column and provide the procedures to be used as outlined in the NBFSPQ Operational Procedures, COA-5. Evaluation methods that are not cognitive, manipulative, portfolio, or project based should be identified in the "Other" column.

OBJECTIVE / JPR, RK, RS		COGNITIVE	MANIPULATIVE			
SECTION	ABBREVIATED TEXT	WRITTEN TEST	SKILLS STATION	PORTFOLIO	PROJECTS	OTHER
6.1.1	Fire Fighter I					Chapter 1, 5 (p 6, 116–130)
6.2.1	Complete a basic incident report					Chapter 5 (p 128)
6.2.1(A)	RK: basic incident reports					Chapter 5 (p 128)
6.2.1(B)	RS: complete reports					Chapter 5, Skill Drill 5–2 (p 128)

OBJECTIVE / JPR, RK, RS		COGNITIVE	MANIPULATIVE			
SECTION	ABBREVIATED TEXT	WRITTEN TEST	SKILLS STATION	PORTFOLIO	PROJECTS	OTHER
6.2.2	Communicate the need for team assistance					Chapter 4 (p 105)
6.2.2(A)	RK: Standard operating procedures					Chapter 4 (p 105)
6.2.2(B)	RS: operate fire department communications equipment					Chapter 4 (p 103)
6.3.1	Extinguish an ignitible liquid fire					Chapter 17 (p 560)
6.3.1(A)	RK: foam					Chapter 17 (p 560–564)
6.3.1(B)	RS: prepare a foam concentrate					Chapter 17, Skill Drill 17–17, 17–18, 17–19 (p 563–565)
6.3.2	Coordinate an interior attack line team's accomplishment of an assignment in a structure fire					Chapter 15, 23 (p 449, 681, 690, 692)
6.3.2(A)	RK: proper nozzle and hose					Chapter 15, 23 (p 453, 684)
6.3.2(B)	RS: assemble a team					Chapter 23, Skill Drill 22–1 (p 685)
6.3.3	Control a flammable gas cylinder fire operating as a member of a team					Chapter 23 (p 700–701)
6.3.3(A)	RK: Characteristics of pressurized flammable gases					Chapter 23 (p 700–701)
6.3.3(B)	RS: advances and retreats					Chapter 23, Skill Drill 22–12 (p 701–702)
6.3.4	Protect evidence of fire cause and origin					Chapter 29 (p 849–850)
6.3.4(A)	RK: assess origin and cause					Chapter 29 (p 849–853, 859–862)
6.3.4(B)	RS: locate the fire's origin area					Chapter 29 (p 852–853, 859–860)
6.4.1	Extricate a victim entrapped in a motor vehicle as part of a team					Chapter 25 (p 741–743, 745–757)
6.4.1(A)	RK: fire department's role at a vehicle accident					Chapter 25 (p 735, 738–741, 745–752)
6.4.1(B)	RS: operate hand and power tools					Chapter 25 Skill Drill 25–3, 25–4, 25–5, 25–6, 25–7, 25–8, 25–9, 25–10 (p 742–757)
6.4.2	Assist rescue operation teams					Chapter 10, 26 (p 260–262, 766–777)
6.4.2(A)	RK: fire fighter's role at a technicial rescue operation					Chapter 9, 26 (p 245, 764–770, 772–783)
6.4.2(B)	RS: identify and retrieve various types of rescue tools					Chapter 9, 26 (p 245, 771, 773)
6.5.1	Perform a fire safety survey in a occupied structure					Chapter 27 (p 797–801)
6.5.1(A)	RK: Organizational policy and procedures					Chapter 27 (p 791–797)
6.5.1(B)	RS: complete forms					Chapter 27 (p 796, 800–801)
6.5.2	Present fire safety information to station visitors or small groups					Chapter 27 (p 790, 793–796)
6.5.2(A)	RK: informational materials					Chapter 27 (p 800–803)
6.5.2(B)	RS: document presentations and using prepared materials					Chapter 27 (p 800–803)
6.5.3	Prepare a preincident survey					Chapter 24 (p 713, 719–726)

OBJECTIVE / JPR, RK, RS		COGNITIVE	MANIPULATIVE			
SECTION	ABBREVIATED TEXT	WRITTEN TEST	SKILLS STATION	PORTFOLIO	PROJECTS	OTHER
6.5.3(A)	RK: sources of water supply					Chapter 24, 28 (713–715, 719–721, 810–840)
6.5.3(B)	RS: components of fire suppression and detection systems					Chapter 28 (p 810–840)
6.5.4	Maintain power plants, power tools, and lighting equipment					Chapter 9, 19 (p 245–248, 601)
6.5.4(A)	RK: Types of cleaning methods					Chapter 9, 19 (p 252–253, 615)
6.5.4(B)	RS: correct tools					Chapter 19 (p 601)
6.5.5	Perform an annual service test on fire hose					Chapter 16 (p 502–503)
6.5.5(A)	RK: hose service testing					Chapter 16 (p 502–503)
6.5.5(B)	RS: hose testing equipment and nozzles and to record results					Chapter 16, Skill Drill 16–13, 16–14 (p 502–503)

Pro Board Assessment Methodology Matrices for NFPA 472

NFPA 472 - HazMat Awareness - 2013 Edition

INSTRUCTIONS: In the column titled "Cognitive/Written Test" place the number of questions from the Test Bank that are used to evaluate the applicable JPR, RK, RS, or objective. In the column titled "Manipulative/Skill Station" identify the skill sheets that are used to evaluate the applicable JPR, RS, or objective. When the Portfolio or Project method is used to evaluate a particular JPR, RK, RS, or objective, identify the applicable section in the appropriate column and provide the procedures to be used as outlined in the NBFSPQ Operational Procedures, COA-5. Evaluation methods that are not cognitive, manipulative, portfolio, or project based should be identified in the "Other" column.

OBJECTIVE / JPR, RK, RS		COGNITIVE	MANIPULATIVE			
SECTION	ABBREVIATED TEXT	WRITTEN TEST	SKILLS STATION	PORTFOLIO	PROJECTS	OTHER
4.2.1	Detecting the Presence of Hazardous Materials					Chapter 31 (p 898)
4.2.1(1)	definition of hazardous materials and WMD					Chapter 30 (p 870)
4.2.1(2)	hazard classes					Chapter 32 (p 934)
4.2.1(3)	primary hazards					Chapter 32 (p 934)
4.2.1(4)	difference between hazardous materials incidents and other emergencies					Chapter 30 (p 873–874)
4.2.1(5)	typical occupancies and locations					Chapter 31, 32 (p 899–902, 925–933)
4.2.1(6)	typical container shapes					Chapter 31, 32 (p 899–902, 925–933)
4.2.1(7)	facility and transportation markings and colors					Chapter 31, 32 (p 899–902, 925–933)
4.2.1(8)	NFPA 704					Chapter 32 (p 899–901)
4.2.1(9)	placards and labels					Chapter 31 (p 899)
4.2.1(10)	information on material safety data sheets					Chapter 31 (p 902)
4.2.1(11)	examples of clues					Chapter 31 (p 898)
4.2.1(12)	limitations of using the senses					Chapter 31 (p 898)
4.2.1(13)	four types of locations					Chapter 32 (p 925)
4.2.1(14)	difference between a chemical and a biological incident					Chapter 30 (p 882–888)
4.2.1(15)	indicators of possible criminal or terrorist activity involving chemical agents					Chapter 31 (p 902)
4.2.1(16)	indicators of possible criminal or terrorist activity involving biological agents					Chapter 31 (p 902–903)
4.2.1(17)	indicators of possible criminal or terrorist activity involving radiological agents					Chapter 31 (p 903)
4.2.1(18)	indicators of possible criminal or terrorist activity involving illicit laboratories					Chapter 31 (p 903)

OBJECTIVE / JPR, RK, RS		COGNITIVE	MANIPULATIVE			
SECTION	ABBREVIATED TEXT	WRITTEN TEST	SKILLS STATION	PORTFOLIO	PROJECTS	OTHER
4.2.1(19)	indicators of possible criminal or terrorist activity involving explosives					Chapter 31 (p 903)
4.2.1(20)	indicators of secondary device					Chapter 31, 32 (p 903, 937)
4.2.2	Surveying the Hazardous Materials Incident from a safe location.					Chapter 32 (p 932)
4.2.2(1)	determining the specific names					Chapter 31 (p 899–901)
4.2.2(2)	obtaining the names of, UN/NA identification numbers					Chapter 31 (p 899–901)
4.2.2(3)	obtaining the names of hazardous materials					Chapter 31 (p 899–901)
4.2.3	Collecting Hazard Information.					Chapter 31 (p 901–902)
4.2.3(1)	determining the appropriate guide page					Chapter 31 (p 899–900)
4.2.3(2)	general types of hazards					Chapter 32 (p 934)
4.4.1	Initiating Protective Actions					Chapter 32 (p 934–936)
4.4.1(1)	location of both the local emergency response plan and the organization's					Chapter 32 (p 934–936)
4.4.1(2)	role of the first responder					Chapter 30 (p 872)
4.4.1(3)	basic precautions					Chapter 30 (p 872–873)
4.4.1(4)	identificiation of response information					Chapter 32 (p 934–935)
4.4.1(5)	identify the recommended personal protective equipment					Chapter 31 (p 903–916)
4.4.1(6)	identify the definitions					Chapter 31 (p 903–916)
4.4.1(7)	identify the shapes of recommended initial isolation and protective action zones					Chapter 32 (p 942–943)
4.4.1(8)	describe the difference between small and large spills					Chapter 32 (p 942–943)
4.4.1(9)	identify circumstances under which distances are used at a haz mat incidents					Chapter 32 (p 942–943)
4.4.1(10)	describe the difference between the isolation distances and protective action distances					Chapter 32 (p 934–939)
4.4.1(11)	identify the techniques used to isolate the hazard area and deny entry					Chapter 32 (p 943)

OBJECTIVE / JPR, RK, RS		COGNITIVE	MANIPULATIVE			
SECTION	ABBREVIATED TEXT	WRITTEN TEST	SKILLS STATION	PORTFOLIO	PROJECTS	OTHER
4.4.1(12)	identify at least four specific actions necessary when an incident is suspected					Chapter 31 (p 916–918)
4.4.2	Initiating the Notification Process					Chapter 32 (p 934–936)

NFPA 472 - HazMat Operational - Core Competencies - 2013 Edition

INSTRUCTIONS: In the column titled "Cognitive/Written Test" place the number of questions from the Test Bank that are used to evaluate the applicable JPR, RK, RS, or objective. In the column titled "Manipulative/Skill Station" identify the skill sheets that are used to evaluate the applicable JPR, RS, or objective. When the Portfolio or Project method is used to evaluate a particular JPR, RK, RS, or objective, identify the applicable section in the appropriate column and provide the procedures to be used as outlined in the NBFSPQ Operational Procedures, COA-5. Evaluation methods that are not cognitive, manipulative, portfolio, or project based should be identified in the "Other" column.

OBJECTIVE / JPR, RK, RS		COGNITIVE	MANIPULATIVE			
SECTION	ABBREVIATED TEXT	WRITTEN TEST	SKILLS STATION	PORTFOLIO	PROJECTS	OTHER
5.1.1.2	Awareness level					Chapter 30 (p 870–872)
5.2.1	survey the incident					Chapter 31 (p 898)
5.2.1.1	identify shapes of containers					Chapter 32 (p 926–933)
5.2.1.1.1	identify each type of tank car					Chapter 32 (p 930–932)
5.2.1.1.2	identify each type of intermodal tank					Chapter 32 (p 930–932)
5.2.1.1.3	identify each type of cargo tank					Chapter 32 (p 929–931)
5.2.1.1.4	identify each type of storage tank					Chapter 32 (p 929–931)
5.2.1.1.5	identify each package by type					Chapter 32 (p 926–929)
5.2.1.1.6	identify each container/ package by type					Chapter 32 (p 926–927)
5.2.1.2	identify the markings that differentiate one container from another					Chapter 32 (p 927)
5.2.1.2.1	identify the vehicle or tank identification markings					Chapter 32 (p 930–932)
5.2.1.2.2	identify the markings indicating container size, product contained, and/or site identification numbers					Chapter 32 (p 926)
5.2.1.3	identify the name(s) of hazmats					Chapter 32 (p 927)
5.2.1.3.1	identify pipeline marker information					Chapter 32 (p 932)
5.2.1.3.2	identify pesticide label information					Chapter 32 (p 928)
5.2.1.3.3	identify radioactive material label information					Chapter 32 (p 933–934)
5.2.1.4	identify conditions surrounding a haz mat incident					Chapter 32 (p 935–936)
5.2.1.5	ways to verify information					Chapter 31 (p 902)
5.2.1.6	additional hazards associated with criminal or terrorist activities					Chapter 31 (p 902–903)
5.2.2	collect hazard and response information					Chapter 30, 31 (p 881–882, 901–902)
5.2.2(1)	definitions associated with the DOT hazard classes					Chapter 31 (p 899–900)
5.2.2(2)	ways to obtain a material safety data sheet					Chapter 31 (p 903)

OBJECTIVE / JPR, RK, RS		COGNITIVE	MANIPULATIVE			
SECTION	ABBREVIATED TEXT	WRITTEN TEST	SKILLS STATION	PORTFOLIO	PROJECTS	OTHER
5.2.2(3)	identify hazard and response information					Chapter 31 (p 903)
5.2.2(4)	Identify type, procedures, and information regarding CHEMTREC					Chapter 31 (p 904)
5.2.2(5)	methods of contacting the manufacturer					Chapter 31 (p 903)
5.2.2(6)	identify type of assistance available with respect to criminal or terrorist activities					Chapter 31 (p 902–903)
5.2.2(7)	identify procedure for contacting local, state, and federal authorities					Chapter 31 (p 902–903)
5.2.2(8)	describe properties or radioactive materials					Chapter 30 (p 880–881)
5.2.3	Predicting the Behavior of a Material and its Container					Chapter 30 (p 875–889)
5.2.3(1)	interpret the hazard and response information					Chapter 30 (p 887–898)
5.2.3(1)(a)	significance of chemical and physical properties on behavior of the container and/or its contents					Chapter 30 (p 875–880)
5.2.3(1)(b)	differences between terms					Chapter 30 (p 875–880)
5.2.3(2)	types of stress					Chapter 30 (p 875–880)
5.2.3(3)	ways in which containers can breach					Chapter 30 (p 875–880)
5.2.3(4)	ways in which containers can release their contents					Chapter 30 (p 875–880)
5.2.3(5)	dispersion patterns					Chapter 30 (p 875–880)
5.2.3(6)	time frames for predicting the length of time that exposures					Chapter 30 (p 889)
5.2.3(7)	health and physical hazards					Chapter 30 (p 880)
5.2.3(8)	health hazards					Chapter 30 (p 880)
5.2.3(9)	warfare agents					Chapter 30 (p 880)
5.2.4	Estimating the Potential Harm					Chapter 31 (p 916–917)
5.2.4(1)	resource for determining the size of an endangered area					Chapter 31 (p 916–917)
5.2.4(2)	number and type of exposures					Chapter 31 (p 916–917)
5.2.4(3)	resources available for determining the concentrations					Chapter 31 (p 916–917)
5.2.4(4)	factors for determining the extent of physical, health, and safety hazards					Chapter 31 (p 916–917)
5.2.4(5)	impact of time, distance and shielding					Chapter 31 (p 916–917)
5.3.1	Describing Response Objectives for Hazardous Materials Incidents					Chapter 32 (p 937–939)
5.3.1(1)	steps for determining the number of exposures					Chapter 32 (p 937–939)
5.3.1(2)	steps for determining defensive response objectives					Chapter 32 (p 935)
5.3.1(3)	assess risk to responder					Chapter 32 (p 937–939)
5.3.1(4)	assess potential for secondary attack					Chapter 32 (p 937–939)

	OBJECTIVE / JPR, RK, RS	COGNITIVE	MANIPULATIVE			
SECTION	ABBREVIATED TEXT	WRITTEN TEST	SKILLS STATION	PORTFOLIO	PROJECTS	OTHER
5.3.2	Identifying Action Options					Chapter 32 (p 937–939)
5.3.2(1)	options to accomplish objectives					Chapter 32 (p 937–939)
5.3.2(2)	prioritization of emergency medical care					Chapter 32 (p 937–939)
5.3.3	Determining Suitablity of Personal Protective Equipment					Chapter 31, 32 (p 903–916, 939–942)
5.3.3(1)	appropriate respiratory protection					Chapter 31, 32 (p 903–916, 939–942)
5.3.3(2)	appropriate personal protective clothing					Chapter 31, 32 (p 903–916, 939–942)
5.3.4	Identifying Decontamination Issues					Chapter 32 (p 951–957)
5.3.4(1)	ways that personnel, personal protective equipment, apparatus, and tools and equipment become contaminated					Chapter 32 (p 951)
5.3.4(2)	potential for secondary contamination					Chapter 32 (p 951)
5.3.4(3)	importance and limitations of decontamination procedures					Chapter 32 (p 951–957)
5.3.4(4)	purpose of emergency decontamination procedures					Chapter 32 (p 951–952)
5.3.4(5)	factors to be considered in emergency decontamination					Chapter 32 (p 951–952)
5.4.1	Establishing and Enforcing Scene Control Procedures					Chapter 31, 32 (916–917, 952)
5.4.1(1)	procedures for establishing scene control					Chapter 32 (p 952)
5.4.1(2)	criteria for determining the locations					Chapter 32 (p 952)
5.4.1(3)	basic techniques for the following protective actions					Chapter 31 (p 916–917)
5.4.1(4)	demonstrate emergency decontamination					Chapter 32 (p 951–952)
5.4.1(5)	identify safety items					Chapter 31, 32 (916–917, 952)
5.4.1(6)	procedures for ensuring coordinated communication					Chapter 31, 32 (916–917, 952)
5.4.2	Preserving Evidence					Chapter 32 (p 950–951)
5.4.3	Initiating ICS					Chapter 32 (p 934–935, 943)
5.4.3(1)	role of the first responder					Chapter 30 (p 872–876)
5.4.3(2)	levels of hazardous materials incidents					Chapter 32 (p 939)
5.4.3(3)	purpose, need, benefits, and elements of an incident management system					Chapter 32 (p 939)
5.4.3(4)	identify duties and responsibilities					Chapter 30 (p 872–876)
5.4.3(5)	considerations for determining location of command post					Chapter 32 (p 939)
5.4.3(6)	procedures for requesting additional resources					Chapter 32 (p 934–935, 943)
5.4.3(7)	Role and objectives of other agencies					Chapter 32 (p 934–935, 943)

OBJECTIVE / JPR, RK, RS		COGNITIVE	MANIPULATIVE			
SECTION	ABBREVIATED TEXT	WRITTEN TEST	SKILLS STATION	PORTFOLIO	PROJECTS	OTHER
5.4.4	Using Personal Protective Equipment					Chapter 32 (p 940–943)
5.4.4(1)	importance of the buddy system					Chapter 32 (p 943)
5.4.4(2)	importance of the backup personnel					Chapter 32 (p 943)
5.4.4(3)	identify safety precautions					Chapter 32 (p 934–935, 943)
5.4.4(4)	symptoms of heat and cold stress					Chapter 32 (p 940–943)
5.4.4(5)	physical capabilities required for, and the limitations of, personnel working in the personal protective equipment					Chapter 32 (p 940–941)
5.4.4(6)	procedures for cleaning, disinfecting, and inspecting personal protective equipment					Chapter 32 (p 957–958)
5.4.4(7)	maintenance, testing, inspections, storage of PPE					Chapter 32 (p 934–935, 943)
5.5.1	Evaluating the Status of Planned Response					Chapter 32 (p 950–951)
5.5.1(1)	considerations for evaluating whether actions were effective					Chapter 32 (p 950–951)
5.5.1(2)	circumstances under which it would be prudent to withdraw					Chapter 32 (p 950–951)
5.5.2	Communicating the Status of the Planned Response					Chapter 32 (p 934–935)
5.5.2(1)	methods for communicating the status of the planned response					Chapter 32 (p 934–935)
5.5.2(2)	methods for immediate notification					Chapter 32 (p 934–935)

NFPA 472 - HazMat Operational w/Personal Protective Equipment Mission Specific Competencies - 2013 Edition

INSTRUCTIONS: In the column titled 'Cognitive/Written Test' place the number of questions from the Test Bank that are used to evaluate the applicable JPR, RK, RS, or objective. In the column titled 'Manipulative/Skill Station' identify the skill sheets that are used to evaluate the applicable JPR, RS, or objective. When the Portfolio or Project method is used to evaluate a particular JPR, RK, RS, or objective, identify the applicable section in the appropriate column and provide the procedures to be used as outlined in the NBFSPQ Operational Procedures, COA-5. Evaluation methods that are not cognitive, manipulative, portfolio, or project based should be identified in the 'Other' column.

OBJECTIVE / JPR, RK, RS		COGNITIVE	MANIPULATIVE			
SECTION	ABBREVIATED TEXT	WRITTEN TEST	SKILLS STATION	PORTFOLIO	PROJECTS	OTHER
6.2.1.1.2	awareness level (Chapter 4), all core competencies at the operations level (Chapter 5), and all competencies in this section.					Chapter 30, 31, 32 (p 870–871, 880–894, 898–914, 934–935, 903–917, 939–943)
6.2.3.1	select PPE					Chapter 31, 32 (p 903–917, 939–943)
6.2.3.1(1)	Describe the types of protective clothing and equipment					Chapter 31, 32 (p 903–917, 939–943)
6.2.3.1(2)	Describe personal protective equipment options for hazards					Chapter 31, 32 (p 903–917, 939–943)
6.2.3.1(3)	Select personal protective equipment for mission-specific tasks at hazardous materials					Chapter 31, 32 (p 903–917, 939–943)
6.2.3.1(3)(a)	Describe terms and explain their impact and significance					Chapter 31, 32 (p 903–917, 939–943)

OBJECTIVE / JPR, RK, RS		COGNITIVE	MANIPULATIVE			
SECTION	ABBREVIATED TEXT	WRITTEN TEST	SKILLS STATION	PORTFOLIO	PROJECTS	OTHER
6.2.3.1(3)(b)	identify three indications of material degradation					Chapter 31, 32 (p 903–917, 939–943)
6.2.3.1(3)(c)	identify the different designs of vapor-protective and splash-protective clothing					Chapter 32 (p 907–908)
6.2.3.1(3)(d)	identify the relative advantages and disadvantages of heat exchange units					Chapter 32 (p 939–943)
6.2.3.1(3)(e)	identify physiological and psychological stresses					Chapter 32 (p 943)
6.2.3.1(3)(f)	describe local procedures for technical decontamination					Reserved
6.2.4.1	Using Protective Clothing and Respiratory Protection					Chapter 31, 32 (p 903–917, 939–943)
6.2.4.1(1)	Describe at least three safety procedures					Chapter 31, 32 (p 903–917, 939–943)
6.2.4.1(2)	Describe at least three emergency procedures					Chapter 31, 32 (p 903–917, 939–943)
6.2.4.1(3)	Demonstrate the ability to don, work in, and doff personal protective equipment					Chapter 32 (p 903–917)
6.2.4.1(4)	Demonstrate local procedures					Reserved
6.2.4.1(5)	Describe the maintenance, testing, inspection, storage, and documentation procedures					Chapter 31, 32 (p 903–917, 939–943)
6.2.5.1	Reporting and Documenting the Incident					Chapter 31 (p 903–908)

NFPA 472 - HazMat Operational w/Product Control Mission Specific Competencies - 2013 Edition

INSTRUCTIONS: In the column titled "Cognitive/Written Test" place the number of questions from the Test Bank that are used to evaluate the applicable JPR, RK, RS, or objective. In the column titled "Manipulative/Skill Station" identify the skill sheets that are used to evaluate the applicable JPR, RS, or objective. When the Portfolio or Project method is used to evaluate a particular JPR, RK, RS, or objective, identify the applicable section in the appropriate column and provide the procedures to be used as outlined in the NBFSPQ Operational Procedures, COA-5. Evaluation methods that are not cognitive, manipulative, portfolio, or project based should be identified in the "Other" column.

OBJECTIVE / JPR, RK, RS		COGNITIVE	MANIPULATIVE			
SECTION	ABBREVIATED TEXT	WRITTEN TEST	SKILLS STATION	PORTFOLIO	PROJECTS	OTHER
6.6.1.1.2	awareness level (Chapter 4), all core competencies at the operations level (Chapter 5), all mission-specific competencies for personal protective equipment (Section 6.2), and all competencies in this section.					Chapter 30, 31, 32 (p 872–873, 874–882, 898, 925–932, 899–902, 933, 934–935, 936–955, 962–963)
6.6.3.1	identify control options					Chapter 31 (p 910–916)
6.6.3.1(1)	Identify the options to accomplish a given response objective.					Chapter 31 (p 910–916)
6.6.3.1(2)	Identify the purpose for and the procedures					Chapter 31 (p 910–916)
6.6.3.2	select personal protective equipment					Chapter 31, 32 (p 903–917, 939–943)
6.6.4.1	perfrom control options					Chapter 31 (p 910–916)
6.6.4.1(1)	Using special purpose or hazard suppressing foams or agents					Chapter 31 (p 910–916)
6.6.4.1(2)	Identify the characteristics and applicability of Class B foams					Chapter 17 (p 561–562)

OBJECTIVE / JPR, RK, RS		COGNITIVE	MANIPULATIVE			
SECTION	ABBREVIATED TEXT	WRITTEN TEST	SKILLS STATION	PORTFOLIO	PROJECTS	OTHER
6.6.4.1(3)	demonstrate how to perform control activities:					Chapter 31 (p 910–916)
6.6.4.1(4)	Identify location and describe use of emergency remote shutoff devices					Chapter 31 (p 910–916)
6.6.4.1(5)	Describe the use of emergency remote shutoff devices					Chapter 32 (p 950)
6.6.4.2	describe local procedures for going through the technical decontamination process					Reserved

Glossary

911 dispatcher/telecommunicator Working from the communications centre, the dispatcher takes the calls from the public, sends appropriate units to the scene, assists callers with treatment instructions until the EMS unit arrives, and assists the IC with needed resources.

"A" posts Vertical support members that form the sides of the windshield of a motor vehicle.

"B" posts Vertical support members located between the front and rear doors of a motor vehicle.

"C" posts Vertical support members located behind the rear doors of a motor vehicle.

A tool A cutting tool with a pry bar built into the cutting part of the tool.

Absorption The process of applying a material (an absorbent) that will soak up and hold a hazardous material in a spongelike manner, for collection and disposal. Also the process by which hazardous materials travel through bodily tissues until they reach the bloodstream.

Accelerator A device that speeds up the removal of the air from a dry-pipe or preaction sprinkler system. An accelerator reduces the time it takes for water to start flowing from sprinkler heads.

Accidental Fire-cause classification that includes fires with a proven cause that does not involve a deliberate human act.

Accordion hose load A method of loading hose onto a vehicle in accordion-like sections. It is achieved by standing the hose on its edge, then placing the next fold on its edge, and so on.

Acid A material with a pH value of less than 7.

Activity logging system A device that keeps a detailed record of every incident and activity that occurs.

Acute health effect A health problem caused by relatively short exposure periods to a harmful substance, which produces observable conditions such as eye irritation, coughing, dizziness, and skin burns.

Adaptor A device that joins hose couplings of the same type, such as male to male or female to female.

Adjustable-gallonage fog nozzle A nozzle that allows the operator to select a desired flow from several settings.

Adsorption A process that occurs when a liquid hazardous material accumulates on the surface of a solid.

Advanced Care Paramedic (ACP) EMS personnel with the highest level of training, including cardiac monitoring, administering drugs, inserting advanced airways, manual defibrillation, and other advanced assessment and treatment skills.

Adz A hand tool constructed of a thin, arched blade set at right angles to the handle. It is used to chop brush for clearing a fire line or to mop up a wildland fire.

Aerating The process of inducing air into the foam solution, which expands and finishes the foam.

Aerial fuels Fuels located more than 1.8 m (6 ft.) off the ground, usually part of or attached to trees.

Aerial ladder A self-supporting, turntable-mounted, power-operated ladder of two or more sections permanently attached to a self-propelled automotive fire apparatus and designed to provide a continuous egress route from an elevated position to the ground. (NFPA 1901)

Agro-terrorism The intentional use of chemical or biological agents against the agricultural industry or food supply.

Air bill Shipping document for materials sent via air transport.

Air cylinder The component of the SCBA that stores the compressed air supply.

Air line The hose through which air flows, either within an SCBA or from an outside source to a supplied air respirator.

Air management The use of a limited air supply in such a way as to ensure that it will last long enough to enter a hazardous area, accomplish needed tasks, and return safely.

Air-purifying respirator (APR) Filtering device that removes particulates and contaminants from the air.

Air-sampling detector A system that captures a sample of air from a room or enclosed space and passes it through a smoke detection or gas analysis device.

Aircraft rescue fire fighter (ARFF) An individual with specialized training in aircraft fires, extrication of victims from aircraft, and extinguishing agents. These fire fighters wear special types of turnout gear and respond in fire apparatus that protect them from high-temperature fires.

Alarm initiation device An automatic or manually operated device in a fire alarm system that, when activated, causes the system to indicate an alarm condition.

Alarm notification device An audible and/or visual device in a fire alarm system that makes occupants or other persons aware of an alarm condition.

Alarm valve A valve that signals an alarm when a sprinkler head is activated and prevents nuisance alarms caused by pressure variations.

Alcohol-resistant foam Used for fighting fires involving water-soluble materials or fuels that are destructive to other types of foams. Some alcohol-resistant foams are capable of forming a vapour-suppressing aqueous film on the surface of hydrocarbon fuels. (NFPA 412)

Alpha particle A positively charged particle emitted by certain radioactive materials that quickly loses energy and can travel only 2.5–5.1 cm (1–2 in.) from its source. Clothing or a sheet of paper can stop this type of energy. Alpha particles are not dangerous to plants, animals, or people unless the alpha-emitting substance enters the body.

Alternative water sources Reliable water supply sources found in rural areas, such as ponds or lakes.

Alternative-powered vehicle A vehicle that is powered by compressed natural gas or other fuel.

Ammonium phosphate An extinguishing agent used in dry chemical fire extinguishers that can be used on Class A, B, and C fires.

Ammonium-nitrate fertilizer and fuel oil (ANFO) An explosive made of commonly available materials.

Annealed The process of forming standard glass.

Anthrax An infectious disease spread by the bacterium *B. anthracis*; it is typically found around farms, infecting livestock.

Arched roof A rounded roof usually associated with a bow-string truss.

Area of origin The room or area where a fire began. (NFPA 921)

Arson The crime of maliciously and intentionally, or recklessly, starting a fire or causing an explosion. (NFPA 921)

Arsonist A person who deliberately sets a fire to destroy property with criminal intent.

Asphyxiant A material that causes a victim to suffocate.

Assistant or division chief A midlevel chief who often has a functional area of responsibility, such as training, and who answers directly to the fire chief.

Atom The smallest particle of an element, which can exist alone or in combination.

Attack engine An engine from which attack lines have been pulled.

Attack hose Hose designed to be used by trained fire fighters and fire brigade members to combat fires beyond the incipient stage; also known as an attack line. (NFPA 1961)

Automatic location identification (ALI) An enhanced 911 service feature that displays where the call originated or where the phone service is billed.

Automatic number identification (ANI) An enhanced 911 service feature that shows the calling party's phone number on the display screen at the dispatcher's terminal.

Automatic sprinklers The working ends of a sprinkler system, which serve to activate the system and apply water to the fire.

Automatic sprinkler system A system of pipes filled with water under pressure that discharge water immediately when a sprinkler head opens.

Automatic-adjusting fog nozzle A nozzle that can deliver a wide range of water stream flows. It operates by means of an internal spring-loaded piston.

Auxiliary system A fire alarm system that sounds an alarm in the building and transmits a signal to the fire department via a public alarm-box system.

Awareness level (29 CFR 1910.12 First Responder at the Awareness Level) Personnel who, in the course of their normal duties, could encounter an emergency involving hazardous materials and weapons of mass destruction (WMDs) and who are expected to recognize the presence of the hazardous materials and WMDs, protect themselves, call for trained personnel, and secure the scene. (NFPA 472)

Awning windows Windows that have one large or two medium-size panels that are operated by a hand crank from the corner of the window.

Backdraft A phenomenon that occurs when a fire takes place in a confined area, such as a sealed aircraft fuselage, and burns undetected until most of the oxygen contained therein is consumed. The heat continues to produce flammable gases, mostly in the form of carbon monoxide. These gases are heated above their ignition temperature, and when a supply of oxygen is introduced (e.g., when normal entry points are opened), the gases ignite with explosive force. (NFPA 402)

Backfiring A planned operation to remove fuel from a wildland fire by burning out large selected areas. This operation should be carried out only under the close supervision of experienced, authorized fire officers.

Backpack pump extinguisher A portable fire extinguisher consisting of a 15- to 30 L (4- to 8-gal.) water tank that is worn on the user's back and features a hand-powered piston pump for discharging the water.

Backpack The harness of the SCBA, which supports the components worn by a fire fighter.

Backup entry team A dedicated team of fully qualified and equipped responders who are ready to enter the hot zone at a moment's notice to rescue any member of the hot-zone entry team.

Backup personnel Individuals who remove or rescue those working in the hot zone in the event of an emergency.

Ball valve A valve used on nozzles, gated wyes, and engine discharge gates. They consist of a ball with a hole in its middle.

Balloon-frame construction An older type of wood-frame construction in which the wall studs extend vertically from the basement of a structure to the roof without any fire stops.

Bam-bam tool A tool with a case-hardened screw, which is secured in the keyway of a lock and used to remove the keyway from the lock.

Bangor ladder A ladder equipped with tormentor poles or staypoles that stabilize the ladder during raising and lowering operations.

Bank To cover a fire to ensure it burns slowly.

Bankshot (or bank-down) method A method that applies the stream onto a nearby object, such as a wall, instead of directly aiming at the fire.

Base (bed) section The lowest or widest section of an extension ladder. (NFPA 1931)

Base A material with a pH value greater than 7.

Base station A stationary radio transceiver with an integral AC power supply. (NFPA 1221)

Batch mixing Pouring foam concentrate directly into the fire apparatus water tank, thereby mixing a large amount of foam at one time.

Battalion chief Typically the first level of fire chief, these chiefs are often in charge of running calls and supervising multiple stations or districts within a city; also called a district chief. A battalion chief is usually the officer in charge of a single-alarm working fire.

Battering ram A tool made of hardened steel with handles on the sides, which is used to force doors and to breach walls. Larger versions may be used by as many as four people; smaller versions are made for one or two people.

Beam The main structural side of a ground ladder. (NFPA 1931)

Beam detector A smoke detection device that projects a narrow beam of light across a large, open area from a sending unit to a receiving unit. When the beam is interrupted by smoke, the receiver detects a reduction in light transmission and activates the fire alarm.

Bend A knot used to join two ropes together.

Beta particles A type of radiation that is capable of travelling 3–4.6 m (10–15 ft.). Heavier materials, such as metal and glass, can stop this type of energy.

Bight A U shape created by bending a rope with the two sides parallel.

Bill of lading Shipping document for materials sent via roads and highways.

Bimetallic strip A thermostatic component made from two distinct metals that respond differently to heat, often found on rate-of-rise heat detectors; when heated, the metals will bend or change shape.

Biological agents Disease-causing bacteria, viruses, and other organisms that attack the human body.

Bite A small opening made to enable better tool access in forcible entry.

Black An area that has already been burned.

Black fire A hot, high-volume, high-velocity, turbulent, ultra-dense black smoke that indicates an impending flashover or autoignition.

Blister agent A chemical that causes the skin to blister.

Blistering agent A chemical that causes the skin to blister.

Block creel construction Rope constructed without knots or splices in the yarns, ply yarns, strands or braids, or rope. (NFPA 1983)

Blood agent A chemical that, when absorbed by the body, interferes with the transfer of oxygen from the blood to the cells.

Boiling liquid/expanding vapour explosion (BLEVE) An explosion that occurs when a tank containing a volatile liquid at the bottom and a flammable gas at the top is heated to the point of rupture.

Boiling point The temperature at which the vapour pressure of a liquid equals the surrounding atmospheric pressure. (NFPA 30)

Bolt cutter A cutting tool used to cut through thick metal objects, such as bolts, locks, and wire fences.

Booster hose (booster line) A noncollapsible hose used under positive pressure having an elastomeric or thermoplastic tube, a braided or spiralled reinforcement, and an outer protective cover. (NFPA 1962)

Booster hose A noncollapsible hose used under positive pressure with an elastomeric or thermoplastic tube, a braided or spiralled reinforcement, and a protective cover; also known as a booster line. (NFPA 1962)

Bowstring truss A truss that is curved on the top and straight on the bottom.

Box A burning structure.

Box-end wrench A hand tool used to tighten or loosen bolts. The end is enclosed, as opposed to an open-end wrench. Each wrench is a specific size, and most have ratchets for easier use.

Braided rope Rope constructed by intertwining strands in the same way that hair is braided.

Branch The organizational level having functional, geographical, or jurisdictional responsibility for major aspects of incident operations. (NFPA 1026)

Branch director A person in a supervisory position in either the operations or logistics function to provide a span of control. (NFPA 1561)

Breakaway type nozzle A nozzle with a tip that can be separated from the shut-off valve.

Bresnan distributor nozzle A nozzle that can be placed in confined spaces. The nozzle spins, spreading water over a large area.

Bulk storage container A large-volume container that has an internal volume greater than 450 L (119 gal.) for liquids or 400 kg (882 lb.) for solids, or a capacity of greater than 400 kg (882 lb.) for gases.

Bulkhead The wall that separates the engine compartment from the passenger compartment in a motor vehicle; also known as a firewall.

Bung A small opening in a closed-head drum.

Bunker coat The protective coat worn by a fire fighter for interior structural firefighting; also called a turnout coat.

Bunker pants The protective trousers worn by a fire fighter for interior structural firefighting; also called turnout pants.

Butt The end of the beam that is placed on the ground, or other lower support surface, when ground ladders are in the raised position. (NFPA 1931)

Butt plate An alternative to a simple butt spur; a swivelling plate with both a spur and a cleat or pad that is attached to the butt of the ladder.

Butt spurs That component of ground ladder support that remains in contact with the lower support surface to reduce slippage. (NFPA 1931)

Butterfly valve A valve found on the large pump intake where the suction hose connects to the inlet of the fire pump.

Call box A system of telephones connected by phone lines, radio equipment, or cellular technology to a communications centre or fire department.

Canadian Transport Emergency Centre (CANUTEC) A national advisory centre operated by the TDG Directorate of Transport Canada. CANUTEC maintains a scientific data bank on chemicals manufactured, stored, and transported in Canada and is staffed by professional scientists who specialize in emergency response and are experienced in interpreting technical information and providing advice.

Captain The second rank of promotion in the fire service (between the lieutenant and the battalion chief). A captain is responsible for managing a fire company and for coordinating the activities of that company with the other shifts.

Carabiner An auxiliary equipment system item; load-bearing connector with a self-closing gate used to join other components of life safety rope. (NFPA 1983)

Carbon monoxide (CO) A toxic gas produced through incomplete combustion.

Carbon-dioxide extinguishing system A fire-suppression system designed to protect either a single room or series of rooms by flooding the area with carbon dioxide.

Carbon-dioxide fire extinguisher A fire extinguisher that uses CO_2 gas as the extinguishing agent. It is rated for use on Class B and C fires.

Carboy A glass, plastic, or steel container ranging in volume from 19 to 57 L (5 to 15 gallons).

Carcinogen A cancer-causing substance identified in one of several published lists, including, but not limited to, the *NIOSH Pocket Guide to Chemical Hazards* (Centers for Disease Control and Prevention 2007), *Hazardous Chemicals Desk Reference*, 6th edition (Wiley-Blackwell, 2008), and *2013 TLVs® and BEIs®* (American Conference of Governmental Industrial Hygeienists 2013). (NFPA 1851)

Carpenter's handsaw A saw designed for cutting wood.

Carryall A piece of heavy canvas with handles, which can be used to tote debris, ash, embers, and burning materials out of a structure.

Cartridge/cylinder-operated fire extinguisher A fire extinguisher that has the expellant gas in a separate container from the extinguishing agent storage container. The storage container is pressurized by a mechanical action that releases the expellant gas.

Cascade system A method of piping air tanks together to allow air to be supplied to the SCBA fill station using a progressive selection of tanks, each with a higher pressure level. (NFPA 1901)

Case-hardened steel Steel created in a process that uses carbon and nitrogen to harden the outer core of a steel component, while the inner core remains soft. Case-hardened steel can be cut only with specialized tools.

Casement windows Windows in a steel or wood frame that open away from the building via a crank mechanism.

Ceiling hook A tool with a long wooden or fibreglass pole that has a metal point with a spur at right angles at one end. It can be used to probe ceilings and pull down plaster–lath material.

Cellar nozzle A nozzle used to fight fires in cellars and other inaccessible places. The device works by spreading water in a wide pattern as the nozzle is lowered through a hole into the cellar.

Census metropolitan area (CMA) An area consisting of one or more adjacent municipalities positioned around a major urban hub.

Central station An off-premises facility that monitors alarm systems and is responsible for notifying the fire department of an alarm. These facilities may be geographically located some distance from the protected building(s).

Chain of command A rank structure—spanning from the fire fighter through to the fire chief—for managing a fire department and fire-ground operations.

Chain of custody A legal term used to describe the paperwork or documentation describing the movement, storage, and custody of evidence, such as the record of possession of a gas can from a fire scene.

Chainsaw A power saw that uses the rotating movement of a chain equipped with sharpened cutting edges. It is typically used to cut through wood.

Chase Open space within walls for wires and pipes.

Chemical change The ability of a chemical to undergo an alteration in its chemical makeup, usually accompanied by the release of some form of energy.

Chemical energy Energy that is created or released by the combination or decomposition of chemical compounds.

Chemical Transportation Emergency Center (CHEMTREC®) A national call centre for basic chemical information, established by the Chemical Manufacturers Association and now operated by the ACC.

Chemical-pellet sprinkler head A sprinkler head activated by a chemical pellet that liquefies at a preset temperature.

Chemical-resistant material A material specifically designed to inhibit or resist the passage of chemicals into and through the material by the processes of penetration, permeation, or degradation.

Chief of the department The top position in the fire department. The fire chief has ultimate responsibility for the fire department and usually answers directly to the mayor or other designated public official.

Chief's trumpet An obsolete amplification device that enabled a chief officer to give orders to fire fighters during an emergency. It was a precursor to the bullhorn and portable radios.

Chisel A metal tool with one sharpened end that is used to break apart material in conjunction with a hammer, mallet, or sledgehammer.

Chlorine A yellowish gas that is approximately two-and-a-half times heavier than air and slightly water-soluble. It has many industrial uses but is also a choking agent and can damage the lungs when inhaled.

Choking agent A chemical designed to inhibit breathing. It is typically intended to incapacitate rather than kill.

Chronic health hazard A health problem that occurs as a result of long-term exposure to a substance; also known as a chronic health effect.

Churning Recirculation of exhausted air that is drawn back into a negative-pressure fan in a circular motion.

Circumstantial evidence The means by which alleged facts are proven by deduction or inference from other facts that were observed directly.

Clapper mechanism A mechanical device installed within a piping system that allows water to flow in only one direction.

Class A fire A fire in ordinary combustible materials such as wood, cloth, paper, rubber, and many plastics. (NFPA 10)

Class A foam Foam intended for use on Class A fires. (NFPA 1150)

Class B fire A fire in flammable liquids, combustible liquids, petroleum greases, tars, oils, oil-based paints, solvents, lacquers, alcohols, and flammable gases. (NFPA 10, 2007)

Class B foam Foam intended for use on Class B fires. (NFPA 1901)

Class C fire A fire that involves energized electrical equipment. (NFPA 10)

Class D fire A fire in combustible metals, such as magnesium, titanium, zirconium, sodium, lithium, and potassium. (NFPA 10)

Class I standpipe A standpipe system designed for use by fire department personnel only. Each outlet should have a valve to control the flow of water and a 64-mm (2.5-in.) male coupling for fire hose.

Class II standpipe A standpipe system designed for use by occupants of a building only. Each outlet is generally equipped with a length of 38-mm (1.5-in.) single-jacket hose and a nozzle, which are preconnected to the system.

Class III standpipe A combination system that has features of both Class I and Class II standpipes.

Class K fire A fire in a cooking appliance that involves combustible cooking media (i.e., vegetable or animal oils and fats). (NFPA 10)

Claw The forked end of a tool.

Claw bar A tool with a pointed claw-hook on one end and a forked- or flat-chisel pry on the other end. It is often used for forcible entry.

Clean agent Electrically nonconducting, volatile, or gaseous fire extinguishant that does not leave a residue upon evaporation. (NFPA 2001)

Clemens hook A multipurpose tool that can be used for several forcible entry and ventilation applications because of its unique head design.

Closed-circuit breathing apparatus SCBA designed to recycle the user's exhaled air. This system removes carbon dioxide and generates fresh oxygen.

Closet hook A type of pike pole intended for use in tight spaces, commonly 0.6–1.2 m (2–4 ft.) in length.

Cockloft The concealed space between the top-floor ceiling and the roof of a building.

Coded system A fire alarm system that divides a building or facility into zones and utilizes audible notification devices to identify the zone in which an alarm originated.

Cold zone (technical rescue incident) The control zone of an incident that contains the command post and such other support functions as are deemed necessary to control the incident. (NFPA 1500)

Combination attack A type of attack employing both direct attack and indirect attack methods.

Combination ladder A ground ladder that is capable of being used both as a stepladder and as a single or extension ladder. (NFPA 1931)

Combustibility A property that describes whether a material will burn and how quickly it will burn.

Combustion A chemical process of oxidation that occurs at a rate fast enough to produce heat and usually light in the form of either a glow or a flame. (NFPA 101)

Come along A hand-operated tool used for dragging or lifting heavy objects that uses pulleys and cables or chains to multiply a pulling or lifting force.

Command The first component of the IMS. It is the only position in the IMS that must always be staffed.

Command staff The PIO, safety officer, and liaison officer, all of whom report directly to the IC and are responsible for functions in the IMS that are not a part of the function of the line organization. (NFPA 1561)

Communications centre A building or portion of a building that is specifically configured for the primary purpose of providing emergency communications services or PSAP services to one or more public safety agencies under the authority or authorities having jurisdiction. (NFPA 1221)

Company officer The officer, or any other position of comparable responsibility in the department, in charge of a fire department company or station. (NFPA 1143)

Compartment A space completely enclosed by walls and a ceiling. The compartment enclosure is permitted to have wall openings to an adjoining space if the openings have a minimum lintel depth of 20 cm (8 in.) from the ceiling and the openings do not exceed 2.4 m (8 ft.) in width. A single opening of 91 cm (36 in.) or less in width without a lintel is permitted when there are no other openings to adjoining spaces. (NFPA 13)

Competent ignition source An ignition source that can ignite a fuel under the existing conditions at the time of the fire. It must have sufficient heat and be in close enough proximity to the fuel for a sufficient amount of time to ignite the fuel.

Compressed air foam (CAF) Class A foam produced by injecting compressed air into a stream of water that has been mixed with 0.1 percent to 1.0 percent foam.

Compressed air foam system (CAFS) A foam system that combines air under pressure with foam solution to create foam. (NFPA 1901)

Compressor A device used for increasing the pressure and density of a gas. (NFPA 853)

Computer-aided dispatch (CAD) A combination of hardware and software that provides data entry, makes resource recommendations, notifies and tracks those resources before, during, and after fire service alarms, preserving records of those alarms and status changes for later analysis. (NFPA 1221)

Conduction Heat transfer to another body, or within a body, by direct contact. (NFPA 921)

Confined space An area large enough and so configured that a member can bodily enter and perform assigned work, but which has limited or restricted means for entry and exit and is not designed for continuous human occupancy. (NFPA 1500)

Confinement Actions taken to keep a material, once released, in a defined or local area. (NFPA 472)

Conflagration A large fire, often involving multiple structures.

Consensus document A code document jointly developed by people representing various organizations and interests. NFPA codes and standards are consensus documents.

Consist A list of every car on a train.

Container Any vessel or receptacle that holds material, including storage vessels, pipelines, and packaging.

Containment Actions taken to keep a material in its container (e.g., stop the release of the material or reduce the amount being released). (NFPA 472)

Contaminated A term used to describe evidence that may have been altered from its original state.

Contamination The process of transferring a hazardous material from its source to people, animals, the environment, or equipment, all of which can act as carriers for the material.

Contemporary construction Buildings constructed since about 1970 that incorporate lightweight-construction techniques and engineered wood components. These buildings exhibit less resistance to fire than older buildings.

Control zone A designated area at a hazardous materials incident, based on safety and the degree of hazard, established to help limit the number of civilian and public service personnel who might be exposed to the hazardous materials.

Convection Heat transfer by circulation within a medium such as a gas or a liquid. (NFPA 921)

Conventional vehicle A vehicle that uses an internal combustion engine and is fuelled by either gasoline or diesel fuel.

Convulsant A chemical capable of causing convulsions or seizures when absorbed by the body.

Coping saw A saw designed to cut curves in wood.

Core temperature The temperature of the central parts of the body (i.e., heart, lungs, and vital organs).

Corrosivity The ability of a material to cause damage (on contact) to skin, eyes, or other parts of the body.

Council rake A long-handled rake constructed with hardened triangular-shaped steel teeth that is used for raking a fire line down to soil with no subsurface fuel, for digging, rolling burning logs, and cutting grass and small brush.

Coupling One set or pair of connection devices attached to a fire hose that allow the hose to be interconnected to additional lengths of hose or adapters and other fire fighting appliances. (NFPA 1963)

Crew A team of two or more fire fighters; also known as a company. (NFPA 1500)

Cribbing Short lengths of timber/composite materials, usually 10 by 10 cm (4 by 4 in.) and 46 by 61 cm (18 by 24 in.) long that are used in various configurations to stabilize loads in place or while a load is moving. (NFPA 1006)

Critical incident stress debriefing (CISD) A postincident meeting designed to assist rescue personnel in dealing with psychological trauma as the result of an emergency. (NFPA 1006)

Critical incident stress management (CISM) A program designed to reduce both acute and chronic effects of stress related to job functions. (NFPA 450)

Cross-zoned system A fire alarm system that requires activation of two separate detection devices before initiating an alarm condition. If a single detection device is activated, the alarm control panel will usually indicate a problem or trouble condition.

Crowbar A straight bar made of steel or iron with a forked chisel on the working end that is suitable for performing forcible entry.

Cryogenic liquid A gaseous substance that has been chilled to the point at which it liquefies; a liquid having a boiling point lower than −101.1°C (−150°F) at an absolute pressure of 101 kPa (14.7 psia); also known as a cryogen.

Curtain wall A non-load-bearing wall that separates the inside and outside of a building but is not part of the building's support structure.

Curved roof A roof with a curved shape.

Cutting tools Tools that are designed to cut into metal or wood.

Cutting torch A torch that produces a high-temperature flame capable of heating metal to its melting point, thereby cutting through an object. Because of the high temperatures (3148°C [5700°F]) that these torches produce, the operator must be specially trained before using this tool.

Cyanide A highly toxic chemical agent that attacks the circulatory system.

Cyberterrorism An intentional electronic attack against government or private computer systems.

Cylinder A portable compressed gas container, such as the body of the fire extinguisher where the extinguishing agent is stored.

Cylindrical locks The most common fixed locks in use today. The locks and handles are placed into predrilled holes in the doors. One side of the door will usually have a key-in-the-knob lock; the other will have a keyway, a button, or some other type of locking/unlocking mechanism.

Damming A process used when liquid is flowing in a natural channel or depression and its progress can be stopped by blocking the channel.

Dangerous cargo manifest Shipping document for materials sent via a marine vessel, typically located in a tube-like container.

Deadbolt Surface- or interior-mounted lock on or in a door with a bolt that provides additional security.

Dead load The weight of all materials of construction incorporated into a building, including, but not limited to, walls, floors, roofs, ceilings, stairways, built-in partitions, finishes, cladding, and other similarly incorporated architectural and structural items, as well as fixed service equipment, such as cranes. (NFPA 5000)

Decay phase The phase of fire development in which the fire has consumed either the available fuel or oxygen and is starting to die down.

Deck gun An apparatus-mounted master stream device that is intended to flow large amounts of water directly onto a fire or exposed building.

Decontamination The physical and/or chemical process of reducing and preventing the spread of contaminants from people, animals, the environment, or equipment involved at hazardous materials/WMD incidents. (NFPA 472)

Decontamination corridor The area where decontamination is performed, usually located within the warm zone. (NFPA 472)

Decontamination team The team responsible for reducing and preventing the spread of contaminants from persons and equipment used at a hazardous materials incident. Members of this team establish the decontamination corridor and conduct all phases of decontamination.

Defend-in-place A strategy in which the victims are protected from the fire without relocating them.

Defensible space An area as defined by the authority having jurisdiction (typically 9.1 m [30 ft.] wide or more) between an improved property and a potential wildland fire where combustible materials and vegetation have been removed or modified to reduce the potential for fire on improved property spreading to wildland fuels or to provide a safe working area for fire fighters protecting life and improved property from wildland fire. (NFPA 1051)

Defensive attack Exterior fire-suppression operations directed at protecting exposures.

Defensive objective A response objective that is primarily defensive in purpose and does not involve stopping the leak or release of a hazardous material.

Degradation The physical destruction or decomposition of a clothing material, owing to chemical exposure, general use, or ambient conditions (e.g., storage in sunlight).

Dehydration A state in which fluid losses are greater than fluid intake into the body. If left untreated, dehydration may lead to shock and even death.

Deluge sprinkler system A sprinkler system in which all sprinklers are open. When an initiation device, such as a smoke or heat detector, is activated, the deluge valve opens and water discharges from all of the open sprinkler heads simultaneously.

Deluge valve A valve assembly designed to release water into a sprinkler system when an external initiation device is activated.

Demonstrative evidence Materials that are used to demonstrate a theory or explain an event.

Department of Transportation (DOT) marking system A unique system of labels and placards that, in combination with the *ERG*, offers guidance for first responders operating at a hazardous materials incident in the United States.

Depressions Indentations felt on a kernmantle rope that indicate damage to the interior (kern) of the rope.

Depth of char The thickness of the layer of a material that has been consumed by a fire. The depth of char on wood may be used to help determine the intensity of a fire at a specific location.

Designated incident facilities Assigned locations where specific functions are always performed.

Dewar container A container designed to preserve the temperature of a cold liquid held inside.

Diking The placement of materials to form a barrier that will keep a hazardous material in liquid form from entering an area or that will hold the material in an area.

Dilution The process of adding some substance (usually water) to a product to weaken its concentration.

Direct attack (structure fire) Firefighting operations involving the application of extinguishing agents directly onto the burning fuel. (NFPA 1145)

Direct attack (wildland and ground fire) A method of fire attack in which fire fighters focus on containing and extinguishing the fire at its burning edge.

Direct evidence Evidence that is reported first-hand, such as statements from an eyewitness who saw or heard something.

Direct line A telephone that connects two predetermined points.

Discipline The guidelines that a department sets for fire fighters to work within.

Disinfection The process used to inactivate virtually all recognized pathogenic micro-organisms but not necessarily all microbial forms (e.g., bacterial endospores). (NFPA 1581)

Dispatch To send out emergency response resources promptly to an address or incident location for a specific purpose. (NFPA 450)

Dispatcher A trained individual who is responsible for answering requests for emergency and non emergency assistance form citizens. This individual assess the need for a response and alerts responders to the incident.

Disposal A two-step removal process for contaminated items that cannot be properly decontaminated. Items are bagged and placed in appropriate containers for transport to a hazardous waste facility.

Distributors Relatively small-diameter underground pipes that deliver water to local users within a neighbourhood.

Diversion Redirecting spilled or leaking material to an area where it will have less impact.

Division A supervisory level established to divide an incident into geographic areas of operations. (NFPA 1561)

Division of labour The breaking down of an incident or task into a series of smaller, more manageable tasks and assignment of personnel to complete those tasks.

Division supervisor A person in a supervisory position who is responsible for a specific geographic area of operations at an incident. (NFPA 1561)

Doff To take off an item of clothing or equipment.

Don To put on an item of clothing or equipment.

Door An entryway; the primary choice for forcing entry into a vehicle or structure.

Double-action pull-station A manual fire alarm activation device that requires two steps to activate the alarm: the user must push in a flap, lift a cover, or break a piece of glass before activating the alarm.

Double-female adaptor A hose adaptor used to join two male hose couplings.

Double-hung windows Windows that have two movable sashes that can go up and down.

Double-jacket hose A hose constructed with two layers of woven fibres.

Double-male adaptor A hose adaptor used to join two female hose couplings.

Double-pane glass A window design that traps air or inert gas between two pieces of glass to help insulate a house.

Drag rescue device (DRD) A component integrated within the turnout coat to aid in the rescue of an incapacitated fire fighter. (NFPA 1851)

Drum A barrel-like container used to store a wide variety of substances, including food-grade materials, corrosives, flammable liquids, and grease. Drums may be constructed of low-carbon steel, polyethylene, cardboard, stainless steel, nickel, or other hybrid materials.

Dry barrel hydrant A type of hydrant used in areas subject to below-freezing weather. The valve that allows water to flow into the hydrant is located underground, and the barrel of the hydrant is normally dry.

Dry bulk cargo tank A tanks designed to carry dry bulk goods, such as powders, pellets, fertilizers, or grain; they are generally V-shaped with rounded sides that funnel toward the bottom.

Dry chemical extinguishing system An automatic fire extinguishing system that discharges a dry chemical agent.

Dry chemical fire extinguisher An extinguisher that uses a mixture of finely divided solid particles to extinguish fires. The agent is usually sodium bicarbonate, potassium bicarbonate, or ammonium phosphate based, with additives being included to provide resistance to packing and moisture absorption and to promote proper flow characteristics. These extinguishers are rated for use on Class B and C fires, although some are also rated for Class A fires.

Dry hydrant An arrangement of pipe permanently connected to a water source other than a piped, pressurized water supply system, which provides a ready water supply for firefighting purposes and utilizes the drafting (suction) capability of a fire department pump. (NFPA 1142)

Dry powder fire extinguisher A fire extinguisher that uses an extinguishing agent in powder or granular form, such as sodium chloride and graphite-based powders. It is designed to extinguish Class D combustible-metal fires by crusting, smothering, or heat-transferring means.

Dry sprinkler system A sprinkler system in which the pipes are normally filled with compressed air. When a sprinkler head is activated, it releases the air from the system, which opens a valve so that pipes fill with water.

Dry-pipe valve The valve assembly on a dry sprinkler system that prevents water from entering the system until the air pressure is released.

Drywall hook A specialized version of a pike pole that can remove drywall more effectively because of its hook design.

Dual-path pressure reducer A feature that automatically provides a backup method for air to be supplied to the regulator of an SCBA if the primary passage malfunctions.

Duck-billed lock breaker A tool with a point that can be inserted into the shackles of a padlock. As the point is driven farther into the lock, it gets larger and forces the shackles apart until they break.

Dump valve A large opening from the water tank of a mobile water supply apparatus, used for unloading purposes. (NFPA 1901)

DuoDote® kit A prefilled auto-injector that provides two medications—atropine and pralidoxime chloride—for use in treating the effects of exposure to nerve agents and insecticide poisoning.

Duplex channel A radio system that uses two frequencies per channel: one frequency transmits and the other receives messages. Such a system uses a repeater site to transmit messages over a greater distance than is possible with a simplex system.

Dutchman A short fold placed in a hose when loading it into the bed, which prevents the coupling from turning in the hose bed.

Dynamic rope A rope generally made from synthetic materials that is designed to be elastic and stretch when loaded. Dynamic rope is often used by mountain climbers.

Early suppression fast-response (ESFR) sprinkler head A sprinkler head designed to react quickly and suppress a fire in its early stages.

Eco-terrorism Terrorism directed against causes that radical environmentalists think would damage the earth or its creatures.

Egress To go or come out; to exit from an area or a building.

Ejectors Also called smoke ejectors, electric fans used in negative-pressure ventilation.

Electric vehicle An automotive-type vehicle for on-road use, such as passenger automobiles, buses, trucks, vans, electric vehicles, and the like, primarily powered by an electric motor that draws current from a rechargeable storage battery, fuel cell, photovoltaic array, or other source of electric current. Electric motorcycles and similar-type vehicles and off-road, self-propelled electric vehicles, such as industrial trucks, hoists, lifts, transports, golf carts, airline ground-support equipment, tractors, boats, and the like, are not included in this definition. (NFPA 70)

Electrical energy Heat that is produced by electricity.

Electrolytes Certain salts and other chemicals that are dissolved in bodily fluids and cells. Proper levels of electrolytes need to be maintained for good health and strength.

Elevated master stream device A nozzle mounted on the end of an aerial device that is capable of delivering large amounts of water onto a fire or exposed building from an elevated position.

Elevated water-storage tower An above ground water-storage tank designed to maintain pressure in a water distribution system.

Elevation pressure The amount of pressure created by gravity; also known as head pressure.

Emergency decontamination The physical process of immediately reducing contamination of individuals in potentially life-threatening situations with or without the formal establishment of a decontamination corridor. (NFPA 472)

Emergency incident rehabilitation A function on the emergency scene that cares for the well-being of the fire fighters. It includes relief from climatic conditions, rest, cooling or warming, rehydration, calorie replacement, medical monitoring, member accountability, and release.

Emergency Medical Responder (EMR) The first trained individual to arrive at the scene of an emergency to provide initial medical care.

Emergency Medical Services (EMS) company A company that may be made up of medical units and first-response vehicles. Members of this company respond to and assist in the transport of medical and trauma victims to medical facilities. They often have medications, defibrillators, and paramedics who can stabilize a critical patient.

Emergency Medical Services (EMS) personnel Personnel who are responsible for administering prehospital care to people who are sick and/or injured. Prehospital calls make up the majority of responses in most fire departments, so fire fighters are often cross-trained as EMS personnel.

Emergency Response Guidebook (ERG) A reference book, written in plain language, to guide emergency responders in their initial actions at the incident scene. (NFPA 472)

Emergency traffic An urgent message, such as a call for help or evacuation, transmitted over a radio that takes precedence over all normal radio traffic.

Employee Assistance Programs (EAPs) Fire service programs that provide confidential help to fire fighters with personal issues.

Emulsification The process of neutralizing or reducing the harmful effects of a hazardous material by changing its chemical properties.

End-of-service-time-indicator (EOSTI) A warning device on an SCBA that alerts the user that the end of the breathing air is approaching.

Endothermic A reaction that absorbs heat or requires heat to be added.

Engine company A group of fire fighters who work as a unit and are equipped with one or more pumping engines with a rated capacity of 2839 litres per minute (750 gallons per minute [gpm]) or more. (NFPA 1410)

Entrapment A condition in which a victim is trapped by debris, soil, or other material and is unable to extricate himself or herself.

Escape rope A system component; a single-purpose, one-time use, emergency self-escape (self-rescue) rope; not classified as a life safety rope. (NFPA 1983)

Evacuation signal A distinctive signal intended to be recognized by the occupants as requiring evacuation of the building. (NFPA 72)

Evaporation The loss of body heat caused by the conversion of water or another fluid (often sweat) from a liquid to a gas.

Exhauster A device that accelerates the removal of the air from a dry-pipe or preaction sprinkler system.

Exit Drills in the Home (EDITH) A public safety program designed to teach occupants how to safely exit a house in the event of a fire or other emergency. It stresses the importance of having a preestablished meeting place for all family members.

Exothermic A reaction that results in the release of energy in the form of heat.

Expansion ratio A description of the volume increase that occurs when a liquid material changes to a gas.

Explosive limits The upper and lower concentration limits, at a specified temperature and pressure, of a flammable gas or vapour in the air, expressed as a percentage of the fuel by volume; also known as flammability limits.

Exposure The process by which people, animals, the environment, and equipment are subjected to or come in contact with a hazardous material. (NFPA 472)

Extension Fire that moves into areas not originally involved in the incident, including walls, ceilings, and attic spaces; also, the movement of fire into uninvolved areas of a structure.

Extension ladder A non-self-supporting ground ladder that consists of two or more sections travelling in guides, brackets, or the equivalent arranged so as to allow length adjustment. (NFPA 1931)

Exterior wall A wall—often made of wood, brick, metal, or masonry—that makes up the outer perimeter of a building. Exterior walls are often load bearing.

Extinguishing agent A material used to stop the combustion process. Extinguishing agents may include liquids, gases, dry chemical compounds, and dry powder compounds.

Extra (high) hazard locations Occupancies where the total amounts of Class A combustibles and Class B flammables are greater than expected in occupancies classed as ordinary (moderate) hazards.

Face piece A component of SCBA that fits over the face.

False alarm The activation of a fire alarm system when no fire or emergency condition exists.

Film-forming fluoroprotein (FFFP) A protein-based foam concentrate incorporating fluorinated surfactants that forms a foam capable of producing a vapour-suppressing aqueous film on the surface of hydrocarbon fuels. This foam can show an acceptable level of compatibility with dry chemicals and might be suitable for use with those agents. (NFPA 412)

Finance/administration section The section responsible for all costs and financial actions of the incident or planned event, including the time unit, procurement unit, compensation/claims unit, and the cost unit. (NFPA 1026)

Fine fuels Fuels that ignite and burn easily, such as dried twigs, leaves, needles, grass, moss, and light brush.

Finger A narrow point of fire whose extension is created by a shift in wind or a change in topography.

Fire A rapid, persistent chemical reaction that releases both heat and light.

Fire alarm annunciator panel Part of the fire alarm system that indicates the source of an alarm within a building.

Fire alarm control panel The component in a fire alarm system that controls the functions of the entire system.

Fire and life-safety education specialist A member of the fire department who deals with the public on education, fire safety, and juvenile fire-safety programs.

Fire apparatus driver/operator A member of the fire department who is authorized by the authority having jurisdiction to drive and/or operate fire department vehicles. (NFPA 1451)

Fire apparatus maintenance personnel People who repair and service fire and EMS vehicles, ensuring that they are always ready to respond to emergencies.

Fire code A set of legally adopted rules and regulations designed to prevent fires and protect lives and property in the event of a fire.

Fire department connection (FDC) A fire hose connection through which the fire department can pump water into a sprinkler system or standpipe system.

Fire Fighter I A person at the first level of progression who has demonstrated the knowledge and skills to function as an integral member of a firefighting team under direct supervision in hazardous conditions. (NFPA 1001)

Fire Fighter II A person at the second level of progression who has demonstrated the skills and depth of knowledge to function under general supervision. (NFPA 1001)

Fire helmet Protective head covering worn by fire fighters to protect the head from falling objects, blunt trauma, and heat.

Fire hook A tool used to pull down a burning structure.

Fire hydraulics The physical science of how water flows through a pipe or hose.

Fire load The weight of combustibles in a fire area (measured in sq. m or sq. ft) or on a floor in buildings and structures, including either contents or building parts, or both. (NFPA 914)

Fire mark Historically, an identifying symbol on a building informing fire fighters that the building was insured by a company that would pay them for extinguishing a fire.

Fire marshal/fire inspector/fire investigator A member of the fire department who inspects businesses and enforces laws that deal with public safety and fire codes. A fire investigator may also respond to a fire scene to help the IC investigate the cause of a fire. Investigators may have full police powers and deal directly with investigations and arrests.

Fire plug A valve installed to control water accessed from wooden pipes.

Fire point The lowest temperature at which a liquid will ignite and achieve sustained burning when exposed to a test flame in accordance with ASTM D92, *Standard Test Method for Flash and Fire Points by Cleveland Open Cup Tester.* (NFPA 704) Also known as flame point.

Fire police Members of the fire department who protect fire fighters by controlling traffic and securing the scene from public access. Many fire police are sworn peace officers as well as fire fighters.

Fire prevention Activities conducted to prevent fires and protect lives and property in the event of a fire. Fire prevention activities include the enactment and enforcement of fire codes, the inspection of properties, the presentation of fire safety education programs, and the investigation of the causes of fires.

Fire protection engineer A member of the fire department who is responsible for reviewing plans and working with building owners to ensure that the design of and systems for fire detection and suppression will meet applicable codes and function as needed.

Fire shelter An item of protective equipment configured as an aluminized tent utilized for protection, by means of reflecting radiant heat, in a fire-entrapment situation. (NFPA 1500)

Fire tetrahedron A geometric shape used to depict the four components required for a fire to occur: fuel, oxygen, heat, and chemical chain reactions.

Fire triangle A geometric shape used to depict the three components of which a fire is composed: fuel, oxygen, and heat.

Fire wall A fire division assembly with a fire-resistance rating of 3 test hours or longer built to permit the complete burn out and collapse of the structure on one side of the wall without the fire extending through the fire wall or the collapse of the fire wall itself. (NFPA 901)

Fire warden An individual charged with enforcing fire regulations in colonial Canada.

Fire window A window assembly rated in accordance with NFPA 257, *Standard on Fire Test for Window and Glass Block Assemblies,* and installed in accordance with NFPA 80, *Standard for Fire Doors and Other Opening Protectives.* (NFPA 5000)

Fire-ground command (FGC) An IMS developed in the 1970s for day-to-day fire department incidents generally handled with fewer than 25 units or companies.

Fire-resistive construction Buildings that have structural components of noncombustible materials with a specified fire resistance. Materials can include concrete, steel beams, and masonry-block walls. Fire-resistive construction is also known as Type I building construction, as defined in NFPA 220, *Standard on Types of Building Construction.*

FIRESCOPE (FIre RESources of California Organized for Potential Emergencies) An organization of agencies established in the early 1970s to develop a standardized system for managing fire resources at large-scale incidents such as wildland fires.

Fixed-gallonage fog nozzle A nozzle that delivers a set number of litres per minute (gallons per minute) as per the nozzle's design, no matter what pressure is applied to the nozzle.

Fixed-temperature heat detector A sensor that responds when its operating element is heated to a predetermined temperature.

Flame detector A sensor that detects the radiant energy emitted by a flame.

Flame point The lowest temperature at which a substance releases enough vapours to ignite and sustain combustion; also known as fire point.

Flameover The condition where unburned fuel (pyrolysate) from the originating fire has accumulated in the ceiling layer to a sufficient concentration (i.e., at or above the lower flammable limit) that it ignites and burns; it can occur without ignition of, or prior to the ignition of, other fuels separate from that which caused the event; also known as rollover. (NFPA 921)

Flammability limits The upper and lower concentration limits, at a specified temperature and pressure, of a flammable gas or vapour in the air, expressed as a percentage of the fuel by volume; also known as explosive limits.

Flammable range The range of concentrations between the UFL and LEL. (NFPA 68)

Flammable vapour A concentration of constituents in air that exceeds 10 percent of its LFL. (NFPA 115)

Flanking attack A direct method of suppressing a wildland or ground fire that involves placing a suppression crew on one flank of a fire.

Flashover A transition phase in the development of a compartment fire in which surfaces exposed to thermal radiation reach their ignition temperature more or less simultaneously and fire spreads rapidly throughout the space, resulting in full-room involvement or total involvement of the compartment or enclosed space. (NFPA 921)

Flash point The minimum temperature of a liquid at which sufficient vapour is given off to form an ignitable mixture with the air, near the surface of the liquid, or within the vessel used. (NFPA 30)

Flat bar A specialized type of prying tool made of flat steel with prying ends suitable for performing forcible entry.

Flat hose load A method of putting a hose on a vehicle in which the hose is laid flat and stacked on top of the previous section.

Flat roof A horizontal roof; often found on commercial or industrial occupancies.

Flat-head axe A tool that has a head with an axe on one side and a flat head on the opposite side.

Floodlight A light that can illuminate a broad area.

Floor runner A piece of canvas or plastic material, usually 91–122 cm (3–4 ft.) wide and available in various lengths, that is used to protect flooring from dropped debris and dirt from shoes and boots.

Flow pressure The amount of pressure created by moving water.

Flow switch An electrical switch that is activated by water moving through a pipe in a sprinkler system.

Fluoroprotein foam A protein-based foam concentrate to which fluorochemical surfactants have been added. This has the effect of giving the foam a measurable degree of compatibility with dry chemical extinguishing agents and an increase in tolerance to contamination by fuel. (NFPA 402)

Fly section Any section of an aerial telescoping device beyond the base section. (NFPA 1901)

Foam concentrate A concentrated liquid foaming agent as received from the manufacturer. (NFPA 11)

Foam eductor A device placed in the hose line that draws foam concentrate from a container and introduces it into the fire stream.

Foam injector A device installed on a fire pump that meters out foam by pumping or injecting it into the fire stream.

Foam proportioner A device or method to add foam concentrate to water to make foam solution. (NFPA 1901)

Foam solution A homogeneous mixture of water and foam concentrate in the proper proportions. (NFPA 1901)

Fog-stream nozzle A nozzle that is placed at the end of a fire hose and separates water into fine droplets to aid in heat absorption.

Folding ladder A single-section ladder with rungs that can be folded or moved to allow the beams to be brought into a position touching or nearly touching each other. (NFPA 1931)

Forcible entry Techniques used by fire personnel to gain entry into buildings, vehicles, aircraft, or other areas of confinement when normal means of entry are locked or blocked. (NFPA 402)

Forward lay A method of laying a supply line where the line starts at the water source and ends at the attack engine.

Forward staging area A strategically placed area, close to the incident site, where personnel and equipment can be held at the ready for rapid response to an emergency event.

Four-way hydrant valve A specialized type of valve that can be placed on a hydrant to allow another engine to increase the supply pressure without interrupting flow.

Frangible-bulb sprinkler A sprinkler with a liquid-filled bulb. The sprinkler becomes activated when the liquid is heated and the glass bulb breaks.

Freelancing The dangerous practice of acting independently of command instructions.

Freight bill Shipping document for materials sent via roads and highways.

Fresno ladder A narrow, two-section extension ladder that has no halyard. Because of its limited length, it can be extended manually.

Friction loss The reduction in pressure caused by the water's contact with the side of the hose. This contact requires additional force to overcome the drag created by the hose wall.

Frostbite A localized condition that occurs when the layers of the skin and deeper tissue freeze. (NFPA 704)

Fuel A material that will maintain combustion under specified environmental conditions. (NFPA 53)

Fuel cell An electrochemical system that consumes fuel to produce an electric current. The main chemical reaction used in a fuel cell for producing electric power is not combustion, but sources of combustion may be used within the overall fuel cell system, such as reformers/fuel processors. (NFPA 70)

Fuel cell stack Many fuel cells assembled into a group to produce larger quantities of electricity.

Fuel compactness The extent to which fuels are tightly packed together.

Fuel continuity The relative closeness of wildland fuels, which affects a fire's ability to spread from one area of fuel to another.

Fuel moisture The amount of moisture present in a fuel, which affects how readily the fuel will ignite and burn.

Fuel volume The amount of fuel present in a given area.

Fully developed phase The phase of fire development in which the fire is free-burning and consuming much of the fuel.

Fully encapsulating suit A protective suit that completely covers the fire fighter, including the breathing apparatus, and does not let any vapour or fluids enter the suit. It is commonly used in hazardous materials emergencies.

Fusible-link sprinkler A sprinkler with an activation mechanism that incorporates two pieces of metal held together by low-melting-point solder. When the solder melts, it releases the link and water begins to flow.

Gamma radiation A type of radiation that can travel significant distances, penetrate most materials, and pass through the body. Gamma rays are the most destructive type of radiation to the human body.

Gas A material that has a vapour pressure greater than 300 kPa (43.5 psi) absolute at 50°C (122°F) or is completely gaseous at 20°C (68°F) at a standard pressure of 101.3 kPa (14.7 psi) absolute. (NFPA 30A)

Gas detector A device that detects and/or measures the concentration of dangerous gases.

Gate valve A valve found on hydrants and sprinkler systems.

Gated wye A valved device that splits a single hose into two separate hoses, so that each hose can be turned on and off independently.

General use life safety rope A life safety rope that is no larger than 16 mm (0.7 m.) and no smaller than 11 mm (0.5 m.), with a minimum breaking strength of 40 kN(8992 lbf).

Generator An electromechanical device for the production of electricity. (NFPA 1901)

Geographic information system (GIS) A system of computer software, hardware, data, and personnel to describe information tied to a spatial location. (NFPA 450)

Glass blocks Thick pieces of glass similar to bricks or tiles.

Glazed A glass or transparent or translucent plastic sheet used in windows, doors, skylights, or curtain walls. [ASCE/SEI 7:6.2] (NFPA 5000)

Glucose The source of energy for the body. One of the basic sugars, it is the body's primary fuel, along with oxygen.

Governance The process by which an organization exercises authority and performs the functions assigned to it.

Grade A measurement of the angle used in road design and expressed as a percentage of elevation change over distance. (NFPA 1901)

Gravity-feed system A water distribution system that depends on gravity to provide the required pressure. The system storage is usually located at a higher elevation than the end-users.

Green An area of unburned fuels.

Gripping pliers A hand tool with a pincer-like working end that can be used to bend wire or hold smaller objects.

Gross decontamination The phase of the decontamination process during which the amount of surface contaminants is significantly reduced. (NFPA 472)

Ground cover fire A fire that burns loose debris on the surface of the ground.

Ground duff Partly decomposed organic material on a forest floor; a type of light fuel.

Group A supervisory level established to divide an incident into functional areas of operation. (NFPA 1561)

Group supervisor A person in a supervisory position who is responsible for a functional area of operation. (NFPA 1561)

Growth phase The phase of fire development in which the fire spreads beyond the point of origin and begins to involve other fuels in the immediate area.

Guide line A rope used for orientation when fire fighters are inside a structure where there is low or no visibility. The line is attached to a fixed object outside the hazardous area.

Guides Strips of metal or wood that serve to guide a fly section during extension. Channels or slots in the bed or fly section may also serve as guides.

Gusset plate A connecting plate used in trusses, typically made of wood or lightweight metal.

Gypsum A naturally occurring material consisting of calcium sulphate and water molecules.

Gypsum board The generic name for a family of sheet products consisting of a noncombustible core made primarily of gypsum with paper surfacing. (NFPA 5000)

Hacksaw A cutting tool designed for use on metal. Different blades can be used for cutting different types of metal.

Halligan tool A prying tool that incorporates a pick and a fork, specifically designed for use in the fire service.

Halogenated agent fire extinguisher An extinguisher that uses a halogenated extinguishing agent.

Halogenated extinguishing agent A liquefied gas extinguishing agent that puts out fires by chemically interrupting the combustion reaction between the fuel and oxygen.

Halon 1211 A halogenated agent whose chemical name is bromochlorodifluoromethane, $CBrClF_2$; it is a multipurpose, Class ABC-rated agent effective against flammable liquid fires. (NFPA 408)

Halon 1301 A liquefied-gas extinguishing agent that puts out a fire by chemically interrupting the combustion reaction between fuel and oxygen. Halon agents leave no residue.

Halyard Rope used on extension ladders for the purpose of raising a fly section(s). (NFPA 1931)

Hammer A striking tool.

Hand light A small, portable light carried by fire fighters to improve visibility at emergency scenes; it is often powered by rechargeable batteries.

Handle The grip used for holding and carrying a portable fire extinguisher.

Handline nozzle A nozzle with a rated discharge of less than 1591 L/min. (350 gpm). (NFPA 1964)

Handsaw A manually powered saw designed to cut different types of materials. Examples include hacksaws, carpenter's handsaws, keyhole saws, and coping saws.

Hard suction hose A hose used for drafting water from static supplies (e.g., lakes, rivers, and wells). It can also be used for supplying fire apparatus pumps from hydrants, if designed for that purpose. The hose contains a semirigid or rigid reinforcement. (NFPA 1963)

Hardware The parts of a door or window that enable it to be locked or opened.

Harness A piece of equipment worn by a rescuer that can be attached to a life safety rope.

Hazard The potential for harm or damage to people, property, or the environment. (NFPA 1710)

Hazardous material A substance that, when released, is capable of creating harm to people, the environment, and property. (NFPA 472)

Hazardous materials branch The function within an overall IMS that deals with the mitigation and control of the hazardous materials/WMD portion of an incident. (NFPA 472)

Hazardous materials branch director/group supervisor Commanders of hazardous materials incidents beyond the operations level.

Hazardous materials company A fire company that responds to and controls scenes where hazardous materials have spilled or leaked. Responders wear special suits and are trained to deal with most chemicals.

Hazardous materials information research team A dedicated team of responders who serve as an information gathering and referral point for the IC as well as the hazardous materials officer.

Hazardous Materials Information System (HMIS) A colour-coded marking system used in the United States by which employers give their personnel the necessary information to work safely around chemicals.

Hazardous materials officer The person responsible for directing and coordinating all operations involving hazardous materials/WMDs as assigned by the incident commander; also known as hazardous materials branch director/group supervisor (NIMS). (NFPA 472)

Hazardous materials safety officer The person who is responsible for ensuring the safety of hazardous materials personnel and ensuring that appropriate hazardous materials/weapons of mass destruction practices are followed.

Hazardous materials technician A person who responds to hazardous materials or weapons of mass destruction incidents using a risk-based response process by which they analyze the problem at hand, select applicable decontamination procedures, and control a release while using specialized protective clothing and control equipment. (NFPA 472)

Hazardous materials technician with specialties A hazardous materials technician with training in areas such as specialized chemicals, containers, and their uses.

Hazardous waste Waste that is potentially damaging to the environment or human health, due to its toxicity, ignitability, corrosivity, or chemical reactivity, or another cause. (NFPA 80)

Head of the fire The main or running edge of a fire; the part of the fire that spreads with the greatest speed.

Heads-up display (HUD) A visual display of information and system conditions status that is visible to the wearer of the SCBA.

Heat detector A fire alarm device that detects an abnormally high temperature, an abnormally high rate of rise in temperature, or both.

Heat exhaustion A mild form of shock caused when the circulatory system begins to fail as a result of the body's inadequate effort to give off excessive heat.

Heat sensor label A label that changes colour at a preset temperature to indicate a specific heat exposure. (NFPA 1931)

Heat stroke A severe and sometimes fatal condition resulting from the failure of the temperature-regulating capacity of the body. It is caused by prolonged exposure to the sun or high temperatures. Reduction or cessation of sweating is an early symptom; a body temperature of 40.5°C (105°F) or higher, a rapid pulse, hot and dry skin, a headache, confusion, unconsciousness, and convulsions can also occur. Heatstroke is a true medical emergency requiring immediate transport to a medical facility.

Heating, ventilation, and air conditioning (HVAC) system A system to manage the internal environment that is often found in large buildings.

Heavy fuels Fuels of a large diameter, such as large brush, heavy timber, snags, stumps, branches, and dead timber on the ground. These fuels ignite and are consumed more slowly than light fuels.

Heel of the fire The side opposite the head of the fire, which is often close to the area of origin.

Higbee indicators Indicators on male and female threaded couplings that identify where the threads begin. These indicators should be aligned before fire fighters start to thread the couplings together.

High temperature–protective equipment Protective equipment designed to safeguard the wearer from high temperatures during short-term exposures.

High-angle operation A rope rescue operation where the angle of the slope is greater than 45 degrees. In this scenario, rescuers depend on life-safety rope rather than a fixed support surface, such as the ground.

High-efficiency particulate air (HEPA) filter A filter capable of catching particles as small as 0.3 micron—much smaller than a typical dust of alpha radiation particle.

Hitch A knot that attaches to or wraps around an object so that when the object is removed, the knot will fall apart. (NFPA 1670, *Standard on Operations and Training for Technical Search and Rescue Incidents*)

Hockey puck lock A type of lock with hidden shackles that cannot be forced open through conventional methods.

Hollow-core door A door made of panels that are honeycombed inside, creating an inexpensive and lightweight design.

Horizontal evacuation Moving occupants from a dangerous area to a safe area on the same floor level.

Horizontal ventilation The process of making openings on the same level as the fire so that smoke, heat, and gases can escape horizontally from a building through openings such as doors and windows.

Horizontal-sliding windows Windows that slide open horizontally.

Horn The tapered discharge nozzle of a CO_2–type fire extinguisher.

Horseshoe hose load A method of loading hose in which the hose is laid into the bed along its three walls, so that it resembles a horseshoe.

Hose appliance A piece of hardware (other than a nozzle) that is generally intended to be connected to a fire hose to control or convey water. (NFPA 1962)

Hose clamp A device used to compress a fire hose so as to stop water flow.

Hose jacket A device used to stop a leak in a fire hose or to join hoses that have damaged couplings.

Hose liner The inner portion of a hose that is in direct contact with the flowing water; also known as the hose inner jacket.

Hose roller A device that is placed on the edge of a roof and used to protect hose as it is being hoisted up and over the roof's edge.

Hot zone (technical rescue incident) The control zone immediately surrounding a hazardous materials spill/incident site that is directly dangerous to life and health. All personnel working in the hot zone must wear complete, appropriate protective clothing and equipment. Entry requires approval by the IC or a designated sector officer. Complete backup, rescue, and decontamination teams must be in place at the perimeter before operations begin.

Hot-zone entry team The team of fire fighters assigned to enter into the designated hot zone.

Hux bar A multipurpose tool that can be used for several forcible entry and ventilation applications because of its unique design. It may also be used as a hydrant wrench.

Hybrid vehicle A vehicle that uses a battery-powered electric motor and an internal combustion engine.

Hydrant wrench A hand tool that is used to operate the valves on a hydrant; it may also be used as a spanner wrench. Some models are plain wrenches, whereas others have a ratchet feature.

Hydraulic shears A lightweight, hand-operated tool that can produce up to 4536 kg (10,000 lb.) of cutting force.

Hydraulic spreader A lightweight, hand-operated tool that can produce up to 4536 kg (10,000 lb.) of prying and spreading force.

Hydraulic ventilation Ventilation that relies on the movement of air caused by a fog stream that is placed 0.6–1.2 m (2–4 ft.) in front of the open window.

Hydrogen cyanide A toxic gas produced by the combustion of materials containing cyanide.

Hydroplaning The loss of steering or braking ability (and thus, control) because of a layer of water on the road that prevents sufficient contact between the vehicle's tires and the road's surface.

Hydrostatic testing Pressure testing of a fire extinguisher to verify its strength against unwanted rupture. (NFPA 10)

Hypothermia A condition in which the internal body temperature falls below 35°C (95°F), usually a result of prolonged exposure to cold or freezing temperatures.

Hypoxia A state of inadequate oxygenation of the blood and tissue sufficient enough to cause impairment of function. (NFPA 99B)

I-beam A ladder beam constructed of one continuous piece of I-shaped metal or fibreglass to which the rungs are attached.

Ignitable liquid Classification for liquid fuels including both flammable and combustible liquids.

Ignition phase The phase of fire development in which the fire is limited to the immediate point of origin.

Ignition point The minimum temperature at which a substance will burn.

Ignition temperature The minimum temperature at which a fuel, when heated, will ignite in air and continue to burn; also known as auto-ignition temperature.

Immediately dangerous to life and health (IDLH) Any condition that would pose an immediate or delayed threat to life, cause irreversible adverse health effects, or interfere with an individual's ability to escape unaided from a hazardous environment. (NFPA 1670, 2004)

Improvised explosive device (IED) An explosive or incendiary device fabricated in an improvised manner.

IMS general staff The chiefs of each of the four major sections of IMS (outside of command): operations, planning, logistics, and finance/administration.

Incendiary A fire that is intentionally ignited under circumstances in which the person knows that the fire should not be ignited.

Incendiary device A device or mechanism used to start a fire or explosion.

Incident action plan (IAP) The objectives reflecting the overall incident strategy, tactics, risk management, and member safety that are developed by the IC. IAPs are updated throughout the incident. (NFPA 1500)

Incident command post (ICP) The field location at which the primary tactical-level, on-scene incident command functions are performed. (NFPA 1026)

Incident commander (IC) The person who is responsible for all decisions relating to the management of an incident and is in charge of the incident site. (NFPA 1500)

Incident Management System (IMS) The combination of facilities, equipment, personnel, procedures, and communications operating within a common organizational structure that has responsibility for the management of assigned resources to effectively accomplish stated objectives pertaining to an incident or training exercise. (NFPA 1670)

Incipient stage The initial or beginning stage of a fire, in which it can be controlled or extinguished by portable extinguishers or small amounts of dry extinguishing agents, without the need for protective clothing or breathing apparatus. (NFPA 484)

Incomplete combustion A burning process in which the fuel is not completely consumed, usually due to a limited supply of oxygen.

Incubation period The time period between a victim's initial infection by an organism and the appearance of symptoms.

Indirect application of water The use of a solid object such as a wall or ceiling to break apart a stream of water, creating more surface area on the water droplets and thereby causing the water to absorb more heat.

Indirect attack (structure fire) Firefighting operations involving the application of extinguishing agents to reduce the buildup of heat released from a fire without applying the agent directly onto the burning fuel. (NFPA 1145). Also a method of attack in which the control line is located along natural fuel breaks, at favourable breaks in the topography, or at considerable distance from the fire, and the intervening fuel is burned out.

Information management Fire fighters or civilians who take care of the computer and networking systems that a fire department needs to operate.

Ingestion Exposure to a hazardous material by swallowing it.

Inhalation Exposure to a hazardous material by breathing it into the lungs.

Injection Exposure to a hazardous material through cuts or other breaches in the skin.

Integrated communications The ability of all appropriate personnel at the emergency scene to communicate with their respective supervisor and subordinates.

Interior attack The assignment of a team of fire fighters to enter a structure and attempt fire suppression.

Interior finish The exposed surfaces of walls, ceilings, and floors within buildings. (NFPA 101)

Interior fire separations Interior walls that extend from the floor to the underside of the rated assembly above. They often enclose fire-rated interior corridors or service rooms or divide a floor area into separate fire compartments. Interior fire separations also enclose vertical shafts, stairwells and elevator shafts.

Interior wall A wall inside a building that divides a large space into smaller areas; also known as a partition.

Intermodal tank A bulk container that can be shipped by all modes of transportation—air, sea, and land.

Inverter Equipment used to change the voltage level or waveform, or both, of electrical energy. Commonly, an inverter is a device that changes DC input to AC output. Inverters may also function as battery chargers that use alternating current from another source and convert it into direct current for charging batteries. Also known as a power conditioning unit (PCU or power conversion system (PCS). (NFPA 70)

Ionization smoke detector A device containing a small amount of radioactive material that ionizes the air between two charged electrodes to sense the presence of smoke particles.

Ionizing radiation Radiation of sufficient energy to alter the atomic structure of materials or cells with which it interacts, including electromagnetic radiation, such as X-rays, gamma rays, and microwaves, and particulate radiation, such as alpha and beta particles. (NFPA 1991)

Irons A combination of tools, usually consisting of a Halligan tool and a flat-head axe, that are commonly used for forcible entry.

Irritant A substance such as mace that can be dispersed to briefly incapacitate a person or groups of people.

Island An unburned area surrounded by fire.

J tool A tool that is designed to fit between double doors equipped with panic bars.

Jalousie windows Windows made of small slats of tempered glass, which overlap each other when the window is closed. Often found in trailers and mobile homes, jalousie windows are held together by a metal frame and operated by a small hand wheel or crank found in the corner of the window.

Jamb The part of a doorway that secures the door to the studs in a building.

Junction box A device that attaches to an electrical cord to provide additional outlets.

K tool A tool that is used to remove lock cylinders from structural doors so the locking mechanism can be unlocked.

Kelly tool A steel bar with two main features: a large pick and a large chisel or fork.

Kerf cut A cut that is only the width and depth of the saw blade. It is used to inspect cockloft spaces from the roof.

Kernmantle rope Rope made of two parts—the kern (interior component) and the mantle (the outside sheath).

Kevlar® A strong synthetic material used in the construction of protective clothing and equipment.

Keyhole saw A saw designed to cut circles in wood for keyholes.

Kilopascal (kPa) The standard unit of measuring pressure.

Knot A fastening made by tying together lengths of rope or webbing in a prescribed way. (NFPA 1670)

Label A smaller version (100-mm [4-in.] diamond-shaped marking) of a placard, placed on any side of an individual box or smaller package.

Ladder belt A compliant equipment item that is intended for use as a positioning device for a person on a ladder. (NFPA 1983)

Ladder gin An A-shaped structure formed with two ladder sections. It can be used as a makeshift lift when raising a trapped person. One form of the device is called an A-frame hoist.

Ladder halyards Rope used on extension ladders to raise a fly section.

Ladder pipe A monitor that is fed by a hose and that holds and directs a nozzle while attached to the rungs of a vehicle-mounted aerial ladder. (NFPA 1965)

Laminar smoke flow The smooth or streamlined movement of smoke, which indicates that the pressure in the building is not excessively high.

Laminated glass A type of safety glass comprised of a thin layer of plastic between two layers of glass, which prevents the glass from shattering and falling apart when broken.

Laminated wood Pieces of wood that are glued together.

Large-diameter hose (LDH) A hose with a diameter of 89 mm (3.5 in.) or more that is designed to move large volumes of water to supply master stream appliances, portable hydrants, manifolds, standpipe and sprinkler systems, and fire department pumpers from hydrants and in relay. (NFPA 1410)

Latch A spring-loaded latch bolt or a gravity-operated steel bar that, after release by physical action, returns to its operating position and automatically engages the strike plate when it is returned to the closed position. (NFPA 80)

Lath Thin strips of wood used to make the supporting structure for roof tiles.

Leap-frogging A condition in which fire spreads from one floor to the other through exterior windows (auto-exposure).

Legacy construction An older type of construction (prior to 1970) that used sawn lumber.

Level A protection A heavy encapsulating suit that provides the highest level of chemical protection for skin, eyes, and respiration.

Level B protection PPE used when the type and atmospheric concentration of identified substances require a high level of respiratory protection but less skin protection.

Level C protection PPE used when the type of airborne substance is known, its concentration is measured, the criteria for using APRs are met, and skin and eye exposure is unlikely.

Level D protection PPE used when the atmosphere contains no known hazard, and when work functions preclude splashes, immersion, or the potential for unexpected inhalation of or contact with hazardous levels of chemicals.

Lewisite An oily, colourless to dark-brown liquid blistering agent with an odour of geraniums.

Liaison officer A member of the command staff who serves as the point of contact for assisting or coordinating agencies. (NFPA 1026)

Lieutenant A company officer who is usually responsible for a single fire company on a single shift; the first in line among company officers.

Life line A rope secured to a fire fighter that enables the fire fighter to retrace his or her steps out of a structure.

Life safety rope Rope dedicated solely for the purpose of supporting people during rescue, firefighting, other emergency operations, or during training evolutions. (NFPA 1983)

Light (low) hazard locations Occupancies where the total amount of combustible materials is less than expected in an ordinary hazard location.

Light-emitting diode (LED) An electronic semiconductor that emits a single-colour light when activated.

Lightweight construction Lightweight materials or advanced engineering, or both practices, which result in a weight savings without any sacrifice of strength or efficiency. (NFPA 414)

Line detector Wire or tubing that can be strung along the ceiling of large, open areas to detect an increase in heat.

Liquid Any material that (1) has a fluidity greater than that of 300 penetration asphalt when tested in accordance with American Society of Testing and Materials (ASTM) D5, *Standard Test Method for Penetration of Bituminous Materials*, or (2) is a viscous substance for which a specific melting point cannot be determined but that is determined to be a liquid in accordance with ASTM D4359, *Standard Test Method for Determining Whether a Material Is a Liquid or a Solid*. (NFPA 30)

Liquid splash–protective clothing The garment portion of a CPE that is designed and configured to protect the wearer against chemical liquid splashes but not against chemical vapours or gases. (NFPA 472)

Liquid surge The force imposed upon a fire apparatus by the contents of a partially filled water or foam-concentrate tank when the vehicle accelerates, decelerates, or turns. (NFPA 1002)

Live load The load produced by the use and occupancy of the building or other structure, which does not include construction or environmental loads, such as wind load, snow load, rain load, earthquake load, flood load, or dead load. Live loads on a roof are those produced during maintenance by workers, equipment, and materials; and during the life of the structure by movable objects such as planters and by people. (NFPA 5000, ASCE/SEI 7-4.1)

Load-bearing wall A wall designed to provide structural support for a building.

Loaded-stream fire extinguisher A water-based fire extinguisher that uses an alkali metal salt as a freezing-point depressant.

Local alarm system A fire alarm system that sounds an alarm only in the building where it was activated; that is, no signal is sent out of the building.

Lock body The part of a padlock that holds the main locking mechanisms and secures the shackles.

Locking mechanism A device that locks an extinguisher's trigger to prevent its accidental discharge. Also, a standard doorknob lock, deadbolt lock, or sliding latch.

Lockout and tagout systems Methods of ensuring that electricity and other utilities have been shut down and switches are locked so that they cannot be switched on, so as to prevent flow of power or gases into the area where rescue is being conducted.

Logistics section The section responsible for providing facilities, services, and materials for the incident or planned event, including the communications unit, medical unit, and food unit within the service branch and the supply unit, facilities unit, and ground support unit within the support branch. (NFPA 1026)

Logistics section chief The general staff position responsible for directing logistical functions. It is generally assigned on complex, resource-intensive, or long-duration incidents.

Loop A piece of rope formed into a circle.

Louvre cut A cut that is made using power saws and axes to cut along and between roof supports so that the sections created can be tilted into the opening.

Low-angle operation A rope rescue operation on a mildly sloping surface (less than 45 degrees) or flat land. In this scenario, fire fighters depend on the ground for their primary support, and the rope system is a secondary means of support.

Low-volume nozzle A nozzle that flows 182 L/min. (40 gpm) or less.

Lower flammable limit (LFL) The concentration of a combustible material in air below which ignition will not occur. Mixtures below this limit are said to be "too lean"; also known as lower explosive limit (LEL). (NFPA 329)

Malicious false alarm A fire alarm signal sent when there is no fire, usually initiated by an individual who wishes to cause a disturbance.

Mallet A short-handled hammer.

Manual pull-station A device with a switch that either opens or closes a circuit, activating the fire alarm; also known as a manual fire alarm box.

Manufactured (mobile) home A factory-assembled structure, or a structure that can be transported in one or more sections, that is built on a permanent chassis and designed to be used as a dwelling without a permanent foundation when connected to the required utilities, including the plumbing, heating, air-conditioning, and electric systems contained therein.

Masonry A built-up unit of construction or combination of materials such as clay, shale, concrete, glass, gypsum, tile, or stone set in mortar. (NFPA 5000)

Mass arson Arson in which an offender sets three or more fires at the same site or location during a limited period of time.

Mass decontamination The physical process of reducing or removing surface contaminants from large numbers of victims in potentially life-threatening situations, in the fastest time possible. (NFPA 472)

Master stream device A large-capacity nozzle that can be supplied by two or more 64-mm (2.5-in.) hose lines or large-diameter hose or fixed piping. It can flow between 1591 and 6818 L/min (350 and 1500 gpm). These devices include deck guns and portable ground monitors.

Master stream nozzle A nozzle with a rated discharge of 1591 L/min (350 gpm) or greater. (NFPA 1964)

Master-coded alarm An alarm system in which audible notification devices can be used for multiple purposes beyond a fire alarm.

Matter A substance made up of atoms and molecules.

Maul A specialized striking tool, weighing 2.7 kg (6 lb.) or more, with an axe on one end and a sledgehammer on the other end.

Mayday A code indicating that a fire fighter is lost, missing, or trapped, and requires immediate assistance.

McLeod tool A hand tool used for constructing fire lines and overhauling wildland fires. One side of the head consists of a five- to seven-toothed fire rake; the other side is a hoe.

Mechanical energy A form of potential energy that can generate heat through friction.

Mechanical saw A saw that is usually powered by an electric motor or a gasoline engine. The three primary types of mechanical saws are chainsaws, rotary saws, and reciprocating saws.

Mechanical ventilation A process of removing heat, smoke, and gases from a fire area by using exhaust fans, blowers, air-conditioning systems, or smoke ejectors. (NFPA 402)

Medium-diameter hose (MDH) Hose with a diameter of 64 or 76 mm (2.5 or 3 in.).

Mildew A fungus that can grow on hose if the hose is stored wet. Mildew can damage the jacket of a hose.

Mobile data terminal Technology that allows fire fighters to receive information while in the fire apparatus or at the station.

Mobile radio A two-way radio that is permanently mounted in a fire apparatus.

Mobile water supply apparatus A vehicle designed primarily for the transportation—pickup, transport, and delivery—of water to fire emergency scenes, where the water will then be applied by other vehicles or pumping equipment. (NFPA 1901)

Mortise locks Door locks with both a latch and a bolt built into the same mechanism; the two locking mechanisms operate independently of each other. Mortise locks are often found in hotel rooms.

Multipurpose dry chemical fire extinguisher A fire extinguisher rated to fight Class A, B, and C fires.

Multipurpose hook A long pole with a wooden or fibreglass handle and a metal hook on one end used for pulling.

Municipal water system A water distribution system that is designed to deliver potable water to end-users for domestic, commercial, industrial, and fire protection purposes.

Mushrooming The process in which rising smoke, heat, and gases encounter a horizontal barrier such as a ceiling and begin to move out and back down.

National Fire Protection Association (NFPA) A private organization that develops and maintains nationally recognized minimum consensus standards on many areas of fire safety as well as specific standards on hazardous materials.

National Incident Management System (NIMS) A system mandated by the United States Department of Homeland Security Presidential Directive 5 (HSPD-5), *Management of Domestic Incidents*, that provides a systematic, proactive approach guiding government agencies at all levels, the private sector, and nongovernmental organizations to work seamlessly to prepare for, prevent, respond to, recover from, and mitigate the effects of incidents, regardless of cause, size, location, or complexity, so as to reduce the loss of life or property and harm to the environment. (NFPA 1026)

National Institute for Occupational Safety and Health (NIOSH) The United States federal agency responsible for research and development on occupational safety and health issues.

National Response Center (NRC) An agency maintained and staffed by the United States Coast Guard; it should always be notified if any spilled material could possibly enter a navigable waterway.

National Terrorism Advisory System (NTAS) A national system to communicate information about terrorist threats in the United States by providing timely information to the public, government agencies, and first responders. It issues imminent threat alerts and elevated threat alerts.

Natural ventilation The flow of air or gases created by the difference in the pressures or gas densities between the outside and inside of a vent, room, or space. (NFPA 853)

Negative-pressure ventilation Ventilation that relies on electric fans to pull or draw the air from a structure or area.

Nerve agent A toxic substance that attacks the central nervous system in humans.

Neutron A penetrating particle found in the nucleus of an atom that is removed through nuclear fusion or fission. Although neutrons are not radioactive, exposure to neutrons can create radiation.

NFPA 704 hazard identification system A hazardous materials marking system designed for fixed-facility use.

Nomex® A fire-resistant synthetic material used in the construction of PPE for firefighting.

Nonambulatory A term describing individuals who cannot move themselves to an area of safety owing to their physical condition, medical treatment, or other factors.

Nonbearing wall Any wall that is not a load-bearing wall. (NFPA 5000, ASCE/SEI 7-11.2)

Nonbulk storage vessel Any container other than a bulk storage container.

Noncoded alarm An alarm system that provides no information at the alarm control panel to indicate where the activated alarm is located.

Normal operating pressure The observed static pressure in a water distribution system during a period of normal demand.

Nose cups An insert inside the face piece of an SCBA that fits over the user's mouth and nose.

Nozzle A constricting appliance attached to the end of a fire hose or monitor to increase the water velocity and form a stream. (NFPA 1965)

Nozzle shut-off A device that enables the fire fighter at the nozzle to start or stop the flow of water.

Nuisance alarm A fire alarm signal caused by the malfunction or improper operation of a fire alarm system or component.

Obscuration rate A measure of the percentage of light transmission blocked between the sender and receiver unit of a beam detector.

Occupancy The purpose for which a building or other structure, or part thereof, is used or intended to be used. (NFPA 5000)

Occupational Safety and Health Administration (OSHA) The United States federal agency that regulates worker safety and, in some cases, responder safety. OSHA is part of the United States Department of Labor.

Offensive attack An advance into the fire building by fire fighters with hose lines or other extinguishing agents that are intended to overpower the fire.

Open-circuit breathing apparatus SCBA in which the exhaled air is released into the atmosphere and is not reused.

Open-end wrench A hand tool that is used to tighten or loosen bolts. The end is open, as opposed to a box-end wrench. Each wrench is a specific size.

Operations level Persons who respond to hazardous materials/weapons of mass destruction incidents for the purpose of implementing or supporting actions to protect nearby persons, the environment, or property from the effects of the release. (NFPA 472)

Operations section The section responsible for all tactical operations at an incident or planned event, including up to 5 branches, 25 divisions/groups, and 125 single resources, task forces, or strike teams. (NFPA 1026)

Operations section chief The general staff position responsible for managing all operations activities. It is usually assigned when complex incidents involve more than 20 single resources or when command cannot be involved in the details of tactical operations.

Operator lever The handle of a door that turns the latch to open it.

Ordinary construction Buildings whose exterior walls are made of noncombustible or limited-combustible materials, but whose interior floors and walls are made of combustible materials. Ordinary construction is also known as Type III building construction, as defined in NFPA 220, *Standard on Types of Building Construction*.

Ordinary (moderate) hazard locations Occupancies that contain more Class A and Class B materials than are found in light hazard locations.

Outside stem and yoke (OS&Y) valve A sprinkler control valve with a valve stem that moves in and out as the valve is opened or closed.

Overhaul The process of final extinguishment after the main body of a fire has been knocked down. All traces of fire must be extinguished at this time. (NFPA 402, 2002)

Oxidation A reaction with oxygen in either its elemental form or the form of one of its compounds. (NFPA 53)

Oxygen deficiency Any atmosphere where the oxygen level is less than 19.5 percent. Low oxygen levels can have serious effects on people, including adverse reactions such as poor judgment and lack of muscle control.

Packaging The process of securing a victim in a transfer device, with regard to existing and potential injuries or illness, so as to prevent further harm during movement. (NFPA 1006)

Padlocks The most common types of locks on the market today, built to provide regular-duty or heavy-duty service. Several types of locking mechanisms are available, including keyways, combination wheels, and combination dials.

Parallel chord truss A truss in which the top and bottom chords are parallel.

Parapet walls Walls on a flat roof that extend above the roofline.

Partition A nonstructural interior wall that spans horizontally or vertically from support to support. The supports may be the basic building frame, subsidiary structural members, or other portions of the partition system. [ASCE/SEI 7:11.2] (NFPA 5000)

Party wall A wall constructed on the line between two properties.

PASS Acronym for the steps involved in operating a portable fire extinguisher: pull pin, aim nozzle, squeeze trigger, sweep across burning fuel.

Pawls Devices attached to a fly section(s) to engage ladder rungs near the beams of the section below for the purpose of anchoring the fly section(s). (NFPA 1931)

PBI® A fire-retardant synthetic material used in the construction of PPE.

Peak cut A ventilation opening that runs along the top of a peaked roof.

Pendant sprinkler A sprinkler designed to be mounted on the underside of sprinkler piping so that the water stream is directed downward; usually marked SSP (standard spray pendant).

Penetration The movement of matter through a suit's closures, including zippers, buttonholes, seams, flaps, or other design features of chemical-protective clothing, as well as through punctures, cuts, and tears. (NFPA 472)

Permeation A chemical action involving the movement of chemicals on a molecular level through intact material. (NFPA 472)

Permissible exposure limit (PEL) The maximum permitted 8-hour, time-weighted average concentration of an airborne contaminant per United States regulations. (NFPA 5000)

Personal alert safety system (PASS) A device that continually monitors for lack of movement of the wearer and, if no movement is detected, automatically activates an alarm signal indicating the wearer is in need of assistance. The device can also be manually activated to trigger the alarm signal. (NFPA 1982)

Personal dosimeter A device that measures the dose of radiation to an individual.

Personal flotation device (PFD) A displacement device worn to keep the wearer afloat in water. (NFPA 1925)

Personal protective equipment (PPE) The basic PPE for wildland fire suppression includes a helmet, protective footwear, gloves, and flame-resistant clothing as defined in NFPA 1977, Standard on Protective Clothing and Equipment for Wildland Fire Fighting. (NFPA 1051)

Personnel accountability report (PAR) One of a series of periodic reports verifying the status of responders assigned to an incident or planned event. (NFPA 1026)

Personnel accountability system A system that readily identifies both the locations and the functions of all members operating at an incident scene. (NFPA 1500)

Personnel accountability tag (PAT) An identification card used to track the location of a fire fighter on an emergency incident.

pH A measure of the acidity or basic nature of a material; more technically, an expression of the concentration of hydrogen ions in the substance.

Phosgene A chemical agent that causes severe pulmonary damage; it is a by-product of incomplete combustion.

Photoelectric smoke detector A device that uses a light source and a photosensitive sensor to detect visible products of combustion.

Physical change A transformation in which a material changes its state of matter—for instance, from a liquid to a solid.

Physical evidence Items that can be observed, photographed, measured, collected, examined in a laboratory, and presented in court to prove or demonstrate a point.

Pick The pointed end of a pick axe, which can be used to make a hole or bite in a door, floor, or wall.

Pick-head axe A tool that has a head with an axe on one side and a pointed end (pick) on the opposite side.

Piercing nozzle A nozzle that can be driven through sheet metal or other material to deliver a water stream to that area.

Pike pole A pole with a sharp point (pike) on one end coupled with a hook. It is used to make openings in ceilings and walls. Pike poles are manufactured in different lengths for use in rooms of different heights.

Pincer attack A direct method of suppressing a wildland or ground fire, in which one suppression crew attacks the right flank of the fire and a second suppression crew attacks the left flank of the fire.

Pipe bomb A device created by filling a section of pipe with an explosive material.

Pipe wrench A wrench having one fixed grip and one movable grip that can be adjusted to fit securely around pipes and other tubular objects.

Pipeline A length of pipe including pumps, valves, flanges, control devices, strainers, and/or similar equipment for conveying fluids. (NFPA 70)

Pipeline right-of-way An area, patch, or roadway that extends about a metre on either side of the pipe itself and that may contain warning and informational signs about hazardous materials carried in the pipeline.

Pitched chord truss A type of truss typically used to support a sloping roof.

Pitched roof A roof with sloping or inclined surfaces.

Pitot gauge A type of gauge used to measure the velocity and pressure of water as it is discharged from an opening. In particular, it is used to determine the flow of water from a hydrant or nozzle.

Placard Signage required to be placed on all four sides of highway transport vehicles, railroad tanker cars, and other forms of hazardous materials transportation that identifies the hazardous contents of the vehicle using a standardized system with 250-mm (10.75-in.) diamond-shaped indicators.

Plague An infectious disease caused by the bacterium *Y. pestis*, which is commonly found on rodents.

Planning section The section responsible for the collection, evaluation, dissemination, and use of information related to the incident situation, resource status, and incident forecast. (NFPA 1026)

Planning section chief The general staff position responsible for planning functions. It is assigned when command needs assistance in managing information.

Plaster hook A long pole with a pointed head and two retractable cutting blades on the side.

Plate glass A type of glass that has additional strength so it can be formed in larger sheets but will still shatter upon impact.

Platform frame A type of vehicle frame resembling a ladder, which is made up of two parallel rails joined by a series of cross members. This kind of construction is typically used for luxury vehicles, SUVs, and all types of trucks.

Platform-frame construction A construction technique for building the frame of a structure one floor at a time. Each floor has a top and bottom plate, which act as fire stops.

Plume The column of hot gases, flames, and smoke rising above a fire; also called a convection column, thermal updraft, or thermal column. (NFPA 921)

Pocket A deep indentation of unburned fuel along the fire's perimeter, often found between a finger and the head of the fire.

Point of origin The exact location where a heat source and a fuel come in contact with each other and a fire begins. (NFPA 921)

Polar solvent A water-soluble flammable liquid, such as alcohol, acetone, ester, and ketone.

Policy A formal statement that provides guidelines for present and future actions. Policies often require personnel to make judgments.

Portable ladder A ladder carried on fire apparatus but designed to be removed from the apparatus and deployed by fire fighters where needed.

Portable monitor A monitor that can be lifted from a vehicle-mounted bracket and moved to an operating position on the ground by not more than two people. (NFPA 1965)

Portable radio A battery-operated, hand-held transceiver. (NFPA 1221)

Portable tanks Folding or collapsible tanks used at the fire scene to hold water for drafting.

Positive-pressure ventilation Ventilation that relies on fans to push or force clean air into a structure.

Post indicator valve (PIV) A sprinkler control valve with an indicator that reads either open or shut, depending on its position.

Post One of the vertical support members, or pillars, of a vehicle that holds up the roof and forms the upright columns of the passenger compartment.

Pounds per square inch (psi) The standard imperial unit for measuring pressure.

Power take-off shaft A supplemental mechanism that enables a fire engine to operate a pump while the engine is still moving.

Powered air-purifying respirator (PAPR) An air-purifying respirator with a hood or helmet, breathing tube, canister, cartridge, filter, and a blower that passes ambient air through the purifying element. The blower can be stationary or portable. It can also be designed and equipped with a full face piece, in which case it is referred to as a full-face-piece powered air-purifying respirator (FFPAPR). (NFPA 1404)

Preaction sprinkler system A dry sprinkler system that uses a deluge valve instead of a dry-pipe valve and requires the activation of a secondary device before the pipes will fill with water.

Preincident plan A written document resulting from the gathering of general and detailed information to be used by public emergency response agencies and private industry for determining the response to reasonable anticipated emergency incidents at a specific facility.

Preincident survey The process used to gather information to develop a preincident plan.

Premixed foam Solution produced by introducing a measured amount of foam concentrate into a given amount of water in a storage tank. (NFPA 11)

Pressure gauge A device that measures and displays pressure readings. In an SCBA, the pressure gauges indicate the quantity of breathing air that is available at any time.

Pressure indicator A gauge on a pressurized portable fire extinguisher that indicates the internal pressure of the expellant.

Primary Care Paramedic (PCP) A member of EMS who can perform limited procedures that usually fall between those provided by an EMR and those provided by an ACP, including IV therapy, interpretation of cardiac rhythms, and defibrillation.

Primary cut The main ventilation opening made in a roof to allow smoke, heat, and gases to escape.

Primary feeders The largest-diameter pipes in a water distribution system, which carry the greatest amounts of water.

Primary search A rapid search to locate living victims.

Private water system A privately owned water system that operates separately from the municipal water system.

Projected windows Windows that project inward or outward on an upper hinge; also called factory windows. They are usually found in older warehouses or commercial buildings.

Proprietary system A fire alarm system that transmits a signal to a monitoring location owned and operated by the facility's owner.

Protection plates Reinforcing material placed on a ladder at chafing and contact points to prevent damage from friction and contact with other surfaces.

Protective hood A part of a fire fighter's PPE that is designed to be worn over the head and under the helmet; it provides thermal protection for the neck and ears.

Protein foam A protein-based foam concentrate that is stabilized with metal salts to make a fire-resistant foam blanket. (NFPA 402)

Pry axe A specially designed hand axe that serves multiple purposes. Similar to a Halligan bar, it can be used to pry, cut, and force doors, windows, and many other types of objects. Also called a multipurpose axe.

Pry bar A specialized prying tool made of a hardened steel rod with a tapered end that can be inserted into a small area.

Public information officer (PIO) A member of the command staff who is responsible for interacting with the public and media or with other agencies with incident-related information requests. (NFPA 1026)

Public information officer An individual who has demonstrated the ability to conduct media interviews and prepare news releases and media advisories. (NFPA 1035)

Public safety answering point (PSAP) A facility in which 911 calls are answered. (NFPA 1221)

Pulaski axe A hand tool that combines an adz and an axe for brush removal.

Pulley A device with a free-turning, grooved metal wheel (sheave) used to reduce rope friction. Side plates are available for a carabiner to be attached. (NFPA 1670)

Pulmonary edema Fluid buildup in the lungs.

Pump tank fire extinguisher A nonpressurized, manually operated water extinguisher that is rated for use on Class A fires. Discharge pressure is provided by a hand-operated, double-acting piston pump.

Pyrolysis The destructive distillation of organic compounds in an oxygen-free environment whereby the organic matter is converted into gases, liquids, and char. (NFPA 820)

Pyromaniac A pathological fire-setter.

Rabbet A type of door frame in which the stop for the door is cut into the frame.

Rabbet tool A hydraulic spreading tool designed to pry open doors that swing inward.

Radiation dispersal device Any device that causes the purposeful dissemination of radioactive material without a nuclear detonation; also known as a dirty bomb.

Radiation The emission and propagation of energy through matter or space by means of electromagnetic disturbances that display both wave- and particle-like behaviour. (NFPA 801). The transfer of heat to colder objects in an environment by radiant energy—for example, heat gain from a fire.

Radio repeater system A radio system that automatically retransmits a radio signal on a different frequency.

Radioactive isotope An atom that has unequal numbers of protons and neutrons in its nucleus and that emits radioactivity.

Radioactivity The spontaneous decay or disintegration of an unstable atomic nucleus, accompanied by the emission of radiation. (NFPA 801)

Radiological agents Materials that emit radiation.

Rafters Joists mounted in an inclined position to support a roof.

Rail The top or bottom piece of a trussed beam assembly used in the construction of a trussed ladder. Also, the top and bottom surfaces of an I-beam ladder. Each beam has two rails.

Rapid intervention crew (RIC) A minimum of two fully equipped personnel who are on site, in a ready state, for immediate rescue of injured or trapped fire fighters. Also known as rapid intervention team (RIT).

Rapid intervention crew/company universal air connection (RIC UAC) A system that allows emergency replenishment of breathing air to the SCBA of disabled or entrapped fire or emergency services personnel. (NFPA 1981)

Rapid oxidation A chemical process that occurs when a fuel is combined with oxygen, resulting in the formation of ash or other waste products and the release of energy as heat and light.

Rate-of-rise heat detector A fire detection device that responds when the temperature increases at a rate that exceeds a predetermined value.

Rear of the fire The side opposite the head of the fire. Also called the heel of the fire.

Reciprocating saw A saw that is powered by an electric motor or a battery motor, and whose blade moves back and forth.

Recommended exposure level (REL) The established standard limit of exposure to a hazardous material. It is based on the maximum time-weighted concentration at which 95 percent of exposed, healthy adults suffer no adverse effects over a 40-hour workweek. RELs are set by the NIOSH but do not have the force of United States federal law.

Reconnaissance report The inspection and exploration of a specific area for the purpose of gathering information for the incident commander.

Recovery phase The stage of a hazardous materials incident after the imminent danger has passed, when cleanup and the return to normalcy have begun.

Reducer A fitting used to connect a small hose line or pipe to a larger hose line or pipe. (NFPA 1142)

Regulation A rule, usually issued by a government or other legally authorized agency, that dictates how something must be done. Regulations are often developed to implement a law.

Rehabilitate To restore someone or something to a condition of health or to a state of useful and constructive activity.

Rehabilitation An intervention designed to mitigate against the physical, physiological, and emotional stress of firefighting so as to sustain a member's energy, improve performance, and decrease the likelihood of on-site injury or death. (NFPA 1521)

Rekindle A return to flaming combustion after apparent but incomplete extinguishment. (NFPA 921)

Relative humidity The ratio between the amount of water vapour in the air (gas) at the time of measurement and the amount of water vapour that could be in the air (gas) when condensation begins, at a given temperature. (NFPA 79)

Remote annunciator A secondary fire alarm control panel in a different location than the main alarm panel; it is usually located near the front door of a building.

Remote shut-off valve A device that allows for the shut-off of a system or the isolation a valve from a remote location.

Remote station system A fire alarm system that sounds an alarm in the building and transmits a signal to the fire department or an off-premises monitoring location.

Remote-controlled hydrant valve A valve attached to a fire hydrant to allow the operator to turn the hydrant on without flowing water into the hose line.

Removal A mode of decontamination that applies specifically to contaminated soil that can be taken away from the scene.

Rescue Those activities directed at locating endangered persons at an emergency incident, removing those persons from danger, treating the injured, and providing for transport to an appropriate health care facility. (NFPA 402)

Rescue company A group of fire fighters who work as a unit and are equipped with one or more rescue vehicles. (NFPA 1410)

Rescue-lift air bags Extrication tools consisting of air bags, filler hoses, air regulators, control valves, and a supply of compressed air. These tools are used to lift vehicles that have been involved in crashes.

Reservoir A water storage facility.

Residential sprinkler system A sprinkler system designed to protect residential dwellings.

Residual pressure The pressure that exists in the distribution system as measured at the residual hydrant at the time the flow readings are taken at the flow hydrants. (NFPA 24)

Resource management A standard system of assigning and tracking the resources involved in an incident. Under NIMS, this includes mutual-aid agreements; the use of special federal, state, local, and tribal teams; and resource-mobilization protocols. (NFPA 1026)

Respiration The loss of body heat as warm air in the lungs is exhaled into the atmosphere and cooler air is inhaled.

Response Immediate and ongoing activities, tasks, programs, and systems to manage the effects of an incident that threatens life, property, operations, or the environment. (NFPA 1600)

Retention The process of creating an area to hold hazardous materials.

Reverse lay A method of laying a supply line where the supply line starts at the attack engine and ends at the water source.

Rim locks Surface- or interior-mounted locks located on or in a door with a bolt that provide additional security.

Risk–benefit analysis An assessment of the risk to rescuers versus the benefits that can be derived from their intended actions. (NFPA 1006)

Rocker lugs Fittings on threaded couplings that aid in coupling the hoses; also known as rocker pins.

Rollover The condition where unburned fuel (pyrolysate) from the originating fire has accumulated in the ceiling layer to a sufficient concentration (i.e., at or above the lower flammable limit) that it ignites and burns; it can occur without ignition of, or prior to the ignition of, other fuels separate from that which caused the event; also known as flameover. (NFPA 921)

Roof covering The membrane, which can also be the roof assembly, that resists fire and provides weather protection to the building against water infiltration, wind, and impact. (NFPA 5000)

Roof decking The rigid component of a roof covering.

Roof hooks The spring-loaded, retractable, curved metal pieces that allow the tip of a roof ladder to be secured to the peak of a pitched roof. The hooks fold outward from each beam at the top of a roof ladder.

Roof ladder A single ladder equipped with hooks at the top end of the ladder. (NFPA 1931)

Roofman's hook A long pole with a solid metal hook used for pulling.

Rope bag A bag used to protect and store rope so that the rope can be easily and rapidly deployed without kinking.

Rope record A record for each piece of rope that includes a history of when the rope was placed in service, when it was inspected, when and how it was used, and which types of loads were placed on it.

Rotary saw A saw that is powered by an electric motor or a gasoline engine, and that uses a large rotating blade to cut through material. The blades can be changed depending on the material being cut.

Round turn A piece of rope looped to form a complete circle with the two ends parallel.

Rubber-covered hose Hose whose outside covering is made of rubber, which is said to be more resistant to damage; also known as rubber-jacket hose.

Run cards Cards used to determine a predetermined response to an emergency.

Rung The ladder crosspieces on which a person steps while ascending or descending. (NFPA 1931)

Running end The part of a rope used for lifting or hoisting.

Safe location A location remote or separated from the effects of a fire so that such effects no longer pose a threat. (NFPA 101)

Safety Data Sheet (SDS) A form, provided by manufacturers and compounders (blenders) of chemicals, containing information about the chemical composition, physical and chemical properties, health and safety hazards, emergency response, and waste disposal of the material.

Safety knot A knot used to secure the leftover working end of the rope.

Safety officer A member of the command staff who is responsible for monitoring and assessing safety hazards and unsafe situations and for developing measures for ensuring personnel safety. (NFPA 1561)

Salvage A firefighting procedure for protecting property from further loss following an aircraft accident or a fire. (NFPA 402)

Salvage cover A large square or rectangular sheet made of heavy canvas or plastic material that is spread over furniture and other items to protect them from water runoff and falling debris.

San Francisco hook A multipurpose tool that can be used for several forcible entry and ventilation applications because of its unique design, which includes a built-in gas shut-off and directional slot.

Saponification The process of converting the fatty acids in cooking oils or fats to soap or foam; the action caused by a Class K fire extinguisher.

Sarin A nerve agent that, when dispersed, sends droplets into the air that, when inhaled, harm intended victims.

SCBA harness The part of SCBA that allows fire fighters to wear it as a backpack.

SCBA regulator The part of the SCBA that reduces the high pressure in the cylinder to a usable lower pressure and controls the flow of air to the user.

Screwdriver A tool used for turning screws.

SCUBA dive rescue technician A responder who is trained to handle water rescues and emergencies, including recovery and search procedures in both water and under-ice situations. (SCUBA stands for *self-contained underwater breathing apparatus*.)

Search The process of locating victims who are in danger.

Search and rescue The process of searching a building for a victim and extricating the victim from the building.

Search rope A guide rope used by fire fighters that allows them to maintain contact with a fixed point.

Seat belt cutter A specialized cutting device that cuts through seat belts.

Seat of the fire The main area of the fire.

Secondary collapse A subsequent collapse in a building or excavation. (NFPA 1006)

Secondary containment Any device or structure that prevents environmental contamination when the primary container or its appurtenances fail. Secondary containment shall be designed and constructed to intercept and contain pesticide spills and leaks and to prevent runoff or leaching of pesticides into the environment. Examples of secondary containment include dikes, curbing, and double-walled tanks. (NFPA 434)

Secondary contamination The process by which a contaminant is carried out of the hot zone and contaminates people, animals, the environment, or equipment.

Secondary cut An additional ventilation opening made for the purpose of creating a larger opening or limiting fire spread.

Secondary device An explosive device designed to injure emergency responders who have responded to an initial event.

Secondary feeders Smaller-diameter pipes that connect the primary feeders to the distributors.

Secondary loss Property damage due to smoke, water, or other measures taken to extinguish a fire.

Secondary search A methodical search designed to locate deceased victims after the fire has been suppressed, searching areas not covered during the primary search.

Sector A designated geographic or functional area within the IMS hierarchy similar to a group or division as defined by NIMS.

Self-contained breathing apparatus (SCBA) A respirator with an independent air supply that is used by fire fighters to enter toxic or otherwise dangerous atmospheres.

Self-contained underwater breathing apparatus (SCUBA) A respirator with an independent air supply that is used by underwater divers.

Self-expelling agents The agents in a fire extinguisher that have sufficient vapour pressure at normal operating temperatures to expel themselves. (NFPA 10)

Self-rescue Escaping or exiting a hazardous area under one's own power. (NFPA 1006)

Sensitizer A chemical that causes a large portion of people (or animals) to develop an allergic reaction after repeated exposure to the substance.

Serial arson A series of fires set by the same offender, with a cooling-off period between fires.

Shackles The U-shaped part of a padlock that runs through a hasp and then is secured back into the lock body.

Shelter-in-place A method of safeguarding people in a hazardous area by keeping them in a safe atmosphere, usually inside structures.

Shipping paper A shipping order, bill of lading, manifest, or other shipping document serving a similar purpose and containing the information required by regulations of Transport Canada. (NFPA 498)

Shock load An instantaneous load that places a rope under extreme tension, such as when a falling load is suddenly stopped as the rope becomes taut.

Shoring A structure such as a metal hydraulic, pneumatic/mechanical, or timber system that supports the sides of an excavation and is designed to prevent cave-ins. (NFPA 1670)

Shut-off valve Any valve that can be used to shut down water flow to a water user or system.

Siamese connection A device that allows two hoses to be joined to flow into a single hose.

Sidewall sprinkler A wall-mounted sprinkler that discharges water horizontally into a room.

Signal words Information on a pesticide label that indicates the relative toxicity of the material.

Simplex channel A radio system that uses one frequency to transmit and receive all messages.

Single resource An individual, a piece of equipment and its personnel, or a crew or team of individuals with an identified supervisor that can be used on an incident or planned event. (NFPA 1026)

Single-action pull-station A manual fire alarm activation device in which the user takes a single step—moving a lever, toggle, or handle—to activate the alarm.

Single-station smoke alarm A single device that both detects visible and invisible products of combustion (namely, smoke) and sounds an alarm.

Size-up The observation and evaluation of existing factors that are used to develop objectives, strategy, and tactics for fire suppression. (NFPA 1051)

Slash Debris resulting from natural events such as wind, fire, snow, or ice breakage, or from human activities such as building or road construction, logging, pruning, thinning, or brush cutting. (NFPA 1144)

Sledgehammer A long, heavy hammer that requires the use of both hands.

Small-diameter hose (SDH) Hose with a diameter of 25–51 mm (1–2 in.).

Smallpox A highly infectious disease caused by the virus *Variola*.

Smoke The airborne solid and liquid particulates and gases evolved when a material undergoes pyrolysis or combustion, together with the quantity of air that is entrained or otherwise mixed into the mass. (NFPA 318)

Smoke colour The attribute of smoke that reflects the stage of burning of a fire and the material that is burning in the fire.

Smoke density The thickness of smoke. (Because it has a high mass per unit volume, smoke is difficult to see through.)

Smoke detector A device that detects visible and invisible products of combustion (namely, smoke) and sends a signal to a fire alarm control panel.

Smoke inversion The condition in which smoke hangs low to the ground because of the presence of cold air.

Smoke particles Airborne solid materials consisting of ash and unburned or partially burned fuel released by a fire.

Smoke velocity The speed of smoke leaving a burning building.

Smoke volume The quantity of smoke, which indicates how much fuel is being heated.

Smooth-bore nozzle/tip A nozzle that produces a straight stream that is a solid column of water.

Socket wrench A wrench that fits over a nut or bolt and uses the ratchet action of an attached handle to tighten or loosen the nut or bolt.

Soft suction hose A large-diameter hose designed to be connected to the large port on a hydrant (steamer connection) and into the engine.

Solid A material that has three dimensions and is firm in substance; one of the three phases of matter.

Solid beam A ladder beam constructed of a solid rectangular piece of material (typically wood), to which the ladder rungs are attached.

Solid stream A stream made by using a smooth-bore nozzle to produce a penetrating stream of water.

Solid-core door A door design that consists of wood filler pieces inside the door. This construction creates a stronger door that may be fire rated.

Solidification The process of chemically treating a hazardous liquid so as to turn it into a solid material, thereby making the material easier to handle.

Soman A nerve gas that is both a contact hazard and a vapour hazard; it has the odour of camphor.

Sounding The process of striking a roof with a tool to determine where the roof supports are located.

Spalling The chipping or pitting of concrete or masonry. (NFPA 921)

Span of control The maximum number of personnel or activities that can be effectively controlled by one individual (usually three to seven). (NFPA 1006)

Spanner wrench A type of tool used to couple or uncouple hoses by turning the rocker lugs on the connections.

Special-use railcars Boxcars, flatcars, cryogenic and corrosive tank cars, and high-pressure compressed gas tube cars.

Specialist level Persons who respond to hazardous materials/weapons of mass destruction incidents who have received more specialized training than hazardous materials technicians. Most of the training that specialist employees receive is either product or transportation mode specific.

Specific gravity The ratio of the weight of a given volume of gas to that of the same volume of air, both measured under the same conditions. (NFPA 54)

Split hose bed A hose bed arranged so as to enable the engine to lay out either a single supply line or two supply lines simultaneously.

Split hose lay A scenario in which the attack engine forward lays a supply line from an intersection to the fire, and the supply engine reverse lays a supply line from the hose left by the attack engine to the water source.

Spoil pile A pile of excavated soil next to the excavation or trench. (NFPA 1006)

Spot detector A single-station heat alarm. These devices are often spaced throughout an area.

Spot fire A new fire that starts outside areas of the main fire, usually caused by flying embers and sparks.

Spotlight A light designed to project a narrow, concentrated beam of light.

Spree arson A series of fires started by an arsonist who sets three or more fires at separate locations, with no emotional cooling-off period between fires.

Spring-loaded centre punch A spring-loaded punch used to break automobile glass.

Sprinkler piping The network of piping in a sprinkler system that delivers water to the sprinkler heads.

Sprinkler stop A mechanical device inserted between the deflector and the orifice of a sprinkler head to stop the flow of water.

Sprinkler system For fire protection purposes, an integrated system of underground and overhead piping designed in accordance with fire protection engineering standards. The installation includes at least one automatic water supply that supplies one or more systems. The portion of the sprinkler system above ground is a network of specially sized or hydraulically designed piping installed in a building, structure, or area, generally overhead, and to which sprinklers are attached in a systematic pattern. Each system has a control valve located in the system riser or its supply piping. Each sprinkler system includes a device for actuating an alarm when the system is in operation. The system is usually activated by heat from a fire and discharges water over the fire area. (NFPA 13, 2007)

Sprinkler wedge A wedge-shaped piece of wood placed between the deflector and the orifice of a sprinkler head to stop the flow of water.

Stack effect The vertical air flow within buildings caused by the temperature-created density differences between the building interior and exterior or between two interior spaces. (NFPA 92B)

Staging area A prearranged, strategically placed area where support response personnel, vehicles, and other equipment can be held in an organized state of readiness for use during an emergency. (NFPA 424)

Standard operating procedure (SOP) A written organizational directive that establishes or prescribes specific operational or administrative methods to be followed routinely for the performance of designated operations or actions. (NFPA 1521)

Standing part The part of a rope between the working end and the running end.

Standpipe system An arrangement of piping, valves, hose connections, and allied equipment installed in a building or structure, with the hose connections located in such a manner that water can be discharged in streams or spray patterns through attached hose and nozzles, for the purpose of extinguishing a fire, thereby protecting a building or structure and its contents in addition to protecting the occupants. (NFPA 14, 2007)

State of matter The physical state of a material—solid, liquid, or gas.

Static pressure The pressure that exists in a distribution system at a given point under normal conditions as measured at the residual hydrant with no hydrants flowing. (NFPA 24)

Static rope A rope generally made out of synthetic material that stretches very little under load.

Static water source A water source that is not under pressure, such as a pond, river, stream, or other body of water.

Staypoles Poles attached to each beam of the base section of extension ladders, which assist in raising the ladder and help provide stability of the raised ladder. (NFPA 1931)

Steamer port The large-diameter port on a hydrant.

Step chocks Specialized cribbing assemblies made out of wood or plastic blocks in a step configuration. They are typically used to stabilize vehicles.

Stop A piece of material that prevents the fly sections of a ladder from becoming overextended, leading to collapse of the ladder.

Stored-pressure fire extinguisher A fire extinguisher in which both the extinguishing material and the expellant gas are kept in a single container and that includes a pressure indicator or gauge. (NFPA 10)

Stored-pressure water-type fire extinguisher A fire extinguisher in which water or a water-based extinguishing agent is stored under pressure.

Storz-type hose coupling A hose coupling that has the properties of both a male and a female coupling. It is connected by engaging the lugs and turning the coupling clockwise one-third of a turn.

Straight stream A stream made by using an adjustable nozzle to provide a straight stream of water.

Strike team leader The person in charge of a strike team. This individual is responsible to the next higher level in the incident organization and serves as the point of contact for said strike team within the organization.

Strike team Specified combinations of the same kind and type of resources with common communications and a leader. (NFPA 1026)

Striking tools Tools designed to strike other tools or objects such as walls, doors, or floors.

Strip cut Another term for a trench cut.

Subsurface fuels Partially decomposed matter that lies beneath the ground, such as roots, moss, duff, and decomposed stumps.

Suggested operating guideline (SOG) Another term for SOP that implies that the fire fighter or officer must use his or her personal judgment in completing the procedure.

Sulphur mustard A clear, yellow, or amber oily liquid blister agent with a faint, sweet odour of mustard or garlic. Sulphur mustard can be dispersed in an aerosol form. It causes exposed skin to blister.

Supplied-air respirator An atmosphere-supplying respirator for which the source of breathing air is not designed to be carried by the user. (NFPA 1404)

Supply hose Hose designed for the purpose of moving water between a pressurized water source and a pump that is supplying attack lines; also known as a supply line. (NFPA 1961)

Surface fuels Fuels that are close to the surface of the ground, such as grass, leaves, twigs, needles, small trees, logging slash, and low brush. Also called ground fuels.

Survivability profiling The process of using available data and observations to determine if there is a reasonable chance that victims trapped inside might still be alive. This process is used to decide whether to send fire fighters into a fire or other hazardous environment.

Sweep (or roll-on) method A method of applying foam that involves sweeping the stream just in front of the target.

Tabun A nerve gas that is both a contact hazard and a vapour hazard; it operates by disabling the chemical connection between the nerves and their target organs.

Talk-around channel A simplex channel used for on-site communications.

Tamper alarm An alarm that sounds when the clear plastic cover of a pull-station is lifted; it is intended to prevent malicious false alarms.

Tamper seal A retaining device that breaks when the locking mechanism is released.

Tamper switch A switch on a sprinkler valve that transmits a signal to the fire alarm control panel if the normal position of the valve is changed.

Target hazard Any occupancy type or facility that presents a high potential for loss of life or serious impact to the community resulting from fire, explosion, or chemical release.

Task force Any combination of single resources assembled for a particular tactical need with common communications and a leader. (NFPA 1051)

Task force leader The person in charge of a task force. This individual is responsible to the next higher level in the incident organization and serves as the point of contact for the task force within the organization.

TC306/MC306/DOT406 flammable-liquid tanker Commonly known as a gasoline tanker; a tanker that typically carries gasoline or other flammable and combustible materials.

TC307/MC307/DOT407 chemical hauler A tanker with a rounded or horseshoe-shaped tank.

TC312/MC312/DOT412 corrosives tanker A tanker that often carries aggressive acids, such as concentrated sulphuric or nitric acid, and features reinforcing rings along the side of the tank.

TC331/MC331 pressure cargo tanker A tank commonly constructed of steel with rounded ends and a single open compartment inside; there are no baffles or other separations inside the tank.

TC341/CGA341/MC338 cryogenics tanker A low-pressure tanker designed to maintain the low temperature required by the cryogens it carries.

TDD/TTY/text phone system User devices that allow speech-and/or hearing-impaired citizens to communicate over a telephone system. TDD stands for telecommunications device for the deaf; TTY stands for teletype; text phones visually display text. The displayed text is the equivalent of a verbal conversation between two hearing persons.

Technical decontamination The planned and systematic process of reducing contamination to a level that is as low as reasonably achievable.

Technical rescue incident (TRI) A complex rescue incident requiring specially trained personnel and special equipment to complete the mission. (NFPA 1670)

Technical rescue team A group of rescuers specially trained in the various disciplines of technical rescue.

Technical rescue technician A fire fighter who is trained in special rescue techniques for incidents involving structural collapse, trench rescue, swiftwater rescue, confined-space rescue, and other unusual rescue situations.

Technical use life safety rope A life safety rope with a diameter that is 9.5 mm (0.4 in.) or greater, but is less than 12.5 mm (0.5 in.), with a minimum breaking strength of 20 kN (4496 lbf). Used by highly trained rescue teams that deploy to very technical environments, such as mountainous and/or wilderness terrain.

Technician level Persons who respond to hazardous materials/weapons of mass destruction incidents and are allowed to enter heavily contaminated areas using the highest levels of protection. This is the most aggressive level of training as identified in NFPA 472.

Telephone interrogation The phase in a 911 call during which the dispatcher asks questions to obtain vital information such as the location of the emergency.

Tempered glass A type of safety glass that is heat treated so that under stress or fire it will break into small pieces that are not as dangerous.

Temporal-3 pattern The standard audible signal made by a fire alarm to alert occupants of a building.

Ten-codes A system of predetermined coded messages, such as "What is your 10-20?", used by responders over the radio.

Tender shuttle A method of transporting water from a source to a fire scene, using a number of mobile water supply apparatus; also known as a tanker shuttle.

Thermal column A cylindrical area above a fire in which heated air and gases rise and travel upward.

Thermal conductivity A property that describes how quickly a material will conduct heat.

Thermal imaging camera An electronic camera that detects differences in temperature based on infrared energy and then generates images based on those data. It is commonly used in smoke-filled environments to locate victims as well as to search for hidden fire during size-up and overhaul.

Thermal layering The stratification (layering) that occurs in a room as a result of a fire.

Thermal radiation The means by which heat is transferred to other objects.

Thermoplastic material A plastic material capable of being repeatedly softened by heating and hardened by cooling and, in its softened state, repeatedly shaped by moulding or forming. (NFPA 5000)

Thermoset material A plastic material that, after having been cured by heat or other means, is substantially infusible and cannot be softened and formed. (NFPA 5000)

Threaded hose coupling A type of coupling that requires a male fitting and a female fitting to be screwed together.

Threshold limit value–ceiling (TLV–C) The concentration of a hazardous material to which a worker should not be exposed, even for an instant.

Threshold limit value–short-term exposure limit (TLV–STEL) The maximum concentration of a hazardous material to which a worker can sustain a 15-minute exposure, not more than four times daily, with a 60-minute rest period required between each exposure; the lower the value, the more toxic the substance.

Threshold limit value–skin A designation indicating that direct or airborne contact with a material could result in significant exposure from absorption through the skin, mucous membranes, and eyes.

Threshold limit value–time-weighted average (TLV–TWA) The airborne concentration of a material to which a worker can be exposed for 8 hours a day, 40 hours a week, and not suffer any ill effects.

Tie rod A metal rod that runs from one beam of the ladder to the other to keep the beams from separating. Tie rods are typically found in wood ladders.

Time marks Status updates provided to the communications centre every 10–20 minutes. Such an update should include the type of operation, the progress of the incident, the anticipated actions, and the need for additional resources.

Tip The very top of the ladder.

Topography The land surface configuration. (NFPA 1051)

Tote A portable tank characterized by its unique style of construction, which usually holds around 1000 L (264 gal.) of product.

Toxic Any material whose properties contain the inherent capacity to produce injury to a biological system, which is dependent on concentration, rate, method, and site of absorption. (NFPA 306)

Toxic inhalation hazard (TIH) material A gas or volatile liquid that is extremely toxic to humans and poses a hazard to health during transportation.

Toxic product of combustion A hazardous chemical compound released when a material decomposes under heat.

Toxicology The study of the adverse effects of chemical or physical agents on living organisms.

Trace (transfer) evidence A minute quantity of physical evidence that is conveyed from one place to another.

TRACEMP An mnemonic to help remember the effects and potential exposures to a hazardous materials incident—that is, thermal, radiation, asphyxiant, chemical, etiologic, mechanical, and psychogenic.

Trailers Combustible materials, such as rolled rags, blankets, and newspapers or ignitable liquid, that are used to spread fire from one point or area to other points or areas, often used in conjunction with an incendiary device.

Training officer The person designated by the fire chief as having authority for overall management and control of the organization's training program. (NFPA 1401)

Transfer of command The formal procedure for transferring the duties of an IC from one person to another at an incident scene. (NFPA 1026)

Trench cut A roof cut that is made from one load-bearing wall to another load-bearing wall and that is intended to prevent horizontal fire spread in a building.

Triage The process of sorting victims based on the severity of their injuries and medical needs to establish treatment and transportation priorities.

Triangular cut A triangle-shaped ventilation cut in the roof decking that is made using a saw or an axe.

Trigger The button or lever used to discharge the agent from a portable fire extinguisher.

Triple-layer load A hose loading method in which the hose is folded back onto itself to reduce the overall length to one-third before loading the hose in the bed. This load method reduces deployment distances.

Truck company A group of fire fighters who work as a unit and are equipped with one or more pieces of aerial fire apparatus. (NFPA 1410)

Trunking system A radio system that uses a shared bank of frequencies to make the most efficient use of radio resources.

Truss A collection of lightweight structural components joined in a triangular configuration that can be used to support either floors or roofs.

Truss block A piece of wood or metal that ties the two rails of a trussed beam ladder together and serves as the attachment point for the rungs.

Trussed beam A ladder beam constructed of top and bottom rails joined by truss blocks that tie the rails together and support the rungs.

Tube trailers High-volume transportation devices made up of several individual compressed gas cylinders banded together and affixed to a trailer.

Turbulent smoke flow Agitated, boiling, aggressive smoke, which indicates great heat in the burning building. It is a precursor to flashover.

Turnout coat Protective coat that is part of a protective clothing ensemble for structural firefighting; also called a bunker coat.

Turnout pants Protective trousers that are part of a protective clothing ensemble for structural firefighting; also called bunker pants.

Twisted rope Rope constructed of fibres twisted into strands, which are then twisted together.

Two-in–two-out rule A safety procedure that requires a minimum of two personnel to enter a hazardous area and a minimum of two backup personnel to remain outside the hazardous area during the initial stages of an incident.

Two-way radio A portable communication device used by fire fighters. Every firefighting team should carry at least one radio to communicate distress, progress, changes in fire conditions, and other pertinent information.

Type I construction Buildings with structural members made of noncombustible materials that have a specified fire resistance; also known as fire-resistive construction.

Type II construction Buildings whose structural members are made of noncombustible materials without fire resistance; also known as noncombustible construction.

Type III construction Buildings with exterior walls made of noncombustible or limited combustible materials, but with interior floors and walls made of combustible materials; also known as ordinary construction.

Type IV construction Buildings constructed with noncombustible or limited combustible exterior walls, and interior walls and floors made of large-dimension combustible materials; also known as heavy timber construction.

Type V construction Buildings with exterior walls, interior walls, floors, and roof structures made of wood; also known as wood-frame construction.

Underventilated fire A fire that has an adequate supply of fuel available but is not receiving an adequate supply of oxygen.

Underwriters Laboratories of Canada (ULC) The Canadian organization that tests and certifies that fire extinguishers (among many other products) meet established standards. The United States equivalent is Underwriters Laboratories, Inc.

Undetermined Fire-cause classification that includes fires for which the cause has not or cannot be proven.

Unibody (unit-body) The frame construction most commonly used in vehicles. The base unit is made of formed sheet metal; structural components are then added to the base to form the passenger compartment. Subframes are attached to each end. This type of construction does away with the rail beams used in platform-frame vehicles; also known as unit body.

Unified command An application of the IMS that allows all agencies with jurisdictional responsibility (geographical or functional) for an incident or planned event to manage said incident or event by establishing a common set of objectives and strategies. (NFPA 1561)

Unity of command The concept that each person within an organization reports to one, and only one, designated person. (NFPA 1026)

Universal precautions Procedures for infection control that treat blood and certain body fluids as capable of transmitting infectious diseases.

Unlocking mechanism A keyway, combination wheel, or combination dial.

Unwanted alarm A fire alarm signal caused by a device reacting properly to a condition that is not a true fire emergency (e.g., cigarette smoke).

Upper flammable limit (UFL) The highest concentration of a combustible substance in a gaseous oxidizer that will propagate a flame. (NFPA 68)

Upright sprinkler A sprinkler designed to be installed on top of the supply piping; usually marked SSU (standard spray upright).

Utility rope Rope used for securing objects, for hoisting equipment, or for securing a scene to prevent bystanders from being injured. Utility rope must never be used in life safety operations.

V-agent An oily liquid nerve agent that can persist for several weeks; primarily a contact hazard.

Vacuuming The process of cleaning up dusts, particles, and some liquids using a vacuum with HEPA filtration to prevent recontamination of an environment.

Vapour The gaseous state of a substance, particularly of those that are normally liquids or solids at ordinary temperatures. (NFPA 92)

Vapour density The weight of an airborne concentration (vapour or gas) as compared to an equal volume of dry air.

Vapour dispersion The process of lowering the concentration of vapours by spreading them out.

Vapour pressure The pressure, measured in kilopascals or psi absolute (psia), exerted by a liquid, as determined by ASTM D323, *Standard Test Method for Vapor Pressure of Petroleum Products* (*Reid Method*). (NFPA 385)

Vapour suppression The process of controlling vapours given off by hazardous materials by covering the product with foam or other material or by reducing the temperature of the material, thereby preventing vapour ignition.

Vapour-protective clothing The garment portion of a CPE that is designed and configured to protect the wearer against chemical vapours or gases. (NFPA 472)

Vent pipes Inverted J-shaped tubes that allow for pressure relief or natural venting of a pipeline for maintenance and repairs.

Ventilation The changing of air within a compartment by natural or mechanical means. Ventilation can be achieved by introduction of fresh air to dilute contaminated air or by local exhaust of contaminated air. (NFPA 402)

Verification system A fire alarm system that does not immediately initiate an alarm condition when a smoke detector activates. The system will wait a preset interval (generally 30–60 seconds) before checking the detector again. If the condition is clear, the system returns to normal status; if the detector still senses smoke, the system activates the fire alarm.

Vertical ventilation The process of making openings so that smoke, heat, and gases can escape vertically from a structure.

Voice recording system Recording devices or computer equipment connected to telephone lines and radio equipment in a communications centre to record telephone calls and radio traffic.

Volatility The ability of a substance to produce combustible vapours.

VX A nerve agent.

Wall post indicator valve A sprinkler control valve that is mounted on an outside wall of a building. The position of the indicator tells whether the valve is open or shut.

Warm zone The zone where the decontamination process takes place in the decontamination corridor, which is set up in between the cold and hot zones.

Water catch-all A salvage cover that has been folded to form a container to hold water until it can be removed.

Water chute A salvage cover that has been folded to direct water flow out of a building or away from sensitive items or areas.

Water curtain nozzle A nozzle used to deliver a flat screen of water that forms a protective sheet of water to protect exposures from fire.

Water hammer The surge of pressure that occurs when a high-velocity flow of water is abruptly shut off. The pressure exerted by the flowing water against the closed system can be seven or more times that of the static pressure. (NFPA 1962)

Water knot A knot used to join the ends of webbing together.

Water main A generic term for any underground water pipe.

Water-motor gong An audible alarm notification device that is powered by water moving through the sprinkler system.

Water supply A source of water for firefighting activities. (NFPA 1144)

Water thief A device that has a 64-mm (2.5-in.) inlet and a 64-mm (2.5-in.) outlet, in addition to two 38-mm (1.5-in.) outlets. It is used to supply multiple hoses from one source.

Water vacuum A device similar to a wet/dry shop vacuum cleaner that can pick up liquids. It is used to remove water from buildings.

Waybill Shipping document for railroad transportation.

Weapons of mass destruction (WMDs) A. Any destructive device, such as any explosive, incendiary, or poison-gas bomb; grenade; rocket having a propellant charge of more than 113 grams (4 ounces); missile having an explosive or incendiary charge of more than 7 g (0.25 oz.); mine; or device similar to the above. B. Any weapon involving toxic or poisonous chemicals. C. Any weapon involving a disease organism. D. Any weapon that is designed to release radiation or radioactivity at a level dangerous to human life. (NFPA 472)

Wedges Material used to tighten or adjust cribbing and shoring systems. (NFPA 1006)

Wet barrel hydrant A hydrant used in areas that are not susceptible to below-freezing temperatures. The barrel of the hydrant is normally filled with water.

Wet chemical extinguishing agent An extinguishing agent for Class K fires. It commonly consists of solutions of water and potassium acetate, potassium carbonate, potassium citrate, or any combination thereof.

Wet chemical extinguishing system An extinguishing system that discharges a proprietary liquid extinguishing agent. It is often installed over stoves and deep fryers in commercial kitchens.

Wet chemical fire extinguisher A fire extinguisher for use on Class K fires that contains a wet chemical extinguishing agent.

Wet sprinkler system A sprinkler system in which the pipes are normally filled with water.

Wetting-agent water-type fire extinguisher An extinguisher that expels water combined with a chemical or chemicals to reduce its surface tension; it is used for Class A fires.

Wheeled fire extinguisher A portable fire extinguisher equipped with a carriage and wheels, which is intended to be transported to the fire by one person. (NFPA 10)

Wilderness search and rescue (SAR) The process of locating and removing a victim from the wilderness.

Wildland Land in an uncultivated, more or less natural state and covered by timber, woodland, brush, and/or grass. (NFPA 901)

Wildland/brush company A fire company that is dispatched to wildland and brush fires where larger engines cannot gain access. Wildland/brush companies have four-wheel drive vehicles and special firefighting equipment.

Wildland fire An unplanned fire burning in vegetative fuels. (NFPA 1051)

Wildland–urban interface An area where undeveloped land with vegetative fuels mixes with human-made structures.

Wired glass A glazing material with embedded wire mesh. (NFPA 80)

Wood-frame construction Buildings whose exterior walls, interior walls, floors, and roofs are made of combustible wood materials. Wood-frame construction is also known as Type V building construction, as defined in NFPA 220, *Standard on Types of Building Construction*.

Wooden beam A load-bearing member assembled from individual wood components.

Wood truss An assembly of small pieces of wood or wood and metal.

Working end The part of the rope used for forming a knot.

Wye A device used to split a single hose into two separate lines.

Zoned coded alarm A fire alarm system that indicates which zone was activated both on the alarm control panel and through a coded audio signal.

Zoned noncoded alarm A fire alarm system that indicates the activated zone on the alarm control panel.

Zoned system A fire alarm system that divides a building or facility into zones so that the area where an alarm originated can be identified.

Index

Note: Page numbers followed by *f* or *t* indicate material in figures or tables, respectively.